GEO5

Global Environment Outlook

Environment for the future we want

United Nations Environment Programme

achieving refuge interface
knowledge affluent abandon beliefs
thaw conditions funds deliver upheaval rivers
terrestrial decline tax
ice GROW catch results damage tide
dynamics
GMOs agreements nest logging robust diminishing
rights deepen precipitation
binding approach alter arctic
projection atmosphere adverse
CSD tundra variability solar polar wetland mitigate marine petrol example
sprawl media overexploit burn business learning drivers beyond source
extinct bleaching less methods integrity permafrost thresholds inclusive flow
cascading technologies vertebrate THREAT erode axis LAKE services earth
opportunities density maize acquisition
convergence unintended reserve consequence sectors
modify cloud fertilizer El Niño negotiate
restore contain CONVENTION work hazardous well-being plankton
adopt biofuels integrated meat governance urgent
industrialization municipal
frequent extent
goods poverty warming
multilateral carbon vehicles science bilateral endemic stocks households protocol
evolve instruments UN decreasing
GHG downturn
fire cattle natural
wildlife emerging health invasive
supply agricultural
spill
literacy debt
humid pests boundaries
controls
TROPICAL
peace success plant methane
uncertain
restoration
persistent widespread
human
AMAZON free gains
BIRDS forests
novel economies
soy pollutants cholera committed eutrophic
pristine
Rio+20
biodiversity disappear sink
integrate local
predict
achieving
Agenda21 corridors clouds
new

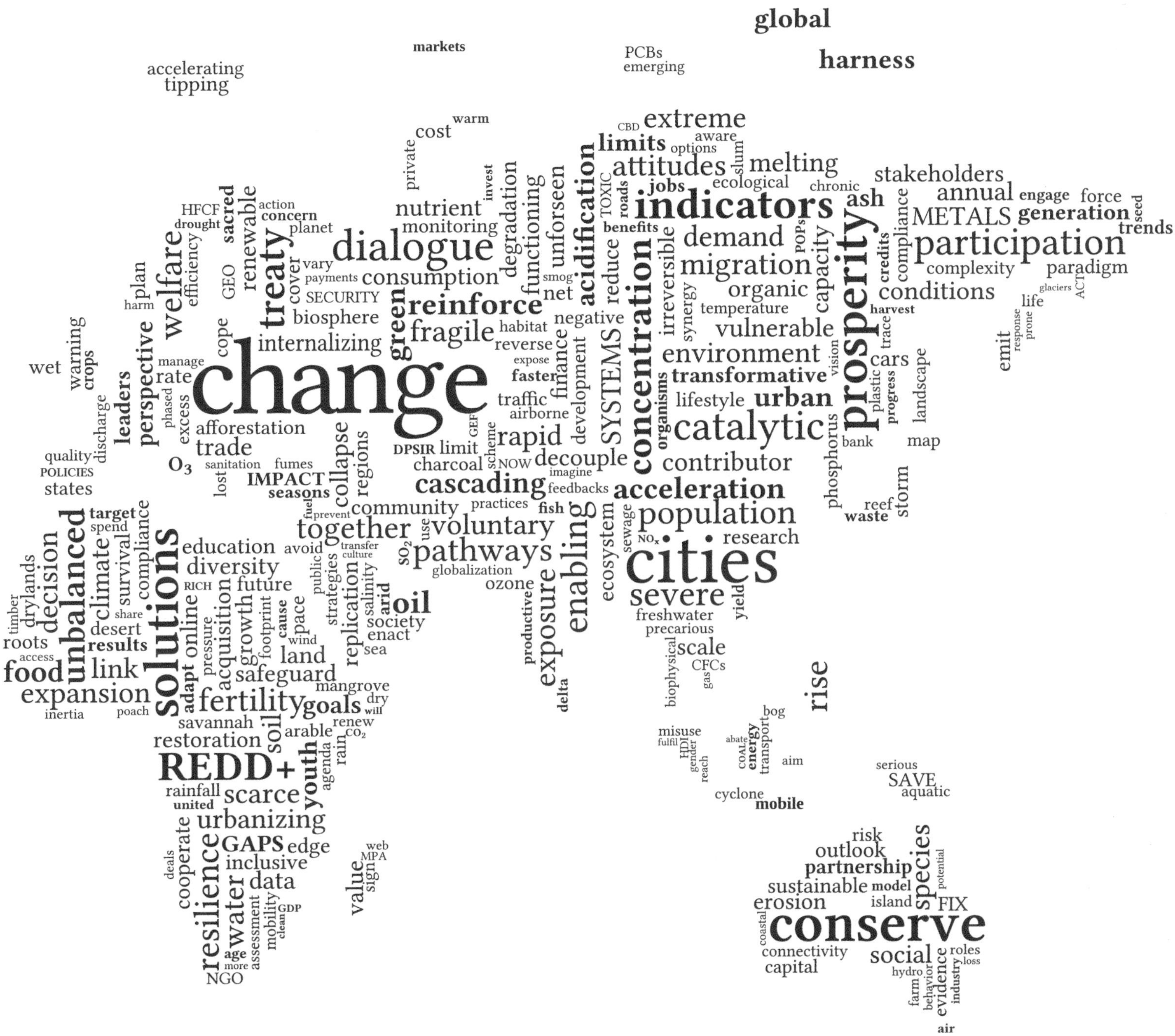
global
markets
harness
accelerating
tipping
PCBs
emerging
warm
cost
CBD
extreme
limits
aware
options
slum
private
invest
attitudes
melting
stakeholders
jobs
ecological
chronic
annual
engage
force
seed
HFCF
action
concern
drought
sacred
renewable
nutrient
monitoring
degradation
functioning
unforseen
acidification
TOXIC
roads
indicators
ash
compliance
METALS
generation
trends
planet
dialogue
benefits
demand
POPs
capacity
prosperity
credits
participation
plan
welfare
efficiency
GEO
treaty
cover
vary
payments
consumption
smog
reduce
concentration
irreversible
migration
complexity
paradigm
harm
SECURITY
reinforce
net
organic
conditions
glaciers
ACT
life
biosphere
green
negative
temperature
harvest
response
prone
warning
perspective
cope
internalizing
fragile
habitat
synergy
vulnerable
trace
wet
crops
manage
reverse
finance
environment
vision
emit
rate
change
expose
SYSTEMS
faster
development
transformative
cars
leaders
phased
excess
traffic
organisms
lifestyle
urban
plastic
landscape
progress
airborne
discharge
afforestation
GEF
scheme
rapid
catalytic
quality
trade
collapse
regions
DPSIR
limit
bank
map
POLICIES
O3
sanitation
fumes
charcoal
NOW
decouple
contributor
phosphorus
states
lost
IMPACT
cascading
imagine
storm
seasons
feedbacks
acceleration
target
community
practices
fish
reef
spend
fuel
prevent
waste
population
unbalanced
together
use
voluntary
climate
survival
compliance
solutions
education
avoid
transfer
culture
NOx
research
drylands
decision
diversity
SO2
pathways
sewage
timber
RICH
future
public
strategies
salinity
globalization
ecosystem
cities
online
acquisition
arid
ozone
enabling
share
pressure
growth
footprint
cause
pace
replication
oil
exposure
severe
yield
desert
society
freshwater
precarious
roots
results
wind
enact
productive
access
land
sea
scale
food
link
safeguard
biophysical
CFCs
adapt
rise
expansion
mangrove
delta
gas
fertility
goals
dry
inertia
poach
will
bog
savannah
renew
CO2
misuse
arable
fulfil
energy
soil
HDI
abate
transport
restoration
rain
COAL
REDD+
youth
agenda
gender
reach
aim
serious
rainfall
scarce
SAVE
united
cyclone
aquatic
mobile
urbanizing
risk
cooperate
GAPS
edge
web
outlook
species
deals
inclusive
MPA
partnership
resilience
data
value
sign
sustainable
model
potential
water
erosion
island
FIX
mobility
GDP
coastal
conserve
assessment
clean
connectivity
social
roles
age
capital
loss
more
hydro
NGO
farm
behavior
evidence
industry
air

Acknowledgements

This fifth *Global Environment Outlook* (*GEO-5*) assessment report is a product of the strong dedication and extraordinary investment of numerous individuals, whose knowledge, expertise and insight helped shape this important body of work. UNEP acknowledges the contributions made by many governments, individuals and institutions to the preparation and publication of this report. A full list of names of individuals and institutions involved in the assessment process is included from pages 498–504. Special thanks are extended to:

High-Level Intergovernmental Advisory Panel

Rajender Ahlawat, Hussein A. Al-Gunied, Mohammed Saif Al-Kalbani, Wahid Al-Shuely, Burcu Bursali, Mantang Cai, Jorge Laguna Celis, Raouf Dabbas, Guilherme da Costa,Martijin Dadema, Sandra De Carlo, Idunn Eidheim, Prudence Galega, Nilkanth Ghosh, Rosario Gomez, Xia Guang, Han Huiskamp, Jos Lubbers, John Michael Matuszak, Samira Nateche, Kim Thi Thuy Ngoc, Van Tai Nguyen, Jose Rafael Almonte Perdomo, Majid Shafie-Pour-Motlagh, Jiang Wei, Albert Williams and Daniel Ziegerer.

Science and Policy Advisory Board

Joseph Alcamo, Asma Ali Abahussain, Pinhas Alpert, Torkil Jonch Clausen, Ahmed Djoghlaf, Susanne Droege, Kejun Jiang, Nicholas King, Filipo Lansigan, Anne Larigauderie, Emilio Lèbre La Rovere, Jacqueline McGlade, Luisa T. Molina, Toral Patel-Weynand, Nicolas Perritaz, Carlos A. Quesada, Chirapol Sintunawa, Sandra Torrusio, George Varughese and Robert Watson.

Data and Indicators Working Group

Asma Ali Abahussain, Ezgi Akpinar-Ferrand, Barbara Clark, Sandra De Carlo, Volodymyr Demkine, Alexander Gorobets, Eszter Horvath, Koffi Kouadio, Murari Lal, Samwiri Musisi-Nkambwe, Ambinistoa Lucie Noasilalaonomenjanahary, Toral Patel-Weynand, Muhammad Munir Sheikh, Ashbindu Singh, Anil Kumar Thanappan, Susan Tumwebaze, Héctor Tuy and Jaap van Woerden.

Coordinating Lead Authors

May Antoniette Ajero, Dolors Armenteras, Jane Barr, Ricardo Barra, Ivar Baste, James Dobrowolski, Nicolai Dronin, Amir El-Sammak, Tom P. Evans, C. Max Finlayson, Erica Brown Gaddis, Keisha Garcia, Nesreen Ghaddar, Paul Roger Glennie, Yi Huang, Carol A. Hunsberger, Maria Ivanova, Jill Jäger, Peter N. King, Johan C.I. Kuylenstierna, Bernice Lee, Marc A. Levy, Lailai Li, Clever Mafuta, Ruben Mnatsakanian, Jennifer Clare Mohamed-Katerere, Alexandra C. Morel, Begum Ozkaynak, Neeyati Patel, Renat Perelet, Laszlo Pinter, Pierre Portas, Walter Rast, Asha Singh, Detlef P. van Vuuren, Roy Victor Watkinson and Joanna Noelia Kamiche Zegarra.

Scientific Peer-Reviewers (Coordinated by the Earth System Science Partnership)

Keigo Akimoto, Mahmoud Ali, Erik Ansink, Masroor Ellahi Babar, David Barkin, Janos Bogardi, Philippe Bourdeau, Josep Canadell, Graciela Ana Canziani, Andrea Birgit Chavez Michaelesen, Kevin Cheung, Antonio Cruzado, Shobhakar Dhakal, Serigne Faye, Marina Fischer-Kowalski, Amadou Thierno Gaye, Mark O. Gessner, Evgeny Gordov, Dagmar Haase, Itsuki Handoh, Nick Harvey, Lars Hein, Gerhard J. Herndl, Shu-Li Huang, Falk Huettmann, Ada Ignaciuk, Muhammad Mohsin Iqbal, Louise Jackson, Sharad Jain, Ian Jenkinson, Rainer Krug, Nelson Lourenco, Angela M. Maharaj, Miyuki Nagashima, Daiju Narita, Isabelle Niang, Patrick Nunn, Jay O'Keeffe, Jean-Pierre Ometto, Ursula Oswald Spring, Claudia Pahl-Wostl, Nirmalie Pallewatta, Henrique M. Pereira, Erika Pires Ramos, Germán Poveda, Francesc Prenafeta, Seema Purushothaman, Dork Sahagian, Galia Selaya, Mika Sillanpaa, Maria Siwek, Erika Techera, Holm Tiessen, Klement Tockner, Aysun Uyar, Emma Archer van Garderen, Tracy Van Holt, Stefano Vignudelli, Hassan Virji, Angela Wagener and Hong Yang.

Outreach Group

Adel Farid Abdel-Kader, Robert Barnes, Matthew Billot, Peter Browne, Bryan Coll, Richard Crompton, Ivica Cvetanovski, Marie Daher, Salif Diop, Silvia Giada, Peter Gilruth, Elisabeth Guilbaud-Cox, Suzanne Howard, Alexander Juras, Satwant Kaur, Fatoumata Keita-Ouane, Fanina Kodre-Alexander, Alejandro Laguna, Thor-Jürgen Greve Løberg, Amos Muema, Nicole Lettington, Michael Logan, Angele Luh, Kelvin Memia, Graciela Metternicht, Waiganjo Njoroge, Nick Nuttall, Neeyati Patel, Audrey Ringler, Stuart Roberts, Andrea Salinas, Ashbindu Singh, Janet Fernandez Skaalvik, Anna Stabrawa, Mia Turner, Frank Turyatunga, Isabelle Valentiny, Ronald Witt, Jinhua Zhang, Laetitia Zobel and Shereen Zorba.

GEO-5 Funding

The Governments of Canada, Norway, Republic of Korea, the Netherlands, Sweden, Switzerland and the Gwangju Metropolitan City, Republic of Korea, together with the UNEP Environment Fund, provided the necessary funding for the production of *GEO-5* and subsequent outreach activities. Contributions were also provided by GRID-Arendal and the Development Bank of Latin America.

Environment Canada Environnement Canada

REGERINGSKANSLIET Government Offices of Sweden

NORWEGIAN MINISTRY OF FOREIGN AFFAIRS

Ministry of Foreign Affairs of the Netherlands

MINISTRY OF ENVIRONMENT

Schweizerische Eidgenossenschaft
Confédération suisse
Confederazione Svizzera
Confederaziun svizra
Federal Office for the Environment FOEN

CAF DEVELOPMENT BANK OF LATIN AMERICA

UNEP GRID ARENDAL *Environmental Knowledge for Change*

Global Environment Outlook 5

GEO-5 Core Team: Matthew Billot (Head of GEO Unit), Ludgarde Coppens, Volodymyr Demkine, Salif Diop, Peter Gilruth, Jason Jabbour, Fatoumata Keita-Ouane, Josephine Nyokabi Mwangi, Brigitte Ohanga, Nalini Sharma

Regional Coordinating Team: Adel Farid Abdel-Kader, Fouad Abousamra, Silvia Giada, Graciela Metternicht, Charles Sebukeera, Ashbindu Singh, Anna Stabrawa, Frank Turyatunga, Jaap van Woerden, Ronald Witt, Jinhua Zhang

Production Coordination: Jason Jabbour

Production Support: Sarah Abdelrahim, Sylvia Adams, Joana Akrofi, Joseph Alcamo, Chris Ambala, Liana Archaia-Atanasova, Suzanne Bech, Charles Davies, Tessa Goverse, Loise Kinuthia, Fanina Kodre, Sunday Leonard, Erick Litswa, Kelvin Memia, Monika G. MacDevette, Patrick M'mayi, Edwin Mwanyika, Trang Nguyen, Thierry De Oliveira, Janak Pathak, Audrey Ringler, Pinya Sarasas, Tunnie Srisakulchairak, Mick Wilson, Shereen Zorba

UNEP World Conservation Monitoring Centre (UNEP-WCMC): Mari Bieri, Satu Glaser, Maxwell Gomera, Abisha Mapendembe, Alison M. Rosser, Jörn Scharlemann, Matt J. Walpole

Global Resources Information Database Centre, Norway (GRID-Arendal): Björn Alfthan, John Crump, Lawrence Hislop, Tiina Kurvits, Thor-Jürgen Greve Løberg, Clever Mafuta, Riccardo Pravettoni, Peter Prokosch, Petter Sevaldsen, Janet Fernandez Skaalvik

GEO-5 E-peer-review System: Herb Caudill, Shane Kunkle

Data Support: Andrea de Bono, Dominique del Pietro, Stefan Schwarzer, Jaap van Woerden

Maps and Graphics: Riccardo Pravettoni (GRID-Arendal), UNEP/GRID-Geneva, Mattias Turini, Nieves López Izquierdo, Audrey Ringler

Editorial Team: Bart Ullstein, Helen de Mattos, Christine Hawkins, Catherine P. McMullen, Jason Jabbour, Jörn Scharlemann

Design and Layout: GRID-Arendal, Ali Cherri

Editorial and Outreach Coordination: Neeyati Patel

Contents

Figures

Tables

Boxes

Foreword

Anyone wishing to understand the pace and scale of environmental change will find UNEP's flagship assessment report – *Global Environment Outlook-5: Environment for the future we want* – compelling reading. Equally, anyone seeking a paradigm shift that can bring us closer to a truly sustainable world will find this latest edition of the GEO series rich in opportunities and policy options.

GEO-5 is designed to be the most comprehensive, impartial and in-depth assessment of its kind. It reflects the collective body of recent scientific knowledge, drawing on the work of leading experts, partner institutions and the vast body of research undertaken within and beyond the United Nations system.

The launch of *GEO-5* coincides with the final stages of preparation for the UN Conference on Sustainable Development (Rio+20), taking place two decades after the Rio Earth Summit that set the agenda for contemporary thinking about sustainable development. The report underlines the reasons why world leaders need to show decisive leadership in Rio and beyond. It highlights the state, trends and trajectories of the planet and its people, and showcases more than 100 initiatives, projects and policies from across the globe that are pioneering positive environmental change.

In a world with a growing population, glaring inequality and a precarious environmental base, it is imperative that Governments collaborate to balance the economic, social and environmental strands of sustainable development. *GEO-5* highlights not just the perils of delaying action, but the options that exist to transform sustainable development from theory to reality. I commend *GEO-5* to all who wish to invest in this generational opportunity to create the future we want.

BAN Ki-moon

Secretary General of the United Nations
United Nations Headquarters, New York

May 2012

Preface

Since the days of the ancient Egyptians, Greeks and Chinese, through the Islamic Golden Age and the Renaissance, philosophers and scientists have sought to make sense of the forces and processes of the natural world and humanity's place within them. In the past half century or so, this endeavour has accelerated as concerns over the impacts of industrialization have emerged and more recently been fuelled by a growing realization that people – once marginal influencers of environmental change – are now its principal drivers, from biodiversity loss to climate change.

The *Global Environment Outlook: Environment for the future we want (GEO-5)* is part of this broad sweep of history, and is a major contribution to the public understanding of the way ecosystems and the atmosphere are responding to patterns of unprecedented consumption and production – patterns taking place on a planet of 7 billion people, rising to more than 9 billion by 2050. Its findings on the state of the planet, globally and regionally, are unsurprisingly sobering and cause for profound concern – they should serve as a reminder to world leaders and delegates attending the Rio+20 Summit in June as to why they are there.

Bridging the science-policy interface remains problematic – translating the findings of science into environmental law and policy making has been a challenge stretching back through Rio 1992 to the Stockholm Conference on the Human Environment of 1972. Encouragingly, a growing scientific understanding and technological progress have not fallen on deaf ears; they have inspired a myriad of treaties and agreements covering such issues as the trade in endangered species, the protection of the ozone layer, climate change, biodiversity loss and the banning of persistent organic pollutants.

GEO-5 adds new dimensions to the discourse through its assessment of progress towards meeting internationally agreed goals and identifying gaps in their achievement. Out of 90 goals and objectives assessed, significant progress could only be shown for four. Of equal concern, progress could not be appraised for 14 goals and objectives simply because data were lacking.

Another *GEO-5* innovation is that it highlights a regional selection of more than a hundred policies and transformational actions that have been tried and tested successfully in countries and communities around the world. These policy options give decision makers tools that could be adapted to their own settings.

Such policy options are part of a broad sweep of emerging work termed the Green Economy, which in the context of sustainable development and poverty eradication is one of the two major themes for Rio+20. The summit is about taking stock and renewing commitments, but it is also about the integration of scientific findings in evidence-based policy making and the re-engagement of society in endeavours to move the world on to a sustainable path.

When nations take stock of sustainable development 20 years after the Rio Earth Summit of 1992, the limited achievements and endemic knowledge divide between North and South should be high on the agenda.

In summary, science must underpin policy making, but as five GEO assessments and reports have shown, it is not enough. Realizing and implementing science-based policies is where the real gap resides, and this can be bridged not by more satellite observations, field monitoring, computations and scenario modeling but by courage, decisiveness and political leadership that matches the reality that *GEO-5* confirms.

Achim Steiner

United Nations Under-Secretary General and Executive Director
United Nations Environment Programme

Introduction

THE EARTH SYSTEM CONTEXT

The Earth System provides the basis for all human societies and their economic activities. People need clean air to breathe, safe water to drink, healthy food to eat, energy to produce and transport goods, and natural resources that provide the raw materials for all these services. However, the 7 billion humans alive today are collectively exploiting the Earth's resources at accelerating rates and intensities that surpass the capacity of its systems to absorb wastes and neutralize the adverse effects on the environment. In fact, the depletion or degradation of several key resources has already constrained conventional development in some parts of the world.

Within the Earth System – which acts as a single, self-regulating system comprised of physical, chemical, biological and human components – the effects of human activities can be detected at a planetary scale (Chapter 7). These have led scientists to define a new geological epoch, the Anthropocene, based on evidence that atmospheric, geological, hydrological, biological and other Earth System processes are being altered by human activity. The most readily recognized changes include a rise in global temperatures and sea levels, and ocean acidification, all associated with the increase in emissions of greenhouse gases, especially carbon dioxide and methane (Chapters 2 and 4). Other human-induced changes include extensive deforestation and land clearance for agriculture and urbanization, causing species extinctions as natural habitats are destroyed (Chapters 3 and 5).

While humans have long been aware of the effects of their activities on the local environment, only in the last few decades has it become apparent that these activities can cumulatively affect the global environment (Chapters 1–7). In the past, anthropogenic pressures on natural resources were less pervasive and the Earth's atmosphere, land and water could carry the load of human consumption and production. However, in the second half of the 20th century the effects of many diverse local changes compounded at accelerating rates to produce global consequences. Globalization allows goods to be produced under circumstances that consumers would refuse to tolerate in their own community, and permits waste to be exported out of sight, enabling people to ignore both its magnitude and its impacts. However, just as waste has – literally – reached the ends of the Earth, environmental concerns have become globalized as well (Chapter 1).

These threats to the Earth System have led the science community and policy makers to work together more closely to meet the challenge in a sustainable and collaborative manner.

THE SCIENCE-POLICY CONTEXT

At the 1972 United Nations Conference on the Human Environment, 119 nations came together for the first time to discuss serious environmental concerns raised by the scientific and conservation communities. As an initial step, the conference established UNEP to catalyse international and UN-wide environmental action. Twenty years on, the United Nations Conference on Environment and Development in Rio de Janeiro approved Agenda 21, a blueprint for the introduction of sustainable development, a concept first articulated as "satisfying the needs of the present generation without compromising the chance for future generations to satisfy theirs" in the World Commission on Environment and Development 1987 report *Our Common Future*. In the second decade of the new century, Agenda 21 remains a vibrant and relevant guide with many of its precepts yet to be applied, particularly in regard to consumption.

The 2000 Millennium Summit, which brought world leaders together to discuss the role of the United Nations at the turn of the 21st century, produced eight Millennium Development Goals (MDGs) to make up for shortcomings that resulted from a focus on economic objectives while international development stalled. The MDGs address the integration of sustainable development principles into country policies and programmes and aim to reverse the impoverishment of human and environmental resources, while setting time-bound targets and establishing metrics. MDG 7, which specifically addresses the environment, set targets to make significant reductions in the rate of biodiversity loss by 2010, to halve the proportion of the population without sustainable access to safe drinking water and basic sanitation by 2015, and to achieve a significant improvement in the lives of at least 100 million slum dwellers by 2020.

As understanding has developed about the relationship between human well-being and environmental change, so have the attempts to make it relevant for policy makers. The dependence of social development and economic activity on environmental services and stability is increasingly understood. An economy functions within a society, or within and between societies, using natural and human resources to produce marketable goods and services. At the same time, societies survive and thrive within the environment determined by the physical limits of atmosphere, land, water, biodiversity and other material resources.

Interacting environmental, social and economic forces produce a complex system that has been the focus of substantial research, but it is only in the last two decades that information and communication technologies have enabled researchers to model and explore the intricate complexities of the whole Earth System.

Insights gained from the ability to appreciate the power and nuance of Earth System complexities demand a new perception of the responsibilities and accountabilities of nation states towards planetary stewardship (Chapter 16 and 17). This not only requires the realization of environment and development goals and targets but also the development of specific goals aimed at global sustainability, addressing the needs of the most vulnerable as well as the wants of the more powerful.

The elaboration of such goals requires scientifically credible indicators and information to guide, track and report progress (Chapter 8). Integrated environmental assessments are tools, within a broad and deep toolkit, that have been developed to meet this need. However, for the most part, policy developments and revisions have failed to adequately incorporate assessment findings and other scientific information into international policy priorities.

BACKGROUND

The main goal of UNEP's *Global Environment Outlook* (GEO) is to keep governments and stakeholders informed of the state and trends of the global environment. Over the past 15 years, the GEO reports have examined a wealth of data, information and knowledge about the global environment; identified potential policy responses; and provided an outlook for the future. The assessments, and their consultative and collaborative processes, have worked to bridge the gap between science and policy by turning the best available scientific knowledge into information relevant for decision makers.

Previous GEO reports focused on an analysis of environmental issues and the identification of responses, using an integrated approach that provided a comprehensive and multidisciplinary overview across different themes. This fifth *Global Environment Outlook* (*GEO-5*) builds on previous reports, continuing to provide analyses of the state, trends and outlook for, and responses to, environmental change. But it also adds new dimensions through its assessment of progress towards meeting internationally agreed goals and identifying gaps in their achievement (Chapters 2–6), on analysing promising response options that have emerged in the regions (Chapters 9–15), and presenting potential responses for the international community (Chapters 16–17). Furthermore, for the first time, *GEO-5* suggests that there should be a fundamental shift in the way environmental issues are analysed, with consideration given to the drivers of global change, rather than merely to the pressures on the environment.

Details of the process followed by the UNEP Secretariat in developing *GEO-5*, including the assemblage of more than 600 scientists guided by governmental, scientific and policy advisory bodies, are presented in the *GEO-5* Process section.

STRUCTURE

The *GEO-5* report is made up of 17 chapters organized into three distinct but linked parts.

Part 1 – State and trends of the global environment

To explore today's rapidly changing socio-economic conditions, Chapter 1 examines the drivers of environmental change – the overarching socio-economic forces that exert varying degrees of influence, or pressures, on the environment. Chapter 1 identifies and describes these major root causes of the environmental challenges and provides some suggestions for policy interventions.

Using the drivers, pressures, state, impacts and responses (DPSIR) analytical framework (Figure 1), the *GEO-5* assessment presents the latest state and trends of the global environment under the themes of atmosphere, land, water, biodiversity and, for the first time in the GEO series, chemicals and waste (Chapters 2–6).

The DPSIR framework is used to identify and evaluate the complex and multidimensional cause-and-effect relationships between society and the environment. The DPSIR framework used in GEO assessments is an extension of the pressure-state-response model developed by the OECD and the European Environment Agency in the mid-1990s. Drivers such as population dynamics, economic demand and unsustainable consumption and production patterns are processes that lead to impacts on the environment. These drivers often directly or indirectly result in environmental pressures including increased emissions of pollutants and wastes and destructive resource extraction. Such pressures cause changes to the environment with concomitant impacts on both humans and ecosystems. The DPSIR analytical framework helps to identify these processes. Finally, it suggests responses, which can take many forms at many scales from community action to international treaties, not only to the underlying drivers, but also to the environmental pressures and their impacts on ecosystems and human health.

Chapters 2–6 evaluate whether a selection of internationally agreed environmental goals are being met for each of the themes; Chapter 7 provides a synthesis of the thematic information from an Earth System perspective. Part 1 concludes with a review of the need to strengthen the collection, analysis and interpretation of data relevant to tracking the state and trends of the environment as a fundamental requirement for further research, for monitoring and evaluation, for scientific assessments, and for effective policy making (Chapter 8).

Part 2 – Policy options from the regions

Part 2 of *GEO-5*, Chapters 9–14, presents an appraisal of policy options from the regions (Figure 2) that show potential for helping to speed up the accomplishment of internationally agreed goals. This was requested by UNEP's Governing Council and provides readers wishing to implement successful policies with promising avenues for exploration.

To direct the policy appraisal, multi-stakeholder consultations were undertaken in each region to identify priority environmental challenges and related internationally agreed goals.

Following a screening exercise, policies or policy clusters that either demonstrated a record of success with respect to their associated goals or featured innovative characteristics combined with promising initial results were retained and analysed in further detail. The policy appraisal was based on literature review, documented case studies and expert opinion. It was not always possible to apply a consistent appraisal methodology due to the multi-faceted and non-quantifiable elements of some of the internationally agreed goals and the multi-dimensional and cross-cutting nature of the co-benefits and trade-offs of the policies. Consistency of approach was also hampered by a lack of underlying data and indicators.

Figure 1 The GEO-5 DPSIR conceptual framework

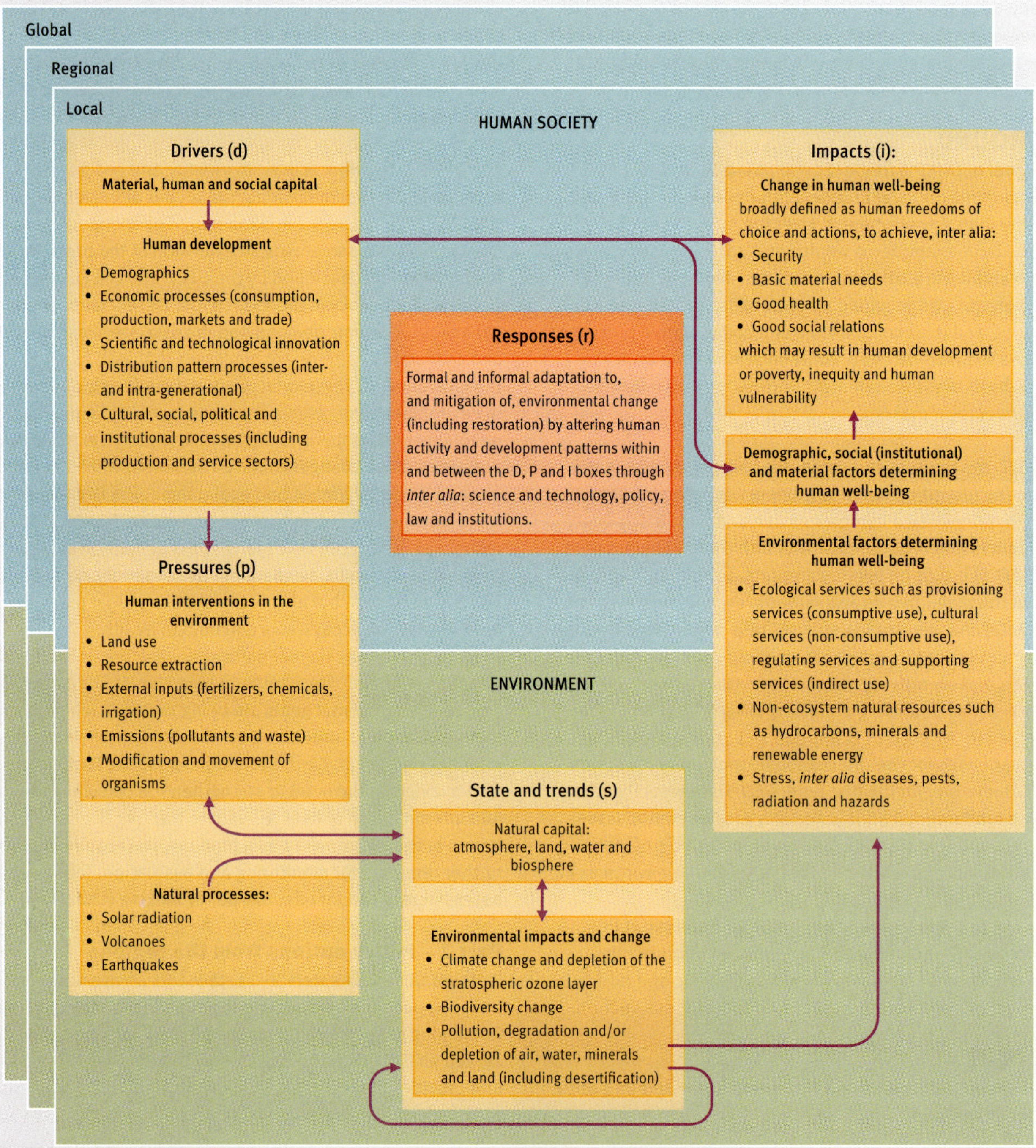

Source: 2012 United Nations Environment Programme DEWA/ GRID-Geneva

The appraisal explored the benefits of the policies and the enabling conditions that facilitated their adoption or success. Other characteristics that were analysed include the monitoring and tracking of environmental, economic or social outcomes; cross-cutting effects on other priority themes and internationally agreed goals; and the potential for their application in new contexts.

Each region identified policy responses that were effective and potentially suitable for replication and/or adoption in other countries. Some highly promising approaches featured in the regional chapters are worthy of closer analysis and possible testing by governments.

The regional summary at the end of Part 2 (Chapter 15) presents an overview of the priority environmental challenges selected by the regions; a discussion on commonalities, challenges, and opportunities; and a summary of the policy options.

Part 3 – Opportunities for a global response

The final part of *GEO-5* begins with an analysis of the type of actions required to reach a sustainable world. It first reviews existing environmental treaties and internationally agreed goals to construct a possible vision for 2050 with specific goals and targets. Next, existing scenario studies are reviewed in the context of two possible categories: conventional world scenarios that depict possible development if present trends continue and, second, global scenarios that aim to achieve a sustainable world. The analysis that follows identifies a range of measures that could enable the world to reach the sustainable development targets identified by *GEO-5*. Achieving these targets, however, requires radical departure from current trends. To account for the interactions of policies across sectors in the dense and interlinked system of global activities, an integrated sustainable world scenario is included in the analysis to examine the extent and complexity of policy changes needed to achieve the vision for 2050 (Chapter 16).

Chapters 16 and 17 review the state of knowledge of how public institutions, the private sector and civil society could generate effective and efficient responses to environmental change. While many responses at national and regional levels have successfully put societies on trajectories that are beginning to address some of these challenges, the analysis confirms that global environmental change cannot be addressed successfully by any single approach.

GEO-5 concludes by identifying action to undertake at the global level, combined with relevant national applications where appropriate, to enable the adoption of truly transformative policies – as well as the legal, institutional and policy frameworks required to make them successful. *GEO-5* will provide the reader not only with an understanding of the complexity of the threats humanity faces, but possible policy solutions and transformative pathways to a sustainable future.

The *GEO-5* process contributes to UNEP's mission of providing leadership and encouraging partnership in caring for the environment by inspiring, informing, and enabling nations and peoples to improve their quality of life without compromising that of future generations. To facilitate its development the Earth was divided into regions which largely reflect the concerns and remits of the six UNEP's Regional Offices, and allowed them to provide regional support to the working teams preparing *GEO-5*. A full breakdown of the regions, sub-regions and their respective nation states can be found on the Environmental Data Explorer (formerly the GEO Data Portal), at www.unep.org/geo/data.

Figure 2 UNEP regions

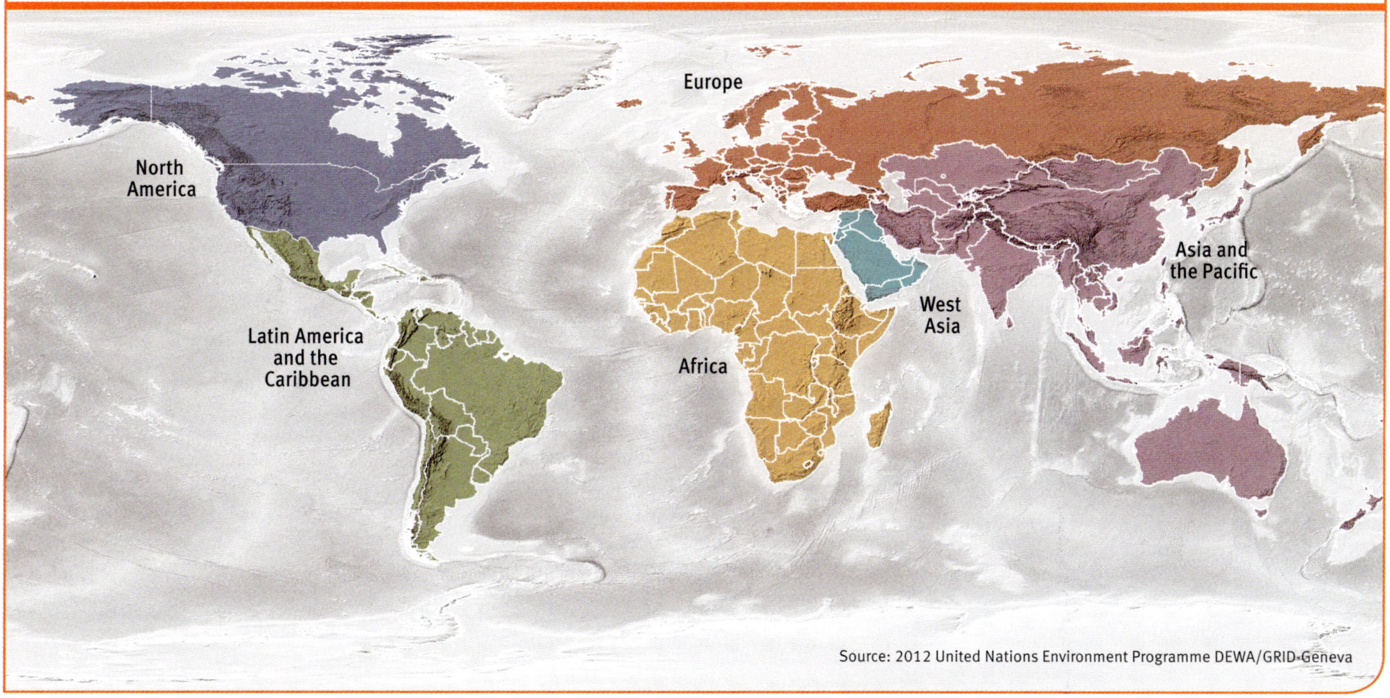

Source: 2012 United Nations Environment Programme DEWA/GRID-Geneva

Part 1: State and Trends of the Environment

"As we watch the sun go down, evening after evening, through the smog across the poisoned waters of our native Earth, we must ask ourselves seriously whether we really wish some future universal historian on another planet to say about us: 'With all their genius and with all their skill, they ran out of foresight and air and food and water and ideas'"

U Thant, UN Secretary General, addressing 7th Session of the General Assembly, New York, 1970

CHAPTER 1

Drivers

Coordinating lead authors: Marc A. Levy and Alexandra C. Morel

Lead authors: Susana B. Adamo, Jane Barr, Catherine P. McMullen, Thomas Dietz, David López-Carr and Eugene A. Rosa

Contributing authors: Alec Crawford, Elizabeth R. Desombre, Matthew Gluschankoff, Konstadinos Goulias, Jason Jabbour, Yeonjoo Kim, David Laborde Debucquet, Ana Rosa Moreno, Siwa Msangi, Matthew Paterson, Batimaa Punsalmaa, Ray Tomalty and Craig Townsend

Principal scientific reviewer: Shobhakar Dhakal

Chapter coordinator: Jason Jabbour

Main Messages

The scale, spread and rate of change of global drivers are without precedent. Burgeoning populations and growing economies are pushing environmental systems to destabilizing limits. The idea that the perturbation of a complex ecological system can trigger sudden feedbacks is not new: significant scientific research has explored thresholds and tipping points that the planetary system may face if humanity does not control carbon emissions. Understanding feedbacks from the perspective of drivers reveals that many of them interact in unpredictable ways. Generally, the rates of change in these drivers are not monitored or managed, and so it is not possible to predict or even perceive dangerous thresholds as they approach. Critically, the bulk of research has been on understanding the effects of drivers on ecosystems, not on the effects of changed ecosystems on the drivers – the feedback loop.

Patterns of globalization – links between trade, finance, technology and communication – have made it possible for trends in drivers to generate intense pressures in concentrated parts of the world very quickly. There has been a rapid rise in the production of biomass-based fuels for transport – from maize, sugar cane, palm oil and rapeseed. In the early years of the 21st century, biodiesel became more widely available, with production growing at around 60 per cent per year, reaching nearly 13 million tonnes of oil equivalent in 2009. However, recent information raises concerns about the direct environmental and social consequences of large-scale biofuel production. These complex issues include, but are not limited to, land clearance and conversion, the introduction of potentially invasive species, the overuse of water, effects on the global food market, and the purchase or leasing of land by foreign investors to produce food and biofuels, typically in developing and sometimes semi-arid countries.

Drivers typically have high inertia and path dependencies, which can act as barriers to effective action. Three-quarters of the agricultural land in the United States is dedicated to just eight commodity crops: maize, wheat, cotton, soybeans, rice, barley, oats and sorghum. This dominance is reinforced by a set of interlocking structural constraints including high levels of producer subsidies, dietary preferences, and a large industrialized food processing economy. For example, of the top 20 sources of industrial pollution in the United States, eight are slaughterhouses, but even with well-understood environmental and health problems associated with this food system, its highly entrenched nature makes it extremely difficult to modify.

Although reducing the drivers of environmental change directly may appear politically difficult, it is possible to accomplish some environmental co-benefits by targeting more expedient objectives, such as international goals on human well-being. Education is recognized as a basic human right, included in the Universal Declaration of Human Rights. Achieving universal primary education is Goal 2 of the Millennium Development Goals, and it is linked to the improvement of gender equality and women's empowerment. Together with access to reproductive health, education is a key determinant of fertility levels. Greater investment in education has been correlated with declining fertility, rising incomes and increasing longevity, and also with an educated citizenry able to express concern about environmental matters.

Surveillance and monitoring get results. Even where policy responses are not immediately possible, awareness of the importance of drivers can justify increased efforts at surveillance and monitoring. Many of the most important drivers identified in this chapter are currently not subject to systematic monitoring, their impacts even less so. The evidence, then, is compelling for the need to enhance the understanding and monitoring of drivers and their links with the environment.

INTRODUCTION

The last 100 years was characterized by exceptional growth both in the human population and in the size of the global economy, with the population quadrupling to 7 billion and global economic output, expressed as gross domestic product (GDP), increasing about 20-fold (Maddison 2009). This expansion has been accompanied by fundamental changes in the scale, intensity and character of society's relationship with the natural world (Steffen *et al.* 2007; MA 2005; McNeill 2000). In tracking and analysing these transformations, a new understanding of the complexities of the Earth's biophysical systems has been developed.

It is four decades since Lovelock (1972) introduced the idea that the Earth's systems were a complex organism. More recently, science has struggled with the realization that many Earth systems are at planetary boundaries that must not be crossed (Rockström *et al.* 2009). These concepts are useful to communicate both the dependence of human development on the environment and the urgency with which the consequences of collective human activity on the biological, physical and chemical processes of the Earth's systems need to be addressed. The impacts of human activities include alteration of the global carbon cycle by carbon dioxide (CO_2) and methane (CH_4) emissions; disruption of the nitrogen, phosphorous and sulphur cycles; interruptions in natural river flows that interfere with the water cycle; destruction of ecosystems that has led to the extinction of countless species; and drastic modification of the planet's land cover (Rockström *et al.* 2009).

FRAMEWORK

The fifth *Global Environment Outlook* (*GEO-5*) is organized using the DPSIR framework consisting of drivers, pressures, states, impacts and responses along a continuum (Stanners *et al.* 2007). Drivers refer to the overarching socio-economic forces that exert pressures on the state of the environment. While *GEO-4* identified drivers within a thematic context, *GEO-5* identifies two major drivers on the continuum – population and economic development – that influence cross-cutting dynamic patterns and generate complex systemic interactions. For example, the pressure of supplying food, feed and fibre to growing urban centres threatens biodiversity, a pressure then exacerbated by climate change.

Pressures can include resource extraction, land-use change and the modification and movement of organisms. For example, as economic growth and the demand for agricultural products rise, so does the conversion of land for agricultural purposes, as well as the use of agrochemicals. Similarly, market demands, trade and globalization patterns can lead to the inadvertent transport of invasive species that may wreak havoc on the natural ecosystems they newly inhabit.

The DPSIR framework asks three questions (Pinter *et al.* 1999):

- What is happening to the environment and why (pressure and state)?
- What is the consequence of the changed environment (impact)?
- If appropriate, what is being done about it and how effective is it (response)?

Questions regarding the role of drivers behind pressures – and the relationship between the two – can lead to persistent theoretical discussions. *GEO-5* assumes that such roles and relationships are fluid, sometimes arbitrary, a stance that should serve the purposes of this assessment.

To facilitate policy-making, this report considers leverage points to be advantageous places to intervene in the complex human interaction with the Earth System (Meadows 1999). In many cases, the most important leverage points for policy may not be the pressures themselves but the drivers. There can be substantial co-benefits, and trade-offs, associated with altering drivers in order to reduce pressure on the environment.

To effectively describe the selected drivers and for a better understanding of the pressures acting on the environment, two questions are asked that focus on why environmental changes are occurring or, more fundamentally, why there is pressure.

- What is the scale or quantity of the driver? This entails both the size of the driver and its growth rate, as well as the extent of its influence and effect on other parameters.
- What is the intensity or quality of the driver? This entails the organization of the driver as well as the various processes it exhibits and influences.

DRIVERS

Population growth and economic development are seen as ubiquitous drivers of environmental change with particular facets exerting pressure: energy, transport, urbanization and globalization. While this list may not be exhaustive, it is useful. Understanding the growth in these drivers and the connections between them will go a long way to address their collective impact and find possible solutions, thereby preserving the environmental benefits on which human societies and economies depend.

Population

Many environmental pressures are proportional to the number of people dependent on natural resources, although technological advances can mitigate individual impacts. When a population of deer, rats or sea urchins grows beyond the carrying capacity of their ecosystem, their populations crash. Sometimes the ecosystem recovers but sometimes it is permanently altered. This has been happening to human populations for millennia as they grow beyond the capacity of their valley, island or landscape to support their society, and they face famine, plague or collapse (Diamond 2005). In the last century, as human numbers grew, people came to exploit most of Earth's surface, but it is not only the scale or quantity of the population that affects the nature of a pressure on the environment. In addition, how human populations are organized – in cities or villages, in nuclear or extended families, as migrants or those that stay behind – makes a difference to the capacity of the environment to support them in their way of life.

Quantity

The human population reached 7 billion in 2011 and is expected to reach 10 billion by 2100 (UN 2011). Using the regions defined by the UN Statistics Division, the Asia and Oceania region has the largest population, Africa is the fastest-growing and most youthful region, and Europe and North America have the slowest-growing populations and the highest proportion of elderly. As of 2012, much of the current growth in global population can be attributed to momentum left from past population increases, shifts in generational composition, and communities with high fertility rates in rural areas of less developed countries and elsewhere (Bongaarts and Bulatao 1999). Population momentum explains the apparent contradiction between a growing population size and declining fertility rates. Higher fertility rates in previous decades have resulted in a large generation of youth now entering or in the reproductive age group. This increase in the reproducing population creates conditions for larger numbers of births overall, even though couples are having fewer children.

Fertility is declining in almost all countries, although rates vary broadly. At the global level, the crude birth rate fell from 37 births per thousand in 1950–1955 to 20 per thousand in 2005–2010, while total fertility, or the number of children per woman, declined from 4.9 in 1950–1955 to 2.6 in 2005–2010 (UN 2011). While the fertility decline was more accentuated in developing countries – from 6.0 to 2.7 children per woman between 1950 and 2010 – fertility levels in the countries of the less developed regions are still spread over a broad range. Among developed countries, fertility levels were already relatively low in 1950 at 2.8 children per woman, but continued to fall to 1.6 children per woman in 2010, which is less than the replacement rate of 2.1 children per woman (Box 1.1) (UN 2011). Although the global growth rate peaked more than 40 years ago, some estimates suggest there will be another billion people by 2025 and a further billion before mid-century (UN 2009a).

Fertility and mortality are closely linked. Fewer pregnancies, for example, translate into a reduction in maternal mortality, which in many countries is still a leading cause of death for women of childbearing age. Further, lower infant and child mortality may lead to lower fertility rates as parents become better able to depend on their children surviving (Palloni and Rafalimanana 1999).

The epidemiological transition closely mirrors the fertility aspect of the demographic transition. In regions that are in an early demographic stage – those with high birth and death rates – death clusters around infants, whose deaths are mostly related to nutritional deficiencies, and those dying of communicable diseases such as influenza, malaria, tuberculosis and HIV/AIDS. In regions that have entered a later demographic stage – those with lower birth and death rates – infant mortality is low and deaths coalesce around the elderly and are associated with obesity and aging, with many deaths due to cancer and heart disease (Murray and Lopez 1997).

Mortality transitions remain distinct between developed and developing countries, despite improvements. Infant mortality has continued to decline and life expectancy to rise everywhere. Global average life expectancy in 1950–1955 was 47 years, while in 2005–2010 it was 65–68 for men and 70 for women (UN 2009a). There are, of course, important regional variations, particularly in terms of infant mortality in the least developed countries, young adult mortality in countries affected by the HIV epidemic, and old-age mortality in developed countries (de Sherbinin *et al.* 2007; Rindfuss and Adamo 2004). Table 1.1 shows notable disparities in mortality rates. Infant mortality rates vary from 74 deaths per 1 000 live births in Africa to 6 deaths per 1 000 in Europe and North America.

Table 1.1 Demographic data, 2011*

	Africa	Asia and Oceania	Europe	Latin America and the Caribbean	North America	World (all countries with data)
Birth rate per 1 000 population	36	18	11	18	13	20
Death rate per 1 000 population	12	7	11	6	8	8
Life expectancy	58	70	76	74	78	70
Total fertility rate per woman	4.7	2.2	1.6	2.2	1.9	2.5
Infant mortality rate per 1 000 live births	74	39	6	19	6	44
Net migration rate per 1 000 population	-1	0.04	2	-1	3	N/A
Internal migration rate 1990–2005, %	15.4	13.2	22.3	19.3	17.8	17.5
Married women aged 15–49 using contraception, all methods, %	29	64	73	74	78	61
Married women aged 15–49 using contraception, modern methods, %	25	59	60	67	73	55

* Unless otherwise stated.

Source: PRB 2011; UNDP 2009

Box 1.1 Facilitating the demographic transition through education

Population levels and growth rates are not subject to international goals and targets, although population is directly relevant to major policy areas, including the Millennium Development Goals (MDGs). The most cost-effective method of reducing population pressures is through meeting the demand for contraception: many countries formulate policy targets around meeting unmet demand while increasing demand through investing in education for girls. Given that approximately 40 per cent of pregnancies remain unintended, great potential exists to meet latent demand for contraception (Singh *et al.* 2010).

Education is recognized as a basic human right included in the Universal Declaration of Human Rights (UNDHR 1948). Achieving universal primary education is MDG 2, linked to the improvement of gender equality and women's empowerment (UN 2000). Together with access to reproductive health (MDG 5b), education is a key determinant of fertility levels. Increasing investment in education has been correlated with declining fertility, rising incomes and greater longevity (Bulled and Sosis 2010), and an educated human population is also able to express greater concern about environmental matters (White and Hunter 2009).

In developing countries, girls' education is critical not only for reducing fertility, but for the associated lower mortality rates and improvements in health (Lutz and Samir 2011). Between 1970 and 2009, more than half of the deaths prevented among children under the age of five could be attributed to increased women's education during their reproductive age (Gakidou *et al.* 2010). In addition, women have been better equipped to resist violence by gaining greater socio-economic standing through education. This empowerment has, for example, helped women avoid HIV/AIDS infection (Bhana *et al.* 2009; Vyas and Watts 2009).

Great opportunities exist for positive interventions in education. An ethical imperative and a social and economic good, universal education for girls would also empower them to make their own choices concerning starting and expanding their families. Globally, girls represent 60 per cent of the 77 million children not attending primary school (CARE 2011). To achieve the MDG of universal primary school enrolment by 2015 it is estimated that an additional US$10–30 billion per year needs to be invested on top of the approximately US$80 billion currently spent annually on primary education (Bruns *et al.* 2003; Devarajan *et al.* 2002).

Migration is another component of the demographic transition and is characterized by shifts from predominantly rural-rural migration in regions at early stages of the transition, to rural-urban and international migration in regions at later stages. The most dynamic of the three population processes, population movements produce local and global environmental consequences. Migration may have any of three direct impacts on the environment:

- rural-rural migration produces direct household impacts on natural resources, often through agricultural expansion;
- rural-urban migration and associated livelihood changes are often accompanied by changing patterns of energy use and increased meat and dairy consumption, which can intensify land pressures in productive rural areas; and
- international migration, with remittances sent home, can have a direct impact through land-use investments or an indirect impact through increased meat, dairy and material consumption.

Africa is increasingly urbanizing, although most of the population remains rural; Asia and Oceania and Latin America and the Caribbean are already largely urbanized and migration streams are increasingly international; and the United States and Europe have high internal migration associated with labour mobility (UNDESA 2011; Zaiceva and Zimmerman 2008).

The sending and receiving areas of rural-urban and international migration remain connected through remittances, with specific characteristics varying considerably across regions. The potential remittance-driven impact on land use change is significant, while remittance-driven consumption may be similar in scale but more diffuse in its environmental impacts (World Bank 2011b).

Figure 1.1 The demographic transition

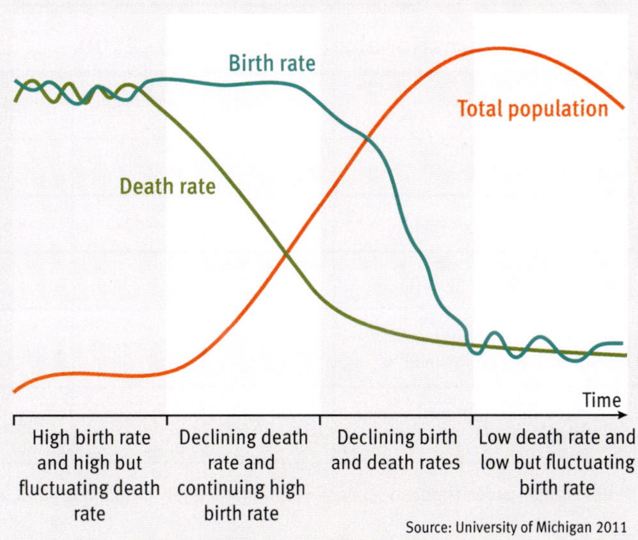

Source: University of Michigan 2011

Figure 1.2 Urban population, 1950–2050

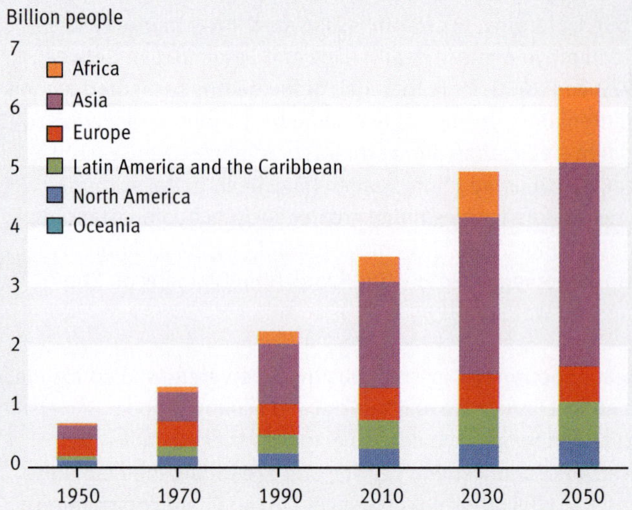

Source: UN 2009b

Internal migration is increasingly dominated by rural-urban flows, a trend that is expected to continue (Sommers 2010; Rindfuss and Adamo 2004; Cohen and Small 1998). However, in some developing countries, a minority of rural-rural migrants has a disproportionate impact on tropical deforestation (Carr 2009; Lambin *et al.* 2003). Increasing migration to coastal areas and small islands can affect the environmental integrity of coastal wetlands and associated fisheries (Rindfuss and Adamo 2004).

World population is unevenly distributed, with densities in 2010 varying from 21 000 people per km^2 in Macao to 0.03 per km^2 in Greenland. This is due to a number of factors including settlement history, regional variations in demographic dynamics such as fertility, mortality and migration, and the fact that some locations are simply less suitable for human occupation (Adamo and de Sherbinin 2011). Population is particularly concentrated at lower elevations and near coasts. An estimate from 1998 suggested that a zone below an altitude of 100 metres, comprising 15 per cent of all inhabited land, houses about 30 per cent of the human population (Cohen and Small 1998). Low-elevation coastal zones are even more concentrated, representing about 2 per cent of total land area but housing 13 per cent of the population, and growing rapidly (McGranahan *et al.* 2007).

In 1950 only 29 per cent of the world population lived in urban settings and only New York and Tokyo, with their populations of more than 10 million people, qualified as megacities. The urban proportion reached 50 per cent in 2010 with 20 megacities, with the bulk of the urban population in Asia and Latin America (Figure 1.2). Urban growth rates are high in both Asia and Africa (Satterthwaite *et al.* 2010), with the highest rates in recent decades in middle-sized cities (Montgomery 2008).

Quality

Beyond the size and growth rates of populations, the way people settle and the way they consume can result in effects on different ecosystems.

While all of the world's net population growth by 2050 is projected to occur in the world's poorest cities (UN 2009b), virtually all land-cover change will take place in rural environments. The greatest human imprint on the Earth's surface has been the conversion of forest to agriculture. Currently, 37.4 per cent of the planet's land surface is used for agricultural production (Foley *et al.* 2011).

Located on only 0.5 per cent of the global terrestrial surface (Schneider *et al.* 2009), urban areas' demand for food is disproportionately large in terms of world land use. At the same time, forest loss is no longer correlated to rural

Table 1.2 International migration, 1950–2100

	1970–1975	1975–1980	1980–1985	1985–1990	1990–1995	1995–2000	2000–2005	2005–2010
More developed regions	6 122	6 076	5 643	7 433	11 895	13 821	17 450	16 558
Less developed regions	-6 122	-6 076	-5 643	-7 433	-11 895	-13 821	-17 450	-16 558
Least developed countries	-4 872	-4 301	-5 735	-3 562	2 563	-3 061	-3 351	-5 559
Less developed regions, excluding least developed countries	-1 250	-1 775	92	-3 871	-14 458	-10 760	-14 099	-10 999
Less developed regions, excluding China	-5 043	-6 210	-5 438	-7 194	-11 068	-13 535	-15 316	-15 107

Note: Figures are in thousands. Positive numbers imply net immigration, negative ones net emigration.

Source: UN 2011

Figure 1.3 Change in population density, 1990–2005

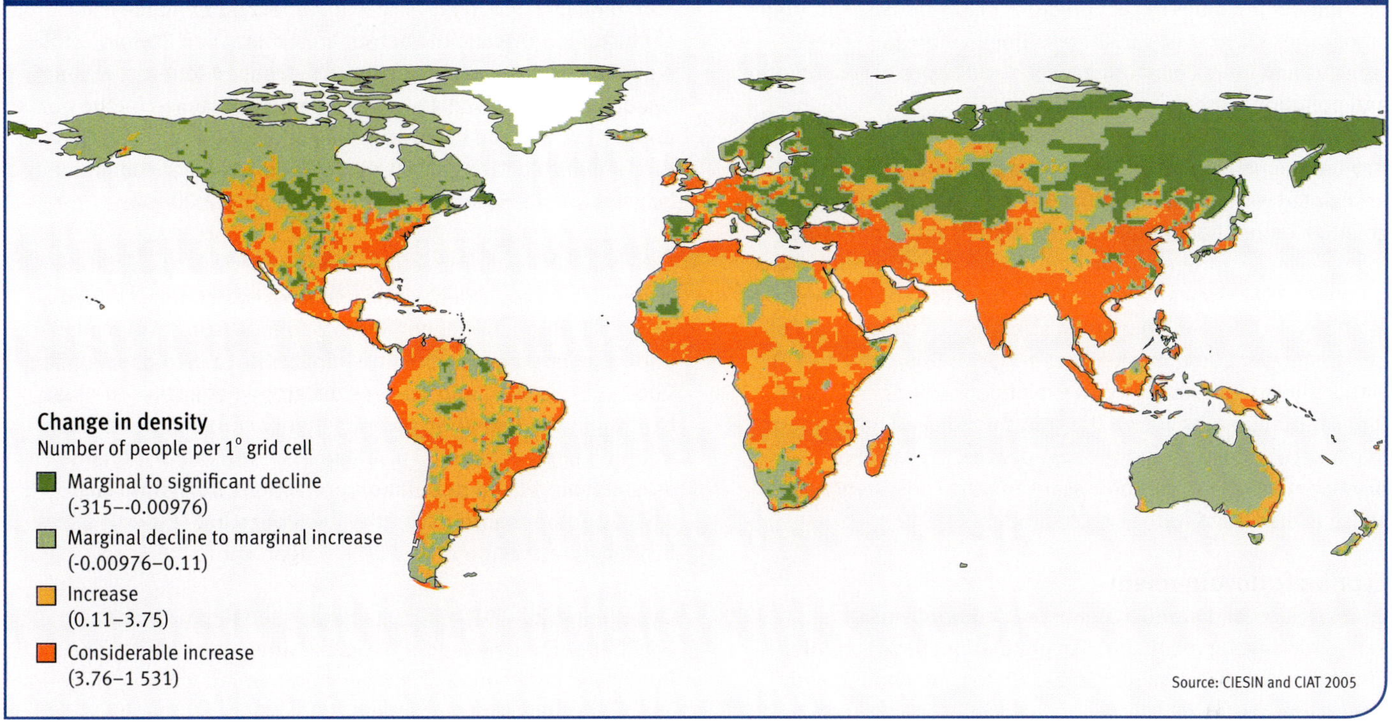

population growth; rather at the national scale, it is linked to the international demand for agricultural products and timber harvesting for urban consumption (DeFries *et al.* 2010).

The world is nearly evenly divided between rural and urban inhabitants. One half includes rural food producers with a direct impact on land in space and time. Their effect on forests is particularly acute and widespread following rural-rural migration and the associated conversion of forests to agricultural land. This very small minority of all migrants is responsible for a significant proportion of tropical deforestation yet remains very little researched (Carr 2009). From a drivers perspective, it is also much more difficult to manage this phenomenon due to the scale and diffuse nature of the activity. The second type is the burgeoning urban population who are concentrated in space but whose impacts on the land are indirect albeit significant.

A rising human population has also been identified as the principal root cause of the water crisis (UNEP 2006). Overall, humans use more than a quarter of terrestrial evapotranspiration for growing crops and more than half of accessible water run-off (Postel *et al.* 1996). While climate change is making some places wetter (Clark and Aide 2011), much of Africa and the Middle East currently suffer a water scarcity that is worsening with the expanding populations (Sowers *et al.* 2010). Population growth has also been implicated in water scarcity in rapidly developing countries such as China, where urban growth has exacerbated a decline in the availability of clean water by overwhelming the water supply and sanitation infrastructure (Jiang 2009).

Population is not the only problem: groundwater use is highly inequitable, for example in India where 10 per cent of large farms consume 90 per cent of groundwater (Aguilar 2011; Kumar *et al.* 1998). Nor is a thirsty populace the only outcome. In the Republic of Tanzania, a diverse series of drivers, including population growth, has led to water conflicts (Mbonile 2005). Water scarcity can also provoke migration, as documented throughout Africa (Mwang'ombe *et al.* 2011; Grote and Warner 2010; Mbonile 2005).

Addressing population as a driver of global environmental change, households can be considered as units for analysing consumption patterns (Jiang and Hardee 2009; UNFPA 2008; Liu *et al.* 2003; MacKellar *et al.* 1995). In the developed world, household size is shrinking as their composition changes from extended families to nuclear ones (Bongaarts 2001). As a consequence, the rise in the number of households has been faster than population growth (Liu *et al.* 2003). Research suggests that this can cause double the rise in energy consumption that would occur from population growth alone (MacKellar *et al.* 1995), as there is an increase in the number of appliances and the level of electricity consumed per person (Zhou *et al.* 2011). Larger households generally use less energy per person than small ones, conforming to the expectations of economies of scale (O'Neill *et al.* 2001; Ironmonger *et al.* 1995). The age composition of a household also has an impact on energy consumption, Lenzen *et al.* (2006), working with data from Australia, Brazil, Denmark, India and Japan, found that the residents' average age is positively related with per-person energy consumption, while household size and urban location are negatively associated. Transport, too, is likely to be more sensitive to the number of households, since an increase in the number of homes occurs primarily in low-density suburban landscapes (Seto *et al.* 2010), resulting in more passenger vehicles and more commuting, which add to petrol consumption and pollution.

Beyond the household unit, studies also identify impacts associated with absolute population size. A study of Californian counties found that population size significantly contributes to increases in nitrogen oxide and carbon monoxide emissions (Cramer 1998). Similarly, researchers have observed a positive relationship between population size and CO_2 emissions (Cole and Neumayer 2004; Mackellar *et al.* 1995; Bongaarts 1992), with an inverted U-shaped curve relation for sulphur dioxide (Cole and Neumayer 2004). How households and populations impact ecosystems is highly dependent on the stage of development, the geographic scale and the ecosystem itself, which is discussed further in Chapters 2–6.

Economic development

Consumption and production are both components of economic development and, like population, have a multiplier effect on environmental pressures. While consumption and production are technically separate socio-economic drivers, they are so inextricably linked that it is difficult to discuss them independently: the consumption of raw materials by the primary industries of mining and forestry leads to the manufacture of products that are in turn consumed by individual customers.

Quantity

The production of goods for consumption requires materials – minerals, water, food, fibre – and energy. During the 20th century, global economic output grew more than 20-fold, while materials extraction grew to almost 60 billion tonnes per year (Maddison 2009). This level of materials consumed by the human population is of the same scale as major global material flows in ecosystems, such as the amount of biomass produced annually by green plants (Krausmann *et al.* 2009; UNEP 2009b).

Consumption and production trends appear to have stabilized in developed countries, while in emerging economies such as Brazil, China, India, and Mexico, per-person resource use and associated environmental impacts have increased since 2000 (SERI 2008), and the less developed countries are just beginning the transition towards higher consumption levels. Should global economic development continue in a business-as-usual mode and population projections persist through 2050, another sharp rise in the level of global resource use is likely (Krausmann *et al.* 2009; SERI 2008).

Over the period 1970–2010, average global growth rates in GDP per person measured in purchasing power parity (ppp) fluctuated between -2 and 5 per cent annually; the average was about 3.1 per cent (World Bank 2011a). Since 2001, however, China has grown at 10 per cent per year, a seven-year doubling time, and India at 8 per cent per year, a nine-year doubling time, with environmental pressures increasing at much the same pace. As a result, China is now the world's largest emitter of

Figure 1.4 Change in economic output, 1990–2005

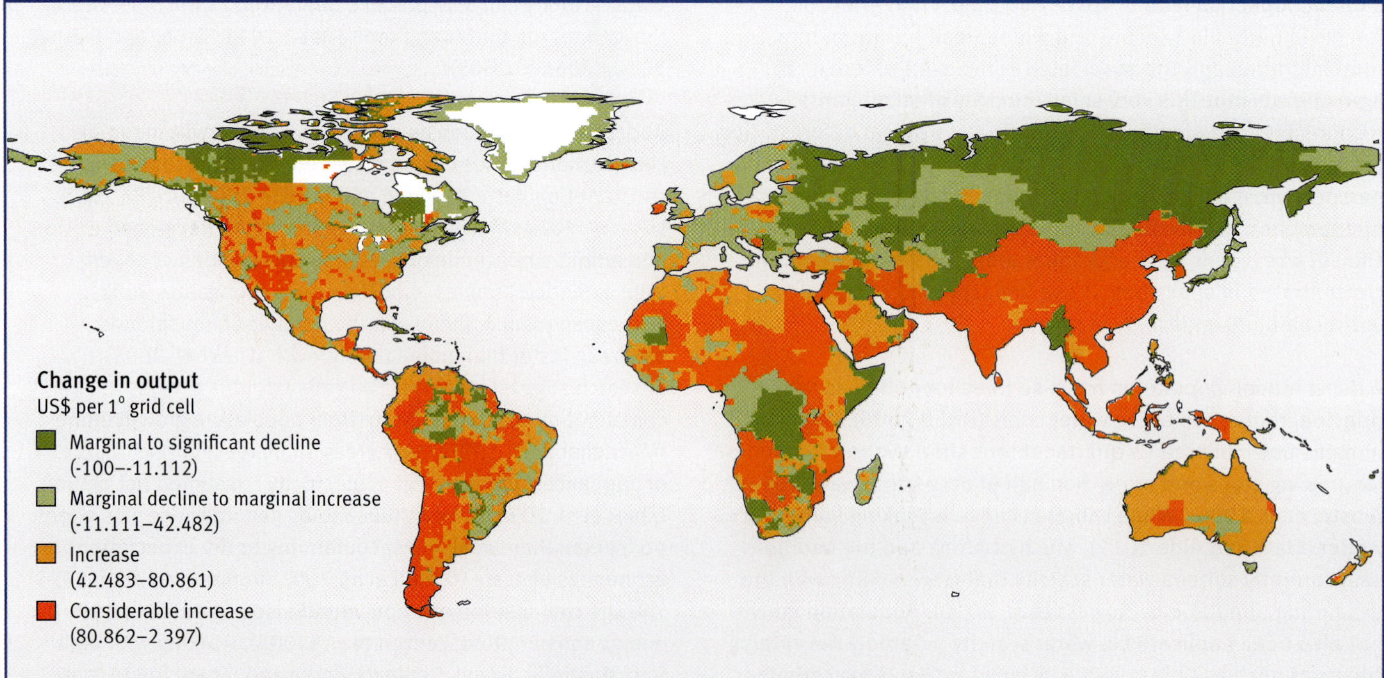

Note: The change in economic output aggregated across all cells within a country's borders equates to the change in GDP.

Source: Nordhaus *et al.* 2008

Box 1.2 Expressing prosperity beyond GDP

Within the traditional accounting framework for benchmarking economic performance, a considerable amount of nature's capital and services is externalized (excluded), thereby ignoring key environmental pressures and the forces driving them. Including those pressures requires alternative metrics to GDP and related benchmarks. Such alternatives can be measured in either monetary or physical units.

An alternative monetary approach seeks to maintain the traditional accounting framework and its reliance on market transactions, but augments it by internalizing (including) environmental costs and pressures. A common approach for accomplishing this is to assign market values to nature's assets and services with the goal of taking full account of both market and non-market costs and benefits (Abraham and Mackie 2005; NRC 2004, 1994; Nordhaus and Kokkelenberg 1999), a procedure that was first attempted by Costanza *et al.* in 1997.

An alternative physical approach, stemming from the industrial metabolism or industrial ecology tradition, seeks to identify the rates and volumes of material flows through the economy. A system such as material flow accounting (MFA) is presumed to reveal more accurately the pressures on resources and the undesirable impacts on the environment from any part of the life cycle of resources – from extraction through combustion or conversion into a usable commodity and consumer consumption, to recycling, disposal or stewardship.

Two leading indicators are used to chart trends in global, national and urban material flows:

- total material extraction per unit of GDP; and
- metabolic rates – the amount of resource use per person.

During the 20th century, total material extraction increased from 7 billion tonnes to almost 60 billion, while GDP increased by a factor of 24 (Krausmann *et al.* 2009). Over the same period per-person resource use doubled from 4.6 tonnes to around 9 tonnes, while per person income increased by a factor of seven (UNEP 2011a; Krausmann *et al.* 2009). At the same time, resource prices were declining or stagnant. Taken together, these data indicate that resource decoupling or dematerialization, both in the aggregate and on a per-person basis, took place during the 20th century. Since there were no overarching policies specifically devoted to decoupling during the period, it appears that it took place spontaneously, perhaps due to forces within the global economic system. However, there is a clear need for further research to identify the responsible factors.

A more serious challenge – due to limitations in the available data – is determining whether material use is increasing or decreasing on a country-by-country basis. In a production-based system of accounts, environmental pressure is allocated to the country where the pressure occurs, while a consumption-based system allocates the pressure to the country where a product is finally consumed.

Furthermore, trade accounts only measure the weight of traded commodities entering a country, ignoring hidden or indirect flows – materials that are extracted or moved but are not traded directly. Finally, industrialized countries tend to be material importers while developing countries tend to be exporters. Due to these data limitations and patterns, the resource intensity of the advanced countries may be grossly understated because their high resource use is actually happening in exporting countries (Caldeira and Davis 2011).

These data limitations may account for the finding that, with the same standard of living, more densely populated areas and regions consume fewer resources per person than do less densely populated ones (Lenzen *et al.* 2006; Larivière and Lafrance 1999; Kenworthy and Laube 1996). The difference is even more pronounced when comparing industrialized high-density areas with low-density ones. Since high-density areas are nearly equivalent to urbanization, these areas – not the hinterland – are the hub of international trade where goods and services are received, while the resource intensity and environmental impacts are felt elsewhere as resource extraction typically takes place in areas of low population (Rosa and Dietz 2009).

greenhouse gases per year and, since 2010, its economy is second in size only to the United States (World Bank 2011a).

Much of China's economic growth has come from its expansion in manufacturing, both for domestic markets and for export. By comparison, the average growth rate is negative for sub-Saharan Africa and less than 1 per cent for the Middle East and North Africa, although Figure 1.4 shows considerable variation across these regions. In addition, since 1995, Russia's annual growth rate has fluctuated between -7.8 per cent and 10.0 per cent, with an average of 3.3 per cent (World Bank 2011c).

It is difficult to project economic growth: during the 1980s and 1990s the Republic of Korea experienced growth spurts at rates similar to China's and India's recent ones, before slowing to more moderate rates (World Bank 2011b). Using the concept of an ecological footprint, which aggregates all environmental pressures into a measure of hypothetical land required to meet current rates of resource use (Wackernagel *et al.* 2002, 1999), China and India are expected to appropriate 37 per cent of the projected increase in global footprint over the period 2001–2015 unless they are able to improve their production efficiency annually by 2.9 and 2.2 per cent, respectively (Dietz *et al.* 2007).

Whether these growth rates are realistic when put in the context of the Earth System's biophysical boundaries remains to be seen (Chapter 7) (Rockström *et al.* 2009).

Quality

Technology is a key factor in the production of goods and services and an important one in terms of environmental impact. It has been argued that over time, factors of intensity or quality, affected by technological innovation, may more than compensate for the adverse effects of the rise in population, so that economic growth eventually leads to environmental improvements. An example of this is greenhouse gas emission rates in developed countries since 1970, where, it is claimed, emissions increased more slowly than economic activity because of shifts towards technologies that have a lower environmental impact (Bruvoll and Medin 2003; Hamilton and Turton 2002). However, it is not certain whether other sectors were so successful – efforts to reduce deforestation at the national level might have shown domestic improvement, but demand may have driven increased deforestation in other countries (Meyfroidt and Lambin 2009).

Figure 1.5 A simple interpretation of the environmental Kuznets curve

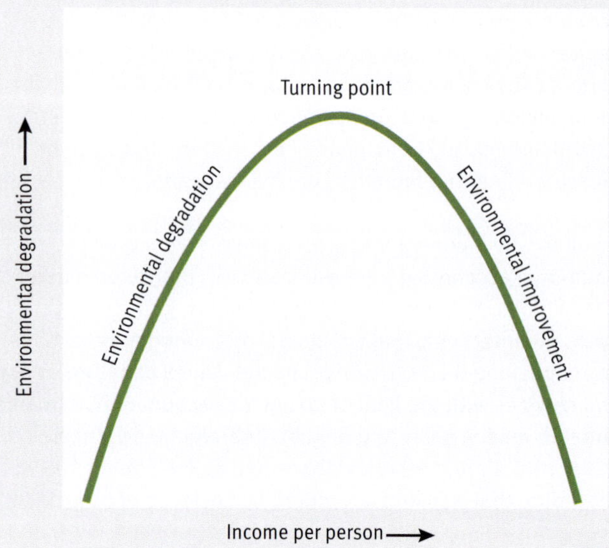

The environmental Kuznets curve (Figure 1.5) (Grossman and Krueger 1995) suggested that as countries become more affluent, concern about the environment increases, leading to policies that protect it. At the same time, preferences shift away from the most environmentally damaging goods and services.

This theory has been extensively examined (Carson 2010; Mol 2010; York *et al.* 2010; Aslanidis and Iranzo 2009; Galeotti *et al.* 2009; Jalil and Mahmud 2009; Lee *et al.* 2009; Roberts and Grimes 1997) and while debate continues, there seems to be clear evidence that some companies and industrial sectors have reduced their environmental impact, as the theory predicts. However, there are many obstacles to a shift towards more environmentally benign technologies: in some cases, these are economic challenges as environmentally sound technologies often have higher overall costs. But in many cases, simple cost/benefit calculations are not sufficient to explain the slow pace of growth in new technologies. For example, although researchers have noted the energy efficiency gap for years (Jaffe and Stavins 1994), whereby economically beneficial investments in energy efficiency have not been made, neither consumers nor industry have made significant investments in closing that gap despite the potentially favourable returns in energy costs saved, particularly when life-cycle costing is applied.

On the other hand, technological change that improves resource efficiency can have a perverse environmental effect by decreasing the costs of resource use and thus increasing demand. If the increased demand is greater than the efficiency gains, the overall consumption of a resource can actually increase, with concomitant increases in environmental impact. This phenomenon is known as the Jevons paradox or the rebound effect (Polimeni and Polimeni 2006; York 2006). The choice of technology, which is shaped by economic factors and individual and public decisions, is critical in determining the overall human impact on the environment. Research to explain the obstacles to adopting more environmentally benign, cost-effective technology is just beginning. One key factor, at least for households, is unfamiliarity with life-cycle costing and a lack of understanding of the energy and cost impacts of commonly used technologies (Attari *et al.* 2010; Carrico *et al.* 2009), and it appears that the same factors may also affect organizational decision making.

Values

It is commonplace to identify values as a key driver of environmental change. At one level, the argument is straightforward: human decisions, especially about consumption, are influenced by values and those decisions have impacts on the environment. However, research on human decision making notes that values are only one element in the cognitive processes, with beliefs and norms also of great importance (Stern 2011). While some decisions reflect a formal weighing of values and beliefs, many are made without much reflection, on the basis of normative expectations, emotions and interpretations of symbols or quick judgements (Kahneman 2003; Jaeger *et al.* 2001).

There is a voluminous canon of literature exploring the social psychology of environmental decision making, in which several generalizations can be discerned (Carrico *et al.* 2011; Schultz and Kaiser 2011; Stern 2011; Stern *et al.* 2010). First, no single factor is sufficient to explain such decisions. Values, beliefs and norms, and trust in others who must also take action or who are providing information, all matter. Second, decisions are often context-specific in the sense that individuals read the context, such as whether to emphasize a gain or a loss, and frame the decision based on that reading. Sometimes individuals act as consumers, sometimes as members of a community, sometimes as citizens. Third, social networks are of immense importance in providing context as well as shaping values, beliefs, norms, trust and other significant factors (Henry 2009; Jackson and Yariv 2007). Fourth, values, beliefs, norms, trust and other individual characteristics interact with the character of the action to be taken in shaping behaviour – for example, social psychological factors may matter little when a pro-environmental action is exceptionally easy or hard to undertake, but may be critical for actions of intermediate difficulty (Guagnano *et al.* 1995).

Social psychology has developed many concepts to explain the factors underlying environmental decision making. Among these, values have been explored the most thoroughly and tested empirically across many national contexts (Dietz *et al.* 2005). In particular, altruism towards other humans, other species and the biosphere has consistently been found to predict pro-environmental attitudes and behaviour. In addition, a willingness to cooperate with others in experimental games, conducted in both laboratory and field settings, varies considerably across individuals and cultures (Henrich *et al.* 2010, 2005). Recently, the propensity to cooperate has been shown to matter in managing forest commons (Rustagi *et al.* 2010; Vollan and Ostrom 2010), with a substantial amount of literature showing the importance of trust in commons dilemmas (Fehr 2009). However, research on trust has not yet been linked to the larger literature on values. Consumer surveys have revealed a range of reasons why an individual is unwilling to pay more for an environmentally sensitive product (WBCSD 2010). The top three reasons involve poor understanding of, or apathy towards, the negative environmental impacts of consumption decisions, while the fourth most common was whether the individual viewed an action as common practice among their peers. This last point reveals the importance of societal pressure on values and by extension how decisions that impact the environment are influenced by it.

Diets

With economic growth comes a change in dietary intensity, which Popkin (2002) describes as the nutrition transition. This happens in three states: decreased occurrence of famine with rising incomes; the emergence of chronic diet-related diseases due to changes in activity and food consumption patterns; and a stage of behavioural change where diet and activity levels are better managed for prolonged healthier lives.

The growth in food consumption and related requirements for animal feed largely determine the pace at which supplies need to grow to keep up with the domestic and export demand for agricultural goods. Urbanization, demographic change and household wealth in a number of fast-evolving regions – Brazil,

Figure 1.6 Change in meat supply by region, 1960–2007

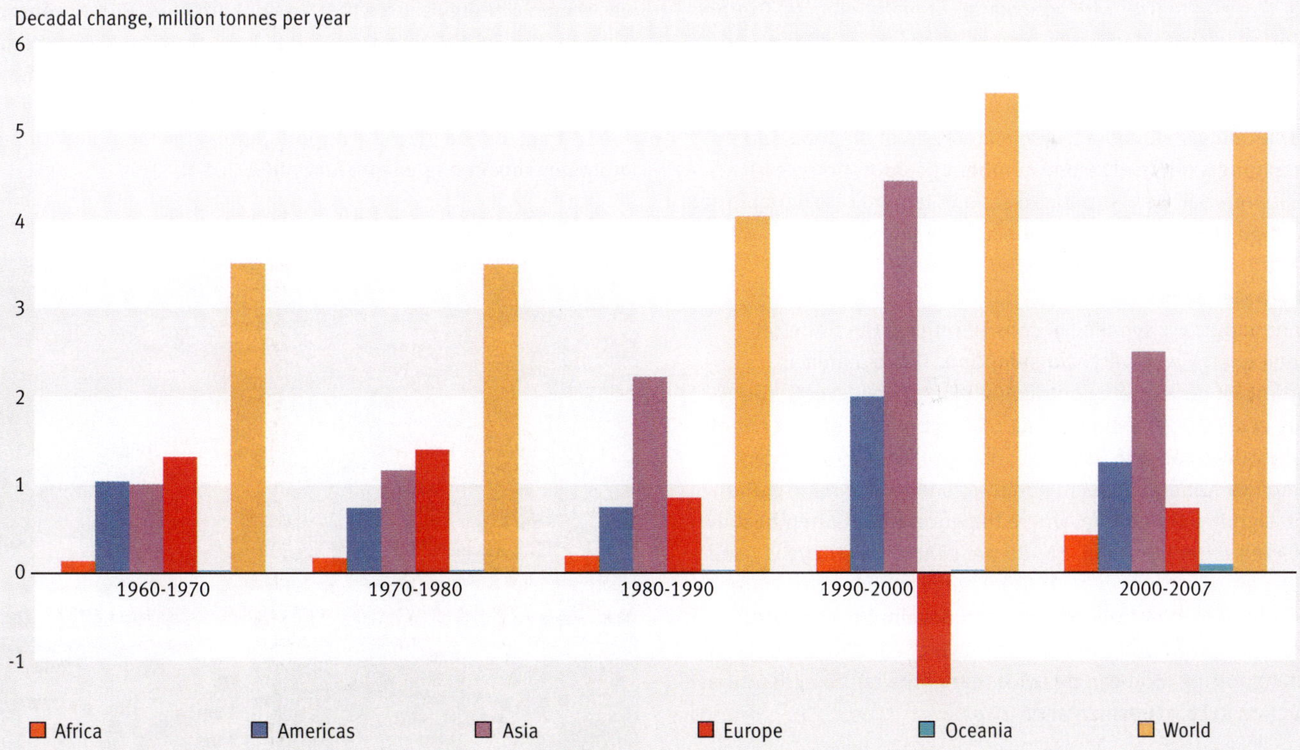

Europe exhibited a significant decline in its meat supply between 1990 and 1999 due to the bovine spongiform encephalopathy (BSE) crisis of the early 1990s.

Source: FAOSTAT 2010

China, India and Indonesia – suggest that changes in food consumption patterns are likely to have profound effects on regional food systems (Satterthwaite *et al.* 2010). These changes in consumption and consumption preferences introduce increased pressures on food and energy systems from the demand side, which forces compensating adjustments to take place on the supply side through market-mediated, price-driven interactions with producers.

As regional economies continue to grow, so, too, does the consumption and production of of meat (Figure 1.6). Livestock production is the largest anthropogenic land use, accounting for 30 per cent of the land surface of the globe and 70 per cent of all agricultural land; 33 per cent of total arable land is used for producing animal feed (Steinfeld *et al.* 2006). Pelletier and Tyedmers (2010) suggest that, by 2050, the livestock sector alone may occupy the majority of, or significantly overshoot, recent estimates of humanity's biophysical limits within three environmental areas: climate change, reactive nitrogen mobilization, and appropriation of plant biomass at planetary scales.

As urban areas are generally wealthier than rural ones, there are considerable differences in dietary composition, with urban diets characterized by higher levels of meat, dairy and vegetable oil. These foods are often imported and require more energy-intensive production (de Haen *et al.* 2003; Popkin 2001). Globalization and urbanization are cited as causing dietary *convergence* and *adaptation*. The former refers to the focusing of caloric intake on a smaller number of staple crops, such as wheat, rice and maize, with concomitant health impacts. Dietary adaptation is characterized by a greater reliance on processed foods due to lifestyle changes, greater exposure to advertising and time constraints on food preparation. This concentration of consumption also favours the concentration of the food supply chain among a relatively small number of corporations, with an implicit preference for supermarkets and larger-scale agricultural production (Kennedy *et al.* 2005).

Energy-water nexus

Another important dynamic of consumption is the trade-off between energy and water consumption. This dynamic is important for both energy production and agriculture. Gerbens-Leenes *et al.* (2009) estimate that 60–80 per cent of water used globally is dedicated to irrigation, rising to nearly 90 per cent in some low-rainfall areas. In addition, energy use for irrigation can be significant. In India, where the government often heavily subsidizes water pumping, 15–20 per cent of electricity is used for this purpose (Shah *et al.* 2004). Energy use for agriculture is considerable in both developed and developing countries, although in developed countries the energy used for processing and transporting food can be twice that of the entire agricultural production sector (Bazilian *et al.* 2011).

Water can also be an important resource for energy production and mineral extraction. However, freshwater pollution is a common side effect of mining, including recent hydraulic fracturing activities (Scott *et al.* 2011). China suffers from water scarcity due to a dwindling supply as well as to industrial pollution; the World Bank (2006) estimates that up to a third of water scarcity in China is due to pollution, the cost of which is equivalent to 1–3 per cent of GDP.

THE DRIVER-PRESSURE CONTINUUM

As population and economic development have continued to grow despite depressions and downturns, technological innovations have enhanced the integration of communities and societies into a global civilization. Technological advances in energy and transport continually generate new opportunities for growth in production and consumption, while ingenuity applied to communication and mobility has created new goods and services that previous generations could not have imagined. The growth and integration of human settlements, societies and relationships is evidenced by rapid urbanization and globalization.

Energy

Quantity

As the world population increases, more people aspire to higher material living standards – creating an ever greater demand for goods and services as well as for the energy required to provide these. From 1992 until 2008, per-person energy consumption increased at a rate of 5 per cent annually. In 2009 total global energy use decreased for the first time in 30 years – by 2.2 per cent – as a result of the financial and economic crisis (Enerdata 2011); half of this occurred in the OECD countries (IEA 2011). Oil, natural gas and nuclear power consumption all decreased while hydroelectric and renewable energy consumption increased. Coal was the only energy source that was not affected. Primary energy consumption in 2010 is estimated to have risen by 4.7 per cent worldwide, easily surpassing the minor reduction in 2009. The rate of growth in the future, however, is expected to decrease due to an assumed levelling of population growth and continued improvements in energy efficiency (IEA 2011).

By 2030, more than 55 per cent of the population of Asia will be urban.

© Klbae Park/UN Photo

The shares of energy inputs are likely to change, with the proportion produced from oil decreasing and natural gas increasing. Coal levels are expected to stay relatively constant and nuclear energy use will increase due to investments in Asia. However, with potential policy changes following the Fukushima disaster in 2011, it is difficult to predict the growth trajectory of nuclear power. If nuclear energy plans are not followed through, more coal is likely to be used, with significant implications for climate change mitigation efforts (IEA 2011). Developing regions show a particularly strong increase in per-person energy consumption between 2005 and 2010, although, as of 2010, this seems to be levelling off. The three major economic sectors in terms of energy consumption (IEA 2011) are:

- manufacturing: 33 per cent;
- households: 29 per cent;
- transport: 26 per cent.

Electricity and heat generation account for more than 40 per cent of all CO_2 emissions (IEA 2010). Between 1992 and 2008, the annual rise in CO_2 emissions of more than 3 per cent and the total rise of 66 per cent – a much greater increase than that of the global population – was primarily the result of growth in industrial production, as well as higher living standards in many developing countries.

Emissions from a coal-fired power plant rise into the atmosphere.

© Sasha Radosavljevic/iStock

On a per-person basis, the largest growth in electricity production occurred in the developed countries, increasing from 8.3 megawatt hours (MWh) in 1992 to nearly 10 MWh in 2008, a difference of 1.7 MWh per person (IEA 2010), though in percentage terms this was the smallest rise at 22 per cent. The global average per-person electricity production grew by 33 per cent, from 2.2 MWh in 1992 to 3.0 MWh in 2008, while that of developing countries grew by 68 per cent, from 1 MWh to 1.7 MWh (IEA 2010).

In 2010, 1.44 billion people globally – around 20 per cent of the world population – were still suffering from energy poverty, without access to reliable electricity or the power grid, and entirely dependent on biomass for cooking and lighting (UNEP 2011b).

The energy commodity that dominates trade volume and value is crude oil, with China continuing to rival the United States in terms of consumption (EIA 2010). The Middle East accounts for about half of all global oil trade (IEA 2008). Coal production increased by 3–5 per cent per year during 2005–2009, with China experiencing a 16 per cent increase in production during 2008–2009 and reaching 44 per cent of the world's total coal production of 3.05 billion tonnes. With rapidly increasing energy demand, however, China became a net importer of coal for the first time in 2007 (Kahrl and Roland-Holst 2008). The United States is the second largest producer of coal at 975 million tonnes per year, followed by India producing 566 million tonnes.

Quality

Renewable energy production is gaining much attention: the amount of energy produced from renewable sources, including sun, wind, water and wood, amounted to 13 per cent of the world supply in 2008, and estimates suggest 16 per cent in 2010 (REN21 2011). However, the largest renewable source is biomass at 10 per cent, with nearly two-thirds of that used in cooking and heating in developing countries (IPCC 2011). Thus, when biomass is excluded, other renewable sources provide only about 3 per cent of world energy.

There has been a 30 000 per cent rise in solar energy supply since 1992, a 6 000 per cent increase in wind energy and a 3 500 per cent rise in biofuel production, all from very low bases. This is mainly due to the decreasing cost of these technologies and the 2010 adoption by 199 countries of policies to promote renewable energy (REN21 2011).

There has been a rapid rise in the production of biomass-based fuels for transport – from maize, sugar cane, oil palm and rapeseed. While ethanol has been widely used in Brazil for two decades, its use accelerated globally at the end of the 1990s, increasing by 20 per cent each year to reach 30 million tonnes of oil equivalent in 2009. In the early years of the 21st century, biodiesel became available, with production growing at around 60 per cent per year, reaching nearly 13 million tonnes of oil equivalent in 2009. However, recent information on biofuel production raises concerns about the direct environmental and social impacts of land clearance and conversion, the introduction of potentially invasive species, the overuse of water and the consequences for the global food market. An additional cause for concern is the purchase or leasing of land by wealthier nations to produce food and biofuels – typically in developing and sometimes semi-arid countries. This trend may have serious impacts on fossil and renewable water resources, as well as on local food security (UNEP 2009a).

Investment in greening the energy sector is setting new records, totalling US$211 billion in 2010, up 32 per cent from 2009, and nearly five and a half times the 2004 figure. For the first time, new investment in utility-scale renewable energy projects in developing countries surpassed that of developed economies (UNEP 2011c).

The number of nuclear power plants, seen by some as an opportunity to meet the growing demand for energy, has increased by more than 20 per cent since 1992, rising to 435 by mid-2012. According to the International Atomic Energy Agency (IAEA 2008), in the 30 countries that have nuclear power, the share of electricity generated ranges from 78 per cent in France to 2 per cent in China, which has 14 operational plants, 25 under construction and more planned (WNA 2011a). Since 1992, energy production from nuclear power sources has grown by almost 30 per cent, although the share of nuclear power in the total supply has fallen from 17.5 per cent in 1992 to 13.5 per cent in 2008. Today, around the world, 60 plants are under construction, 155 planned and 339 proposed (WNA 2011b).

Global energy consumption is expected to continue to grow. Though China's energy intensity decreased by 66 per cent between 1980 and 2002 (IEA 2008; Polimeni and Polimeni 2006), India's energy use per unit of GDP remained relatively constant over the same period and, due to its growing economy, the country is expected to contribute 8 per cent of the world's projected growth in emissions by 2030 (World Bank 2008). If the international community continues to have difficulty in addressing climate change in the near future, temperatures could increase by 3.5–6°C by the end of the century (IEA 2011). To stem the rise in global GHG emissions, the Kyoto Protocol encouraged the transfer of cleaner technologies from developed to developing economies. Trade was assumed to be the means of distributing these technologies, but without a significant reduction in existing trade barriers, this route will have limited impact (World Bank 2008).

Serious inequities remain in meeting global demand for access to energy. Today, 1.3 billion people are lacking electricity and 2.7 billion people still rely on the traditional use of biomass for food preparation, with concomitant impacts on deforestation rates, soil erosion and human health (IEA 2011). The reliance on fuelwood also has a demographic aspect, as per-person fuelwood consumption is shown to increase with decreasing household size but to decrease with urbanization, indicating a wealth effect (Knight and Rosa 2011). In order to achieve universal access to primary energy by 2030, an annual investment of US$48 billion is needed (IEA 2011).

Transport

Quantity

Transport serves people, production and consumption and is an important facilitator of trade. The global economy is currently recovering from a severe recession, with global industrial production and trade climbing back to pre-crisis levels, albeit with marked geographic differences: GDP is growing fastest in China, by 10.3 per cent per year, and India, by 9.7 per cent, in 2010. Data published by Global Insight (2010) suggest that in the next 40 years Brazil, Russia, India and China (the BRIC countries) will start to approach the United States in terms of GDP, surpassing Germany, the United Kingdom, France and Italy, with the distinct possibility that China will have the world's highest GDP by 2050. This unequal growth has implications for world trade and the flow of goods, posing considerable challenges and opportunities in terms of logistics and supply chains.

Countries and entire regions appear to be specializing in their attempts to become competitive, creating even greater demand for transport. For instance, Europe, the United States, Canada and Japan are dependent on fruit exports from Central and South America, some Western European countries, many Eastern European countries, and portions of Africa. Similar differential production-consumption trends happen with all products, pushing the demand for transport even higher and making freight inelastic to fuel prices. An evolving trend to manage this ever increasing world trade is containerization, which by many in the industry is considered a major revolution in handling goods, using larger ships to achieve economies of scale. It is estimated that from 80 to 90 per cent of world trade is by sea (UNCTAD 2011).

In the United States, the Bureau of Transportation Statistics (BTS 2011) reports that container trade in 2005 and 2006 was double that of the previous decade, increasing to 46.3 million 20-foot-equivalent units (TEUs, 19–43 cubic metres). At the global scale, container trade tripled during the same period. The European Union (EU), the world's largest trading bloc, carries out 90 per cent of its external trade and 40 per cent of its internal trade by sea, totalling 3.5 billion tonnes (Reynaud 2009; Goulias 2008). However, studies in major ports show that any environmental benefits of seafaring cargo require significant attention at the place of loading and unloading. The Port of Los Angeles in California, a major hub, has, for example, implemented a variety of policies including the introduction of cleaner trucks with refuelling stations for natural gas, performance standards for cargo handlers and harbour craft, modernized and cleaner rail locomotives, and reduced vessel speeds (Port of Los Angeles 2010).

After a slump in 2008 and 2009, air freight began to return to its pre-economic-crisis levels, with annual international growth of 21 per cent in 2010, although 2011 growth is expected to be heavily dependent on consumer spending (IATA 2011). Data from the International Transport Forum (ITF) show some recovery for rail freight but it is still suffering from the economic crisis with unknown implications for the long term; exceptionally, India continues to increase its rail freight. Similarly, recovery of road freight is very slow at the national and international levels for many OECD and ITF countries.

For passenger travel China, India and Brazil recorded 7.1 per cent growth in 2010 relative to 2009. According to the International Air Transport Association, there were 2.4 billion domestic and international passengers in 2010, approximately 6.4 per cent more than ever before, with a similar trend observed in passenger

In 2011, the Beijing rapid transit subway system delivered over 2.18 billion rides. *© Niclas Mäkelä*

kilometres travelled. Rail passenger travel continued to decline, providing space for possible substitution by freight. Data on passenger kilometres travelled in private cars suffers poor harmonization, yet it is clear that the economic crisis reduced overall travel. Moreover, possible saturation of passenger travel by car is observed in developed economies that exhibit non-significant increases in passenger kilometres, hovering at around one digit percentage growth per year.

Quality

While transport enables human interactions that contribute to development, the infrastructure for fast, motorized means of travel also creates displacement and barriers that can divide communities and reduce well-being. Roads and the enormous amount of parking to store the world's 1 billion cars are the commonest barriers, but airports and seaports for container ships are also significant.

In societies with extremely high levels of mobility, inequities in the social distribution of related environmental pressures and benefits are of increasing concern (Adams 1999). Because most human settlements are located close to supplies of water and agricultural land, transport infrastructure displaces food production while also fragmenting landscapes that are then less able to support wildlife (Huijser *et al.* 2008). Transport also has secondary environmental impacts through expanded human access to land, as the infrastructure promotes economic activities such as mining, forestry or power generation in new locations. In addition, transport enables more extensive permanent human settlement, particularly suburban and urban growth.

Most energy for transport comes from fossil fuels, and the rise of the car has produced various specific environmental impacts, from urban health problems through land and water degradation to contributing to climate change. Many people are optimistic about the long-term prospects for shifting to cars powered by fuel cells and electric motors, but a near-term change will be difficult, and the car is noticeably more intensive in its environmental impacts than its competitor technologies, exhibiting the highest levels of energy consumption and greenhouse gas emissions (Chester and Horvath 2009). Private car ownership can also impact patterns of urbanization by permitting dispersed and low-density sprawl, which in many contexts reflects individual household dissatisfaction with urban environments, but collectively degrades environmental quality. Like the transport infrastructure that makes them possible, these new or expanded built areas impinge on natural landscapes and amplify the direct environmental impacts of transport.

There may have been a temporary decrease in transport activity in, for example, the United Kingdom and United States due to the economic recession (Millard-Ball and Schipper 2011; Metz 2010). However, these declines are likely to be outweighed by increases in private vehicle ownership in rapidly developing low- and middle-income nations. At present, the number of motor vehicles in the world is growing much faster than the number of people (World Bank 2012). While it is unlikely that the levels of hypermobility reached in the United States will ever be reached in many other nations, there is still massive potential for growth in the level of travel and shifts towards individual motorized vehicles, especially as incomes increase. In developing nations including China and India, the ownership and use of highly

The worldwide motor-vehicle industry now produces more than 220 000 cars a day. © *Josemoraes/iStock*

polluting motorcycles is increasing faster than cars (Pucher *et al.* 2007). Even when more fuel-efficient vehicles are introduced, rising numbers may outweigh efficiency benefits.

However, with aggressive moves by governments and advocacy groups in the creation of green markets, two related phenomena could emerge. The first is an offset trade market in which companies can buy offsets, as futures and options, to counterbalance their inability to manage and decrease CO_2 production (Lequet and Bellasen 2008). The second is an attempt to develop carbon-neutral supply chains in which the amount of CO_2 produced is offset by a variety of mitigating actions that include partnerships with the local supply chains. From a policy perspective, these could deliver some development benefit by encouraging small local producers to partner with multinational companies, helping reach carbon neutrality. Similarly, new markets are developing around a lifestyle based on promoting health, the environment, social justice, personal justice and sustainable living. Such developments offer new policy opportunities for more sustainable development worldwide that incorporates green transport policies across all sectors.

Urbanization

Quantity

Urbanization exhibits complex interactions with food, discussed earlier, and energy. Urban areas, which house half the world's population, utilize two-thirds of global energy and produce 70 per cent of global carbon emissions (IEA 2008). The amount of energy an urban area consumes is largely dependent on the built environment – whether residential and commercial buildings or transport infrastructure. Beijing and Shanghai's rapid economic growth, for example, has been accompanied by a decrease in the proportion of emissions due to industrial activities since 1985. With the increase in personal vehicle ownership, however, emissions from transport have increased significantly, sevenfold for Beijing and eightfold for Shanghai between 1985 and 2006 (Dhakal 2009). This increase may, in part, have been offset by an energy-efficiency labelling programme implemented by the Chinese government, credited with avoiding 1.4 billion tonnes of CO_2 emissions for 2006–2010 (Zhan *et al.* 2011).

In general, urban populations in developing countries generate higher greenhouse gas emissions per person than surrounding rural populations, while the opposite is true for developed countries (Dhakal 2010). Energy consumption in urban areas, much like food consumption, can be far removed from where environmental impacts occur, with populations remaining oblivious of the greenhouse gas and water pollution impacts of their consumption (Scott *et al.* 2011).

Due to the links between them, it is difficult to reliably project rates of spatial expansion in urban areas without accurate projections of population growth and GDP. The challenge is magnified by recent research suggesting that the relationship between these three factors can vary significantly across regions. Assessing changing urban spatial spread using satellites shows urban areas to be growing at an average rate of 3–7 per cent per year, with China exhibiting the highest rates. The contribution of population and GDP growth to this expansion has been found to be 28 and 72 per cent respectively for North America and 23 and 30 per cent respectively for India. In the same study, African city growth showed no relationship to GDP, although there is a recognition that in many developing countries there is significant informal economic activity that is not captured by GDP statistics (Seto *et al.* 2010).

In terms of the spatial distribution of people in growing cities, the defining feature, perhaps most common in East Asia, is peripheral development (Seto *et al.* 2010). Quantifying this phenomenon using satellite imagery for 2000 shows a range of estimates of the total spatial spread of urban areas of 0.2–2.4 per cent of the terrestrial land surface, due partially to differing definitions of urban land cover (Potere and Schneider 2007). In developed countries such as the United States and Canada, about half the urban population lives in suburbs, while in the developing world squatter settlements or slums host more than one-third of urban populations (UN-Habitat 2003).

The spatial distribution of cities demonstrates the complex interactions between urbanization and transport. For instance, when comparing per-person greenhouse gas emissions, Bangkok is dominated by transport emissions, while New York and London have significantly larger contributions from residential and commercial buildings (Croci *et al.* 2011). The ability to travel within a city is extremely important both in terms of the environmental impact and of economic productivity (Bertaud *et al.* 2011). In developing countries the majority of trips are taken by commuters but, as incomes increase, individuals are likely to make more personal trips. This preference often precipitates the acquisition of personal vehicles, as the locations of shopping or entertainment centres, schools or hospitals are widely spread and less easily connected by a public transport system (Bertaud *et al.* 2011). Finally, the type of fuel used is an important factor affecting the environmental impact of urban areas. Many trains already run on electricity, but should electric vehicle use increase, more electricity will be needed and – unless energy sources are priced according to their carbon intensity – an increase in electricity production using coal is likely, leading to significant increases in greenhouse gas emissions (Bertaud *et al.* 2011).

Quality

Cities have been seen as an opportunity for developing more sustainable resource management and reducing greenhouse gas emissions. While per-person emissions are generally lower in the cities of developed countries than in surrounding rural areas, the sources are much more diffuse and therefore difficult to manage with one overarching policy tool (Bertaud *et al.* 2011). Beyond mitigation activities, cities, particularly in developing countries, need to evolve climate adaptation measures (World Bank 2011d). Several cities across South America, Africa and Asia have shown significant leadership in developing innovative adaptation strategies (Heinrichs *et al.* 2011).

Developing cities are being encouraged to achieve zero waste, the principles of which include a reduction in waste incineration, the recycling of greater volumes of paper and plastics and the mining of precious metals and rare earth elements from existing landfills (Zaman and Lehmann 2011).

The question remains whether the Earth can support several billion additional people with a direct impact on land through subsistence farming, or additional urban billions with indirect impacts through consumer demand for fats and proteins from meats that are mostly produced on large corporate farms. The answer to this question will ultimately reveal how much land will be converted to livestock rearing, feedstock production and agriculture. Not evident in the short term is whether an accelerated or delayed demographic transition is more or less taxing on land systems. But if the living standards of the poorest are raised to more equitably match those of the developed world, then population growth should slow and the related environmental impact should begin to diminish. Demographic and health transitions will continue to be major predictors of environmental change in general and of land-use and land-cover change in particular. Fundamental to facilitating demographic and health transitions will be investments in maternal and child health and education.

Globalization

Quantity

Trade in food, fuels and minerals has increased dramatically over recent decades and shows few signs of slowing. International trade has grown rapidly since 1990, by 12 per cent per year, doubling in six years (Figure 1.7) (Peters *et al.* 2011). In addition, annual emissions from exports have grown at 4.3 per cent, often due to production moving from developed

Figure 1.7 Growth in population, GDP, trade and CO_2 emissions, 1990–2008

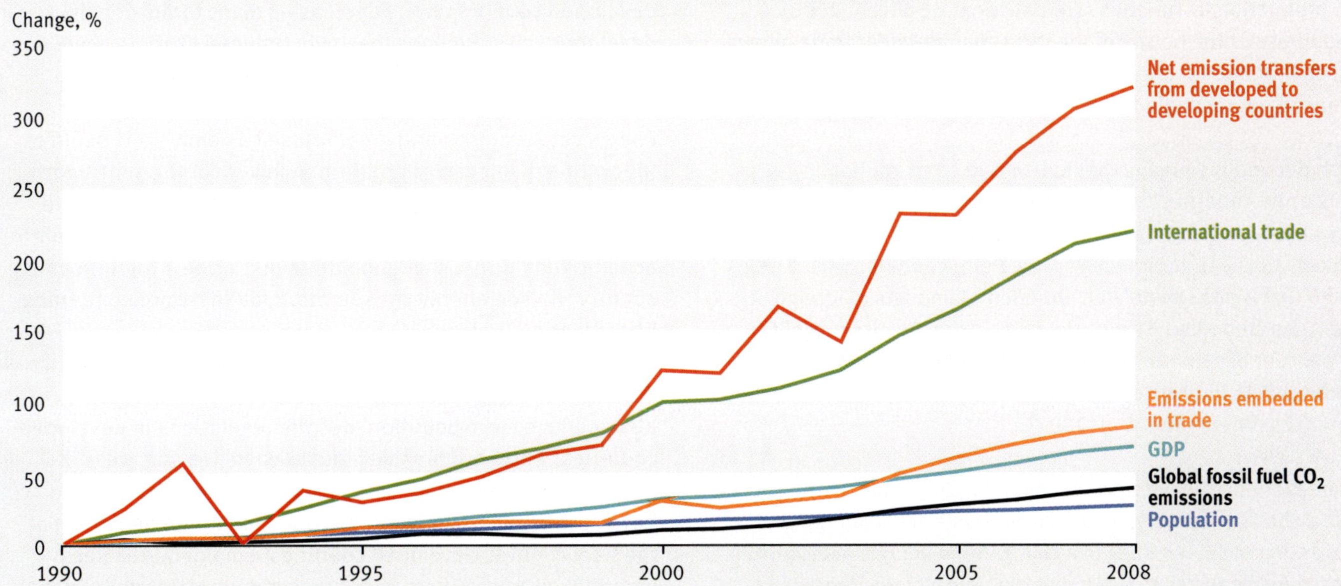

Source: Peters *et al.* 2011

The amount of energy being produced globally from renewable sources, including solar, is on the rise. *© Fernando Alonso Herrero/iStock*

countries to sites with less sophisticated technology in developing countries (Peters *et al.* 2011).

Greater liberalization of trade can exert pressure on the environment in any of three ways:

- increasing economic activity and by extension natural resource extraction, a scale effect;
- changing the type of economic activity to either more or less polluting industries, affecting intensity; and
- changing the technology or intensity of production that can sometimes encourage more environmentally friendly production techniques (Kirkpatrick and Scrieciu 2008).

Regardless of the nature of the local change, wider trade allows the environmental impacts of production to be completely removed, or decoupled, from the site of consumption.

Such decoupling means that household consumption in developed countries can have significant environmental impacts elsewhere, particularly in developing nations. Tracing the impacts of consumption in Norway, Peters and Hertwich (2006) found that a household's environmental impacts in foreign countries embodied 61 per cent of its indirect emissions of CO_2, 87 per cent of sulphur dioxide, and 34 per cent of nitrogen oxides, while imports only represented 22 per cent of household expenses (Wiedmann *et al.* 2007).

China is an instructive case for understanding trade. In the second half of the 20th century, it rapidly shifted its economy towards a processing base, resulting in a change from being a net exporter of primary resources to a net importer. Much of this processed merchandise is exported directly, with China's environment absorbing the pollution (Ma *et al.* 2006). Between 2002 and 2007, for example, 8–12 per cent of China's CO_2 emissions were attributable to exports to the United States (Xu *et al.* 2009).

Quality

Globalization is confounding the expected effect of the environmental Kuznets curve in countries with emerging economies. With affluence should come improvement in environmental conditions, but the link is proving difficult to confirm. In the case of China, nitrogen oxides and sulphur dioxide emissions have shown a complicated relationship with increasing income, suggesting that the reliance on coal-fired power may be negating improvements in other manufacturing technology (Brajer *et al.* 2011).

Some invoke a traditional economic dynamic at work – a regulatory race to the bottom, where deregulation is expected to attract economic activity and create a comparative advantage over competitors. This notion suggests that concern for the environment and increasing environmental regulation in the developed countries result in migration of the most polluting industries to less affluent nations, although explicit evidence of this is inconclusive (Kirkpatrick and Scrieciu 2008). A different explanation has also been offered – that the pattern is more akin to the rapidly industrializing countries being stuck at the bottom, since there were no regulations to begin with (Porter 1999). A related argument has also been made over the environmental effects of trade (Jorgenson 2007; Cole 2003).

Either way, the consequence is the same – the creation of centres of pollution in developing countries. This suggests that the environmental Kuznets curve, relevant to a national context, has been disguising the displacement of pollution across national borders, with consumption in the most affluent nations driving environmentally polluting production and consumption to less affluent ones. For example, Cole (2006, 2004, 2003) has shown that trade increases environmental damage in the least developed countries while decreasing many forms of pollution in developed ones. Perhaps the environmental Kuznets curve does not work when all borders have been crossed by pollution.

Energy consumption and greenhouse gas emissions seem to follow this displacement pattern. A low-income country with less stringent regulations will find that an increase in trade openness increases energy consumption as its comparative advantage in dirty production deepens, while a high-income country will see energy consumption fall in response to trade liberalization (Cole 2006).

So, will future goods produced for consumption inevitably also produce more pollution, despite regulations in developed countries? Carbon-intensive industries are leaving areas of stricter carbon regulation and moving to those that do not have such regulations (World Bank 2008). At the beginning of the 21st century, developed countries remained the largest greenhouse gas emitters in per-person terms. However, in the next few decades, the growth of emissions will come primarily from developing countries. So, despite 20 years of negotiations

Box 1.3 Greenhouse gas emissions and international trade

Recently developed analytic methodologies allow the representation of carbon emissions embodied in goods and services that are internationally produced, consumed and traded (Peters and Hertwich 2006). Plotting these data over time illustrates changes in trade balances and the transfer of emissions (Caldeira and Davis 2011). The most recent emission and trade data reveal the effects of the global financial crisis that started in 2008 (Peters *et al.* 2012).

Figure 1.8 tracks economic activity and CO_2 emissions in developed and developing countries for 1990–2010. The tinted areas represent relative trade balances, with consumption lower than production in developing countries, but higher than production in developed countries. In developing countries, the total emissions embodied in the production and consumption of goods and services rose steeply, especially after 2002, with the trade balance increasing slowly as production and consumption diverged. In contrast, the emissions embodied in production and consumption in developed countries were more horizontal until about 2002, after which they rose steeply, peaking in 2008. Their negative trade balance increased over the decades. As represented by embodied carbon emissions, developed countries seem to be back to business as usual by 2010, while emissions in developing countries have passed them with hardly a pause. On a per-person basis a large disparity persists between CO_2 emissions from developed and developing countries, as shown on the right.

Although the global financial crisis could have presented an opportunity to establish the decoupling of economic development from carbon emissions, the return of high emissions growth in 2010 may mark the passing of the opportunity. The effects of environmentally sound and low-carbon economic stimulus packages are not yet evident, but the persistent implementation of low-carbon economic plans oriented towards resource-efficiency could show positive effects in future tracking of embodied emissions (Peters *et al.* 2012).

Figure 1.8 The transfer of CO_2 emissions between developed and developing countries, 1990–2010

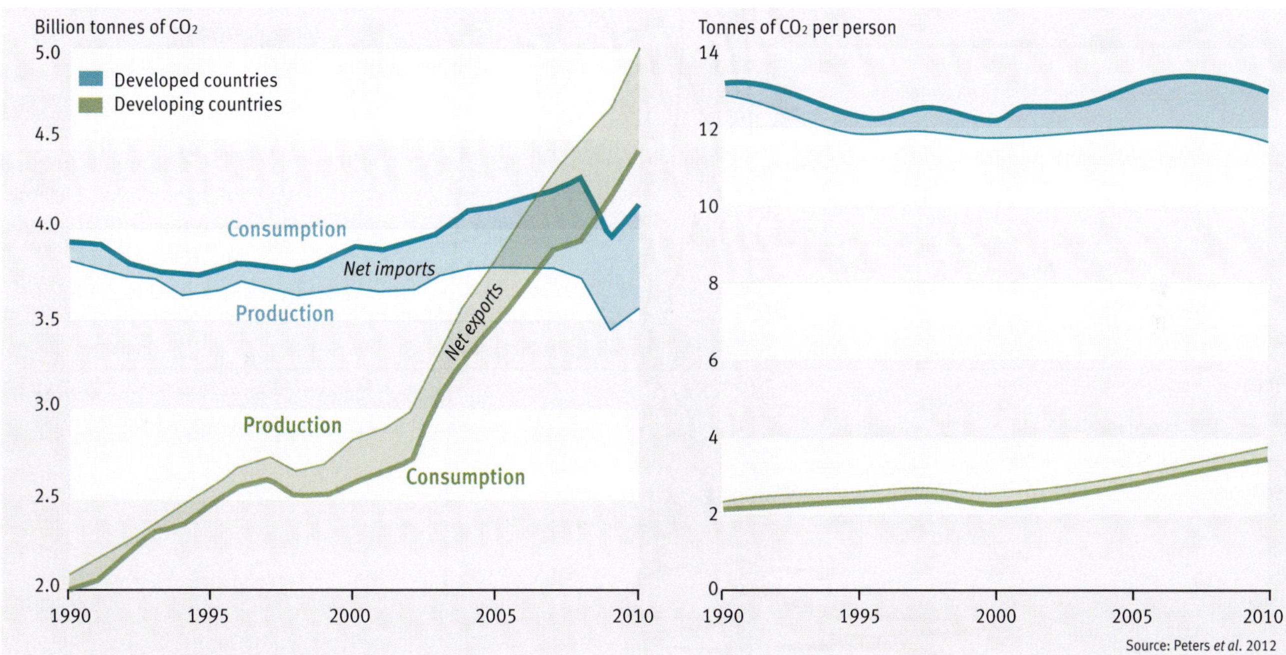

Source: Peters *et al.* 2012

to avoid this outcome, developing countries will be following the same energy- and carbon-intensive development path as their developed counterparts have done (World Bank 2008).

DISCUSSION

Drivers are interacting in unpredictable ways, resulting in some surprising consequences. This section, which links the drivers with a number of pressures on the environment, is intended to illustrate the complexity and provide some methods with which policy makers might be able to work to ameliorate the effects.

Critical thresholds

Critical thresholds are being approached or even crossed. Ecosystems and the biosphere are systems that may change in a

Figure 1.9 The great acceleration after the Second World War

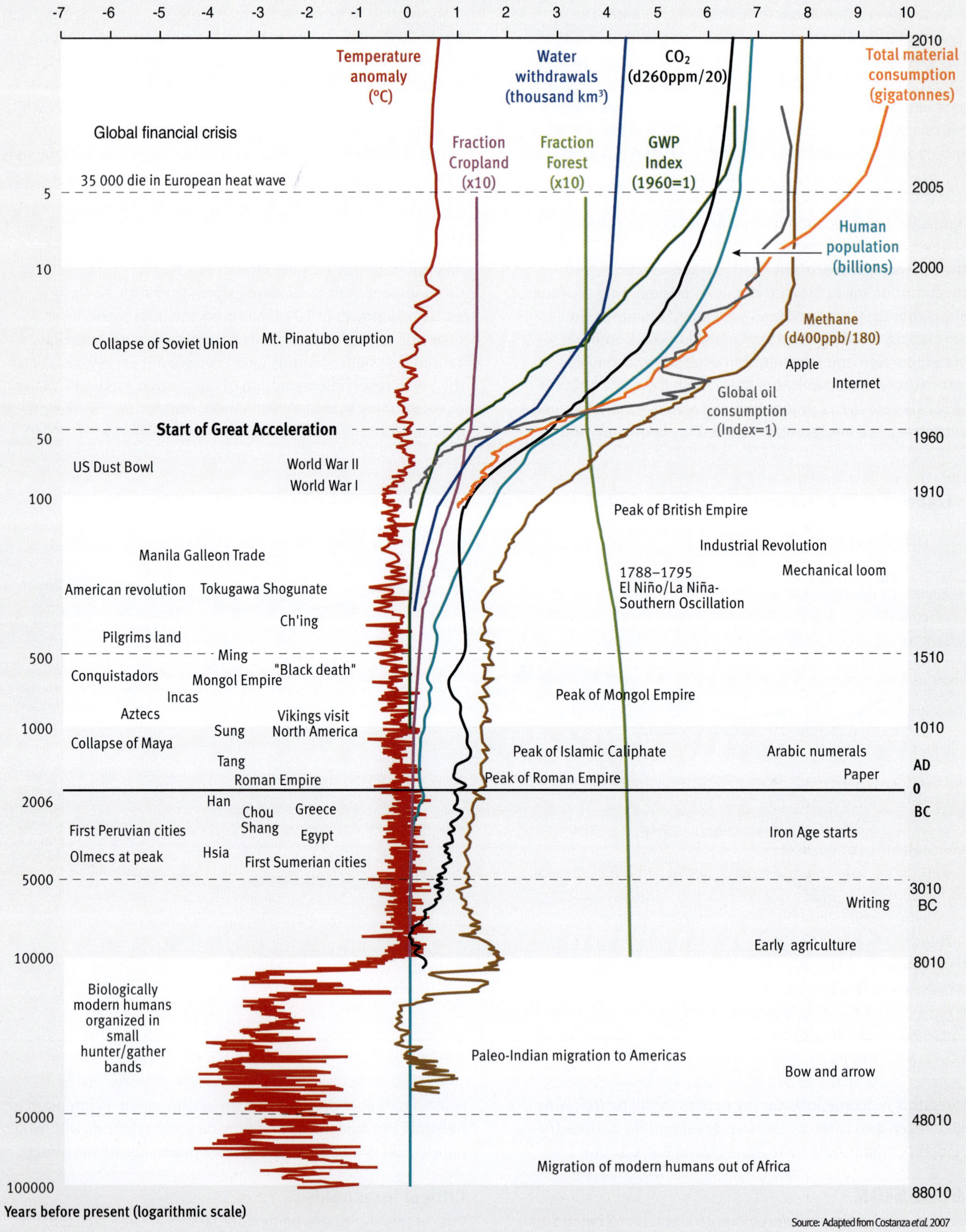

Source: Adapted from Costanza *et al.* 2007

direct and linear way as a result of human stresses, or that may have more complicated dynamics (Levin 1998). Although some can absorb a substantial amount of stress before they exhibit any response, change can take place abruptly and irrevocably when a threshold is exceeded, leaving little opportunity for human adaptation (Carpenter *et al.* 2011; Folke *et al.* 2004).

To understand the dynamics of a complex system, analysts seek out leverage points. The study of leverage points in complex systems suggests that indirect interventions can have great power and direct interventions can be used to enhance co-benefits, that both probable and possible outcomes should be addressed, and that difficult challenges can be broken down to manageable portions. The system must be monitored for both intended and unintended change (Meadows 1999).

The idea that the perturbation of a complex ecological system can trigger sudden feedbacks is not new: significant scientific research has explored thresholds and tipping points that the planetary system may face if humanity does not control carbon emissions. From the perspective of drivers, understanding feedbacks reveals that many of them interact in unpredictable ways. Generally, the rates of change in these drivers are not monitored or controlled, and so it is not possible to predict or even perceive the thresholds as they approach. Critically, the bulk of research has been on understanding the effects of drivers on ecosystems, not on the effects of changed ecosystems on the drivers – the feedback loop.

From Figure 1.9, it is evident that the rate of these changes and the anthropogenic drivers acting on them are accelerating. In fact, Costanza *et al.* (2007) argues that this "great acceleration" began after the Second World War, with the scale of population growth and economic consumption and production increasing at rates that are orders of magnitude greater than in previous eras. It is this scale and speed that makes redirecting humanity's trajectory toward more sustainable development within the limits of planetary boundaries an extremely daunting challenge but one that we cannot afford to delay.

Overexploitation of natural resources

Considering that 14–16 per cent of animal protein consumed globally comes from the sea, overfishing of marine resources offers a useful example of overexploitation of natural resources. At the global level, overfishing has been widespread but far from universal, and in those parts of the world with the capacity to manage fisheries, there is evidence that overfishing can be stopped and that previously overfished stocks can recover (Worm *et al.* 2009). There remain, however, a number of cases where overfishing continues despite the efforts of the international community, emphasizing the need for capacity building for both policy formulation and effective management.

The greatest expansion in fishing fleets and harvesting occurred after the Second World War, as governments provided significant subsidies to encourage more investment in harvesting technologies, which massively increased yields. In many cases the increased yield proved to be unsustainable, and fishery declines were widespread by the 1970s (Pauly 2009). Extension of jurisdiction with the United Nations Convention on the Law of the Sea (UNCLOS) resulted in improved management practices in many coastal areas, but a second round of expansion of fishing capacity resulted in a second round of declines (FAO 2010). Overcapacity remains a serious problem in global fisheries despite an international agreement to address it, the 1999 International Plan of Action for the Management of Fishing Capacity (FAO 2010).

Part of the problem in sustainably managing fisheries is the difficulty of monitoring the state of fish populations, especially in areas outside the jurisdiction of national or international authorities where biological information and even basic catch data may be unavailable or unreliable. Moreover in many fisheries, data are not recorded on species taken as by-catch – unwanted fish caught inadvertently, often returned to the sea dead or dying – so their status and the impacts of fishing are unknown and unmanaged (Myers and Worm 2005). More generally, poor monitoring means that there is little knowledge about the dynamics of many fish populations, making it difficult to discern whether the observed populations are showing signs of natural variability or imminent collapse (Carpenter *et al.* 2011). Chapters 4 and 5 discuss the environmental impacts of these collapses in more detail.

Driver combinations and feedbacks on human health

Looking specifically at food production, human and ecosystem exposure to chemicals increased dramatically with the industrialization of agriculture (Wallinga 2009). There has been limited research on the human and environmental health impacts of long-term exposure to these chemicals, but it is known that the risks are much higher in developing countries where 99 per cent of current global deaths from pesticide exposure occur, both from occupational exposure and from casual exposure resulting from lax or absent health and safety controls (De Silva *et al.* 2006).

Nitrate pollution from both crop cultivation and livestock production is among the most destructive impacts of food production, with the scale of meat production having serious ramifications for local pollution levels. In the United States, for example, of the top 20 sources of industrial pollution, eight are slaughterhouses (Hamerschlag 2011; EPA 2009). In addition, the country's Concentrated Animal Feeding Operations (CAFOs) produced 500 million tonnes of manure in 2007: three times the United States' 2007 total amount of human waste (Hamerschlag 2011; EPA 2009). A further problem from centralized meat production facilities involves how bacteria convert excess nitrate in such waste into nitrous oxide, a potent greenhouse gas, or it can leach into waterways and groundwater (Wallinga 2009).

Generating intense pressures

Drivers of environmental change are growing, evolving and combining at such an accelerating pace, at such a large scale and with such widespread reach that they are exerting unprecedented pressure on the environment. Most forms of consumption and production use the environment as a source of raw materials

Box 1.4 Information and communication technologies: a vicious cycle?

Democratic Republic of Congo

The rising pace and scale of drivers of environmental change are related to the process of globalization, which has enhanced the rapidity and reach with which people, ideas, and technologies move. The explosive demand for mobile telephones and the resources with which they are made has concentrated impacts in producing countries. Since 1994, more than 10 billion mobile telephones have been produced, and as of mid-2010, there were an estimated 5 billion users worldwide (ITU 2010). This growth has led to an accelerating demand for tantalum, extracted from coltan ore, a key component of consumer electronics. Most coltan is mined in Australia, but approximately 8–9 per cent of the global coltan supply comes from the eastern Democratic Republic of the Congo (DRC) (Global Witness 2010). The environmental impacts are likely to be significant for a number of reasons: among other things, illegal mining operations are carried out with few environmental safeguards, often within the borders of national parks; land clearance and pollution from the mines contribute to erosion and the degradation of streams and water tables; and mining operations typically lead to an increase in poaching and the local bushmeat trade, threatening wildlife (Hayes 2002). In addition, since most mining operations in the eastern DRC are outside government control, lucrative revenues from the extraction and trade in coltan and other minerals have often been used to finance violence and other human rights violations.

Pearl River basin, China

In 2008, a quarter of the world's electronic equipment was manufactured in China and more specifically in the Pearl River basin of southern China (Yunjie *et al.* 2010). China's GDP growth was 9 per cent in 2009, while the Guangdong region of the basin exhibited growth levels 2–3 percentage points above national averages (World Bank 2011e). In the past decade, the region constituted one-fifth of China's area, contained a third of its population and produced 40 per cent of national GDP (Barak 2009). The environmental impact of this economic growth has been poorly monitored, with estimates of tens of thousands of tonnes of heavy metals, nitrates and fuel being dumped untreated into the ocean each year (AsiaNews 2005). Without better coordination of water treatment, farmers have suffered severe crop losses from using the heavily polluted water for irrigation. The information technology industry has been blamed for much of the heavy metal dumped in the region, with the Pearl River basin named the most polluted river system in the country in 2004 and 2005 (Xu 2010).

Agbogbloshie, Ghana

A huge dumping site for electronic waste is located in the suburbs of Ghana's capital city, Accra. The Agbogbloshie slum, populated by domestic migrants from the northern reaches of Ghana, has witnessed an explosion of discarded computers, screens, hard drives and mobile telephones over the last ten years. What was once a productive wetland has become a hazardous chemical zone, home to approximately 40,000 people (Safo 2011). The local economy depends on recycling this waste, with the majority of the workforce young boys aged 11–18 earning roughly US$8 per day. The sources of much of this waste appear to be Parties to the Basel Convention, although a significant proportion also seems to come from the United States, which along with only Afghanistan and Haiti has not signed this treaty.

To date there has been little study of the effects of this trade, but toxins have been discovered in soil and food samples due to chemicals accumulating in the food chain (Dogbevi 2011; Monbiot 2011), and the local toll could be considerable. Exposure to chemical fumes can inhibit development of the reproductive and nervous systems, particularly in children with high lead levels, while mercury, cadmium and lead may all retard the cognitive and immunological development of the young workforce. The story of Abogbloshie gives an initial snapshot of the very real, localized environmental and health impacts of rapidly emerging global phenomena such as the shift to information technology – replete with its disposable approach to obsolete equipment. It is a cautionary tale of how technological innovation can have both an extraordinary effect on the global economy and society itself while, nearly invisibly, wreaking havoc on the more vulnerable, especially where the necessary institutional oversight is absent. It is this disconnection between the global and local that the current economic paradigm has created, and researchers must work backwards through the supply chain if the present situation is to be understood.

and as a sink for wastes. The impacts can be highly concentrated in some parts of the world – such as nuclear waste storage facilities and residual accumulation of toxic compounds at e-waste recycling sites (Box 1.4) – or systemically spread over the entire globe – such as PCBs delivered along the food chain from equator to poles – and they can quickly create new and potentially dangerous situations. In many instances their impacts can be so deep, rapid and unpredictable that they risk exceeding environmental thresholds and societal capacity to monitor them or respond adequately.

The combination and scale of some drivers can create dynamic patterns that, in turn, generate complex systemic interactions. One example is the rise in greenhouse gas emissions, the scale of which has defied global efforts to stimulate the necessary action to stem emissions. In addition to rising global temperatures and sea levels, scientists predict that the pace and scale of climate change could eventually exceed certain ecological limits or thresholds, leading to surprising and dangerous consequences such as the alteration of the world ocean's chemical composition with increasing proportions of acidifying carbon, the global loss of coral reef ecosystems, or the collapse of the West Antarctic ice sheet (Fabry *et al.* 2008; Lenton *et al.* 2008).

One driver can trigger a series of drivers and pressures that act in a domino fashion. For example, concerns about climate change impacts, including crop vulnerability and food insecurity, gave rise to policies that included mandates to increase biofuel production, such as legislation introduced in 2003 in the EU and in 2008 in the United States. The resulting demand generated a cascading set of pressures including crop diversion to biofuels. This diversion of cropland then contributed to higher food prices in 2008 and 2010, increasing worries about food insecurity.

Inertia and path dependencies

As global ecological and institutional systems are extremely complex and slow to change, decisions made today have long-term and far-reaching impacts. Without addressing the drivers behind the current trajectory, it will be difficult to move to an environmentally sustainable suite of choices and outcomes. At the same time the need for urgency must be recognized. Finally, due to the inertia in the system and an unwillingness to address these drivers in the past, future generations are committed to a range of impacts that could have been avoided. The most daunting of these problems is climate change, where a coalescing of several drivers has made reducing carbon emissions a very complicated task. For instance, current fossil-fuel-dependent energy and transport infrastructures are estimated to have committed the planet to emitting 496 billion tonnes of CO_2 from now until 2060 (Davis *et al.* 2010). These calculations do not include currently uncommitted transport network extensions, additional fossil-fuel-based power plants or the complex economy of refuelling stations or factories dependent on combustion energy, all of which are entirely reliant on the current model of energy generation and transport. The issue is not solely about the existing physical infrastructure that would be costly to replace, but the millions of jobs, processing facilities and entire sub-industries that have developed as a result of the *status quo*.

The case for investments in transport infrastructure has been made before. However, the institutionalization of global food production offers similar barriers to change. United States farm policy provides an illustrative example of this phenomenon, although it is by no means the only country where it occurs. Currently, 74 per cent of agricultural land in the United States is dedicated to eight commodity crops: maize, wheat, cotton, soybeans, rice, barley, oats and sorghum, supported by 70–80 per cent of government agricultural subsidies (Jackson *et al.* 2009), while the farming industry has consolidated to become an industrialized food production system. Unfortunately, the emphasis on producing these eight crop commodities has resulted in a food system where healthier food options, such as vegetables and fruits, increased in price by more than 100 per cent between 1985 and 2000, while the price of unhealthy fats and oils derived from these basic foodstuffs rose by only 35 per cent (Jackson *et al.* 2009). With many of the country's consumers making daily consumption decisions based on cost, decades of investment in this vertically integrated and politically powerful industry make fundamental changes in the health outcomes of the food system extremely challenging.

However, not all health effects are diet related, but can be linked to such atmospheric pollution as nitrate formation and chemical pollution resulting from enhanced pesticide use, amongst other sources. For instance, in the United States, a high proportion of maize and soybean crops are genetically modified to resist the effects of the herbicide glyphosate, applied in vast quantities to eradicate weeds. Within the supply chain, maize and soy make up 83–91 per cent of livestock feed grains. Ongoing research raises the question of the endocrine-disrupting potential of glyphosate (Daniel and Margareta 2009; Gasnier *et al.* 2009). The residence time of glyphosate in the environment is difficult to model, as it is dependent on a number of biophysical factors (Vereecken 2005) and monitoring capability is only recently catching up with its widespread use. However, in communities located near agricultural fields, evidence of glyphosate and its most common degradate aminomethylphosphonic acid (AMPA) can be found in the atmosphere, rain and local water bodies (Chang *et al.* 2011).

Organic, pesticide-free maize stalks, in Santa Cruz, California.

© David Gomez/iStock

Box 1.5 Conclusions of driver-centred thinking

A display at the UN Conference on Environment and Development, in June 1992, registered increases in world population, and decreases in the amount of productive land. *© Michos Tzovaras/UN Photo*

Focus on causes, rather than effects. It has not conventionally been popular to think about drivers – the causes – as a focus for environmental policy. Rather, policy responses typically concentrate on reducing pressures – the effects. There are, however, two compelling reasons to take a fresh look at drivers as an appropriate focus for policy. Firstly, unprecedented rates of change are being experienced and even where coping with one set of pressures is successful, others are around the corner. Secondly the global community has embraced a set of international environmental goals that are designed to tackle the drivers of environmental change more directly than previous efforts. The major legal agreements of the 1992 United Nations Conference on Environment and Development – on climate change, biodiversity and land degradation – recognize that long-term progress requires an ability to manage the evolution of underlying drivers. A relevant set of insights is available, providing policy makers with a menu of leverage points from which to choose driver-focused options for managing environmental problems.

The relationship between human well-being and environmental sustainability is synergistic. MDG 1 to end poverty and hunger, MDG 2 to achieve universal education, and MDGs 3–5 on gender equality and child and maternal health are all synergistic with MDG 7 on environmental sustainability. For example, approximately three-quarters of all human land use is for meat and dairy production. Red meat is several times more demanding of land and water than poultry or vegetarian foods, and is also linked to cancer and heart disease. Policies encouraging lower consumption of red meat would contribute to the MDGs related to human health and environmental sustainability. Similarly, universal education and enhanced gender equality are mutually synergistic. Improvement in both of these areas increases demand for maternal and child health services, reducing unwanted births which in turn reduces population impacts on the environment.

Indirect interventions can go a long way. Sometimes policy interventions targeted directly at drivers are not practical. Policies that set specific targets for population growth, for example, are seldom politically viable and have been called into question on moral and humanitarian grounds. However, there are often policy options that can reduce a driver indirectly in ways that are more acceptable. Fertility rates, for example, have been shown to be very responsive to levels of women's education and to access to family planning programmes, consistent with two key MDGs as well as imperatives of ethical human justice.

Direct interventions can be targeted at many different entry points. Even where indirect interventions are not practical, the fully disaggregated representation of key drivers opens up opportunities for effective intervention. For example, economic growth is generally considered a positive outcome across the world, so policies aimed at reducing growth, whether directly or indirectly, are not well received. However, that does not mean that driver-oriented policies are impossible. In China, for example, recognition of the problems associated with growth has led to ambitious targets aimed at energy efficiency.

Unintended consequences matter. Policies intended to bring about improvement in one environmental domain may result in unintended consequences in another. Negative consequences may take the form of cross-systemic links, the effects on food security of biofuel promotion, for example, or of path dependence such as policies that favour one type of infrastructure and make a switch to more favourable infrastructures more difficult. Policy makers seeking to manage drivers need to find ways of designing policies to minimize such negative consequences.

Even intractable drivers can be reframed. A core tenet of conflict resolution is to break down seemingly intractable elements into separate parts, which can then be subject to effective bargaining. Recent discussions around alternative measures of well-being have elements in common with this. Whereas GDP per person is treated as a proxy for well-being and as a universal policy objective, recent explorations have promoted alternative formulations where GDP is analytically separated from well-being. This opens up investigations into a broader range of proxies for well-being that could be pursued.

Surveillance and monitoring get results. Even where policy responses are not immediately possible, awareness of the importance of drivers justifies increased surveillance and monitoring. Many of the most important drivers are currently not subject to systematic monitoring, their impacts even less so. The evidence, then, is compelling for the need to enhance the collection and monitoring of anthropogenic drivers and their links with the environment.

REFERENCES

Abraham, K.G. and Mackie, C. (2005). *Beyond the Market: Designing Non-Market Accounts for the United States.* National Academy Press, Washington, DC

Adamo, S. and De Sherbinin, A. (2011). The impact of climate change on the spatial distribution of populations and migration. In *Population Distribution, Urbanization, Internal Migration and Development: An International Perspective* (ed. UN Population Division). United Nations, New York. http://www.un.org/esa/population/publications/PopDistribUrbanization/PopulationDistributionUrbanization.pdf

Adams, J. (1999). *The Social Implications of Hypermobility.* OECD Environmental Directorate, Paris

Aguilar, D. (2011). Groundwater reform in India: an equity and sustainability dilemma. *Texas International Law Journal* 46(3), 623–653

AsiaNews (2005). *Pearl River Pollution a Serious Concern.* http://www.asianews.it/news-en/Pearl-River-pollution-a-serious-concern-3264.html (accessed 5 September 2011)

Aslanidis, N. and Iranzo, S. (2009). Environment and development: is there a Kuznets curve for CO_2 emissions? *Applied Economics* 41(6), 803–810

Attari, S.Z., Dekay, M.L., Davidson, C.I. and De Bruien, W.B. (2010). Public perceptions of energy consumption and savings. *Proceedings of the National Academy of Sciences of the United States of America* 107(37), 16054–16059

Barak, R. (2009). Fighting pollution on the Pearl River. *China Dialogue* (online). http://www.chinadialogue.net/article/show/single/en/3266-Fighting-pollution-on-the-Pearl-River (accessed 5 September 2011)

Bazilian, M., Rogner, H., Howells, M., Hermann, S., Arent, D., Gielen, D., Steduto, P., Mueller, A., Komor, P., Tol, R.S.J. and Yumkella, K.K. (2011). Considering the energy, water and food nexus: towards an integrated modelling approach. *Energy Policy* 39, 7896–7906

Bertaud, A., Lefèvre, B. and Yuen, B. (2011). GHG emissions, urban mobility, and morphology: a hypothesis. In *Cities and Climate Change: Responding to an Urgent Agenda* (eds. Hoornweg, D., Freire, M., Lee, M.J., Bhada-Tata, P. and Yuen, B.). World Bank, Washington, DC

Bhana, D., Morrell, R. and Pattman, R. (2009). Gender and education in developing contexts: postcolonial reflections on Africa. In *International Handbook of Comparative Education* (eds. Cowen, R. and Kazamias, A.M.). pp.703–713. Springer, Netherlands

Bongaarts, J. (2001). *Household Size and Composition in the Developing World.* Population Council, New York

Bongaarts, J. (1992). Population growth and global warming. *Population and Development Review* 18(2), 299–319

Bongaarts, J. and Bulatao, R.A. (1999). Completing the demographic transition. *Population and Development Review* 25(3), 515–529

Brajer, V., Mead, R.W. and Xiao, F. (2011). Searching for an environmental Kuznets curve in China's air pollution. *China Economic Review* 22(3), 383–397

Bruns, B., Mingat, A., and Rakotomalala, R. (2003). *Achieving Universal Primary Education by 2015 – A Chance for Every Child.* Washington, DC: The World Bank.

Bruvoll, A. and Medin, H. (2003). Factors behind the environmental Kuznets curve: a decomposition of the changes in air pollution. *Environmental and Resource Economics* 24(1), 27–48

BTS (2011). *America's Container Ports: Linking Markets at Home and Abroad.* Bureau of Transportation Statistics, Washington, DC

Bulled, N. and Sosis, R. (2010). Examining the relationship between life expectancy, reproduction, and educational attainment. A cross-country analysis. *Human Nature* 21, 269–289

Caldeira, K. and Davis, S.J. (2011). Accounting for carbon dioxide emissions: a matter of time. *Proceedings of the National Academy of Sciences of the United States of America* 108(21), 8903–8908

CARE (2011). *White Paper: Women's Empowerment.* CARE USA

Carpenter, S.R., Cole, J.J., Pace, M.L., Batt, R., Brock, W.A., Cline, T., Coloso, J., Hodgson, J.R., Kitchell, J.F., Seekell, D.A., Smith, L. and Weidel, B. (2011). Early warnings of regime shifts: a whole-ecosystem experiment. *Science* 332, 1079–1082

Carr, D. (2009). Population and deforestation: why rural migration matters. *Progress in Human Geography* 33(3), 355–378

Carrico, A., Vandenbergh, M.P., Stern, P.C., Gardner, G.T., Dietz, T. and Gilligan, J.M. (2011). Energy and climate change: key lessons for implementing the behavioral wedge. *George Washington Journal of Energy and Environmental Law* 2, 61–67

Carrico, A.R., Padgett, P., Vandenbergh, M.P., Gilligan, J. and Walston, K.A. (2009). Costly myths: an analysis of idling beliefs and behavior in personal motor vehicles. *Energy Policy* 37(8), 2881–2888

Carson, R.T. (2010). The environmental Kuznets curve: seeking empirical regularity and theoretical structure. *Review of Environmental Economics and Policy* 4(1), 3–23

Chang, F.C., Simcik, M.F. and Capel, P.D. (2011). Occurrence and fate of the herbicide glyphosate and its degradate aminomethylphosphonic acid in the atmosphere. *Environmental Toxicology and Chemistry* 30(3), 548–555

Chester, M.V. and Horvath, A. (2009). Environmental assessment of passenger transportation should include infrastructure and supply chains. *Environmental Research Letters* 4, 1–8

CIESIN and CIAT (2005). Gridded population of the world, version 3 (GPWv3). Center for International Earth Science Information Network, Columbia University and Centro Internacional de Agricultura Tropica. Socioeconomic Data and Applications Center (SEDAC), Columbia University, Palisades, NY. http://sedac.ciesin.columbia.edu/gpw

Clark, M.L. and Aide, T.M. (2011). *An analysis of decadal land change in Latin America and the Caribbean mapped from 250-m MODIS data.* 34th International Symposium on Remote Sensing of Environment, 10–15 April 2011, Sydney

Cohen, J. and Small, C. (1998). Hypsographic demography: the distribution of human population by altitude. *Proceedings of the National Academy of Sciences of the United States of America* 95, 14009–14014

Cole, M.A. (2006). Does trade liberalization increase national energy use? *Economics Letters* 92(1), 108–112

Cole, M.A. (2004). Trade, the pollution haven hypothesis and the environmental Kuznets curve: examining the linkages. *Ecological Economics* 48(1), 71–81

Cole, M.A. (2003). Development, trade, and the environment: how robust is the environmental Kuznets curve? *Environment and Development Economics* 8(04), 557–580

Cole, M.A. and Neumayer, E. (2004). Examining the impact of demographic factors on air pollution. *Population and Environment* 26(1), 5–21

Costanza, R., d'Arge, R., De Groot, R., Farber, S., Grasso, M., Hannon, B., Limburg, K., Naeem, S., O'Neill, R.V., Paruelo, J., Raskin, R.G., Sutton, P. and Van Den Belt, M. (1997). The value of the world's ecosystem services and natural capital. *Nature* 387(6630), 253–260

Costanza, R., Graumlich, L., Steffen, W., Crumley, C., Dearing, J., Hibbard, K., Leemans, R., Redman, C. and Schimel, D. (2007). Sustainability or collapse: what can we learn from integrating the history of humans and the rest of nature? *Ambio* 36(7), 522–527

Cramer, J.C. (1998). Population growth and air quality in California. *Demography* 35(1), 45–56

CRI (2009). *Research Report on China's Cigarette Industry, 2009.* China Research and Intelligence, Shanghai

Croci, E., Melandri, S. and Molteni, T. (2011). Comparing mitigation policies in five large cities: London, New York City, Milan, Mexico City and Bangkok. In *Cities and Climate Change: Responding to an Urgent Agenda* (eds. Hoornweg, D., Freire, M., Lee, M.J., Bhada-Tata, P. and Yuen, B.). World Bank, Washington, DC

Daniel, H. and Margareta, W. (2009). Effects of Roundup and glyphosate formulations on intracellular transport, microtubules and actin filaments in *Xenopus laevis* melanophores. *Toxicology in Vitro* 24(3), 795

Davis, S.J., Caldeira, K. and Matthews, H.D. (2010). Future CO_2 emissions and climate change from existing energy infrastructure. *Science* 329(5997), 1330–1333

DeFries, R.S., Rudel, T., Uriarte, M. and Hansen, M. (2010). Deforestation driven by urban population growth and agricultural trade in the twenty-first century. *Nature Geoscience* 3, 178–181

De Haen, H., Stamoulis, K., Shetty, P. and Pingali, P. (2003). The world food economy in the twenty-first century: challenges for international co-operation. *Development Policy Review* 21(5–6), 683–696

De Sherbinin, A., Carr, D., Cassels, S. and Jiang, L. (2007). Population and environment. *Annual Review of Environment and Resources* 32, 345–73

De Silva, H.J., Samarawickrema, N.A. and Wickremasinghe, A.R. (2006). Toxicity due to organophosphorus compounds: what about chronic exposure? *Transactions of the Royal Society of Tropical Medicine and Hygiene* 100(9), 803–806

Devarajan, S., M.J. Miller & E. V. Swanson (2002). *Goals for Development: History, Prospects and Costs.* Policy Research Working Paper 2819. Washington, DC: The World Bank.

Dhakal, S. (2010). GHG emissions from urbanization and opportunities for urban carbon mitigation. *Current Opinion in Environmental Sustainability* 2(4), 277–283

Dhakal, S. (2009). Urban energy use and carbon emissions from cities in China and policy implications. *Energy Policy* 37(11), 4208–4219

Diamond, J. (2005). *Collapse: How Societies Choose to Fail or Succeed.* Viking Press.

Dietz, T., Rosa, E.A. and York, R. (2007). Driving the human ecological footprint. *Frontiers in Ecology and the Environment* 5(1), 13–18

Dietz, T., Fitzgerald, A. and Shwom, R. (2005). Environmental values. *Annual Review of Environment and Resources* 30, 335–372

Dogbevi, E.K. (2011). E-waste in Ghana – how many children are dying from lead poisoning? *Ghana Business News*, 7 June 2010

EIA (2010). *World Energy Projection System Plus.* US Energy Information Administration. Washington, DC

Enerdata (2011). *Global Energy Statistical Yearbook.* Enerdata, Grenoble

EPA (2009). *National Water Quality Inventory: Report 2000.* US Environmental Protection Agency, Washington, DC

Fabry, V.J., Seibel, B.A., Feely, R.A. and Orr, J.C. (2008). Impacts of ocean acidification on marine fauna and ecosystem processes. *ICES Journal of Marine Science* 65, 414–432

FAO (2010). *The State of World Fisheries and Aquaculture.* Food and Agriculture Organization, Rome

FAOSTAT (2010). *Food Supply: Livestock and Fish Primary Equivalent.* 2 June 2010. UN Food and Agriculture Organization, Rome

Fehr, E. (2009). On the economics and biology of trust. *Journal of the European Economics Association* 7(2–3), 235–266

Foley, J.A., Ramankutty, N., Brauman, K.A., Cassidy, E.S., Gerber, J.S., Johnston, M., Mueller, N.S., O'Connell, C., Ray, D.K., West, P.C., Balzer, C., Bennett, E.M., Carpenter, S.R., Hill, F., Monfreda, C., Polasky, S., Rockström, J., Sheehan, J., Siebert, S., Tilman, D. and Zaks, D.P.M. (2011). Solutions for a cultivated planet. *Nature* 478, 337–342

Folke, C., Carpenter, S., Walker, B., Scheffer, M., Elmqvist, T., Gunderson, L. and Holling, C.S. (2004). Regime shifts, resilience, and biodiversity in ecosystem management. *Annual Review of Ecology and Systematics* 35, 557–581

Gakidou, E., Cowling, K., Lozano, R. and Murray, C.J. (2010). Increased educational attainment and its effect on child mortality in 175 countries between 1970 and 2009: a systematic analysis. *The Lancet* 376(9745), 959–974

Galeotti, M., Manera, M. and Lanza, A. (2009). On the robustness of robustness checks of the environmental Kuznets curve hypothesis. *Environmental and Resource Economics* 42, 551–574

Gasnier, C., Dumont, C., Benachour, N., Clair, E., Chagnon, M.C., and Séralini, G.E. (2009). Glyphosate-based herbicides are toxic and endocrine disruptors in human cell lines. *Toxicology* 262(3), 184–191

Gerbens-Leenes, P.W., Hoekstra, A.Y. and Van Der Meer, T. (2009). The water footprint of energy from biomass: a quantitative assessment and consequences of an increasing share of bio-energy in energy supply. *Ecological Economics* 68(4), 1052–1060

Global Insight. 2010. Economic Outlook 2010. IHS Global Insight, Englewood. http://www.globalinsight.com

Global Witness (2010). *The Hill Belongs to Them: The Need for International Action on Congo's Conflict Minerals Trade.* Global Witness, London. http://www.globalwitness.org/sites/default/files/library/The%20hill%20belongs%20to%20them141210.pdf

Goulias, K.G. (2008). Supply chain and transportation: a smorgasbord of issues. In *Agri-food Logistics in the Mediterranean Area* (ed. Gattuso, D.). Franco Angeli, Milan

Grossman, G. and Krueger, A. (1995). Economic growth and the environment. *Quarterly Journal of Economics* 110, 353–377

Grote, U. and Warner, K. (2010). Environmental change and migration in sub-Saharan Africa. *International Journal of Global Warming* 2(1), 17–47

Guagnano, G.A., Stern, P.C. and Dietz, T. (1995). Influences on attitude-behavior relationships: a natural experiment with curbside recycling. *Environment and Behavior* 27, 699–718

Hamerschlag, K. (2011). *Meat Eater's Guide to Climate Change and Health.* Environmental Working Group, Washington, DC. http://static.ewg.org/reports/2011/meateaters/pdf/report_ewg_meat_eaters_guide_to_health_and_climate_2011.pdf

Hamilton, C. and Turton, H. (2002). Determinants of emissions growth in OECD countries. *Energy Policy* 30, 63–71

Hayes, K. (2002). Update on coltan mining in the Democratic Republic of Congo. *Oryx* 36, 12–13

Heinrichs, D., Aggarwal, R., Barton, J., Bharucha, E., Butsch, C., Fragkias, M., Johnston, P., Kraas, F., Krellenberg, K., Lampis, A., Ling, O.G. and Vogel, J. (2011). Adapting cities to climate change: opportunities and constraints. In *Cities and Climate Change: Responding to an Urgent Agenda.* (eds. Hoornweg, D., Freire, M., Lee, M.J., Bhada-Tata, P. and Yuen, B.). World Bank, Washington, DC

Henrich, J., Ensminger, J., McElreath, R., Barr, A., Barrett, C., Bolyanatz, A., Cardenas, J.C., Gurven, M., Gwako, E., Henrich, N., Lesorogol, C., Marlowe, F., Tracer, D. and Ziker, J. (2010). Markets, religion, community size, and the evolution of fairness and punishment. *Science* 327(5972), 1480–1484

Henrich, J., Boyd, R., Bowles, S., Camerer, C., Fehr, E., Gintis, H., McElreath, R., Alvard, M., Barr, A., Ensminger, J., Henrich, N.S., Hill, K., Gil-White, F., Gurven, M., Marlowe, F.W., Patton, J.Q. and Tracer, D. (2005). "Economic man" in cross-cultural perspective: behavioral experiments in 15 small scale societies. *Behavioral and Brain Sciences* 28, 795–855

Henry, A.D. (2009). The challenge of learning for sustainability: a prolegomenon to theory. *Human Ecology Review* 16(2), 131–140

Huijser, M.P., McGowen, P., Fuller, J., Hardy, A., Kociolek, A., Clevenger, A.P., Smith, D. and Ament, R. (2008). *Wildlife-Vehicle Collision Reduction Study: Report to Congress.* United States Department of Transportation, Washington, DC

IAEA (2008). *Nuclear Power Global Status.* International Atomic Energy Agency, Vienna

IATA (2011). *Cargo E-Chartbook Q1 2011.* International Air Transport Association, Geneva

IEA (2011). *World Energy Outlook 2011.* International Energy Agency, OECD, Paris

IEA (2010). *CO_2 Emissions from Fossil Fuel Combustion.* International Energy Agency, Paris

IEA (2008). *World Energy Outlook 2008.* International Energy Agency, OECD, Paris

IPCC (2011). Summary for policymakers. In *IPCC Special Report on Renewable Energy Sources and Climate Change Mitigation* (eds. Edenhofer, O., Pichs-Madruga, R., Sokona, Y., Seyboth, K., Matschoss, P., Kadner, S., Zwickel, T., Eickemeier, P., Hansen, G., Schlomer, S., von Stechow, C.). Cambridge University Press, Cambridge and New York

Ironmonger, D.S., Aitken, C.K. and Erbas, B. (1995). Economies of scale in energy use in adult-only households. *Energy Economics* 17(4), 301–310

ITU (2010). ITU sees 5 billion mobile subscriptions globally in 2010. Press release, 15 February 2010. International Telecommunication Union, Barcelona

Jackson, M.O. and Yariv, L. (2007). Diffusion of behavior and equilibrium properties in network games. *American Economic Review* 97(2), 92–98

Jackson, R.J., Minjares, R., Naumoff, K.S., Shrimali, B.P. and Martin, L.K. (2009). Agriculture policy is health policy. *Journal of Hunger and Environmental Nutrition* 4(3–4), 393–408

Jaeger, C., Renn, O., Rosa, E.A. and Webler, T. (2001). *Risk, Uncertainty and Rational Action.* Earthscan, London

Jaffe, A.B. and Stavins, R.N. (1994). The energy-efficiency gap: what does it mean? *Energy Policy* 22, 804–810

Jalil, A. and Mahmud, S.F. (2009). Environment Kuznets curve for CO_2 emissions: a cointegration analysis for China. *Energy Policy* 37, 5167–5172

Jiang, Y. (2009). China's water scarcity. *Journal of Environmental Management* 90(11), 3185–3196

Jiang, L. and Hardee, K. (2009). How do recent population trends matter to climate change? *Population Research and Policy Review* 30(2), 287–312

Jorgenson, A.K. (2007). The effects of primary sector foreign investment on carbon dioxide emissions for agricultural production in less-developed countries, 1980–99. *International Journal of Comparative Sociology* 48, 29–42

Kahneman, D. (2003). A perspective on judgment and choice. *American Psychologist* 58(9), 697–720

Kahrl, F. and Roland-Holst, D. (2008). China's water-energy nexus. *Water Policy* 10(S1), 51–65

Kennedy, G., Nantel, G. and Shetty, P. (2005). *Globalization of Food Systems in Developing Countries: Impact on Food Security and Nutrition.* Food and Agriculture Organization of the United Nations, Rome. http://www.fao.org/docrep/007/y5736e/y5736e00.htm

Kenworthy, J.R. and Laube, F.B. (1996). Automobile dependence in cities: an international comparison of urban transport and land use patterns with implications for sustainability. *Environmental Impact Assessment Review* 16(4–6), 279–308

Kirkpatrick, C. and Scrieciu, S.S. (2008). Is trade liberalisation bad for the environment? A review of the economic evidence. *Journal of Environmental Planning and Management* 51(4), 497–510

Knight, K.W. and Rosa, E.A. (2011). Household dynamics and fuelwood consumption in developing countries: a cross-national analysis. Population and Environment, 1-14.

Krausmann, F., Gingrich, S., Eisenmenger, N., Erb, K.-H., Haberl, H. and Fischer-Kowalski, M. (2009). Growth in global materials use, GDP and population during the 20th century. *Ecological Economics* 68(10), 2696–2705

Kumar, C., Malhotra, K., Raghuram, S. and Pais, M. (1998). Case study: India. Water and population dynamics in a rural area of Tumkur district, Karnataka. In *Water and Population Dynamics: Case Studies and Policy Implications* (eds. Sherbinin, A.D. and Dompka, V.). American Association for the Advancement of Science (AAAS), Washington, DC

Lambin, E.F., Geist, H.J. and Lepers, E. (2003). Dynamics of land-use and land-cover change in tropical regions. *Annual Review of Environment and Resources* 28, 205–241

Larivière, I. and Lafrance, G. (1999). Modelling the electricity consumption of cities: effect of urban density. *Energy Economics* 21(1), 53–66

Lee, C.-C., Chiu, Y.-B. and Sun, C.-H. (2009). Does one size fit all? A reexamination of the environmental Kuznets curve using the dynamic panel data approach. *Review of Agricultural Economics* 31(4), 751–778

Lenton, T.M., Held, H., Kriegler, E., Hall, J.W., Lucht, W., Rahmstorf, S. and Schellnhuber, H.J. (2008). Tipping elements in the Earth's climate system. *Proceedings of the National Academy of Sciences of the United States of America* 105, 1786–1793

Lenzen, M., Wier, M., Cohen, C., Hayami, H., Pachauri, S. and Schaeffer, R. (2006). A comparative multivariate analysis of household energy requirements in Australia, Brazil, Denmark, India and Japan. *Energy* 31(2–3), 181–207

Lequet, B. and Bellasen, V. (2008). *Comprendre la compensation carbone.* Pearson Education, Paris

Levin, S.A. (1998). Ecosystems and the biosphere as complex adapative systems. *Ecosystems* 1, 431–436

Liu, J., Daily, G.C., Ehrlich, P.R. and Luck, G.W. (2003). Effects of household dynamics on resource consumption and biodiversity. *Nature* 421, 530–533

Lovelock, J.E. (1972). Gaia as seen through the atmosphere. *Atmospheric Environment* 6(8), 579–580

Lutz, W. and Samir, K.C. (2011). Global human capital: integrating education and population. *Science* 333, 587–592

MA (2005). *Ecosystems and Human Well-Being: Synthesis*. Millennium Ecosystem Assessment. Island Press, Washington, DC. http://www.millenniumassessment.org/documents/document.356.aspx.pdf

Ma, T., Li, B., Fang, C., Zhao, B., Luo, Y. and Chen, J. (2006). Analysis of physical flows in primary commodity trade: a case study in China. *Resources, Conservation and Recycling* 47(1), 73–81

MacKellar, F.L., Lutz, W., Prinz, C. and Goujon, A. (1995). Population, households, and CO_2 emissions. *Population and Development Review* 21(4), 849–865

Maddison, A. (2009). *Historical Statistics for the World Economy: 1–2001 AD*. http://www.ggdc.net/maddison/

Mbonile, M.J. (2005). Migration and intensification of water conflicts in the Pangani Basin, Tanzania. *Habitat International* 29(1), 41–67

McGranahan, G., Balk, D. and Anderson, B. (2007). The rising tide: assessing the risks of climate change and human settlements in low elevation coastal zones. *Environment and Urbanization* 19, 17–37

McNeill , J.R. (2000). *Something New Under the Sun: An Environmental History of the Twentieth Century*. Norton, New York

Meadows, D. (1999). *Leverage Points: Places to Intervene in a System*. Sustainability Institute, Hartland, VT

Metz, D. (2010). Saturation of demand for daily travel. *Transport Reviews* 30(5), 659–674

Meyfroidt, P. and Lambin, E.F. (2009). Forest transition in Vietnam and displacement of deforestation abroad. *Proceedings of the National Academy of Sciences of the United States of America* 106(38), 16139–16144

Millard-Ball, A. and Schipper, L. (2011). Are we reaching peak travel? Trends in passenger transport in eight industrialized countries. *Transport Reviews* 31(3), 357–378

Mol, A.P.J. (2010). Ecological modernization as a social theory of environmental reform. In *The International Handbook of Environmental Sociology* (eds. Redclift, M.R. and Woodgate, G.). Edward Elgar Publishing, Cheltenham

Monbiot, G. (2011). From toxic waste to toxic assets, the same people always get dumped on. *Guardian*, 21 September 2009. http://www.guardian.co.uk/commentisfree/cif-green/2009/sep/21/global-fly-tipping-toxic-waste

Montgomery, M.R. (2008). The urban transformation of the developing world. *Science* 319(5864), 761–764

Murray, C.J.L. and Lopez, A.D. (1997). Global mortality, disability, and the contribution of risk factors: Global Burden of Disease Study. *The Lancet* 349(9063), 1436–1442

Mwang'ombe, A.W., Ekaya, W.N., Muiru, V.M., Wasonga, V.O., Mnene, W.M., Mongare, P.N. and Chege, S.W. (2011). Livelihoods under climate variability and change: an analysis of the adaptive capacity of rural poor to water scarcity in Kenya's drylands. *Journal of Environmental Science and Technology* 4(4), 403–410

Myers, R.A. and Worm, B. (2005). Extinction, survival or recovery of large predatory fishes. *Philosophical Transactions of the Royal Society B: Biological Sciences* 360(1453), 13–20

Nordhaus, W. (2008) New metrics for environmental economics: gridded economic data. *Integrated Assessment* 8(1), 73–84

Nordhaus, W.D. and Kokkelenberg, E.C. (1999). *Nature's Numbers: Expanding the National Economic Accounts to Include the Environment*. National Academy Press, Washington, DC

NRC (2004). *Valuing Ecosystem Services: Toward Better Environmental Decision-Making*. National Research Council. National Academy Press, Washington, DC

NRC (1994). *Assigning Economic Value to Natural Resources*. National Research Council. National Academy Press, Washington, DC

O'Neill, B.C., MacKellar, F.L. and Lutz, W. (2001). *Population and Climate Change*. Cambridge University Press, Cambridge

Palloni, A. and Rafalimanana, H. (1999). The effects of infant mortality on fertility revisited: new evidence from Latin America. *Demography* 36(1), 41–58

Pauly, D. (2009). Beyond duplicity and ignorance in global fisheries. *Scientia Marina* 73(2), 215–224

Pelletier, N. and Tyedmers, P. (2010). Forecasting potential global environmental costs of livestock production 2000–2050. *Proceedings of the National Academy of Sciences of the United States of America* 107(43), 18371–18374

Peters, G.P. and Hertwich, E.G. (2006). The importance of import for household environmental impacts. *Journal of Industrial Ecology* 10(3), 89–110

Peters, G.P., Marland, G., Quéré, C.L., Boden, T., Canadell, J.G. and Raupach, M.R. (2012). Rapid growth in CO_2 emissions after the 2008–2009 global financial crisis. *Nature Climate Change* 2, 2–4

Peters, G.P., Minx, J.C., Weber, C.L. and Edenhofer, O. (2011). Growth in emission transfers via international trade from 1990 to 2008. *Proceedings of the National Academy of Sciences of the United States of America* 108(21), 8903–8908

Pinter, L., Cressman, D.R. and Zahedi, K. (1999). *Capacity Building for Integrated Environmental Assessment and Reporting – Training Manual*. International Institute for Sustainable Development and United Nations Environment Programme, Winnipeg

Polimeni, J.M. and Polimeni, R.I. (2006). Jevons' paradox and the myth of technological liberation. *Ecological Complexity* 3(4), 344–353

Popkin, B.M. (2002). An overview on the nutrition transition and its health implications: the Bellagio meeting. *Public Health Nutrition* 5(1A), 93–103

Popkin, B.M. (2001). The nutrition transition and obesity in the developing world. *Journal of Nutrition* 131(3), 871S–873S

Porter, G. (1999). Trade competition and pollution standards: "race to the bottom" or "stuck at the bottom". *The Journal of Environment and Development* 8(2), 133–151

Port of Los Angeles (2010). *Port of Los Angeles Annual Budget Fiscal Year 2010/2011*. Los Angeles

Postel, S.L., Daily, G.C. and Ehrlich, P.R. (1996). Human appropriation of renewable fresh water. *Science* 271(5250), 785–788

Potere, D. and Schneider, A. (2007). A critical look at representations of urban areas in global maps. *GeoJournal* 69, 55–80

PRB (2011). *World at 7 Billion: World Population Data Sheet 2011*. Population Reference Bureau, Washington, DC. http://www.prb.org/Publications/Datasheets/2011/world-population-data-sheet/data-sheet.aspx

Pucher, J., Peng, Z.-R., Mittal, N., Zhu, Y. and Korattyswaroopam, N. (2007). Urban transport trends and policies in China and India: impacts of rapid economic growth. *Transport Reviews* 27(4), 379–410

REN21 (2011). *Renewables 2011 Global Status Report*. Renewable Energy Policy Network for the 21st Century, Paris

Reynaud, C. (2009). *Globalization and its Impacts on Inland and Intermodal Transport*. OECD/ITF, Paris

Rindfuss, R. and Adamo, S. (2004). Population trends: implications for global environmental change. *IHDP Update* 3, 1–3

Roberts, J.T. and Grimes, P.E. (1997). Carbon intensity and economic development 1962–1971: a brief exploration of the environmental Kuznets curve. *World Development* 25, 191–198

Rockström, J., Steffen, W., Noone, K., Persson, Ö., Chapin, F.S., Lambin, E.F., Lenton, T.M., Scheffer, M., Folke, C., Schellnhuber, H.J., Nykvist, B., De Wit, C.A., Hughes, T., Van Der Leeuw, S., Rodhe, H., Sörlin, S., Snyder, P.K., Costanza, R., Svedin, U., Falkenmark, M., Karlberg, L., Corell, R.W., Fabry, V.J., Hansen, J., Walker, B., Liverman, D., Richardson, K., Crutzen, P. and Foley, J.A. (2009). A safe operating space for humanity. *Nature* 461(7263), 472–475

Rosa, E.A. and Dietz, T. (2009). Global transformations: passage to a new ecological era. In *Human Footprints on the Global Environment: Threats to Sustainability* (eds. Rosa, E.A., Diekmann, A., Dietz, T. and Jaeger, C.). The MIT Press, Cambridge, MA

Rustagi, D., Engel, S. and Kosfeld, M. (2010). Conditional cooperation and costly monitoring explain success in forest commons management. *Science* 330(6006), 961–965

Safo, A. (2011). End of the road for "Sodom and Gomorrah" squatters. *News from Africa* 10 March 2011. http://www.newsfromafrica.org/newsfromafrica/articles/art_827.html

Satterthwaite, D., McGranahan, G. and Tacoli, C. (2010). Urbanization and its implications for food and farming. *Philosophical Transactions of the Royal Society B: Biological Sciences* 365(1554), 2809–2820

Schneider, A., Friedl, M.A. and Potere, D. (2009). A new map of global urban extent from MODIS data. *Environmental Research Letters* 4, article 044003

Schultz, P.W. and Kaiser, F.G. (2011). Promoting pro-environmental behavior. In *Handbook of Environmental and Conservation Psychology* (ed. Clayton, S.). Oxford University Press, Oxford

Scott, C.A., Pierce, S.A., Pasqualetti, M.J., Jones, A.L., Montz, B.E. and Hoover, J.H. (2011). Policy and institutional dimensions of the water-energy nexus. *Energy Policy* 39(10), 6622–6630

SERI (2008). *Global Resource Extraction 1980 to 2005*. Sustainable Europe Research Institute, Vienna

Seto, K.C., Sánchez-Rodríguez, R. and Fragkias, M. (2010). The new geography of contemporary urbanization and the environment. *Annual Review of Environment and Resources* 35, 167–194

Shah, T., Scott, C., Kishore, A. and Sharma, A. (2004). *Energy-Irrigation Nexus in South Asia: Improving Groundwater Conservation and Power Sector Viability*. International Water Management Institute, Colombo

Singh, S., Sedgh, G. and Hussain, R. (2010). Unintended pregnancy: worldwide levels, trends, and outcomes. *Studies in Family Planning* 41(4), 241–250

Sommers, M. (2010). Urban youth in Africa. *Environment and Urbanization* 22(2), 317–332

Sowers, J., Vengosh, A., and Weinthal, E. (2010). Climate change, water resources, and the politics of adaptation in the Middle East and North Africa. *Climatic Change* 104(3), 599–627

Stanners, D., Bosch, P., Dom, A., Gabrielsen, P., Gee, D., Martin, J., Rickard, L. and Weber, J.-L. (2007). Frameworks for Environmental Assessment and Indicators at the EEA. In *Sustainability Indicators – A Scientific Assessment* (eds. Hák, T., Moldan, B. and Dahl, A.). Island Press, Washington, DC

Steffen, W., Crutzen, P.J. and McNeill, J.R. (2007). The Anthropocene: are humans now overwhelming the great forces of nature? *Ambio* 36(8), 614–621

Steinfeld, H., Gerber, P., Wassenaar, T.D., Castel, V., Rosales, M. and Haan, C.D. (2006). *Livestock's Long Shadow: Environmental Issues and Options.* FAO Press, Rome

Stern, P.C. (2011). Contributions of psychology to limiting climate change. *American Psychologist* 66(4), 303–314

Stern, P.C., Gardner, G.T., Vandenbergh, M.P., Dietz, T. and Gilligan, J.M. (2010). Design principles for carbon emissions reduction programs. *Environmental Science and Technology* 44(13), 4847–4848

UN (2011). *World Population Prospects: The 2010 Revision.* Population Division, Department of Economic and Social Affairs, United Nations, New York

UN (2009a). *World Mortality.* Population Division, Department of Economic and Social Affairs, United Nations, New York

UN (2009b). *World Urbanization Prospects: The 2009 Revision.* Population Division, Department of Economic and Social Affairs, United Nations, New York. http://esa.un.org/unpd/wup/index.htm

UN (2000). *Millennium Development Goals.* http://www.un.org/millenniumgoals/

UNCTAD (2011) *United Nations Conference on Trade and Development,* Review of Maritime Transport, UNCTAD/RMT/2011

UNDESA (2011). *World Urbanization Prospects, the 2009 Revision.* United Nations Department of Economic and Social Affairs http://esa.un.org/unpd/wup/Analytical-Figures/Fig_10.htm

UNDHR (1948). Article 26. In *The Universal Declaration of Human Rights.* United Nations. http://www.un.org/en/documents/udhr/

UNDP (2009). *Human Development Report.* United Nations Development Programme, New York

UNDP (1998). *Human Development Report 1998: Consumption for Human Development.* United Nations Development Programme, New York

UNEP (2011a). *Decoupling Natural Resource Use and Environmental Impacts from Economic Growth.* United Nations Environment Programme, Nairobi

UNEP (2011b). *Towards a Green Economy: Pathways to Sustainable Development and Poverty Eradication – A Synthesis for Policy Makers.* United Nations Environment Programme, St-Martin Bellevue

UNEP (2011c). *UNEP Global Trends in Renewable Energy Investment 2011: Analysis in Trends and Issues in the Financing of Renewable Energy.* United Nations Environment Programme, Frankfurt

UNEP (2009a). *Towards Sustainable Production and Use of Resources: Assessing Biofuels.* United Nations Environment Programme, Paris

UNEP (2009b). *UNEP Year Book: Resource Efficiency.* United Nations Environment Programme, Nairobi

UNEP (2006). *Challenges to International Waters: Regional Assessments in a Global Perspective.* United Nations Environment Programme, Nairobi

UNFPA (2008). *Population and Climate Change: Framework of UNFPA's Agenda.* http://www.unfpa.org/pds/climate/docs/climate_change_unfpa.pdf

UN-Habitat (2003). *The Challenge of Slums: Global Report on Human Settlements 2003.* Earthscan, London

University of Michigan (2011). Population Growth over Human History. Michigan Population Studies Centre. http://www.globalchange.umich.edu/globalchange2/current/lectures/human_pop/human_pop.html

Vereecken, H. (2005). Mobility and leaching of glyphosate: a review. *Pest Management Science* 61(12), 1139–1151

Vollan, B. and Ostrom, E. (2010). Cooperation and the commons. *Science* 330(6006), 923–924

Vyas, S. and Watts, C. (2009). How does economic empowerment affect women's risk of intimate partner violence in low and middle income countries? A systematic review of published evidence. *Journal of International Development* 21(5), 577–602

Wackernagel, M., Schulz, N.B., Deumling, D., Linares, A.C., Jenkins, M., Kapos, V., Monfreda, C., Loh, J., Myers, N., Norgaard, R. and Randers, J. (2002). Tracking the ecological overshoot of the human economy. *Proceedings of the National Academy of Sciences of the United States of America* 99(14), 9266–9271

Wackernagel, M., Onisto, L., Bello, P., Linares, A.C., Falfán, I.S.L., García, J.M., Guerrero, A.I.S. and Guerrero, M.G.S. (1999). National natural capital accounting with the ecological footprint concept. *Ecological Economics* 29(3), 375–390

Wallinga, D. (2009). Today's food system: how healthy is it? *Journal of Hunger and Environmental Nutrition* 4(3-4), 251–281

WBCSD (2010). *Sustainable Consumption: Facts and Trends.* World Business Council for Sustainable Development.

White, M. and Hunter, L. (2009). Public perception of environmental issues in a developing setting: environmental concern in coastal Ghana. *Social Science Quarterly* 90(4), 960–982

Wiedmann, T., Lenzen, M., Turner, K. and Barrett, J. (2007). Examining the global environmental impact of regional consumption activities – Part 2: Review of input-output models for the assessment of environmental impacts embodied in trade. *Ecological Economics* 61(1), 15–26

WNA (2011a). *Nuclear Power in China.* World Nuclear Association. http://www.world-nuclear.org/info/inf63.html

WNA (2011b). *World Nuclear Power Reactors and Uranium Requirements.* World Nuclear Association. http://www.world-nuclear.org/info/reactors.html

World Bank (2012). *World Development Indicators.* http://data.worldbank.org/indicator/IS.VEH.NVEH.P3/countries/1W?display=graph

World Bank (2011a). *Data Indicators: GDP growth (annual %).* World Bank, Washington, DC

World Bank (2011b). *Migration and Remittances Factbook 2011.* 2nd ed. World Bank, Washington, DC

World Bank (2011c). *World Development Indicators.* http://data.worldbank.org/indicator/ (accessed 9 January 2012)

World Bank (2011d). Introduction: cities and the urgent challenges of climate change. In *Cities and Climate Change: Responding to an Urgent Agenda* (eds. Hoornweg, D., Freire, M., Lee, M.J., Bhada-Tata, P. and Yuen, B.). World Bank, Washington, DC

World Bank (2011e). *World Development Indicators 2011: Part 2.* World Bank, Washington, DC

World Bank (2008). *International Trade and Climate Change: Economic, Legal and Institutional Perspectives.* World Bank, Washington, DC

World Bank (2006). *China Water Quality Management – Policy and Institutional Considerations.* World Bank, Washington, DC. http://www-wds.worldbank.org/external/default/WDSContentServer/WDSP/IB/2006/10/18/000310607_20061018111318/Rendered/PDF/377520CHA01Wat1management001PUBLIC1.pdf

Worm, B., Hilborn, R., Baum, J.K., Branch, T.A., Collie, J.S., Costello, C., Fogarty, M.J., Fulton, E.A., Hutchings, J.A., Jennings, S., Jensen, O.P., Lotze, H.K., Mace, P.M., McClanahan, T.R., Minto, C., Palumbi, S.R., Parma, A.M., Ricard, D., Rosenberg, A.A., Watson, R. and Zeller, D. (2009). Rebuilding global fisheries. *Science* 325(5940), 578–585

Xu, J. (2010). IT pollution threatens Pearl River delta. *Chinadaily.com.cn* (online). http://www.chinadaily.com.cn/china/2010-05/31/content_9913000.htm (accessed 5 September 2011)

Xu, M., Allenby, B. and Chen, W. (2009). Energy and air emissions embodied in China – US trade: eastbound assessment using adjusted bilateral trade data. *Environmental Science and Technology* 43(9), 3378–3384

York, R. (2006). Ecological paradoxes: William Stanley Jevons and the paperless office. *Human Ecology Review* 13(2), 143–147

York, R., Rosa, E.A. and Dietz, T. (2010). Ecological modernization theory: theoretical and empirical challenges. In *The International Handbook of Environmental Sociology.* 2nd ed. (eds. Redclift, M.R. and Woodgate, G.). Edward Elgar Publishing, Cheltenham

Yunjie, L., Shumin, C. and Wen, L. (2010). The sustainable develoment of ICT in China. The rise and future development of the internet. In *Global Information Technology Report 2009–2010: ICT for Sustainability* (eds. Dutta, S. and Mia, I.). World Economic Forum, Geneva

Zaiceva, A. and Zimmerman, K.F. (2008). Scale, diversity, and determinants of labour migration in Europe. *Oxford Review of Economic Policy* 24(3), 427–451

Zaman, A.U. and Lehmann, S. (2011). Challenges and opportunities in transforming a city into a "zero waste city". *Challenges* 2(4), 73–93

Zhan, L., Ju, M. and Liu, J. (2011). Improvement of China energy label system to promote sustainable energy consumption. *Energy Procedia* 5, 2308–2315.

Zhang, Z., Lohr, L. Escalante, C. and Wetzstein, M. (2010). Food versus fuel: what do prices tell us? *Energy Policy* 38(1), 445–451

Zhou, W., Zhu, B., Chen, D., Griffy-Brown, C., Ma, Y. and Fei, W. (2011). Energy consumption patterns in the process of China's urbanization. *Population and Environment* 29 March

CHAPTER 2

Atmosphere

Coordinating lead authors: Johan C.I. Kuylenstierna and May Antoniette Ajero

Lead authors: Drew Shindell, Eric Zusman, Frank Murray, Geir Braathen, Kevin Hicks, Linn Persson, Lisa Emberson, Martha Barata, Sara Feresu, Sara Terry, T.S. Panwar, Yousef Meslmani and Nguyen Thi Kim Oanh

Contributing authors: Luis Abdón Cifuentes, Msafiri Jackson, Nicholas Muller, Paulo Artaxo, Seydi Ababacar Ndiaye, Susan Casper Anenberg and Emily Nyaboke (GEO Fellow)

Principal scientific reviewer: Evgeny Gordov

Chapter coordinator: Volodymyr Demkine

Main Messages

The global atmosphere is at a critical stage, particularly in relation to climate change. There is considerable scientific evidence of the causes and solutions that could protect human health and ecosystems, and effective action has resulted in the achievement of some internationally agreed goals. The phase-out of ozone-depleting substances and lead in petrol by implementing relatively simple and cost-effective solutions demonstrates that, when most major stakeholders agree, significant progress is possible.

Progress in achieving environmental goals for particulate matter (PM) and tropospheric ozone is mixed, despite the high levels of concern about their impacts, as solutions are complex and can be costly. Much of the developed world has successfully reduced concentrations of indoor and outdoor PM, sulphur and nitrogen compounds to levels close to or within World Health Organization guidelines. However, more remains to be done. Concern is high in Africa, Asia and Latin America, where levels of PM in many cities remain far in excess of the guidelines. Existing solutions can be relatively costly and the time it will take to meet guideline or target concentrations will depend on the priority given to the issue. Tropospheric ozone also remains a significant problem and is proving difficult to address despite some progress with regard to peak concentrations in Europe and North America.

Climate change is the most important atmospheric issue. While there is considerable concern about this complex problem, progress has been slow due to varying levels of motivation and because some low-carbon technological solutions are considered expensive. Despite attempts to develop low-carbon economies in a number of countries, atmospheric concentrations of greenhouse gases continue to increase to levels likely to push global temperatures beyond the internationally agreed limit of 2°C above the pre-industrial average temperature. Application of current low-carbon technologies and existing policy options would reduce the risks posed by climate change, but a gap of several billion tonnes of carbon dioxide (CO_2) equivalent remains between present emission reduction pledges and those necessary to reach climate targets.

Complementary strategies to address short-lived climate forcers (SLCFs) – black carbon, methane and tropospheric ozone – could, if implemented widely, significantly reduce the rate of temperature increase in the near term while delivering substantial co-benefits for human health and food security. Given that policy tools and technological solutions already exist, progress in reducing SLCFs could be rapid. Nevertheless, this has to be seen as a complementary strategy to the reductions in anthropogenic CO_2 emissions necessary to protect the Earth from exceeding the 2°C limit.

Climate change, air quality and stratospheric ozone depletion are increasingly seen as closely related issues but governments are not addressing them in an integrated manner. An integrated approach to atmospheric protection could support economic development and, by addressing key sectors, policy makers could accomplish multiple goals. Addressing sources of pollution can affect the different gases and particles they emit and deliver multiple climate and air quality benefits. The challenge is to find those solutions that maximize the benefits and lend themselves to widespread implementation.

Investment in achieving atmospheric goals is expected to be cost-effective. The benefits of these policies include reducing anthropogenic radiative forcing, saving millions of lives and significantly enhancing quality of life. Achievement of these benefits and climate and air quality goals requires widespread implementation of currently available technology and proven policy – but it is likely that transformative changes affecting major drivers of emissions are required as well.

INTRODUCTION

Substances emitted to the atmosphere as a result of human activities are a challenge to both the environment and development: millions of people die prematurely each year from indoor and outdoor air pollution; ozone-depleting substances (ODS) have thinned the ozone layer and created seasonal holes in the stratospheric ozone layer over polar regions; and climate change is happening now, and atmospheric concentrations of greenhouse gases and other substances that affect climate continue to increase. Climate change threatens, amongst other things, food security and biodiversity, and it is likely to increase storm damage on all parts of the globe. People in many of the developing regions are especially vulnerable.

These atmospheric issues are addressed by several global and regional agreements including Agenda 21 (UNCED 1992) and the Johannesburg Plan of Implementation (WSSD 2002). Internationally agreed goals and, in some cases, targets have been established. In addition, there are some internationally agreed guidelines related to human health and ecosystems that are used to monitor progress in addressing atmospheric issues.

This chapter uses key indicators to assess progress in relation to goals set at global and regional levels for atmospheric issues. It considers whether progress is on track to achieve these goals using existing policies and measures, and whether they are sufficient to address the key issues important to human well-being and development. The chapter then considers the outlook for the different issues and what more needs to be done. Where existing policies fall short, the need for more transformative change is highlighted, a concept that is expanded in Chapter 16.

Figure 2.1 Impacts of and links between selected substances emitted to the atmosphere

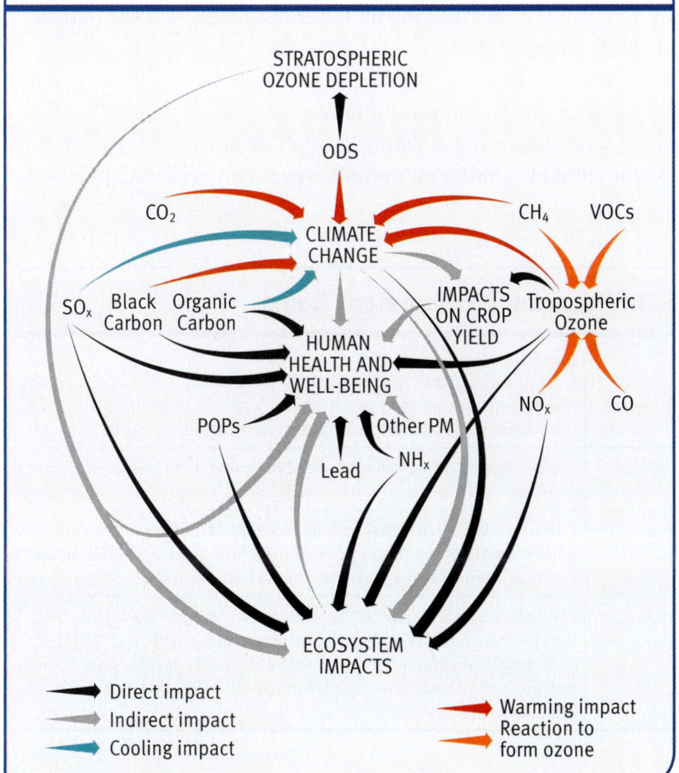

The scientific basis for the development of air pollution policy has greatly improved and there is increasing understanding of the socio-economic aspects of atmospheric issues (Stern 2007). Recently, science has pointed to new challenges such as near-term climate change and short-lived climate forcers (SLCFs) (Shindell *et al.* 2012; UNEP and WMO 2011), and knowledge about thresholds and tipping points has improved (Lenton *et al.* 2008).

Climate change, air quality and stratospheric ozone depletion are closely related, as individual pollutants can have multiple impacts on health, crop yields, ecosystems, cooling or heating of the atmosphere and stratospheric ozone depletion, all with the potential to affect human well-being (Figure 2.1). Many sources also emit multiple pollutants that can both affect air quality and cause climate change. Yet, despite these links, most governments address these issues separately, in part because goals were set in this way 20 years ago. Depending on which measures are implemented, there could be co-beneficial or antagonistic outcomes and, unless a more integrated approach is developed, there is a risk that different atmospheric policies could work against each other.

INTERNATIONAL GOALS AND TARGETS

Major goals to protect the environment and human well-being from the impact of substances emitted to the atmosphere were established in Agenda 21 (UNCED 1992) and the Johannesburg Plan of Implementation (WSSD 2002). These emphasized the need to identify threshold levels of pollutants and greenhouse gases that cause "dangerous anthropogenic interference with the climatic system and environment" (Agenda 21 Chapter 9). Meeting the objectives to phase out chlorofluorocarbons (CFCs) and other ozone-depleting substances – as defined in the 1985 Vienna Convention for the Protection of the Ozone Layer (UNEP 1985) and its 1987 Montreal Protocol on Substances that Deplete the Ozone Layer (UNEP 1987) – was considered essential. These also recognized the importance of the 1979 Convention on Long-Range Transboundary Air Pollution (CLRTAP) and its protocols to reduce regional air pollution, and recommended that these programmes be continued and enhanced, and their experience shared with other regions.

The Johannesburg Plan of Implementation went on to consider air quality as a part of overall development, promoting an integrated approach to policy making. It stressed the need to reduce respiratory diseases and other health impacts resulting from air pollution, paying particular attention to women and children. It supported the phasing out of lead in petrol, measures to prevent children's exposure to lead, and efforts to strengthen the monitoring, surveillance and treatment of lead poisoning. Another focus was to assist developing countries in providing affordable energy to rural communities, particularly to reduce dependence on traditional fuels for cooking and heating.

Delegates at the 2011 UN Climate Change Conference in Durban, South Africa. *© UNFCCC/Jan Golinski*

Atmospheric issues are closely linked to the Millennium Development Goals (MDGs) (UN 2000), as shown in Table 2.1. Other non-atmosphere-related conventions such as the 1992 Convention on Biological Diversity (CBD) also have links with the impacts of atmospheric pollution. The Aichi Biodiversity Targets (CBD 2010a) include two atmosphere-related targets:

- Target 8: by 2020, pollution, including from excess nutrients, will have been brought to levels that are not detrimental to ecosystem function and biodiversity; and
- Target 10: by 2015, the multiple anthropogenic pressures on coral reefs, and other vulnerable ecosystems impacted by climate change or ocean acidification will have been minimized, so as to maintain their integrity and functioning.

Atmospheric goals and targets are supported by both legally and non-legally binding environmental agreements (Table 2.2), most of which contain globally agreed quantitative targets and timelines for implementation that have catalysed the development and implementation of national regulation. The goals and targets refer to different aspects of control, including:

- control of drivers, for example the total ban – with a few exceptions – of the production and consumption of ozone-depleting substances, and the phase-out of leaded petrol;
- reducing pressures, for example emission reductions of carbon dioxide (CO_2) and other greenhouse gases; and
- targeting concentrations of, for example, particulate matter (PM) and CO_2.

For outdoor and indoor air pollution there are no global targets as such, but the World Health Organization (WHO) has established air quality guidelines, based on scientific research,

Table 2.1 Atmospheric issues affecting achievement of the Millennium Development Goals

Goal	Target	Impact
Eradicate extreme poverty and hunger	Halve, between 1990 and 2015, the proportion of people suffering from hunger	Climate variability and change (trends and extremes) affect crop production; tropospheric ozone directly affects crop yields
Achieve universal primary education	Ensure that, by 2015, children everywhere, boys and girls alike, will be able to complete a full course of primary schooling	Lead exposure affects cognitive development and function in young children
Reduce child mortality	Reduce by two-thirds, between 1990 and 2015, the under-five mortality rate	Children are the most susceptible to health impacts of air pollution and lead poisoning; indoor air pollution from cooking with biomass fuels disproportionately affects women and small children
Ensure environmental sustainability	Reduce biodiversity loss, achieving, by 2010, a significant reduction in the rate of loss	Climate change is one of the greatest threats to biodiversity; eutrophication from nitrogen deposition affects the diversity of terrestrial vegetation; tropospheric ozone affects vegetation composition in sensitive ecosystems; ocean acidification and warming affect marine biodiversity

Source: (goals and targets) UN 2000

to help assess progress towards reducing risks from air pollution (WHO 2006). The limit for global temperature increase at the end of the century – the agreed 2°C limit – was set on the basis of scientific discussion of the potential impacts, but also political realities and the likelihood that it could be achieved (Hare *et al.* 2011). Countries set national air-quality standards and even greenhouse gas commitments or targets specific to their international obligations, development situation and institutional capacities. The Copenhagen Accord (UNFCCC 2009) invited developed countries to submit economy-wide emission reduction targets for 2020 and developing countries to submit nationally appropriate mitigation actions (NAMAs). The Cancun Agreements (UNFCCC 2010) legally recognized these pledged targets and actions, formally anchoring them in the United Nations Framework Convention on Climate Change (UNFCCC). The CLRTAP remains the only regional agreement on transboundary air pollution that sets targets for many different pollutants. Some regions, and sub-regions – Africa, Asia and South America – have cooperation agreements that show intent to reduce emissions, but these are not binding, and in some cases have not been implemented due to lack of human and financial resources.

Table 2.2 Selected internationally agreed goals and themes related to atmospheric issues

Major themes from internationally agreed goals		Numerical target	Coverage
Stratospheric ozone depletion			
Vienna Convention for the Protection of the Ozone Layer (UNEP 1985)	Protect the ozone layer		Global
Montreal Protocol on Substances that Deplete the Ozone Layer (UNEP 1987)	Eliminate ozone-depleting substances	Zero production and consumption of ozone-depleting substances, with stated exceptions	Global
Climate change			
United Nations Framework Convention on Climate Change (UNFCCC 1992)	Stabilize greenhouse gas concentrations in the atmosphere at a level that would prevent dangerous anthropogenic interference with the climate system		Global
Kyoto Protocol to the UNFCCC (UNFCCC 1998)	Reduce greenhouse gas emissions from industrialized countries	At least 5% reduction in emissions from Annex 1 (developed) countries by 2012 compared to 1990; specific national reduction commitments	Annex 1 countries
Cancun Agreements (UNFCCC 2010)	Reduce global emissions to limit the global average temperature increase to 2°C above pre-industrial levels	Quantified economy-wide emission pledges for 2020 for Annex 1 Parties and nationally appropriate mitigation actions for non-Annex 1 Parties (developing countries)	Global
EU 20-20-20 targets (EEA 2009)	Reduce greenhouse gas emissions from EU countries by 2020	20% reduction in emissions from 1990 levels; 20% energy consumption to come from renewable sources; 20% cut in primary energy use compared to projected levels	EU Member States
Lead pollution			
Agenda 21 (UNCED 1992); Johannesburg Plan of Implementation (WSSD 2002)	Prevent exposure to lead	Zero emission of lead from transport	Global
Air quality for health and ecosystems			
WHO guidelines (WHO 2006)	Reduce the health effects of air pollution	Guidelines set for $PM_{2.5}$, PM_{10}, SO_2, NO_2, Pb, CO and O_3, e.g. $PM_{2.5}$ – 10 µg per m³ annual mean; PM_{10} – 20 µg per m³ annual mean WHO guidelines for ecosystems set on the basis of critical levels and loads as set for CLRTAP	Globally recommended
Air pollution			
EU directives for air quality, vehicles, stationary sources and national emissions (EC 2008)	Improve human health and environmental quality up to 2020	Guidelines set for $PM_{2.5}$, PM_{10}, SO_2, NO_2, Pb, CO and O_3, e.g. $PM_{2.5}$ – 25 µg per m³ annual mean, PM_{10} – 40 µg per m³ annual mean; critical loads and levels also set for ecosystems; national emissions ceilings set for each EU country for SO_2, NO_x, VOC and NH_3	EU Member States
Transboundary air pollution			
Convention on Long-Range Transboundary Air Pollution (CLRTAP) (UNECE 1979)	Protect humans and the environment from air pollution, to be implemented through objectives set in the convention protocols	The Gothenburg Protocol (UNECE 2005) sets reductions for all Parties – a multi-pollutant/multi-effect protocol with reduction targets for emissions compared to 1990 to be achieved in 2010 (being revised with 2020 targets): refer to Annex II of the protocol for specific country emission ceilings	UNECE countries of Europe, Central Asia and North America
ASEAN Haze Agreement (ASEAN 2002)	Monitor and prevent transboundary haze pollution as a result of land and/or forest fires	Agreed to adopt the policy of zero burning	ASEAN countries of South East Asia

PROGRESS IN ACHIEVING ATMOSPHERIC GOALS

This section examines progress in addressing concerns about the atmosphere over the 20 years or so since Agenda 21 (UNCED 1992) identified key priorities. The goals and targets set globally and regionally for a number of atmospheric issues are compared to the current situation, examining whether they have been met and determining the size of the gap between the current situation and the goals and targets.

Progress is described against key indicators by considering atmospheric issues in three main categories:

- examples of where targets are not being met and the situation remains far from sustainable;
- examples of mixed progress, with some regions having met targets and others remaining far from them; and
- examples of good progress, where targets have been set and are largely met.

Box 2.1 Climate change

Related goals
Prevent dangerous anthropogenic interference with the climate system (UNFCCC)

Indicators
Temperature trends; rainfall changes; sea ice extent; CO_2 concentrations; greenhouse gas emissions

Global trends
Slow progress; not on track to avoid crossing UNFCCC's agreed temperature rise limit of 2°C

Climate change: goals far from being met

There is broad scientific consensus that anthropogenic emissions of CO_2 and other greenhouse gases are the leading cause of contemporary climatic changes (IPCC 2007). Four independent analyses show that 2000–2009 was the warmest decade on record with atmospheric concentrations of CO_2 also increasing (Figure 2.2). A look at regional temperature changes shows that the greatest warming over the past century is at high latitudes (Figure 2.3).

Climate change threatens human well-being in many ways, from a greater frequency of heat waves and severe storms, to shifts in rainfall patterns and rising sea levels (IPCC 2007). Changes in the frequency of tropical cyclones are uncertain, but it is likely that their intensity will increase with rising temperatures (IPCC 2011).

Humans are directly affected, for example through alterations in freshwater supply, agricultural productivity and health, and indirectly by the economic and social impacts of loss of biodiversity and ecosystem services. Climate change is therefore considered to be the most significant problem related to changes in the atmosphere being faced by humankind. A review of the economics literature suggests that damage from climate impacts could amount annually to 1–2 per cent of world gross domestic product (GDP) by 2100 if temperatures increase by 2.5°C above pre-industrial levels. These damage estimates rise to 2–4 per cent of world GDP for a 4°C increase (Aldy *et al.* 2010). The few studies that estimate damages for extreme warming find that annual losses to world GDP by 2100 could range from 10.2 per cent for a 6°C warming increase (Nordhaus and Boyer 2000) to 11.3 per cent for a 7.4°C warming increase (Stern 2007). While valuation is sensitive to underlying assumptions about discount rates and catastrophic effects, it is clear that the socio-economic impacts of climate change are likely to be very large.

More than 22 000 homes were inundated in the disastrous Brisbane Flood of 2011, the worst flood in Australia's history. *© On-Air/iStock*

Impacts will probably be especially significant in the Arctic where warming is likely to be the greatest. Large portions of the Arctic have experienced temperature increases of more than 2°C compared to 1890–1910 (Figure 2.3), and Arctic sea ice cover has decreased dramatically, with declines in both the autumn and the winter (Figure 2.4). The Greenland and Antarctic ice sheets both show rapidly increasing melt rates and the Greenland melt area has expanded markedly (Rignot *et al.* 2011). Other areas where sizeable climate change impacts are projected include the sub-tropics, where arid regions are expected to expand, and low-lying areas, where sea-level rise is likely to cause the most damage. Less developed countries that have limited capacity to adapt to these changes are at risk of not meeting their development goals.

Extreme weather events are expected to change in frequency and intensity, becoming more common as the climate warms, including regional heat waves and both wet and dry extremes of precipitation (IPCC 2007). Europe experienced two extremely hot summers in 2003 and 2010, and research shows that the probability of extreme summer heat, including mega-heat waves, will increase by a factor of 5–10 within the next 40 years (Barriopedro *et al.* 2011). The frequency of heavy precipitation events has increased over most of the world's land area (Chapter 4), and more intense and longer droughts have been observed since the 1970s, particularly in the tropics and sub-tropics (IPCC 2007). Long-term trends show a tendency towards drier conditions in the Sahel and northern India (Figure 2.5).

Figure 2.2 Trends in temperature change and atmospheric CO_2 concentrations, 1850–2010

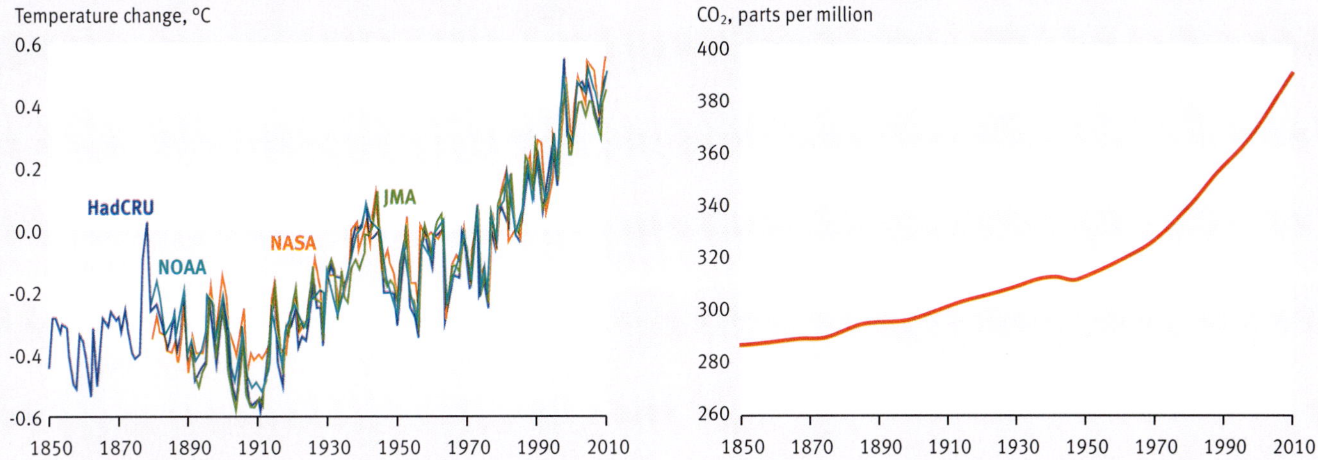

Note: 0 = 1961–1990 global mean

Source: NOAA NCDC; NASA GISS; Hadley Climatic Research Unit at the University of East Anglia (HadCRU); Japan Meteorological Agency (JMA)

Source: Scripps Institute of Oceanography, NOAA

There is growing concern that inaction will lead to changes that are irreversible at human timescales – so-called tipping points. Increased release of carbon stored in permafrost, as CO_2 or methane, is an example of a change that could give rise to a cycle of further warming and further releases of greenhouse gases (Schaefer *et al.* 2011; Lawrence and Slater 2005).

Figure 2.3 Temperature change over the 20th century

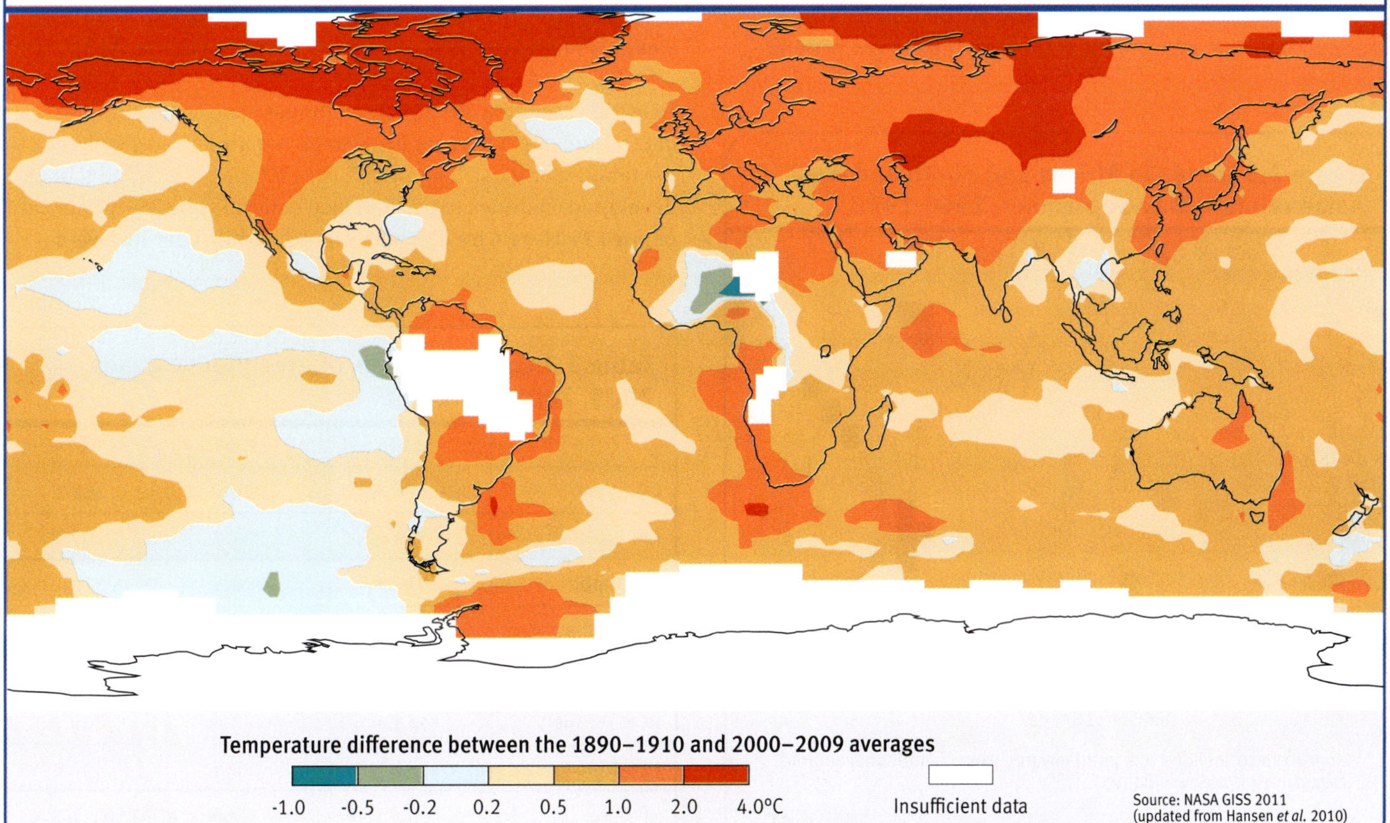

Source: NASA GISS 2011 (updated from Hansen *et al.* 2010)

Figure 2.4 Trends in Arctic sea ice extent in winter and autumn, 1979–2010

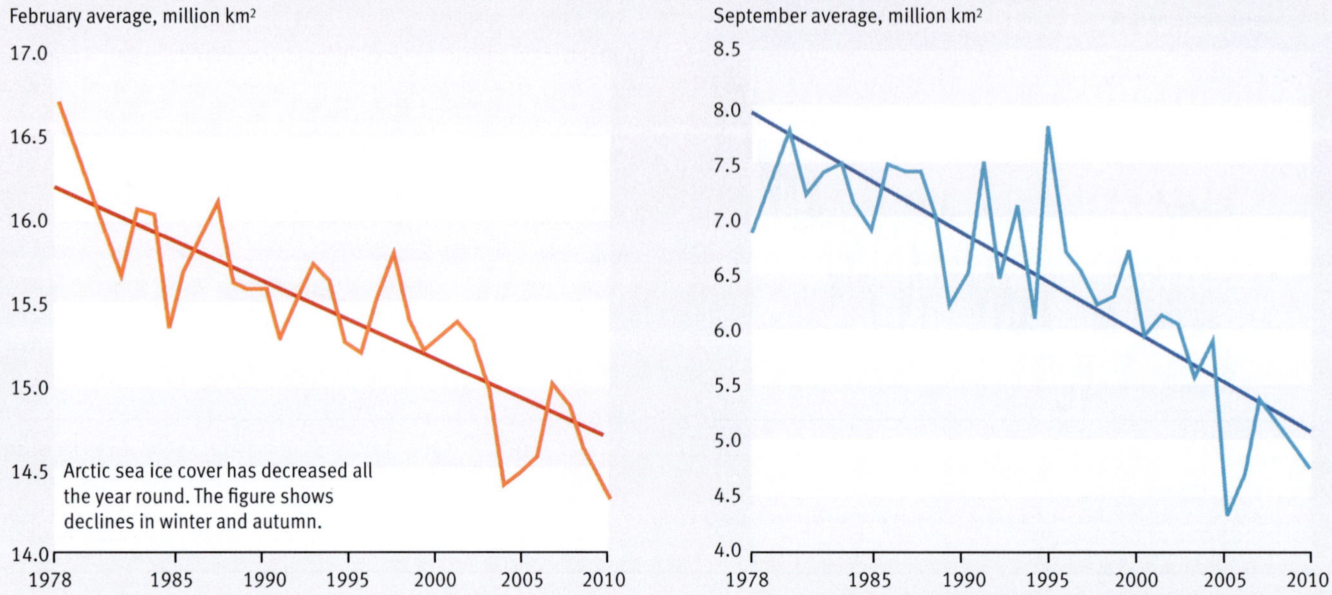

Source: NSIDC 2011

Concentrations and emissions of most anthropogenic greenhouse gases have increased during recent years (Table 2.3 and Figure 2.6). Growth rates have been especially high for concentrations of several hydrofluorocarbons (HFCs) while emissions of CO_2 from fossil fuel consumption have followed the more pessimistic of the widely used projections of the Intergovernmental Panel on Climate Change *Special Report on Emission Scenarios* (SRES) (IPCC 2000) during the last decade, despite a brief downturn in global emissions in 2009 associated with the economic recession (Figure 2.6). The rapid growth of CO_2 concentrations is also associated with similarly rapid increases in ocean acidification (Chapter 4).

Figure 2.5 Trends in African and South and West Asian rainfall, May-September, 1960-1998

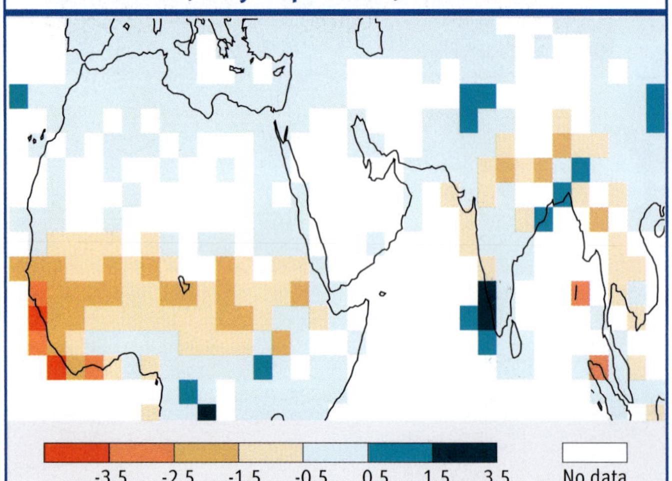

Blue areas correspond to a trend towards wetter conditions, and red areas towards a drier climate.

Source: Hulme *et al.* 1998

To avoid exceeding the 450 ppm atmospheric concentrations of CO_2-equivalent that are likely to be required to stay within the temperature rise limit of 2°C, the IPCC has concluded that developed countries need to reduce emissions by 25–40 per cent below 1990 levels by 2020 (IPCC 2007), while peer-reviewed

Table 2.3 Concentrations of greenhouse gases, 2005, 2009 and 2010

	2005	2009	2010
CO_2 (ppm)	378.7	386.3	388.5
CH_4 (ppb)	1 774.5	1 794.2	1 799.1
N_2O (ppb)	319.2	322.5	323.1
CFC-11 (ppt)	251.5	243.1	240.5
CFC-12 (ppt)	541.5	532.6	530.8
HCFC-22 (ppt)	168.3	198.4	206.2
HFC-134a (ppt)	34.4	52.4	57.8

Source: NOAA GMD 2011a

Figure 2.6 Trends in fossil fuel emissions, calculated and IPCC scenarios, 1990–2015

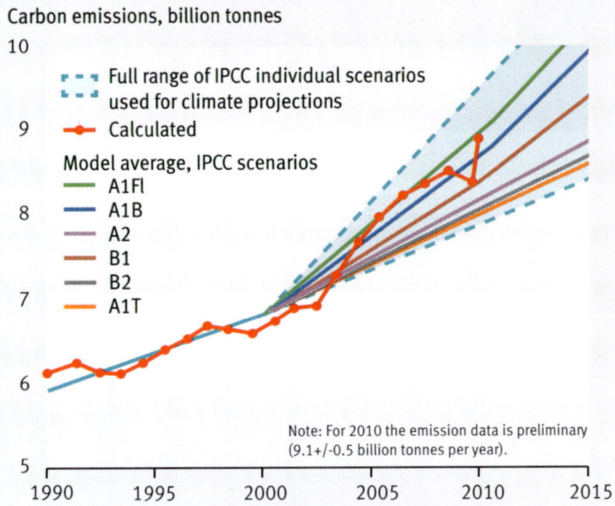

Source: adapted from Raupach and Canadell 2010, with observed emissions data from the Carbon Dioxide Information Analysis Center (CDIAC) and International Energy Agency (IEA), and preliminary data for 2010 from Peters *et al.* 2011b

literature has concluded that developing countries need to reduce emissions by 15–30 per cent relative to business-as-usual by 2020 (den Elzen and Höhne 2010, 2008). Further reductions are then required beyond 2020 to achieve the target. While some countries have reduced CO_2 emissions since the Kyoto Protocol entered into force in 2005, many appear unlikely to reach their Kyoto targets. Further, many of the same countries reporting reductions have increased imports of carbon-intensive products – so-called carbon leakage. Accounting for the foreign CO_2 emissions embedded within imported products, emissions have in fact increased in many developed nations and the net domestic-plus-embedded emissions are far greater than the Kyoto targets (Peters *et al.* 2011a).

The years since the Bali Action Plan (UNFCCC 2008) have seen 42 developed countries pledge quantified economy-wide emission targets up to 2020, while 44 developing countries have pledged nationally appropriate mitigation actions. Nonetheless, these commitments fall short – by about 6 billion tonnes of CO_2-equivalent – of levels likely to keep temperatures within safe limits as shown in Figure 2.7. This figure compares the expected emissions in 2020, resulting from four possible interpretations of pledges of mitigation action from 86 countries, with the range of emission levels that would providea greater than 66 per cent chance of staying below the agreed 2°C temperature rise limit. The gap between expected emissions and the agreed UNFCCC 2°C limit lies between 6 billion and 11 billion tonnes of CO_2-equivalent. The size of the gap depends on the extent to which the pledges are implemented and how they are applied (UNEP 2011a).

Addressing sources of pollution can deliver both climate and air quality benefits. *© Morten Madsen/iStock*

Figure 2.7 The emissions gap

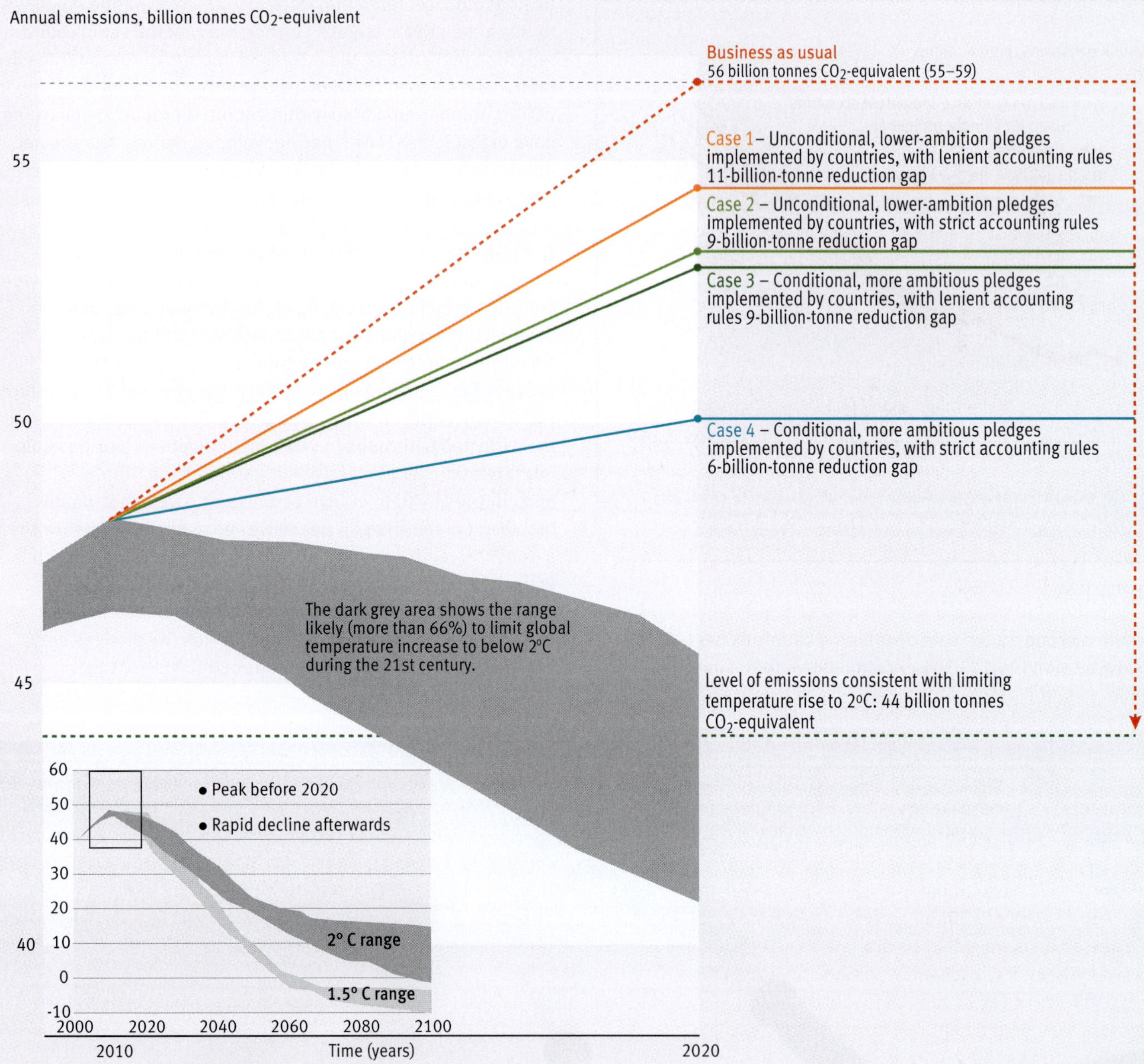

A comparison of the gap between expected emissions in 2020 according to submitted country pledges, formally recognized in Cancun, and the range of emissions likely to limit temperature increase to the agreed UNFCCC 2°C limit.

Source: UNEP 2011a

An historical tendency to underestimate rates of climate change suggests that non-linear changes and material losses at the higher end of estimated ranges are also possible (Smith *et al.* 2009; Stern 2007). Overall, prospects for long-term climate change look bleak if there is no demonstrable progress at both international and national levels.

Even if negotiations take longer than expected at the international level, national actions should continue to move forward. A growing body of low-carbon research has shown that in countries ranging from the United Kingdom (Strachan *et al.* 2008) and Japan (Fujino *et al.* 2008) to Thailand (Shrestha *et al.* 2008), it would be economically and technically feasible to cut emissions in half by 2050. The results of these studies are based on placing a price on carbon through, for instance, an emissions trading scheme. It is important to note, however, that market-based instruments such as emissions trading schemes or the Clean Development Mechanism (CDM) may not work in all

The boom in global trade has led to significant emissions of CO_2 and key pollutants including SO_2, NO_x and black carbon from international shipping. © *Mark Wragg/iStock*

contexts or benefit all regions equally. For example, in the CDM market, Latin America and Asia and the Pacific account for more than 87 per cent of all projects while Africa accounts for less than 3 per cent (UNFCCC 2012).

Other studies suggest that mainstreaming climate change into existing development plans could provide a more promising alternative to market-based instruments, especially for those developing countries that constrain their development with an emissions cap (Shukla *et al.* 2008). This is reinforced by research that shows that because of the relatively greater value of co-benefits, such as improved local air quality, low-income countries have the most to gain from mitigating greenhouse gases in a manner consistent with development priorities (Nemet *et al.* 2010). Capturing these co-benefits not only requires that policy makers become adept at mainstreaming climate change in development plans, but also necessitates decision-making frameworks that explicitly acknowledge synergies between climate change and other atmospheric issues. Such an integrated approach can readily be put in place at local and city levels, where a considerable amount of climate mitigation and air quality control has already been implemented.

Mixed progress

There are examples of improvement in some regions while large difficulties remain in many others, and global targets are still far from being met. Four major atmospheric issues are described below: sulphur, nitrogen, small particulate matter (typically described as PM_{10} and $PM_{2.5}$) and tropospheric ozone.

Sulphur pollution

Sulphur dioxide (SO_2) emissions, predominantly from fossil fuel use in power generation, industry and transport, have detrimental effects on human health by contributing to $PM_{2.5}$, on terrestrial and freshwater ecosystems by acidification (Rodhe *et al.* 1995), on man-made materials and cultural heritage by corrosion (Kucera *et al.* 2007), and on biodiversity (Bobbink *et al.* 1998) and forestry (Menz and Seip 2004). Sulphate aerosols also cool the atmosphere (Forster *et al.* 2007), which makes it important to track them in order to assess the overall benefits of greenhouse gas reduction strategies.

Since issues of transboundary air pollution were highlighted in Agenda 21 (UNCED 1992) there have been considerable

Box 2.2 Sulphur pollution

Related goals
CLRTAP, EU Directives and WHO guidelines for human health and ecosystems

Indicators
Sulphur emissions; exceedance of critical loads/levels (thresholds above which harmful effects are observed)

Global trends
Mixed regional progress

Figure 2.8 Regional trends in sulphur dioxide emissions, 1850–2050

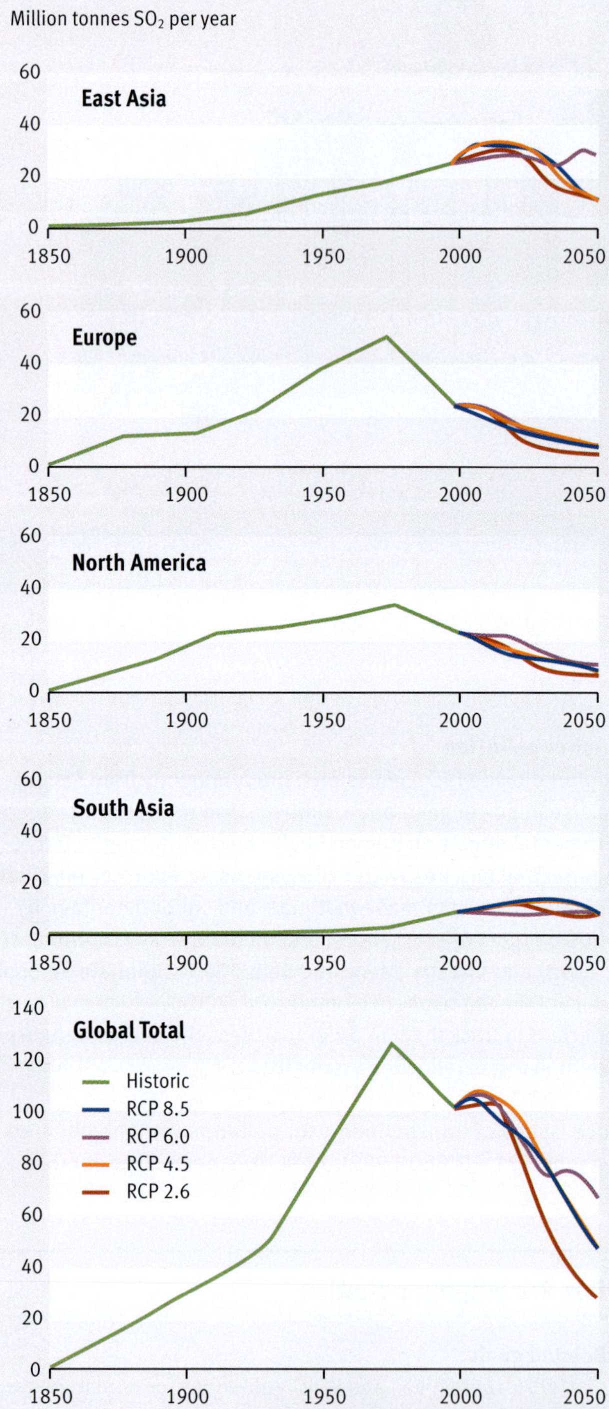

Emission trends from 1850–2000 and four Representative Concentration Pathway (RCP) scenarios from 2000–2050, developed to contribute to the Fifth Assessment of the IPCC, are shown for the four source regions and for the global total from the hemispheric transport of air pollution (HTAP) multi-model experiments.

Source: HTAP 2010

reductions in sulphur dioxide emissions in Europe and North America, achieving the targets of the CLRTAP protocols, the National Emission Ceiling (NEC) Directives of the European Union (EU) and clean air legislation in Canada and the United States (Figure 2.8). Key to the development of country targets in Europe was the use of critical loads (deposition thresholds above which harmful effects are observed) (Nilsson and Grennfelt 1988). Successful implementation of legislation brought about a drop of around 20 per cent in global emissions between 1980 and 2000. Emissions from Europe and North America were dominant until about 2000, when East Asian emissions started to dominate. According to the Representative Concentration Pathway (RCP) scenarios (Figure 2.8), global sulphur dioxide emissions were projected to decline steadily after 2005, and by 2050 to be 30, 50 or 70 per cent lower than 2000 levels. This set of four new pathways was developed for the climate modelling community as a basis for near- and long-term modelling experiments (van Vuuren *et al.* 2011; Moss *et al.* 2010).

As sulphur deposition has abated in Europe and North America, acidification has also diminished and some freshwater ecosystems have recovered, although critical loads are still exceeded in some areas (Wright *et al.* 2005; Stoddard *et al.* 1999). In Asia, the increase in emissions has put sensitive ecosystems at risk from the effects of soil acidification (Figure 2.9). However, the large-scale acidification of natural lakes experienced in Europe and North America has not been seen in Asia, and may be unlikely due to the nature of the region's soil and geology (Hicks *et al.* 2008). In 2005 it was estimated that the critical load for soils in China was exceeded by sulphur deposition across 28 per cent of the country's territory, mainly in eastern and south-central China. The area in exceedance is projected to decrease to 20 per cent in 2020, given the implementation of current plans for emission reductions (Zhao *et al.* 2009).

Further action on sulphur emissions is being taken through the revision of the Gothenburg Protocol in Europe. In Asia, action is also being taken to improve the efficiency of energy use and reduce sulphur dioxide emissions. For example, as part of its five-year plans, China implemented flue-gas desulphurization and the phasing out of small, inefficient units in the power sector in a move to achieve the national goal of a 10 per cent reduction in sulphur dioxide emissions between 2005 and 2010 (Zhang 2010).

Global efforts are also being made to reduce sulphur emissions from key sectors, including transport and shipping. The human health effects of particulate matter measuring 2.5 micrometres or less in diameter ($PM_{2.5}$) are being tackled by lowering the sulphur content of diesel fuels – for example, UNEP's Partnership for Clean Fuels and Vehicles (PCFV) is promoting the reduction of sulphur in vehicle fuels to 50ppm or below worldwide (UNEP 2012). Sulphur emissions from shipping have become an important policy issue in Europe, while the International Convention for the Prevention of Marine Pollution from Ships (MARPOL) is committed to the progressive global reduction in

Figure 2.9 Areas at risk and timeframe for acidification damage in Asia

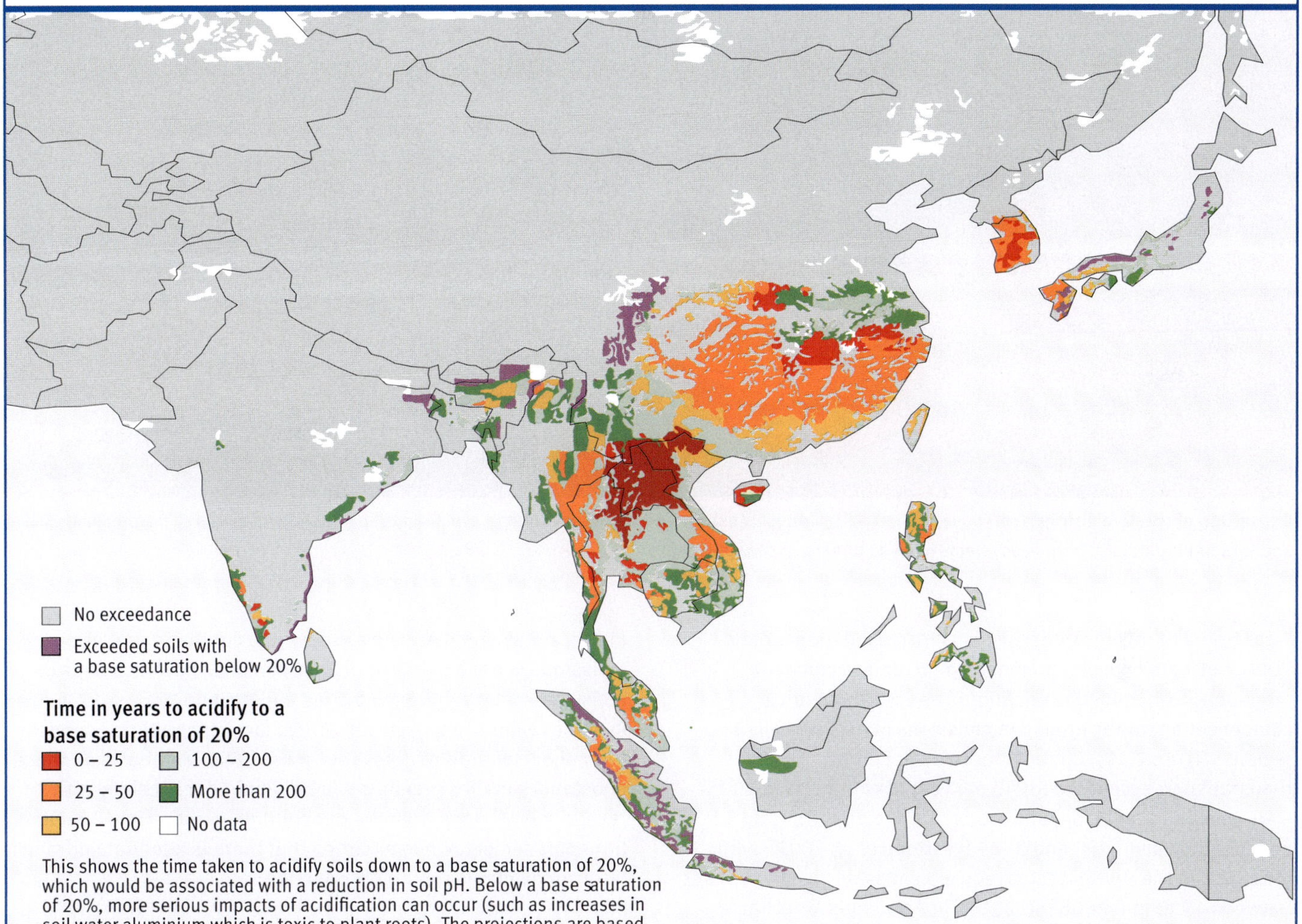

This shows the time taken to acidify soils down to a base saturation of 20%, which would be associated with a reduction in soil pH. Below a base saturation of 20%, more serious impacts of acidification can occur (such as increases in soil water aluminium which is toxic to plant roots). The projections are based on emissions according to the IPCC SRES A2 scenario for 2050.

Source: Hicks *et al.* 2008.

emissions of sulphur oxides, nitrogen oxides and particulate matter (MARPOL 2011 Annex VI).

Nitrogen compounds

Human activity linked to energy use and food production has more than doubled the amount of reactive nitrogen circulating in the environment over the past century (ENA 2011). This is emitted to the atmosphere as nitrogen oxides (NO_x), mainly from the transport and industry sectors, and ammonia (NH_3) and nitrous oxide (N_2O), mainly from agriculture. They have multiple effects on the atmosphere, terrestrial ecosystems, freshwater and marine systems, and on human health, a phenomenon known as the nitrogen cascade (Galloway *et al.* 2003). Nitrogen compounds are precursors of atmospheric $PM_{2.5}$, which has impacts on human health, while nitrogen oxide is a precursor of tropospheric ozone, which has impacts on health, crop yields, ecosystems and climate. Nitrous oxide and tropospheric ozone are also important greenhouse gases. Nitrogen deposition drives biodiversity loss through eutrophication and acidification in terrestrial and aquatic ecosystems (Bobbink *et al.* 1998). However, it can also be of benefit to crop yields, and can increase carbon sequestration through the stimulation of forest growth (ENA 2011).

Box 2.3 Atmospheric nitrogen pollution

Related goals
CBD, CLRTAP, EU directives and WHO guidelines for human health and ecosystems

Indicators
Nitrogen oxides and ammonia emissions; nitrogen deposition; exceedance of critical loads/levels – thresholds above which harmful effects are observed

Global trends
Mixed: reductions in nitrogen oxides in some regions; ammonia emissions set to increase in all regions

Deposition of reactive nitrogen compounds from agriculture, transport and industrial sources can lead to increased emissions of nitrous oxide (N_2O) and loss of biodiversity from ecosystems such as forests. *© Orchidpoet/iStock*

Total global nitrogen oxide emissions increased until around 2000, but were expected to remain more or less constant thereafter, with reductions in Europe and North America compensating for the growth in emissions in Asia and all other regions (Figure 2.10). Control measures in Europe – where road transport accounted for 40 per cent of emissions in 2005 – succeeded in reducing total nitrogen oxide emissions by 32 per cent between 1990 and 2005 (Vestreng *et al.* 2009), while measures in the United States reduced emissions by 36 per cent between 1990 and 2008 (IJC 2010). In Asia, emissions have continued to increase over the past two decades, with the growth rate itself accelerating during this period (Figure 2.10). Emissions from international shipping are estimated to have risen from 16 million tonnes in 2000 to 20 million tonnes of nitrogen dioxide (NO_2) in 2007 (IMO 2009).

Global ammonia emissions, largely from the agricultural sector, have increased fivefold since the middle of the last century and are projected to continue to climb in all regions with the possible exception of Europe, where they have decreased slightly and may stabilize (Figure 2.10) (EEA 2009). Nevertheless, there is a lack of concern and focus on this issue in Europe, and there is often resistance to major changes from the farming community. In most other regions ammonia is not regulated under major emission control laws. However, the Gothenburg Protocol of CLRTAP is being revised with more stringent targets and is likely to lead to a further reduction in emissions in Europe.

Despite these improvements, nitrogen-based air pollution from agriculture, industry and traffic in urban areas contributes significantly to $PM_{2.5}$ concentrations as secondary nitrate and ammonium particles, which are reducing people's life expectancy by several months across much of Central Europe (ENA 2011).

In Africa, Asia and Latin America, where control of nitrogen emissions is not a high priority, projections show increases in emissions of both nitrogen oxides and ammonia (Figure 2.10). In some regions, especially in Africa, lack of monitoring capacity is a major issue. To address this, more policy emphasis on these substances in these regions will be required, especially with regard to emissions from the agricultural, energy, industrial and transport sectors, while ensuring that there is adequate nitrogen fertilizer available for food production.

Current technology can deliver significant reductions in emissions of nitrogen oxide, but growth in certain sectors, particularly transport, can counteract control measures. Changed management practices will be needed to reduce ammonia emissions, and more fundamental consideration of agricultural policy and practice, as well as changes in consumption patterns of meat and dairy products, are required if large reductions are to be achieved.

Rising atmospheric nitrogen deposition will lead to environmental effects associated with the nitrogen cascade (Galloway *et al.* 2003), including impacts on plant diversity. The Convention on Biological Diversity has recognized nitrogen deposition as an indicator of the threat to biodiversity (CBD 2010b) and especially to sensitive ecosystems that receive a total nitrogen deposition above 10 kg per hectare per year (Figure 2.11). However, the full impact is difficult to estimate as there is little quantification of the effects on biodiversity outside Europe and North America.

Designing effective policies to balance the positive impacts of nitrogen deposition, such as increases in crop yields and carbon sequestration, and the negative impacts, such as loss of biodiversity and increased greenhouse gas emissions, underlines the necessity for a truly integrated approach to nitrogen management in the environment.

Figure 2.10 Regional trends in emissions of nitrogen oxides and ammonia, 1850–2050

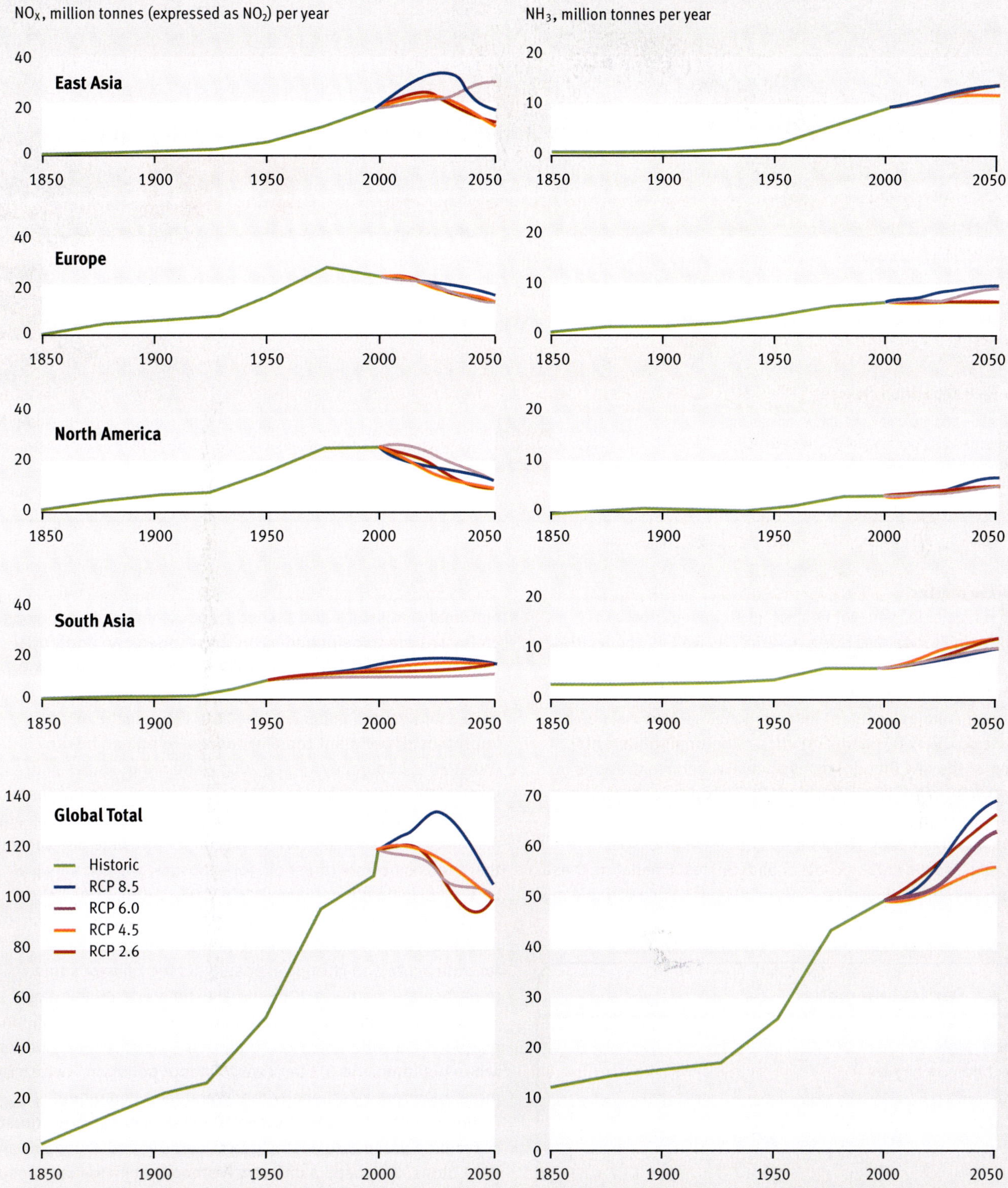

Emission trends from 1850–2000 and four RCP scenarios from 2000–2050, developed to contribute to the Fifth Assessment of the IPCC, are shown for the four regions and for the global total from the HTAP multi-model experiments. Total nitrogen oxide emissions (left) remain more or less constant at global scale while global ammonia emissions (right) are projected to increase in most scenarios.

Source: HTAP 2010

Figure 2.11 Trends in nitrogen deposition to protected areas, 2000–2030

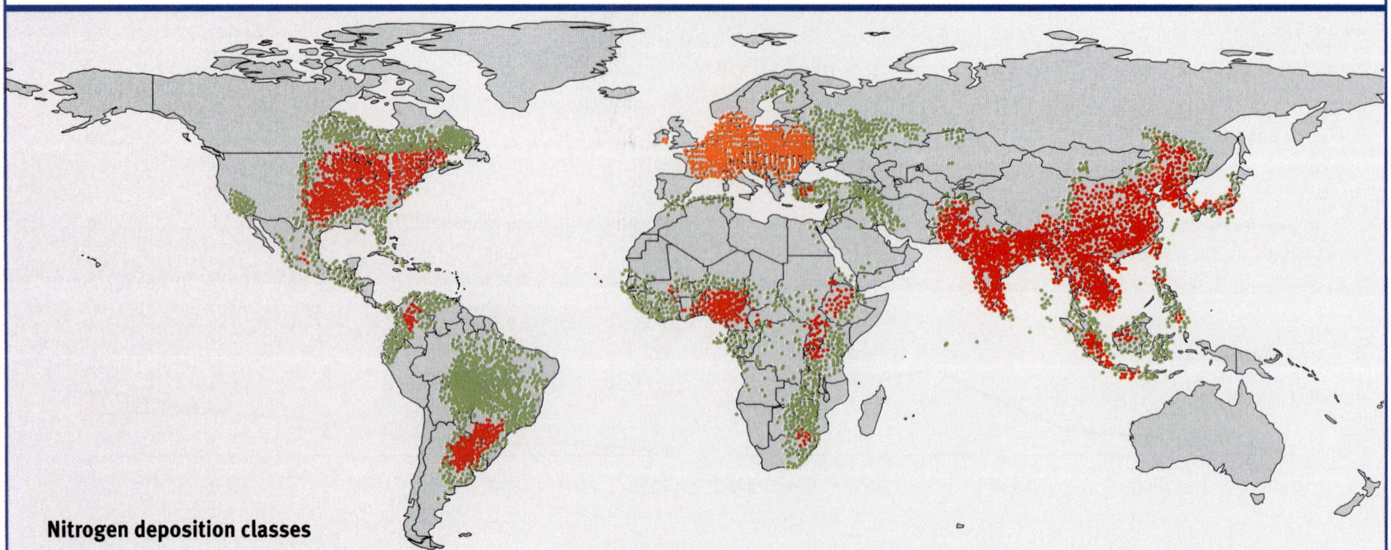

Nitrogen deposition classes

- Exceeds 10 kg per hectare and is increasing
- Exceeds 10 kg per hectare and is decreasing
- Between 5 and 10 kg per hectare and increasing

Note: Red indicates areas protected under the Convention on Biological Diversity where total nitrogen deposition is relatively high and is estimated to increase by 2030, green areas where it is still relatively low but increasing, and orange where it is high but is decreasing.

Source: Bleeker *et al.* 2011.

Particulate matter

Control of particulate matter has achieved mixed progress worldwide. In Europe and North America, as well as some cities in Latin America and Asia, emissions of PM_{10} – particles of 10 micrometres in diameter or less – have been reduced, but they remain a major pollutant in many other cities in Asia and Latin America. Very few cities in Africa monitor air pollutants; however, of the few that do, many show PM_{10} concentrations in excess of WHO guidelines (WHO 2012). Outdoor concentrations in high-income countries come close to the WHO PM_{10} guideline of 20 micrograms per m^3 (Figure 2.12). In Africa, the most widespread issue is indoor levels of particulates. Regulating these pollutants is complex because they are composed of a variable mix of primary emissions and secondary pollutants, where the original emissions are transformed in the atmosphere. An additional challenge for cities is the elimination of particulate hotspots.

Box 2.4 Particulate matter

Related goals
Protect human health

Indicators
PM levels

Global trends
Mixed progress in relation to WHO guidelines, with significant reductions in the EU and North America and some Latin American and Asian cities, but mostly high concentrations in urban areas in Asia and Latin America; data for Africa is insufficient, but some cities have high PM levels

Particulate matter, especially the finer $PM_{2.5}$, is the most important air pollutant causing damage to human health (WHO 2011; Carnelley and Le 2001). The prime sources of particulate matter relate to the energy, transport and industry sectors, but open burning of solid waste and crop residues are also important sources. Health research worldwide has shown that there is no safe threshold for exposure, as even very low levels cause health damage (WHO 2006, 1999). Impacts on health are predominantly associated with respiratory and cardiovascular illnesses, but the range of effects is broad for both acute and chronic exposure. Based on exposures to particulate matter in 2004, WHO estimated that annually 5.3 per cent of premature deaths worldwide, about 3.1 million people, are attributable to air pollution – 2 per cent to outdoor urban pollution and 3.3 per cent to indoor pollution – which is more than from all other environmental risks combined (Table 2.4) (WHO 2009). However, a more recent study estimated 3.7 million premature deaths due to outdoor anthropogenic $PM_{2.5}$ alone, as it used a different method that includes exposure in rural areas, does not have a low-concentration threshold, and used updated concentration-response relationships (Annenberg *et al.* 2010). Worldwide, approximately 41 million disability-adjusted life years (DALYs) – the sum of potential healthy life years lost due to illness – are attributed to solid fuel and methods use with about 18 million, or 44 per cent of the total, occurring in sub-Saharan Africa (UNDP and WHO 2009). Household energy

Table 2.4 Global burden of disease due to particulate air pollution

Air pollution type	Premature deaths	Morbidity (DALYs)
Urban outdoor	1.15 million = 2.0% of deaths worldwide 0.61 million males and 0.54 million females 8% of lung cancer deaths 5% of cardiopulmonary deaths 3% of respiratory infection deaths	8.7 million DALYs
Indoor	1.97 million = 3.3% of deaths worldwide 0.89 million males and 1.08 million females 21% of lower respiratory infection deaths 35% of chronic obstructive pulmonary deaths 3% of lung cancer deaths 0.9 million deaths due to pneumonia among children younger than five years	41 million DALYs
Total air pollution	3.12 million = 5.3% of deaths worldwide	49.7 million DALYs

Note: DALYs – disability adjusted life years: the sum of potential healthy life years lost due to illness.

Source: WHO 2009

interventions, which reduce dependence on traditional fuels and methods for cooking and heating, clearly have the potential to improve health and promote achievement of the MDGs. Even in high-income countries such as the United Kingdom, $PM_{2.5}$ is estimated to have contributed to 29 000 premature deaths and the loss of 340 000 life-years in 2008 (COMEAP 2010), despite considerable progress in reducing concentrations.

Recent assessments of the long-range transport of air pollution indicate that intercontinental transport of particulate matter is contributing to exceedances of public health standards and visibility targets. Long-range transport of particulates may be responsible for 380 000 premature deaths worldwide, of which 75 per cent are attributable to (mineral) dust $PM_{2.5}$ (HTAP 2010). The impacts of air pollution from natural sources are an emerging atmospheric issue requiring attention, and will be discussed in the subsequent section on atmospheric governance.

Various measures, including technological improvements to vehicles, increased transport and energy efficiencies and cleaner fuels and filters, have been successful in developed countries and to some extent in developing ones. However, while the latter are catching up in the use of cleaner technologies, such efficiencies are being compromised by a rapid increase in emission sources, for example fuel use for energy and transport. Where indoor particulates are concerned, global partnerships are promoting cleaner energy and improved cooking stoves.

Most developed and developing countries have adopted ambient air quality standards (Figure 2.12), but concentrations of particulates in most cities exceed the levels recommended by WHO's ambient air quality guidelines for protecting human health and ecosystems (Figure 2.13). Most of the PM_{10} standards in developing countries are less stringent than the interim targets set by WHO to promote a progressive reduction in air pollution. WHO has also recommended $PM_{2.5}$ guidelines, but many countries have yet to adopt standards and monitoring practices. In Asia in 2010, for example, only four out of the 22 countries have standards for $PM_{2.5}$ supported by monitoring. There is growing interest in the health impacts of micrometre and sub-micrometre particles, as discussed below in the section on emerging issues.

Projections of a 20 per cent reduction in $PM_{2.5}$ emissions by 2020 in Europe are expected to lead to a 40 per cent fall in the associated years of life lost compared to the year 2000; nonetheless, $PM_{2.5}$ air pollution is still expected to shorten statistical life expectancy by 4.6 months (Amann *et al.* 2011). However, if the new National Emission Ceilings in Europe are implemented the benefits will outweigh the costs 12–37 times depending on the valuation method (AEA 2010), and PM emissions could be reduced by 35–50 per cent depending on the portfolio of measures. Meanwhile, the US Environmental Protection Agency reports that lowered levels of ambient $PM_{2.5}$ and ozone due to the US Clean Air Act are projected to result in avoided mortalities valued at US$1.2 trillion in 2010 and US$1.8 trillion in 2020 (2006 dollars). Reductions in exposure to particulate matter account for more than 90 per cent of these projected and realised benefits (USEPA 2010).

Using traditional cooking methods with biomass as a fuel causes severe indoor pollution by particulates and significant outdoor concentrations of black carbon and other particles. *© Stillpictures/nbsp*

Figure 2.12 National ambient air quality standards and WHO guidelines for PM_{10}

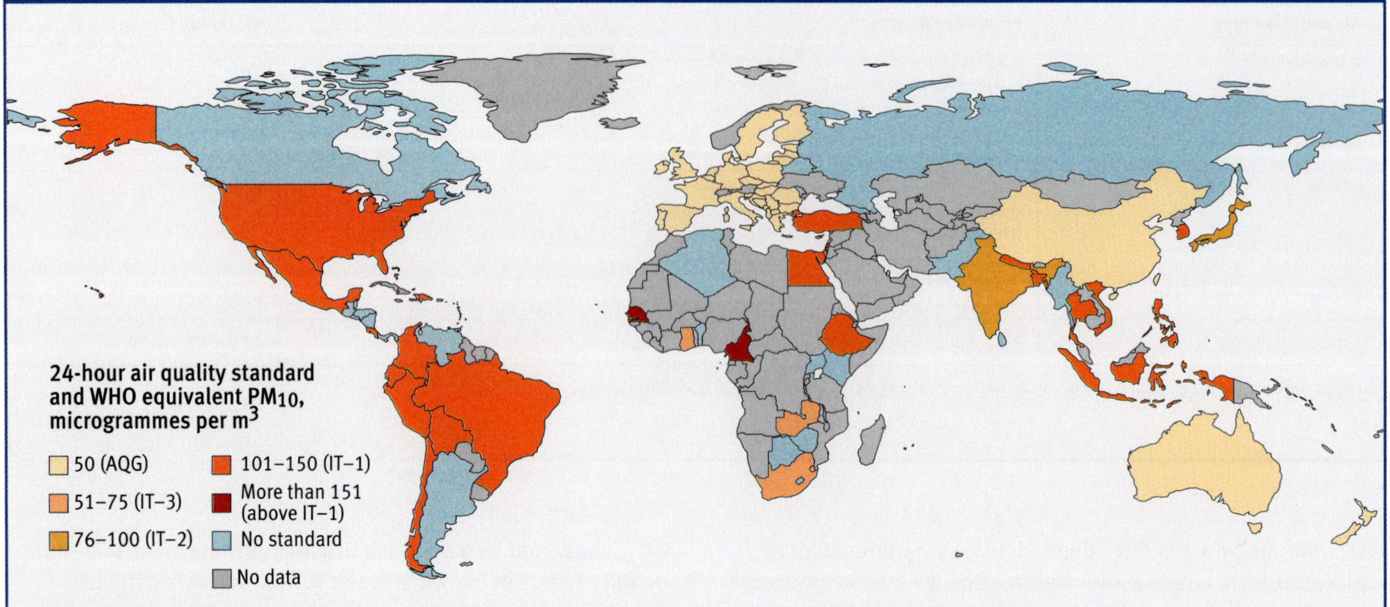

Note: AQG–WHO 24-hour Air Quality Guidelines.
IT–WHO 24-hour Interim Target.

Source: Vahlsing and Smith 2010

Figure 2.13 Urban PM_{10} trends in selected regions and cities, 1993–2009

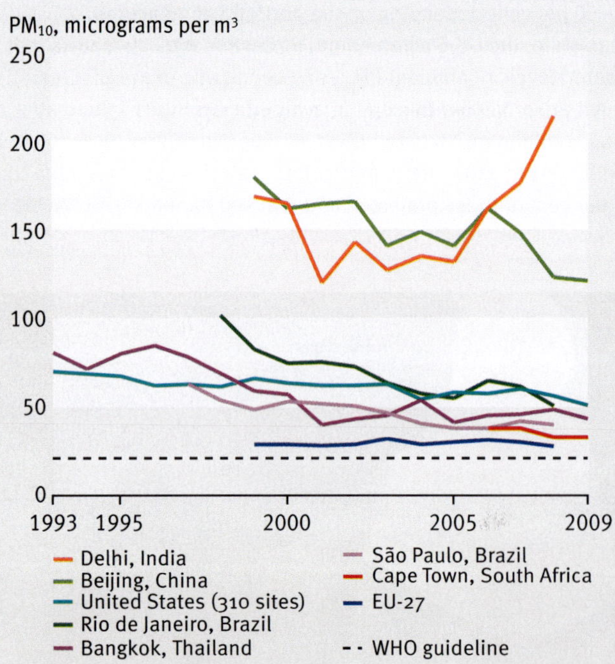

Note: Air quality data come from a variable number of monitoring sites.

Source: Government of NCT of Delhi 2010; Beijing Statistical Yearbook; USEPA; Brazilian Statistic and Geographic Institute; Companhia de Tecnologia de Saneamento Ambiental; Fundação Estadual de Engenharia e de Meio Ambiente; City of Cape Town; EC 2011; WHO 2006

There are some uncertainties that need to be resolved to inform better policy making for particulate matter and health. These include the concentration and impact of particle sizes and a better understanding of the nature of primary and secondary PM pollution in different locations through monitoring, emission inventories and modelling, as well as through the use of source apportionment and estimation of the economic value of health impacts. Efforts to harmonize ambient air quality standards and the building of capacity have the potential to fast-track PM reduction in developing countries, amplifying the successful policies and technologies applied in Europe and North America and in some Asian and Latin American cities.

Tropospheric and surface ozone

Tropospheric ozone (O_3) in the lower atmosphere, from 0–10 up to 20 km above the Earth's surface, is responsible for ozone's impact on warming. Ground-level or surface ozone refers to concentrations at ground level that affect both human health and ecosystems. There is mixed progress in controlling tropospheric ozone: peak concentrations have decreased in Europe and North America, and North America, while background concentrations have increased. In rapidly industrializing regions both background and peak concentrations have been steadily rising (Royal Society 2008).

Ozone causes harm in three main ways. Firstly, surface ozone damages human health and its impact is considered second only to particulate matter. It is responsible for an estimated 0.7 million respiratory deaths globally each year (Anenberg *et al.* 2010), more than 75 per cent of which are in Asia. Ozone

Box 2.5 Tropospheric ozone

Related goals
Protect human health, crop yields, ecosystems and climate

Indicators
Precursor emissions; ozone levels

Global trends
Mixed progress in relation to CLRTAP targets: some reductions in EU and North America and mostly increasing concentrations in Asia; data for Africa is insufficient

can also have chronic health effects resulting in permanent lung damage (Royal Society 2008).

Secondly, surface ozone is the most important air pollutant causing damage to vegetation (Emberson *et al.* 2009; Ashmore 2005), diminishing crop yields and forest productivity and altering net primary productivity. Estimates suggest, for example, that ozone-induced yield losses range between 3 and 16 per cent for four staple crops – maize, wheat, soybean and rice – which translate into annual global economic losses of US$14–26 billion (HTAP 2010).

Lastly, ozone is the third most important greenhouse gas after CO_2 and methane (IPCC 2007), but is classified as a short-lived climate forcer due to its residence time in the atmosphere of just days to weeks. Tropospheric ozone is estimated to have been responsible for a change in radiative forcing of +0.35 (-0.1, +0.3) watts per m^2 since pre-industrial times, compared to a combined anthropogenic radiative forcing of +1.6 (-1.0, +0.8) watts per m^2 (IPCC 2007). These ozone-induced changes are thought to be responsible for 5–16 per cent of the global temperature change since pre-industrial times (Forster *et al.* 2007). Reductions in biomass caused by ozone also influence the amount of carbon sequestered within terrestrial ecosystems. This effect is estimated to increase atmospheric CO_2 concentrations such that the additional radiative forcing could exceed warming due to the direct radiative effect of tropospheric ozone in the atmosphere (Sitch *et al.* 2007).

Ozone is not directly emitted into the atmosphere but rather is formed when precursor pollutants – nitrogen oxides and volatile organic compounds, including methane, and carbon monoxide – react in the presence of sunlight. As such, ozone concentrations tend to be higher at some distance – tens to thousands of kilometres – downwind of precursor pollutant sources, causing ozone to pollute at the local, regional and hemispheric scale.

Photochemical reactions account for approximately 90 per cent of the ozone in the troposphere, with the remaining 10 per cent directly transported from the stratosphere. Around 30 per cent of tropospheric ozone is due to anthropogenic emissions, with 40 per cent of the change in the global ozone burden since pre-industrial times due to increases in methane, and the remainder to increases in emissions of nitrogen oxides, carbon monoxide and volatile organic compounds other than methane (HTAP 2010). The origin of ground-level or surface ozone relevant to effects on human health and ecosystems across the polluted regions of the northern hemisphere is 20–25 per cent from the stratosphere and a similar proportion from natural precursor sources including lightning and emissions from soils, vegetation and fire, along with oxidation of naturally occurring methane. The anthropogenic contribution thus typically exceeds 50 per cent over these regions (Figure 2.14).

Figure 2.14 Sources of ozone over polluted regions of the northern hemisphere, 1850 and 2000

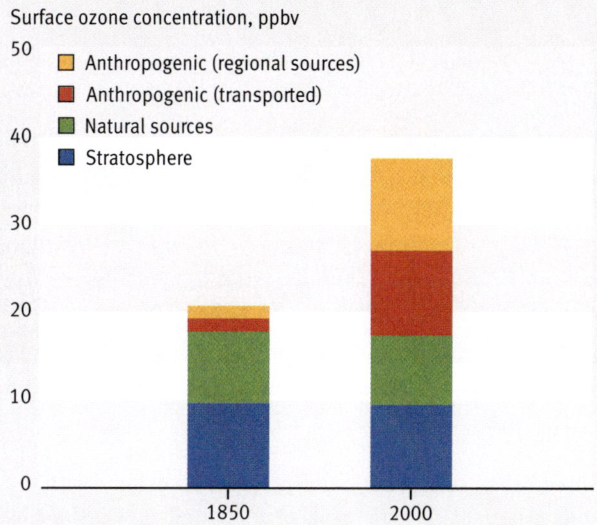

Note: ppbv - parts per billion by volume.

Source: HTAP 2010

Surface ozone does more damage to food crops than any other air pollutant. *© Evgeny Kuklev/iStock*

Ground-level ozone is one of the major contributors to city smog. © *T. Kimura*

Elevated ozone concentrations tend to be associated with regions experiencing high levels of uncontrolled emissions from industrial and urban centres as well as seasonal periods with high solar radiation. This causes high variability in global and seasonal concentrations. Regions in North America, Europe and Asia have been identified as having a high anthropogenic ozone load (Figure 2.15).

The targets defined for tropospheric ozone in the United Nations Economic Commission for Europe (UNECE) region are now being exceeded at many locations. Coordinated action in Europe, however, has resulted in emissions of nitrogen oxide and volatile organic compounds currently being 30 and 35 per cent lower than in 1990, leading to reductions in short-term peak ozone concentrations to daily peak values of around 60 micrograms per m^3 of air. In contrast, mean ozone concentrations at many locations have been increasing due to a variety of different factors. For example, local emission reductions of nitrogen oxides, and hence nitric oxide, remove a key mechanism of ozone destruction and can result in increases in concentrations in urban areas (Royal Society 2008). There is also evidence of background ozone concentrations increasing by up to 10 micrograms per m^3 of air per decade since the 1970s (Royal Society 2008) due to changes in stratospheric ozone incursions, hemispheric transport and ozone formation responding to climate change. This will increase both mean and peak ozone levels.

Global photochemical modelling studies performed for the HTAP (2010) assessment provide estimates of changes in surface ozone concentration for those regions that currently show the highest concentrations. These data indicate recent reductions in surface ozone in North America and Europe, which are likely to be due to effective controls of nitrogen oxides and volatile organic compounds over the past two decades in response to the US Clean Air Act and CLRTAP and EU targets in Europe. In contrast, trends in Asia are continuing upwards due to continued rapid industrialization across the region (Figure 2.15). However, these regional trends may hide large local variations.

Future changes in tropospheric ozone concentrations have been explored using a number of different global photochemical models for a variety of emission scenarios and provide variable results (Figure 2.16). The HTAP (2010) assessment used a mean from six global models to assess the implications of emission changes between 2000 and 2050 following the RCP emission

Figure 2.15 Regional changes in concentrations of surface ozone, 1960–2000

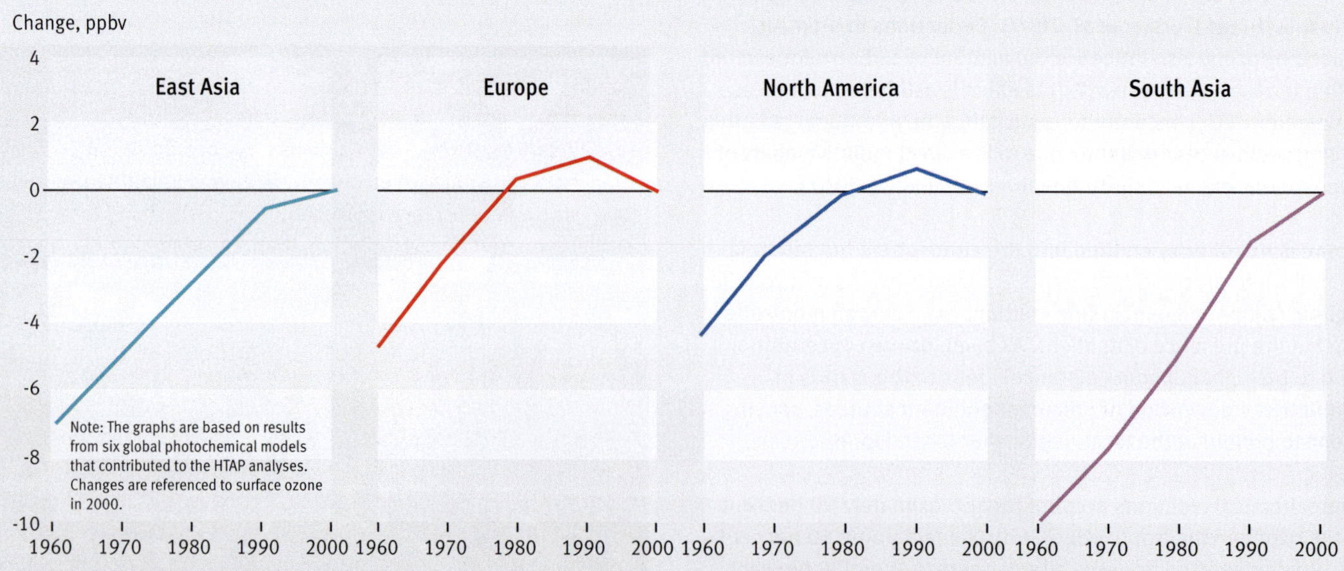

Source: HTAP 2010

Figure 2.16 Projected changes in surface ozone concentrations over polluted regions of the northern hemisphere, 2000–2050

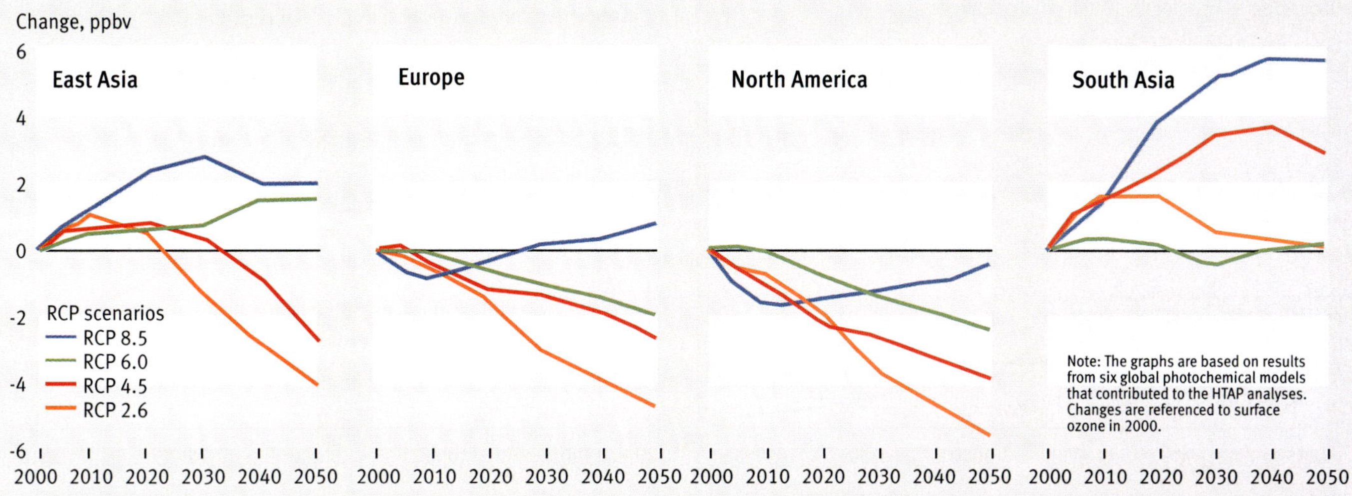

Source: HTAP 2010

scenarios. The outlook for ozone concentrations is heavily dependent on global and regional emission pathways.

Assessment of the effectiveness of policies introduced to control ozone requires an extended global monitoring network covering rural as well as urban locations. Improved understanding of ozone's impact on human health and ecosystems and how climate change will affect its formation, as well as how ozone acts in combination with other stressors such as global warming and excessive nitrogen deposition, will also be important. The growing interest in ozone as a short-lived climate forcer and the associated human health, arable-agriculture and ecosystem benefits that its reduction might bring (UNEP/WMO 2011) make this a pollutant of particular interest for policy intervention.

Box 2.6 Stratospheric ozone

Related goals
Protection of the stratospheric ozone layer

Indicators
Consumption of ozone-depleting substances; atmospheric burden; annual extent of the Antarctic ozone hole

Global trends
Significant progress

Progress on internationally agreed goals

There are two examples of substantial progress in solving issues and achieving targets: protection of the stratospheric ozone layer and the removal of lead from petrol.

Stratospheric ozone layer

Global regimes to address stratospheric ozone depletion include the 1985 Vienna Convention for the Protection of the Ozone Layer and its 1987 Montreal Protocol on Substances that Deplete the Ozone Layer. The latest scientific assessments confirm the success of the action taken under the Montreal Protocol to eliminate consumption of ozone-depleting substances (Figure 2.17) (WMO 2011; UNEP 2010).

Stratospheric ozone protects humans and other organisms because it absorbs ultraviolet-B (UV-B) radiation from the sun. In humans, heightened exposure to UV-B radiation increases the risk of skin cancer, cataracts and suppression of the immune system. Excessive UV-B exposure can also damage terrestrial plant life, single-cell organisms and aquatic ecosystems. In the mid-1970s, it was discovered that the thinning of the stratospheric ozone layer was linked to the steady increase of chlorofluorocarbons (CFCs) – used for refrigeration and air conditioning, foam blowing and industrial cleaning – in the atmosphere.

The most severe and surprising ozone loss – which came to be known as the ozone hole – was discovered to be recurring in springtime over the Antarctic. Thinning of the ozone layer has also been observed over other regions, such as the Arctic (Manney *et al.* 2011) and northern and southern mid-latitudes.

Despite the drastic reduction in consumption of ozone-depleting substances (Figure 2.17), their concentrations in the stratosphere remain high (as shown by the Ozone Depleting Gas Index (ODGI), Figure 2.18) as they have long atmospheric lifetimes. Figure 2.18

Figure 2.17 Consumption of ozone-depleting substances, 1986–2009

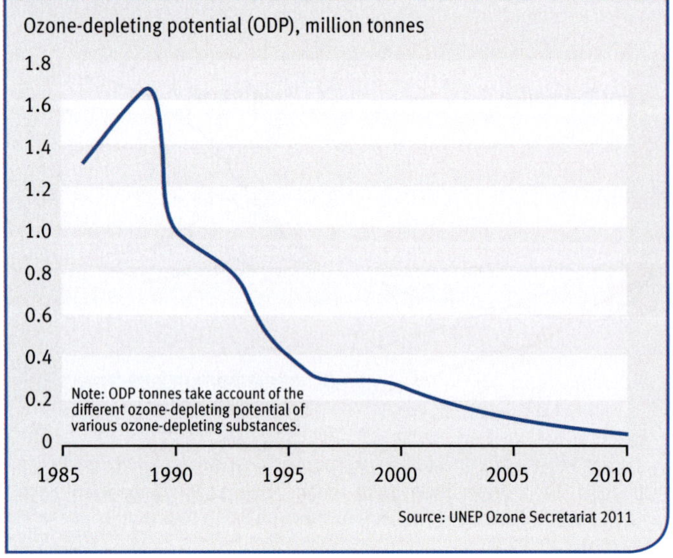

Figure 2.18 Reduction of ozone-depleting substances in the stratosphere, 1994–2009

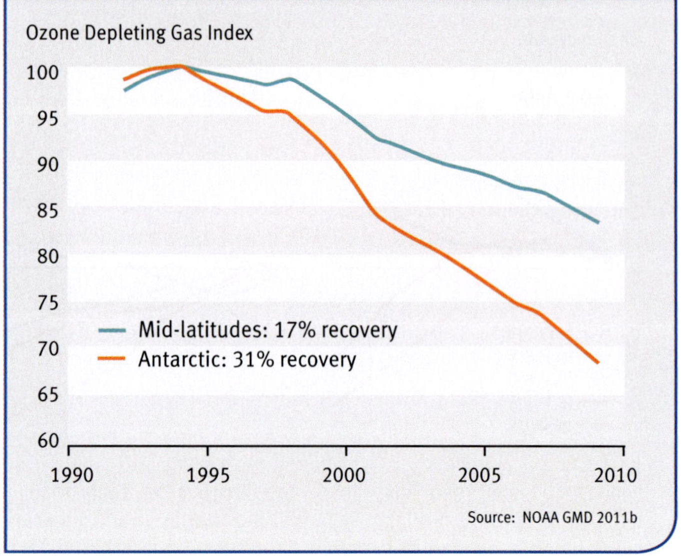

shows a 31 per cent recovery from a peak in 1994 at mid-latitudes and a 17 per cent recovery in the Antarctic.

The hole in the Antarctic ozone layer is the clearest manifestation of the effect of ozone-depleting substances: springtime Antarctic total column ozone losses continue to occur each year, with the extent affected by meteorological conditions. Figure 2.19 shows the development of Antarctic ozone depletion over the last three decades, measured from 19 July to 1 December each year. The largest ozone hole on record occurred in 2006 (WMO 2011).

Model simulations for an ozone depletion scenario without the Montreal Protocol, the "World Avoided" scenario, show that there would have been a 300 per cent increase in the UV radiation at

Figure 2.19 Antarctic ozone hole extent, 1980–2010

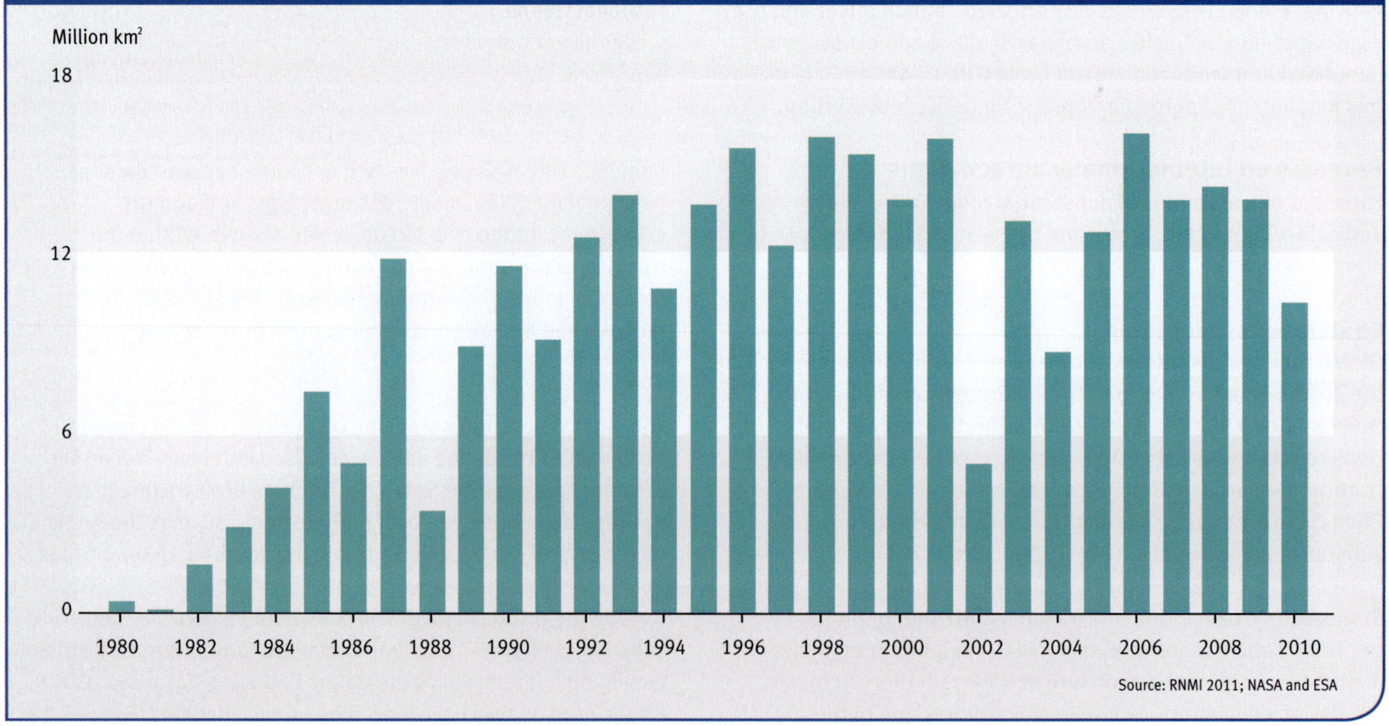

Figure 2.20 The World Avoided modelled UV index, 1975, 2020 and 2065

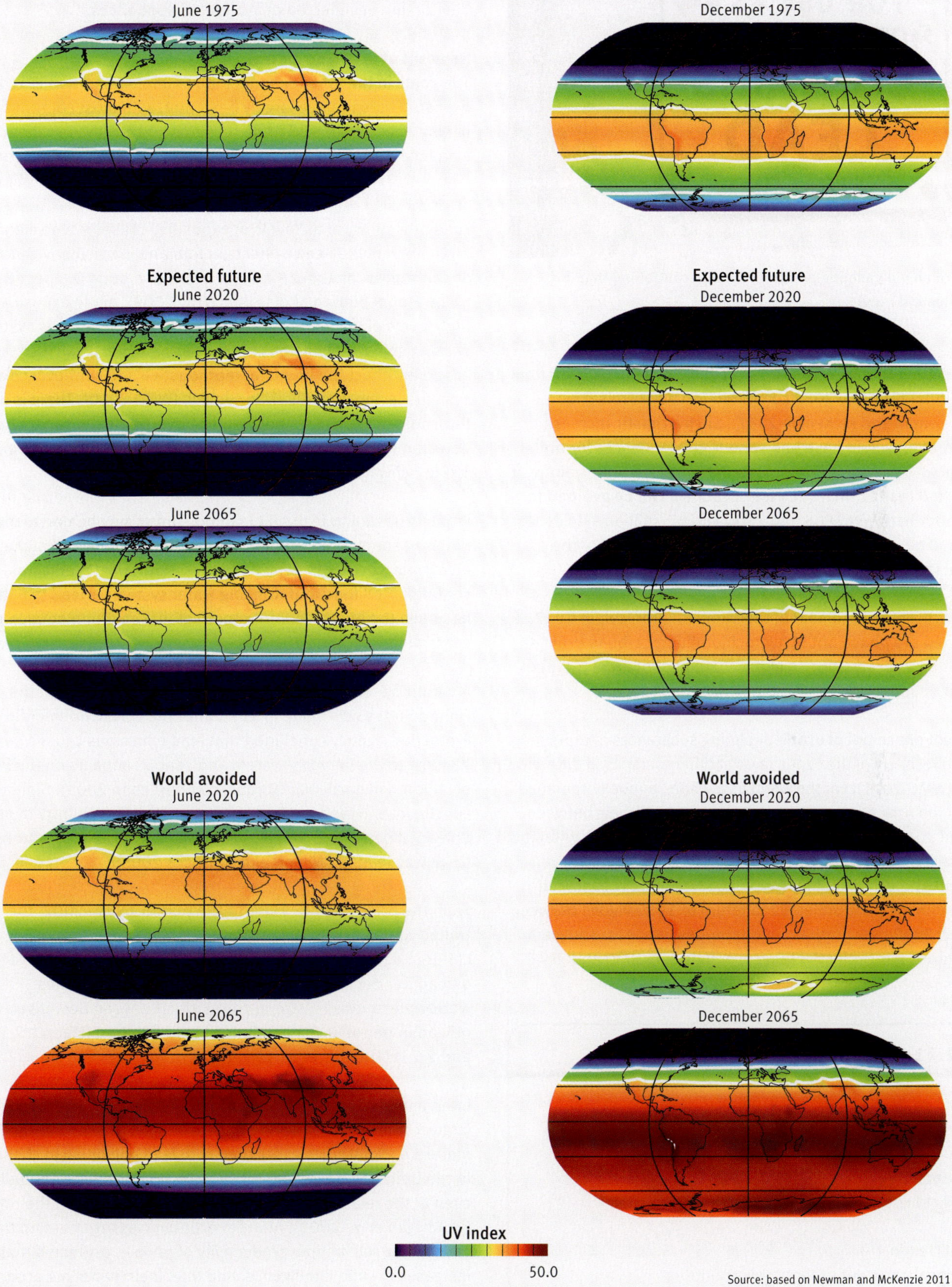

Source: based on Newman and McKenzie 2011

The side of an old gasoline pump with a sign cautioning that it contains lead. © *Tim Messick*

mid-northern latitudes or a 550 per cent increase in UV radiation at the mid-latitudes by 2065 compared to 1980 levels (Figure 2.20) (Newman and McKenzie 2011). Such a drastic increase in UV-radiation would have had serious consequences for both human health and the environment. In the United States alone, it is estimated that 22 million cases of cataract will be avoided for people born between 1985 and 2100 and 6.3 million skin cancer deaths will be avoided up to the year 2165 as a result of the Montreal Protocol (USEPA 2010).

The most recent amendment of the Montreal Protocol, in 2007, accelerated the phase-out of hydrochlorofluorocarbons (HCFCs), contributing to a reduction in global warming potential (GWP) of about 18 billion tonnes of CO_2-equivalent emissions.

The current phase-out of ozone-depleting substances is expected to lead to recovery of the ozone layer at different times in different regions (WMO 2011). For the world as a whole, annually averaged total column ozone is projected to return to 1980 levels between 2025 and 2040, but this will take until mid-century in the Antarctic, with small episodic Antarctic ozone holes likely to persist even at the end of the 21st century (WMO 2011). Annually averaged total column ozone is projected to return to 1980 values between 2015 and 2030 over northern mid-latitudes, while for southern mid-latitudes it is projected to recover between 2030 and 2040.

Box 2.7 Lead in petrol

Related goals
Prevention of exposure to lead

Indicators
Number of countries with leaded petrol

Global trends
Globally phased out except in six countries

Despite the successful implementation of certain Montreal Protocol provisions, some issues remain regarding the capture of ozone-depleting substances in old equipment and the destruction of collected or stockpiled appliances.

Removal of lead from petrol

The goals in the Johannesburg Plan of Implementation to reduce exposure to lead have largely been met, with most countries having phased out lead in petrol since 2002, although there is evidence that leaded petrol is still sold in at least six countries (Figure 2.21).

Lead poisoning, at all levels of exposure, causes adverse and often irreversible health effects in humans, particularly children, and accounts for about 9 million DALYs or around 0.6 per cent of the global burden of disease (WHO 2009). Acute exposure to high levels of lead affects the brain and central nervous system, causing coma, convulsions and even death. Lead can also adversely affect the immune, reproductive and cardiovascular systems, even at relatively low levels (WHO 2010). There is no threshold of exposure under which adverse effects cannot be detected (Lanphear *et al.* 2005; Schneider *et al.* 2003; Lovei 1998; Schwartz 1994).

While exposure to lead and lead poisoning may be due to many different sources and products – including paints and pigments, electronic waste, cosmetics and toys, traditional medicines, contaminated food and drinking water systems – lead in petrol has been the biggest contributor to global environmental lead contamination (WHO 2010).

A health-based regulation to remove lead from petrol in the United States was made in 1973 after the US Environmental Protection Agency concluded that lead emissions caused serious damage to the nervous system and serious impacts on the health of children in particular (Bridbord and Hanson 2009). Similar conclusions in Japan led to it becoming the first country to market unleaded petrol and, by 1981, less than 3 per cent of petrol sold there was leaded (Wilson and Horrocks 2008).

From the period 1976–1980 to the period 1999–2002, the United States saw a 98 per cent reduction in the proportion of children aged one to five with levels of more than 10 micrograms of lead per decilitre of blood (CDC 2005). Other studies worldwide showed a strong correlation between decreased use of lead in petrol and reductions of lead in blood (Figures 2.22 and 2.23) (Thomas *et al.* 1999).

Interventions to prevent lead poisoning have demonstrated very large economic benefits. An analysis of the direct medical and indirect societal costs associated with lead poisoning in children in the United States found these to be US$43 billion annually, even at the relatively low levels of exposure to lead at the time (Landrigan *et al.* 2002). Another economic assessment, in this case for the full lifetime productivity of people, estimated that the increases in child intelligence, and thus their in-lifetime economic productivity, resulting from the removal of lead from petrol produced a benefit of between US$110 billion and US$319 billion in each birth cohort in the United States (Grosse *et al.* 2002).

According to Gould (2009), for every US$1 spent to reduce lead hazards, there would be a benefit of US$17–220, a better cost-benefit ratio than that for vaccines, which have long been described as the single most cost-beneficial medical or public health intervention. Another methodology that uses GDP extrapolation applied to published literature showed the global benefits of the phase-out of lead in petrol to be US$1–6 trillion per year, with a best estimate of US$2.45 trillion per year, roughly 4 per cent of global GDP (Tsai and Hatfield 2011).

Recent evidence on health impacts has led the US Environmental Protection Agency to tighten its rolling three-month average air quality standard for lead from 1.5 micrograms of lead per m^3 of air in 1978 to 0.15 micrograms per m^3 in 2008 (USEPA 2008). The WHO annual ambient air guideline for lead remains at

Figure 2.21 Leaded petrol phase-out, 2002 and 2011

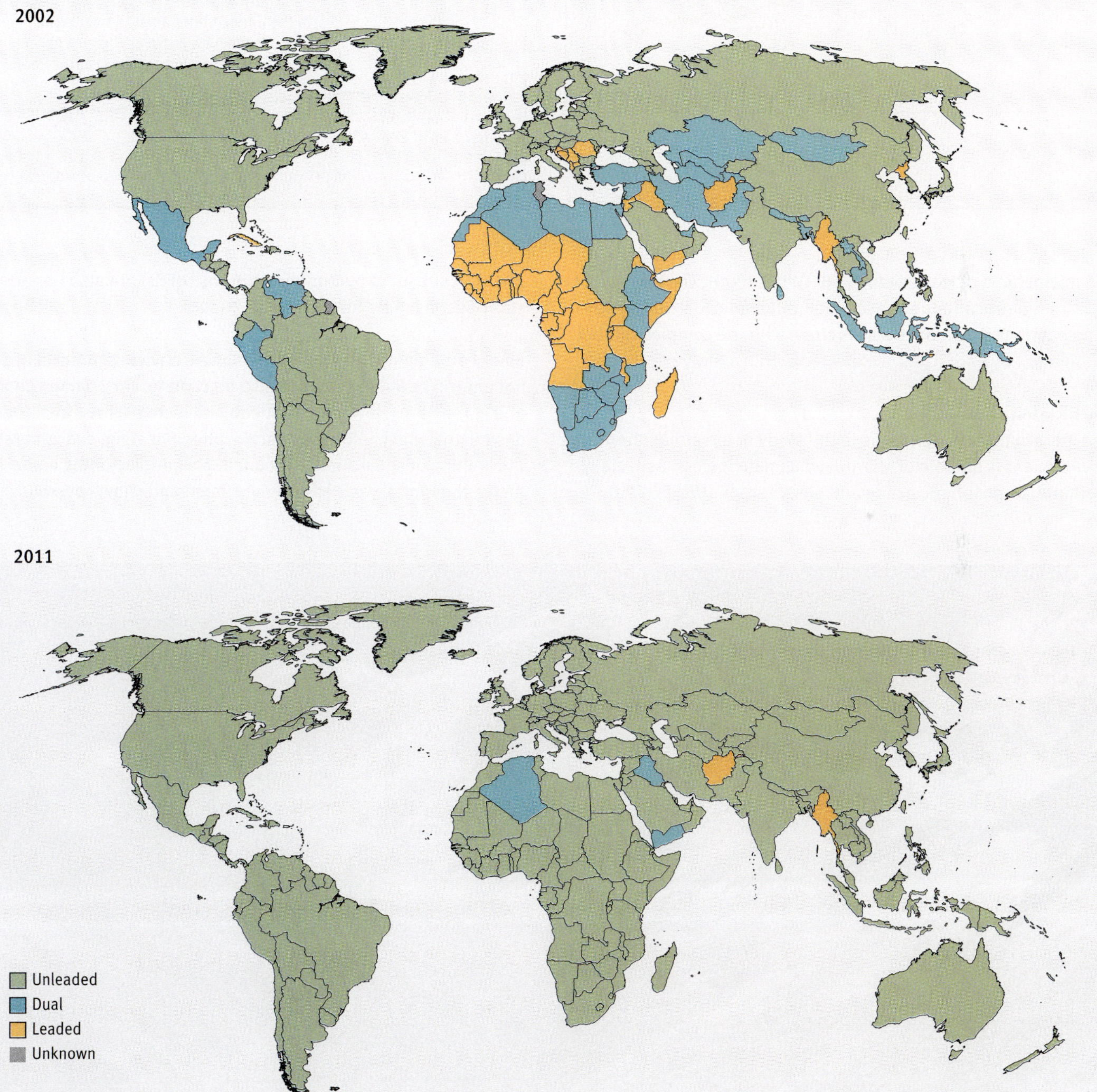

Source: UNEP 2011b

Figure 2.22 Petrol and blood lead levels in Sweden following the phase-out of lead in petrol, 1976–2004

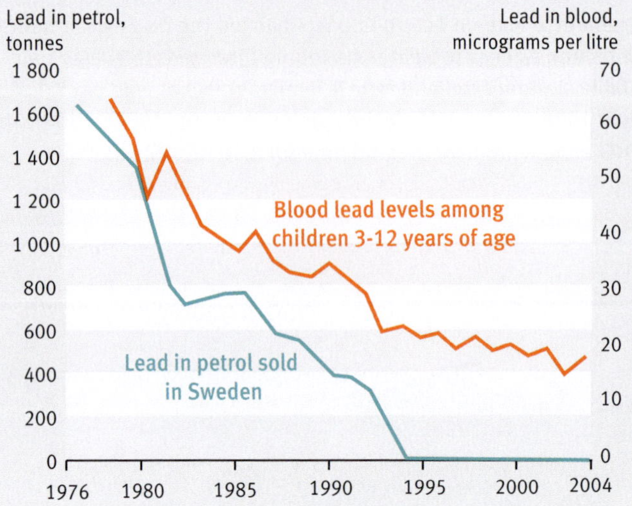

Source: EEA 2011

Figure 2.23 Blood lead levels in the United States following the phase-out of lead in petrol, 1976–2008

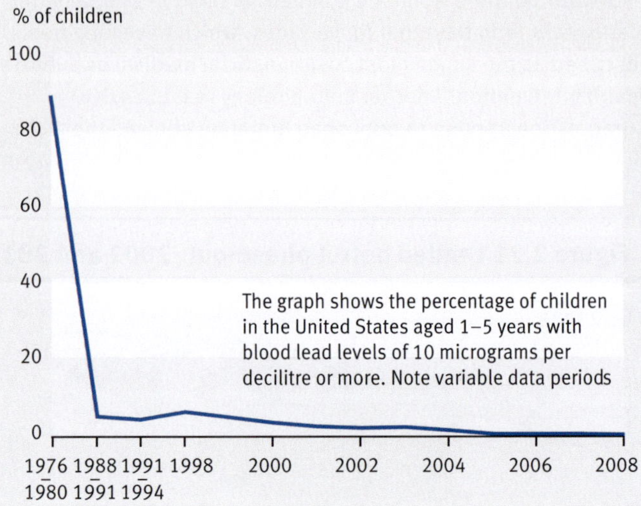

Source: CDC 2012, 2003

0.5 micrograms of lead per m^3 of air (WHO 2000). The removal of lead from petrol and the consequent reductions in health risks is an outstanding global success story, with the complete global elimination of lead in petrol expected within a few years.

EMERGING ISSUES

The most important new issue in the study of the atmospheric environment is the role of short-lived climate forcers, especially methane, tropospheric ozone and black carbon (UNEP/WMO 2011). A subset of hydrofluorocarbons (HFCs) are also important short-lived climate forcers (UNEP 2011c).

Black carbon particles in the atmosphere have a significant impact not only on human health but also on climate. They darken snow and ice surfaces, lower their albedo and increase their absorption of sunlight, which, along with atmospheric heating, exacerbates the melting of snow and ice around the world, including in the Arctic, the Himalayas and other glaciated and snow-covered

Snow and ice cover in the Himalayas is affected by aerosols, including black carbon. *© Arsgera*

The production of bricks in traditional kilns is a locally significant source of black carbon in South Asia. *© Alexander Kataytsev/iStock*

regions. This affects the water cycle and may increase risks of flooding. Methane is a powerful greenhouse gas and an important precursor for ozone generation. Methane, black carbon and tropospheric ozone are fundamentally different from the longer-lived greenhouse gases as they remain in the atmosphere for a relatively short time. Reducing black carbon and methane emissions now will slow the rate of climate change within the first half of this century (Shindell *et al.* 2012; UNEP/WMO 2011).

A second major emerging issue is the health effect of fine particles of natural origin. Every year, very large amounts of soil-derived dust and particles from wildfires engulf major populated areas. These can include soil particles from arid regions being deposited on coastal cities of China, Saharan dust reaching cities in Africa and the Mediterranean, and dust from drought-affected inland areas being deposited on cities in the United States and Australia. In addition, smoke from wildfires commonly raises particulate concentrations in Africa, Siberia, the Mediterranean, the United States, South East Asia and Australia. These particles can have major impacts on human health and a recent study suggests that almost 300 000 excess deaths per year can be attributed to fine particles of natural origin (Liu *et al.* 2009a, 2009b). The sources can, however, be at least partially controlled (Chapter 3). Major interventions to re-vegetate degraded landscapes are continuing in several countries and the 2003 ASEAN Agreement on Transboundary Haze Pollution is an example of an international agreement aimed at addressing the international transport of particles generated from forest burning.

As understanding of the relationship between particle size, the number of particles and health impacts has improved, concern has grown about the impacts of fine particles (smaller than 2.5 micrometres in diameter) and ultrafine particles (1 micrometre and sub-micrometre sizes) on respiratory and cardiovascular health (Schmid *et al.* 2009; Valavanidis *et al.* 2008). As the volume of evidence is rapidly growing, it is likely that in the next few years air quality standards and guidelines to protect health by controlling exposure to ultrafine particles will be developed and become a focus of air quality policy, monitoring and management.

A number of new approaches to address the challenges of climate change have been proposed, including carbon capture and storage and geo-engineering (IPCC 2005; Rasch *et al.* 2008).

ATMOSPHERIC GOVERNANCE AND AN INTEGRATED APPROACH TO MANAGEMENT

In this section a simple analytical framework based on the level of concern for the different issues, the relative cost of addressing the problems, and the complexity of managing atmospheric issues is used to evaluate the governance of atmospheric problems. It suggests there is no one-size-fits-all solution to most atmospheric problems (Levy *et al.* 1993). The use of targets and timetables that worked for stratospheric ozone depletion may stall negotiations over climate change (Sunstein 2007). Emissions trading schemes that worked well for the reduction of sulphur dioxide in some developed countries may need to be complemented by other measures in developing countries (Chang and Wang 2010). Many emission sources release both greenhouse gases and air pollutants; some air pollutants have an additional effect on the climate; and reducing consumption of ozone depleting substances also reduces their impact on climate. There will be a growing need for decision-making frameworks and enabling environments that explicitly recognize the integrated nature of the atmosphere.

The elimination of lead from petrol was made easier by a cost-effective alternative that proved easy to communicate to politicians and other stakeholders. With the well-timed support of international initiatives such as the UNEP Partnership for Clean Fuels and Vehicles, one country after another introduced lead-free fuels (Hilton 2006).

Though there was no global binding agreement on the phase-out of lead in petrol, there are parallels with the phase-out of ozone-depleting substances as a relatively manageable problem with cost-effective solutions and a high level of concern. For the elimination of substances that harm the ozone layer, governments agreed to the Vienna Convention, setting in motion an international negotiation process that culminated in the Montreal Protocol. The protocol became the model for other international agreements that called for a series of targets and timetables to eliminate ozone-depleting substances in the developed world, and for the creation of a multilateral fund to finance replacement technologies for developing countries beginning to manufacture CFCs (Benedick 1998). The process of arriving at this agreement helped raise concern, lower costs and clarify complexities.

Progress with other pollutants has been more uneven. For example, in the case of sulphur dioxide, existing technology, affordable abatement costs and growing understanding have made the issue increasingly manageable in much of the developed world. However, while target setting and the installation of flue-gas desulphurization have become common, growth in the number of coal-fired power plants has overwhelmed efforts to reduce emissions. Consequently, levels of acid deposition in East Asia remain high.

The health impacts of particulate matter make its control the highest priority. However, measures can be costly and complex due to a multiplicity of industrial, transport, energy, commercial, domestic and natural sources, especially in developing countries. Various measures, such as technological improvements

A compact electric car, intended for urban use, charges its battery in a special charging station. © *iStock/code6d*

in vehicles, increased engine efficiency, cleaner fuels and particulate filters, have been successfully applied in different cities. Across the developed world urban levels of particulate matter began to fall sharply in the 1950s and 1960s. Cleaner technologies have enjoyed some success in reducing emissions in developing countries, but have not been sustained in rapidly growing cities where high demands for motorization, energy and industrial products have increased aggregate emissions. Both the complexity of the issue and the costs have hampered progress. To reduce exposure to indoor particulate matter national policies, including on rural development and energy, need to be mainstreamed into overall development policy.

Governance issues related to climate change have a high level of complexity, mixed levels of concern and long lead times between actions and benefits, often beyond political timescales. The governance approach for climate change has in many ways followed a similar approach to that of the ozone layer, but with different results due to differences in the nature of the issues. A growing degree of concern led to a global agreement, the UNFCCC, which allowed negotiation of the Kyoto Protocol. This was intended to initiate the process of reducing anthropogenic greenhouse gas emissions, but even if fully implemented was never designed to be enough to stay within the UNFCCC agreed limit of a 2°C temperature increase.

The approach of developing binding national targets within a global framework has so far not delivered emission reductions that would help achieve climate targets and internationally agreed goals. In the medium term, a promising approach appears to be the development of nationally appropriate mitigation actions – or NAMAs – to encourage countries to contribute to actions in a national context.

The achievement of global climate goals will most likely require a transformational change addressing the main drivers of emissions (Chapter 16) such as in the way electricity is generated, the efficiency with which energy and resources are used, and the management of terrestrial ecosystems (Chapters 3 and Chapter 12). Consumption levels and production processes may require the introduction of approaches such as a circular economy in which material flows are either made up of biological nutrients designed to re-enter the biosphere, or materials designed to circulate without entering the biosphere (Braungart *et al.* 2007). Such changes take time and, in the short term, existing cost-effective options need to be rolled out as far and as rapidly as possible to deliver early emission reductions that put the world on a path towards achieving established climate goals.

Mitigating near-term climate change – the warming likely to be experienced over the next two to four decades – is important to prevent damage to vulnerable ecosystems such as the Arctic, and to vulnerable societies such as those in drought- or flood-prone regions. Addressing CO_2 is not likely to be sufficient to mitigate warming in these timescales, partly because it is long-lived. Fortunately, near-term warming can be addressed by additional complementary policy action that reduces concentrations of short-lived climate forcers, including black carbon, methane and tropospheric ozone (Box 2.8) (Shindell *et al.* 2012; UNEP/WMO 2011). Addressing these short-lived substances is an example of an integrated approach to atmospheric governance that provides an opportunity for policy development to meet multiple goals in a cost-effective way. The improved awareness of atmospheric brown clouds emphasizes the integration of different atmospheric issues (Box 2.9). There is increasing evidence that the Antarctic ozone hole has affected the surface climate in the southern hemisphere (Polvani *et al.* 2011), and further links also exist between climate change and ozone-depleting substances as many are also very strong greenhouse gases. Indeed, avoiding emissions of CFCs has brought about a significant contribution to climate change mitigation (Velders *et al.* 2007).

These interactions and links between different atmospheric issues provide opportunities to improve progress in reaching internationally set goals, maximizing benefits and avoiding policy conflicts. To make rapid progress, scientific knowledge needs to be properly presented to policy makers to allow them to deal more effectively with the complexity of the issues. The way in which options are analysed, costs and benefits weighed up, and evidence-based policies developed, requires considerable improvement. This will require closer interaction between the science and policy communities, increased stakeholder participation, capacity enhancement and technology transfer.

CONCLUSIONS, GAPS AND OUTLOOK

Concern for the impacts of atmospheric issues at global, regional and national scales has led to considerable efforts to control emissions to meet internationally agreed goals and targets. Some issues have been effectively dealt with. For others there has only been partial success, with improvements in some regions but problems remaining in others.

Targets to protect the global atmosphere from air pollution are being met for stratospheric ozone depletion and lead in petrol.

Box 2.8 Complementary actions to limit near-term climate change and improve air quality

Implementing a limited number of measures targeting black carbon, tropospheric ozone and methane has the potential to reduce the rise in global temperature projected for 2050 by 0.5°C – roughly half of the warming in the reference scenario (Figure 2.24) – substantially reducing the rate at which the world warms over the next few decades (Shindell *et al.* 2012; UNEP/WMO 2011). About half of this reduction could be attributed to reductions in methane emissions and about half to measures addressing incomplete combustion and targeting black carbon emissions. The projected reduced warming in the Arctic, about 0.7°C lower than the reference scenario in 2050 according to this research, would be likely to be greater than that seen globally (UNEP/WMO 2011). There are further benefits for regional climate as several detailed studies of the Asian monsoon suggest that regional forcing by absorbing particles substantially alters precipitation patterns (UNEP/WMO 2011). As the reductions in atmospheric forcing are largest over the Indian sub-continent and other parts of Asia, the emission reductions might have a substantial effect on the Asian monsoon, mitigating disruption of rainfall patterns.

Full implementation of the identified measures would also substantially improve air quality, reducing premature deaths globally due to significant reductions in indoor and outdoor air pollution, and improving crop yields. The reductions in $PM_{2.5}$ and tropospheric ozone concentrations resulting from the implementation of the measures could, by 2030, avoid 2.4 million premature deaths (within a range of 0.7–4.6 million deaths) and the loss of 52 million tonnes of the global production of maize, rice, soybean and wheat each year (within a range of 30–140 million tonnes), or 1–4 per cent (UNEP/WMO 2011).

Figure 2.24 Projected effects of measures to reduce CO_2, methane and black carbon emissions in relation to a reference scenario

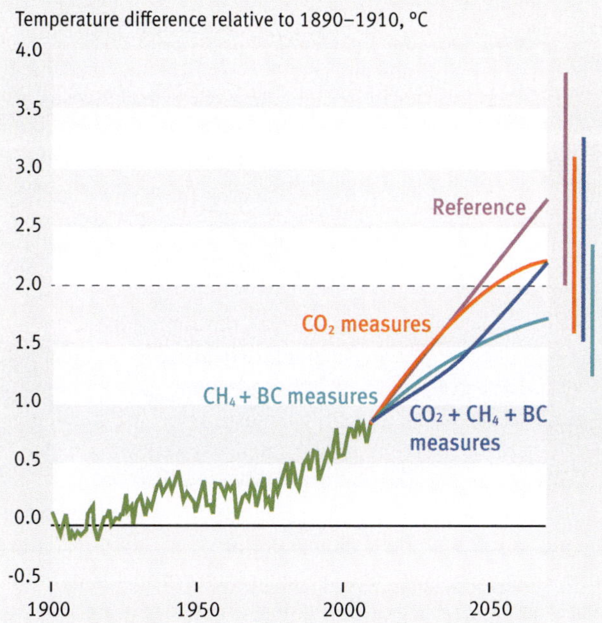

Implementation by 2030 of black carbon (BC) and methane (CH_4) measures identified in the UNEP/WMO Integrated Assessment of Black Carbon and Tropospheric Ozone, together with measures to reduce CO_2 emissions, starting now, would improve the chances of keeping the Earth's temperature increase to less than 2°C relative to pre-industrial levels. The bulk of the benefits of methane and black carbon measures are realized by 2040. The lines on the right-hand side of the graph represent uncertainty in the estimates. The projections are added to the observed deviation of temperature from the 1890–1910 mean (up to 2010).

Source: UNEP/WMO 2011

For the majority of the world, however, most air quality guidelines are not being met as there is insufficient implementation of policies. Meanwhile, important ecosystems are experiencing pollution loads in excess of critical thresholds. In the near term, atmospheric issues such as particulate matter and other pollutants could, with adequate commitment and resources, be effectively addressed by the wider implementation of existing policies and technologies.

The current development trajectory, based on existing models of international governance, is unlikely to meet internationally agreed atmospheric goals, especially those for mitigating climate change and reducing the health impacts of pollutants. Carefully selected approaches at national and regional scales need to be encouraged and facilitated by global coordination to increase the chance of reaching the targets in the near term.

Box 2.9 Atmospheric brown clouds

Atmospheric brown clouds, which have been observed as widespread layers of brownish haze, particularly in South Asia (Figure 2.25), are regional-scale plumes of air pollutants consisting mainly of aerosol particles such as black carbon and precursor gases that produce aerosols and ozone. These clouds significantly affect the regional climate, the hydrological cycle and glacial melting. The pollutants can be transported by long-range and regional transport phenomena that push this haze towards the Himalayan ridge, where plain-to-mountain wind systems favour air-mass transport to high altitudes (Bonasoni *et al.* 2010). The widespread nature of atmospheric brown clouds and findings related to their diverse and adverse effects have increased the need for the development of science, capacity and emission-reduction measures within an integrated framework.

Figure 2.25 Atmospheric brown cloud over part of South Asia

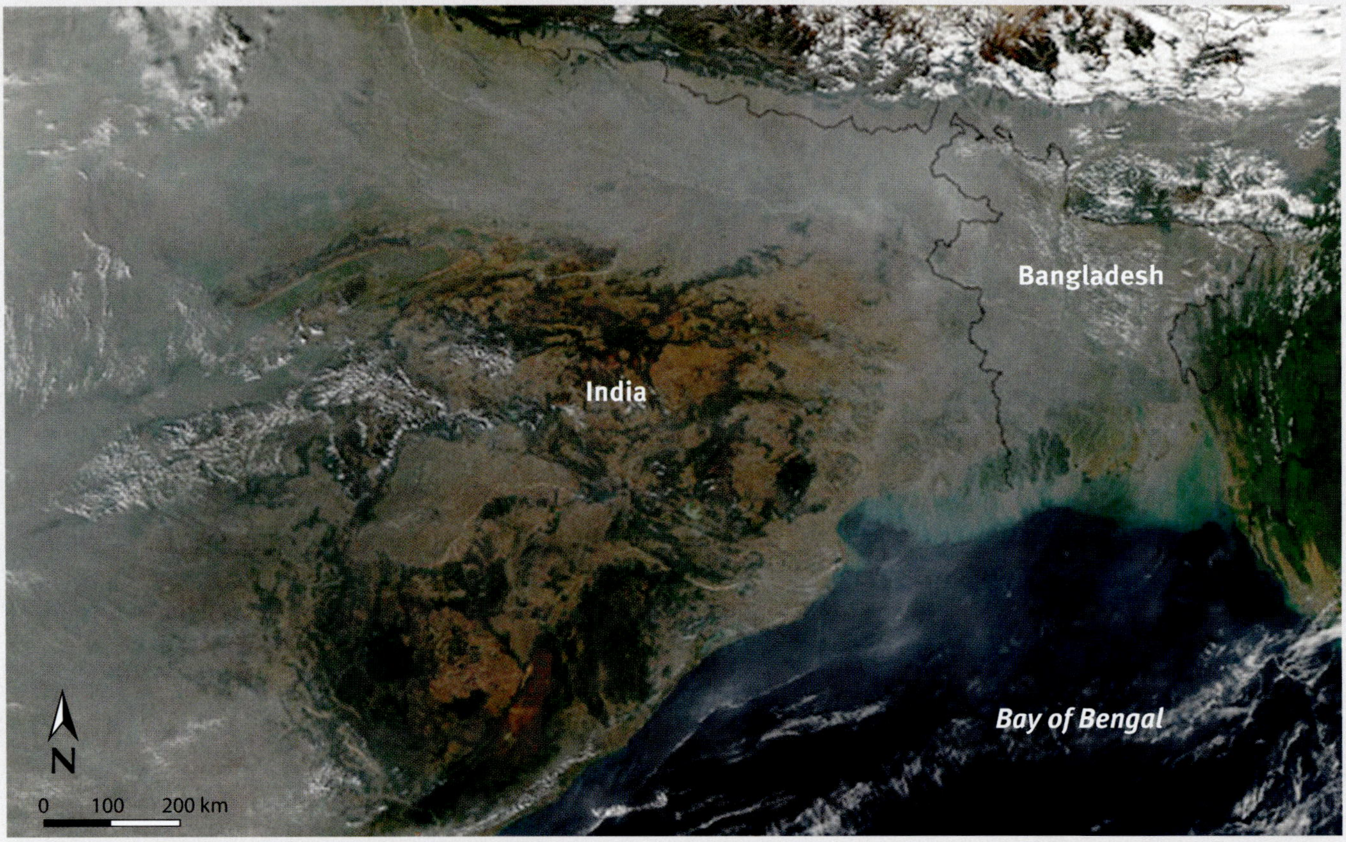

Source: NASA-MODIS

Climate change presents the global community with one of the most serious challenges to achieving development goals. Serious impacts from climate change are unlikely to be avoided on the basis of current emission reduction pledges. In the medium term, progress could be made by encouraging further national pledges, taking into account individual country circumstances, and by the wide application of current technological and policy approaches.

Measures to reduce emissions of short-lived climate forcers could contribute to reducing the temperature increase in the near term but, ultimately, a transformation in the way energy is provided and the efficiency with which electricity and other resources are used, coupled with a shift in consumption and production patterns and investment in innovation, will be required to achieve long-term climate goals. Such transformative change will also affect other atmospheric problems. But action should start now with the measures that are currently available while transformation takes place. Such action would bring significant benefits, particularly if the atmospheric issues and required policies were considered in an integrated way.

Table 2.5 provides a summary of progress and a prognosis for the development of key atmospheric issues in their relation to goals and targets.

Table 2.5 Progress towards goals (see Table 2.2)

A: Significant progress	C: Very little to no progress	X: Too soon to assess progress
B: Some progress	D: Deteriorating	?: Insufficient data

Key issues and goals		State and trends	Outlook	Gaps
1. Stabilization of greenhouse gas concentrations in the atmosphere at a level that would prevent dangerous anthropogenic interference with the climate system				
Climate change limit the increase in global average temperature to less than 2°C above pre-industrial levels	C	Rising CO_2 and other greenhouse gas emissions, and increasing concentrations; SLCF concentrations remain high and some are increasing Temperature increases observed globally and regionally over last decades	Efficiency improvements and some progress towards meeting Kyoto targets Likely to breach the 2°C limit without further commitment and action	Improvement in monitoring and reporting of pledged actions; financial and technical support to developing countries; policy integration of climate change and other atmospheric issues
2. Protect the ozone layer by taking precautionary measures to control total global production and consumption of ozone-depleting substances (ODS) equitably, with the ultimate objective of eliminating them				
Stratospheric ozone depletion Zero consumption of ozone-depleting substances	A	About 98% achievement (in 2009) in reducing the production and consumption of substances covered by the Montreal Protocol Decreasing atmospheric concentrations Stabilization of the Antarctic ozone hole	Continued decrease in atmospheric concentrations of ozone-depleting substances; recovery of ozone layer by mid-century	Recovery and destruction of ozone-depleting substances; from equipment, chemical stockpiles, foams, and other products not yet released into the atmosphere
3. Reduce respiratory diseases and other health impacts resulting from air pollution, with particular focus on women and children				
Particulate matter (urban/ outdoor) WHO guidelines and national targets	B	Outdoor concentrations of particulate matter in most parts of Europe and North America are within or are approaching WHO and EU guidelines; concentrations in Africa and Asia remain high	Slow progress in developing countries in Africa and Asia, as any efficiencies are likely to be offset by increased consumption and levels of activity	Monitoring, mainly in developing countries; standards for particulate matter, political will and awareness of the issues in some developing countries
Particulate matter (indoor) Households cooking using biomass	C	In poor rural areas of the world, for example in parts of Africa and Asia, there is little access to cleaner cookstoves and fuels, and indoor particulate matter is very high; significant health impacts, especially for women and children	Continued poverty and other barriers will prevent the transition to modern fuels or use of improved cooking facilities	Monitoring and associated technology in developing countries; mechanisms to enable purchase of efficient cooking stoves, institutional strengthening, and political will to address the issue
Tropospheric ozone WHO guidelines for health	B	Peak tropospheric ozone concentrations decreasing in Europe and North America except in ozone hotspots	Further reductions in Europe and North America will lead to decreased ozone, but there will be increases in precursors and ozone elsewhere	More ozone and precursor monitoring in developing countries; awareness of the issue
4. Enhanced cooperation at international, regional and national levels to reduce air pollution, including transboundary air pollution and acid deposition				
Tropospheric ozone CLRTAP goals	B	Peak concentrations of ozone are decreasing due to diminishing precursor emissions (nitrogen oxides, volatile organic compounds, methane and carbon monoxide) in Europe and North America except in ozone hotspots; increasing concentrations elsewhere; background concentrations increasing	Improvements in some regions are being offset by an increase in background ozone	Technology to minimize emissions of ozone precursors; monitoring in rural settings; implementation of policies in different sectors for different precursor emissions; regional and inter-regional cooperation
Sulphur dioxide WHO guidelines CLRTAP emission targets	B	Sulphur dioxide emissions and concentrations have been significantly reduced in Europe and North America	Overall sulphur dioxide emissions will decrease due to global desulphurization but increasing emissions are expected in some rapidly developing countries in Asia	Further sulphur dioxide emission reductions, especially in Asia
Nitrogen WHO guidelines CLRTAP emission targets	B	Nitrogen dioxide concentrations globally remained constant due to reductions in North America and Europe off-setting slight increases in Africa, Asia and Latin America	In Africa, Asia and Latin America where nitrogen emissions are not high priority, increases in both nitrogen oxide and ammonia emissions are expected, especially from agriculture and motorization	Awareness of issues and policy focus; improved technology to minimize emissions of nitrogen; understanding of long-range transport and impacts in all regions
5. Prevention of children's exposure to lead				
Lead Eliminate lead in petrol	A	Lead phased out in petrol globally except in six countries; lead blood levels in children have gone down	Lead from other sources, such as paint, has to be tackled globally	Policies and studies on lead in paint from developing countries

REFERENCES

AEA (2010). *Cost Benefit Analysis for the Revision of the National Emission Ceilings Directive.* http://ec.europa.eu/environment/air/pollutants/pdf/necd_cba.pdf

Aldy, J.E., Krupnick, A.J., Newell, R.G., Parry, I.W.H. and Pizer, W.A. (2010). Designing Climate Mitigation Policy. *Journal of Economic Literature* 48(4), 903–934

Amann, M., Bertok, I., Borken-Kleefeld, J., Cofala, J., Heyes, C., Höglund-Isaksson, L., Klimont, Z., Nguyen, B., Posch, M., Rafaj, P., Sandler, R., Schöpp, W., Wagner, F. and Winiwarter, W. (2011). Cost-effective control of air quality and greenhouse gases in Europe: modeling and policy applications. *Environmental Modelling and Software* (in press). doi:10.1016/j.envsoft.2011.07.012

Anenberg, S.C., Horowitz, L.W., Tong, D.Q. and West, J.J. (2010). An estimate of the global burden of anthropogenic ozone and fine particulate matter on premature human mortality using atmospheric modeling. *Environmental Health Perspectives* 118(9), 1189–1195

ASEAN (2002). *ASEAN Agreement on Transboundary Haze Pollution.* http://www.aseansec.org/pdf/agr_haze.pdf

Ashmore, M.R. (2005). Assessing the future global impact of ozone on vegetation. *Plant, Cell and Environment* 28, 949–964

Barriopedro, D., Fischer, E.M., Luterbacher, J., Trigo, R.M. and García-Herrera, R. (2011). The hot summer of 2010: redrawing the temperature record map of Europe. *Science* 332(6026), 220–4

Benedick, R.E. (1998). *Ozone Diplomacy: New Directions in Safeguarding the Planet.* Harvard University Press, Cambridge, MA

Bleeker, A., Hicks, W.K., Dentener, F., Galloway, J. and Erisman, J.W. (2011). Nitrogen deposition as a threat to the world's protected areas under the Convention on Biological Diversity. *Environmental Pollution* 159, 2280–2288

Bobbink, R., Hornung, M. and Roelofs, J.G.M. (1998). The effects of air-borne nitrogen pollutants on species diversity in natural and semi-natural European vegetation. *Journal of Ecology* 86, 738

Bonasoni, P., Laj, P., Marinoni, A., Sprenger, M., Angelini, F., Arduini, J., Bonafè, U., Calzolari, F., Colombo, T., Decesari, S., Di Biagio, C., di Sarra, A.G., Evangelisti, F., Duchi, R., Facchini, M.C., Fuzzi, S., Gobbi, G.P., Maione, M., Panday, A., Roccato, F., Sellegri, K., Venzac, H., Verza, G.P., Villani, P., Vuillermoz, E. and Cristofanelli, P. (2010). Atmospheric brown clouds in the Himalayas: first two years of continuous observations at the Nepal Climate Observatory-Pyramid (5079 m). *Atmospheric Chemistry and Physics* 10, 7515–7531

Braungart, M., McDonough, W. and Bollinger, A. (2007). Cradle-to-cradle design: creating healthy emissions – a strategy for eco-effective product and system design. *Journal of Cleaner Production* 15(13–14), 1337–1348

Bridbord, K. and Hanson, D. (2009). A personal perspective on the initial federal health based regulation to remove lead from gasoline. *Environmental Health Perspectives* 117(8), 1195–1201

Carnelley, T. and Le, X.C. (2001). *Correlation Between Chemical Characteristics and Biological Reactivity of Particulate Matter in Ambient Air.* Alberta. http://environment.gov.ab.ca/info/library/6646.pdf

CBD (2010a). *Aichi Biodiversity Targets.* Secretariat of the Convention on Biological Diversity, Montreal. http://www.cbd.int/sp/targets/

CBD (2010b). *Global Biodiversity Outlook 3.* Secretariat of the Convention on Biological Diversity, Montreal. http://www.cbd.int/gbo3/ebook/

CDC (2012) *CDC's National Surveillance Data (1997–2009).* US Centers for Disease Control and Prevention. http://www.cdc.gov/nceh/lead/data/national.htm

CDC (2005). Blood lead levels: United States 1999–2002. *Morbidity and Mortality Weekly Report*, 54(20), 513–516

CDC (2003). Blood lead levels: United States 1999–2002. *Morbidity and Mortality Weekly Report*, 52(SS-10)

CDIAC (2010). Carbon Dioxide Information Analysis Center. http://cdiac.ornl.gov/

Chang, Y.-C. and Wang, N. (2010). Environmental regulations and emissions trading in China. *Energy Policy* 38(7), 3356–3364

COMEAP (2010). *The Mortality Effects of Long-Term Exposure to Particulate Air Pollution in the United Kingdom.* Committee on the Medical Effects of Air Pollutants. Health Protection Agency, United Kingdom

Den Elzen, M. and Höhne, N. (2010). Sharing the reduction effort to limit global warming to 2° C. *Climate Policy* 10, 247–260

Den Elzen, M. and Höhne, N. (2008). Reductions of greenhouse gas emissions in Annex I and non-Annex I countries for meeting concentration stabilisation targets. *Climatic Change* 91, 249–274

EC (2011). Eurostat. http://epp.eurostat.ec.europa.eu/tgm/table.do?tab=table&init=1&plugin=1&language=en&pcode=tsien110

EC (2008). *Directive 2008/50/EC on Ambient Air Quality and Cleaner Air for Europe.* http://eur-lex.europa.eu/LexUriServ/LexUriServ.do?uri=OJ:L:2008:152:0001:0044:EN:PDF

EEA (2009). *NEC Directive Status Report of 2008.* EEA technical report 11/2009. European Environment Agency. http://www.eea.europa.eu/publications/

Emberson, L.D., Büker, P., Ashmore, M.R., Mills, G., Jackson, L., Agrawal, M., Atikuzzaman, M.D., Cinderby, S., Engardt, M., Jamir, C., Kobayashi, K., Oanh, K., Quadir, Q.F. and Wahid, A. (2009). A comparison of North American and Asian exposure-response data for ozone effects on crop yields. *Atmospheric Environment* 43(12), 1945–1953. doi:10.1016/j.atmosenv.2009.01.005

ENA (2011). *The European Nitrogen Assessment: Sources, Effects and Policy Perspectives* (eds. Sutton, M.A., Howard, C.M., Erisman, J.W., Billen, G., Bleeker, A., Grennfelt, P., Van Grinsven, H. and Grizzetti, B.) Cambridge University Press. http://www.nine-esf.org/ENA-Book

Forster, P., Ramaswamy, V., Artaxo, P., Berntsen, T., Betts, R., Fahey, D.W., Haywood, J., Lean, J., Lowe, D.C., Myhre, G., Nganga, J., Prinn, R., Raga, G., Schulz, M. and Van Dorland, R. (2007). Changes in atmospheric constituents and in radiative forcing. *In Climate Change 2007: The Physical Science Basis. Contribution of Working Group I to the Fourth Assessment Report of the Intergovernmental Panel on Climate Change* (eds. Solomon, S., Qin, D., Manning, M., Chen, Z., Marquis, M., Averyt, K.B., Tignor, M. and Miller, H.L.). Cambridge University Press, Cambridge, and New York

Fujino, J., Hibino, G., Ehara, T., Matsuoka, Y., Masui, T. and Kainuma, M. (2008). Back-casting analysis for 70% emission reduction in Japan by 2050. *Climate Policy* 8, S108–S124

Galloway, J.N., Aber, J.D., Erisman, J.W., Seitzinger, S.P., Howarth, R.W., Cowling, E.B. and Cosby, B.J. (2003). The nitrogen cascade. *BioScience* 53(4), 341–356

Gould, E. (2009). Childhood lead poisoning: conservative estimates of the social and economic benefits of lead hazard control. *Environmental Health Perspectives* 117, 1162–1167

Government of NCT of Delhi (2010). *State of the Environment Report for Delhi, 2010.* http://www.delhi.gov.in/wps/wcm/connect/9e24b08042c37602aaafaa6c8168d2a2/SoE+Delhi+2010.pdf?MOD=AJPERES&lmod=301990690&CACHEID=9e24b08042c37602aaafaa6c8168d2a2

Grosse, S.D., Matte, T.D., Schwartz, J. and Jackson, R.J. (2002). Economic gains resulting from the reduction in children's exposure to lead in the United States. *Environmental Health Perspectives* 110(6), 563–569

Hansen, J., Ruedy, R., Sato, M. and Lo, K. (2010). Global surface temperature change. *Reviews of Geophysics.* 48, RG4004. doi:10.1029/2010RG000345

Hare, W.L., Cramer, W., Schaeffer, M., Battaglini, A. and Jaeger, C.C. (2011). Climate hotspots: key vulnerable regions, climate change and limits to warming. *Regional Environmental Change* 11, S1–S13. doi:10.1007/s10113-010-0195-4

Hicks, W.K., Kuylenstierna, J.C.I., Owen, A., Dentener, F., Seip, H.M. and Rodhe, H. (2008). Soil sensitivity to acidification in Asia: status and prospects. *Ambio* 37, 295–303

Hilton, F.G. (2006). Poverty and pollution abatement: evidence from lead phase-out. *Ecological Economics* 56(1), 125–131

HTAP (2010). *Hemispheric Transport of Air Pollution, 2010. Part A: Ozone and Particulate Matter.* Air Pollution Studies No. 17. (eds. Dentener, F., Keating T. and Akimoto, H. Prepared by the Task Force on Hemispheric Transport of Air Pollution (HTAP) acting within the framework of the Convention on Long-range Transboundary Air Pollution (LRTAP) of the United Nations Economic Commission for Europe (UNECE). United Nations, New York and Geneva

Hulme, M., Osborn, T.J. and Johns, T.C. (1998). Precipitation sensitivity to global warming: comparison of observations with HadCM2 simulations. *Geophysical Research Letters* 25, 3379–3382

IJC (2010). *US and Canada Air Quality Agreement Progress Report.* International Joint Commission. www.ijc.org

IMO (2009). *Second IMO GHG Study 2009* (eds. Buhaug, Ø., Corbett, J.J., Endresen, Ø., Eyring, V., Faber, J., Hanayama, S., Lee, D.S., Lee, D., Lindstad, H., Markowska, A.Z., Mjelde, A., Nelissen, D., Nilsen, J., Pålsson, C., Winebrake, J.J., Wu, W., Yoshida, K.). International Maritime Organization, London

IPCC (2011). Summary for Policymakers. In: *Intergovernmental Panel on Climate Change Special Report on Managing the Risks of Extreme Events and Disasters to Advance Climate Change Adaptation* (eds. Field, C.B., Barros, V., Stocker, T.F., Qin, D., Dokken, D., Ebi, K.L., Mastrandrea, M. D., Mach, K.J., Plattner, G.-K., Allen, S.K., Tignor, M. and P.M. Midgley). Cambridge University Press, Cambridge and New York

IPCC (2007). *Climate Change 2007: Synthesis Report.* Contribution of Working Groups I, II and III to the Fourth Assessment Report of the Intergovernmental Panel on Climate Change. Geneva

IPCC (2005). *Carbon Dioxide Capture and Storage* (eds. Metz, B., Davidson, O., de Coninck, H., Loos, M. and Meyer, L.). IPCC Special Report. Cambridge University Press, Cambridge

IPCC (2000). Summary for Policymakers: Emissions Scenarios. Special Report of IPCC Working Group III. Intergovernmental Panel on Climate Change. http://www.ipcc.ch/pdf/special-reports/spm/sres-en.pdf

Kucera, V., Tidblad, J, Kreislova, K., Knotkova, D., Faller, M., Reiss, D., Snethlage, R., Yates, T., Henriksen, J., Schreiner, M., Melcher, M., Ferm, M., Lefèvre, R.-A. and Kobus J. (2007). UN/ECE ICP materials dose-response functions for the multi-pollutant situation. *Water, Air and Soil Pollution Focus* 7, 249–258. doi:10.1007/s11267-006-9080-z

Landrigan, P.J., Schechter, C.B., Lipton, J.M., Fahs, M.C. and Schwartz, J. (2002). Environmental pollutants and disease in American children: estimates of morbidity, mortality, and costs for lead poisoning, asthma, cancer, and developmental disabilities. *Environmental Health Perspectives* 110(7), 721–728

Lanphear B.P., Hornung R., Khoury J., Yolton, K., Baghurst, P., Bellinger, D.C., Canfield, R.L., Dietrich, K.N., Bornschein, R., Greene, T., Rothenberg, S.J., Needleman, H.L., Schnaas, L., Wasserman, G., Graziano, J. and Roberts, R. (2005). Low-level environmental lead exposure and children's intellectual function: an international pooled analysis. *Environmental Health Perspectives* 113(7), 894–899

Lawrence, D.M. and Slater, A.G. (2005). A projection of severe near-surface permafrost degradation during the 21st century, *Geophysical Research Letters* 32, L24401. doi:10.1029/2005GL025080

Lenton, T.M., Held, H., Kriegler, E., Hall, J.W., Lucht, W., Rahmstorf, S. and Schellnhuber, H.J. (2008). Tipping elements in the Earth's climate system. *Proceedings of the National Academy of Sciences of the United States of America* 105(6), 1786–1793. cgi/doi/10.1073/pnas.0705414105

Levy, M.A., Haas, P.M. and Keohane, R.O. (1993). Improving the effectiveness of international environmental institutions. In *Institutions for the Earth: Sources of Effective International Environmental Protection* (eds. Haas, P.M., Keohane, R.O. and Levy, M.A.). MIT Press, Cambridge, MA

Liu, J., Mauzerall, D.L., Horowitz, L.W., Ginoux, P. and Fiore, A.M. (2009a). Evaluating inter-continental transport of fine aerosols: (1) Methodology, global aerosol distribution and optical depth. *Atmospheric Environment.* doi:10.1016/j.atmosenv.2009.03.054

Liu, J., Mauzerall, D.L. and Horowitz, L.W. (2009b). Evaluating inter-continental transport of fine aerosols: (2) Global health impacts. *Atmospheric Environment.* doi:10.1016/j.atmosenv.2009.05.032

Lovei, M. (1998). *Phasing out Lead from Gasoline: Worldwide Experience and Policy Implications.* World Bank, Washington, DC

Manney, G.L., Santee, M.L., Rex, M., Livesey, N.J., Pitts, M.C., Veefkind, P., Nash, E.R., Wohltmann, I., Lehmann, R., Froidevaux, L., Poole, L.R. Schoeberl, M.R., Haffner, D.P., Davies, J., Dorokhov, V., Gernandt, H., Johnson, B., Kivi, R., Kyrö, E., Larsen, N., Levelt, P.F., Makshtas, A., McElroy, C.T., Nakajima, H., Parrondo, M.C., Tarasick, D.W., von der Gathen, P., Walker, K.A. and Zinoviev, N.S. (2011). Unprecedented Arctic ozone loss in 2011. *Nature* 478, 469–475. doi:10.1038/nature10556

MARPOL (2011). *International Convention for the Prevention of Pollution from Ships (MARPOL).* http://www.imo.org/about/conventions/listofconventions/pages/international-convention-for-the-prevention-of-pollution-from-ships-%28marpol%29.aspx

Menz, F.C. and Seip, H.-M. (2004). Acid rain in Europe and the United States: an update. *Environmental Science and Policy* 7(4), 253–265

Moss, R.H., Edmonds, J.A., Hibbard, K.A., Manning, M.R., Rose, S.K., van Vuuren, D.P., Carter, T.R., Emori, S., Kainuma, M., Kram, T., Meehl, G.A., Mitchell, J.F.B. Nakicenovic, N., Riahi, K., Smith, S.J., Stouffer, R.J., Thomson, A.M., Weyant, J.P. and Wilbanks, T.J. (2010). The next generation of scenarios for climate change research and assessment. *Nature* 463(7282), 747–756. doi:10.1038/nature08823

NASA GISS (2011). *GISS Surface Temperature Analysis (GISTEMP).* National Aeronautics and Space Administration Goddard Institute for Space Studies. http://data.giss.nasa.gov/gistemp/

Nemet, G.F., Holloway T., and Meier, P. (2010). Implications of incorporating air-quality co-benefits into climate change policymaking. *Environmental Research Letters* 5, 014007. doi:10.1088/1748-9326/5/1/014007

Newman P.A. and McKenzie, R (2011). UV impacts avoided by the Montreal Protocol. *Photochemical and Photobiological Sciences* 10, 1152–1160, doi:10.1039/c0pp00387e

Nilsson, J. and Grennfelt, P. (1988). *Critical Loads for Sulphur and Nitrogen.* Nordic Council of Ministers, Copenhagen

NOAA GMD (2011a). Carbon Cycle Greenhouse Gases Group (CCGG). National Oceanic and Atmospheric Administration Global Monitoring Division (GMD). www.esrl.noaa.gov/gmd/ccgg

NOAA GMD (2011b). NOAA Ozone Depleting Gas Index. National Oceanic and Atmospheric Administration Global Monitoring Division (GMD). http://www.esrl.noaa.gov/gmd/odgi/

Nordhaus, W.D and Boyer, J. (2000). *Warming the World: Economic Models of Global Warming.* MIT Press

NSIDC (2011). *NSIDC News.* National Snow and Ice Data Center, University of Colorado, Boulder. http://nsidc.org/arcticseaicenews/

Peters, G.P., Minx, J.C., Weber, C.L. and Edenhofer, O. (2011a). Growth in emission transfers via international trade from 1990 to 2008. *Proceedings of the National Academy of Sciences of the United States of America* 108(21), 8903–8908

Peters, G.L., Marland, G., Le Quéré, C., Boden, T., Canadell, J.G. and Raupach, M.R. (2011b). Rapid growth in CO_2 emissions afte the 2008–2009 global financial crisis. *Opinion and Comment, Nature Climate Change* 2, 2-4

Polvani, L.M., Waugh, D.W., Correa, G.J.P. and Son, S.-W. (2011). Stratospheric ozone depletion: the main driver of 20th century atmospheric circulation changes in the southern hemisphere. *Journal of Climate* 24, 795–812

Rasch, P.J., Crutzen, P.J. and Coleman, D.B. (2008). Exploring the geoengineering of climate using stratospheric sulfate aerosols: the role of particle size. *Geophysical Research Letters* 35, L02809

Raupach, M.R. and Canadell, J.G. (2010). Carbon and the Anthropocene. Current Opinion in Environmental Sustainability 2, 210–218

Raupach, M.R., Marland, G., Ciais, P., Le Quéré, C., Canadell, J.G., Klepper, G. and Field, C.B. (2007). Global and regional drivers of accelerating CO_2 emissions. *Proceedings of the National Academy of Sciences of the United States of America* 104(24), 10288–10293

Rignot, E.I., Velicogna, M.R., van den Broeke, A., Monaghan, A. and Lenaerts, J. (2011). Acceleration of the contribution of the Greenland and Antarctic ice sheets to sea level rise. *Geophysical Research Letters*, 38, L05503. doi:10.1029/2011GL046583

RNMI (2010). *Monitoring Atmospheric Composition and Climate – Interim Implementation.* Royal Netherlands Meteorological Institute. http://www.temis.nl/macc/index.php?link=o3_msr_intro.html

Rodhe, H., Langner, J., Gallardo, L. and Kjellstrom, E. (1995). Global scale transport of acidifying pollutants. *Water, Air, Soil Pollution* 85(1), 37–50

Royal Society (2008). *Ground-level Ozone in the 21st Century: Future Trends, Impacts and Policy Implications.* Science Policy Report. http://royalsociety.org

Schaefer, K., Zhang, T., Bruhwiler, L. and Barrett, A.P. (2011). Amount and timing of permafrost carbon release in response to climate warming. *Tellus B* 63(2), 165–180

Schmid, O., Möller, W., Semmler-Behnke, M., Ferron, G.A., Karg, E., Lipka, J., Schulz, H., Kreyling, W.G., Stoeger, T. (2009). Dosimetry and toxicology of inhaled ultrafine particles. *Biomarkers* 14 Suppl 1:67-73. http://www.ncbi.nlm.nih.gov/pubmed/19604063

Schneider, J.S., Huang, F.N., Vemuri, M.C. (2003). Effects of low-level lead exposure on cell survival and neurite length in primary mesencephalic cultures. *Neurotoxicology and Teratology* 25, 555–555

Schwartz, J. (1994). Low-level lead exposure and children's IQ: a meta-analysis and search for a threshold. *Environmental* Research 65, 42–55

Shindell, D., Kuylenstierna, J.C.I., Vignati, E., Van Dingenen, R., Amann, M., Klimont, Z., Anenberg, S.C., Muller, N., Janssens-Maenhout, G., Raes, F., Schwartz, J., Faluvegi, G., Pozzoli, L., Kupiainen, K., Höglund-Isaksson, L., Emberson, L., Streets, D., Ramanathan, V., Hicks, K., Oanh, K., Milly, G., Williams, M., Demkine, V. and Fowler, D. (2012). Simultaneously mitigating near-term climate change and improving human health and food security. *Science* 335(6065), 183–189. doi:10.1126/science.1210026

Shrestha, R.M., Pradhan S. and Liyanage, M. (2008). Effects of a carbon tax on greenhouse gas mitigation in Thailand. *Climate Policy* 8, S140–S155

Shukla, P.R., Dhar, S. and Diptiranjan, M. (2008). Low-carbon society scenarios for India. *Climate Policy* 8, S156–S176

Sitch, S., Cox, P.M., Collins, W.J. and Huntingford, C. (2007). Indirect radiative forcing of climate change through ozone effects on the land carbon sink. *Nature* 448(16), 791–795

Smith, J.B., Schneider, S.H., Oppenheimer, M., Yohe, G.W., Hare, W., Mastrandrea, M.D., Patwardhan, A., Burton, I., Corfee-Morlot, J., Magazda, C.H.D., Füssel, H-M., Pittock, A.B., Rahman, A., Suarez, A. and van Ypersele, J.-P. (2009). Assessing dangerous climate change through an update of the Intergovernmental Panel on Climate Change (IPCC) "reasons for concern". *Proceedings of the National Academy of Sciences* 106, 4133–413

Stern, N. (2007). *The Economics of Climate Change: The Stern Review.* Cambridge University Press, Cambridge and New York

Stoddard, J.L., Jeffries, D.S., Lukewille, A., Clair, T.A., Dillon, P.J., Driscoll, C.T., Forsius, M., Johannessen, M., Kahl, J.S., Kellogg, J.H., Kemp, A., Mannio, J., Monteith, D.T., Murdoch, P.S., Patrick, S., Rebsdorf, A., Skjelkvale, B.L., Stainton, M.P., Traaen, T., van Dam, H., Webster, K.E., Wieting, J. and Wilander, A. (1999). Regional trends in aquatic recovery from acidification in North America and Europe. *Nature* 401(6753), 575–578

Strachan, N., Foxon, T. and Fujino, J. (2008). Policy implications from the Low-Carbon Society (LCS) modelling project. *Climate Policy* 8, S17–S29

Sunstein, C. (2007). Of Montreal and Kyoto: a tale of two protocols. *Harvard Environmental Law Review* 31(1), 1–66

Thomas, V.M., Robert, H.S., James, J. and Thomas, G. (1999). Effects of reducing lead in gasoline: an analysis of the international experience. *Environmental Science and Technology* 33(22), 3942–3948

Tsai, P.L. and Hatfield, T.H. (2011). Global benefits from the phaseout of leaded fuel – going unleaded. *Journal of Environmental Health* 74(5), 8–14

UN (2000). *Millennium Development Goals.* http://www.un.org/millenniumgoals/

UNCED (1992). *Agenda 21.* http://www.un.org/esa/sustdev/documents/agenda21/english/Agenda21.pdf

UNDP/WHO (2009). *The Energy Access Situation in Developing Countries: A Review Focusing on the Least Developed Countries (LDCs) and Sub-Saharan Africa (SSA).* UNDP, New York. http://content.undp.org/go/cms-service/stream/asset/?asset_id=2205620

UNECE (2005). *The 1999 Gothenburg Protocol to Abate Acidification, Eutrophication and Ground-level Ozone.* Amended 2005. United Nations Economic Commission for Europe. http://www.unece.org/fileadmin/DAM/env/lrtap/full%20text/1999%20Multi.E.Amended.2005.pdf

UNECE (1979). *Convention on Long-Range Transboundary Air Pollution (CLRTAP).* http://www.unece.org/fileadmin/DAM/env/lrtap/full%20text/1979.CLRTAP.e.pdf

UNEP (2012). *Reduction in Sulphur in Fuels. Partnership for Clean Fuels and Vehicles.* United Nations Environment Programme, Nairobi. http://www.unep.org/transport/pcfv/corecampaigns/ campaigns.asp#sulphur (accessed 23 March 2012)

UNEP (2011a). *Bridging the Emissions Gap.* United Nations Environment Programme, Nairobi

UNEP (2011b). *Global Status of Leaded Petrol Phase-Out.* United Nations Environment Programme, Nairobi. http://www.unep.org/transport/PCFV/PDF/MapWorldLead_January2011.pdf and http://unep.org/transport/pcfv/PDF/leadprogress.pdf (accessed 26 May 2011)

UNEP (2011c). *HFCs: A Critical Link in Protecting Climate and the Ozone Layer.* United Nations Environment Programme, Nairobi. http://www.unep.org/climatechange/Publications/Publication/tabid/429/language/en-US/Default.aspx?ID=6224)

UNEP (2010). *Environmental Effects of Ozone Depletion: 2010 Assessment.* United Nations Environment Programme, Nairobi

UNEP (1987). *Montreal Protocol on Substances that Deplete the Ozone Layer.* Ozone Secretariat, United Nations Environment Programme, Nairobi. http://ozone.unep.org/pdfs/Montreal-Protocol2000.pdf

UNEP (1985). *Vienna Convention for the Protection of the Ozone Layer.* Ozone Secretariat, United Nations Environment Programme, Nairobi. http://ozone.unep.org/pdfs/viennaconvention2002.pdf

UNEP Ozone Secretariat (2011). Data Access Centre. http://ozone.unep.org/new_site/en/ozone_data_tools_access.php

UNEP/WMO (2011). *Integrated Assessment of Black Carbon and Tropospheric Ozone: Summary for Decision Makers.* UNON/Publishing Services Section/Nairobi, ISO 14001:2004. http://www.unep.org/dewa/Portals/67/pdf/BlackCarbon_SDM.pdf

UNFCCC (2012) *CDM in Numbers: Registration.* United Nations Framework Convention on Climate Change. http://cdm.unfccc.int/Statistics/Registration/RegisteredProjByRegionPieChart.html

UNFCCC (2010). *Cancun Agreements.* http://cancun.unfccc.int

UNFCCC (2009). *The Copenhagen Accord.* http://unfccc.int/resource/docs/2009/cop15/eng/l07.pdf

UNFCCC (2008). *The Bali Action Plan.* http://unfccc.int/resource/docs/2007/cop13/eng/06a01.pdf

UNFCCC (1998). *Kyoto Protocol to the United Nations Framework Convention on Climate Change.* http://unfccc.int/resource/docs/convkp/kpeng.pdf

UNFCCC (1992). *United Nations Framework Convention on Climate Change.* FCCC/INFORMAL/84 GE.05-62220 (E) 200705. http://unfccc.int/resource/docs/convkp/conveng.pdf

USEPA (2010). *Protecting the Ozone Layer Protects Eyesights: A Report on Cataract Incidence in the United States Using the Atmospheric and Health Effects Framework Model.* US Environmental Protection Agency, Washington, DC. http://www.epa.gov/ozone/science/effects/AHEFCataractReport.pdf

USEPA (2008). *National Ambient Air Quality Standards for Lead (Final Rule).* US Environmental Protection Agency, Washington, DC. http://www.epa.gov/oaqps001/lead/fr/20081112.pdf

Vahlsing, C. and Smith, K.R. (2010). Global review of national ambient air quality standards for PM_{10} and SO_2 (24h). *Air Quality Atmosphere and Health.* doi:10.1007/s11869-010-0131-2

Valavanidis, A., Fiotakis, K., Vlachogianni, T. (2008) Airborne particulate matter and human health: toxicological assessment and importance of size and composition of particles for oxidative damage and carcinogenic mechanisms. *J Environ Sci Health C Environ Carcinog Ecotoxicol Rev* 26(4):339-62 http://www.ncbi.nlm.nih.gov/pubmed/19034792

van Vuuren, D.P., Edmonds, J., Kainuma, M., Riahi, K., Thomson, A., Hibbard, K., Hurtt, G.C. Kram, T., Krey, V., Lamarque, J.-F., Masui, T., Meinshausen, M., Nakicenovic, N., Smith, S.J. and Rose, S.K. (2011). The representative concentration pathways: an overview. *Climatic Change* 109, 5–31

Velders, G.J.M., Andersen, S.O., Daniel, J.S., Fahey, D.W. and McFarland, M. (2007). The importance of the Montreal Protocol in protecting climate. *Proceedings of the National Academy of Sciences of the United States of America* 104(12), 4814–4819

Vestreng, V., Ntziachristos, L, Semb, A., Reis, S., Isaksen, I.S.A., and Tarrason, L. (2009). Evolution of NOx emissions in Europe with focus on road transport control measures. *Atmospheric Chemistry and Physics* 9, 1503–1520

WHO (2012). Database: outdoor air pollution in cities. http://www.who.int/phe/health_topics/outdoorair/databases/en/index.html

WHO (2011). *Health in the Green Economy: Health Co-benefits of Climate Change Mitigation – Housing Sector.* World Health Organization, Geneva

WHO (2010). *Childhood Lead Poisoning.* World Health Organization, Geneva. http://www.who.int/ceh/publications/leadguidance.pdf

WHO (2009). *Global Health Risks: Mortality and Burden of Disease Attributable to Selected Major Risks.* World Health Organization, Geneva. http://whqlibdoc.who.int/publications/2009/9789241563871_eng.pdf

WHO (2006). *WHO Air Quality Guidelines for Particulate Matter, Ozone, Nitrogen Dioxide and Sulfur Dioxide: Global Update 2005.* World Health Organization, Geneva

WHO (2000). *Air Quality Guidelines for Europe.* Second Edition. WHO Regional Publications European Series No. 91. World Health Organization Regional Office for Europe, Copenhagen

WHO (1999). *Air Quality Guidelines.* World Health Organization, Geneva

Wilson, N. and Horrocks, J. (2008). Lessons from the removal of lead from gasoline for controlling other environmental pollutants: a case study from New Zealand. *Environmental Health* 7, 1. doi:10.1186/1476-069X-7-1

WMO (2011). *Scientific Assessment of Ozone Depletion: 2010.* World Meteorological Organization Global Ozone Research and Monitoring Project Report No. 52. World Meteorological Organization, Geneva

Wright, R.F., Larssen, T., Camarero, L., Cosby, B.J., Ferrier, R.C., Helliwell, R., Forsius, M., Jenkins, A., Kopacek, J., Majer, V., Moldan, F., Posch, M., Rogora, M. and Schopp, W. (2005). Recovery of acidified European surface waters. *Environmental Science & Technology* 39(3), 64A–72A

WSSD (2002). *Johannesburg Plan of Implementation.* http://www.un.org/esa/sustdev/documents/ WSSD_POI_PD/English/POIToc.htm

Zhang, Z.X. (2010). *China in the Transition to a Low Carbon Economy.* East-West Centre Working Papers. Economics Series 109

Zhao, Y., Duan, L., Xing, J., Larssen, T., Nielsen, C.P. and Hao, J.M. (2009). Soil acidification in China: is controlling SO_2 emissions enough? *Environmental Science & Technology* 43(21), 8021–8026

CHAPTER 3

Land

Coordinating lead authors: Carol Hunsberger and Tom P. Evans

Lead authors: T. Mitchell Aide, Juan Albaladejo Montoro, Saturnino (Jun) M. Borras Jr., Héctor Francisco del Valle, Tahia Devisscher, Jason Jabbour, Shashi Kant, David López-Carr, Hillary Masundire, Narcisa G. Pricope (GEO Fellow) and Roberto Sánchez-Rodriguez

Contributing authors: Magdi T. Abdelhamid, Björn Alfthan, Fethi Ayache, Asmeret Asefaw Berhe, Chizoba Chinweze, Jana Frélichová, Lawrence Hislop, William K. Pan, Björn Schulte-Herbrüggen, Jessica Smith, Carlos Souza Jr., Tracy L. Timmins (GEO Fellow) and Leo C. Zulu

Principal scientific reviewer: Jean-Pierre Ometto

Chapter coordinator: Jason Jabbour

Main Messages

Pressure on land resources has increased during recent years despite international goals to improve their management. The fourth *Global Environment Outlook* (UNEP 2007) highlighted the unprecedented land-use changes created by a burgeoning population, economic development and global markets. The outcome of those drivers continues to cause resource depletion and ecosystem degradation.

Economic growth has come at the expense of natural resources and ecosystems. Many terrestrial ecosystems are being seriously degraded because land-use decisions often fail to recognize non-economic ecosystem functions and biophysical limits to productivity. For example, deforestation and forest degradation alone are likely to cost the global economy more than the losses of the 2008 financial crises. The current economic system, built on the idea of perpetual growth, sits uneasily within an ecological system that is bound by biophysical limits. However, some market-based approaches that attach value to ecosystem services offer incentives to reduce environmental damage.

Competing demands for food, feed, fuel, fibre and raw materials are intensifying pressures on land. Demands for food and livestock feed are increasing rapidly due to human population growth and changing diets. Demand for biofuels and raw materials have also risen sharply, driven by the increased population, greater consumption and biofuel-friendly policies. This simultaneous growth is causing land conversion, land degradation and pressure on protected areas. Climate change is placing additional stress on productive areas. One result is heightened tension between goals related to production and those related to conservation.

Globalization and urbanization are aggravating competing demands on land. These processes expand and intensify the pressure on land systems by increasing the distances between places where products originate and where they are consumed. The greater distances can obscure the drivers of resource depletion and ecosystem degradation, produce higher environmental costs due to transport and infrastructure, and complicate the negotiation of sustainable land management practices. Large-scale international land deals are both an emerging outcome of and a contributor to this trend. Internationally coordinated responses are needed to address related social and environmental pressures.

Improved governance and capacity building are crucial to achieving sustainable land management. Many interventions meant to protect ecosystems have failed because they were created without recognizing local values or engaging local communities in their design and implementation. Capacity building across spatial and temporal scales is needed to improve land management. Current governance approaches include market-based strategies such as the collaborative UN programme for Reducing Emissions from Deforestation and Forest Degradation (REDD), centralized institutional strategies such as certification, and decentralized strategies such as community-based resource management. All offer both opportunities and challenges for improving land governance.

Potential exists to create more sustainable land systems. To solve these complex problems, it is critical to understand how diverse social and ecological drivers affect land systems at local, regional, national and global scales. A concerted effort by international organizations, the scientific community, and national and local institutions to coordinate their actions can create the policy options needed to achieve this goal.

INTRODUCTION

Changing climate patterns, economic globalization, population growth, increasing use of natural resources and rapid urbanization are putting pressure on terrestrial ecosystems as never before, and virtually all of them are under stress. Biophysical limits on what is available for human use are real and there are strong signals that these limits are close to being reached or have already been exceeded (Rockström *et al.* 2009). Even so, the fact that some areas show recent gains in forested area or land reclamation (Lambin and Meyfroidt 2010; Nepstad *et al.* 2009; Bai *et al.* 2008) suggests that declines are not inevitable, and indeed that recovery may be possible – even though original ecosystem functions may be modified or pressure on ecosystems may shift elsewhere (Meyfroidt *et al.* 2010).

Growing demands for food, feed, fuel, fibre and raw materials create local and distant pressures for land-use change (Lambin and Meyfroidt 2011). The cascade of outcomes resulting from these demands is complicated by urbanization and globalization, which separate the production of goods from their consumption over vast distances (Barles 2010; Kissinger and Rees 2010). The central question is how these demands can be met – or managed – in ways that recognize the joint imperatives of human well-being and environmental sustainability. Addressing this requires careful examination of the social relations and biophysical processes involved in managing terrestrial ecosystems, setting priorities for policies and policy instruments, and considering the likely distribution of implications, both positive and negative.

The fourth *Global Environment Outlook* (*GEO-4*) (UNEP 2007) noted that increased demand for water, waste disposal and food had led to unsustainable patterns of land use and land degradation. It identified forest cover and composition, cropland expansion, intensification of agriculture, desertification and urban development as key topics in land-use change. *GEO-4* concluded that continued inaction on land stewardship, combined with increased climate change, would reduce social resilience, making recovery from future stresses difficult or impossible. This chapter provides an update on the state and trends of global land systems including wetlands, explores major and emerging issues influencing changes in land use, examines the implications of recent changes for achieving international accords, and suggests some broad responses.

INTERNATIONAL GOALS

The international goals selected to guide this chapter cover vital targets related to food security, poverty reduction and environmental sustainability (Table 3.1). This chapter identifies biophysical, social, economic and political factors that may enable or constrain their attainment.

Table 3.1 Selected internationally agreed goals and themes related to land

Major themes from internationally agreed goals	Johannesburg Plan of Implementation (WSSD 2002) Paragraph 40b	Millennium Development Goal 1 (UN 2000)	Millennium Development Goal 7 (UN 2000)	World Food Summit Plan of Action (FAO 1996) Paragraph 33g	Ramsar Convention on Wetlands (1971)	United Nations Convention to Combat Desertification (UNCCD 1994) Article 2
Promote food security		X		X		
Reduce the proportion of people who suffer from hunger		X				
Improve access to food		X		X		
Increase food production		X				
Reverse the loss of environmental resources			X	X	X	X
Reduce the deforestation rate and increase forest coverage				X		
Halt the destruction of tropical forests				X		
Stem the loss of wetlands					X	
Combat desertification and mitigate the effects of drought						X
Practise integrated land-use planning and management	X		X	X	X	X
Integrate the principles of sustainable development into country policies and programmes	X		X			X
Recognize, maintain and develop the multiple benefits of ecosystem services (in addition to their economic value)				X	X	

The stakes are high: as Chapter 16 demonstrates, failure to achieve these targets could have severe impacts on human well-being and environmental integrity.

STATE AND TRENDS

This section uses selected indicators to gauge the current state of agricultural land, forests, drylands, wetlands, polar areas and human settlements, and recent changes to these land covers and uses.

Agriculture

Demands for food and livestock feed are rising rapidly due to population growth, urbanization and changing diets that include more animal products. One of the consequences of these changes is the widespread expansion of agricultural land allocated to livestock, both directly and indirectly through cropland dedicated to animal feed production (Rudel *et al.* 2009; Naylor *et al.* 2005). At a time when water shortage and land degradation remain threats to food security, accelerated interest in biofuel, feeds and fibre in recent years imposes competing demands on how agricultural land is used.

Agricultural land and production trends

In 2009, there were approximately 3.3 billion hectares of pasture and 1.5 billion hectares of cropland globally, with the extent and proportion of total land area varying greatly across regions (Figure 3.1) (FAO 2012). In 2009, all regions except Europe had a greater proportion of land area devoted to pasture than to cropland. Although there has been only a slight increase in total cropland extent over the past decade, there has been a considerable change in the crops grown (Figure 3.2) (FAO 2012). Maize is an important crop in all regions other than West Asia, with the area harvested increasing by more than 25 per cent across Africa and Asia and the Pacific between 2001 and 2010. In total, approximately 160 million hectares of maize were harvested in 2010. Asia and the Pacific have the largest area of rice, but Europe and Africa experienced the greatest percentage increases between 2001 and 2010 – about 30 and 20 per cent respectively. The dominant soybean-producing regions are Latin

Box 3.1 Eradicating hunger

Related goals
Eradicate extreme poverty and hunger

Indicators
Proportion of malnourished people

Global trends
Proportion decreasing, but absolute number increasing

Most vulnerable communities
People who are food insecure due to chronic poverty, climate variation or food price fluctuations

Regions of greatest concern
Africa, Asia and the Pacific

Figure 3.1 Area in use for cropland and pasture in 2009, by region, and global change between 1960 and 2010

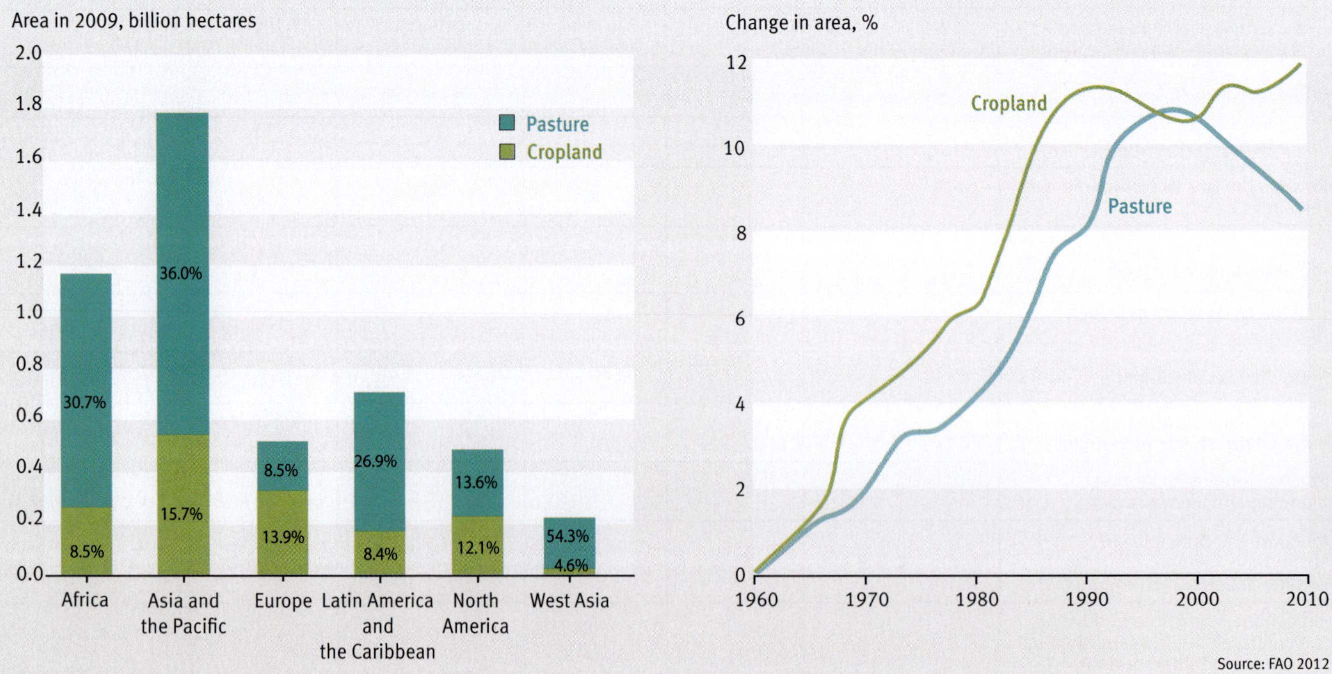

Source: FAO 2012

Maize field in the foreground of an ethanol plant in Midwest United States, where the most common feedstock used for ethanol production continues to be maize. *© iStock/SimplyCreativePhotography*

America and the Caribbean and North America, with the United States, Brazil and Argentina the three largest producers. Asia and the Pacific and Europe are the primary producers of wheat.

Increases in the area used for these crops have been accompanied by overall growth in yields (FAO 2012). Globally, the current yields of wheat, maize and rice have been estimated at 64, 50 and 64 per cent of their potential respectively, but the size of the yield gap varies greatly from region to region under the influence of different factors (Neumann *et al.* 2010). Larger gaps between actual and potential yields tend to occur where low-input agriculture is practised (Licker *et al.* 2010). Africa and Latin America and the Caribbean – two regions where crop area has expanded since 2001– still have relatively low yields compared to North America and Europe; if region-specific constraints can be assessed and overcome (Neumann *et al.* 2010), there may be potential to increase food production in these regions while minimizing cropland expansion.

Agricultural productivity is limited by biophysical and other factors. Extending conventional agriculture into uncultivated lands requires mechanization to modify the surface, and supplements in the form of fertilizers, herbicides, pesticides and irrigation water. Excessive use of machinery and chemical supplements, however, breaks up soil structure, increases erosion, chemically pollutes soil, contaminates groundwater and surface water, changes greenhouse gas fluxes, destroys habitat and builds genetic resistance to chemical supplements (Blanco-Canqui and Lal 2010; Foley *et al.* 2005; Buol 1995). With widespread adoption of intensive, mechanized, high-input agricultural practices, the rate of soil erosion has greatly increased. Erosion in conventional agricultural systems is now over three times higher than in systems practising conservation agriculture, and over 75 times higher than in systems with natural vegetation (Montgomery 2007). Globally, soil erosion is contributing to the decline in agricultural land available per person (Boardman 2006) as degraded land is abandoned (Bakker *et al.* 2005; Lal 1996). Thus, the yield gains achieved by these methods come with ecological costs.

In continuously cultivated, low-input agricultural systems, rapid declines in soil fertility and yield, together with international commodity price changes, continue to impact human well-being in agricultural communities (Koning and Smaling 2005). Sustainable intensification techniques offer the potential to improve soil fertility and yields in some situations while avoiding some of the problems of high-input agriculture just presented.

While the future impact of climate change on global food production is difficult to specify, substantial evidence suggests that an increasing number of people will be directly affected by climate change impacts on agricultural areas (World Bank 2010).

Figure 3.2 Area harvested in 2010 and the change between 2001 and 2010, selected crops

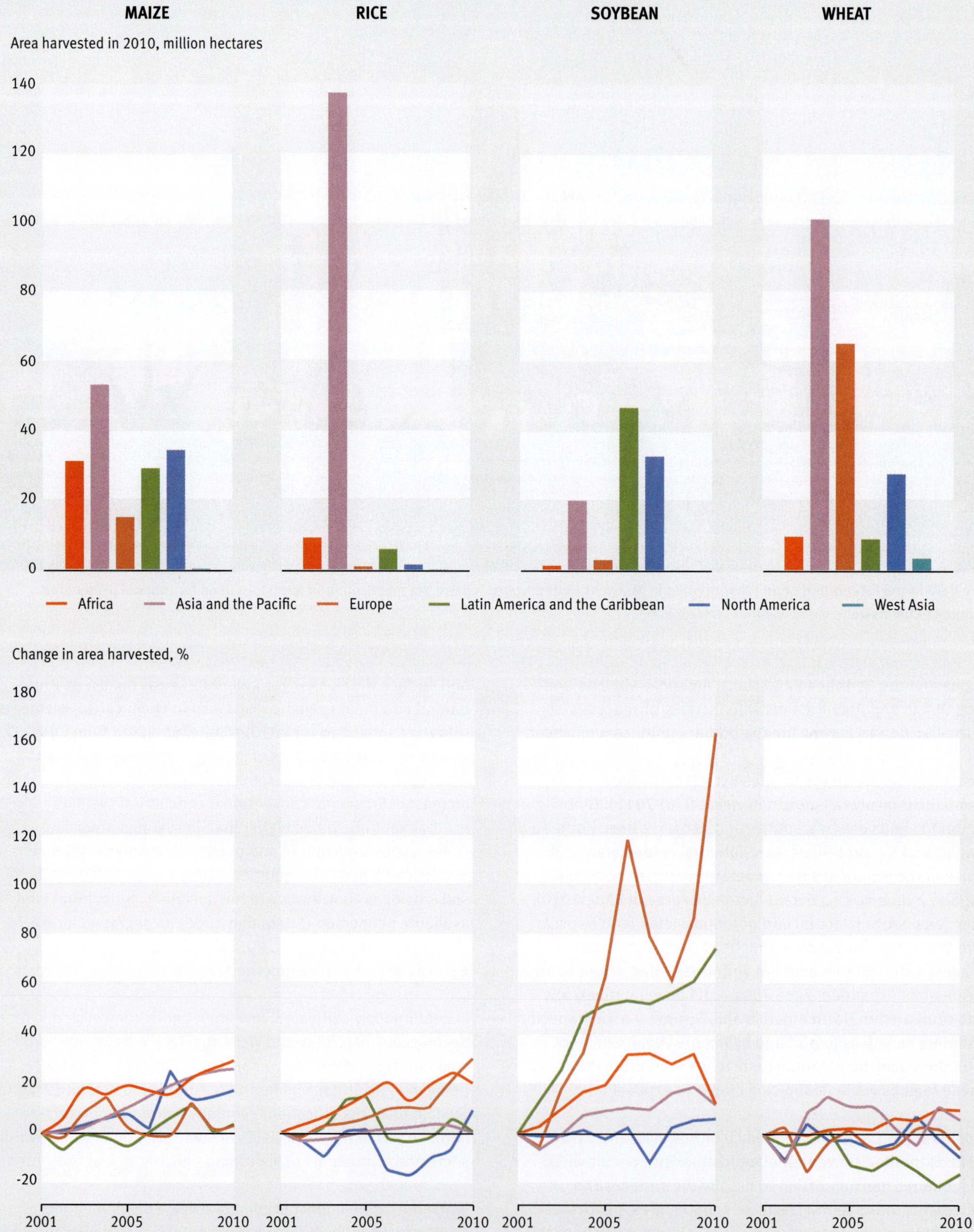

Source: FAO 2012

Figure 3.3 Average food supply in 2007 and the change between 1998 and 2007, by region

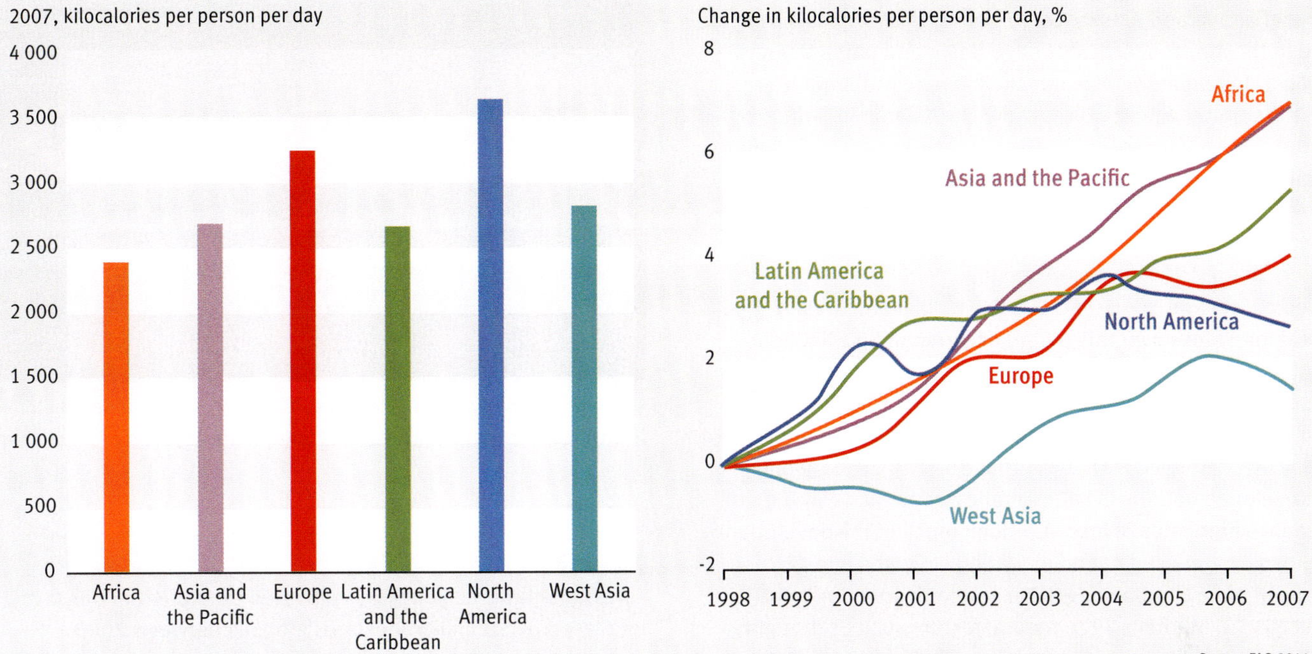

Source: FAO 2011

Consumption trends

While the proportion of undernourished people has been declining – from 14 per cent of world population in 1995–1997 to 13 per cent in 2010 – the absolute number rose over the same period from 788 million to an estimated 925 million due to population growth (Box 3.1) (FAO 2010b). Areas with chronic food insecurity face many obstacles, including regional conflicts, weak governance structures and a breakdown of local institutions, all of which affect access to and distribution of food (FAO 2010a). Many of the world's undernourished people live in areas that are also particularly vulnerable to climate variability. Africa and Asia and the Pacific were the regions with the lowest average food consumption in 2007 (Figure 3.3) (FAO 2012), but they were also the regions that had experienced the highest percentage increase. While the Asia and Pacific region is home to the largest number of undernourished people, at 578 million, sub-Saharan Africa has the highest proportion of undernourished people – about 30 per cent of its population in 2010 (FAO 2010b).

Forests

Forests play a crucial role in terrestrial ecosystems and provide a multitude of services such as shelter, habitats, fuel, food, fodder, fibre, timber, medicines, security and employment; regulating freshwater supplies; storing carbon and cycling nutrients; and helping to stabilize the global climate. Historically, forests have been under pressure due to increasing demands for shelter, agricultural land, meat production, and fuel and timber extraction, but in recent decades this pressure has increased due to competing demands for agricultural expansion and biofuel production, rapid urbanization and infrastructure development, and increased global demand for forest products. Forests are also under increasing stress from changes in mean annual temperatures, altered precipitation patterns, and more frequent and extreme weather events (Allen *et al.* 2010; Tiwari 2009).

Box 3.2 Forests

Related goals
Reduce deforestation and increase forest cover

Indicators
Net forest change

Global trends
Some forest gains in temperate areas; deforestation slowing in some tropical countries; overall tropical deforestation remains high

Most vulnerable communities
Forest-dependent people in tropical countries

Regions of greatest concern
Africa, Latin America and the Caribbean

Forest area

Forests cover just over 4 billion hectares, 31 per cent of the world's total land area (FAO 2011). The majority of these are boreal forests extending across northern and central Russia and much of Canada and Alaska. Large expanses of tropical forest are found in the Amazon, Africa's Congo Basin and parts of South East Asia. Temperate forests remain in a patchy distribution across the United States, Europe and the Asian mid-latitudes.

The rate of forest loss from both deforestation and natural causes is slowing, but remains alarmingly high (Box 3.2). At the global level, annual forest loss decreased from 16 million hectares in the 1990s to approximately 13 million hectares between 2000 and 2010 (FAO 2011). The highest rates of tropical forest loss over this period occurred in South America and Africa (Figure 3.4). Some rapidly developing countries that suffered extensive deforestation in the 1990s, including Brazil and Indonesia, have significantly reduced their rates of tropical forest loss (FAO 2011; Ometto *et al.* 2011), while less developed nations in Latin America and Africa continue to experience high rates of loss. Although much of the developed world has experienced net reforestation since the late 1800s as a result of rural-urban migration and farm abandonment (Walker 1993; Mather 1992), natural factors such as drought, forest fire and insect attacks have exacerbated forest loss in recent decades. However, the key drivers of forest loss are population growth, poverty, economic growth, land pricing, international demand for timber and other forest products, insecurity of the rights of local people, and incomplete valuation of forest ecosystems (Carr *et al.* 2005; Lambin *et al.* 2001).

Clearance in the Amazon, where a substantial portion of deforestation is attributed to cattle ranching and large-scale soybean production.

© iStock/luoman

Plantations

Forest plantations, generally cultivated for industrial purposes, increased by 50 million hectares globally between 2000 and 2010, reaching 264 million hectares or 7 per cent of the total forest area (Table 3.2) (FAO 2011). Asia accounted for 28 million hectares, or 58 per cent of this increase. Generally, monoculture plantations tend not to enrich local biodiversity, but they do provide ecosystem services including timber, carbon and water storage and soil stabilization.

Figure 3.4 Change in forest area by region, 1990–2010

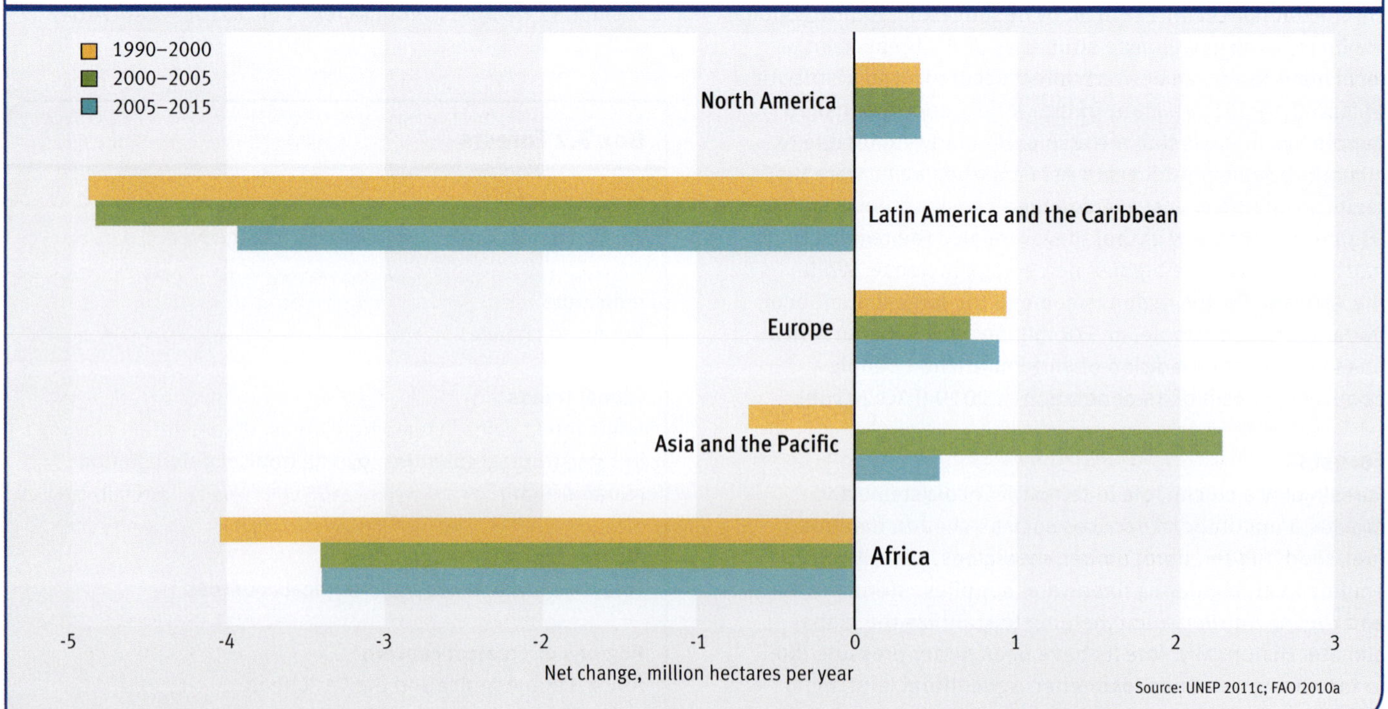

Source: UNEP 2011c; FAO 2010a

Table 3.2 Plantation area in 2010 and the increase between 2000 and 2010, by region

	Africa	Asia and the Pacific	Europe	Latin America and the Caribbean	North America	West Asia	World
Plantation area 2010, thousand hectares	15 409	121 802	69 318	14 952	37 529	5 073	264 084
Annual increase, thousand hectares	245	2 948	401	407	809	115	4 925
Annual increase, %	1.75	2.82	0.6	3.23	2.46	2.6	2.09

Note: FAO data has been applied to GEO regional categories, except for Afghanistan, Turkey and Iran, which are included in West Asia.

Source: FAO 2011

Productive and protective forest area

The global forest area designated for the production of timber and non-timber products declined from about 1.16 billion hectares in 2000 to about 1.13 billion hectares in 2010, an annual decrease of about 2.91 million hectares or 0.25 per cent (FAO 2011). However, the global forest area designated for protection of soil and water increased from about 272 million hectares in 2000 to about 299 million hectares in 2010, an annual increase of some 2.77 million hectares or 0.97 per cent (FAO 2011). Similarly, the global forest area designated for biodiversity conservation has increased from around 303 million hectares to about 366 million hectares, an annual increase of about 6.33 million hectares or 1.92 per cent (FAO 2011). The main reason for the decrease in forest area designated for production is deforestation, and for the increase in protective forest area is afforestation (FAO 2010a).

Forest management and certification

The Forest Stewardship Council (FSC) and the Programme for the Endorsement of Forest Certification (PEFC) are the two main forest management certification organizations. There was an increase in certified forests of about 20 per cent per year between 2002 and 2010 under these two agencies (UNEP 2011c). However, in 2010 about 10 per cent of the total forest area was under FSC- or PEFC-certified forest management (UNEP 2011c). These trends indicate that while there is improvement in forest management, much work remains to be done.

Forest carbon stock

Forests are considered an important sink for atmospheric carbon dioxide (CO_2) because of their ability to store carbon in their biomass and soil (Anderson *et al.* 2011). More than 75 per cent of the total terrestrial biomass carbon stock and more than 40 per cent of the soil organic carbon stock are found in forest ecosystems (Jandl *et al.* 2007). In the 1990s, forest carbon sequestration was equivalent to approximately one-third of carbon emissions from fossil fuel combustion and land-use change (Bonan 2008). Boreal forests store more carbon in their soils than tropical forests, while tropical forests store much more carbon in their plant biomass (Prentice *et al.* 2001). Pan *et al.* (2011) estimate that global forest systems constituted a total carbon sink of 2.4±0.4 billion tonnes of carbon per year from 1990 to 2007.

Fires are a major source of greenhouse gas emissions from forests (van der Werf *et al.* 2010). Boreal forest ecosystems are prone to frequent and severe wildfires leading to large carbon emissions. Amiro *et al.* (2001) estimated that during the period 1949–1999, on average 2 million hectares of the Canadian boreal forest burnt annually (ranging from 0.3 to 7.5 million hectares in any given year), emitting a yearly average of 27±6 million tonnes of carbon (ranging from 3 to 115 million tonnes in any given year). Sukhinin *et al.* (2004) estimated that on average 7.7 million hectares of area burnt annually between 1995 and 2002 in eastern Russia and that 55 per cent, 4.2 million hectares, of that area was forest. Gillett *et al.* (2004) found that recent increases in area burnt in Canada are a result of anthropogenic climate change. In the future more fires, more area burnt and longer fire seasons may be expected in temperate and boreal regions (Flannigan *et al.* 2009).

Drylands, grasslands and savannahs

Drylands, grasslands and savannahs experience high spatial and temporal variability in rainfall, resulting in dramatic differences in plant growth, habitats and human livelihoods. Drylands cover approximately 40 per cent of the world's land surface and are home to more than 2 billion people, 90 per cent of whom are in developing countries (UNEP 2007). However, the spatial extent of drylands remains uncertain due to variations in ecosystem sub-types, data variability and the different classes and thresholds applied to remotely sensed data, making global comparisons challenging (Reynolds *et al.* 2007). Grasslands range from very dry, almost desert-like, to humid types. Savannahs are mixed tree-grass ecosystems, ranging from almost treeless grasslands to closed-canopy woodlands that occupy large areas in the tropics and sub-tropics, particularly in Africa, Latin America and Australia (Mistry 2000).

Trends in drylands, grasslands and savannahs

Fluctuations in precipitation are a major driver of change in plant cover, but grazing intensity has also been directly linked to long-term dryland degradation (Miehe *et al.* 2010). Transformation of rangelands to cultivated croplands is leading to a significant, persistent decrease in overall dryland plant productivity. Sietz *et al.* (2011) indicated that the most important factors causing vulnerability in drylands are water stress, poverty, soil degradation, natural agronomic constraints and isolation from political centres.

Net primary productivity (NPP) is the net amount of carbon captured by vegetation through photosynthesis each year (Melillo *et al.* 1993). Approximately 2 per cent of global terrestrial NPP is lost yearly due to dryland degradation, equivalent to 4–10 per cent of dryland potential NPP (Zika and Erb 2009). Figure 3.5 shows how dryland degradation, measured in terms of NPP loss, is most widespread in the Sahelian and Chinese arid and semi-arid regions, followed by the Iranian and Middle Eastern drylands and to a lesser extent the Australian and Southern African regions. Sustainable development in drylands will rely on techniques that improve soil fertility, conserve soil and water and increase agricultural efficiency, such as mulch farming, conservation tillage and diverse cropping systems (Mortimore *et al.* 2009).

Figure 3.5 Global extent of drylands and human-induced dryland degradation

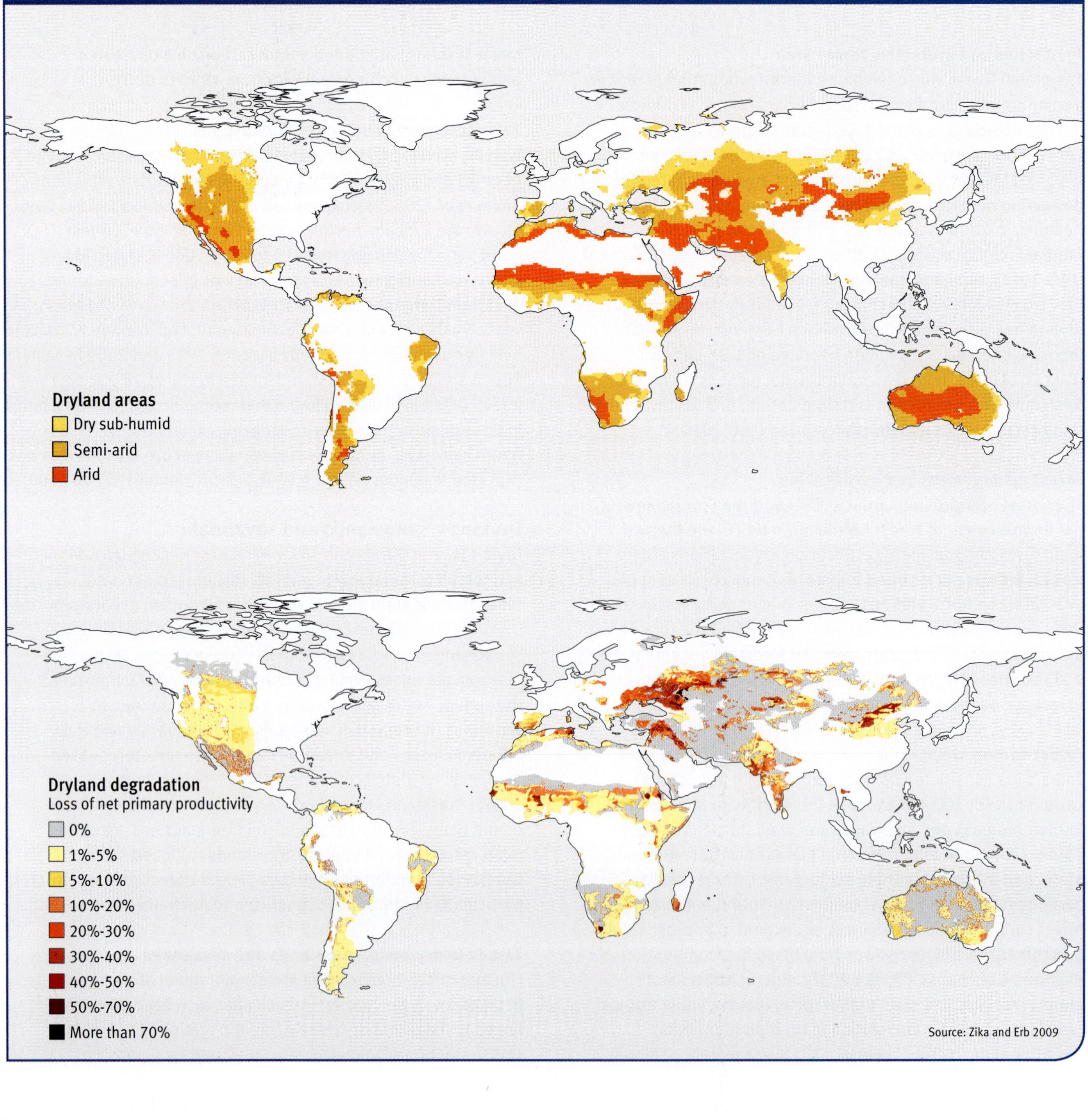

Source: Zika and Erb 2009

Figure 3.6 UNCCD operational objectives and achievements, 2010

Operational objective	Performance indicator	Current achievement level	Overall target	Target due
Advocacy, awareness and education	Information and awareness	25%	**30%** of global population informed about desertification, land degradation and drought and/or synergies with climate change and biodiversity	2018
Policy framework	National action plan alignment	5%	**80%** of affected country Parties with a formulated/revised national action plan aligned to the 2008–2018 Strategic Plan	2014
	Joint planning of the Rio conventions*	72%	**100%** of affected country Parties with joint national action plans in place or functional mechanisms to ensure synergies between the three Rio conventions	2014
Science, technology knowledge	Dryland monitoring	38%	**60%** affected country Parties with established and supported national dryland monitoring systems	2018
Capacity building	Dryland capacity-building	71%	**90%** of affected country Parties implementing dryland-specific capacity-building initiatives	2014
Finance and technology transfer	Integrated investment framework	15%	**50%** of affected country Parties with integrated investment frameworks	2014

* Convention on Biological Diversity (CBD), United Nations Framework Convention on Climate Change (UNFCCC) and United Nations Convention to Combat Desertification (UNCCD)

The figure evaluates progress towards the UNCCD targets, showing substantial progress in some areas and highlighting a need for improvement in others. Particularly encouraging is the high level of awareness about dryland degradation globally. Challenges have been encountered in aligning national action plans and developing integrated investment frameworks. The assessment process has also revealed problems in data availability and reporting methods (UNEP-WCMC 2011), potentially enabling UNCCD to address these lessons learnt prior to the next reporting cycle in 2012.

Source: Prepared by UNEP-WCMC

As an international response to desertification, land degradation and drought in drylands, the United Nations Convention to Combat Desertification (UNCCD) was adopted in 1995 and has since been signed by 194 Parties – 193 countries plus the European Union. Following mixed results in its initial implementation phase (UNCCD 2007), Parties to the Convention adopted a ten-year strategic plan for 2008–2018 to revitalize it. The plan includes a results-based management approach built on a set of specific objectives and indicators, and a new monitoring, assessment and reporting process – the performance review and assessment of implementation system.

Wetlands

In 2003, the European Space Agency, in collaboration with the secretariat of the Convention on Wetlands of International Importance (Ramsar Convention), launched the GlobWetland project to demonstrate the current capabilities of Earth observation technology to support inventorying, monitoring and assessment of wetland ecosystems. The project revealed a major gap between the findings of the Earth observation and wetland communities (Jones *et al.* 2009), with considerable inconsistency in global wetlands estimations (Table 3.3).

The conversion of wetlands continues. For both inland and coastal wetlands, the most salient drivers of change are population growth and increasing economic development, which in turn promote infrastructure development and land conversion including agricultural expansion (Wood and van Halsema 2008). Other direct drivers affecting wetlands are deforestation, increased withdrawal of freshwater, diversion of freshwater flows, disruption and fragmentation of the landscape, nitrogen loading, overharvesting, siltation, changes in water temperatures and invasion by alien species (Fraser and Keddy 2005). In 14 deltas analysed by Coleman *et al.* (2008), over

Table 3.3 Estimates of global wetland area

Region	**Global review of wetlands resources** (MA 2005b; Finlayson *et al.* 1999)		**Global lakes and wetlands database** (Lehner and Döll 2004)	
	Million hectares	% of global wetland area	Million hectares	% of global wetland area
Africa	125	10	131	14
Asia	204	16	286	32
Europe	258	20	26	3
Neotropics	415	32	159	17
North America	242	19	287	31
Oceania	36	3	28	3
Total	1 280	100	917	100

half of the studied wetland area of 1.6 million hectares had been irretrievably lost over a 14-year period due to natural causes, conversion to agriculture or industrial use. Global climate change may exacerbate the loss and degradation of coastal wetlands. For example, Syvitski *et al.* (2009) analysed the effects of human activities on delta subsidence, susceptibility to flooding and vulnerability to sea level rise, concluding that the area of deltas at risk of flooding could increase by more than 50 per cent by the end of this century.

The deforestation, drainage and conversion to agriculture of peatland results in substantial emissions of CO_2 and nitrous oxide (Mitra *et al.* 2005). Globally, peatlands cover 3 per cent of the world's land surface, about 400 million hectares, of which 50 million hectares are being drained and degraded, producing the equivalent of 6 per cent of all global CO_2 emissions (Crooks *et al.* 2011). Avoiding further wetland degradation could result in significant climate change mitigation (Wetlands International 2011).

Because of increasing demand for land for food, feed, biofuels and materials, the loss of wetlands and associated ecosystem services is likely to continue (CA 2007). Globally, coastal wetlands such as mangroves are continuing to decline by more than 100 000 hectares, over 0.7 per cent, per year, but that rate of loss has slowed relative to the 1 per cent per year of the 1980s. Although, in most regions, rates of loss have decreased compared to the 1980s and 1990s, mangrove losses in Asia accelerated again during 2000–2005 (UNEP-WCMC 2010). Despite these losses, the Asia and Pacific region holds the largest spatial extent of mangrove systems – more than 50 per cent of the global total. Other major mangrove areas are in northern Latin America, Eastern and Western Africa, and the Red Sea.

Polar regions

The Arctic's permafrost – the top 3.5 metres of soil that remains permanently frozen for 24 months or more – contains the largest deposits of organic carbon on Earth. But due to some of the most rapid warming on the planet (McGuire *et al.* 2009; Tarnocai *et al.* 2009), with temperatures in the permafrost having already risen by up to 2°C over the past two to three decades (AMAP 2011), this is likely to become a substantial source of carbon emissions over the next century (Schuur *et al.* 2008). The Arctic's tundra and boreal forest ecosystems currently act as a carbon sink (McGuire *et al.* 2009), but it is possible that the Arctic region will become a net emitter over the course of the 21st century (Schuur *et al.* 2008; Zimov *et al.* 2006) as up to 90 per cent of near-surface permafrost is expected to disappear due to thawing by 2100 (Lawrence *et al.* 2008).

Methane emissions, primarily from wetlands, also play an important role in the carbon balance of the Arctic (O'Connor *et al.* 2010). Although just 2 per cent of global methane emissions originate in the Arctic, this region has seen the largest proportional increase in emissions, rising by nearly a third between 2003 and 2007 (Bloom *et al.* 2010). Some of these emissions originate in the escape of methane from sequestration within hydrate crystals frozen beneath permafrost. These methyl hydrates also occur in abundance beneath the deep ocean floor and within continental shelves (O'Connor *et al.* 2010). Methane is 25 times more effective at trapping heat in the atmosphere than CO_2 over a 100-year horizon (IPCC 2007).

Other climate-related land changes occurring in the Arctic include the northward movement of tree lines, woody vegetation encroachment into the tundra, and a longer growing season – resulting in an increase in plant productivity (Figure 3.7) (Epstein *et al.* 2012; Walker *et al.* 2012; Callaghan *et al.* 2011; Wang and Overland 2004; Zhou *et al.* 2001; Myneni *et al.* 1998). Whilst these processes remove CO_2 from the atmosphere, it is likely that the release of carbon from thawing permafrost and other processes will outpace CO_2 sequestration by vegetation (Schuur *et al.* 2008; Zimov *et al.* 2006).

Environmental changes such as the northward advance of tree lines, combined with rapid industrial development, create challenges for traditional livelihoods such as reindeer herding.

Figure 3.7: Changes in Arctic vegetation, 1982–2005

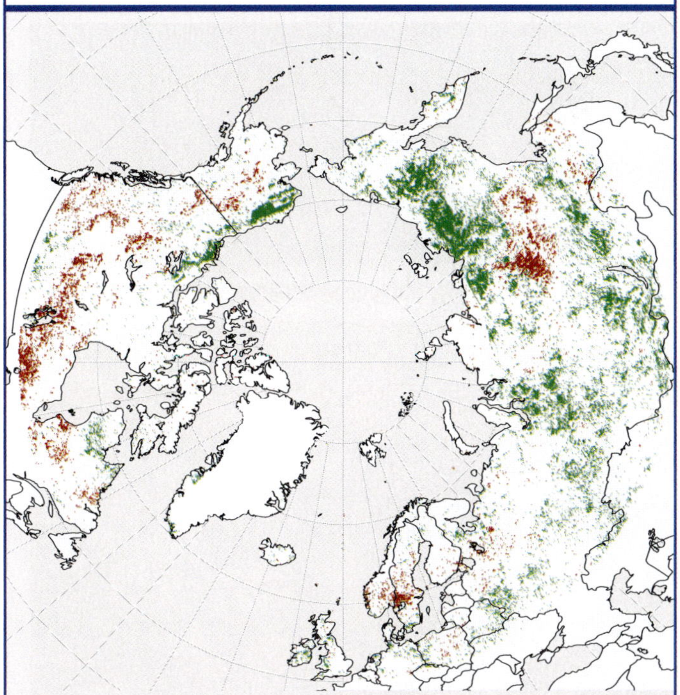

Source: CAFF 2010

Access to many areas on land, especially in northern Canada and Russia, is becoming more difficult as ice roads melt earlier and freeze later, severely affecting communities and industrial development (AMAP 2011; Stephenson *et al.* 2011). At the same time, because the seasonal Arctic Ocean ice cover is decreasing in area, volume and duration, new economic opportunities are presenting themselves, including increased tourism, forestry, agriculture, and expanding oil, gas and mining developments. Nonetheless, some communities in the Arctic most affected by thawing permafrost and/or coastal erosion are being forced to relocate (ACIA 2005), and further research is needed to foresee how living conditions are likely to change and to evaluate possible adaptation options, taking the region's indigenous peoples into particular consideration (AMAP 2011).

At the southern pole, the landmass of Antarctica also has a profound effect on the Earth's climate and ocean systems. However, in contrast to the Arctic, the Antarctic land mass is 99 per cent covered by glacial ice. The changes occurring in this region are discussed in greater detail in Chapters 4 and 7.

Urban areas and human infrastructure

Urbanization has progressed at an extraordinary rate in recent decades and this growth is projected to continue throughout the century. Urban areas are the hubs of social processes, driving many changes through material demands that affect land use and cover, biodiversity and water resources, locally to globally. Nevertheless, if well planned, urban areas can reduce the overall pressure on land resources of a growing population.

Satellite-based studies calculate urban land cover at less than 1 per cent of the planet's total land surface (Schneider *et al.* 2009). However, the impact of urban areas on the global environment cannot be measured only by their physical expansion. Some studies estimate that 60–70 per cent of total anthropogenic greenhouse gas emissions are directly or indirectly related to urban areas, with a few wealthy cities contributing the majority of emissions (Dodman 2009). It is the concentration of population, economic activities and wealth generation in urban areas that drives their impact on the global environment, with demands for food, energy, water and production materials that have significant consequences for land-use change around the world (Grimm *et al.* 2008).

Most of the understanding of urbanization as a land-change process is based on individual case studies (Seto *et al.* 2010) that reveal significant differences in urbanization processes between regions and countries, and even within countries. Ecological footprint analyses of cities provide a symbolic parameter illustrating the impacts of those differences on the local and global environment. For example, the inhabitants of a typical city of 650 000 inhabitants in the United States collectively require 3 million hectares of land to meet their domestic needs, while those of a similar-sized city in India require just 280 000 hectares (Newman 2006).

Urban trends

The UN Population Division projects that between 2007 and 2050, the world's urban population will increase by more than 3 billion, with almost all future population growth expected to take place in the cities and towns of developing countries (Montgomery 2008). By 2050, more than 70 per cent of China's population and 50 per cent of India's is likely to be urban, with China expected to have 30 additional cities of more than 1 million inhabitants and India 26 (Seto *et al.* 2010).

Urbanization is not a homogeneous process (Seto *et al.* 2010). Recent studies suggest a significant increase in land requirements for urban uses in the next 40 years – potentially an additional 100–200 million hectares (Bettencourt *et al.* 2007) (Figure 3.8). This increase is expected to occur primarily in sprawled patterns and to have major effects on greenhouse gas emissions, air pollution and waste management (Lobo *et al.* 2009).

Very large cities exert local and global impacts on the environment, for example the emission of greenhouse gases or aerosols that have a dimming effect in the atmosphere. Small and medium cities, despite their own environmental impacts, may have better opportunities to improve their relationship with the environment and social well-being, particularly in low-income and middle-income countries, where population will concentrate in the future (Seto *et al.* 2010, Martine *et al.* 2008). Only 12 per cent of the total urban population in developing countries lives in very large urban areas of more than 10 million people, while 40 per cent lives in cities of less than 1 million (Figure 3.9) (Montgomery 2008).

Figure 3.8 Urban expansion in the Pearl River Delta, China, 1990–2009

The upper delta area shown in the left-hand image had over 7 million people in 1990, but has since more than tripled to over 25 million, with the cities of Dongguan, Foshan, Guangzhou and Shenzhen beginning to merge into one continuous city. This intense urbanization has led to the loss of productive farmland and natural areas, as well as creating a variety of environmental problems.

USGS EROS Data Center 2010 and UNEP 2011c

Figure 3.9 Distribution of the urban population of developing countries, by city size

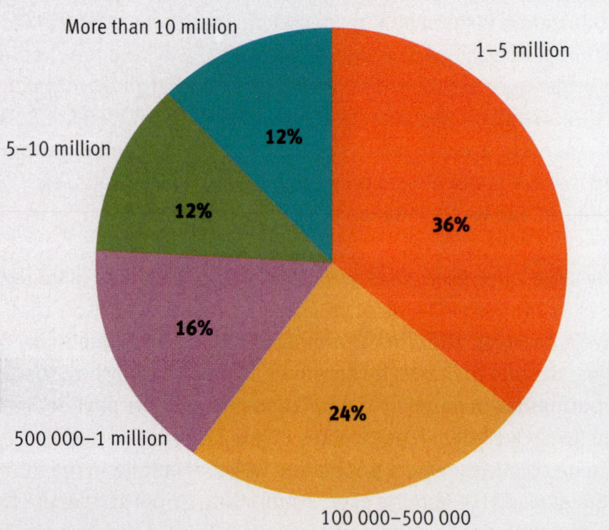

Source: Montgomery 2008

MAJOR ISSUES IN LAND CHANGE

The changes in land use presented in this chapter are a product of complex interactions between human actions and biophysical processes. International goals provide one set of guidelines for land management, but these are often overshadowed by other pressures and competing demands. Here, four major themes are explored that help to explain the apparent movement away from achieving land-related goals:

- economic growth at the expense of natural capital;
- competing demands for land;
- increased separation of production from consumption; and
- governance challenges related to sustainable land management.

Each is illustrated with examples of impacts on land resulting from these pressures, as well as opportunities to move land management decisions towards social and ecological outcomes in line with international goals.

Economic growth and natural capital

The global economic system is based on the pursuit of perpetual and unsustainable growth. Distorted incentives have reduced natural capital, while often rendering attempts to curtail resource or energy use politically problematic (Chapter 1) (Daly and Farley 2010; Dasgupta 2009). Simply put, economic growth has come at the expense of natural capital.

Today, many terrestrial ecosystems exhibit degradation and reduced resilience. This can be linked to the failure to account for the vital functions of these ecosystems in economic cost-benefit analyses. For example, financial pressures have encouraged the irrigation and subsequent salinization of vast dryland areas, making them very difficult to rehabilitate (Sakadevan and Nguyen 2010). Wetlands continue to be drained for agriculture and urban development, destroying their ability to regulate water quantity and quality and buffer against extreme weather events (Box 3.3). Deforestation and forest degradation produce financially attractive short-term returns, but global natural capital losses have been recently estimated at between US$2 trillion and US$ 4.5 trillion each year (Kumar 2010).

Ecosystems have priceless spiritual, aesthetic and cultural dimensions. They are also the cornerstones of economies, but their real value remains effectively invisible in national profit and loss accounts (TEEB 2010). Allowing the privatization of profits from the extraction of natural capital at the expense of more innovative and equitable land management approaches is a pervasive problem across all land covers and uses. Incentives that are narrowly focused on economic growth often encourage land management that degrades ecosystem services,

Box 3.3 Restoring wetlands along the Mississippi

Wetlands can help control floods by absorbing and storing high levels of precipitation. However, the Mississippi River basin in the United States has historically been managed by draining wetlands for agriculture and building dams and levees to contain floodwaters, a strategy that has worsened the impacts of flood events (Hey and Philippi 1995). The coastal wetlands of the Mississippi Delta have likewise been replaced with artificial flood control structures, compromising ecosystem services such as soil formation, provision of habitat for fish and crustaceans, and protection against severe storms (Twilley and Rivera-Monroy 2009).

In 2005, Hurricanes Katrina and Rita brought into focus the importance of maintaining wetlands as buffers against natural hazards. The State of Louisiana has since assigned 37 per cent of revenues from new oil and gas projects to coastal protection and restoration; combined with other funds, this could provide up to US$1 billion per year over the next 30 years (Day *et al.* 2007). Research suggests that an investment of US$10–15 billion in restoring the Mississippi Delta could generate the equivalent of US$62 billion by avoiding losses from storm damage and reduced ecosystem functions while gaining additional ecological benefits (Batker *et al.* 2010).

Box 3.4 The Mau Forests complex, Kenya

The Mau Forests complex in Kenya provides goods and services worth US$1.5 billion a year through water for hydroelectricity, agriculture, tourism and urban and industrial use, as well as erosion control and carbon sequestration (TEEB 2010). Alternative accounting has helped spur the government of Kenya to invest in rehabilitating the area and its vital ecological services, though challenges remain in addressing the interests of people living there (UNEP 2011a).

© Christian Lambrechts

while including and valuing ecosystem services in accounting systems can help protect and enhance them. Successful strategies rest on improving understanding of ecosystem functions and building that understanding into policies and institutions (Daily *et al.* 2009). Indeed, recognition of the multiple uses and multiple values of ecosystems can be used to leverage resources for their protection (Boxes 3.3 and 3.4).

Over the past two decades, payment for ecosystem services (PES) has gained attention as a mechanism with the potential to account for services provided by ecosystems in market transactions, build bridges and balance interests between the users and providers of these services, and deal with the linked challenges of conservation and poverty alleviation (Pascual and Corbera 2011; Engel *et al.* 2008). Payment for ecosystem services involves a suite of approaches linked to a broad central idea: "the transfer of resources among social actors with the objective of creating incentives to align individual and/or collective land-use decisions with the social interest in the management of natural resources" (Muradian *et al.* 2010).

The concept of PES offers several advantages over conventional conservation approaches: it complements command-and-control and polluter-pays principles with more flexible, incentive-based approaches; it is conditional and voluntary, with the potential to promote equity, accountability and cost effectiveness; and it can produce co-benefits for livelihoods and contribute to poverty alleviation (Borner *et al.* 2010; van Hecken and Bastiansen 2010). Positive land-use outcomes have been achieved through some PES initiatives in, for example, Colombia, Costa Rica and Nicaragua, where tree cover has increased and degraded pasture decreased due to a regionally integrated PES project (Chapter 12).

However, groups who oppose the idea of nature being commoditized or traded have criticized the concept (Pascual and Corbera 2011; Corbera *et al.* 2007). Furthermore, despite promising initial benefits such as increased land-tenure security, current evidence of PES's cost effectiveness and the conditions under which it has positive environmental and socio-economic impacts remains inconclusive, particularly in developing countries with weak governance (Pattanayak *et al.* 2010; Wunder *et al.* 2008).

Challenges ahead for PES focus on cost effectiveness, monitoring capacity, enforcement, transparency and accountability, and clear boundaries to land access and tenure rights (Borner *et al.* 2010). Taking into account social norms and culture, building trust between actors and dealing with power relations will ultimately define benefit allocation strategies and successful long-term implementation of PES (Bille 2010; van Hecken *et al.* 2010).

Competing demands for land

The challenge of feeding a growing human population has been compounded by rising affluence in some regions. Changing diets and increasing demand for biofuels and other industrial materials such as timber have intensified competition for land and pressures on terrestrial ecosystems.

Food security

To meet the Millennium Development Goal (MDG) 1c on reducing hunger, global food production will have to increase and food distribution improve. To meet MDG 7 and other environmental goals, agriculture needs to reduce its current environmental impacts (Figure 3.10).

Although estimates vary, the Food and Agriculture Organization of the United Nations (FAO) projects that to reduce the proportion of developing countries' populations that are chronically undernourished to 4 per cent in the year 2050, world food production will need to increase by 70 per cent from 2005 levels (Bruinsma 2009). Although food consumption per person is increasing across all regions, it is unevenly distributed and the number of malnourished people continues to rise as more grain is diverted to produce meat for those who can afford it. Livestock and poultry can serve as an important source of protein in areas of chronic food insecurity and provide an important buffer in times of crop failure, but a disproportionate share of agricultural land is dedicated to meat and dairy production for consumption in developed countries. Such land use is less efficient in meeting global food needs and comes with greater environmental consequences than cropland (Steinfeld *et al.* 2006). For

Figure 3.10 Food security and environmental goals for agriculture by 2050

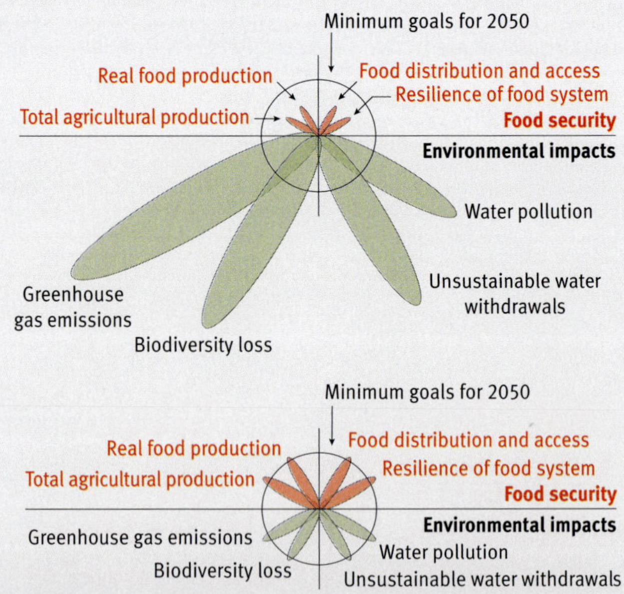

Each lobe represents the status of a particular theme, with the circle defining the balance required to meet the goals. The upper figure shows the current situation, reflecting a shortfall on food security goals and excessive environmental impacts; the lower figure shows a hypothetical situation in which all goals for 2050 are achieved.

Source: Adapted from Foley *et al.* 2011

example, it is estimated that the amount of grain fed to livestock in the United States is more than seven times that consumed directly by the population (Pimentel and Pimentel 2003).

Meanwhile, about one-third of all food that is produced for human consumption is wasted or lost – approximately 1 300 million tonnes annually (Toulmin *et al.* 2011). The concept of food security moves beyond the question of whether adequate food is available and considers whether people have physical and economic access to food (FAO 2008). This draws attention to a broad set of social and political issues related to food distribution.

It will be challenging to meet future global demand for food while avoiding, or at least mitigating, negative impacts on forests, wetlands and other ecosystems – and at the same time reducing poverty, supporting livelihoods, and ensuring food safety and animal welfare. There is little debate that more land will have to be allocated to agriculture, but this will not be sufficient without increasing yields and reducing losses along the food supply chain. Climate change is likely to complicate matters further by affecting crop yields in many areas (Figure 3.11) (Ringler *et al.* 2010; Lobell *et al.* 2008).

A variety of agricultural approaches is likely to provide the best outcomes for food security and environmental well-being. High-input, intensive agricultural methods undeniably increase agricultural yields, though these gains may come at the expense of long-term soil fertility (Foley *et al.* 2005). Location-specific approaches are also needed in order to achieve sustainable land use based on biophysical as well as socio-economic considerations (Chapter 12) (DeFries and Rosenzweig 2010), while agroecology and urban agriculture can contribute to the global food supply (Perfecto and Vandermeer 2010; Zezza and Tasciottia 2010). Agricultural practices that conserve soils and nutrients such as no-till farming (Chapter 12) can complement efforts to restore degraded and abandoned agricultural land.

The world's food system faces increasingly complex and interconnected challenges. *© Ralf Hettler/iStock*

Figure 3.11 Projected changes in sub-Saharan African crop yields due to climate change, 2050

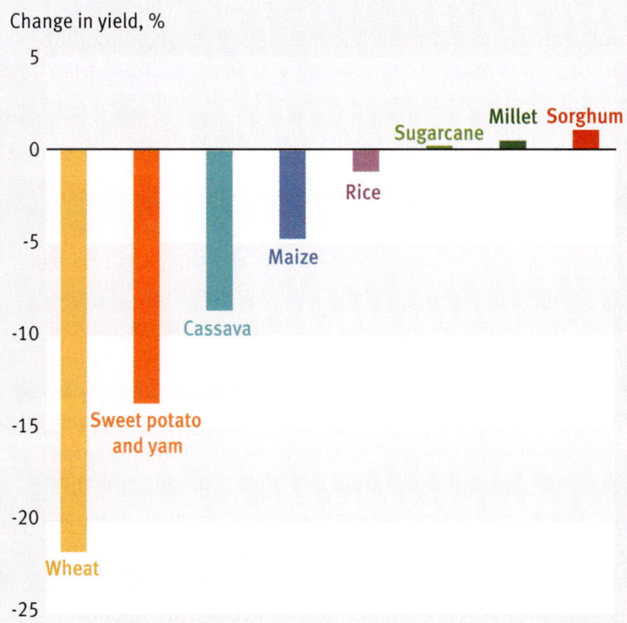

Source: Ringler *et al.* 2010

Meeting the global need for food will be one of the most important challenges of this century, and a portfolio of solutions including conservation agriculture, high-yielding cultivars, and efficient and carefully managed use of fertilizers is needed rather than promotion of a single strategy. Advocates of genetically modified crops point out their potential to increase yields while reducing the use of agricultural chemicals (Brookes and Barfoot 2010; Fedoroff *et al.* 2010), although resistance to their use remains, in part, due to the uncertainty of potential risks to human health and further loss of agricultural biodiversity (Chapter 5).

Meat production

Meat production has increased significantly during the past two decades, outpacing the rate of population growth over the same period (Figure 3.12). Large differences in meat consumption exist both within and between countries, ranging from an average of 83 kg per person per year in North America and Europe to 11 kg per person per year in Africa (FAO 2009). Population growth, urbanization and increasing incomes are expected to continue to raise demand for meat, particularly in developing countries (Delgado 2010).

Figure 3.12 Change in global population and in meat, fish and seafood supplies, 1992–2007

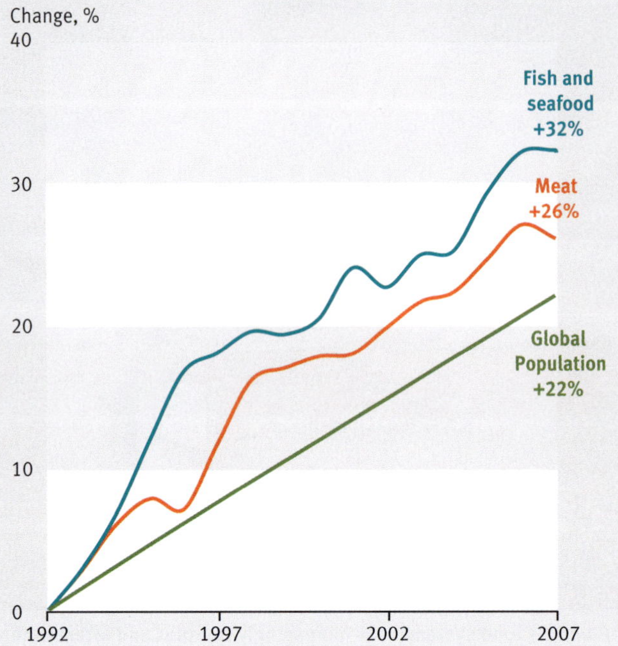

Source: UNEP 2011c

The environmental impacts of meat production depend on intensity, extent and management. Nonetheless, growing demand for meat worldwide has been an important driver of deforestation in South America, as forest is cleared to plant soy for livestock feed (Box 3.5). As meat production has grown, so has the area harvested for soybean crops, which expanded to 98.8 million hectares in 2009 from 74.3 million hectares in 2000, and 50.4 million hectares 30 years ago (FAO 2012). An increasing demand for meat has the potential to compound rangeland degradation. Livestock production accounts for over 8 per cent of global freshwater use and is among the largest sources of water pollution leading to eutrophication,

Meat and dairy production systems account for a large proportion of the global land area. *© Anna Kontorov*

algal blooms, coral reef degradation, human health issues, antibiotic resistance and disruption of nutrient cycling (Steinfeld *et al.* 2006). Considering the entire commodity chain, including deforestation for grazing and forage production, meat production accounts for 18–25 per cent of the world's greenhouse gas emissions, which is more than global transport (UNEP 2009b; Fiala 2008; Steinfeld *et al.* 2006). Reducing meat consumption in regions where it is relatively high could thus bring a range of environmental benefits (Marlow *et al.* 2009).

Biofuels

An urgent search for renewable sources of energy has resulted in policies that promote the use of biofuels. Increased production of crops that can be used for multiple purposes including food, feed or fuel – such as oil palm, soy, maize and sugar cane – is indicative of this trend (Figure 3.14). However, subsidies that promote biofuels have been linked to distortions in the world food system, leading to increases in food prices (Pimentel *et al.* 2009). Recent changes in the linked production of food, feed and fuel have far-reaching impacts for ecology as well as for social relations and vulnerability (Bernstein and Woodhouse 2010; McMichael and Scoones 2010). While no energy source is completely problem-free, biofuels present particular challenges to land use and terrestrial ecosystems. This, combined with the recent rapid increase in their production, is the reason for examining them here.

While a major motivation for promoting and investing in biofuels has been the desire to reduce greenhouse gas emissions, recent research shows that their emissions balance varies widely depending on which crops are grown, where, and which production methods are used (Cerri *et al.* 2011; Johnston *et al.* 2009; Pimentel *et al.* 2009). Biofuel crops have been linked to deforestation, for example in Indonesia (Box 3.6), and to encroachment into conservation lands. Once these land-use changes are taken into account, the biofuel carbon balance can become negative, meaning that more carbon is released producing and using biofuels than the equivalent amount of energy from fossil fuels (Melillo *et al.* 2009; Fargione *et al.* 2008; Searchinger *et al.* 2008).

Crop-use changes stemming from demand for biofuels have already been observed. For example, in 2007, the United States converted 24 per cent of its corn to ethanol, supported by government subsidies. The US Renewable Fuels Standard of 2007 mandated an increase in biofuel production from around 6.5 billion litres (1.7 billion US gallons) per year in 2001 to 136 billion litres (36 billion US gallons) per year by 2022 (US Government 2007). Also in 2007, US farmers planted the largest area in maize since 1944: 37.8 million hectares, an area 20 per cent bigger than in 2006 (Gillon 2010). This crop change, which was subsidized, resulted in calling back into production many set-aside lands in the Conservation Reserve Program (CRP) that used to help check surpluses, maintain price levels and promote an ecological balance. Between late 2007 and March 2009, the total area of CRP land in the United States dropped from 14.9 million to 13.6 million

Box 3.5 Brazil's forest policy and soy moratorium

While most Amazonian deforestation is linked to cattle pasture and ranching, forest conversion for cropland – especially soy – increased in Mato Grosso during 2000–2004 (Morton *et al.* 2006), and evidence suggests that by displacing pastures, soy production may also drive deforestation (Barona *et al.* 2010). A sharp decline in annual deforestation during 2004–2009 (Figure 3.13) coincided with the introduction of new policies as part of the Action Plan for Prevention and Control of Deforestation in the Amazon (PPCDAm). These include:

- creating new protected areas in deforestation hot spots;
- establishing a deforestation monitoring programme using satellite imagery;
- an assertive law enforcement strategy allowing for property apprehension, forfeiture or even destruction;
- withholding public rural credit from producers who break environmental regulations; and
- an obligation on municipalities to reduce deforestation rates below a certain threshold and register protected areas in a GIS database to make illegal deforestation promptly apparent (BRASIL 2009).

Pressure from consumers in Europe and a Greenpeace campaign against illegal deforestation also led the Brazilian Vegetable Oil Industry Association (ABIOVE) and National Cereal Exporters' Association (ANEC) to sign an agreement in July 2006 in which members pledge not to acquire soybeans from newly deforested areas in the Amazon. The success of this moratorium has prompted efforts to persuade the beef industry to make its own commercial agreement.

Despite the apparent success of these and other policies and agreements in reducing deforestation, challenges remain. For example, many are concerned that proposed changes to Brazil's forest code may reduce forest protection (Tollefson 2011). The rise of deforestation in other biomes and countries is also a concern, which has led the Brazilian government to launch an action plan for the Cerrado biome (BRASIL 2010) and disseminate lessons learned to neighbouring Amazonian countries.

Figure 3.13 Clear-cut deforestation in the Brazilian Amazon, 1988–2011

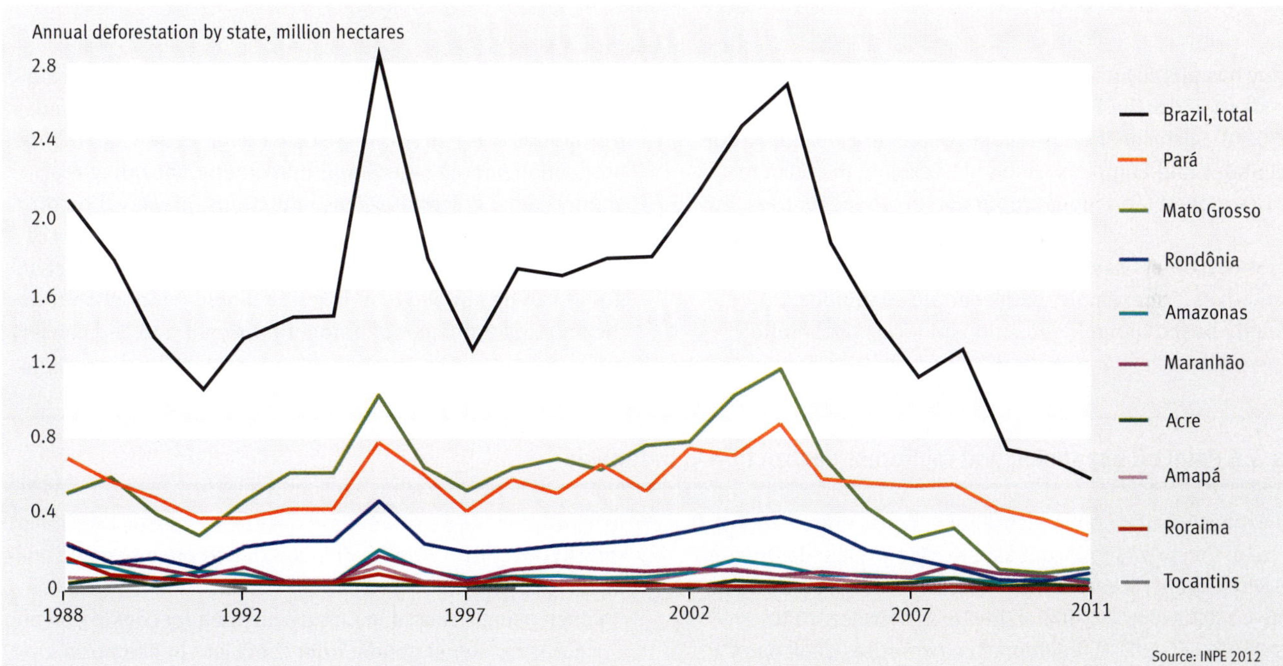

hectares (Gillon 2010). In other words, close to 1.3 million hectares of conservation lands were lost in just over a year.

A similar trend can be seen in the European Union (EU), particularly Germany, whose production capacity for biodiesel increased fivefold between 2004 and 2008 (Franco *et al.* 2010). Although Germany's rapeseed cultivation reached 1.53 million hectares in 2007, a little over half of which was used for fuel to meet its EU mandatory biodiesel blending target, Germany needs an additional 1.8 million hectares of rapeseed, which can be done only by increasing the conversion of permanent grassland – similar to the US CRP. However,

Figure 3.14 Area under cultivation for selected crops in humid tropical countries, 1960–2010

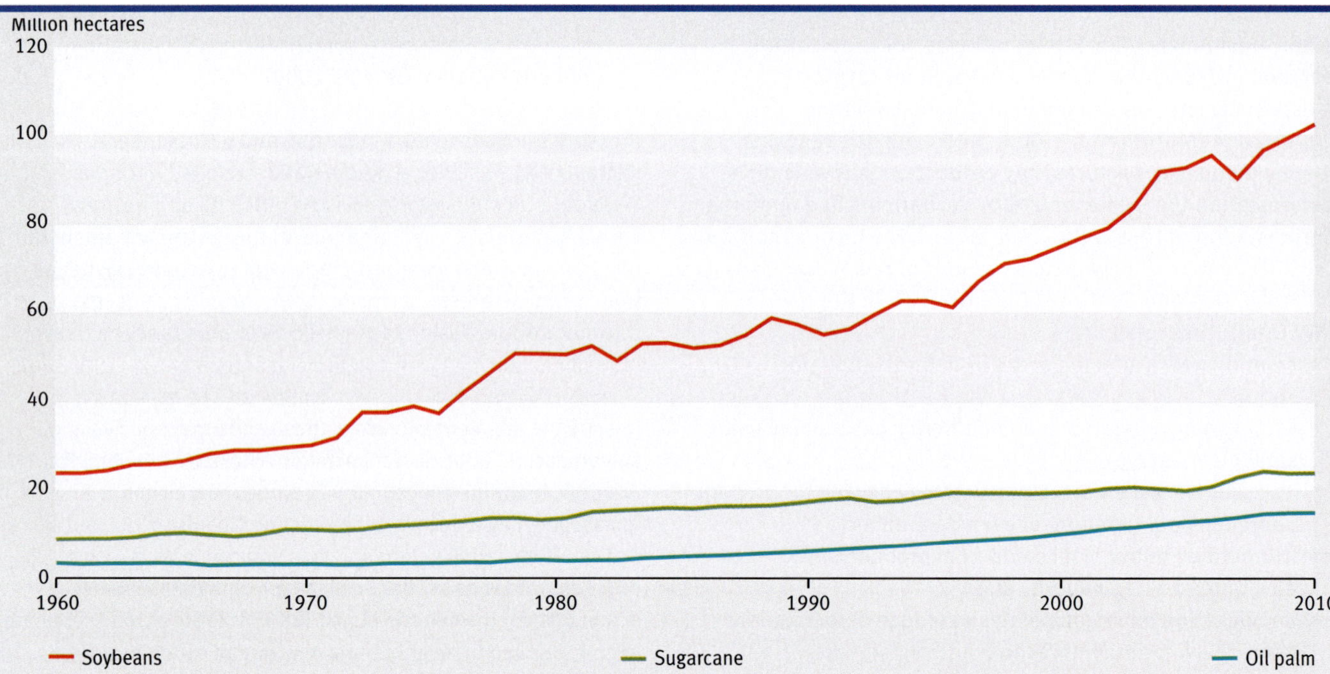

Source: UNEP 2011c

Germany has already used its maximum allowable 5 per cent of grassland under the EU Common Agricultural Policy (Franco *et al.* 2010). Such constraints on agricultural expansion in the United States and European Union help explain the push to outsource biofuel (and food) production to other countries.

Critiques of biofuels have been accompanied by the emergence of alternatives. For example, under certain conditions, community-based biofuel production for local consumption can be desirable, such as in Brazil where some small-scale farmers produce fuel for their own vehicles and equipment (Fernandes *et al.* 2010). To be considered beneficial, biofuel production should satisfy multiple criteria, including real energy gains, greenhouse gas reductions, preservation of biodiversity and maintenance of food security (Tilman *et al.* 2009). Indeed, the principles of ecoagriculture (Milder *et al.* 2008) can be applied to help guide biofuel production towards the mutual attainment of production, livelihood and conservation

Box 3.6 Palm oil expansion and rainforest destruction in Indonesia

The expansion of oil palm plantations, both for food and fuel, is one of the most significant causes of rainforest destruction in South East Asia, where the area under oil palm increased from 4.2 million to 7.1 million hectares between 2000 and 2009 (FAO 2012). In Indonesia, two-thirds of oil palm expansion has occurred by converting rainforest (UNEP 2009a). Clearing tropical forests produces a carbon debt that lasts from decades to centuries, contradicting one of the main reasons for pursuing biofuels in the first place (Gibbs *et al.* 2008). It also compromises vital ecosystem functions provided by rainforests that cannot be replaced by plantations.

In 2009 the Indonesian government projected a dramatic increase in the area planted to oil palm during the next one or two decades – up to 20 million hectares – mostly on cleared forest land (UNEP 2009a). This target was based on two linked assumptions:

- increasing demand in China and India for cooking oil and other consumer goods, from chocolate to shampoo, that use palm oil; and
- increasing demand for biofuels in Europe and elsewhere (McCarthy 2010; White and Dasgupta 2010).

In May 2011 the President of Indonesia signed a two-year moratorium on new permits to clear primary forests and peatlands, potentially slowing oil palm expansion; however, secondary forests and existing contracts remain exempt (USDA 2011).

Table 3.4 Timber and fibre consumption, 2002 and 2008

Type	2002 Million m³	2008 Million m³	2002–2008 % change
Fuelwood	1 795	1 867	+4
Industrial wood	1 595	1 544	-3
Wood-based panels	197	263	+34
Pulp for paper	185	191	+3
Paper and paper board	324	388	+20

Source: FAO 2011b, 2005

objectives. While such systems represent only a tiny portion of overall biofuel production, they provide an opportunity for equitably distributed alternative fuels to benefit land-based ecosystems, for example by reducing charcoal production.

Timber and wood products

Forests are the main source of timber for fuel, industry, pulp, paper and wood-based composites (Table 3.4). Key factors contributing to the rise in consumption are population and economic growth (FAO 2011). In addition, an increase in the absolute number of people living in poverty, especially in rural areas, and continued urbanization contribute to the growth in consumption of wood fuels, while enhanced economic growth in emerging economies contributes to the increase in consumption of paper and paper products.

Protected areas

Protected areas are an important mechanism for the conservation of vulnerable environmental resources, although there are controversies as to whether they sometimes come at the expense of the livelihoods of local people. Rates of deforestation are much lower within reserves than outside them (Scharlemann *et al.* 2010; Nagendra 2008), and some research cites the positive benefits that protected areas have on the conservation of ecosystem services (Stolton and Dudley 2010). But when the underlying pressures imposed by local populations are not adequately considered, substantial monitoring and enforcement are needed to enforce rules designed to sustain natural resources, and governance has been found to be most effective when local users participate in the design and implementation of natural resource governance. There is also some evidence of spill-over effects in countries that enact conservation policies, for example by increasing cereal imports from elsewhere (Rudel *et al.* 2009). Protection in a given area has also been found to contribute to deforestation on the adjacent land to which displaced human populations have moved (Wittemyer *et al.* 2008). Despite the growing area of land with protected status – currently almost 13 per cent of the planet's terrestrial area is under some degree of protection (Chapter 5) – policy makers should not rely solely on this mechanism to preserve natural resources (Ostrom and Cox 2010). Instead, they should develop capacity for adaptive management strategies that produce the best institutional fit for natural resource problems while taking into consideration the need to protect local property rights and local livelihoods.

The separation of consumption from the impacts of production

Urbanization and globalization contribute to the separation between places where resources and goods originate and where products are consumed. Recent research suggests that the spatial distance between production and consumption is both significant and growing (Erb *et al.* 2009). As a result, many of the ecological costs of consumption are borne by people and places increasingly far from consumption sites. While urbanization draws people into densely populated spaces and concentrates demand for food, materials and consumer products, globalization and trade facilitate the movement of people and goods, making both regional and international transfers of resources and finished products possible. Large-scale land acquisitions to supply food, fodder and other forest products as well as other natural resources to markets in distant countries are both a recent outcome of and a contributor to the separation of production and consumption (Toulmin *et al.* 2011). If carefully planned and managed, urbanization and globalization can present opportunities to increase efficiency of resource use.

Drivers of increased separation

Urbanization affects land use and land cover, water use and biodiversity at local and regional scales through social processes that drive consumption patterns and material demands. Higher purchasing power among many urban workers contributes to improved quality of life, but at the cost of new challenges for natural resources and environmental management. For example, Western-style diets are increasingly being adopted globally in urban areas (Pingali 2006). Similarly, improved urban lifestyles are accompanied by higher consumption of water and energy and increased carbon emissions. These urban consumption patterns intensify stresses on distant as well as local ecosystems.

Globalization is not new, but its current iteration has some distinct features (Chapter 1). Lower trade barriers, improved communication technologies and relatively cheap transport have all encouraged countries to become increasingly specialized in their economic activities and reliant on international trade to connect products and services with distant markets (Gibbon *et al.* 2008). While international trade can make use of strategic advantages to produce goods in an efficient way, it also makes it easier to externalize both environmental and social costs. The well-being of individuals in one place is often based on the degradation of the environment elsewhere, for example by non-renewable resource extraction. Meanwhile, both resources and pollution are embedded in trade (Chapter 4), and countries that place greater emphasis on free-market economic policies have been linked to higher levels of environmental degradation (Özler and Obach 2009). The challenge for the global economy

Yasuni National Park on the fringes of the Amazon Basin in Ecuador – believed to be the single most biodiverse place on the planet – has come under severe threat following the discovery of rich oil deposits beneath the park's rivers. In December 2011, US$116 million payment for ecosystem services was raised by crowd-sourcing, temporarily halting ecological devastation and the release of more than 400 million tonnes of carbon dioxide (CO_2). *© Sebastian Liste*

is to encourage the best of what it can offer in terms of efficient resource use while taking measures to reduce the occurrence, concentration and transfer of environmental and social costs.

Land deals

Recent changes in production patterns can be linked to the convergence of food, energy, environmental and financial crises, and a continuing surge in the mineral and timber industries (Tables 3.2 and 3.4; Chapter 1). These interactions have brought corporations and some national governments, based in the global North and South, to forge widespread land deals, sometimes referred to as land grabs, in distant countries. The UN Committee on Food Security suggests that such large-scale land acquisition now involves close to 100 million hectares (Toulmin *et al.* 2011). Concentrated in the global South, these land deals are intended to produce food, feed, biofuels, timber and minerals, usually for export. This ongoing global rush for land is altering land-use patterns and social relations, and involves a new combination of people and pressures. Given the rapid pace of recent developments and projected growth in demands for food, feed, biofuels and materials, it is likely to have major impacts on future land use.

The 2007–2008 food price spike inspired multi-sectoral investors to purchase or lease land for food production and export (Toulmin *et al.* 2011). At the same time, biofuel blending requirements in the EU and many other countries have provided another impetus for external land deals and land-use change. This has directly and indirectly inspired the expansion of oil palm plantations in Colombia, Guatemala, Indonesia and Malaysia, sugar cane ethanol production in Brazil and Southern Africa, soy cultivation in Argentina and Brazil, and the planting of jatropha in Ghana and India, amongst other developments (Franco *et al.* 2010). The emerging pattern of production in these newly opened sites is large-scale, industrial monoculture (Novo *et al.* 2010; Richardson 2010). Even in cases where contract growing with smallholders is promoted as a key component of new enterprises, monoculture and industrial production methods are adopted, for example in the oil palm sector in Indonesia (McCarthy 2010).

In theory the term marginal lands, often applied to land deals, refers to lands that are far from road networks, are not irrigated, and are not used for intensive commercial agriculture. However, in practice there are indications that land deals have encroached on prime agricultural lands, suggesting that investors do not want to invest in lands with little access to water sources or transport infrastructure.

Displacement of local, including indigenous, people is a potential outcome of these land deals. This becomes a problem if people have nowhere go to seek employment or construct livelihoods (Li 2011). This has happened in several sites of current land deals, pushing people to further crowd urban spaces or into more fragile environments such as remaining forest, higher slopes or river banks. For example, in the Democratic Republic of the Congo, large-scale agricultural investment has reportedly pushed local farmers into a national park (Deininger *et al.* 2011). But not all land deals have led, or will lead, to dispossession. Different outcomes of land deals for the rural poor are illustrated by McCarthy (2010) in Jambi, Indonesia, where three villages showed three broad trajectories: dispossession, relatively successful incorporation into the oil palm enclave, and adverse incorporation with precarious employment and livelihoods.

There are competing views on how to respond. One position argues that land deals offer both opportunities and threats, and that opportunities can be harnessed and threats managed by promoting a voluntary land-deal code of conduct (Deininger 2011). In contrast, proponents of minimum human rights principles argue that voluntary codes may be insufficient to ensure that agricultural investment "benefits the poor in the South, rather than leading to a transfer of resources to the rich in the North" (De Schutter 2011). An in-between position is reflected in the Voluntary Guidelines for the Democratic Governance of Natural Resources promoted by FAO, which, unlike corporate-led codes of conduct, bind member states to mandatory reporting. How these viewpoints unfold remains to be seen.

Land governance

Many of the challenges for sustainable land management stem from underlying weaknesses in land governance systems. Generally, there are three components of a governance system: actors and organizations, institutions, and practices (GFI 2009). Incompatibility between these is one of the most common reasons for the lack of successful transition from resource-extractive to sustainable management of land resources. For example, various countries have redirected their policies and management rules towards sustainable forest management, but due to structural and cultural resistance in forestry organizations, management practices have not changed to the expected level (Kumar and Kant 2005). Other common features of poor land governance are low levels of transparency, accountability and participation in decision making, and a lack of capacity amongst the actors and organizations responsible for land management.

Land governance includes structures ranging from totally centralized to completely decentralized. A major challenge is to find the best governance system, which depends on existing governance alongside the social, economic and environmental conditions and their dynamics (Kant 2000).

Market-based approaches

Heightened interest in carbon sequestration has inspired new incentives and financing for ecosystem protection. Local and global initiatives have started to invest in market-based climate approaches that attach a financial value to the carbon stored in forests, offering incentives for developing countries to invest in low-carbon development. One such opportunity – Reducing Emissions from Deforestation and Forest Degradation (REDD) in developing countries – has emerged as an important component of a global strategy to reduce emissions while generating financial flows from North to South (Scharlemann *et al.* 2010; Angelsen 2009). Since its inception, REDD has evolved into REDD+, which now goes beyond deforestation and forest degradation to include conservation, sustainable forest management and enhancement of forest carbon stocks. Evidence of the potential for carbon sequestration in drylands and grasslands is accumulating in support of REDD+ programmes for these ecosystems as well as forests (Neely *et al.* 2009).

At this stage, REDD+ has not been incorporated into any formal international carbon market, but it is likely to form a key element of a post-Kyoto climate change treaty by promoting the avoidance of deforestation and allied measures as eligible activities for countries seeking to meet their obligations. Carbon offset payments would encourage developing countries to reduce national deforestation rates, while REDD+ could include incentives to promote afforestation, reforestation and improved forest management. Research suggests that when appropriate techniques are used, forest restoration is a cost-effective means of sequestering carbon while providing abundant social and ecological benefits (Sasaki *et al.* 2011).

Proponents from both science and policy believe that REDD+ will not just conserve forests; they also consider it one of the most cost-effective carbon abatement options worldwide (Corbera *et al.* 2010; Dickson and Osti 2010; Sikor *et al.* 2010; UN-REDD 2010; Kindermann *et al.* 2008; Thoms 2008). With the right safeguards in place, REDD+ could offer crucial new incentives for achieving sustainable development goals – which have proved elusive since the 1992 Rio Earth Summit – by simultaneously enabling biodiversity conservation, watershed protection, capacity building in tropical forest nations and poverty alleviation for rural communities (Sikor *et al.* 2010).

Much of the debate around REDD+ has focused on its international aspects. However, its success will largely depend on allocating benefits at the local to national levels and creating domestic safeguards to prevent perverse incentives and the marginalization of forest-dependent communities (Phelps *et al.* 2010; Cotula and Mayers 2009; Daniel and Mittal 2009). To this end, some stakeholders are concerned that REDD+ could pose new risks to already vulnerable populations through restricted access to land, tenure insecurity, conflict over resources, centralization of power, and distortion effects in local economic systems. These observers caution that REDD+ will only achieve lasting results if it is suitable for adaptation to the particular circumstances of relevant countries and can meet the needs of local people while building their capacity (IUCN 2010/11; Mayers *et al.* 2010; Preskett *et al.* 2008).

The risks and opportunities for REDD+ will depend on several factors, including how it will be financed and implemented. Many challenges are shared by forest countries, but responses and solutions will often have to be developed according to country-specific and local characteristics. Ultimately, if REDD+ is to be successful, it must generate substantial revenues to implement conservation and sustainable forest management while supporting rural poverty reduction and livelihoods. At the same time, it must recognize the dynamic complexity of global systems, where cause and effect are often distant in time and space.

Land management and decentralization

Governance plays a major role in how land resources are monitored and used and how environmental protection is enforced. Proponents of decentralized natural resource management suggest that giving local-level officials greater responsibilities should result in more efficient, flexible, equitable, accountable and participatory governance (Blair 2000). Local-level decision makers often know more about local conditions and are therefore well positioned to develop new management solutions. This is important from the perspective of adaptive management and providing decision makers with the flexibility to quickly develop solutions to unforeseen problems (Ostrom 2007). But decentralization is only effective if local governments have the financial resources and technical capacity to monitor environmental change (Andersson 2004). Positive outcomes from decentralized environmental governance are also unlikely in the absence of public participation in local government decision making (Larson 2002; Blair 2000); this emphasizes the importance of developing the capacity of local-level stakeholders in the sustainable management of land systems.

Capacity building for sustainable land management

Capacity building recognizes the knowledge systems, perspectives and values of all stakeholders and uses an in-depth understanding of how a resource system functions. As sustainable land management requires a different set of organizational, technical, economic, environmental and managerial skills from that of many land managers, building the capacity of all actors and organizations can be central to its successful integration.

Box 3.7 Sustainable dryland management

Promising management strategies for dryland ecosystems across the world include afforestation to counteract chronic carbon loss due to land degradation, with successful examples in Israel (Tal and Gordon 2010), Iran (Amiraslani and Dragovich 2011) and eastern Uganda (Buyinza *et al.* 2010). Other progressive strategies for adaptively managing drylands include planting resilient nitrogen-fixing crops (Saxena *et al.* 2010), dune stabilization measures, run-off control, improved range management and integrated land management, for example Iran's National Plan to Combat Desertification. Programmes that build community resilience through watershed restoration in drylands, such as the Watershed Organization Trusts in India, are also promising, as are models of polycentric adaptive governance increasingly adopted in Australia (Marshall and Smith 2010; Smith *et al.* 2010). Enhanced monitoring programmes based on vegetation indices and real-time climatic data are also important in allowing for early-warning and management interventions (Veron and Paruelo 2010).

Land degradation in dryland ecosystems provides an example where the lack of capacity – scientific, technical and collaborative – limits success in addressing environmental problems. Degradation in dryland systems is driven by multiple causes and characterized by complex feedbacks that are made worse by global climate change (Ravi *et al.* 2010; Verstraete *et al.* 2009). Despite concerted efforts and a wide array of initiatives (Box 3.7), drylands continue to be threatened because of lack of agreement on the underlying driving mechanisms, characteristics and consequences of degradation (Reynolds *et al.* 2007). Long-term harmonized data are necessary not only to understand the root causes of observed changes, but also to forecast and disentangle those, possibly irrevocable, impacts of global change from the often more temporary or local variability induced by other human activity. These data gaps, and the subsequent lack of capacity and common strategies among dryland nations, can severely hamper progress towards internationally agreed goals on dryland conservation and rehabilitation.

OUTLOOK

Complex forces are affecting land resources, some at dramatic rates of change and with diverse regional and national characteristics. Certainly some land conversion trends are on an unsustainable trajectory, as global population growth and rising consumption exert ever greater pressures on land. Continued deforestation, wetland conversion and dryland degradation are of particular concern. An increasing portion of the pressure on tropical forests is shifting from the activities of small household farms to large industrial plantations producing soy, meat and dairy products, palm oil, sugar cane and other products destined for global markets (DeFries *et al.* 2010, 2008). Land degradation continues to hamper soil productivity and ecological functions in many regions. At the same time, there is significant potential to reduce greenhouse gas emissions from agricultural production (Smith *et al.* 2007). Two phenomena that have arisen since *GEO-4* are the expansion of biofuel production and a growing number of land deals in developing countries. These and other processes are unfolding rapidly. While their longer-term implications remain uncertain, early evidence of their social and environmental consequences should be closely considered. In combination, these processes are seriously affecting the environment in several regions and require urgent attention.

Data and monitoring gaps

One key to avoiding environmental damage is to effectively monitor environmental trends, yet major data gaps limit the ability to avert unwanted outcomes. Global data on land degradation have not been updated for a long time, although new estimates using satellite material are being developed. Datasets exist for land cover but do not always adequately represent areas that have experienced selective cutting or other types of modification. Forest cover losses in boreal and temperate forests are not as well studied as those in tropical forests, while evidence is still emerging of the significant carbon sequestration potential of rangelands and grasslands. Records of ecosystem change are improving, mainly through remote sensing, but reliable data on land-use change are still fragmented and often not comparable – the extent of drylands, for example, is uncertain because of the classifications and methodologies used by different programmes. Similarly, there are discrepancies between a number of wetland inventories (Ramsar Convention Secretariat 2007) and there is no comprehensive global wetlands database.

Satellite remote sensing is an essential tool for monitoring global land resources, but no such technology exists for population patterns. National census efforts, the best current technique, are sporadic and underfunded in many countries, and there is a significant data gap for population changes in rural areas. Further, it is critical to track the consequences for the environment of rapid and extensive urbanization, with its uncertain implications for land resources.

Data on biofuels – including the extent of production and use – are incomplete at the global level, although national datasets can be found for some countries. Similarly, there is a need for improved national and global monitoring of land transactions including large-scale land deals. There are also few standard indicators that governments can use to monitor the environmental impacts of different patterns of land tenure. Finally, standard methodologies for the badly needed valuation of ecosystem services are at an early stage of development.

Goal gaps

Table 3.5 summarizes progress toward the themes expressed in internationally agreed goals on land use and conservation. However, some important topics are not reflected in them. For example, there are no goals or targets that reflect the vulnerabilities and challenges specific to the polar regions.

Table 3.5 Progress towards goals (see Table 3.1)

A: Significant progress B: Some progress	C: Very little to no progress D: Deteriorating	X: Too soon to assess progress ?: Insufficient data

Key issues and goals	State and trends		Outlook	Gaps
1. Promote food security				
Reduce proportion of people who suffer from hunger	B	Proportion of malnourished people decreasing, but absolute number increasing	Depends on up-coming policy decisions and interventions	See following entries on increasing food production and access
Improve household economic access to food	C	Food per person is increasing overall, but a large gap remains between and within regions, particularly for rural poor households who now spend more than half of their income on food; one-third of food produced for human consumption is lost or wasted; land and food price volatility is influenced by rising demands for biofuels, among other economic forces	Drivers remain in place for land and food price volatility to continue; without interventions, the gap in food per person is likely to persist	Interventions to reduce post-harvest food waste; stimulate smallholder farmer-centred agricultural growth – promoting affordable access to land, water and tenure rights for poor households; coordinate domestic and regional biofuel policies to avoid worsening global food insecurity
Increase food production	C	Agricultural yields are generally increasing but a large gap remains between regions	Yields are unlikely to improve much more in developed countries; with efforts focusing on decreasing the yield gap in developing countries, much depends on how this is accomplished	Location-specific approaches to increase yields and achieve sustainable land use, for example smallholder farmer-centred agricultural growth; increased nutrient-use efficiency; improved temporal and spatial matching of nutrient supply with plant demand
2. Reverse loss of environmental resources				
Reduce deforestation rate and increase forest coverage	B	Slight slowing of deforestation but rate is still high; deforestation is concentrated in the tropics; temperate areas are experiencing some forest regrowth	Demand for timber and fibre is likely to rise; clearing for agricultural expansion, including biofuels, is likely to continue without a change in policies	Improved understanding of forest degradation; regional policy coordination to avoid leakage shifting deforestation from regulated to unregulated areas
Halt the destruction of tropical forests	B	Deforestation rate has slowed in some tropical countries, but net forest loss in Latin America and the Caribbean and Africa remains close to 7 million hectares per year	The area under the REDD+ programme and schemes for payment for ecosystem services is likely to increase, providing new incentives to protect tropical forests and their ecosystem services	Data and monitoring on carbon stocks/flux; number and area of community-managed REDD+ areas; national adaptation strategies with ecosystem-based components
Stem the loss of wetlands	C/D	Continued conversion of wetlands for agriculture, aquaculture and human infrastructure	Pressure on wetlands is likely to continue or increase as demand for agricultural land and urban expansion continues	Improved inventory and monitoring of global wetlands; renewed commitment to the Ramsar Convention at the national level
Combat desertification and mitigate the effects of drought	C	Net primary productivity is decreasing in drylands	Pressure on drylands is likely to continue	Improved inventory and monitoring of global drylands
3. Practise integrated land-use planning and management				
Integrate principles of sustainable development into country policies and programmes	B	Good progress in countries affected by the UNCCD in establishing mechanisms to ensure synergy between conventions on desertification, biodiversity and climate change, but few countries have integrated investment frameworks	Depends on upcoming policy decisions and interventions	Greater integration/collaboration between sectors
Recognize, maintain and develop the multiple benefits of ecosystem services, for example for biodiversity, and for their cultural, scientific, and recreational value in addition to their economic value	C	Some examples of valuing multiple benefits of ecosystem services, but overall still largely externalized	Depends on upcoming policy decisions and interventions	Improved non-market valuation techniques; capacity building to include multiple and local values in land-use decision making

Coon Creek Watershed in southwest Wisconsin, once one of the most heavily eroded regions in the United States, is now an impressive and integrated farmland mosaic thanks to advances in soil and farmland restoration. *© Jim Richardson*

Issues of capacity building and stakeholder participation are also inadequately represented in international goals. Several of the land-related goals that do exist lack quantifiable targets, complicating the task of assessing progress towards their achievement. A particular challenge is to acknowledge the interactions between different components of social-ecological systems at different scales.

Goals cannot be considered in isolation. Due to tensions and synergies, progress towards one goal must be viewed in light of implications for others. For example, Figure 3.10 highlights friction between MDG 1 on reducing hunger and MDG 7 on environmental sustainability: if food production is increased through agricultural expansion, it directly compromises the protection of forests, wetlands and other ecosystems. Meanwhile, efforts to address the education and health issues expressed in MDGs 2–6 can indirectly help achieve MDGs 1 and 7 in the long term. Thus, an integrated perspective on goal achievement is crucial.

Discussion of key issues

Economic growth and land resources

The global economy has quadrupled during the last 25 years (IMF 2006), but 60 per cent of the world's major ecosystem goods and services underpinning livelihoods have been degraded or used unsustainably (MA 2005a). This means that traditional economic growth cannot be the foundation of sustainable development. A new paradigm of economic welfare is required – one that is focused on improving human welfare and social equity, and reducing environmental risks and ecological scarcities. One such approach, the green economy proposed by UNEP in 2010, includes:

- valuation of natural resources and environmental assets;
- pricing policies and regulatory mechanisms that translate these values into market and non-market incentives; and
- measures of economic welfare that are responsive to the use, degradation and loss of ecosystem goods and services (UNEP 2011b).

The transition from traditional economic growth to the green economy will require changes to national regulations, policies, subsidies, incentives and accounting systems, as well as to global legal and market infrastructures, an appropriate international trade structure and targeted development aid.

Meeting the growing demand for food

Both global population and per-person consumption continue to grow. The achievement of MDG 1, the eradication of extreme hunger and poverty, will require getting more food to more people. How this is accomplished will have important implications for MDG 7 – environmental sustainability. Population growth is an important part of this complex interaction, but changing lifestyles and consumption patterns, particularly the increasing global demand for animal products, are also significant. Friction between these two MDG goals could be reduced by:

- improving efficiency along the whole food chain by increasing crop yields through research and extension, and reducing food waste and spoilage by improving transport, storage and distribution infrastructure in developing countries and changing behaviour in wealthier societies, where much food waste occurs in food retail markets and homes;
- implementing full-cost accounting for food products that reflects the environmental and social costs of their production in order to facilitate a shift in consumption patterns;
- encouraging, where appropriate, innovative approaches to food production to shorten food supply chains and enhance food security;
- evaluating the ecosystem service and carbon balance implications of potential biofuel production to inform land-use planning and management, and reducing competition between food and biofuel production, particularly in areas with the highest crop production potential.

Growing demand for non-food resources

Crop- and plantation-based biofuel production has increased rapidly in recent years and the land-use transitions associated with this could have strong environmental and social impacts. Fuel-blending targets in numerous countries mandate the continued expansion of biofuel production. Next-generation biofuels – from, for example, algae or cellulose – are still under development and are not likely to contribute a significant share of biofuel production in the near future. Governments should recognize that targets for biofuel production have both direct and indirect implications for land use at national and global scales.

Large-scale land acquisitions are growing with potentially major impacts on land-use change and social relations. Recent reports have advocated the establishment of an observatory of land tenure and rights to food to monitor access to land and ensure that land investments result in decreased hunger and poverty in host communities and countries (Toulmin *et al.* 2011). United Nations organizations could play an important role in creating precedents that could help improve food access in developing countries.

Complexity and policy challenges

An important step towards addressing these challenges is to monitor, study and understand how social and biophysical drivers interact, and the diversity of social, economic and environmental consequences they generate at local, regional and global levels. A concerted effort by international organizations, the scientific community, and national and local institutions could create the comprehensive monitoring network needed to achieve this goal – but to be effective there needs to be strong coordination between these actors.

Limitations in the assessment of land-change processes cannot and should not delay action to address their driving forces, with the precautionary principle being applied to reduce their negative impacts. Current evidence of their consequences highlights the need to act in the short term to avoid potentially irreversible negative outcomes in the long term. There are no easy answers to these complex problems, and single and isolated actions might achieve only limited positive outcomes rather than broad solutions. New governance approaches to land management could help incorporate adaptive management, capacity building and more efficient valuation of ecosystem services and natural resources by combining market-based tools with a bigger role for community agency and bottom-up approaches. New governance approaches could also help foster the changes in consumption patterns needed to reduce pressure on land systems and create better knowledge and awareness of the multiple values of ecosystems. While the leadership of UN organizations and other international institutions is a central element in these efforts, governments have a crucial role, responsibility and opportunity to act as agents of change.

New governance approaches could foster the changes in consumption patterns needed to reduce pressure on land systems and create better knowledge and awareness of the multiple values of ecosystems.

© Frank van den Bergh/iStock

REFERENCES

ACIA (2005). *Arctic Climate Impact Assessment*. Cambridge University Press, Cambridge

Allen, C.D., Macalady, A.K., Chenchouni, H., Bachelet, D., McDowell, N., Vennetier, M., Kitzberger, T., Rigling, A., Breshears, D.D., Hogg, E.H., Gonzalez, P., Fensham, R., Zhang, Z., Castro, J., Demidova, N., Lim, J.-H., Allard, G., Running, S.W., Semerci, A. and Cobb, N. (2010). A global overview of drought and heat-induced tree mortality reveals emerging climate change risks for forests. *Forest Ecology and Management* 259, 660–684

AMAP (2011). *Snow, Water, Ice, Permafrost in the Arctic (SWIPA): Executive Summary*. Arctic Monitoring and Assessment Secretariat, Oslo

Amiraslani, F. and Dragovitch, D. (2011). Combating desertification in Iran over the last 50 years: an overview of changing approaches. *Journal of Environmental Management* 92(1), 1–13

Amiro, B.D., Todd, J.B., Wotton, B.M., Logan, K.A., Flannigan, M.D., Stocks, B.J., Mason, J.A., Martell, D.L. and Hirsch, K.G. (2001). Direct carbon emissions from Canadian forest fires, 1949–1999. *Canadian Journal of Forest Research* 31, 512–525

Anderson, R.G., Canadell, J.G., Randerson, J.T., Jackson, R.B., Hungate, B.A., Baldocchi, D.D., Ban-Weiss, G.A., Bonan, G.B., Caldeira, K., Cao, L., Diffenbaugh, N.S., Gurney, K.R., Kueppers, L.M., Law, B.E., Luyssaert, S. and O'Halloran, T.L. (2011). Biophysical considerations in forestry for climate protection. *Frontiers in Ecology and the Environment* 9(3), 174–182. doi:10.1890/090179

Andersson, K. (2004). Who talks with whom? The role of repeated interactions in decentralized forest governance. *World Development* 32(2), 233–249

Angelsen, A. (ed.) (2009). *Realising REDD+*. Centre for International Forestry Research, Bogor

Bai, Z.G., Dent, D.L., Olsson, L. and Schaepman, M.E. (2008). *Global Assessment of Land Degradation and Improvement: 1. Identification by Remote Sensing*. GLADA Report 5. ISRIC – World Soil Information, Wageningen

Bakker, M.M., Govers, G., Kosmas, C., Vanacker, V., van Oost, K. and Rounsevell, M. (2005). Soil erosion as a driver of land-use change. *Agriculture, Ecosystems and Environment* 105(3), 467–481

Barles, S. (2010). Society, energy and materials: the contribution of urban metabolism studies to sustainable urban development issues. *Journal of Environmental and Planning Management* 53(4), 439–455

Barona, E., Ramankutty, N., Hyman, G. and Coomes, O.T. (2010). The role of pasture and soybean in deforestation of the Brazilian Amazon. *Environmental Research Letters* 5, 124002–124009

Batker, D., de la Torre, I., Costanza, R., Swedeen, P., Day, J., Boumans, R. and Bagstad, K. (2010). *Gaining Ground: Wetlands, Hurricanes, and the Economy: the Value of Restoring the Mississippi River Delta*. Earth Economics, Tacoma

Bernstein, H. and Woodhouse, P. (eds.) (2010). Productive forces in capitalist agriculture: political economy and political ecology. Special issue of *Journal of Agrarian Change* 10(3)

Bettencourt, L.M., Lobo, J., Helbing, D., Kuhnert, C. and West, G.B. (2007). Growth, innovation, scaling, and the pace of life in cities. *Proceedings of the National Academy of Sciences of the United States of America* 104(17), 7301–7306

Bille, R. (2010). Action without change? On the use and usefulness of pilot experiments in environmental management. *Veolia Environment* 3, 1–6

Blair, H. (2000). Participation and accountability at the periphery: democratic local governance in six countries. *World Development* 28(1), 21–39

Blanco-Canqui, H. and Lal, R. (2010). *Principles of Soil Conservation and Management*. pp.493–512. Springer

Bloom, A., Palmer, P.I., Fraser, A.D., Reay, S. and Frankenberg, C. (2010). Large-scale controls of methanogenesis inferred from methane and gravity spaceborne data. *Science* 327(5963), 322–325

Boardman, J. (2006). Soil erosion science: reflections on the limitations of current approaches. *Catena* 68, 73–86

Bonan, G. (2008). Forests and climate change: forcings, feedbacks, and the climate benefits of forests. *Science* 320, 1444–1449

Borner, J., Wunder, S.,Wertz-Kanounnikoff, S., Tito, M.R., Pereira, L. (2010). Direct conservation payments in the Brazilian Amazon: scope and equity implications. *Ecological Economics* 69, 1272–1282

BRASIL (2010). *Plano de ação para prevenção e controle do desmatamento e das queimadas no Cerrado –PPCerrado*. Presidência da República. Casa Civil. Brasília. http://www.casacivil.gov.br/.arquivos/101116%20-%20PPCerrado_Vfinal.pdf

BRASIL (2009). *Plano de ação para a prevenção e o controle do desmatamento na Amazônia Legal – PPCDAm, 2ª fase (2000–2011) Rumo ao desmatamento ilegal zero*. Presidência da República. Casa Civil. Brasília. http://www.mma.gov.br/estruturas/168/_publicacao/168_publicacao02052011030251.pdf

Brookes, G. and Barfoot, P. (2010). Global impact of biotech crops: environmental effects, 1996–2008. *AgBioForum* 13(1), 76–94

Bruinsma, J. (2009). The resource outlook to 2050: by how much do land, water and crop yields need to increase by 2050? In *How to Feed the World in 2050: Proceedings of the Expert Meeting on How to Feed the World in 2050 24–26 June 2009, FAO Headquarters, Rome*. Food and Agriculture Organization of the United Nations, Rome. http://www.fao.org/docrep/012/ak542e/ak542e00.htm

Buol, S.W. (1995). Sustainability of soil use. *Annual Review of Ecology and Systematics* 26, 25–44

Buyinza, M., Senjonga, M. and Lusiba, B. (2010). Economic valuation of a tamarind (*Tamarindus indica* L.) production system: green money from drylands of eastern Uganda. *Small-Scale Forestry* 9(3), 317–329

CA (2007). *Water for Food, Water for Life: A Comprehensive Assessment of Water Management in Agriculture*. Comprehensive Assessment of Water Management in Agriculture, Battaramulla

CAFF (2010). Trends in Arctic vegetation productivity 1982–2005 (Greening of the Arctic). Figure produced by Ahlenius, H., using data from Goetz *et al.* 2007. In *Arctic Biodiversity Trends 2010: Selected Indicators of Change*. CAFF International Secretariat, Akureyri. http://maps.grida.no/go/graphic/trends-in-arctic-vegetation-productivity-1982-2005-greening-of-the-arctic

Callaghan, T.V., Tweedie, C.E., Åkerman, J., Andrews, C., Bergstedt, J., Butler, M.G., Christensen, T.R., Cooley, D., Dahlberg, U., Danby, R.K., Daniëls, F.J.A., de Molenaar, J.G., Dick, J., Mortensen, C.E., Ebert-May, D., Emanuelsson, U., Eriksson, H., Hedenås, H., Henry, G.H.R., Hik, D.S., Hobbie, J.E., Jantze, E.J., Jaspers, C., Johansson, C., Johansson, M., Johnson, D.R., Johnstone, J.F., Jonasson, C., Kennedy, C., Kenney, A.J., Keuper, F., Koh, S., Krebs, C.J., Lantuit, H., Lara, M.J., Vanessa D.L., Lougheed, L., Madsen, J., Matveyeva, N., McEwen, D.C., Myers-Smith, I.H., Narozhniy, Y.K., Olsson, H., Pohjola, V.A., Price, L.W., Rigét, F., Rundqvist, S., Sandström, A., Tamstorf, M., Bogaert, R.V., Villarreal, S., Webber, P.J., Zemtsov. V.A. (2011). Multi-decadal changes in tundra environments and ecosystems: synthesis of the International Polar Year – Back to the Future project (IPYBTF). *Ambio* 40, 705–716

Carr, D.L., Suter, L. and Barbieri, A. (2005). Population dynamics and tropical deforestation: state of the debate and conceptual challenges. *Population and Environment* 27(1), 89–113

Cerri, C.C., Galdos, M.V., Maia, S.M.F., Bernoux, M., Feigl, B.J., Powlson, D. and Cerri, C.E.P. (2011). Effect of sugarcane harvesting systems on soil carbon stocks in Brazil: a review. *European Journal of Soil Science* 62, 23–28

Coleman, J.M., Huh, O.K. and Braud, D.J. (2008). Wetland loss in world deltas. *Journal of Coastal Research* 24(1A), 1–14

Corbera, E., Estrada, M. and Brown, K. (2010). Reducing greenhouse gas emissions from deforestation and forest degradation in developing countries: revisiting the assumptions. *Climatic Change* 100, 355–388

Corbera, E., Brown, K. and Adger, W.N. (2007). The equity and legitimacy of markets for ecosystem services. *Development and Change* 38(4), 587–613

Cotula, L. and Mayers, J. (2009). *Tenure in REDD: Start-point or Afterthought?* International Institute for Environment and Development, London

Crooks, S., Herr, D., Tamelander, J., Laffoley, D. and Vandever, J. (2011). *Mitigating Climate Change through Restoration and Management of Coastal Wetlands and Near-shore Marine Ecosystems: Challenges and Opportunities*. Environment Department Paper 121. World Bank, Washington, DC

Daily, G.C., Polasky, S., Goldstein, J., Kareiva, P.M., Mooney, H.A., Pejchar, L., Ricketts, T.H., Salzman, J. and Shallenberger, R. (2009). Ecosystem services in decision making: time to deliver. *Frontiers in Ecology and the Environment* 7(1), 21–28

Daly, H. and Farley, J. (2010). *Ecological Economics: Principles and Applications*, 2nd ed. Island Press, Washington, DC

Daniel, S. and Mittal, A. (2009). *The Great Land Grab: Rush for World's Farmland Threatens Food Security for the Poor*. The Oakland Institute, Oakland, CA

Dasgupta, P. (2009). The place of nature in economic development. In *Development Economics* (eds. Rodrik, D. and Rosenzweig, M.) 5, 4977–5046. Handbooks in Economics series (eds. Arrow, K.J. and Intriligator, M.D.). North-Holland, Amsterdam

Day, J.W. Jr., Boesch, D.F., Ellis, J., Clairain, E.J., Kemp, G.P., Shirley, B., Laska, S.B., Mitsch, W.J., Orth, K., Hassan Mashriqui, H., Reed, D.J., Shabman, L., Simenstad, C.A., Streever, B.J., Twilley, R.R., Watson, C.C., Wells, J.T. and Whigham, D.F. (2007). Restoration of the Mississippi delta: lessons from hurricanes Katrina and Rita. *Science* 315(5819), 1679–1684

DeFries, R. and Rosenzweig, C. (2010). Toward a whole-landscape approach for sustainable land use in the tropics. *Proceedings of the National Academy of Sciences of the United States of America* 107(46), 19627–19632

DeFries, R.S., Rudel, T., Uriarte, M. and Hansen, M. (2010). Deforestation driven by urban population growth and agricultural trade in the twenty-first century. *Nature Geoscience* 3, 178–181

DeFries, R.S., Morton, D.C., van der Werf, G.R., Giglio, L., Collatz, G.J., Randerson, J.T., Houghton, R.A., Kasibhatla, P.K. and Shimabukuro, Y. (2008). Fire-related carbon emissions from land use transitions in southern Amazonia. *Geophysical Research Letters* 35, L22705

Deininger, K. (2011). Challenges posed by the new wave of farmland investment. *The Journal of Peasant Studies* 38(2), 217–247

Deininger, K., Byerlee, D., Lindsay, J., Norton, A., Selod, H. and Stickler, M. (2011). *Rising Global Interest in Farmland: Can it Yield Sustainable and Equitable Benefits?* World Bank, Washington, DC

Delgado, C.L. (2010). Future of animal agriculture: demand. In *Encyclopedia of Animal Science*, 2nd ed. (eds. Pond, W.G. and Bell, A.W.). Marcel Dekker, New York

De Schutter, O. (2011). How not to think of land-grabbing: three critiques of large-scale investments in farmland. *The Journal of Peasant Studies* 38(2), 249–279

Dickson, B. and Osti, M. (2010). *What are the Ecosystem-Derived Benefits of REDD+ and Why do they Matter?* Multiple Benefits Series 1. UN-REDD Programme, Nairobi

Dodman, D. (2009). Blaming cities for climate change? An analysis of urban greenhouse gas emissions inventories. *Environment and Urbanization* 21(1), 185–201

Engel, S., Pagiola, S. and Wunder, S. (2008). Designing payments for environmental services in theory and practice: an overview of the issue. *Ecological Economics* 65, 663–674

Epstein, H.E., Raynolds, M.K., Walker, D.A., Bhatt, U.S., Tucker, C.J., and Pinzon, J.E. (2012). Dynamics of aboveground phytomass of the circumpolar Arctic tundra during the past three decades. *Environmental Research Letters* 7(1)

Erb, K.-H., Krausmann, F., Lucht, W. and Haberl, H. (2009). Embodied HANPP: mapping the spatial disconnect between global biomass production and consumption. *Ecological Economics* 69(2), 328–334

FAO (2012). *FAO Statistics*. Food and Agriculture Organization of the United Nations, Rome

FAO (2011). *2011: State of the World's Forests*. Food and Agriculture Organization of the United Nations, Rome

FAO (2010a). *Global Forest Resources Assessment 2010*. FAO Forestry Paper No. 163. Food and Agriculture Organization of the United Nations, Rome. http://www.fao.org/docrep/013/i1757e/i1757e.pdf

FAO (2010b). *The State of Food Insecurity in the World: Addressing Food Insecurity in Protracted Crises*. Food and Agriculture Organization of the United Nations, Rome

FAO (2009). *The State of Food and Agriculture 2009: Livestock in the Balance*. Food and Agriculture Organization of the United Nations, Rome. http://www.fao.org/docrep/012/i0680e/i0680e.pdf

FAO (2008). *An Introduction to the Basic Concepts of Food Security*. Practical Guides series. Food Security Programme, Food and Agriculture Organization of the United Nations, Rome. http://www.fao.org/docrep/013/al936e/al936e00.pdf

FAO (2005). State of the world's forests 2005. Food and Agriculture Organization of the United Nations, Rome

FAO (1996). *World Food Summit Plan of Action*. Food and Agriculture Organization of the United Nations, Rome. http://www.fao.org/docrep/003/w3613e/w3613e00.htm

Fargione, J., Hill, J., Tilman, D., Polasky, S. and Hawthorne, P. (2008). Land clearing and the biofuel carbon debt. *Science* 319, 1235–1238

Fedoroff, N.V., Battisti, D.S., Beachy, R.N., Cooper, P.J.M., Fischhoff, D.A. and Hodges, C.N. (2010). Radically rethinking agriculture for the 21st century. *Science* 327(5967), 833–834

Fernandes, B.M., Welch, C.A. and Gonçalves, E.C. (2010). Agrofuel policies in Brazil: paradigmatic and territorial disputes. *The Journal of Peasant Studies* 37(4), 793–819

Fiala, N. (2008). Meeting the demand: an estimation of potential future greenhouse gas emissions from meat production. *Ecological Economics* 67(3), 412–419

Finlayson, C.M., Davidson, N.C., Spiers, A.G. and Stephenson, N.J. (1999). Global wetland inventory – current status and future priorities. *Marine and Freshwater Research* 50, 717–727

Flannigan, M.D., Krawchuk, M.A., de Groot, W.J., Wotton, B.M. and Gowman, L.M. (2009). Implications of changing climate for global wildland fire. *International Journal of Wildland Fire* 18, 483–507

Foley, J.A., Ramankutty, N., Brauman, K.A., Cassidy, E.S., Gerber, J.S., Johnston, M., Mueller, N.S., O'Connell, C., Ray, D.K., West, P.C., Balzer, C., Bennett, E.M., Carpenter, S.R., Hill, F., Monfreda, C., Polasky, S., Rockström, J., Sheehan, J., Siebert, S., Tilman, D. and Zaks, D.P.M. (2011). Solutions for a cultivated planet. *Nature* 478, 337–342

Foley, J., DeFries, R., Asner, G., Barford, C., Bonan, G., Carpenter, S.R., Chapin, F.S., Coe, M.T., Daily, G.C., Gibbs, H.K., Helkowski, J.H., Holloway, T., Howard, E.A., Kucharik, C.J., Monfreda, C., Patz, J.A., Prentice, I.C., Ramankutty, N. and Snyder, P.K. (2005). Global consequences of land use. *Science* 309(5734), 570–574

Franco, J., Levidow, L., Fig, D., Goldfarb, L., Honicke, M. and Mendonça, M.L. (2010). Assumptions in the European Union biofuels policy: frictions with experiences in Germany, Brazil and Mozambique. *The Journal of Peasant Studies* 37(4), 661–698

Fraser, L.H. and Keddy, P.A. (eds.) (2005). *The World's Largest Wetlands: Ecology and Conservation*. Cambridge University Press, New York

GFI (2009). The governance of forests tool kit (version 1). http://www.wri.org/gfi (accessed 6 September 2011)

Gibbon, P., Bair, J. and Ponte, S. (2008). Governing global value chains: an introduction. *Economics and Society* 37(3), 315–338

Gibbs, H.K., Johnston, M., Foley, J., Holloway, T., Monfreda, C., Ramankutty, N. and Zaks, D. (2008). Carbon payback times for crop-based biofuel expansion in the tropics: the effects of changing yield and technology. *Environmental Research Letters* 3, 034001

Gillett, N.P., Weaver, A.J., Zwiers, F.W. and Flannigan, M.D. (2004). Detecting the effect of climate change on Canadian forest fires. *Geophysical Research Letters* 31(18), L18211. doi:10.1029/2004GL020876

Gillon, S. (2010). Fields of dreams: negotiating an ethanol agenda in the Midwest United States. *The Journal of Peasant Studies* 37(4), 723–748

Goetz, S.J., Mack, M.C., Gurney, K.R., Randerson, J.T. and Houghton, R.A. (2007). Ecosystem responses to recent climate change and fire disturbance at northern high latitudes: observations and model results contrasting northern Eurasia and North America. *Environmental Research Letters* 2(4), 045031

Grimm, N., Faeth, S., Golubiewski, N., Redman, C., Wu, J., Bai, X. and Briggs, J. (2008). Global change and the ecology of cities. *Science* 319, 756–760

Hey, D.L. and Philippi, N.S. (1995). Flood reduction through wetland restoration: the Upper Mississippi River basin as a case history. *Restoration Ecology* 3(1), 4–17

IMF (2006). World economic outlook database. International Monetary Fund, Washington, DC. http://www.imf.org/external/pubs/ft/weo/2006/02/data/download.aspx

INPE (2012). *Prodes Project: Monitoring the Brazilian Amazon Forest by Satellite* (in Portuguese). National Institute for Space Research, São José dos Campos. http://www.obt.inpe.br/prodes/

IPCC (2007). *Climate Change 2007: The Physical Science Basis*. Working Group I contribution to the Fourth assessment report of the Intergovernmental Panel on Climate Change. Cambridge University Press, Cambridge

IUCN (2010/11). *IUCN's Thematic Work on REDD: Community Forest Resource Planning – The Building of a Community of REDD Practitioners*. International Union for the Conservation of Nature, Gland. http://www.iucn.org/about/work/programmes/forest/fp_our_work/fp_our_work_thematic/redd/iucn_s_thematic_work_on_redd?/

Jandl, R., Lindner, M., Vesterdal, L., Bauwens, B., Baritz, R., Hagedorn, F., Johnson, D.W., Minkkinen, K. and Bryne, K.A. (2007). How strongly can forest management influence soil carbon sequestration? *Geoderma* 137, 253–268

Johnston, M., Foley, J.A., Holloway, T., Kucharik, C. and Monfreda, C. (2009). Resetting global expectations from agricultural biofuels. *Environmental Research Letters* 4(1), 014004. doi:10.1088/1748-9326/4/1/014004

Jones, K., Lanthier, Y., van der Voet, P., van Valkengoed, E., Taylor, D. and Fernández-Prieto, D. (2009). Monitoring and assessment of wetlands using earth observation: the GlobWetland project. *Journal of Environmental Management* 90(7), 2154–2169

Kant, S. (2000). A dynamic approach to forest regimes in developing countries. *Ecological Economics* 32(2), 287–300

Kindermann, G., Obersteiner, M., Sohngen, B., Sathaye, J., Andrasko, K., Ewald, R., Schlamadinger, B., Wunder, S. and Beach, R. (2008). Global cost estimates of reducing carbon emissions through avoided deforestation. *Proceedings of the National Academy of Sciences of the United States of America* 105(30), 10302–10307

Kissinger, M. and Rees, W. (2010). An interregional ecological approach for modelling sustainability in a globalizing world: reviewing existing approaches and emerging directions. *Ecological Modelling* 221, 2615–2623

Koning, N. and Smaling, E.M.A. (2005). Environmental crisis or "lie of the land"? The debate on soil degradation in Africa. *Land Use Policy* 22(1), 3–11

Kumar, P. (ed.) 2010. *The Economics of Ecosystems and Biodiversity: Ecological and Economic Foundations*. Earthscan, Washington

Kumar, S. and Kant, S. (2005). Bureaucracy and new management paradigms: modeling foresters' perceptions regarding community-based forest management in India. *Forest Policy and Economics* 7(4), 651–669

Lal, R. (1996). Deforestation and land-use effects on soil degradation and rehabilitation in western Nigeria. III. Runoff, soil erosion and nutrient loss. *Land Degradation and Development* 7, 99–119

Lambin, E. and Meyfroidt, P. (2011). Global land use change, economic globalization, and the looming land scarcity. *Proceedings of the National Academy of Sciences of the United States of America* 108(9), 3465–3472

Lambin, E. and Meyfroidt, P. (2010). Land use transitions: socio-ecological feedback versus socio-economic change. *Land Use Policy* 27, 108–118

Lambin, E.F., Turner, B.L., Geist, H.J., Agbola, S.B., Angelsen, A., Bruce, J.W., Coomes, O.T., Dirzo, R., Fischer, G., Folke, C., George, P.S., Homewood, K., Imbernon, J., Leemans, R., Li, X., Moran, E.F., Mortimore, M., Ramakrishnan, P.S., Richards, J.F., Skånes, H., Steffen, W., Stone, G.D., Svedin, U., Veldkamp, T.A., Vogel, C. and Xu, J. (2001). The causes of land-use and land-cover change: moving beyond the myths. *Global Environmental Change* 11(4), 261–269

Larson, A.M. (2002). Natural resources and decentralization in Nicaragua: are local governments up to the job? *World Development* 30(1), 17–31

Lawrence, D.M., Slater, A.G., Tomas, R.A., Holland, M.M. and Deser, C. (2008). Accelerated Arctic land warming and permafrost degradation during rapid sea ice loss. *Geophysical Research Letters* 35, L11506. doi:10.1029/2008GL033985

Lehner, B. and Döll, P. (2004). Development and validation of a global database of lakes, reservoirs and wetlands. *Journal of Hydrology* 296, 1–22

Li, M.T. (2011). Forum on global land grabbing: centering labour in the land grab debate. *The Journal of Peasant Studies* 38(2), 281–298

Licker, R., Johnston, M., Barford, C., Foley, J.A., Kucharik, C.J., Monfreda, C. and Ramankutty, N. (2010). Mind the gap: how do climate and agricultural management explain the 'yield gap' of croplands around the world? *Global Ecology and Biogeography* 19(6), 769–782

Lobell, D.B., Burke, M.B., Tebaldi, C., Mastrandrea, M.D., Falcon, W.P. and Naylor, R.L. (2008). Prioritizing climate change adaptation needs for food security in 2030. *Science* 319, 607–610

Lobo, J., Strumsky, D. and Bettencourt, L.M.A. (2009). *Metropolitan Areas and CO_2 Emissions: Large is Beautiful.* Rotman School of Management, University of Toronto, Toronto

MA (2005a). *Ecosystems and Human Well-being: Synthesis.* Millennium Ecosystem Assessment. Island Press, Washington, DC. http://www.millenniumassessment.org/documents/document.356.aspx.pdf

MA (2005b). *Ecosystems and Human Well-being: Wetlands and Water: Synthesis.* Millennium Ecosystem Assessment. World Resources Institute, Washington, DC. http://www.maweb.org/documents/document.358.aspx.pdf

Marlow, H.J., Hayes, W.K., Soret, S., Carter, R.L., Schwab, E.R. and Sabaté, J. (2009). Diet and the environment: does what you eat matter? *American Journal of Clinical Nutrition* (89)5, 1699S–1703S

Marshall, G.R. and Smith, D.M.S. (2010). Natural resources governance for the drylands of the Murray-Darling basin. *Rangeland Journal* 32(3), 267–282

Martine, G., McGranahan, G., Montgomery, M. and Fernandez-Castilla, R. (2008). Introduction. In The New Global Frontier: Urbanization, Poverty and Environment in the 21st Century (ed. Martine, G., McGranahan, G., Montgomery, M. and Fernandez-Castilla, R.) pp.1-16. Earthscan

Mather, A.S. (1992). The forest transition. *Area* 24, 367–379

Mayers, J., Maginnis, S. and Arthur, E. (2010). *REDD Readiness Requires Radical Reform: Prospects for Making the Big Changes Needed to Prepare for REDD-Plus in Ghana.* TFD Publication No. 1. The Forests Dialogue, Yale University, New Haven, CT. http://cmsdata.iucn.org/downloads/tfd_reddreadiness_ghana_report_lo_res__1_.pdf

McCarthy, J. (2010). Processes of inclusion and adverse incorporation: oil palm and agrarian change in Sumatra, Indonesia. *The Journal of Peasant Studies* 37(4), 821–850

McGuire, A.D., Anderson L.G., Christensen, T.R., Dallimore, S., Guo, L., Hayes, D.J., Heimann, M., Lorenson, T.D., Macdonald, R.W. and Roulet, N. (2009). Sensitivity of the carbon cycle in the Arctic to climate change. *Ecological Monographs* 79(4), 523–555

McMichael, P. and Scoones, I. (eds.) (2010). Special issue on biofuels, land and agrarian change. *The Journal of Peasant Studies* 37(4), 575–962

Melillo, J.M., Reilly, J.M., Kicklighter, D.W., Gurgel, A.C., Cronin, T.W., Paltsev, S., Felzer, B.S., Wang, X., Sokolov, A.P. and Schlosser, C.A. (2009). Indirect emissions from biofuels: how important? *Science* 326, 1397–1399

Melillo, J.M., McGuire, A.D., Kicklighter, D.W., Moore, B., Vorosmarty, C.J., Schloss, A.L. (1993). Global climate change and terrestrial net primary production. *Global Change Biology* 363, 234–240

Meyfroidt, P., Rudel, T.K. and Lambin, E.F. (2010). Forest transitions, trade, and the global displacement of land use. *Proceedings of the National Academy of Sciences of the United States of America* 107(49), 20917–20922

Miehe, S., Kluge, J., von Wehrden, H. and Retzer, V. (2010). Long-term degradation of Sahelian rangeland detected by 27 years of field study in Senegal. *Journal of Applied Ecology* 47(3), 692–700

Milder, J.C., McNeely, J.A., Shames, S.A. and Scherr, S.J. (2008). Biofuels and ecoagriculture: can bioenergy production enhance landscape-scale ecosystem conservation and rural livelihoods? *International Journal of Agricultural Sustainability* 6(2), 105–121

Mistry, J. (2000). *World Savannas: Ecology and Human Use.* Pearson Education Limited, Harlow

Mitra, S., Wassmann, R. and Vlek, P.L.G. (2005). An appraisal of global wetland area and its organic carbon stock. *Current Science* 88(1), 25–35

Montgomery, M. (2008). The urban transformation of the developing world. *Science* 319, 761–764

Montgomery, D.R. (2007). Soil erosion and agricultural sustainability. *Proceedings of the National Academy of Sciences of the United States of America* 104(33), 13268–13272

Mortimore, M., Anderson, S., Cotula, L., Davies, J., Faccer, K., Hesse, C., Morton, J., Nyangena, W., Skinner, J. and Wolfangel, C. (2009). *Dryland Opportunities: A New Paradigm for People, Ecosystems and Development.* International Union for the Conservation of Nature, Gland. http://pubs.iied.org/pdfs/G02572.pdf

Morton, D.C., DeFries, R.S., Shimabukuro, Y.E., Anderson, L.O., Arai, E., del Bon Espirito-Santo, F., Freitas, R. and Morisette, J. (2006). Cropland expansion changes deforestation dynamics in the southern Brazilian Amazon. *Proceedings of the National Academy of Sciences* 103(39), 14637–14641.

Muradian, R., Corbera, E., Pascual, U., Kosoy, N. and May, P.H. (2010). Reconciling theory and practice: an alternative conceptual framework for understanding payments for environmental services. *Ecological Economics* 69, 1202–1208

Myneni, R.B., Tucker, C.J., Asrar, G. and Keeling, C.D. (1998). Interannual variations in satellite sensed vegetation index data from 1981 to 1991. *Journal of Geophysical Research* 103, 6145–6160

Nagendra, H. (2008). Do parks work? Impact of protected areas on land cover clearing. *Ambio* 37, 330–337

Naylor, R., Steinfeld, H., Falcon, W., Galloway, J., Smil, V., Bradford, E., Alder, J. and Mooney, H. (2005). Losing the links between livestock and land. *Science* 310, 1621–1622

Neely, C., Running, S. and Wilkes, A. (eds.) (2009). *Review of Evidence on Drylands Pastoral Systems and Climate Change: Implications and Opportunities for Mitigation and Adaptation.* Land and Water Discussion Paper No. 8. Food and Agriculture Organization of the United Nations, Rome. ftp://ftp.fao.org/docrep/fao/012/i1135e/i1135e00.pdf

Nepstad, D., Soares-Filho, B.S., Merry, F., Lima, A., Moutinho, P., Carter, J., Bowman, M., Cattaneo, A., Rodrigues, H., Schwartzman, S., McGrath, D.G., Stickler, C.M., Lubowski, R., Piris-Cabezas, P., Rivero, S., Alencar, A., Almeida, O. and Stella, O. (2009). The end of deforestation in the Brazilian Amazon. *Science* 326, 1350–1351

Neumann, K., Verburg, P.H., Stehfest, E. and Müller, C. (2010). The yield gap of global grain production: a spatial analysis. *Agricultural Systems* 103(5), 316–326

Newman, P. (2006). The environmental impact of cities. *Environment and Urbanization* 18(2), 275–295

Novo, A., Jansen, K., Slingerland, M. and Giller, K. (2010). Biofuel, dairy production and beef in Brazil: competing claims on land use in Sao Paulo state. *The Journal of Peasant Studies* 37(4), 769–792

O'Connor, F.M., Boucher, O., Gedney, N., Jones, C.D., Folberth, G.A., Coppell, R., Friedlingstein, P., Collins, W.J., Chappellaz, J., Ridley, J. and Johnson C.E. (2010). Possible role of wetlands, permafrost, and methane hydrates in the methane cycle under future climate change: a review. *Reviews of Geophysics* 48, RG4005. doi:10.1029/2010RG000326

Ometto, J.P., Aguiar, A.P.D. and Martinelli, L.A. (2011). Amazon deforestation in Brazil: effects, drivers and challenges. *Carbon Management* 2(5), 575–585

Ostrom, E. (2007). A diagnostic approach for going beyond panaceas. *Proceedings of the National Academy of Sciences of the United States of America* 104(39), 15181–15187

Ostrom, E. and Cox, M. (2010). Moving beyond panaceas: a multi-tiered diagnostic approach for social-ecological analysis. *Environmental Conservation* 37, 451–463

Özler, Ş.İ. and Obach, B.K. (2009). Capitalism, state economic policy and ecological footprint: an international comparative analysis. *Global Environmental Politics* 9(1), 79–108

Pan, Y., Birdsey, R.A., Fang, J., Houghton, R., Kauppi, P.E., Kurz, W.A., Phillips, O.L., Shvidenko, A., Lewis, S.L., Canadell, J.G., Ciais, P., Jackson, R.B., Pacala, S.W., McGuire, A.D., Piao, S., Rautiainen, A., Sitch, S. and Hayes, D. (2011). A large and persistent carbon sink in the world's forests. *Science* 333, 988–993

Pascual, U. and Corbera, E. (2011). Pagos por servicios ambientales: perspectivas y experiencias innovadoras para la conservación de la naturaleza y el desarrollo rural (Payment for ecosystem services: perspectives and experiences for conservation and rural development). *Revista Española de Estudios Agrosociales y Pesqueros* 228, 11–29

Pattanayak, S.K., Wunder, S. and Ferraro, P.J. (2010). Show me the money: do payments supply environmental services in developing countries? *Review of Environmental Economics and Policy* 4(2), 254–274

Perfecto, I. and Vandermeer, J. (2010). The agroecological matrix as alternative to the land-sparing/agriculture intensification model. *Proceedings of the National Academy of Sciences of the United States of America* 107(13), 5786–5791

Phelps, J., Webb, E.L. and Agrawal, A. (2010). Does REDD+ threaten to recentralize forest governance? *Science* 328(5976), 312–313

Pimentel, D. and Pimentel, M. (2003). Sustainability of meat-based and plant-based diets and the environment. *American Journal of Clinical Nutrition* 78(3), 6605–6635

Pimentel, D., Marklein, A., Toth, M.A., Karpoff, M.N., Paul, G.S., McCormack, R., Kyriazis, J. and Krueger, T. (2009). Food versus biofuels: environmental and economic costs. *Human Ecology* 37(1), 1–12

Pingali, P. (2006). Westernization of Asian diets and the transformation of food systems: implications for research and policy. *Food Policy* 32, 281–298

Prentice, I.C., Farquhar, G.D., Fasham, M.J.R., Goulden, M.L., Heimann, M., Jaramillo, V.J., Kheshgi, H.S., Le Quéré, C., Scholes, R.J. and Wallace, D.W.R. (2001). The carbon cycle and atmospheric carbon dioxide. In *Climate Change 2001: The Scientific Basis* (ed. Houghton, J.T., Ding, Y., Griggs, D.J., Noguer, M., van der Linden, P.J., Dai, X., Maskell, K. and Johnson, C.A.). pp.183–237. Contribution of Working Group I to the Third assessment report of the Intergovernmental Panel on Climate Change. Cambridge University Press, Cambridge http://unfccc.int/resource/cd_roms/na1/mitigation/Resource_materials/IPCC_TAR_Climate_Change_2001_Scientific_Basis/TAR-03.pdf

Preskett, L., Huberman, D., Bowen-Jones, E., Edwards, G. and Brown, J. (2008). *Making REDD Work for the Poor.* A Poverty Environment Partnership (PEP) report. http://www.cbd.int/doc/meetings/for/wscb-fbdcc-01/other/wscb-fbdcc-01-oth-10-en.pdf

Ramsar Convention Secretariat (2007). *Wetland Inventory: A Ramsar Framework for Wetland Inventory.* Ramsar Handbooks for the Wise Use of Wetlands, 3rd ed. vol. 12. Ramsar Convention Secretariat, Gland. http://www.ramsar.org/pdf/lib/lib_handbooks2006_e12.pdf

Ramsar Convention Secretariat (1971). *Convention on Wetlands of International Importance especially as Waterfowl Habitat* (as amended in 1982 and 1987). Ramsar Convention Secretariat, Gland. http://www.ramsar.org/cda/en/ramsar-documents-texts-convention-on/main/ramsar/1-31-38%5E20671_4000_0

Ravi, S., Breshears, D.D., Huxman, T.E. and D'Odorico, P. (2010). Land degradation in drylands: interactions among hydrologic-aeolian erosion and vegetation dynamics. *Geomorphology* 116, 236–245

Reynolds, J.F., Stafford Smith, M., Lambin, E.F., Turner, B.L. II, Mortimore, M., Batterbury, S.P.J., Downing, T.E., Dowlatabadi, H., Fernández, R.J., Herrick, J.E., Huber-Sannwald, E., Jiang, H., Leemans, R., Lynam, T., Maestre, F.T., Ayarza, M. and Walker, B. (2007). Global desertification: building a science for dryland development. *Science* 316, 847–851

Richardson, B. (2010). Big sugar in southern Africa: rural development and the perverted potential of sugar/ethanol exports. *The Journal of Peasant Studies* 37(4), 917–938

Ringler, C., Zhu, T., Cai, X., Koo, J. and Wang, D. (2010). *Climate Change Impacts on Food Security in Sub-Saharan Africa: Insights from Comprehensive Climate Change Scenarios*. IFPRI Discussion Paper No. 1042. International Food Policy Research Institute, Washington, DC

Rockström, J., Steffen, W., Noone, K., Persson, Å., Chapin, F.S. III, Lambin, E., Lenton, T.M., Scheffer, M., Folke, C., Schellnhuber, H.J., Nykvist, B., de Wit, C.A., Hughes, T., van der Leeuw, S., Rodhe, H., Sörlin, S., Snyder, P.K., Costanza, R., Svedin, U., Falkenmark, M., Karlberg, L., Corell, R.W., Fabry, V.J., Hansen, J., Walker, B., Liverman, D., Richardson, K., Crutzen, P. and Foley, J. (2009). Planetary boundaries: exploring the safe operating space for humanity. *Ecology and Society* 14(2), 32. http://www.ecologyandsociety.org/vol14/iss2/art32/

Rudel, T.K., Schneider, L., Uriarte, M., Turner, B.L. II, DeFries, R., Lawrence, D., Geoghegan, J., Hecht, S., Ickowitz, A., Lambin, E.F., Birkenholtz, T., Baptista, S. and Grau, R. (2009). Agricultural intensification and changes in cultivated areas, 1970–2005. *Proceedings of the National Academy of Sciences of the United States of America* 106, 20675–20680

Sakadevan, K. and Nguyen, M.-L. (2010). Extent, impact, and response to soil and water salinity in arid and semiarid regions. In *Advances in Agronomy* (ed. Sparks, D.L.) 109, 55–74. Academic Press, San Diego, CA

Sasaki, N., Asner, G.P., Knorr, W., Durst, P.B. and Piriyadi, H.R. (2011). Approaches to classifying and restoring degraded tropical forests for the anticipated REDD+ climate change mitigation mechanism. *iForest - Biogeosciences and Forestry* 4, 1–6. http://www.sisef.it/iforest/pdf/Sasaki_556.pdf

Saxena, K.B., Mula, M.G., Sugui, F.P., Layaoen, H.L., Domoguen, R.L., Pascua, M.E., Mula, R.P., Dar, W.D., Gowda, C.L.L., Kumar, R.V. and Eusebio, J.E. (2010). *Pigeonpea: A Resilient Crop for the Philippine Drylands*. Information Bulletin No. 85. International Crops Research Institute for the Semi-Arid Tropics, Andhra Pradesh. http://openaccess.icrisat.org/bitstream/10731/3590/1/Pigeonpea-resilient-crop.pdf

Scharlemann, J.P.W., Kapos, V., Campbell, A., Lysenko, I., Burgess, N.D., Hansen, M.C., Gibbs, H.K., Dickson, B. and Miles, L. (2010). Securing tropical forest carbon: the contribution of protected areas to REDD. *Oryx* 44(3), 352–357

Schneider, A., Friedl, M.A. and Potere, D. (2009). A new map of global urban extent from MODIS satellite data. *Environmental Research Letters* 4(4), 044003. doi:10.1088/1748-9326/4/4/044003

Schuur, E.A.G., Bockheim, J., Canadell, J.G., Euskirchen, E., Field, C.B., Goryachkin, S.V., Hagemann, S., Kuhry, P., Lafleur, P., Lee, H., Mazhitova, G., Nelson, F.E., Rinke, A., Romanovsky, V., Shiklomanov, N., Tarnocai, C., Venevsky, S., Vogel, J.G. and Zimov, J.G. (2008). Vulnerability of permafrost carbon to climate change: implications for the global carbon cycle. *BioScience* 58, 701–714

Searchinger, T., Heimlich, R., Houghton, R.A., Dong, F., Elobeid, A. and Fabiosa, J. (2008). Use of US croplands for biofuels increases greenhouse gases through emissions from land-use change. *Science* 319, 1238–1240

Seto, K., Sanchez-Rodriguez, R. and Fragkias, M. (2010). The new geography of contemporary urbanization and the environment. *Annual Review of Environment and Resources* 35, 167–194

Sietz, D., Lüdeke, M.K.B. and Walther, C. (2011). Categorisation of typical vulnerability patterns in global drylands. *Global Environmental Change* 21, 431–440

Sikor, T., Stahl, J., Enters, T., Ribot, J.C., Singh, N., Sunderlin, W.D. and Wollenberg, L. (2010). REDD-Plus, forest people's rights and nested climate governance. *Global Environmental Change* 20, 423–425

Smith, P., Gregory, P.J., van Vuuren, D., Obersteiner, M., Havlík, P., Rounsevell, M., Woods, J., Stehfest, E. and Bellarby, J. (2010). Competition for land. *Philosophical Transactions of the Royal Society B* 365, 2941–2957

Smith, P., Martino, D., Cai, Z., Gwary, D., Janzen, H.H., Kumar, P., McCarl, B., Ogle, S., O'Mara, F., Rice, C., Scholes, R.J., Sirotenko, O., Howden, M., McAllister, T., Pan, G., Romanenkov, V., Schneider, U. and Towprayoon, S. (2007). Policy and technological constraints to implementation of greenhouse gas mitigation options in agriculture. *Agriculture Ecosystem Environment* 118, 6–28

Steinfeld, H., Gerber, P., Wassenaar, T.D., Castel, V., Rosales, M. and de Haan. C. (2006). *Livestock's Long Shadow: Environmental Issues and Options*. Food and Agriculture Organization of the United Nations, Rome

Stephenson, S.R., Smith, L.C. and Agnew, J.A. (2011). Divergent long-term trajectories of human access to the Arctic. *Nature Climate Change* 1, 156–160

Stolton, S. and Dudley, N. (eds.) (2010). *Arguments for Protected Areas: Multiple Benefits for Conservation and Use*. Earthscan, London

Sukhinin, A.I., French, N.H.F., Kasischke, E.S., Hewson, J.H., Soja, A.J., Csiszar, I.A., Hyer, E.J., Laboda, T., Conard, S.G., Romasko, V.I., Pavlichenko, E.A., Miskiv, S.I. and Slinkina, O.A. (2004). AVHRR-based mapping of fires in Russia: new products for fire management and carbon cycle studies. *Remote Sensing of Environment* 93, 546–564

Syvitski, J.P.M., Kettner, A.J., Overeem, I., Hutton, E.W.H., Hannon, M.T., Brakenridge, G.R., Day, J., Vörösmarty, C., Saito, Y., Giosan, L. and Nicholls, R.J. (2009). Sinking deltas due to human activities. *Nature Geoscience* 2, 681–686

Tal, A. and Gordon, J. (2010). Carbon cautious: Israel's afforestation experience and approach to sequestration. *Small-Scale Forestry* 9(4), 409–428

Tarnocai, C., Canadell, J.G., Schuur, E.A.G., Kuhry, P., Mazhitova, G. and Zimov, S. (2009). Soil organic carbon pools in the northern circumpolar permafrost region. *Global Biogeochemical Cycles* 23, GB2023. doi:10.1029/2008GB003327

TEEB (2010). *TEEB for Local and Regional Policy Makers*. The Economics of Ecosystems and Biodiversity, Bonn. http://www.teebweb.org/ForLocalandRegionalPolicy/tabid/1020/Default.aspx

Thoms, C.A. (2008). Community control of resources and the challenge of improving local livelihoods: a critical examination of community forestry in Nepal. *Geoforum* 39(3), 1452–1465

Tilman, D., Socolow, R., Foley, J.A., Hill, J., Larson, E., Lynd, L., Pacala, S., Reilly, J., Searchinger, T., Somerville, C. and Williams, R. (2009). Beneficial biofuels: the food, energy, and environment trilemma. *Science* 325(5938), 270–271

Tiwari, P.C. (2009). Sustainable land use for adaptation to long term impacts of climate change in Himalaya. In *IOP Conference Series: Earth and Environmental Science* 6, 342007. doi:10.1088/1755-1307/6/4/342007

Tollefson, J. (2011). Brazil revisits forest code. *Nature* 476, 259–260

Toulmin, C., Borras, S., Bindraban, P., Mwangi, E. and Sauer, S. (2011). *Land Tenure and International Investments in Agriculture: A Report by the UN Committee on Food Security High Level Panel of Experts*. Food and Agriculture Organization of the United Nations, Rome

Twilley, R.R. and Rivera-Monroy, V. (2009). Sediment and nutrient tradeoffs in restoring Mississippi River delta: restoration versus eutrophication. *Journal of Contemporary Water Research and Education* 141(1), 39–44

UN (2000). *Millennium Development Goals*. http://www.un.org/millenniumgoals/

UNCCD (2010). *Fostering Evidence-based Decision-Making in UNCCD Implementation: Initial Results from PRAIS Reports in 2010*. United Nations Convention to Combat Desertification, Bonn. http://www.mediaterre.org/docactu,dW5pc2ZlcmEvZG9jcy9wcmFpcy1icmllZmluZ3BhcG VyMnJlc3VsdHM=,1.pdf

UNCCD (2007). *Follow-up to the Joint Inspection Unit Report and Strategy Development to Foster Implementation of the Convention. Situational Analysis*. ICCD/COP(8)/INF.5 Prepared by Unisféra International Centre (Unisféra, Canada) and Integrated Environmental Consultants Namibia (IECN), Namibia. http://www.unccd.int/cop/officialdocs/cop8/pdf/inf5eng.pdf

UNCCD (1994). *United Nations Convention to Combat Desertification in those Countries Experiencing Serious Drought and/or Desertification, Especially in Africa*. http://www.unccd.int/convention/text/pdf/conv-eng.pdf and http://www.unccd.int/convention/text/convention.php

UNEP (2011a). *European Commission and UNEP Announce New Partnership to Catalyze Green Economy: Support for Kenya's Mau Forest Restoration Project Spotlighted*. Press Release. United Nations Environment Programme, New York. http://hqweb.unep.org/Documents.Multilingual/Default.asp?DocumentID=659&ArticleID=6911&l=en&t=long

UNEP (2011b). *Green Economy Report: Towards a Green Economy – Pathways to Sustainable Development and Poverty Eradication*. United Nations Environment Programme, New York. http://www.unep.org/greeneconomy/GreenEconomyReport/tabid/29846/Default.aspx

UNEP (2011c). *Keeping Track of our Changing Environment: from Rio to Rio+20 (1992–2012)*. United Nations Environment Programme, Nairobi

UNEP (2009a). *Towards Sustainable Production and Use of Resources: Assessing Biofuels*. International Panel for Sustainable Resource Management, United Nations Environment Programme, Nairobi

UNEP (2009b). (eds. Nellemann, C., MacDevette, M., Manders, T., Eickhout, B., Svihus, B., Prins, A.G. and Kaltenborn, B.P.). *The Environmental Food Crisis – The Environment's Role in Averting Future Food Crises*. United Nations Environment Programme, GRID-Arendal, Arendal

UNEP (2007). *Global Environment Outlook GEO-4: Environment for Development*. United Nations Environment Programme, Nairobi

UNEP-WCMC (2011). *PRAIS Briefing Paper: 3. Lessons*. UNEP World Conservation Monitoring Centre, Cambridge. http://www.unep-wcmc.org/medialibrary/2011/12/08/a2df8f9a/3.%20LESSONS.pdf

UNEP-WCMC (2010). *The Ramsar Convention on Wetlands and its Indicators of Effectiveness*. International Expert Workshop on the 2010 Biodiversity Indicators and Post-2010 Indicator Development. A workshop convened by the UNEP World Conservation Monitoring Centre (UNEP-WCMC), in cooperation with the Convention on Biological Diversity (CBD), 6–8 July 2009. UNEP World Conservation Monitoring Centre, Cambridge

UN-REDD (2010). *Perspectives on REDD+*. United Nations Collaborative Programme on Reducing Emissions from Deforestation and Forest Degradation in Developing Countries, Geneva

USDA Foreign Agricultural Service (2011). *Indonesia Forest Moratorium 2011*. Global Agricultural Information Network Report Number ID1127. http://www.usdaindonesia.org/public/uploaded/Indonesia%20Forest%20Moratorium_Jakarta_Indonesia_6-8-2011.pdf

US Government (2007). *Energy Independence and Security Act of 2007*. 110th Congress, USA

USGS EROS (2010). U.S. Geological Survey - EROS Data Center 2010 http://landsat.usgs.gov/index.php

van der Werf, G.R., Randerson, J.T., Giglio, L., Collatz, G.J., Mu, M., Kasibhatla, P.S., Morton, D.C., DeFries, R.S., Jin, Y. and van Leeuwen, T.T. (2010). Global fire emissions and the contribution of deforestation, savanna, forest, agricultural, and peat fires (1997–2009). *Atmospheric Chemistry and Physics* 16153–16230

van Hecken, G. and Bastiansen, J. (2010). Payments for ecosystem services in Nicaragua: do market-based approaches work? *Development and Change* 41(3), 421–444

van Hecken, G., Bastiansen, J. and Vasquez, W.F. (2010). *Institutional Embeddedness of Local Willingness to Pay for Environmental Services: Evidence from Matiguás, Nicaragua*. IDPM-UA Discussion Paper 2010-04. Institute of Development Policy and Management, University of Antwerp

Veron, S.R. and Paruelo, J.M. (2010). Desertification alters the response of vegetation to changes in precipitation. *Journal of Applied Ecology* 47(6), 1233–1241

Verstraete, M., Scholes, R. and Stafford Smith, M. (2009). Climate and desertification: looking at an old problem through new lenses. *Frontiers in Ecology and the Environment* 7(8), 421–428

Walker, R. (1993). Deforestation and economic development. *Canadian Journal of Regional Science* XVI (3), 481–497

Walker, D.A., Epstein, H.E., Raynolds, M.K., Kuss, P., Kopecky, M.A., Frost, G.V., Daniëls, F.J.A, Leibman, M.O., Moskalenko, N.G., Matyshak, G.V., Khitun, O.V., Khomutov, A.V., Forbes, B.C., Bhatt, U.S., Kade, A.N., Vonlanthen C.M. and Tichý, L. (2012). Environment, vegetation and greenness (NDVI) along the North America and Eurasia Arctic transects. *Environmental Research Letters* 7(1)

Wang, M. and Overland, J.E. (2004). Detecting Arctic climate change using KoÅNppen climate classification. *Climatic Change* 67, 43–62

Wetlands International (2011). *Association and Foundation Wetlands International: annual plan and budget 2011*. Wetlands International, Wageningen

White, B. and Dasgupta, A. (2010). Agrofuels capitalism: a view from political economy. *The Journal of Peasant Studies* 37(4), 593–607

Wittemyer, G., Elsen, P., Bean, W.T., Burton, A.C. and Brashares, J.S. (2008). Accelerated human population growth at protected area edges. *Science* 321, 123–126

Wood, A. and van Halsema, G.E. (eds.) (2008). *Scoping Agriculture-Wetland Interactions: Towards a Sustainable Multiple-Response Strategy*. FAO Water Report 33. Food and Agriculture Organization of the United Nations, Rome

World Bank (2010). *World Development Report 2010: Development and Climate Change*. The World Bank, Washington, DC

WSSD (2002). *Johannesburg Plan of Implementation*. World Summit on Sustainable Development. http://www.un.org/esa/sustdev/documents/WSSD_POI_PD/English/POIToc.htm [Accessed 20 May 2012]

Wunder, S., Engel, S. and Pagiola, S. (2008). Taking stock: a comparative analysis of payments for environmental services programs in developed and developing countries. *Ecological Economics* 65, 834–852

Zezza, A. and Tasciottia, L. (2010). Urban agriculture, poverty, and food security: empirical evidence from a sample of developing countries. *Food Policy* 35(4), 265–273

Zhou, L.M., Tucker, C.J., Kaufmann, R.K., Slayback, D., Shabanov, N.V. and Myneni, R.B. (2001). Variations in northern vegetation activity inferred from satellite data of vegetation index during 1981 to 1999. *Journal of Geophysical Research – Atmospheres* 106, 20069–20083

Zika, M. and Erb, K.H. (2009). The global loss of net primary production resulting from human-induced soil degradation in drylands. *Ecological Economics* 69, 310–318

Zimov, S.A., Schuur, E.A.G. and Chapin, F.S. III (2006). Permafrost and the global carbon budget. *Science* 312, 1612–1613

CHAPTER 4

Water

Coordinating lead authors: Erica Brown Gaddis, Paul Roger Glennie, Yi Huang and Walter Rast

Lead authors: Magdi Abdelhamid, Maite Aldaya, Björn Alfthan, Peter Koefoed Bjørnsen, Mariele Evers, Gensuo Jia, Ljubomir Jeftic, Alioune Kane, Santiago Reyna and Judith Weis

Contributing authors: Hermanni Backer, Hans Günter Brauch, Eberhard Braune, Salif Diop, Carlo Giupponi, Sherry Heileman, Lawrence Hislop, Tiina Kurvits, Robin Mahon, Liana Talaue-McManus, Lisa Speer and Jaap van Woerden

Principal scientific reviewer: Úrsula Oswald Spring

Chapter coordinator: Salif Diop

Main Messages

Increasing water-use efficiency in all sectors is vital to ensure sustainable water resources for all uses. Human water demands, with only limited improvements in efficiency, are increasing and are already unsustainable in many regions. Nevertheless, potential exists for efficiency gains: irrigation efficiency, for example, could be increased by approximately one-third simply by implementing existing technology. At the local level, integrated demand and supply strategies are critical. At a river-basin level, more efficient and fair water allocation systems are needed. More broadly, virtual water trade can ease water demands in some locations.

Recognition of ecosystem water needs within allocation systems will help protect life-supporting ecosystem services. Freshwater and marine ecosystem services are critical to human development and integral to the transition to a green economy. Inadequately articulated objectives and lack of data, however, make it difficult to evaluate progress in meeting environmental water requirements. Better strategies and tools are needed for efficient, equitable water allocation between users, including the environment. Full implementation of international commitments and enforcement of legally binding agreements, and due consideration of customary water-use arrangements, will facilitate sustainable human and ecosystem use.

Reducing both point and non-point pollution is imperative to improve ecosystem health and provide safe water for humans. Substantial achievements in reducing some pollutants have occurred since 1992, although many water bodies are still affected, and many new contaminants have poorly understood effects. Treating municipal and industrial wastewater is achievable with existing technology, but requires better regulatory oversight, infrastructure investment and capacity building, especially in developing countries. Integrated land-water management and stakeholder participation are necessary to reduce non-point pollution of both freshwater and marine systems.

Improved water supply and sanitation is probably the single most cost-effective means of reducing water-related death and disease globally. Although the Millennium Development Goal (MDG) target on water supply was met in 2010, more than 600 million people will still lack access to safe drinking water in 2015. The MDG target on sanitation is unlikely to be met, with 2.5 billion people currently without improved sanitation facilities; poor rural populations are most affected. Meeting the water supply and sanitation MDGs would reduce the water-related global disease burden by about 10 per cent. Increased investment in infrastructure, capacity building and regulation are needed, and the participation of women is crucial for water management and the prevention of water-borne disease.

Climate-sensitive policies across all water-related sectors are essential to address extreme events and increased climatic variability. Floods and droughts still cause losses of billions of dollars annually. Climate change is altering the hydrologic cycle, threatening freshwater and marine ecosystems as well as human water security in many regions. Open oceans play a major role in regulating global climate and weather patterns, with climate change impacts manifested in warmer surface waters and rising sea levels. Ocean warming and acidification threaten tropical coral reef ecosystems, with rapid contraction predicted by 2050. Mitigation and adaptation to climate change impacts must be considered within the context of other drivers and pressures. Those related to energy production are likely to require trade-offs between human energy needs, water demands and ecosystem protection.

The pace of increasing demands on freshwater and ocean resources must be matched by improved governance. Freshwater systems integrate human activities and land management across nations and regions. The open oceans are a major global commons and require effective international cooperation and governance. Most human and environmental water problems result from inadequate governance involving policy, institutional, financial and/or stakeholder issues. Integrated management approaches for addressing these constraints require time and resources to be successful. They need enhanced integration of policies and institutions between sectors and governance levels, implementation and enforcement of relevant agreements and goals, improved monitoring and resolution of transboundary issues. Good governance, including stakeholder and private-sector participation and gender considerations, is critical to increasing societal and environmental resilience and sustainability.

INTRODUCTION

Aquatic ecosystems are major integrators of natural and anthropogenic processes. As the ultimate sink for pollutants, freshwater and marine ecosystems are among the most sensitive indicators of the environmental impacts of human activities. They support a wide diversity of life (Chapter 5), providing important goods and services that directly or indirectly support and sustain human existence and livelihoods. Adequate freshwater supplies of acceptable quality are recognized as a human right by the UN General Assembly's declaration on clean water and sanitation.

Mangroves are significant breeding grounds for marine life, and protect coastal areas from storm surges and other natural hazards. © *Jeremy Sterk*

As highlighted in the *Millennium Ecosystem Assessment* (MA 2005), freshwater and marine ecosystems provide various services, including provisioning (food, water, fibre, fuel), regulating (climate, hydrological, purification), cultural (spiritual, recreational), and supporting (sediment transport, nutrient cycling). Such ecosystem services are a function of water, land, biodiversity and atmospheric links. Healthy aquatic ecosystems not only provide goods and services, but also enhance resilience against the negative impacts of environmental perturbations or disasters. Aquatic systems also drive major global bio-geochemical cycles; the open oceans play a major role in regulating global climate and weather patterns.

This chapter addresses freshwater and marine systems as distinct but linked hydrological components of the water environment. It assesses progress towards achieving water-related goals in major multilateral environmental agreements identified by the *GEO-5* High-Level Intergovernmental Advisory Panel and regional consultations. Based on the drivers-pressures-states-impacts-responses framework (DPSIR) (Stanners *et al.* 2007) used for the *GEO-5* assessment, this chapter focuses on the state, trends and impacts of the water environment, with references to drivers (Chapter 1) and responses (Parts 2 and 3), and other environmental sectors (Chapters 2, 3, 5 and 6) where appropriate.

Although freshwater ecosystem goods and services are extensive, competition and multi-sectoral demands for water have resulted in overexploitation and contamination of resources in many regions. Competing water uses and their impacts on sustainable aquatic resources, including quantity and quality issues, are discussed, including the water needs of ecosystems. The chapter also addresses inequitable and unsustainable water demands in many countries. Pollution from land- and marine-based activities (Chapter 6) continues to degrade coastal areas and open oceans. Water quality trends are discussed. Continuing overfishing severely impacts many fish stocks, particularly marine species (Chapter 5).

Many predicted global climate change impacts will be manifested in changes to the hydrologic cycle. How they could affect the water environment is highlighted, including increased frequency, duration and severity of droughts and floods. Predicted climate change impacts and their uncertainties are discussed, including the vulnerabilities and adaptation needs of many communities.

Watersheds comprise a group of linked water systems that can include rivers, lakes, reservoirs, wetlands, underlying aquifers and downstream marine systems, although these links are often not considered in developing water management plans. This is a significant omission, since simultaneously managing for human health and social concerns – including disease and poverty, economic development and sustainable environmental integrity – within complex global connectivity often requires environmental and economic trade-offs, some very difficult. Because many water-related problems result from policy, institutional, financial or other governance inadequacies, this chapter also discusses both freshwater and marine governance elements identified in the multilateral environmental agreements. It concludes by identifying major policy and data gaps for achieving water-related goals. Policy options to address the issues raised in this chapter are addressed throughout Part 2 of *GEO-5*.

INTERNATIONALLY AGREED GOALS

Freshwater was selected as a priority issue in all UNEP *GEO-5* regional scoping consultations, with most regions identifying Paragraph 26c of the Johannesburg Plan of Implementation (Box 4.1) as the most important freshwater goal, with water availability and marine issues also identified in several regions. Although limited by gaps in global-scale data and specific targets, the degree to which water-related multilateral environmental agreements have been addressed is a focus of the current chapter.

Goals were identified on the basis of their policy relevance and ability to illustrate intergovernmental cooperation since the United Nations Conference on Environment and Development (UNCED) in 1992 and earlier (Table 4.1).

Box 4.1 Johannesburg Plan of Implementation Paragraph 26c

Improve the efficient use of water resources and promote their allocation among competing uses in a way that gives priority to the satisfaction of basic human needs and balances the requirement of preserving or restoring ecosystems and their functions, in particular in fragile environments, with human domestic, industrial and agriculture needs, including safeguarding drinking water quality.

Source: WSSD 2002

Table 4.1 Selected internationally agreed goals and themes related to water

Major themes from internationally agreed goals		Johannesburg Plan of Implementation (JPOI) (WSSD 2002)**							Millennium Development Goals (MDGs) (UN 2000)*			UN Millennium Declaration (2000) (UN 2000)**	United Nations Convention on the Law of the Sea (UNCLOS 1982)**	London Convention on Marine Pollution (1972)*	MARPOL Convention on marine pollution (1973)	Convention on Biological Diversity (1992)		Ramsar Convention on Wetlands (1972)*	International Watercourses Convention (1997)	United Nations Framework Convention on Climate Change (UNFCCC 1992)	Ballast Water Management Convention (2004)	Global Programme of Action for the Protection of the Marine Environment from Land-based Activities (GPA) (1995)	FAO Responsible Fisheries (1995)*	UN Fish Stocks Agreement (2001)	Barbados Programme of Action for Small Island Developing States (1994)	Dublin Principles on Water and Sustainable Development (1992)	Regional seas conventions and programmes	Multilateral freshwater agreements
		7*	25*	26**	30*	31*	32	40–45	1*	4	7*	55/2				1992 targets	Aichi Targets											
Ecosystems	Protect and restore freshwater ecosystems and their services		X	X				X			X					X	X	X	X	X	X	X	X		X	X	X	X
	Protect and restore marine ecosystems and their services				X	X	X	X				X	X	X	X	X		X	X	X	X	X	X	X	X		X	
	Conserve and improve management of wetlands					X		X								X	X	X	X							X	X	X
	Ensure environmental water needs			X	X							X					X	X	X						X	X		X
Human well-being	Reduce water-related human health hazards	X	X	X			X		X	X	X	X		X	X				X		X	X			X		X	X
	Ensure equitable access to improved drinking water supply		X								X	X							X							X		X
	Secure adequate sustainable freshwater supply	X		X				X			X	X							X						X	X		X
	Develop programmes for mitigating effects of extreme water-related events			X																X					X			X
	Mitigate and adapt to adverse effects of climate change on the water environment			X													X			X					X			X
Water-use efficiency	Improve the efficient use of water resources			X				X				X							X						X	X	X	X
Water quality	Reduce and control freshwater pollution	X	X	X				X		X		X					X	X	X			X			X	X	X	X
	Reduce and control marine pollution				X		X	X					X	X	X		X	X			X	X		X	X		X	
	Improve sanitation coverage including sewage collection, treatment and disposal		X							X	X								X							X		X
Institutional and legal issues	Recognize the economic value of water																		X						X	X		X
	Develop and enforce effective legal frameworks and regulations		X	X	X							X	X	X	X	X				X		X	X	X	X		X	X
	Strengthen institutional coordination mechanisms			X	X	X	X					X		X	X	X	X		X	X	X	X	X	X	X		X	X
Water resources management	Develop and implement integrated management strategies and plans		X	X	X		X				X	X	X	X	X	X		X	X	X	X	X	X	X	X	X	X	X
	Develop adequate monitoring systems (national, regional and global)		X	X		X		X					X	X	X	X			X	X		X		X	X		X	X
	Improve stakeholder participation and mainstream gender in water management		X					X	X																X	X	X	X
	Improve groundwater management		X	X															X						X	X	X	X

Note: Numbers at the top of the columns under JPOI, MDG and UN Millennium Declaration represent specific paragraphs, goals or articles.

* Selected by the GEO-5 High-Level Intergovernmental Advisory Panel (HLIAP). ** Selected at regional consultations.

STATE AND TRENDS

Water scarcity

Human-environment competition for scarce water resources

Water scarcity is a significant and increasing threat to the environment, human health, development, energy security and the global food supply (Pereira *et al.* 2009). Ecosystems, which provide life-supporting goods and services (Chapter 5), suffer from multiple pressures, including the need for water of adequate quantity and quality as well as appropriate timing (environmental flows). The indicator used here is blue water scarcity (Figure 4.1), the proportion of groundwater and surface water consumed relative to the sustainable water available for human use, after accounting for environmental flows (Hoekstra and Mekonnen 2011). Water scarcity is a significant factor in human water security, with a fifth of the global population living in areas with physical water scarcity (Comprehensive Assessment of Water Management in Agriculture 2007).

Falkenmark and Rockström (2004) estimated the water required to maintain ecosystem goods and services as 75 per cent of the total water use, while direct human water use represented 25 per cent of the total. These figures include both blue (groundwater and surface water) and green water (water stored in the soil). Water is overcommitted in many places, leaving insufficient resources for both human and environmental needs (Gleick and Palaniappan 2010). In a study of 424 of the world's major river basins, containing a population of 3.9 billion people, environmental flow requirements were violated in 223 basins, containing 2.67 billion people facing severe water scarcity during at least one month of the year (Figure 4.1) (Hoekstra and Mekonnen 2011). Although arid regions of Northern Africa and the Middle East are not included in this analysis, other data suggest that the proportion of renewable water withdrawn in those regions exceeds 50–75 per cent, leaving little environmental flow (FAO 2008).

Although many goals in the Johannesburg Plan of Implementation acknowledge the importance of marine and coastal ecosystems (WSSD 2002), there is less recognition of water needs to support freshwater ecosystems, which are themselves legitimate water users (Chapter 5). Although the importance of formally recognizing the environment as a legitimate water user is increasing, it remains on a relatively small scale in practice, with many aquatic ecosystems still at risk (Garrick *et al.* 2009).

Box 4.2 Water scarcity

Goals
Ensure environmental water needs; conserve and improve management of wetlands

Indicators
Blue water scarcity

Global trends
Deteriorating

Most vulnerable communities
Poor communities highly dependent on ecosystem services

Regions of greatest concern
West Asia, South Asia, Mesoamerica, Australia

Figure 4.1 Annual average water scarcity in major river basins, 1996–2005

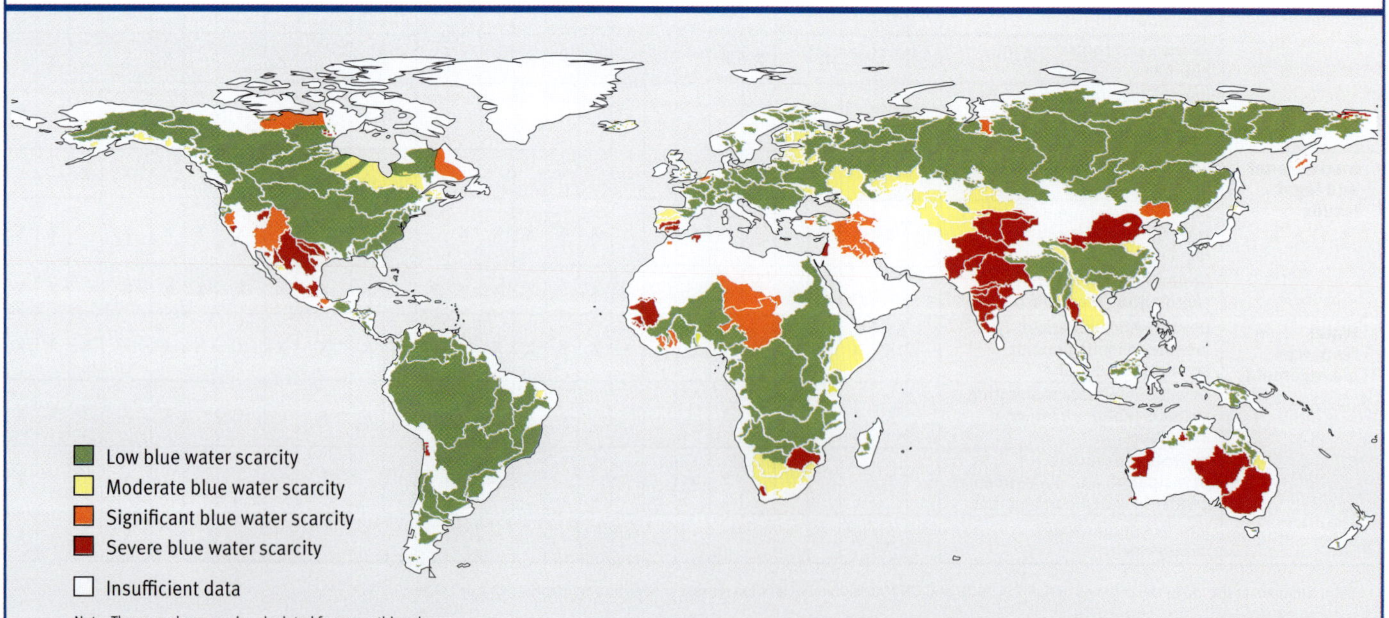

Note: The annual average is calculated from monthly values.

Source: Hoekstra and Mekonnen 2011

Water demand

Global withdrawals have tripled over the last 50 years (UNESCO 2009) to meet the demands of a growing population with increasing wealth and consumption levels. While water supply over this period has remained relatively constant, demand now exceeds sustainable supply in many places, with serious long-term implications (2030 Water Resources Group 2009). The planetary boundary for human consumptive blue water use – when used groundwater and surface water is not made available for reuse in the same basin – is estimated to be 4 000 km³ per year, with current consumptive blue water use estimated at approximately 2 600 km³ per year. Projected water demands may to reach planetary boundaries in the coming decades (Rockström *et al.* 2009).

Agricultural, industrial and domestic water withdrawals have steadily increased. Agriculture is by far the largest global water user (Figure 4.2), with withdrawals for this purpose being unsustainable in many places due to unbalanced long-term irrigation water budgets (MA 2005), as evidenced by the mining of aquifers and reliance on large water diversion projects. These withdrawals are projected to continue increasing, placing further pressure on aquatic ecosystems, which themselves also require water of adequate quantity, quality and timing for sustained health.

Box 4.3 Water demand

Goals
Secure adequate sustainable freshwater supply

Indicators
Water withdrawals; groundwater withdrawals; net water footprint

Global trends
Deteriorating

Most vulnerable communities
Developing countries with increasing water demand; communities dependent on groundwater-irrigated agriculture

Regions of greatest concern
Groundwater withdrawals: Asia and Pacific, parts of North America
Water footprint: North America, Latin America and the Caribbean, Europe

Many communities are dependent on unsustainable groundwater withdrawals (aquifer mining) to meet agricultural and domestic water demands, further threatening water security in many regions. Between 1960 and 2000, global groundwater

Figure 4.2 Current and projected water withdrawals by sector, 2000–2050

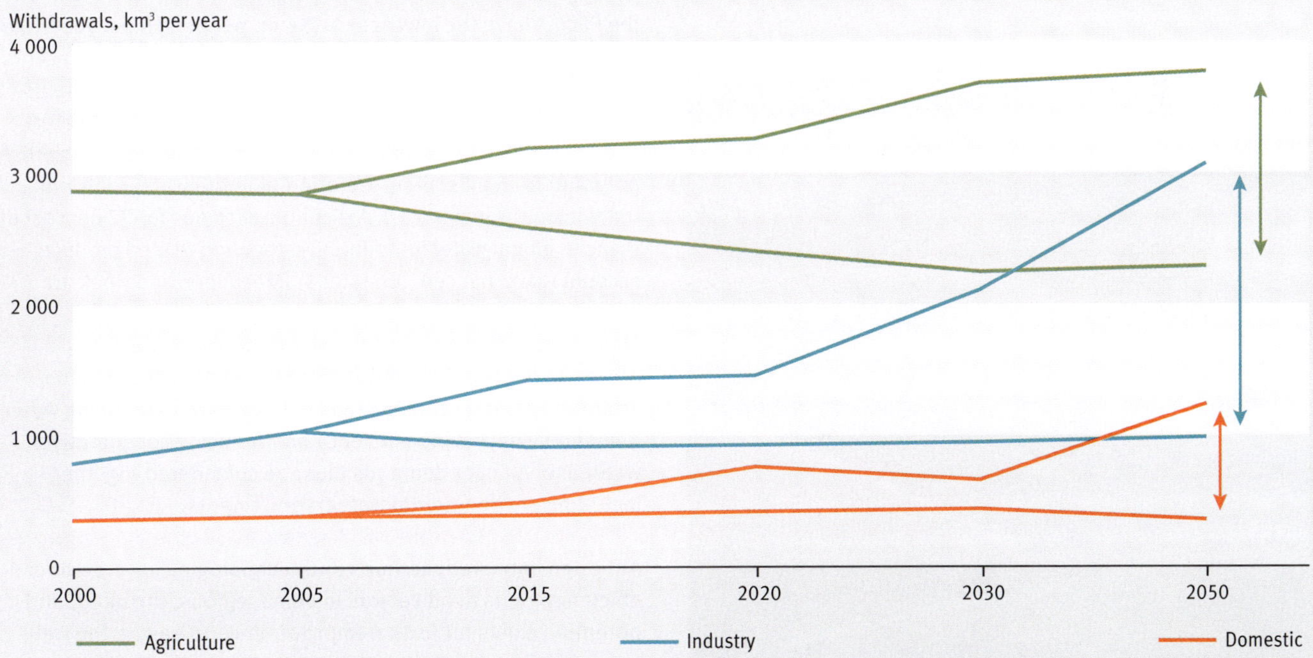

Note: The graph includes three different groups of scenarios (Millennium Ecosystem Assessment, *GEO-4* and OECD) and shows the range between their maximum and minimum values at each five-year interval.

Source: CESR, University of Kassel, Germany; Alcamo *et al.* 2005a, 2005b (MA scenarios); Rothman *et al.* 2007 (*GEO-4* scenarios); Bakkes *et al.* 2008 and OECD 2008 (OECD scenarios)

Figure 4.3 Global annual groundwater depletion, 2000

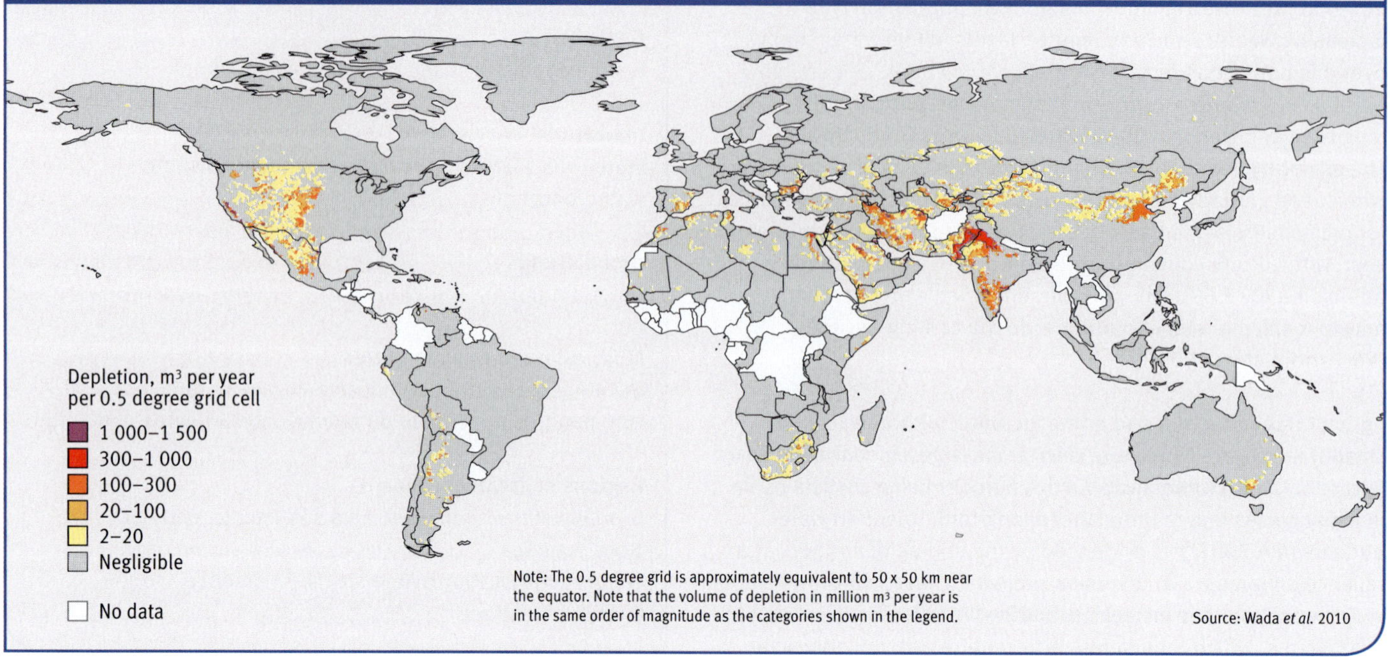

withdrawal increased from 312 km^3 to 734 km^3 per year, resulting in groundwater depletion increasing from 126 km^3 to 283 km^3 per year (Wada *et al.* 2010). Many globally important agricultural centres are particularly dependent on groundwater, including northwest India, northeastern China, northeast Pakistan, California's Central Valley, and the western United States (Figure 4.3) (Wada *et al.* 2010).

Sprinkler irrigation systems are usually more efficient than flood systems. *© Pgiam/iStock*

Not all water withdrawals result in consumptive water use, since much withdrawn water is returned in the form of wastewater or irrigation return flows. Rain-fed agriculture also represents significant human water use without direct water withdrawals. Global water consumption per person, as measured by the water footprint, averages 1 387 m^3 per year. North America has the highest water footprint at 2 798 m^3 per person per year, while Asia and the Pacific have the lowest at 1 156 m^3 per person per year (Figure 4.4). Of the total global water footprint, 74 per cent represents rainwater stored in soil (green water), 11 per cent represents the consumptive use of surface and groundwater (blue water), and 15 per cent represents the freshwater required to assimilate pollution from all sources (referred to in the water footprint terminology as grey water). Agriculture accounts for 92 per cent of the total global water footprint; livestock and related products alone account for 27 per cent (Chapter 1) (Mekonnen and Hoekstra 2011).

Water use efficiency and the virtual water trade

Since the renewable supply of water is relatively constant, addressing water scarcity relies in large part on reducing water demand by improving efficiency and reducing consumptive water use. All user demands must be considered together, including environmental water requirements.

Although improved methods and technologies have produced efficiency gains in all sectors in some regions, the need, and potential, exists for further improvements to ensure the well-being of a growing world population while minimizing the impacts on ecosystems and their goods and services.

The need and potential for improvement is greatest in the agricultural sector (Figure 4.5), since approximately 70 per

Box 4.4 Water-use efficiency

Goals
Improve the efficient use of water resources

Indicators
Irrigation efficiency; net virtual water trade

Global trends
Some progress

Most vulnerable communities
Those dependent on irrigated agriculture in arid areas; poor communities in net virtual-water exporting countries

Regions of greatest concern
West Asia, Asia and the Pacific, parts of Africa and Central America

cent more food will be needed by 2050 to cope with a growing population and dietary changes (Chapter 1) (Boelee 2011). Improvements in the application, conveyance, distribution and management of irrigation can raise the overall efficiency of the water getting to the crops from approximately 35 per cent to 75 per cent or more (Rohwer *et al.* 2007). Broader agricultural water-use efficiency strategies include land-water management and reuse (Ali 2010), while food supply chains beyond the farm can also be more water efficient.

There is inadequate global data to evaluate the overall state and trends of industrial and domestic water-use efficiency. Nonetheless, opportunities do exist for significant improvement in these sectors, particularly where there are major withdrawals and/or rapid urbanization is occurring (Chapter 1). Water allocation efficiency is also required at the river-basin level to ensure sustainable, equitable and economic water use.

At national, regional and global scales, the virtual water trade – the water embedded in traded products ranging from crops to manufactured goods – can be a tool for improving overall efficiency by capitalizing on the comparative advantages of certain water uses in particular regions. About one-fifth of the global water footprint is related to production for export (Figure 4.6). The global virtual water trade for agricultural and industrial products totalled 2 320 km^3 per year between 1996 and 2005, with crops contributing 76 per cent and animal and industrial products each contributing 12 per cent (Mekonnen and Hoekstra 2011). The virtual water trade can efficiently redistribute water and partially help to address the disconnection between consumption and the impacts of production (Chapter 3). Water-scarce basins, countries or

Figure 4.4 Annual global and regional water footprint, 1996–2005

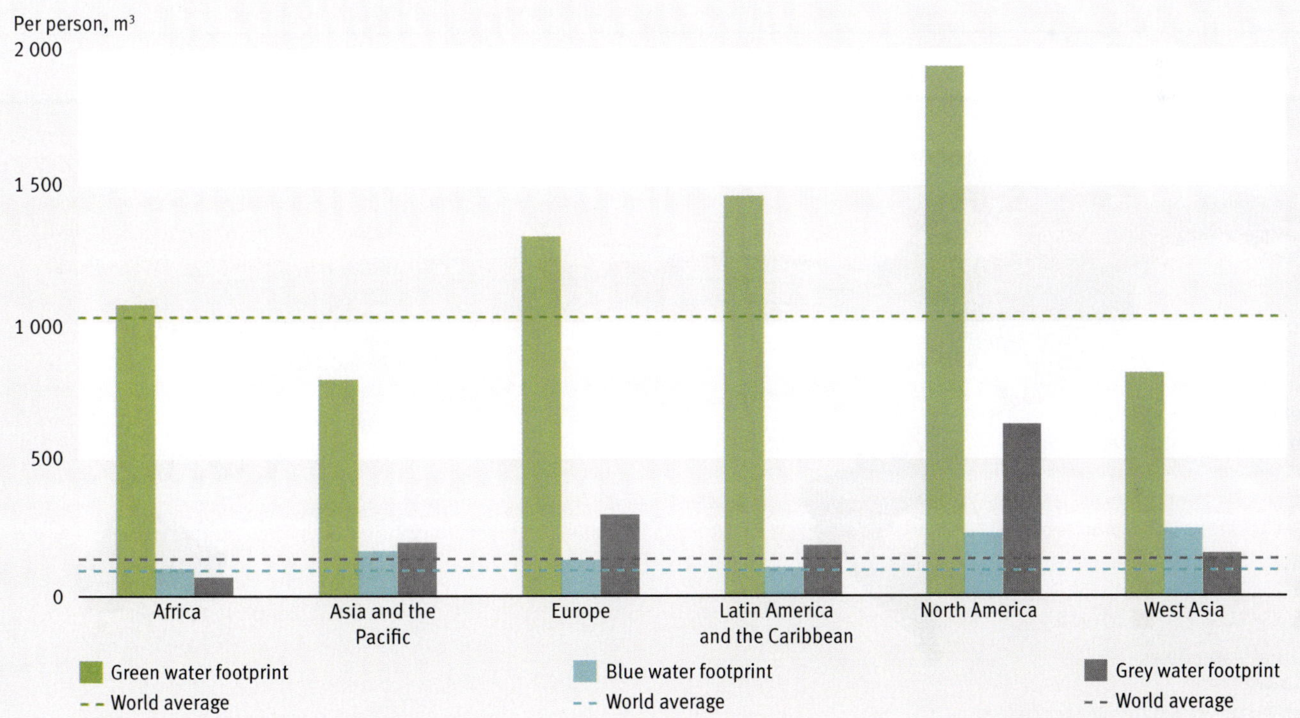

Source: Mekonnen and Hoekstra 2011

Figure 4.5 Global irrigation efficiencies, 2000

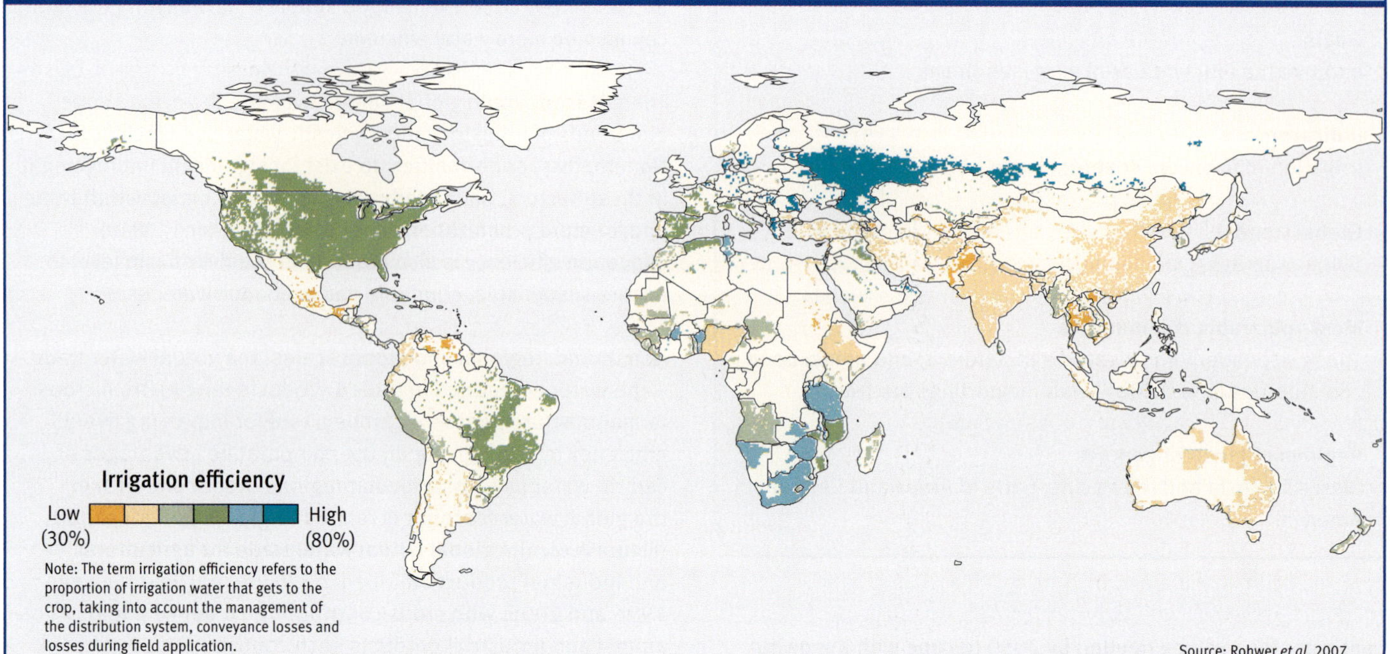

Note: The term irrigation efficiency refers to the proportion of irrigation water that gets to the crop, taking into account the management of the distribution system, conveyance losses and losses during field application.

Source: Rohwer *et al.* 2007

regions, for example, can import water-intensive products through trade, preserving scarce water resources for more valuable purposes. However, it can also lead to overexploitation of water resources in net exporter countries, prioritizing commodity water needs over basic local needs, especially where strong economic drivers promote commodity exports (Chapter 1). Another characteristic is that some net virtual water exporters, such as Australia or South Asia, are also water scarce, while some net importers may have abundant supplies, as is the case for Central Europe.

Figure 4.6 Virtual water imports, exports and flows around the world, 1996–2005

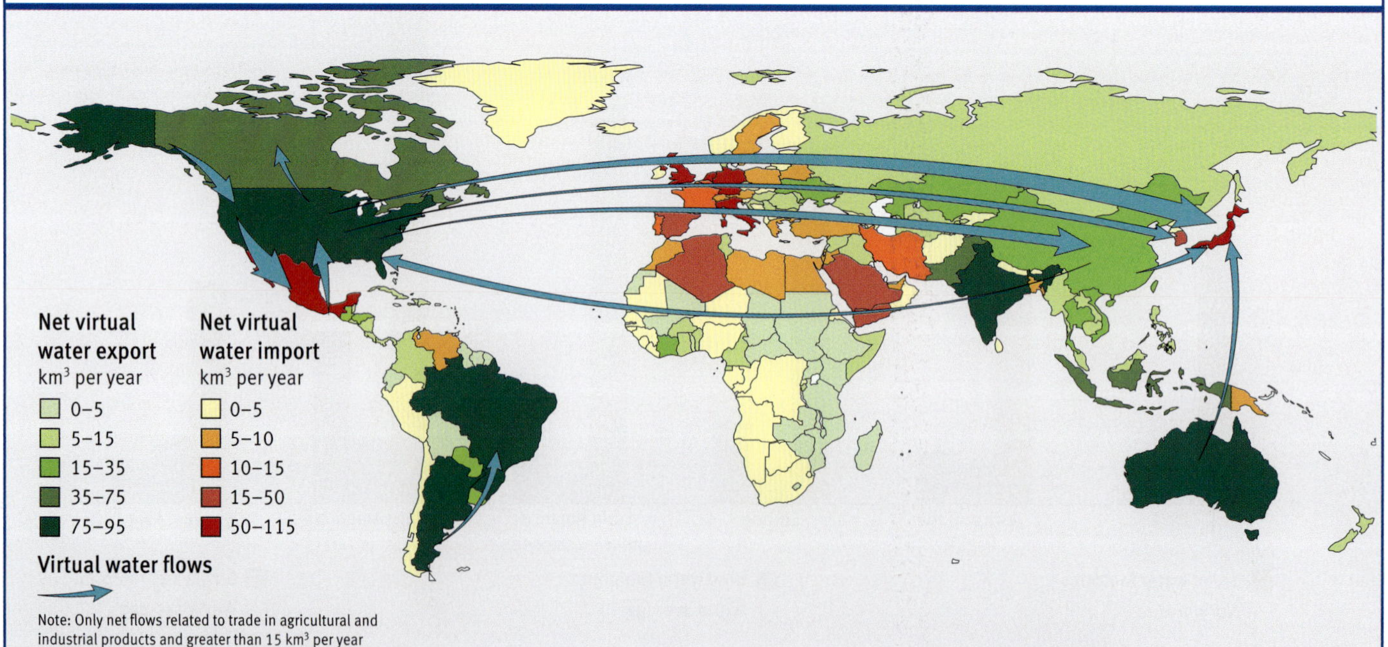

Note: Only net flows related to trade in agricultural and industrial products and greater than 15 km³ per year are shown. The larger the arrow, the greater the flow.

Source: Adapted from Mekonnen and Hoekstra 2011

Changes to the hydrologic regime

Extreme events: floods and droughts

The number of flood and drought events classified as disasters – when ten or more people are killed, 100 are affected, a state of emergency is declared or international assistance is requested (EM-DAT 2011) – has risen since the 1980s, as have the total area and number of people affected and the level of damages (EM-DAT 2011; Rosenfeld *et al.* 2008; Kleinen and Petschel-Held 2007). River channelization, floodplain loss, urbanization, particularly in coastal areas, and changing land use are major reasons for the increasing impacts of floods and droughts as well as growing vulnerability to those impacts (Chapter 1). The number of people affected and total damages vary significantly, making it hard to identify trends with confidence (Lugeri 2010). Vulnerability depends on the preparedness and capacity to anticipate and react to extreme events. There are varying preparedness levels on a regional basis for dealing with sudden-onset (floods) and gradual-onset (droughts) disasters (IOM 2010).

Floods cause loss of life and billions of dollars of damage annually (Figure 4.7), with the economic losses higher in developed countries due to the financial valuation and insuring of assets. Between the 1980s and the 2000s, a 230 per cent rise in the number of flood disasters was accompanied by increasing levels of damages (Figure 4.7) (EM-DAT 2011). In addition, the number of people exposed to floods increased by 114 per cent (UNISDR 2011). Over 95 per cent of deaths related to natural disasters between 1970 and 2008 occurred in developing countries (IPCC 2011) and although governments in South and East Asia, for example, increased their disaster preparedness levels, the capacity of communities to cope with such extreme events is weakening because of inadequate social capacity and greater flood severity (Osti *et al.* 2011). Looking to the future, higher precipitation intensity is forecast for the northern

Box 4.5 Extreme events

Goals
Develop programmes for mitigating the effects of extreme water-related events

Indicators
Number of people affected by floods and droughts; total damages from floods and droughts

Global trends
Modest progress in some years or regions and a deteriorating situation in others

Most vulnerable communities
Deltas, low-lying areas, development in river floodplains, islands, and urban areas with inadequate drainage infrastructure are vulnerable to floods; communities directly dependent on rain-fed agriculture are vulnerable to droughts

Regions of greatest concern
South East Asia, North America (Mississippi Basin) and Latin America (Amazon Basin) for floods; small island developing states (SIDS), West Asia, Northern and Western Africa, Australia and South and Central Asia for droughts

Figure 4.7 People affected by and damages associated with floods and droughts, 1980–2010

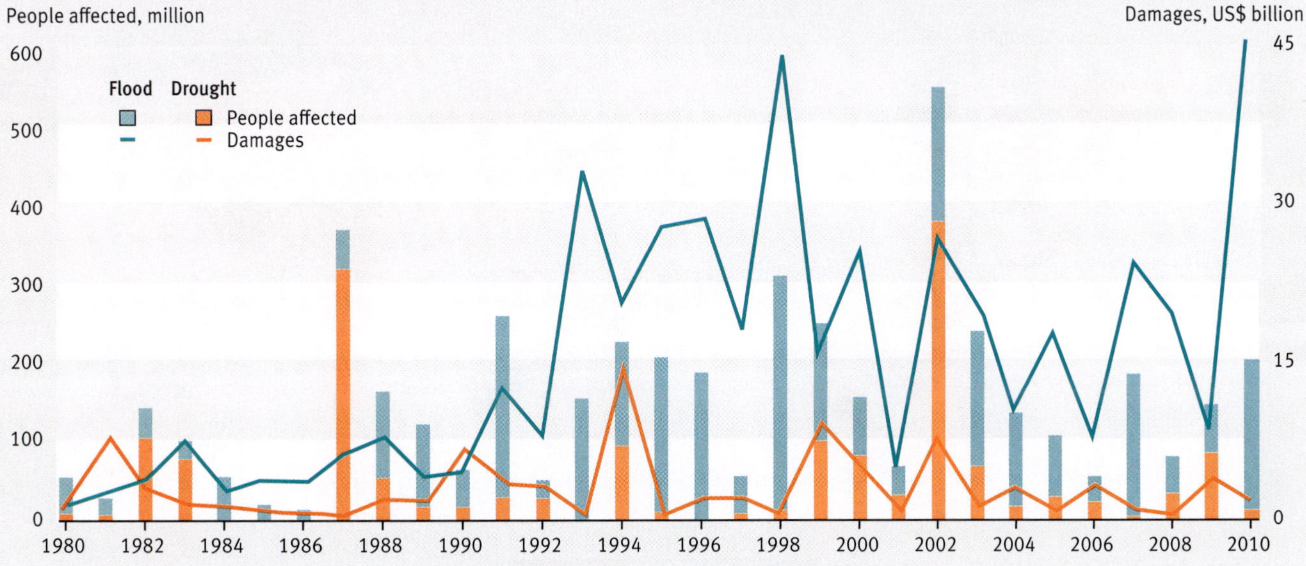

Note: The costs of the large floods in Thailand and Pakistan in 2011 are not included here.

Source: EM-DAT 2011

hemisphere and equatorial areas, with many already arid and semi-arid areas expected to get drier (IPCC 2007a).

The number of drought disasters rose by 38 per cent between the 1980s and the 2000s, the number of people affected increased and related damages also increased (EM-DAT 2011). Droughts disrupt sustainable social and economic development, hindering achievement of the Millennium Development Goals (MDGs), and place additional stresses on ecosystems. Communities dependent on rain-fed crops, which represent approximately 70 per cent of global crop production, often have few alternative food sources beyond international aid (Portmann *et al.* 2010). This is evidenced by the severe ongoing drought in Eastern Africa and the reduction of net primary production in Latin America, Africa and South East Asia (Zhao and Running 2010). Droughts also affect irrigation and can exacerbate water resource conflicts, with arid and semi-arid areas being particularly vulnerable, especially in the context of climate change.

Dams and river fragmentation

Dam building and river control significantly benefit humans, providing flood protection, reliable water supplies and hydroelectric power. But dams can also have detrimental impacts on ecosystems, including channel fragmentation and flow modification, altering ecosystem processes and affecting aquatic organisms, particularly migratory species. Improved management of existing dams to ensure environmental flows and retain or create fish passes is important to mitigate conflicts, although such measures often fall short as a full remedy (Gleick 2003). Careful trade-off analyses are necessary to ensure that the design, location and operation of new dams minimize environmental impacts (Matthews *et al.* 2011).

Box 4.6 Dams and river fragmentation

Goals
Secure an adequate and sustainable freshwater supply; reduce water-related human health hazards (flood protection); protect and restore freshwater ecosystems and their services (often conflicting)

Indicators
Dam density

Global trends
Dam density is increasing; there is some progress on adequate supply of sustainable freshwater; freshwater ecosystems and their services are deteriorating

Most vulnerable communities
Populations displaced by dam construction; populations dependent on dams for water supply

Regions of greatest concern
Developing countries, Asia, Southern Africa

Dam density is highest in industrialized countries (Figure 4.8), although construction in developed regions has slowed because most suitable locations have already been used, and because recent legislation and public pressure do not support dam construction. However, dam building is being actively pursued in many developing countries to secure water and electricity

Figure 4.8 Global density of medium to large dams

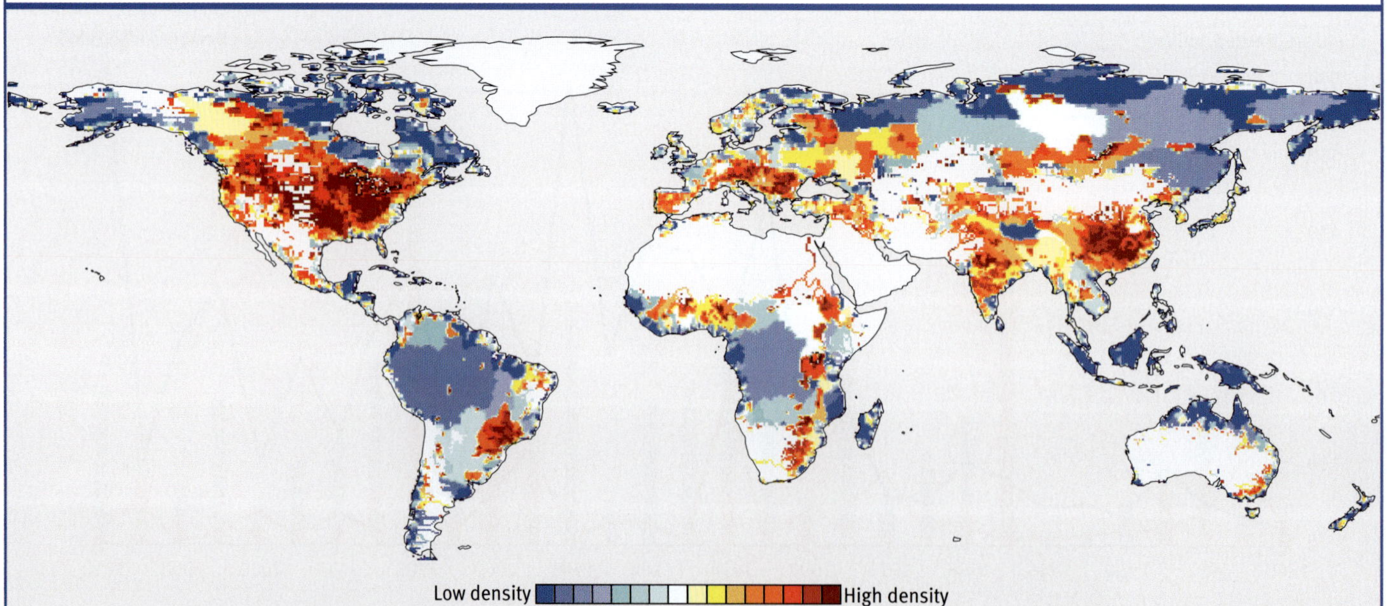

Source: Vörösmarty *et al.* 2010

supplies. As this trend is likely to continue (Chapter 1), dam planning should consider any predicted increase in flow variability associated with climate change.

Freshwater and marine water quality

Groundwater contamination

Groundwater around the world is threatened by pollution from agricultural and urban areas, solid waste, on-site wastewater treatment, oil and gas extraction and refining, mining, manufacturing and other industrial sources. The primary causes are inadequate control of these activities and exceedance of the natural attenuation capacity of underlying soils and strata (Foster *et al.* 2006). Salinization of overexploited aquifers, especially in coastal areas, is another serious concern, particularly for communities dependent on groundwater for drinking.

Groundwater nitrate concentrations are increasing, especially in areas of rapid urbanization, inadequate sanitation and/or heavy agricultural fertilizer use. Nitrate in groundwater contributes

Mining and mineral extraction can significantly reduce river or groundwater levels. © *BanksPhotos/iStock*

Box 4.7 Groundwater contamination

Goals
Mitigate effects of groundwater contamination

Indicators
Arsenic, nitrate and salinization

Global trends
Very little progress in some areas; deterioration in others

Most vulnerable communities
Populations in rapidly urbanizing areas with inadequate sanitation

Regions of greatest concern
Arsenic is of particular concern in Bangladesh, India, highly populated river deltas in South East Asia, North America and Eastern Europe

Figure 4.9 Estimated risk of arsenic in drinking water, based on hydrogeological conditions

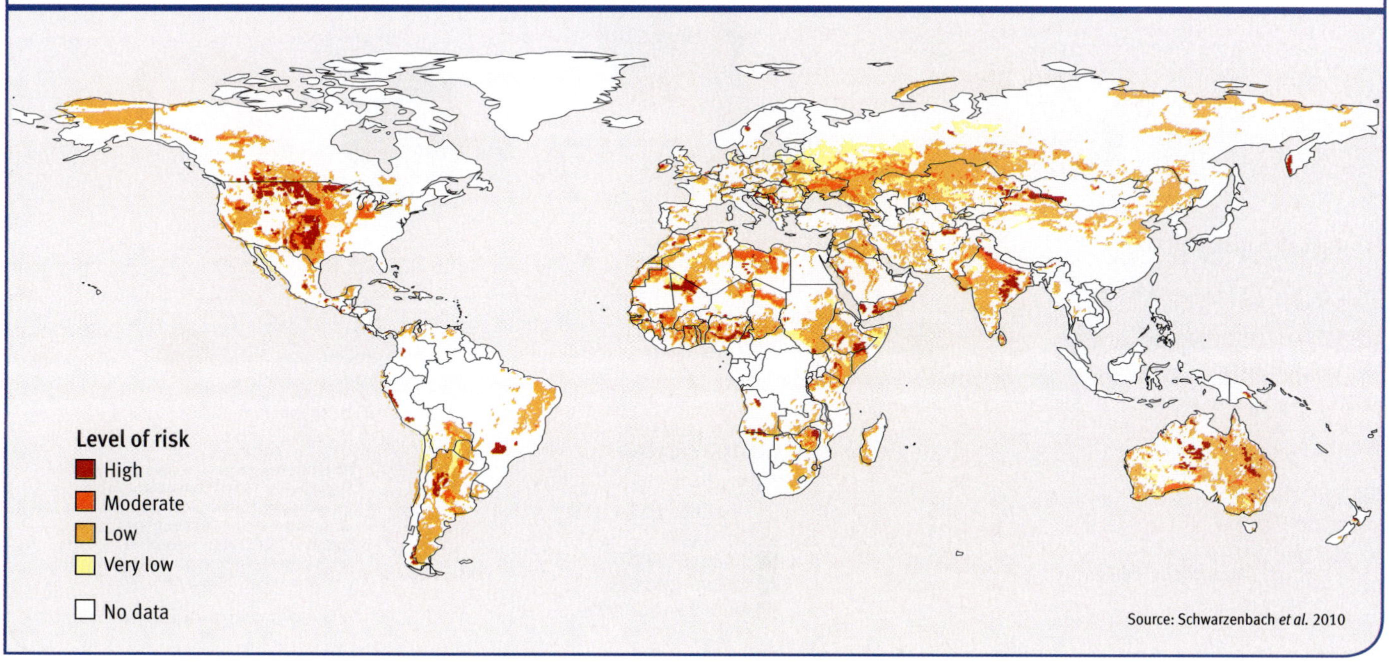

Source: Schwarzenbach *et al.* 2010

to eutrophication and has direct human health impacts. Both naturally occurring arsenic and that mobilized by human activities threaten drinking water quality in many countries (Figure 4.9). Groundwater contaminated with arsenic from natural geologic sources affects 35–75 million people. Surface water pollution in some regions has led to the development of groundwater as a source of drinking water, resulting in inadvertently exposing people to these natural sources of arsenic (Schwarzenbach *et al.* 2010; Brunt *et al.* 2004).

Pathogenic contamination

Pathogenic contamination of surface and groundwater is a critical threat to human health in many areas and contributes to water treatment costs in many communities. Using domestic sewage collection and treatment as a proxy, microbial contamination has decreased over past decades in most developed countries. In contrast, microbial pathogens are often the most pressing water quality issue in many developing countries (Figure 4.10).

Because human and animal faeces are the primary pathogenic sources of water contamination, achieving MDG Goal 7c of halving the population without basic sanitation by 2015 will help reduce such pollution. Nevertheless, although some regions have made significant progress, the world is currently not on track to attain this goal (Figure 4.11). Improved sanitation continues to bypass the poorest communities and individuals, especially in Africa and South Asia (WHO 2012). Unless achieving the MDG sanitation goal in the future includes the provision of wastewater collection and treatment facilities, increasing access to improved sanitation could have

Box 4.8 Pathogenic contamination

Goals
Improve sanitation coverage including sewage collection, treatment and disposal; reduce and control freshwater and marine pollution

Indicator
Faecal coliform concentration; population without access to improved sanitation

Global trends
Some progress

Most vulnerable communities
Poorest and most rural communities

Regions of greatest concern
Africa, South Asia, South Pacific

Figure 4.10 Faecal coliform concentrations in rivers near major cities – an indicator of waterborne pathogens, 1990–2011

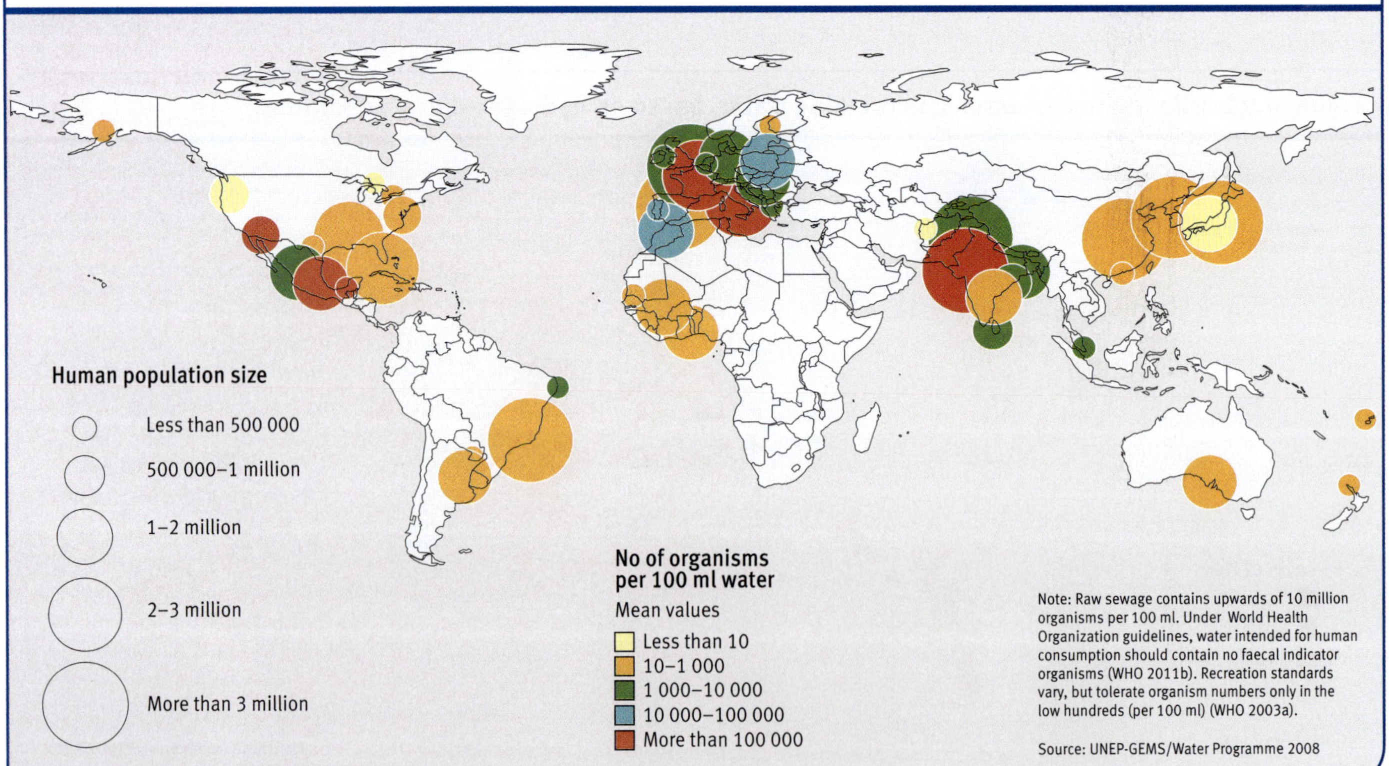

Note: Raw sewage contains upwards of 10 million organisms per 100 ml. Under World Health Organization guidelines, water intended for human consumption should contain no faecal indicator organisms (WHO 2011b). Recreation standards vary, but tolerate organism numbers only in the low hundreds (per 100 ml) (WHO 2003a).

Source: UNEP-GEMS/Water Programme 2008

Figure 4.11 Population without access to improved sanitation compared to MDG target, 1990–2015

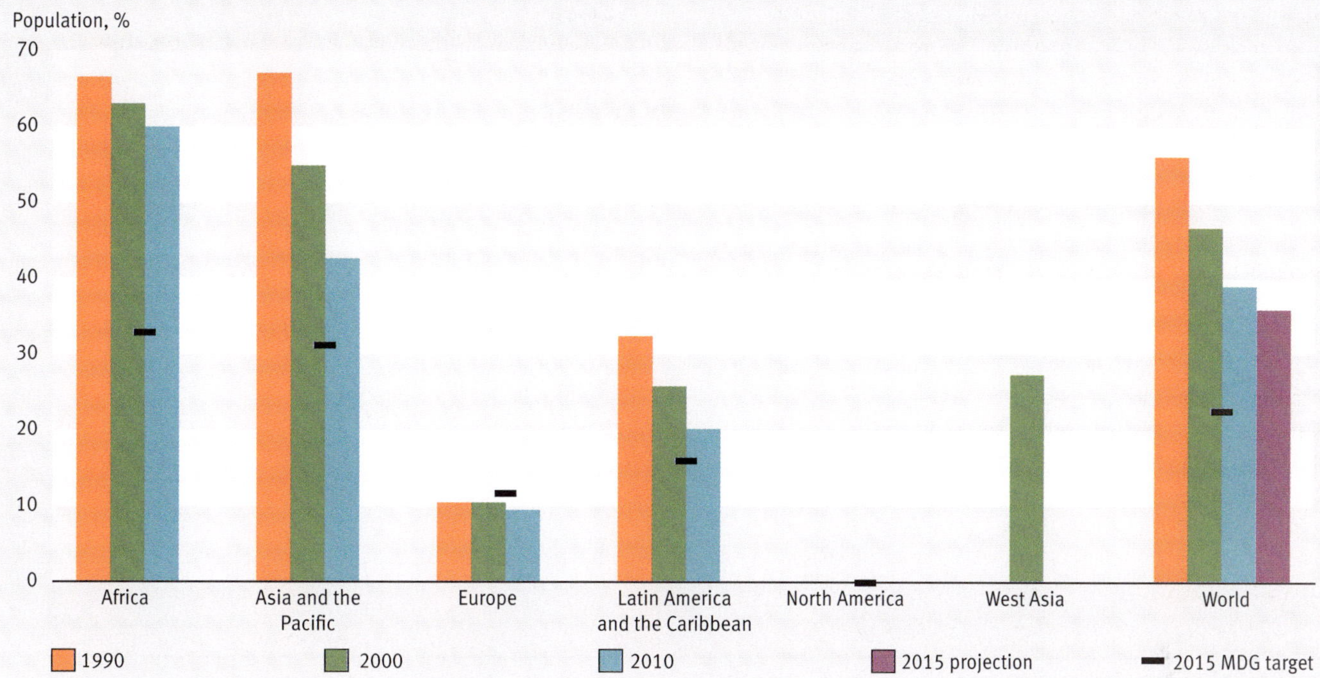

Note: For MDG monitoring, WHO/UNICEF defines an improved sanitation facility as one that hygienically separates human excreta from human contact (WHO 2012). Source: WHO 2012

the unintended negative impact of delivering more untreated wastewater to water bodies, further degrading downstream water quality (Biswas and Tortajada 2011).

Nutrient pollution and eutrophication

Eutrophication, resulting from excessive nutrient pollution from human sewage, livestock wastes, fertilizers, atmospheric deposition and erosion (Chapter 3), is a continuing, pervasive water quality problem. Although there has been increased sewage treatment in many areas, much less progress has been made in reducing nutrient loads from non-point sources, including agricultural and urban run-off and atmospheric deposition to freshwater and marine systems. Interference with global nutrient cycles may be reaching planetary boundaries, beyond which marine and freshwater ecosystems might not recover, although specific thresholds for these processes remain uncertain (Rockström *et al.* 2009).

Box 4.9 Nutrient pollution and eutrophication

Goals
Reduce and control freshwater and marine pollution

Indicators
Marine: prevalence of coastal dead zones; frequency and intensity of harmful algal blooms
Freshwater: global river exports of nitrogen and phosphorous

Global trends
Very little progress or deteriorating

Regions of greatest concern
South East Asia, Europe, eastern North America

Global river nutrient export has increased nutrient export has increased by approximately 15 per cent since 1970, by approximately 15 per cent, since 1970, with South Asia accounting for at least half of the increase (Seitzinger *et al.* 2010). There has been a 74 per cent increase in algal and macrophyte gross productivity in lakes since 1970 (Lewis 2011), and a dramatic increase in the number of eutrophic coastal areas since 1990. Under severe eutrophic conditions, algal blooms can produce hypoxic conditions, causing fish kills in lakes, and dead zones in coastal areas. Hypoxia has become a significant and increasing problem in lakes and rivers, estuaries and coastal areas around the world (Diaz *et al.* 2010; Rabalais *et al.* 2010; Diaz and Rosenberg 2008). At least 169 coastal areas are considered hypoxic, with dead zones especially prevalent in the seas around South East Asia, Europe and eastern North America (Figure 4.12). Only 13 coastal areas appear to be recovering (Diaz *et al.* 2010; Rabalais *et al.* 2010), most in North America and northern Europe. Whereas phosphorus loads are projected to level off, global river nitrogen loads are likely to

Figure 4.12 World hypoxic and eutrophic coastal areas, 2010

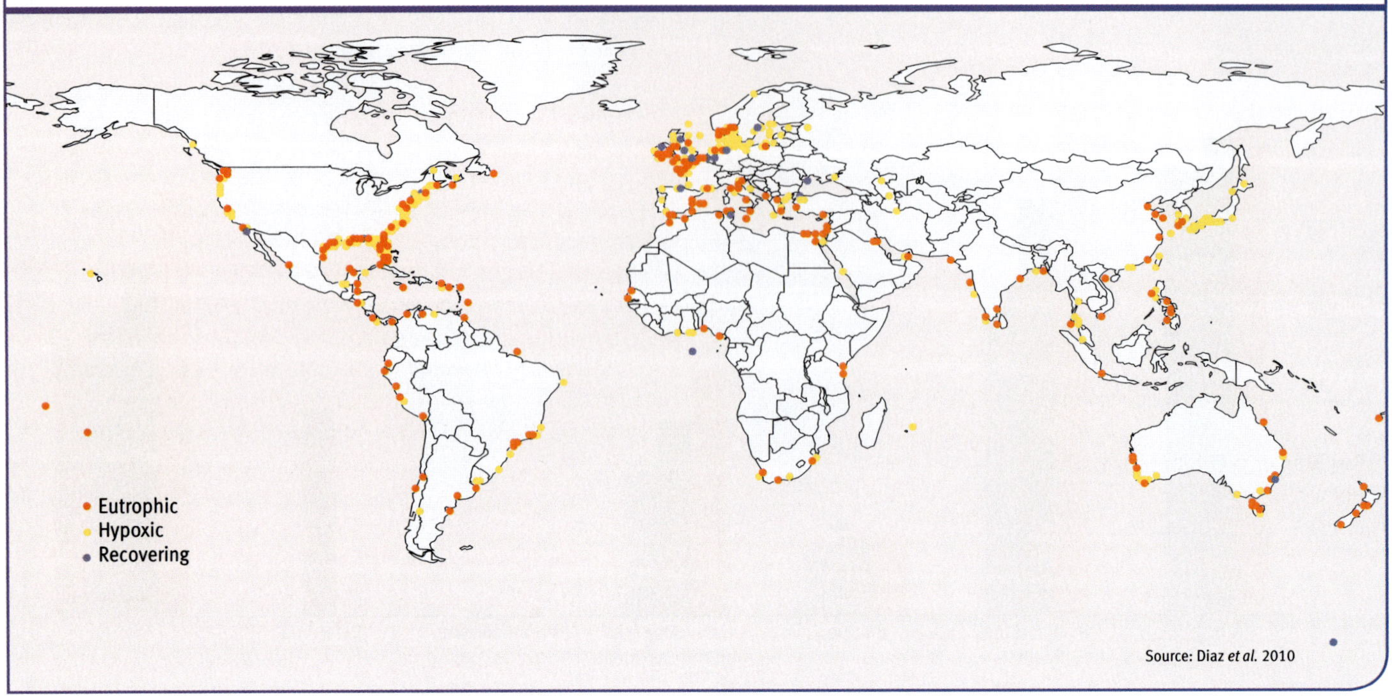

increase by an additional 5 per cent by 2030, mostly in South Asia (Seitzinger *et al.* 2010).

Nutrients can also cause harmful algal blooms in freshwaters and coastal areas, some releasing algal toxins that directly affect human health (WHO 2003a), aquatic organisms and livestock. The number of reported outbreaks of paralytic shellfish poison, a harmful algal toxin found in eutrophic waters, increased from fewer than 20 in 1970, to more than 100 in 2009 (Anderson *et al.* 2010).

Box 4.10 Marine litter

Goals
Reduce marine pollution

Indicator
Levels of litter at the shoreline; levels on the sea bottom and in marine gyres

Global trends
Little to no progress

Most vulnerable communities
Coastal populations

Regions of greatest concern
Unknown

Marine litter

Litter is found in all the world's oceans because of poor solid waste management and the increased use of plastic (UNEP 2009). It damages wildlife, fisheries and boats, contaminates coastal areas, and presents safety and human health risks. Marine litter accumulates on coastal beaches, on the sea bottom (Galgani *et al.* 2000) and large marine gyres in both the Atlantic and Pacific Oceans (Law et al. 2010; Martinez et al. 2009).

Of the 12 seas surveyed between 2005 and 2007, the South East Pacific, North Pacific, East Asian Sea and Wider Caribbean coasts contained the most marine litter (UNEP 2009), and the Caspian, Mediterranean and Red Seas the least. Regional studies of the Baltic Sea (HELCOM 2009), Northeast Atlantic (OSPAR 2009), US coastline (Sheavly 2007) and North Atlantic Subtropical Gyre indicated no statistically significant changes in litter quantity between 1986 and 2008, while data from the Mid-Atlantic indicated an increase in land-based and general-source marine litter during 1997–2007 (Ribic *et al.* 2010).

Persistent toxic chemicals

Toxic pollutants include the trace metals cadmium, lead and mercury, pesticides and their by-products such as dichlorodiphenyltrichloroethane (DDT) and chlordecone, industrial chemicals and combustion by-products. They are still used in many places and thus continue to accumulate in aquatic systems, leaving a legacy of sediment contamination; they are found in 90 per cent of water bodies. The pollutants of greatest concern are persistent, toxic and bioaccumulative (Chapter 6). Organisms can accumulate contaminants from water, sediment and food, acquiring tissue contaminant levels much higher

Box 4.11 Toxic chemicals

Goals
Reduce marine and freshwater pollution

Indicators
Concentration of organochlorines in predatory fish species; concentration of persistent organic pollutants in Arctic air

Global trends
Some progress

Most vulnerable communities
Coastal populations; populations dependent on fish for food

Regions of greatest concern
Polar regions

than those in the surrounding environment. Organochlorine compounds such as polychlorinated biphenyl (PCB) or DDT concentrate in fatty tissues, remain for long periods and biomagnify up the food chain, with the highest concentrations found in top predators.

Concentrations of many persistent organic pollutants (POPs), which tend to accumulate in the Arctic (Hung *et al.* 2010), have decreased in Arctic air samples since the early 1990s (Chapter 2). Tissue concentrations of at least three organochlorine chemicals in 12 deep-sea fish species in the Western Pacific (Figure 4.13) (Oshihoi *et al.* 2009) and PCB concentrations in at least four fish species in San Francisco Bay (Davis *et al.* 2003) have also fallen since the mid-1990s (Chapter 6).

Emerging water quality concerns

Although conventional toxic pollutants are declining in many industrialized areas, additional contaminants are raising new concerns, for example the use of flame retardants such as polybrominated diphenyl ethers (PBDEs), a type of POP, has increased exponentially over the past 30 years in Europe, North America and Japan (Schwarzenbach *et al.* 2010). There are also mounting concerns about pharmaceuticals and personal care products that are not removed by most sewage systems, and thus enter the environment after use. The long-term risks to aquatic organisms and humans are largely unknown, although it is clear that pharmaceuticals and endocrine-disrupting compounds can have biological effects at very low concentrations (Schwarzenbach *et al.* 2010).

Nanoparticles and microplastics are relatively new water pollutants (Chapter 6). Nanoparticles – particles measuring 1–100 nanometres, or billionths of a metre – are increasingly used in modern life. An emerging field of nanoecotoxicology is examining their environmental fate and potential impacts on aquatic ecosystems (Hassellöv *et al.* 2008; Navarro *et al.* 2008). Microplastics, from the deterioration of plastic objects, may contain additives that accumulate in aquatic organisms (GESAMP 2010; Ryan *et al.* 2009), and their concentrations, especially in marine systems, are expected to follow increases in global plastic consumption. Additional types of pollutants about which little is

Figure 4.13 Trends in organochlorine contamination in selected deep-sea fish species, 1995–2005

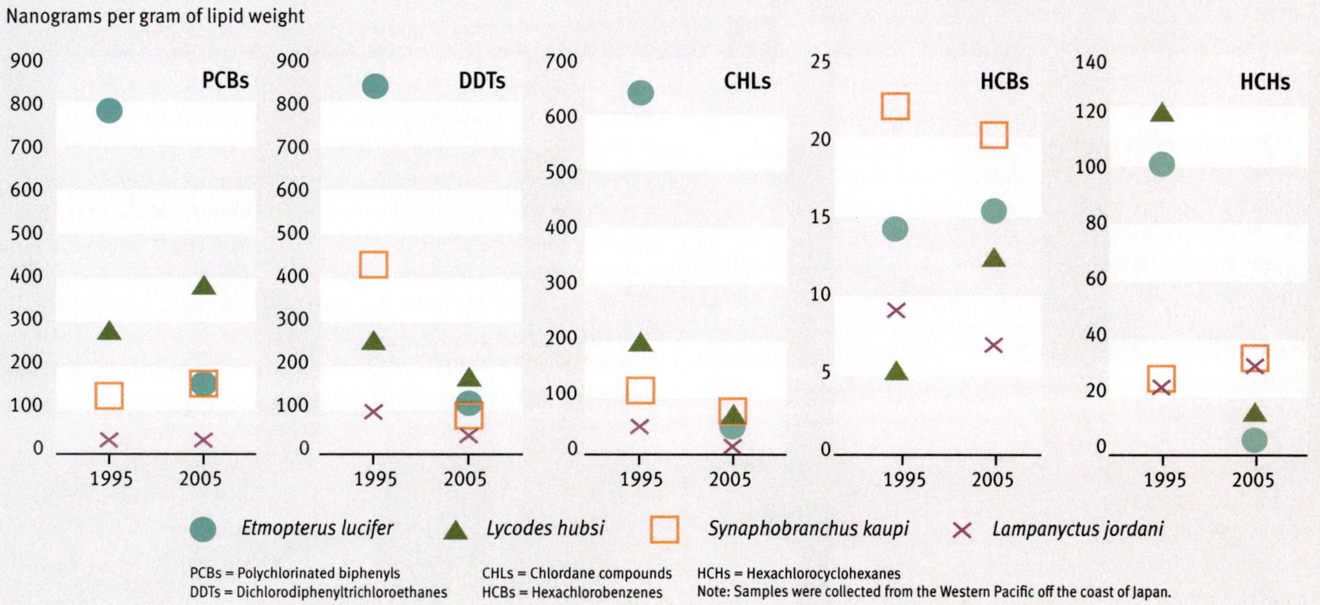

Source: Oshihoi *et al.* 2009

Box 4.12 Ballast water and invasive species

Invasive species, a form of biological pollution, pose great threats to aquatic ecosystems and can cause severe environmental and economic damage. Ballast water is a major vector for transporting species around the world. The Ballast Water Convention of 2004 required the implementation of management plans, with open-ocean ballast exchange commonly used to reduce introductions. Since this is unfeasible in many shipping routes, some countries, including Denmark and Australia, have instituted regulations requiring ballast water treatment to kill resident organisms.

currently known will doubtless continue to be identified. Although not new, industrial, medical, military and accidental releases of radioactive substances are of renewed concern, as illustrated by the water contamination after the 2011 tsunami damaged Japanese nuclear power plants. Invasive alien species also remain a problem for many coastal areas (Box 4.12; Chapter 5).

CROSS-CUTTING ISSUES

Water security and human health

As previously noted, regional differences exist regarding both absolute water availability and the limitations placed on it by inadequate infrastructure. Both relate to water security, as does water pollution, because they can all affect the range of human and environmental water uses. Despite improvements, lack of access to drinking water of adequate quality and quantity remains one of the largest human health problems globally. Inadequate water supply is an inherently regional phenomenon, however, caused by basin-level water scarcity, regional water quality, inadequacies of infrastructure and governance, cultural perspectives and inequitable water pricing.

Box 4.13 Water security

Goals
Secure adequate sustainable freshwater supply

Indicators
Human water security threat

Global trends
Deteriorating

Most vulnerable communities
Developing countries with increasing water demand

Regions of greatest concern
Africa, West Asia, Asia and the Pacific, Latin America and the Caribbean

Although the MDG target on water supply was met in 2011, more than 600 million people will lack safe water supplies in 2015. *© Kibae Park/UN Photo*

Figure 4.14 Threats to water security with and without infrastructure investment, 2000

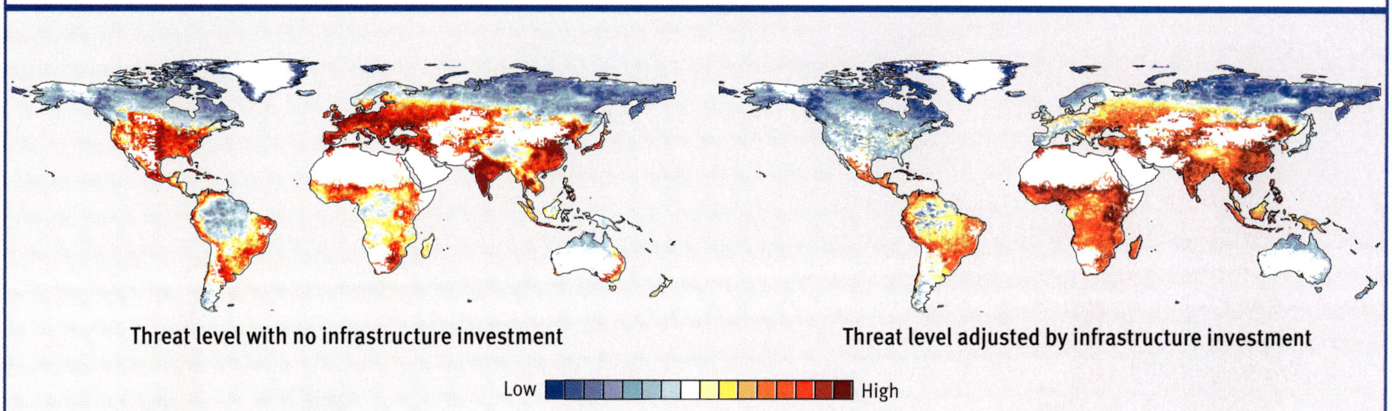

Source: Vörösmarty *et al.* 2010

Water security

Although several definitions for water security have been proposed since the 1992 Rio Earth Summit, none has been universally accepted (Oswald Spring and Brauch 2009). Varying definitions, leading to numerous indices based on different criteria, make it difficult to generate trend data. The Ministerial Declaration of the Hague broadly defines water security to include the protection and improvement of freshwater and marine ecosystems, sustainable development and political stability, with the aim of providing every person with access to "enough safe water at an affordable cost to lead a healthy and productive life", as well as protecting vulnerable communities from water-related risks and hazards (World Water Council 2000).

About 80 per cent of the world's population lives in areas with high water security threats, the most severe category encompassing 3.4 billion people, almost all in developing countries. Water security threat here refers to the cumulative effect of 23 drivers that have an impact on water resources, categorized into watershed disturbance, pollution, water resource development and biotic factors (Vörösmarty *et al.* 2010). More people are likely to experience severe water stresses in the coming decades because of increased demands (Chapter 1) in addition to altered precipitation patterns associated with climate change.

Figure 4.14 highlights the global threat to human water security and compares it with the magnitude of threat after adjusting for the effects of previous and current infrastructure investment. With higher investments in infrastructure in the industrialized countries, the figures show that human water security can be increased, overcoming the various threats to water resources (Vörösmarty *et al.* 2010), while low investments in developing countries means their water security remains poor. Investments must be coupled with adequate institutional capacity, and because infrastructure development often occurs at the expense of aquatic biodiversity and environmental quality, it is imperative that environmental risks related to investments are considered and appropriately mitigated.

Box 4.14 Access to improved water

Goals
Ensure equitable access to improved drinking water supply

Indicators
Proportion of population without an improved drinking water source; rural-urban equity

Global trends
Significant progress on improved supply; modest progress on rural-urban equity

Most vulnerable communities
Poor in developing countries and rural areas

Regions of greatest concern
South Pacific sub-region; most of Africa, especially the West Indian Ocean sub-region

Equitable access to improved drinking water

Although water security is an increasing problem in many regions of the world, significant progress in access to improved drinking water has been made since 1990. However, several regions, including most of Africa and other rural areas in developing countries, still lack access to improved drinking water sources (UNDESA 2010). The UN General Assembly declared access to clean water and sanitation as a human right in July 2010, although the right is not yet recognized or applied in many countries.

Recent data suggest that the MDG drinking water target was met in 2010 (Figure 4.15). However, there are important inequities in this improvement. Whereas only 4 per cent of people in urban areas lacked access to improved drinking water in 2010, in rural

Figure 4.15 Population without access to improved drinking water, 1990–2015

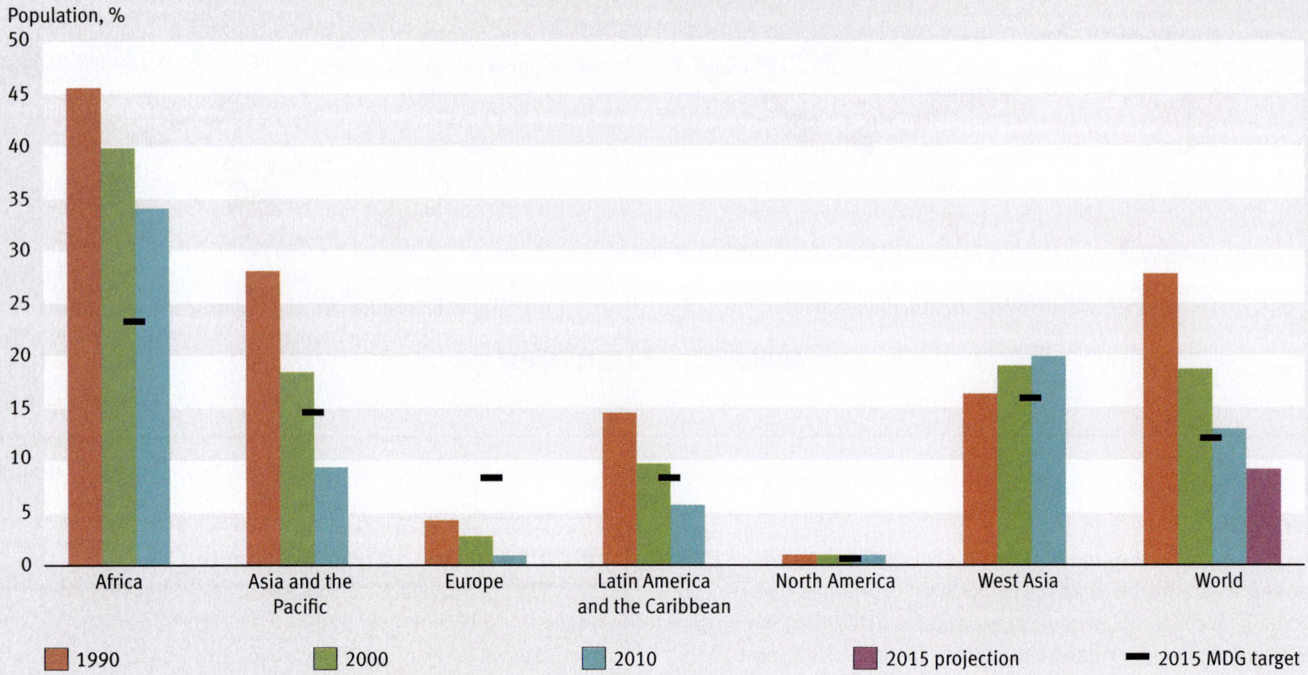

Note: As defined by WHO/UNICEF, an improved drinking-water source is one that, by nature of its construction or through active intervention, is protected from outside contamination, particularly from faecal matter.

Source: WHO 2012

areas 19 per cent of residents lacked such access. Progress towards achieving MDG 7c primarily reflects increased use of technology and infrastructure to overcome poor water quality or water scarcity (WHO 2012).

Water-related diseases

Water-related diseases, as defined by the World Health Organization (WHO), include those caused by microorganisms and chemicals in drinking water; diseases like schistosomiasis,

Box 4.15 Water-related diseases

Goals
Reduce water-related human health hazards

Indicators
Water-related-disease deaths measured as disability-adjusted life years (DALYs); number of reported cholera cases

Global trends
Some progress

Most vulnerable communities
Poor in developing countries and rural areas; communities that have experienced natural disasters

Regions of greatest concern
Africa

Dengue fever and malaria, both diseases transmitted by mosquitoes, are most problematic where there is stagnant water in which mosquitoes can breed. © *Salem Alkait/iStock*

Figure 4.16 Cholera cases by region, 1989–2009

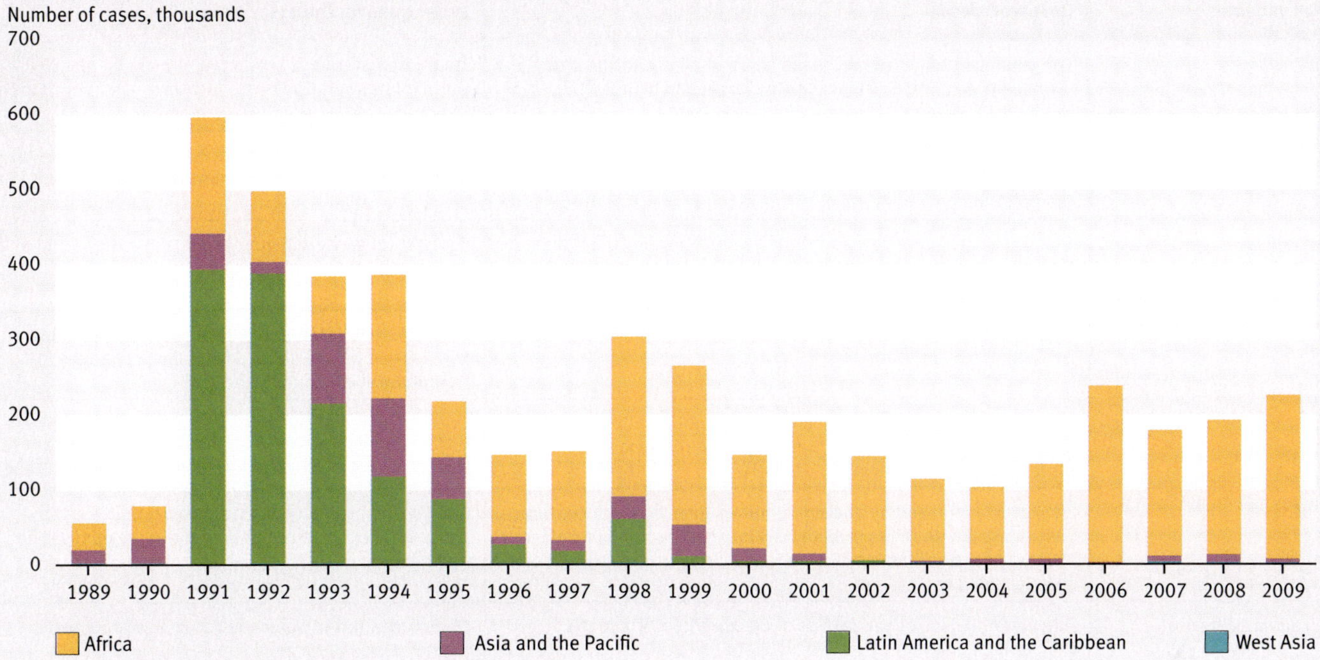

Note: Reported cases are an underestimate of total cases because many go unreported.

Source: WHO 2010

Box 4.16 Diarrhoea in children in Africa

At any given time, over half the world's hospital beds are filled with people suffering from water-related diseases (UNDP 2006). Diarrhoeal diseases make up more than 4 per cent of the global disease burden, 90 per cent being linked to environmental pollution and lack of access to safe drinking water and sanitation (Prüss-Üstün *et al.* 2008). Africa has the highest burden of diarrhoea-related childhood deaths, accounting for 70 per cent of the 1.3 million deaths of children less than five years old in 2008. Not surprisingly, access to basic sanitation is also poorest in sub-Saharan Africa, with 330 million people lacking access to proper sanitation (WHO 2011a).

whose vector spends part of its life cycle in water; diseases like malaria with water-related vectors; and others such as legionellosis carried by aerosols containing certain micro-organisms.

Such diseases are a major public health concern, especially in Africa. Globally, diarrhoea related to inadequate sanitation and water supply was the second largest contributor to the 2004 global disease burden, claiming more than 70 million disability-adjusted life years (DALYs), years lost due to ill-health, disability or early death (Box 4.15) (Prüss-Üstün *et al.* 2008). Global health statistics indicate that Africa and South Asia contain the areas most severely affected by waterborne disease (WHO 2004).

The WHO is focusing on reducing 25 different water-related diseases (WHO 2011a). There have been some notable successes in the reduction of onchocerciasis, malaria, schistomiasis and cholera. However, globally reported cholera incidence – which serves as a proxy where complete data on water-related disease trends are lacking – has increased in recent years, mainly in Africa (Figure 4.16). In 2009, 45 countries from all continents reported 221 226 cases of cholera (Figure 4.16). Water-related diseases are a continuing public health problem in developing countries lacking access to adequate drinking water and sanitation, as further evidenced by the cholera epidemic in Haiti following the 2010 earthquake (Walton and Ivers 2011).

Water-energy-climate nexus

Water, energy, economic development and climatic change are interdependent issues. Increases in human population and per-person consumption related to economic development drive energy demands. Meanwhile the use of fossil fuel energy produces greenhouse gas emissions that contribute to climate change, which has effects on water, including extreme weather events, loss of ice cover, water scarcity and sea level rise. In turn, responses to climate change have implications for the water environment. Some

Table 4.2 Observed and projected impacts of climate change on key hydrological variables

Key variables	Observed trends	Projections for the 21st century
Precipitation	The trend is unclear; increases in general precipitation over land from 30° N to 85° N; notable decreases from 10° S to 30° N	Total precipitation is projected to increase (by 1–3% per °C of temperature rise), with varying changes at the regional scale
Precipitation intensity	Disproportionate increase in volume of precipitation in heavy or extreme precipitation events; intensification in extreme precipitation on global scale	Heavy precipitation is projected to increase by approximately 7% per °C of temperature rise
Droughts	Drought increased in the 20th century measured by the Palmer Drought Severity Index, although some areas became wetter and/or drought intensity has lessened	Drought intensity will increase in some areas and seasons; patterns are complex and difficult to predict
Tropical cyclones	High degree of uncertainty about any detectable change being related to climate change	Likely increase in average tropical cyclone maximum wind speed but decreased frequency; changes in frequency and track are uncertain
Glaciers and snow cover	There is a decrease in glacial masses, but not in all regions, and decreased snow cover in northern hemisphere regions; peak run-off from glacier and snowmelt is earlier	Continued decrease in glacial mass and snow cover
Sea level	Sea levels increased by about 0.2 metres over the 20th century; a rise equivalent to 0.3 metres per century has been recorded since the early 1990s, although it is not clear if this is an acceleration in long-term sea level rise	Sea level is projected to rise by 0.2–0.6 metres by 2100, although the upper end of the range could be much higher
Ocean acidification	The mean surface ocean pH has decreased from 8.2 to 8.1	pH is projected to decrease to 7.7 or 7.8 by 2100 if present trend persists
Sea surface temperature	Increased by 0.5° C since 1980	Continued increase

Source: IPCC 2011; Feely *et al.* 2009; World Bank 2009; IPCC 2007c

forms of solar energy consume significant quantities of water, often in arid regions. With increasing water scarcity, some regions also rely on desalinization of marine water, requiring large energy inputs (World Bank 2009). In addition, droughts have the potential to decrease hydropower production (Box 4.21).

Climate change impacts on the water cycle and ocean warming

The hydrologic cycle refers to the continuous movement of water through the oceans, atmosphere and over and under land surfaces. There is strong evidence that climate change is altering global and regional hydrologic cycles (Bates *et al.* 2008; IPCC 2007a; Kundzewicz *et al.* 2007), with impacts predicted to be manifested as changing precipitation patterns, increased intensity of extreme weather events and consequent natural disasters, retreating glaciers resulting in altered river discharge regimes, and more intense droughts in semi-arid regions (Table 4.2) (IPCC 2007b).

Although there is considerable uncertainty regarding projected impacts on specific water systems, climate change has the potential to seriously affect water management (Bates *et al.* 2008). Nonetheless, the global impacts of other human activities on the hydrologic cycle – urbanization, industrialization, water resources development – are likely to exceed those related to climate change, at least for the next two to three decades (Gordon *et al.* 2005). If climate change impacts are to be addressed, the cost of the additional water infrastructure needed by 2030 to provide a sufficient quantity of water for all countries is estimated at US$9–11 billion per year (UNFCCC 2007), 85 per cent of this in developing countries. There are additional costs associated with flood

Box 4.17 Climate change impacts on human security

Goals
Mitigate and adapt to adverse effects of climate change on the water environment

Indicators
Extreme precipitation; glacial retreat; drought intensity; water sector costs of climate change adaptation

Global trends
Some progress on adaptation and mitigation strategies; little or no progress on funding and implementation

Most vulnerable communities
People dependent on rain-fed agriculture and/or glacial melt; those relying on non-renewable groundwater in the long term

Regions of greatest concern
Arid regions, tropics, and coastal areas that experience cyclones and hurricanes

risk management and water quality protection (Parry *et al.* 2009). There are signs of increased awareness of mitigation and adaptation needs: of 191 water projects funded by the World Bank between 2006 and 2008, 35 per cent incorporated mitigation and adaptation measures for climate change impacts (World Bank 2009). At the same time, however, local and regional efforts to increase protection against floods and other extreme events are likely to have significant negative impacts on aquatic ecosystems themselves.

The most direct climate change impact on oceans is increased sea surface temperature (SST), which has risen by 0.5ºC globally since the 1980s and is predicted to continue increasing throughout the 21st century (IPCC 2007a). Global precipitation is predicted to increase at a rate of 1–3 per cent per degree of surface warming (Wentz *et al.* 2007), with more extreme precipitation events predicted for many tropical and temperate regions (IPCC 2011; Gorman and Schneider 2009).

Melting ice sheets and sea level rise

Sea level rise is caused by ocean thermal expansion and by melting glaciers and ice sheets (IPCC 2007a). Although average global sea level has remained relatively constant for almost 3 000 years, it increased by approximately 170 mm during the 20th century (IPCC 2007b), and is projected to rise by at least another 400 mm (+/-200 mm) by 2100 (IPCC 2007a). Measurements from 1993 to 2008 indicate that sea levels are already rising twice as fast as in previous decades (Cazenave and Llovel 2010) and are exceeding the rise predicted by climate models.

Although there is considerably variability associated with these and other estimates of sea level rise (Levitus *et al.* 2009; Ishii and Kimoto 2009), 25–50 per cent of sea level rise observed since 1960 has been attributed to thermal expansion (Cazenave and Llovel 2010; Antonov *et al.* 2005; Willis *et al.* 2004). Some variability may result from water impounded in reservoirs, which is estimated to have reduced sea level rise by 30–55 mm over the past 50 years (Chao *et al.* 2008). Small glaciers and ice caps exhibited significant mass losses over the 20th century (Dyurgerov and Meier 2005) and freshwater run-off from melting land-based ice sources will increase in the future. However, with losses accelerating over the past 20 years, melting Greenland and Antarctic ice sheets have become the biggest contributors to sea level rise, and will remain the dominant contributor to sea level rise in the 21st century if current trends continue (Rignot *et al.* 2011; Rignot 2008).

Box 4.18 Sea level rise

Goals
Mitigate and adapt to adverse effects of climate change on the water environment

Indicators
Sea level rise; cost of adaptation to sea level rise

Global trends
Very little to no progress

Most vulnerable communities
Coastal areas, island communities, high-density populations in deltas

Regions of greatest concern
Coastal areas (deltas and African coast), small island developing states, the Arctic, Antarctica and high mountain regions

Box 4.19 Ocean acidification

Goals
Protect and restore marine ecosystems and their services

Indicators
Ocean pH

Global trends
Deteriorating

Most vulnerable communities
Communities dependent on tropical fisheries that rely on coral reef ecosystems and other calcareous primary producers

Regions of greatest concern
Tropical oceans

Because of the high concentrations of human populations and infrastructure in coastal zones (McGranahan *et al.* 2007), many countries are vulnerable to sea level rise and associated coastal and low-lying community flooding (Chapter 7). Developing countries, particularly small island developing states (SIDS) and deltaic areas, are especially vulnerable (IPCC 2007c), many with limited capacity to adapt to rising sea levels or recover from associated losses. The estimated costs of coastal adaptation range from US$26 billion to US$89 billion per year by the 2040s, depending on the magnitude of sea-level rise (World Bank 2010).

Ocean acidification

The oceans annually absorb a substantial proportion of anthropogenic carbon dioxide (CO_2), which reacts with water to form carbonic acid, thereby making the ocean more acidic. The mean surface ocean pH has already decreased from a pre-industrial average of about 8.2 to a present value of 8.1, though there are regional differences (Figure 4.17; Chapter 2), and Feely *et al.* (2009) project a pH decrease to a mean of about 7.8 by 2100. Ocean acidification may be approaching the planetary boundary (Rockström *et al.* 2009).

Figure 4.17 CO_2 concentrations and ocean acidification in the North Pacific, 1960–2010

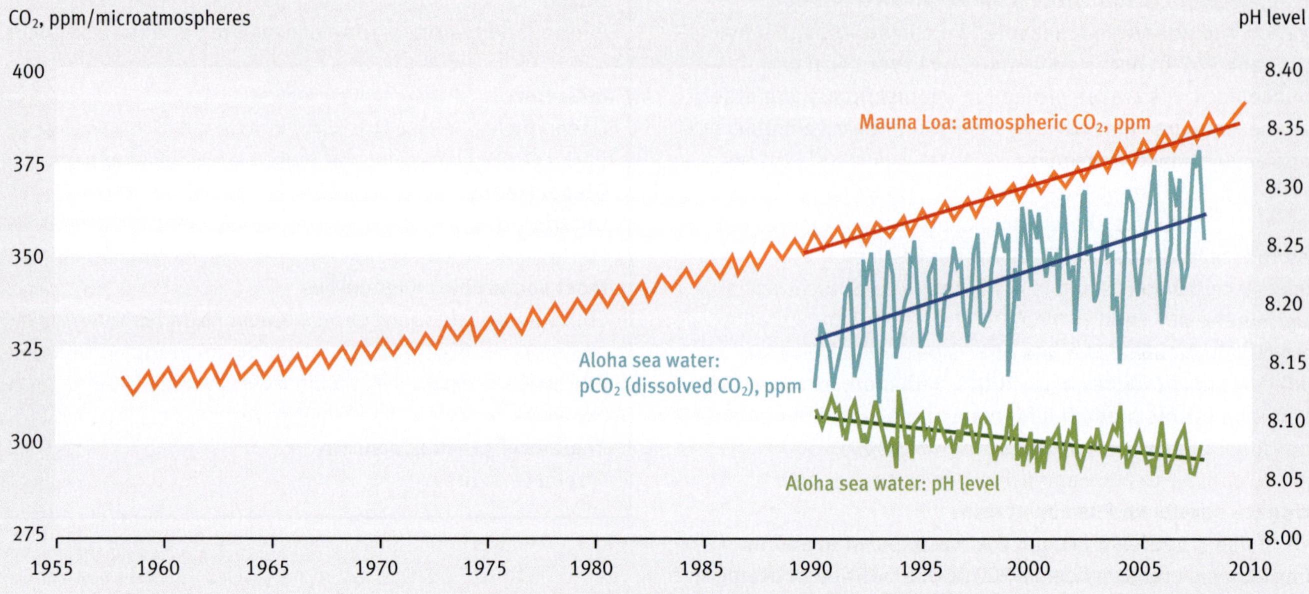

Source: Feely *et al.* 2009

Increased ocean acidity affects marine animals with carbonate shells and skeletons, calcareous algae and other organisms (Langdon and Atkinson 2005). Affected organisms include reef-building corals as well as animals critical to ocean food webs, including several important human food sources such as crabs and molluscs. Combined with higher water temperatures, ocean acidification is thought to be a major cause of coral bleaching, destroying coral reef ecosystems around the world (Hoegh-Guldberg *et al.* 2007), with some studies projecting a rapid contraction of tropical coral reefs by 2050 (Chapter 5) (Logan 2010). Coral reefs provide important ecosystem services, such as spawning and nursery grounds for some commercially important fish species. Impairment of these ecosystems and their services are becoming increasingly evident and illustrate the need for governance to enhance their protection.

Ocean acidification is threatening marine life, particularly corals and shellfish. It could have a devastating effect on communities dependent on fishing and aquaculture. *© Extreme-photographer/iStockv*

Impacts of energy development on water resources

While global data are lacking, the energy sector is believed to account for approximately 40 per cent of total water withdrawals in the United States and European Union (EU) (Glennie *et al.* 2010). Water demands for energy range from extraction and processing of raw materials to driving hydropower turbines and cooling thermoelectric plants, including nuclear. Fossil fuel extraction can also have serious impacts on water quality.

Oil and gas exploration and production can affect both freshwater and marine ecosystems. Newly proven technologies are accelerating the expansion of new natural gas wells in shale gas basins (EIA 2011). Associated water resource impacts are currently being researched, including aquifer contamination with potentially explosive methane levels (Osborn *et al.* 2011), surface and groundwater contamination, streams receiving water discharges (Johnson *et al.* 2007), and high consumptive water use for well drilling and completion (Chapter 7). Oil sand exploitation also requires large water volumes and can produce severe water pollution (Kelly *et al.* 2010).

Box 4.20 The Deepwater Horizon oil spill

The *Deepwater Horizon* oil spill of 4.9 million barrels of crude oil into the Gulf of Mexico in 2010 was the largest accidental marine oil spill in history. Although the economic and ecological costs have not yet been fully quantified, it caused extensive damage to marine life, wildlife habitats, fishing and tourism. Unlike previous oil spills, in which most of the oil dissipated or evaporated, immense underwater plumes and thick layers of dissolved oil remained on the seafloor in spring 2011, tar balls continued to wash ashore and wetlands marsh grass continued to foul and die.

Source: National Commission on the BP Deepwater Horizon Oil Spill and Offshore Drilling 2011

Oil spills continue to pose an environmental threat, particularly to marine ecosystems. Although the number of oil tanker spills has decreased significantly since the 1970s and 1980s (ITOPF 2010), the recent large spill associated with offshore oil and gas exploration in the Gulf of Mexico is evidence of ongoing risks to marine ecosystems (Box 4.20). Nevertheless, with increasing global oil and gas demands, such offshore activity is expected to increase over the next two decades, facilitated by the resolution of maritime boundaries and improved access to previously inaccessible areas as Arctic ice melts. The Arctic contains approximately 20 per cent of the world's undiscovered but technically recoverable oil and gas resources (Bird *et al.* 2008; AMAP 2007), but the region is uniquely vulnerable to oil spills because of its remoteness, harsh physical environment, the aggregation of large numbers of marine mammals and the slow rate of oil degradation in cold water.

The most water-intensive form of electricity production is biomass, followed by hydropower, oil, coal and nuclear, gas, some concentrated solar power systems and geothermal, solar photovoltaics, and wind. Exact values vary greatly, depending on electricity generation type and location (Glennie *et al.* 2010). Many forms of concentrated solar power, for example, which may be most effective in arid areas exhibiting high solar energy levels, also require significant quantities of water for cooling, sometimes as much as fossil-fuel-powered plants. There are cases in which water scarcity is already affecting energy production. More than half of existing or planned capacity for major power companies in South and South East Asia, for example, is located in water-scarce or water-stressed areas (Figure 4.18) (WRI 2010).

Figure 4.18 Thermal power and hydropower plant locations and water stress levels in five countries of South and South East Asia

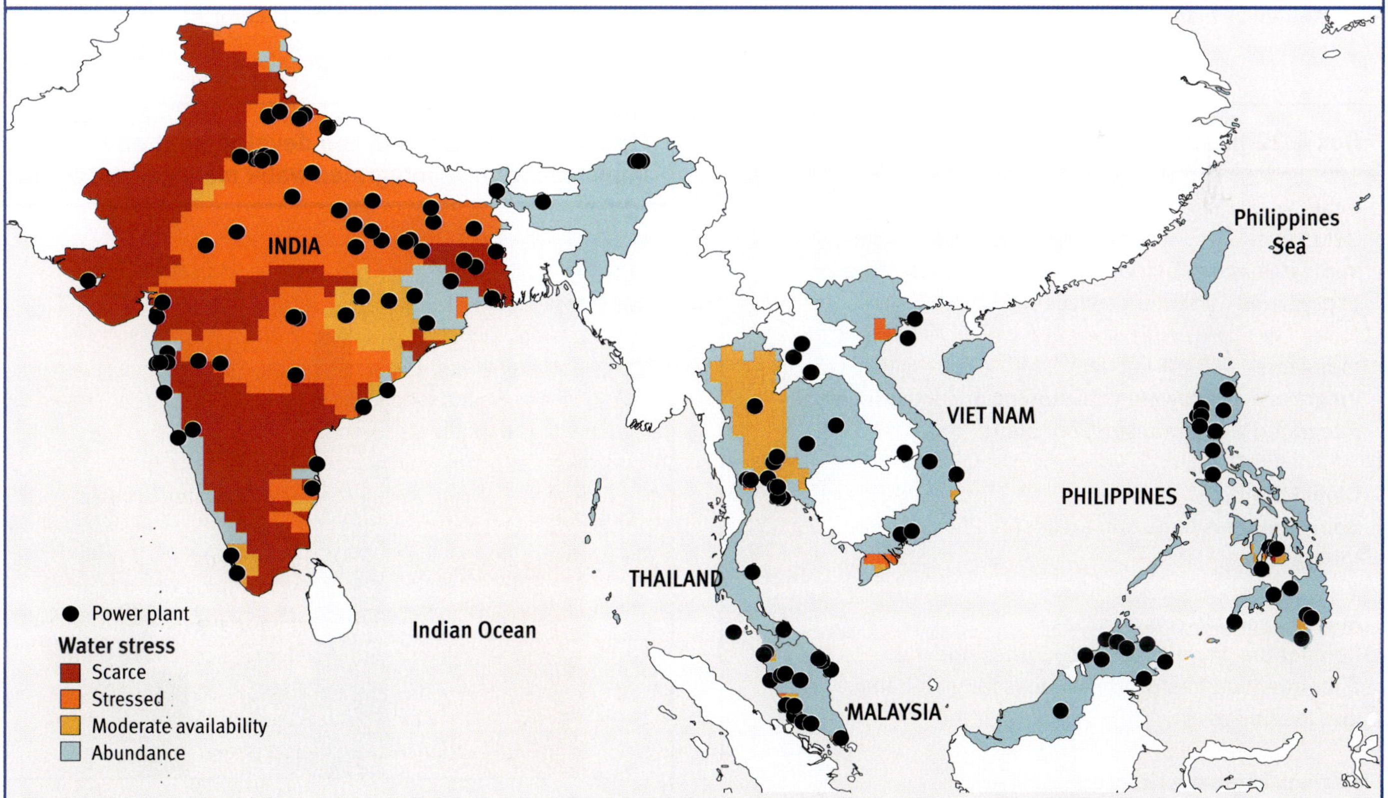

Note: Water stress indicators represent the ratio of total withdrawals to utilizable water, but do not reflect anticipated demographic or climate changes such as the timing or quantity of precipitation.

Source: WRI

Box 4.21 The impacts of drought on hydropower production

Droughts have significantly decreased hydropower output in Eastern Africa over the past ten years, adversely impacting national economies. Low water levels in Lake Victoria between 2004 and 2006, for example, reduced hydropower output in Uganda by 50 megawatts, contributing to a fall in the economic growth rate from 6.2 per cent to 4.9 per cent over this period (Karakezi *et al.* 2009).

Climate change mitigation policies can also affect water demands for electricity production. Capture and storage of carbon emissions from coal-fired plants, for example, can increase water consumption by 45–90 per cent (Glennie *et al.* 2010). Further, increasing the proportion of electricity generated from biomass or some types of concentrated solar power is likely to have significant negative impacts on water availability, highlighting the need for selecting power generation types that use less water and more efficient technologies (Chapter 12).

Water governance

Water problems frequently translate into inadequacies of water governance (RCSE-SU and ILEC 2011; UNESCO 2006), as illustrated by many of the water goals laid out in Table 4.1.

Adaptive freshwater management and integrated planning

Agenda 21 of UNCED called for "integrated approaches to the development, management and use of water resources" (UNCED 1992), subsequently leading to the development of several integrated management paradigms, including integrated water resources management (Global Water Partnership 2000), integrated lake basin management (International Lake Environment Committee 2006), and integrated coastal zone management, as mentioned in the Jakarta Mandate on Marine and Coastal Biodiversity (CBD 1997) and other outputs of the Convention on Biological Diversity (CBD). Integrated management approaches also offer a degree of protection against the negative impacts of natural disasters such as the devastating earthquake and tsunami that struck Japan in 2011.

The need for integrated approaches was formalized in Paragraph 26 of the Johannesburg Plan of Implementation, which states that governments should develop integrated water resources management and water efficiency plans by 2005 through actions at all levels (WSSD 2002). This overall target has not been met. However, data from studies in 2003 and 2005, primarily involving developing countries, and for 2008 and 2012 for all countries, do suggest significant headway – from the development of plans to their implementation – particularly in developed countries (Figure 4.19). Progress appears to have slowed, however, in developing countries (UN-Water 2012).

Some water professionals and policy makers suggest that in some cases the integrated management concept is insufficiently specific for practical implementation (Placht 2007; Watson *et al.*

Box 4.22 Integrated water management

Goals
Develop and implement integrated management strategies and plans; protect and restore freshwater ecosystems and their services

Indicators
Progress made towards developing and implementing integrated water management plans

Global trends
Some progress in certain areas; insufficient data for others

Most vulnerable communities
Populations in developing countries directly dependent on freshwater systems for well-being and livelihoods

Regions of greatest concern
Developing regions, particularly those with water shortages and/or water quality degradation

Figure 4.19 Progress in the development and implementation of integrated water management plans

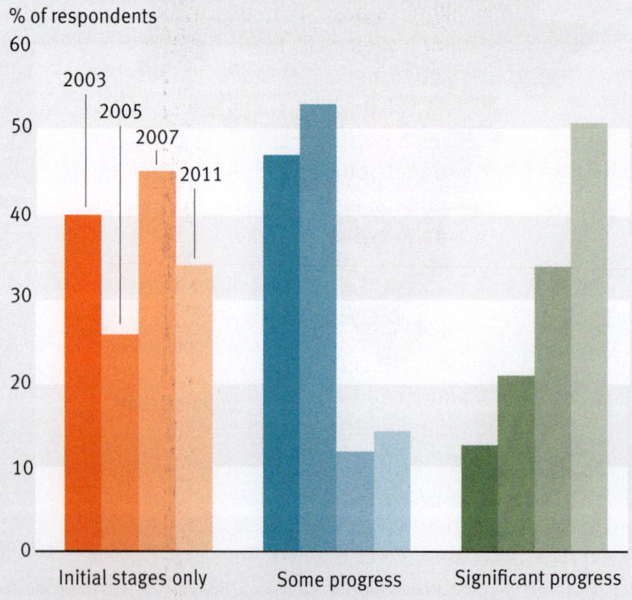

Source: Global Water Partnership 2006; UN-Water 2012

Coastal sand dunes buffer the coastline against wave damage and protect the land from saltwater intrusion. *© Rui Miguel da Costa Neves Saraiva*

2007), and is slow to get under way owing to many institutional, economic, political and resource constraints (Brauch *et al.* 2009; Lansky and Uitto 2005). Further, although integrated water management proposes cross-sectoral coordination, all relevant government agencies and key water stakeholders may not be in agreement (Biswas 2004). Because participatory approaches do not always ensure consideration of gender perspectives, a systematic assessment of the differing impacts of economic development on women and men is also necessary to ensure that water issues affecting both genders are part of programme planning, implementation and evaluation, including ensuring institutional and organizational changes for gender equality as an ongoing commitment (Bennett *et al.* 2005). Further, although integrated management can be applied at many levels, from village to basin to national to transboundary, there are management issues particular to each of these levels (Lenton and Muller 2009), necessitating both bottom-up and top-down approaches. Existing evidence, however, suggests that integrated policies have focused largely on higher-level activities such as national policy reforms or the establishment of river basin organizations, rather than on on-the-ground implementation of integrated management activities at the local level (Perret *et al.* 2006).

The European Commission (EC) applied integrated water resources management principles in its Water Framework Directive in 2000, and a flood risk management directive in 2007. Further, although implicit in goals such as Paragraph 26 of the Johannesburg Plan of Implementation, there is no global multilateral environmental agreement specifically directed at aquifer conservation. There are, however, several regional groundwater initiatives, including the 2008 establishment of the Africa Groundwater Commission (AMCOW 2008). Because poor groundwater governance is a major issue, recognizing groundwater systems in national laws would be a first step towards improved groundwater governance, followed by establishing sustainable institutions and financing.

While social science literature on regional experiences with integrated water management and international river basin management is increasing, there is little data on the state and trends of such approaches, particularly their long-term benefits and impacts. Research has focused more on the concept and its application, and less on relevant policy implementation, highlighting the need for better progress indicators as well as continued monitoring frameworks to assess effectiveness (RCSE-SU and ILEC 2011; UN-Water 2012). Certain policy initiatives have supported, *inter alia*, international river basin regimes, and include the Transboundary Water Assessment Programme (TWAP), to be implemented by UNEP for the purpose of developing a methodology for monitoring and assessing trends regarding, among other things, environmental and human water stresses, pollution, population density and water system resilience.

Marine governance

Marine systems are a major food source, a means of transport for international shipping, a tourism attraction and a climate change regulator. Coastal dunes and tidal wetlands are important buffers against tidal flooding. A number of international conventions have been established to protect the marine environment, demonstrating significant levels of international cooperation, although a common limitation is their dependence on national legislation that may reflect other agendas.

Regarding international agreements, the 1972 Convention on the Prevention of Marine Pollution by Dumping of Wastes and Other Matter (the London Convention) and the 1973 International Convention for the Prevention of Pollution from Ships (MARPOL) address marine pollution. The United Nations Convention on the Law of the Sea (UNCLOS 1982), ratified by 160 countries and in force since 1994, represents a unified approach towards shared use of the oceans and their resources, addressing navigation, economic rights, pollution, marine conservation and scientific exploration. Notwithstanding the concern with increasing marine litter, the conventions are generally viewed as positive frameworks for controlling and preventing marine pollution. The 2004 International Convention for the Control and Management of Ships' Ballast Water and Sediments exemplifies collaborative actions to address the introduction of alien invasive species, which can cause significant environmental and economic damage.

Another noteworthy international effort is the Global Programme of Action for the Protection of the Marine Environment from Land-based Activities (GPA), adopted by 108 governments and the EC in 1995. Although not enforceable, the GPA was designed to guide national and regional authorities in undertaking sustained action to prevent, reduce and/or eliminate marine degradation from land-based activities. Many countries profess to subscribe to its goals, providing a means for developing collaborative strategies to address coastal and offshore water degradation from influent freshwaters. Marine spatial planning, similar to land planning or zoning on public lands, is another emerging area of possibilities for marine governance.

The regional seas conventions (UNEP and independent conventions), other action plans, and the large marine ecosystem concept promulgated by the US National Oceanic and Atmospheric Administration (NOAA), also represent integrated management approaches (Figure 4.20). Development and implementation of these plans differ, however, depending on the countries involved, with some programme guidelines being binding on the participating states while others are not.

Figure 4.20 Map of 18 regional seas and 64 large marine ecosystems, 2011

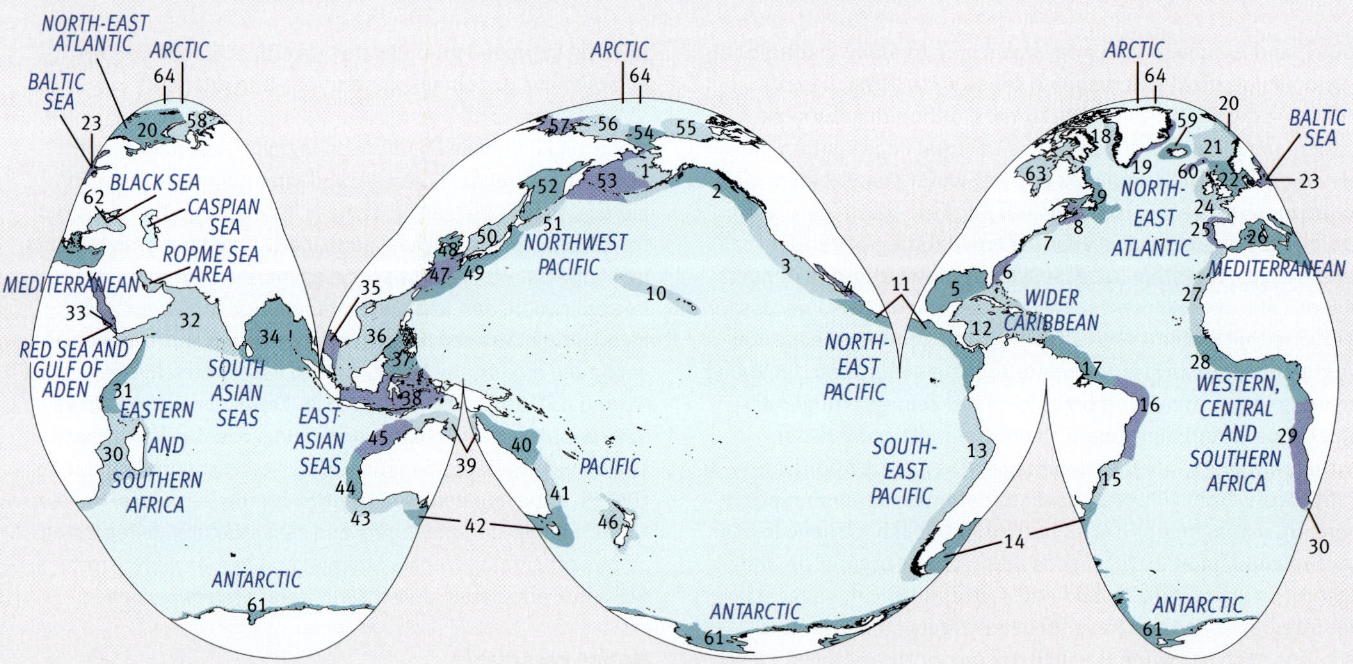

Large marine ecosystems of the world

1. East Bering Sea
2. Gulf of Alaska
3. California Current
4. Gulf of California
5. Gulf of Mexico
6. Southeast US Continental Shelf
7. Northeast US Continental Shelf
8. Scotian Shelf
9. Newfoundland-Labrador Shelf
10. Insular Pacific-Hawaiian
11. Pacific Central-American Coastal
12. Caribbean Sea
13. Humboldt Current
14. Patagonian Shelf
15. South Brazil Shelf
16. East Brazil Shelf
17. North Brazil Shelf
18. West Greenland Shelf
19. East Greenland Shelf
20. Barents Sea
21. Norwegian Shelf
22. North Sea
23. Baltic Sea
24. Celtic-Biscay Shelf
25. Iberian Coastal
26. Mediterranean Sea
27. Canary Current
28. Guinea Current
29. Benguela Current
30. Agulhas Current
31. Somali Coastal Current
32. Arabian Sea
33. Red Sea
34. Bay of Bengal
35. Gulf of Thailand
36. South China Sea
37. Sulu-Celebes Sea
38. Indonesian Sea
39. North Australian Shelf
40. Northeast Australian Shelf-Great Barrier Reef
41. East-Central Australian Shelf
42. Southeast Australian Shelf
43. Southwest Australian Shelf
44. West-Central Australian Shelf
45. Northwest Australian Shelf
46. New Zealand Shelf
47. East China Sea
48. Yellow Sea
49. Kuroshio Current
50. Sea of Japan
51. Oyashio Current
52. Okhotsk Sea
53. West Bering Sea
54. Chukchi Sea
55. Beaufort Sea
56. East Siberian Sea
57. Laptev Sea
58. Kara Sea
59. Iceland Shelf
60. Faroe Plateau
61. Antarctic
62. Black Sea
63. Hudson Bay
64. Arctic Ocean

Note: ROPME is the Regional Organization for the Protection of the Marine Environment, one of the regional seas programmes.

Source: UNEP/DEWA/GRID-Geneva 2011

Box 4.23 Competition and conflict

Goals
Strengthen institutional coordination mechanisms

Indicators
Number of conflictive and cooperative events; number of institutions and treaties

Global trends
Some progress

Most vulnerable communities
Communities in transboundary basins with inadequate institutional frameworks

Regions of greatest concern
Those with water stress and undergoing rapid development

The open oceans beyond national jurisdiction comprise almost half the planet's surface, with rapidly advancing technology opening major new oceanic frontiers for commercial uses including fishing, shipping, resource exploration and for potential marine engineering such as deep-ocean CO_2 sequestration. The open oceans and deep-sea ecosystems, including seamounts, trenches and canyons, cold water corals and hydrothermal vents, exhibit a wealth of biodiversity. Larger, slow-growing, long-lived and heterogeneously distributed species are adapted to stable conditions in these environments, and are particularly sensitive to environmental stresses. Governance of areas beyond national boundaries is weak and fragmented, however, and requires strengthening in preparation of increasing human activities and their impacts on areas within national jurisdictions, as well as to ensure conservation and sustainable use of the open oceans.

Water as a basis for conflict and cooperation

Competition for shared water resources can cause conflicts, particularly at the local level, with water needs usually immediate, and resources often inadequate to address all competing needs. Intrastate conflicts occur between rural and urban sectors – such as agricultural, industrial and municipal – and between water-reliant livelihood activities – such as fishing, agriculture and livestock grazing. Population growth, economic development and climate change can exacerbate management issues. Further, about 40 per cent of the global population lives in transboundary river basins that cover nearly half of the Earth's land surface and provide more than 60 per cent of global freshwater flows (Figure 4.21), imparting an additional management difficulty.

There also are increasing incidents of deliberate or threatened poisoning of water supplies (Pacific Institute 2011; Greenberg 2009). Sixty-nine water conflicts were documented for 2000–2010 in the Water Conflict Chronology List maintained by the Pacific Institute, compared to 54 recorded for 1975–1999. Although specific incidents were not described in detail, De Stefano *et al.* (2010) found about 67 per cent of 1 831 reported water events for 1948–1999 were cooperative, with only 28 per cent being conflictive in nature; by 2000–2008, the proportion of conflictive events had increased slightly, to 33 per cent. Infrastructure and water quantity were consistently the major issues likely to generate conflict (Figure 4.22). Although

Figure 4.21 International river basins, 2000

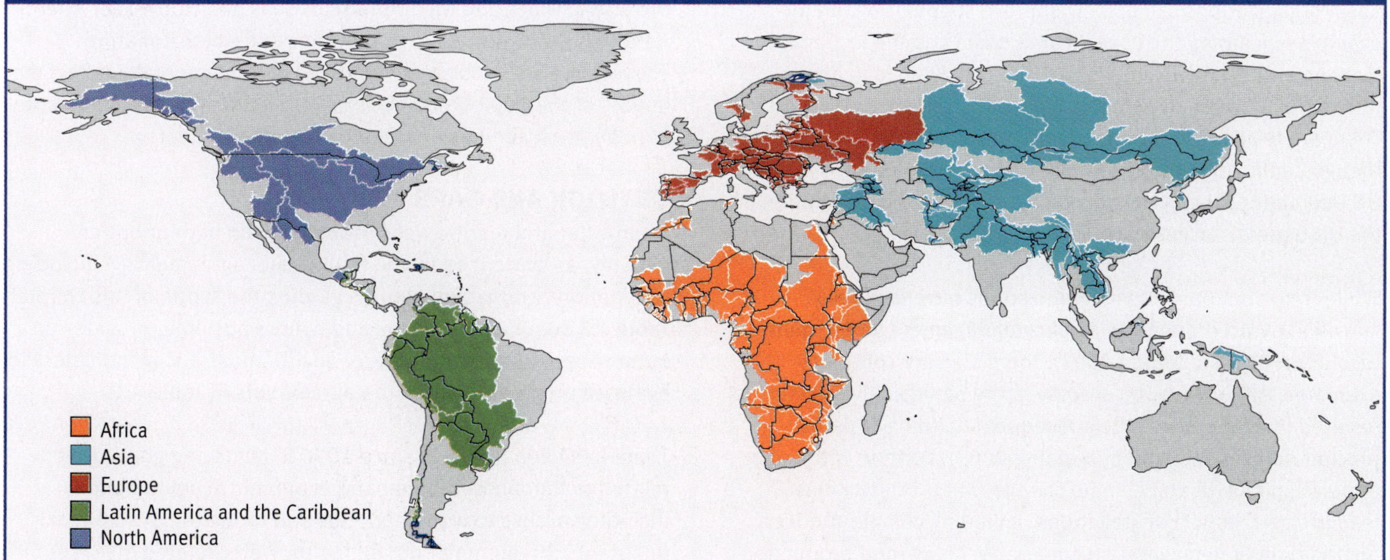

Source: Wolf 2007

Figure 4.22 Freshwater conflict by type of issue, 1948–1999 and 2000–2008

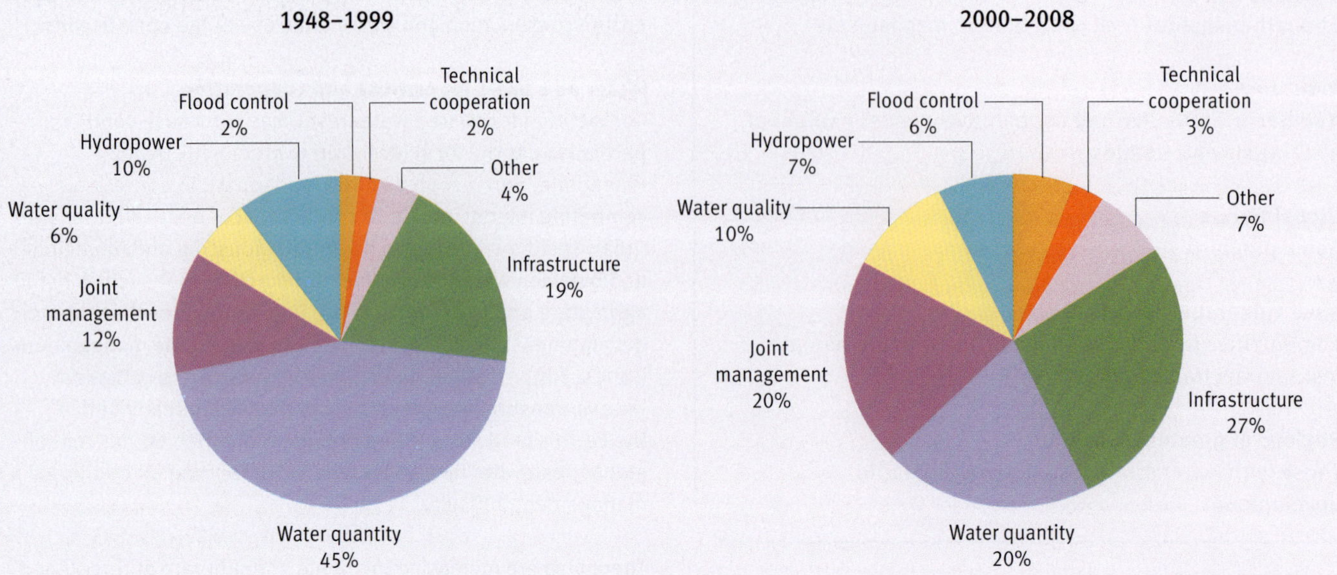

Source: De Stefano *et al.* 2010

water conflicts have occurred in many locations, and disputes could increase in the future (Kundzewicz and Kowalczak 2009; Greenberg 2009), current evidence suggests greater potential for cooperation than for conflict, particularly at the international level (De Stefano *et al.* 2010).

About 158 of the 263 international freshwater basins still lack cooperative management frameworks, while less than 20 per cent of the 106 basins with water institutions have multilateral agreements in effect (De Stefano *et al.* 2010). Evidence suggests, however, that freshwater systems with established transboundary basin organizations can usually improve cooperation, major examples including the Lake Victoria basin and the La Plata, Mekong and Senegal River basins (Chapter 8). In fact, about 295 international water agreements have been signed since 1948. There are relatively few transboundary groundwater institutions, though codification of the law of transboundary aquifers by the UN International Law Commission (ILC), adopted by resolution of the UN General Assembly in 2008, is a major advance.

While transboundary basin organizations have facilitated so-called hydro-diplomacy, conflict management and dispute resolution (Oswald Spring 2007), there also are contrary examples. Although water scarcity in the Senegal River basin resulted in cooperation, the subsequent building of dams precipitated violent conflict (Kipping 2009). Further, with population growth and climate change, water scarcity may lead to new conflict constellations, including climate-induced degradation of freshwater resources, declining food production and increased storm and flood disasters, which may further undermine food security (WBGU 2008).

No similar analysis exists for potential conflicts related to overfishing and deep-sea mineral exploration in the open oceans, although several international agreements identified in the section on marine governance address them to varying degrees. The sustainable use of coastal areas and ocean resources requires effective coordination and cooperation at regional and global levels, with examples including UNEP's 13 regional seas programmes and the 64 large marine ecosystems (Figure 4.20). The EU Marine Strategy Framework Directive is another regional instrument, applicable in European waters under the jurisdiction of EU Member States that border the Baltic, Black and Mediterranean Seas and North-East Atlantic. Although the regional seas programmes and large marine ecosystems are consistent with UNCLOS and generally reflect the targets of the 2002 World Summit on Sustainable Development (WSSD), their achievement status remains unclear.

OUTLOOK AND GAPS

Freshwater and marine water issues remain high priorities globally, as evidenced by the multilateral agreements – including conventions and action plans – guiding the scope of this chapter. Table 4.3 summarizes trends and, where possible, provides an outlook of the state of the water environment, using indicators to evaluate progress towards the agreements in Table 4.1.

There has been progress since 1990 in achieving goals directly related to human well-being and economic development, including access to water supplies and reduction of some toxic pollutants threatening human health. Water-related diseases and water supply in the rural areas of developing countries, however, require increased attention. There has also been progress on water governance with the development of integrated water resource management plans and transboundary water agreements. However, these plans must now be implemented, adequately funded and enforced to improve aquatic ecosystems and the sustainability of their life-supporting goods and services.

Improving water security and ensuring equitable access to water resources remains a challenge. Against a background of continuing water degradation and overexploitation, the need for sustainable water supplies remains one of humanity's most critical resource needs. There also has been little to no progress in most regions in reducing nutrient loads to freshwaters and coastal areas, or for governance beyond national jurisdictions.

The complexity of the drivers and associated pressures on aquatic ecosystems is a key barrier to attaining internationally agreed goals directed at addressing their root causes. Lack of appropriate indicators or targets for many environmental, socio-economic and governance goals makes assessing progress towards achieving water-related goals and sustainable aquatic ecosystems especially problematic. Other major barriers include inadequate capacity, limited access to technology and funding, information and data gaps, and lack of quantifiable targets. More emphasis, including enhanced monitoring efforts, should be directed to acquiring reliable data on the impacts of climate change and extreme weather events on human health and well-being, and on environmental integrity. Unfortunately, monitoring of water quality, quantity and ecosystem health has been reduced in many regions. As a result, there are increasing uncertainties regarding assessment and management of the water environment, due both to data gaps and to the rapidly changing nature of water issues, including those related to climate change.

Men pulling a raft packed with their belongings through flood waters on a road in Pathum Thani, Thailand, in October 2011. *© ruchos/iStock*

Finally, analysis of the state and trends regarding the multilateral environmental agreements identified in Table 4.1 indicates a continuing major need for research, policy development and implementation on the national and international level. Data collection, including of gender-specific data, also requires greater attention, especially regarding the impacts of extreme weather events – storms, floods and droughts – on affected people. This should form a basis for future policy development, adoption and implementation to enhance the security and livelihoods of all affected by such events, including women, children and the elderly (Bennett *et al.* 2005). Although this assessment is limited by many data and information gaps, sufficient information exists to develop useful policy action to address the water and related land issues identified in this chapter.

Table 4.3 Progress towards goals (see Table 4.1)

A: Significant progress	C: Very little to no progress	X: Too soon to assess progress
B: Some progress	D: Deteriorating	?: Insufficient data

Key issues and goals		State and trends	Outlook	Gaps
1. Ecosystem				
Protect and restore freshwater ecosystems and their services	?	Half of countries have made significant progress towards developing and implementing integrated water resources management plans. It is unclear, however, how many reflect improved management of freshwater ecosystems. Many medium and large-scale dams have been constructed since 1990, especially in developing countries, disrupting the hydrologic regime integral to freshwater ecosystem function *See Chapter 5*	Pressure to build more dams and irrigation infrastructure will continue to be driven by increasing demand for energy and food	Global data on the state of freshwater ecosystems; quantifiable targets for ecosystem preservation and restoration from acute and chronic impacts
Protect and restore marine ecosystems and their services	D/B	D: Ocean warming and acidification is accelerating and stressing marine ecosystems, particularly coral reefs; 415 coastal areas are eutrophic of which 169 have hypoxic dead zones B: There are 18 regional seas conventions and action plans involving 143 countries that aim to improve ecosystem health, among other goals; 64 large marine ecosystems cover the world's coastal regions, some being effectively managed, while others lack funding and commitment of participating countries, resulting in slow progress B: See Chapter 5 regarding protected areas	Many tropical coral reefs could rapidly die by 2050 due to ocean acidification and warming; other significant threats to marine ecosystems include land-based pollution and lack of governance of high seas	pH target for oceans

Table 4.3 Progress towards goals (see Table 4.1) *continued*

1. Ecosystem *continued*				
Conserve and improve management of wetlands	D	See Chapters 3 and 5		
Ensure environmental water needs	D	Human water consumption jeopardizes ecosystems by utilizing environmental flows in nearly a third of major river basins	Expected to get worse as water demands increase	Data on monthly environmental flows required to maintain ecosystem services at the basin level; legal recognition of environmental water needs (Part 2); target to define and ensure that minimum environmental water requirements are met at the basin level; incorporation of environmental flows into basin allocation schemes
2. Human well-being				
Reduce water-related human health hazards	B	Increased access to improved water supply and sanitation has reduced water-related human health hazards globally and there have been notable successes in reducing some water related diseases; nevertheless, 3.5 million people still die each year from water-related diseases as of 2004; the frequency of paralytic shellfish poisoning has increased by a factor of five since 1970	Projected to continue improving access to water supply and sanitation. Africa projected to lag behind the rest of the world	Updated water-related disease and hazards data; mechanism for strict implementation of MDG at local scales
Ensure equitable access to improved drinking water supply	A/B	A: Population without access to improved drinking water supply has been reduced from 23% in 1990 to 13% in 2008 and is projected to be 9% by 2015 B: More improvement has been made in urban than rural communities, leaving large inequities in access; the reliability and quality of water supplies are of concern in many areas	Population without access to improved drinking water supply is projected to be 9% by 2015, meeting the related MDG	Data on safe (not just improved) drinking water access by region; mechanism for strict implementation of MDG; agreed definition of equitable
Secure adequate and sustainable freshwater supply	D/B	D: Global water withdrawals have tripled over the last 50 years to meet increasing demands, with groundwater particularly at risk; 80% of people live in areas with high levels of threats to water security, including 3.4 billion people in the most severe threat category B: Construction of dams is improving access to freshwater supply in many developing countries	More people are likely to experience more severe water stresses in coming decades; planetary boundaries for freshwater use are expected to be reached in the coming decades	Water security metric defined and data developed to allow tracking of trends over time (groundwater recharge; global withdrawals and consumption from the energy sector; global overlays of water scarcity and demand from energy); agreed definition of water security and related metrics
Develop programmes for mitigating the effects of extreme water-related events	B/D	B: Many governments report major progress towards implementation of disaster risk reduction strategies D: The number of drought and flood disasters increased by 38% and 230% respectively from the 1980s to the 2000s, while the number of people exposed to floods increased by 114%	Increased precipitation intensity and aridity is expected to accentuate extreme water-related events in many parts of the world	Holistic cost-benefit analyses of various adaptation and mitigation measures and impact analyses of mitigation efforts; policy integration, both horizontal (e.g. between sectors) and vertical (e.g. from global to regional to local); risk management strategies for vulnerable communities
Mitigate and adapt to adverse effects of climate change on the water environment	B/C	B: Broad adaptation tools, scenario-based approaches and adaptive management are being formulated at multiple scales; planned interventions in the water sector can be also be found in national adaptation programmes of action (NAPAs) of the least developed countries; 35% of World Bank water projects during 2006–2008 included mitigation and adaptation measures for climate change C: The costs of adapting to climate change are additional to those required to meet current MDG targets on water and sanitation, which are themselves underfunded	As scientific uncertainty is reduced at regional and local levels and awareness increases, mitigation and adaptation measures are expected to increase; climate change adaptation costs for the water sector and sea level rise will be at least US$35–100 billion per year	Reporting of mitigation and adaptation outcomes; monitoring and early warning for water-related climate extremes; long-term observatories for monitoring changes to the hydrologic cycle as a result of climate change
3. Water use efficiency				
Improve the efficient use of water resources	B	Irrigation efficiency is poor in many regions; irrigation technologies have become more efficient but have not been widely applied; some efficiency improvements have occurred through virtual water trade	The rate of implementation of water efficiency is not on track to keep pace with growing demand; virtual water trade could help efficiently redistribute water	Water resource efficiency trend data by sector (including energy sector) and country; virtual water trade trend data; efficiency impacts of virtual water trade; quantitative efficiency targets by sector; water allocation efficiency including environmental flows
4. Water quality				
Reduce and control freshwater pollution	?/C	No global datasets of freshwater quality have been available to assess overall trends; there have been some local water quality improvements but faecal coliforms in at least parts of most major river systems exceed WHO standards for drinking; gross algal and macrophyte productivity in lakes has increased by 74% globally	No outlook data identified	Global and regional data on sediment, nutrients, marine litter, toxic chemicals and emerging contaminants; rigorous global and regional water quality index based on comprehensive long-term data; water quality standards and targets for emerging contaminants

Table 4.3 Progress towards goals (see Table 4.1) *continued*

4. Water quality *continued*				
Reduce and control marine pollution	D/C/B	D: At least 415 coastal areas with serious eutrophication. Global nutrient run-off increased by approximately 15% since 1970. C: No statistically significant changes in the quantity of coastal or marine litter, although data are scarce for many regions B: Reduction of many contaminants in fish tissue; notable recent contamination events include the Fukushima nuclear crisis in Japan and the Deepwater Horizon oil spill in the Gulf of Mexico	Nitrogen loads to oceans are projected to increase from 43.2 million tonnes per year in 2000 to 45.5 million tonnes per year in 2030	Global and regional data on sediment, nutrients, marine litter, toxic chemicals and emerging contaminants
Improve sanitation coverage, including sewage collection, treatment and disposal	B	The population with access to improved sanitation increased from 54% to 61% during 1990–2008, though improvements are bypassing the poorest and most rural communities; 2.6 billion people (1 in 2.5) were without access to improved sanitation in 2008	Not globally on track to meet the MDG target of halving the proportion of people without access to improved sanitation	Water security metric defined and data developed to allow tracking of trends over time (groundwater recharge; global withdrawals and consumption from the energy sector; global overlays of water scarcity and demand from energy); agreed definition of water security and related metrics
5. Institutional and legal				
Recognize the economic value of water	?	See Chapter 5 for discussion of ecosystem services; Chapters 10, 11 and 12 give examples of water pricing schemes and market-based solutions that reflect the value of water and aquatic ecosystems	No outlook data identified	Data on scope, magnitude and value of water-related ecosystem services (e.g. value of wetlands as buffers against extreme events); goals and targets recognizing, protecting and valuing ecosystem services for human and environmental health and well-being
Develop and enforce legal frameworks and regulations	B	UNCLOS was ratified by 160 countries and the Global Programme of Action (GPA) adopted by 108 countries; legal frameworks for industrial and municipal wastewater discharge exist in most developed countries although non-point-source regulations lag behind; governance of areas beyond national boundaries is weak and fragmented; enforcement remains an issue in many regions.	No outlook data identified	Capacity to effectively assess and regulate environmental impacts beyond national jurisdictions
Strengthen institutional coordination mechanisms	B	Two-thirds of transboundary water-related events are cooperative, although the number of water conflicts has increased since the 1970s; 295 international water agreements have been signed since 1948; less than 20% of the 106 basins with water institutions have multilateral agreements in effect; 143 countries participate in 18 regional seas programmes, and the large marine ecosystem approach has delineated 64 management units globally	No outlook data identified	Metrics of coordination effectiveness
6. Water resources management				
Develop and implement integrated management strategies and plans	B/?	There has been an increased recognition of the need for integrated approaches for freshwater and marine system management; about half of countries have made significant progress towards developing and implementing integrated approaches to water resources management and water efficiency, but the 2002 WSSD target is far from being met; implementation is slowed by financial, legal and/or capacity barriers; there is insufficient data to evaluate the long-term effectiveness of integrated water resource management	Developing countries in particular will face difficulties implementing integrated management approaches due to lack of funding, capacity and governance	Reporting mechanism and meaningful governance indicators for countries' progress towards integrated water resources management, including the effectiveness of such approaches; implementation of policy goals
Develop adequate monitoring systems (national, regional and global)	C/D	Data are fragmented, lacks complete global coverage or is not regularly updated; marine monitoring and remote sensing data acquisition has increased, but global freshwater monitoring has declined and is now inadequate; modelling and remote sensing are complementing monitoring in many instances, but still rely on adequate data	Comprehensive monitoring systems will continue to be limited by financing and capacity	Metadata on existing data; agreed quantitative targets on comprehensive monitoring and reporting systems
Improve stakeholder participation and mainstream gender in water management	?	No quantitative global data are available to assess this goal; stakeholder engagement and gender mainstreaming is becoming more common globally, but is still lacking in many regions	No outlook data available	Data to assess stakeholder participation, including roles of women and men, and separating data by sex; institutionalized stakeholder participation; systematic gender impact assessment
Improve groundwater management	C/D/?	C: Arsenic and nitrates threaten aquifers in many countries D: Many aquifers are being drawn down at unsustainable rates; efficient management requires more data for quantitative assessment of the problem ?: Transboundary groundwater systems have been largely ignored due largely to insufficient data and lack of agreement	No outlook data available	Global level datasets on groundwater contamination, availability and withdrawal; transboundary management of groundwater resources (precluded by the data gap)

REFERENCES

2030 Water Resources Group (2009). *Charting our Water Future: Economic Frameworks to Inform Decision-Making.* http://www.mckinsey.com/App_Media/Reports/Water/Charting_Our_Water_Future_Full_Report_001.pdf

Alcamo, J., van Vuuren, D.P. and Cramer, W. (2005a). Change in ecosystem services and their drivers across the scenarios. In *Ecosystems and Human Well-being: Scenarios.* Volume 2 (eds. Carpenter, S.R., Pingali, P., Bennett, E.M. and Zurek, M.B.). Island Press, Washington, DC

Alcamo, J., van Vuuren, D., Ringler, C., Cramer, W., Masui, T., Alder, J. and Schulze, K. (2005b). Changes in nature's balance sheet: model-based estimates of future worldwide ecosystem services. *Ecology and Society* 10(2), 19

Ali, M.H. (2010). *Fundamentals of Irrigation and On-Farm Water Management* Volume 1, and *Practices of Irrigation and On-Farm Water Management* Volume 2. Springer Science+Business Media, New York, NY

AMAP (2007). *Arctic Oil and Gas 2007: Overview Report of the Assessment of Oil and Gas Activities in the Arctic.* Arctic Monitoring and Assessment Programme, Oslo. http://www.amap.no/oga/

AMCOW (2008). *Roadmap for the Africa Groundwater Commission.* African Ministers' Council on Water. UNEP/UNESCO/UWC, Nairobi

Anderson, D.M., Reguera, B., Pitcher, G.C. and Enevoldsen, H.O. (2010). The IOC international harmful algal bloom program: history and science of impacts. *Oceanography* 23, 72–85

Antonov, J.L., Levitus, S. and Boyer, T.P. (2005). Thermostatic sea level rise, 1955–2003. *Geophysical Research Letters* 32, L12602

Bakkes, J.A. and Boschet, P.R. (eds.) (2008). *Background Report to the OECD Environmental Outlook to 2030: Overviews, Details, and Methodology of Model-based Analysis.* MNP Report 500113001. Netherlands Environmental Assessment Agency (Milieu-en Natuurplanbureau) and Organisation for Economic Co-operation and Development, Paris

Bates, B.C., Kundzewicz, Z.W., Wu, S. and Palutikof, J.P (eds.) (2008). *Climate Change and Water.* Technical paper of Intergovernmental Panel on Climate Change. IPCC Secretariat, Geneva

Bennett, V., Dávila-Poblete, S. and Rico, M.N. (2005). *Opposing Currents: The Politics of Water and Gender in Latin America.* University of Pittsburg Press, Pittsburg, PA

Bird, K.J., Charpentier, R.R., Gautier, D.L., Houseknecht, D.W., Klett, T.R., Pitman, J.K., Moore, T.E., Schenk, C.J., Tennyson, M.E. and Wandrey, C.J. (2008). *Circum-Arctic Resource Appraisal: Estimates of Undiscovered Oil and Gas North of the Arctic Circle.* US Geological Survey Fact Sheet 2008-3049. http://pubs.usgs.gov/fs/2008/3049/

Biswas, A.K. (2004). Integrated water resources management: a re-assessment. *Water International* 29(2), 248–256

Biswas, A. and Tortajada, C. (2011). Water quality management: an introductory framework. *Water Resources Development* 27(1), 5–11

Boelee, E. (ed.) (2011). *Ecosystems for Water and Food Security.* United Nations Environment Programme, Nairobi and International Water Management Institute, Colombo

Brauch, H.G., Oswald Spring, U., Grin, J., Mesjasz, C., Kameri-Mbote, P., Behera, N.C., Chourou, B. and Krummenacher, H. (eds.) (2009). *Facing Global Environmental Change: Environmental, Human, Energy, Food, Health and Water Security Concepts.* Springer-Verlag, Berlin; Heidelberg; New York

Brunt, R., Vasak, L. and Griffioen, J. (2004). *Arsenic in Groundwater: Probability of Occurrence of Excessive Concentration on Global Scale.* Report SP 2004-1. International Groundwater Resource Centre (IGRAC), Delft

Cazenave, A. and Llovel, W. (2010). Contemporary sea level rise. *Annual Review of Marine Science* 2, 145–173

CBD (1997). *Jakarta Mandate on Marine and Coastal Biological Diversity.* Secretariat of the Convention on Biological Diversity. http://www.cbd.int/doc/meetings/mar/jmem-01/official/jmem-01-02-en.pdf

Chapagain, A.K. and Hoekstra, A.Y. (2008). The global component of freshwater demand and supply: an assessment of virtual water flows between nations as a result of trade in agricultural and industrial products. *Water International* 33(1), 19–32

Chao, B.F., Wu, Y.H. and Li, Y.S. (2008). Impact of artificial reservoir water impoundment on global sea level. *Science* 320(5), 212–214

Comprehensive Assessment of Water Management in Agriculture (2007). *Water for Food, Water for Life: A Comprehensive Assessment of Water Management in Agriculture.* Earthscan, London and International Water Management Institute, Colombo

Davis, J.A., Hunt, J.A., Greenfield, B.K., Fairey, R., Sigala, M., Crane, D.B., Regalado, K. and Bonnema, A. (2003). *Contaminants in Fish from the San Francisco Bay 2003.* SFEI Contribution 432. San Francisco Estuary Institute, Oakland, CA

De Stefano, L., Edwards, P., de Silva, L. and Wolf, A.T. (2010). Tracking cooperation and conflict in international basins: historic and recent trends. *Water Policy* 12, 871–884

Diaz, R.J., Selman, M. and Chique-Canache, C. (2010). *Global Eutrophic and Hypoxic Coastal Systems: Eutrophication and Hypoxia – Nutrient Pollution in Coastal Waters.* World Resources Institute, Washington, DC. http://www.wri.org/project/eutrophication

Dyurgerov, M.B. and Meier, M.F. (2005). *Glaciers and the Changing Earth System: A 2004 Snapshot.* Occasional Paper 58. Institute of Arctic and Alpine Research, University of Colorado, Boulder, CO

EIA (2011). *World Shale Gas Resources: An Initial Assessment of 14 Regions Outside the United States.* http://www.eia.gov/analysis/studies/worldshalegas/pdf/fullreport.pdf

EM-DAT (2011). *EM-DAT: The OFDA/CRED International Disaster Database.* Université Catholique de Louvain, Brussels. www.emdat.be

Falkenmark, M. and Rockström, J. (2004). *Balancing Water for Humans and Nature: The New Approach in Ecohydrology.* Earthscan, London

FAO (2008). *FAO-Aquastat: Proportion of Renewable Water Resources Withdrawn (MDG Water Indicator).* Food and Agriculture Organization of the United Nations, Rome. http://www.fao.org/nr/water/aquastat/globalmaps/index.stm (accessed May 2011)

Feely, R.A., Doney, S.C. and Sarah, R. (2009). Ocean acidification: present conditions and future changes in a high-CO_2 world. *Oceanography* 22(4), 36–47

Foster, S., Garduno, H., Kemper, K., Tuinhof, A., Nanni, M. and Dumars, C. (2006). *Groundwater Quality Protection: Defining Strategy and Setting Priorities.* Briefing Note Series 8. World Bank, Washington, DC

Galgani, F., Leaute, J.P., Moguedet, P., Souplet, A., Verin, Y., Carpentier, A., Goraguer, H., Latrouite, D., Andral, B., Cadiou, Y., Mahe, J.C., Poular, J.C. and Nerisson, P. (2000). Litter on the sea floor along European coasts. *Marine Pollution Bulletin* 40(6), 516–527. http://www.sciencedirect.com/science/article/pii/S0025326X99002349

Garrick, D., Siebentritt, M.A., Aylward, B., Bauer, D.C.J. and Purkey, A. (2009). Water markets and freshwater services: policy reform and implementation in the Columbia and Murray-Darling Basins. *Ecological Economics* 69, 366–379

GESAMP (2010). *Proceedings of the GESAMP International Workshop on Plastic Particles as a Vector in Transporting Persistent, Bio-accumulating and Toxic Substances in the Oceans.* GESAMP Rep. Stud. No. 82 (eds. Bowmer, T. and Kershaw, P.J.). IMO/FAO/UNESCO-IOC/UNIDO/WMO/IAEA/UN/UNEP Joint Group of Experts on the Scientific Aspects of Marine Environmental Protection

Gleick, P.H. (2003). Global freshwater resources: soft-path solutions for the 21st century. *Science* 302, 1524–1528

Gleick P.H. and Palaniappan, M. (2010). Peak water limits to freshwater withdrawal and use. *Proceedings of the National Academy of Sciences of the United States of America* 107, 11155–11162

Glennie, P., Lloyd, G.J. and Larsen, H. (2010). *The Water-Energy Nexus: The Water Demands of Renewable and Non-Renewable Electricity Sources.* DHI, Hørsholm

Global Water Partnership (2006). *Setting the Stage for Change: Second Informal Survey by the GWP Network g*
Giving the Status of the 2005 WSSD Target on National Integrated Water Resources Management and Water Efficiency Plans. Global Water Partnership, Stockholm

Global Water Partnership (2000). *Integrated Water Resources Management.* Background Paper No. 4. Technical Advisory Committee, Global Water Partnership, Stockholm

Gordon, L.J., Steffen, W., Jonsson, B.F., Folke, C., Falkenmark, M. and Johannessen, A. (2005). Human modification of global water vapor flows from the land surface. *Proceedings of the National Academy of Sciences of the United States of America* 102, 7612–7617

Gorman, P.A. and Schneider,T. (2009). The physical basis for increases in precipitation extremes in simulations of 21st century climate change. *Proceedings of the National Academy of Sciences of the United States of America* 106(35), 14773–14777

Greenberg, M.R. (2009). Water, conflict, and hope. *American Journal of Public Health* 99(11), 1928–1930

Hassellöv, M., Readman, J.W., Ranville, J.F. and Tiede, K. (2008). Nanoparticle analysis and characterization methodologies in environmental risk assessment of engineered nanoparticles. *Ecotoxicology* 17(5), 344–361

HELCOM (2009). *Marine Litter in the Baltic Sea Region: Assessment and Priorities for Response.* Helsinki Commission, Baltic Marine Environment Protection Commission.

Hoegh-Guldberg, O., Mumby, P.J., Hooten, A.J., Steneck, R.S., Greenfield P., Gomez, E., Harvell, C.D., Sale, P.F., Edwards, A.J., Caldeira, K., Knowlton, N., Eakin, C.M., Iglesias-Prieto, R., Muthiga, N., Bradbury, R.H., Dubi, A. and Hatziolos, M.E. (2007). Coral reefs under rapid climate change and ocean acidification. *Science* 318, 1737–1742

Hoekstra, A.Y. and Mekonnen, M.M. (2011). *Global Water Scarcity: Monthly Blue Water Footprint Compared to Blue Water Availability for the World's Major River Basins.* Value of Water Research Report Series No.53. UNESCO-IHE, Delft

International Lake Environment Committee (2006). *Managing Lakes and their Basins for Sustainable Use: A Report for Lake Basin Managers and Stakeholders.* International Lake Environment Committee Foundation, Kusatsu

IMO (1972) *Convention on the Prevention of Marine Pollution by Dumping of Wastes and Other Matter.* International Maritime Organization. http://www.ecolex.org/server2.php/libcat/docs/TRE/Multilateral/En/TRE000420.txt

IOM (2010). *Disaster Risk Reduction, Climate Change Adaptation and Environmental Migration: A Policy Perspective.* International Organization for Migration, Geneva

IPCC (2011). Summary for policymakers. In *Intergovernmental Panel on Climate Change Special Report on Managing the Risks of Extreme Events and Disasters to Advance Climate Change Adaptation* (eds. Field, C.B., Barros, V., Stocker, T.F., Qin, D., Dokken, D., Ebi, K.L., Mastrandrea, M.D., Mach, K.J., Plattner, G.-K., Allen, S., Tignor, M., Midgley, P.M.). Cambridge University Press, Cambridge

IPCC (2007a). *Climate Change 2007: The Physical Science Basis.* Working Group I contribution to the Fourth assessment report of the Intergovernmental Panel on Climate Change. Cambridge University Press, Cambridge

IPCC (2007b). *Climate Change 2007: Synthesis Report.* Contribution of Working Groups I, II and III to the Fourth assessment report of the Intergovernmental Panel on Climate Change (eds. Pachauri, R.K. and Reisinger, A.). IPCC, Geneva

IPCC (2007c). *Climate Change 2007: Impacts, Adaptation and Vulnerability.* Working Group II contribution to the Fourth assessment report of the Intergovernmental Panel on Climate Change. Cambridge University Press, Cambridge

Ishii, M. and Kimoto, M. (2009). Re-evaluation of historical ocean heat content variations with varying XBT and MBT depth bias corrections. *Journal of Oceanography* 65(3), 287–299. doi:10.1007/s10872-009-0027-7

ITOPF (2010). Oil tanker spill statistics. International Tanker Owners Pollution Federation Ltd. http://www.itopf.com/information-services/data-and-statistics/statistics/index.html

Johnson, B.M., Kanagy, L.E., Rodgers, J.H. and Castle, J.W. (2007). Chemical, physical, and risk characterization of natural gas storage produced waters. *Water, Air and Soil Pollution* 191, 33–54

Karakezi, S., Kimani, J., Onguru, O. and Kithyoma, W. (2009). *Large Scale Hydropower, Renewable Energy and Adaptation to Climate Change: Climate Change and Energy Security in East and Horn of Africa.* Energy, Environment and Development Network for Africa (AFREPEN/FWD), Nairobi. http://www.boell.or.ke/downloads/RenewableEnergyandAdaptationtoClimateChangePublication.pdf (accessed 1 September 2010) and www.afrepren.org/Pubs/Occasional_Papers/pdfs/OP33.pdf

Kelly, E.N., Schindler, D.W., Rodson, P.V., Short, J.W., Radmanovich, R. and Nielsen, C.C. (2010). Oil sands development contributes elements toxic at low concentrations to the Athabasca River and its tributaries. *Proceedings of the National Academy of Sciences of the United States of America* 107(37), 16178–16183

Kipping, M. (2009). Water security in the Senegal River basin: water cooperation and water conflicts. In *Facing Global Environmental Change: Environmental, Human, Energy, Food, Health and Water Security* (eds. Brauch, H.G., Oswald Spring, U., Grin, J., Mesjasz, C., Kameri-Mbote, P., Behera, N.C., Chourou, B. and Krummenacher, H.). pp. 675–684. Springer-Verlag, Berlin; Heidelberg; New York

Kleinen, T. and Petschel-Held, G. (2007). Integrated assessment of changes in flooding probabilities due to climate change. *Climatic Change* 81, 283–312

Kundzewicz, Z.W. and Kowalczak, P. (2009). The potential for water conflict is on the increase. *Nature* 459, 31

Kundzewicz, Z.W., Mata, I.J., Arnell, N.W., Döll, P., Kabat, P., Jiménez, B., Miller, K.A, Oki, T., Sen, Z. and Shiklomanov, I.A. (2007). Freshwater resources and their management. In *Climate change 2007: impacts, adaptation and vulnerability.* Working Group II contribution to the Fourth assessment report of the Intergovernmental Panel on Climate Change (eds. Parry, M.I., Canziani, O.F., Palutikoj, J.P., van der Lindenand, P.J. and Hanson, C.E.). pp.173–210. Cambridge University Press, Cambridge

Langdon C. and Atkinson, M.J. (2005). Effect of elevated pCO_2 on photosynthesis and calcification of corals and interactions with seasonal change in temperature/irradiance and nutrient enrichment. *Journal of Geophysical Research* 110, C09S07

Lansky, L. and Uitto, J.I. (eds.) (2005). *Enhancing participation and governance in water resources management: conventional approaches and information technology.* United Nations University Press, Tokyo; New York; Paris

Law, K.L., Morét-Ferguson, K., Maximenko, S., Proskurowski, N.A., Peacock, E.E., Hafner, J. and Reddy, C.M. (2010). Plastic accumulation in the North Atlantic subtropical gyre. *Science* 329(5996), 1185–1188

Lenton, R. and Muller, M. (2009). *Integrated Water Resources Management in Practice: Better Water Management for Development.* Earthscan, London

Levitus, S., Antonov, J.L., Boyer, T.P., Locarnini, R.A., Garcia, H.E. and Mishonov, A.V. (2009). Global ocean heat content 1955–2008 in light of recently revealed instrumentation. *Geophysical Research Letters,* 36

Lewis, W.M. (2011). Global primary production of lakes: 19th Baldi Memorial Lecture. *Inland Waters* (in press)

Logan, C.A. (2010). A review of ocean acidification and America's response. *Bioscience* 60, 819–828

Lugeri, N., Kundzewicz, Z.W., Genovese, E., Hochrainer, S. and Radziejewski, M. (2010). River flood risk and adaptation in Europe – assessment of the present status. *Mitigation and Adaptation Strategies for Global Change* 15, 621–639
MA (2005). *Ecosystems and Human Well-Being: Wetlands and Water Synthesis.* Millennium Ecosystem Assessment. World Resources Institute, Washington, DC

MARPOL (2011). *International Convention for the Prevention of Pollution from Ships (MARPOL).* http://www.imo.org/about/conventions/listofconventions/pages/international-convention-for-the-prevention-of-pollution-from-ships-%28marpol%29.aspx

Martinez, E., Maamaatuaiahutapu, K. and Taillandier, V. (2009). Floating marine debris surface drift: convergence and accumulation toward the South Pacific subtropical gyre. *Marine Pollution Bulletin*, 58(9), 1347–1355

Matthews, J., Wickel, B. and Freeman, S. (2011). Converging currents in climate-relevant conservation: water, infrastructure, and institutions. *PLOS Biology* 9(9), e1001159

McGranahan, G., Balk, D. and Anderson, B. (2007). The rising tide: assessing the risks of climate change and human settlements in low elevation coastal zones. *Environment and Urbanization* 19, 17–37

Mekonnen, M.M. and Hoekstra, A.Y. (2011). *National Water Footprint Accounts: The Green, Blue and Grey Water Footprint of Production and Consumption.* Value of Water Research Report Series No. 50. UNESCO-IHE, Delft

National Commission on the BP Deepwater Horizon Oil Spill and Offshore Drilling (2011). *Deep Water: The Gulf Oil Disaster and the Future of Offshore Drilling.* Report to the President, United States of America

Navarro, E., Baun, A., Behra, R., Hartmann, N.B., Filser, J., Miao, A.J., Quigg, A., Santshi, P.H. and Sigg, L. (2008). Environmental behaviour and ecotoxicity of engineered nanoparticles to algae, plants, and fungi. *Ecotoxicology* 17, 372–386

OECD (2008). *Environmental Outlook to 2030.* Organisation for Economic Co-operation and Development, Paris

Osborn, S.G., Vengosh, A., Warnder, N.R. and Jackson, R.B. (2011). Methane contamination of drinking water accompanying gas-well drilling and hydraulic fracturing. *Proceedings of the National Academy of Sciences of the United States of America* 108(20), 8172–8176. http://www.pnas.org/content/early/2011/05/02/1100682108

Oshihoi, T., Isobe, T., Takahashi, S., Kubodera, T. and Tanabe, S. (2009). Contamination status of organohalogen compounds in deep-sea fishes in northwest Pacific ocean off Tohoku, Japan. In *Interdisciplinary Studies on Environmental Chemistry – Environmental Research in Asia* (eds. Obayashi, Y., Isobe, T., Subramanian, A., Suzuki, S. and Tanabe S.). pp. 67–72. Terrapub, Tokyo

OSPAR (2009). *Marine Litter in the North-East Atlantic Region: Assessment and Priorities for Response.* OSPAR Commission, London.

Osti, R., Hishinuma, S., Miyake, K. and Inomata, H. (2011). Lessons learned from statistical comparison of flood impact factors among southern and eastern Asian countries. *Journal of Flood Risk Management* 4(3), 203–215

Oswald Spring, U. (2007). Hydro-diplomacy: opportunities for learning from an interregional process. In *Integrated Water Resources Management and Security in the Middle East* (eds. Lipchin, C., Pellant, E., Saranga, D. and Amster, A.). pp.163–200. Springer, Dordrecht

Oswald Spring, U. and Brauch, H.G. (2009). Securitizing water. In *Facing Global Environmental Change: Environmental, Human, Energy, Food, Health and Water Security Concepts* (eds. Brauch, H.G., Oswald Spring, U., Grin, J., Mesjasz, C., Kameri-Mbote, P., Behera, N.C., Chourou, B. and Krummenacher, H.). Springer-Verlag, Ebook at SpringerLink

Pacific Institute (2011). *Water Conflict Chronology List.* http://www.worldwater.org/conflict/list/

Parry, M., Arnell, N., Berry, P., Dodman, D., Fankhauser, S., Hope, C., Kovats, S., Nicholls, R., Satterthwaite, D., Tiffin, R. and Wheeler, T. (2009). *Assessing the Costs of Adaptation to Climate Change: A Review of the UNFCCC and Other Recent Estimates.* International Institute for Environment and Development and Grantham Institute for Climate Change, London

Pereira, L.A.S., Cordery, I. and Iacovides, I. (2009). *Coping with Water Scarcity: Addressing the Challenges.* Springer Science

Perret, S., Stefano, F. and Rashid, H. (eds.) (2006). *Water Governance for Sustainable Development: Approaches and Lessons from Developing and Transitional Countries.* Earthscan, London

Placht, M. (2007). Integrated water resource management: incorporating integration, equity, and efficiency to achieve sustainability. *International Development, Environment and Sustainability* 3. http://fletcher.tufts.edu/ierp/ideas/issue3.html

Portmann, F.T., Siebert, S. and Döll, P. (2010). MIRCA 2000 – Global monthly irrigated and rainfed crop areas around the year 2000: a new high-resolution data set for agricultural and hydrological modeling. *Global Biogeochemical Cycles* 24, GB1011. doi:10.1029/2008GB003435

Prüss-Üstün, A., Bos, R., Gore, F. and Bartram, J. (2008). *Safer Water, Better Health: Costs, Benefits and Sustainability of Interventions to Protect and Promote Health.* World Health Organization, Geneva. http://www.who.int/quantifying_ehimpacts/publications/saferwater/en/index.html

Rabalais, N.N., Diaz, R.J., Levin, L.A., Turner, R.E., Gilbert, D. and Zhang, J. (2010). Dynamics and distribution of natural and human-caused hypoxia. *Biogeosciences* 7, 585–619

RCSE-SU and ILEC (2011). *Development of ILBM Platform Process: Evolving Guidelines through Participatory Improvement.* Research Center for Sustainability and Environment, Shiga University and International Lake Environment Committee, Kusatsu

Ribic, C.A., Sheavly, S.B., Rugg, D.J. and Erdmann, E.S. (2010). Trends and drivers of marine debris on the Atlantic coast of the United States 1997–2007. *Marine Pollution Bulletin* 60, 1231–1242

Rignot, E. (2008). Changes in West Antarctic ice dynamics observed with ALOS PALSAR. *Geophysical Research Letters* 35, L12505

Rignot, E., Velicogna, I., van den Broeke, M.R., Monaghan, A. and Lenaerts, J. (2011). Acceleration of the contribution of the Greenland and Antarctic ice sheets to sea level rise. *Geophysical Research Letters* 38, L05503

Rockström, J., Stefen, W., Noone, K., Persson, A., Chapin, F.S., Lambin, E.F., Lenton, T.M., Scheffer, M., Folke, C., Schellnhuber, H.J., Nykvist, B., de Wit, C.A., Hughes, T., van der Leeuw, S., Rodhe, H., Sörlin, S., Snyder, P.K., Costanza, R., Svedin, U., Falkenmark, M., Karlberg, L., Corell, R.W., Fabry, V.J., Hansen, J., Walker, B., Liverman, D., Richardson, K., Crutzen, P. and Foley, J.A. (2009). A safe operating space for humanity. *Nature* 461, 472–475

Rohwer, J., Gerten, D. and Lucht, W. (2007). *Development of Functional Irrigation Types for Improved Global Crop Modelling*. PIK Report No. 104. Potsdam Institute for Climate Impact Research, Potsdam

Rosenfeld, D., Lohmann, U., Raga, G.B., O'Dowd, C.D., Kulmala, M., Fuzzi, S., Reissell, A. and Andreae, M.O. (2008). Flood or drought: how do aerosols affect precipitation? *Science* 321(5894), 1309–1313

Rothman, D., Agard, J. and Alcamo, J. (2007). The future today. In *Global Environment Outlook-4 (GEO-4)*. pp.395–454. United Nations Environment Programme, EarthPrint, Stevenage

Ryan, P.G., Moore, C.J., van Franeker, J.A. and Moloney, C.L. (2009). Monitoring the abundance of plastic debris in the marine environment. *Philosophical Transactions of the Royal Society* B 364, 1999–2012

Sauer, A., Klop, P. and Agrawal S. (2010). *Over Heating: Financial Risks from Water Constraints on Power Generation in Asia: India, Malaysia, Philippines, Thailand, Vietnam*. World Resources Institute, Washington, DC

Schwarzenbach, R.P., Egli, T., Hofstetter, T.B., von Gunten, U. and Wehrli, B. (2010). Global water pollution and human health. *Annual Review of Environment and Resources* 35, 109–136

Seitzinger, S.P., Mayorga, E., Bouwman, A.F., Kroeze, C., Beusen, A.H.W., Billen, G., Van Drecht, G., Dumont, E., Fekete, B.M., Garnier, J. and Harrison, J.A. (2010). Global river nutrient export: a scenario analysis of past and future trends. *Global Biogeochemical Cycles* 24, GBOA08

Sheavly, S.B. (2007). *National Marine Debris Monitoring Program: Final Program Report, Data Analysis and Summary*. Ocean Conservancy, Washington, DC

Stanners, D., Bosch, P., Dom, A., Gabrielsen, P., Gee, D., Martin, J., Rickard, L. and Weber, J.-L. (2007). Frameworks for environmental assessment and indicators at the EEA. In *Sustainability Indicators – A Scientific Assessment* (eds. Hák, T., Moldan, B. and Dahl, A.). Island Press, Washington, DC.

UNCED (1992). *Agenda 21* (Chapter 18). United Nations Conference on Environment and Development. http://www.un.org/esa/sustdev/documents/agenda21/english/Agenda21.pdf

UNCLOS (1982). *United Nations Convention on the Law of the Sea*. http://www.un.org/depts/los/convention_agreements/texts/unclos/unclos_e.pdf

UNDESA (2010). *Millennium Development Goals Report*. United Nations Department of Economic and Social Affairs, New York. http://mdgs.un.org/unsd/mdg/Resources/Static/Products/Progress2010/MDG_Report_2010_En.pdf

UNDP (2006). *Human Development Report 2006. Beyond Scarcity: Power, Poverty and the Global Water Crisis*. United Nations Development Programme, New York. http://undp.org/en/media/HDR06-complete.pdf (accessed February 2010)

UNEP (2009). *Marine Litter: A Global Challenge*. United Nations Environment Programme, Nairobi

UNEP-GEMS/Water Programme (2008). *Water Quality for Ecosystem and Human Health*, 2nd ed. United Nations Environment Programme GEMS/Water Programme, Burlington. http://www.gemswater.org/publications/pdfs/water_quality_human_health.pdf

UNESCO (2009). *Water in a Changing World*. 3rd United Nations World Water Development Report. United Nations Educational, Scientific and Cultural Organization, Paris. http://webworld.unesco.org/water/wwap/wwdr/wwdr3/tableofcontents.shtml

UNESCO (2006). *Water: A Shared Responsibility*. 2nd United Nations World Water Development Report. United Nations Educational, Scientific and Cultural Organization, Paris. http://www.unesco.org/water/wwap/wwdr/wwdr2/

UNFCCC (2007). *Investment and Financial Flows to Address Climate Change*. Climate Change Secretariat, United Nations Framework Convention on Climate Change, Bonn

UNFCCC (1992). *United Nations Framework Convention on Climate Change*. http://unfccc.int/key_documents/the_convention/items/2853.php

UNISDR (2011). *Revealing Risk, Redefining Development*. 2011 Global Assessment Report on Disaster Risk Reduction. United Nations International Strategy for Disaster Risk Reduction, Geneva

UN-Water (2012). Status Report on the Application of Integrated Approaches to Water Resources Management. http://www.unwater.org/rio2012/

Vörösmarty, C.J., McIntyre, P.B., Gessner, M.O., Dudgeon, D., Prusevich, A., Green, P., Glidden, S., Bunn, S.E., Sullivan, C.A., Liermann, C.R. and Davies, P.M. (2010). Global threats to human water security and river biodiversity. *Nature*, 467(7315), 555–561

Wada, Y., van Beek, L.P.H., van Kempen, C.M., Reckman, J.W.T.M., Vasak, S. and Bierkens, M.F.P. (2010). Global depletion of groundwater resources. *Geophysical Research Letters* 37, L20402

Walton, D.A. and Ivers, L.C. (2011). Responding to cholera in post-earthquake Haiti. *New England Journal of Medicine* 364, 3–5

Watson, N., Walker, G. and Medd, W. (2007). Critical perspectives on integrated water management. *The Geographical Journal* 173(4), 297–299

WBGU (2008). *World in Transition – Climate Change as a Security Risk*. Earthscan, London. http://www.wbgu.de/wbgu_jg2007_engl.html

Wentz, F.J., Ricciardulli, L., Hilburn, K. and Mears, C. (2007). How much more rain will global warming bring? *Science* 317, 233–235

WHO (2012). WHO/UNICEF Joint Monitoring Programme (JMP) for water supply and sanitation: data resources and estimates. World Health Organization, Geneva. http://www.wssinfo.org/data-estimates/introduction

WHO (2011a). *Water-Related Diseases: Information Sheets*. Water, sanitation and health. World Health Organization, Geneva. http://www.who.int/water_sanitation_health/diseases/diseasefact/en/index.html

WHO (2011b). *Guidelines for Drinking-Water Quality*. Fourth edition. World Health Organization, Geneva. http://www.who.int/water_sanitation_health/publications/2011/9789241548151_ch07.pdf

WHO (2010). *Weekly Epidemiological Record* 85(31), 293–308. World Health Organization, Geneva

WHO (2004). *Burden of Disease (in DALYs) Attributable to Water, Sanitation and Hygiene*. World Health Organization, Geneva

WHO (2003a). Algae and cyanobacteria in fresh water. In *Guidelines for Safe Recreational Waters Volume 1: Coastal and Fresh Waters*. World Health Organization, Geneva. http://www.who.int/water_sanitation_health/bathing/srwe1-chap8.pdf

WHO (2003b). Faecal pollution and water quality. In *Guidelines for Safe Recreational Waters Volume 1: Coastal and Fresh Waters*. World Health Organization, Geneva. http://www.who.int/water_sanitation_health/bathing/srwe1/en/index.html

Willis, J., Roemmich, D. and Cornuelle, B. (2004). Interannual variability in upper-ocean heat content, temperature and thermosteric expansion on global scales. *Journal of Geophysical Research* 109, C12037

Wolf, A.T. (2007). Shared waters: conflict and cooperation. *Annual Review of Environment and Resources* 32, 3.1–3.29

World Bank (2010). *The Cost to Developing Countries of Adapting to Climate Change: New Methods and Estimates*. The Global Report of the Economics of Adaptation to Climate Change Study Consultation Draft. World Bank, Washington, DC

World Bank (2009). *Water and Climate Change: Understanding the Risks and Making Climate-Smart Investment Decisions*. World Bank, Washington, DC http://siteresources.worldbank.org/EXTNTFPSI/Resources/DPWaterClimateChangeweblarge.pdf

World Water Council (2000). *Ministerial Declaration of The Hague on Water Security in the 21st Century*. http://www.worldwatercouncil.org/fileadmin/wwc/Library/Official_Declarations/The_Hague_Declaration.pdf

WSSD (2002). *Johannesburg Plan of Implementation* (JPOI). World Summit on Sustainable Development. http://www.un.org/esa/sustdev/documents/WSSD_POI_PD/English/POIToc.htm

Zhao, M. and Running, S.W. (2010). Drought-induced reduction in global terrestrial net primary production from 2000 through 2009. *Science* 329(5994), 940–943

CHAPTER 5

Biodiversity

Coordinating lead authors: Dolors Armenteras and C. Max Finlayson

Lead authors: John Agard, Stuart H.M. Butchart, Joji Carino, William W.L. Cheung, Ben Collen, Leslie G. Firbank, Simon Hales, Marc Hockings, Robert Hoft, Justin Kitzes, Melodie A. McGeoch, Christian Prip, Thomasina E.E. Oldfield, Kent H. Redford and Heikki Toivonen

Contributing authors: Neil Burgess, Mario Baudoin, Bastian Bertzky, Nigel Dudley, Rod Fuentes, Alessandro Galli, Valerie Kapos, Linda Krueger, Yogesh Gokhale, Ashish Kothari, J. Carter Ingram, Camilo Garcia Ramirez, Dan Laffoley, Jörn P.W. Scharlemann, Damon Stanwell-Smith, John G. Robinson, Bas Verschuuren, Johanna von Braun, Kabir Bavikatte, Holly Shrumm and Monica Morales Rivas (GEO Fellow)

Principal scientific reviewer: Klement Tockner

Chapter coordinator: Alison M. Rosser and Matt J. Walpole

Main Messages

The pressure on biodiversity continues to increase. Habitat loss and degradation from agriculture and infrastructure development, overexploitation, pollution and invasive alien species remain the predominant threats. Climate change is increasing in importance and will have profound impacts, particularly in combination with other threats. Greater integration of policies and institutional responses, including effective engagement of local communities, is required to stop and reverse current trends. The world lost over 100 million hectares of forest from 2000 to 2005, and has lost 20 per cent of its seagrass and mangrove habitats since 1970 and 1980 respectively. In some regions, 95 per cent of wetlands have been lost. The condition of coral reefs globally has declined by 38 per cent since 1980. Two-thirds of the world's largest rivers are now moderately to severely fragmented by dams and reservoirs.

The state of global biodiversity is continuing to decline, with substantial and ongoing losses of populations, species and habitats. For instance, vertebrate populations have declined on average by 30 per cent since 1970, and up to two-thirds of species in some taxa are now threatened with extinction. Declines are most rapid in the tropics, in freshwater habitats and for marine species utilized by humans. Conversion and degradation of natural habitats is ongoing, with some having experienced declines of 20 per cent since 1980. Limited successes, such as saving particular species from extinction, reversing the decline of some populations, and restoring some habitats, are outweighed by continuing declines.

The benefits humans obtain from biodiversity are at risk. Conversion of natural habitats to large-scale, commercial agriculture has resulted in net benefits for human well-being. However, this has often been accompanied by reductions in other services, such as carbon sequestration and flood regulation. Continuing ecological degradation, unsustainable levels of consumption and inequities in sharing of the benefits from biodiversity threaten the improvements in human well-being and health that have been achieved in recent decades.

There has been an increase in responses to the loss and degradation of biodiversity, although these have failed to reduce the decline, and more effort is needed. Successful responses include: increases in the designation of protected areas, now covering nearly 13 per cent of land area, and increasing recognition of indigenous and local community-managed areas; and adoption of policies and actions for managing invasive alien species and genetically modified organisms (GMOs). About 55 per cent of countries have legislation to prevent the introduction of new alien species and control existing invasives, but less than 20 per cent are estimated to have comprehensive strategies and management plans, and there is a lack of data on their effectiveness. Successful responses also include regulations that support sustainable harvesting and reduced pollution; successful species recoveries and habitat restoration; and some progress towards equitable access to and benefit sharing of genetic resources. International financing for biodiversity conservation is estimated to have grown by about 38 per cent in real terms since 1992 and now stands at US$3.1 billion per year. But less than 1.5 per cent of the marine area is covered by protected areas.

An opportunity to develop a concerted global approach to stop and reverse the decline of biodiversity is provided by the recent adoption of the Strategic Plan for Biodiversity (2011–2020) including the Aichi Biodiversity Targets, and acceptance of the Nagoya Protocol on Access and Benefit Sharing.

INTRODUCTION

Biodiversity is formally defined by the Convention on Biological Diversity (CBD) as: "the variability among living organisms from all sources including, among others, terrestrial, marine and other aquatic ecosystems and the ecological complexes of which they are part; this includes diversity within species, between species and of ecosystems" (UN 1992 Article 2).

In recent years the links between biodiversity and ecosystem services and the benefits people derive from these have received increasing attention (CBD 2010b; TEEB 2010; Sutherland *et al.* 2009; UNEP 2007; MA 2005a; 2005b). There is growing evidence that biodiversity has a vital role in attaining the Millennium Development Goals: it contributes to poverty reduction and to sustaining human livelihoods and well-being through, for example, underpinning food security and human health, providing clean air and water, and supporting economic development (UNEP 2007; MA 2005a). Given the importance of biodiversity and evidence of its ongoing decline (CBD 2010b), it is essential to chart progress in reducing and, as far as possible, reversing the rate of decline.

Recent assessments of the status of biodiversity have shown little evidence of improvement. The third *Global Biodiversity Outlook* (*GBO-3*), was launched in May 2010 (CBD 2010b), and showed that biodiversity has continued to decline since publication of the Millennium Ecosystem Assessment (MA 2005a) and the last *Global Environment Outlook* (*GEO-4*) (UNEP 2007). This chapter builds on these recent assessments. The three objectives of the CBD, namely, the conservation of biological diversity, the sustainable use of its components, and fair and equitable sharing of the benefits arising out of the utilization of genetic resources, as well as the missions and objectives of other biodiversity-related conventions are all considered.

The current chapter presents globally agreed indicators and goals for biodiversity, in particular the Aichi Biodiversity Targets (Box 5.1). The implications for human well-being of not achieving these targets are examined and gaps in achieving internationally agreed goals for biodiversity are identified, so as to frame key messages for the international community. Current knowledge of the pressures, state and trends affecting biodiversity and of the benefits of biodiversity to people is synthesized from past assessments and recent publications. Management responses that address these pressures are also examined so as to chart progress in safeguarding biodiversity. In particular, cross-boundary issues are tackled from both an ecological and an equity perspective. The links between biodiversity and traditional knowledge and cultural diversity are also considered before concluding with a look to the future.

INTERNATIONALLY AGREED GOALS

Goals and targets are one aspect of the policy agenda for assessing progress in meeting global commitments for biodiversity. Eighteen goals related to biodiversity have been identified (Table 5.1 and Box 5.2). These range from the Millennium Development Goal 7 to ensure environmental sustainability, to the most recent five strategic goals and 20 Aichi Biodiversity Targets of the Strategic Plan for Biodiversity 2011–2020 (Box 5.1). These biodiversity goals and targets have been clustered into themes and prioritized by taking into account the links between them and by reference to key biodiversity issues (Table 5.1 and Box 5.2).

Lake Nakuru National Park, Kenya, renowned as a sanctuary for more than 400 bird species, also offers refuge to large ungulates including waterbucks.

© Jason Jabbour

Box 5.1 Strategic Plan for Biodiversity 2011–2020 and the Aichi Biodiversity Targets

Actor Edward Norton, UN Goodwill Ambassador for Biodiversity, addresses a press conference on the dangers of global biodiversity loss. *© Rick Bajornas/UN Photo*

The Strategic Plan for Biodiversity 2011–2020 (CBD 2010c), including the Aichi Biodiversity Targets (CBD 2010a), was adopted by the Parties to the Convention on Biological Diversity (CBD) in October 2010, following many regional consultations, expert workshops and high-level events organized in collaboration with numerous partners. The plan contains five strategic goals and establishes targets for achieving the vision of "a world living in harmony with nature and where, by 2050, biodiversity is valued, conserved, restored and wisely used, maintaining ecosystem services, sustaining a healthy planet and delivering benefits essential for all people" (CBD 2010c Decision X/2).

It is envisaged that the plan will be implemented primarily through activities at the national or sub-national level, with supporting action at the regional and global levels. Countries have committed to developing national and regional targets, using the plan and its Aichi Targets as a flexible framework to integrate these targets into national biodiversity strategies and action plans, and to develop indicators to report on progress in 2014 and 2018.

Strategic goal A: Address the underlying causes of biodiversity loss by mainstreaming biodiversity across government and society

Target 1: By 2020, at the latest, people are aware of the values of biodiversity and the steps they can take to conserve and use it sustainably.

Target 2: By 2020, at the latest, biodiversity values have been integrated into national and local development and poverty reduction strategies and planning processes and are being incorporated into national accounting, as appropriate, and reporting systems.

Target 3: By 2020, at the latest, incentives, including subsidies, harmful to biodiversity are eliminated, phased out or reformed in order to minimize or avoid negative impacts, and positive incentives for the conservation and sustainable use of biodiversity are developed and applied, consistent and in harmony with the Convention and other relevant international obligations, taking into account national socio-economic conditions.

Target 4: By 2020, at the latest, governments, business and stakeholders at all levels have taken steps to achieve or have implemented plans for sustainable production and consumption and have kept the impacts of use of natural resources well within safe ecological limits.

Strategic goal B: Reduce the direct pressures on biodiversity and promote sustainable use

Target 5: By 2020, the rate of loss of all natural habitats, including forests, is at least halved and where feasible brought close to zero, and degradation and fragmentation is significantly reduced.

Target 6: By 2020, all fish and invertebrate stocks and aquatic plants are managed and harvested sustainably, legally and applying ecosystem-based approaches, so that overfishing is avoided, recovery plans and measures are in place for all depleted species, fisheries have no significant adverse impacts on threatened species and vulnerable ecosystems and the impacts of fisheries on stocks, species and ecosystems are within safe ecological limits.

Target 7: By 2020, areas under agriculture, aquaculture and forestry are managed sustainably, ensuring conservation of biodiversity.

Target 8: By 2020, pollution, including from excess nutrients, has been brought to levels that are not detrimental to ecosystem function and biodiversity.

Target 9: By 2020, invasive alien species and pathways are identified and prioritized, priority species are controlled or eradicated, and measures are in place to manage pathways to prevent their introduction and establishment.

Target 10: By 2015, the multiple anthropogenic pressures on coral reefs, and other vulnerable ecosystems impacted by climate change or ocean acidification are minimized, so as to maintain their integrity and functioning.

Strategic goal C: Improve the status of biodiversity by safeguarding ecosystems, species and genetic diversity

Target 11: By 2020, at least 17 per cent of terrestrial and inland water areas, and 10 per cent of coastal and marine areas, especially areas of particular importance for biodiversity and ecosystem services, are conserved through effectively and equitably managed, ecologically representative and well-connected systems of protected areas and other effective area-based conservation measures, and integrated into the wider landscapes and seascapes.

Target 12: By 2020, the extinction of known threatened species has been prevented and their conservation status, particularly of those most in decline, has been improved and sustained.

Target 13: By 2020, the genetic diversity of cultivated plants and farmed and domesticated animals and of wild relatives, including other socio-economically as well as culturally valuable species, is maintained, and strategies have been developed and implemented for minimizing genetic erosion and safeguarding their genetic diversity.

Strategic goal D: Enhance the benefits to all from biodiversity and ecosystem services

Target 14: By 2020, ecosystems that provide essential services, including services related to water, and contribute to health, livelihoods and well-being, are restored and safeguarded, taking into account the needs of women, indigenous and local communities, and the poor and vulnerable.

Target 15: By 2020, ecosystem resilience and the contribution of biodiversity to carbon stocks has been enhanced, through conservation and restoration, including restoration of at least 15 per cent of degraded ecosystems, thereby contributing to climate change mitigation and adaptation and to combating desertification.

Target 16: By 2015, the Nagoya Protocol on Access to Genetic Resources and the Fair and Equitable Sharing of Benefits Arising from their Utilization is in force and operational, consistent with national legislation.

Strategic goal E: Enhance implementation through participatory planning, knowledge management and capacity building

Target 17: By 2015, each Party has developed, adopted as a policy instrument, and has commenced implementing an effective, participatory and updated national biodiversity strategy and action plan.

Target 18: By 2020, the traditional knowledge, innovations and practices of indigenous and local communities relevant for the conservation and sustainable use of biodiversity, and their customary use of biological resources, are respected, subject to national legislation and relevant international obligations, and fully integrated and reflected in the implementation of the Convention with the full and effective participation of indigenous and local communities, at all relevant levels.

Target 19: By 2020, knowledge, the science base and technologies relating to biodiversity, its values, functioning, status and trends, and the consequences of its loss, are improved, widely shared and transferred, and applied.

Target 20: By 2020, at the latest, the mobilization of financial resources for effectively implementing the Strategic Plan for Biodiversity 2011–2020 from all sources, and in accordance with the consolidated and agreed process in the Strategy for Resource Mobilization, should increase substantially from the current levels. This target will be subject to changes contingent to resource needs assessments to be developed and reported by Parties.

Delegates in Nagoya, Japan, during the tenth Conference of the Parties to the Convention on Biological Diversity, where signatories adopted a new UN Strategic Plan, including the Aichi Biodiversity Targets. *© IISD/Earth Negotiations Bulletin*

Table 5.1 Selected internationally agreed goals and themes related to biodiversity

Major themes from internationally agreed goals		Biodiversity			
		Pressures	State and trends	Benefits	Responses
		Aichi Biodiversity Targets			
		5,6,7,8,9,10	*11,12,13*	*14,15,16*	*1–20*
International Plant Protection Convention (IPPC) (FAO 1951) Article 1	Measures to prevent the introduction and spread of plant pests and to promote appropriate measures for their control				X
Convention on International Trade in Endangered Species of Wild Fauna and Flora (CITES 1973) Preamble	International cooperation for the protection of species of wild fauna and flora against overexploitation through international trade			X	X
Ramsar Convention on Wetlands (UN 1973) Article 3	Promote conservation of wetlands included in the list and wise use of other wetlands in national territory		X	X	X
Convention on the Conservation of Migratory Species of Wild Animals (CMS 1979) Preamble	Concerted action by states within the national jurisdictional boundaries of which migratory species spend any part of their life cycle to conserve and effectively manage such species		X		X
Agenda 21 (UNCED 1992) Chapter 17 Paragraph 86	Identify priority marine ecosystems and limit use in these areas, through, inter alia, designation of protected areas	X	X		X
Convention on Biological Diversity (CBD 1992) Article 1	Conservation and sustainable use of biodiversity and the fair and equitable sharing of benefits arising from the utilization of genetic resources		X	X	X
Article 6	National strategies for the conservation and sustainable use of biodiversity and integration of such into relevant plans, programmes and policies		X	X	X
Article 8j	Maintain knowledge of indigenous communities relevant for the conservation and sustainable use of biological diversity, promote their wider application and encourage the equitable sharing of resulting benefits		X	X	X
Article 10	Sustainable use of components of biological diversity and encourage relevant cooperation, protect traditional cultural practices and support remedial action where biological diversity has been reduced	X	X	X	X
Decision VII/28 Paragraph 1.2.3	Sustainable use of components of biological diversity and encourage relevant cooperation, protect traditional cultural practices and support remedial action where biological diversity has been reduced	X	X	X	X
CBD COP 7 (2004) Decision VII/30 Annex II	Integrate protected areas into broader landscapes and seascapes through ecological networks, ecological corridors and/or buffer zones to maintain ecological processes and take into account the needs of migratory species	X			X
2011–2050 Vision (CBD 2010c)	Control threats from invasive alien species	X	X	X	X
Millennium Summit (2000) Millennium Development Goal (MDG) 7 (UN 2000)	Living in harmony with nature and where, by 2050, biodiversity is valued, conserved, restored and wisely used, maintaining ecosystem services, sustaining a healthy planet and delivering benefits essential for all people	X	X	X	X
Johannesburg Plan of Implementation (JPOI) (WSSD 2002) Paragraph 44	Ensure environmental sustainability	X	X	X	X
Cartagena Protocol on Biosafety to the CBD (CBD 2000) Article 1	Sustainable use of biological diversity and fair and equitable sharing of benefits arising from the use of genetic resources	X			
International Treaty on Plant Genetic Resources for Food and Agriculture (ITPGRFA) (FAO 2001)	Ensuring an adequate level of protection in transfer, handling and use of living modified organisms resulting from modern biotechnology		X	X	X
Article 1 Paragraph 1.1	Sustainable agriculture and food security through the conservation and sustainable use of plant genetic resources and the fair and equitable sharing of the benefits arising out of their use	X	X	X	X
World Summit Outcome (UNGA 2005)	Promote and safeguard the fair and equitable sharing of benefits arising out of the utilization of genetic resources; significantly reduce the rate of biodiversity loss by 2010	X	X	X	X

STATE AND TRENDS

Biodiversity is affected by multiple drivers and pressures that modify its ability to provide ecosystem services to people. The interaction of multiple drivers, including demographic, economic, socio-political, scientific and technological ones, as discussed in Chapter 1, is known to increase pressures on biodiversity, leading to further decline, degradation and loss. However, the mechanisms associated with such loss require further research.

Pressures

The principal pressures on biodiversity include habitat loss and degradation, overexploitation, alien invasive species, climate change and pollution (Figure 5.1) (Baillie *et al.* 2010; Vié *et al.* 2009; MA 2005a). These pressures are continuing to increase (Box 5.3) (Butchart *et al.* 2010; CBD 2010b).

Habitat loss

Habitat loss in the terrestrial domain has been caused largely by the expansion of agriculture: more than 30 per cent of land has been converted for agricultural production (Foley *et al.* 2011). Large-scale commercial agriculture has adversely affected biodiversity, particularly agro-biodiversity (Belfrage 2006; Rosset 1999). Moreover, the growing demand for biofuels has taken a toll, with expanses of forests and natural lands in South East Asia being converted into mono-crop plantations (Danielsen *et al.* 2009; Fitzherbert *et al.* 2008).

Direct habitat loss is a major threat to coastal ecosystems through aquaculture (Valiela *et al.* 2004). Wetlands in particular have faced a 50 per cent loss in the 20th century (MA 2005b). Freshwater ecosystems are severely affected by fragmentation (Nilsson *et al.* 2005) and floodplain ecosystems

> **Box 5.2 Biodiversity vision: a world in harmony with nature**
>
> **Related goals**
> Reduce direct pressures on biodiversity; improve the status of biodiversity; enhance benefits from biodiversity; strengthen responses to safeguard biodiversity
>
> **Indicators**
> Trends in: invasive species and pollutants such as nitrogen deposition; extinction risk of species; extent, condition and integrity of biomes, habitats and ecosystems; status of species harvested for food and medicine; development and effectiveness of protected areas, indigenous and community-conserved areas, sustainable use management and payment for ecosystem services programmes; and in the number of languages and speakers as a proxy for traditional knowledge supporting sustainable resource use and conservation
>
> **Global status and trend**
> Pressures on biodiversity are expected to increase, and the status of biodiversity is expected to decrease, but encouragingly, responses are starting to increase

Figure 5.1 Major threats to vertebrates listed as critically endangered, endangered or vulnerable on the IUCN Red List

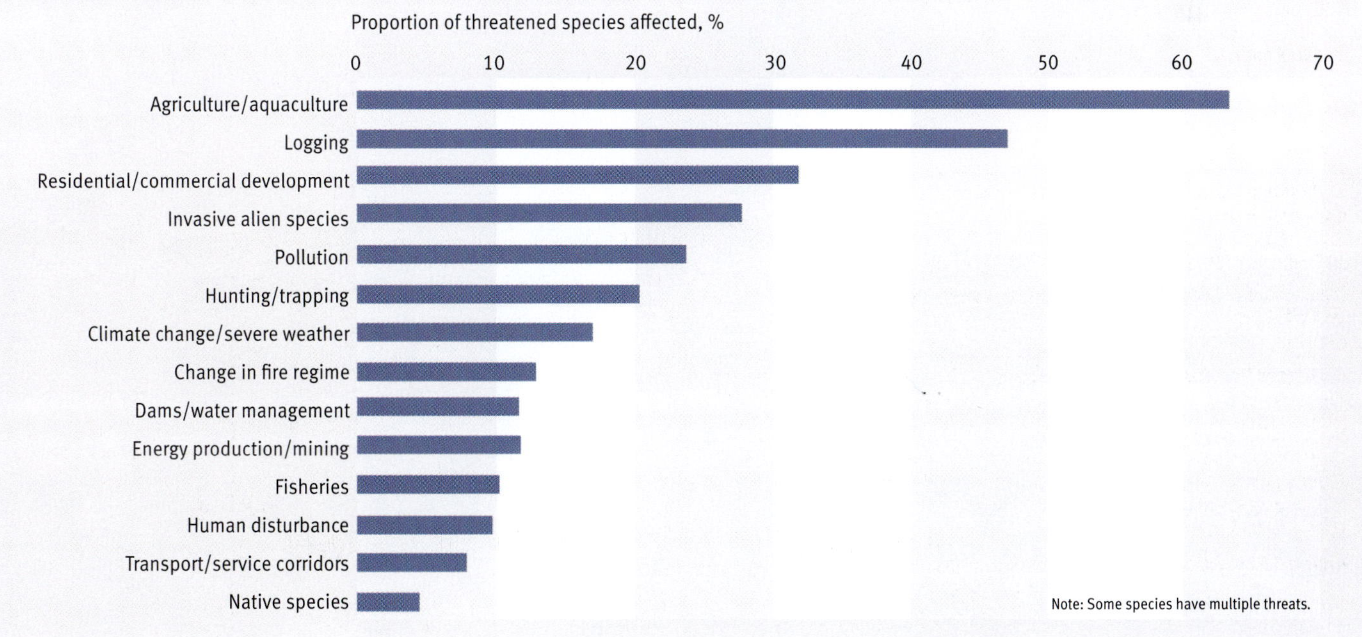

Source: Baillie *et al.* 2010

Box 5.3 Global Biodiversity Outlook

The *Global Biodiversity Outlook* is a periodic report prepared by the Secretariat of the Convention on Biological Diversity. The third edition (*GBO-3*) was one of the main assessments of progress towards the 2010 Biodiversity Target of significantly reducing the current rate of biodiversity loss at global, regional and national levels, and was an important source of information in the development of the Strategic Plan for Biodiversity 2011–2020 and the Aichi Biodiversity Targets.

The main conclusion of *GBO-3* was that the 2010 Biodiversity Target had not been met. Specifically, the underlying causes of biodiversity loss have not been addressed despite increasing responses by governments. Pressures on biodiversity have remained high or continued to increase, leading to ongoing degradation of ecosystems, reductions in species populations and increasing extinction risks, as well as erosion of genetic variety (Figure 5.2).

Most future scenarios of biodiversity change project continuing high levels of population and species extinctions and loss of habitats, with associated decline of some ecosystem services important to human well-being. There is a high risk of degradation of a broad range of such services if ecosystems are pushed beyond threshold levels.

While the conclusions of *GBO-3* give cause for concern, the report also provides a message of hope. Many actions in support of biodiversity have been taken, and have had significant and measurable results in particular areas and amongst targeted species and ecosystems. This suggests that, with adequate resources and political will, the tools exist for reducing the erosion of biodiversity. Preventing further loss in the near term will be extremely challenging, but it can be achieved in the longer term if effective action is initiated now in support of an agreed long-term vision. Initiating action to address the underlying causes of biodiversity decline is paramount. Failure to use this opportunity will result in many ecosystems moving into new, unprecedented states in which the capacity to provide for the needs of present and future generations is highly uncertain.

are also threatened (Tockner *et al.* 2008; Tockner and Stanford 2002). Benthic habitats have been degraded as a consequence of bottom trawling and other destructive fishing methods (Thrush and Dayton 2002).

Overexploitation

Overexploitation of wild species to meet consumer demand threatens biodiversity, with unregulated overconsumption contributing to declines in terrestrial, marine and freshwater

Land conversion to oil palm plantations in Sabah, Malaysia, has encroached on the natural habitat of the orangutan, significantly threatening the species. © *Johannes Refisch/UNEP*

Figure 5.2 Biodiversity indicator trends

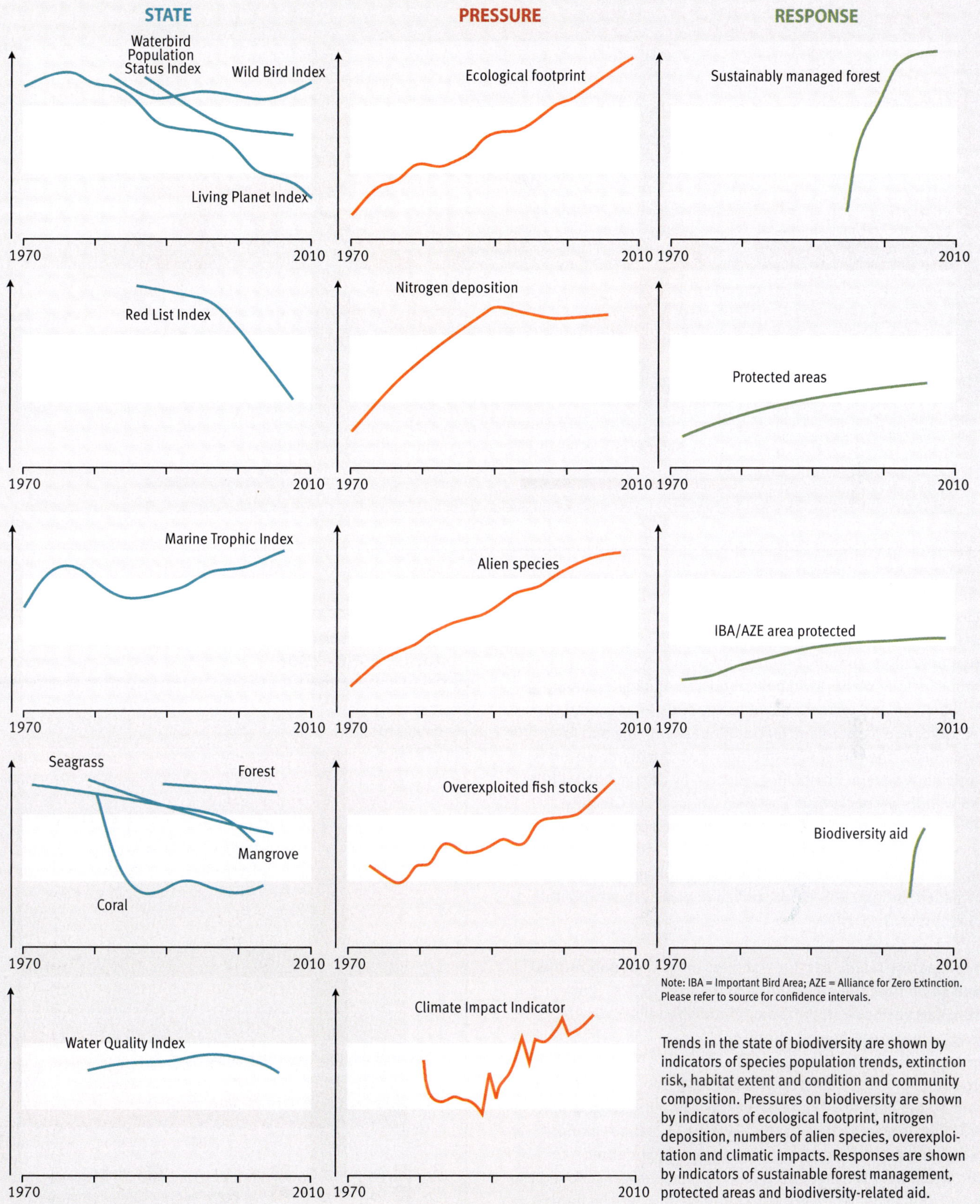

Note: IBA = Important Bird Area; AZE = Alliance for Zero Extinction. Please refer to source for confidence intervals.

Trends in the state of biodiversity are shown by indicators of species population trends, extinction risk, habitat extent and condition and community composition. Pressures on biodiversity are shown by indicators of ecological footprint, nitrogen deposition, numbers of alien species, overexploitation and climatic impacts. Responses are shown by indicators of sustainable forest management, protected areas and biodiversity-related aid.

Vertical scales and units vary - see source for details

Source: Adapted from Butchart *et al.* 2010

Figure 5.3 Numbers of vertebrates globally threatened by overexploitation, 2010

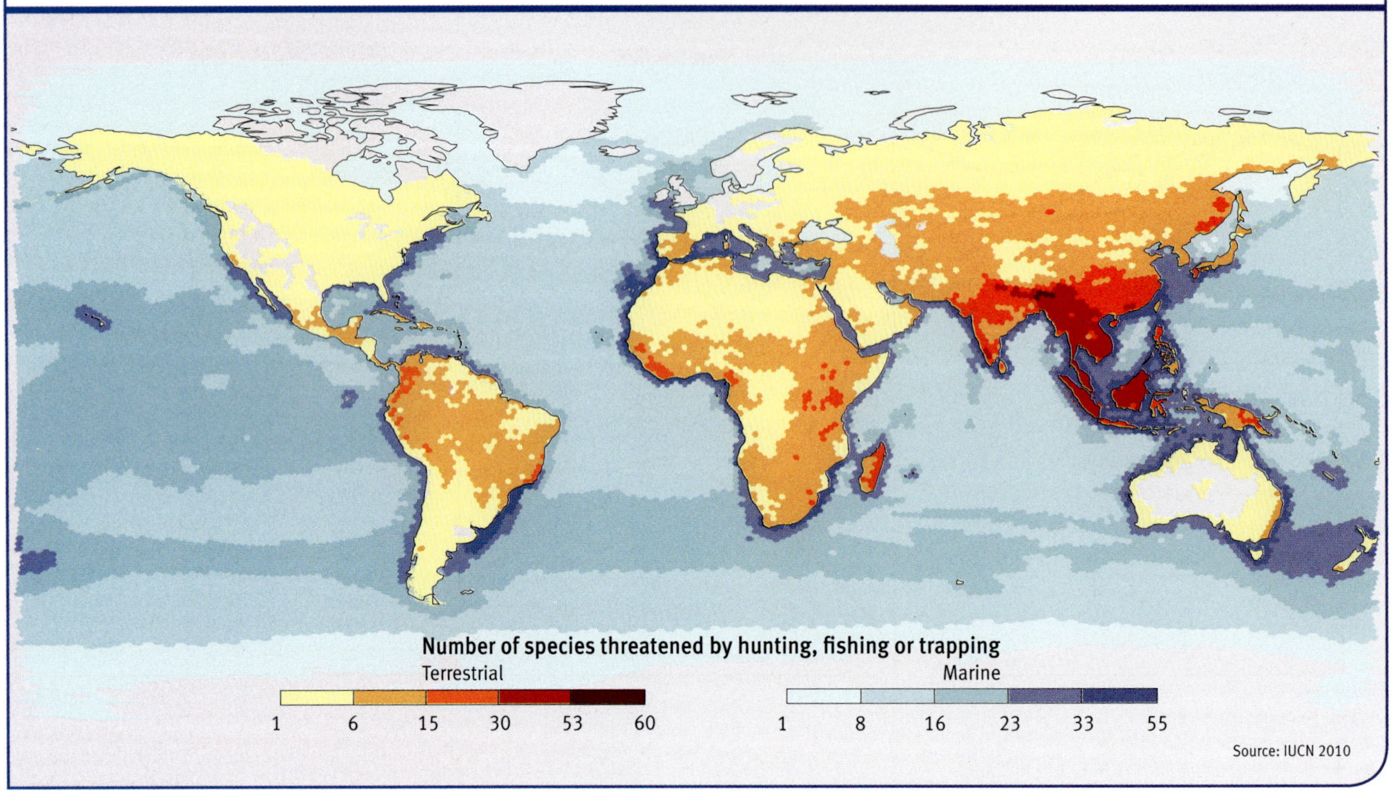

Source: IUCN 2010

ecosystems (Figure 5.3) (Peres 2010; Vorosmarty *et al.* 2010; Kura *et al.* 2004; Dulvy *et al.* 2003). Although overexploitation is often difficult to quantify in terrestrial systems, major exploited groups include plants for timber, food and medicine; mammals for wild meat and recreational hunting; birds for food and the pet trade; and amphibians for traditional medicine and food (Vié *et al.* 2009). The threat to vertebrates from overexploitation is particularly severe, driven, in particular, by demand for wildlife and wildlife products from East Asia (Figure 5.3). Globally, utilized vertebrate populations have declined by 15 per cent since 1970 as indicated by the Living Planet Index (Butchart *et al.* 2010). Similarly, the extinction risk of utilized bird species increased during 1988–2008, driven in part by overexploitation (Butchart *et al.* 2010).

In the marine realm, capture fisheries more than quadrupled their catch from the early 1950s to the mid-1990s. Since then, catches have stabilized or diminished (FAO 2010b), despite increased fishing effort (Anticamara *et al.* 2011; Swartz *et al.* 2010). The proportion of marine fish stocks that are overexploited, depleted or recovering from depletion rose from 10 per cent in 1974 to 32 per cent in 2008 (Figure 5.4) (FAO 2010b; Worm *et al.* 2009). Of the 133 local, regional and global extinctions of marine species documented worldwide over the last 200 years, 55 per cent were caused by overexploitation, while the remainder were driven by habitat loss and other threats (Dulvy *et al.* 2003). Commercial fisheries are the principal threat to fish stocks, but overexploitation

Figure 5.4 Trends in the state of global fishery stocks, 1950–2006

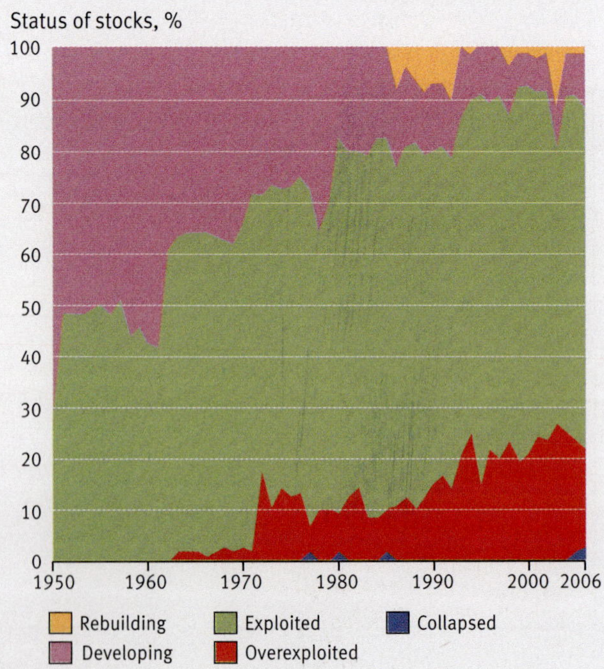

Source: Kleisner and Pauly 2011

in artisanal fisheries also occurs (Garcia and Rozenberg 2010). Such practices can ultimately lead to major shifts in community composition. For example, coral communities have been transformed into algal-dominated systems because of overfishing of herbivores (Mumby 2009).

The use of destructive fishing practices further amplifies the impacts of unsustainable fishing on marine biodiversity and habitats (FAO and UNEP 2009). Technology can enhance the intensity and range of human impacts on marine biodiversity although it can also play a significant role in making fishing practices less destructive. Moreover, abandoned and lost fishing gear is having negative ecological consequences on marine biodiversity (also known as ghost fishing) (Brown and Macfadyen 2007).

Overfishing is also a problem in freshwater wetlands, although in many cases adequate data are not available to quantify the extent of the loss (Kura *et al.* 2004). Recreational fishery practices such as stocking and selective take can also have important evolutionary impacts on freshwater fish stocks (Jorgensen *et al.* 2007). By-catch from fisheries can be a major threat to groups such as sharks, turtles and albatrosses.

Invasive alien species

Invasive alien species threaten native biodiversity and are spreading through both deliberate and unintentional introductions as a consequence of increasing levels of global travel and trade. Poorly planned economic introductions, air transport, hull-fouling and ballast water from ships, as well as trade in pets, garden plants and aquarium species, are significant pathways for the dispersal of invasive species (Reise *et al.* 2006; Bax *et al.* 2003). Invasive alien species affect native species principally through predation, competition and habitat modification (McGeoch *et al.* 2010; Vié *et al.* 2009; Strayer *et al.* 2006). Invasive species have major economic costs, estimated by one study to total US$1.4 trillion annually (Pimentel *et al.* 2004). They are found in nearly all countries and habitats, including marine ecosystems – for example the red lionfish *Pterois volitans* affects coral reef fish in the Caribbean (González *et al.* 2009) – and freshwater ecosystems: the Nile Perch *Lates niloticus*, for instance, has an impact on native fish in Lake Victoria (Balirwa *et al.* 2003). Invasive species have particularly acute effects on the terrestrial biodiversity of small islands (McGeoch *et al.* 2010). Data from Europe show that the number of alien species has increased by 76 per cent since 1970 (Butchart *et al.* 2010), a pattern that is likely to be similar in other places. In another study, invasive alien species were a factor in more than 50 per cent of vertebrate extinctions where the cause was known, and were the sole cause of 20 per cent of extinctions (Clavero and García-Berthou 2005).

Climate change

Climate change is an increasingly important threat to species and natural habitats. There is widespread evidence that changes in phenology, including the timing of reproduction and migration, physiology, behaviour, morphology, population density and distributions of many different types of species are driven by climate change (Rosenzweig *et al.* 2007). For example, trends in European bird populations since 1990 show a growing impact: populations are increasing among the species projected to benefit from climate change while population decline is documented for those projected to undergo range contraction (Gregory *et al.* 2009). In the Arctic, tundra habitats are shrinking owing to tree-line advance (Callaghan *et al.* 2005). In the marine realm, climate change is causing widespread die-off of coral reefs through rising temperatures and ocean acidification (Baker *et al.* 2008; Carpenter *et al.* 2008; Hoegh-Guldberg *et al.* 2007). The Arctic ice cap is also shrinking rapidly, with likely impacts on ice-dependent species (McRae *et al.* 2010; IPCC 2007), as well as shifts in phenology and distribution of marine species (Dulvy *et al.* 2008; Hiddink and Ter Hofstede 2008; Richardson 2008; Perry *et al.* 2005). Recent studies have also projected distribution shifts of 1 066 marine fish and invertebrate species polewards at an average rate of 40 km per decade (Cheung *et al.* 2009), leading to likely disruption of community composition and local extinctions.

For many wetlands, changes in rainfall and evaporation are expected to have major impacts on water regimes, affecting both migratory and residential species (Finlayson *et al.* 2006), while changes in flow in both the short and long term will impact many aquatic species (Bates *et al.* 2008; Xenopoulos and Lodge 2006). Climate change will also act synergistically with other threats, such as the spread of diseases and invasive alien species (Benning *et al.* 2002). However, in many instances it may be difficult to differentiate the effects of these different threats, as has been outlined for wetlands and rivers in Australia (Finlayson *et al.* 2011).

Pollution

Pollutants such as pesticide and fertilizer effluents from agriculture and forestry, industry including mining and oil or gas extraction, sewage plants, run-off from urban and suburban areas, and oil spills, harm biodiversity directly through mortality and reduced reproductive success, and also indirectly through habitat degradation (MA 2005a). Inland wetlands and coastal marine habitats face a major threat from waterborne pollutants (Chapter 6) (Finlayson and D'Cruz 2005). Meanwhile, atmospheric pollution in terrestrial systems, particularly the deposition of eutrophying and acidifying compounds such as nitrogen and sulphur (Chapter 2), is also important. Rates of nitrogen deposition increased sharply after 1940 but have levelled out since 1990, probably owing to an overall decrease in biomass burning, though there is regional variation (Butchart *et al.* 2010). Nevertheless, nitrogen deposition continues to be a significant threat to biodiversity, especially for species that have adapted to low-nitrogen habitats (Dise *et al.* 2011).

Additional threats

Additional threats to biodiversity include changes in fire regimes, problematic native species (Figure 5.1) and negative influences from human activities. Influences from human activities that may be harmful to biodiversity include artificial illumination,

Box 5.4 The ecological footprint: an indicator of the pressures on biodiversity

The ecological footprint is a resource accounting tool that measures how much biologically productive land and sea area – crop and grazing land, forests, fishing grounds and built-up land – is demanded by a given population or activity, and compares this to how much land and sea is available (Kitzes and Wackernagel 2009; Wackernagel *et al.* 2002; Wackernagel and Rees 1996). It has become an increasingly popular headline indicator of broad human pressures on the environment, although its methodology and application continue to be debated (Kitzes *et al.* 2009; Best *et al.* 2008; Fiala 2008).

Ecological footprint analysis shows that the global demand for biologically productive areas has approximately doubled since the 1960s (WWF 2010). In 2007, global society demanded more than 1.5 planets' worth of productive biological capacity, a deficit that can only be met through the depletion of stocks of renewable resources or the accumulation of waste product, most importantly carbon dioxide (CO_2) in the atmosphere (Figure 5.5). Together with other indicators (Butchart *et al.* 2010), this trend provides evidence of an overall increase in pressures on biodiversity. The continued growth of these pressures is likely to increase the difficulty of halting or reversing global loss.

Figure 5.5 The ecological footprint, 1961–2007

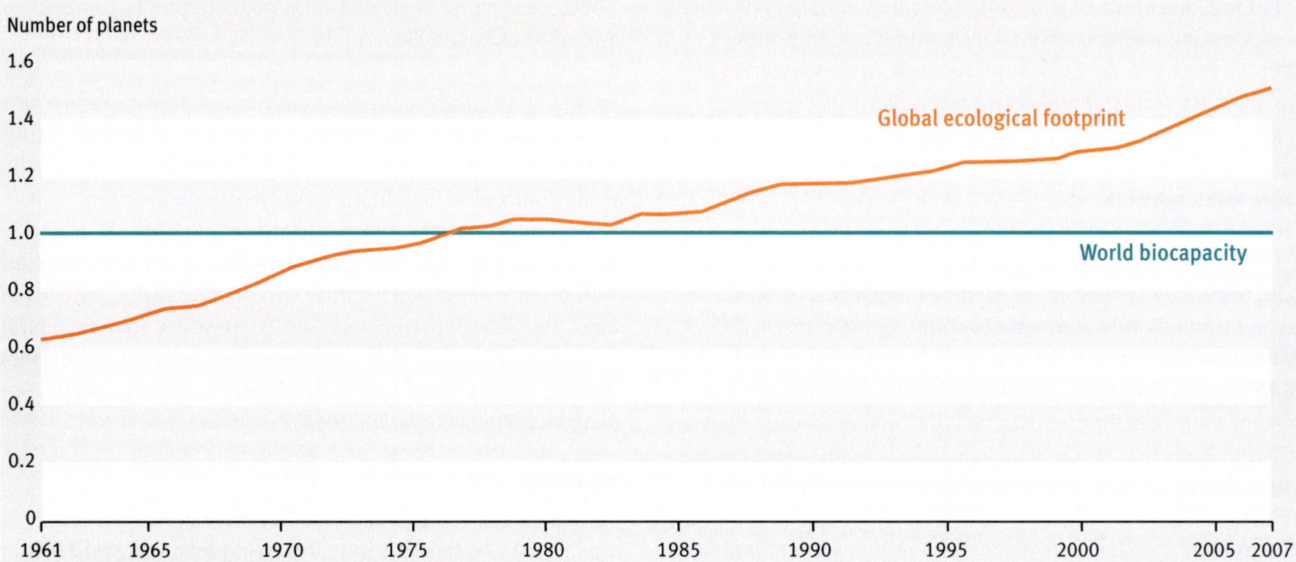

Source: WWF 2010

genetically modified organisms (Box 5.5), microplastics, nanotechnology, geo-engineering, and high levels of human appropriation of net primary productivity (Box 5.4) (Cole 2011; Gough 2011; Galgani *et al.* 2010; Hölker *et al.* 2010; Sutherland *et al.* 2009, 2008). Scientific understanding of the specific nature of threats to biodiversity from these influences is building. Meanwhile, the causes of some recent biodiversity declines remain unclear, and further research is required to elucidate the problems and identify solutions, for example for mammals in northern Australia (Woinarski *et al.* 2011) or trans-Saharan migrant birds (Moeller *et al.* 2008).

Patterns of biodiversity change

Biodiversity is deteriorating at the level of populations, species and ecosystems, and genetic diversity is also suspected to be declining, although trends remain largely unknown (Box 5.3) (Butchart *et al.* 2010; CBD 2010b; Vié *et al.* 2009). Populations of vertebrate species recorded in the Living Planet Index have declined on average by 30 per cent since 1970 (Figure 5.6) (Loh 2010; Collen *et al.* 2008a). Declines in freshwater populations are steeper, at 35 per cent since 1970, than those for terrestrial populations, which have fallen by 25 per cent and marine populations by 24 per cent; those in the tropics are steeper than those in temperate latitudes. Habitat-specific trends are available for some regions for birds and show, for example, that European farmland bird populations have declined by 48 per cent on average since 1980 (Gregory *et al.* 2005). North American grassland and dryland species have declined by 28 per cent and 27 per cent respectively since 1968; but North American wetland bird species have increased by 40 per cent (Butchart *et al.* 2010; NABCI US Committee 2009).

At the species level, the proportion of species threatened with extinction – classified as critically endangered, endangered or vulnerable on the IUCN Red List – ranges from 13 per cent for birds to 63 per cent for cycads, and averaging almost 20 per cent for vertebrates (Baillie *et al.* 2010; Hoffmann *et al.* 2010). Furthermore, Red List Indices for mammals, birds, amphibians and corals show that considerably more species have become more threatened with extinction over recent decades than have become less threatened, and declines have been steepest for corals (Figure 5.7) (Butchart *et al.* 2010; Hoffmann *et al.* 2010). The composition of biological communities is increasingly disrupted by human activities, in particular through overexploitation. For example, in some oceans the community structure appears to have shifted to lower trophic levels owing to fisheries targeting predators and larger fish species (Branch *et al.* 2010; Pauly and Watson 2005). This phenomenon of fishing down the food web has been reported widely in many parts of the ocean, such as in Canada (Pauly *et al.* 1998), Brazil (Freire and Pauly 2010), India (Bhathal and Pauly 2008), Thailand (Pauly and Chuenpagdee 2003), the North Sea (Heath 2005) and the Caribbean (Wing and Wing 2001). However, the use of catch data to indicate fishing down the food web may be confounded by data quality and factors such as the spatial expansion of fisheries (Swartz *et al.* 2010), and may need careful interpretation if independent data on stock levels are unavailable (Branch *et al.* 2011). Other indicators, such as the Fishing-In-Balance (FIB) index, may be preferable in future (Kleisner and Pauly 2010; Bhathal and Pauly 2008).

At the level of habitats, losses include more than 100 million hectares of forest globally during 2000–2005, or 3 per cent of the 3.2 billion hectares in existence in 2000 (Hansen *et al.* 2010);

Figure 5.6 Living Planet Index, 1970–2007

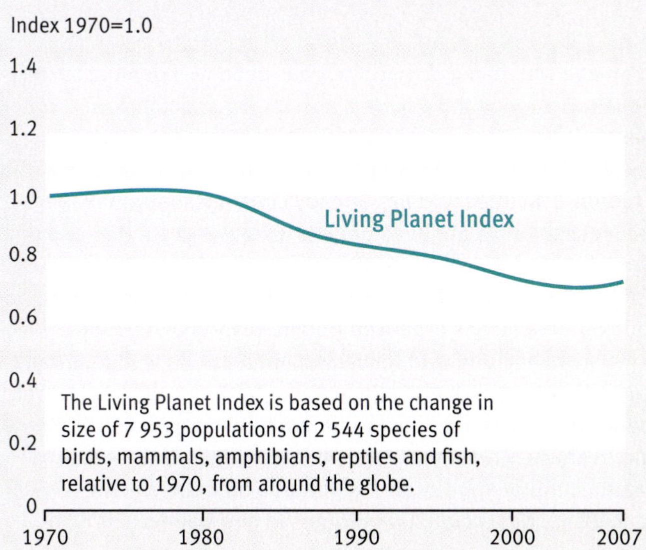

Source: WWF 2010

Figure 5.7 Red List Indices of species survival for all species of birds, mammals, amphibians and corals, 1980–2010

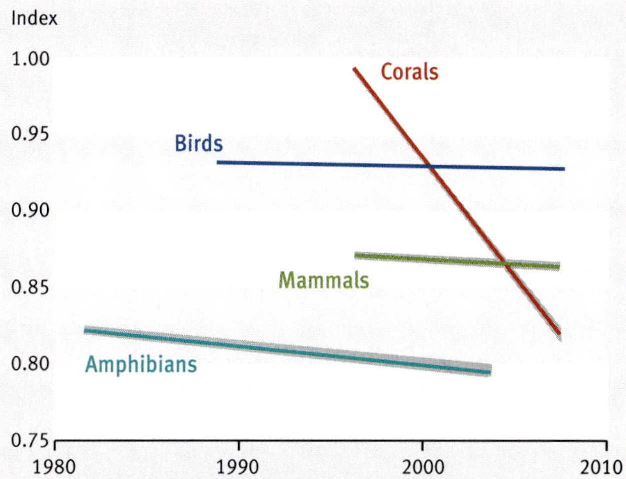

Note: Shaded areas show 95 per cent confidence intervals. The numbers of data-sufficient extant species in year of first assessment were: 9 785 birds, 4 555 mammals, 4 416 amphibians and 704 corals (warm water reef-building species only). An index of 1 equates to all species being classified as Least Concern while an index of 0 equates to all species being classified as Extinct

Source: Adapted from Vié *et al.* 2009

20 per cent of mangroves since 1980; and 20 per cent of seagrasses since 1970 (Butchart *et al.* 2010; Waycott *et al.* 2009). Remaining habitats are increasingly degraded – measures of net primary productivity, for example, show that around one-quarter of the terrestrial land area is degraded, including around 30 per cent of all forests, 20 per cent of cultivated zones and 10 per cent of grasslands (Bai *et al.* 2008). Similarly, coral reefs have declined globally by 38 per cent since 1980 (Butchart *et al.* 2010; Spalding *et al.* 2003). Natural habitats are also becoming increasingly fragmented – 80 per cent of remaining forest fragments in the Brazilian Atlantic Forest are now smaller than 50 hectares (Ribeiro *et al.* 2009), while two-thirds of the world's largest rivers are now moderately to severely fragmented by dams and reservoirs (Nilsson *et al.* 2005).

CROSS-CUTTING ISSUES

Benefits to people from biodiversity

Biodiversity underpins the ecosystem services that supply benefits to people (UNEP 2007; MA 2005a). The deterioration or loss of biodiversity and ecosystem services tends to affect poor people most directly as they are the most dependent on local ecosystems and often live in places that are most vulnerable to ecosystem change (UNEP 2007). Because the precise mechanisms of human dependence on biodiversity are not fully understood and biodiversity is undervalued – especially for regulating services – maintenance of biodiversity is rarely fully integrated into policy. Progress has been made since the Millennium Ecosystem Assessment (TEEB 2010; MA 2005a), which strongly supported the concept of ecosystem services

Figure 5.8 Relationships between biodiversity, ecosystem services and human well-being

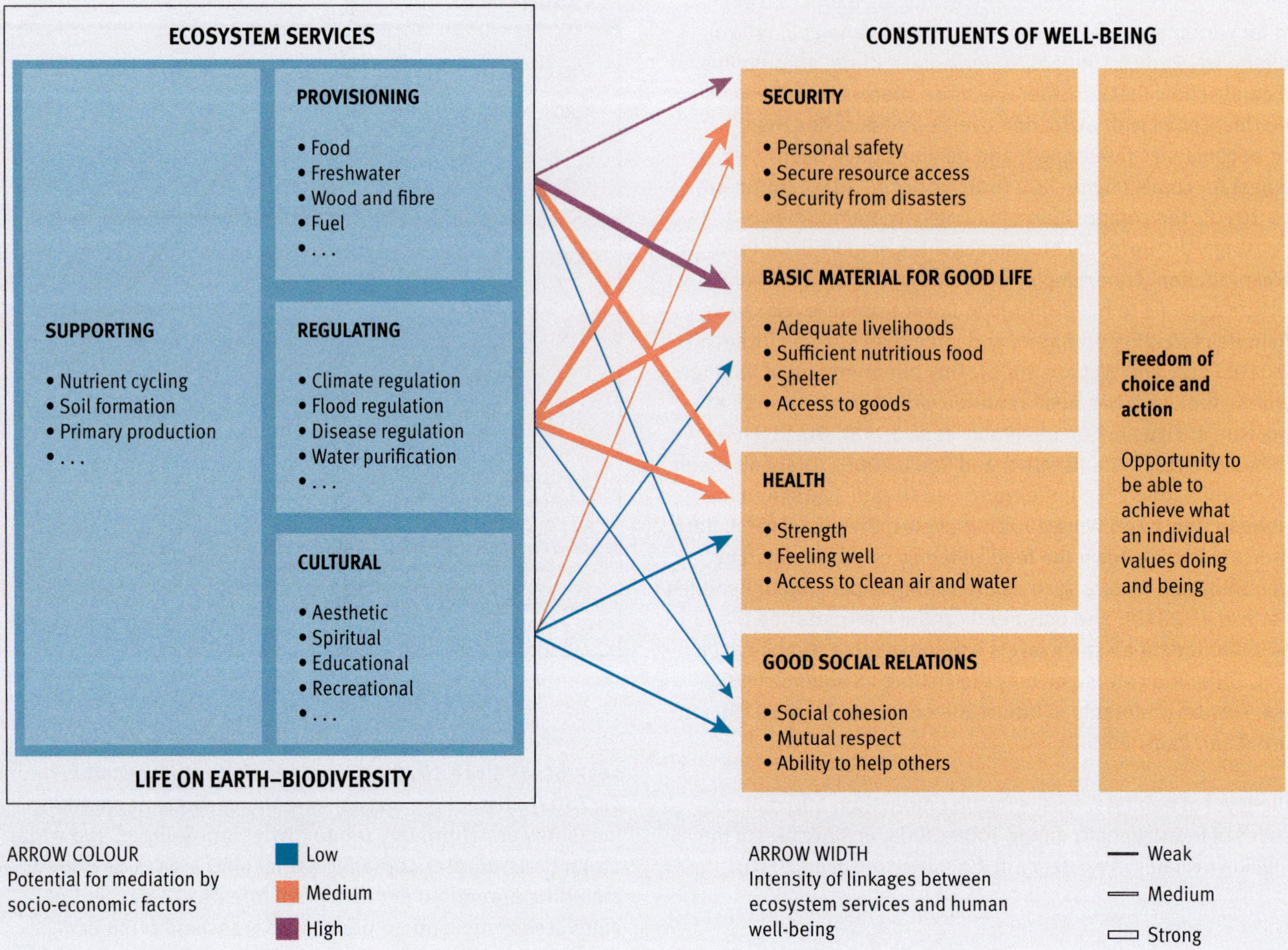

Source: UNEP/GRID-Arendal 2005

and their role in providing supporting, provisioning, regulating and cultural services (Figure 5.8). More recently, The Economics of Ecosystem Services and Biodiversity (TEEB 2010) and green economy approaches have quantified the value of biodiversity and ecosystem services (UNEP 2011). The Ramsar Convention on Wetlands has further outlined the direct links that exist between ecosystem services from wetlands and human health (Horwitz and Finlayson 2011; Horwitz *et al.* 2011).

Biodiversity and human well-being

Biodiversity and ecosystem services provide food, medicines, fish and timber products as well as biomass, energy and water-related services that people need for their livelihoods and well-being. Too often, the use and management of these provisioning services has failed to focus on conserving the ecosystems providing them. This has resulted in the degradation of regulating and supporting services that are important for overall system functioning and long-term resilience to change and therefore to human well-being, a point that has been well demonstrated when considering the effects of expanding agriculture and water management (Gordon *et al.* 2010; Falkenmark *et al.* 2007). Decreases in provisioning services may be a definitive signal that a biophysical threshold has already been passed with respect to an ecosystem's ability to provide a service, as in the case of a number of fishery collapses (Westley *et al.* 2011).

Food and medicines produced from terrestrial and aquatic ecosystems include wild-harvested products, as well as farmed crops, livestock, fish and aquaculture products. Wild-harvested foods, such as wild meat, non-timber forest products, wild fruits and freshwater resources, remain important for food security, health, cultural identity and adaptation for many people (Golden *et al.* 2011; Nasi *et al.* 2008; Robinson and Bennett 2000). Likewise, in some Asian and African countries, up to 80 per cent of the population depend on traditional medicines (WHO 2003). Assessment of the status of birds and mammals used

Figure 5.9 Red List Indices of species survival for birds and mammals used for food and medicine, 1988–2008

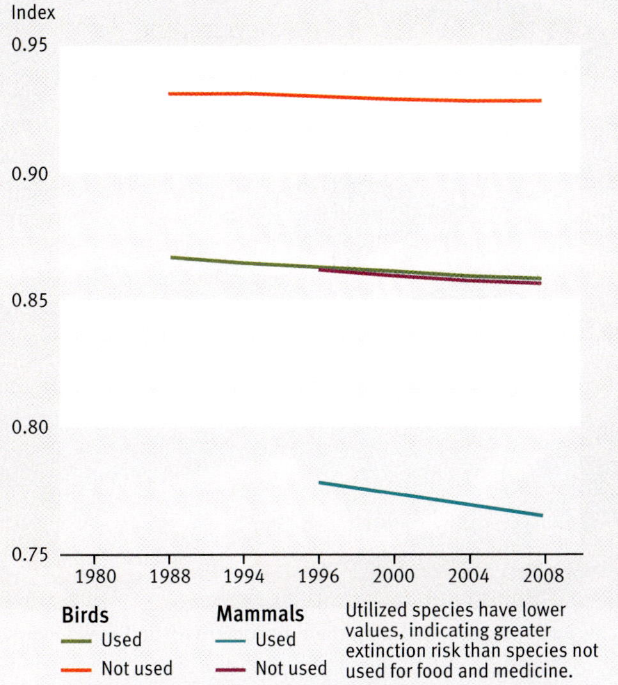

Note: An index of 1 equates to all species being classified as Least Concern while an index of 0 equates to all species being classified as Extinct

Source: IUCN 2010

for these purposes indicates that they are on average facing a greater extinction risk than other species (Figures 5.9 and 5.10). Although global data for plants are not available, medicinal plants face a high risk of extinction in those parts of the world where people are most dependent on them. This emphasizes the threat posed by biodiversity loss to the health and well-being of people directly dependent on the availability of wild species.

Fisheries provide a major source of food, revenues and employment with globally over 80 million tonnes of biomass being captured annually from the ocean (Sumaila *et al.* 2010) and large amounts from inland waters (Kura *et al.* 2004). However, as fish stocks are depleted, this supply is becoming increasingly dependent on aquaculture, which itself can have many negative environmental and social impacts such as pollution, introduction of exotic species and displacement of small-scale fishing practices (Barnhizer 2001; Naylor *et al.* 2000; Emerson 1999). Recent estimates suggest that in 2000 alone, the potential global catch losses due to overfishing amounted to 7–36 per cent of the actual tonnage landed that year, resulting in a landed value loss of US$6.4–36.0 billion. This amount could have helped prevent around 20 million people worldwide suffering from undernourishment (Srinivasan *et al.* 2010).

Agricultural production is also supported by biodiversity and ecosystem services (Altieri 1999), and agricultural diversity can in turn contribute to food security by supporting adaptation to a changing climate (Thrupp 2000). Small-scale livestock husbandry

Figure 5.10 Distribution and conservation status of medicinal plant species assessed for the IUCN Red List, by region, 2009

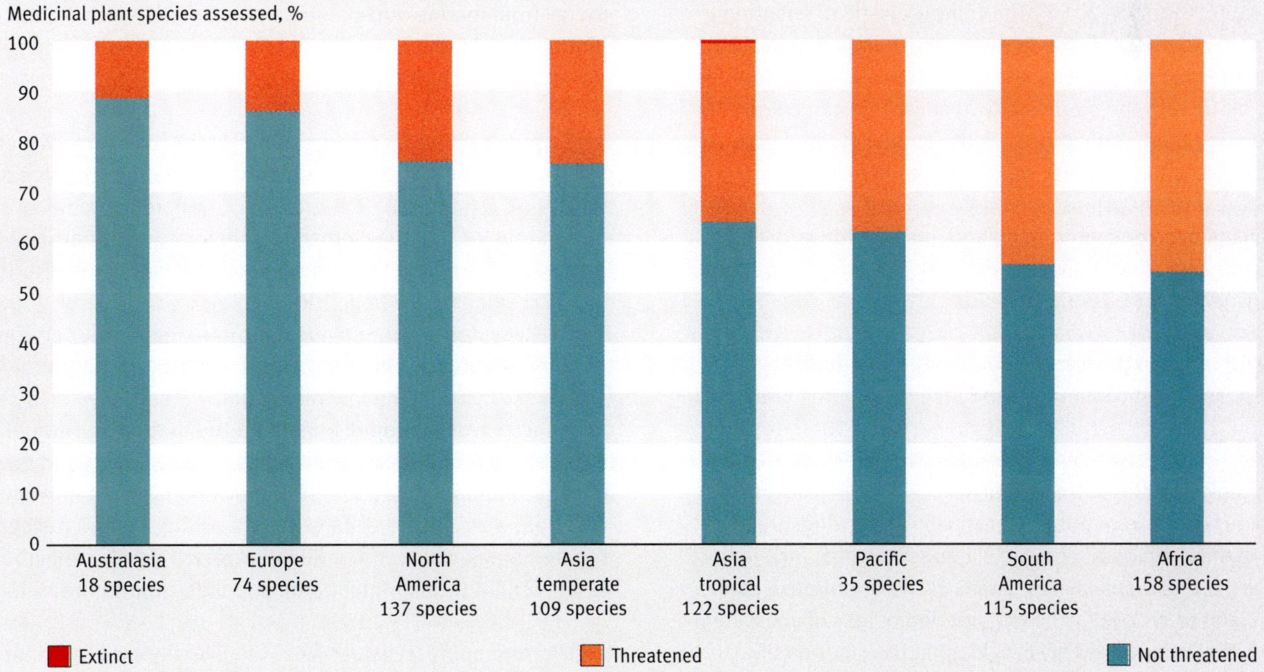

Source: TRAFFIC and IUCN SSC Medicinal Plants Specialist Group 2009

Large-scale dams and their associated reservoirs affect biodiversity by displacing species and restricting their movement up and down rivers.
© Nikola Miljkovic/iStock

and pastoralism can both contribute to maintaining biodiversity and to sustainable local economies, adaptation to climate change, resistance to disease and cultural diversity (FAO 2009). Equally, overgrazing can cause soil erosion and desertification, thereby decreasing provisioning services. Threats from livestock production to biodiversity are likely to grow as demand for meat and dairy increases, requiring more livestock feed and more water (Thornton 2010). The complex issue of ensuring a sustainable food supply for an expanding human population has been addressed in recent assessments (IAASTD 2009; Molden 2007), along with the biodiversity benefits that can be obtained by balancing food production with the supply of other ecosystem services. Pressures on land, water and biodiversity from agriculture and aquaculture could be reduced in some countries by reducing overconsumption of food, shifting towards diets comprising less meat/fish, and reducing crop losses and food waste (Godfray *et al.* 2010; WHO 2005).

Energy for much of the world's population is derived from biomass. The most commonly used fuels for heating and cooking are wood, charcoal and plant and animal waste (Berndes *et al.* 2003). Hydropower depends on high volumes and regular rates of flow of water to dams from functioning ecosystems in the catchment area, but often contributes to widespread negative environmental and social impacts, particularly loss of biodiversity and displacement (WHO 2009; Greathouse *et al.* 2006; Ligon *et al.* 1995). The degradation or loss of ecosystem services that provide energy is evident in the siltation of reservoirs and loss of water volume associated with deteriorating catchment areas (Nilsson *et al.* 2005); in the deforestation created by the overharvesting of woody vegetation; and in the overuse of agricultural waste and animal manure. The loss of ecosystem services associated with overharvesting, poor management, climate change and, for example, an increase in forest fires, is often felt by already marginalized groups who have to collect fuelwood and/or other forms of biomass for household energy needs (CBD 2010b). The development of renewable energy from marine and coastal environments, such as that from offshore wind farms, may result in trade-offs between energy production and habitat loss.

Freshwater from surface and groundwater ecosystems is a critical provisioning service used for drinking, sanitation, cooking and agriculture (Chapter 4). Wetlands and rivers regulate flows and material cycles that play indispensible roles in supporting human life systems and benefiting many sectors of society (Arthurton *et al.* 2007; Falkenmark *et al.* 2007; Finlayson and D'Cruz 2005). These ecosystems also provide important regulatory services in the form of water purification, erosion control and storm buffering (Morris *et al.* 2003). Meanwhile, groundwater ecosystems provide great social and economic benefits through the provision of low-cost, high-quality water supplies for both urban and rural areas (Bjorklund *et al.* 2009). Groundwater is also important for irrigation, with Siebert *et al.* (2010) reporting that 40 per cent of irrigated areas, some 300 million hectares representing about 20 per cent of total farmland, are supplied by groundwater.

Cultural and spiritual values from biodiversity are important to many communities (Posey 1999). Many have benefited from exploiting the recreational and cultural value of biodiversity for ecotourism (Ehrlich and Ehrlich 1992). For example, lakes, wetlands, rivers and coastal ecosystems offer significant ecotourism potential with, for example, coral reef tourism in Belize estimated to be worth US$150–196 million per year (Cooper *et al.* 2009). These aquatic ecosystems also supply water that is integral to many social, spiritual and religious activities. Examples include the sacred status of water sources and riparian zones for the Bantu-speaking peoples of Southern Africa (Bernard 2003) and the duty of care exercised by Maori in New Zealand for the life force exhibited by water (Williams 2006).

The wildlife and timber trade, comprising the sale or exchange of wild animal and plant resources, is prevalent within national borders. However, significant volumes can also be traded internationally, for example highly prized products such as caviar and medicines. The primary motivating factor for wildlife traders is financial, ranging from small-scale local income generation to major profit-oriented business, such as marine fisheries and logging companies. In some cases, harvest and trade of species can provide a significant proportion of local or national income. Overall, the legal trade in wildlife including live animals, animal products for clothing and food, ornamental and medicinal plants, fish and timber was estimated at over US$300 billion in 2009 (TRAFFIC in prep.; Roe 2008). Furthermore, illegal trade is believed to be substantial, possibly worth US$10 billion (Haken 2011). Timber and seafood are the most important categories of international wildlife trade in terms of both volume and value: around 90 million tonnes of fish were captured in 2008 with trade valued at more than US$100 billion (FAO 2010b), while the trade in primary wood products in 2009 was valued at US$189 billion (FAO 2010a).

Biodiversity and climate change

Biodiversity plays an important role both in supporting efforts to mitigate climate change and in enabling societal adaptation to its effects. Ecosystems store and sequester carbon through biological and biophysical processes that are underpinned by biodiversity. About 2 500 gigatonnes (billion tonnes) of carbon are stored in terrestrial ecosystems, compared to approximately 750 gigatonnes in the atmosphere (Chapter 3) (Ravindranath and Oswald 2008). Almost 38 000 gigatonnes are stored in the oceans, of which about 37 000 gigatonnes are in deep ocean layers that will only feed back to atmospheric processes over very long time scales (Sabine *et al.* 2004). Around 1 150 gigatonnes are stored in forests, with 30–40 per cent in biomass and 60–70 per cent in soil. Significant carbon stocks are also found in other terrestrial ecosystems including wetlands and peatlands. Indeed, the latter cover only 3 per cent of the land area, but reputedly contain nearly 30 per cent of all global soil carbon (Parish *et al.* 2008). Marine ecosystems on average take up an additional 2.2 gigatonnes of carbon per year (Le Quéré *et al.* 2009; Canadell *et al.* 2007). The critical role of freshwaters in the global carbon cycle has only recently been demonstrated (Battin *et al.* 2009; Cole *et al.* 2007).

The importance of forests in storing almost half of all terrestrial carbon, and sequestering carbon from the atmosphere, means that they play a major role in climate change mitigation. Primary forests are more biologically diverse, and also more carbon dense, than other forest ecosystems. Modified natural forests and plantations have less biodiversity and lower carbon stocks than primary forests under similar environmental conditions (CBD 2009a). Efforts to maintain forest health, for example through incentive mechanisms such as Reducing Emissions from Deforestation and Degradation (REDD+), have the potential to help mitigate climate change. These can also have multiple benefits for biodiversity if interventions ensure that environmental and social safeguards are respected, such as full and effective participation of indigenous and local communities (Cotula and Mathieu 2008), and if they avoid displacing deforestation and degradation from areas of lower conservation value to those of higher biodiversity value, or exerting pressures on other native ecosystems.

Many of the options available to help society adapt to the effects of climate change depend on and are enhanced by biodiversity. Ecosystem-based adaptation uses the range of opportunities for the sustainable management, conservation and restoration of ecosystems to provide services that enable people to adapt to the impacts of climate change. For example, intact, well-functioning ecosystems, with natural levels of species diversity, are usually more able to continue to provide ecosystem services, and resist and recover more readily from extreme

The tropical forests of Panama, which contain some 1 569 known species of amphibians, birds, mammals and reptiles, are also a valuable carbon sink. *© Jason Jabbour*

weather events than degraded, impoverished ones (CBD 2009a). Healthy ecosystems also play an important role in protecting infrastructure and enhancing human security, and therefore in reducing the risk from disasters (ISDR 2009). Ecosystem-based adaptation options are often more accessible to the rural poor than interventions based on infrastructure and engineering, and there can be multiple social, economic and environmental co-benefits for local communities from its use when designed and managed appropriately.

Responses to the threats to biodiversity

Managing agriculture and biodiversity

Successful management of agricultural landscapes requires a reduction of habitat loss and degradation whilst providing an adequate supply of food for a growing human population. Sustainable agriculture has received increasing attention because expanding agriculture is globally the principal driver of biodiversity decline (Brussaard *et al.* 2010; IAASTD 2009; MA 2005b). In recent years attention has been given to a new paradigm of ecoagriculture or integrated conservation-agriculture, which seeks to integrate biodiversity conservation with rural development. This paradigm is being explicitly considered in shaping conservation strategies with clearly identified economic and ecological relationships that include ecosystem services (IAASTD 2009; Scherr and McNeely 2008). The extensification of agriculture may require more land than intensive agriculture to achieve the same production levels (Godfray *et al.* 2010; Phalan *et al.* 2011), but it may be more sustainable in the long term and have fewer impacts on wildlife and human health (Perfecto and Vandermeer 2010). New approaches that combine the most effective, least harmful practices from intensive and extensive farming, sometimes termed sustainable intensification, will be needed (Royal Society 2009). In this context the use of GMOs in agriculture and also in aquaculture potentially presents both threats and opportunities for biodiversity (Box 5.5).

Managing invasive species

Successful management of invasive species relies on preventing the introduction and spread of species to new areas, as well as controlling and eradicating established invaders. Ten different international agreements and organizations have some relevance, including the International Plant Protection Convention, the World Trade Organization, the International Maritime Organization, the Convention on International Civil Aviation and the Convention on Biological Diversity. Since 1970 there has been a significant increase in the number of parties to these agreements (Figure 5.11), with 81 per cent of the world's countries acceding to them (McGeoch *et al.* 2010). Although this signifies an international intent to manage biological invasions, no international agreement currently deals exclusively with the trade, transport or control of alien and invasive species (Stoett

Box 5.5 Genetic modification

Genetic modification (GM) remains controversial, both a potential threat and an opportunity for biodiversity conservation, depending on the context. The technology is widely used in pharmaceuticals and crop production, but many consider it an unwarranted risk to the environment and human health. A genetically modified organism (GMO) is defined by the Cartagena Protocol on Biosafety as any living organism that possesses a novel combination of genetic material obtained through the use of modern biotechnology (CBD 2000); generally by the transfer of genetic material from one species to another. The vast proportion of GM crops has been modified to be tolerant of broad-spectrum herbicides to allow more efficient weed control and/or to express a toxin (Bt) that acts against the caterpillars of butterflies and moths that live and feed inside the crop plants.

Genetically modified crops were first planted commercially in 1996, and by 2010 they covered 148 million hectares. Although the largest areas were in the United States, Canada, Brazil, China and Argentina, the greatest number of adopters – 14.4 million out of an estimated total of 15.4 million – were small farmers in developing countries (James 2010).

Genetic modification technologies are being developed to control malaria, both by making wild mosquito populations less capable of carrying the malaria parasite (malERA 2011; Sinkins and Gould 2006) and by reducing mosquito numbers through introducing sterility, replacing the use of radiation (Bax and Thresher 2009).

Several environmental risks from GMOs have been identified, including the loss of genetic diversity of agricultural species and their wild relatives through gene flow, although this also occurs with non-GM crops (Piñeyro-Nelson *et al.* 2009). Another concern is the effects on organisms that are not the target of the GM trait, although Bt crops have few toxic effects on non-target species as the Bt toxins produced are highly specific and only expressed in the plant itself. Further, the effects tend to be outweighed by overall increases in invertebrate numbers because of lower levels of pesticide use (Marvier *et al.* 2007). Lower pesticide use also has benefits for human health in some areas (Raybould and Quemada 2010). In contrast, GM crops tolerant of broad-spectrum herbicides such as glyphosate often result in fewer weeds than conventional crops, and therefore make less food available to farmland birds (Gibbons *et al.* 2006). In addition, species are evolving resistance to both glyphosate and Bt (Powles 2010; Liu *et al.* 2010). These latter outcomes are examples that raise concern over the complexities of the environmental implications of GMOs.

Figure 5.11 Commitments to manage alien invasive species, 1970–2010

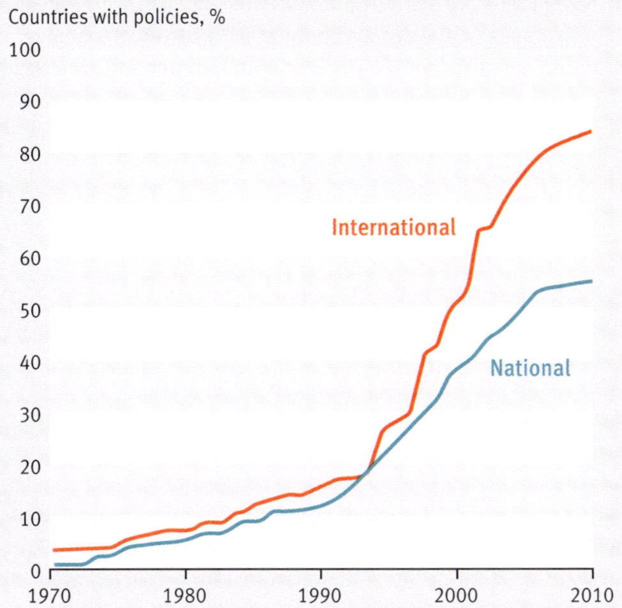

Source: McGeoch *et al.* 2010

2010). At a national level, only 55 per cent of countries have legislation to prevent the introduction of new ones and to control existing ones, and less than 20 per cent are estimated to have comprehensive strategies and management plans. In many cases, information on existing management activities either does not exist or is not readily available (Stoett 2010).

To control threats from invasive alien species, the following actions are seen as necessary:

- integrated planning to prevent further introductions by managing priority pathways;
- focus on controlling established species and priority invaders with significant impacts on biodiversity (Hulme 2009); and
- investment in the knowledge generation, data collation and research needed for risk assessments (McGeoch *et al.* 2010).

Managing wildlife trade and use

Wildlife use and trade can be managed through a variety of measures, including regulatory measures such as policies and laws and voluntary ones such as certification schemes; formal measures such as positive and economic incentives, and informal ones such as influencing sustainable consumer behaviour; direct measures such as customs inspections and other enforcement actions, and indirect ones such as economic influences. These measures can be applied at a variety of levels from the local, such as delineating resource extraction zones in protected areas or establishing community-based natural resource management, to the global, such as through the Convention on International Trade in Endangered Species of Wild Fauna and Flora (CITES) (Roe 2008).

Managing the impacts of climate change on biodiversity through mitigation and adaptation

Managing the impacts of climate change will be important, as recent studies show that the range shifts of terrestrial organisms towards the poles and higher altitudes as a result of climate change may be significantly faster than previously thought (Tewksbury *et al.* 2011). Minimizing the adverse impacts of climate change on biodiversity is dependent on:

- efforts to mitigate climate change itself (Chapter 3);
- measures to ensure that those activities and societal adaptation efforts do not themselves have adverse impacts on biodiversity; and
- application of best practice in conserving and restoring biodiversity in the face of climatic change.

Of the wide range of approaches, many are dependent on the conservation and sustainable use of healthy ecosystems, and offer opportunities for synergies in terms of climate change mitigation and maintenance of biodiversity. In particular, this concerns intact forests and wetlands, but also natural and semi-natural grasslands and many agricultural ecosystems. For example, some agricultural approaches, such as conservation tillage and agroforestry, can result in the maintenance and enhancement of terrestrial carbon stocks and also contribute to the conservation and sustainable use of biodiversity (CBD 2009a). Traditional knowledge and systems of small-scale livestock husbandry, farming, and forest product collection can greatly enable local mitigation and adaptation in culturally appropriate ways (RECOFTC 2010; IUCN 2008). However, ecosystem-based approaches also carry risks and these need

The rehabilitation of wetlands is an important tool for restoring biodiversity and building resilience to the impacts of climate change.

© J. Smith/Still Pictures

Considerable efforts have been made over the past decade to promote marine protected areas in Eastern Africa. *© J Tamelander/IUCN*

to be assessed and addressed. In the case of forests, the United Nations Framework Convention on Climate Change (UNFCCC 1992) and CBD have recognized a need for safeguards, particularly for biodiversity and human communities, to minimize risks associated with REDD+. There is also a risk of conflict between the goals of the Kyoto Protocol's Clean Development Mechanism for carbon sequestration and biodiversity conservation (Heiskanen 2009; Kneteman and Green 2009).

Other concerns about the impacts of mitigation activities include those related to artificial ocean fertilization through using nutrients such as iron or nitrogen to increase the uptake and sequestration of atmospheric carbon. The effectiveness of this approach is highly uncertain and increasingly thought to be quite limited. Potential negative environmental effects include increased production of methane and nitrous oxide and changes in phytoplankton community composition, which may lead to toxic algal blooms (CBD 2009b). Alternative sources of energy production, such as biofuels, hydropower, wind farms and oceanic tide generators have all been documented to have impacts on biodiversity if safeguards are not developed (Keder and McIntyre Galt 2009; McDonald *et al.* 2009). The most fundamental strategy for conserving biodiversity in the face of climate change will continue to be promoting the conservation of intact and functioning ecosystems supported by restoration initiatives wherever possible (CBD 2009a).

Managing area-based conservation

Protected areas are seen by many as the core means of preventing ongoing losses of species and habitats. Protected areas have expanded over the past 20 years in both number and area (Figures 5.12 and 5.13) and now cover 13 per cent of the world's land area (IUCN and UNEP-WCMC 2011). However, coverage is uneven, and 6 of the 14 global biomes and half of the 821 terrestrial ecoregions do not meet the CBD target for 10 per cent of their area to be protected by 2010 (Jenkins and Joppa 2009). Furthermore, the expansion of the world's protected area network needs to be targeted at the most important sites for biodiversity. Some 51 per cent of the 587 sites identified by the Alliance for Zero Extinction as critical for the survival of hundreds of highly threatened species, and 49 per cent of the more than 10 000 important bird areas are still entirely outside the protected area network (Butchart *et al.* 2012). Even more importantly, the performance of protected areas in maintaining populations of their key species is poorly documented. Although some studies have found wildlife declines within some protected areas (Woinarski *et al.* 2011; Craigie *et al.* 2010), others demonstrate that protected areas have been effective in maintaining species that would otherwise have disappeared (Bruner *et al.* 2001). However, not all species may require protected areas to ensure their survival (Pereira and Daly 2006), and protected areas require complementary broad-scale conservation measures (Boyd *et al.* 2008).

Uneven protected area coverage of biomes is most evident in the marine realm, despite a CBD target to protect 10 per cent of the ocean by 2012. By the end of 2010 marine protected

Figure 5.12 Extent of nationally designated protected areas, 1990–2010

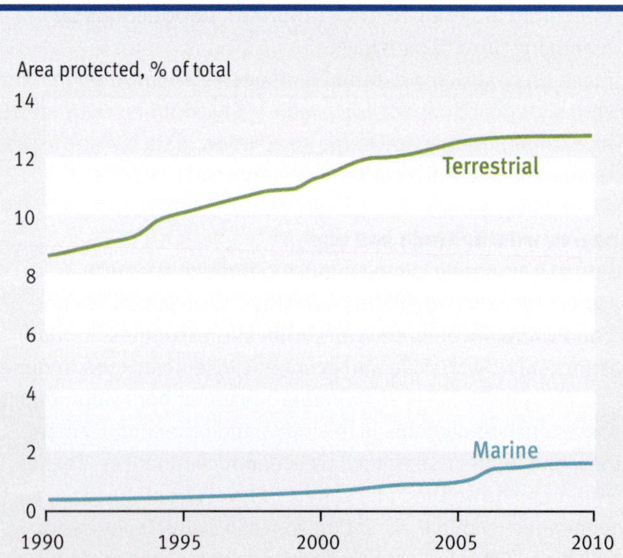

Source: IUCN and UNEP-WCMC 2011

areas covered 1.6 per cent of the ocean area (IUCN and UNEP-WCMC 2011). Indeed, at the end of 2010, only 12 countries had designated more than 10 per cent of their waters, often through large areas, while 121 countries had yet to designate more than 0.5 per cent of their marine jurisdiction (Toropova *et al.* 2010). In response, the CBD has retained the 10 per cent target, with a revised achievement date of 2020.

Marine protected areas can be designated at a variety of levels of protection, but those with complete protection provide the greatest biodiversity benefits. A review of 112 independent studies in 80 different protected areas found significantly higher fish populations inside the reserves than in surrounding areas or in the same place before protection was established. Relative to reference sites, population densities were 91 per cent higher, biomass 192 per cent higher and average organism size and diversity 20–30 per cent higher, usually between one and three years after establishment of a reserve. These trends occurred even in small marine protected areas (Halpern 2003).

Protected areas can also play a key role in climate change mitigation and adaptation, preventing the conversion of natural habitats to other land uses and hence avoiding significant release of carbon (Dudley *et al.* 2010b). Emissions from land-use change, mainly forest loss, contribute up to 17 per cent of all anthropogenic greenhouse gas emissions (IPCC 2007). It has been estimated that about 15 per cent of the global terrestrial carbon stock is stored in the world's protected area network (Campbell *et al.* 2008), and the role that this can play in climate mitigation is underlined by the fact that between 2000 and 2005, protected areas in humid tropical forests lost about half as much carbon as the same area of unprotected forest (Scharlemann *et al.* 2010).

Indigenous and community-conserved areas

Protected areas can be effectively managed by many groups, from government agencies to local communities and indigenous peoples, and from non-governmental organizations (NGOs) to private individuals. Recently, the full range of IUCN protected area categories has been brought into use for designating protected areas (Dudley *et al.* 2010a). For example, in Australia, protected areas established and managed by indigenous communities comprise nearly a quarter of Australia's national reserve system by area. Indigenous and community-conserved areas (ICCAs) and sacred natural sites (SNSs) have proven successful in conserving a rich biological and biocultural diversity by supporting the maintenance of traditional environmental knowledge and practices (Porter-Bolland *et al.* 2012; Sobrevila 2008). These community areas are extremely diverse, manifesting myriad ethical, economic, cultural, spiritual and political dimensions (Brown and Kothari 2011; Borrini-Feyerabend *et al.* 2010a, 2010b; Kothari 2006; Posey 1999). They include waterfowl nesting wetlands, roosting sites or other critical wild animal habitats, and landscapes with mosaics of natural and agricultural ecosystems such as the Potato Park in the Andean Highlands of Peru and the rice terraces of the Philippines. A number of studies demonstrate the wide range of values they provide (Box 5.6) (Mallarach *et al.* 2012; Verschuuren *et al.* 2010; ICCA 2009).

Figure 5.13 Proportion of each terrestrial ecoregion covered by protected areas, 2011

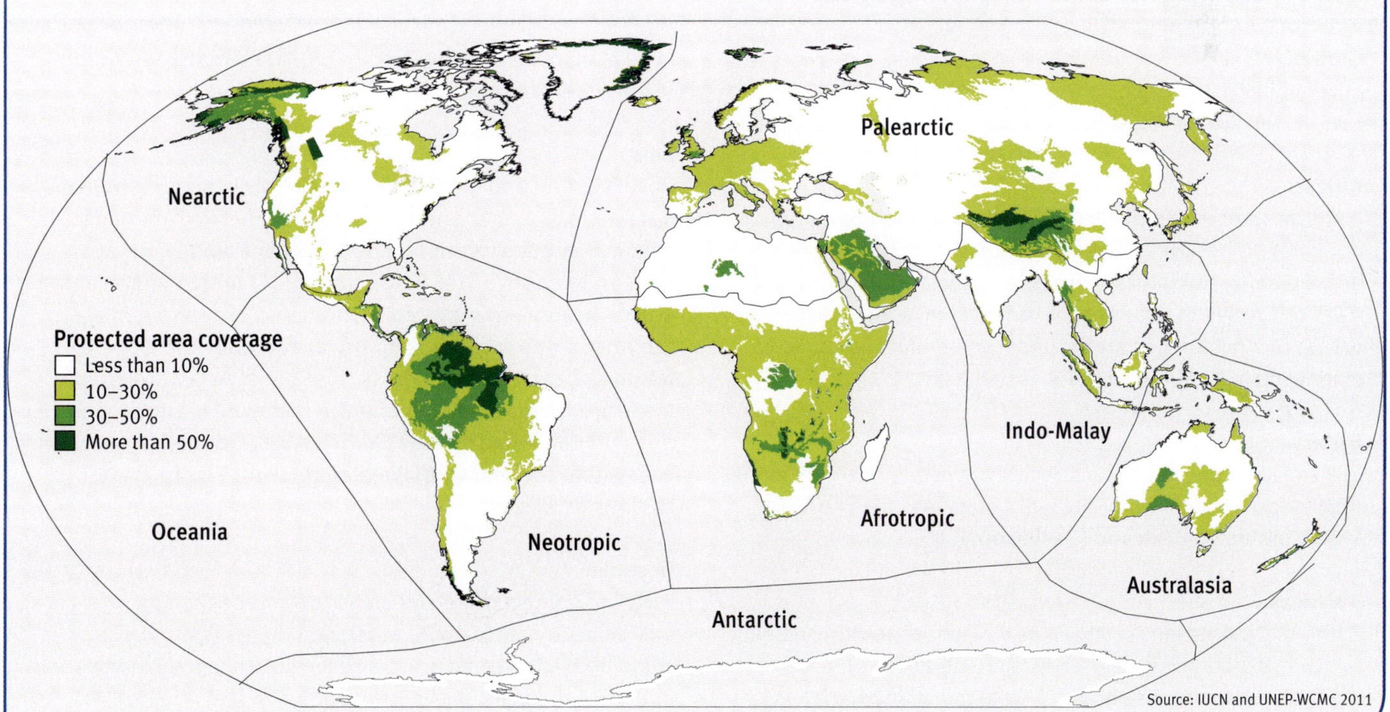

Source: IUCN and UNEP-WCMC 2011

The number and extent of ICCAs and SNSs have not been comprehensively estimated. It has been, nevertheless, suggested that in some parts of the world their area is similar to that currently under government-managed protection (Box 5.6) (Molnar *et al.* 2004). Furthermore, it has been estimated that communities own or manage 22 per cent of all forests in 18 developing countries (White and Martin 2002). Recent analyses highlight the potential effectiveness of indigenous and community-managed areas in tropical forest conservation. For example, such areas can be more effective in reducing tropical deforestation than forest protected areas (Porter-Bolland *et al.* 2012), and indigenous and multiple-use protected areas can reduce the incidence of tropical forest fires as effectively as strict protection (Nelson and Chomitz 2011).

ICCAs and SNSs are increasingly recognized as legitimate and powerful tools for the security of both their custodians and the biodiversity they encompass, supported by a range of conservation, human-rights and development instruments. A preliminary survey of the laws and policies of 27 countries and one sub-national region showed that progress in national recognition of ICCAs and SNSs is patchy: some countries are moving rapidly, others slowly, and some not at all (Kothari *et al.* 2010). The biggest challenge, now that ICCAs and SNSs have global attention, is in gaining appropriate national recognition and support, particularly for tenure, customary practices and decision-making institutions, and other fundamental human rights (Stevens 2010). Activities relating to governance, participation, equity and benefit sharing in relation to protected areas merit increasing consideration.

Recognizing the value of cultural diversity and traditional knowledge

The recognition of human and natural systems as unified social-ecological systems is increasingly important for safeguarding biodiversity (Ostrom 2007). This growing understanding underscores the links between biological and cultural diversity and the role of local and indigenous peoples in the sustainable governance and management of biodiversity (Sutherland 2003; Moore *et al.* 2002). The Strategic Plan for Biodiversity and the Aichi Biodiversity Targets support greater respect of traditional knowledge and its full integration and reflection in CBD implementation at all levels, with the full and effective participation of indigenous and local communities (Aichi Target 18, Box 5.1). Information on the status and trends of linguistic diversity (Figure 5.14) has been used as a proxy indicator for traditional knowledge, innovations and practices, including those about biodiversity. Traditional knowledge is an invaluable and irreplaceable source of information about biodiversity and human relationships; its loss entails a loss of collective cultural heritage and capacity to adapt to and live sustainably within specific ecosystems and areas (Maffi and Woodley 2010; Swiderska 2009).

Access and benefit sharing of genetic resources and associated traditional knowledge

The fair and equitable sharing of the benefits of exploiting genetic resources is one of the three CBD objectives (Article 1), recognized as critical for biodiversity conservation. Through the recently adopted Nagoya Protocol on Access to Genetic Resources and the Fair and Equitable Sharing of Benefits Arising from their Utilization, standards are established for regulating access to genetic resources

Box 5.6 Examples of community management

Globally:
Community-controlled or managed forests total some 400–800 million hectares (Molnar *et al.* 2004; White *et al.* 2004).

Africa:
Forty-seven of the approximately 70 Kaya forests, totalling about 6 000 hectares, have been legally recognized in Kenya and are being cared for in collaboration with local communities (Githitho 2003). In the Republic of Tanzania, a total area in excess of 2 million hectares is under community-based forest management (Blomley and Iddi 2009).

Europe:
In a small country like Estonia, there are estimated to be more than 7 000 sacred natural sites, although less than 500 are legally protected (Valk and Kaasik 2007).

Americas:
A fifth of the Amazon is classed as indigenous territory helping to achieve biodiversity conservation (Oviedo 2006), and more than 800 000 hectares of boreal forest and wetlands have been declared protected traditional territory in Canada (Government of Manitoba 2011).

Asia:
Several thousand natural ecosystem sites, ranging from a hectare to several hundred square kilometres, are under community conservation in South Asia (Kalpavriksh 2011; Jana and Paudel 2010; Pathak 2009). At least 13 720 sacred groves have been reported in India and experts estimate the total number for the country at 100 000–150 000 (Malhotra *et al.* 2001). Across South East Asia and Japan, there are hundreds of community-managed marine areas oriented towards sustainable fisheries and coastal/marine ecosystem conservation (Yagi *et al.* 2010; Ferrari 2006; Lavides *et al.* 2006).

Oceania:
Forty indigenous protected areas cover more than 23 million hectares (DSEWPC 2011), and there are hundreds of community-conserved areas and locally managed marine areas in the South Pacific (Govan *et al.* 2009).

A Kenyan fisherman sets out in a traditional wooden boat to fish beyond the coral reef almost a kilometre offshore.

© Cheryl-Samantha Owen/ samowenphotography.com

and the distribution of benefits from their use, as well as the associated traditional knowledge. The principle underlying CBD recognizes that states have a sovereign right to exploit their own resources pursuant to their own environmental policies (Article 3).

Figure 5.14 Language endangerment as a share of all languages, 2010

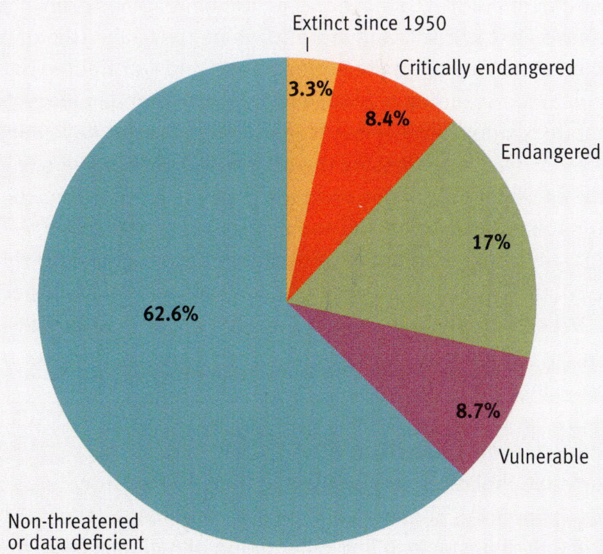

Source: Moseley 2010

Access to genetic resources has emerged as a major political rallying point in international negotiations. Much of the world's biodiversity is concentrated in the forests of developing countries in the tropics, but much of the technology and financial capital that can convert elements of biodiversity into commercial products rests with the developed countries. Hence, while unprecedented biodiversity loss is a global concern, commercial use and the associated issues of intellectual property fundamentally alter the nature of biodiversity as a global public good (Giraud 2008; Gupta 2006; Schuler 2004). The impetus behind the Nagoya Protocol arose from growing discontent amongst developing countries and indigenous and local communities regarding the lack of implementation of the benefit-sharing provisions of CBD since it came into force in 1993. This was compounded by only a handful of user countries undertaking any compliance measures to prevent bio-piracy despite the adoption of guidelines in 2002.

The Nagoya Protocol is an important milestone for rectifying the issues of equity associated with the commercial use of genetic resources and associated traditional knowledge. The protocol is also unprecedented in its recognition of the right of indigenous and local communities to regulate access to traditional knowledge associated with genetic resources in accordance with their customary laws and procedures. The protocol opened for signature in February 2011 and will not enter into force until 90 days after 50 countries have signed. A number of countries

Figure 5.15 The number and type of access and benefit-sharing measures, 2011

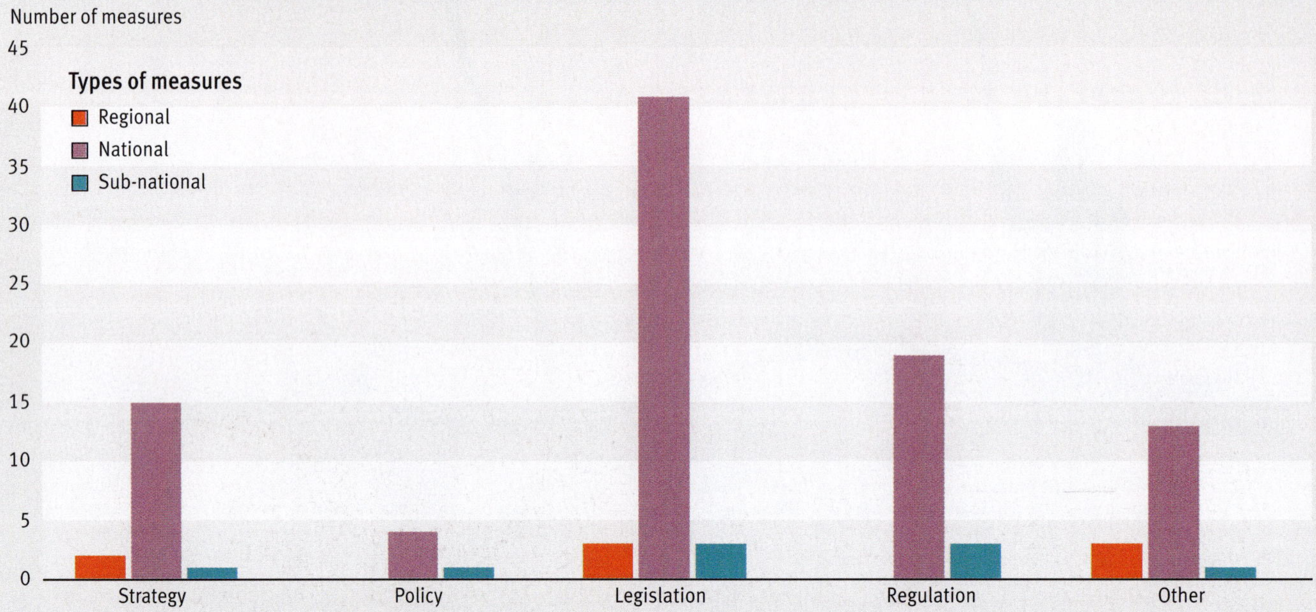

Note: The category Other includes resolutions, guidelines and declarations.

Source: CBD 2012

already have national legislation and regulations pertaining to issues of access and benefit sharing, and monitoring the further development of such regulations could provide a useful indicator of progress (Figure 5.15).

In the marine realm, ten countries own 90 per cent of the patents deposited with marine genes – with 70 per cent belonging to just three – but account for only about 20 per cent of the world's coastline. These nations benefit from access to the advanced technologies required to explore the vast genetic reservoir of the oceans, leading to a call for policies targeting capacity building to improve access for other countries (Arnaud-Haond *et al.* 2011).

PROGRESS, GAPS AND OUTLOOK

Assessment of progress and gaps

Conservation strategies

Protected areas are one of the primary responses for maintaining biodiversity, particularly on land, but are generally deemed to be insufficient (Rodrigues *et al.* 2004). The exclusion of local communities from many state and private protected areas along with the failure to fully acknowledge their role in safeguarding biodiversity remains a challenge to real progress. Outside protected areas the proportion of sustainably managed production landscapes – for agriculture, forestry, fisheries and aquaculture, amongst others – is increasing, but only slowly. For example, the area of forest certified by the Forest Stewardship Council (FSC) as sustainably managed continues to grow, reaching 149 million hectares in 2012 (FSC 2012), and there are additional areas managed under the Programme for the Endorsement of Forest Certification (PEFC), but this remains a small fraction of the global total of managed forests. Similarly, fish products certified by the Marine Stewardship Council (MSC) constituted only 7 per cent of global fisheries in 2007 (Jacquet *et al.* 2009).

National biodiversity strategies and action plans

The CBD requires all member states to develop a national biodiversity strategy and action plan as the primary mechanism for the implementation of its strategic plan. To date, 172 of the 193 signatory countries have adopted their plans or equivalent instruments (CBD 2011). The large number of plans is an achievement in itself, and more so where they have stimulated conservation action at the national level and contributed to a better understanding of biodiversity, its value and management. In spite of these achievements, national strategies have not been fully effective in addressing the main drivers of biodiversity loss. Only a few countries have used the plans as mechanisms for mainstreaming biodiversity and ecosystem services, and there is generally poor coordination with other relevant policies (Prip *et al.* 2010; CBD 2010c). However, Parties to the CBD are expected by 2014 to revise their plans in line with the new Strategic Plan for Biodiversity 2011–2020, which includes reference to improving mainstreaming.

Resource mobilization

Many national reports submitted to the CBD have identified the lack of financial, human and technical resources as the most widespread obstacle to implementation of national strategies and the CBD in general. Thus, the fulfilment of the Aichi Target to substantially increase resource mobilization will be crucial to enable the other targets to be achieved.

While documentation is lacking for both the current and the required level of financing to safeguard biodiversity, there is no doubt that the gap between the two is substantial. Estimates suggest that existing financing is in the order of tens of billions of dollars a year, while total needs are of the order of hundreds of billions a year (Rands *et al.* 2010; Berry 2007; James *et al.* 2001). International financing for biodiversity is estimated to have grown by approximately 38 per cent in real terms since 1992 and now stands at US$3.1 billion annually (OECD 2010; Gutman and Davidson 2008). The Global Environmental Facility (GEF) will provide US$1.2 billion for CBD implementation from 2010 to 2014, an increase of 29 per cent compared to the previous four years.

Increasingly, innovative financial mechanisms are considered essential tools to mobilize additional resources for biodiversity. These include payment for ecosystem services, biodiversity offsets, ecological fiscal reforms, markets for green products and biodiversity in new sources of international development finance. For example, further information on schemes such as Reducing Emissions from Deforestation and Degradation (REDD+) is available in Chapter 3.

Madagascar – a global biodiversity hotspot with many endemic species including lemurs – is using payment-for-ecosystem-services schemes to attract new funding for the protection of biodiversity and ecosystems. *© Tdhster/iStock*

Knowledge gaps for biodiversity monitoring

Although indicators of the state of biodiversity are predominantly showing declines (Butchart *et al.* 2010; CBD 2010b), there are considerable gaps in their geographic, taxonomic and temporal coverage (Pereira *et al.* 2010a, 2010b; Walpole *et al.* 2010; Collen *et al.* 2008a, 2008b). While biodiversity loss is a global phenomenon, its impact may be greatest in the tropics where available indicators and data coverage are the least complete.

Particular gaps in knowledge for state indicators include: grassland and wetland extent, habitat condition, primary productivity, genetic diversity of wild species, freshwater and terrestrial trophic integrity, ecosystem functioning and ocean acidification. Pressure indicators lack data on pollution, exploitation in terrestrial and freshwater ecosystems, wildlife disease incidence and freshwater extraction. The principal gaps in response indicators include sustainable management of agriculture and freshwater fisheries, and management of invasive alien species.

A prominent gap in knowledge concerns ecosystem services (UNEP-WCMC 2011; TEEB 2010). Indicators of the biodiversity that underpins these services should be tailored to the scales at which ecological processes that produce the services occur, such as the landscape scale for agriculture and biomass production, and the watershed for direct water use and hydroelectricity generation.

Other responses to biodiversity loss include policy action to tackle an array of issues including hunting and pollution, and enforcement of environmental impact assessments and mitigation measures for infrastructure development; however, global trend data are unavailable for these. Given that most global biodiversity targets, such as the Aichi Targets, require action at the national scale, national biodiversity data are crucial for tracking progress towards global biodiversity targets, and to inform national strategies. National Red Lists of threatened species are one of the many examples of nationally relevant biodiversity data that may provide suitable input for reporting on progress towards these goals and for informing national conservation priority setting (Zamin *et al.* 2010), although there are others which are also suitable (Jones *et al.* 2011). The Group on Earth Observations Biodiversity Observation Network (GEO BON) is expected to make an important contribution to future monitoring efforts (GEO BON 2011), whilst the Biodiversity Indicators Partnership (BIP 2011) is supporting global and national biodiversity indicator development for the Aichi Targets and for national biodiversity strategies and action plans.

Projections, scenarios and horizon scans

While recognizing a time frame of increasing uncertainty, this section synthesizes biodiversity studies from short-term projections through to longer-term scenarios with a view to distilling relatively short-term policy implications. This relies heavily on the *GBO-3* analysis of biodiversity scenarios (Leadley *et al.* 2010; Pereira *et al.* 2010a), for which scientists from a wide range of disciplines came together to seek consensus on projections and scenarios for biodiversity change during the 21st century.

Although quantitative projections and scenario methods are well advanced, the range of projected changes reported by the studies reviewed is rather broad, partially because there are significant opportunities to intervene through better policies, but also because of large uncertainties in the projections. The projections of global change impacts on biodiversity show continuing and, in many cases, accelerating species extinctions (Figure 5.16), loss of natural habitat, and changes in the distribution and abundance of species and biomes over the 21st century. Possible thresholds, amplifying feedbacks and time-lag effects leading to tipping points appear to be widespread and make the impacts of global change on biodiversity hard to predict, difficult to control once they begin, and slow and expensive to reverse once they have occurred. For many important cases, the degradation of ecosystem services goes hand-in-hand with species extinctions, declining species abundance or widespread shifts in species and biome distributions; however, the conservation of biodiversity and of some services, especially provisioning services, is often at odds. Strong action at international, national and local levels to mitigate the drivers of biodiversity change and to develop adaptive management strategies could significantly reduce or reverse undesirable and dangerous biodiversity transformations if urgently, comprehensively and appropriately applied.

Policy implications

The accumulated evidence, cited above, suggests that there is greater success in halting the negative changes in biodiversity and ecosystem services when a proactive attitude in support of a sustainable environment is adopted. Overall, the above synthesis, coupled with inputs from UNEP's Foresight Initiative (Peduzzi *et al.* 2011), suggests that:

- land must be used more efficiently to decrease the rate of habitat loss;
- mitigation of climate change is urgent and there is a significant risk of tipping points near or even before the 2°C mean global surface temperature target agreed at the UNFCCC meeting in Cancun in 2010;
- payments for ecosystem services and the greening of national accounts can help to protect biodiversity if appropriately applied;
- protected areas by themselves have not been adequate to achieve the target of reducing the rate of biodiversity loss by 2010;
- potential collapse of oceanic ecosystems requires an integrated and ecosystem-based approach to ocean governance; and
- recognizing the importance of local participation and community support, it is crucial to ensure that policies are integrated, sensitive and inclusive of local communities. This applies to conservation strategies, preservation of local cultures and languages, and access and benefit sharing of genetic resources and traditional knowledge.

Outlook summary

A summary of progress in achieving the main biodiversity goals is provided in Table 5.2. It also outlines gaps in data and policy and is based on expert opinion. The International Platform on Biodiversity and Ecosystem Services (IPBES) is expected to play an important role at the science-policy interface in future (Perrings *et al.* 2011).

Figure 5.16 Scenarios of species change

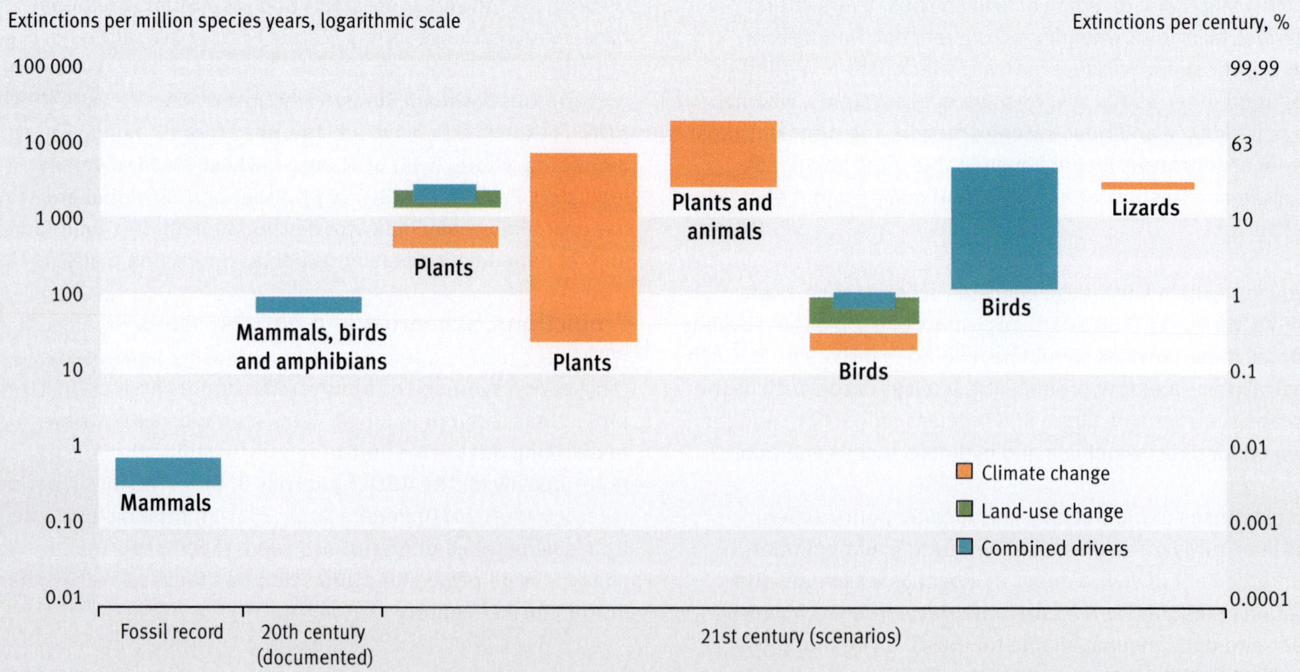

Source: Pereira *et al.* 2010a

Table 5.2 Progress towards goals (see Table 5.1)

A: Significant progress B: Some progress	C: Very little to no progress D: Deteriorating	X: Too soon to assess progress ?: Insufficient data

Key issues and goals	State and trends		Outlook	Gaps
1. Reduce the direct pressures on biodiversity *(Notes 4, 6, 7, 13; CBD Targets 5–10)*				
Drivers of habitat loss and degradation	C	Continuing increases in pressures from, for example, agriculture and infrastructure development	Increasing pressure	Quantification of trends in habitat extent and condition resulting from different drivers
Levels of exploitation	C	Significant proportion of species is threatened by overexploitation; legal international trade is successfully managed for a small number of species	Increasing pressure	More systematic measures of exploitation levels, particularly at local/national scale, including illegal trade
Spread and impact of invasive alien species	B/C	Numbers and extent of invasive alien species are increasing where quantified; impacts have been successfully mitigated and the spread limited in some cases	Continuing spread and impact, with local exceptions	Numbers/impacts in developing countries, policy implementation and effectiveness at local/national scale
Pressure from pollutants	B	Generally increasing pressures from pollution, but nitrogen deposition since the 1990s may be levelling off	Increasing pressure, with local exceptions for certain pollutants	Trends in levels of pollutants other than nitrogen
Impacts of climate change	C	Increasing impacts on phenology, abundance, distribution and community composition in all ecosystems	Increasing pressure	Impacts on population trends and interactions with other threats
2. Improve the status of biodiversity *(Notes 1, 2, 3, 4, 7, 8, 9, 11, 12; CBD Targets 11–13)*				
Genetic diversity of wild species	?	Genetic diversity of cultivated crops and domesticated animals has declined and, while un-quantified in wild species, is likely to be declining	Continuing decline	Data collection on genetic diversity of wild populations
Population abundance of species	C	Declining at the global scale, most rapidly in the tropics, freshwater habitats and for utilized marine species; there are some exceptions due to effective conservation action, for example North American waterbirds	Continuing decline	Trends for plants and invertebrates; tropical coverage patchy; systematic monitoring largely confined to birds in developed countries
Extinction risk of species	C	13–63 % of species in different groups are threatened with extinction; trends, where known, are declining (most rapidly for corals)	Continuing decline	Trends for plants, invertebrates and remaining vertebrate classes; national scale extinction risk trends
Extent, condition and integrity of biomes, habitats and ecosystems	C	Declines in all natural habitats with known trends, for example forests, mangroves, seagrasses and coral reefs; some exceptions, for example reforestation in some temperate countries	Continuing decline	Consistent and repeated remote-sensing monitoring, including for non-forest areas; metrics of condition and fragmentation
3. Enhance sustainable benefits (ecosystem services) from biodiversity *(Notes 1, 2, 3, 4, 9, 11, 12; CBD Targets 14–16)*				
Status of species harvested for food and medicine	C	Extinction risk trends are worse for species harvested for food and medicine than for other species	Benefits currently unsustainable and likely to decline	Trends for plants and invertebrates; disaggregation of all data into small-scale subsistence use or large-scale and/or commercial use
Equitable use of natural resources	C	For some countries the per-person ecological footprint is high and/or increasing relative to life expectancy, indicating inefficiency and often unsustainability in resource use	Potential for the global ecological footprint to be reduced while enhancing human well-being, requiring major adjustments in benefit sharing	Suitable data for footprint analysis, including spatial and temporal resolution data on the intensity and magnitude of natural resource use at global level

Table 5.2 Progress towards goals (see Table 5.1) *continued*

4. Strengthen responses to safeguard biodiversity *(Notes 1, 2, 3, 4, 6, 7, 8, 9, 10, 12, 13; CBD Targets 1–20)*				
Extent, biodiversity coverage and integrity of protected areas	B	Terrestrial coverage has reached nearly 13%, but marine coverage is less than 1.5%; representativeness at the scale of ecoregions is fairly high, but the proportion of fully protected key biodiversity sites is low	Protected area extent is likely to increase if governments fulfil their commitments; more careful site selection and better management will be required to protect biodiversity; jurisdictional uncertainties and conflicts need resolution	Data on trends in the effectiveness of protected areas and on jurisdictional uncertainties and conflict
Extent, biodiversity coverage and integrity of indigenous and community-conserved areas (ICCAs), sacred natural sites (SNSs) and other community-managed natural areas	B	Community-based governance and management approaches exist largely without state recognition or are newly developing; external drivers of biodiversity loss and/ or other factors undermine the capacity of ICCAs, SNSs, and other such areas to conserve biodiversity	Likely to increase in importance; empowerment of local communities in decision making is needed, plus greater awareness amongst government protected-area officials	Data on the location, extent, legal status and effectiveness of these areas for biodiversity conservation; possible forms and modes of appropriate state recognition and support
Schemes such as REDD+ or payment for ecosystem services (PES), where biodiversity supports mitigation of, and adaptation to, climate change	B	The development of REDD+ and PES schemes is increasing	The area under REDD+ and PES schemes is likely to increase, providing both opportunities and potential threats for biodiversity conservation	Potential indicators such as number and area of community-managed REDD+ areas or number of national adaptation strategies with ecosystem-based components
Proportion of sustainably managed production areas	C	Area certified as sustainably managed increasing, but the proportional area remains minimal, with an uneven global distribution	Area of certified production increasing, especially in developed countries	Effectiveness for biodiversity conservation; impacts of these approaches in non-certified areas
Policy responses addressing invasive alien species	B	Proportion of countries with relevant legislation increasing, but implementation and transboundary cooperation are poor	Policy responses increasing but ineffective without considerably improved implementation	More data needed on implementation and effectiveness
Action for species recovery, site safeguarding and habitat restoration	B	Numerous local examples show that successful conservation programmes prevent extinctions, restore habitats and conserve sites; however, the scale of these efforts remains inadequate	Improvements in coordination and integration are expected, but on their own will remain insufficient	More data on species recovery and restoration needed
Number of countries with national mechanisms addressing access and benefit sharing	B	Agreement of Nagoya Protocol on access and benefit sharing is a significant step forward, with increasing numbers of signatories and countries with relevant legislation	Implementation of the Nagoya Protocol could address this issue effectively	Data required on access and benefit-sharing agreements and beneficiaries, and on the benefits and sustainability of utilizing genetic resources
Number of languages and speakers as a proxy for traditional knowledge supporting sustainable resource use and conservation	C	Number of languages and speakers is declining, suggesting less traditional knowledge in support of sustainable use and conservation	Appropriate mechanisms, including support for customary sustainable use of biodiversity and secure tenure, may help to halt the decline in traditional knowledge	Indicators to capture intergenerational transfer of traditional knowledge and provision of incentives; indicators on the retention of traditional knowledge to assess social-ecological resilience

Notes: 1. CBD Article 1; **2.** CBD Article 6; **3.** CBD Article 8j; **4.** CBD Article 10; **5.** CBD COP 7 Decision VII/28 Paragraph 1.2.3; **6.** CBD COP 7 Decision VII/30 Annex II; **7.** Agenda 21 Chapter 17 Paragraph 86; **8.** CMS 1979 Preamble; **9.** CITES 1973 Preamble; **10.** ICPP Article 1; **11.** Ramsar Article 3; **12.** ITPGRFA Article 1 Paragraph 1.1; **13.** Cartagena Protocol on Biosafety Article 1.

REFERENCES

Altieri, M.A. (1999). The ecological role of biodiversity in agroecosystems. *Agriculture, Ecosystems and Environment* 74, 19–31

Anticamara, J.A., Watson, R., Gelchu, A. and Pauly, D. (2011). Global fishing effort (1950–2010): trends, gaps, and implications. *Fisheries Research* 107, 131–136

Arnaud-Haond, S., Arrieta, J.M. and Duarte, C.M. (2011). Marine biodiversity and gene patents. *Science* 331(6024), 1521–1522

Arthurton, R., Barker, S., Rast, W., Huber, M., Alder, J., Chilton, J., Gaddis, E., Pietersen, K., Zöckler, C., Al-Droubi, A., Dyhr-Nielsen, M., Finlayson, M., Fortnam, M., Kirk, E., Heileman, S., Rieu-Clark, A., Schäfer, M., Snoussi, M., Danling Tang, L., Tharme, R., Vadas, R. and Wagner, G. (2007). Water. In *Global Environment Outlook-4: Environment for Development*. pp.115–156. United Nations Environment Programme, Nairobi

Bai, Z.G., Dent, D.L., Olsson, L. and Schaepman, M.E. (2008). Proxy global assessment of land degradation. *Soil Use and Management* 24(3), 223–234

Baillie, J.E.M., Griffiths, J., Turvey, S.T., Loh J. and Collen, B. (2010). *Evolution Lost: Status and Trends of the World's Vertebrates*. Zoological Society of London, London

Baker, A.C., Glynn, P.W. and Riegl, B. (2008). Climate change and coral reef bleaching: an ecological assessment of long-term impacts, recovery trends and future outlook. *Estuarine Coastal and Shelf Science* 80(4), 435–471

Balirwa, J.S., Chapman, C.A., Chapman, L.J., Cowx, I.G., Geheb, K., Kaufman, L., Lowe-McConnell, R.H., Seehausen, O., Wanink, J.H., Welcomme, R. and Witte, F. (2003). Biodiversity and fishery sustainability in the Lake Victoria basin: an unexpected marriage? *BioScience* 53(8), 703–716

Barnhizer, D. (2001). Trade, environment, and human rights: the paradigm case of industrial aquaculture and the exploitation of traditional communities. In *Effective Strategies for Protecting Human Rights: Economic Sanctions, Use of National Courts and International Fora, and Coercive Power* (ed. Barnhizer, D.). pp.137–155. Ashgate, Burlington, VT

Bates, B., Kundzewica, Z.W., Wu, S. and Palutikof, J. (eds.) (2008). *Climate Change and Water*. IPCC Technical Paper VI. IPCC Secretariat, Intergovernmental Panel on Climate Change, Geneva

Battin, T.J., Luyssaert, S., Kaplan, L.A., Aufdenkampe, A.K., Richter, A. and Tranvik, L.J. (2009). The boundless carbon cycle. *Nature Geoscience* 2, 598–600

Bax, N.J. and Thresher, R.E. (2009). Ecological, behavioral, and genetic factors influencing the recombinant control of invasive pests. *Ecological Applications* 19(4), 873–888

Bax, N., Williamson, A., Aguero, M., Gonzalez, E. and Geeves, W. (2003). Marine invasive alien species: a threat to global biodiversity. *Marine Policy* 27, 313–323

Belfrage, K. (2006). The effects of farm size and organic farming on diversity of birds, pollinators and plants in Swedish landscape. *Ambio* 34(8), 582–588

Benning, T.L., LaPointe, D., Atkinson, C.T. and Vitousek, P.M. (2002). Interactions of climate change with biological invasions and land use in Hawaiian Islands: modelling the fate of endemic birds using a geographic information system. *Proceedings of the National Academy of Sciences of the United States of America* 99, 14246–14249

Bernard, P.S. (2003). Ecological implications of the water spirit beliefs in southern Africa: the need to protect knowledge, nature, and resource rights. In *Science and Stewardship to Protect and Sustain Wilderness Values* (eds. Watson, A. and Sproull, J.). 7th World Wilderness Congress Symposium, Port Elizabeth, South Africa, 2–8 November 2001

Berndes, G., Hoogwijk, M. and van den Broek, R. (2003). The contribution of biomass in the future global energy supply: a review of 17 studies. *Biomass and Bioenergy* 25(1), 1–28

Berry, P. (2007). *Adaptation Options on Natural Ecosystems*. A report to the United Nations Framework Convention on Climate Change Secretariat, Financial and Technical Support Division. Environmental Change Unit, University of Oxford, Oxford

Best, A., Giljum, S., Simmons, C., Blobel, D., Lewis, K., Hammer, M., Cavalieri, S., Lutter, S. and Maguire, C. (2008). *Potential of the Ecological Footprint for Monitoring Environmental Impacts from Natural Resource Use: Analysis of the Potential of the Ecological Footprint and Related Assessment Tools for Use in the EU's Thematic Strategy on the Sustainable Use of Natural Resources*. Report to the European Commission, Directorate General for the Environment

Bhathal, B. and Pauly, D. (2008). "Fishing down marine food webs" and spatial expansion of coastal fisheries in India, 1950–2000. *Fisheries Research* 91, 26–34

BIP (2011). *Biodiversity Indicators Partnership*. http://www.bipindicators.net (accessed 30 November 2011)

Björklund, G., Bullock, A., Hellmuth, M., Rast, W., Vallée, D. and Winpenny, J. (2009). Water's many benefits. In *United Nations World Water Development Report 3: Water in a Changing World*. World Water Assessment Programme, pp 80–95. UNESCO, Paris and Earthscan, London

Blomley, T. and Iddi, S. (2009). *Participatory Forest Management in Tanzania 1993–2009, Lessons Learned and Experiences to Date*. Forestry and Beekeeping Division, United Republic of Tanzania Ministry of Natural Resources and Tourism

Borrini-Feyerabend, G., Kothari, A., Alcorn, J., Amaya, C., Bo, L., Campese, J., Carroll, M., Chapela, F., Chatelain, C., Corrigan, C., Crawhall, N., de Vera, D., Dudley, N., Hoole, A., Farvar, M.T., Ferguson, M., Ferrari, M.F., Finger, A., Foggin, M., Hausser, Y., Ironside, J., Jallo, B., Jonas, H., Jones, M., Lasimbang, J., Lassen, B., Lovera, S., Martin, G., Morris, J., Nelson, F., Okuta, J.S., Oviedo, G., Pathak, N., Ramirez, R., Rasoarimanana, V., Riascos de la Peña, J.C., Royo, N., Sandwith, T., Shrumm, H., Smyth, D., Stevens, S., Surkin, J. and Wild, R. (2010a). *Strengthening What Works – Recognising and Supporting the Conservation Achievements of Indigenous Peoples and Local Communities*. Briefing Note 10. IUCN Commission on Environmental, Economic and Social Policy, International Union for Conservation of Nature, Gland

Borrini-Feyerabend, G., Lassen, B., Stevens, S., Martin, G., Riascos de la Peña, J.C., Ráez-Luna, E.F. and Farvar, M.T. (2010b). *Bio-cultural Diversity Conserved by Indigenous Peoples and Local Communities – Examples and Analysis*. Companion document to Briefing Note 10. IUCN Commission on Environmental, Economic and Social Policy, International Union for Conservation of Nature, Gland

Boyd, C., Brooks, T.M., Butchart, S.H.M., Edgar, G.J., da Fonseca, G.A.B., Hawkins, F., Hoffmann, M., Sechrest, W., Stuart, S.N. and van Dijk, P.P. (2008). Spatial scale and the conservation of threatened species. *Conservation Letters* 1, 37–43

Branch, T.A., Jensen, O.P., Ricard, D., Ye, Y. and Hilborn, R. (2011). Contrasting global trends in marine fishery status obtained from catches and from stock assessments. *Conservation Biology* 25, 777–786

Branch, T.A., Watson, R., Fulton, E.A., Jennings, S., McGilliard, C.R., Pablico, G.T., Ricard, D. and Tracey, S.R. (2010). The trophic fingerprint of marine fisheries. *Nature* 468(7322), 431–435

Brown, J. and Kothari, A. (2011). Traditional agricultural landscapes and community conserved areas: an overview. *Management of Environmental Quality: An International Journal* 22(2), 139–153

Brown, J. and MacFadyn, G. (2007). Ghost fishing in European water: impacts and management responses. *Marine Policy* 31, 488–504

Bruner, A.G., Gullison, R.E., Rice, R.E and da Fonseca, G.A.B. (2001). Effectiveness of parks in protecting tropical biodiversity. *Science* 291(550), 125–128

Brussaard, L., Caron, P., Campbell, B., Lipper, L., Mainka, S., Rabbinge, R., Babin, D. and Pulleman, M. (2010). Reconciling biodiversity conservation and food security: scientific challenges for a new agriculture. *Current Opinion in Environmental Sustainability* 2(1–2), 34–42

Butchart, S.H.M., Scharlemann, J.P.W., Evans, M.I., Quader, S., Aricò, S., Arinaitwe, J., Balman, M., Bennun, L.A., Besançon, C., Boucher ,T.M., Bertzky, B., Brooks, T.M., Burfield, I.J., Burgess, N.D., Chan, S., Clay, R.P., Crosby, M.J., Davidson, N.C., De Silva, N., Devenish, C., Dutson, G.C.L., Díaz Fernández, D.F., Fishpool, L.D.C., Fitzgerald, C., Foster, M., Heath, M.F., Hockings, M., Hoffmann, M., Knox, D., Larsen, F.W., Lamoreux, J.F., Loucks, C., May, I., Millett, J., Molloy, D., Morling, P., Parr, M., Ricketts, T.H., Seddon, N., Skolnik, B., Stuart, S.N., Upgren, A. and Woodley, S. (2012). Protecting important sites for biodiversity contributes to meeting global conservation targets, *PLoS ONE* 7(3): e32529

Butchart, S.H.M., Walpole, M., Collen, B., van Strien, A., Scharlemann, J.P.W., Almond, R.E.A., Baillie, J.E.M., Bomhard, B., Brown, C., Bruno, J., Carpenter, K.E., Carr, G.M., Chanson, J., Chenery, A.M., Csirke, J., Davidson, N.C., Dentener, F., Foster, M., Galli, A., Galloway, J.N., Genovesi, P., Gregory, R.D., Hockings, M., Kapos, V., Lamarque, J.-F., Leverington, F., Loh, J., McGeoch, M.A., McRae, L., Minasyan, A., Hernández Morcillo, M., Oldfield, T.E.E., Pauly, D., Quader, S., Revenga, C., Sauer, J.R., Skolnik, B., Spear, D., Stanwell-Smith, D., Stuart, S.N., Symes, A., Tierney, M., Tyrrell, T.D., Vié, J.-C. and Watson, R. (2010). Global biodiversity: indicators of recent declines. *Science* 328(5892), 1164–1168

Callaghan, T.V., Björn, L., Chernov, Y.I., Chapin III, F.S., Christensen, T.R., Huntley, B., Ims, R., Johansson, M., Jolly, D., Matveyeva, N.V., Panikov, N., Oechel, W.C. and Shaver, G.R. (2005). Arctic tundra and polar ecosystems. In *Arctic Climate Impact Assessment* (eds. Symon, C., Arris, L. and Heal, B.). pp.243–235. Cambridge University Press, Cambridge

Campbell, A., Kapos, V., Lysenko, I., Scharlemann, J.P.W., Dickson, B., Gibbs, H.K., Hansen, M. and Miles, L. (2008). *Carbon Emissions from Forest Loss in Protected Areas*. United Nations Environment Programme World Conservation Monitoring Centre (UNEP-WCMC), Cambridge

Canadell, J.G., Le Quéré, C., Raupach, M.R., Field, C.B., Buitenhuis, E.T., Ciais, P., Conway, T.J., Gillett, N.P., Houghton, R.A. and Marland, G. (2007). Contributions to accelerating atmospheric CO_2 growth from economic activity, carbon intensity, and efficiency of natural sinks. *Proceedings of the National Academy of Sciences of the United States of America* 114(47), 18866–18870

Carpenter, K.E., Abrar, M., Aeby, G., Aronson, R.B., Banks, S., Bruckner, A., Chiriboga, A., Cortés, J., Delbeek, J.C., DeVantier, L., Edgar, G.J., Edwards, A.J., Fenner, D., Guzmán, H.M., Hoeksema, B.W., Hodgson, G., Johan, O., Licuanan, W.Y., Livingstone, S.R., Lovell, E.R., Moore, J.A., Obura, D.O., Ochavillo, D., Polidoro, B.A., Precht, W.F., Quibilan, M.C., Reboton, C., Richards, Z.T., Rogers, A.D., Sanciangco, J., Sheppard, A., Sheppard, C., Smith, J., Stuart, S., Turak, E., Veron, J.E.N., Wallace, C., Weil, E. and Wood, E. (2008). One-third of reef-building corals face elevated extinction risk from climate change and local impacts. *Science* 321(5888), 560–563

CBD (2012). ABS Measures Database. Convention on Biological Diversity. http://www.cbd.int/abs/measures/

CBD (2011). *National Biodiversity Strategies and Action Plans (NBSAPs)*. Secretariat of the Convention on Biological Diversity, Montreal. http://www.cbd.int/nbsap (accessed 22 November 2011)

CBD (2010a). *Aichi Biodiversity Targets*. Secretariat of the Convention on Biological Diversity, Montreal. http://www.cbd.int/sp/targets/

CBD (2010b). *Global Biodiversity Outlook 3*. Secretariat of the Convention on Biological Diversity, Montreal

CBD (2010c). *Strategic Plan for Biodiversity 2011–2020*. Secretariat of the Convention on Biological Diversity, Montreal. http://www.cbd.int/decision/cop/?id=12268 (accessed 14 November 2011)

CBD (2009a). *Connecting Biodiversity and Climate Change Mitigation and Adaptation: Report of the Second Ad Hoc Technical Expert Group on Biodiversity and Climate Change*. CBD Technical Series 41. Secretariat of the Convention on Biological Diversity, Montreal

CBD (2009b). *Scientific Synthesis of the Impacts of Ocean Fertilization on Marine Biodiversity*. CBD Technical Series 45. Secretariat of the Convention on Biological Diversity, Montreal

CBD (2008). *Access and Benefit-Sharing in Practice: Trends in Partnerships Across Sectors*. CBD Technical Series No. 38. Secretariat of the Convention on Biological Diversity, Montreal

CBD (2000). *Cartagena Protocol on Biosafety to the Convention on Biological Diversity: Text and Annexes*. Secretariat of the Convention on Biological Diversity, Montreal. http://bch.cbd.int/protocol/text/

CBD (1992). *Convention on Biological Diversity*. http://www.cbd.int/ (accessed 30 November 2011)

Cheung, W.W.L., Lam, V.W.Y., Sarmiento, J.L., Kearney, K., Watson, R. and Pauly, D. (2009). Projections of global marine biodiversity impacts under climate change scenarios. *Fish and Fisheries* 10(3), 235–251

CITES (1973). *Convention on International Trade in Endangered Species of Wild Fauna and Flora*. (Amended in 1979) http://www.cites.org/eng/disc/E-Text.pdf

Clavero, M. and García-Berthou, E. (2005). Invasive species are a leading cause of animal extinctions. *Trends in Ecology and Evolution* 20(3), 110

CMS (1979). *Convention on the Conservation of Migratory Species of Wild Animals*. http://www.cms.int/documents/convtxt/cms_convtxt.htm

Cole, M., Lindeque, P., Halsband, C and Galloway, T.S. (2011). Microplastics as contaminants in the marine environment: a review. *Marine Pollution Bulletin* 62, 2588–2597

Cole, J.J., Prairie, Y.T., Caraco, N.F., McDowell, W.H., Tranvik, L.J., Striegl, R.G., Duarte, C.M., Kortelainen, P., Downing, J.A., Middelburg, J.J. and Melack, J. (2007). Plumbing the global carbon cycle: integrating inland waters into the terrestrial carbon budget. *Ecosystems* 10, 171–184

Collen, B., Loh, J., Whitmee, S., McRae, L., Amin, R. and Baillie, J.E.M. (2008a). Monitoring change in vertebrate abundance: the Living Planet Index. *Conservation Biology* 23, 317–327

Collen, B., Ram, M., Zamin, T. and McRae, L. (2008b). The tropical biodiversity data gap: addressing disparity in global monitoring. *Tropical Conservation Science* 1(2), 75–88

Cooper, E., Burke, L. and Bood, N. (2009). *Coastal Capital: Belize. The Economic Contribution of Belize's Coral Reefs and Mangroves*. WRI Working Paper. World Resources Institute, Washington, DC

Cotula, L. and Mathieu, P. (eds.). (2008). *Legal Empowerment in Practice: Using Legal Tools to Secure Land Rights in Africa*. International Institute for Environment and Development (IIED), London

Craigie, I., Baillie, J., Balmford, A., Carbone, C., Collen, B., Green, R.E. and Hutton, J.H. (2010). Large mammal population declines in Africa's protected areas. *Biological Conservation* 143, 2221–2228

Danielsen, F., Beukema, H., Burgess, N.D., Parish, F., Brühl, C.A., Donald, P.F., Murdiyarso, D., Phalan, B., Reihnders, L., Struebig, M. and Fitzherbert, E.B. (2009). Biofuel plantations on forested lands: double jeopardy for biodiversity and climate. *Conservation Biology* 23, 348–358

Dise, N.B., Ashmore, M., Belyazid, S., Bleeker, A., Bobbink, R., de Vries, W., Erisman, J.W., Spranger, T., Stevens, C.J. and van den Berg, L. (2011). Nitrogen as a threat to European terrestrial biodiversity. In *The European Nitrogen Assessment* (eds. Sutton, M.A., Howard, C.M., Erisman, J.W., Billen, G., Bleeker, A., Grennfelt, P., van Grinsven, H. and Grizzetti, B.). *pp.463–494*. Cambridge University Press, Cambridge

DSEWPC (2011). *Declared Indigenous Protected Areas – Case Studies*. Australian Government Department of Sustainability, Environment, Water, Population and Communities. http://www.environment.gov.au/indigenous/ipa/declared/index.html (accessed 11 November 2011)

Dudley, N., Bhagwat, S., Higgins-Zogin, L., Lassen, B., Verschuuren, B. and Wild, R. (2010a). Conservation of biodiversity in sacred natural sites in Asia and Africa: a review of the scientific literature. In *Sacred Natural Sites, Conserving Nature and Culture* (eds. Verschuuren, B., Wild, R., McNeely, J. and Oviedo, G.). pp.19–32. Earthscan, London and Washington, DC

Dudley, N., Stolton, S., Belokurov, A., Krueger, L., Lopoukhine, N., MacKinnon, K., Sandwith, T. and Sekhran, N. (eds.) (2010b). *Natural Solutions: Protected Areas Helping People Cope with Climate Change*. International Union for Conservation of Nature World Commission on Protected Areas (IUCN-WCPA), Gland, The Nature Conservancy (TNC), Arlington, VA, United Nations Development Programme (UNDP), New York, Wildlife Conservation Society (WCS), New York, World Bank, Washington, DC and WWF–World Wildlife Fund for Nature, Gland

Dulvy, N.K., Rogers, S.I., Jennings, S., Stelzenmüller, V., Dye, S.R. and Skjoldal, H.R. (2008). Climate change and deepening of the North Sea fish assemblage: a biotic indicator of regional warming. *Journal of Applied Ecology* 45(4), 1029–1039

Dulvy, N.K., Sadovy, Y. and Reynolds, J.D. (2003). Extinction vulnerability in marine populations. *Fish and Fisheries* 4, 25–64

Ehrlich, P.R. and Ehrlich, A.H. (1992). The value of biodiversity. *Ambio* 21(3), 219–226

Emerson, C. (1999). *Aquaculture Impacts on the Environment*. Cambridge Scientific Abstracts. http://www.csa.com (accessed 17 January 2012)

Falkenmark, M., Finlayson, C.M. and Gordon, L. (2007). Agriculture, water, and ecosystems: avoiding the costs of going too far. In *Water For Food, Water For Life: A Comprehensive Assessment of Water Management in Agriculture* (ed. Molden, D.). pp.234–277. Earthscan, London

FAO (2010a). *The Global Forest Resources Assessment 2010. Main Report*. FAO Forestry Paper 163. Food and Agriculture Organization of the United Nations, Rome

FAO (2010b). *The State of World Fisheries and Aquaculture 2010*. Food and Agriculture Organization of the United Nations, Rome

FAO (2009). *Livestock Keepers: Guardians of Biodiversity*. FAO Animal Production and Health Paper 167. Food and Agriculture Organization of the United Nations, Rome

FAO (2001). *International Treaty on Plant Genetic Resources for Food and Agriculture*. Food and Agriculture Organization of the United Nations. http://www.planttreaty.org/content/texts-treaty-official-versions

FAO (1951). *International Plant Protection Convention*. (Amended 1979 and 1997) http://www.fao.org/Legal/TREATIES/004t-e.htm

FAO and UNEP (2009). *Report of the FAO/UNEP Expert Meeting on Impacts of Destructive Fishing Practices, Unsustainable Fishing, and Illegal, Unreported and Unregulated (IUU) Fishing on Marine Biodiversity and Habitats*. FAO Fisheries and Aquaculture Report No. 932. Food and Agriculture Organization of the United Nations, Rome

Ferrari, M.F. (2006). Rediscovering community conserved areas in South-East Asia: peoples' initiative to reverse biodiversity loss. *Parks* 16(1), 43–48

Fiala, N. (2008). Measuring sustainability: why the ecological footprint is bad economics and bad environmental science. *Ecological Economics* 67(4), 519–525

Finlayson, C.M. and D'Cruz, R. (2005). Inland water systems. In *Ecosystems and Human Well-being: Current State and Trends: Findings of the Condition and Trends Working Group* (eds. Hassan, R., Scholes, R. and Ash, N.). pp.551–583. Island Press, Washington, DC

Finlayson, C.M., Davis, J.A., Gell, P.A., Kingsford, R.T. and Parton, K.A. (2011). The status of wetlands and the predicted effects of global climate change: the situation in Australia. *Aquatic Sciences* 1–21

Finlayson, C.M., Gitay, H., Bellio, M.G., van Dam, R.A. and Taylor, I. (2006). Climate variability and change and other pressures on wetlands and waterbirds – impacts and adaptation. In *Water Birds Around the World* (eds. Boere, G., Gailbraith, C. and Stroud, D.). pp.88–89. Scottish Natural Heritage, Edinburgh

Fitzherbert, E.B., Struebig, M.J., Morel, A., Danielsen, F., Brühl, C.A., Donald, P.F. and Phalan, B. (2008). How will oil palm expansion affect biodiversity. *Trends in Ecology and Evolution* 23(10), 538–545

Foley, J.A., Ramankutty, N., Brauman, K.A., Cassidy, E.S., Gerber, J.S., Johnston, M., Mueller, N.D., O'Connell, C., Ray, D.K., West, P.C., Balzer, C., Bennett, E.M., Carpenter, S.R., Hill, J., Monfreda, C., Polasky, S., Rockström, J., Sheehan, J., Siebert, S., Tilman, D. and Zaks, D.P.M. (2011). Solutions for a cultivated planet. *Nature* 478, 337–342

Freire, K. and Pauly, D. (2010). Fishing down Brazilian marine food webs, with emphasis on the East Brazil Large Marine Ecosystem. *Fisheries Research* 105, 57–62

FSC (2012). *Global FSC Certificates: Type and Distribution*. Forest Stewardship Council, Bonn

Galgani, F., Fleet, D., van Franeker, J., Katsanevakis, S., Maes, T., Mouat, J., Oosterbaan, L., Poitou, I. Hanke, G., Thompson, R., Amato, E., Birkun, A. and Janssen, C. (2010). *Marine Strategy Framework Directive Task Team 10 Report: Marine Litter*. JRC (EC Joint Research Centre) Scientific and Technical Reports

Garcia, S.M. and Rosenberg, A.A. (2010). Food security and marine capture fisheries: characteristics, trends, drivers and future perspectives. *Philosophical Transactions of the Royal Society B* 365(1554), 2869–2880

GEO BON (2011). *Adequacy of Biodiversity Observation Systems to support the CBD 2020 Targets*. A report prepared by the Group on Earth Observations Biodiversity Observation Network (GEO BON) for the Convention on Biological Diversity. GEO BON, Pretoria

Gibbons, D.W., Bohan, D.A., Rothery, P., Stuart, R.C., Haughton, A.J., Scott, R.J., Wilson, J.D., Perry, J.N., Clark, S.J., Dawson, R.J.G. and Firbank, L.G. (2006). Weed seed resources for birds in fields with contrasting conventional and genetically modified herbicide-tolerant crops. *Proceedings of the Royal Society B* 273(1596), 1921–1928

Giraud, G. (2008). Range and limit of geographical indication scheme: the case of basmati rice from Punjab, Pakistan. *International Food and Agribusiness Management Review* 11(1), 51–76

Githitho, A. (2003). The sacred Mijikenda Kaya forests of coastal Kenya and biodiversity conservation. In *The Importance of Sacred Natural Sites for Biodiversity Conservation* (eds. Lee, C. and Schaaf, T.). Proceedings of the International Workshop held in Kumming and Xishuangbanna Biosphere Reserve, People's Republic of China, 2003. pp.27–35. United Nation Educational, Scientific and Cultural Organization, Paris

Godfray, H.C.J., Beddington, J.R., Crute, I.R., Haddad, L., Lawrence, D., Muir, J.F., Pretty, J., Robinson, S., Thomas, S.M. and Toulmin, C. (2010). Food security: the challenge of feeding 9 billion people. *Science* 327(5967), 812–818

Golden, C.D., Fernald, L.C.H., Brashares, J.S., Rasolofoniaina, B.J.R. and Kremen, C. (2011). Benefits of wildlife consumption to child nutrition in a biodiversity hotspot. *Proceedings of the National Academy of Sciences of the United States of America* (in press)

González, J., Grijalba-Bendeck, M., Acero-P., A. and Betancur-R., R. (2009). The invasive red lionfish, *Pterois volitans* (Linnaeus 1758), in the southwestern Caribbean Sea. *Aquatic Invasions* 4(3), 507–510

Gordon, L.J., Finlayson, C.M. and Falkenberg, M. (2010). Managing water in agriculture for food production and other ecosystem services. *Agricultural Water Management* 97(2010), 512–519

Gough, C.M. (2011). Terrestrial primary production: fuel for life. *Nature Education Knowledge* 2(2), 1

Govan, H., Tawake, A., Tabunakawai, K., Jenkins, A., Lasgorceix, A., Techera, E., Tafea, H., Kinch, J., Feehely, J., Ifopo, P., Hills, R., Alefaio, S., Meo, S., Troniak, S., Malimali, S., George, S., Tauaefa, T. and Obed, T. (2009). *Community Conserved Areas: A Review of Status and Needs in Melanesia and Polynesia*. Indigenous and Community Conserved Areas (ICCA) regional review for the Centre for Sustainable Development (CENESTA)/Theme on Indigenous and Local Communities, Equity and Protected Areas (TILCEPA)/Theme on Governance, Equity and Rights (TGER)/International Union for Conservation of Nature (IUCN)/Global Environment Fund- Small Grants Programme (GEF-SGP)

Government of Manitoba (2011). *Province Permanently Designates Largest Area of Protected Land in More Than a Decade*. http://news.gov.mb.ca/news/index.html?archive=&item=11766 (accessed 21 November 2011).

Greathouse, E.A., Pringle, C.M., McDowell, W.H. and Holmquist, J.G. (2006). Indirect upstream effects of dams: consequences of migratory consumer extirpation in Puerto Rico. *Ecological Applications* 16, 339–352

Gregory, R.D., Willis, S.G., Jiguet, F., Voříšek, P., Klvaňová, A., Huntley, B., Collingham, Y.C., Couvet, D. and Green, R.E. (2009). An indicator of the impact of climatic change on European bird populations. *PLoS ONE* 4(3), e4678

Gregory, R.D., van Strien, A., Vorisek, P., Gmelig Meyling, A.W., Noble, D.G., Foppen, R.P.B. and Gibbons, D.W. (2005). Developing indicators for European birds. *Philosophical Transactions of the Royal Society B* 360(1454), 269–288

Gupta, V.K. (2006). Protection of traditional knowledge. In *Perspectives on Biodiversity: A Vision for Megadiverse Countries* (eds. Verma, D.D., Arora, S. and Rai, R.K.). pp.243–258. Ministry of Environment and Forests, Government of India, New Delhi

Gutman, P. and Davidson, S. (2008). *A Review of Innovative International Financial Mechanisms for Biodiversity Conservation with a Special Focus on the International Financing of Developing Countries' Protected Areas*. WWF–World Wide Fund for Nature, Gland

Haken, J. (2011). *Transnational Crime in the Developing World*. Global Financial Integrity, Washington, DC

Halpern, B.S. (2003). The impact of marine reserves: do reserves work and does reserve size matter? *Ecological Applications* 13, 117–137

Hansen, M.C., Stehman, S.V. and Potapov, P.V. (2010). Quantification of global gross forest cover loss. *Proceedings of the National Academy of Sciences of the United States of America* 107, 8650–8655

Heath, M.R. (2005). Changes in the structure and function of the North Sea fish foodweb, 1973–2000, and the impacts of fishing and climate. *ICES Journal of Marine Science* 62, 847–868

Heiskanen, M. (2009). *The Regulatory Development Case of the CDM Forests – Seeking a Vital Balance between the Goals of Carbon Sequestration and Biodiversity Conservation through the New Biodiversitical Concepts*. XIII World Forestry Congress. Buenos Aires, Argentina, 18–23 October 2009. Food and Agriculture Organization of the United Nations, Rome

Hiddink, J.G. and Ter Hofstede, R. (2008). Climate induced increases in species richness of marine fishes. *Global Change Biology* 14(3), 453–460

HLIAP (2010). *Report of the First Meeting of the High-Level Intergovernmental Advisory Panel on the Selection of Internationally Agreed Goals for GEO-5*. 1st High-level Intergovernmental Advisory Panel, Geneva, 28-30 June 2010. United Nations Environment Programme, Nairobi

Hoegh-Guldberg, O., Mumby, P.J., Hooten, A.J., Steneck, R.S., Greenfield, P., Gomez, E., Harvell, C.D., Sale, P.F., Edwards, A.J. , Caldeira, K., Knowlton, N., Eakin, C.M., Iglesias-Prieto, R., Muthiga, N., Bradbury, R.H., Dubi, A. and Hatziolos, M.E. (2007). Coral reefs under rapid climate change and ocean acidification. *Science* 318(5857), 1737–1742

Hoffmann, M., Hilton-Taylor, C., Angulo, A., Boehm, M., Brooks, T.M., Butchart, S.H., Carpenter, K.E., Chanson, J., Collen, B., Cox, N.A., Darwall, W.R., Dulvy, N.K., Harrison, L.R., Katariya, V., Pollock, C.M., Quader, S., Richman, N.I., Rodrigues, A.S., Tognelli, M.F., Vie, J.C., Aguiar, J.M., Allen, D.J., Allen, G.R., Amori, G., Ananjeva, N.B., Andreone, F., Andrew, P., Aquino Ortiz, A.L., Baillie, J.E., Baldi, R., Bell, B.D., Biju, S., Bird, J.P., Black-Decima, P., Blanc, J., Bolanos, F., Bolivar, G., Burfield, I.J., Burton, J.A., Capper, D.R., Castro, F., Catullo, G., Cavanagh, R.D., Channing, A., Chao, N.L., Chenery, A.M., Chiozza, F., Clausnitzer, V., Collar, N.J., Collett, L.C., Collette, B.B., Fernandez, C.F., Craig, M.T., Crosby, M.J., Cumberlidge, N., Cuttelod, A., Derocher, A.E., Diesmos, A.C., Donaldson, J.S., Duckworth, J., Dutson, G., Dutta, S., Emslie, R.H., Farjon, A., Fowler, S., Freyhof, J., Garshelis, D.L., Gerlach, J., Gower, D.J., Grant, T.D., Hammerson, G.A., Harris, R.B., Heaney, L.R., Hedges, S.B., Hero, J.M., Hughes, B., Hussain, S.A., Icochea, M., Inger, R.F., Ishii, N., Iskandar, D.T., Jenkins, R.K.B., Kaneko, Y., Kottelat, M., Kovacs, K.M., Kuzmin, S.L., La Marca, E., Lamoreux, J.F., Lau, M.W.N., Lavilla, E.O., Leus, K., Lewison, R.L., Lichtenstein, G., Livingstone, S.R., Lukoschek, V., Mallon, D.P., McGowan, P.J.K., McIvor, A., Moehlman, P.D., Molur, S., Munoz Alonso, A., Musick, J.A., Nowell, K., Nussbaum, R.A., Olech, W., Orlov, N.L., Papenfuss, T.J., Parra-Olea, G., Perrin, W.F., Polidoro, B.A., Pourkazemi, M., Racey, P.A., Ragle, J.S., Ram, M., Rathbun, G., Reynolds, R.P., Rhodin, A.G.J., Richards, S.J., Rodriguez, L.O., Ron, S.R., Rondinini, C., Rylands, A.B., de Mitcheson, Y.S., Sanciangco, J.C., Sanders, K.L., Santos-Barrera, G., Schipper, J., Self-Sullivan, C., Shi, Y., Shoemaker, A., Short, F.T., Sillero-Zubiri, C., Silvano, D.L., Smith, K.G., Smith, A.T., Snoeks, J., Stattersfield, A.J., Symes, A.J., Taber, A.B., Talukdar, B.K., Temple, H.J., Timmins, R., Tobias, J.A., Tsytsulina, K., Tweddle, D., Ubeda, C., Valenti, S.V., van Dijk, P.P., Veiga, L.M., Veloso, A., Wege, D.C., Wilkinson, M., Williamson, E.A., Xie, F., Young, B.E., Akcakaya, H.R., Bennun, L., Blackburn, T.M., Boitani, L., Dublin, H.T., da Fonseca, G.A.B., Gascon, C., Lacher Jr., T.E., Mace, G.M., Mainka, S.A., McNeely, J.A., Mittermeier, R.A., Reid, G.M., Paul Rodriguez, J., Rosenberg, A.A., Samways, M.J., Smart, J., Stein, B.A. and Stuart, S.N. (2010). The impact of conservation on the status of the world's vertebrates. *Science* 330(6010), 1503–1509

Hölker, F., Wolter, C., Perkin, E.K. and Tockner, K. (2010). Light pollution as a biodiversity threat. *Trends in Ecology and Evolution* 25(12), 681–682

Horwitz, P. and Finlayson, C.M. (2011). Wetlands as settings for human health: incorporating ecosystem services and health impact assessment into water resource management. *Bioscience* 61, 678–688

Horwitz, P., Finlayson, C.M. and Weinstein, P. (2011). *Healthy Wetlands, Healthy People: A Review of Wetlands and Human Health Interactions*. Ramsar Technical Report No. 6. Secretariat of the Ramsar Convention on Wetlands, Gland, and the World Health Organization, Geneva

Hulme, P.E. (2009). Trade, transport and trouble: managing invasive species pathways in an era of globalization. *Journal of Applied Ecology* 46, 10–18

IAASTD (2009). *Agriculture at a Crossroads. Synthesis Report* (eds. McIntyre, B.D., Herren, H.R., Wakhungu, J. and Watson, R.T.). International Assessment of Agricultural Knowledge, Science and Technology for Development. Island Press, Washington, DC

ICCA (2009). *Indigenous People's Conserved Territories and Areas Conserved by Indigenous Peoples and Local Communities. A Bold New Frontier for Conservation*. http://www.iccaforum.org (accessed 21 November 2011).

IPCC (2007). *Fourth Assessment Report of the Intergovernmental Panel on Climate Change*. Intergovernmental Panel on Climate Change, Geneva

ISDR (2009). *Global Assessment Report on Disaster Risk Reduction*. United Nations International Strategy for Disaster Reduction, Geneva

IUCN (2010). *The IUCN Red List of Threatened Species*. http://www.iucnredlist.org/ (accessed 23 November 2011).

IUCN (2008). *Indigenous and Traditional Peoples and Climate Change*. Issues Paper. International Union for Conservation of Nature, Gland

IUCN and UNEP-WCMC (2011). *The World Database on Protected Areas (WDPA)*. International Union for Conservation of Nature, Gland and United Nations Environment Programme World Conservation Monitoring Centre, Cambridge. http://www.wdpa.org/ (January 2011)

Jacquet, J., Hocevar, J., Lai, S., Majluf, P., Pelletier, N., Pitcher, T., Sala, E., Sumaila, R. and Pauly, D. (2009). Conserving wild fish in a sea of market-based efforts. *Oryx* 44(1), 45–56

James, C. (2010). *Global Status of Commercialised Biotech/GM crops: 2010*. ISAAA Brief No. 42. International Service for the Acquisition of Agri-biotech Applications, Ithaca, NY

James, A., Gaston, K.J. and Balmford, A. (2001). Can we afford to conserve biodiversity? *Bioscience* 51(1), 43–52

Jana, S. and Paudel, N.S. (2010). *Rediscovering Indigenous Peoples' and Community Conserved Areas in Nepal*. Forest Action, Kathmandu

Jenkins, C.N. and Joppa, L. (2009). Expansion of the global terrestrial protected area system. *Biological Conservation* 142(10), 2166–2174

Jones, J., Collen, B., Atkinson, G., Baxter, P., Bubb, P., Illian, J., Katzner, T., Keane, A., Loh, J., McDonald-Madden, E., Nicholson, E., Pereira, H., Possingham, H., Pullin, A., Rodrigues, A., Ruiz-Gutierrez, V., Sommerville, M. and Milner-Gulland, E. (2011). The why, what, and how of global biodiversity indicators beyond the 2010 target. *Conservation Biology* 25(3), 450–457

Jorgensen, C., Enberg, K., Dunlop, E.S., Arlinghaus, R., Boukal, D.S., Brander, K., Ernande, B., Gardmark, A., Johnston, F., Matsumura, S., Pardoe, H., Raab, K., Silva, A., Vainikka, A., Dieckmann, U., Heino, M. and Rijnsdorp, A.D. (2007). Managing evolving fish stocks. *Science* 318, 1247–1248

Kalpavriksh (2011). *Recognising and Supporting Indigenous and Community Conserved Areas (ICCAs) in South Asia and Globally. Final Report, February 2011*. Kalpavriksh Environment Action Group. http://www.kalpavriksh.org/community-conserved-areas/research-and-documentation/ccas-in-southasia/148-undp-final-report-feb-2011 (accessed 11 November 2011).

Keder, G. and McIntyre Galt, R. (2009). *Impacts of Climate Change and Selected Renewable Energy Infrastructures on EU Biodiversity and the Natura 2000 Network: Task 4 – Wind, Hydro and Marine Renewable Energy Infrastructures in the EU: Biodiversity Impacts, Mitigation and Policy Recommendations*. European Commission and International Union for Conservation of Nature

Kitzes, J. and Wackernagel, M. (2009). Answers to common questions in Ecological Footprint accounting. *Ecological Indicators* 9(4), 812–817

Kitzes, J., Moran, D., Galli, A.,Wada, Y. and Wackernagel, M. (2009). Interpretation and application of the Ecological Footprint: a reply to Fiala (2008). *Ecological Economics* 68(4), 929–930

Kleisner, K. and Pauly, D. (2011). Stock-catch status plots of fisheries for Regional Seas. In *The State of Biodiversity and Fisheries in Regional Seas* (eds. Christensen, V., Lai, S., Palomares, M.L.D., Zeller, D. and Pauly, D.). pp.37–40. Fisheries Centre Research Reports 19(3)

Kleisner, K. and Pauly, D. (2010). The Marine Trophic Index (MTI), the Fishing in Balance (FiB) Index and the spatial expansion of fisheries. In *The State of Biodiversity and Fisheries in Regional Seas* (eds. Christensen, V., Lai, S., Palomares, M.L.D., Zeller, D. and Pauly, D.). pp.41–44. Fisheries Centre Research Reports 19(3)

Kneteman, C. and Green, A. (2009). The twin failures of the CDM: recommendations for the "Copenhagen Protocol". *The Law and Development Review* 2(1), 9

Kothari, A. (2006). Community conserved areas. In *Managing Protected Areas: A Global Guide* (eds. Lockwood, M.L., Worboys, G. and Kothari, A.). pp.549–573. Earthscan, London

Kothari, A., Menon, M. and O'Reilly, S. (2010). *Territories and Areas Conserved by Indigenous Peoples and Local Communities (ICCAs): How Far Do National Laws and Policies Recognize Them?* International Union for Conservation of Nature (IUCN) Commission on Environmental, Economic and Social Policy-World Commission on Protected Areas (CEESP-WCPA), Theme on Indigenous and Local Communities, Equity, and Protected Areas (TILCEPA) and Kalpavriksh, Pune

Kura, Y., Revenga, C., Hoshino, E. and Mock, G. (2004). *Fishing for Answers*. World Resources Institute, Washington, DC

Lavides, M.N., Pajaro, M.G. and Nozawa, C.M.C. (2006). *Atlas of Community-Based Marine Protected Areas in the Philippines*. Haribon Foundation for the Conservation of Natural Resources, Inc. and Panama KaSaPilipinas

Leadley, P., Pereira, H.M., Alkemade, R., Fernandez- Manjarrés, J.F., Proença, V., Scharlemann, J.P.W. and Walpole, M.J. (2010). *Biodiversity Scenarios: Projections of 21st Century Change in Biodiversity and Associated Ecosystem Services: A Technical Report for the Global Biodiversity Outlook 3*. Convention on Biological Diversity Technical Series No 50. Secretariat of the Convention on Biological Diversity, Montreal

Le Quéré, C., Raupach, M.R., Canadell, J.G, Marland, G., Bopp, K., Ciais, P., Conway, T.J., Doney, S.C., Feely, R.A., Foster, P., Friedlingstein, P., Gurney, K., Houghton, R.A., House, J.I., Huntingford, C., Levy, P.E., Lomas, M.R., Majkut, J., Metzl, N., Ometto, J.P., Peters, I.C., Randerson, J.T., Running, S.W., Sarmiento, J.L., Schuster, U., Sitch, S., Takahashi, T., Viovy, N., van der Werf, G. and Woodward, F.I. (2009). Trends in the sources and sinks of carbon dioxide. *Nature Geoscience* 2, 831–836

Ligon, F.K., Dietrich, W.E. and Trush, W.J. (1995). Downstream ecological effects of dams. *BioScience* 45(3), 183–192

Liu, F., Xu, Z., Zhu, Y.C., Huang, F., Wang, Y., Li, H., Li, H., Gao, C., Zhou, W. and Shen, J. (2010). Evidence of field-evolved resistance to Cry1Ac-expressing Bt cotton in *Helicoverpa armigera* (Lepidoptera: Noctuidae) in northern China. *Pest Management Science* 66, 155–161. doi:10.1002/ps.1849

Loh, J. (ed.). (2010). *2010 and Beyond: Rising to the Biodiversity Challenge*. WWF–World Wide Fund for Nature, Gland

MA (2005a). *Ecosystems and Human Well-being: Synthesis*. Millennium Ecosystem Assessment. World Resources Institute. Island Press, Washington, DC

MA (2005b). *Ecosystems and Human Well-being: Wetlands and Water Synthesis*. Millennium Ecosystem Assessment. World Resources Institute. Island Press, Washington, DC

Maffi, L. and Woodley, E. (2010). *Biocultural Diversity Conservation: A Global Sourcebook*. Earthscan, London

malERA Consultative Group on Vector Control (2011). A research agenda for malaria eradication: vector control. *PLoS Medicine* 8(1), 34–41.

Malhotra, K.C., Gokhale, Y., Chatterjee, S. and Srivastava, S. (2007). *Sacred Groves in India: An Overview*. Aryan Books International, New Delhi and Indira Gandhi Rashtriya Manav Sangrahalaya, Bhopal

Malhotra, K.C., Gokhale, Y., Chatterjee, S. and Srivastava, S. (2001). *Cultural and Ecological Dimensions of Sacred Groves in India*. Indian National Science Academy, New Delhi and Indira Gandhi Rashtriya Manav Sangrahalaya, Bhopal

Mallarach, J.-M., Papayannis, T. and Väisänen, R. (eds.) (2012). *Sacred Natural Sites in European Protected Areas. Proceedings of the Third Workshop of the Delos Initiative, Inari 2010*. International Union for Conservation of Nature, Gland

Marvier, M., McCreedy, C., Regetz, J. and Kareiva, P. (2007). A meta-analysis of effects of Bt cotton and maize on nontarget invertebrates. *Science* 316(5830), 1475–1477

McDonald, R.I., Fargione, J., Kiesecker, J., Miller, W.M. and Powell, J. (2009). Energy sprawl or energy efficiency: climate policy impacts on natural habitat for the United States of America. *PLoS ONE* 4(8), e6802

McGeoch, M.A., Butchart, S.H.M., Spear, D., Marais, E., Kleynhans, E.J., Symes, A., Chanson, J. and Hoffmann, M. (2010). Global indicators of biological invasion: species numbers, biodiversity impact and policy responses. *Diversity and Distributions* 16(1), 95–108

McRae, L., Zöckler, C., Gill, M., Loh, J., Latham, J., Harrison, N., Martin, J. and Collen, B. (2010). *Arctic Species Trend Index 2010: Tracking Trends in Arctic Wildlife*. CAFF CBMP Report No. 20. Conservation of Arctic Flora and Fauna International Secretariat, Akureyri

Moeller, A.P., Rubolini, D. and Lehikoinen, E. (2008). Populations of migratory bird species that did not show a phenological response to climate change are declining. *Proceedings of the National Academy of Sciences of the United States of America* 105(42), 16195–16200

Molden, D. (ed). (2007). *Water For Food, Water For Life: A Comprehensive Assessment of Water Management in Agriculture*. Earthscan, London and Water Management Institute, Colombo

Molnar, A., Scherr, S. and Khare, A. (2004). *Who Conserves the World's Forests: Community Driven Strategies to Protect Forests and Respect Rights*. Forest Trends and Eco-agriculture Partners, Washington, DC

Moore, J.L., Manne, L., Brooks, T., Burgess, N.L., Davies, R., Rahbek, C., Williams, P. and Balmford, A. (2002). The distribution of cultural and biological diversity in Africa. *Proceedings of the Royal Society B* 269(1501), 1645–1653

Morris, B.L., Lawrence, A.R., Chilton, P.J., Adams, B., Calow, R. and Klinck, B.A. (2003). *Groundwater and its Susceptibility to Degradation: A Global Assessment of the Problems and Options for Management*. Early Warning and Assessment Report Series, RS, 03-3. United Nations Environment Programme, Nairobi

Moseley, C. (ed.) (2010). *Atlas of the World's Languages in Danger*. UNESCO Publishing, Paris

Mumby, P.J. (2009). Phase shifts and the stability of macroalgal communities on Caribbean coral reefs. *Coral Reefs* 28, 761–773

NABCI US Committee (2009). *The State of the Birds: United States of America, 2009*. North American Bird Conservation Initiative, US Department of Interior, Washington, DC

Nagoya Protocol (2011). *Access and Benefit-sharing*. ABS Measures Search Page. http://www.cbd.int/abs/measures/ (accessed 8 September 2011)

Nasi, R., Brown, D., Wilkie, D., Bennett, E., Tutin, C., van Tol, G. and Christophersen, T. (2008). *Conservation and Use of Wildlife Based Resources: The Bushmeat Crisis*. Technical Series No. 33. Secretariat of the Convention on Biological Diversity, Montreal and Center for International Forestry Research, Bogor

Naylor, R.L., Goldburg, R.J., Primavera, J.H., Kautsky, N., Beveridge, M.C.M., Clay, J., Folke, C., Lubchenco, J., Mooney, H. and Troell, M. (2000). Effect of aquaculture on world fish supplies. *Nature* 405, 1017–1024

Nelson, A. and Chomitz, K.M. (2011). Effectiveness of strict vs. multiple use protected areas in reducing tropical forest fires: a global analysis using matching methods. *PLoS ONE* 6(8), e22722

Nijar, G.S. (2011). *The Nagoya Protocol on Access and Benefit Sharing of Genetic Resources: An Analysis*. Centre of Excellence for Biodiversity Law (Ceblaw), Kuala Lumpur

Nilsson, C., Reidy, C.A., Dynesius, M. and Revenga, C. (2005). Fragmentation and flow regulation of the world's large river systems. *Science* 308(5720), 405–408

OECD (2010). *Paying for Biodiversity: Enhancing the Cost-Effectiveness of Payments for Ecosystem Service*. Organisation for Economic Co-operation and Development, Paris

Ostrom, E.A. (2007). A diagnostic approach for going beyond panaceas. *Proceedings of the National Academy of Sciences of the United States of America* 104, 15181

Oviedo, G. (2006). Community conserved areas in South America. *Parks* 16(1), 49–55

Parish, F., Sirin, A., Charman, D., Joosten, H., Minayeva, T., Silvius, M. and Stringer, L. (eds.) (2008). *Assessment on Peatlands, Biodiversity and Climate Change: Main Report*. Global Environment Centre, Kuala Lumpur and Wetlands International, Wageningen

Pathak, N. (ed.). (2009). *Community Conserved Areas in India: A Directory*. Kalpavriksh, Pune, Delhi. http://www.kalpavriksh.org/community-conserved-areas/cca-directory (accessed 07 November 2011)

Pauly, D. and Chuenpagdee, R. (2003). Fisheries and coastal systems: the need for integrated management. *Journal of Business Administration and Policy Analysis* 30–31, 1–18

Pauly, D. and Watson, R. (2005). Background and interpretation of the 'marine trophic index' as a measure of biodiversity. *Philosophical Transactions of the Royal Society B* 360(1454), 415–423

Pauly, D., Christensen, V., Dalsgaard, J., Froese, R. and Torres, F.C. (1998). Fishing down marine food webs. *Science* 279, 860–863

Peduzzi, P., Harding, R., Richard, J., Kluser, S., Duquesnoy, L. and Boudol. 2011. *UNEP Foresight Process: Phase I: Results of the UNEP consultation*. United Nations Environment Programme, Nairobi

Pereira, H.M. and Daily, G.C. (2006). Modeling biodiversity dynamics in countryside landscapes. *Ecology* 87, 1877–1885

Pereira, H.M., Belnap, J., Brummitt, N., Collen, B., Ding, H., Gonzalez-Espinosa, M., Gregory, R.D., Honrado, J., Jongman, R.H., Julliard, R., McRae, L., Proença, V., Rodrigues, P., Opige, M., Rodriguez, J.P., Schmeller, D.S., van Swaay, C. and Vieira, C. (2010a). Global biodiversity monitoring. *Frontiers in Ecology and the Environment* 8, 459–460

Pereira, H.M., Leadley, P.W., Proença, V., Alkemade, R., Scharlemann, J.P.W., Fernandez-Manjarrés, J.F., Araújo, M.B, Balvanera, P., Biggs, R., Cheung, W.W.L., Chini, L., Cooper, H.D., Gilman, E.L., Guénette, S., Hurtt, G.C., Huntington, H.P., Mace, G.M., Oberdorff, T., Revenga, C., Scholes, R.J., Sumaila U.R. and Walpole, M. (2010b). Scenarios for global biodiversity in the 21st century. *Science* 330(6010), 1496–1501

Peres, C.A. (2010). Overexploitation. In *Conservation Biology for All* (eds. Sodhi, N.S. and Ehrlich, P.R.). pp.107–131. Oxford Scholarship Online Monographs. http://www.oxfordscholarship.com (accessed 17 January 2012)

Perfecto, I. and Vandermeer, J. (2010). The agroecological matrix as alternative to the land-sparing/agriculture intensification model. *Proceedings of the National Academy of Sciences of the United States of America* 107(13), 5786–5791

Perrings, C., Duraiappah, A., Larigauderi, A. and Mooney, H. (2011). The biodiversity and ecosystem services science-policy interface. *Science* 331(6021), 1139–1140

Perry, A.L., Low, P.J., Ellis, J.R. and Reynolds, J.D. (2005). Climate change and distribution shifts in marine fishes. *Science* 308(5730), 1912–1915

Phalan, B., Balmford, A., Green, R.E. and Scharlemann, J.P.W. (2011). Minimising the harm to biodiversity of producing more food globally. *Food Policy* 36(supplement 1), S62–S71

Pimentel, D., Zuniga, R. and Morrison, D. (2004). Update on the environmental and economic costs associated with alien-invasive species in the United States. *Ecological Economics* 52(3), 273–288

Piñeyro-Nelson, A., van Heerwaarden, J., Perales, H.R., Serratos-Hernandez, J.A., Rangel, A., Hufford, M.B., Gepts, P., Garay-Arroyo, A., Rivera-Bustamante, R. and Alvarez-Buylla, R. (2009). Transgenes in Mexican maize: molecular evidence and methodological considerations for GMO detection in landrace populations. *Molecular Ecology* 18(4), 750–761

Porter-Bolland, L., Ellis, E.A., Guariguata, M.R., Ruiz-Mallen, I., Negrete-Yankelvich, S. and Reyes-Garciam, V. (2012). Community managed forests and forest protected areas: an assessment of their conservation effectiveness across the tropics. *Forest Ecology and Management* 268, 6–17.

Posey, D.A. (ed.). (1999). *Cultural and Spiritual Values of Biodiversity*. United Nations Environmental Programme and Intermediate Technology Publications, London

Powles, S. (2010). Gene amplification delivers glyphosate-resistant weed evolution. *PNAS* 107(3), 955–956. doi:10.1073/pnas.0913433107

Prip, C., Gross, T., Johnston, S. and Vierros, M. (2010). *Biodiversity Planning: An Assessment of National Biodiversity Strategies and Action Plans*. United Nations University Institute of Advanced Studies, Yokohama

Rands, M.R.W., Adams, W.M., Bennun, L., Butchart, S.H.M., Clements, A., Coomes, D., Entwistle, A., Hodge, I., Kapos, V., Scharlemann, J.P.W., Sutherland, W.J. and Vira, B. (2010). Biodiversity conservation: challenges beyond 2010. *Science* 329(5997), 1298–1303

Ravindranath, N.H. and Ostwald, M. (2008). *Carbon Inventory Methods Handbook for Greenhouse Gas Inventory, Carbon Mitigation and Roundwood Production Projects*. Advances in Global Change Research. vol. 29. Springer Verlag, New York

Raybould, A. and Quemada, H. (2010). *Bt* crops and food security in developing countries: realised benefits, sustainable use and lowering barriers to adoption. *Food Security* 2, 247–259

RECOFTC (2010). *The Role of Social Forestry in Climate Change Mitigation and Adaptation in the ASEAN Region*. The Center for People and Forests (RECOFTC), ASEAN Social Forestry Network (ASFN) and Swiss Agency for Development and Cooperation (SDS), Bangkok

Reise, K., Olenin, S. and Thieltges, D.W. (2006). Are aliens threatening European aquatic coastal ecosystems? *Helgoland Marine Research* 60, 77–83

Ribeiro, M.C., Metzger, J.P., Martensen, A.C., Ponzoni, F.J. and Hirota, M.M. (2009). Brazilian Atlantic forest: how much is left and how is the remaining forest distributed? Implications for conservation. *Biological Conservation* 142(6), 1141–1153

Richardson, A.J. (2008). In hot water: zooplankton and climate change. *ICES Journal of Marine Science* 65(3), 279–295

Robinson, J.G. and Bennett, E.L. (eds.). (2000). *Hunting for Sustainability in Tropical Forests*. Columbia University Press, New York

Rodrigues, A.S.L., Akçakaya, A.R., Andelman, S.J., Bakarr, M.I., Boitani, L., Brooks, T.M., Chanson, J.S., Fishpool, L.D.C., Da Fonseca, G.A.B., Gaston, K.J., Hoffmann, M., Marquet, P.A., Pilgrim, J.D., Pressey, R.L., Schipper, J., Sechrest, W., Stuart, S.N., Underhill, L.G., Waller, W., Watts, M.E.J. and Yan, X. (2004). Global gap analysis: priority regions for expanding the global protected-area network. *BioScience* 54(12), 1092–1100

Roe, D. (2008). *Trading Nature. A Report, with Case Studies, on the Contribution of Wildlife Trade Management to Sustainable Livelihoods and the Millennium Development Goals*. TRAFFIC International, Cambridge and WWF–World Wide Fund for Nature, Gland

Rosenzweig, C., Casassa, G., Karoly, D.J., Imeson, A., Liu, C., Menzel, A., Rawlins, S., Root, T.L., Seguin, B. and Tryjanowski, P. (2007). Assessment of observed changes and responses in natural and managed systems. In *Climate Change 2007: Impacts, Adaptation and Vulnerability. Contribution of Working Group II to the Fourth Assessment Report of the Intergovernmental Panel on Climate Change* (eds. Parry, M.L., Canziani, O.F., Palutikof, J.P., van der Linden, P.J. and Hanson, C.E.). pp.79–131. Cambridge University Press, Cambridge

Rosset, P.M. (1999). *The Multiple Functions and Benefits of Small Farm Agriculture*. Policy Brief. Institute for Food and Development Policy, Oakland and Transnational Institute, Amsterdam

Royal Society (2009). *Reaping the Benefits: Science and the Sustainable Intensification of Global Agriculture*. The Royal Society, London

Sabine, C.L., Feely, R.A., Gruber, N., Key, R.M., Lee, K., Bullister, J.L., Wanninkhof, R., Wong, C.S., Wallace, D.W.R., Tilbrook, B., Millero, F.J., Peng, T.-H., Kozyr, A., Ono, T. and Rios, A.F. (2004). The oceanic sink for anthropogenic CO_2. *Science* 305(5682), 367–371

Scharlemann, J.P.W., Kapos, V., Campbell, A., Lysenko, I., Burgess, N.D., Hansen, M.C., Gibbs, H.K., Dickson, B. and Miles, L. (2010). Securing tropical forest carbon: the contribution of protected areas to REDD. *Oryx* 44(3), 352–357

Scherr, S.J. and McNeely, J.A. (2008). Biodiversity conservation and agricultural sustainability: towards a new paradigm of 'ecoagriculture' landscapes. *Philosophical Transactions of the Royal Society B* 363(1491), 477–494

Schuler, P. (2004). Biopiracy and commercialization of ethnobotanical knowledge. In *Poor People's Knowledge: Promoting Intellectual Property in Developing Countries* (eds. Finger, J.M. and Schuler, P.). pp.159–181. World Bank, Washington, DC and Oxford University Press, Oxford

Siebert, S., Burke, J., Faures, J.M., Frenken, K., Hoogeveen, J., Doll, P. and Portmann, F.V. (2010). Groundwater use for irrigation – a global inventory. *Hydrology and Earth System Sciences* 14, 1863–1880.

Sinkins, S.P. and Gould, F. (2006). Gene drive systems for insect disease vectors. *Nature Reviews Genetics* 7, 427–435

Sobrevila, C. (2008). *The Role of Indigenous Peoples in Biodiversity Conservation: The Natural but Often Forgotten Partners*. World Bank, Washington, DC

Spalding, M., Taylor, M., Ravilious, C., Short, F. and Green, E. (2003). Global overview: the distribution and status of seagrasses. In *World Atlas of Seagrasses* (eds. Green, E.P. and Short, F.T.). pp.5–25. University of California Press, Berkeley, CA

Srinivasan, U.T., Cheung, W.W.L., Watson, R. and Sumaila, U.R. (2010). Food security implications of global marine catch losses due to overfishing. *Journal of Bioeconomics* 12, 183–200

Stevens, S. (2010). Implementing the UN Declaration on the Rights of Indigenous Peoples and International Human Rights Law through the recognition of ICCAs. *Policy Matters* 17(3), 181–194

Stoett, P. (2010). Framing bioinvasion: biodiversity, climate change, security, trade, and global governance. *Global Governance* 16, 103–120

Strayer, D.L., Eviner, V.T., Jeschke, J.M. and Pace, M.L. (2006). Understanding the long-term effects of species invasions. *Trends in Ecology and Evolution* 21(11), 645–661

Sumaila, U.R., Khan, A.S., Dyck, A.J., Watson, R., Munro, G., Tydemers, P. and Pauly, D. (2010). A bottom-up re-estimation of global fisheries subsidies. *Journal of Bioeconomics* 12, 201–225

Sutherland, W.J. (2003). Parallel extinction risk and global distribution of languages and species. *Nature* 423, 276–279

Sutherland, W.J., Adams, W.M., Aronson, R.B., Aveling, R., Blackburn, T.M., Broad, S., Ceballos, G., Côté, I.M., Cowling, R.M., Da Fonseca, G.A.B., Dinerstein, E., Ferraro, P.J., Fleishman, E., Gascon, C., Hunter Jr., M., Hutton, J., Kareiva, P., Kuria, A., Macdonald, D.W., MacKinnon, K., Madgwick, F.J., Mascia, M.B., McNeely, J., Milner-Gulland, E.J., Moon, S., Morley, C.G., Nelson, S., Osborn, D., Pai, M., Parsons, E.C.M., Peck, L.S., Possingham, H., Prior, S.V., Pullin, A.S., Rands, M.R.W., Ranganathan, J., Redford, K.H., Rodriguez, J.P., Seymour, F., Sobel, J., Sodhi, N.S., Stott, A., Vance-Borland, K. and Watkinson, A.R. (2009). One hundred questions of importance to the conservation of global biological diversity. *Conservation Biology* 23, 557–567

Sutherland, W.J., Bailey, M.J., Bainbridge, I.P., Brereton, T., Dick, J.T.A., Drewitt, J., Dulvy, N.K., Dusic, N.R., Freckleton, R.P., Gaston, K.J., Gilder, P.M., Green, R.E., Heathwaite, A.L., Johnson, S.M., Macdonald, D.W., Mitchell, R., Osborn, D., Owen, R.P., Pretty, J., Prior, S.V., Prosser, H., Pullin, A.S., Rose, P., Stott, A., Tew, T., Thomas, C.D., Thompson, D.B.A., Vickery, J.A., Walker, M., Walmsley, C., Warrington, S., Watkinson, A.R., Williams, R.J., Woodroffe, R. and Woodroof, H.J. (2008). Future novel threats and opportunities facing UK biodiversity identified by horizon scanning. *Journal of Applied Ecology* 45, 821–833

Swartz, W., Sala, E., Tracey, S., Watson, R. and Pauly, D. (2010). The spatial expansion and ecological footprint of fisheries (1950 to present). *PLoS ONE* 5(12): e15143.

Swiderska, K. (2009). *Protecting Community Rights over Traditional Knowledge: Implications of Customary Law and Practices. Key Findings and Recommendations 2005–2009*. International Institute for Environment and Development (IIED), London

TEEB (2010). *The Economics of Ecosystems and Biodiversity: Mainstreaming the Economics of Nature: A Synthesis of the Approach, Conclusions and Recommendations of TEEB*. The Economics of Ecosystems and Biodiversity. Progress Press, Malta

Tewksbury, J.J., Sheldon, K.S. and Ettinger, A.K. (2011). Ecology: moving farther and faster. *Nature Climate Change* 1, 396–397

Thornton, P.K. (2010). Livestock production: recent trends, future prospects. *Philosophical Transactions of the Royal Society B* 365(1554), 2853–2867

Thrupp, L.A. (2000). Linking agricultural biodiversity and food security: the valuable role of agrobiodiversity for sustainable agriculture. *International Affairs* 76(2), 265–281

Thrush, S.F. and Dayton, P.K. (2002). Disturbance to marine benthic habitats by trawling and dredging: implications for marine biodiversity. *Annual Review of Ecology and Systematics* 33, 449–473

Tockner, K. and Stanford, J.A. (2002). Riverine floodplains: present state and future trends. *Environmental Conservation* 29, 308–330

Tockner, K., Bunn, S.E., Quinn, G., Naiman, R., Stanford, J.A. and Gordon, C. (2008). Floodplains: critically threatened ecosystems. In *Aquatic Ecosystems* (ed. Polunin, N.C.). pp.45–61. Cambridge University Press, Cambridge

Toropova, C., Meliane, I., Laffoley, D., Matthews, E. and Spalding, M. (eds.) (2010). *Global Ocean Protection: Present Status and Future Possibilities.* Agence des aires marines protégées, Brest, International Union for Conservation of Nature World Commission on Protected Areas (IUCN WCPA), Gland, Washington, DC and New York, United Nations Environment Programme World Conservation Monitoring Centre (UNEP-WCMC), Cambridge, The Nature Conservancy (TNC), Arlington, VA, United Nations University (UNU), Tokyo and Wildlife Conservation Society (WCS), New York

TRAFFIC (in prep.). *Global Values of Wildlife Trade.* The Wildlife Trade Monitoring Network, Cambridge.

TRAFFIC and IUCN SSC Medicinal Plants Specialist Group (2009). *Biodiversity for Food and Medicine Indicator – Biannual Substantive Report to the Biodiversity Indicators Partnership.* http://www.traffic.org/trade/

UN (2000). *Millennium Development Goals.* http://www.un.org/millenniumgoals/

UN (1971). *Convention on Wetlands of International Importance especially as Waterfowl Habitat.* Ramsar (Iran), 2 February 1971. UN Treaty Series No. 14583. (Amended 1982 and 1987). http://www.ramsar.org/cda/en/ramsar-documents-texts/main/ramsar/1-31-38_4000_0__

UNCED (1992). *Agenda 21.* United Nations Conference on Environment and Development. http://www.un.org/esa/sustdev/documents/agenda21/english/Agenda21.pdf

UNEP (2011). *Towards a Green Economy: Pathways to Sustainable Development and Poverty Eradication.* United Nations Environment Programme, Nairobi

UNEP (2007). *Global Environment Outlook 4: Environment for Development.* United Nations Environment Programme. Progress Press, Valletta

UNEP/GRID-Arendal (2008). *Major Pathways and Origins of Invasive Species Infestations in the Marine Environment.* UNEP/GRID-Arendal Maps and Graphics Library. http://maps.grida.no/go/graphic/major-pathways-and-origins-of-invasive-species-infestations-in-the-marine-environment (accessed 3 September 2011)

UNEP/GRID-Arendal (2005). *Linkages between Ecosystem Services and Human Well-being.* UNEP/GRID-Arendal Maps and Graphics Library. http://maps.grida.no/go/graphic/linkages-between-ecosystem-services-and-human-well-being (accessed 22 November 2011)

UNEP-WCMC (2011). *Developing Ecosystem Service Indicators: Experiences and Lessons Learned from Sub-global Assessments and Other Initiatives.* Technical Series No. 58. Secretariat of the Convention on Biological Diversity, Montreal

UNFCCC (1992). *United Nations Framework Convention on Climate Change.* FCCC/INFORMAL/84 GE.05-62220 (E) 200705. http://unfccc.int/resource/docs/convkp/conveng.pdf

UNGA (2005). *World Summit Outcome 2005.* United Nations General Assembly. http://daccess-dds-ny.un.org/doc/UNDOC/GEN/N05/487/60/PDF/N0548760.pdf?OpenElement

Valiela, I., Rutecki, D. and Fox, S. (2004). Saltmarshes: biological controls of foodwebs in a diminishing environment. *Journal of Experimental Marine Biology and Ecology* 300(1–2), 131–159

Valk, H. and Kaasik, A. (2007). *Looduslikud pühapaigad: väärtused ja kaitse.* Õpetatud Eesti Seltsi. Toimetised. Verhandlungen der Gelehrten Estnischen Gesellschaft. *Looduslikud pühapaigad: Väärtused ja kaitse.* Õpetatud Eesti Seltsi

Verschuuren, B., Wild, R., McNeely, J. and Oviedo, G. (eds.) (2010). *Sacred Natural Sites, Conserving Culture and Nature.* Earthscan, Oxford

Vié, J.-C., Hilton-Taylor, C. and Stuart, S.N. (eds.) (2009). *Wildlife in a Changing World. An Analysis of the 2008 IUCN Red List of Threatened Species.* International Union for Conservation of Nature, Gland

Vorosmarty, C.J., McIntyre, P.B., Gessner, M.O., Dudgeon, D., Prusevich, A., Green, P., Glidden, S., Bunn, S.E., Sullivan, C.A., Reidy Liermann, C. and Davies, P.M. (2010). Global threats to human water security and river biodiversity. *Nature* 467, 555–561

Wackernagel, M. and Rees, W. (1996). *Our Ecological Footprint: Reducing Human Impact on the Earth.* New Society Publishers, Gabriola Island, BC

Wackernagel, M., Schulz, N.B., Deumling, D., Linares, A.C., Jenkins, M., Kapos, V., Monfreda, C. and Loh, J. (2002). Tracking the ecological overshoot of the human economy. *Proceedings of the National Academy of Sciences of the United States of America* 99(14), 9266–9271

Walpole, M., Almond, R.E.A., Besançon, C., Butchart, S.H.M., Campbell-Lendrum, D., Carr, G.M., Collen, B., Collette, L., Davidson, N.C., Dulloo, E., Fazel, A.M., Galloway, J.N., Gill, M., Goverse, T., Hockings, M., Leaman, D.J., Morgan, D.H.W., Revenga, C., Rickwood, C.J., Schutyser, F., Simons, S., Stattersfield, A.J., Tyrrell, T.D., Vié, J.-C. and Zimsky, M. (2010). Tracking progress toward the 2010 biodiversity target and beyond. *Science* 325(5947), 1503–1504

Waycott, M., Duarte, C.M., Carruthers, T.J.B., Orth, R.J., Dennison, W.C., Olyarnik, S., Calladine, A., Fourqurean, J.W., Heck, K.L., Hughes, A.R., Kendrick, G.A., Kenworthy, W.J., Short, F.T. and Williams, S.L. (2009). Accelerating loss of seagrasses across the globe threatens coastal ecosystems. *Proceedings of the National Academy of Sciences of the United States of America* 106(30), 12377–12381

Westley, F., Olsson, P., Folke, C., Homer-Dixon, T., Vredenburg, H., Loorbach, D., Thompson, J., Nilsson, M., Lambin, E., Sendzimir, J., Banarjee, B., Galaz, V. and van der Leeuw, S. (2011). Tipping towards sustainability: emergent pathways of transformation. Working Paper No 3. In *3rd Nobel Laureate Symposium on Global Sustainability: Transforming the World in an Era of Global Change.* Stockholm, Sweden, 16–19 May 2011

White, A. and Martin, A. (2002). *Who Owns the World's Forests? Forest Tenure and Public Forests in Transition.* Forest Trends and Center for International Environmental Law, Washington, DC

White, A., Molnar, A. and Khare, A. (2004). *Who Owns, Who Conserves, and Why it Matters.* Forest Trends Association, Washington, DC

WHO (2009). *Health Impact Assessment (HIA) – Health and Social Impacts of Large Dams.* http://www.who.int/hia/examples/energy/whohia020/en/index.html (accessed 07 November 2011)

WHO (2005). *Ecosystems and Human Well-being: Health Synthesis.* World Health Organization, Geneva

WHO (2003). *Traditional Medicine.* WHO Fact Sheet No.134 revised May 2003. http://www.who.int/mediacentre/factsheets/2003/fs134/en/ (accessed 18 September 2011)

Williams, J. (2006). Resource management and Maori attitudes to water in southern New Zealand. *New Zealand Geographer* 62, 73–80

Wing, S.R. and Wing, E.S. (2001). Prehistoric fisheries in the Caribbean. *Coral Reefs* 20, 1–8

Woinarski, J.C.Z, Legge, S., Fitzsimons, J.A., Traill, B.J., Burbidge, A., Fisher, A., Firth, R.S.C., Gordon, I.J., Griffiths, A.D., Johnson, C.D., McKenzie, L., Palmer, C., Radford, I., Rankmore, B., Ritchie, E.G., Ward, S. and Ziembicki, M. (2011). The disappearing mammal fauna of northern Australia: context, cause, and response. *Conservation Letters* 4(3), 192–201

Worm, B., Hilborn, R., Baum, J.K., Branch, T.A., Collie, J.S., Costello, C., Fogarty, M.J., Fulton, E.A., Hutch ings, J.A., Jennings, S., Jensen, O.P., Lotze, H.K., Mace, P.M., McClanahan, T.R., Minto, C., Palumbi, S.R., Parma, A.M., Ricard, D., Rosenberg, A.A., Watson, R. and Zeller, D. (2009). Rebuilding global fisheries. *Science* 325(5940), 578–585

WSSD (2002). *Johannesburg Plan of Implementation.* World Summit on Sustainable Development. http://www.un.org/esa/sustdev/documents/WSSD_POI_PD/English/POIToc.htm

WWF (2010). *Living Planet Report 2010. Biodiversity, Biocapacity and Development.* WWF–World Wide Fund For Nature, Gland

Xenopoulos, M.A. and Lodge, D.M. (2006). Going with the flow: using species-discharge relationships to forecast losses in fish biodiversity. *Ecology* 87, 1907–1914

Yagi, N., Takagi, A.P., Takada, Y. and Kurokura, H. (2010). Marine protected areas in Japan: institutional background and management framework. *Marine Policy* 34(6), 1300–1306

Zamin, T., Baillie, J.E.M., Miller, R.M., Rodrigues, J.P., Ardid, A. and Collen, B. (2010). National Red Listing beyond the 2010 target. *Conservation Biology* 24(4), 1012–1020

CHAPTER 6

Chemicals and Waste

Coordinating lead authors: Ricardo Barra, Pierre Portas and Roy Victor Watkinson

Lead authors: Oladele Osibanjo, Ian Rae, Martin Scheringer and Claudia ten Have

Contributing authors: Borislava Batandjieva, Walter Giger, Ivan Holoubek, Heather Jones-Otazo, Liu Lili, Philip Edward Metcalf, Karina Silvia Beatriz Miglioranza, Arthur Russell Flegal, Adebola A. Oketola (GEO Fellow) and Monica Montory (GEO Fellow)

Principal scientific reviewer: Mika Sillanpaa

Chapter coordinator: Ludgarde Coppens

Main Messages

There is an extensive but incomplete body of scientific knowledge on the impacts of chemicals and wastes on humans and the environment, with particular information and data gaps on the uses, emissions, exposure pathways and effects of chemicals. Global understanding of the complexity of properties and environmental impact of chemicals and wastes is therefore markedly deficient. The fourth *Global Environment Outlook* (2007) indicated that data were incomplete globally and that, for many regions, it was important to evaluate the magnitude of chemical contamination and its impacts on the environment and human health. But little has occurred since then. The UN Secretary-General, in his May 2011 report on policy options for waste management to the Commission on Sustainable Development, stated that: "the barriers to effective management and minimization include lack of data, information, and knowledge on waste scenarios". And the UN-Habitat report on waste management in cities stated that "waste reduction is desirable but, typically, it is not monitored anywhere" (UN-Habitat 2010).

Over the last decade chemical production has shifted from the countries of the Organization for Economic Co-operation and Development (OECD) to the BRIC countries (Brazil, Russia, India and China) and other developing countries, accompanied by a doubling of sales and the development of many new types of chemical. The OECD's share of world production is now 9 per cent less than in 1970. Much of this shift has been due to major emerging economies. In 2004, China accounted for the largest share of BRIC production at 48 per cent, followed by Brazil and India at 20 per cent each, and Russia at 12 per cent (OECD 2008b). Chemical consumption in developing countries is likewise growing much faster than in the developed world and could account for a third of global consumption by 2020.

Chemicals play an important role in human life, economic development and prosperity, yet they can also have adverse impacts on the environment and human health. The diversity and potential consequences of such impacts, combined with limited capacity in developing countries and economies in transition to manage these impacts, make the sound management of chemicals and waste a key cross-cutting issue. A recent study by the World Health Organization (WHO) (Prüss-Ustün *et al.* 2011) indicated that 4.9 million deaths were attributable to environmental exposure to chemicals in 2004. In many regions, hazardous waste streams are mixed with municipal or solid wastes and then either dumped or burned in the open air (UN-Habitat 2010).

Global chemical pollution is a serious threat to sustainable development and livelihoods. The problem has impacts on both humanity and ecosystems, and includes adverse effects from long-term exposure to low or sub-lethal concentrations of single chemicals or to mixtures of chemicals. Currently, more than 90 per cent of water and fish samples from aquatic environments are contaminated by pesticides. Estimates indicate that about 3 per cent of exposed agricultural workers suffer from an episode of acute pesticide poisoning every year (Thunduyil *et al.* 2008). Pollution with persistent organic pollutants (POPs) is widespread, in particular affecting remote areas such as the Arctic and Antarctic.

Emerging issues requiring better understanding and prompt action to prevent harm to health and

the environment include the sound management of electronic and electrical waste (e-waste), endocrine-disrupting chemicals, plastics in the environment, open burning, and the manufacture and use of nanomaterials. E-waste has become one of the major environmental challenges of the 21st century: it is the fastest-growing waste stream in the world, estimated at 20–50 million tonnes per year (Schwarzer *et al.* 2005). It is of particular interest because it contains not only hazardous substances – such as heavy metals including mercury and lead, and endocrine-disrupting substances such as brominated flame retardants (BFRs) – but also many strategic metals such as gold, palladium and rare earth metals that can be recovered and recycled. Very little is known about whether nanomaterials or nanoparticles are released from products when they are incinerated, buried or degraded over time, so it is possible that they will pose a serious waste disposal challenge. Sound decision making on nanotechnology has provoked much debate among developed country regulators, and increasingly among the regulators of developing countries (Morris *et al.* 2010).

Effective management of these issues requires better information gathering and integrated approaches to chemicals, radioactive materials and waste management, supported where appropriate by improved environmental governance. The process for greater cooperation and coordination between the chemicals and waste conventions (Basel, Rotterdam and Stockholm) provides an opportunity to enhance awareness raising, knowledge transfer, capacity building and national implementation that should be further explored.

INTRODUCTION

More than 248 000 chemical products are commercially available (CAS 2011) and subject to regulatory and inventory systems. Chemicals provide valuable benefits to humanity including in agriculture, medicine, industrial manufacturing, energy extraction and generation, and public health and disease vector control. Chemicals play an important role in achieving developmental and social goals, especially for improving maternal health, reducing child mortality and ensuring food security, and advances in their production and management have increased their safe application. Nonetheless, because of their intrinsic hazardous properties, some pose risks to the environment and human health. Simultaneous exposure to many chemicals – the cocktail or synergistic effect – is likely to exacerbate the impacts.

Chemicals are released at many steps in their life cycle, from the extraction of raw materials, through production chains, transport and consumption, to final waste disposal. They are distributed through indoor environments, food and drinking water, and through soils, rivers and lakes. Certain long-lived chemicals such as persistent organic pollutants (POPs) and heavy metals are transported globally, reaching otherwise pristine environments such as rain forests, deep oceans or polar regions, and can quickly pass along the food chain, bioaccumulating to cause toxic effects in humans and wildlife.

Products derived from chemicals often become hazardous wastes in their end-of-life phase, generating additional pollution risks that can devalue their initial benefits and counteract development advantages. Pollution from dumping and uncontrolled open burning is common (UN-Habitat 2010), and is even increasing in some parts of the world, though some progress has been made in recent decades. The causes of mismanagement often lie in such factors as deficiencies in institutional and regulatory frameworks. Such shortcomings also have an impact on the growing transboundary movement of hazardous wastes from developed to developing countries, where compliance, monitoring and enforcement of regulations tend to be weak, and the financial and technical capacity to implement improved waste management practices is limited. This leads to a risk of rapidly increasing exposure for greater portions of the population and to related, often serious, health problems, in particular for women and children.

Broadly, a two-speed situation exists, with developed countries generally having comprehensive systems for chemical and hazardous waste management, while developing countries generally do not. Developing countries and economies in transition struggle with basic landfill co-disposal of many types of wastes, with little capacity for their separation and sound management.

While many developing countries have ratified the multilateral environmental agreements on chemicals and wastes – such as the Basel Convention on the Control of Transboundary Movement of Hazardous Wastes and their Disposal (Basel Convention 1989) – these are not always transposed into national legislation in a comprehensive manner. In addition, given the cross-sectoral nature of the issue, the regulation and management of chemicals in most developing countries is spread over several ministries – including agriculture, industry, labour, environment and health – and between several agencies within each ministry.

In most countries, it is the poorest members of the population that are at particular risk of exposure. This may be due to occupational exposure, poor living conditions, lack of access to clean water and food, domestic proximity to polluting activities, or a lack of knowledge about the detrimental impacts of chemicals – or a combination of these factors.

Radioactive contamination is another source of potential environmental and health hazards, both from controlled emissions and waste management, and from accidental release. The controlled release of radionuclides to the atmospheric and aquatic environments may occur as authorized effluent discharge, while uncontrolled release may occur as a result of accidents and at legacy sites left by nuclear weapons testing. The management and disposal of radioactive waste from industry, research and medicine, as well as from nuclear power, is relevant to almost all countries, requiring different approaches according to the volume, radioactivity and other properties of the waste.

Initially, governance instruments for chemicals and wastes could be considered to have been reactive, piecemeal and isolated, and with mixed success – the Montreal Protocol on

Waste treatment plant, Los Angeles, United States. *© John Crall/iStock*

Substances that Deplete the Ozone Layer (UNEP 1987), for example, being effective in reducing the impact of ozone-depleting substances, while the Basel Convention (1989) has struggled to reduce the transboundary movement of hazardous waste. There have been significant advances over the past decade, however, and regulatory instruments are now improving with the better and more widespread understanding of the life cycle of chemicals and their association with the generation and processing of wastes. Efforts to bring the work of the Basel, Rotterdam and Stockholm Conventions together constitute a first step towards addressing the entire life cycle of chemicals. This also applies to the establishment of the Strategic Approach to International Chemicals Management (SAICM) and the current negotiation for an international agreement on mercury. Similarly, the Joint Convention on the Safety of Radioactive Waste Management and the Safety of Spent Nuclear Fuel Management is a significant step forward. However, ensuring that these efforts are sustained and fully anchored at the national level requires further investment in better science-based understanding of chemicals and wastes, policy creativity to balance development and sustainability imperatives, public-private partnerships to link technological innovation and societal responsibility, and allocation of funds for comprehensive capacity building.

INTERNATIONALLY AGREED GOALS

This chapter evaluates progress towards internationally agreed goals relevant to chemicals and wastes. The goals are those identified by the *GEO-5* High-Level Intergovernmental Advisory Panel from key multilateral environmental agreements and related agreements and declarations, as further considered and prioritized in regional consultations. The current lack of data, a key constraint on many aspects of chemical and waste management, has not been seen as a reason to preclude the selection of a goal. The goals evaluated are reflected in Table 6.1.

In the 1970s and 1980s the human health and environmental impacts of chemicals and waste led to the creation of a number of key international agreements. These, along with other related goal-based international agreements and declarations such as those emanating from the 2002 World Summit on Sustainable Development (WSSD) in Johannesburg, constitute a framework for organizing and implementing specific goals for the environmentally sound design, production, consumption and recycling or disposal of chemicals and hazardous waste (Box 6.1). These goals are also considered against the background of the Millennium Development Goals (MDGs), specifically MDG 1 for eradicating extreme poverty and hunger, and MDG 7 for ensuring environmental sustainability. MDG 7 includes specific targets for ozone-depleting substances, as well as for improved access to safe drinking water and sanitation facilities.

The broad set of principles pivotal to the development of international agreements comprises prior informed consent for the transboundary movement of hazardous waste and certain hazardous chemicals; transparency through national reporting; the environmentally sound management of chemicals and waste; waste prevention; the precautionary approach; and the polluter-pays principle. These are addressed through specific obligations such as the implementation of control measures, monitoring of the state of the environment, and compliance regimes with supportive delivery mechanisms including capacity building and training, international cooperation, synergies and partnerships.

Goals relevant to the sound management of chemicals and waste aim to protect human health and the environment while improving resource efficiency. They can be grouped into six themes:

- sound management of chemicals throughout their life cycle, including persistent organic pollutants and heavy metals, and of waste;
- control of the transboundary movement of hazardous wastes as well as responsible trade in hazardous chemicals;
- transparent science-based risk assessment and risk management procedures, as well as monitoring systems at the national, regional and global levels;
- support for countries to strengthen their capacity for the sound management of chemicals and waste;
- protection and preservation of the marine environment from all sources of pollution;
- safe radioactive and nuclear waste management.

Box 6.1 Multilateral environmental agreements and the sound management of chemicals

The sound management of chemicals is addressed by 17 different multilateral agreements including the 1998 Rotterdam Convention on the Prior Informed Consent Procedure for Certain Hazardous Chemicals and Pesticides in International Trade and the 2001 Stockholm Convention on Persistent Organic Pollutants (both effective since 2004). In addition, in 2006 the First International Conference on Chemicals Management established the Strategic Approach to International Chemicals Management (SAICM), a multi-stakeholder policy framework for achieving the safe management of chemicals worldwide by 2020 (SAICM 2009). So far, more than 300 activities have been conducted under the SAICM Global Plan of Action. Pollutant release and transfer registers have been promoted and currently just 23 countries have established a functioning national register. The globally harmonized system of classification and labelling of chemicals, containing all criteria necessary for classification of chemicals according to their intrinsic hazardous properties, has been established, as have provisions for hazard communication. However, many challenges remain and the lack of prioritization of sound management of chemicals, the limitations of legislation, the lack of information and the lack of adequate financial resources, including for the funding of activities concerning the remediation of contaminated sites, are still major obstacles to achieving the 2020 goal (CSD 2010).

Table 6.1 Selected internationally agreed goals related to chemicals and waste

Johannesburg Plan of Implementation (JPOI) (WSSD 2002) Paragraph 23 *(Prioritized during regional consultations in Europe and the Asia and Pacific region)*	Renew the commitment, as advanced in Agenda 21, to sound management of chemicals throughout their life cycle and of hazardous wastes for sustainable development as well as for the protection of human health and the environment, *inter alia*, aiming to achieve, by 2020, that chemicals are used and produced in ways that lead to the minimization of significant adverse effects on human health and the environment, using transparent science-based risk assessment procedures and science-based risk management procedures, taking into account the precautionary approach, as set out in principle 15 of the Rio Declaration on Environment and Development (UNCED 1992a), and support developing countries in strengthening their capacity for the sound management of chemicals and hazardous wastes by providing technical and financial assistance.
Paragraph 22 *(Prioritized during regional consultations in the Asia and Pacific region)*	Prevent and minimize waste and maximize reuse, recycling and use of environmentally friendly alternative materials, with the participation of government authorities and all stakeholders, in order to minimize adverse effects on the environment and improve resource efficiency, with financial, technical and other assistance for developing countries.
Paragraph 23g	Promote reduction of the risks posed by heavy metals that are harmful to human health and the environment, including through a review of relevant studies, such as the United Nations Environment Programme global assessment of mercury and its compounds.
Stockholm Convention on Persistent Organic Pollutants (Stockholm Convention 2001) Article 1	... to protect human health and the environment from persistent organic pollutants.
Rotterdam Convention on the Prior Informed Consent Procedure for Certain Hazardous Chemicals and Pesticides in International Trade (Rotterdam Convention 2001) Article 1	... to promote shared responsibility and cooperative efforts among Parties in the international trade of certain hazardous chemicals in order to protect human health and the environment from potential harm and to contribute to their environmentally sound use, by facilitating information exchange about their characteristics, by providing for a national decision-making process on their import and export and by disseminating these decisions to Parties.
Basel Convention on the Control of Transboundary Movements of Hazardous Wastes and their Disposal (Basel Convention 1989) Preamble	... to protect, by strict control, human health and the environment against the adverse effects which may result from the generation and management of hazardous waste and other wastes.
International Convention for the Prevention of Pollution from Ships 1973, as modified by the Protocol of 1978 (MARPOL 1973/78) Article 17	The Parties to the Convention accept the obligation to promote, in consultation with other international bodies and with assistance from UNEP and coordination with the Executive Director of UNEP, the necessary support for Parties who may require technical assistance in the following areas: a) Training scientific and technical staff; b) obtaining equipment and monitoring installations where needed; c) ease the adoption of additional measures and conditions which seek to prevent or reduce pollution in the marine environment by ships; and d) encourage research; preferably within the concerned countries, in order to promote realization of the aims and objectives of this convention.
International Convention on the Prevention of Marine Pollution by the Dumping of Wastes and Other Matter (London Convention 1972) Article 2	Contracting Parties shall individually and collectively protect and preserve the marine environment from all sources of pollution and take effective measures, according to their scientific, technical and economic capabilities, to prevent, reduce and where practicable eliminate pollution caused by dumping or incineration at sea of wastes or other matter.
Article 12	The Contracting Parties pledge themselves to promote, within the competent specialized agencies and other international bodies, measures to protect the marine environment against pollution caused by: (a) hydrocarbons, including oil and their wastes.
Agenda 21 (UNCED 1992b) Chapter 22 Paragraph 3	The objective of this programme area is to ensure that radioactive wastes are safely managed, transported, stored and disposed of, with a view to protecting human health and the environment, within a wider framework of an interactive and integrated approach to radioactive waste management and safety.
Joint Convention on the Safety of Spent Fuel Management and on the Safety of Radioactive Waste Management (IAEA 1997) Article 1	The objectives of this Convention are: (i) to achieve and maintain a high level of safety worldwide in spent fuel and radioactive waste management [...]; (ii) to ensure that during all stages of spent fuel and radioactive waste management there are effective defenses against potential hazards so that individuals, society and the environment are protected from harmful effects of ionizing radiation, now and in the future, in such a way that the needs and aspirations of the present generation are met without compromising the ability of future generations to meet their needs and aspirations; (iii) to prevent accidents with radiological consequences and to mitigate their consequences should they occur during any stage of spent fuel or radioactive waste management.

STATE AND TRENDS

The fourth *Global Environment Outlook* (2007) indicated that data were incomplete at the global level, making it a challenge for many regions to evaluate the magnitude of chemical contamination and its impacts on the environment and human health. Little has changed in five years and a worldwide effort is needed to fill this gap. The forthcoming UNEP Global Chemicals Outlook report should assist: it aims to provide a framework for assessing and setting priorities to stimulate further international action on sound management of chemicals.

Chemicals and wastes: data and indicators

The lack of data on existing chemicals, and the rapid technological changes that bring new chemicals to the market, have hindered the production of an established set of indicators with time-series data that can be used to identify the state and trends of chemicals and wastes. Several possible indicators to fill this gap are proposed below. In addition, extensive investment in collating the required data and solidifying the knowledge base is required to construct long-term time series.

Underlying data on waste generation, treatment and recycling are difficult to obtain. Some are available on hazardous waste through reports to the Secretariat of the Basel Convention (Figure 6.1), providing information on the quantity, characteristics, destination and mode of treatment or disposal of hazardous waste that is subject to international movement, but even this is incomplete and unverified – as was reported in 2011 to the tenth Conference of the Parties to the Convention (UNEP 2011a). Global data on non-hazardous waste generation and disposal have not been systematically reported and are therefore unsatisfactory. As stated by the UN Secretary-General in his May 2011 report to the Commission on Sustainable Development: "*The barriers to effective management and minimization include lack of data, information, and knowledge on waste scenarios, lack of comprehensive regulations and weak enforcement of existing legislation, weak technical and organizational capacities, poor public awareness and cooperation, and lack of funds.*" (UNCSD 2011)

There is an urgent need to improve the availability and quality of these basic datasets, with a focus on comparability between countries, timeliness and coherence over time, and interpretability. As waste is increasingly seen as a potential resource, waste data and indicators should be more closely linked to economic and social information systems and material flow accounting. The measurability issue is critical to assess waste generation, including municipal, industrial, agricultural, mining, military, radioactive and nuclear.

Box 6.2 Johannesburg Plan of Implementation (JPOI) (WSSD 2002) Paragraph 23

Issue
The sound management of chemicals throughout their life cycle for the protection of human health and the environment

Related goals
To ensure, by 2020, that chemicals are used and produced in ways that lead to the minimization of significant adverse effects on human health and the environment

Indicators
Number of signatory countries to the three conventions on chemicals and wastes (Basel, Rotterdam, Stockholm); number of implementation plans being put in place by these countries

Global trends
Some progress

Most vulnerable communities
Labour force, women and children in developing countries, consumers worldwide

Regions of greatest concern
Africa, Latin America and Asia

Figure 6.1 Transmission of national reports by Parties to the Basel Convention, 1999–2009

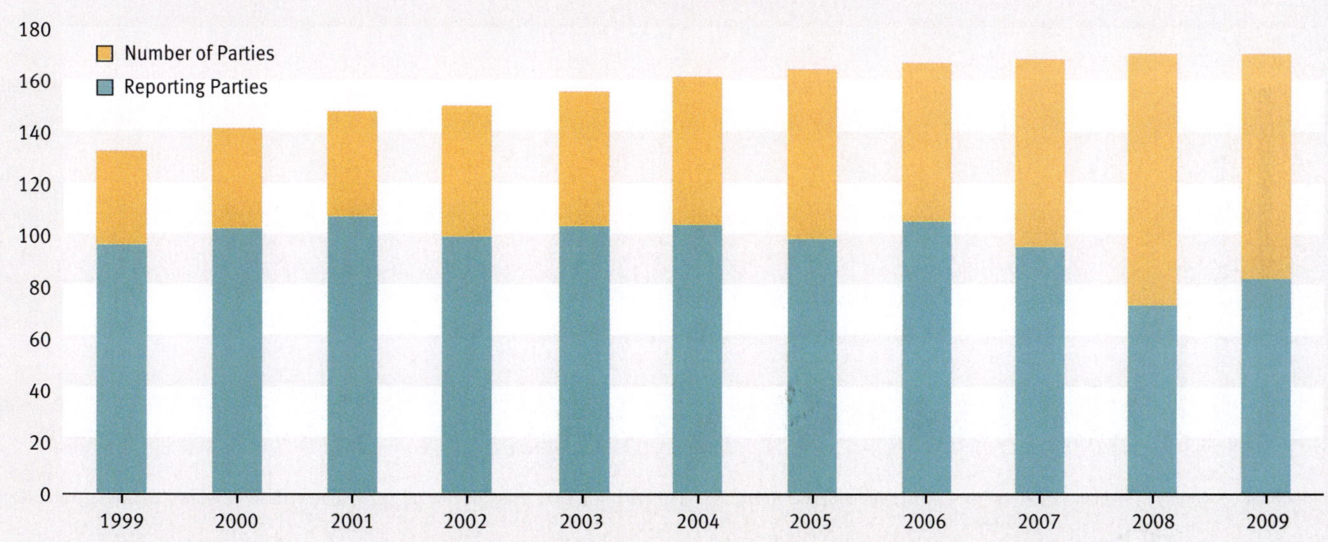

Source: Basel Convention 2011

Three indicators to help inform governments and municipalities of industry performance and progress are highlighted here. It is imperative that data for these indicators are generated to guide decision making on sound global management of wastes. The key indicators proposed are:

- quantity and types of waste – solid, organic, hazardous and non-hazardous – managed or finally disposed of;
- waste and hazardous waste generation per person; and
- the amount of municipal or household waste, industrial solid waste and hazardous waste that is recycled.

Large-scale chemical plant at night. *© Tetsuo Morita/iStock*

Status and trends of the chemical industry

The chemical industry is a major driver of economic growth and its performance is a leading indicator of economic development. In 2008 the global chemicals industry had an estimated turnover of about US$3.7 trillion (OECD 2010a) and was growing at 3.5 per cent per year. More than 20 million people around the globe are employed by it directly or indirectly, and it is an intensive energy consumer and a ubiquitous generator of emissions.

While companies in the countries of the Organisation for Economic Co-operation and Development (OECD) continue to account for the bulk of world production (74.5 per cent in 2004), the OECD's share is now 9 per cent less than in 1970. Much of this shift has been caused by the major emerging economies, particularly the BRIC countries (Brazil, Russia, India and China). In 2004, China accounted for most BRIC production (48 per cent), followed by Brazil and India (20 per cent each), and Russia (12 per cent) (OECD 2008b). Chemical consumption in developing countries is likewise growing much faster than in developed countries, and could account for a third of global consumption by 2020. At the same time, some data show that developed countries are reducing chemical use. For example, overall use of pesticides in OECD countries declined by 5 per cent during 1990–2002, although trends vary from country to country (OECD 2008a). Total releases and transfers of the 152 pesticides that are common to the United States and Canada dropped by 18 per cent and the production of ozone-depleting substances almost stopped; emissions of acid rain precursors dropped by 48 per cent, ozone precursors by 38 per cent and non-methane volatile organic compounds by 26 per cent. Nonetheless, international cooperation between all governments is needed to build capacity, share information and promote effective chemicals management globally (OECD 2008b). Figure 6.2 shows the sales data of the world's major chemical-producing countries.

Figure 6.2 Chemical sales by country, 2009

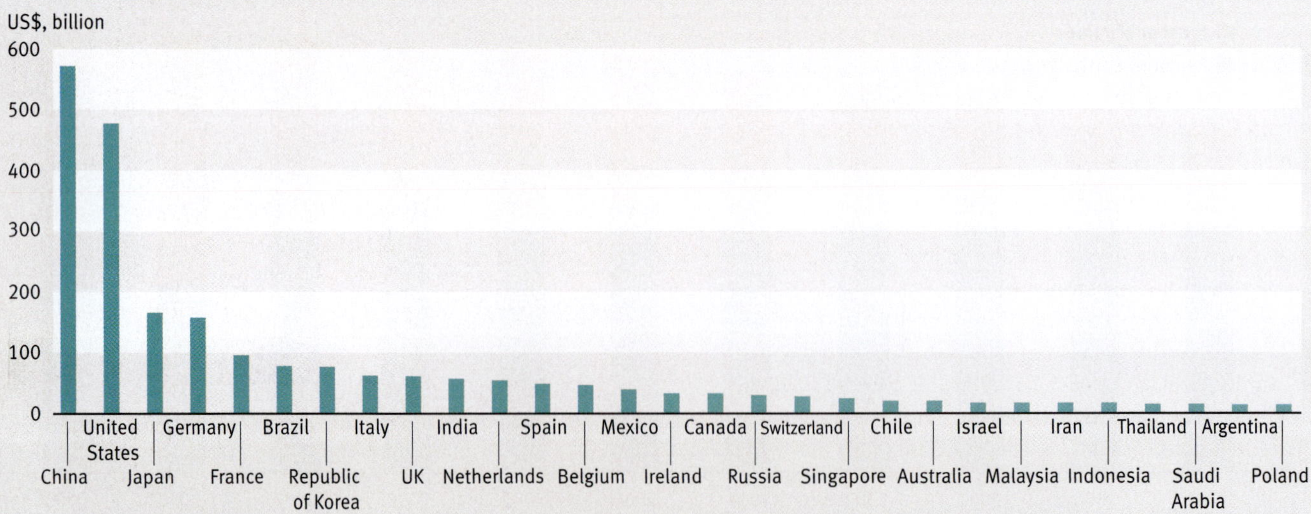

Source: OECD 2010b

Waste as an issue of global significance

The growing interdependence of the global economy along with the increasing generation and complexity of waste worldwide can lead countries to unsound waste management and disposal operations, and there may come a point when related costs are such that the economy and public services fail to keep up. Integrated policies are required to support sustainable economic development through recycling, recovery, reuse and other operations aimed at reducing both the use of natural resources and the quantities of waste, as it is inevitable that some resource inputs to industrial production are returned to the environment as waste, and may be hazardous. A critical issue is to reverse current trends in waste generation, which would require a high level of commitment to minimize both quantities and levels of hazard. Furthermore, unsound recycling comes with the risk of pollution and increased human exposure to toxic substances. Recycling can also be misused as a disguise for criminal operations.

The introduction of many new chemical substances to the market leads to the production of new kinds of wastes. In many regions, hazardous waste streams are mixed with municipal or solid wastes and then either dumped or burned in the open (UN-Habitat 2010). This raises issues of environmental and social justice, as the people most affected by such precarious practices are usually the poor who live and work adjacent to dump sites.

Through globalization, materials may be produced in one country or region, used in another and managed as waste in a third. Electrical and electronic equipment provides a case in point (Schluepa *et al.* 2009; Cui and Forssberg 2003). The treatment of end-of-life electronics, including toxic substances and plastics with associated flame-retardants as well as precious metals, exemplifies the two sides of this business. The original equipment has the potential to contribute to protecting human health, supporting livelihoods and creating jobs, while also promoting a shift from waste to resources that supports economic development, energy efficiency and the conservation of natural resources. However, unsound or inadequate waste management can have profound human health impacts and cause serious harm to the environment. Extending the useful life of electrical and electronic equipment and using less harmful substances in these products is one way to reduce the waste burden and its accompanying hazards.

Municipal waste

The unsound management of waste can lead to mutually reinforcing undesirable effects. It can pollute and contaminate the environment, pose a threat to human health and represent a loss of resources in the form of both materials and energy. The recent UN-Habitat report on solid waste management in cities refers to the escalating challenge of managing it across the globe, and amply demonstrates the complexity and variety of issues faced, including the difficulty of achieving objectives when progress goes unrecorded, stating for example that "waste reduction is desirable but, typically, it is not monitored anywhere" (UN-Habitat 2010).

Box 6.3 Waste in the OECD

The quantity of municipal waste generated in OECD countries has risen steeply since 1980, exceeding an estimated 650 million tonnes in 2007 (556 kg per person). In most countries for which data are available, increased affluence associated with economic growth and changes in consumption patterns tend to generate higher levels of waste per person. Over the past 20 years, however, waste generation has risen less rapidly than either GDP or expenditure on private consumption, with a slow-down in recent years. The amount and composition of municipal waste going to final disposal depend on national waste management practices. Despite improvements in these practices, only a few countries have succeeded in reducing the quantity of solid waste for disposal (OECD 2010b).

Municipal waste constitutes a significant percentage of the total waste a country generates (OECD 2008b), with annual figures ranging from 0.4 to 0.8 tonnes per person, and solid waste generation increasing at an estimated rate of about 0.5–0.7 per cent per year. Waste complexity is also increasing with the co-disposal of assorted waste types: biodegradable components currently account for almost 50 per cent of municipal solid waste and electronic waste (e-waste) for 5–15 per cent. The management of waste is further complicated by the range and diversity of waste generators, from mining and a wide variety of manufacturers through agricultural and medical waste to household rubbish. In addition, the sound management of municipal waste constitutes a sizable and continuous part of a municipality's budget.

Many countries do not have the infrastructure to deal with ever more complex waste streams. Nor do many have the regulatory and physical infrastructure to derive some rebate from the recyclable materials that are inevitably part of municipal waste.

Municipal truck with robotic arm collecting residential waste for recycling. *© Paul Vasarhelyi/iStock*

Life-cycle thinking: identifying the range of impacts from chemicals and wastes

What ultimately determines how humans and ecosystems are exposed to toxic chemicals is defined by their life-cycle characteristics. Releases of substances not only occur during chemical production but also during the use of products containing chemicals (Figure 6.3), and finally at their disposal. Life-cycle thinking promotes an integrated approach to the sustainable production and consumption of such substances.

The entire life cycle of resource use, from extraction and production/manufacture through consumption/use to post-consumption disposal, produces undesirable environmental impacts from emissions and wastes. These impacts can include unintended side effects such as endocrine disruption, which directly interferes with growth and development in most animals, and can also affect people (WHO 2002). Life-cycle analysis helps understand such impacts, but, while a useful tool, it can be extremely complex. Too often, when problems are identified, shifts to alternative chemicals that have the same intended properties may result in further unexpected or undesirable outcomes (Muir and Howard 2010).

The latest materials to raise concern are those arising from synthetic biology and engineered nanomaterials. With the accelerated pace at which new technologies and chemicals are being deployed (Poliakoff *et al.* 2002), a different approach is needed in which their implications are systematically and comprehensively assessed before they reach production. The use of green chemistry principles in chemical design and the adoption of clean production processes may help to prevent problems at a later stage. While this is happening in some parts of the world through the use of exposure models – for example by the Canadian Centre for Environmental Modelling and Chemistry (CEMC 2012) – for some technologies and chemicals, life-cycle analysis has yet to become a universal systematic approach. This may well require new forms of international governance (Finnveden *et al.* 2009).

The high number and diversity of chemicals and the complexity of their life cycles inevitably lead to a situation where the scientific understanding of the impacts of chemicals, and the regulatory schemes used to manage them, lag behind technological and economic developments.

Poverty and exposure to chemicals: vulnerable groups

The overwhelming majority of impacts from unsafe chemical use and unsound waste disposal – including death, impairment of health and ecosystem degradation – occur in situations of poverty (Sexton *et al.* 2011). Increased risks of exposure to toxic and hazardous chemicals and wastes predominantly affect the poor, who routinely face such risks because of their occupation, poor living standards and lack of knowledge about the detrimental impacts of exposure to these chemicals and wastes. Many of the poor enter the informal sector of the

Figure 6.3 Life-cycle analysis of chemicals

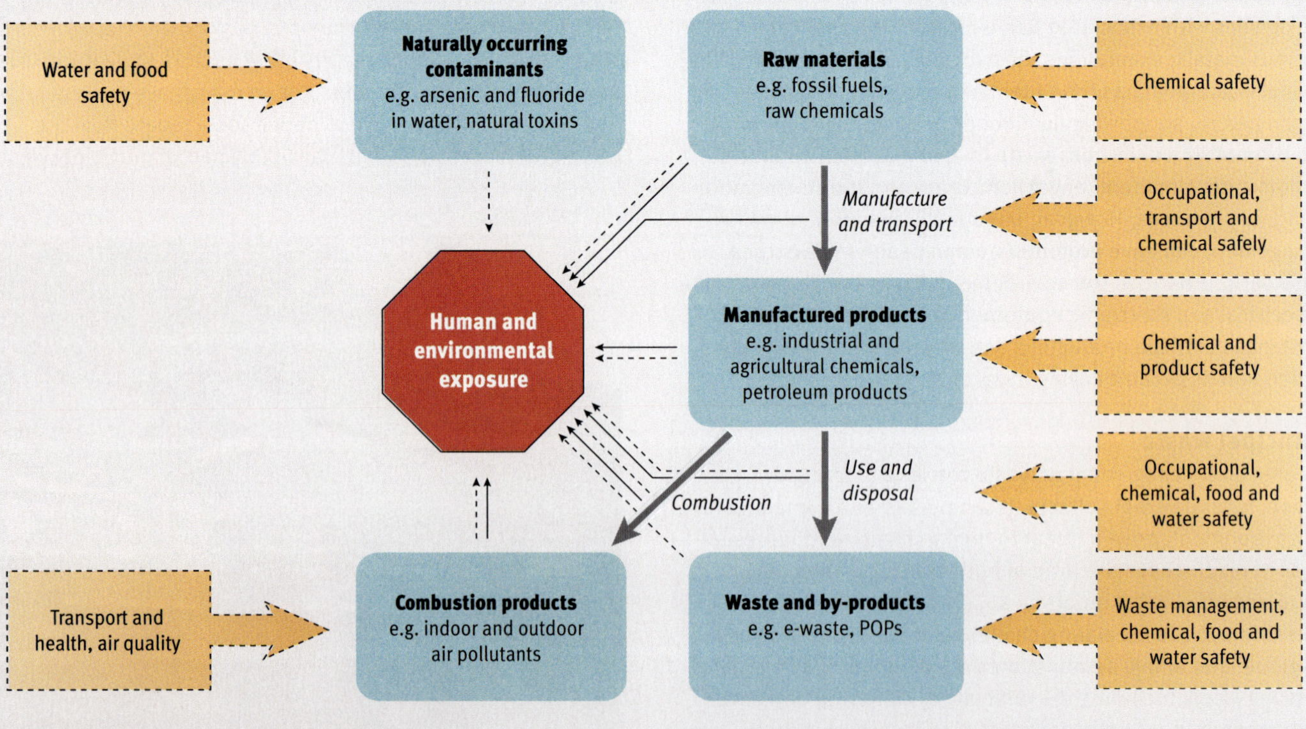

Note: Dashed lines are management options; solid lines correspond to the life cycle.

Source: Adapted from Prüss-Ustün *et al.* 2011

Shacks along a polluted waterway in Manila, Philippines.

© Marcus Lindström/iStock

economy where they may encounter new kinds of toxic hazards such as electronic and electrical waste (e-waste). Risk is not only related to the dose they receive from such exposure, but also to important factors such as age, nutritional status and co-exposure to other chemicals. Children are particularly susceptible to the negative health impacts of chemicals due to their rapid growth and development and greater exposure relative to body weight (Sheffield and Landrigan 2011).

A recent study by the World Health Organization (WHO) (Prüss-Ustün *et al.* 2011) indicated that 4.9 million deaths were attributable to environmental exposure to chemicals in 2004. Indoor smoke from the use of solid fuels, outdoor air pollution and second-hand smoke are among the most critical causes. The study concluded that the known burden of chemicals, while considerable, is an underestimate because data on many chemicals are scarce.

Changes in the global production, trade and use of chemicals and the concomitant production of hazardous wastes are not always accompanied by corresponding control measures, thus increasing the risk of releasing hazardous chemicals into the environment. It is estimated that there are 2 million contaminated sites in Europe, the United States and the Russian Federation alone. Data for developing countries and economies in transition are more difficult to obtain, but indications are worrying. The Global Inventory Project – which involves the Blacksmith Institute together with the United Nations Industrial Development Organization (UNIDO), the Green Cross and the European Commission – is currently assessing the state of contaminated areas in 80 countries worldwide, with trace metal and pesticide pollution among the ten most problematic types of contamination (Blacksmith Institute 2011). This is the first such attempt to give governments, international organizations and affected communities aggregated data for decision making.

Marine pollution

The oceans cover 71 per cent of the Earth's surface and are polluted to varying degrees, threatening marine life, fisheries, mangroves, coral reefs, and estuarine and coastal zones, with

Figure 6.4 PCBs in beached plastics

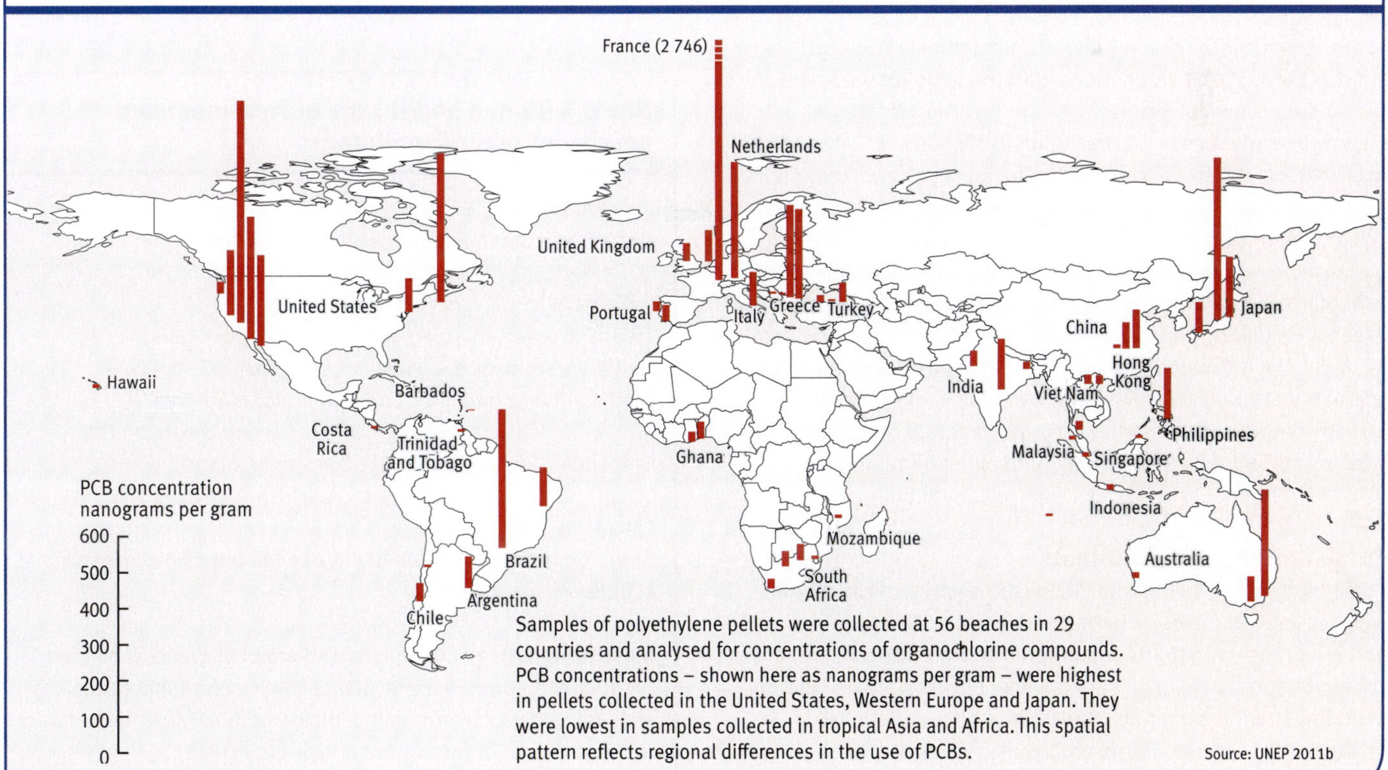

Samples of polyethylene pellets were collected at 56 beaches in 29 countries and analysed for concentrations of organochlorine compounds. PCB concentrations – shown here as nanograms per gram – were highest in pellets collected in the United States, Western Europe and Japan. They were lowest in samples collected in tropical Asia and Africa. This spatial pattern reflects regional differences in the use of PCBs.

Source: UNEP 2011b

Box 6.4 Waste generated on board ship

The world fleet comprises more than 80 000 vessels of which around 50 000 merchant ships carry out 90 per cent of international trade. Every ship generates waste during its operation or when transporting cargo, including sludge, oily tank washings known as slops, rubbish from the crew, and cargo residues. Depending on its size, a ship can generate a few hundred tonnes of slops during a voyage. With 50 000 ships of more than 500GT (gross tonnage) in the world fleet, and assuming an average of ten port calls per ship, half a million port calls take place annually (Mikelis 2010). Port states are required by the International Convention for the Prevention of Pollution from Ships 1973, as modified by its 1978 Protocol (MARPOL 73/78), to provide adequate port reception facilities to collect waste generated on board ships. Illicit discharges of slops represent a major source of marine pollution. For instance, according to the Regional Marine Pollution Emergency Response Centre for the Mediterranean Sea (REMPEC), there are more than 2 500 illicit discharges of ship waste in the Mediterranean Sea annually. The 1972 London Convention on the Prevention of Marine Pollution by Dumping of Wastes and Other Matter, along with its 1996 Protocol, is one of the first global conventions to protect the marine environment from the effects of human activities and has been in force since 1975. Its objective is to promote the effective control of all sources of marine pollution and to take all practicable steps to prevent pollution of the sea by dumping of wastes and other matter.

some 80 per cent of the pollution coming from land-based sources (UNEP 2011b). Common man-made pollutants include pesticides, chemical fertilizers, heavy metals, detergents, oil, sewage, plastics and other solids (UNEP 2011b). Many of these pollutants accumulate in the deep oceans and sediments (Jacobsen *et al.* 2010; Zarfl and Matthies 2010; Wania and Daly 2002), where they are consumed by small marine organisms and may be reintroduced to the global food chain. Some 20 per cent of marine pollution originates from direct disposal into the oceans: regular discharges of oily wastes from ships, accidental oil spills and untreated sewage disposal in enclosed areas such as the Mediterranean are threats to marine ecosystems (UNEP 2011b). Figure 6.4 shows the occurrence of PCB in beached plastics around the globe. Some of the most harmful pollutants also come from diffuse sources such as air pollution.

Persistent organic pollutants

Persistent organic pollutants (POPs) are a group of chemicals with common features including persistence, bioaccumulation and long-range transport. Combined with their toxicity, these characteristics have significant adverse effects both on wildlife, including marine mammals, and on human populations, in particular such vulnerable groups as nursing mothers and infants. The health effects of exposure to POPs include neuro-developmental disorders, endocrine disruption and carcinogenicity (Diamanti-Kandarakis *et al.* 2009).

The Stockholm Convention on POPs was adopted in 2001 in response to an urgent need for global action, and entered into force in 2004. It currently has 177 Parties and calls for documentation of the amounts of POPs that are still present in different countries and for global monitoring of these substances in human tissue (blood and milk). This is one of two indicators proposed for monitoring and assessing the status and trends of POPs in the environment and their impact on human health. The Stockholm Convention established a Global Monitoring Plan as a source of globally consistent and reliable data. Collection of data is at an early stage and more will become available in the coming years, but individual studies already provide historical and regional trends for some substances. An example is DDT, for which Ritter *et al.* (2011) report global time series of concentrations in human tissue from many individual measurements (Figure 6.5). In general, DDT body burdens have fallen over recent decades, but are still considerably higher in tropical than northern regions. Where DDT is used for malaria control, concentrations are still very high and the decrease is less pronounced than elsewhere.

The other indicator for POPs is trends of selected atmospheric POPs in both urban/industrialized and remote regions. Concentrations of these substances in the air follow changes in emissions more closely than concentrations in food and human tissue, and reflect the effect of atmospheric long-range transport. Hung *et al.* (2010) provide a summary of long-term time trends of various POPs measured at Arctic monitoring stations. In general, concentrations of most substances in Arctic air show a falling trend, but

Box 6.5 Human health, the environment and persistent organic pollutants

Related goals
To protect human health and the environment from POPs

Indicators
Trends in levels of selected POPs in human tissue; trends in atmospheric levels of selected POPs such as PCBs (conventional POPs, regulated for many years) and endosulfan (emerging POPs, added to the list in the Stockholm Convention in 2010)

Global trends
Some progress; it is too early to use the above indicators for evaluation

Most vulnerable communities and areas of greatest concern
Arctic communities, in particular children; communities in areas with indoor residual spraying of dichlorodiphenyltrichloroethane (DDT); children of the world exposed to POPs

Figure 6.5 DDT levels in humans, 1960–2008

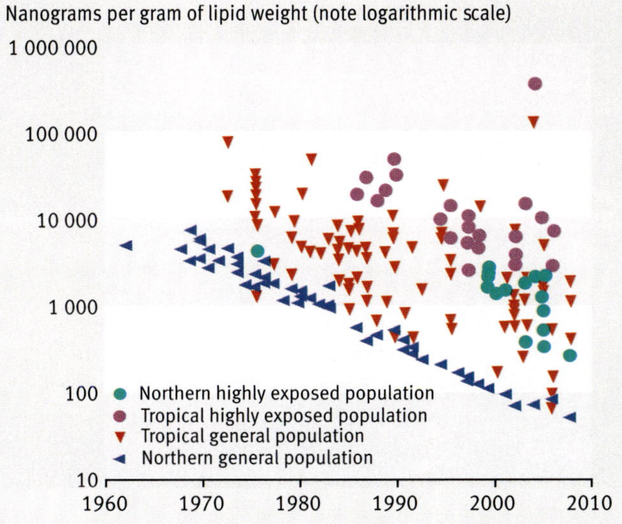

Source: Ritter *et al.* 2011

half-lives are often long – five to ten years and sometimes even longer. In recent years the decrease has come to a halt for several compounds, and some concentrations have been observed to be on the rise, for example polychlorinated biphenyls (PCBs), chlordane and DDT. Long-term trends for two PCBs are shown in Figure 6.6.

The environmental behaviour of POPs is strongly affected by temperature and other climate-related factors (UNEP/AMAP 2010; Macleod *et al.* 2005), including precipitation patterns, wind fields and extreme weather events. In general, climate change is expected to cause greater mobilization of POPs from primary and secondary sources as well as increased airborne transport (Lamon *et al.* 2009). It is unclear to what extent higher temperatures will accelerate the degradation of POPs, but the melting of ice in which they have been held for decades contributes to rising amounts of POPs and other pollutants in the environment (Bogdal *et al.* 2010).

Pesticides including POPs

Pesticides are compounds designed to kill specific pests but often reach non-target organisms as well. In one study, more than 90 per cent of sampled water and fish were found to be contaminated by several pesticides and estimates indicated that

Figure 6.6 Trends in two PCBs from air monitoring data at two sites in the northern hemisphere, 1995–2005

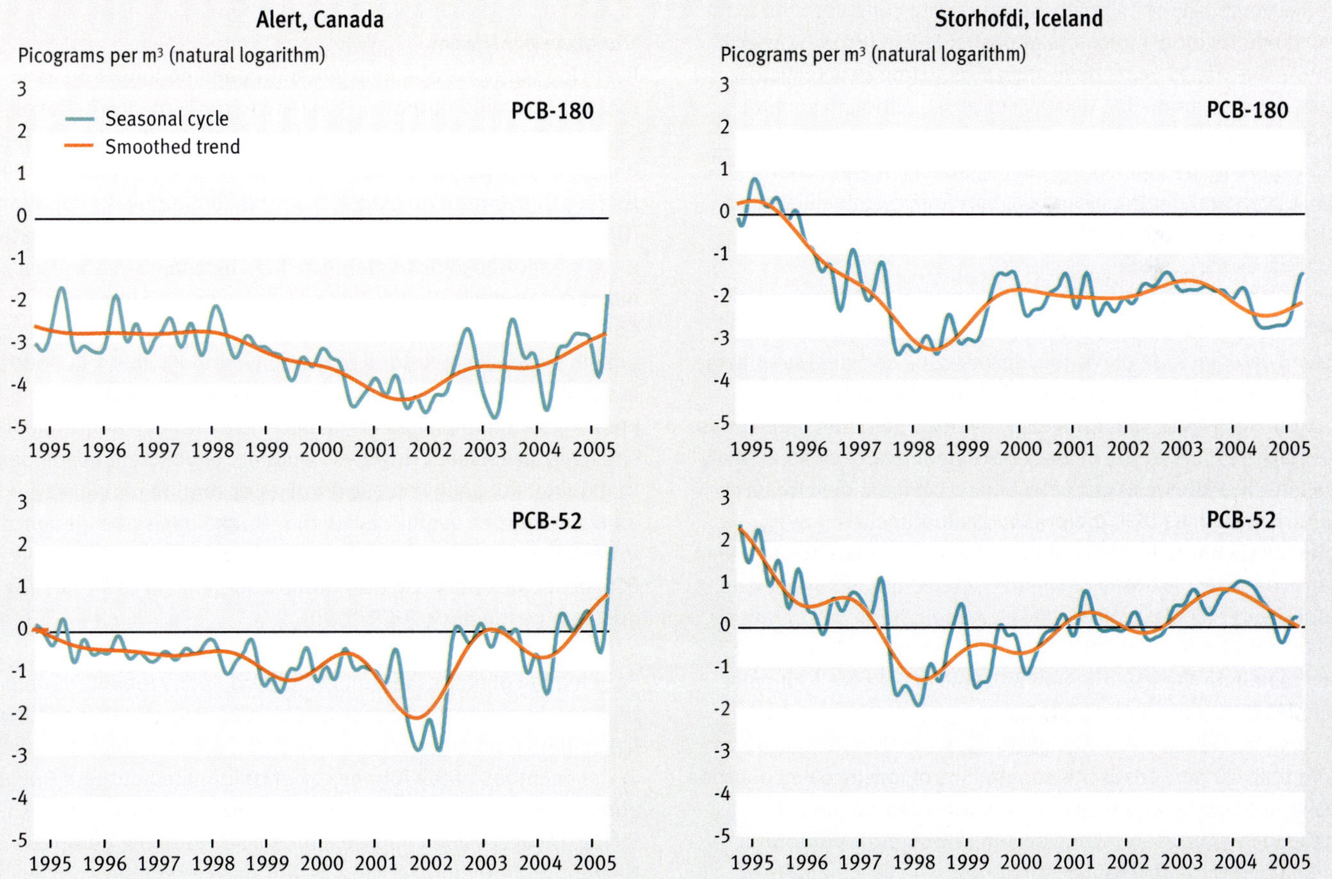

Source: Adapted from Hung *et al.* 2010

A farmer, wearing no protective equipment, sprays his vines with pesticide. *© Alistair Scott/iStock*

about 3 per cent of exposed agricultural workers suffer from an episode of acute pesticide poisoning every year (Thunduyil *et al.* 2008). It is therefore imperative to know the nature of exposure and causes of contamination, and to identify action that can be taken to reduce pesticide levels in terrestrial and aquatic ecosystems. Long-term pesticide sales data constitute the main global and regional indicators of pesticide use (Brodesser *et al.* 2006). The last 25 years have seen a reduction in insecticide sales due to mammalian toxicity concerns, although general pesticide sales increased from US$5.4 billion in 2004 to US$7.5 billion by 2009 in the Latin American region, with 2,4-D, paraquat, methamidophos, methomyl, endosulfan and chlorpyrifos accounting for a high proportion of these sales (Brodesser *et al.* 2006).

Globally, the main 15 pesticides found in streams and groundwater include the herbicides atrazine and di-ethylatrazine, metolachlor, cyanazine and alachlor, and the insecticide diazinon. However, regarding fish, riverbed sediments and soils, the main pesticides still include persistent insecticides, heavily used in the 1960s and currently banned in most developed countries, such as DDT, dieldrin and chlordane. Moreover, endosulfan sulphate, the metabolite of endosulfan still in use in many countries, is a very common contaminant of surface and groundwater (Ondarza *et al.* 2011). Although the use of most organochlorine insecticides came to an end 10–25 years ago, they remain in the environment at levels of concern (Gonzalez *et al.* 2010; Ondarza *et al.* 2010).

More than 70 per cent of the populations of low-income countries live in rural areas, and 97 per cent of rural populations are engaged in agriculture. While developing countries account for just one-third of global pesticide use, the vast majority of pesticide poisonings occur in these countries (Brodesser *et al.* 2006).

The extent of human exposure and the health effects of pesticides under future climate change conditions will depend on the adoption of less toxic practices that take account of changes in factors such as temperature and precipitation (Boxall *et al.* 2009).

Obsolete pesticides

Pesticides become obsolete when they can no longer be used for their intended purpose. There are four major international agreements for their regulation: the Stockholm, Rotterdam and Basel Conventions and the 1998 Protocol on POPs to the 1979 Geneva Convention on Long-Range Transboundary Air Pollution (UNECE Geneva Convention 1979/98). It is difficult to estimate exact quantities of obsolete pesticides because many are very old and documentation is scarce. Parties to the Stockholm Convention are in the process of collecting information on nine POPs that were added to the convention's annex in 2009, including hexachlorocyclohexane (HCH), and a good deal is known about dump sites of the latter even though some smaller sites may be missed. However, amounts of obsolete pesticides that do not fall under the Stockholm Convention remain vague and can only be roughly calculated. On the basis of experience in Africa and the Middle East, UNEP estimates that on average POP pesticides make up only around 30 per cent of all existing obsolete pesticides (UNEP 2000).

Country-by-country assessments carried out by the International HCH and Pesticide Association (IHPA 2009) suggest that obsolete pesticides could amount to between 256 000 and 263 000 tonnes in the countries of the former Soviet Union, the Southern Balkans and new Member States of the European Union (defined as EU-12, EU accession countries, the countries of the European Neighbourhood Policy (ENP), the Russian Federation and Central Asia together), costing approximately US$780 million to dispose of, while some estimates for Africa by UNEP Chemicals suggested

Table 6.2 Quantities of obsolete pesticides

Region	Estimated quantities (tonnes)	Estimated disposal costs at US$3 000–5 000 per tonne (US$ million)
Africa	27 395	82.2–137.0
Asia	6 463	19.4–32.3
Eastern Europe	240 998	722.9–1 204.9
Latin America and the Caribbean	11 284	33.9–56.4
Near East	4 528	13.6–22.6
Total	**290 668**	**872–1 453**

Note: The latest updates given by FAO vary from 1994 to 2006.

Source: FAO 2012

that there may be as much as 120 000 tonnes remaining (UNEP 2002), costing some US$200–250 million to dispose of, applying UN Food and Agriculture Organization cost estimates (FAO 2002). These assessments alone identify 376 000–383 000 tonnes for disposal at a cost of US$968–1 040 million. The most recent FAO figures, shown in Table 6.2, indicate that there are some 290 000 tonnes of pesticide stockpiles with estimated disposal costs of US$3 000–5 000 per tonne (FAO 2012).

The Africa Stockpiles Programme (ASP), launched in 2005, aimed to clear all obsolete pesticides and contaminated waste in Africa within 10–15 years and to promote prevention measures and capacity building. It is very likely that the costs of inaction by far exceed the costs of cleaning up. As underlined by the European Environment Agency (EEA), downplaying the costs of inaction is a frequent phenomenon (Koppe and Keys 2001) and analysis suggests that the costs of inaction are high (OECD 2008c).

Metals, metalloids and heavy metals

Inorganic pollutants, including metals and metalloids, also adversely impact human populations on a global scale (Blacksmith Institute 2011). Unlike organic chemicals, metallic elements do not degrade and may accumulate in the environment and become increasingly bio-available over time. Their impacts are often most severe in the developing countries where they are mined, processed, used and recycled with limited environmental control and regulation. Populations in more developed countries also suffer from historic and on-going industrial emissions of pollutants, as well as from associated releases of other pollutants such as sulphur oxides, which cause acid rain, and acid mine drainage (Carn *et al.* 2007). Contamination even extends to Antarctica, as industrial pollutant emissions are carried there by long-range atmospheric transport from other continents (Caroli *et al.* 2001). Pollutants can also be re-released after decades as glaciers melt (Geisz *et al.* 2008).

Poisoning by naturally occurring arsenic is a global problem (Ravenscroft *et al.* 2009). More than a decade ago it was estimated that 130 million people around the world have been exposed to toxic levels of arsenic in drinking water, above the WHO recommended limit of 10 parts per billion (Smith and Lingus 2000), but there is mounting evidence that arsenic toxicity occurs at levels below that standard (Wasserman *et al.* 2004). There are also many unexplored sources of arsenic and the total number of people affected may be higher (Huang *et al.* 2011). Associated toxicities include diabetes and skin, kidney, lung, neurological and vascular diseases – most notably blackfoot disease which leads to gangrene – and bladder cancer. These diseases are most prevalent in vulnerable populations living on subsistence diets of arsenic-contaminated foods and with limited access to clean water, minerals and nutrients, which partially counteract that toxicity. Arsenic pollution in Bangladesh, which resulted from drilling wells to protect the population from surface waters contaminated with pathogens (Lokuge *et al.* 2004), has been described as "the largest poisoning of a population in history" (Smith and Lingus 2000). Populations in both developed and developing countries may be exposed to arsenic at contaminated sites left behind by the formerly widespread use of arsenic as a pesticide.

Lead is among the most prominent of the global contaminants (Rauch and Pacyna 2009), with several activities being responsible for acute lead poisoning. There are on-going human health problems at previous mining and smelting sites, including in Kabwe, Nigeria (Nweke and Sanders 2009) and the Rudnaya River Valley, Russia (von Braun *et al.* 2002), where high lead levels in children persisted after the smelters in both areas were closed, and in La Oroya, Peru, where 99.7 per cent of the children living nearest the smelter were found to have dangerously high levels of lead in their systems (Fraser 2009). On a global scale, some 85 per cent of lead-acid batteries are recycled, but there are recycling sites such as in Dakar,

The Lavender red copper open pit mine in Bisbee, Arizona, United States.

Senegal (Haefliger *et al.* 2009), where the average blood-lead concentration of children was at 130 micrograms per decilitre, enough to cause acute toxicity or even death (ATSDR 2007). Children may also be exposed to lead in paints, which has been phased out in developed countries but persists in some developing ones (Lanphear *et al.* 1998). Electronic waste recycling can also involve exposure to lead in solder, and there are sites such as Guiyu, China (Huo *et al.* 2007), where 82 per cent of the children tested in the village had blood-lead concentrations above the US Centers for Disease Control action level of 10 micrograms per decilitre (ATSDR 2007). Although that level is two orders above the estimated natural lead level, there is no established lower threshold for lead toxicity in humans (Flegal and Smith 1992).

Most coals contain tiny proportions of mercury, so industrial mercury fluxes to the biosphere are projected to increase with greater fossil fuel combustion (Soerensen *et al.* 2010). While large amounts of mercury are thus released into the environment from numerous industrial activities, reports of acute neurotoxicity from mercury poisoning are now primarily associated with its use to amalgamate gold in artisanal mining, which is practised in more than 50 countries (Bose-O'Reilly *et al.* 2008). In Indonesia and Zimbabwe, all of the children tested in two mining areas were found to have elevated mercury levels and corresponding signs of mercury intoxication, whether they were directly involved in mining or not (Bose-O'Reilly *et al.* 2008). This poisoning of children is of special concern because mercury, even at sub-lethal levels, is a neurotoxin that can permanently impair development and – like some other toxins – foster auto-immune resistance, making children and adults more vulnerable to infection and disease, as has been found with artisanal gold miners in Brazil (Feingold *et al.* 2010). Currently, UNEP is convening an intergovernmental negotiating committee to prepare a global legally binding instrument on mercury: more than 100 countries are participating and a global treaty text is expected to be ready for adoption in late 2013 (Selin and Selin 2006).

A number of other metals such as zinc, copper and manganese could have harmful human and environmental impacts at certain levels. Cadmium, which was once used in pigments and for electroplating, is the most toxic, and contaminated sites may remain. Its main uses now are in rechargeable nickel-cadmium batteries, and collection and recycling of these items must be efficient if it is not to be released to the environment. Cadmium is also released into the environment by some fossil fuel combustion, and is in addition a natural contaminant in phosphate deposits, so may be transferred in fertilizers and taken up by root vegetables (Jarup and Akesson 2009).

Radioactive material

Radioactive material has been in use since the 1890s, increasing significantly with the advent of nuclear energy in the 1940s and its exploitation in weapons, with a concomitant increase in the generation of radioactive waste and contaminated sites. In addition, the use of radioactive materials in industry, research and medicine continues and increases, as do the mining and processing of minerals containing elevated concentrations of naturally occurring radionuclides. Some contaminated sites have been remediated at significant cost, while others remain to be addressed. The rising cost and reduced availability of fossil fuels have from time to time favoured the adoption of nuclear power, as have recent concerns over greenhouse gas emissions. However, social attitudes to nuclear accidents such as those at Three Mile Island and Chernobyl – which are rare but can have a very high impact – have exerted a restraining influence. It was predicted in 2008 that the use of nuclear energy would increase by 15–45 per cent by 2020 and by 25–95 per cent by 2030 (IAEA 2008a), but future activity is likely to be affected by responses to the more recent Fukushima disaster.

Radioactive waste takes many physical and chemical forms and has differing radioactive properties. The international system of classification (IAEA 2009a) links waste classes (exempt, very short lived, very low level, low level, intermediate level, high level) to options for management and disposal. Disposal is the final

Table 6.3 Global inventory of radioactive waste, 2004

Waste source	Low- and intermediate-level waste		Spent fuel		High-level waste		Mining	
	Volume (million m³)	Activity (million TBq)	Mass (million MTHM)	Activity (million TBq)	Volume (million m³)	Activity (million TBq)	Volume (million m³)	Activity (TBq million)
Nuclear power	2	1.2	0.17	28 000	0.034	42	1 600	0.028
Industrial or medical use	2	1.2						
Weapons	4	0.7			0.8	31	250	0.0046
Total	**8**	**3.1**	**0.17**	**28 000**	**0.8**	**73**	**1 850**	**0.033**

Note: MTHM – metric tonne of heavy metal; TBq – Tera-Becquerel.

Source: IAEA 2008b

Crates for storing radioactive material. © *Clearviewimages/iStock*

step in the management of radioactive waste, generally in near-surface or deep land-based facilities. Apart from high-level and some intermediate-level waste, the majority has been disposed of in such facilities. Table 6.3 presents an estimate of the global inventory of radioactive waste (IAEA 2008b).

About a hundred near-surface facilities exist, and others for disposal of waste of various levels are under development in a number of countries, although the process of selecting and designing a site is often contentious. Many nuclear reactors are ageing and will need to be decommissioned in the near future, resulting in radioactive waste and signalling the need for disposal facilities and trained professionals to operate them. As of 2 February 2012, 435 nuclear power reactors with a combined capacity of about 368 gigawatts are in operation in 30 countries, of which around 75 per cent are more than 20 years old, and 63 plants with a combined capacity of 61 gigawatts are under construction in 14 countries (European Nuclear Society 2012).

Contracting Parties to the Joint Convention on Radioactive Waste and Spent Fuel increased steadily after its establishment in 1997 to number 58 in April 2011, and are committed to ensuring a high level of safety in radioactive waste management. At the 2009 triennial review meeting, the reports of 45 Contracting Parties were reviewed with the conclusion that there is a commitment to improve safety, make progress in building, maintaining and implementing legal/regulatory frameworks, and observe good practices in national radioactive waste management strategies and policies (IAEA 2009b). Despite progress since the 2006 review meeting, however, the 2009 meeting concluded that much still needed to be done to meet the following challenges:

- implementation of national policies for the long-term management of spent fuel, including disposal;
- siting, construction and operation of spent fuel and radioactive waste disposal facilities;
- management of legacy wastes;
- monitoring of disused sealed sources and recovery of orphan sources;
- knowledge management and human resources development; and
- provision of financial resources for liabilities.

There has been a growing trend for disposal facilities to undergo international peer-reviewed safety demonstration (IAEA 2006). In addition, the 2010 General Conference of the International Atomic Energy Agency (IAEA) created an International Working Forum on Regulatory Supervision of Legacy Sites (IAEA 2010), aimed at enhancing regulatory regimes, the professional development of regulators and the application of safety and environmental assessment.

EMERGING ISSUES

Policy making and regulatory processes are naturally prone to lag behind rapid changes taking place in the global production and distribution of chemicals and wastes. The challenge is to protect human health and the environment from the undesirable effects of chemicals and wastes even when there are inadequate quantitative data and the potential life-cycle hazards of both old and new materials are incompletely understood.

Nanomaterial and nanoparticles

Many new materials are produced as minute particles of a nanometre – or one-billionth of a metre – in size, and they exhibit chemical and biological properties that are quite different from those of the corresponding bulk materials. Commercial applications of nanomaterials include, for example, food packaging, personal care products, cosmetics and pharmaceuticals. Their unique properties make nanomaterials useful in cancer therapies, the neutralization of pollution or improvement of energy efficiency. However, safety testing is in its infancy and governments have been slow to adapt existing regulations to these new materials, even though they are widely marketed and some potential for human exposure has been identified (Morris *et al.* 2010). More research is needed for a better understanding of workplace and consumer exposure and related impacts on human health, especially as some of these materials are known to pass through the skin and are small enough to penetrate cell membranes and cause toxic effects at cellular and sub-cellular scales. Furthermore, very little is known about whether nanomaterials or nanoparticles are released from products when they are incinerated, buried or degrade over time, so it is possible that they will pose a serious waste disposal challenge. Sound decision making on nanotechnology has provoked much debate among developed-country regulators, and is increasingly doing so among the regulators of developing countries (Morris *et al.* 2010).

Plastics in the environment

Plastics are ubiquitous in the environment. They are widely used in many products and have many formulations. The simple plastic bag is a prime example of how a utilitarian object can become an environmental hazard. More than 500 billion plastic bags are used every year but many are improperly disposed of, ending up as marine litter. This significant problem was highlighted in the *UNEP Yearbook 2011* (UNEP 2011b), showing that discarded plastic debris forms a major component of marine litter, degrading into micro-pollutants in ocean gyres, fouling beaches in coastal waters, and entering the food chain where it is consumed by marine fauna such as turtles and sea birds, weakening or killing them by affecting their digestion, respiration and reproduction. There is concern that these plastics also act as transport vectors of persistent organic pollutants such as PCBs and similar compounds, with chronic effects on wildlife. The solution is sound management, preventing the escape or discharge of this material, yet rates of plastic recycling and reuse vary greatly, from more than 80 per cent in some EU countries to only a small percentage in many developing ones. The Global Programme of Action (GPA) for the protection of the Marine Environment from Land-based Activities and other local and regional initiatives are seeking to address this issue (Astudillo *et al.* 2009; Young *et al.* 2009).

Electronic waste

The high turnover of equipment in the information and communication technology industry has caused an increase in obsolete electrical and electronic products, which in turn has generated almost uncontrollable volumes of end-of-life products driving a global trade in e-waste. As the fastest growing waste stream in the world, estimated at 20-50 million tonnes per year, e-waste has become one of the major environmental challenges of the 21st century (Schwarzer *et al.* 2005). Generated by a wide range of electrical products, it is of particular interest because it contains not only hazardous substances including heavy metals such as mercury and lead, and endocrine-disrupting substances such as brominated flame retardants (BFRs), but also many strategic metals such as gold, palladium and rare earth metals that can be recovered and recycled. E-waste can thereby serve as a valuable source of secondary raw materials, reducing pressure on scarce natural resources and the environmental footprint of the mining industry.

Discarded computer circuit boards. © *roccomontoya/iStock*

Developing countries nonetheless remain the destination of most of the e-waste exported from developed countries as second-hand or used equipment. Yet these countries often lack the infrastructure, capacity and resources for its sound management (UNEP 2009), with the informal sector and vulnerable groups employing crude processing methods such as open-air burning or acid leaching to recover valuable metals like copper and gold. In the process, toxic substances in the waste may be released into the environment, posing a high risk to ecological and human health. Recent studies have revealed that by 2016, developing countries will generate twice as much e-waste as developed countries (Zoeteman *et al.* 2010), but while electronic equipment has positive impacts on development and progress, it can have negative impacts on both human health and environmental integrity as end-of-life e-waste. This is a growing environmental and public health issue that threatens attainment of the Millennium Development Goals (MDGs) in developing countries and economies in transition.

Endocrine disruptors

Endocrine disruption is the term given to the alteration of hormonal signals in living systems when they are exposed to chemical substances. A considerable number of chemicals have been shown to be endocrine disruptors, affecting the growth and reproductive and neurological development of many species, including humans (Waye and Trudeau 2011; Gore and Patisaul 2010; Toppari *et al.* 1996; Colborn *et al.* 1993). In addition, the numerous chemical substances, both natural and anthropogenic, that are present in the environment in low concentrations come together to exacerbate exposure of both humans and wildlife. Many investigations have been conducted since publication of the *Global Assessment of the State-of-the-Science of Endocrine Disruptors* (WHO 2002), and it is clear that both inorganic and organic substances can affect hormonal signalling. UNEP has proposed listing this as an emerging policy issue for listing under the Strategic Approach to International Chemicals Management (SAICM).

Open burning

In open burning, the pollutants produced by combustion are released directly to the air and so enter the environment in uncontrolled ways. Open burning can include forest wildfires, planned combustion activities such as burning stubble in preparation for a subsequent grain crop, irresponsible burning of waste such as domestic rubbish and e-waste, arson-initiated combustion of scrap tyres, and even public detonation of fireworks (Lemieux *et al.* 2004). Polycyclic aromatic hydrocarbons (PAHs) are always released in these processes, and (in the case of fireworks) heavy metals such as lead and copper are also released. PAHs are widespread in the environment in both developed and developing countries (Barra *et al.* 2007), and concern about their carcinogenic properties has led to their classification as primary pollutants by agencies such as the US Environmental Protection Agency.

Gaps in the understanding of chemical toxicity

Since humans are continually exposed to a multitude of manufactured chemicals, there is a need to understand the behaviour of these chemicals and their interaction with human health and the environment. Previously unsuspected properties of widely used chemicals present legacy problems that raise concern in the scientific community and among the public. For example, of the many chemicals that have been found to have endocrine-disrupting properties, bisphenol-A is present in many plastic baby bottles and food can liners, and phthalate esters in various flexible plastics including some children's toys (Hengstler *et al.* 2011). Consumer vigilance is not enough to prevent exposure in such cases because the presence of these chemicals is usually not evident to the non-expert. This places a heavy responsibility on public authorities to inform people about potential risks associated with manufactured chemicals, and on manufacturers to exercise the extended or individual producer responsibility approach and to search for alternatives.

Most existing chemical regulations worldwide address the effects of individual substances. Managing single chemicals is difficult enough, but there is also concern about gaps in understanding human exposure to mixtures of chemicals (Rajapakse *et al.* 2002; Silva *et al.* 2002). As mentioned, little has been done to study the toxicology of mixtures. There is an urgent need for further risk assessment of the combined exposure of multiple chemicals – the chemical cocktail or synergistic effects – to human health and the environment. Integrated environmental risk assessments based on state-of-the-art dynamic pollutant modelling and toxicological experiments on chemical cocktails will help quantify planetary boundaries for chemical pollution (Handoh and Kawai 2011; Rockström *et al.* 2009).

GAPS AND OUTLOOK

Chemical properties, patterns of use and the environment

There is a lack of information about the health and environmental effects of many chemical substances and about the products in which different types of chemicals are used (OECD 2008b). Huge gaps in the assessment of chemicals arise from two causes. First, many were introduced and became established items of commerce before systematic assessments began. Where there is mounting evidence of harm or potential harm, action can lead to regional controls and eventual listing under global conventions, but most industrial chemicals remain unassessed. Second, concerns have arisen over hitherto unsuspected properties such as the endocrine activity of phthalates and bisphenol A, for example, or long-range transport coupled with bioaccumulation. Furthermore, academic assessment suggests the potential for further industrial chemicals and pesticides to qualify as POPs (Muir and Howard 2010, 2006). It should also be noted that wastes are often mixed, which makes it extremely difficult to assess the risks of any chemicals present. In addition, residues arising from the recycling of hazardous waste may contain a higher concentration of toxic materials than the recyclable materials themselves.

Long-term monitoring programmes for POPs in the environment as well as in human tissue need to be maintained and expanded, in particular in the southern hemisphere. They are essential for a better understanding of trends in global chemical pollution and for the Stockholm Convention's evaluation of effectiveness.

More extensive work on chemical toxicity inventories is aiming to fill a significant gap. An example is the European legislation on Registration, Evaluation, Authorisation and Restriction of Chemicals (REACH). This has extended the number of chemicals covered by regulations, notably those that were on the market before 1981 and previously exempt (Chapter 11).

Limited information on chemicals in products makes it difficult to document the extent of the risk posed to human health and the environment. Initiatives such as the ongoing UNEP Global Chemicals Outlook and the Cost of Inaction (UNEP Mainstreaming of Chemicals) will help to fill some important knowledge gaps.

In addition to scientific knowledge gaps, sound chemicals and waste management is also hampered by a lack of resources, capacity and compliance monitoring. A lack of education and training also limits appropriate management of chemicals and wastes in many developing countries. Increased trade resulting from free trade agreements can complicate this picture (Vogel 1997), as such agreements may well exert even more pressure on emerging economies with respect to regulating or restricting chemical use.

Chemicals, wastes and drinking water

At the global level, about 1.1 billion people do not have access to a safe water supply and 2.6 billion people do not have access to adequate sanitation facilities. The associated health impacts are alarming: 1.7 million deaths per year, of which 90 per cent are children under five years of age (WHO/UNICEF 2005). The costs of water pollution may represent between 0.3 and 1.9 per cent of rural gross domestic product (GDP) (WHO/UNICEF 2005). Industrial sectors with the potential for significant water pollution include the chemicals sector, food and beverage sector, textile and mining industry and pulp and paper sector. The policy framework for regulating the industrial point sources of water pollution is well developed in most OECD countries, although some pollutants such as heavy metals and chlorinated solvents remain a concern. Increasing attention is being paid to non-point sources, such as agricultural run-off, which are more difficult to regulate but can lead to nitrate pollution of water bodies. In addition to efforts to reduce the run-off of organic pollutants from fertilizers and manure, organophosphates from pesticides are also a concern. Studies reviewed by the OECD (2008a, 2008b) suggest that national measures to reduce agricultural run-off and manage storm water, including targeted measures to reduce a variety of different pollutants such as arsenic and nitrates, could yield health benefits in excess of US$100 million in large OECD economies (Hammer *et al.* 2011). In non-OECD countries, the cost of inaction with respect to unsafe water supply and sanitation is particularly acute.

Box 6.6 Funding: an ongoing challenge

Much effort at the intergovernmental level goes into identifying the funding and support needs of capacity building, technical assistance and institutional strengthening for the sound management of chemicals and waste in developing countries and economies in transition. This is reflected in the decisions of the Conferences of the Parties to the Basel, Rotterdam and Stockholm Conventions, especially for national implementation plans. International funding for implementation of the chemicals and wastes agenda is currently managed and channelled through the World Bank, Global Environment Facility (GEF), United Nations Development Programme (UNDP), UNEP, the United Nations Industrial Development Organization (UNIDO), the United Nations Institute for Training and Research (UNITAR), FAO, WHO and the Quick Start Programme of SAICM, as well as the OECD and regional development banks. Some funding is also available through private-sector bodies. In addition, SAICM, the Inter-Organization Programme for the Sound Management of Chemicals (IOMC), Intergovernmental Forum on Chemical Safety (IFCS), and Organisation for the Prohibition of Chemical Weapons (OPCW) play supportive and coordinating roles.

The existing approach is hampered by fragmentation, disconnections and insufficient coordination, and adequate funding remains a fundamental challenge. For example, lack of agreement on funding has played a significant part in delaying establishment of a compliance mechanism for the Stockholm Convention. As a result, in 2009 the Executive Director of UNEP launched the Consultative Process on Financing Options for Chemicals and Wastes to look at overall funding needs and possibilities. Between 2009 and 2011 participants discussed four tracks:

- mainstreaming of sound management of chemicals and hazardous wastes;
- industry involvement, including private-public partnerships and the use of economic instruments at the national and international levels;
- a new trust fund similar to the Multilateral Fund; and
- introducing safe chemicals and waste management as a new GEF focal area, expanding the existing POPs focal area under GEF or establishing a new trust fund under the GEF.

The final meeting of the Consultative Process in October 2011 resulted in a document outlining an integrated approach to financing the sound management of chemicals and wastes (UNEP 2012). This formed the basis of a report by the UNEP Executive Director to the UNEP Governing Council Special Session in February 2012, which in turn resulted in governments requesting the Executive Director for a fully-fledged proposal on an integrated approach ensuring optimal funding for the chemicals and wastes sector. A decision on this is expected at the third SAICM International Conference on Chemicals Management in September 2012 and the UNEP Governing Council in 2013.

The follow-up to the Consultative Process is an important opportunity to raise the profile of financing for sound management of chemicals and wastes and links to human health and development, the environment and carbon. It is an intrinsic component of development and a necessary objective to bring lasting social, environmental and economic benefits. Without adequate infrastructure in the key sectors of health, water, sanitation, energy, transport, information and communication technology, and disaster management, there is little hope of protecting people from the risks of exposure to harmful chemicals, hazardous or radioactive waste and other waste streams contaminating the environment.

Reinforcing a global response

The Basel, Rotterdam and Stockholm Conventions and other instruments that address chemicals and wastes – including the Montreal Protocol on Ozone-Depleting Substances, MARPOL, the London Convention, and regional treaties like the Bamako, Waigani or Mediterranean Conventions, as well as the future Minamata Convention on Mercury – represent the foundation on which to build and consolidate a global response to protect human health and the environment from the adverse effects of chemicals and waste. Discussions conducted under the auspices of these global instruments enable emerging problems to be foreseen and facilitate the formulation of ways to manage issues soundly and collectively on a sustainable basis. All these global legally binding instruments, as well as regional agreements such as those agreed by the OECD and the European Commission, share the universal principle of an environmentally sound management of chemicals and waste. A key feature of this global architecture is transparency in the collection and dissemination of information. The EU chemicals legislation, REACH, is exemplary of such efforts (Hartung and Rovida, 2009). But large gaps remain, both in addressing the number of chemicals and nanomaterials present in the market, and in the fact that many countries are unable to manage hazardous chemicals and waste in an environmentally sound way.

With the Basel, Rotterdam and Stockholm Conventions sharing the common objective of protecting human health and the environment from hazardous chemicals and wastes, the Parties to these agreements have embarked on rationalizing their operations to improve assistance to countries in managing chemicals at different stages of their life cycle. This has been exemplified by the establishment of the International Panel on Chemical Pollution (IPCP) in 2008, enhanced cooperation and coordination between the three conventions during their respective Conferences of the Parties in 2008 and 2009, and their simultaneous extraordinary meetings in Bali, Indonesia, in February 2010. Since early 2011, the convention secretariats have been working under a joint Executive Secretary, opening up the possibility of a more holistic approach to the sound management of chemicals and waste (Basel Convention 2012).

Outlook

Table 6.4 summarizes the main goals into key themes and uses the indicators described in this chapter to illustrate progress towards their achievement. It also makes recommendations for consideration alongside those from other chapters in Part 1 when developing policy options and responses as outlined in Parts 2 and 3.

Table 6.4 Progress towards goals (see Table 6.1)

A: Significant progress B: Some progress	C: Very little to no progress D: Deteriorating	X: Too soon to assess progress ?: Insufficient data

Key issues and goals		State and trends	Outlook	Gaps
1. Provide sound management of chemicals throughout their life cycle and of waste				
Sound management of chemicals	B	17 multilateral agreements are in place and over 300 activities have been conducted under the SAICM Global Plan of Action; 23 countries have a functioning national chemical register; a globally harmonized system of classification and labelling of chemicals has been established	The number of developing countries implementing sound chemicals management is increasing	Strengthening of the life-cycle approach; a more integrated global framework for risk assessment and risk management of chemicals; plans for the implementation of sound management of chemicals, most notably for developing countries heavily involved in the production, trade and use of chemicals, particularly in Asia and the Pacific and Latin America
Sound management of waste	B	Objective standards of environmentally sound waste management are not optimal; practices vary widely according to local norms and conditions; cities face increasing problems with management of municipal waste, poor monitoring and open burning, and illegal traffic in waste	Waste production will increase according to current trends in consumption and trade	Data on waste
Improve resource efficiency	C	Efficient ways of transforming waste into energy, applicable in developing countries, are lacking	Environmentally sound energy recovery will benefit some situations if balanced with emphasis on waste recovery for reuse and recycling rather than competing with energy solutions	Technology transfer and capacity building for the long term to ensure the performance of any facility is maintained
Prevent and minimize waste by maximizing reuse, recycling and environmentally friendly alternative materials	?	Many initiatives and some regional and national programmes exist, including Reduce, Reuse, Recycle; EU Producer Responsibility Directives; Basel Mobile Phone Partnership; and Basel Partnership Action on Computer Equipment	Potential to raise many of these efforts to a global level	Global measurement methodology and data on waste minimization; reliable data and trend information
Strict control of the generation and management of hazardous and other waste	?	Data from national reporting to the Basel Convention Secretariat is sparse and difficult to interpret; reporting by Parties is declining	Trend may continue if Parties are not supported in improving compliance and changing direction	Additional effort on awareness raising and capacity building
2. Protect human health and the environment from POPs				
Eliminate or restrict the production, use, import and export of POPs	B/B/B	B: Some progress, but it is still too early to evaluate headway on the basis of the indicators; B: long-term records show a decrease in atmospheric POP concentrations during the 1980s and 1990s, but trends have levelled off since 2000; B: in urban regions of Western industrialized countries PCB emissions are ongoing and of the order of 0.1–1.0 gram per person per year	Continuing exposure to POPs in all parts of the world is likely; climate change may increase exposure due to greater mobilization of POPs	Support to developing countries for their national implementation plans under the Stockholm Convention; greater attention to DDT exposure from malaria control, and exposure to polybrominated diphenyl ethers (among other chemicals) from handling e-waste
3. Reduction of the risks posed by heavy metals				
Restrict production and use of heavy metals	B	Progress in developed countries has resulted in a less frequent occurrence of acute toxicity, but exposure still occurs at industrial and legacy sites, and there is increasing concern regarding possible subtle developmental effects of chronic, low-level exposure; significant problems remain in developing countries, where heavy metals are often mined, processed, used and recycled with limited control, and where most cases of acute toxicity occur, particularly for lead, mercury and arsenic	Ongoing global negotiations on mercury are positive, and further efforts are required to include heavy metals (lead, cadmium, arsenic) in international agreements	Additional research into alternative, more benign chemicals for use in consumer goods to help reduce the heavy metal burden to the environment; more stringent occupational, human health and environmental standards

Table 6.4 Progress towards goals (see Table 6.1) *continued*

3. Reduction of the risks posed by heavy metals *continued*				
Restrict import and export of heavy metals, and waste containing heavy metals	B	The rapid turnover of e-products and high cost of eliminating the hazardous substances within them have led to growth in the trade and movement of e-waste to developing countries where labour costs and health and environmental standards are lower	EU directive on the restriction of certain hazardous substances in e-waste is a positive step; global initiatives are also required	Global initiatives such as that on mercury would be beneficial
Improve waste disposal techniques for waste containing heavy metals	?	Much controlled disposal takes place in developed countries, but work is still needed in many developing countries to improve disposal following mining, smelting, battery and e-waste recycling	Internationally agreed goals for lead, mercury and possibly other heavy metals exist or are in development; further efforts are required	More stringent occupational, human health and environmental standards, as well as more stringent regulations surrounding disposal
4. Promote shared responsibility and cooperative efforts between Parties in the international trade of certain hazardous chemicals				
Develop national decision-making processes for the import and export of hazardous chemicals	C	Many developing countries lack policies on sound management, with weak institutional and regulatory frameworks and poor enforcement of existing laws; there is no coherent approach for national decision-making, with a multiplicity of government agencies having mandates for the import and export of hazardous chemicals, causing jurisdictional conflict and weakening decision-making	Foreseeable improvement if the rate of notifications under the Rotterdam Convention continues or is enhanced	Improvement of mechanisms, governance and a regulatory framework for effective decision making at regional and national levels, including promoting synergy in the implementation of international chemical and waste agreements
Facilitate information exchange about chemical characteristics	?	There is a lack of datasets and effective mechanisms available to national, regional and international stakeholders for the retrieval and dissemination of information about chemicals' characteristics; poor information exchange between government agencies at the national level is hindering informed decision making		Development and implementation of functional international, regional and national networks for the exchange of information on the characteristics of hazardous chemicals and wastes
5. Use transparent science-based risk assessment and risk management procedures				
Sound management of chemicals throughout their life cycle	?	Risk assessment is being used at the international level (Stockholm Convention), but is constrained in developing countries by the lack of data on exposure and effects caused by chemicals and wastes, and also by capacity constraints Under the Montreal Protocol, training has been provided for ozone officers to detect illegal shipments; the Globally Harmonized System (GHS) for labelling of chemicals in the workplace is hazard-based and can be used as input to risk assessment There is uncertainty about both the hazardousness of certain chemicals and the risk they pose; chemicals in products are often not identified, sometimes for reasons of commercial confidentiality	The situation may improve through the activities of the convention review committees, the intergovernmental process on mercury, EU REACH and national reassessments of chemicals	Data on children's vulnerability to chemical risk (average risk assessment only uses adult data); training in chemicals identification and risk management under the auspices of SAICM; disclosure of product composition
Encourage research in order to prevent, eliminate and reduce pollution of the marine environment	A	Historically, investment in marine pollution research has been more active in the northern hemisphere; more recently, important efforts are being made in the developing world to protect marine resources – often an important food source – from pollution		Scientifically sound data on contamination
6. Develop adequate monitoring systems (national, regional and global)				
Develop sound science-based monitoring programmes	?	For POPs, the Global Monitoring Plan is in place; for a wide range of additional chemicals, biomonitoring programmes are inadequate in most countries and human exposure is incompletely documented; hazardous waste reporting systems are available for Parties to the Basel Convention but are not fully taken up or reported; the impacts of waste due to unsound disposal are difficult to quantify	Global monitoring programmes involving chemicals are being developed and harmonization and global coverage are expected to be reached in the coming years	Comprehensive regional and global monitoring programmes for building spatial and temporal trends of key chemicals and wastes, as well as datasets and indicators to enable monitoring of change; biomarkers and bioindicators to assist the assessment of chemical exposure and effects; training and appropriate laboratory facilities in developing countries, and support to build capacity for monitoring hazardous residues imported from developed countries (storage, disposal or reprocessing)

Table 6.4 Progress towards goals (see Table 6.1) *continued*

7. Capacity development				
Sound management of chemicals and hazardous wastes	C	In spite of efforts by international agencies, capacity development for sound management in developing countries is still lacking; Basel and Stockholm Convention Regional Centres have been established to enhance the capacity of governments and stakeholders in developing countries, but adequate financing mechanisms are not yet in place	Situation likely to improve if Stockholm and Basel Conventions, SAICM and the GHS can be supported through innovative financing mechanisms	An adequate financing mechanism, as well as information and knowledge sharing between north and south
Improve resource efficiency	B	Waste disposal rather than integrated waste management is practised without resource and materials recovery, and national waste policy and legislations on integrated waste management as well as infrastructure for collection are inadequate; crude and resource-inefficient recycling exists in the informal economy	Environmentally sound management of waste instead of indiscriminate disposal and uncontrolled open burning	Promotion of regional and national initiatives on waste-to-energy and waste-to-organic-fertilizers, along with waste recycling and materials recovery through pilot/demonstration projects
Control of transboundary movements of hazardous waste	?	The control system available via the Basel Convention's prior informed consent notification process can work very well when fully utilized, though the process is vulnerable to circumvention and illegal traffic	Full national implementation of the Basel Convention and measures under SAICM, along with further stimulus, would improve the rate of progress	Improved capacity building and funding mechanisms to sustain implementation and compliance of the Basel Convention; elaboration of synergistic convention initiatives at the regional and national levels; improved cooperation between international, regional and national networks for controlling the transboundary movement of hazardous waste Greater cooperation, for example through the European IMPEL and global INECE networks, to improve compliance and enforcement, together with Interpol Pollution Crimes Working Group initiatives; better reporting as well as active collaboration and cooperation with Basel Convention focal points in developing countries
8. Protect and preserve the marine environment from all sources of pollution				
Pollution from ships	B	Progress has been made under the MARPOL Convention – 150 countries have ratified although many are yet to comply; control of greenhouse gas emissions from international shipping, and a draft of the reduction mechanisms themselves are under further consideration by the International Maritime Organization's Marine Environment Protection Committee (MEPC)	Likely to improve with the development of new mechanisms under MEPC	Development and implementation of functional international networks for the control of ships' wastes, including disposal facilities in ports
Protect the marine environment	X	Protection of the marine environment has not always been given priority by institutional arrangements or by environmental regulations in developing countries; UNEP's regional seas conventions have not all been transposed into law or implemented; many countries have not ratified or implemented MARPOL; the level of coastal and marine pollution continues to increase, with a lack of control of land-based sources of pollution at the regional and national levels; unsustainable exploitation of marine resources and the marine environment is widespread	Mixed	International action to promote the ratification, transposition into national law and implementation of MARPOL, the regional seas conventions and the London Convention at the regional and national levels, as well as to develop multilateral agreements on chemicals and waste
9. Radioactive waste management and safety				
Ensure that radioactive wastes are safely managed, transported, stored and disposed of	B	Radioactive waste from the operation of nuclear facilities and uses of radioactive material in medicine, industry and research are generally controlled according to international standards and reported at meetings of the Joint Convention on Spent Fuel and Radioactive Waste; some legacy sites remain from nuclear weapons production and testing; some uranium mining legacy sites remain in Africa and Central Asia	Radioactive waste will continue to be generated by the nuclear industry, medical and industrial uses and mining and mineral exploitation, with elevated levels of naturally occurring radionuclides; management and disposal facilities will be needed in the foreseeable future	A closer link between the Joint Convention (already an important global instrument for safe management of radioactive waste) and other international instruments on hazardous materials to develop beneficial synergies; support for international efforts to assist with remediation of uranium mining legacy sites
Prevent accidents with radiological consequences and mitigate consequences of accidental releases	B	The Fukushima accident illustrates that nuclear accidents can still occur despite improvements since Chernobyl, and the Convention on Nuclear Safety (sister convention to the Joint Convention) is intended to ensure countries maintain a high level of safety	A number of countries have decided to phase out their nuclear programmes following the Fukushima accident while others continue to develop their programmes; it is too early to say what the overall impact will be	More emphasis to ensure that the objectives of the Convention on Nuclear Safety and the Joint Convention are achieved

REFERENCES

Astudillo, J.C., Bravo, M., Dumont, C.P. and Thiel, M. (2009). Detached aquaculture buoys in the SE Pacific: potential dispersal vehicles for associated organisms. *Aquatic Biology* 5, 219–231

ATSDR (2007). *Toxicological Profile for Lead*. Agency for Toxic Substances and Disease Registry. US Department of Health and Human Services, Public Health Service, Atlanta, GA

Barra, R., Castillo, C. and Torres, J.P.M. (2007). Polycyclic aromatic hydrocarbons in the South American environment. *Reviews of Environmental Contamination and Toxicology* 191, 1–22

Basel Convention (2012). *The Synergies Process*. http://www.basel.int/TheConvention/Synergies/tabid/1320/Default.aspx

Basel Convention (2011). Basel Convention website. http://www.basel.int/Countries/NationalReporting/StatusCompilations/tabid/1497/Default.aspx

Basel Convention (1989). *The Basel Convention on the Control of Transboundary Movement of Hazardous Wastes and their Disposal*. http://www.basel.int/

Blacksmith Institute (2011). *Top Ten of the Toxic Twenty. The World's Worst Toxic Pollution Problems Report 2011*. Blacksmith Institute, New York and Green Cross Switzerland, Zurich. http://www.worstpolluted.org

Bogdal, C., Nikolic, D., Lüthi, M.P., Schenker, U., Scheringer, M. and Hungerbühler, K. (2010). Release of legacy pollutants from melting glaciers: model evidence and conceptual understanding, *Environmental Science and Technology* 44(11): 4063-4069

Bose-O'Reilly, S.B., Lettmeier, R.M., Gothe, R.M., Beinhoff, C., Siebert, U. and Drasch, G. (2008). Mercury as a serious hazard for children in gold mining areas. *Environmental Research* 107(1), 89–97

Boxall, A., Hardy, A., Beulke, S., Boucard, T., Burgin, L., Falloon, P., Haygarth, P., Hutchinson, P., Kovats, S., Leonardi, G., Levy, L., Nichols, G., Parsons, S., Potts, L., Stone, D., Topp, E., Turley, D., Walsh, K., Wellington, E. and Williams, R. (2009). Impacts of climate change on indirect human exposure to pathogens and chemicals from agriculture. *Environmental Health Perspectives* 117(4), 508–514

Brodesser, J., Byron, D.H., Cannavan, A., Ferris, I.G., Gross-Helmert, K., Hendrichs, J., Maestroni, B.M., Unsworth, J., Vaagt, G. and Zapata, F. (2006). *Pesticides in developing countries and the International Code of Conduct on the Distribution and the Use of Pesticides*. Austrian Agency for Health and Food Safety (AGES) Meeting on Risks and Benefits of Pesticides, Vienna, Austria, 30 March 2006. http://www-naweb.iaea.org/nafa/fep/public/2006-AGES-CoC.pdf

Carn, S.A., Krueger, A.J., Krotkov, N.A., Yang, K. and Levelt, P.F. (2007). Sulfur dioxide emissions from Peruvian copper smelters detected by the ozone-monitoring instrument. *Geophysical Research Letters* 34(09801) L09801, doi:10.1029/2006GL029020

Caroli, S., Cescon, P. and Walton, D.W.H. (eds.) (2001). *Environmental Contamination in Antarctica: A Challenge to Analytical Chemistry*. Elsevier Science, Oxford

CAS (2011). Chemicals Abstract Service. www.cas.org (accessed July 2011)

CEMC (2012) Canadian Centre for Environmental Modelling and Chemistry website. www.trentu.ca/academic/aminss/envmodel

Colborn, T., vom Saal, F.S. and Soto, A.M. (1993). Developmental effects of endocrine-disrupting chemicals in wildlife and humans. *Environmental Health Perspectives* 101(5), 378–384

CSD (2010). *Review of implementation of Agenda 21 and the Johannesburg Plan of Implementation: Chemicals*. Report of the Secretary-General. Commission on Sustainable Development, 18th session. http://daccess-dds-ny.un.org/doc/UNDOC/GEN/N10/245/37/PDF/N1024537.pdf?OpenElement

Cui, J. and Forssberg, E. (2003). Mechanical recycling of waste electric and electronic equipment: a review. *Journal of Hazardous Materials* 99(3), 243–263

Diamanti-Kandarakis, E., Bourguignon, J.P., Giudice, L.C., Hauser, R., Prins, G.S., Soto, A.M., Zoeller, T. and Gore, A.C. (2009). Endocrine-disrupting chemicals: an Endocrine Society scientific statement. *Endocrine Reviews* 30(4), 293–342

European Nuclear Society (2012). http://www.euronuclear.org/info/ (accessed February 2012)

FAO (2012) *Prevention and Disposal of Obsolete Pesticides*. Food and Agriculture Organization of the United Nations, Rome. http://www.fao.org/agriculture/crops/obsolete-pesticides/where-stocks/zh/ (accessed March 2012)

FAO (2002) *Stockpiles of Obsolete Pesticides in Africa Higher than Expected*. Food and Agriculture Organization of the United Nations, Rome. http://www.fao.org/english/newsroom/news/2002/9109-en.html

Feingold, B.J., Vegosen, L., Davis, M., Leibler, J., Peterson, A. and Silbergeld, E.K. (2010). A niche for infectious disease in environmental health: rethinking the toxicological paradigm. *Environmental Health Perspectives* 118(8), 1165–1172

Finnveden, G., Hauschild, M.Z., Ekvall, T., Guine´e, J., Heijungs, R., Hellweg, S., Koehler, A., Pennington, D. and Suh, S. (2009). Recent developments in Life Cycle Assessment. *Journal of Environmental Management* 91, 1–21

Flegal, A.R. and Smith, D.R. (1992). Lead levels in preindustrial humans. *New England Journal of Medicine* 326, 1293–1294

Fraser, B. (2009). La Oroya's legacy of lead. *Environmental Science and Technology* 43(15), 5555–5557

Geisz, H.N., Dickhut, R.M., Cochran, M.A., Fraser, W.R. and Ducklow, H.W. (2008). Melting glaciers: a probable source of DDT to the Antarctic Marine Ecosystem. *Environmental Science and Technology* 42, 3958–3962

Gonzalez, M., Miglioranza, K.S.B., Aizpún, J.E., Isla, F.I. and Peña, A. (2010). Assessing pesticide leaching and desorption in soils with different agricultural activities from Argentina (Pampa and Patagonia). *Chemosphere* 81(3), 351–356

Gore, A.C. and Patisaul, H.B. (2010). Neuroendocrine disruption: historical roots, current progress, questions for the future. Front. *Neuroendocrinology* 31, 395–399

Haefliger, P., Mathieu-Nolf, M., Lociciro, S., Ndiaye, C., Coly, M., Diouf, A., Faye, A.L., Sow, A., Tempowski, J., Pronczuk, J., Filipe Junior, A.P., Bertollini, R. and Neira, M. (2009). Mass lead intoxication from informal used lead-acid battery recycling in Dakar, Senegal. *Environmental Health Perspectives* 117(10), 1535–1540

Hammer, S. Kamal-Chaoui, L., Robert, A. and Plouin, M. (2011). *Cities and Green Growth: A Conceptual Framework*. OECD Regional Development Working Papers 2011/08, OECD Publishing. http://dx.doi.org/10.1787/5kg0tflmzx34-en

Handoh, I.C. and Kawai, T. (2011). Bayesian uncertainty analysis of the global dynamics of persistent organic pollutants: towards quantifying the planetary boundaries for chemical pollution. In *Interdisciplinary Studies on Environmental Chemistry – Marine Environmental Modeling and Analysis* (eds. Omori, K., Guo, X., Yoshie, N., Fujii, N., Handoh, I.C., Isobe, A. and Tanabe, S.). pp.179–187. Terrapub, Tokyo

Hartung, T. and Rovida, C. (2009). Chemicals regulators have overreached. *Nature* 460, 1080–1081

Hengstler, J.G., Foth, H., Gebel, T., Kramer, P.J., Lilienblum, W., Schweinfurth, H., Völkel, W., Wollin, K.M. and Gundert-Remy, U. (2011). Critical evaluation of key evidence on the human health hazards of exposure to bisphenol A. *Critical Reviews in Toxicology* 41, 263–291

Huang, X., Sillampaa ,T., Gjessing, E.T., Peraniemi, S. and Vogt, R.D. (2011). Water quality in the southern Tibetan Plateau: chemical evaluation of the River Yarlung Tsangpo (Brahmaputra). *River Research and Applications* 27, 113–121

Hung, H., Kallenborn, R., Breivik, K., Su, Y., Brorström-Lundén, E., Olafsdottier, K., Thorlacius, J.M., Leppänen, S., Bossi, R., Skov, H., Manö, S., Patton, G.W., Stern, G., Sverko, E. and Fellin, P. (2010). Atmospheric monitoring of organic pollutants in the Arctic under the Arctic Monitoring and Assessment Programme (AMAP): 1993–2006. *Science of the Total Environment* 408, 2854–2873

Huo, X., Peng, L., Xu, X.J., Zheng, L., Qiu, B., Qi, Z., Zhang, B., Han, D. and Piao, Z. (2007). Elevated blood lead levels of children in Guiyu, an electronic waste recycling town in China. *Environmental Health Perspectives* 115(7), 1113–1117

IAEA (2010). *Measures to Strengthen International Cooperation in Nuclear, Radiation, Transport and Waste Safety*. General Conference Resolution GC (54)/RES/7 adopted 24 September 2010. International Atomic Energy Agency, Vienna

IAEA (2009a). *Classification of Radioactive Waste General Safety Guide*. Series No. GSG-1. International Atomic Energy Agency, Vienna

IAEA (2009b). *Summary Report. Joint Convention on the Safety of Spent Fuel Management and on the Safety of Radioactive Waste Management*. Third Review Meeting of the Contracting Parties, 11–20 May, Vienna. JC/RM3/02/Rev2. International Atomic Energy Agency, Vienna

IAEA (2008a). *20/20 Vision for the Future. Background Report by the Director General for the Commission of Eminent Persons*. International Atomic Energy Agency, Vienna

IAEA (2008b). *Estimation of Global Inventories of Radioactive Waste and Other Radioactive Material*. TECDOC-1591. International Atomic Energy Agency, Vienna

IAEA (2006). *An International Peer Review of the Programme for Evaluating Sites for Near Surface Disposal of Radioactive Waste in Lithuania*. Report of the IAEA International Review Team. International Atomic Energy Agency, Vienna

IAEA (1997). *Joint Convention on the Safety of Spent Fuel Management and on the Safety of Radioactive Waste Management*. International Atomic Energy Agency, Vienna. http://www.iaea.org/Publications/Documents/Infcircs/1997/infcirc546.pdf

IHPA (2009) *Obsolete (Lethal) Pesticides: A Ticking Time Bomb and Why We Have to Act Now*. International HCH and Pesticides Association. http://www.ihpa.info/docs/library/reports/timeBomb_Obsolete_Pesticides.pdf

IPCP (undated). International Panel on Chemical Pollution, Zurich. http://www.ipcp.ch/

Jarup, L. and Akesson, A. (2009). Current status of cadmium as an environmental health problem *Toxicology and Applied Pharmacology* 238, 201–208.

Jacobsen, J.K., Massey, L. and Gulland, F. (2010). Fatal ingestion of floating net debris by two sperm whales (*Physeter macrocephalus*). *Marine Pollution Bulletin* 60(15), 765–767

Koppe, J.G. and Keys, J. (2001). PCBs and the precautionary principle. In: *Late Lessons from Early Warnings: The Precautionary Principle 1896–2000* (eds. Harremoes, P., Gee, D., MacGarvin, M., Stirling, A., Keys, J., Wynne, B. and Vaz, S.G.). pp.64–72. Environmental Issue Report No. 22. European Environment Agency, Copenhagen

Lamon, L., Valle, M.D., Critto, A. and Marcomini, A. (2009). Introducing an integrated climate change perspective in POPs modelling, monitoring and regulation. *Environmental Pollution* 157(7), 1971–1980

Lanphear, B., Matte, T., Rogers, J., Clickner, R., Dietz, B., Bornschein, R., Succop, P., Mahaffey, K., Dixon, S., Galke, W., Rabinowitz, M., Farfel, M., Rohde, C., Schwartz, J. Ashley, P. and Jacobs, D. (1998). The contribution of lead-contaminated house dust and residential soil to children's blood lead levels: a pooled analysis of 12 epidemiologic studies. *Environmental Research* 79(1), 51–68

Lemieux, P.L., Lutes, C.C. and Santoianni, D.A. (2004). Emissions of organic air toxics from open burning: a comprehensive review. *Progress in Energy and Combustion Science* 30, 1–32

Lokuge, K.M., Smith, W., Caldwell, B., Dear, K. and Milton, A.H. (2004). The effect of arsenic mitigation interventions on disease burden in Bangladesh. *Environmental Health Perspectives* 112, 1172–1177

London Convention (1972/96). *Convention on the Prevention of Marine Pollution by Dumping of Wastes and Other Matter.* Adoption 1972; 1996 Protocol. http://www.imo.org/About/Conventions/ListOfConventions/Pages/Convention-on-the-Prevention-of-Marine-Pollution-by-Dumping-of-Wastes-and-Other-Matter.aspx

MacLeod, M., Riley,W.J. and McKone, T.E. (2005). Assessing the influence of climate variability on atmospheric concentrations of polychlorinated biphenyls using a global-scale mass balanced model (BETR-Global). *Environmental Science and Technology* 39, 6749–6756

MARPOL (1973/78). *International Convention for the Prevention of Pollution from Ships.* Adoption 1973, 1978 Protocol. International Maritime Organization (IMO), London. http://www.imo.org/about/conventions/listofconventions/pages/international-convention-for-the-prevention-of-pollution-from-ships-(marpol).aspx

Mikelis, N. (2010). *IMO's Action Plan on Tackling the Inadequacy of Port Reception Facilities.* Ships' Waste: Time for action! Conference organised by EUROSHORE and FEBEM-FEGE, Brussels, 14 October 2010. International Maritime Organization, London

Morris, J., Willis, J., De Martinis, D., Hansen, B., Laursen, H., Sintes, J.R., Kearns, P. and Gonzalez, M. (2010). Science policy considerations for responsible nanotechnology decisions. *Nature Nanotechnology* 6, 73–77. doi:10.1038/nnano.2010.191

Muir, D. and Howard, P. (2010). Identifying new persistent and bioaccumulative organics among chemicals in commerce. *Environmental Science and Technology* 44, 2277–2285

Muir, D. and Howard, P. (2006). Are there other persistent organic pollutants? A challenge for environmental chemists. *Environmental Science and Technology* 40, 7157–7166

Nweke, O.C. and Sanders, W.H. (2009). Modern environmental health hazards: a public health issue of increasing significance in Africa. *Environmental Health Perspectives* 117(6), 863–870

OECD (2010a). *Cutting Costs in Chemicals Management: How OECD helps Governments and Industry.* Organisation for Economic Co-operation and Development, Paris

OECD (2010b). *OECD Factbook: Economic, Environmental and Social Statistics.* Organisation for Economic Co-operation and Development, Paris

OECD (2008a). *OECD Environmental Data: Compendium 2008.* Organisation for Economic Co-operation and Development, Paris. http://www.oecd.org/dataoecd/56/45/41255417.pdf

OECD (2008b). *OECD Environmental Outlook to 2030.* Organisation for Economic Co-operation and Development, Paris

OECD (2008c) *Costs of Inaction on Key Environmental Challenges.* Organisation for Economic Co-operation and Development, Paris

Ondarza P.M., Gonzalez, M., Fillmann, G. and Miglioranza, K.S.B. (2011). Polybrominated diphenyl ethers and organochlorine compound levels in brown trout (*Salmo trutta*) from Argentinean Patagonia. *Chemosphere* 83, 1597–1602

Ondarza, P.M., Miglioranza, K.S.B., Gonzalez, M., Shimabukuro, V.M., Aizpún, J.E. and Moreno, V.J. (2010). Organochlorine compounds (OCCs) in common carp (*Cyprinus carpio*) from Patagonia Argentina. *Journal of the Brazilian Society of Ecotoxicology* 5, 41–46

Poliakoff, M., Fitzpatrick, J.M., Farren, T.R. and Anastas, P.T. (2002). Green chemistry: the science and policy of change. *Science* 297, 807–810

Prüss-Ustün, A., Vickers, C., Haefliger, P. and Bertollini, R. (2011). Knowns and unknowns on burden of disease due to chemicals: a systematic review. *Environmental Health* 10, 9–24

Rajapakse, N., Silva, E. and Kortenkamp, A. (2002). Combining xenoestrogens at levels below individual no-observed-effect concentrations dramatically enhances steroid hormone action. *Environmental Health Perspectives* 110, 917–921

Rauch, J.N. and Pacyna, J.M. (2009). Earth's global Ag, Al, Cr, Cu, Fe, Ni, Pb, and Zn cycles. *Global Biogeochemical Cycles* 23, GB2001

Ravenscroft, O., Brammer, H. and Richards, K. (2009). *Arsenic Pollution: A Global Synthesis.* Wiley-Blackwell, Chichester

Ritter, R., Scheringer, M., MacLeod, M. and Hungerbühler, K. (2011). Assessment of nonoccupational exposure to DDT in the tropics and the north: relevance of uptake via inhalation from indoor residual spraying. *Environmental Health Perspectives* 119, 707–712

Rockström, J., Steffen, W., Noone, K., Persson, Å., Chapin III, S.F., Lambin, E.F., Lenton, T.M., Scheffer, M., Folke, C., Schellnhuber, H.J., Nykvist, B., de Wit, C.A., Hughes, T., van der Leeuw, S., Rodhe, H., Sörlin, S., Snyder, P.K, Costanza, R., Svedin, U., Falkenmark, M., Karlberg, L., Corell, R.W., Fabry, V.J., Hansen, J., Walker, B., Liverman, D., Richardson, K., Crutzen, P. and Foley, J.A (2009). A safe operating space for humanity. *Nature* 461, 472–475

Rotterdam Convention (2001). *Rotterdam Convention on the Prior Informed Consent Procedure for Certain Hazardous Chemicals and Pesticides in International Trade.* Revised in 2011. http://www.pic.int/TheConvention/Overview/TextoftheConvention/tabid/1048/language/en-US/Default.aspx

SAICM (2009). Background information in relation to the emerging policy issue of electronic waste. *Implementation of the Strategic Approach to International Chemicals Management: Emerging Policy Issues.* International Conference on Chemicals Management, Geneva, 11–15 May. SAICM/ICCM.2/INF/36. Strategic Approach to International Chemicals Management

Schluepa, M., Hageluekenn., C., Kuehr, R., Magalini, F., Maurer, C., Meskers, C., Mueller, E. and Wang. F. (2009). *Recycling from E-waste to Resources: Sustainable Innovation and Technology Transfer.* UNEP/DTIE

Schwarzer, S., De Bono, A., Giuliani, G., Kluser, S. and Peduzzi, P. (2005). *E-Waste, the Hidden Side of IT Equipment's Manufacturing and Use.* UNEP Early Warning on Emerging Environmental Threats No. 5. United Nations Environment Programme/GRID Europe. http://www.grid.unep.ch/products/3_Reports/ew_ewaste.en.pdf

Selin, N.E. and Selin, H. (2006). Global politics of mercury pollution: the need for multi-scale governance. *Review of European Community and International Environmental Law* 15(3), 258–269

Sexton, K., Ryan, A.D., Adgate, J.L., Barr, D.B. and Needham, L.L. (2011). Biomarker measurements of concurrent exposure to multiple environmental chemicals and chemical classes in children. *Journal of Toxicology and Environmental Health, Part A* 74(14), 927–942

Sheffield, P.E. and Landrigan, P.J. (2011). Global climate change and children's health: threats and strategies for prevention. *Environmental Health Perspectives* 119(3), 291–298

Silva, E., Rajapakse, N. and Kortenkamp, N. (2002). Something from "nothing" – eight weak estrogenic chemicals combined at concentrations below NOEC produce significant mixture effect. *Environmental Science and Technology* 36(8), 1751–1756

Smith, A.H. and Lingus, E.O. (2000). Contamination of drinking-water by arsenic in Bangladesh: a public health emergency. *Bulletin of the World Health Organization* 78(9), 1093–1103

Soerensen, A.L., Sunderland, E.M., Holmes, C.D., Jacob, D.J.,Yantosca, R.M., Skov, H., Christensen, J.H., Strode, S.A. and Mason, R.P. (2010). An improved global model for air-sea exchange of mercury: high concentrations over the North Atlantic. *Environmental Science and Technology* 44(22), 8574–8580

Stockholm Convention (2001). *Stockholm Convention on Persistent Organic Pollutants.* Adopted 2001. Secretariat of the Stockholm Convention, Châtelaine. http://chm.pops.int/default.aspx

Thundiyil, J.G., Stober, J., Besbelli, N. and Pronczuk, J. (2008). Acute pesticide poisoning: a proposed classification tool. *Bulletin of the World Health Organization* 86(3), 205–209

Toppari, J., Larsen, J.C., Christiansen, P., Giwercman, A., Grandjean, P., Guillette, L.J., Jegou, B., Jensen, T.K., Jouannet, P., Keiding, N., Leffers, H., McLachlan, J.A., Meyer, O., Muller, J., Rajpert-De Meyts, E., Scheike, T., Sharpe, R., Sumpter, J. and Skakkebaek, N.E. (1996). Male reproductive health and environmental xenoestrogens. *Environmental Health Perspectives* 104(4), 741–803

UNCED (1992a). *Rio Declaration on Environment and Development.* United Nations Convention on Environment and Development, Rio de Janeiro

UNCED (1992b). *Agenda 21.* United Nations Convention on Environment and Development. http://www.un.org/esa/sustdev/documents/agenda21/english/Agenda21.pdf

UNCSD (2011). *Report of the Secretary-General: Policy Options and Actions for Expediting Progress in Implementation: Waste Management.* Commission on Sustainable Development 19th Session, 2–13 May. Doc. E/CN.17/2011/6. United Nations Economic and Social Council. http://www.un.org/esa/dsd/csd/csd_pdfs/csd-19/sg-reports/CSD-19-SG-report-waste-management-final-single-spaced.pdf

UNECE Geneva Convention (1979/98). Convention on Long-range Transboundary Air Pollution, 1998 Aarhus Protocol on Persistent Organic Pollutants (POPs). http://www.unece.org/fileadmin/DAM/env/lrtap/full%20text/1998.POPs.e.pdf

UNEP (2012). 12th Special Session GC/GMEF Website: UNEP/GCSS.XII/8 and UNEP/GCSS.XII/7. United Nations Environment Progamme, Nairobi. http://www.unep.org/gc/gcss-xii/docs/info_docs.asp

UNEP (2011a). *Selected Documents Relevant to the Work of the Implementation and Compliance Committee, Conference of the Parties to the Basel Convention on the Control of Transboundary Movements of Hazardous Wastes and Their Disposal.* Tenth meeting, Cartagena, Colombia, 17–21 October 2011. UNEP/CHW.10/INF/11. Implementation and Compliance Committee, United Nations Environment Programme, Nairobi

UNEP (2011b). *UNEP Yearbook 2011: Emerging Issues in Our Global Environment.* United Nations Environment Programme, Nairobi

UNEP (2010). *Report of the First Meeting of the Global Alliance to Eliminate Lead in Paints.* United Nations Environment Programme, Nairobi. http://www.unep.org/hazardoussubstances/Portals/9/Lead_Cadmium/docs/GAELP/FirstMeeting/GAELP_8_Meeting_report.pdf

UNEP (2009). *Recycling from E-Waste to Resources.* Sustainable Innovation and Technology Transfer Industrial Sector Studies DTI /1192/PA. United Nations Environment Programme, Nairobi
UNEP (2007). *Global Environment Outlook 4: Environment for Development.* United Nations Environment Programme and Earthscan, Nairobi

UNEP (2002). *Proceedings: Subregional Workshop on Support for the Implementation of the Stockholm Convention on Persistent Organic Pollutants (POPs).* Port of Spain, Trinidad and Tobago, 4–8 June. United Nations Environment Programme – Chemicals, Geneva. http://www.pops.int/documents/implementation/gef/TT_Proceedings.pdf

UNEP (2000). *Related Work on Persistent Organic Pollutants under the Food and Agriculture Organization of the United Nations, Review of Ongoing International Activities Relating to the Work of the Committee.* Intergovernmental Negotiating Committee for an International Legally Binding Instrument for Implementing International Action on Certain Persistent Organic Pollutants, 5th Session, Johannesburg, 4–9 December. United Nations Environment Programme, Nairobi. http://www.pops.int/documents/meetings/inc5/Fr/inf5-4/inf4.doc

UNEP (1987). *Montreal Protocol on Substances that Deplete the Ozone Layer.* Ozone Secretariat, United Nations Environment Programme, Nairobi. http://ozone.unep.org/pdfs/Montreal-Protocol2000.pdf

UNEP/AMAP (2010). *Climate Change and POPs. Predicting the Impacts.* Report of a UNEP/AMAP expert group. Secretariat of the Stockholm Convention, Geneva

UN-Habitat (2010). *Solid Waste Management in the World's Cities: Water and Sanitation in the World's Cities 2010.* United Nations Human Settlements Programme and Earthscan, London and Washington, DC

Vogel, D. (1997). Trading up and governing across: transnational governance and environmental protection. *Journal of European Public Policy* 4, 556–571

von Braun, M.C., von Lindern, I.H., Khristoforova, N.K., Kachur A.H., Yelpatyevsky, P.V., Elpatyevskaya, V.P. and Spalinger, S.M. (2002). Environmental lead contamination in the Rudnaya Pristan–Dalnegorsk Mining and Smelter District, Russian Far East. *Environmental Research* 88(3), 164–173

Wania, F. and Daly, G.L. (2002). Estimating the contribution of degradation in air and deposition to the deep sea to the global loss of PCBs. *Atmospheric Environment* 36–37, 5581–5593

Wasserman, G.A., Xinhua, L., Parvez, F., Ahsan, H., Factor-Litvak, P., van Geen, A., Slavkovich, V., Lolacono, N.J., Cheng, Z., Hussain, I., Momotaj, H. and Graziano, J.H. (2004). Water arsenic exposure and children's intellectual function in Araihazar, Bangladesh. *Environmental Health Perspectives* 112, 1329–1333

Waye, A. and Trudeau, V.J. (2011). Neuroendocrine disruption: more than hormones are upset. *Journal of Toxicology and Environmental Health Part B: Critical Reviews* 14(5-7)

WHO (2002). *Global Assessment of the State-of-the-Science of Endocrine Disruptors* (eds. Damstra, T., Barlow, S., Kavlock, R., Bergman, A. and Van Der Kraak, G.). International Programme on Chemical Safety, World Health Organization, Geneva. http://www.who.int/ipcs/publications/new_issues/endocrine_disruptors/en/

WHO/UNICEF (2005). *Water for Life; Making It Happen.* WHO/UNICEF Joint Monitoring Programme for Water Supply and Sanitation. http://www.who.int/water_sanitation_health/waterfor life.pdf; http://www.who.int/water_sanitation_health/waterforlife.pdf

WSSD (2002). Johannesburg Plan of Implementation. World Summit on Sustainable Development. http://www.un.org/esa/sustdev/documents/WSSD_POI_PD/English/POIToc.htm

Young, L.C., Vanderlip, C., Duffy, D.C., Afanasyev, V. and Shaffer, S.A. (2009). Bringing home the trash: do colony-based differences in foraging distribution lead to increased plastic ingestion in Laysan albatrosses? *PloS ONE* 4, 10

Zarfl, C. and Matthies, M. (2010). Are marine plastic particles transport vectors for organic pollutants to the Arctic? *Marine Pollution Bulletin* 60(10), 1810–1840

Zoeteman, B.C.J., Krikke, H.R. and Venselaar, J. (2010). Handling WEEE waste flows: on the effectiveness of producer responsibility in a globalizing world. *International Journal of Advanced Manufacturing Technology* 47, 415–436

CHAPTER 7

An Earth System Perspective

Coordinating lead authors: Jill Jäger and Neeyati Patel

Lead authors: Vladimir Ryabinin, Pushker Kharecha, James Reynolds, Lawrence Hislop and Johan Rockström

Contributing authors: Andrew Githeko, Pauline Dube, Niki Frantzeskaki, Derk Loorbach, Jan Rotmans, Genrikh Alekseev, Benjamin Gaddis and Jiansheng Ye (GEO Fellow)

Principal scientific reviewer: Gerhard J. Herndl

Chapter coordinator: Neeyati Patel

Main Messages

The changes discussed in Chapters 2–6 are taking place within an integrated, interconnected whole that is the Earth System. Humans are an integral part of that system.

The Earth System is complex and composed of interacting components. Non-linear interactions within and among these components, supplemented by the inherent difficulties in anticipating human behaviour, impose limits on the predictability of the Earth System.

As human pressures within the Earth System increase, several critical thresholds are approaching or have been exceeded, beyond which abrupt and non-linear changes to the life-support functions of the planet could occur. This has significant implications for human well-being now and in the future. For example: climate variability and extreme weather influence food security; crossing of thresholds leads to significant health impacts, as shown by the increase in malaria in response to rising temperatures; increased frequency and severity of climatic events affect both natural assets and human security; and accelerating changes such as of temperature and sea level rise affect the social cohesion of indigenous communities: in Alaska, for example, permafrost thawing and increased flooding are forcing villages to relocate.

Traditional expert-driven, top-down approaches to problem solving are not flexible enough to address complex, non-linear changes in the Earth System effectively. After more than 20 years of tackling problems of unsustainable development as more or less isolated issues, an integrated Earth Systems approach is needed for informed and effective decision-making.

There is an urgent need to address the underlying drivers of the human pressures on the Earth System. At the same time, it is necessary to adopt approaches that can deal better with the complexities and inherent uncertainties of the Earth System. This must include three elements: basic research to understand interactions and feedbacks; sustained long-term monitoring and observation to underpin basic research; and regular evaluation of progress to allow the adjustment of responses when observations indicate that this is necessary.

INTRODUCTION

The first pictures of the Earth from space stimulated an immediate and profound appreciation of its finite boundaries. Scientific advancements continue to enable a better view of the Earth as a whole. This includes a combination of surface and remote-sensing global observation systems that can document global-scale phenomena, advances in the ability to reconstruct past states of the environment, and enhanced computing power to conduct global-scale simulation experiments (Steffen *et al.* 2004b). Evidence shows that human activities are now so pervasive and profound in their consequences that they affect the Earth at a planetary scale.

Following on from Chapter 1, which highlighted major drivers of change, and Chapters 2–6, which illustrated environmental changes and their impacts at the local, regional and even global levels, this chapter discusses change from an Earth System perspective and provides a context for steering transitions in the way we live, work and govern the planet.

THE EARTH SYSTEM

A system is a collection of component parts that interact with one another within a defined boundary. The Earth System is a complex social-environmental system, including the vast collection of interacting physical, chemical, biological and social components and processes that determine the state and evolution of the planet and life on it. The biophysical components of the Earth System are often referred to as spheres: atmosphere, biosphere, hydrosphere and geosphere. They provide environmental processes that regulate the functioning of the Earth, such as the climate system, the ecological services generated by the living biosphere, including food production, and natural resources like fossil fuels and minerals. Humans are an integral part of the Earth System. All spheres include countless subsystems and levels of organization. The interactions within and between these spheres are complex and the predictability of future states of the Earth System is limited.

Unprecedented changes

Some experts suggest that the Earth has entered a new geological epoch, the Anthropocene (Zalasiewicz *et al.* 2011, 2010). The word was coined by Nobel Laureate Paul Crutzen to capture the idea that humans are now overwhelming the forces of nature. An implication of entering the Anthropocene would be the leaving of the Holocene, the interglacial period that has provided humanity, over the past 10 000 years, with extraordinarily good living conditions, enabling the development of modern societies and a world with 7 billion people (Folke *et al.* 2011).

Crutzen (2002) suggests that the Industrial Revolution 250 years ago saw the beginning of the Anthropocene. The unprecedented rise in human population since the early 19th century, from less than a billion to 7 billion at present, is inherent to the Anthropocene as it unfolds (Zalasiewicz *et al.* 2010). Many societal changes have accompanied this proliferation of the human population, such as increased consumption of natural resources and an enormous expansion of dependence on fossil fuels (Chapter 1).

The Earth System demonstrates complexity in its natural variability independently of, and previous to, human influence.

Figure 7.1 Changes in atmospheric CO_2 concentrations

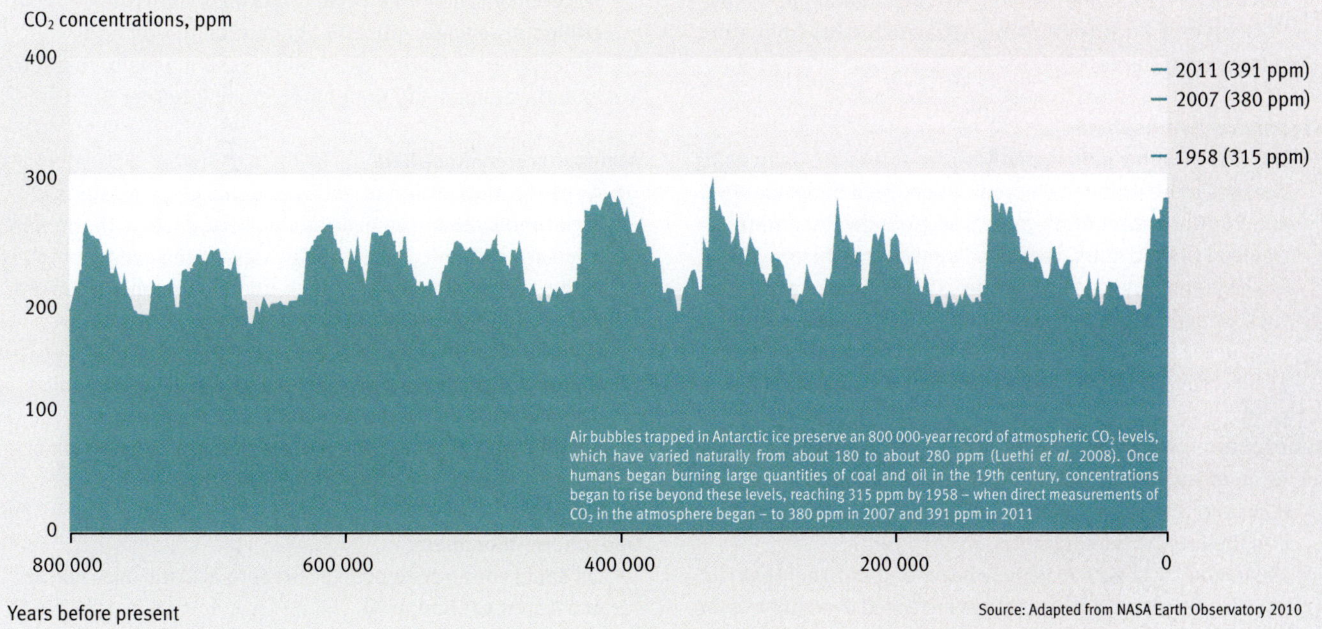

Source: Adapted from NASA Earth Observatory 2010

Ice cores in Antarctica have shown that during the past 800 000 years, air temperature and carbon dioxide (CO_2) concentrations have oscillated within a relatively limited range (Luethi *et al.* 2008), with variations that could be largely linked to factors such as the irregularities of the Earth's rotation and motion along its orbit around the Sun (Hays *et al.* 1976). Current concentrations of atmospheric CO_2 are, however, well outside the range of the past (Figure 7.1), having risen from 310 parts per million (ppm) in 1950 to 391 ppm in 2011 (NOAA 2011), with half the total rise in atmospheric CO_2 since the pre-industrial era having occurred in the last 30 years (Steffen *et al.* 2007).

Biodiversity, the variety of life on Earth, has evolved over the last 3.8 billion years or so of the planet's approximately 5-billion-year history. Five major extinction events have been recorded over this period, but, unlike the previous events – which were due to natural upheavals and planetary change – the current loss of biodiversity is mainly due to human activities and is often referred to as the sixth global extinction (Barnosky *et al.* 2011; Eldredge 2001). According to the *Global Biodiversity Outlook 3* (CBD 2010), the abundance of some vertebrate populations fell by nearly one-third on average between 1970 and 2006 and continues to fall globally. Many biologists consider that coming decades will see the loss of large numbers of species (Leadley *et al.* 2010), increasing the risk of abrupt change in landscapes and seascapes (Estes *et al.* 2011). Fewer scientists appear to have recognized that, in the longer term, these extinctions will alter not only biological diversity but also the evolutionary processes by which diversity is generated (Myers and Knoll 2001).

Earth System interactions

Given the interconnections between the different spheres of the Earth System, changes in one part of the system have effects in one or more of the others. Box 7.1 shows examples from Chapters 2–6 that illustrate Earth System interactions between spheres and how they are being changed by human activities.

EARTH SYSTEM COMPLEXITIES

The complexity of the Earth System is associated with its countless interacting processes, at many scales and levels of system organization. Importantly, these interactions mean that changes rarely occur in linear and incremental ways. Instead, the dominant behaviour when the various systems on Earth undergo change is for it to happen in a non-linear way, driven by feedbacks that either dampen change (negative feedbacks) or reinforce it (positive feedbacks) (Steffen *et al.* 2004a). Many such feedbacks shape the Earth System.

Positive feedbacks are increases in system reaction that may destabilize the system and move it into another state – a regime

Box 7.1 Examples of Earth System interactions influenced by human activities

Atmosphere–biosphere

- Altered atmospheric concentrations of sulphur dioxide affect terrestrial and freshwater ecosystems through acid rain (Chapter 2), with impacts including significant losses of fish stocks and other sensitive aquatic species, and effects on biodiversity and forestry.
- The biosphere in polar regions has been contaminated by the long-range transport of industrial pollutants from other continents (Chapter 6).

Geosphere–hydrosphere

- The rate at which global groundwater stocks are decreasing because of abstractions more than doubled between 1960 and 2000 (Chapter 4). Depletion of groundwater aquifers can lead to land subsidence and saltwater intrusion into freshwater supplies. Furthermore, due to human activities such as agriculture, nutrient mobilization in watersheds around the world, including of phosphorous and nitrogen, has increased significantly since 1960 (Chapter 4).

Atmosphere–geosphere

- As much as 90 per cent of near-surface permafrost may thaw and disappear by 2100, releasing CO_2 and methane into the atmosphere (Chapter 3).
- The frequency of both extremely heavy and extremely light or absent precipitation (drought) has increased over much of the world's land area. Long-term trends show a tendency towards drier conditions in the Sahel and northern India (Chapter 2).

Biosphere–hydrosphere

- Dam building and the control of rivers and their floodplains affect ecosystems and biodiversity (Chapter 4 and 5).
- Water pollutants from waste disposal of industrial effluents, sewage, rubbish, agricultural run-off and atmospheric pollution (acid rain) present a major threat to inland wetlands and their biodiversity (Chapter 5).

Atmosphere–hydrosphere

- A substantial portion of anthropogenic CO_2 emissions is absorbed by the oceans annually. This reacts with the water to create carbonic acid, thereby making the ocean more acidic. The mean surface ocean pH has already decreased from 8.2 to 8.1 and is projected to fall to 7.7 by 2100 (Chapter 4).
- Certain long-lived chemicals such as persistent organic pollutants (POPs) and heavy metals reach the marine environment and are transported globally, causing toxic effects in humans and wildlife (Chapter 6).

Geosphere–biosphere

- Oil spills continue to pose a threat to aquatic and marine ecosystems (Chapter 4).

shift (Box 7.2). An example of a positive feedback is the effect of black carbon deposition in the Arctic (McConnell *et al.* 2007). Black carbon particles are emitted into the atmosphere from the incomplete combustion of biomass and fossil fuels (Chapter 2). The Arctic climate is especially vulnerable to black carbon deposition because of its impact on the albedo (reflectivity) of snow, glaciers and sea ice. Black carbon makes the surface darker, thus reflecting less radiation, which leads to an increase in warming and ice/snow melt. Ramanathan and Carmichael (2008) report that, at high elevations in the Himalayas, positive feedback from the increased absorption of solar radiation by black carbon may be as important in the melting of snowpack and glaciers as are the rising temperatures that result from increased atmospheric CO_2.

An important relationship between the temperature and carbon content of the Earth's atmosphere manifests itself both on relatively short and on geological timescales (Pagini *et al.* 2010), and is the result of many contributing feedbacks in the atmosphere and other components of the Earth System. For example, with the higher temperature and greater acidity of ocean waters, the ability of the ocean to act as a carbon sink weakens (Steffen *et al.* 2004b). This positive feedback is one that increases system reaction and is thus destabilizing.

Another destabilizing feedback, increasingly discussed in climate science, is related to the carbon reservoirs in the Arctic permafrost. If rising temperatures lead to permafrost thaw, this will release carbon and lead to further increases in temperature and consequently to even more permafrost thaw and more releases of carbon (Krey *et al.* 2009).

Forests are a valuable carbon sink and provide a negative feedback response to anthropogenic CO_2 emissions. *© Eugenio Opitz*

The eutrophication of this river is evident from the bright green water, caused by a dense bloom of bluegreen alga Microcystis. *© Heike Kampe/iStock*

The role of biodiversity in such feedback processes is not currently well understood because of the complexity of interactions in physical, chemical and biological processes. It is, however, well known that a positive feedback, which could enhance climate change, will occur if the carbon stored below ground is released to the atmosphere by accelerated respiration induced by soil warming (Rustad *et al.* 2001).

A negative feedback is when the initial response is suppressed, which tends to be stabilizing. For example, if increased water in the atmosphere leads to greater cloud cover, this raises the percentage of sunlight reflected away from the Earth (albedo), which leads to a fall in the temperature of the atmosphere and a decrease in the rate of evaporation (Schmidt *et al.* 2010).

So far, the dominant aggregate response of the Earth System to human pressure has been to dampen its impacts (Steffen *et al.* 2004b). This is explained by the inherent resilience of the Earth System, where the biosphere interacts with the climate system, in particular, to buffer disturbances including some induced by humans. As a result, as a negative feedback response to CO_2 emissions from human activities, the global carbon sink in the biosphere has increased from approximately 2 billion tonnes of carbon per year in the 1960s to approximately 4 billion in 2005 (Canadell *et al.* 2007). However, there are indications that the capacity of the biosphere to buffer pressures from global environmental change is declining (Le Quéré *et al.* 2009), and there is growing evidence of positive feedbacks occurring at the local level, for example eutrophication of lakes (Qin *et al.* 2007), through to the regional level, such as accelerated melting of Arctic ice cover due to a regional amplification of warming (Serreze and Barry 2011).

Box 7.2 Regime shifts

Ecological regime shifts that occur when a certain threshold is passed are usually abrupt, with the change in response far greater than the change in forcing (change in drivers). Regime shifts can also be long-lasting, with concomitant impacts on human economies and societies (Briggs *et al.* 2009). Some ecosystems may be highly vulnerable to change when subjected to the synergistic effects of two or more disturbances with a combined impact greater than the sum of their separate effects (Folke *et al.* 2004). An example is a pastureland that is exposed to the combined effects of drought and overgrazing, leading to alterations in soils, biodiversity and productivity, and resulting in a new ecosystem with different structural and functional characteristics. This is an example of gradual changes in feedbacks (Wysham and Hastings 2008; Levin 1998). In contrast, the shift of an ecosystem from one state to another can also result from catastrophic or extreme external perturbations (van Nes and Scheffer 2007), as illustrated below.

Three examples of regime shifts are shown in Figure 7.2. The first shows how the synergistic effect of drought and bark beetle infestation in northern New Mexico, United States, led to a high mortality of ponderosa pine (*Pinus ponderosa*), which was replaced by a *piñon*-juniper woodland and was thus manifested as a loss of forested area. This is one of the most rapid landscape-scale regime shifts recorded (Allen and Breshears 1998).

The second example is also rapid, but the underlying cause is quite different. de Young *et al.* (2008) illustrated the effects of interactions between organisms. A species-specific pathogen on Caribbean coral reefs caused mass mortalities of sea urchins in the early 1980s. Urchin densities dropped to 1 per cent of their original level, which in turn enabled brown fleshy algae, no longer subject to grazing, to overgrow the reef. The reef community had undergone a regime shift. The new state persisted in some areas for more than 20 years, although the shift had occurred within one or two years for both the trigger (the pathogen) and the algal dominance.

Brown *et al.* (1997) described a shift from a grass- to a shrub-dominated ecosystem over a period of about ten years in a semi-arid site in southern Arizona, United States. The single factor driving this was a slowly changing driver: seasonal precipitation. Winter precipitation favoured the expansion of woody shrubs over grass species, giving rise to the extinction of several animal species and increases in other previously rare ones.

Figure 7.2 Examples of regime shifts resulting from different drivers and feedbacks

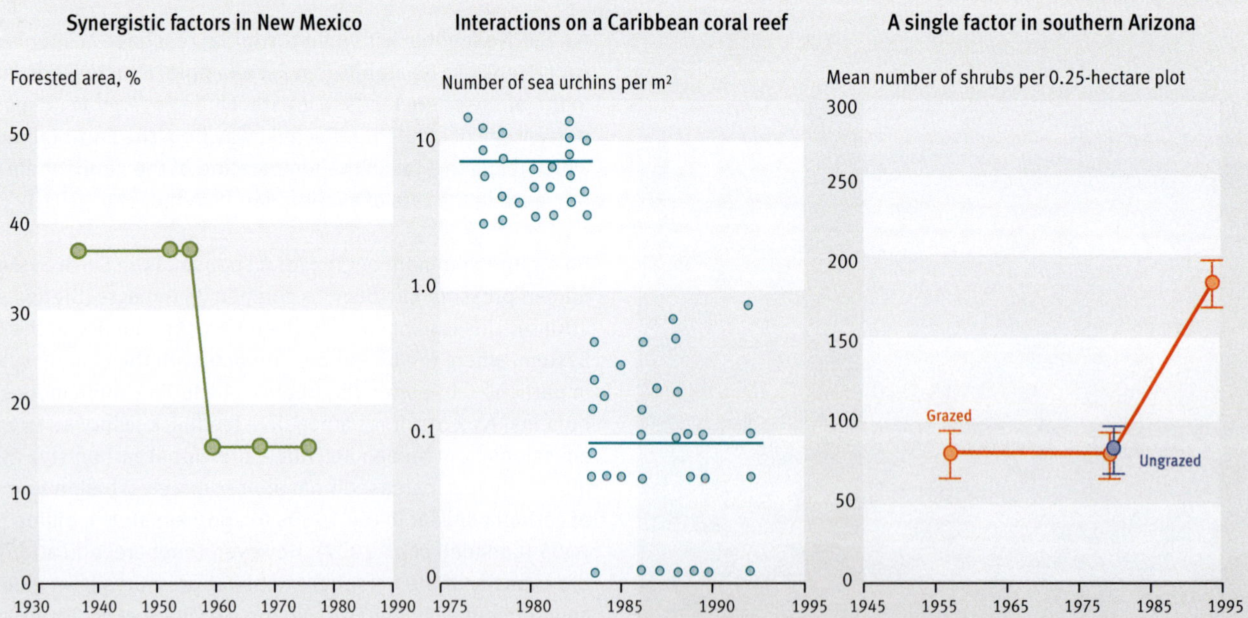

Note: Establishing that the grazed and ungrazed areas in southern Arizona were equally affected by changes in precipitation patterns helped to eliminate grazing as a factor influencing the regime shift.

Sources: de Young *et al.* 2008 (Caribbean); Allen and Breshears 1998 (New Mexico); Brown *et al.* 1997 (Arizona)

EARTH SYSTEM CHANGES AND IMPLICATIONS FOR HUMAN WELL-BEING

The key Earth System changes discussed above have impacts on the environment, on economies and on societies. Illustrative and by no means comprehensive examples of these impacts follow, showing the interconnectedness of the Earth System and the effects that human activities and environmental change have at all scales.

Polar regions

Many of the complex changes in global environmental conditions tend to amplify in the polar regions. For example, heat flux from the lower latitudes leads to accelerating melt of sea ice as well as a loss of mass in Arctic glaciers and the Greenland and Antarctic ice sheets, all of which contribute to global sea level rise. There are many ways in which the polar regions affect the lower latitudes and the whole globe.

The Arctic

As shown in Figure 7.3, the Arctic region has experienced the largest increase in surface temperatures in recent decades (shown in red). This amplification of global warming has been confirmed by instrumental records and reconstruction of past climates, and has been demonstrated in climate model simulations (Serreze and Barry 2011).

The amplification is caused by several factors including heat transport to the Arctic (Graversen *et al.* 2008); melt of sea ice (Screen and Simmonds 2010) enhanced by black carbon depositions on snow (Hansen and Nazarenko 2004) and a corresponding reduction in albedo; an increase in long-wave (infrared) energy emitted downward by the atmosphere (Francis and Hunter 2006); and an increase in heat-absorbing black carbon aerosols in the atmosphere (Chapter 2) (Shindell and Faluvegi 2009). The rapid shrinking of Arctic sea-ice cover is part of the positive climate feedback. In addition to the reduction in the area covered by ice over the past 30 years, as shown by satellite data, substantial loss is also taking place in the oldest and thickest ice (Maslanik *et al.* 2011). The disappearance of sea ice, which acts as a thermal insulator between the ocean and atmosphere, results in an enhanced upward heat flux that warms the lower troposphere in the Arctic, affecting the general atmospheric circulation in a significant part of the northern hemisphere (Serreze and Barry 2011; WWF 2010). This alters storm tracks, precipitation patterns and the conditions that lead to heat waves and cold spells. For example, the emerging atmospheric pattern of a warm Arctic Ocean and cold continents favours more frequent and severe Arctic air outbreaks during the cold season (Petoukhov and Semenov 2010), affecting the well-being of hundreds of millions of people living in the mid-latitudes of the northern hemisphere.

Figure 7.3 Observed change in annual mean surface air temperature, 1960–2009

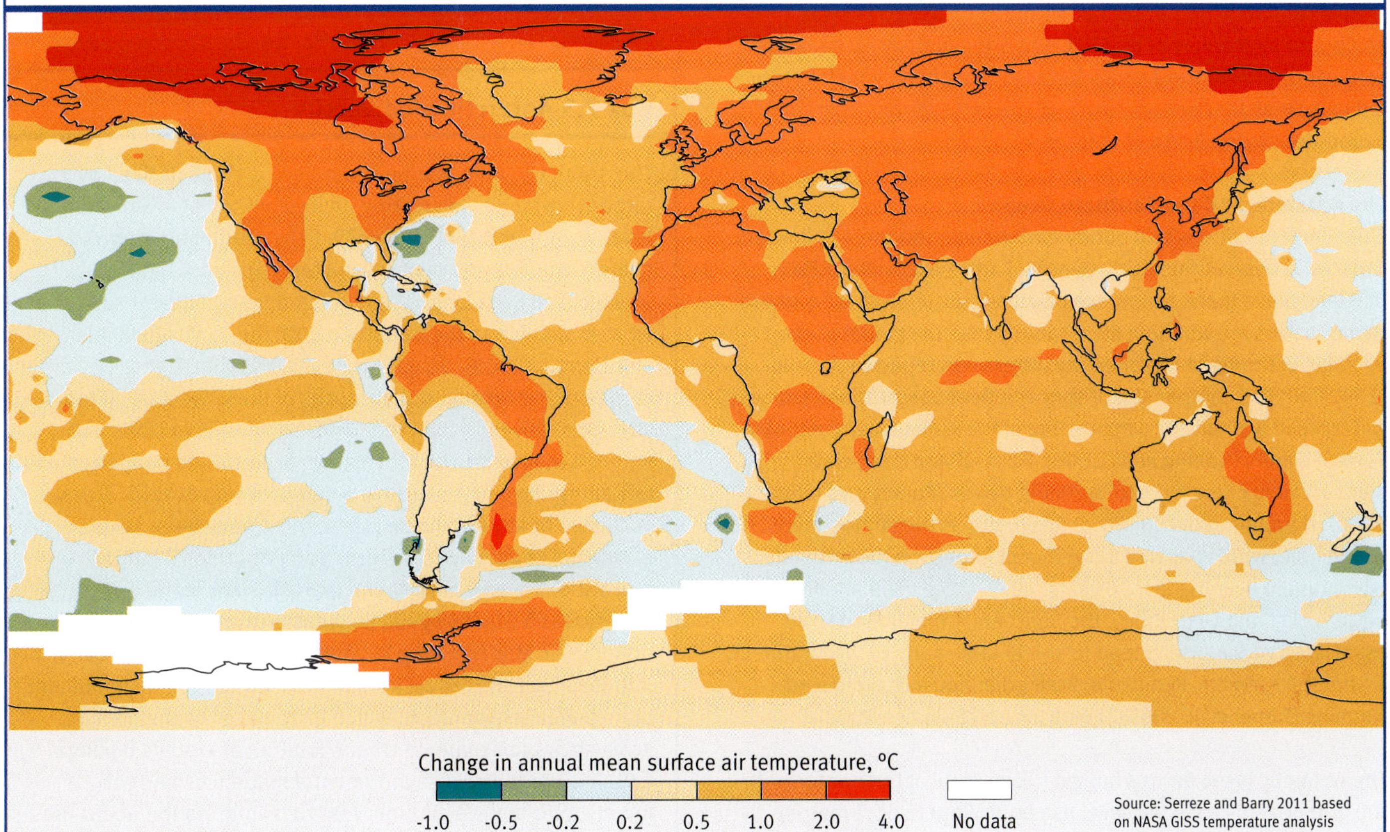

Source: Serreze and Barry 2011 based on NASA GISS temperature analysis (http://data.giss.nasa.gov/gistemp).

The Arctic Council's *Snow, Water, Ice and Permafrost in the Arctic* assessment (SWIPA 2011) shows that temperatures in the permafrost have risen by as much as 2°C over the past two to three decades, particularly at colder sites. Warming in the Arctic causes permafrost thaw and loss (Lawrence *et al.* 2008), as seen in the increased depth of seasonally thawing soil above the permafrost in Scandinavia, Arctic Russia west of the Urals, and inland Alaska; the 30–80 km northward retreat of the southern limit of permafrost in Russia between 1970 and 2005; and the 130 km retreat in Quebec, Canada, during the past 50 years (ACIA 2004). A regional process of permafrost thawing supports increased microbial activity and is likely to lead to the release of carbon that is currently locked up in frozen soils (Tarnocai *et al.* 2009), initiating a global positive climate feedback (Schaefer *et al.* 2011). It is possible that by 2030 the Arctic will become a carbon source rather than a carbon sink (Schaefer *et al.* 2011).

The warming and opening of Arctic waters also has implications for the accessibility of hydrocarbons and other natural resources (Stephenson *et al.* 2011). The production of oil and gas and increased shipping could turn the Arctic into an area of rapid industrial development, leading to additional anthropogenic emissions of carbon. This is another example of a positive climate feedback, involving both natural influences (the greenhouse effect) and social ones (human activities).

Another unexpected manifestation of the global and regional links in the Earth System was observed in the Arctic in boreal spring 2011. An unprecedented stratospheric ozone loss of approximately 80 per cent at altitudes of 18–20 km was attributed by Manney *et al.* (2011) to anomalously long-lasting cold conditions in the Arctic's lower stratosphere, which in turn led to a persistent enhancement of the atmospheric content of ozone-destroying forms of chlorine.

The Antarctic and the Southern Ocean

This remote region is still poorly understood and there is limited capacity to observe the highly complex Earth System interactions that take place there. Numerous observations show that Southern Ocean waters are warming more rapidly than the global ocean average. Warming of intermediate waters was reported by Gille (2002), while a comparison of ship and float observations showed widespread warming and freshening of the Antarctic Circumpolar Current waters (Böning *et al.* 2008). Abyssal and deep-water measurements also indicate warming trends (Purkey and Johnson 2010). A profound peculiarity of this region is the stratospheric ozone hole, which has had a significant impact on the Antarctic environment over the last 30 years, altering the main regional pattern of climate variability, the Southern Annular Mode and associated winds, which tend to shield large parts of the continent, except the Antarctic Peninsula, from greenhouse-gas induced warming (Turner *et al.* 2009; Thompson and Solomon 2002).

The Antarctic is the Earth's largest frozen store of freshwater, with the potential to cause sea level rise equivalent to 61.1 metres (IPCC 2001). While significant parts of the Antarctic ice sheet rest on land, these areas are still below the current mean sea level. For example, the ice body of the West Antarctic ice sheet is in many places more than 1 000 metres below the ocean surface. Recent estimates indicate a potential contribution to global sea level of 3.3 metres by this ice (Bamber *et al.* 2009). Recent airborne geophysical measurements over previously unexplored areas of the East Antarctic ice sheet (Young *et al.* 2011) have shown that it too is largely resting below sea level. There are therefore concerns about marine ice sheet stability in a rapidly warming climate. While recent regional temperature trends in Antarctica have not been very significant and in some locations negative, the Faraday/Vernadsky Station in the northwestern part of the Antarctic Peninsula has observed an increase of 0.53°C per decade for the period 1951–2006 (Turner *et al.* 2009). This local warming and the accompanying changes in winds are considered to be the main causes of the collapse of the Larsen ice shelves A in 1995 and B in 2002. The potential destruction or accelerated melt of the West Antarctic ice sheet under current warming is a subject of intensive research (Huybrechts 2009; Pollard and DeConto 2009).

As a general rule, when atmospheric concentrations of CO_2 increase, the oceans tend to absorb more of it. However, in the Southern Ocean, which represents a significant fraction of the global ocean carbon sink (Takahashi *et al.* 2009), there is a declining ability to absorb CO_2 (Le Quéré *et al.* 2007). One of the likely reasons for this is a 15–20 per cent intensification of the circumpolar westerly winds over the Southern Ocean since the 1970s, which can be partially attributed to the effects of the stratospheric ozone hole (Thompson and Solomon 2002). This phenomenon also has important implications for Antarctic biodiversity (Box 7.3).

Box 7.3 Antarctic biodiversity

The Antarctic Large Marine Ecosystem is characterized by a rather short food web linking phytoplankton at lower trophic levels to Antarctic krill, on which fish, squid, baleen whales, seals, penguins and seabirds all feed (Hill *et al.* 2006). There was uncontrolled, or poorly regulated, harvesting of seals, whales and fish in the period up until the 1980s (Sherman and Hempel 2008). Seals, whales and seabirds were the initial targets of harvesting, but as stocks of these were depleted attention moved to fish, then krill, and finally to crabs and squid. Because much of the earlier exploitation happened rather rapidly and resulted in a series of catastrophic stock collapses of key organisms, there must have been severe ecological consequences (Nicol and Robertson 2006). A precautionary approach to the management of the krill fishery was adopted by the Convention for the Conservation of Antarctic Marine Living Resources (CCAMLR), which came into force in 1982, as an interim measure during development of its preferred approach, which includes enhancement of ecosystem monitoring; further evolution of models linking krill, their predators, environmental influences and the fishery; and high-resolution, real-time information about the behaviour of fishing vessels (Hewitt *et al.* 2001).

The ensemble of climate-chemistry models of the World Climate Research Programme Chemistry-Climate Model Validation project (CCMVal2) indicates that complete recovery of the stratospheric ozone layer resulting from implementation of the Montreal Protocol should occur around the middle of this century (Eyring *et al.* 2010). However, the ozone layer restoration may affect the Southern Annual Mode and associated winds, weakening the currently existing restraint on greenhouse gas-induced warming in Antarctica and the Southern Ocean (Turner *et al.* 2009), and potentially leading to other significant local-to-global changes.

Implications for human well-being

The patterns of change in the polar regions and their links with activities and impacts elsewhere on the globe illustrate the archetypical pattern of vulnerability described in *GEO-4* (UNEP 2007) as "misusing the global commons". This misuse leads to the exposure of people and the environment to resource depletion, for example dwindling fisheries or land in the case of sea level rise, and to environmental transformations such as climate change and sea level rise. Those who are most vulnerable to changes resulting from misuse of the global commons are usually not responsible for the misuse itself.

From an Earth System perspective, the ongoing and potential future changes of the polar regions – given the sometimes long timescales, the interconnections with the rest of the globe, the interactions between problem areas like stratospheric ozone recovery and global warming, and possible catastrophic events such as the melting of the West Antarctic ice sheet – point to the need for holistic responses to managing the global commons in order to reduce the vulnerability of people and the environment to potentially very large pressures.

The Hindu Kush-Himalaya

The Hindu Kush-Himalaya (Figure 7.4), sometimes referred to as the Third Pole, is one of the most dynamic and complex mountain systems in the world. It contains the largest amount of snow and ice found outside the polar regions, including more than 100 000 km^2 of glacier cover, and the sources of ten of the largest rivers in Asia. This mountain system, stretching 3 500 km through some of the world's wettest and driest environments, rising 8 km vertically through nearly every life zone existing on Earth, and at the geographical centre of the largest and densest concentration of humans, is recognized as an extremely fragile environment and particularly vulnerable to global warming (Bates *et al.* 2008; Xu *et al.* 2007).

Extreme vulnerability to natural hazards in the countries of South Asia is cyclical and repeatedly causes major setbacks in the socio-economic and equitable development of the region. Climate change is expected to increase both the frequency and magnitude of extreme weather events that lead to disasters, and calls for speedy action (Cruz *et al.* 2007).

Uncertainties about the rate and magnitude of climate change and its potential impacts prevail, but there is no question that climate change is one of the many pressures that are gradually and powerfully changing the ecological and socio-economic landscape in the Himalayan region. This is particularly true in relation to water and ecosystem services, with significant implications for mountain communities and livelihoods, as well as downstream users, especially women who, for instance, often need to trek longer distances for potable water and fuel (UNEP 2011b). In mountain areas, however, the influence of the changing climate has to be understood in the frame of overall changes due to modernization – communication, transport, infrastructure, monetization and others – and migration (CIDA 2002), which alters traditional gender relations.

Victims of the worst floods to hit Pakistan in several years walk through water-filled streets in the northwestern city of Nowshera.

© Amjad Jamal/UN Photo

Implications for human well-being

Climate change impacts in the Hindu Kush-Himalaya region and downstream are particularly severe due to the large number of people depending on climate-sensitive livelihoods such as agriculture. Here, more than 20 per cent live below the poverty line, amounting to around 260 million people. The International Food Policy Research Institute concluded that the negative impact of climate change on world cereal production may vary from 0.6 per cent to 0.9 per cent per year, but that in South Asia the impact could be as high as 18.2–22.1 per cent by 2080 (von Braun 2007). Recent studies conclude that the Himalayan region and its downstream areas, including the Indo-Gangetic plains, the grain basket of South Asia, are also particularly vulnerable to climate change (Ma *et al.* 2009; Xu *et al.* 2009; Bates *et al.* 2008; Cruz *et al.* 2007; Beniston 2003; Nijssen *et al.* 2001).

Poor and marginalized groups such as mountain populations and the inhabitants of downstream flood plains are particularly vulnerable to climate change. The rough topography of the

Figure 7.4 Recent flood events in the Hindu Kush-Himalaya region

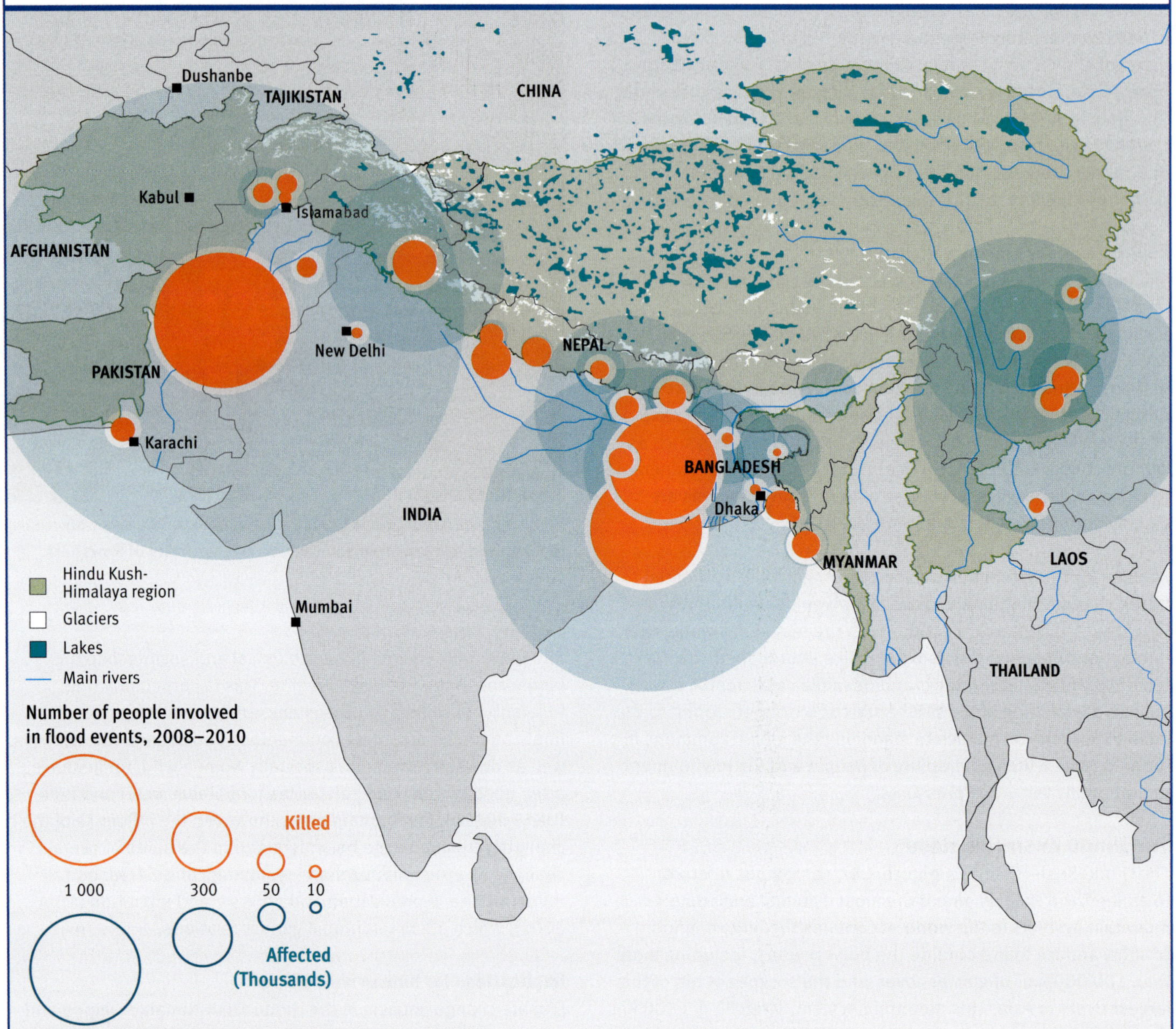

Note: Data for people killed or affected may not have been available for all flood events.

Source: ICIMOD 2010a

Himalayas combined with the precariousness of many homesteads on low incomes, makes the region a particularly flood-sensitive area, with mudslides and unstable ground (Figure 7.4) serious threats to settlement areas. Moreover, the risks of death and destruction are increased by the fact that people, after floods, often rebuild on the same risk-prone areas.

Mountain livelihoods are much more susceptible to environmental and economic upheaval than are livelihoods in the plains, and poverty in the mountains is exacerbated by climate change (ICIMOD 2010b). Women, in particular, are vulnerable to the impacts of climate change and environmental degradation (ICIMOD 2009; Plümper and Neemayer 2007).

The Amazon

The Amazon forest is an extremely important component of the Earth System. It is the repository of the greatest terrestrial diversity of organisms on Earth (Cochrane and Barber 2009; Foley *et al.* 2007), it exchanges vast amounts of water and energy with the atmosphere and thus affects local and regional climates (da Rocha *et al.* 2009), and it is a major carbon sink and reservoir containing 90 billion tonnes of carbon (Chao *et al.* 2009). This is about a fifth of the total carbon contained in the world's tropical forests (Pan *et al.* 2011; Chao *et al.* 2009).

The Amazon has recently experienced two once-in-a-century droughts in the space of five years – 2005 and 2010. Both events caused rapid, widespread tree mortality leading to large increases in carbon emissions in undisturbed regions that are normally net carbon sinks (Lewis *et al.* 2011; Phillips *et al.* 2009). Even under declining deforestation rates, droughts increase vulnerability to fires by enhancing forest flammability and fire spread (Aragão *et al.* 2007). According to model simulations by Vergara and Scholz (2010), in a zero-deforestation scenario the effect of climate change alone, which is predominantly driven by fossil-fuel emissions, could reduce the extent of the Amazon biome by one-third by 2100. But when the effects of deforestation and fire are combined with those of climate change, the models show a much greater reduction.

Deforestation in general, and in the Amazon in particular, is a product of the interactions of multiple socio-economic factors (Chapter 3) in addition to the natural factors discussed above. The largest of these by far is conversion to cattle pasture to satisfy growing international demand for beef (Zaks *et al.* 2009). As discussed in Chapter 3, there remains a basis for optimism regarding the future of the Amazon. For example, Brazil's governmental Action Plan for Prevention and Control of Deforestation in the Amazon (PPCDAm) coordinates a set of government initiatives focused on three goals: monitoring and law enforcement, land tenure regularization, and fostering alternatives for sustainable land use.

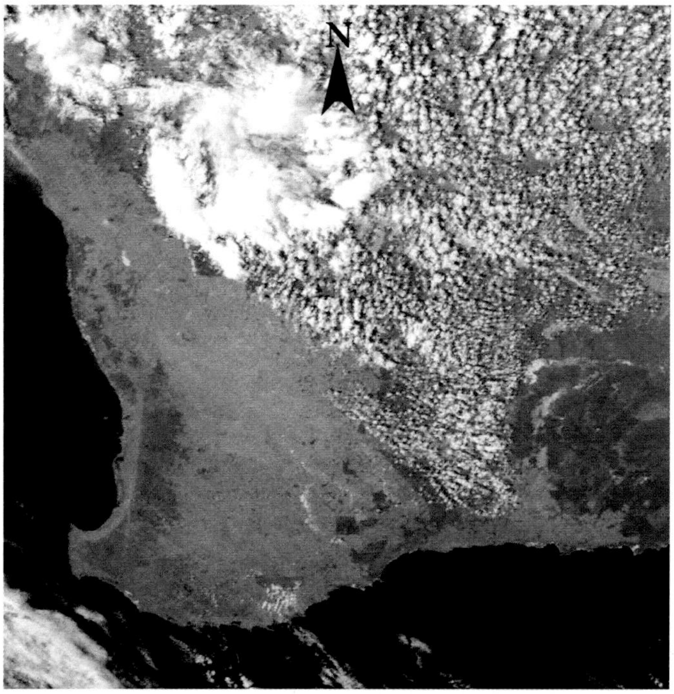

Western Australia's rabbit-proof fence separates native vegetation (east) from farmland and pastures (west). There are more clouds on the darker eastern side of the fence where native vegetation grows.

Source: Pielke *et al.* 2011

Implications for human well-being

In addition to its importance in regulating the global climate through acting as a carbon sink, the Amazon also provides livelihoods for both indigenous peoples and recent settlers (Parry 2008). Foley *et al.* (2007) note that the Amazon system regulates freshwater and river flows, modulates regional climate patterns and controls the spread of vector-borne and water-borne diseases, all of which are crucial to human well-being. Farmers in the Amazon are vulnerable to climate change through the effects of drought, flood and fire on planting times, the spread of disease, and impacts on food, water and human security (Brondizio and Moran 2008).

Drylands

Desertification or land degradation in drylands is one of the greatest environmental challenges facing human society (Chapter 3). The world's drylands – arid, semi-arid and dry sub-humid climatic regions, ranging from deserts to steppes and savannahs – cover approximately 40 per cent of the world's land surface area and are home to nearly 2 billion people (Ezcurra 2006). Degradation of a landscape is a particularly complex problem because it involves a tight coupling of socio-economic, meteorological and ecological processes (Reynolds and Stafford Smith 2002).

Western Australia

One example of how changes in land cover can affect the regional climate in an arid zone is the rabbit-proof fence in Western Australia, built to prevent rabbits from damaging cropland and pastures. The fence spans more than 750 km and separates native vegetation to the east from 13 million hectares of cropland to the west. It did little to protect the crops from rabbits, but it illustrates how vegetation affects climate: there are more clouds and it rains more frequently on the east side of the fence where native vegetation remains (see photo). Several plausible explanations are provided by Nair *et al.* (2011), who measured numerous physical and biological variables on both sides of the fence. They found that these variables differed substantially throughout the year on the agricultural side of the fence whereas only small seasonal variations were found over the native vegetation. Nair and colleagues concluded that the darker surface and greater roughness of the native vegetation resulted in an enhanced heat flux into the atmosphere, which increases the chances of cloud formation. Since measurements began in the 1970s, rainfall observations show about a 20 per cent decline in winter rainfall, confined mainly to agricultural areas.

The Sahel

The Sahel is a large semi-arid region that runs east-west across Africa south of the Sahara, and extends through ten countries. Rainfall is extremely variable and predominantly driven by two major, and undoubtedly interacting, factors: patterns in global sea surface temperature (Biasutti *et al.* 2008) and large-scale changes in land cover that impact land-atmosphere interactions (Huber *et al.* 2011). The role of rainfall variability and vegetation dynamics in the Sahel has been the subject of many high-profile studies, and is particularly important because the population of Sahelian countries is projected to quadruple by 2020 relative to its population of 19 million in 1960 (Brown and Crawford 2009).

Almost 6 000 years ago the Sahel was covered by grasslands and shrublands (Prentice and Jolly 2000; Hoelzmann *et al.* 1998), with records of marine sediments and archaeological evidence showing a switch to arid conditions thereafter (Foley *et al.* 2003; de Menocal *et al.* 2000). More recently, there has been a marked shift from relatively wetter conditions with higher rainfall in the 1950s and 1960s to drier conditions in the 1970s and 1980s, followed by a general trend in increasing precipitation throughout the Sahelian region over the past 30 years (Huber *et al.* 2011), leading to what is generally referred to as a greening trend. Huber *et al.* (2011), however, demonstrate the complexity of this trend, since vegetation changes are not always directly related to precipitation changes.

Implications for human well-being

In the case of Western Australia, land-use change brought unintended consequences. As well as the decrease in rainfall, the removal of deep-rooted native vegetation also led to a rise in the water table, increasing the surface salinity of the farmland and, hence, further decreasing agricultural productivity. As humans continue to clear land for agriculture a paradox is in the making: while food production may increase in the short run, it may be seriously decreased in the long run (Noticewala 2007).

Another consequence of the widespread clearance of native vegetation in Australia for cattle raising and farming was its impact on indigenous peoples who had relied on previously abundant wildlife for their traditional diets. Many groups had little choice but to work on cattle stations and adapt to European foods (Kouris-Blazos and Wahlqvist 2000). This has had a detrimental impact on their nutritional status and well-being, leading to chronic diseases associated with obesity (Wolfenden *et al.* 2011).

Subsistence agriculture is the main source of household livelihoods in many parts of Africa, especially in dryland regions such as the Sahel (Kumssa and Jones 2010). This constitutes a serious food security risk given the complex feedbacks between human activities, land cover and climate. The African Partnership Forum (APF 2007) estimates that 75–250 million people living in the African drylands will be affected by climate change.

Although a greening of the Sahel region is observed, rainfall in the Western part of the region has not increased (Huber *et al.* 2011). A study by Mertz *et al.* (2010) of 1 249 households in five Sahelian countries with annual rainfall ranging from 400 to 900 mm, found that climate factors, mainly inadequate rainfall, are believed by 30–50 per cent of households to be a cause of decreasing rain-fed crop production, whereas a wide range of other factors, such as changes of land tenure, was held responsible for the remaining 50–70 per cent. The differences between the rain-fed crop and livestock sectors, as well as between the driest and wettest zones studied by Mertz *et al.* (2010), illustrate the difficulty faced by people on the agricultural margins in the driest part of the Sahel who are trying to develop their rain-fed agriculture. Adaptation to climate change in the drylands will have to take these complex interactions into account.

Fires

The majority of global biomass burning occurs in the tropics where there are cycles of drought and exceptionally wet years (Liu *et al.* 2010; van der Werf *et al.* 2008; Goldammer and de Ronde 2004). The African continent has the highest occurrence of vegetation fires, accounting for an estimated 30–50 per cent of the total annual biomass burned globally (Roberts and Wooster 2008; Dwyer *et al.* 2000).

Large and uncontrolled fires have increased in the recent past on all vegetated continents and have caused tens of billions of US dollars of damage (Bowman *et al.* 2009). Evidence from the United States and from Canada (Figure 7.5) shows that fire extent in both countries increased significantly over the 20th century. In the western United States, the frequency of large fires has increased nearly fourfold and the extent more than sixfold since the mid-1980s (USGCRP 2009). Recent fires in the Canadian and Alaskan tundra are unprecedented in the last 5 000 years (Hessl 2011). Satellite observations demonstrate a strongly non-linear relationship between climate and human activity: droughts result in more rapid deforestation while also reducing the fire buffering effect of peatland water tables, thereby increasing ecosystem vulnerability to fire (van der Werf *et al.* 2008). Projecting future fire dynamics is challenging because of the non-linearities in the various causal factors (Hessl 2011; Flannigan *et al.* 2009; van der Werf *et al.* 2008) and the unresolved question of whether direct human activity or climate change plays a more dominant role in general (Bowman *et al.* 2009).

Implications for human well-being

Among the significant effects of fires on human well-being are the destruction of assets such as homes (Bowman *et al.* 2009); effects on human health and mortality, as demonstrated by the fires in Russia in 2010; and the lost livelihoods of rural resource-dependent communities, as experienced, for example, in Lebanon in 2007 (IUCN 2008). In addition, the continuation or exacerbation of current global fire trends would have serious

Vermilion River, British Columbia, Canada, the aftermath of a forest fire that devastated the area three years prior. *© Bruce Smith/iStock*

Figure 7.5 Forest fires in Canada, 1920-1999

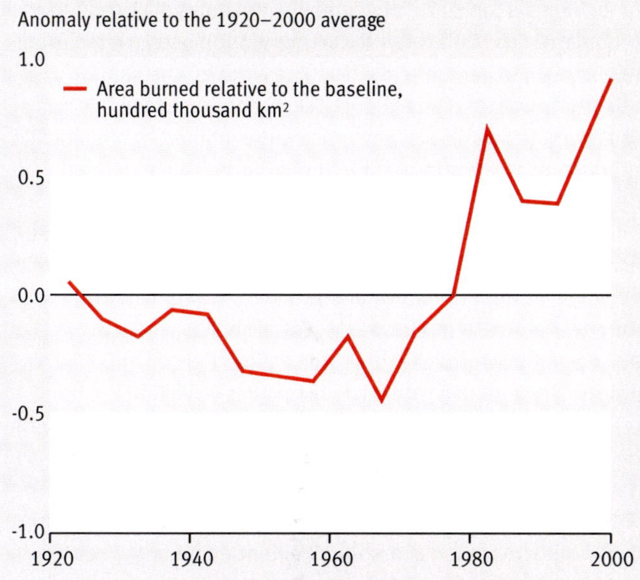

Source: Gillett *et al.* 2004/American Geophysical Union

consequences for the enormous amounts of carbon stored in forests and other ecosystems, potentially unleashing an amplifying climate-carbon feedback and increasing the risk of dangerous climate change.

Shale gas basins

New proven technologies such as directional drilling and hydraulic fracturing have made the extraction of natural gas from low permeability geologic formations (shale formations) economically viable. These practices have accelerated the construction of new natural gas wells and their accompanying infrastructure – pipelines, roads, compressor stations and evaporation ponds – and led to extensive land fragmentation and disturbance, degraded air quality, and degraded surface- and groundwater quality. In parts of both the eastern and western United States, there has been rapid growth in the pace of development as new geologic targets become economical to drill using new technologies. Although to date such shale gas extraction has primarily taken place in the United States, it is expected to reach other parts of the world (Figure 7.6) as use of the new technology expands, and as changing access, profitability compared to other gas resources, and shale gas characteristics make it viable (Kuuskraa and Stevens 2009).

While substituting the burning of coal with natural gas leads to emission reductions and may have some local air quality benefits (Howarth *et al.* 2011), the impacts on air quality near concentrated natural gas development can be quite severe due to the release of hazardous air pollutants such as benzene, ozone-forming precursors and fugitive dust, among others. More broadly, such fuel switching, along with the continuing large-scale development and use of unconventional fossil fuels like shale gas, is likely to exacerbate human-induced climate change because the methane emissions are at least 30 per cent higher than from conventional gas (Howarth *et al.* 2011; Wigley 2011). Furthermore, evaporation ponds used for

Figure 7.6 World shale gas basins identified by the US Energy Information Agency

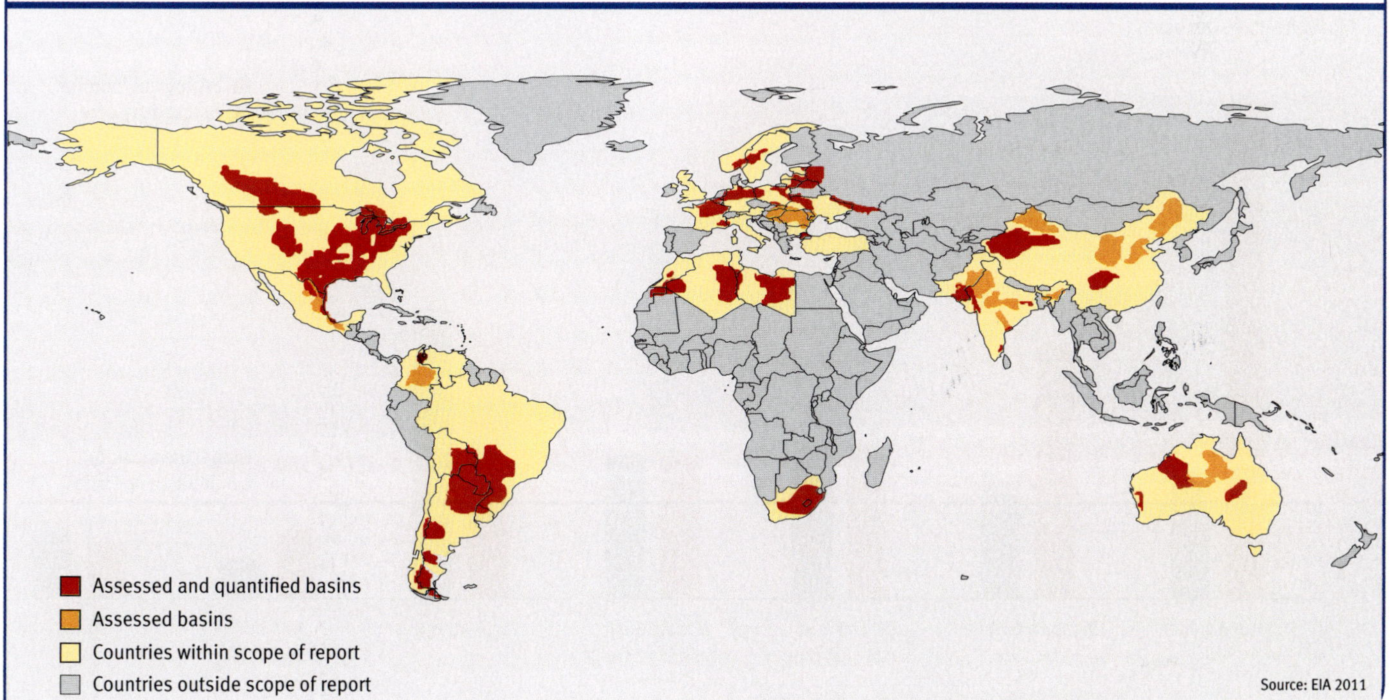

Source: EIA 2011

the disposal of produced water – or coal seam gas water – in the western United States have recently been found to be major sources of volatile organic and hazardous air pollutants (USEPA 2009). The impacts of natural gas development on water resources are also broad-ranging, including the contamination of groundwater aquifers with potentially explosive levels of methane (Osborn *et al.* 2011), of surface- and groundwater with chlorides, metals and organic compounds, and of streams where produced water is discharged (Johnson *et al.* 2008), alongside high rates of consumptive water use for the drilling and completion of wells. The complex nature of many of the geologic formations from which natural gas is recovered may lead to many unknown impacts on groundwater resources. This is of particular concern because of the huge volume of shale gas deposits worldwide (EIA 2011; IEA 2011).

Impacts on human well-being

The chemicals used in hydraulic fracturing, as well as related surface water contamination and air pollution, are thought to be harmful to human health (Finkel and Law 2011).

OVERSHOOT

Scientific understanding of the functioning of the Earth System and recent changes in it indicate a risk of crossing thresholds or tipping points, which would lead to fundamental state changes with major implications for human societies. Such shifts might include the transformation of rainforest to savannah or of hard to soft coral reefs, and changes in rainfall patterns (Box 7.2). The risk of abrupt changes causing regional to global impacts – captured in such concepts as tipping elements (Schellnhuber 2009) and planetary boundaries (Rockström *et al.* 2009a) – is a relatively recent insight from Earth System science. These frameworks

Box 7.4 The ecological footprint

An ecological footprint measures the area of biologically productive land and water that a population uses, with current technology, to generate the resources it consumes and to absorb its wastes. Kitzes *et al.* (2008) compared the ecological footprint with available biocapacity, the amount of biologically productive area available to a population within a defined geographical area. Figure 7.7 shows that in North America and Western Europe the footprint exceeds biocapacity. Kitzes *et al.* (2008) concluded that if everyone in the world had an ecological footprint equivalent to that of the typical North American or Western European, global society would overshoot the planet's biocapacity three- to fivefold. Central and Eastern Europe together live within the biocapacity available in that region, but with a per-person level of consumption that cannot be sustainably adopted at a global scale. Conversely, the Asia and Pacific region lives beyond the biocapacity available within its borders, but with a per-person ecological footprint that would not cause overshoot if extended globally. Residents of Africa, on average, use less than the biocapacity available per person either regionally or globally.

Figure 7.7 The ecological footprint and biocapacity of regions, 2002

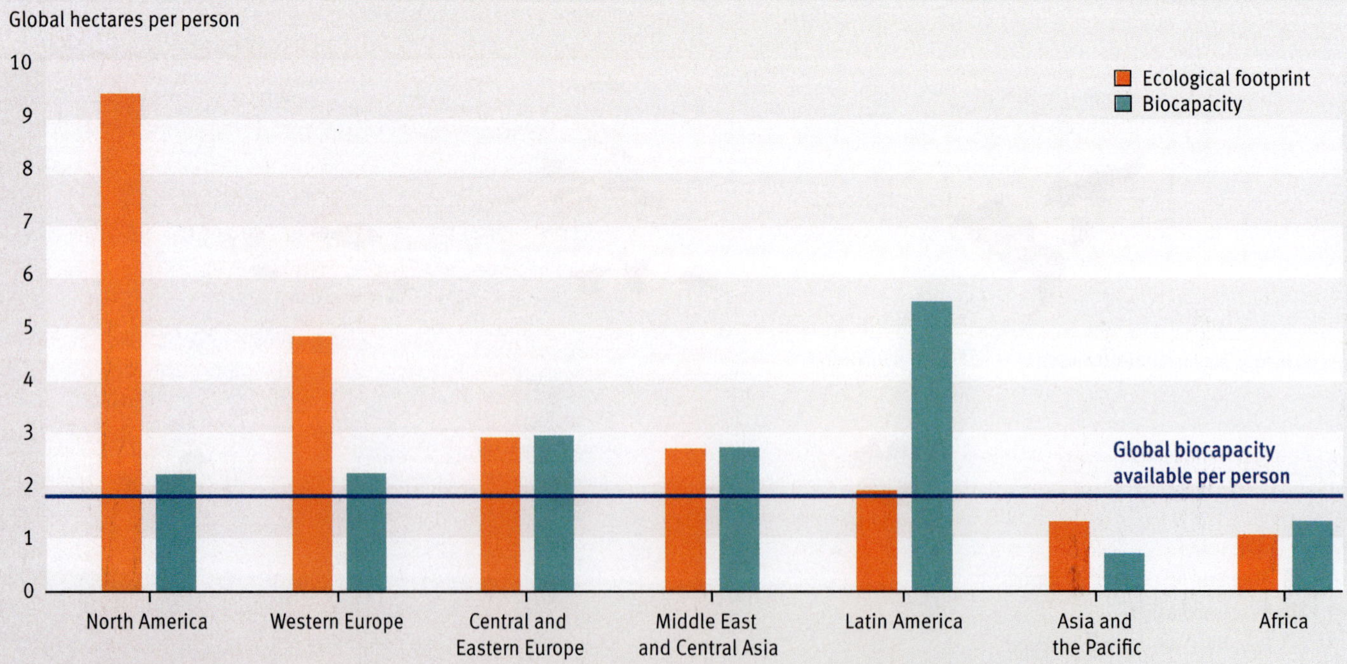

Figure 7.8 Global material extraction, 1900–2005

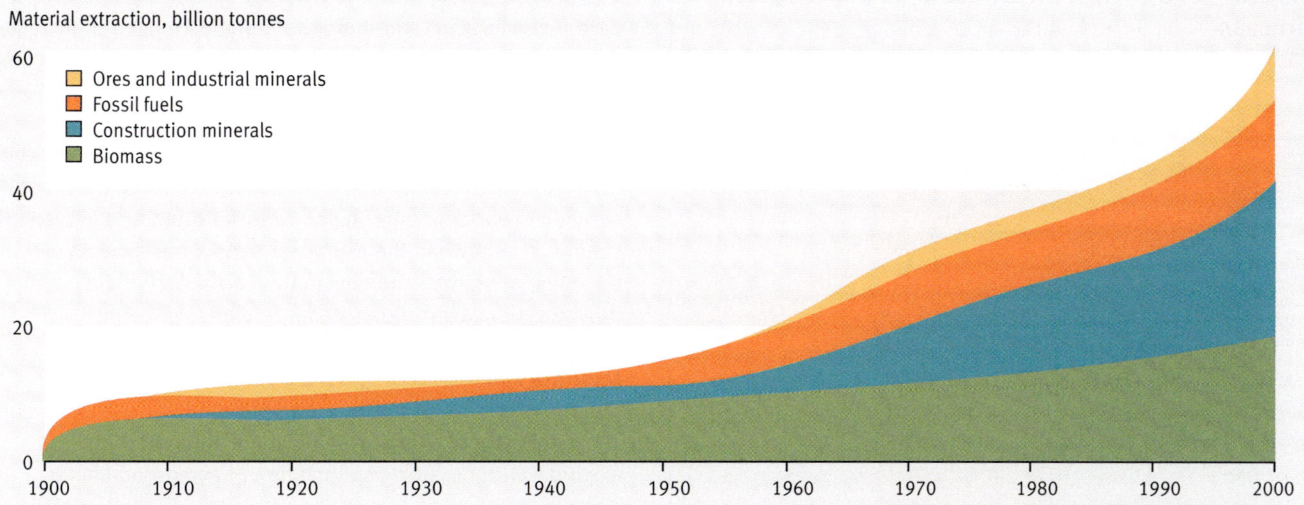

Source: Krausmann *et al.* 2009

for global sustainability complement previous concepts such as limits to growth, carrying capacity, ecological footprint and overshoot, which have in common an estimate of stocks of natural resources and critical loads of different pollutants predominantly in relation to health. Underlying all of these descriptors are different methods and assumptions for identifying points at which the ability of the Earth System to absorb anthropogenic changes is exceeded. While these methods and assumptions are still being discussed in the scientific literature, their conclusions all point in the same direction: thresholds in the Earth System are being reached and the consequences are significant.

Forty years ago, Meadows *et al.* (1972) argued in *The Limits to Growth* that unchecked consumption and economic growth on a finite planet were leading the Earth towards overshoot of its carrying capacity, which would be followed by major impacts on the global economy. Hall and Day (2009) looked back at the conclusions of this study and found that its warnings were generally correct. Turner (2008) compared historical data for 1970–2000 with scenarios presented in *The Limits to Growth* and found that 30 years of historical data compared favourably with key features of the business-as-usual scenario, which results in collapse of the global system midway through the 21st century.

The ecological footprint (Chapter 5) is used to understand human demand on the biosphere and the Earth's biocapacity. While there is still a need for improved data, humanity's ecological footprint overall has doubled since 1966, with major regional disparities (WWF 2010). Box 7.4 and Figure 7.7 show the significant regional differences in both footprint and available biocapacity, indicating which regions use more biocapacity than is available. Of significance with regard to overshoot is the ecological footprint of megacities.

Another approach that points to Earth System limits looks at resource use (UNEP 2011a). Material flow accounting, which quantifies all materials used in economic activities, accounts for the total material mobilized during extraction and for the materials actually used in economic processes measured in terms of their mass (tonnes). At the beginning of the 21st century, estimates of the quantity of global raw materials extracted ranged between 47 and 59 billion tonnes per year (Figure 7.8 shows the higher estimates), with global annual material extraction having increased by a factor of eight during the 20th century (UNEP 2011a). Three scenarios developed by UNEP (2011a) all demonstrate that without significant improvements in resource productivity, it will not be possible to meet the needs of 9 billion people by 2050.

Recognizing interactions and non-linear dynamics within the Earth System, the concept of planetary boundaries was introduced by Rockström *et al.* (2009a) to identify those key environmental processes that provide humanity with a safe operating space for well-being. Rockström *et al.* (2009b) identified nine planetary processes and proposed safe boundaries for seven of them – climate change, rate of biodiversity loss, the nitrogen and phosphorous cycle, stratospheric ozone depletion, ocean acidification, global freshwater use and change in land use. The proposed boundary position was placed at what was considered to be a safe distance from the risk of critical feedbacks and non-linear shifts that could trigger deleterious changes in critical environmental systems (Figure 7.9). The safe position for a process was based on an assessment of the current state of science, acknowledging that there is and will always be an uncertainty range for environmental risks. The safe boundary level for each environmental process was selected for the lower end of this scientific uncertainty range to reflect a precautionary approach (Rockström *et al.* 2009a,

Figure 7.9 Conceptual description of planetary boundaries where the boundary is set to avoid the crossing of a critical threshold in an Earth System process

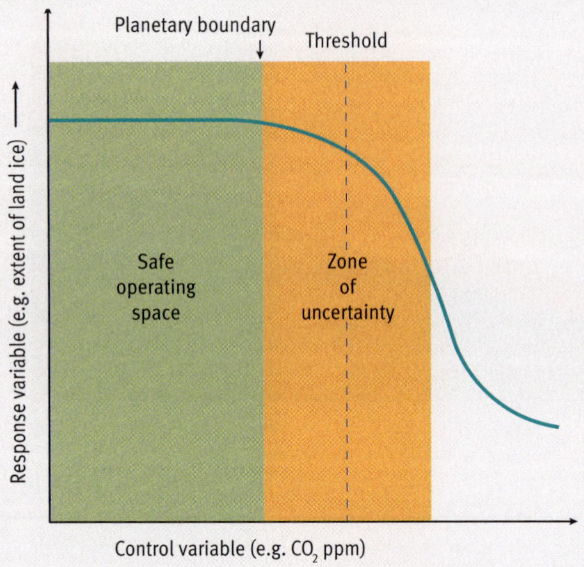

Source: Stockholm Resilience Centre 2009

2009b). The boundaries relate to rates and processes driven by human activities and not to resource scarcities.

The planetary boundaries framework indicates that environmental challenges at the global scale go well beyond climate change. Furthermore, evidence indicates that the processes analysed interact: transgressing one safe boundary may affect the distance from others. For example, expanding agricultural land may undermine the climate change boundary by increasing carbon emissions from terrestrial ecosystems. While the specific numbers used in the planetary boundaries analysis may be challenged (Nature 2009), the approach provides input to the debate about overshoot: early analyses indicate that humanity has already transgressed three boundaries – climate change, the rate of biodiversity loss, and global interference with the nitrogen cycle (Rockstrom *et al.* 2009b). A recent follow-up on human interference with the phosphorous cycle indicates that the phosphorous boundary has also been exceeded in freshwater systems (Carpenter and Bennett 2011).

Implications for human well-being

Ecosystems are essential for human well-being through their provisioning, regulating, supporting and cultural services (TEEB 2010; MA 2005). Human well-being refers to the extent to which individuals have the ability to live the kind of lives they value and the opportunities to achieve their potential (UNEP 2007), and is determined by a range of factors including access to resources, not just financial ones, security, good health and social relations (Introduction). All of these factors are affected by changes in the Earth System. Global interconnectedness in the human-environment system also means that well-being in one place may be affected by practices elsewhere. Chapters 2–6 provide examples of how changes in subsystems of the Earth System affect well-being, although they generally refer to aspects such as food and water security rather than assets, social cohesion and personal security.

From an Earth System perspective, it is important to consider the consequences for human well-being of exceeding the planet's carrying capacity or entering periods of abrupt and irreversible change. As shown in the examples below, the impacts of complex, non-linear changes in the Earth System are already having serious consequences for human well-being.

Multiple and interacting drivers affect human security

Climate variability and extreme weather influence food security. These drivers are complex and involve different pathways (regional water scarcity, salinization of agricultural lands, destruction of crops through flood events, disruption of food logistics through disasters, and increased burden of infectious plant diseases or pests) (IPCC 2007).

Crossing thresholds: significant health impacts

Land-use change and deforestation alter habitats by raising local temperatures and removing shade, changes that can facilitate the rapid development of malaria vectors (da Silva-Nunes *et al.* 2008; Afrane *et al.* 2005). Pascual *et al.* (2006) demonstrated the importance of the well-recognized non-linear and threshold responses of malaria (a biological system) to the effect of regional temperature change.

Unprecedented events affect assets and human security

Climate change already undermines human security and will do so increasingly in the future, by reducing access to, and the quality of, natural resources that are important to sustain livelihoods (Barnett and Adger 2007). In Bangladesh, for example, a significant number of people are affected every year by riverbank erosion and floods that lead to loss of agricultural land, infrastructure and communication systems. These assets are essential for maintaining livelihoods (Poncelet *et al.* 2010).

Rapid change and indigenous communities

As discussed, the Arctic is warming faster than elsewhere on the planet. Since 1975, temperatures in Alaska have increased by an average of 2.0–3.5°C. Approximately 200 indigenous villages along the navigable waters of Alaska's coasts and rivers are threatened by accelerated rates of erosion or flooding, and five communities have concluded that relocation is the only solution. Studies show that displacement has considerable cultural, social, economic and psychological impacts (Bronen 2010).

TRANSITIONS AND SYSTEMIC RESPONSES TO EARTH SYSTEM CHALLENGES

Earth System challenges have been characterized as "persistent problems of unsustainability" that are "… complex, ill-structured, involve many stakeholders, are surrounded by structural uncertainties, and are hard to manage" (Rotmans 2006). Persistent problems tend to reappear when only their symptoms are treated or when the measures taken are only marginal and incremental, and thus inadequate to deal with root causes. For each of the different issues discussed in Chapters 2–6 – climate change, land degradation, biodiversity loss, water security and chemical pollution – symptoms of unsustainability mask deeper underlying problems in societal structures and institutions.

The persistence of the problems is due to what Rotmans (2006) refers to as system failures:

- institutional system failures – a predominance of institutions that block innovation;
- economic system failures – inadequate market development or lack of investment capital;
- social system failures – entrenched behaviour;
- ecological system failures – regime shifts described earlier in this chapter.

Transition management

Addressing these system failures requires new, innovative forms of governance, including transition management (Chapter 16) (Grin *et al.* 2010). Ultimately, ignoring these systemic failures will result in non-linear, systemic and fundamental changes in the composition and functioning of the societal system, with changes in structures, cultures and practices (Loorbach and Rotmans 2010).

Transitional change is different from normal or gradual societal change, as shown in Figure 7.10, and occurs in four phases: pre-development, take-off, acceleration and stabilization. Between the pre-development and acceleration phases, a moment of take-off is often recognized, tipping the system over a point of no return (Frantzeskaki and de Haan 2009).

Figure 7.10 Transition phases

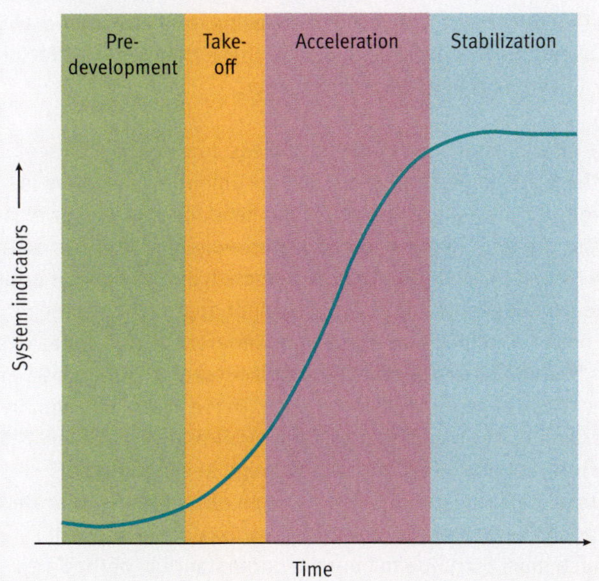

Source: Frantzeskaki and de Haan 2009

Reducing large vessels in certain areas in Chile provided an innovative solution to fisheries management. *© Joris Van Ostaeyen*

Historical transitions, such as the post-1950 emergence of personal mobility, intensive agriculture or fossil energy infrastructure, were partly driven by the promise of solving societal problems such as poverty, inequality, lack of education and so on. However, these transitions have, in turn, produced their own problems. While individuals might now have access to cheap energy and mobility, the results are pollution, resource exploitation and congestion. The challenge in dealing with complex and persistent modern problems is to find new ways of dealing with them in a more anticipatory and exploratory manner. It is necessary to improve understanding of the dynamics of complex processes of change and try to influence their pace and direction.

A more thorough understanding of the forces that drive societal transitions is essential in policy making for the Earth System (Frantzeskaki and de Haan 2009). While current mainstream policy and research approaches predominantly seek to improve existing systems, leading to gradual improvement, transitions thinking requires a fundamental shift. The current trend of marginal improvements and optimization of existing systems leads to lock-ins not only in technological systems but also in policy and consequently social systems (Frantzeskaki and Loorbach 2010) that divert society from sustainability. Escaping such lock-ins requires radical shifts – transformative changes (Chapters 16 and 17) – that fundamentally alter structures, cultures and practices to achieve sustainability in the long term.

As transitions are increasingly likely to occur given the instabilities in both socio-economic and ecological systems, it is crucial that strategies be developed to influence these transitions effectively in terms of their speed and direction (Loorbach *et al.* 2011; Loorbach and Rotmans 2006). While steering a transition in a command-and-control manner is not workable, it is possible to influence a transition by using various approaches, including coordinating existing social movements, niche-innovations and new practices in general. The importance

Box 7.5 Innovative response to a crisis

In Chile, a combination of fisheries collapse and the move to democracy provided the opportunity to try some new arrangements for managing fisheries based on informal partnerships and trust between fishers, scientists and managers. There was a general recognition that Chile's fish stocks were in trouble and people were looking for answers, while at the same time societal change had begun. This supported openness to new approaches. Scientific understanding of coastal ecosystems in the region was good and provided the basis for a new management plan and the testing of new cooperative models for fishery management. The result was a national system of marine tenure that allocates exclusive ocean territories to local and small-scale fisheries. By cutting the number of large vessels in distinct territories, fishing pressure was reduced.

Source: Gelcich *et al.* 2010

of long-term stable and systemic monitoring for managing transitions is illustrated by the example in Box 7.5.

Understanding the inevitability of transitions, and learning to govern and manage transition processes, is particularly important given the evidence of Earth System changes (Loorbach *et al.* 2011). New kinds of multi-level change processes are required that involve a dynamic interplay between gradually introduced, top-down changes and self-organizing bottom-up processes of social innovation, because traditional expert-driven, top-down approaches to problem solving are not flexible enough to address complex, non-linear and rapidly changing situations effectively.

Box 7.6 The transition to improved governance of the Great Barrier Reef

Olsson *et al.* (2008) found that in the case of the Great Barrier Reef, management had to be flexible, adaptive and responsive to continual scientific monitoring. This flexibility enabled new interactions and ways of working, with leadership and consensus building also being important. The Great Barrier Reef Marine Park Authority and its Chair played an essential role in seeking and obtaining the support of the public, industry and governments at all levels for putting the management of the world's largest coral reef system on to an ecological footing. A critical step in the process was to enlist public support for managing the reef more flexibly. One of the most visible and controversial initiatives under the new regime was to extend the area closed to all forms of fishing from 6 per cent to 33 per cent of the total reef area, creating the largest no-take zone in the world. The Barrier Reef example illustrates a shift in thinking towards an integrated view of humans and nature based on active stewardship of marine ecosystems for human well-being.

Source: Westley *et al.* 2011

These change processes require the active involvement of agents from science, policy, civil society and business, both in the development of new knowledge and in its application (O'Riordan 2008). The processes are necessarily iterative and involve developing a joint framing of a problem, a shared vision of the future, experimenting with solutions, evaluation and learning. Bottom-up solutions developed in this way should contribute to improving local sustainability and also reinforce the initial top-down changes and support their further extension (Weaver 2011). This is also proposed by the German Advisory Council on Global Change (WBGU 2011), which points to the need to empower the state to determine priorities and underline them with clear signals, while at the same time giving citizens more extensive opportunities to have a voice, to get involved in decision making and to take a more active role in politics.

With regard to bottom-up responses, Westley *et al.* (2011) point out that there are enormous reservoirs for learning and innovation that are often revealed in moments of crisis. Success involves listening to local communities for ideas, informing local populations of the resources and possibilities available, trusting them and allowing a diversity of innovative responses to emerge, as opposed to insisting on a top-down planning process. One of the examples cited by Westley *et al.* (2011) is summarized in Box 7.6.

REFLECTIONS

The Earth System is complex, with many interactions between and within subsystems, feedbacks and non-linearity. Humans, as an integral part of the Earth System, are changing it through their sheer numbers and their activities, although the impacts of these changes are not uniformly distributed, with some people and places more affected than others. As a result of the enormous complexity of the system as a whole, it is not possible to predict the outcomes of rapidly increasing human pressures on the Earth System, but it is clear that thresholds have been or are being reached, beyond which abrupt and irreversible changes occur. These changes will affect the basic life-support functions of the planet.

While efforts have been made to address some of the changes, and there are some success stories documented elsewhere in this report, this analysis points to the need for approaches that address the underlying drivers of anthropogenic pressures on the Earth System, especially population growth and overconsumption. At the same time, it is necessary to adopt approaches that can deal better with the complexities and inherent uncertainties of the Earth System. Such approaches are discussed in Chapters 15 and 16 as adaptive governance processes. These must, however, be underpinned by sustained long-term monitoring and observation of all relevant aspects of the system, regular evaluation of progress, and adjustment of goals when observations indicate that this is necessary. At the same time, basic and applied research must continue to improve understanding of the Earth System and make this knowledge available in the search for solutions to persistent problems of unsustainability.

REFERENCES

ACIA (2004). *Impacts of a Warming Arctic: Arctic Climate Impact Assessment.* Cambridge University Press, Cambridge

Afrane, Y.A., Lawson, B.W., Githeko, A.K. and Yan, G. (2005). Effects of microclimatic changes caused by land use and land cover on duration of gonotrophic cycles of *Anopheles gambiae* (Diptera: Culicidae) in Western Kenya Highlands. *Journal of Medical Entomology* 42, 974–980

Allen, C.D. and Breshears, D.D. (1998). Drought-induced shift of a forest-woodland ecotone: rapid landscape response to climate variation. *Proceedings of the National Academy of Sciences of the United States of America* 95, 14839–14842

APF (2007). *Climate Change and Africa.* Document prepared jointly by the African Partnership Forum (APF) and the Secretariat of the New Partnership for Africa's Development (NEPAD) for the 8th APF Meeting in Berlin, 22–23 May, 2007

Aragão, L.E.O.C., Malhi, Y., Roman-Cuesta, R.M., Saatchi, S., Anderson, L.O. and Shimabukuro, Y.E. (2007). Spatial patterns and fire response of recent Amazonian droughts. *Geophysical Research Letters* 34, L07701

Bamber, J.L., Riva, R.E.M., Vermeersen, B.L.A. and LeBrocq, A.M. (2009). Reassessment of the potential sea-level rise from a collapse of the West Antarctic ice sheet. *Science* 324, 901–903

Barnett, J. and Adger, W.N. (2007). Climate change, human security and violent conflict. *Political Geography* 26, 639–655

Barnosky, A.D., Matzke, N., Tomiya, S., Wogan, G.O.U., Swartz, B., Quental, T.B., Marshall, C., McGuire, J.L., Lindsey, E.L., Maguire, K.C., Mersey, B. and Ferrer, E.A. (2011). Has the Earth's sixth mass extinction already arrived? *Nature* 471, 5–57

Bates, B.C., Kundzewicz, Z.W., Wu, S. and Palutikof, J.P. (eds.) (2008). *Climate Change and Water.* IPCC Technical Paper VI. IPCC Secretariat, Geneva

Beniston, M. (2003). Climatic change in mountain regions: a review of possible impacts. *Climatic Change* 59, 5–31

Biasutti, M., Held, I.M., Sobel, A.H. and Giannini, A. (2008). SST forcings and Sahel rainfall variability in simulations of the twentieth and twenty-first centuries. *Journal of Climate* 21, 3471–3486

Böning, C.W., Dispert, A., Visbeck, M., Rintoul, S.R. and Schwarzkopf, F. (2008). The response of the Antarctic Circumpolar Current to recent climate change. *Nature Geoscience* 1, 864–869. doi: 10.1038/ngeo362

Bowman, D.M.J.S., Balch, J.K., Artaxo, P., Bond, W.J., Carlson, J.M., Cochrane, M.A., D'Antonio, C.M., DeFries, R.S., Doyle, J.C., Harrison, S.P., Johnston, F.H., Keeley, J.E., Krawchuk, M.A., Kull, C.A., Marston, J.B., Moritz, M.A., Prentice, I.C., Roos, C.I., Scott, A.C., Swetnam, T.W., van der Werf, G.R. and Pyne, S.J. (2009). Fire in the Earth system. *Science* 324, 481–484

Briggs, R., Carpenter, S.R. and Brock, W.A. (2009). Turning back from the brink: detecting an impending regime shift in time to avert it. *Proceedings of the National Academy of Sciences of the United States of America* 106(3), 826–831

Brondizio, E.S. and Moran, E.F. (2008). Human dimensions of climate change: the vulnerability of small farmers in the Amazon. *Philosophical Transactions of the Royal Society B* 363, 1803–1809

Bronen, R. (2010). Forced migration of Alaskan indigenous communities due to climate change. In *Environment, Forced Migration and Social Vulnerability* (eds. Afifi, T. and Jäger, J.). pp.87–98. Springer Verlag, Berlin

Brown, O. and Crawford, A. (2009*). Climate Change and Security in Africa.* A study for the Nordic-African Ministers of Foreign Affairs Forum, 2009. International Institute for Sustainable Development (IISD), Winnipeg

Brown, J.H., Valone, T.J. and Curtin, C.G. (1997). Reorganization of an arid ecosystem in response to recent climate change. *Proceedings of the National Academy of Sciences of the United States of America* 94, 9729–9733

Canadell, J.G., Le Quéré, D., Raupach, M.R., Field, C.R., Buitenhuis, E., Ciais, P., Conway, T.J., Gillett, N.P., Houghton, R.A. and Marland, G. (2007). Contributions to accelerating atmospheric CO2 growth from economic activity, carbon intensity, and efficiency of natural sinks. PNAS 104, 18866–18870

Carpenter, S.R. and Bennett, E.M. (2011). Reconsideration of the planetary boundary for phosphorus. *Environmental Research Letters* 6, 014009. doi: 014010.011088/011748-019326/014006/014001/014009

CBD (2010). *Global Biodiversity Outlook 3.* Secretariat of the Convention on Biological Diversity, Montreal

Chao, K.J., Phillips, O.L., Baker, T.R., Peacock, J., Lopez-Gonzalez, G., Vásquez Martínez, R., Monteagudo, A. and Torres-Lezama, A. (2009). After trees die: quantities and determinants of necromass across Amazonia. *Biogeosciences* 6, 1615–1626

CIDA (2002). *Gender Equality and Climate Change: Why Consider Gender Equality when Taking Action on Climate Change?* Canadian International Development Agency (CIDA), Hull

Cochrane, M.A. and Barber, C.P. (2009). Climate change, human land use and future fires in the Amazon. *Global Change Biology* 15, 601–612

Crutzen, P.J. (2002). Geology of mankind. *Nature* 415, 23–23

Cruz, R.V., Harasawa, H., Lal, M., Wu, S., Anokhin, Y., Punsalmaa, B., Honda, Y., Jafari, M., Li, C. and Huu Ninh, N. (2007). Asia. In *Climate Change 2007: Impacts, Adaptation and Vulnerability. Contribution of Working Group II to the Fourth Assessment Report of the Intergovernmental Panel on Climate Change* (eds. Parry, M.L., Canziani, O.F., Paulutkof, J.P., van de Linden, P.J. and Hanson, C.E.). pp.469–506. Cambridge University Press, Cambridge

da Rocha, H.R., Manzi, A.O., Cabral, O.M., Miller, S.D., Goulden, M.L., Saleska, S.R., R.-Coupe, N., Wofsy, S.C., Borma, L.S., Artaxo, P., Vourlitis, G., Nogueira, J.S., Cardoso, F.L., Nobre, A.D., Kruijt, B., Freitas, H.C., von Randow, C., Aguiar, R.G. and Maia, J.F. (2009). Patterns of water and heat flux across a biome gradient from tropical forest to savanna in Brazil. *Journal of Geophysical Research* 114, G00B12

da Silva-Nunes, M., Codeço, C.T., Malafronte, R.S., da Silva, N.S., Juncansen, C., Muniz, P.T. and Ferreira, M.U. (2008). Malaria on the Amazonian frontier: transmission dynamics, risk factors, spatial distribution, and prospects for control. *American Journal of Tropical Medicine and Hygiene* 79(4), 624–35

de Menocal, P., Ortiz, J., Guilderson, T., Adkins, J., Sarnthein, M., Baker, L. and Yarusinsky, M. (2000). Abrupt onset and termination of the African humid period: rapid climate responses to gradual insolation forcing. *Quarternary Science Reviews* 19, 347–61

de Young, B., Barange, M., Beaugrand, G., Harris, R., Perry, R.I., Scheffer, M. and Werner, F. (2008). Regime shifts in marine ecosystems: detection, prediction and management. *Trends in Ecology and Evolution* 23, 402–409

Dwyer, E., Pinnock, S., Grégoire, J.-M. and Pereira, J.M.C. (2000). Global spatial and temporal distribution of vegetation fire as determined from satellite observations. *International Journal of Remote Sensing* 21(6/7), 1289–1302

EIA (2011). *World Shale Gas Resources: An Initial Assessment of 14 Regions outside the United States.* US Energy Information Administration, Washington, DC

Eldredge, N. (2001). *The Sixth Extinction.* American Institute of Biological Sciences. http://www.actionbioscience.org/newfrontiers/eldredge2.html (accessed 16 September 2011)

Estes, J.A., Terborgh, J., Brashares, J.S., Power, M.E., Berger, J., Bond, W.J., Carpenter, S.R., Essington, T.E., Holt, R.D., Jackson, J.B.C., Marquis, R.J., Oksanen, L., Oksanen, T., Paine, R.T., Pikitch, E.K., Ripple, W.J., Sandin, S.A., Scheffer, M., Schoener, T.W., Shurin, J.B., Sinclair, A.R.E., Soulé, M.E., Virtanen, R. and Wardle, D.A. (2011). Trophic downgrading of planet Earth. *Science* 333, 301–306

Eyring, V., Shepherd, T.G. and Waugh, D.W. (2010). *SPARC Report on Evaluation of Chemistry-Climate Models.* SPARC Report No. 5. Stratospheric Processes And Their Role In Climate. WCRP-132, WMO/TD-No. 1526

Ezcurra, E. (ed.) (2006). *Global Deserts Outlook.* Division of Early Warning and Assessment, United Nations Environment Programme, Nairobi

Finkel, M.L. and Law, A. (2011). The rush to drill for natural gas: a public health cautionary tale. *American Journal of Public Health* 101, 784–785

Flannigan, M.D., Krawchuk, M.A., de Groot, W.J., Wotton, B.M. and Gowman, L.M. (2009). Implications of changing climate for global wildland fire. *International Journal of Wildland Fire* 18, 483–507

Foley, J., Asner, G., Costa, M., Coe, M., Defries, R., Gibbs, H., Howard, E., Olson, S., Patz, J., Ramankutty, N. and Snyderf, P. (2007). Amazonia revealed: forest degradation and loss of ecosystem goods and services in the Amazon Basin. *Frontiers in Ecology and the Environment* 5, 25–32

Foley, J.A., Coe, M.T., Scheffer, M. and Wang, G.L. (2003). Regime shifts in the Sahara and Sahel: interactions between ecological and climatic systems in northern Africa. *Ecosystems* 6(6), 524–539

Folke, C., Jansson, Å., Rockström, J., Olsson, P., Carpenter, S.R., Chapin, F.S., Crépin, A.-S., Daily, G., Danell, K. and Ebbesson, J. (2011). Reconnecting to the biosphere. *Ambio.* doi: 10.1007/s13280-011-0184-y

Folke, C., Carpenter, S., Walker, B., Scheffer, M., Elmqvist, T., Gunderson, L. and Holling, C.S. (2004). Regime shifts, resilience, and biodiversity in ecosystem management. *Annual Review of Ecology Evolution and Systematics* 35, 557–581

Francis, J.A. and Hunter, E. (2006). New insight into the disappearing Arctic sea ice. *Eos, Transactions, American Geophysical Union* 87(46)

Frantzeskaki, N. and de Haan, H. (2009). Transitions: two steps from theory to policy. *Futures* 41, 593–606

Frantzeskaki, N. and Loorbach, D. (2010). Towards governing infrasystem transitions: reinforcing lock-in or facilitating change? *Technological Forecasting and Social Change* 77, 1292–1301

Gelcich, S,. Hughes, T.P., Olsson, P., Folke, C., Defeo, O., Fernández, M., Foale, S., Gunderson, L.H., Rodríguez-Sickert, C., Scheffer, M., Steneck, R.S. and Castilla, J.C. (2010). Navigating transformations in governance of Chilean marine coastal resources. *Proceedings of the National Academy of Sciences of the United States of America* 107(39), 16794–16799. doi:10.1073/pnas.1012021107

Gille, S.T. (2002). Warming of the Southern Ocean since the 1950s. *Science* 295(5558), 1275–1277. doi:10.1126/science.1065863

Gillett, N.P., Weaver, A.J., Zwiers, F.W. and Flannigan, M.D. (2004). Detecting the effect of climate change on Canadian forest fires. *Geophysical Research Letters* 31, L18211. doi:10.1029/2004GL020876

Goldammer , J.G. and de Ronde, C. (eds.) (2004). *Wildland Fire Management Handbook for Sub-Sahara Africa.* Global Fire Monitoring Centre (GFMC), Freiburg

Graversen, R.G., Mauritsen, T., Tjernstrom, M., Kallen, E. and Svensson, G. (2008). Vertical structure of recent Arctic warming. *Nature* 451, 53–56

Grimm, N.B., Faeth, S.H., Golubiewski, N.E., Redman, C.L., Wu, J., Bai, X. and Briggs, J.M. (2008). Global change and the ecology of cities. *Science* 319, 756–760

Grin, J., Rotmans, J. and Schot, J. (2010). *Transitions to Sustainable Development: New Directions in the Study of Long-Term for Transformative Change.* Routledge, New York

Hall, C.A.S. and Day, J.W. (2009). Revisiting the limits to growth after peak oil. *American Scientist* 97(3), 230

Hansen, J. and Nazarenko, L. (2004). Soot climate forcing via snow and ice albedos. *Proceedings of the National Academy of Sciences of the United States of America* 101, 423–428

Hays, J.D., Imbrie, J. and Shackleton, N.J. (1976). Variations in the Earth's orbit: pacemaker of the ice ages. *Science* 194, 1121–1132

Hessl, A.E. (2011). Pathways for climate change effects on fire: models, data, and uncertainties. *Progress in Physical Geography* 35, 393–407

Hewitt, R.P., Watkins J.L., Naganobu, M., Tshernyshkov, P., Brierley, A.S., Demer, D.A., Kasatkina, S., Takao, Y., Goss, C., Malyshko, A., Brandon, M.A., Kawaguchi, S., Siegel, V., Trathan, P.N., Emery, J.H., Everson, I. and Miller, D.G.M. (2001). Setting a precautionary catch limit for Antarctic krill. *Oceanography* 15(3), 26–33

Hill, S.L., Murphy, E.J., Reid, K., Trathan, P.N. and Constable, A.J. (2006). Modelling Southern Ocean ecosystems: krill, the food-web, and the impacts of harvesting. *Biological Reviews* 81, 581–608

Hoelzmann, P., Jolly, D., Harrison, S.P., Laarif, F., Bonnefille, R. and Pachur, H.-J. (1998). Mid-Holocene land-surface conditions in northern Africa and the Arabian Peninsula: a data set for the analysis of biogeophysical feedbacks in the climate system. *Global Biogeochemical Cycles* 12, 35–52

Howarth, R., Santoro, R. and Ingraffea, A. (2011). Methane and the greenhouse-gas footprint of natural gas from shale formations. *Climatic Change* 106, 679–690

Huber, S., Fensholt, R. and Rasmussen, K. (2011). Water availability as the driver of vegetation dynamics in the African Sahel from 1982 to 2007. *Global and Planetary Change* 76, 186–195

Huybrechts, P. (2009). Global change: west-side story of Antarctic ice. *Nature* 458, 295–296

ICIMOD (2010a). *Mountain GeoPortal.* International Center for Integrated Mountain Development, Kathmandu. http://geoportal.icimod.org/Downloads/FreeDataDownloads.aspx

ICIMOD (2010b). *Understanding Mountain Poverty: Exploring the Specificities of Poverty in the Mountain Areas of the Greater Himalayan Region.* International Center for Integrated Mountain Development, Kathmandu

ICIMOD (2009). *Local Responses to Too Much and Too Little Water in the Greater Himalayan Region.* International Center for Integrated Mountain Development, Kathmandu

IEA (2011). *World Energy Outlook 2011 Special Report: Are We Entering a Golden Age of Gas?* International Energy Authority, Paris

IPCC (2007). *Fourth Assessment Report of the Intergovernmental Panel on Climate Change.* Cambridge University Press, Cambridge, United Kingdom and New York, NY

IPCC (2001). *Climate Change 2001: The Scientific Basis.* Intergovernmental Panel on Climate Change. Cambridge University Press, Cambridge

IUCN (2008). *Lebanon's National Forest Fire Management Strategy, Second Draft.* http://cmsdata.iucn.org/downloads/forest_strategy_english_final_may09_1.pdf (accessed 11 August 2011)

Johnson, B., Kanagy, L., Rodgers, J. and Castle, J. (2008). Chemical, physical, and risk characterization of natural gas storage produced waters. *Water, Air, and Soil Pollution* 191, 33–54

Kitzes, J., Wackernagel, M., Loh, J., Peller, A., Goldfinger, S., Cheng, D. and Tea, K. (2008). Shrink and share: humanity's present and future Ecological Footprint. *Philosophical Transactions of the Royal Society B* 363, 467–475

Kouris-Blazos, A. and Wahlqvist, M. (2000). Indigenous Australian food culture on cattle stations prior to the 1960s and food intake of older Aborigines in a community studied in 1988. *Asia Pacific Journal of Clinical Nutrition* 9, 224–231

Krausmann, F., Gingrich, S., Eisenmenger, N., Erb, K.H., Haberl, H. and Fischer-Kowalski, M. (2009). Growth in global materials use, GDP and population during the 20th century. *Ecological Economics* 68(10), 2696–2705

Krey, V., Canadell, J., Nakicenovic, N., Abe, Y., Andruleit, H., Archer, D., Grubler, A., Hamilton, N.T.M., Johnson, A., Kostov, V., Lamarque, J., Langhorne, N., Nisbet, E., O'Neill, B., Riahi, K., Riedel, M., Wang, W. and Yakushev, V. (2009). Gas hydrates: entrance to a methane age or climate threat? *Environmental Research Letters* 4, 034007. doi:034010.031088/031748-039326/034004/034003/034007

Kumssa, A. and Jones, J.F. (2010). Climate change and human security in Africa. *International Journal of Sustainable Development and World Ecology* 17, 453–461

Kuuskraa, V.A. and Stevens, S.H. (2009). *Worldwide Gas Shales and Unconventional Gas: A Status Report.* Advanced Resources International, Inc., Arlington, VA

Lawrence, D.M., Slater, A.G., Tomas, R.A., Holland, M.M. and Deser, C. (2008). Accelerated Arctic land warming and permafrost degradation during rapid sea ice loss. *Geophysical Research Letters* 35, L11506

Leadley, P., Pereira, H.M., Alkemade, R., Fernandez-Manjarrés, J.F., Proença, V., Scharlemann, J.P.W. and Walpole, M.J. (2010). *Biodiversity Scenarios: Projections of 21st Century Change in Biodiversity and Associated Ecosystem Services.* Technical Series No. 50. Secretariat of the Convention on Biological Diversity, Montreal

Le Quéré, C., Raupach, M.R., Canadell, J.G., Marland, G., Bopp, L., Ciais, P., Conway, T.J., Doney, S.C., Feely, R.A., Foster, P., Friedlingstein, P., Gurney, K., Houghton, R.A., House, J.I., Huntingford, C., Levy, P.E., Lomas, M.R, Majkut, J., Metzl, N., Ometto, J.P., Peters, G.P., Prentice, I.C., Randerson, J.T., Running, S.W., Sarmiento, J.L., Schuster, U., Sitch, S., Takahashi, T., Viovy, N., van der Werf, G.R. and Woodward, F.I. (2009). Trends in the sources and sinks of carbon dioxide. *Nature Geoscience* 2, 831–836. doi: 10.1038/ngeo689

Le Quéré, C.L., Rödenbeck, C., Buitenhuis, E.T., Conway, T.J., Langenfelds, R., Gomez, A., Labuschagne, C., Ramonet, M., Nakazawa, T., Metzl, N., Gillett, N. and Heimann, M. (2007). Saturation of the Southern Ocean CO_2 sink due to recent climate change. *Science* 316(5832), 1735–1738

Levin, S.A. (1998). Ecosystems and the biosphere as complex adaptive systems. *Ecosystems* 1, 431–436

Lewis, S.L., Brando, P.M., Phillips, O.L., van der Heijden, G.M.F. and Nepstad, D. (2011). The 2010 Amazon drought. *Science* 331, 554

Liu, Y., Stanturf, J. and Goodrick, S. (2010). Trends in global wildfire potential in a changing climate. *Forest Ecology and Management* 259 (4), 685–697

Loorbach, D. and Rotmans, J. (2010). The practice of transition management: examples and lessons from four distinct cases. *Futures* 42, 237–246

Loorbach, D. and Rotmans, J. (2006). Managing transitions for sustainable development. In *Understanding Industrial Transformation: Views from Different Disciplines* (eds. Olshoorn, X. and Wieczorek, A.J.). Springer, Dordrecht

Loorbach, D., Frantzeskaki, N. and Thissen, W. (2011). A transition research perspective on governance for sustainability. In *European Research on Sustainable Development* (eds. Jaeger, C.C., Tàbara, J.D. and Jaeger, J.). pp.73–89. Springer, Berlin-Heidelberg

Luethi, D., Le Floch, M., Bereiter, B., Blunier, T., Barnola, J.-M., Siegenthaler, U., Raynaud, D., Jouzel, J., Fischer, H., Kawamura, K. and Stocker, T.F. (2008). High-resolution carbon dioxide concentration record 650,000–800,000 years before present. *Nature* 453, 379–382

MA (2005). *Ecosystems and Human Well-being: Synthesis.* Millennium Ecosystem Assessment. Island Press, Washington, DC

Ma, X., Xu, J.C., Luo, Y., Aggarwal, S.P. and Li., J.T. (2009). Response of hydrological processes to land cover and climate change in Kejie watershed, southwest China. *Hydrological Processes.* doi:10.1002/hyp.7233

Manney, G.L., Santee, M.L., Rex, M., Livesey, N.J., Pitts, M.C., Veefkind, P., Nash, E.R., Wohltmann, I., Lehmann, R., Froidevaux, L., Poole, L.R., Schoeberl, M.R., Haffner, D.P., Davies, J., Dorokhov, V., Gernandt, H., Johnson, B., Kivi, R., Kyrö, E., Larsen, N., Levelt, P.F., Makshtas, A., McElroy, C.T., Nakajima, H., Parrondo, M.C., Tarasick, D.W., von der Gathen, P., Walker, K.A. and Zinoviev, N.S. (2011). Unprecedented Arctic ozone loss in 2011. *Nature* 478(7370), 469–475. doi:10.1038/nature10556

Maslanik, J., Stroeve, J., Fowler, C. and Emery, W. (2011). Distribution and trends in Arctic sea ice age through spring 2011. *Geophysical Research Letters* 38, L13502. doi:10.1029/2011GL047735

McConnell, J.R., Edwards, R., Kok, G.L., Flanner, M.G., Zender, C.S., Saltzman, E.S., Banta, J.R., Pasteris, D.R., Carter, M.M. and Kahl, J.D.W. (2007). 20th-century industrial black carbon emissions altered Arctic climate forcing. *Science* 317, 1381–1384

Meadows, D.H., Meadows, D.L., Randers, J. and Behrens III, W.W. (1972). *The Limits to Growth.* Universe Books, New York

Mertz, O., Mbow, C., Nielsen, J.O., Maiga, A., Diallo, D., Reenberg, A., Diouf, A., Barbier, B., Moussa, I.B., Zorom, M., Ouattara, I. and Dabi, D. (2010). Climate factors play a limited role for past adaptation strategies in West Africa. *Ecology and Society* 15(4), 25. http://www.ecologyandsociety.org/vol15/iss4/art25/

Myers, N. and Knoll, A.H. (2001). The biotic crisis and the future of evolution. *Proceedings of the National Academy of Sciences of the United States of America* 98, 5389–5392

Nair, U.S., Wu, Y., Kala, J., Lyons, T.J., Pielke, R.A. and Hacker, J.M. (2011). The role of land use change on the development and evolution of the west coast trough, convective clouds, and precipitation in southwest Australia. *Journal of Geophysical Research* 116, D07103. doi:07110.01029/02010JD014950

Nature (2009). Earth's boundaries? *Nature* 461, 447–448. doi:10.1038/461447b

NASA Earth Observatory (2010). *If Earth has Warmed and Cooled throughout History, What Makes Scientists Think that Humans are Causing Global Warming Now?* National Aeronautics and Space Administration. http://earthobservatory.nasa.gov/blogs/climateqa/if-earth-has-warmed-and-cooled-throughout-history-what-makes-scientists-think-that-humans-are-causing-global-warming-now/

NASA GISS (2011). Surface Temperature Analysis. National Aeronautics and Space Administration Goddard Institute for Space Studies. http://data.giss.nasa.gov/gistemp/

Nicol, S. and Robertson, G. (2006). Ecological consequences of Southern Ocean harvesting. *Books Online 2006*, 48–61. http://www.publish.csiro.au/paper/9780643090712_03 (accessed 19 November 2011)

Nijssen, B., O'Donnell, G.M., Hamlet, A. and Letternmaier, D.P. (2001). Hydrological sensitivity of global rivers to climate change. *Climate Change* 50, 143–175

NOAA (2011). *Current Trends in Atmospheric Carbon Dioxide*. Earth System Research Laboratory, Global Monitoring Division. U.S. Department of Commerce, National Oceanic and Atmospheric Administration. http://www.esrl.noaa.gov/gmd/dv/

Noticewala, S. (2007). At Australia's bunny fence, variable cloudiness prompts climate study. http://www.nytimes.com/2007/08/14/science/earth/14fenc.html?pagewanted=print (accessed 10 September 2011)

Olsson, P., Folke, C. and Hughes, T.P. (2008). Navigating the transition to ecosystem-based management of the Great Barrier Reef, Australia. *Proceedings of the National Academy of Sciences of the United States of America* 105, 9489–9494

O'Riordan, T. (2008). Some reflections on the conditions for favouring integrated sustainability assessment. *International Journal of Innovation and Sustainable Development* 3(1–2), 153–162

Osborn, S.G., Vengosh, A., Warner, N.R. and Jackson, R.B. (2011). Methane contamination of drinking water accompanying gas-well drilling and hydraulic fracturing. *Proceedings of the National Academy of Sciences of the United States of America* 108, 8172–8176

Pagini, M., Liu, Z., LaRiviere, J. and Ravelo, A.C. (2010). High Earth-system climate sensitivity determined from Pliocene carbon dioxide concentrations. *Nature Geoscience* 3, 27–30

Pan, Y., Birdsey, R.A., Fang, J., Houghton, R., Kauppi, P.E., Kurz, W.A., Phillips, O.L., Shvidenko, A., Lewis, S.L., Canadell, J.G., Ciais, P., Jackson., R.B., Pacala1, S.W., McGuire, A.D., Piao, S., Rautiainen, A., Sitch, S. and Hayes, D. (2011). A large and persistent carbon sink in the world's forests. *Science* 333, 988–993

Parry, B. (2008). *Amazon: An Extraordinary Journey Down The Greatest River On Earth*. Penguin Books, London

Pascual, M., Ahumada, J.A., Chaves L.F., Rodó X., and Bouma, M. (2006). Malaria resurgence in the East African highlands: temperature trends revisited. *Proceedings of the National Academy of Sciences of the United States of America* 103(15), 5829–5834. doi:10.1073/pnas.0508929103

Petoukhov, V.A. and Semenov, V.A. (2010). A link between reduced Barents-Kara sea ice and cold winter extremes over northern continents. *Journal of Geophysical Research* 115, D21111

Phillips, O.L., Aragão, L.E.O.C., Lewis, S.L., Fisher, J.B., Lloyd, J., López-González, G., Malhi, Y., Monteagudo, A., Peacock, J., Quesada, C.A., van der Heijden, G., Almeida, S., Amaral, I., Arroyo, L., Aymard, G., Baker, T.R., Bánki, O., Blanc, L., Bonal, D., Brando, P., Chave, J., de Oliveira, Á.C.A., Cardozo, N.D., Czimczik, C.I., Feldpausch, T.R., Freitas, M.A., Gloor, E., Higuchi, N., Jiménez, E., Lloyd, G., Meir, P., Mendoza, C., Morel, A., Neill, D.A., Nepstad, D., Patiño, S., Peñuela, M.C., Prieto, A., Ramírez, F., Schwarz, M., Silva, J., Silveira, M., Thomas, A.S., ter Steege, H., Stropp, J., Vásquez, R., Zelazowski, P., Dávila, E.A., Andelman, S., Andrade, A., Chao, K.-J., Erwin, T., di Fiore, A., Honorio C., E., Keeling, H., Killeen, T.J., Laurance, W.F., Cruz, A.P., Pitman, N.C.A., Vargas, P.N., Ramírez-Angulo, H., Rudas, A., Salamão, R., Silva, N., Terborgh, J. and Torres-Lezama, A. (2009). Drought sensitivity of the Amazon rainforest. *Science* 323, 1344–1347

Pielke, R.A., Pitman, A., Niyogi, D. Mahmoud, R., McAlpine, C., Hossain, F., Goldewijk, K.K., Nair, U., Betts, R., Fall, S., Reichstein, M., Kabat, P. and de. Noblet, N. (2011). Land use/land cover changes and climate: modeling analysis and observational evidence. *WIREs Climate Change* 2, 828-850. doi:10.1002/wcc.144 http://wires.wiley.com/WileyCDA/WiresArticle/wisId-WCC144.html

Plümper, T. and Neemayer, E. (2007). The gendered nature of natural disasters: the impact of catastrophic events on the gender gap in life expectancy, 1981–2002. *Annals of the Association of American Geographers* 97(3), 551–566

Pollard, D. and DeConto, R.M. (2009). Modelling West Antarctic ice sheet growth and collapse through the past five million years. *Nature* 458, 329–332

Poncelet, A., Gemenne, F., Bousetta, H. and Martiniello, M. (2010). A country made for disasters: environmental vulnerability and forced migration in Bangladesh. In *Environment, Forced Migration and Social Vulnerability*. (eds. Afifi, T. and Jaeger, J.), Springer, Berlin

Prentice, I.C. and Jolly, D. (2000). Mid-Holocene and glacial-maximum vegetation geography of the northern continents and Africa. *Journal of Biogeography* 27, 507–19

Purkey, S.G. and Johnson, G.C. (2010). Warming of global abyssal and deep Southern Ocean waters between the 1990s and 2000s: contributions to global heat and sea level rise budgets. *Journal of Climate* 23, 6336–6351. doi:10.1175/2010JCLI3682.1

Qin, B., Liu, Z. and Havens, K. (2007). *Eutrophication of Shallow Lakes with Special Reference to Lake Taihu, China*. Springer, Dordrecht

Ramanathan, V. and Carmichael, G. (2008). Global and regional climate changes due to black carbon. *Nature Geoscience* 1, 221–227

Reynolds, J.F. and Stafford Smith, D.M. (eds.) (2002). *Global Desertification: Do Humans Cause Deserts?* Dahlem Workshop Report 88. Dahlem University Press, Berlin

Roberts, C.J. and Wooster, M.J. (2008). Fire detection and fire characterization over Africa using Meteosat SEVIRI. *IEEE Transactions on Geoscience and Remote Sensing* 46(4), 1200–1218

Rockström, J., Steffen, W., Noone, K., Persson, Å., Chapin III, F.S., Lambin, E., Lenton, T.M., Scheffer, M., Folke, C., Schellnhuber, H., Nykvist, B., de Wit, C.A., Hughes, T., van der Leeuw, S., Rodhe, H., Sörlin, S., Snyder, P.K., Costanza, R., Svedin, U., Falkenmark, M., Karlberg, L., Corell, R.W., Fabry, V.J., Hansen, J., Walker, B., Liverman, D., Richardson, K., Crutzen, P. and Foley, J. (2009a). Planetary boundaries: exploring the safe operating space for humanity. *Ecology and Society* 14, 32. http://www.ecologyandsociety.org/vol14/iss32/art32/

Rockstrom, J., Steffen, W., Noone, K., Persson, Å., Chapin, F.S., Lambin, E.F., Lenton, T.M., Scheffer, M., Folke, C., Schellnhuber, H.J., Nykvist, B., de Wit, C.A., Hughes, T., van der Leeuw, S., Rodhe, H., Sorlin, S., Snyder, P.K., Costanza, R., Svedin, U., Falkenmark, M., Karlberg, L., Corell, R.W., Fabry, V.J., Hansen, J., Walker, B., Liverman, D., Richardson, K., Crutzen, P. and Foley, J.A. (2009b). A safe operating space for humanity. *Nature* 461, 472–475

Rotmans, J. (2006). Tools for integrated sustainability assessment: a two-track approach. *The Integrated Assessment Journal* 6, 35–57

Rustad, L.E, Campbell, J.L., Marion, G.M., Norby, R.J., Mitchell, M.J., Hartley, A.E., Cornelissen, J.H.C., Gurevitch, J. and GCTE-NEWS (2001). A meta-analysis of the response of soil respiration, nitrogen mineralization, and aboveground plant growth to experimental ecosystem warming. *Oecologia* 126, 543–562

Schaefer, K., Zhang, T., Bruhwiler, L. and Barrett A.P. (2011). Amount and timing of permafrost carbon release in response to climate warming. *Tellus B* 63(2), 165–180

Schellnhuber, H.-J. (2009). Tipping elements in the Earth system. *Proceedings of the National Academy of Sciences of the United States of America* 106(49), 20561–20563. doi:10.1073/pnas.0911106106

Schmidt, G.A., Ruedy, R.A., Miller, R.L. and Lacis, A.A. (2010). Attribution of the present-day total greenhouse effect. *Journal of Geophysical Research* 115, D20106

Screen, J.A. and Simmonds, I. (2010). The central role of diminishing sea ice in recent Arctic temperature amplification. *Nature* 464, 1334–1337

Serreze, M.C. and Barry, R.G. (2011). Processes and impacts of Arctic amplification: a research synthesis. *Global and Planetary Change* 77, 85–96

Sherman, K. and Hempel, G. (2008). *The UNEP Large Marine Ecosystem Report: A Perspective on Changing Conditions in LMEs of the World's Regional Seas.* United Nations Environment Programme, Nairobi

Shindell, D. and Faluvegi, G. (2009). Climate response to regional radiative forcing during the twentieth century. *Nature Geoscience* 2, 294–300

Steffen, W., Crutzen, P.J. and McNeill, J.R. (2007). The Anthropocene: are humans now overwhelming the great forces of Nature? *Ambio* 36, 614–621

Steffen, W., Andreae, M.O., Bolin, B., Cox, P.M., Crutzen, P.J., Cubasch, U., Held, H., Nakidenovic, N., Scholes, R.J., Talaue-McManus, L. and Turner, B.L. (2004a). Abrupt changes: the Achilles' heels of the Earth system. *Environment: Science and Policy for Sustainable Development* 46, 8–20

Steffen, W., Sanderson, A., Tyson, P.D., Jäger, J., Matson, P.A., Moore III, b., Oldfield, F., Richardson, K., Schellnhuber, H.J, Turner II, B.L., Wasson, R.J. (2004b). *Global Change and the Earth System.* Springer, Berlin

Stephenson, S., Smith, L. and Agnew, J. (2011). Divergent long-term trajectories of human access to the Arctic. *Nature Climate Change* 1, 156–160

Stockholm Resilience Centre (2009). *Tipping Towards the Unknown.* University of Stockholm. http://www.stockholmresilience.org/research/researchnews/tippingtowardstheunknown.5.7cf9c5aa121e17bab42800021543.html 20

SWIPA (2011). *Snow, Water, Ice and Permafrost in the Arctic (SWIPA) – Executive Summary 2011.* Arctic Monitoring and Assessment Programme

Takahashi, T., Sutherland, S.C., Wanninkhof, R., Sweeney, C., Feely, R.A., Chipman, D.W., Hales, B., Friederich, G., Chavez, F., Sabine, C., Watson, A., Bakker, D.C.E., Schuster, U., Metzl, N., Yoshikawa-Inoue, H., Ishii, M., Midorikawa, T., Nojiri, Y., Körtzinger, A., Steinhoff, T., Hoppema, M., Olafsson, J., Arnarson, T.S., Tilbrook, B., Johannessen, T., Olsen, A., Bellerby, R., Wong, C.S., Delille, B., Bates, N.R. and de Baar, H.J.W. (2009). Climatological mean and decadal change in surface ocean p CO_2, and net sea–air CO_2 flux over the global oceans. *Deep Sea Research Part II: Topical Studies in Oceanography* 56, 554–577

Tarnocai, C., Canadell, J., Schuur, E., Kuhry, P., Mazhitova, G. and Zimov, S. (2009). Soil organic carbon pools in the northern circumpolar permafrost region. *Global Biogeochemical Cycles* 23

TEEB (2010). *The Economics of Ecosystems and Biodiversity: Mainstreaming the Economics of Nature: A synthesis of the Approach, Conclusions and Recommendations of TEEB.* http://www.teebweb.org/TEEBSynthesisReport/tabid/29410/Default.aspx

Thompson, D.W.J. and Solomon, S. (2002). Interpretation of recent Southern Hemisphere climate change. *Science* 296(5569), 895–899. doi:10.1126/science.1069270

Turner, G.M. (2008). A comparison of *The Limits to Growth* with 30 years of reality. *Global Environmental Change* 18, 397–411

Turner, J., Bindschadler, R., Convey, P., di Prisco, G., Fahrbach, E., Gutt, J., Hodgson, D., Mayewski, P. and Summerhayes, C. (2009). *Antarctic Climate Change and the Environment.* Scar and Scott Polar Research Institute, Cambridge

UNEP (2011a). *Decoupling Natural Resource Use And Environmental Impacts From Economic Growth. A Report of the Working Group on Decoupling to the International Resource Panel.* United Nations Development Programme, Nairobi

UNEP (2011b). Nellemann, C., Verma, R., and Hislop, L. (eds). *Women at the Frontline of Climate Change: Gender Risks and Hopes.* A Rapid Response Assessment. United Nations Environment Programme and GRID-Arendal

UNEP (2007). *Global Environment Outlook 4: Environment for Development.* United Nations Development Programme, Nairobi

USEPA (2009). *Measurement of Emissions from Produced Water Ponds: Upstream Oil and Gas Study 1.* National Risk Management Research Laboratory, US Environmental Protection Agency, Cincinnati

USGCRP (2009). *Global Climate Change Impacts in the United States* (eds. Karl, T., Melillo, J.M., and Peterson T.C.). US Global Change Research Program. Cambridge University Press, Cambridge. 188 pp. http://www.globalchange.gov/what-we-do/assessment/previous-assessments/global-climate-change-impacts-in-the-us-2009

van der Werf, G.R., Dempewolf, J., Trigg, S.N., Randerson, J.T., Kasibhatla, P.S., Giglio, L., Murdiyarso, D., Peters, W., Morton, D.C., Collatz, G.J., Dolman, A.J. and DeFries, R.S. (2008). Climate regulation of fire emissions and deforestation in equatorial Asia. *Proceedings of the National Academy of Sciences of the United States of America* 105, 20350–20355

van Nes, E.H. and Scheffer, M. (2007). Slow recovery from perturbations as a generic indicator of a nearby catastrophic shift. *American Naturalist* 169, 738–747

Vergara, W. and Scholz, M.S. (2010). *Assessment of the Risk of Amazon Dieback.* World Bank, Washington, DC

von Braun (2007). *The World Food Situation. New Driving Forces and Required Actions.* International Food Policy Research Institute, Washington, DC

WBGU (2011). *World In Transition. A Social Contract for Sustainability.* German Advisory Council on Global Change, Berlin

Weaver, P.M. (2011). Pragmatism and pluralism: creating clumsy and context-specific approaches to sustainability science. In *European Research on Sustainable Development* (eds. Jaeger, C.C., Tabara, J.D. and Jaeger, J.). pp.173–186. Springer-Verlag, Berlin

Westley, F., Olsson, P., Folke, C., Homer-Dixon, T., Vredenburg, H., Loorbach, D., Thompson, J., Nilsson, M., Lambin, E., Sendzimir, J., Banarjee, B., Galaz, V. and van der Leeuw, S. (2011). *Tipping towards Sustainability: Emergent Pathways of Transformation.* Prepared for the 3rd Nobel Laureate Symposium on Global Sustainability: Transforming the World in an Era of Global Change, Stockholm. http://www.stockholmresilience.org/seminarandevents/seminarandeventvideos/

Wigley, T. (2011). Coal to gas: the influence of methane leakage. *Climatic Change.* doi:10.1007/s10584-10011-10217-10583

Wolfenden, L., Hardy, L.L., Wiggers, J., Milat, A.J., Bell, C. and Sutherland, R. (2011). Prevalence and socio-demographic associations of overweight and obesity among children attending child-care services in rural and regional Australia. *Nutrition and Dietetics* 68, 15–20

WWF (2010). *Living Planet Report 2010.* WWF–World Wide Fund For Nature, Gland

Wysham, D.B. and Hastings, A. (2008). Sudden shifts in ecological systems: intermittency and transients in the coupled Ricker population model. *Bulletin of Mathematical Biology* 70, 1013–1031

Xu, J., Grumbine, R.E., Shrestha, A., Eriksson, M., Yang, X., Wang, Y.U.N. and Wilkes, A. (2009). The melting Himalayas: cascading effects of climate change on water, biodiversity, and livelihoods. *Conservation Biology* 23, 520–530

Xu, J.C., Shrestha, A.B., Vaidya, R., Eriksson, M. and Hewitt, K. (2007). *The Melting Himalayas: Regional Challenges and Local Impacts of Climate Change on Mountain Ecosystems and Livelihoods.* Technical paper. International Center for Integrated Mountain Development, Kathmandu

Young, D.A., Wright, A.P., Roberts, J.L., Warner, R.C., Young, N.W., Greenbaum, J.S., Schroeder, D.M., Holt, J.W., Sugden, D.E., Blankenship, D.D., van Ommen, T.D. and Siegert, M.J. (2011). A dynamic early East Antarctic ice sheet suggested by ice-covered fjord landscapes. *Nature* 474, 72–75

Zaks, D.P.M., Barford, C.C., Ramankutty, N. and Foley, J.A. (2009). Producer and consumer responsibility for greenhouse gas emissions from agricultural production—a perspective from the Brazilian Amazon. *Environmental Research Letters* 4, 044010

Zalasiewicz, J., Williams, M., Haywood, A. and Ellis, M. (2011). The Anthropocene: a new epoch of geological time? *Philosophical Transactions of the Royal Society A* 369, 835–841

Zalasiewicz, J., Williams, M., Steffen, W. and Crutzen, P. (2010). The new world of the Anthropocene. *Environmental* Science and Technology 44, 2228–2231

CHAPTER 8

Review of Data Needs

Contributing authors: Charles Davies, Ashbindu Singh and Jaap van Woerden

Principal scientific reviewer: Rainer M. Krug

This chapter has been prepared by the Secretariat

Main Messages

Global research programmes and rapidly improving technologies for collecting environmental information and presenting it in engaging ways are informing the debate about present and future environmental challenges. Nevertheless, deficiencies in scientifically credible data on the environment – in particular time series on such issues as freshwater quantity and quality, groundwater depletion, ecosystem services, loss of natural habitat, land degradation, and chemicals and waste – are a major handicap in developing evidence-based policies.

Official environment statistics is still an emerging field, with poor availability and quality of data in many countries. Environment statistics, mostly collected or compiled by national statistical offices, are one of the most important sources of information for assessment reports like *GEO-5*, but global and regional reports from the United Nations and other agencies regularly show gaps, or use old data or estimates.

Capacity development to support environmental information, especially in developing countries, needs to be stepped up significantly. Different countries often use different approaches to produce data on the same issue, making comparisons difficult. This highlights the need for regular monitoring and, to allow comparisons across nations and regions, the harmonization of approaches to those that follow international standards. Also of high priority are the coordination at country level of existing – if fragmented – scientific and environmental data; the provision of easy access to a range of potential users, for example on the internet; and linking this data with official statistics that are used for policy making.

International cooperation is essential, since environmental problems do not follow national boundaries. Some of the many global and regional initiatives supporting environmental information are mentioned in this review. International cooperation and sharing of comparable data are especially important in addressing global issues such as climate change, and tackling environmental problems related to transboundary watercourses, oceans and seas, and polar regions. At present, cooperation mechanisms are much stronger in some areas than others.

Adequate information does exist to develop effective environmental policies; data gaps rarely justify inaction. However, more systematic data collection efforts can help governments to assess their progress towards international goals, improve the focus of their policies and monitor their impact, and direct scarce resources to address the most critical environmental challenges.

BACKGROUND

This review provides a snapshot of the data on which *GEO-5* is based, highlights some of the limitations in the data that are currently available, describes some global and regional programmes that are supporting environmental information, and identifies some of the highest priorities for promoting more effective environmental monitoring in countries and regions.

The review is intended to have a practical, rather than a detailed, technical focus. In line with the request of the Global Intergovernmental and Multi-Stakeholder Consultation, it focuses on data relevant to track the state and trends of the environment (Part 1), and touches more briefly on data needs relating to policy responses (Parts 2 and 3).

DEFINITIONS

Data: "facts and statistics used for reference or analysis" (COD 2003). Data is used here to mean points of information, normally collected through some kind of scientific method. An example of a point of data or "datum" is a measurement of the temperature in downtown Cairo at 8:00am.

Datasets: collections of data on a particular issue, for example historical temperature records for downtown Cairo.

Information: "facts or knowledge provided or learned as a result of research or study" (COD 2003). Information is used here as a broader term, including facts, data, anecdotes and the results of analysis that are understood, correctly or incorrectly, by the person using them; for example, the best time to visit Cairo is in the cooler months between November and March.

Statistics: is used here to describe official data collected by national statistical offices.

Environment statistics: statistics that describe the state of and trends in the environment, covering the media of the natural environment (air and climate, water, land and soil), the biota within the media and human settlements (OECD 2007).

INTRODUCTION

The factual and scientific quality of an assessment such as *GEO-5* relies to a large extent on what data are available on the state and trends of the environment. Economic and social data are important to analyse the drivers and socio-economic impacts of environmental change (Box 8.1) and consider possible responses and scenarios. Indices, obtained by combining and packaging data on a number of variables, can be used to summarize information and make it easier to communicate and understand, and a variety of other tools can be used to visualize, present and disseminate data and information.

The availability of information relies on activities such as data collection, for example measurements of pollution levels in air and water, sea surface temperature, or images from satellite remote sensing that can be used to produce land-cover maps; monitoring programmes involving regular, comparable measurements or time series; analysis of data to produce information that policy makers can use, such as comparisons of trends over time or progress towards established targets; and the interpretation of results to explain patterns and trends. Existing gaps and capacity-building needs relate to all these areas.

Rapid advances in information technology, remote sensing, geographic information systems (GIS), global positioning systems (GPS), database management, measuring instruments, data visualization tools, social media and the internet offer unprecedented opportunities to collect and disseminate information. The trend towards digital data has opened up new possibilities for state-of-the-environment reporting, allowing users to directly access and download data, maps and other information, in addition to the more traditional periodic, analytical reports such as *GEO*. That said, many national statistical offices have not yet been able to exploit this potential to modernize their statistical systems, an area that could benefit from international coordination and support.

Internationally comparable data are necessary to track environmental changes at the regional and global level. Most data to track the state and trends of the environment are collected at the country level, but both availability and quality remain poor in a large number of countries. Many do not produce internationally comparable data because they follow their own national guidelines or a modified version of international guidelines.

Data are produced by a wide range of public and private sources but these are often scattered and difficult to compare globally. In addition, privately produced data may be protected by intellectual property rights and available only at cost; although in the absence of property rights, data may not have been collected at all.

Box 8.1 The three principal data gaps on drivers of global environmental change

Human migration
Data are needed on where migrants come from and where they go, with respect to both international migration and internal migration, and also regarding both permanent and temporary migration. Data would ideally include timing, numbers and geographic locations.

Agricultural systems
Basic information is needed on inflows and outflows of nutrients and water, as well as other important resource flows.

Environmental footprints for economic production
At the level of country and product, information is needed on energy and water inputs and key pollution outputs in order to understand how patterns of production and consumption affect environmental systems.

In 2009, the United Nations reported: "*Environment statistics frequently lack one or more of the standard attributes of high-quality statistics, namely, relevance, accuracy, timeliness, accessibility, interpretability and coherence. The fact that environment statistics are* ad hoc, *widely dispersed and of varying degrees of quality clearly underlines the need for a framework, that is, a basic organizing structure to guide environment statistics* (UN 2009)."

INTERNATIONAL PROGRAMMES SUPPORTING ENVIRONMENTAL INFORMATION

A number of international programmes support data collection and analysis, and coordinate, compile, disseminate and present data and information on various issues relating to the state of the environment. It is important to note that environmental data and information from UN sources often rely on statistics collected by national governments.

The United Nations has developed a core set of Indicators of Sustainable Development in response to the 2002 World Summit on Sustainable Development (WSSD) and decisions of the UN Commission on Sustainable Development (CSD). The latest revision of these indicators, published in October 2007, contains a core set of 50 indicators covering poverty, governance, health, education, demographics, natural hazards, atmosphere, land, oceans, seas and coasts, freshwater, biodiversity, economic development, global economic partnership, and consumption and production patterns (UN 2007b).

The *Framework for the Development of Environment Statistics* (UN 1984) serves as a template and guide for countries to develop and organize environmental and related socio-economic data. The United Nations Statistics Division (UNSD) has endorsed a work programme to update this framework with a view to transforming it into a hub for a broader range of producers of environment statistics, including not only traditional statistical data-collection instruments of national statistical systems but also information from scientific monitoring. The update aims to improve coordination of environmental data within countries, and improve coordination of environmental, economic and social data (UN 2009).

Data collection, dissemination, training and capacity-building programmes in environment statistics are coordinated by an Intersecretariat Working Group on Environment Statistics (IWG-ENV), convened by the UNSD. The UNSD/UNEP Questionnaire on Environment Statistics covers the themes of water, air, land and waste (UN 2011). The UNSD also collects specific data on the status of national environmental-economic accounting (UN 2007a) and has developed a System of Environmental-Economic Accounting (SEEA) under the United Nations Committee of Experts on Environmental-Economic Accounting (UNCEEA). The Committee for the Coordination of Statistical Activities (CCSA) coordinates the statistical data work of UN entities and international partners, such as the Organisation for Economic Co-operation and Development (OECD) and the Statistical Office of the European Union (Eurostat),

Mount Gwanaksan weather station, Republic of Korea. © *Matteusus/iStock*

and exchange of data between UN entities is facilitated by the UN data mechanism (data.un.org).

Prominent sets of indicators and indices relate to the Millennium Development Goals (MDGs), in particular ten indicators related to MDG 7 on ensuring environmental sustainability; the United Nations Development Programme (UNDP) Human Development Index (HDI); Yale University's Environmental Performance Index and the OECD's Core and Key Environmental Indicators as well as the Core Set of Indicators of the European Environment Agency (EEA). A number of environment-related, global and regional conventions have monitoring and reporting programmes, and many of the convention secretariats, such as that for the Montreal Protocol, support countries with the collection, monitoring, interpretation and analysis of data in their area of responsibility (UNEP 1999).

Geospatial data, from technologies such as satellite remote sensing and networks of weather stations and ocean buoys, are another important source of environmental information. Large, international global-change research programmes and initiatives such as the Group on Earth Observations System of Systems (GEOSS) offer new opportunities to access this type of information. GEOSS was set up to provide decision-making support tools to a wide variety of users by linking existing and planned observation systems around the world, and supporting the development of new systems where gaps exist. It also promotes common technical standards that make it possible to combine data from many different observation instruments into coherent datasets. In addition, the GEOSS Data Sharing Principles have encouraged satellite operators to enable greater access to data (GEO 2010).

Table 8.1 Environmental Data Explorer: data providers

Updated information and web links to data providers can be found on the Environmental Data Explorer website (geodata.grid.unep.ch).

Biodiversity Indicators Partnership (BIP)
Bureau de Recherches Géologiques et Minières (BRGM)
Carbon Dioxide Information Analysis Center (CDIAC)
Center for Environmental Systems Research (CESR)
Center for International Earth Science Information Network (CIESIN)
Centre for Research on the Epidemiology of Disasters (CRED)
Colorado Center for Astrodynamics Research
Commonwealth Scientific and Industrial Research Organisation (CSIRO)
Conservation International (CI)
Environmental Systems Research Institute (ESRI)
European Commission Joint Research Centre (JRC) – Institute for Environment and Sustainability (IES)
European Space Agency (ESA)
Food and Agriculture Organization of the United Nations (FAO)
Forest Stewardship Council (FSC)
Global Footprint Network (GFN)
Global Land Cover Facility (GLCF)
Intergovernmental Panel on Climate Change (IPCC)
International Centre for Tropical Agriculture (CIAT)
International Energy Agency (IEA)
International Groundwater Resources Assessment Centre (IGRAC)
International Labour Organization (ILO)
International Organization for Standardization (ISO)
International Union for the Conservation of Nature (IUCN)
IUCN and UNEP World Conservation Monitoring Centre – World Database on Protected Areas (WDPA)
National Aeronautics and Space Administration (NASA), United States
NASA Goddard Space Flight Center (GSFC)
National Center for Ecological Analysis and Synthesis (NCEAS), United States
National Marine Fisheries Service, United States
National Geophysical Data Center (NGDC), United States
National Oceanic and Atmospheric Administration (NOAA), United States
Netherlands Environment Assessment Agency (PBL)
Nuclear Energy Agency (NEA)
Organisation for Economic Co-operation and Development (OECD)
Programme for the Endorsement of Forest Certification (PEFC) International
Ramsar Convention Bureau
Secretariat of the Basel Convention
Secretariat of the Convention on Biological Diversity (CBD)
Secretariat of the United Nations Convention to Combat Desertification (UNCCD)
Secretariat of the United Nations Framework Convention on Climate Change (UNFCCC)
United Nations Children's Fund (UNICEF)
United Nations Development Programme (UNDP)
United Nations Educational, Scientific and Cultural Organization (UNESCO)
United Nations Environment Programme (UNEP)
UNEP/GRID-Arendal
UNEP Global Environmental Monitoring System – Water Programme (GEMS Water)
UNEP World Conservation Monitoring Centre (UNEP-WCMC)
United Nations Refugee Agency (UNHCR) (formerly United Nations High Commissioner for Refugees)
United Nations Office of Legal Affairs (OLA)
United Nations Population Division (UNPD)
United Nations Statistics Division (UNSD)
United States Geological Survey (USGS)
University of California at Berkeley Museum of Vertebrate Zoology
University of Maryland (UMD)
University of Montana Department of Ecosystem and Conservation Sciences
Water Footprint Network (WFN)
World Development Indicators (WDI), World Bank
World Energy Council (WEC)
World Glacier Monitoring Service (WGMS)
World Health Organization (WHO)
WHO – Second Administrative Level Boundaries dataset project (SALB)
WHO/UNICEF – Joint Monitoring Programme (JMP) for Water Supply and Sanitation
WHO/UNICEF – Roll Back Malaria
World Bank

UNEP's Environmental Data Explorer (geodata.grid.unep.ch) compiles and presents a wide range of economic, social and environmental data covering more than 500 variables, which can be used to track the state and trends of the environment in support of GEO and other environmental assessments. Updated lists of data providers and variables can be accessed on the Data Explorer (Table 8.1).

THEMATIC GAPS

Many global programmes focus on supporting and compiling data collected at the national level. In almost all thematic areas, data availability is geographically unbalanced and data are generally scarcer in developing countries. Data collected at sub-national levels – for example on urban air quality – tend to be even more fragmented. Variables that are linked to industrial activities and organized sectors of the economy – including some sources of carbon dioxide (CO_2) emissions – are easier to measure and monitor. Other issues, including forest cover, can be assessed on a broad scale using satellite remote sensing. However, the effects of environmental change, such as air or water pollution eroding human health, may be spread over wide areas and be difficult both to measure and to attribute to a particular cause. These limitations present a serious challenge to measuring the consequences of environmental change.

Research on both scientific and policy aspects of the state of the environment is continuously developing and countries face greater challenges in collecting data on newly emerging issues. Data collection and monitoring in the polar regions, on the high seas and in the upper atmosphere rely on international cooperation programmes.

Atmosphere

- Climate data remain limited in some regions, with marked scarcity in developing countries. Understanding temperature changes at finer resolution than continental scales – and attributing them to human or natural causes – remains difficult, and is complicated by factors such as land-use change and pollution (IPCC 2007). The impacts of climate change and extreme events depend on a range of economic, social, geographic, cultural, institutional, governance and environmental factors, such as levels of wealth and education, disability and health status, as well as gender, age and social status. In general, data on disasters and measures to reduce disaster risk are lacking at the local level (IPCC 2011).

- Data relating to emissions of greenhouse gases, ozone-depleting substances and many other pollutants have improved in recent years as they are required and supported by a number of international programmes and conventions, including the United Nations Framework Convention on Climate Change (UNFCCC) and the Montreal Protocol. The policy relevance of emissions data can be increased by disaggregating them by sector, for example transport, and by such sub-sectors as road, air and water transport and type of fuel and engine. Data availability is more limited for countries that are not Parties to the relevant conventions; gaps are often filled through the use of model estimates.

- There are many gaps in air quality data relating to pollutants that are not the focus of global conventions – nitrates, sulphates, tropospheric ozone, particulate matter and black carbon – especially in developing countries, and, even where data are available, they can be fragmented and difficult to access. Indoor air pollution is a leading cause of death, especially in low-income countries, with a disproportionate impact on women; WHO's programme to assess the global burden of disease uses estimates of the effects of indoor air pollution, because it is impractical to monitor inside people's homes (WHO 2010, 2009).

Sugar cane, one of the world's major biofuel crops. Important gaps in data on the production and use of biofuels remain. *© Wendy Townrow/iStock*

Land

Different methods of assessing land cover and use continue to yield very different results. In general, there are many deficiencies in the available data on these issues.

- The extent of drylands in the world is uncertain because of different classifications and methodologies used by different programmes (ICTSD 2007).
- The limited globally comparable data on land degradation – an essential information base for dryland nations to address the problem – date from the Global Assessment of Human Induced Soil Degradation (GLASOD) in 1990 (UN 2004), although new estimates using satellite data are being developed.
- There is no comprehensive and complete global database of wetlands, and different estimates of global wetlands extent are very inconsistent (Lehner and Döll 2004; Finlayson *et al.* 1999).
- Remote sensing has advanced knowledge of land cover and land use, but reliable information on changes is limited as data from different points in time are often not comparable because of changing sensor technology, insufficient ground truthing and a lack of agreement on ecosystem delineations. There are multiple definitions for forest, for example.
- Satellite-derived estimates of urban area, less than 0.5 per cent of global land cover, are four- to sixfold lower than previous estimates based on global urban population maps (Schneider *et al.* 2009).
- Two recent FAO reviews of deforestation showed very different trends for 2000–2005, with one based on national reports showing a slower rate of deforestation than in the past, while the other, based on remote sensing, suggests a faster rate (FAO and EC-JRC 2011; Hansen *et al.* 2010).
- Baseline data and monitoring of changes in carbon stocks are needed, and evidence is still emerging of the significant carbon sequestration potential of rangelands and grasslands.
- Data on biofuels – including the extent of production and use – are fragmented and incomplete at the global level, although datasets can be found at the national level in some countries.
- The International Fund for Agricultural Development (IFAD) is monitoring a set of indicators to assess country performance in ensuring that poor people have access to land and tenure security (IFAD 2008). Governments could apply these and other social, economic and environmental indicators (Bach *et al.* 2009) to evaluate the impacts of land-use changes and large-scale international land deals in Africa or elsewhere.

Water

- Comprehensive data on water quality and quantity remain a priority; mapping and the compilation of an inventory of transboundary aquifer systems will be a major task for the Transboundary Water Assessment Programme (UNEP 2011a).
- In general, data on groundwater, including availability, quality, extraction, uses, management and legislation, are more limited than data on surface waters. Rectifying this should be a priority since groundwater is being extracted unsustainably in many regions. Additionally, limited data are available on groundwater contamination from substances such as nitrates and arsenic.
- Information on access to drinking water and sanitation has improved to track the MDG 7 target; sex-disaggregated data on these issues remain a high priority.
- No global datasets are available to evaluate trends in all water-related diseases, although global trend data on cholera are used as a proxy.
- UN-Water compiles information on the state and trends of integrated water management approaches (UN-Water 2008a).
- Exchange of comparable information, as well as joint monitoring and assessment, is necessary for long-term, sustainable and reliable cooperation in managing transboundary waters (UN-Water 2008b).

A Nepali girl drinking from a city fountain on Patan Durbar Square, Kathmandu, where local residents sometimes queue for hours waiting for their turn to collect clean drinking water. *© Wendy Townrow/iStock*

Box 8.2 Glacier monitoring in the Himalayas

Glaciers in the Himalayas and other high mountains in Asia are the source of the continent's major rivers, supporting vulnerable, densely populated river basins downstream. Understanding fluctuations in the mass of glaciers is essential for decision making on water resources, agriculture and disaster risk reduction in downstream areas. Changes in the mass of glaciers depend on a multitude of complex elements, including the glacier's size, the microclimate, local topography, altitude range, aspect in relation to the sun, and variations in influences of the Indian monsoons and the deserts of Central Asia and western China. There is also huge uncertainty about how snow and glacial melting in the Himalayan region will continue to respond to climate change, and how such change will affect ecosystems and human well-being.

The ability to measure the area and length of glaciers has been improved through remote sensing, although there are still limitations, including access to appropriate satellite images and confirmation of results through field surveys (ground truthing). Glacier length and area alone, however, are not the only significant factors: thickness is also important but is much more difficult to measure. Lack of sufficient Himalayan meteorological data and monitoring stations has also been a major handicap in drawing conclusions on the impact of climate change on glaciers, snow cover and associated phenomena such as glacial lake outburst floods. Another challenge is that the Himalayas are spread across eight countries with different financial capabilities and socio-economic objectives, meaning that international coordination may be required to strengthen a long-term programme in the region.

"*We have anecdotal evidence that glaciers may be receding, but we need precise and carefully vetted data, both through satellite imaging and ground surveys.*" Indian Prime Minister Manmohan Singh

Source: Jacob *et al.* 2012; UNEP 2009b; Haeberli 2008; Zemp *et al.* 2008

- Comparable data are needed on water resource efficiency by sector and country, water footprint, and movements of virtual water in traded products.

- A body of research and a number of initiatives are improving data on emerging issues relating to oceans and marine ecosystems, including ocean acidification, marine litter, water quality for bathing, carbon sequestered by marine organisms, the status of coral reefs and algal blooms. All regions have at least some information on fisheries status and trends but, in general, there are major gaps in global coverage of data on the marine environment as a whole, especially in areas beyond national jurisdiction, and consistent time series are rarely maintained (UNEP and IOC-UNESCO 2009).

- The availability of data on glaciers and ice coverage is improving through observation networks and remote sensing (Box 8.2), but information on permafrost is mostly still at the research stage, with monitoring taking place in only a few areas.

Biodiversity

Data on the state of biodiversity, such as protected areas and threatened species, is improving but uneven (BIP 2010). For instance, there is better data on birds and mammals than on invertebrates and plants. In general, monitoring is least extensive in tropical regions, although they contain the greatest share of global biodiversity (UNEP 2011a).

- Data on invasive species populations, when available, are probably substantial underestimates, especially for many developing countries. Addressing this gap is a high priority for small islands, which suffer heavily from invasive alien species.

- Countries will report on the Aichi Biodiversity Targets 2011–2020 under the Convention on Biological Diversity (CBD). A recent review evaluates adequacy of existing observation systems to support reporting on these targets and identifies a number of data gaps (GEO BON 2011).

- Under the previous 2010 Biodiversity Indicators Partnership (BIP 2010), indicators for themes such as the status of access and benefit sharing, as well as traditional knowledge, could not be fully developed because of a lack of comprehensive data.

- A substantial effort has been made in the past to gather data on natural resources that are consumed directly, such as fish and timber, but data quality is not adequate to monitor changes in the composition of fish catches that result from fishing down the food web.

- There is a lack of data allowing the linking of trends in the state of biodiversity with the drivers of biodiversity loss, such as changes in habitat extent, ocean acidification, overfishing and chemicals.

- There is no comprehensive assessment of the number and extent of community-managed protected areas.

- Initiatives such as the Millennium Ecosystem Assessment (MA) and The Economics of Ecosystems and Biodiversity (TEEB) have pioneered approaches to assess and value a broader range of ecosystem goods and services – for example regulating services including the value of ecosystems for disaster risk reduction, and cultural services – but the ability of most national statistical systems to support these approaches is still limited.

Chemicals and waste

- The effects of chemicals on human health and the environment have only been assessed against modern standards (USEPA 2005) for a small number of substances. Data on the effects at different dosages or concentrations, or the effects of combined exposure to multiple chemicals, are at an even earlier stage of research or are absent. In addition, risk assessment procedures often use average adult data, so risks to children also need to be considered.

- Many chemicals became established items of commerce before systematic assessments were made (Lowell Center for Sustainable Production 2003), and concerns have arisen over unsuspected properties such as endocrine disruption, which can damage the hormonal and reproductive system of people and animals (UNEP 2010).

- Chemicals legislation in a number of developed countries, such as the European Commission's REACH programme, has established data inventories that are greatly improving access to information on issues such as chemical toxicity and socio-economic impacts (EC 2012).

- Many new materials containing nanoparticles are produced and widely marketed, but safety testing has been limited even though some potential for human exposure has been identified (Morris *et al.* 2011; Sass *et al.* 2006).

- Data on hazardous wastes at the international level are mainly provided through reports submitted to the Secretariat of the Basel Convention on the Control of Transboundary Movements of Hazardous Wastes and their Disposal, but the number of national reports is declining and the data they contain can be sparse and difficult to interpret. Data on transboundary movements of hazardous wastes is satisfactory according to a recent analysis report, in part since data from Parties that report also include information on transboundary movements involving Parties that do not. Nonetheless, more data are needed about the generation of hazardous wastes and their treatment by importing states. Concerns have also been expressed about the quality of some of the data, and data do not cover illegal movements or generation and disposal in the informal sector (Basel Convention 2010).

- Reliable data about waste generation, collection and management worldwide are lacking, especially in most developing regions. Municipal and household waste should be a particular priority together with industrial and hazardous waste. Data collection on municipal and household waste is complex and time-consuming, and there is a risk of

Rusty fuel and chemical drums in the Arctic. *© Vladimir Melnik/iStock*

double-counting from different sources, for example waste collection companies and disposal facilities.

- In most high-income countries, severely polluted sites have already been identified and remediated, but in many low- and middle-income countries, pollution hotspots are poorly documented, and sometimes are completely unknown to local and national governments (Blacksmith Institute 2011).

- Long-term monitoring programmes for persistent organic pollutants (POPs) in the environment and in human tissue need to be maintained and expanded, in particular in the southern hemisphere (UNEP 2009a).

- The long-term monitoring of marine pollution faces financial and institutional capacity constraints, especially in developing countries (UNEP/GPA 2006).

Policy and responses

- Quantitative targets can spur the collection of data to track whether the targets are being met, but only a few international goals – such as the MDG 7 targets for water and sanitation – include quantitative environmental targets.

- Some information is available on the implementation status of ecolabelling, certification and similar programmes; on protected areas; and on ratification and implementation of conventions. This comes from convention secretariats or compilations such as ECOLEX (www.ecolex.org).

- Limited data are available on issues such as environmental expenditures, green investments, green gross domestic product accounting, trends in and effectiveness of payment for ecosystem service (PES) schemes, environmental crimes and environmental policy effectiveness.

- Governments and other stakeholders can monitor environmental policies by looking at their implementation status, such as the extent of protected areas or of compliance with vehicle emissions regulations, or at their impact, such as trends in species extinction risk or in air quality.

Social and economic issues

- Social and economic data and indices – for example census data and conventional gross domestic product (GDP) – have a solid history that is difficult to match in the environmental domain. A number of initiatives, such as the System of Environmental-Economic Accounting (SEEA) (UN *et al.* 2003) and UNEP's Green Economy Initiative (UNEP 2011c), have supported the development of environmental and social indicators to complement GDP, and are starting to be applied in different countries worldwide.

- Basic data on population trends and distribution are improving. Socio-economic data relating more closely to the state of the environment – on issues such as poverty-environment and environment-security links – are still based mainly on proxy data and case studies. Tools such as the T-21 model of the Millennium Institute (UNEP 2011b) combine environmental and socio-economic data and produce policy-relevant information, for example to demonstrate how resource depletion could affect GDP.

A team of divers conducting coral reef resilience surveys off the northwestern coast of Pemba Island, Tanzania. *© J Tamelander/IUCN*

- Sex-disaggregated data on issues relating to the environment are generally lacking, especially for developing countries, making it difficult to analyse and understand disparities in natural resource use and management structures.

- Data on resource use and efficiencies as well as material flows are improving, but solid baseline data on issues such as resource stocks are often lacking.

- The availability of data on energy production and consumption, including renewable energy, are also improving through the International Energy Agency and other organizations (IEA 2011).

REGIONAL INITIATIVES AND PRIORITIES

Many regional programmes aim at strengthening environmental information on the basis of the specific needs, development status and priority environmental issues of countries in each region. Table 8.2 lists a selection of these programmes and regional priority needs.

Table 8.2: Selected regional initiatives and priorities for environmental information

Africa	• United Nations Statistics Division (UNSD), the African Centre for Statistics and UNEP are supporting countries in developing a core list of indicators for Africa, based on the Commission on Sustainable Development (CSD) indicators, the Millennium Development Goals (MDGs) and the New Partnership for Africa's Development (NEPAD) • Sub-regional organizations like the Economic Community for West African States (ECOWAS) are increasing their involvement, based on programmes in related areas like food security and economic development • In general, Africa is a high-priority region for capacity development in environment statistics
Asia and the Pacific	• Overall, Asian countries have a relatively high response rate to the UNSD/UNEP Questionnaire on Environment Statistics, whereas the response rate from Pacific countries is low (UN 2011); development in the region varies widely between countries • Regional organizations provide support on certain issues, such as compilation of statistics from various sources by the Economic and Social Commission for Asia and the Pacific (ESCAP) and the Asian Development Bank • A number of programmes exist on specific themes and/or covering different sub-regions, for instance: - Commonwealth Scientific and Industrial Research Organisation (CSIRO)/UNEP Asia-Pacific Material Flows - Association of Southeast Asian Nations (ASEAN) Centre for Biodiversity - UNEP/ South Asian Association for Regional Cooperation (SAARC) South Asia Environment Outlook - Acid Deposition Monitoring Network in East Asia (EANET) - Clean Air Initiative – Asia (data on issues such as air quality, energy and transport in Asian cities) - Mekong River Commission (for example water level data)
Europe	• The European Commission, Eurostat and the European Environment Agency (EEA) are the main sources of environmental data in Western and Central Europe, and they are extending their related capacity-building efforts into Central Asia • Eurostat collects, produces, analyses and disseminates statistics on the state of and pressures on the environment – as well as a range of related economic and social data – based largely on data provided by national statistical offices of its member states (Eurostat 2010) • EEA and its member and cooperating countries operate the European Environment Information and Observation Network (EIONET) for multi-country data collection, organization and dissemination • Eurostat and OECD circulate a Joint Questionnaire on the State of the Environment covering member states of both organizations • Environmental data and information are more sparse in European countries that are not members of either organization, including a number of countries in Eastern and South Eastern Europe; one priority in many of these countries is to restore monitoring networks and related data time series that were discontinued in the 1990s (UNECE 2003) • The Helsinki Commission for the Baltic Sea, the OSPAR Commission for the North-East Atlantic, the Barcelona Convention for the Mediterranean Sea and the Black Sea Commission operate data programmes covering their respective marine areas • The pan-European scientific monitoring network of the UNECE Convention on Long Range Transboundary Air Pollution (CLRTAP) has been pivotal in providing the evidence identifying the most important European air quality issues
Latin America and the Caribbean	• A number of regional initiatives are promoting and coordinating environment statistics in the region, including: - the Latin American and Caribbean Initiative for Sustainable Development (ILAC, 32 agencies from 24 countries as of end 2011) - the Working Group on Environmental Statistics of the Statistical Conference of the Americas (15 institutions from 10 countries) - sub-regional initiatives such as the Andean Community Indicators and the Caribbean Community (CARICOM) Indicators • According to a recent survey of national statistical offices and environment ministries in the region, 81% of the participating institutions had an environment statistics programme, although only 36% had a specific budget dedicated to it; other significant institutional challenges remain in many countries (ECLAC 2011)
North America	• Data and information to track the state of the environment is collected and analysed by a variety of government agencies as well as academic and other institutions, including Environment Canada, the US Environmental Protection Agency (EPA) and the National Oceanic and Atmospheric Administration (NOAA) • US government and academic institutions also collect and provide access to global environmental data on a number of issues including sea level, surface temperature, land cover and coral bleaching (Table 8.1)
West Asia	• The Abu Dhabi Global Environmental Data Initiative (AGEDI), in partnership with UNEP, promotes enhanced collection, dissemination and use of environmental data and information. These and other organizations are co-sponsoring the Global Network of Networks initiative, aimed at effective access to the world's expanding pool of environmental data • A core set of environmental indicators for West Asia, developed by the League of Arab States (LAS) in partnership with the Economic and Social Commission for West Asia (ESCWA) and UNEP, has been established by countries of the region on a voluntary basis • An Arab Environmental Information Network is being developed under the umbrella of LAS, with support from UNEP and in collaboration with ESCWA, AGEDI and other organizations • In many countries of the region, official statistics on the environment are rarely generated, difficult to access and scattered among different institutions, and reporting is fragmented (UNEP 2006); major thematic gaps and priorities include data on land salinization, coastal and marine pollution, disasters, waste management and transport (UNEP 2006)

Source: (UN 2011)

NATIONAL CAPACITY NEEDS

Data collected at the national level are some of the most important sources of information to track the state and trends of the global environment. Environment statistics is an emerging field in most countries, and many have only scattered data (UN 2011). Most developing countries currently have no comprehensive environmental observation system. Data may exist but are often discontinuous, making it difficult to establish a baseline to measure change over time or progress against targets. The recent UNDP/UNEP/GEF synthesis of National Capacity Self-Assessments noted that more than 90 per cent of the 119 participating countries identified "information management and knowledge" as a capacity need. While selecting a list of environmental indicators and collecting information was more straightforward in many countries, the main challenge involved managing this information and coordination of the organizations involved, including research institutions and programmes. National environmental management information systems need to be strengthened, as well as the skill sets of associated staff. Measures to address this include application of standards, use of communication

Figure 8.1 Example of a country snapshot on environment statistics, from Uganda

Uganda

Air and climate

Emissions of:		Year
SO_2 (1000t)	...	
SO_2 per capita (kg)	...	
NOx (1000t)	...	
NOx per capita (kg)	59	1994
CO_2 (million tonnes)	3	2007
CO_2 per capita (tonnes)	0	2007
GHG (million tonnes CO_2 eq.)	42	1994
GHG per capita (tonnes CO_2 eq.)	2.0	1994
Ozone depleting CFCs (ODP tonnes)	0	2008
GHG from energy (%)	9	1994

Biodiversity

Proportion of terrestrial and marine areas protected (%)	10	2009
Number of threatened species	166	2010
Fish catch (tonnes)	450 000	2008
Change in fish catch from previous year (%)	-11	2008

Economy

GDP growth rate from previous year (%)	10	2008
GDP per capita ($US)	523	2009
% Value added agriculture, hunting, forestry, fishing	23	2009
% Value added mining, manufacturing, utilities	12	2009
% Value added other	65	2009

Energy

Energy consumption (1000t oil eq.)	1 064	2007
Energy consumption per capita (kg oil eq.)	38	2007
Energy use intensity (kg oil eq.) per $1,000 (PPP) GDP	...	
Renewable electricity production (%)	72	2007

Land and agriculture

Total area (sq km)	241 550	2008
Agricultural land (sq km)	130 120	2008
Arable land (% of agric. land)	43	2008
Permanent crops (% of agric.land)	17.0	2008
Permanent meadows and pastures (% of agric. land)	39	2008
Change in agricultural land area since 1990 (%)	9	2008
Forest area (sq km)	29 880	2010
Change in forest area since 1990 (%)	-37	2010

Population

Population (1000)	33 425	2010
Population growth rate from previous year (%)	3	2010

Waste

Total population served by municipal waste collection (%)	...	
Municipal waste collected (1000t)	224	2006
Hazardous waste generated (tonnes)	...	

Note: The boundaries, the names shown, and the designations used on this map do not imply official endorsement or acceptance by the United Nations.

Water and sanitation

Long-term average renewable freshwater resources (mio m^3/yr)	66 000	N / A
Urban population with access to improved drinking water source (%)	91	2008
Rural population with access to improved drinking water source (%)	64	2008
Urban population with access to improved sanitation (%)	38	2008
Rural population with access to improved sanitation (%)	49	2008

Source: UN 2011

Flooded houses in the aftermath of tropical storm "Hanna", Haiti. Better information on disaster risk is a growing priority in most regions.

© Marco Dormino/UN Photo

technologies and networks, as well as capacity development, public awareness activities and environmental education. In addition, while many stakeholders recognized the value of traditional knowledge for environmental management, few countries felt it was captured well, if at all, and used to develop environmental policies and programmes (UNDP *et al.* 2010).

Recent assessments of the status of national environmental information include the National Capacity Self-Assessments, the UNSD/UNEP Questionnaire on Environment Statistics, and various regional and bilateral needs assessments. UNSD country snapshots (Figure 8.1) summarize the environment statistics that are available, and indicate those that are not, from each country. The specific needs to strengthen environmental information vary in each country, but normally relate to the following issues.

- Collection of high-quality data that adequately cover a full range of established core indicators that can be used to monitor the state and trends of the environment, such as the UN Sustainable Development Indicators (UN 2007b), and that are coherent and comparable. Figure 8.2 provides general information on the existing and planned thematic coverage of national environment statistics programmes.

- Establishment of long-term monitoring programmes in priority areas – based on consistent support in terms of funding and personnel, amongst others – collecting data that are reliable and are available in comparable time series (UNECE 2003).

- Developing the necessary in-country expertise and capacity for data collection, quality assessment, analysis and interpretation on different themes.

- Strengthening institutional arrangements and expertise for coordination of environmental and scientific information that exist in-country but may be fragmented, based on clear roles and responsibilities of different agencies, and the incorporation of economic, social and environmental data into national statistical systems.

- Promoting the easy accessibility of data and information (UNCED 1992) for a range of users and the general public, addressing language barriers as well as cost, security and intellectual property concerns, providing on-line access and harnessing other new technologies to disseminate and present environmental information.

- Supporting institutional and other arrangements that increase the use of national data, indicators and information, for instance for environmental assessments; policy making; convention reporting; and educational, scientific and awareness-raising purposes.

> *"The challenges posed by environment statistics are generally greater than for most other types of statistics... Most significant perhaps is the fact that a national statistical office must rely heavily on other agencies to collect and supply the bulk of the primary data. Such a high degree of interdependence between different government bodies demands close cooperation and collaboration."*
>
> ADB 2002

Scale is very important with respect to environmental data. Some environmental problems are global, such as climate change; others are best addressed and monitored at a regional level, or

Figure 8.2 National environment statistics programmes and thematic coverage, 2007

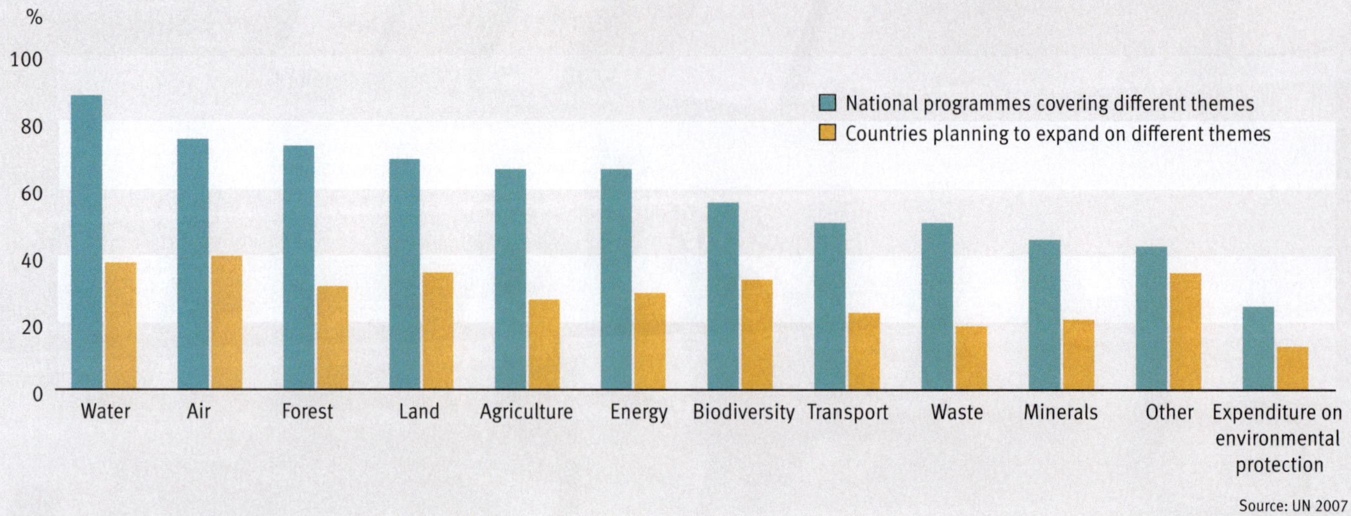

Source: UN 2007

at the level of maritime area (for example fish stocks) or watershed (for example water availability and quality). Many types of air pollution are best monitored at the municipal level. Other issues affect ecosystems, for instance Amazon deforestation affects parts of several countries in South America. The boundaries of an ecological zone, and the scale on which environmental problems occur, will rarely if ever coincide with national or provincial borders or other official geographic delineations (ADB 2002). Nevertheless, government programmes – national statistics in particular – are important for ensuring consistent, long-term support to data collection and dissemination, and the use of scientifically credible information for policy making. In strengthening these programmes, it is useful to consider how environment statistics can be collected at scales relevant to the problems concerned.

Wildebeest crossing the Mara River during the great migration between Kenya and Tanzania. *© SimplyCreativePhotography/iStock*

Many of the constraints on environmental data at the national level are strongly linked with the availability of financial and human resources. The cost of national environmental information systems may vary greatly in different countries, and it is important to ensure that methods for data collection, analysis and dissemination are clear and cost effective (ADB 2002). In order to strengthen the collection of environmental data in countries and, where possible, to improve consistency with international standards it is also important that environmental information supports national policy goals and that policy makers understand its value to their constituency, with international financial support, capacity building and technical assistance available when needed.

CONCLUSIONS

Sound environmental assessments must be underpinned by high-quality scientific environmental, social and economic data. Environmental data are also important to monitor the impacts of environmental policies and programmes. As described in this review, a large number of initiatives are collecting, supporting and improving access to environmental information at the global, regional and national levels. Ongoing and future priorities include promoting, where possible, the use of common standards for data collection and analysis, increased data sharing, consistent time series of environmental observations, capacity building to strengthen environment statistics in a wider range of countries, and harnessing new technologies to communicate environmental information effectively to policy makers and the public.

REFERENCES

ADB (2002). *Handbook on Environment Statistics*. Development Indicators and Policy Research Division, Economics and Research Department, Asian Development Bank, Manila

Bach, H., Bakker, M., Farrington, J., Drillet, Z., Duray, B., Frederiksen, P., Gyuró, E.K., Henrichs, T., Jansson, K., Jensen, T.S., Jombach, S., Jones, L., Kaae, B., Lindner, M., Lopatka, A., Kohlheb, N., Kuhlman, T., Petit, S., Paracchini, M.L., Petersen, L.K., Reid, L., Rothman, D., Scholefield, P., Schulp, N., Stuczynski, T., van Eupen, M., Verburg, P., Verkerk, H., Vogt, J., Vinther, F.P. and Wilson, C. (2009). Indicators – methodology and descriptions. In *SENSOR Report Series 2008/09* (eds. Helming, K. and Wiggering, H.). ZALF, Germany

Basel Convention (2010). *Waste Without Frontiers – Global Trends in Generation and Transboundary Movements of Hazardous Wastes and Other Wastes: Analysis of the Data from National Reporting to the Secretariat of the Basel Convention for the Years 2004–2006*. Secretariat of the Basel Convention, Geneva

BIP (2010). *Biodiversity Indicators and the 2010 Target: Outputs, Experiences and Lessons Learnt from the 2010 Biodiversity Indicators Partnership.* CBD Technical Series No. 53. Secretariat of the Convention on Biological Diversity, Montreal

Blacksmith Institute (2011). *Top Ten of the Toxic Twenty. The World's Worst Toxic Pollution Problems Report 2011*. Blacksmith Institute, New York and Green Cross Switzerland, Zurich. http://www.worstpolluted.org

COD (2003). *Concise Oxford English Dictionary, Tenth Edition*. (ed. Pearsall, J.) Oxford University Press, Oxford

EC (2012). Regulation on Registration, Evaluation, Authorisation and Restriction of Chemical (REACH) substances. European Commission, Brussels http://ec.europa.eu/environment/chemicals/reach/reach_intro.htm

ECLAC (2011). *Report on the Coordination of International Statistical Activities in the Area of the Environment*. Tenth meeting of the Executive Committee of the Statistical Conference of the Americas of ECLAC (Havana, 6–8 April 2011). LC/L.3288(CE.10/7). United Nations Economic Commission for Latin America and the Caribbean, Santiago

Eurostat (2010). *Environmental Statistics and Accounts in Europe: 2010 Edition*. European Commission, Luxemburg

FAO (2007). *The State of Food and Agriculture 2007 – Paying Farmers for Environmental Services*. Food and Agriculture Organization of the United Nations, Rome

FAO and EC-JRC (2011). *Global Forest Land-use Change from 1990 to 2005 – Initial Results from a Global Remote Sensing Survey*. Food and Agriculture Organization of the United Nations, Rome and European Commission Joint Research Centre, Brussels

Finlayson, C.M., Davidson, N.C., Spiers, A.G. and Stevenson, N.J. (1999). Global wetland inventory – current status and future priorities. *Marine and Freshwater Research* 50, 717–27

GEO (2010). *Report on Progress, Beijing Ministerial Summit: Observe, Share, Inform*. Group on Earth Observations. GEO Secretariat, Geneva

GEO BON (2011). *Adequacy of Biodiversity Observation Systems to support the CBD 2020 Targets*. Group on Earth Observations Biodiversity Observation Network. http://www.earthobservations.org/documents/cop/bi_geobon/2011_cbd_adequacy_report.pdf

Haeberli, W. (2008). Changing views of changing glaciers. In Orlove, B., Wiegandt, E. and Luckman, B.H. (eds.), *Darkening Peaks: Glacier Retreat, Science, and Society*. University of California Press, Berkeley, Los Angeles and London

Hansen, M.C., Stehman, S.V. and Potapov, P.V. (2010). Quantification of global gross forest cover loss. *Proceedings of the National Academy of Sciences of the United States of America* 107(19), 8650–8655

ICTSD (2007). *Trade and Sustainable Land Management in the Context of Drylands*. ICTSD Project on Trade and Sustainable Land Management, Selected Issue Briefs. International Centre for Trade and Sustainable Development, Geneva

IEA (2011). *Key World Energy Statistics 2011*. International Energy Agency, Paris

IFAD (2008). *Policy – Improving Access to Land and Tenure Security.* International Fund for Agricultural Development, Rome

IPCC (2011). Summary for policymakers. In *Intergovernmental Panel on Climate Change Special Report on Managing the Risks of Extreme Events and Disasters to Advance Climate Change Adaptation* (eds. Field, C.B., Barros, V., Stocker, T.F., Qin, D., Dokken, D., Ebi, K.L., Mastrandrea, M.D., Mach, K.J., Plattner, G.-K., Allen, S.K., Tignor, M. and Midgley, P.M.). Cambridge University Press, Cambridge and New York

IPCC (2007). *Climate Change 2007: Synthesis Report* (eds. Pachauri, R.K. and Reisinger, A.). Contribution of Working Groups I, II and III to the Fourth Assessment Report of the Intergovernmental Panel on Climate Change. Intergovernmental Panel on Climate Change, Geneva

Jacob, T., Wahr, J., Pfeffer, W. and Swenson, S. (2012). Recent contributions of glaciers and ice caps to sea level rise. *Nature* 482, 514–518

Lehner, B. and Döll, P. (2004). Development and validation of a global database of lakes, reservoirs and wetlands. *Journal of Hydrology* 296, 1–22

Lowell Center for Sustainable Production (2003). *Chemicals Policies in Europe Set New Worldwide Standard for Registration, Evaluation and Authorization of Chemicals (REACH)*. Lowell Center for Sustainable Production, University of Massachusetts Lowell, Lowell, MA

Morris, J., Willis, J., de Martinis, D., Hansen, B., Laursen, H., Sintes, J.R., Kearns, P. and Gonzalez, M. (2011). Science policy considerations for responsible nanotechnology decisions. *Nature Nanotechnology* 6(2), 73–77

OECD (2007). *Glossary of Statistical Terms*. Organisation for Economic Co-operation and Development, Paris. http://stats.oecd.org/glossary (accessed 15 April 2012)

Sass, J., Patrice, S. and Elliott, N. (2006). Nanotechnologies: the promise and the peril. *Sustainable Development Law and Policy* Spring 2006, 11–14, 74

Schneider, A., Friedl, M.A. and Potere, D. (2009). A new map of global urban extent from MODIS satellite data. *Environmental Research Letters* 4, 044003

UN (2011). *United Nations Brochure on Environment Statistics*. Statistics Division, United Nations. http://unstats.un.org/unsd/environment (accessed 20 December 2011).

UN (2009). *Framework for the Development of Environment Statistics*. Report of the Secretary-General for the forty-first session of the Statistical Commission, 23–26 February 2010. Document E/CN.3/2010/9. United Nations Economic and Social Council, New York

UN (2007a). *Global Assessment of Environment Statistics and Environmental-Economic Accounting*. Background document for the thirty-eighth session of the Statistical Commission, 27 February–2 March 2007. Statistics Division, United Nations, New York

UN (2007b). *Indicators of Sustainable Development: Guidelines and Methodologies*. Third Edition. Division for Sustainable Development, United Nations, New York

UN (2004). *Land Degradation and Land Use/Cover Data Sources*. Statistics Division, Department of Economic and Social Affairs, United Nations, New York

UN (1984). *A Framework for the Development of Environment Statistics*. Statistical Papers, Series M, No. 78. Statistics Division, United Nations, New York

UNCED (1992). *Agenda 21*. United Nations Conference on Environment and Development. http://www.un.org/esa/sustdev/documents/agenda21/english/Agenda21.pdf

UNDP, UNEP and GEF (2010). *Capacity Self-Assessments: Results and Lessons Learned for Global Environmental Sustainability*. United Nations Development Programme, United Nations Environment Programme and Global Environment Facility

UNECE (2003). *Environmental Monitoring and Reporting – Eastern Europe, the Caucasus and Central Asia*. United Nations Economic Commission for Europe, New York and Geneva

UN, EC, IMF, OECD and World Bank (2003). *Handbook of National Accounting – System of Integrated Environmental and Economic Accounting 2003*. Final Draft. United Nations, European Commission, International Monetary Fund, Organisation for Co-operation and Development and World Bank

UNEP (2011a). *Methodology for the GEF Transboundary Waters Assessment Programme. Volume 1. Methodology for the Assessment of Transboundary Aquifers, River Basins, Large Marine Ecosystems, and the Open Ocean* (eds. Jeftic, L., Glennie, P., Talaue-MacManus, L. and Thornton, J.A.). United Nations Environment Programme, Nairobi

UNEP (2011b). *Modelling Global Green Investment Scenarios: Supporting the Transition to a Global Green Economy*. United Nations Environment Programme, Nairobi

UNEP (2011c). *Towards a Green Economy – Pathways to Sustainable Development and Poverty Eradication: A Synthesis for Policy Makers*. United Nations Environment Programme, Nairobi

UNEP (2010). *UNEP Yearbook 2010: New Science and Developments in Our Changing Environment*. Division of Early Warning and Assessment (DEWA), United Nations Environment Programme, Nairobi

UNEP (2009a). *Global Monitoring Report under the Global Monitoring Plan for Effectiveness Evaluation*. Note by the Secretariat for the Conference of the Parties of the Stockholm Convention on Persistent Organic Pollutants, fourth meeting, Geneva, 4–8 May 2009. UNEP/POPS/COP.4/33. United Nations Environment Programme, Nairobi

UNEP (2009b). *Recent Trends in Melting Glaciers, Tropospheric Temperatures over the Himalayas and Summer Monsoon Rainfall over India*. Division of Early Warning and Assessment, United Nations Environment Programme, Nairobi

UNEP (2006). *Multi-Scale Databases Comparison for West Asia*. Unpublished. United Nations Environment Programme, Nairobi

UNEP (1999). *Handbook on Data Reporting under the Montreal Protocol*. OzonAction Programme under the Multilateral Fund, Division of Technology, Industry and Economics, United Nations Environment Programme, Paris and Multilateral Fund for the Implementation of the Montreal Protocol, Montreal

UNEP/GPA (2006). *The State of the Marine Environment: Trends and Processes*. United Nations Environment Programme Global Programme of Action for the Protection of the Marine Environment from Land-based Activities. UNEP/GPA Coordination Office, Nairobi

UNEP and IOC-UNESCO (2009). *An Assessment of Assessments: Findings of the Group of Experts Pursuant to UNGA Resolution 60/30. Summary for Decision Makers*. United Nations Environment Programme and Intergovernmental Oceanographic Commission of United Nations Educational, Scientific and Cultural Organization

UN-Water (2008a). *Status Report on Integrated Water Resources Management and Water Efficiency Plans*. Prepared for the 16th session of the Commission on Sustainable Development. UN-Water Report

UN-Water (2008b). *Transboundary Waters: Sharing Benefits, Sharing Responsibilities*. Thematic Paper. Task Force on Transboundary Waters, UN-Water

USEPA (2005). *Guidelines for Carcinogen Risk Assessment*. Document EPA/630/P-03/001F. United States Environmental Protection Agency, Washington, DC

WHO (2010). *WHO Guidelines for Indoor Air Quality: Selected Pollutants*. World Health Organization, Geneva

WHO (2009). *Global Health Risks: Mortality and Burden of Disease Attributable to Selected Major Risks*. World Health Organization, Geneva

Zemp, M., Roer, I., Kaab, A., Hoelzle, M., Paul, F. and Haeberli, W. (2008). *Global Glacier Changes: Facts and Figures*. World Glacier Monitoring Service, Zurich, Switzerland

Part 2: Policy Options

"Parents should be able to comfort their children and say 'everything will be all right... it's not the end of the world, and we are doing the best we can' but I don't think you can say that anymore."

Severn Suzuki, aged 12, addressing UNCED, Rio de Janeiro, June, 1992

"The best way to use hard facts as a motivator is to partner scary information with options for action: 'What can we do? Let's find a way to fix it.' The young, particularly, can handle bad news if there's an avenue for change."

Severn Suzuki, aged 32, Perspectives on Rio+20

CHAPTER 9

Africa

Coordinating lead authors: Jennifer Clare Mohamed-Katerere and Clever Mafuta

Lead authors: Ameer Abdulla, Osman Mirghani M. Ali, Francis Mwaura and Bevlyne Sithole

Contributing authors: Sachooda Ragoonaden, A. Adnan Awad, Kerry W. Bowman, Valerie Rabesahala, Gabriel Grimsditch, Charlotte Karibuhoye, Habtemariam Kassa Belay, Rannveig K. Formo, Marina Gomei, Winnie Lau and Masego Madzawamuse

Principal scientific reviewer: Emma Archer van Garderen

Chapter coordinator: Frank Turyatunga

Main Messages

In Africa, population growth, rapid urbanization, climate change, unsustainable development choices and weak governance persist as critical challenges to achieving both the environmental and the social aspects of important regional goals.

Embarking on mutually reinforcing policies works to Africa's advantage. For nations faced with limited resources, maximizing policy synergies helps deliver social, ecological and economic benefits, reduces trade-offs, and provides multiple paths for addressing common drivers and pressures. For example, sustainable land management policies support positive outcomes in diverse environmental domains including marine ecosystems, biodiversity and freshwater. This will, however, require the harmonization of policies between the local, national, regional and global levels to avoid adverse environmental and social impacts. For example, global policies – such as those for food, biofuels and climate mitigation – should not undercut local livelihoods and adaptive capacity.

Transboundary natural resource policies strengthen the integrated management of shared terrestrial and marine ecosystems. The popularity of this approach in Africa underscores its validity as a useful strategy with wide applicability.

Networks of marine managed areas rather than individual, strictly protected marine areas provide more opportunity for comprehensive management of marine resources. Benefits include fisheries recovery, improved migratory species conservation, reduced land- and sea-based pollution, and enhanced social and ecological resilience to climate change.

The recognition and up-scaling of human rights, including the rights of indigenous people and women, support improved human and environmental well-being, as demonstrated by the recognition of the right to water at national levels and by the African Commission on Human and Peoples' Rights.

Adequate capacity is critical for effective natural resource management and governance. Needs differ across levels, making capacity assessment at multiple scales necessary to identify solutions. Innovative institutional arrangements for pooling knowledge, capacity and financial resources help build long-term collaboration to achieve environmental goals, as approaches for pollution management demonstrate.

Management that considers the whole ecosystem, including people, is more likely to be effective in maintaining environmental goods and services and human well-being, as sustainable land management and integrated coastal zone management show. Africa's extensive use of community-based environmental management has helped poor rural communities generate revenues from wildlife-based enterprises and enabled nations to improve their conservation of large mammals and related ecosystems. The success of this approach hinges on the extent to which governments have devolved authority and rights over natural resources.

Strong accountability with secure participation, access to information and consent ensure that decisions are environmentally and socially sustainable. Africa's growing reliance on extractive resource use makes securing accountability important to avoid special-interest decisions. Monitoring government performance and tracking environmental trends supports effective and timely response to ongoing environmental change, including extreme natural events, and sets the basis for further policy development.

INTRODUCTION

Over the last generation, Africa has built a substantial record of responding to environmental challenges while addressing human well-being, and this provides a point of departure for strengthening policy and implementation. This chapter considers 12 promising policy options identified as contributing to achieving a set of regionally selected international goals (Table 9.1) (see Introduction to *GEO-5* for methodology).

As the policy appraisal shows, innovation by and partnerships between African people and their governments have underpinned this success, while the support of donors has been crucial to the implementation of some policies. The principles of the Paris Declaration on Aid Effectiveness – ownership, harmonization, alignment, management for results and mutual accountability – define collaboration with donors and are designed to ensure that aid supports agreed government priorities and uses, and strengthens government systems rather than developing parallel institutions.

Despite tremendous progress, significant challenges remain, including population growth, rapid urbanization, climate change, unsustainable development choices and weak governance.

Africa's population – 1 billion in 2009 – is growing at close to 2.15 per cent per year (UN 2011), placing increasing demand on environmental resources. In 2010, some 395 million people, or 40 per cent of the total population, lived in urban areas. By 2040, the urban population is likely to be 1 billion and by 2050, 1.23 billion – 60 per cent of the total (UN-Habitat 2010). In Africa's cities – characterized by extremes of prosperous centres and poor, informal settlements – many governments struggle to provide social services including access to water, to achieve food and energy security, and to manage environmental risks (Figures 9.1 and 9.2). Climate change and other adverse environmental change may increase urbanization

Figure 9.1 Exposure and vulnerability to floods in sub-Saharan Africa, 1980–2010

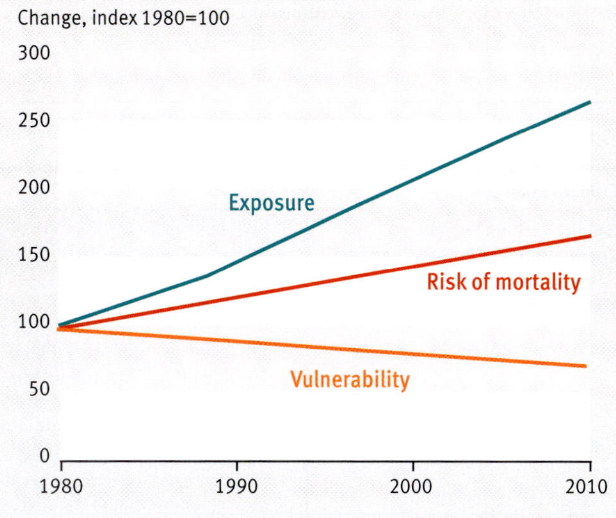

Source: UNISDR 2011

Figure 9.2 Food insecurity in selected Southern African cities, 2008–2009

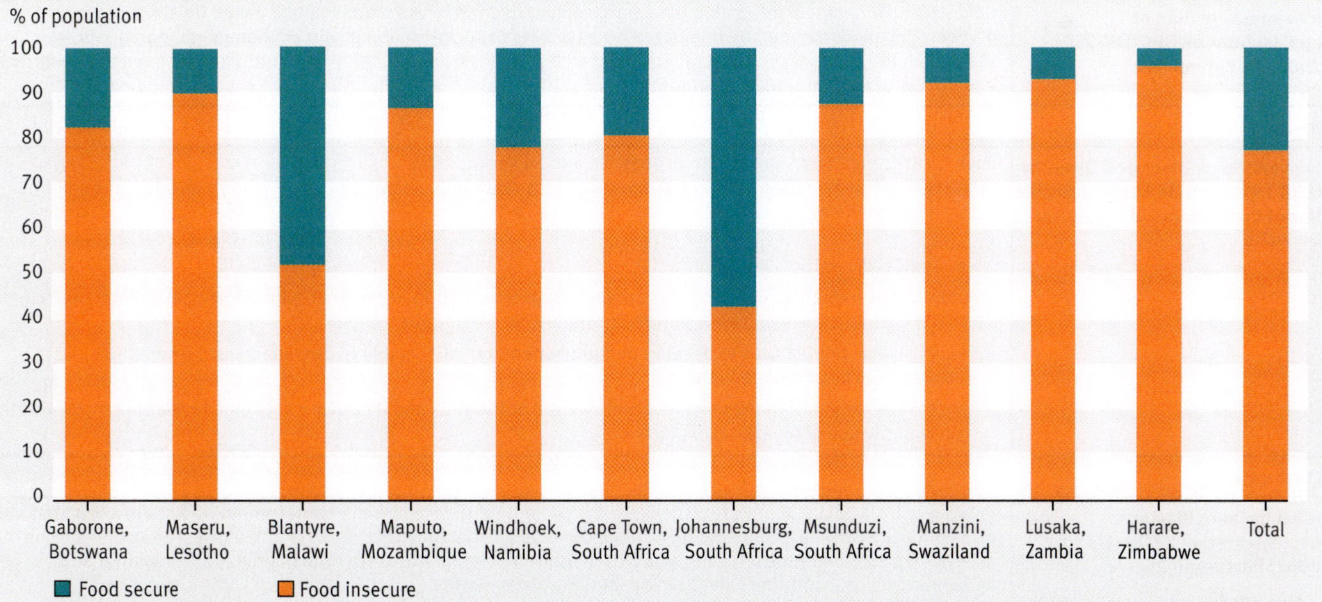

The 1996 World Food Summit in Rome agreed that "food security exists when all people, at all times, have physical and economic access to sufficient, safe and nutritious food to meet their dietary needs and food preferences for an active and healthy life" (FAO 1996). This relies on food availability, accessibility, reliability and quality, as well as acknowledging preference (Frayne et al. 2010; Sen 1981).

Source: Frayne *et al.* 2010

(UN-Habitat 2010) and further strain governments' ability to cope, leading to greater instability (Mohamed-Katerere 2009).

Climate change – by exerting extreme pressure on ecological systems – is likely to increase the stress of vulnerable populations in urban and rural areas (Boko *et al.* 2007). More intense rainfall events contribute to more run-off and floods, threatening food security and settlements, while longer periods between rains and changing seasonal patterns contribute to crop loss. Sea level rise is likely to have significant impacts on coastal settlements, given the high population in potentially hazardous areas (Nicholls 2004), including Akpakpa in Benin and Lagos in Nigeria, making climate-sensitive policies critical for conservation and adaptation. However, most existing policy lacks the framework to address the complex challenges of human vulnerability to climate change (Madzwamuse 2010). The adoption of risk reduction strategies, improved well-being, and better preparedness can result in reduced vulnerability even in these circumstances. Prospects are, however, currently limited; Figure 9.1 shows that exposure to flood and the risk of mortality are outpacing reductions in vulnerability (UNISDR 2011).

Policies and practices that originate outside the region can contribute to environmental change. External investments in

Table 9.1 Regionally selected policy goals

Climate change	
United Nations Framework Convention on Climate Change (UNFCCC 1992) Article 3 Paragraphs 1–3	In their actions to achieve the objective of the Convention and to implement its provisions, the Parties shall be guided, *inter alia*, by the following: 1. The Parties should protect the climate system for the benefit of present and future generations of humankind, on the basis of equity and in accordance with their common but differentiated responsibilities and respective capabilities. Accordingly, the developed country Parties should take the lead in combating climate change and the adverse effects thereof. 2. The specific needs and special circumstances of developing country Parties, especially those that are particularly vulnerable to the adverse effects of climate change, and of those Parties, especially developing country Parties, that would have to bear a disproportionate or abnormal burden under the Convention, should be given full consideration. 3. The Parties should take precautionary measures to anticipate, prevent or minimize the causes of climate change and mitigate its adverse effects. Where there are threats of serious or irreversible damage, lack of full scientific certainty should not be used as a reason for postponing such measures, taking into account that policies and measures to deal with climate change should be cost-effective so as to ensure global benefits at the lowest possible cost. To achieve this, such policies and measures should take into account different socio-economic contexts, be comprehensive, cover all relevant sources, sinks and reservoirs of greenhouse gases and adaptation, and comprise all economic sectors. Efforts to address climate change may be carried out cooperatively by interested Parties.
Land	
Johannesburg Plan of Implementation (JPOI) (WSSD 2002) Paragraph 40b	Develop and implement integrated land management and water-use plans that are based on sustainable use of renewable resources and on integrated assessments of socio-economic and environmental potentials and strengthen the capacity of Governments, local authorities and communities to monitor and manage the quantity and quality of land and water resources.
Biodiversity	
Convention on Biological Diversity (CBD 1992) Article 10: Sustainable Use of Components of Biological Diversity	Each Contracting Party shall, as far as possible and as appropriate: (a) integrate consideration of the conservation and sustainable use of biological resources into national decision-making; (b) adopt measures relating to the use of biological resources to avoid or minimize adverse impacts on biological diversity; (c) protect and encourage customary use of biological resources in accordance with traditional cultural practices that are compatible with conservation or sustainable use requirements; (d) support local populations to develop and implement remedial action in degraded areas where biological diversity has been reduced; and (e) encourage cooperation between its governmental authorities and its private sector in developing methods for sustainable use of biological resources.
Freshwater	
Johannesburg Plan of Implementation (JPOI) (WSSD 2002) Paragraph 26c	Improve the efficient use of water resources and promote their allocation among competing uses in a way that gives priority to the satisfaction of basic human needs and balances the requirement of preserving or restoring ecosystems and their functions, in particular in fragile environments, with human domestic, industrial and agriculture needs, including safeguarding drinking water quality.
Oceans and seas	
Jakarta Mandate on Marine and Coastal Biodiversity of the CBD (CBD 1997)	Promote the conservation and sustainable use of coastal and marine ecosystems as well as their natural resources.

A woman carries firewood across a stream in southern Sudan where floodwaters washed away the bridge. *© Tim McKulka/UN Photo*

land deals have increased rapidly as global demands for food and biofuels have risen – some 45 million hectares of such land investments, 70 per cent of the global total, are in Africa (Deininger *et al.* 2009). Frequently, these investments have adverse impacts on land resources and livelihoods (Cotula 2008). Other examples of external drivers include climate change and waste disposal practices that have adverse impacts on land and water quality, and consequently on human health and food security. This makes the harmonization of policies between different countries as well as different regions a priority.

Over the last ten years, Africa's development approach has focused on securing growth through resource extraction, especially in the oil and mining sectors, and expansion of infrastructure. In the absence of strategic and integrated environmental assessment and robust accountability systems, this has led to environmental degradation. For example, weak accountability in decision-making systems has contributed to forest loss (FAO 2010).

While governance and institutional arrangements vary, some challenges are of regional significance given the similarities across nations. Conflicting laws, values and interests in and between countries adversely affect the ability to develop the collaborative institutional systems that are essential for managing ecosystems and responding to common challenges such as drought (Mohamed-Katerere 2001). At times, this disparity has resulted in resource conflicts, including over the fair allocation of land and water (Mohamed-Katerere 2009; Ashton 2000). Inequitable sharing of the benefits and losses of natural resource management also reduces social-ecological resilience, often creating conflict as deteriorating conditions in one sphere affect another (Mohamed-Katerere 2009). Weak land tenure, insufficient accountability and poor transparency compound this adverse reality. Sectoral planning that treats the environment as a set of separate resources rather than as a composite system further undermines environmental management. In this complex and unsatisfactory governance context, it is often vulnerable groups that suffer most (Jäger *et al.* 2007).

POLICY APPRAISAL

This section appraises the selected policy options and shows that they are often mutually reinforcing, with positive impacts in more than one thematic domain. For example, developing more effective accountability measures or strategies for collaboration are shown to have positive outcomes in diverse policies, including for marine managed areas and natural solutions for adaptation and mitigation. Further, the policy options address a common set of drivers and pressures, as identified in the section above. Table 9.2 provides insight into these links and also shows benefits in social, ecological and economic domains. For African nations under increasing strain and faced with limited resources, maximizing benefits and synergies among these domains while reducing trade-offs can help place environmental and development choices on a path of sustainability.

The appraisal provides broad indicative lessons about conditions for replication and achieving the policy goals. Simple attribution of positive outcomes to any one policy is problematic as multiple factors contribute to success. Weak systems for monitoring and tracking policy outcomes in social, environmental, economic and political domains mean that the appraisal relies primarily on qualitative analysis from peer-reviewed literature and documented project experience. Some of the identified policy options are in the early stages of implementation and consequently show limited impact; nevertheless, current results suggest potential for up-scaling and replication.

Table 9.2 Mutually reinforcing outcomes through effective implementation of selected policy options

	Policy themes and goals				
Policy options	**Biodiversity** Convention on Biological Diversity (CBD 1992) Article 10	**Freshwater** Johannesburg Plan of Implementation (JPOI) (WSSD 2002) Paragraph 26c	**Land** Johannesburg Plan of Implementation (JPOI) (WSSD 2002) Paragraph 40b	**Oceans and seas** Jakarta Mandate of the CBD (CBD 1997)	**Climate change** United Nations Framework Convention on Climate Change (UNFCCC 1992) Paragraphs 1–3
Transboundary natural resource management	Improved conservation through pooling of management, financial and human resources; management improved by harmonizing approaches	Equitable sharing of water resources is ensured, reducing resource conflict	Livelihood and economic benefits reduce reliance on use options that degrade or deplete resources	Harmonized cross-sector regimes secure fisheries while maintaining conservation	Conservation creates new mitigation opportunities and greater availability of environmental resources for adaptation
Marine managed areas	Improved regulation of use regimes; marine networks support joint management			Protected breeding grounds improve fish stocks; multiple needs are reconciled with conservation	Improved ecosystem services and goods support adaptation and mitigation
Regional approaches for marine pollution management	Improved ecosystem quality and resilience enhance biodiversity			Ecosystem restoration helps maintain social and environmental benefits	The reduction of pollution secures livelihood resources that support adaptation
Payment for ecosystem services and biodiversity offsets	Improved local incomes while strengthening conservation perspectives; conservation of offset sites	Better valuation of water resources leading to incentives to protect wetlands and water catchments	Improved social, ecological and economic benefits; diversified income opportunities reduce land pressures	Recognition of the value of the ocean as a carbon sink and for its tourism value helps protect coastal ecosystems	Diverse carbon sinks are secured; enhanced ecosystem services support adaptation and disaster risk reduction
Reducing Emissions from Deforestation and Forest Degradation (REDD+)	Enhanced biodiversity conservation as forest ecosystems are restored and maintained	Water catchments are protected; water quality is maintained or improved	Diverse income opportunities reduce pressure on land	An extended REDD+ (mangroves and seagrass beds) leads to fisheries recovery	Increased earnings and improved ecosystem services support adaptation; mitigation is enhanced
Integrated coastal zone management	Improved conservation is achieved	Saltwater intrusion is reduced	Saltwater intrusion is reduced	Fisheries improve as coastal ecosystems are restored	Increased availability of ecosystem goods and services for adaptation
Sustainable land management	Biodiversity outside protected areas is used sustainably	Available water is more efficiently used; water sources are protected and quality is enhanced	Land productivity is improved through input-use efficiency and rainfall-use efficiency	Coastal systems are restored as agricultural pollutants are reduced	The conservation of natural systems and more efficient land/water use secures valuable resources for adaptation
Human rights	Opportunities are enhanced for resource custodians to protect valuable ecosystems from pollution and non-sustainable use	Water security is improved, providing a fairer distribution mechanism	Tenure and other rights support improved livelihoods	Citizens protect marine resources through action against polluters	Local resilience and adaptation are enhanced by securing access to resources
Local, inclusive and participatory approaches	Local and indigenous knowledge supports restoration and conservation; improved social benefits encourage long-term perspectives		Additional livelihood resources are available, reducing land pressure; long-term perspectives inform local decisions	Coastal resources are more effectively used to support local livelihoods	Adaptation is strengthened as local knowledge and perspectives help define solutions
Water harvesting	The rehabilitation of water catchments supports ecosystem and habitat recovery	Water security is enhanced as supply is improved	Land productivity and food security is improved		Coping capacity is enhanced through improved access to water; flood risk is reduced as run-off is captured
Natural solutions for adaptation to and mitigation of climate change	Restored ecosystems support biodiversity recovery, including of wild crops	Reduced saltwater intrusion from mangrove restoration secures freshwater	Grassland restoration provides pasture and crop biodiversity and reduces likelihood of conversion	Mangrove restoration leads to recovery of coastal systems and fisheries	Crop biodiversity, pasture, fisheries and other ecosystem goods support adaptation; risk is reduced as ecosystems are more secure
Stakeholder pollution management	Biodiversity recovers as pollution is reduced	Improved human health; river integrity is restored; water resources are better valued by users		Coastal and marine systems recover as pollutants are reduced	Reduced flood risk and improved human health contribute to resilience and capacity to cope with climate change

Transboundary natural resource management

Transboundary approaches to environmental and natural resource management contribute to achieving the agreed goals (Table 9.1) in all thematic areas (Table 9.2). There are many examples of successful transboundary initiatives throughout Africa (Huggins *et al.* 2006; Jones and Chonguiça 2001; Wilkie *et al.* 2001), although there are significant variations in their focus, structure, delivery and scope.

These approaches demonstrate success in minimizing biodiversity loss, supporting integrated land and water management, improving local benefits, contributing to fairer and more equitable resource sharing including of ground- and surface water, and improving climate mitigation and the availability of resources for adaptation (Boxes 9.1 and 9.2) (Dudley *et al.* 2010). Importantly, transboundary approaches often enhance cooperation and reduce conflict by facilitating dialogue, establishing networks – including of marine protected areas – and encouraging learning and knowledge sharing (Abdulla *et al.* 2009; Huggins *et al.* 2006; Mohamed-Katerere 2001; Rodgers *et al.* 2001). This helps create the political stability needed for economic and development cooperation. There are, however, many challenges.

Transboundary approaches are inherently complex processes involving many actors, issues and agendas. Strengthening dialogue can contribute to achieving consensus (Box 9.2) (Conca and Dabelko 2002), but reaching it is nonetheless challenging (Ervin *et al.* 2010). Efforts to bring different countries and sectors of society together can create rifts and alienate some communities (Muboko 2011), and scaled-up management may marginalize local users from decision making and reduce access to valuable livelihood resources (Whande 2010). Implementation can also be impeded by ill-defined rights to land and resources, weak governance processes, and conflicting interests and goals (Katerere *et al.* 2001). Consequently, the development and harmonization of laws and policy are essential (Mohamed-Katerere 2001).

The rapid increase in transboundary natural resource management demonstrates that this policy, despite some challenges, has high potential for replication and for managing Africa's diverse shared ecosystems. For example, given that 75 per cent of African countries are coastal and that 70 per cent of river basins are shared by two or more countries, collaborative governance is essential for sustainable approaches.

Box 9.1 The Sangha Tri-National Landscape

The Sangha Tri-National Landscape (TNS) consists of three national parks – Lobéké (Cameroon), Nouabalé-Ndoki (Congo) and Dzanga-Ndoki (Central African Republic) – covering 4.52 million hectares. This area includes logging concessions, community use zones and hunting areas.

The TNS economy is based on the extractive use of timber, bushmeat, palm wine and fish, which provide support to indigenous communities. The goal is to ensure that all use is sustainable. However, the actual impact of this extractive economy on the socio-economic conditions of local people is not easy to quantify because communities are somewhat diffuse.

Biological surveys show healthy populations of endangered wild species, notably elephants and great apes. The TNS also helps maintain the integrity of species' migratory patterns. Experience to date indicates that harmonization is required in transnational legislation for forestry, indigenous land use and hunting. True and just compensation for the area's indigenous communities needs greater attention and analysis.

Source: Usongo 2010; Breuer 2009

Box 9.2 Collaborative water management: Organization for the Development of the Senegal River Basin

In 1974, the Organization for the Development of the Senegal River Basin (OMVS) was established by Mali, Mauritania and Senegal as a supra-national authority to agree water allocation and dam management principles (Varis *et al.* 2006). Water entitlements of riparian countries are made on the basis of demand from three sectors – navigation, energy and irrigation – as negotiated by the parties, and not on the basis of volume demand.

Successes include irrigation of some 375 000 hectares in Senegal, Mali and Mauritania, the provision of 200 megawatts of electricity to the three countries (Madamombe 2005), and securing year-round navigation of a 900-km stretch of the river from Kayes to Saint-Louis, supporting livelihoods for many. A saltwater prevention barrage near the estuary, the Diama Dam, and flood control measures provide support for farmers practising flood-recession agriculture near the dam. In addition, the OMVS has contributed to political and regional cooperation, reducing the potential for conflict and increasing investment in basin resource management.

Despite this success, the complicated institutional set-up of OMVS has not always been able to deal effectively with conflicts (Varis *et al.* 2006). Furthermore, lack of harmonization of OMVS and national action presents challenges for basin management.

Marine managed areas

Marine managed areas are part of a suite of approaches applied in Africa that contribute directly to achieving the oceans and seas goal, and also to the biodiversity and climate change goals by securing coastal ecosystems and environmental services.

The objectives of marine managed areas – which often include strictly protected no-take zones or other marine protected

Fish drying. The successful implementation of policies that enhance fish stocks improves the food security of millions of Africans who rely on fish for protein. *© Jacoline Schoonees*

areas in addition to multi-use areas – complement a broad range of national development and economic goals other than biodiversity conservation. These goals include improved food security, better livelihoods, effective governance and sustained economic growth. Managed areas complement other regulatory policies such as fisheries and water quality management. For example, five-year rotational harvesting in managed areas off the east coast of South Africa contributes to the rapid recovery of oyster populations during fallow years (de Bruyn *et al.* 2009).

Although the establishment of marine protected areas has often been relied on to improve marine conservation, they face multiple challenges. Disparities in governance, institutional capacities, wealth distribution, social capital and availability of ecological data can affect both their establishment and their effectiveness (Abdulla *et al.* 2009, 2008). In some cases, marine protected areas face opposition from adversely impacted sectors of society. For example, tourism operators resisted the establishment of marine protected areas in Kenya because they could not afford the licensing fees, protective clothing, insurance and equipment required by new regulations (Weru 2004). Local fishers who are excluded from their previous fishing zones may also oppose protected area development (Apostolaki *et al.* 2002). In addition, many countries cannot afford comprehensive research on all marine habitats within their jurisdiction, making the identification and development of marine protected areas challenging. Consequently they are often smaller and wider apart than is ecologically viable (Abdulla *et al.* 2009).

Establishing managed marine areas can be an effective alternative, as they include multiple management zones and protected no-take zones. The information required for the design of such no-take zones can be obtained through rigorous quantitative research in a few representative sites combined with comprehensive surveys of traditional knowledge (Johannes 1998). Once designated, many managed marine areas face a lack of adequate resources for proper enforcement of regulations. However, alternative enforcement approaches can be used, including local community guards (Andrews 1998). An added advantage of managed marine areas is that they can control any unsustainable use that is displaced by fully protected no-take zones. Individually, however, countries may be unable to address this as use may occur in areas beyond their legislative jurisdiction, such as the high seas, all of which suggests a need for more collaborative and transboundary approaches. For example, Africa's widest network of managed marine areas, which stretches over 23 sites in six countries of Western Africa – Cape Verde, the Gambia, Guinea, Guinea Bissau, Mauritania and Senegal – has had considerable success in ensuring that fisheries, tourism, and oil and gas development do not adversely affect the marine ecosystem and its biological resources (Karibuhoye 2008).

Africa's 45 649-km coastline (Vafeidis *et al.* 2005), which encompasses 33 of the region's 48 mainland countries and

Box 9.3 A network of managers in the Mediterranean

The establishment of an ecological network of effective marine managed areas requires a multidisciplinary and multi-level approach, with strong commitment from governmental and non-governmental organizations as well as the scientific community. Although governance and legal challenges have halted the implementation of transboundary environmental policies in Northern Africa (Abdulla *et al.* 2008), international cooperation platforms provide a significant opportunity to achieve ecosystem conservation targets. In the Mediterranean, UNEP and the conservation organization WWF partnered to create MedPAN, a network of marine protected area managers that connects more than 40 marine managed areas and supports them to work on a number of coordinated and common initiatives (MedPAN 2011). The network is a tool and a neutral platform for non-governmental and local governmental agencies to deliver services in an efficient way. The partnership has proved successful for the transfer of knowledge, and improved management capacity, data collection, and monitoring and evaluation. This ultimately supports a bottom-up approach, which contributes to building a constituency of practitioners able to influence decision makers on the conservation of marine resources.

six island nations, illustrates the importance of strengthening marine management. Up-scaling managed areas and establishing a network is a step beyond the more traditional approach of establishing them opportunistically as single independent entities. Through interconnections and interdependencies, the individual elements of the network contribute positively to each other's integrity by decreasing overall vulnerability. Marine food webs extend beyond the boundaries of individual areas, and fishers are dependent on different species and geographic regions at different times of the year. Tourism revenues from an accessible managed area with charismatic species can help subsidize the maintenance costs of more remote places with no other values that can be easily captured through current market mechanisms. Many biophysical and socio-economic connections overlap national boundaries, and regional cooperation can promote national interests. Currently unmanaged areas merit priority attention within a larger managed area framework (Abdulla *et al.* 2009). As part of a regional marine conservation agenda, the formation of managed networks in those areas where coverage is minimal and urgently needed – including Northern Africa (Mediterranean Sea), northeast Africa (the Red Sea), the Gulf of Guinea and Southern Africa – has seen some initial successes (Box 9.3).

Regional approaches to marine pollution management

Regional approaches that include a mix of self-regulated, state-enforced and collaborative management are effective in addressing the multiple drivers and diverse scales of marine pollution, and hence in achieving the selected goal for oceans and seas. A reduction in marine pollution also contributes to the biodiversity and climate change goals (Table 9.2).

Coastal urban growth contributes to residential effluent, industrial discharges, storm-water run-off, agricultural and mining leaching, contaminated groundwater seepage, and industrial and vehicle exhaust fumes that enter the marine environment. The coastal cities of Accra in Ghana, Douala in Cameroon, Lagos and Port Harcourt in Nigeria and Luanda in Angola, for example, are all adversely affected by industrial pollutants (Ibe and Sherman 2002). Oil spillage and discharge from marine transport present major management and regulatory challenges, especially for oil-producing countries such as Libya and Nigeria, where problems are severe (Golik *et al.* 1988). Offshore exploration, especially for oil (GEF *et al.* 2006), contributes to pollution from dumping at sea, accidental and intentional oil spills, engine leaks and noise (Abdulla and Linden 2008).

Comprehensive regional marine pollution conventions govern the four major African coastal regions. The Convention for Co-operation in the Protection and Development of the Marine and Coastal Environment of the West and Central African Region (Abidjan Convention) and the Regional Convention for the Conservation of the Red Sea and Gulf of Aden Environment (Jeddah Convention), and their associated protocols, provide important regulatory mechanisms for high-use areas and employ a self-regulatory approach. The benefit of self-regulation is that it is quick to respond, flexible, and sensitive to market circumstances (Osborn and Datta 2006). The primary drawback is that the responsibility falls on the industry to control pollution (Buckley 1994), and the incentives to do so may be insufficient. Waste exchange programmes, initiated under the Guinea Current Large Marine Ecosystem programme, have effectively supported waste reduction and ecosystem recovery (Ukwe and Ibe 2010). In Ghana, this has focused on using the waste from one industry as raw material for another.

Action plans to increase the capacity of port waste reception facilities have been developed regionally, although on-the-ground progress has been limited. For example, the Benguela Current Large Marine Ecosystem programme has promoted the sharing of facilities between ports in the Benguela and Guinea Current regions through assessment of reception facilities, technical training needs and regional capacity requirements as related to the International Convention for the Prevention of Pollution from Ships 1973, as modified by its 1978 Protocol (MARPOL 73/78) (Awad 2008). This process has increased the involvement of key regional stakeholders and advanced some key operational areas of the convention. The Port Management Association of West and Central Africa incorporates marine pollution compliance and requires further investment from ports and industry partners. The need to address the shortage of technical management capacity has been tackled by the International Ocean Institute–Southern Africa through targeted regional training for countries of Western Africa. The existing network of large marine ecosystem programmes, port management associations and regional conventions provides further opportunity for replication of this model. Tax-based policies can complement this approach by extending a company's liability for environmental damage, although a common argument against tax-based policies is that they give the right to pollute if they are not punitive enough. When citizens have standing in the courts they can serve as an important check on industrial practice, as the Niger Delta Ogoni case demonstrates, a case that is also pertinent to the policy option on human rights. This in turn provides an incentive for improved environmental performance.

Although the regional conventions and their protocols are relatively comprehensive in addressing the various marine pollution issues, significant risks remain due to the lack of implementation of these regimes in some countries. But successes exist – as demonstrated by developments under the Convention for the Protection, Management and Development of the Marine and Coastal Environment of the Eastern African Region (Nairobi Convention) (Box 9.4).

With a network consisting of existing regional International Maritime Organization (IMO) offices in Western and Eastern Africa, various regional conventions, and the regional seas and large marine ecosystem programmes, it is clear that the appropriate policy platforms are in place to combat marine pollution. However, there is a paucity of capacity for management in terms of equipment, technical training and institutional support, as well as for implementing existing policies, making investment in these areas a priority.

Box 9.4 Successful pollution management in the Western Indian Ocean

The Nairobi Convention (UNEP 1985) has been instrumental in developing and implementing marine pollution management projects in the Western Indian Ocean, which includes Eastern Africa and the region's island countries.

An operational platform provided by the convention – which brought global investment for the development of policy and management frameworks – was the key strength that led to successful policy initiation and implementation. Other critical features are the IMO regional office in Nairobi and the Agulhas and Somali Currents Large Marine Ecosystem project, which provide additional technical support, including assistance in developing national contingency and action plans.

Project development and implementation underpin the convention's success. For example, the World Bank/GEF-funded Western Indian Ocean Highway Development and Coastal Marine Contamination Prevention Project, implemented through the Indian Ocean Commission and the South African Maritime Safety Authority, has the following goals:

- development of an electronically supported marine highway system for ship guidance and monitoring within the region;
- further extension and implementation of the Indian Ocean Memorandum of Understanding on Port State Control; and
- building capacity for coastal sensitivity mapping and oil-spill response.

Current project success includes the development of a joint draft regional contingency plan. A Regional Coordination Centre for Marine Pollution Preparedness and Response is envisaged to develop a secretariat that would implement the plan.

Source: Jackson 2011

Payment for ecosystem services and biodiversity offsets

Innovative mechanisms such as payment for ecosystem services and biodiversity offsets contribute to achieving all the identified goals (Table 9.1) by encouraging, compensating and rewarding environmental custodians for maintaining or restoring valued environmental services (Swallow *et al.* 2009).

A growing portfolio of payment for ecosystem services in Africa demonstrates benefits for both nature and people, including for watershed services in Eastern and Southern Africa (Stanton *et al.* 2010). Biodiversity offset programmes have been adopted in Ghana, Guinea, Madagascar and South Africa (Madsen *et al.* 2010). Payment for ecosystem services and offset approaches have also been used to support eco-labelling and community tourism, to protect fragile and valued habitats including forests, mangroves and coral reefs (Box 9.5) (Swallow *et al.* 2009), and to sequester carbon (REDD+).

Box 9.5 The Ambatovy Business and Biodiversity Offsets Programme (BBOP), Madagascar

Biodiversity offsets – conservation activities designed to deliver biodiversity benefits in compensation for losses, in a measurable way – are included in the Madagascar Action Plan 2007–2012 as priority projects.

In 2004, the Ambatovy Mining Project adopted the BBOP project, which includes an off-site offset area of 11 600 hectares of endangered forest, 4 900 hectares of on-site conservation zones and a forest corridor to ensure connectivity with remaining eastern rainforests. Support is also given to an adjacent Ramsar wetland and reforestation. The BBOP process led to the integration of the project into local, regional and national plans. Support for capacity building has increased the potential for successful replication.

The project has, however, faced multiple challenges. Since 2006, agreements with the local communities not to expand agriculture and environmentally degrading activities have been breached. This demonstrates the difficulty of achieving consensus when livelihood opportunities are lost. Developing a third-party mechanism to verify claims could ensure fairness and equity in making trade-offs. Other barriers to success relate to difficulties in identifying sites with similar environmental characteristics and comparable biological diversity as offset sites.

Source: Ambatovy Project 2009; Republic of Madagascar 2006

Despite some positive outcomes from these approaches, barriers to success remain (Wunder 2008; Landell-Mills and Porras 2002). Opportunities for local communities continue to be limited: for example, large landowners or companies supply most biodiversity offsets (Box 9.5) even though low-income communities could be competitive suppliers of biodiversity compensation (Milder *et al.* 2010). The weak negotiating capacity of communities makes it difficult for them to participate and secure livelihood benefits that exceed their opportunity costs (Swallow *et al.* 2009; Wunder 2005).

There is considerable potential for expanding payment for ecosystem services in Africa (Table 9.3), as the region lags behind others in developing such approaches (Dillaha *et al.* 2007). In the global carbon offset market for 2011, for example, Africa accounted for less than 3 per cent of emission reduction projects, albeit the region has seen a strong growth trend in the past few years (UNEP Risoe Centre 2011). Enabling factors include agreeing to a set of principles; strengthening the legal framework including certification and capacity building of buyers

Table 9.3 Estimated numbers of low-income households likely to benefit from payment for ecosystem services in developing countries within the next two decades

Ecosystem service	Buyer				
	Public sector	Private, regulated	Private, voluntary	Consumers of eco-certified products	Total number of beneficiaries
Biodiversity conservation	Hundreds of thousands	Millions	Hundreds of thousands	Millions	10–15 million
Watershed protection	Tens of millions	Hundreds of thousands	Hundreds of thousands	Fewer than 100 000	80–100 million
Carbon sequestration	Fewer than 100 000	Tens of millions	Millions	Fewer than 100 000	25–50 million
Landscape beauty or recreation	Hundreds of thousands	Only a few	Millions	Fewer than 100 000	5–8 million

Source: Milder *et al.* 2010

and sellers; encouraging participation by small-scale players by granting them land title or use, access or co-management rights; focusing on long-term livelihood assets rather than on short-term benefits; reducing corruption and "rent-seeking"; creating a more transparent business framework; and facilitating bi- and multilateral knowledge-sharing initiatives (Milder *et al.* 2010; Swallow *et al.* 2009; Wunder 2005).

Reducing Emissions from Deforestation and Forest Degradation

Reducing Emissions from Deforestation and Forest Degradation (REDD+), including the role of conservation, sustainable management of forests and enhancement of forest carbon stocks, is a payment for ecosystem services mechanism currently being negotiated under the United Nations Framework Convention on Climate Change (UNFCCC). Various multilateral processes are underway to support countries in preparing for REDD+ implementation.

With the right safeguards in place, REDD+ could support climate mitigation through carbon sequestration – the climate change goal – and could also address important social and environmental dimensions that could lead to improved livelihoods. Depending on design, REDD+ initiatives may also offer new incentives for addressing the biodiversity and freshwater goals by enhancing forests and the land goal by reducing economic reliance on land-degrading activities. If the current focus is extended beyond terrestrial forests to include mangroves (Crooks *et al.* 2011), REDD+ equivalents could also support realization of the oceans and seas and land goals (Table 9.2).

Although REDD+ is in its preparatory phase, there are some readiness activities, pilot projects and bi-lateral initiatives as well as carbon sequestration projects from which lessons can be garnered. Preliminary evidence suggests benefits for climate mitigation and the environment as well as for people, primarily through supplementary income (Box 9.6) (Bond *et al.* 2010; Madeira 2009).

Early lessons from carbon sequestration projects suggest that unless several challenges facing REDD+ are resolved, market-based approaches could fail to achieve positive outcomes, or might even increase global emissions (Davis 2011; Horta 2009).

Box 9.6 Mozambique: A pilot project in the voluntary carbon market

A voluntary carbon credit project established in 2003 in Sofala Province of Mozambique has helped reduce poverty in the region, which is still suffering from the impacts of the civil war.

By late 2009, the project included 1 510 farmers who rely on subsistence farming, wood gathering and hunting. Between 2003 and 2009, carbon credits totalling US$1.3 million were sold, corresponding to 156 000 tonnes of carbon dioxide (CO_2) at an average price of US$9.0 per tonne. Farmers received a third of the income, the initiating company received a third and its local non-profit subsidiary received a third for project monitoring and evaluation. The project increased rural employment from 8.6 per cent to 32 per cent, whilst 73 per cent of households raised commercial crops compared with 23 per cent previously. There was a measurable increase in literacy and the development of a business ethos and skills.

The main difficulties revolved around measuring and evaluating carbon sequestration, including establishing a baseline and assessing increases in stocks. Existing satellite data were found to be insufficient, and community management and governance presented further challenges. Earnings were reduced by the relatively high costs of carbon sequestration, at US$3.4 per tonne of CO_2, and by the inability to sell all the credits.

Source: Grace *et al.* 2010

Africa's dryland forests are still poorly incorporated in REDD+. *© Yemi Katerere*

For success, REDD+ needs to address the enabling factors identified under the payment for ecosystem services policy option, and must also ensure that:

- earnings exceed opportunities foregone from agriculture and the fuelwood market (Bond *et al.* 2010);
- secure carbon rights that encourage equitable benefit distribution, reduce the potential for conflict, and discourage forest conversion are adopted (Makhado *et al.* 2011);
- enforceable social and environmental safeguards, such as free, prior and informed consent measures, are effective in reducing adverse impacts;
- systems for accurate measurement, monitoring and reporting of emissions (Makhado *et al.* 2011) are implemented;
- reduced implementation costs are achieved;
- effective intersectoral cooperation is established.

A potentially important limitation of REDD+ is that the current UNFCCC forest definition excludes vast areas of open forests, generally in the dry tropics, and therefore overlooks important carbon stocks, for example in much of Eastern and Southern Africa where significant deforestation is taking place (FAO 2011). The inclusion of these dry forests and woodlands would expand the relevance and impact of a post-Kyoto REDD+ mechanism (Hansen *et al.* 2010).

As significant carbon stocks are held in coastal systems (Crooks *et al.* 2011) and soils, carbon credit schemes could be built into the design of new marine protected areas as public-private partnerships to enhance management and financing. Addressing the specific circumstances of communities and organizations running these REDD+ initiatives is also important, as in coastal zones their investments may be at risk from natural disasters. A further challenge for REDD+ is that climate mitigation activities are poorly integrated with adaptation and development. This is particularly problematic given Africa's high levels of poverty and vulnerability to climate change.

Integrated coastal zone management

Integrated coastal zone management provides a management framework that takes into account complex, non-linear interactions between and within human and ecological systems and across temporal and spatial domains, and consequently takes a significant step towards coherent management of entire ecosystems. It prioritizes the land-sea interface with the objective of balancing economic development and environmental protection, consequently contributing to all five selected goals (Tables 9.1 and 9.2).

The number of African coastal countries adopting integrated coastal zone management increased from five in 1993 to 13 in 2000 (Gustavson *et al.* 2008); this is supported by specific commitments to integrated management in regional agreements (Table 9.4).

As a cross-sectoral approach, integrated coastal zone management involves all levels of governance and encourages the involvement of all stakeholders (Hewawasam 2000; Post and Lundin 1996). This is well illustrated for Eastern Africa's coastal countries in the operation of the Secretariat for Eastern Africa

Table 9.4 Selected regional approaches

Agreement	Africa region	Significant feature
Integrated coastal zone management protocol of the Barcelona Convention (UNEP 1976)	Northern Africa	Parties committed to introducing integrated coastal zone management in national and regional policies and adopting regional and national action plans
Nairobi Convention (UNEP 1985)	Eastern Africa and Indian Ocean island countries	National approaches ensure economic growth amongst coastal communities through the sustainable use of coastal resources (World Bank 2011; Gustavson *et al.* 2008)
Regional Coastal Management Programme of the Indian Ocean countries (ReCoMaP)	Eastern Africa and Indian Ocean island countries	Provides assistance to seven countries in applying integrated management principles to national policies and practice, reducing poverty amongst the coastal population (ReCoMaP 2011)
Accra Declaration (1998)	Southern and Western Africa	Water pollution control and biodiversity conservation policies integrated into the Gulf of Guinea Large Marine Ecosystem project

Source: Milder *et al.* 2010

Box 9.7 Action and commitment at regional and national levels

Although action and commitment to integrated coastal management is growing, multiple institutional challenges persist. The Toliary region of southwestern Madagascar suffers from a range of human-induced environmental problems. While these have been reduced using integrated coastal zone management, greater success has been hindered by a lack of regional coordination (Billê and Rochette 2010; Billê 2008). Specific challenges relate to the lack of clearly defined work programmes, procedures and regular coordination meetings.

On the other hand, the Coastal Area Management Programme (CAMP) in the Mediterranean, which includes projects in Algeria, Egypt, Morocco and Tunisia, applies integrated management principles at multiple scales. This includes engagement at the local level, integration in policies and strategies at national and regional levels, and engaging at wider international and Mediterranean basin levels through cooperation and exchange with the Convention for the Protection of the Marine Environment and the Coastal Region of the Mediterranean (Barcelona Convention). Challenges include the lack of continuous financial commitment that has resulted in many projects being halted, inadequate public participation, poor visibility of the projects and an absence of appropriate national legal frameworks (Gonzàlez-Riancho *et al.* 2009; SMAP III 2009).

Coastal Area Management and in South Africa (DEAT 2011; MCM/DEAT 2000). Success requires action and commitment at regional and national levels (Box 9.7).

Experience demonstrates that the integrated coastal zone management protocol of the Barcelona Convention could be strengthened substantially through the use of spatial planning tools. Although these have been used terrestrially for decades, the broader marine community has only recently adopted them. New technologies, including remote sensing, geographic information systems (GIS) and spatial modelling, provide vastly improved capacity to replicate the spatial structure of nature in models of human-environment interactions, and support a strategic decision-making process that creates a blueprint for ocean use. A key strength of these technologies is that they explicitly recognize that there are valid competing demands on natural resources and that ecosystem-based management solutions must work within the capacity of the local communities. As a result, these tools encourage the development of equitable and viable solutions to the conservation of socio-ecological systems (Bode *et al.* 2008).

Sustainable land management

Sustainable land management can strengthen the management of water and land while incorporating social and economic values. Consequently, it supports achievement of the land and freshwater goals and contributes to the biodiversity, oceans and seas, and climate change goals (Tables 9.1 and 9.2).

An example of this approach is the TerrAfrica initiative. This multi-partner platform for consultation and action includes intergovernmental and civil society organizations. In partnership

Box 9.8 Sustainable land management in Burkina Faso and Ethiopia

Experience in Burkina Faso and Ethiopia under the Global Mechanism of the United Nations Convention to Combat Desertification (UNCCD) suggests that small-scale investments and improved farm and community practices can form the basis for scaled-up national sustainable land management programmes.

In Burkina Faso, the *Gestion des Terroirs* approach involves community-based land management, creates awareness of environmental degradation and land conservation, supports development by local government of new regulations for natural resource management, and encourages the adoption of sustainable land management. This includes soil fertility management, mixed farming, the use of organic manure and other agricultural inputs, and water and soil conservation techniques. Traditional approaches for soil restoration and fertilization include, respectively, the use of stone cordons (*diguettes*) and organic farming methods. The data show that the use of agricultural inputs such as fertilizers without investment in sustainable management is often inefficient and uneconomical, suggesting that stronger links need to be made between agricultural production and sustainable land management.

The Ethiopia Strategic Investment Framework for sustainable land management presents a strategy for scaling-up such activities based on best practice. The National Sustainable Land Management Platform has been established and will be replicated at regional level. The programme will cover 177 watersheds in eight regions over five years, based on the model of participatory watershed management. Local involvement in design and priority setting, along with improvements in water availability and food production, can catalyse the spread of existing sustainable land management technologies as farmers learn from each other.

Source: UNCCD/FAO 2010, 2009; TerrAfrica 2009

with the governments of Burkina Faso, Ghana, Namibia and Uganda, TerrAfrica supports country-level approaches. Dialogue on sustainable land management has been initiated in several countries including Eritrea, the Gambia, Malawi, Mali, Niger, Nigeria and Senegal. The success of TerrAfrica, including in Burkina Faso, Ethiopia, Ghana, Mozambique and Uganda, suggests a high potential for integrated and participatory land management approaches to be replicated in other countries (Box 9.8).

Recognizing that addressing climate change as part of sustainable land management is essential to ensure adaptation and address climate-related land changes (Pender *et al.* 2009), TerrAfrica established climate change as a core priority in 2009. Adaptation policies should complement farmers' responses to climate change, including water harvesting and natural solutions such as ecosystem restoration (Table 9.2) (Below *et al.* 2010).

Sustainable land management approaches appear to be most successful where there is high-level political support and when they build upon local knowledge and practice. This contributes to creating effective stakeholder coalitions and platforms, improving the development, management and dissemination of knowledge, and more effectively leveraging the investments required for sustainable management activities.

An on-going challenge in establishing sustainable management is land tenure insecurity. Many governments are rectifying this through land tenure reform: for example, Niger's Rural Code establishes a framework for protecting and revitalizing pastoralism, where previous policies favoured crop farmers (Jamart 2011). It has promoted the preservation of pastoral areas and protected herders' rights to collective use since 1993, including rights to move their livestock in search of water and pasture (Jamart 2011). In 2010 the Rural Code was modified to address outstanding ambiguities. For example, although the code established land commissions as representative organs in which all stakeholders participate, people still turn to religious and customary leaders first to resolve any land issue. A remaining challenge is to stop the encroachment of pastoral areas by croplands as farmers migrate north under demographic pressure.

Box 9.9 The land rights challenge in Mozambique

Mozambique's 1997 Land Law recognizes individual and collective tenure rights and embraces customary African law. Local authorities have control over the delimitation and allocation of land-use rights, the resolution of disputes and resource management (Kanji *et al.* 2006; Burr 2005). The law also protects various human rights including women's rights, customary land claims to uncultivated fields and rights of way, and the rights of internally displaced people – all consistent with rights in international law. The Land Law also expands consultation and the bargaining power of communities. Outside investors, for example, are required to negotiate with customary rights holders to obtain leases within the customary area (CTC 2003; Norfolk and Liversage 2001). Such arrangements, however, are not easy for communities to negotiate and tend to be manipulated by politicians and other influential people (Brown 2003; Hanlon 2002). In some instances, communities are not fully aware of the provisions of the Land Law. An added challenge is that government officials responsible for implementing the law often have little awareness either of the rights or of the procedures for securing them (Serra and Tanner 2008).

Mozambique's experience (Box 9.9) demonstrates that in replicating sustainable land management and land tenure, greater attention must be given to community empowerment and the capacity of state implementing agencies. Given the similarity in land-use systems across Africa, these approaches could be replicated in other countries.

Human rights

Policy approaches that incorporate human rights contribute to the selected freshwater goal (Box 9.10), the land goal through better recognition of local tenure (Box 9.9), and the biodiversity and oceans and seas goals by holding decision makers accountable for decisions that adversely affect the environment. Importantly, these approaches support the achievement of the Millennium Development Goals (MDGs) at the same time as delivering benefits to the environment (Campese *et al.* 2009). Conversely, a lack of rights is often synonymous with high levels of vulnerability (ICHRP 2008; Jäger *et al.* 2007), as experience with external investments in land deals in Africa illustrates (Cotula 2011; Locher 2011).

Human rights are important for protecting people and the environment when there are strong incentives for natural resource exploitation (Bond and Dugard 2007), as is the case in much of Africa. Governance rights, including participation and free prior informed consent, help ensure that local people's rights are taken into account. Human rights provide a benchmark for making sustainable choices and encouraging equitable and non-discriminatory outcomes (ICHRP 2008). Once decisions have been made, litigation can provide a basis for their evaluation. In Nigeria, communities have used human rights law to oppose oil exploration that has adversely affected agricultural land and biodiversity, including for example in the court case *Kenule Beeson Saro-Wiwa, President of the Movement for the Survival of the Ogoni People (MOSOP) and Eight Others,* unreported 1995 (Frynas 1999; Idowu 1999). In 2002, the African Commission on Human and People's rights found that under the African Charter the Nigerian government has an obligation to protect the well-being of the Ogoni People (*Social and Economic Rights Centre v Nigeria).* Giving effect to this decision would limit the way in which oil exploration takes place and ensure protection of the environment, health and livelihoods.

Governance based on human rights may appear to be cumbersome, but it encourages rigour in decision making and ensures that multiple issues and values are taken into account.

Box 9.10 Recognizing a human right to water can promote fairer access

The South African constitution provides a right of access to sufficient water, implemented through the Free Basic Water Policy 42/2001. Many impoverished households benefit from secure access (within 200 metres of the household) to at least 25 litres of water per person per day for domestic use (Mehta 2005). This is equivalent to the World Health Organization's recommendation for minimum consumption, though it does not cover broader health and livelihood needs.

Positive outcomes include a saving of the time and effort women and girls spend collecting water, freeing them to engage in other activities, and less need to resort to unprotected water sources, thus reducing vulnerability to water-borne disease (Mehta 2005). In addition, citizens attribute such policies directly to good governance, and this in turn can support long-term political stability.

A major challenge for the policy is to strike a balance between the human benefits and the cost implications (DWAF 2002a). However, improvements in human well-being are seen as outweighing the associated costs (Stalk 2004). Decentralization of the responsibility for water provisioning to the district level has encouraged the municipalities to become more innovative (Stalk 2004), although some find it costly (DWAF 2002b).

Failure to provide the legally guaranteed quantity has resulted in citizens instituting litigation. In the 2009 Constitutional Court case *Mazibuko v City of Johannesburg*, the court found that the state is required to take reasonable legislative and other measures, within available resources, to achieve the right to water (Larson 2010). Given cost and other barriers, the policy is still to be implemented in rural areas.

Other institutional and organizational challenges include the lack of community access to information and capacity. This case study demonstrates that critical enabling factors include:

- addressing the principle of cost recovery;
- identifying target groups;
- ensuring financing;
- managing demand;
- building capacity;
- improving access to information; and
- facilitating the expansion of infrastructure.

In the long term, human rights approaches encourage political stability and good social relations. However, they can be severely limited by costs (Box 9.10) (Larson 2010), and the capacity of rights holders to claim, protect and enjoy their rights is adversely affected by lack of access to information and knowledge, of access to justice and of capacity.

Notwithstanding these and other challenges, there is potential for replication. Human rights perspectives are increasingly recognized in conservation and governance (Boxes 9.9 and 9.10). The United Nations recognized a right to water in 2010 (UNGA 2010). The African Commission on Human and People's Rights found that failing to provide basic services such as water is a violation of the environmental right in the African Charter. Several African countries including the Democratic Republic of the Congo, South Africa and Uganda now recognize this right in their constitutions (Winkler 2008). For many African nations, the fast pace of urbanization and climate change will increase the challenge of supplying water, making water rights a part of the solution. The extension of a rights-based approach to other resources, such as land (Box 9.9), can support other environmental goals.

Strengthening existing regional institutions could encourage replication and better utilization of these approaches. The work of the African Commission on Human and Peoples' Rights – Africa's main human rights monitoring body – has been limited by state reluctance to give effect to its decisions (Wachira 2008). The African Court on Human and People's Rights, established in 1998, was designed to complement this role but is poorly utilized. For many, the lack of rights of individuals, groups, and non-governmental organizations to initiate legal action is a critical limitation on the effective use of the court by the public. Mali and Burkina Faso have granted individuals and non-governmental organizations direct access to the court (Wachira 2008).

In Sao Tomé, securing access to water is a policy priority. *© Andrew Mohamed*

Local, inclusive and participatory approaches

Policies that reinforce local rights to participate in environmental management help to strengthen stewardship, contributing to the biodiversity, land, water, oceans and seas, and climate change goals (Tables 9.1 and 9.2). These approaches can be incorporated across different conservation policies, such as sustainable land management, integrated coastal zone management, and natural solutions for adaptation to and mitigation of climate change. They broaden the livelihood base for millions of people, for example through transboundary natural resource management, marine managed areas and REDD+ (Box 9.11), strengthen local resilience, including through policies on water harvesting and natural solutions, and encourage learning across levels (Box 9.12).

Since the 1990s, there has been a rise in the number of countries using local, inclusive and participatory approaches, and growth in the extent of land under this type of management (Koech *et al.* 2009; Roe *et al.* 2009). For example, the percentage of forests under community tenure in Africa's ten most forested countries increased during 2002–2008 from 1.2 million hectares to 6.1 million hectares (Sunderlin *et al.* 2008). Several countries including Cameroon, Ethiopia, Ghana, Kenya and Senegal have policies that recognize sacred sites (Dudley *et al.* 2005; Lee and Schaaf 2003). Sites conserved by indigenous peoples and local communities can be successful in strengthening ecosystem management and restoring and maintaining biodiversity, and complement state protected areas (Lee and Schaaf 2003).

A key challenge in sustaining these approaches has been the relatively low level of earnings from environmental management relative to agriculture (Murombedzi 2010). Nevertheless, earnings are improving. In Namibia, conservancies' revenues from wildlife-based enterprises increased from US$73 600 in 1999 to US$4.3 million in 2009, while the Namibian economy earned more than US$32.5 million from community-based natural resource management (NASCO 2010). Other benefits associated with community-based approaches include inclusive governance, infrastructural development and reduced natural resource conflicts (Nelson 2010).

Community-based sustainable wildlife management areas maintain buffalo herds in Zimbabwe's national parks. *© Jennifer Mohamed-Katerere*

Box 9.11 Butterfly farming in Arabuko Forest Reserve

Kenya's 42 000-hectare Arabuko forest is home to the *Kipepeo* (Swahili for butterfly) community-based project, which has earned more than US$80 000 annually from the export of farmed butterfly pupae, thereby decreasing the use of wild biodiversity.

This project demonstrates the tangible link between conservation and sustainable livelihoods. By shifting livelihood strategies from unsustainable use of wood products (firewood, charcoal and timber) to the commercial farming of forest insects, sustainability has been enhanced. This initiative has increased the awareness of communities and national institutions of the ecological and economic importance of insects and their forest habitats. Potential for replication of this kind of project is high, with butterfly farming now being practised in three other areas in Kenya and in the Usambara Forest of the United Republic of Tanzania.

Source: Gordon and Ayiemba 2003

Assessing the overall effectiveness of community-based approaches is challenging, as there has been little empirical monitoring of the impacts on natural resources (Jones 2008). Where there has been monitoring, as in the Democratic Republic of the Congo, Namibia and South Africa, increases in wildlife populations have been recorded (Mehlman *et al.* 2006; Child 2004; Jones 2004). In the Tayna Community Reserve in the Democratic Republic of the Congo there has been a tenfold increase in the elephant encounter rate, a threefold increase in the chimpanzee encounter rate and a twofold increase in the gorilla encounter rate, while during the same period signs of poaching activities decreased sevenfold (Mehlman *et al.* 2006). The West Africa Pilot Community-Based Natural Resource and Wildlife Management Project in Côte d'Ivoire and Burkina Faso reported a reduction in agricultural encroachment within the conservation zones (World Bank 2008). Strengthened community security over the local forest commons has made these landholdings less vulnerable to appropriation by others or to conversion – leading to improved community earnings, biodiversity gains and enhanced forest condition (Sunderlin *et al.* 2008; Banana and Ssembaijwe 2000). These successes suggest a strong basis for replicating this approach.

Box 9.12 Mapping landscapes in southern Cameroon

The Forest Land Oriented Resources Envisioning System (FLORES) is a participatory mapping initiative in eight communities in Akok near Ebolowa, the capital of Cameroon's South Region. Geographic information system (GIS) methods were used in conjunction with focus groups and individual discussions with women, elders and men to explain social realities, perceptions and historical changes in land use. Community-developed base maps in combination with social/cultural data created a new understanding of defined landscape units, ownership, roads and historical land use as well as hunting and fishing rights (Robiglio *et al.* 2003).

This approach enables the identification of social factors that influence land-use dynamics and the alignment of researchers' perceptions with the reality of the communities using the land. Underlying challenges relate to the accuracy of defining spatial boundaries, poor understanding of local languages and high cost and time factors.

Nevertheless, indications are that this approach provides data often lacking in environmental decision making and can be replicated in multiple geographic and social/cultural regions to support better links between environmental planning and social values and priorities (Robiglio *et al.* 2003).

Wheat fields in the Northern Highlands of Ethiopia, where improved land management practices have contributed to a reduction in soil erosion.

© William Davies

Significant barriers to successful implementation remain for many local and participatory approaches. The inadequate enforcement and implementation of local rights remains a challenge: for example, government authorities have often been slow in allocating community entitlement to forests designated as community forests (see section on human rights) (Sunderlin *et al.* 2008). Conflicts between local and state laws, as well as suspicion about communities' capacity to achieve sustainable management, affect government willingness to transfer authority. Better understanding of the multiple meanings and values attributed to forests by local communities can establish a basis for locally appropriate institutional arrangements (Box 9.12). Other barriers include the limited use of markets due to insufficient finance, poor information and technology flows, inadequate market links, and communities' inability to exploit economies of scale (Scherr *et al.* 2004). Enhancing capacity and entitlements, as envisaged by the Convention on Biological Diversity (CBD), will be critical to improving environmental and social outcomes.

Water harvesting

Water harvesting is used to collect run-off or floodwater for storage in the soil or in tanks so that it can be used for the production of crops, trees or fodder and for domestic use. Water harvesting therefore supports the realization of the climate change goal (Table 9.1) in strengthening adaptation by ensuring access to freshwater and by reducing the run-off impacts of extreme rainfall events in tropical, sub-tropical and dryland conditions; in addition it is appropriate for both rural and urban communities. Rainwater harvesting also contributes to achieving the freshwater and land goals (see policy on sustainable land management) and to the biodiversity goal through the restoration of water catchments.

The importance of this policy option is underpinned by climate change and the understanding that by 2020 some 75–250 million

Box 9.13 Rainwater harvesting in Ethiopia

The lack of water for human consumption, livestock and crop farming has been a major constraint in the arid and semi-arid areas of Ethiopia. Nearly 80 per cent of the population lack access to a domestic water supply and an estimated 46 per cent suffer hunger. The government has been engaged in promoting run-off and roof harvesting structures to help address this challenge. Ethiopia has a potential rainwater harvest equivalent to the needs of more than 520 million people (Mati *et al.* 2006). Farmers who harvested water improved their access to water for a longer period and were able to produce vegetables during the dry seasons, so had higher incomes than those who did not harvest rainwater. In Minjar Shenkora district of central Ethiopia, farmers who used harvested water for supplemental irrigation of onions and onion seedlings obtained average net incomes of US$155 per 100 m^2 plot (Akalu and Adgo 2010). In areas where run-off was channelled into micro-catchments, greater plant growth improved fodder production and carrying capacity of the drylands (Abdelkdair and Schultz 2005). However, the initial construction costs of storage structures and inadequate build quality have undermined their wider adoption.

Africans will live in water-stressed areas (Boko *et al.* 2007), while an increase in extreme rainfall events will adversely affect soils and settlements, including cities. The potential of this approach is evident from diverse contexts: Box 9.13 shows its value in the Ethiopian context. Across the Sahel, innovative rainwater harvesting has been applied to hundreds of thousands of hectares, enhancing agricultural productivity and reducing human susceptibility to climatic variability (Box 9.14) (Reij *et al.* 2009). In Mali, research has quantified the impacts of rainwater harvesting on crop yield increases and ground water recharge (Doumbia *et al.* 2008; Kablan *et al.* 2008).

Establishing effective water harvesting can be challenging, with access to resources, labour and skills being limiting factors (Boxes 9.13 and 9.14) (Saico and Kunene 2010). Families may not be able to afford storage facilities that cater to household size (Saico and Kunene 2010). The returns from water-harvesting investments can be long term, so weak land security for smallholders, and particularly for women, may make them reluctant to invest in such technologies.

Nevertheless, the potential for rainwater harvesting is significant and can be replicated in many countries (Mati *et al.* 2006). Integrating water management into national adaptation planning can support the uptake of such technology by addressing legal and policy constraints and increasing community access to financial resources and skills. Several countries, including Togo, recognize water harvesting as a priority in their national adaptation programmes. Support for local knowledge, practice and innovation can empower communities to act and results in the diffusion of water harvesting through farmer-to-farmer learning. Box 9.14 demonstrates how traditional farming knowledge, which has evolved over hundreds of years in response to rainfall variability, has yielded multiple successes in managing scarce water resources and improving food production.

Expanding opportunities for water harvesting can include the rehabilitation of degraded dams, restoration of watersheds, and conservation of existing forests that contribute to water provisioning. These strategies can improve year-round supply of water, soil conservation, and the expansion of livelihood activities including in the agro-pastoral sector (Box 9.13).

Natural solutions for adaptation to and mitigation of climate change

The restoration and maintenance of ecosystems can provide valuable resources for climate adaptation, disaster risk reduction and mitigation (see also section on REDD+, above), and thus help achieve the climate change goal. By enhancing environmental goods and services, ecosystem restoration can also contribute to the realization of the land, oceans and seas, water, and biodiversity goals (Tables 9.1 and 9.2).

Restoration can involve diverse actors at transboundary, national or community level and includes the maintenance of protected areas. By restoring or maintaining ecosystems, natural

Box 9.14 Enhancing traditional water harvesting practices in Burkina Faso

Zaï planting, planting in a shallow pit, in Burkina Faso demonstrates that investment in water and soil conservation improves crop yields. In the Yatenga Province, for example, average sorghum yields increased from 594 kg per hectare during the period 1984–1988 to 733 kg per hectare in the period 1995–2001 as a result of the adoption of *zaï* techniques. Millet yields rose from 473 kg per hectare to 688 kg per hectare for the same periods (Reij and Thiombiano 2003). These improvements have resulted in a reduction of poverty. In Ranawa village, for example, the number of poor families decreased by 50 per cent between 1980 and 2001 (Hien and Ouédraogo 2001).

While these technologies are rooted in local practice and can be mastered by all farmers, indications are that the better-off and medium-income farmers use this technology more than poor farmers simply because they have the resources to pay for labour when needed (Kaboré and Reij 2004). The main disadvantage of the *zaï* techniques is that they require considerable physical effort and good health, especially where digging is required over a large area.

Source: Barry *et al.* 2008

***Zaï* agriculture helps capture water.** © *Jennifer Mohamed-Katerere*

solutions provide opportunities for adaptation and mitigation. Mangrove restoration, for example, can enhance coping capacity by stabilizing coastlines (Duke *et al.* 2007; Mcleod and Salm 2006). Mangrove restoration also supports adaptation through the provisioning of environmental goods such as food, fuel and wood. For example, Nigeria's mangrove forests provide breeding grounds for more than 60 per cent of the fish caught between the Gulf of Guinea and Angola (Carrere 2009). In Sudan, the restoration of rangelands – achieved through rotational grazing and a shift in livestock composition – helped improve livestock pasture and food security (Buffle and Elasha 2011). An unexpected consequence of this effort was that pastoral nomads were attracted to the area, with conflict being avoided by using traditional local institutions and values to negotiate access. Protected areas, including in Niger, have been found to support the *in situ* conservation of wild crop relatives, which are often more drought resistant than domesticated crops and can be used to strengthen agriculture and food security (Dudley *et al.* 2010).

Ecosystem restoration frequently requires a coherent but cross-cutting multisectoral approach as drivers and pressures exist at multiple levels. Large-scale or global drivers include oil exploration, agricultural expansion and pollution, infrastructure and transport development, population growth and settlements, and coastal development (Adger *et al.* 2005). At the same time, local livelihoods can place pressure on resources where governance and management are weak, for example through unsustainable fuelwood harvesting in mangroves (Ajonina *et al.*2005; Ajonina and Usongo 2001). Establishing integrated approaches that address drivers at multiple levels is often challenging, especially where coordination and collaboration between policy development agencies and policy implementing agencies are weak. Poor data collection, monitoring and information further constrain adaptive management. Inadequate legislation that is sector-based, conflicting, deficient and unenforceable provides a weak basis for planning and management (Madzwamuse 2010; Gordon *et al.* 2009).

In addition, enhancing the conservation of ecosystems and their capacity to regenerate requires a better understanding of the links between different ecosystem components (Abdulla *et al.* 2011; Davis *et al.* 2011) as well as of social-ecological resilience (Johnson and Welch 2010; Adger *et al.* 2005). Investing in and generating ecological knowledge and translating it into information that can be used in governance and policy development is essential for management success (Adger *et al.* 2005), and requires a better interface between science, policy makers and communities. Regional cooperation, community-driven strategies and public-private partnerships (Box 9.15) can support learning, improve sustainability and encourage ecosystem approaches. The recently adopted Mangrove Charter for West Africa, which is complemented by country-specific action plans, is an example of this.

Given that adaptation is about local capacity, it is important that strategies and projects enjoy shared understanding between policy makers, technical agencies and communities (Box 9.16) (Patt and Schroeter 2005). Unless this is achieved, there is a risk that adaptation strategies will run counter to local livelihoods, values and cultures, and that uptake will be low, as in the government-initiated resettlement scheme following Cyclone Eline in 2000 in Mozambique (Patt and Schroeter 2005). The loss of easy access to resources and social support were key obstacles to support for resettlement. A second major challenge was the conflicting perception of climate risk severity held by the government and the communities. These results highlight the need for active dialogue across stakeholder groups as a necessary condition for formulating and successfully implementing policies (Patt and Schroeter 2005). Ongoing dialogue creates the basis for reassessing strategies and responding to change (Box 9.16)

Stakeholder pollution management

Pollution management is important for restoring ecosystems and realizing human health goals. It contributes to achieving the social

Box 9.15 Mangrove restoration in Mauritius

In 2008, in response to declining mangrove forests in Mauritius, the non-governmental organization *Association pour le Développement Durable*, with support from the European Union and the Ministry of Finance, planted about 10 000 mangrove seedlings at Le Morne, a small fishing village in the south. The local community was actively involved. Cooperation included training on planting techniques by the Albion Fisheries Research Centre of the Ministry of Fisheries and Rodrigues. In 2011, multiple levels of cooperation and funding from a commercial bank under a corporate social responsibility scheme led to an additional 40 000 seedlings being planted. An island-wide survey has been undertaken to identify potential areas for replication.

Source: ADD 2011

Mangrove planting at Le Morne. © *Subash Chacowry/ADD*

Box 9.16 Social learning and knowledge in community-based adaptation strategies

The relative success of a community-based mangrove management project in Cameroon demonstrates the value of participation and learning for successful adaptation (Ajonina *et al.* 2009).

Communities of the Campo Beach raised over 4 000 mangrove seedlings in community-run nurseries and planted them as a green shield to protect Campo Beach from coastal erosion and wind. This project was a response to collapsing concrete walls along the beach. Dialogue, learning and inclusion in project development led to active community involvement in different aspects of the project, including seedling development, the demarcation of mangrove wood-gathering zones, locally based enforcement and continuous involvement in monitoring and evaluation. Locally appropriate technologies, including energy-efficient fish-smoking houses, have been adopted.

Source: Ajonina *et al.* 2009

Smallholder farmers of Eastern Africa play an important part in the global agenda for the conservation and sustainable use of natural resources and the equitable sharing of related benefits. *© Guenter Guni/iStock*

and environmental aspects of the selected goals for biodiversity, freshwater, oceans and seas, and climate (Tables 9.1 and 9.2).

Africa has relied primarily on regulatory approaches to achieve pollution targets. These approaches influence environmental outcomes by regulating processes or products, limiting the discharge of specified pollutants, and restricting certain polluting activities to specific times or areas (Bernstein 1997). However, regulatory instruments are often inefficient in achieving pollution control objectives, especially where resources for monitoring pollution and compliance are lacking. The level of expenditure required for ensuring compliance with increasingly stringent environmental laws is an unmanageable cost for many governments. In contrast, stakeholder-driven management approaches have the potential to make pollution control economically advantageous to commercial organizations. These approaches can involve varying degrees of incentives, information and administrative capacity for effective implementation and enforcement. The principal types of economic instruments used for controlling pollution include pricing, pollution charging and marketable permits (Bernstein 1997).

Box 9.17 Managing acid mine drainage in the Olifants catchment

The upper Olifants catchment lies in the Gauteng and Mpumalanga provinces of South Africa, where coal mining, mineral processing and agriculture are key economic activities (Hob *et al.* 2008). The Olifants' waters are polluted by acid mine drainage, particularly from coal mining.

The controlled discharge scheme, introduced in the upper Olifants catchment in 1997 with the support of industrial stakeholders, takes advantage of the natural assimilative capacity of the river system during high flow conditions to control the discharge of acid mine drainage (Hob *et al.* 2008). The upper catchment is divided into management units, each with a distinct waste load allocation based on the assimilative capacity of the unit. Participating industries are allowed to discharge poor-quality water to the host management unit in proportion to the assimilative capacity of the unit and their share in the scheme (Limpitlaw *et al.* 2005).

The scheme has succeeded in reducing sulphate concentrations at the Witbank Dam (World Coal Institute 2002), which is expected to contribute to the ecological integrity of the river in the long run. Significantly, discharge during low-flow periods is reduced. Costs are borne by the polluter, ensuring that general tax can be put to other uses. Industries in the area including mines and power stations made significant capital and operational investments towards this project, with one company investing over R100 million (US$13 million) in December 2007 for drainage, storage and treatment systems to improve the quality and quantity of its discharges (World Coal Institute 2002). The flood risk has been reduced and a healthier community is envisaged.

Unfortunately, water quality problems in the Olifants catchment persist. Success for similar initiatives depends on strong institutional capacity, economic stability, government recognition of innovative ideas and the committed involvement of stakeholders.

Box 9.17 on pollution control in the Olifants illustrates how stakeholder involvement can be successfully implemented and lead to a shift in established assumptions about where responsibility for pollution management lies.

CONCLUSION: BUILDING ON SUCCESS

This environmental policy appraisal suggests that opportunities for building on existing success can be effectively harnessed to ensure better implementation and positive outcomes for people and the environment.

Replicating and up-scaling effective approaches is important, but policies should not be blindly replicated, and should be modified to achieve good fit with local, national and regional conditions. As amply demonstrated in the policy options detailed above, it is important to maximize opportunities by focusing on options that are mutually reinforcing and cross-cutting (Table 9.2). Finding and developing synergies is cost effective when financial and human resources are limited. Ensuring that policies are not in conflict with each other and that they do not lead to an externalization of adverse impacts is important.

As the policy appraisal shows, effective policy implementation requires reducing or removing barriers and strengthening enabling conditions. Insufficient monitoring, special-interest decision making, weak governance and rights, and a lack of adequate capacity have undermined policy success.

Policies that have in-built flexibility are needed to address environmental change. Investing in monitoring and evaluation, as well as social learning, supports revision and modification of policy responses, as illustrated in many of the policy options discussed here, including, for example, natural solutions for adaptation and mitigation.

Decision making that is strategic and takes account of how changes in environmental use and governance affect the resilience of the social-ecological system has been shown to be effective in securing economic, social and environmental benefits. Integrating human and ecological understanding and priorities in environmental management can help ensure that choices do not destroy or undermine the environmental resources that underpin future options. Such approaches – including ecosystem-based management – prioritize the interface between people and nature and do not favour just one ecosystem component, industry sector, community or socio-economic group (Davis *et al.* 2011). Ecosystem-based management is one way to maintain the Earth System's ability to adapt to change, as compared to other approaches that focus on fixed targets and state systems, or on hard-engineering solutions that often interfere with natural processes (Abdulla *et al.* 2011).

Strong accountability helps secure government and private-sector commitment to implementation and to achieve agreed outcomes (Najam and Halle 2010) (see also the sections on local, inclusive and participatory approaches, human rights, and stakeholder pollution management). For countries to be better able to demonstrate results, systems for monitoring progress need to be established. Developing performance indicators rather than effort-based indicators, such as the number of meetings held, improves clarity about how and to what extent the purpose of the policy is being achieved (Najam and Halle 2010). Strong and effective national and sub-regional reporting systems help hold implementing agencies to account and provide an opportunity to document successes, which in turn set the basis for up-scaling and replication.

Cooperation has been shown to be effective for achieving sustainable management, including policy options for transboundary coastal and land-based resource management, and where there are multiple stakeholders. This has improved equity, enhanced skills sharing and reduced conflict. In some cases, external support and collaboration with donors have helped establish effective platforms for engagement, learning and sharing knowledge and skills, including in TerrAfrica and under the Nairobi Convention. Partnerships with the private sector and environmental managers or custodians have been shown to be effective in securing benefits in many of the policy options, including payment for ecosystem services, and in mangrove restoration. Several of the options presented, including sustainable land management, show that a high degree of participation at local and government levels helps to ensure relevance, with good outcomes for strengthening sustainability. Decentralization and devolution policies, including in community-based resource management, have achieved positive outcomes for communities and for the environment.

Strengthening the governance and institutional regime for more equitable benefit sharing is critical, given that ecological and social resilience are tightly inter-twined, as South Africa's basic water policy shows. Weak tenure and entitlements stand out as key

The central Namib Desert, Namibia. © *Lucyna Koch/iStock*

Figure 9.3 Selected strategies from the policy options for strengthening key components of capacity

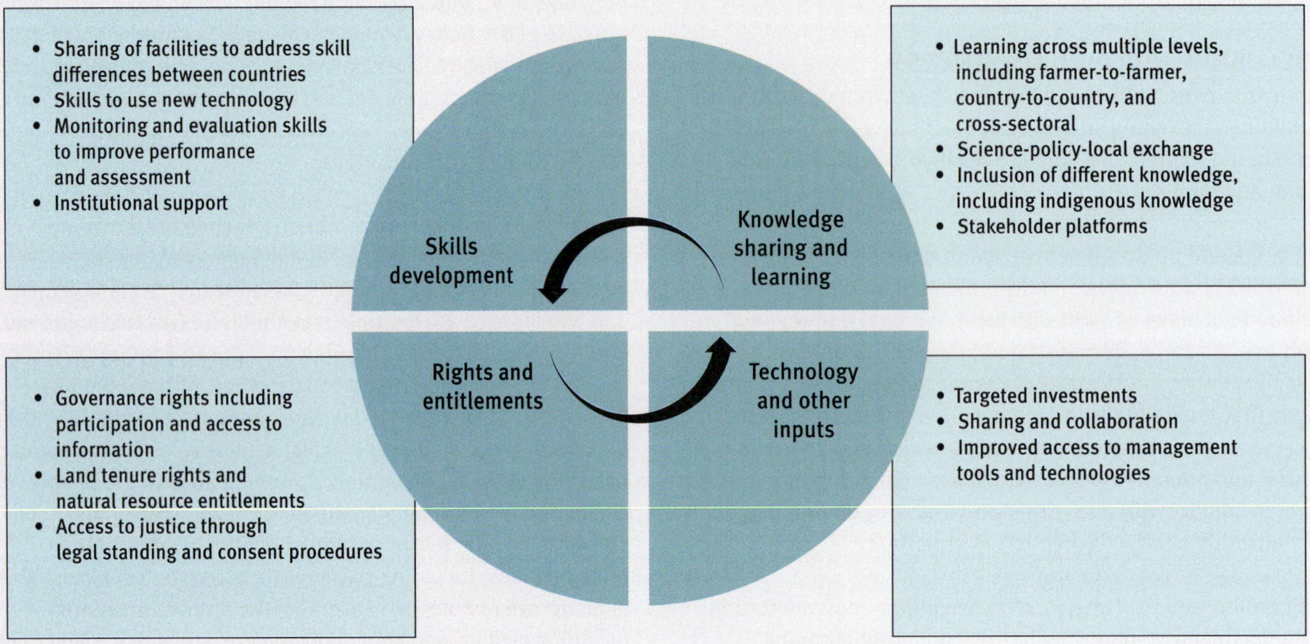

barriers to achieving equitable benefits for payment for ecosystem services including REDD+, community-based management and other policy options. While these are national problems, the scale and commonality of the challenges suggest that developing and adopting regional or global protocols for cooperation and sharing could provide the basis for more effective engagement and management of benefits and losses. Strengthening and integrating human rights perspectives in environmental management frameworks at national and regional levels supports more inclusive, long-term approaches by protecting livelihood rights, ensuring inclusion and reducing conflict. Regional human rights bodies can play an important role in solidifying the piecemeal environmental benefits that human rights recognition has already brought, especially where the mandate of regional courts and the rights of citizens to bring actions are strengthened, as shown in the human rights policy option.

Cloud forests of Nyungwe Forest National Park, in southwestern Rwanda, hold potential for REDD+. *© Guenter Guni/iStock*

Environmental policy is often out of step with realities on the ground, with governments acting alone often unable to effect the necessary change. Innovative institutional arrangements for pooling financial resources, knowledge and capacity, however, can contribute to achieving environmental goals. Improving capacity and equity among diverse communities, including governments, is essential to support collaboration and to secure rights. The policy options demonstrate the potential of various strategies to enhance capacity (Figure 9.3). At regional and sub-regional levels, for example, mechanisms for sharing information and knowledge, as in the management of marine pollution, could be better utilized.

Addressing the barriers to sustainability and achieving the agreed environmental goals addressed in this chapter can create a springboard to improve environmental performance and to move from promising policies to successful policies. Strengthening environmental management can help protect the ecosystem goods and services on which development and human well-being opportunities are based and provide a basis for dealing with key challenges including food security, poverty, urbanization and climate change impacts.

REFERENCES

Abdelkdair, A. and Schultz, R. (2005). Water harvesting in a 'runoff–catchment' agroforestry system in the dry lands of Ethiopia. *Agroforestry Systems* 63(3), 291–298

Abdulla, A. and Linden, O. (eds.) (2008). *Maritime Traffic Effects on Biodiversity in the Mediterranean Sea: Review of Impacts, Priority Areas and Mitigation Measures.* IUCN Technical Series. IUCN Centre for Mediterranean Cooperation, Malaga

Abdulla, A., Game, E., Grimsditch, G., Obura, D., Purkis, S., Rowlands, G. and Rouphael, T. (2011). *Integrating Resilience to Climate Change into Marine Spatial Planning.* UNEP Marine and Coastal Division Series. United Nations Environment Programme, Nairobi

Abdulla, A., Gomei, M., Hyrenbach, D., Notarbartolo-di-Sciara, G. and Agardy, T. (2009). Challenges facing a network of representative marine protected areas in the Mediterranean: prioritizing the protection of underrepresented habitats. *ICES Journal of Marine Science* 66, 22–28

Abdulla, A., Gomei, M., Maison, E. and Piante, C. (2008). *Status of Marine Protected Areas in the Mediterranean Sea.* IUCN, Malaga and WWF, France

Adger, W.N., Huges, T.P., Folke, C., Carpenter, S.R. and Rockstrom, J. (2005). Social-ecological resilience to coastal disasters. *Science* 309, 1036–1039

Ajonina, G.N. and Usongo, L. (2001). Preliminary quantitative impact assessment of wood extraction on the mangroves of Douala-Edea forest reserve Cameroon. *Tropical Biodiversity* 7(2)3, 137–149

Ajonina, G., Tchikangwa, B., Chuyong, G. and Tchamba, M. (2009). The challenges and prospects of developing a community based generalizable method to assess mangrove ecosystems vulnerability and adaptation to climate change impacts: experience from Cameroon. In *The Relevance of Mangrove Forests to African Fisheries, Wildlife and Water Resources. Nature and Faune* (eds. Bojang, F. and Ndeso-Atanga, A.). vol. 24 pp.16–25. Food and Agriculture Organization of the United Nations, Accra. ftp://ftp.fao.org/docrep/fao/012/ak995e/ak995e00.pdf (accessed 29 November 2011)

Ajonina, P.U., Ajonina, G.N., Jin, E., Mekongo, F., Ayissi, I. and Usongo, L. (2005). Gender roles and economics of exploitation, processing and marketing of bivalves and impacts on forest resources in the Douala-Edaa Wildlife Reserve, Cameroon. *International Journal of Sustainable Development and World Ecology* 12(2005), 161–172

Akalu, T.F., and Adgo, E.T. (2010). Water harvesting with geo-membrane lined ponds: impacts on household incomes and rural livelihoods in Minjar Shenkora district of Ethiopia. In Mati, B.M., *Agricultural Water Management Interventions Delivers Returns on Investment in Africa: A Compendium of 18 Case Studies from Six Countries in Eastern and Southern Africa.* VDM Verlag

Ambatovy Project (2009). *BBOP Pilot Project Case Study: The Ambatovy Project.* Business and Biodiversity Offsets Program. http://bbop.forest-trends.org/guidelines/low_ambatovy-case-study.pdf (accessed 29 November 2011)

Andrews, G. (1998). *Mafia Island Marine Park, Tanzania: Implications of Applying a Marine Park Paradigm in a Developing Country.* Proceedings of the International Tropical Marine Ecosystem Management Symposium 1998. Great Barrier Reef Marine Park Authority, Townsville

Apostolaki, P., Milner-Gulland, E.J., McAllister, M.K. and Kirkwood, G.P. (2002). Modeling the effects of establishing a marine reserve for mobile fish species. *Canadian Journal of Fisheries and Aquatic Sciences* 59, 405–415

Ashton, P. (2000). Southern African water conflicts: are they inevitable or preventable? In *Green Cross International: Water for Peace in the Middle East and Southern Africa.* pp.94–98. Green Cross International, Geneva

ADD (2011). *Mangrove Propagation at Le Morne with the Active Participation of the Vulnerable Local Communities and Preparation of a GIS Map Highlighting Potential Sites for an Island-wide Mangrove Restoration Programme.* ADD/MCB-FF Project Third Interim Quarterly Report. Association pour le Developpement Durable, Mauritius. http://www.addmauritius.org/GEO%205%20Third%20MCB%20FF%20report.doc (accessed 11 November 2011)

Association pour le Développement Durable (2009). *Improving the Livelihood and Welfare of Artisanal Fishermen and Other Coastal Communities in Le Morne Village.* ADD/DCP/EU Project Final Report. Association pour le Développement Durable, Mauritius. http://www.addmauritius.org/FINAL%20NARRATIVE%20REPORT_sgw%201.doc (accessed 11 November 2011)

Awad, A.A. (2008). *Assessment Report and Action Plan for Developing Port Waste Reception Facilities in the BCLME Region in Accordance with MARPOL 73/78.* Report for the Benguela Current Large Marine Ecosystem Programme, Windhoek

Banana, A.Y. and Ssembajjwe, W.G. (2000). Successful forestry management: the importance of security of tenure and rule enforcement in Ugandan forests. In *People and Forests: Communities, Institutions and Governance* (eds. Clark, G., McKean, M. and Ostrom, E.). MIT Press, Cambridge, MA

Barry, B., Olaleye, A.O., Zougmoré, R. and Fatondji, D. (2008). *Rainwater Harvesting Technologies in the Sahelian Zone of West Africa and the Potential for Outscaling.* IWMI Working Paper 126. International Water Management Institute, Colombo

Below, T., Artner, A., Siebert, R. and Sieber, S. (2010). Micro level practices to adapt to climate change for African small scale farmers. *Sustainable Land Management* 953. IFPRI, Washington, DC

Bernstein, J.D. (1997). Economic instruments. In *Water Pollution Control – A Guide to the Use of Water Quality Management Principles* (eds. Helmer, R. and Hespanhol, I.). Weinham, Melbourne

Billé, R. (2008). Integrated coastal zone management: four entrenched illusions. *Surveys and Perspectives Integrating Environment and Society* 1(2), 75–86

Billé, R. and Rochette, J. (2010). *Feasibility Assessment of an ICZM Protocol to the Nairobi Convention.* Regional Programme for the Sustainable Management of the Coastal Zone of the Countries of the Indian Ocean, Nairobi

Bode, M., Wilson, K.A., Brooks, T.M., Turner, W.R., Mittermeier, R.A., McBride, M.F., Underwood, E.C. and Possingham, H.P. (2008). Cost-effective global conservation spending is robust to taxonomic group. *Proceedings of the National Academy of Sciences of the United States of America* 105(17), 6498–6501

Boko, M., Niang, I., Nyong, A., Vogel, C., Githeko, A., Medany, M., Osman-Elasha, B., Tabo, R. and Yanda, P. (2007). Africa. In *Climate Change 2007: Impacts, Adaptation and Vulnerability* (eds. Parry M.L., Canziani, O.F., Palutikof, J.P., van der Linden, P.J. and Hanson, C.E.). Contribution of Working Group II to the Fourth Assessment Report of the Intergovernmental Panel on Climate Change. pp.433–467. Cambridge University Press, Cambridge

Bond, P. and Dugard, J. (2007). Water, human rights and social conflict: South African experiences. *Law, Social Justice and Global Development Journal* 2007(1). http://go.warwick.ac.uk/elj/lgd/2008_1/bond_dugard/ (accessed 31 May 2011)

Bond, I., Chambwera, M., Jones, B., Chundama, M. and Nhantumbo, I. (2010). REDD+ in dryland forests: issues and prospects for pro-poor REDD in the miombo woodlands of southern Africa. *Natural Resource Issues* 21. International Institute for Environment and Development, London

Breuer, T. (2009). *Best of the Wild: Wildlife Conservation Society and the Ndoki Landscape.* Wildlife Conservation Society – Congo Program. www.wcs.org/about-us/~/media/Files/.../Ndoki_prospectus.pdf (accessed 29 November 2011)

Brown, T. (2003). *Contestation, Confusion and Corruption: Market-based Land Reform and Local Politics in Zambia.* Paper presented at International Conference on Competing Jurisdictions: Settling Land Claims in Africa, 24–27 September, Vrije Universiteit, Amsterdam

Buckley, R. (1994). Environmental self-regulation in industry. *Environment and Planning Law Journal* 11(1), 3–5C

Buffle, P. and Elasha, B. (2011). *Community-based Rangeland Rehabilitation for Adaptation To Climate Change and Carbon Sequestration.* Ecosystems and Livelihoods Adaptation Network. http://elanadapt.net/sites/default/files/siteimages/6.sudan.pdf (accessed 15 October 2011)

Burr, K. (2005). The evolution of the international law of alienability – the 1997 Land Law of Mozambique as a case study. *Columbia Journal of Transnational Law* 43(3), 961–998

Campese, J., Sunderland, T., Greiber, T. and Oviedo, G. (2009). *Rights-based Approaches. Exploring Issues and Opportunities for Conservation.* International Union for Conservation of Nature (IUCN), Gland and Center for International Forestry Research (CIFOR), Bogor

Carrere, R. (2009). African mangroves: their importance for people and biodiversity. In *The Relevance of Mangrove Forests to African Fisheries, Wildlife and Water Resources. Nature and Faune* (eds. Bojang, F. and Ndeso-Atanga, A.). vol 24 pp.3–7. Food and Agriculture Organization of the United Nations, Accra. ftp://ftp.fao.org/docrep/fao/012/ak995e/ak995e00.pdf (accessed 1 June 2011)

CBD (1997). *Jakarta Mandate on Marine and Coastal Biological Diversity.* Secretariat of the Convention on Biological Diversity. http://www.cbd.int/doc/meetings/mar/jmem-01/official/jmem-01-02-en.pdf

CBD (1992). *Convention on Biological Diversity.* Secretariat of the Convention on Biological Diversity. http://www.cbd.int

Child, B. (ed.) (2004). *Parks in Transition: Biodiversity, Rural Development and the Bottom Line.* Earthscan, London

Conca, K. and Dabelko, G.D. (2002). The problems and possibilities of environmental peacemaking. In *Environmental Peacemaking* (eds. Conca, K. and Dabelko, G.D.). Woodrow Wilson Institute, Washington, DC

Cotula, L. (2011). *Land Deals in Africa. What's in the Contracts?* International Institute for Environment and Development, London

Cotula, L., Dyer, N. and Vermeulen, S. (2008). *Fuelling Exclusion: The Biofuels Boom and Poor People's Access to Land.* Food and Agriculture Organization of the United Nations, Rome and the International Institute for Environment and Development, London

Crooks, S., Herr, D., Tamelander, J., Laffoley, D. and Vandever, J. (2011). *Mitigating Climate Change through Restoration and Management of Coastal Wetlands and Near-shore Marine Ecosystems: Challenges and Opportunities.* Environment Department Paper 121. World Bank, Washington, DC

CTC (2003). *Appraisal of the Potential for a Community Land Registration Negotiation and Planning Support Programme in Mozambique.* Report for UK Department for International Development. CTC Consulting, St. Ives, Cambridge

Davis, C. (2011). *Protecting Forests to Save the Climate: REDD Challenges and Opportunities.* EarthTrends, World Resources Institute. http://earthtrends.wri.org/updates/node/303 (accessed 1 September 2011)

Davis, J., Agardy, T. and Sherwood, K. (2011). *Taking Steps toward Marine and Coastal Ecosystem-based Management – An Introductory Guide.* UNEP Regional Seas Reports and

Studies No. 189. United Nations Environment Programme, Nairobi. http://www.unep.org/ecosystemmanagement (accessed 29 November 2011)

DEAT (2011). *Working for the Environment*. Department of Environmental Affairs and Tourism, Pretoria

De Bruyn, P.A., Moloney, C.L. and Schleyer, M.H. (2009). Application of age-structured production models to assess oyster *Striostrea margaritacea* populations managed by rotational harvesting in KwaZulu-Natal, South Africa. *ICES Journal of Marine Science* 66, 408–419

Deininger, K., Byerlee, D., Lindsay, J., Norton, A., Selod, H. and Stickler, M. (2009). *Rising Global Interest in Agricultural Land*. World Bank, Washington, DC

Dillaha, T., Ferraro, P., Huang, M., Southgate, D., Upadhyaya, S. and Wunder, S. (2007). Payment for watershed services. Regional synthesis. In *USAID PES Sourcebook. Lessons and Best Practices for Pro-poor Payment for Ecosystem Services* (ed. United States Agency for International Development). http://www.katoombagroup.org/~katoomba/documents/tools/PES.Sourcebook.PDF.pdf (accessed 1 July 2011)

Doumbia, M., Jarju, A., Sene, M., Traore, K., Yost, R., Kablan, R., Brannan, K., Berthe, A., Yamoah, C., Querido, A., Traore, P.C.S. and Ballo, A. (2008). Sequestration of organic carbon in West African soils by Aménagement en Courbes de Niveau. *Agronomy for Sustainable Development* 29, 267–275

Dudley, N., Stolton, S., Belokurov, A., Krueger, L., Lopoukhine, N., MacKinnon, K., Sandwith, T. and Sekhran, N. (eds.) (2010). *Natural Solutions: Protected Areas Helping People Cope with Climate Change*. IUCN World Commission on Protected Areas, Gland

Dudley, N., Higgins-Zogib, L. and Mansourian, S. (2005). *Beyond Belief: Linking Faiths and Protected Areas to Support Biodiversity Conservation*. Research report by WWF, Equilibrium and the Alliance of Religions and Conservation (ARC). WWF – World Wide Fund for Nature, Gland

Duke, N.C., Meynecke, J.O., Dittmann, S., Ellison, A.M., Anger, K., Berger, U., Cannicci, S., Diele, K., Ewel, K.C., Field, C.D., Koedam, N., Lee, S.Y., Marchand, C., Nordhaus, I. and Dahdouh-Guebas, F. (2007). A world without mangroves? *Science* 317, 41–42

DWAF. (2002a). *Free Basic Water: Tap into Life. Regulations and guidelines*. Department of Water Affairs and Forestry, Directorate of Interventions and Operations Support, Pretoria

DWAF. (2002b). *Free Basic Water: Tap into life. Prepayment Water Meters and Management Systems*. Department of Water Affairs and Forestry, Directorate of Interventions and Operations Support, Pretoria

Ervin, J., Sekhran, N., Dinu, A., Gidda, S., Vergeichik, M. and Mee, J. (2010). *Protected Areas for the 21st Century: Lessons from UNDP/GEF's Portfolio*. United Nations Development Programme, New York and Convention on Biological Diversity, Montreal

FAO (2011). *State of the World's Forests*. Food and Agriculture Organization of the United Nations, Rome

FAO (2010). *Global Forest Resources Assessment*. Food and Agriculture Organization of the United Nations, Rome

FAO (1996). *Declaration of The World Food Summit*. Food and Agricultural Organization of the United Nations, Rome http://www.fao.org/fileadmin/templates/wsfs/Summit/Docs/Final_Declaration/WSFS09_Declaration.pdf

Frayne, B., Pendleton, W., Crush, J., Acquah, B., Battersby-Lennard, J., Bras, E., Chiweza, A., Dlamini, T., Fincham, R., Kroll, F., Leduka, C., Mosha, A., Mulenga, C., Ruysenaa, S., Nomcebo, S., Tevera, D., Tsoka, M., Tawodzera, G. and Zanamwe, L. (2010). *The State of Urban Food Insecurity in Southern Africa*. Urban Food Security Series 2. Queens University, Kingston and African Food Security Urban Network, Cape Town

Frynas, J.G. (1999). Legal change in Africa: evidence from oil-related litigation in Nigeria. *Journal of African Law* 43(2), 121–150

GEF, UNIDO, UNDP, UNEP, NOAA and NEPAD (2006). *The Transboundary Diagnostic Analysis for the Guinea Current Large Marine Ecosystem*. Programme of the Governments of the GCLME countries with assistance from the Global Environment Facility, United Nations Industrial Development Organization, United Nations Development Programme, United Nations Environment Programme, US National Oceanic and Atmospheric Administration and New Partnership for Africa's Development. Interim Guinea Current Commission, Accra

Golik, A., Weber, K., Salihoglu, I., Yilmaz, A. and Loizides, L. (1988). Pelagic tar in the Mediterranean Sea. *Marine Pollution Bulletin* 19(11), 567–572

Gonzàlez-Riancho, P., Sanò, M., Medina, R., Garcià-Aguilar, O. and Areizaga, J. (2009). A contribution to the implementation of ICZM in the Mediterranean developing countries. *Ocean and Coastal Management* 52, 545–558

Gordon, I. and Ayiemba, W. (2003). Harnessing butterfly biodiversity for improving livelihoods and forest conservation: the Kipepeo project. *Journal of Environment and Development* 12, 82–98

Gordon, C., Tweneboah, E., Mensah, A.M. and Ayivor, J.S. (2009). The application of the ecosystem approach to mangrove management: lessons for Ghana. In *The Relevance of Mangrove Forests to African Fisheries, Wildlife and Water Resources. Nature and Faune* (eds. Bojang, F. and Ndeso-Atanga, A.). vol. 24 pp.16–25. Food and Agriculture Organization of the United Nations, Accra. ftp://ftp.fao.org/docrep/fao/012/ak995e/ak995e00.pdf (accessed 31 May 2011)

Grace, J., Ryan, C.M., Williams, M., Powell, P., Goodman, L. and Tipper, R. (2010). A pilot project to store carbon as biomass in African woodlands. *Carbon Management* 1(2), 227–235
Gustavson, K., Kroeker, Z., Walmsley, J. and Juma, S. (2008). A process framework for coastal zone management in Tanzania. *Ocean and Coastal Management* 52,78–88

Hanlon, J. (2002). *The Land Debate in Mozambique: Will Foreign Investors, the Urban Elite, Advanced Peasants or Family Farmers Drive Rural Development?* Oxfam GB, Pretoria

Hansen M.C., Stehman S.V., and Potapov P.V. (2010). Quantification of global gross forest cover loss. *Proceedings of the National Academy of Sciences of the United States of America* 107, 8650–8655

Hewawasam, I. (2000). Advancing knowledge: a key element of the World Bank's integrated coastal management strategic agenda in sub-Saharan Africa. *Ocean and Coastal Management* 43, 361–377

Hien, F. and Ouédraogo, A. (2001). Joint analysis of the sustainability of a local SWC technique in Burkina Faso. In *Farmer Innovation in Africa: A Source of Inspiration for Agricultural Development* (eds. Reij, C. and Waters-Bayer, A.). Earthscan, London

Hob, H., Oelofse, S.H. and Rascher, J. (2008). Management of environmental impact from coal mining in the upper Olifants river catchment as a function of age and scale. *International Journal of Water Resources Development* 24(30), 417–431

Horta, K. (2009). *Global Climate Politics in the Congo Basin. Unprecedented Opportunity or High-risk Gamble?* International Finance, Development and Environment, Washington, DC and Heinrich-Böll-Stiftung, Lisbon

Huggins, C., Chenje, M. and Mohamed-Katerere, J.C. (2006). Environment for peace and regional cooperation. In *Africa Environment Outlook 2: Our Environment, Our Wealth*. United Nations Environment Programme, Nairobi

Ibe, A.C. and Sherman, K. (2002). The Gulf of Guinea large marine ecosystem project: turning challenges into achievements. In *The Gulf of Guinea Large Marine Ecosystem: Environmental Forcing and Sustainable Development of Marine Resources* (eds. MacGlade, J.M., Cury, P., Koranteng, K.A. and Hardman-Mountford, N.J.). pp.27–39. Elsevier Science, Amsterdam

ICHRP (2008). *Climate Change and Human Rights: A Rough Guide*. International Council on Human Rights Policy, Versoix

Idowu, A.A. (1999). Human rights, environmental degradation and oil multinational companies in Nigeria: the Ogoniland episode. *Netherlands Quarterly of Human Rights* 17(2), 161–184

Jackson, L.J. (2011). *Marine Pollution in the Agulhas and Somali Currents Large Marine Ecosystem*. Report for the ASCLME project. Rhodes University, Grahamstown

Jäger, J., Kok, M., Mohamed-Katerere, J.C., Karlsson, S., Lüdeke, M., Dabelko, G.D., Thomalla, F., de Soysa, I., Chenje, M., Filcak, R., Koshy, L., Long Martello, M., Mathur, V., Moreno, A.R., Narain, V. and Sietz, D. (2007). Vulnerability of people and the environment: challenges and opportunities. In *Global Environment Outlook-4: Environment for Development*. United Nations Environment Programme, Nairobi

Jamart, C. (2011). *Shortcomings of Niger's Rural Code and Challenges for the Future*. Lessons Learned from Niger's Rural Code Paper #6. http://www.agter.org/bdf/en/corpus_chemin/fiche-chemin-93.html (accessed 11 September 2011)

Johannes, R.E. (1998). The case for data-less marine resource management: example from tropical nearshore fisheries. *Trends in Ecology and Evolution* 13, 243–246

Johnson, J. and Welch, D.J. (2010). Marine fisheries management in a changing climate: a review of vulnerability and future options. *Reviews in Fisheries Science* 18(1), 106–124

Jones, B. (2008). *Community Wildlife Management in Southern Africa: A Review of Current Research Activity in the Region and of Recent Literature*. International Institute for Environment and Development, London

Jones, B. (2004). *CBNRM, Poverty Reduction and Sustainable Livelihoods: Developing Criteria for Evaluating the Contribution of CBNRM to Poverty Reduction and Alleviation in Southern Africa*. Commons Southern Africa Occasional Paper Series Number 7. Centre for Applied Social Sciences and Poverty, Land and Agrarian Studies, Harare and Cape Town

Jones, B. and Chonguiça, E. (2001). *Review and Analysis of Specific Transboundary Natural Resource Management Initiatives in the Southern Africa Region*. IUCN-ROSA Series on Transboundary Natural Resource Management Paper 2. International Union for Conservation of Nature, Regional Office for Southern Africa, Harare

Kablan, R., Yost, R.S., Brannan, K., Doumbia, M., Traore, K., Yorote, A.,Toloba, Y., Sissoo, S., Samake, O., Vaksman, M., Dioni, L. and Sissoko, M. (2008). "Aménagement en courbes de niveau", increasing rainfall capture, storage, and drainage in soils of Mali. *Arid Land Research and Management* 22, 62–80

Kaboré, D. and Reij, C. (2004). *The Emergence and Spreading of an Improved Traditional Soil and Water Conservation Practice in Burkina Faso*. EPTD Discussion Paper 114. Environment and Production Technology Division, International Food Policy Research Institute, Washington, DC

Kanji, N., Toulmin, C., Mitlin, D., Cotula, L., Taoli, C. and Hesse, C. (2006). *Innovation in Securing Land Rights in Africa: Lessons from Experience*. International Institute for Environment and Development, London

Karibuhoye, C. (2008). Mise en place du réseau régional d'aires marines protégées en Afrique de l'Ouest (RAMPAO). Une stratégie régionale pour les AMP en Afrique de l'Ouest. In *Actes du 1er colloque national sur les aires marines protégées: Quelle stratégie pour quels objectifs?* 20–22 novembre 2007, Boulogne-sur-Mer. Comité Français UICN, Union mondiale pour la nature, Paris

Katerere, Y., Hill, R. and Moyo, S. (2001). *A Critique of Transboundary Natural Resource Management in Southern Africa*. IUCN-ROSA Series on Transboundary Natural Resource

Management Paper 1. International Union for Conservation of Nature, Regional Office for Southern Africa, Harare

Koech, C.K., Ongugo, P.O., Mbuvi, M.T.E. and Maua, J.O. (2009). *Community Forest Associations in Kenya: Challenges and Opportunities*. Kenya Forestry Research Institute, Nairobi

Landell-Mills, N. and Porras, I.T. (2002). *Silver Bullet or Fool's Gold? A Global Review of Markets for Forest Environmental Services and Their Impact on the Poor.* International Institute for Environment and Development, London

Larson, E.A. (2010). At the intersection of neoliberal development, scarce resources, and human rights: enforcing the right to water in South Africa. *Honors Projects*. Paper 10. http://digitalcommons.macalester.edu/intlstudies_honors/10 (accessed 29 November 2011)

Lee, C. and Schaaf, T. (eds.) (2003). *The Importance of Sacred Natural Sites for Biodiversity Conservation*. Proceedings of an international workshop, Kunming, China, February 2003. United Nations Educational, Scientific and Cultural Organization, Paris

Limpitlaw, D., Aken, M., Lodewijks, H. and Viljoen, J. (2005). *Post-mining Rehabilitation, Land Use and Pollution at Collieries in South Africa*. Paper presented at the Sustainable Development in the Life of Coal Mining colloquium, South African Institute of Mining and Metallurgy, Boksburg, 13 July 2005

Locher, M. (2011). *How Come that Others are Selling our Land? Customary Land Rights, Rural Livelihoods and Foreign Land Acquisition in the Case of a UK-based Forestry Company in Tanzania*. Paper presented at the Global Land Grabbing Conference, Institute of Development Studies, Brighton, 6–8 April 2011

Madamombe, I. (2005). Energy key to Africa's prosperity: challenges in West Africa's quest for electricity. *Africa Renewal* 18(4), 6. http://www.un.org/ecosocdev/geninfo/afrec/vol18no4/184electric.htm (accessed 14 December 2011)

Madeira, E.M. (2009). *REDD in Design: Assessment of Planned First Generation Activities in Indonesia to Reduce Emissions from Deforestation and Degradation (REDD)*. Discussion Paper 09–49. Resources for the Future, Washington, DC

Madsen, B., Carroll, N. and Moore Brands, K. (2010). *State of Biodiversity Markets Report: Offset and Compensation Programs Worldwide*. http://www.ecosystemmarketplace.com/pages/dynamic/resources.library.page.php?page_id=7491&section=our_publications&eod=1 (accessed 29 November 2011)

Madzwamuse, M. (2010). *Climate Governance in Africa: Adaptation Strategies and Institutions*. Heinrich Böll Stiftung. Unity Press, Cape Town

Makhado, R.A., Saidi, T.A, Mantlana, B.K. and Mwayafu, D.M. (2011). Challenges of reducing emissions from deforestation and forest degradation (REDD+) on the African continent. *South African Journal of Science* 107(9–10)

Mati, B., de Bock, T., Malesu, M., Khaka, E., Oduor, A., Nyabenge, M. and Oduor, V. (2006). *Mapping the Potential of Rainwater Harvesting Technologies in Africa: A GIS Overview on Development Domains for the Continent and Ten Selected Countries*. Technical Manual No. 6. World Agroforestry Centre (ICRAF), Nairobi and Netherlands Ministry of Foreign Affairs, The Hague

McLeod, E. and Salm, R.V. (2006). *Managing Mangroves for Resilience to Climate Change*. IUCN Resilience Science Group Working Paper Series No. 2. International Union for Conservation of Nature, Gland

MCM/DEAT (2000). *White Paper for Sustainable Coastal Development in South Africa*. Marine and Coastal Management, Department of Environmental Affairs and Tourism, Pretoria

MedPAN (2011). *The Network of Managers of Marine Protected Areas in the Mediterranean*. http://www.medpan.org/?arbo=reseau (accessed 11 November 2011)

Mehlman, P., Kernan, C. and Bonilla, J.C. (2006). *Conservation International CARPE USAID Final Technical Report. Monte Alen Segmet, Equatorial Guinea, Monte Alen – Monts de Cristal Landscape (1) ad Maiko Tayna Kahuzi-Biega Landscape (10)*. Conservation International, Democratic Republic of Congo, Central African Regional Program for the Environment and United States Agency for International Development

Mehta, L. (2005). *Unpacking Rights and Wrongs: Do Human Rights Make a Difference? The Case of Water Rights in India and South Africa*. IDS Working Paper 260. Institute of Development Studies, Brighton

Milder, J.C., Scherr, S.J. and Bracer, C. (2010). Trends and future potential of payment for ecosystem services to alleviate rural poverty in developing countries. *Ecology and Society* 15(2), 4. http://www.ecologyandsociety.org/vol15/iss2/art4/ (accessed 14 December 2011)

Mohamed-Katerere, J.C. (2009). Climate change, natural resource governance and human security in Africa. Charting new paths. In *Natural Resource Governance and Human Security in Africa. Emerging Issues and Trends* (eds. Kesselman, B., Hughes, T., Kabemba, C., Matose, F. and Rocha, J.). Pax-Africa, Johannesburg

Mohamed-Katerere, J.C. (2001). *Review of the Legal and Policy Framework for Transboundary Natural Resource Management in Southern Africa*. IUCN-ROSA Series on Transboundary Natural Resource Management. International Union for Conservation of Nature, Regional Office for Southern Africa, Harare

Muboko, N. (2011). *Conflict and Sustainable Development: The Case of the Great Limpopo Transfrontier Park (GLTP), Southern Africa*. PhD thesis. Nelson Mandela University, Port Elizabeth

Murombedzi, J.C. (2010). Agrarian social change and post-colonial natural resource management interventions in southern Africa's communal tenure regimes. In *Community Rights, Conservation and Contested Land. The Politics of Natural Resource Governance in Africa* (ed. Nelson, F.). Earthscan, London

Najam, A. and Halle, M. (2010). Global environmental governance: the challenge of accountability. *Sustainable Development Insights* 005. Frederick S. Pardee Center for the Study of the Longer-Range Future, Boston University

NASCO (2010). *Namibia's Communal Conservancies: A Review of Progress and Challenges in 2009.* Namibia Association of CBNRM Support Organisations, Windhoek

Nelson, F. (2010). *Community Rights, Conservation and Contested Land. The Politics of Natural Resource Governance in Africa*. Earthscan, London

Nicholls, R.J. (2004). Coastal flooding and wetland loss in the 21st century: changes under the SRES climate and socio-economic scenarios. *Global Environmental Change* 14(1), 69–86

Norfolk, S. and Liversage, H. (2001). *Land Reform and Poverty Alleviation in Mozambique*. Paper for the Southern African Poverty Relief Network. Human Sciences Research Council, Pretoria

Osborn, D. and Datta, A. (2006). Institutional and policy cocktails for protecting coastal and marine environments from land-based sources of pollution. *Ocean and Coastal Management* 49(9–10), 576–596

Patt, A.G. and Schroter, D. (2005). Perceptions of climate risk in Mozambique: implications for the success of adaptation strategies. *Global Environmental Change* 18, 458–467

Pender, J., Ringler, C. and Magalhaes, M. (2009). *Land and Climate: The Role of Sustainable Land Management for Climate Change Adaptation and Mitigation in Sub-Saharan Africa*. Issues Paper. TerrAfrica Regional Sustainable Land Management. http://www.nepad-caadp.net/pdf/Land&Climate_Paper_English.pdf (accessed 29 November 2011)

Post, J.C. and Lundin, C.G. (eds.) (1996). *Guidelines for Integrated Coastal Zone Management*. Environmentally Sustainable Development Studies and Monograph Series No. 9. World Bank, Washington, DC

ReCoMaP (2011). *Regional Coastal Management Programme of the Indian Ocean*. http://recomap-io.org/home/ (accessed May 2011)

Reij, C. and Thiombiano, T. (2003). *Développement rural et environnement au Burkina Faso: la réhabilitation de la capacité productive des terroirs sur la partie nord du Plateau Central entre 1980 et 2001*. Ambassade des Pays-Bas, GTZ-PATECORE and USAID, Ouagadougou

Reij, C., Tappan, G. and Smale, M. (2009). *Agroenvironmental Transformation in the Sahel: Another Kind of "Green Revolution"*. IFPRI Discussion Paper 00914 for the project on Millions Fed: Proven Successes in Agricultural Development. International Food Policy Research Institute, Washington, DC

Republic of Madagascar (2006). *Madagascar Action Plan 2007–2012*. http://www.madagascar.gov.mg/MAP (accessed 29 November 2011)

Robiglio, V., Mala, W.A. and Diaw, M.C. (2003). Mapping landscapes: integrating GIS and social science methods to model human-nature relationships in southern Cameroon. *Small-scale Forest Economics, Management and Policy* 2(2), 171–220

Rodgers, A., Mugabe, J. and Mathenge, C. (2001). *Beyond Boundaries: Regional Overview of Transboundary Natural Resource Management in Eastern Africa*. Food and Agriculture Organization of the United Nations-United Nations Development Programme (FAO-UNDP), Eastern Africa and African Centre for Technology Studies (ACTS), Nairobi. http://www.worldwildlife.org/bsp/publications/africa/121/121/chap4

Roe, D., Nelson, F. and Sandbrook, C. (eds.) (2009). *Community Management of Natural Resources in Africa: Impacts, Experiences and Future Directions.* Natural Resource Issues No. 18. International Institute for Environment and Development, London

Saico, S.S. and Kunene, S.G. (2010). Viability of rainwater harvesting in supplying domestic water in rural areas of Swaziland: a case of Mpaka community. *Journal of Sustainable Development in Africa* 12(2), 96–109

Scherr, S., White, A. and Kaimowitz, D. (2004). *A New Agenda For Forest Conservation and Poverty Reduction: Making Markets Work for Low-Income Producers*. Forest Trends, Washington, DC

Sen, A. (1981). *Poverty and Famines: An Essay on Entitlement and Deprivation*. Clarendon, Oxford

Serra, C. and Tanner, C. (2008). Legal empowerment to secure and use land and resource rights in Mozambique. In *Legal Empowerment in Practice: Using Legal Tools to Secure Land Rights in Africa* (eds. Cotula, L. and Mathieu, P.). International Institute for Environment and Development, London and Food and Agriculture Organization of the United Nations, Rome

SMAP III (2009). *The Way Forward for the Mediterranean Coast: A Framework for Implementing Regional ICZM Policy at the National and Local Level*. Priority Actions Programme Regional Activity Centre, Split

Stalk, A. (2004). *Management of the Free Basic Water Policy in South Africa*. Master project. Roskilde University, Roskilde

Stanton, T., Echavarria, M., Hamilton, K. and Ott, C. (2010). *State of Watershed Payments: An Emerging Marketplace*. Ecosystem Marketplace. http://www.foresttrends.org/documents/files/doc_2438.pdf (accessed 29 November 2011)

Sunderlin, W.D., Hatcher, J. and Liddle, M. (2008). *From Exclusion to Ownership? Challenges and Opportunities in Advancing Forest Tenure Reform*. Rights and Resources Initiative, Washington, DC

Swallow, B.M., Kallesoe, M.F., Iftikhar, U.A., van Noordwijk, M., Bracer, C., Scherr, S.J., Raju, K.V., Poats, S.V., Kumar Duraiappah, A., Ochieng, B.O., Mallee, H. and Rumley, R. (2009). Compensation and rewards for environmental services in the developing world: framing pan-tropical analysis and comparison. *Ecology and Society* 14(2), 26. http://www.ecologyandsociety.org/vol14/iss2/art26/ (accessed 14 December 2011)

TerrAfrica (2009). Enhancing the TerrAfrica Partnership. http://www.unep.org/south-south-cooperation/exchangeplatform/Publications/GlobalMechanismTeamPublications/EnhancingtheTerrAfricapartnership/tabid/5780/Default.aspx (accessed 11 September 2011)

Ukwe, C.N. and Ibe, C.A. (2010). A regional collaborative approach in transboundary pollution management in the Guinea current region of western Africa. *Ocean and Coastal Management* 53(9), 493–506

UN (2011). *World Population Prospects: The 2010 Revision*. CD-ROM Edition. UN Department of Economic and Social Affairs, Population Division, United Nations, Geneva

UNCCD/FAO (2010). *Policy and Financing for Sustainable Land Management in Africa: The Challenge, Lessons from Experience and Guidance for Action*. Global Mechanism of the United Nations Convention to Combat Desertification, Bonn and the Food and Agriculture Organization of the United Nations, Rome. http://global-mechanism.org/dynamic/documents/document_file/financeactionbox_en.pdf (accessed 11 September 2011)

UNCCD/FAO (2009). *Policy and Financing for Sustainable Land Management in Sub-Saharan Africa: Lessons and Guidance for Action*. Global Mechanism of the United Nations Convention to Combat Desertification, Bonn and the Food and Agriculture Organization of the United Nations, Rome. http://www.caadp.net/pdf/Policy%20and%20Financing%20for%20SLM%20in%20Sub-Saharan%20Africa%201.0.pdf (accessed 11 September 2011)

UNEP Risoe Centre (2011). *Capacity Development for the Clean Development Mechanism*. http://cdmpipeline.org/cdm-projects-region.htm#7 (accessed March 2012)

UNEP (1985). *Convention for the Protection, Management and Development of the Marine and Coastal Environment of the Eastern African Region/Western Indian Ocean* (amended in Nairobi in 2010). United Nations Environment Programme, Nairobi. http://www.unep.org/NairobiConvention/The_Convention/Nairobi_Convention_Text/index.asp

UNEP (1976). *Convention for the Protection of The Mediterranean Sea Against Pollution* (revised in Barcelona in 1995 as the *Convention for the Protection of the Marine Environment and the Coastal Region of the Mediterranean*). United Nations Environment Programme, Nairobi. http://www.unep.ch/regionalseas/regions/med/t_barcel.htm

UNFCCC (1992). *United Nations Framework Convention on Climate Change*. FCCC/INFORMAL/84 GE.05-62220 (E) 200705. http://unfccc.int/resource/docs/convkp/conveng.pdf

UNGA (2010). General Assembly adopts resolution recognizing access to clean water, sanitation as a human right. GA/10967. United Nations General Assembly. http://www.un.org/News/Press/docs/2010/ga10967.doc.htm (accessed 29 November 2011)

UN-Habitat (2010). *The State of African Cities 2010: Governance, Inequality and Urban Land Markets*. United Nations Human Settlements Programme, Nairobi

UNISDR (2011). *Global Assessment Report on Disaster Risk Reduction*. United Nations International Strategy for Disaster Reduction, Geneva

Usongo, L. (2010). Land use planning. In *Landscape-scale Conservation in the Congo Basin: Lessons Learned from the Central Africa Regional Program for the Environment (CARPE)* (eds. Yanggen, D., Angu, K. and Tchamou, N.). International Union for Conservation of Nature (IUCN), Central African Regional Program for the Environment (CARPE) and United States Agency for International Development (USAID). http://cmsdata.iucn.org/downloads/the_book_lessons_learned_from_the_carpe_1.pdf (accessed 29 November 2011)

Vafeidis, A.T., Boot, G., Cox, J., Maatens, R., McFadden, L., Nicholls, R.J., Spencer, T. and Tol, R.S.J. (2005). The DIVA Database Documentation. On DIVA CD and at http://www.dinas-coast.net

Varis, O., Stucki, V. and Fraboulet-Jussila, S. (2006). The Senegal river case. In *Human Development Report 2006. Beyond Scarcity: Power, Poverty and the Global Water Crisis*. United Nations Development Programme, New York. http://www.hdr.undp.org/en/reports/.../olli_varis_senegalriver_casestudy.pdf (accessed 29 November 2011)

Wachira, G.M. (2008). *African Court on Human and Peoples' Rights: Ten Years On and Still No Justice*. Minority Rights Group, London. http://www.unhcr.org/refworld/pdfid/48e4763c2.pdf (accessed 29 November 2011)

Weru, S. (2004). Policy implications in the management of Kenya's marine protected areas. In *Economic Valuation and Policy Priorities for Sustainable Management of Coral Reefs* (eds. Ahmed, M., Chong, C.K. and Cesar, H.). pp.192–197. WorldFish Center, Penang

Whande, W. (2010). Windows of opportunity or exclusion? Local communities in the Great Limpopo Transfrontier Conservation Area, South Africa. In *Community Rights, Conservation and Contested Land. The Politics of Natural Resource Governance in Africa* (ed. Nelson, F.). Earthscan, London

Wilkie, D.S., Hakizumwami, E., Gami, N. and Diafra, B. (2001). *Beyond Boundaries: Regional Overview of Transboundary Natural Resource Management in Central Africa*. Biodiversity Support Program, Washington, DC

Winkler, I. (2008). Judicial enforcement of the human right to water – case law from South Africa, Argentina, and India. *Law, Social Justice and Global Development* 1,4. http://www.go.warwick.ac.uk/elj/lgd/2008_1/winkler (accessed 29 November 2011)

World Bank (2011). *Tanzania Marine and Coastal Environmental Management Project*. http://web.worldbank.org/WBSITE/EXTERNAL/COUNTRIES/AFRICAEXT/TANZANIAEXTN/0,,contentMDK:20992192~menuPK:287367~pagePK:1497618~piPK:217854~theSitePK:258799,00.html (accessed May 2011)

World Bank (2008). *Burkina Faso at a Glance*. World Bank, Washington, DC

World Coal Institute (2002). Water management initiatives in the upper Olifants river catchment. *Good News from Coal* August 2002. http://www.iccwbo.org/uploadedfiles/wbcsd/olifants.pdf (accessed 24 November 2012)

WSSD (2002). *Johannesburg Plan of Implementation*. World Summit on Sustainable Development. http://www.un.org/esa/sustdev/documents/WSSD_POI_PD/English/POIToc.htm

Wunder, S. (2008). Payments for environmental services and the poor: concepts and preliminary evidence. *Environment and Development Economics* 13(3), 279–297

Wunder, S. (2005). *Payments for Environmental Services: Some Nuts and Bolts*. CIFOR Occasional Paper Number 42. Center for International Forestry Research, Bogor

CHAPTER 10

Asia and the Pacific

Coordinating lead authors: Peter King and Lailai Li

Lead authors: Iskandar Abdullaev, Raquibul Amin, Yumiko Asayama, Magnus Bengtsson, Robert Dobias, Mark Elder, Rodrigo Fuentes, Anirban Ganguly, Guibin Jiang, Mikiko Kainuma, Yatsuka Kataoka, Simon Hoiberg Olsen (GEO Fellow) and Diana Suhardiman

Contributing authors: Prodipto Ghosh, Yi Huang, Robert Kipp, Marie Leroy, Keeping Ma, Vishal Narain, Shavkat Rakhmatullaev, Nilapha Ratanavong (GEO Fellow), Jianbo Shi, Poh Poh Wong and Shiqiu Zhang

Principal scientific reviewer: Patrick Nunn

Chapter coordinators: Anna Stabrawa and Jinhua Zhang

Main Messages

Achievement of global environmental goals substantially depends on coordinated policies and action in the Asia and Pacific region, often identified as the global engine of economic growth. The global drivers identified in Chapter 1 – in particular unsustainable economic growth, population increase, mass consumption and urbanization – pose clear challenges to the region's sustainable development. It is therefore important that policy responses are designed to enable the best possible adaptation to pressures and impacts deriving from these drivers.

Asia and the Pacific is the fastest growing region in the world with the most rapidly rising emissions of greenhouse gases, and efforts to combat climate change must accelerate across the region if global efforts are to succeed. Under a business-as-usual scenario, the region will contribute approximately 45 per cent of global energy-related carbon dioxide (CO_2) emissions by 2030. However, intra-regional diversity is great, with China the world's largest emitter, while most Pacific island nations are among the smallest. People from this region have the most to lose from global inaction as many of the countries most at risk from climate change are here. Mainstreaming adaptation concerns into development policies and plans, integrating climate change adaptation and disaster risk reduction, climate proofing infrastructure, and promoting ecosystem-based adaptation are key actions. Significant steps have been taken on both mitigation and adaptation, but much more needs to be done – and urgently – to achieve widespread low-carbon and climate-resilient societies.

Water endowments range from the highly arid temperate zones and water-stressed small island states to Himalayan snowfields and abundant tropics, often alternating between drought and flood. Balancing water supply and demand through coordination between users, and improved water quality management, are essential to achieve global freshwater goals. Successful implementation of policies requires establishing a planning framework for adaptive and integrated management of water resources, under which appropriate pricing and multi-stakeholder participation are essential.

The threat of species extinction is only partly addressed by global goals targeting a significantly reduced rate of biodiversity loss. Despite progress in expanding protected areas, conserving some species, addressing some direct drivers of biodiversity loss, and implementing community-based management and innovative financing, the scale of efforts remains insufficient. In light of the recent Nagoya Protocol on Access to Genetic Resources and the Fair and Equitable Sharing of Benefits Arising from their Utilization, access and benefit-sharing regimes also need to be developed.

As the Asia and Pacific region becomes wealthier it is facing a rapid increase in consumption and its principal side effect – waste. Effective implementation of the 3Rs approach – reduce, reuse, recycle – remains a key goal, although a mix of policies may be necessary to achieve the most cost-efficient outcomes. Changing consumption patterns and behaviour, which reduces waste from the outset, lies at the core of an effective policy mix.

Appropriate controls on chemical production and use, and the provision of safe alternatives as well as appropriate treatment facilities, are key policy concerns. As their use is increasing while their impacts remain poorly monitored and little understood, management of registration, monitoring, export and import as well as information sharing should be strengthened. Proactive measures on emerging contaminants are also necessary.

Governance improvements are critical to enhanced accountability as a means of achieving sustainable development. Integrating sustainability concerns across all policy areas, increased multi-stakeholder participation and capacity improvement can all enhance governance. Additionally, allocating authority to appropriate levels of government, improved monitoring and data collection, access to information and legal redress, as well as greening fiscal policy, have the potential to alter the drivers of environmental change and unsustainable development.

Recommendation of policies to accelerate achievement of the selected global goals remains difficult. There have been some successes in the region but gaps remain. Policy responses are beginning to shift from a focus on environmental impacts to addressing the key drivers through market- and information-based approaches. As many of the policy successes are due to the context in which they are implemented, transferring policies from one country to another, while often practised, requires careful analysis. Creating the necessary enabling environment may be as important as selecting the right mix of policies.

INTRODUCTION

The global drivers identified in Chapter 1 – in particular unsustainable economic growth, population increase, mass consumption and urbanization – pose clear challenges to sustainable development in Asia and the Pacific. It is therefore important that policy responses are designed to enable the best possible adaptation to the pressures and impacts deriving from these drivers.

The objectives of this chapter are to:

- document the chosen priority themes and goals;
- identify the wide range of policies that have been applied in the region to address these goals;
- screen these options according to their effectiveness and compile a shortlist of the most promising policies for further analysis;
- document successful cases where these policies have been implemented and appear to have made a contribution to achieving the global goals;
- analyse the social, environmental, economic and political impacts of these priority policies;
- examine the possibilities and prospects for replication across national borders; and
- conclude which of these policies and/or which combination of these policies should be implemented to accelerate achievement of the global goals.

The chapter concludes with a synthesis of the benefits and limitations of the combined policy packages for each thematic area, an analysis of the enabling environment that needs to be created to allow the selected policies to flourish, and a set of conclusions directed towards the region's decision makers.

POLICY APPRAISAL

While Part 1 of *GEO-5* describes the state and trends of a multitude of environmental issues and challenges, the regional consultation selected five priority themes for their overarching characteristics. The five priority themes and related global goals for the Asia and Pacific region were chosen at the first regional consultation, held in Bangkok, Thailand in September 2010. Generally, the broadest goal was selected to allow all issues covered by other goals to be considered. This means that quantitative targets were not chosen, thus making any quantitative assessment of policy choices more difficult. The themes are relevant for all countries in the region, but specific countries may need to prioritize additional |environmental challenges. Policy responses regarding the selected themes provide insights on how country-specific challenges can be addressed.

Priority themes

Climate change

The priority concern for most countries in the region is how to build resilience, especially in the most vulnerable communities, to climate change impacts already set in motion by past greenhouse gas emissions. Parts of low-lying Pacific island countries may disappear entirely due to sea level rise (Nicholls *et al.* 2011; Nunn 2009; Barnett and Adger 2003), extreme weather events are likely to become more frequent, and marine habitats such as coral reefs and mangroves are threatened by increased temperature and ocean acidification.

Under a business-as-usual scenario, the region will contribute approximately 45 per cent of global energy-related carbon dioxide (CO_2) emissions by 2030 (IEA 2010) and, by one estimate, more than 60 per cent of total global CO_2 emissions by 2100 (Masui *et al.* 2011). However, intra-regional diversity is great – China is the world's largest CO_2 emitter while most Pacific island nations are among the smallest. Transport-related emissions are expected to increase by 57 per cent worldwide between 2005 and 2030, with China and India accounting for more than half of that increase (Leather *et al.* 2009). Nonetheless, there are encouraging signs on mitigation. At least ten countries in the region have voluntarily pledged greenhouse gas emission reductions, including Indonesia's promise of a 26 per cent CO_2 reduction compared to business-as-usual by 2020 (DNPI 2010), and China's of a 40–45 per cent reduction

Box 10.1 Selected climate change goal: United Nations Framework Convention on Climate Change Article 3 Paragraphs 1–3

The Parties should protect the climate system for the benefit of present and future generations of humankind, on the basis of equity and in accordance with their common but differentiated responsibilities and respective capabilities. Accordingly, the developed country Parties should take the lead in combating climate change and the adverse effects thereof.

The specific needs and special circumstances of developing country Parties, especially those that are particularly vulnerable to the adverse effects of climate change, and of those Parties, especially developing country Parties, that would have to bear a disproportionate or abnormal burden under the Convention, should be given full consideration.

The Parties should take precautionary measures to anticipate, prevent or minimize the causes of climate change and mitigate its adverse effects. Where there are threats of serious or irreversible damage, lack of full scientific certainty should not be used as a reason for postponing such measures, taking into account that policies and measures to deal with climate change should be cost-effective so as to ensure global benefits at the lowest possible cost. To achieve this, such policies and measures should take into account different socio-economic contexts, be comprehensive, cover all relevant sources, sinks and reservoirs of greenhouse gases and adaptation, and comprise all economic sectors. Efforts to address climate change may be carried out cooperatively by interested Parties.

Source: UNFCCC 1992

The Mekong River Delta is one of the most important rice granaries in Viet Nam, but being a low-lying coastal region, it is particularly susceptible to floods. © *Bartosz Hadyniak/iStock*

in CO_2 per unit of gross domestic product (GDP) compared to the 2005 level by 2020 (Lommen 2011). With one of the world's greatest potentials for mitigating CO_2 emissions being reduced deforestation alongside improved land-use management (ADB 2009a), Asia and the Pacific can make significant contributions to global efforts at climate change mitigation. Accessing climate funds to enable these contributions, however, is a major concern for developing countries in the region.

Although UNFCCC Article 3 was selected by the regional consultation, three other goals (UNFCCC Article 2, the Bali Action Plan and the Delhi Declaration) selected by the GEO High-Level Intergovernmental Advisory Panel were also considered because adaptation, mitigation, capacity building and financing need to be considered as an integrated package of policy measures.

Biodiversity

The imminent threat of mass extinction of species, brought about by continuing habitat fragmentation, degradation and loss, overexploitation of resources, invasive alien species, illegal wildlife trade, pollution and climate change are priority environmental concerns in the Asia and Pacific region (Box 10.2). The *Global Biodiversity Outlook 3* concluded that the 2010 goal of reversing biodiversity loss had not been achieved (CBD 2010). The Strategic Plan for Biodiversity for 2011–2020 embodying the Aichi Biodiversity Targets now provides the general framework for biodiversity conservation.

Links with the Johannesburg Plan of Implementation (JPOI) Paragraph 44 and its provisions (WSSD 2002) also need to be established.

Freshwater

As outlined in Part 1, the key environmental priorities in the water sector being faced by the region are the quantity and quality of water resources (Box 10.3), climate change, access to safe drinking water and transboundary issues. All of these key challenges are reflected in the selected goal.

The regional consultation also noted that JPOI Paragraphs 25d and 7a should be included in the assessment, as well as taking an innovative approach to links with other themes.

Box 10.2 Selected biodiversity goal: Convention on Biological Diversity Article 1

The objectives of this Convention, to be pursued in accordance with its relevant provisions, are the conservation of biological diversity, the sustainable use of its components and the fair and equitable sharing of the benefits arising out of the utilization of genetic resources, including by appropriate access to genetic resources and by appropriate transfer of relevant technologies, taking into account all rights over those resources and to technologies, and by appropriate funding.

Source: CBD 1992

Box 10.3 Selected freshwater goal: Johannesburg Plan of Implementation Paragraph 26c

Improve the efficient use of water resources and promote their allocation among competing uses in a way that gives priority to the satisfaction of basic human needs and balances the requirement of preserving or restoring ecosystems and their functions, in particular in fragile environments, with human domestic, industrial and agriculture needs, including safeguarding the quality of drinking water.

Source: WSSD 2002

A motorcycle connected to a rickshaw used for the transport of paper which then sold to be recycled in Cambodia. *© Laurent/ iStock*

Chemicals and waste

The chemicals and waste theme encompasses a range of interrelated issues, including the production and use of chemicals, hazardous waste, electronic waste, transboundary movement, product reuse, materials recycling and municipal waste management. During the regional consultations, JPOI Paragraph 23 was selected as the overarching goal for the theme, although JPOI Paragraph 22 was considered equally relevant (Box 10.4).

Box 10.4 Selected goal for chemicals and waste: Johannesburg Plan of Implementation Paragraphs 22 and 23

Prevent and minimize waste and maximize reuse, recycling and use of environmentally friendly alternative materials, with the participation of government authorities and all stakeholders, in order to minimize adverse effects on the environment and improve resource efficiency, with financial, technical and other assistance for developing countries.

Renew the commitment, as advanced in Agenda 21, to sound management of chemicals throughout their life cycle and of hazardous wastes for sustainable development as well as for the protection of human health and the environment, *inter alia*, aiming to achieve, by 2020, that chemicals are used and produced in ways that lead to the minimization of significant adverse effects on human health and the environment, using transparent science-based risk assessment procedures and science-based risk management procedures, taking into account the precautionary approach, as set out in principle 15 of the Rio Declaration on Environment and Development, and support developing countries in strengthening their capacity for the sound management of chemicals and hazardous wastes by providing technical and financial assistance.

Source: WSSD 2002

Box 10.5 Selected governance goal: Johannesburg Declaration on Sustainable Development Paragraph 5

Accordingly, we assume a collective responsibility to advance and strengthen the interdependent and mutually reinforcing pillars of sustainable development – economic development, social development and environmental protection – at the local, national, regional and global levels.

Source: WSSD 2002

The selected global goals are built around the concept of life-cycle thinking. Hence the starting point for effective policies is to use demand management and resource efficiency to minimize waste generation and the use of hazardous chemicals. The region's political recognition of the need to prioritize waste minimization and resource efficiency is not matched by policy implementation (APO 2007). Only weak efforts have been made to address the escalating use of resources and hazardous substances that eventually end up as waste and pollutants (UNEP 2011; Shekdar 2009).

Environmental governance

Environmental governance functions through institutions, laws, norms and processes for collective decision making (Young 1992), and the region has a wide diversity of systems and mechanisms. However, many "remain centralized, expert-driven, compartmentalized, and inflexible" (ESCAP/ADB/UNEP 2012). A persistent problem is that, "many environmental laws, regulations, action plans and programmes [have not been] effectively implemented", making greater progress necessary to achieve the selected global goal of good governance at local, national, regional and global levels (JPOI Paragraph 5).

Policy screening

As indicated in the *GEO-5* Introduction, the first step in the policy analysis framework was to formulate a long list of policy options with the potential to accelerate achievement of the selected global goals, and then identify a few priority policies or policy clusters for further analysis.

In some cases, the long list of policy options was clustered into groups of policies with a common intent prior to screening, for ease of assessment and acknowledging that most policies are implemented as part of a complementary package rather than alone. While all the policies considered may have important contributions to make in certain circumstances, the priority policies in Table 10.1 are believed to be able to accelerate achievement of the selected global goals if implemented consistently across all countries in the region, bearing in mind specific national circumstances. Here, priority means that the policy or cluster of policies was selected for more detailed policy analysis rather than implying high priority for a specific country or sub-region.

Policy analysis

The priority policies in Table 10.1 were subjected to further analysis of their environmental, social and economic benefits and limitations, drawing from the literature and experience of the experts, as well as from 18 case studies of policy implementation, a limited number of which are summarized here. The analysis of limitations is included because even successful policies may have side effects or unintended consequences that need to be understood and addressed during implementation and that may impede replication elsewhere. To illustrate how policy packages may be introduced in coordinated steps, a series of graphic representations (Figures 10.1–10.5) show:

- possible time frames: short term, 1–5 years; medium term, 6–15 years; long term, 16 years or more; and
- direct policy measures that are intended to target the immediate cause of the problem, ranging to indirect policy measures that help to achieve the selected goals by tackling related issues.

Climate change

The key element of the selected global goal for climate change (Box 10.1) is to take a precautionary approach to anticipate, prevent or minimize the causes of climate change and to mitigate its adverse effects (Figure 10.1).

The clean energy policy cluster includes a renewable energy mandate and potentially carbon capture and storage, which, if and when the technology is proven, could contain the greenhouse gas emissions from the largest source – globally 12 billion tonnes of CO_2 per year by 2020 (McKinsey and Company 2009). This cluster also has significant co-benefits such as improved air quality and health improvements, avoidance of environmental damage from mining and exploration for fossil fuels, improved energy security and new green job opportunities (Hughes 2011; Renner 2008), and may offer households and businesses opportunities to generate their own energy and supply the surplus to the grid (Palit and Chaurey 2011; USEPA 2010). Potential limitations include non-climate-related negative

Table 10.1 Policies selected for analysis

Climate change
• **Clean energy:** promote clean energy – renewable energy, energy efficiency, carbon capture and storage • **Energy efficiency:** reduce energy demand – energy efficiency, transport systems • **Technology:** promote technology transfer and diffusion • **Financial policies:** enable economic instruments and innovative financing – carbon tax, emissions trading, eliminating energy subsidies, feed-in tariffs, REDD+ (the UN programme for Reducing Emissions from Deforestation and Forest Degradation in Developing Countries) • **Adaptation:** integrate climate change adaptation into development policies and strategies, and with disaster management • **Land management for carbon sequestration:** REDD+, low-tillage agriculture
Biodiversity
• **Conservation of biological diversity**: increase habitat management including improved effectiveness of protected area management and minimize destructive land-use change, especially deforestation • **Targeted species conservation**: address species conservation and invasive alien species management • **Illegal wildlife trade:** enhance illegal wildlife trade control at national level and through regional cooperation • **Community management:** encourage community-based management of wetlands, forests and coastal areas, including coral reefs and mangroves • **Innovative financing mechanisms**: apply innovative financing mechanisms, such as payment for ecosystem services and REDD+, for biodiversity management • **Access and benefit sharing**: improve access and benefit sharing regimes following the Nagoya Protocol
Freshwater
• **Framework:** apply adaptive and integrated water resources management planning • **Water allocation and cooperation**: promote community-based management for better allocation of water resources • **Basic human needs:** encourage rainwater harvesting/storm water management and support the construction and/or renovation of farm dams for increased and improved water storage • **Water-use efficiency:** promote industrial and domestic wastewater treatment; utilize economic instruments and approaches to enhance efficient use of water • **Water environment:** strengthen water quality legislation and implementation to secure water quality; incorporate the ecosystem approach/environmental flow concept into water resources management
Chemicals and waste
• **Framework:** adopt policy frameworks promoting waste avoidance and reduction of the production and use of hazardous chemicals • **Collection systems and treatment facilities**: establish systems and infrastructure for product reuse and materials recycling, and stimulate markets for recycled materials including both industrial by-products and post-consumer waste; set up safe disposal facilities for hazardous wastes and chemicals that cannot be recycled, either nationally or sub-regionally, paying special attention to the needs and circumstances of developing countries and economies in transition • **International collaboration:** strengthen international collaboration, including technology transfer and financial support as well as information sharing and policy transfer; reinforce control of inappropriate export and import of hazardous chemicals and waste
Environmental governance
• **Policy integration and mainstreaming:** ensure policy integration and coherence and eliminate policy conflicts; build capacity • **Strengthening incentive structures:** greening fiscal policies matched by innovative financing mechanisms • **Accountability and stakeholder participation:** decentralization and devolution of environmental management to the lowest practical level; multi-stakeholder input on all major planning decisions • **Compliance and enforcement:** establish environmental judiciary; combat corruption and inequitable power relationships; base environmental standards on the best available technology

Figure 10.1 Selected climate change policies

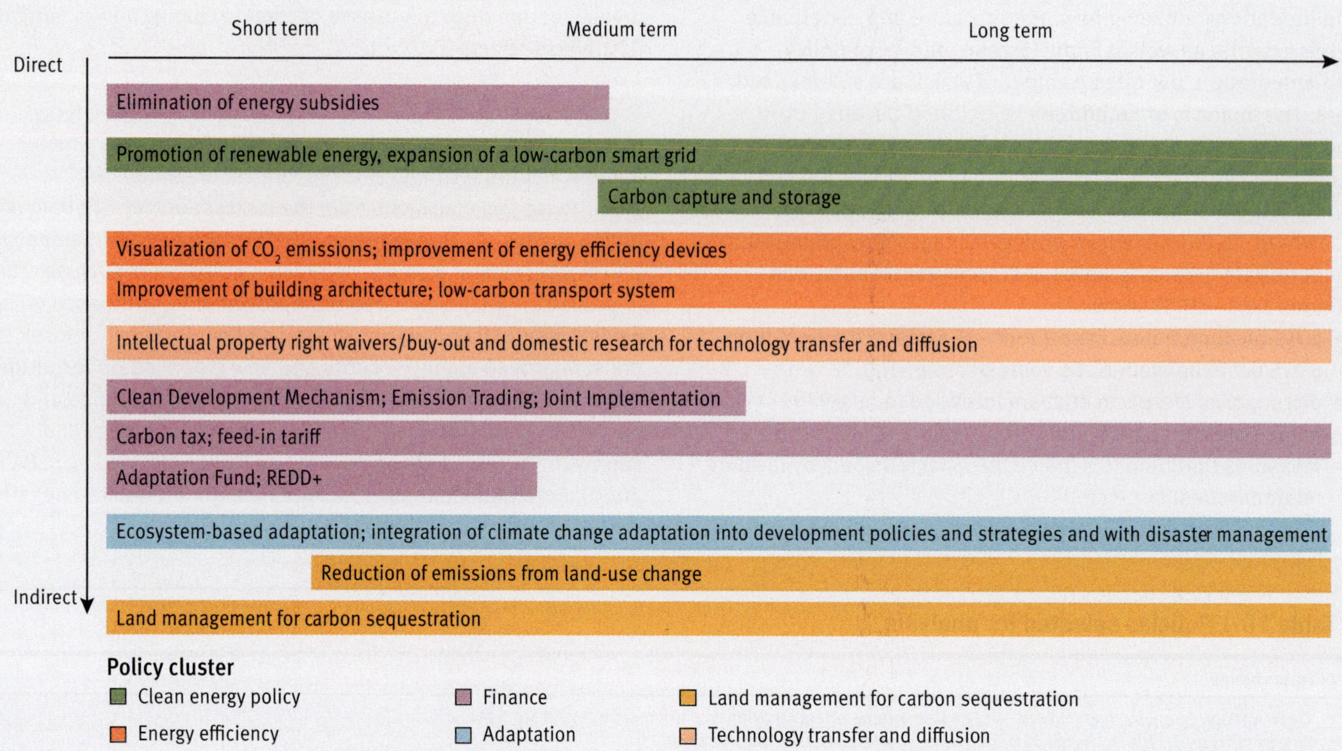

environmental impacts such as mining for rare earth metals; the competition between biofuels and food production and impacts on biodiversity; higher costs to end users; unproven technologies such as carbon capture and storage; and the impacts on businesses and employees in traditional fossil fuel energy production.

The energy efficiency policy cluster is aimed at reducing energy demand through targeted efficiency improvements in buildings, transport and agriculture, globally representing 14 billion tonnes of CO_2 per year by 2030 (McKinsey and Company 2009). The principal benefits are reduced operating and travel costs, health benefits from cleaner air, reduced traffic congestion following the growth of mass transit systems, and lower environmental impacts from low-tillage and other energy-saving agricultural practices. Limitations include the high initial costs of retrofitting buildings and introducing or expanding mass transit systems, high resettlement costs when people are displaced by the building of above-ground mass transit systems, and the potential lack of uptake of mass transit by car-owning urban households. Furthermore, energy efficiency improvements such as hybrid cars, while desirable, may be only temporary or partial due to widespread rebound effects whereby part of the efficiency gain is offset by increased energy use (UNEP 2011; Timilsina and Shrestha 2009). Inadequate public awareness and incomplete markets for energy efficiency are also barriers to greater market penetration (IEA 2007).

The technology policy cluster includes policies that promote technology transfer and diffusion such as technology transfer agreements, intellectual property rights waivers or buy-outs, and domestic research, all of which will also contribute to achieving a precautionary approach to combating climate change. This policy cluster will allow developing countries to leapfrog development stages and avoid the carbon-intensive trajectory of developed countries. It will improve human well-being, assist national development budgets by avoiding a lock-in to fossil-fuel-intensive approaches, and build national capacity for research and development. As a political strategy, such an approach has few limitations, although some intellectual property holders may be resistant to waivers to maintain their global competitiveness, and may be disadvantaged by being pressured to release their rights at less than market value. Indeed, some research claims that stronger intellectual property rights increase international technology transfer (Branstetter *et al.* 2006). Technology transfer from developed to developing countries may not be effective if the latter have inadequate domestic capacity to apply the technology in terms of infrastructure, human and financial capital, and the appropriate institutional environment (Tan and Zhang 2010).

The financial policies cluster includes greenhouse gas emissions trading, the Kyoto Protocol's Clean Development Mechanism, the Joint Implementation mechanism and the Adaptation Fund, amended tariffs and taxes such as feed-in tariffs, a carbon tax and an aviation tax, or elimination of subsidies that promote fossil fuel use (Box 10.6) and of financial incentives that encourage inappropriate land use and forest loss. A carbon tax may allow governments to reduce other taxes or generate additional revenue streams to invest in sustainable development. Earmarked financing provides incentives for scaled up investment in low-carbon technologies by the private sector. Penalizing fossil-fuel-based industries through taxes and tariffs levels the playing field for emerging technologies. Potential limitations include intermediaries benefiting from carbon market initiatives rather than achieving low-carbon objectives. Global schemes such as the Clean Development Mechanism tend to be bureaucratic and cumbersome, and to benefit too few countries (de Lopez *et al.* 2009). Social impacts, particularly for the poor, may be high due to more costly goods and services unless these are offset by lifeline support measures or tax rebates. The distributional impacts of all financial measures need to be carefully analysed prior to adoption.

The adaptation policy cluster anticipates the likely impacts already embedded in the climate system due to historical levels of greenhouse gas emissions and ensures that communities can adapt to inevitable changes (Box 10.7). Policies that facilitate adaptation include mandatory infrastructure design for future climates, or climate proofing; ecosystem-based adaptation in planning and zoning schemes; building climate resilience in agriculture, forestry and fisheries; and integrating climate change adaptation and disaster risk reduction (Srivastava 2011; Mimura *et al.* 2007). These policies have a range of co-benefits such as

Several countries in the region are developing ecosystem-based adaptation strategies for enhancing adaptive capacity and resilience to climate change and variability, including extreme events.

© GYINSEA/ iStock

biodiversity conservation, improved recreational opportunities and managed natural resource harvesting (ADB 2010). The main benefits are a lower incidence of death or injury from extreme storm events and droughts; reduced future economic and social costs; additional economic opportunities for the

Box 10.6 Removing fossil fuel subsidies in Asia and the Pacific

Several countries have begun to remove fossil fuel subsidies. China, for example, has attempted to bring domestic energy prices closer to global market levels as it moves from being largely self-sufficient in energy to being a major importer. In 2007, the country removed price controls for coal, and prices are now negotiated between coal producers and power companies. Crude oil prices and refined oil products now match international levels. In 2010, onshore natural gas benchmark prices rose by 25 per cent following increased gas transmission fees. Preferential tariffs for energy-intensive industries were eliminated and a three-tiered electricity pricing mechanism was introduced for residential use.

In 2010, India announced that petrol pricing would be market driven. Immediate price increases were announced for diesel, liquefied petroleum gas (LPG) and kerosene. Natural gas pricing reform in 2010 allowed state-run producers to sell natural gas from new fields at market prices instead of regulated rates, and the price of natural gas more than doubled. In the coal industry, price reforms are expected to bring domestic prices in line with imports, allowing for quality differences, which will increase electricity prices.

Indonesia has long subsidized energy prices as a poverty support policy, with 19 per cent of the state budget devoted to energy subsidies. Subsidies are becoming increasingly targeted at specific groups and the range of subsidized fuels has declined. In 2010, Indonesia announced plans to eliminate energy subsidies by 2014; to phase out the use of kerosene in favour of LPG and to restrict subsidized fuel to motorcycles, public transport vehicles and older cars; and raised power tariffs by about 10 per cent. In 2010, Malaysia announced plans to reduce subsidies for petrol, diesel and LPG, while Pakistan plans to phase out electricity subsidies and has implemented a 20 per cent tariff increase.

The expected benefits of these policies are to:
- reduce the burden on state budgets;
- prevent use of public funds to support the wealthiest and largest energy consumers;
- ensure that alternative energy sources are at least equally attractive to consumers; and
- reduce environmental damage and contributions to climate change from excessive use of fossil fuels.

Source: IEA/OECD/World Bank 2010

Box 10.7 Adaptation policies in the Maldives

Although they are the smallest contributors of greenhouse gas emissions, small island states are among the most vulnerable to climate change impacts, especially sea level rise, ocean acidification and increased storm severity and frequency. For example, with more than half the settlements and most of the critical infrastructure of the Maldives within 100 metres of the coast, there is serious concern about projections that 85 per cent of the country could be below sea level by 2100 (Khan *et al.* 2002). Recognizing this threat, the Maldives was the first country to declare its intention to be carbon neutral by 2019 (UNEP 2009a) and to view climate change as a critical national development challenge.

In response, the Seventh National Development Plan adopted a policy of identifying ten safer islands, future refuges for people displaced as a result of climate change. This policy includes high-cost infrastructure such as sea walls and desalination plants, and even artificial islands such as Hulhumalé in Malé Atoll. The government has also implemented a softer set of policy measures under the Integrating Climate Change Risks into Resilient Island Planning in the Maldives programme (GEF 2009), which involves working with nature to increase resilience, including coastal afforestation, replenishing natural ridges, climate proofing drainage, coral reef propagation, mangrove planting and beach nourishment. Each island community is involved in choosing which measures are most appropriate.

construction sector; higher property values in secure areas; and increased security and resilience of affected communities. Limitations include the environmental costs associated with large-scale infrastructure such as higher flood levees; social costs if communities or infrastructure need to be relocated away from vulnerable zones; investment costs in climate proofing and possible compensation costs for affected properties and companies; and the political costs of diverting funds to retrofit old and new infrastructure.

The policy cluster on land management for carbon sequestration aims at reducing greenhouse gas emissions from unsustainable land-use practices, including forest loss, biomass decomposition after harvesting, peat fires and decaying drained peat soils, which may contribute 15–20 per cent of total global emissions (van der Werf *et al.* 2009; IPCC 2007; WRI 2005). In South East Asia, emissions from inappropriate land use and forest loss may account for as much as 75 per cent of the total from that sub-region, mostly from forest loss in Indonesia (ADB 2010). Halving deforestation rates by 2050 and maintaining that level until 2100 would account for 12 per cent of the total emission reductions needed to stabilize atmospheric CO_2 at 450 ppm (FAO 2010; Gullison *et al.* 2007). Protection of coastal wetlands and marine ecosystems can also mitigate emissions (Crooks *et al.* 2011). Principal benefits include the conservation and supply of ecosystem services such as biodiversity and water supply and quality; maintenance of indigenous cultural practices; soil conservation; and promotion of local livelihoods. Limitations include possible conflicts with other development objectives; impingement of local economic aspirations due to restrictions applied by protected area managers; and more costly land management practices.

Bunaken National Park, a locally managed marine area in Indonesia, where tourism revenues have contributed to reducing local poverty.

© Piero Malaer/iStock

Biodiversity

The selected biodiversity goal contains elements of conservation of biological diversity, sustainable use of its components, and the fair and equitable sharing of the benefits of using genetic resources (Box 10.2).

The policy cluster for conservation of biodiversity promotes the creation of protected areas, including areas that connect landscapes and seascapes, through identifying areas of high but threatened biodiversity value and biodiversity corridors that link protected areas as a system. The notable progress in the establishment of terrestrial and marine protected areas reported in the *Global Biodiversity Outlook 3* (CBD 2010) may be attributed to explicit policies on protected areas, with many countries in Asia and the Pacific using legislation to establish protected areas (CBD 2010). Existing policies on protected areas may need further improvement, yet they provide a good foundation for attaining the global objective of biodiversity conservation. Commitments like the Convention on Biological Diversity (CBD), the Ramsar Convention on Wetlands, the World Heritage Convention and new funding mechanisms often drive both the establishment and improved effectiveness of protected areas. A shift from revenue generation to a conservation policy mandate has effectively reduced associated illegal land-use change.

Box 10.8 Pacific islands: locally managed marine areas

Defying the current trend of habitat destruction, an outstanding example of community-based management of coastal resources comes from the South Pacific. In the past decade, more than 12 000 km^2 have been brought under a community-based system of marine resource management known as locally managed marine areas. The initiative involves 500 communities in 15 Pacific island states and has helped achieve widespread livelihood and conservation objectives based on traditional knowledge, customary tenure and governance, combined with local awareness of the need for action and the likely benefits, including the recovery of natural resources, greater food security, and improved governance and health. In Fiji, for example, the results of implementing locally managed marine areas since 1997 have included a 20-fold increase in clam density in areas where fishing is banned, an average 200–300 per cent increase in harvests in adjacent areas, a tripling of fish catches, and a 35–45 per cent rise in household incomes. Such initiatives have the potential to be widely replicated wherever the socio-cultural environment is appropriate.

Source: Govan *et al.* 2008

Challenges remain in ensuring that protected areas are part of an ecologically representative network, on land and at sea, and to provide effective protection for threatened and endemic species (ACB 2010). Many protected areas in Asia and the Pacific have communities within or on their periphery for whom formal recognition of stewardship, traditional livelihoods and conservation traditions is necessary. Formal recognition of indigenous and local community-conserved areas (Box 10.8) would increase the ecological coverage of legally protected areas and support community rights in protecting the sites. Possible limitations include competition with development objectives, difficulties in measuring the real values of conservation against development activities yielding shorter-term economic returns, and institutional and individual capacity to effectively enforce conservation laws.

Also covered under this policy cluster are regional cooperation initiatives on transboundary protected areas and biodiversity

Figure 10.2 Selected biodiversity policies

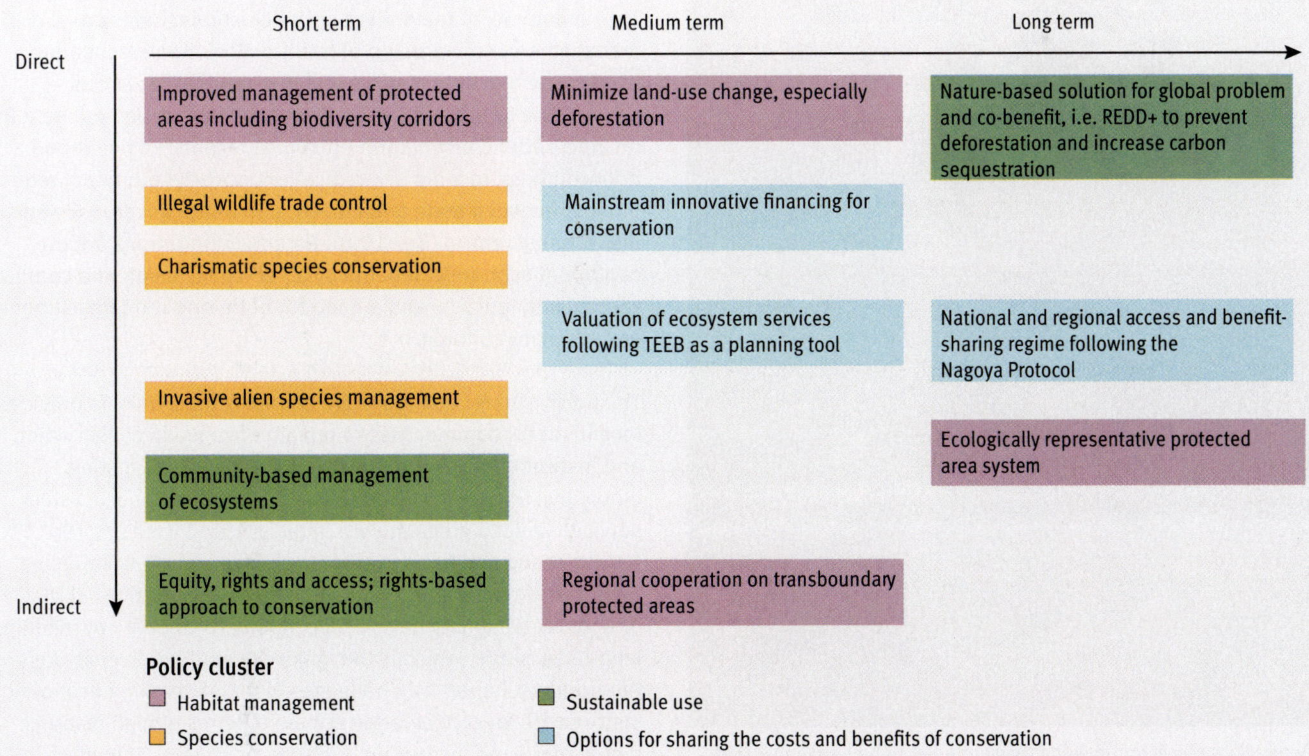

Note: TEEB – The Economics of Ecosystems and Biodiversity study (TEEB 2012).

corridors. Transboundary collaboration fosters the cooperation of national institutions to benefit multiple countries, as demonstrated by several examples involving cross-boundary interest in protecting areas with high levels of biodiversity such as the Greater Mekong sub-region, Terai Arc landscape in India and Nepal, Sulu-Sulawesi marine areas and the Coral Triangle. The benefits from this cooperation are an augmentation of national efforts, transfer of capacities across countries and twin conservation efforts involving several stakeholders across borders. The key challenges are sustainability, differing capacities of institutions involved and the political nature of the cooperation whenever sensitive sovereign issues arise.

The targeted species conservation policy cluster is intended to protect species such as tigers, elephants, pandas, saola – an extremely rare antelope discovered in Viet Nam and Laos in 1994 (Schaller and Vrba 1996) – or other species of biological, economic, spiritual and cultural importance, including species covered by the Convention on International Trade in Endangered Species of Wild Fauna and Flora (CITES). These policies protect species from capture and hunting, being kept as pets or being traded for medicinal or food purposes, as well as promoting their well-being in captivity, such as in wildlife parks or zoos. The policies not only protect valuable species but also help to spread a broader conservation message of the need to conserve species and their habitats. Charismatic species also act as rallying points for political support, promote nature-based tourism and help leverage resources for broader institutional support. The main limitations are that there may be too much attention on such individual species, drawing support away from others in their native habitat.

The Bengal Tigers left in the wild is estimated at to be between 3 000 to 3,900 individuals, with most populations occurring in small groups isolated from each other. *© Neal McClimon/iStock*

Invasive alien species, especially in island ecosystems with high endemism, are a major threat to species conservation. Apart from national quarantine processes and regional networks like the Asia-Pacific Forest Invasive Species Network, control mechanisms are limited. However, there have been a few cases of successful eradication of invasive alien species in the region (GISP 2009).

The illegal wildlife trade policy cluster, based around CITES, is intended to eliminate the illegal wildlife trade, including by strengthening border controls and training customs officials to recognize endangered species, and through campaigns to raise awareness and publicity surrounding successful enforcement cases (ASEAN-WEN 2009). Benefits include the protection of natural assets and saving certain species from extinction. There are also significant co-benefits in the form of improving law and order generally (as wildlife crime is often associated with other criminal activities), improved governance structures and better response mechanisms for wildlife protection within national borders. The main limitations are the need for heavy investment in law enforcement, possibly denying the customary rights of indigenous communities to access non-timber forest products, and possible social conflict where bushmeat is an accepted social norm and a significant source of protein (van Vliet 2011).

The community management policy cluster includes co-management, stewardship of traditional owners, recognition of property-use rights, and policies establishing various sustainable forest and fisheries management schemes. Benefits include better conservation outcomes, expanded livelihood opportunities, income diversification, poverty reduction, reduced tension between state and citizen, and better governance and institutional reform (Box 10.8). Potential limitations are the capture of economic benefits by an elite, exclusion and conflicts within communities, and a need for both long-term investment and capacity building.

The policy cluster on innovative financing mechanisms provides incentives for communities to remain engaged in conservation and institutionalizes their involvement. Countries testing innovative financing mechanisms like payment for ecosystem services under a green growth paradigm are increasing (Box 10.9). Nature-based solutions to climate change – such as Reducing Emissions from Deforestation and Forest Degradation (REDD+) – offer potential for social and biodiversity co-benefits by avoiding land-use change, which is the major driver of biodiversity loss in the tropics (Chapter 1). Challenges to mainstreaming economic instruments for conservation remain in terms of high resource degradation and increasing pressure, as well as difficulties in streamlining supportive policy and legal regimes, institutional mechanisms, and equity and rights of the community.

Box 10.9 Promoting sustainable use of biodiversity: payment for ecosystem services in China and Viet Nam

China: A national environmental policy framework that promotes eco-compensation as a key principle aims to promote sustainable use of natural resources and more balanced growth across regions. China has been implementing some of the largest payment for ecosystem services schemes in the world. For example, more than US$15 billion has been spent since 1999 on the conversion of cropland to forest and grasslands. This programme pays farmers to withdraw land from agricultural use and afforest or plant grasses on sloping and marginal cropland covering more than 9 million hectares. Similarly, almost US$2 billion has been invested in a forest ecosystems compensation fund, which pays households, communities and local governments to protect key forest areas, now covering 44 million hectares. The success of these schemes has created a healthy debate in the government on how to make improvements and has provided the impetus for exploring and developing other market-based tools and innovations to address the country's challenge of balancing growth with environmental concerns. Recently, the government has called for emission-fee reforms for key natural resources, improved resource taxation and fees for mineral resource use, which will expand the horizon for eco-compensation mechanisms. These policy experiences are contributing to the drafting of a national law on the eco-compensation framework.

Viet Nam: Under funding from the US Agency for International Development's Regional Development Mission for Asia, the Asia Regional Biodiversity Conservation Programme has implemented a successful pilot project on payment for forest environmental services in Lam Dong Province, which has improved the livelihoods of 40 000 rural poor and promoted biodiversity conservation while informing the design and subsequent issue of a national decree on such schemes. The decree creates the legal framework for integrating the value of ecosystem services in two pilot areas – Lam Dong and Son La provinces. The policy facilitates payment for forest management while improving the incomes of the communities that provide those management services.

The buyers of the forest services are the electric and water utilities paying for water regulation and soil conservation, and tourism operators paying for the landscape amenity. The Lam Dong pilot area involves high-priority conservation areas in Cat Tien National Park and Bi Doup-Nui Ba National Park, as well as providing a link to the Dong Nai River Basin Conservation Landscape. By December 2010, payments of more than US$4 million for the protection of 210 000 hectares of forest had been made to 22 forest management boards and forestry businesses as well as to 9 870 predominantly ethnic-minority households, each of which receives US$540–615 annually. Forest protection patrols supported by the scheme resulted in half the number of cases of illegal logging and wildlife poaching in one priority watershed area. Replication of this approach throughout Viet Nam would make a significant difference in the incentives to retain forest habitats and protect biodiversity, especially if the pool of buyers of ecosystem services can be expanded – for example to those investing in carbon sequestration through carbon offsets.

Source: (China) Zhang *et al.* 2010; SDPC 2000; (Viet Nam) Winrock International 2011

The policy cluster on access and benefit sharing for the equitable use of genetic resources includes recognition of the rights of indigenous stewards of ecosystems, intellectual property rights protection and regulations preventing biopiracy. The policy cluster draws heavily on the outcome of CBD negotiations on access and benefit sharing, notably the Bonn Guidelines (CBD 2002) and the subsequent international regime. Adoption of the Nagoya Protocol on Access to Genetic Resources and the Fair and Equitable Sharing of Benefits Arising from their Utilization (CBD 2011) will guide the ongoing effort for developing national and regional agreements. The Association of Southeast Asian Nations (ASEAN) draft agreement on Access to Biological and Genetic Resources, combined with draft policies and laws on access to genetic resources, benefit sharing and traditional knowledge in Bangladesh, Cambodia, Mongolia, Nepal and Sri Lanka, will provide additional incentives for implementation.

The principal benefits of these policies include providing an additional incentive for indigenous communities dependent on natural resources to maintain a full range of biological diversity, a fair and equitable return from investors that stand to make a profit from capitalizing on indigenous knowledge and/or stewardship, and a means for governments to protect a national heritage. The main limitations are the difficulties in identifying communities practising traditional knowledge, and its objective validation both for effective rewards and for overcoming potential constraints on voluntary research.

Freshwater

The global goal on freshwater selected for Asia and the Pacific targets improvement of water allocation, conservation of both the quantity and quality of water resources, and the safeguarding of ecosystems (Box 10.3; Figure 10.3). These could be achieved by systematic promotion and application of integrated water resources management as a policy-planning framework.

The concept of integrated water resources management has been in place for almost 30 years and many countries have tried to set up institutional frameworks for it. In practice, however, only a handful of countries in the region have established the necessary legal and institutional capacities for its implementation (UNESCO-WWAP 2006). In most countries, organizational

Figure 10.3 Selected freshwater policies

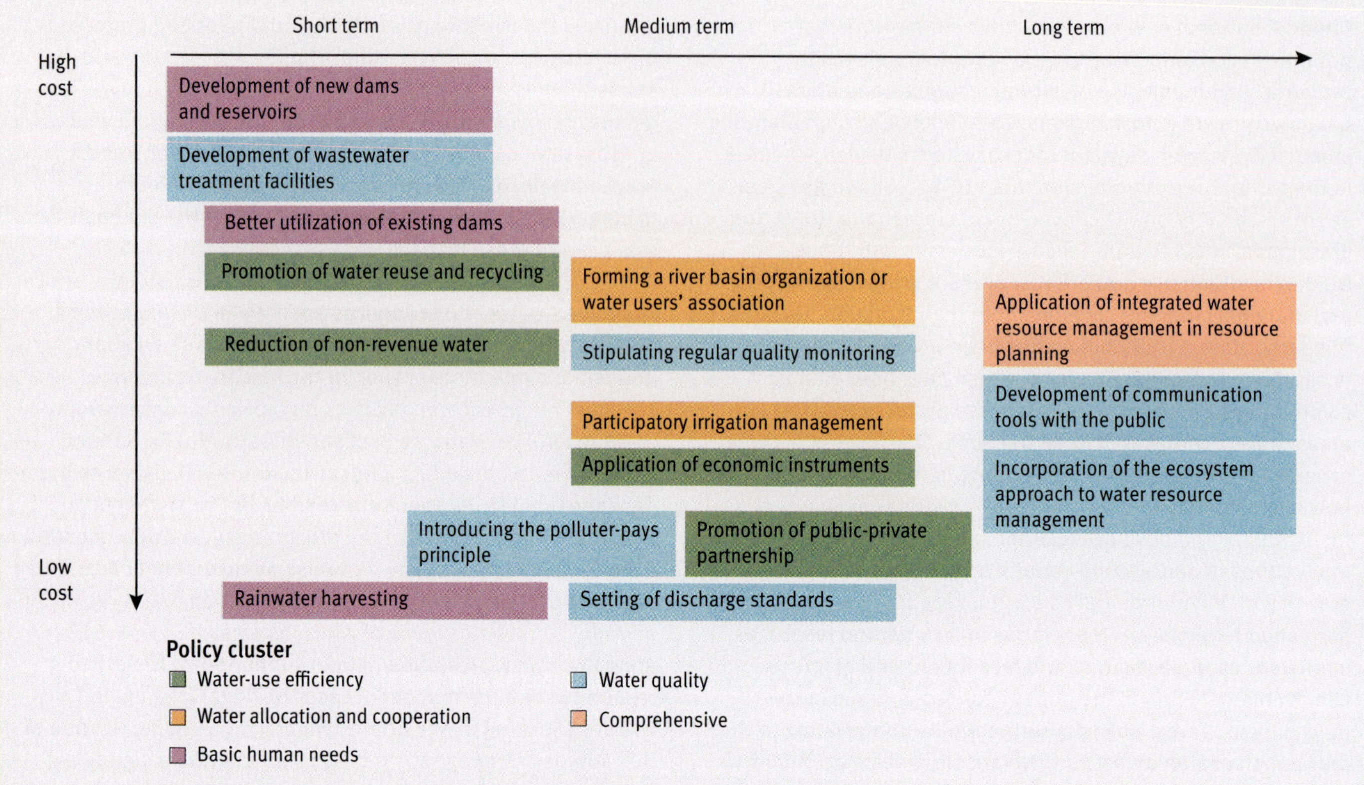

reform has involved setting up apex water bodies and river basin organizations to implement the integrated approach. However, water resources are still managed primarily through a sectoral approach in which policy is shaped by each agency in charge, often without cross-sectoral coordination and with occasional tensions (Molle and Hoanh 2009; Bandaragoda 2006). The application of integrated management can contribute to sustainable water resource use by integrating and balancing human needs with conservation and the restoration of ecosystems, and will enable the region to cope with complex and unpredictable challenges such as future climate change impacts related to changing patterns of extreme events including droughts and tropical cyclones (typhoons) (Chapter 4).

The water allocation and cooperation policy cluster aims to achieve an acceptable balance between potentially competing water uses and involves enhanced conflict management and cooperation capacities. Typical policies may involve assigning existing local communities to implement water resources management, establishing water users' associations or river basin organizations with explicit mandates to manage potential conflicts, using multi-stakeholder participation to draw up basin development plans, or setting up conflict resolution panels. Cooperation institutions established in river basins have helped to reduce potential water related conflicts in Central Asia (Abdullaev and Atabaeva 2011). For example, the inclusion of water users in the canal water management system in the Fergana Valley of Central Asia improved transparency and more equitable water distribution (Abdullaev *et al.* 2009a, 2009b; Dukhovny *et al.* 2008; Abdullaev *et al.* 2006). In Australia, the Murray-Darling agreement and its implementing agency is another example of basin-level institutional arrangements that have been emulated in other countries.

The pioneering of community-based participatory irrigation management in Andhra Pradesh, India shows the importance of water users' associations in leading such initiatives (Gupta 2010; Narain 2003; Ballabh 2002; Mollinga 2001; Parthasarathy 2000; Shashidharan 2000). While the policy has not been an unqualified success, a state-level federated structure of users' associations was created, followed by district- and farmer-level organization structures with well-defined mandates, transparent systems and people-centred decision making. At the local scale, these policies have contributed to increased financial capacity through revenue generation from the group members and the incorporation of a watershed management approach. The main limitations have been difficulties in scaling up from a pilot project level due to resistance from those who were benefiting from the *status quo*, prolonged water scarcity, and unsustainable institutional arrangements once external project funding came to an end.

The basic human needs policy cluster relates to satisfying the need for water and is oriented towards improving and/or

Box 10.10 Uzbekistan: improving the capacity of existing reservoirs in Central Asia

Irrigation is the primary consumer of water in arid countries. In Central Asia it underpins economic development, employment and food security, with reservoirs playing an important role in making it possible. More than 250 large reservoirs with a total volume of more than 163 km^3 were constructed in the region from 1950 to 1990, but most are now silted up. To improve capacity, a state agency in Uzbekistan formulated a policy option to study and de-silt existing reservoirs. This has so far allowed savings of US$250 000 and helped to increase reservoir volume by up to 10 per cent. The policy has now been adopted as standard procedure by the Ministry of Agriculture and Water Resources.

Source: Rakhmatullaev *et al.* 2010; White 2010; Vörösmarty *et al.* 2003; Yang 2003; WCD 2000; Mahmood 1987

increasing the volume of stored water along with the resulting flexibility in use. Policies include planning guidelines to store urban storm water run-off for watering public gardens or for street cleaning, or requiring buildings above a certain size to store water collected on the roof. Rainwater harvesting is already included in national and local policies in some countries such as Sri Lanka (Government of Sri Lanka 2007), Australia (Meinzen 2009) and in some states of India (Rainwater Harvesting Organization 2011). Support for construction of farm dams and renovating existing dams to extract greater volumes of water through application of new technologies are feasible policy options (Box 10.10). National policies on the construction of large water storage facilities for flood control, hydropower and irrigation have potential immediate benefits, but often induce concerns over environmental and social impacts (WCD 2000). The water-use efficiency policy cluster promotes the application of economic instruments, especially in urban areas, including graduated pricing based on the type of water use and volume, pollution charges and combined payments for water and sanitation. The benefits of economic instruments include behavioural change inducing water saving by users, increased revenue for social measures and the operation and maintenance of the water supply. Experience from Cambodia and the Philippines shows that it is possible to improve access to safe drinking water significantly in cities and rural areas (IFAD 2011; ADB 2009c). The principal limitations are the difficulties of making water a priced service rather than a free good, higher management costs, potential cost burdens on some users, shifts to non-priced water resources such as groundwater, unpopularity with politicians who are reluctant to impose new charges and delays in effective institutionalization. Reducing the use of un-priced water is a way of improving water-use efficiency but it needs investment and staff capacity development (Frauendorfer and Liemberger 2010). Manila Water and the Phnom Penh Water Supply Authority have significantly reduced un-priced water use through management improvements (ADB 2009c).

The water environment policy cluster includes strengthening legislation and the implementation of water quality management by ensuring regular monitoring and reporting, applying the polluter-pays principle through pollution fees, incorporating total pollution load control systems, and mandating treatment levels for different categories of wastewater. The promotion of industrial and domestic wastewater treatment also contributes to improving water quality. Public communication strategies such as a water quality index should also be considered as a way of drawing attention to changes in water quality. The main benefit is that sufficient water of the desired quality becomes available, thus reversing current trends. Improved water quality reduces health risks and helps achieve one of the Millennium

The historic Iwabuchi floodgate, in Tokyo, remains crucial for communities living along two of Japan's rapidly-flowing, flood-prone rivers. © *Juergen Sack*

Box 10.11 The Yellow River, China: balancing environmental and human needs through quotas and pricing reform

Several provinces share water from northern China's Yellow (Huang) River. The river began exhibiting a partial failure to reach the sea in 1972 and continuous interruption was observed after 1987. The annual frequency of cut-off days reached its peak of 226 days in 1997, and the severe reduction in flow impaired the ecosystem health of the river basin as well as its services to society.

In 1998, the National Development and Reform Commission, formerly the State Development Planning Commission, and the Ministry of Water Resources issued annual water-use quotas and a distribution scheme for the river, as well as the Implementation Regulation for Water Resource Allocation among Provinces in the Huang River Basin. These management policies determined total water withdrawals on the basis of hydrology, the need for sediment transport and other ecological factors, and established annual provincial water withdrawals including a seasonal distribution plan for greater withdrawal in the rainy season than in the dry season.

In March 1999, the Yellow River Water Conservancy Commission issued the first water withdrawal quota directive and started the water withdrawal control plan for the whole basin. This policy was extended from the main Yellow River to its tributaries in 2006. In the same year, the State Council issued Water Withdrawal and Water Resource Fee Collection Rules, marking a new chapter in water management policy based on economic measures such as water pricing and resource fees. Trade in water user rights between various sectors was observed in some provinces.

Implementation of these policies has ensured uninterrupted flow of the river to the sea since 2000 and improved the water resource and ecological health of the whole basin. Ecosystem integrity and biological diversity have improved greatly. Rare species have reappeared, with the number of bird species in Shandong Yellow River Delta National Nature Reserve rising from 187 in 2000 to 283 by 2006. Populations of rare and endangered plant and animal species in the Shell and Wetland System Nature Reserve doubled over the same five-year period.

Source: Wang and Zhang 2010; UNEP 2008b; State Council 2006a, 2006b; NDRC 1998a, 1998b

Development Goals. The benefits to industry and the water supply sector are lower water treatment costs. Limitations include the high wastewater treatment costs and the difficulty in convincing polluters, especially marginal small to medium-sized enterprises, to undertake the necessary control measures voluntarily, thus requiring resolute implementation of the polluter-pays principle and command-and-control regulations. Ensuring environmental flow, which requires a high level of political commitment, is also important for the health of water bodies (Box 10.11).

Due to industrial waste and urban sewage, Nanchuan is now one of China's most contaminated rivers. *© Sinopictures*

Chemicals and waste

The global goal for chemicals and waste focuses on life-cycle analysis, transparency and risk assessment for a participatory approach to minimize risks to human health and the environment (Box 10.4; Figure 10.4).

The Asia and Pacific region is facing rapidly growing challenges in waste and chemicals management, fuelled by the region's combination of strong economic growth and population increase, and its rapid industrialization and urbanization. In low- and middle-income countries in particular, volumes of waste are growing and waste streams are becoming increasingly complex, containing ever larger amounts of hazardous substances (Harhay *et al.* 2009). The capacity to collect and properly treat this waste is lagging, with consequent impacts on human health and the environment. Similarly, the use of agricultural and industrial chemicals and the unintended generation of hazardous substances are escalating, leading to impacts that are currently poorly monitored and therefore little understood.

The policy cluster on product redesign and sustainable consumption addresses these problems at source, but it will

Box 10.12 Phase-out of ozone-depleting substances in India

Over the past two decades, the government of India has undertaken a number of policy measures to encourage technologies that do no damage to the ozone layer. These include licensing the export and import of ozone-depleting substances (ODS); granting duty exemptions for goods needed to comply with the Montreal Protocol's Multilateral Fund, under which India is entitled to assistance for phasing out ODS and switching to technologies that do not depend on such substances; and stopping investment in new projects that make use of them.

As a producer and user of seven of the 20 substances controlled under the Montreal Protocol, India acceded to the treaty in 1992. In 1993, India prepared a detailed country programme to phase out ODS and subsequently updated the programme in 2006 in consultation with the Confederation of Indian Industry and other stakeholders. In 1997, India was producing almost 40 000 tonnes of ODS, primarily the chlorofluorocarbon CFC-12 and carbon tetrachloride (CTC), with domestic consumption estimated at 14 000 tonnes. By 1999, some 226 projects had been approved by the Multilateral Fund at a cost of US$58 million and estimated to remove 7 682 tonnes of ODS from the aerosol, foam, halon, air conditioning and solvent industries. As of 2010, India had successfully phased out the production and consumption of CFCs, CTC and halons as specified by the Montreal Protocol and excepting the use of pharmacy-grade CFCs for respiratory ailments. A complete, step-wise phase-out was proposed for all ODS by 2040.

The successful phase-out of ODS in India, as well as being attributed to the above policy provisions, is also associated with the creation of a national ozone cell and an empowered steering committee supported by a technology and finance standing committee and a standing committee on monitoring. Other factors included involving Indian industry from the beginning and awareness campaigns for the general public and affected industries. A detailed monitoring mechanism was set up and the ozone cell in the national environment agency conducted site inspections and ensured that funds were used appropriately, timely progress reports were submitted, and the intended impacts were being achieved. Similar success stories from other countries in the Asia and Pacific region suggest that India's policy approaches for implementing the Montreal Protocol are readily transferable.

Source: UNEP 2010; WMO 2010; Ozone Cell 1999

take a relatively long time to realize its full potential. Effective solutions to the challenges of waste and chemicals require preventive approaches. Substances with hazardous properties need to be phased out and replaced with safer options as far as possible. Production systems and products need to be redesigned with their full life cycle in mind, minimizing resource consumption, chemical hazards and waste generation. Overall, there is a need to encourage more sustainable patterns of consumption that can deliver quality of life with the minimum environmental burden.

Several actions towards this objective are already being taken. Green public procurement has proven an effective tool for creating a market for improved products, including those made of recycled materials (FOEN 2008). Extended producer responsibility for electronics has promoted design changes and fostered closer collaboration between manufacturers and recyclers. A good example of chemical substitution has been the phasing out of ozone-depleting substances under the Montreal Protocol on Substances that Deplete the Ozone Layer (Box 10.12). Waste avoidance has had some success for industrial waste but has proven more challenging for household waste.

The principal benefits of such policies include minimizing the need for waste treatment and chemical safety measures, and a reduction in both illegal waste dumping and pollution. The main limitation is that some companies may have to internalize costs that they were previously able to shift to the public domain, although evidence suggests that taking a life-cycle approach can often also reduce production costs (Barringer 2003).

Decentralized wastewater treatment system in Nadu, India, for small volume usage. *© Chinch Gryniewicz/SpecialistStock*

Figure 10.4 Selected chemical and waste policies

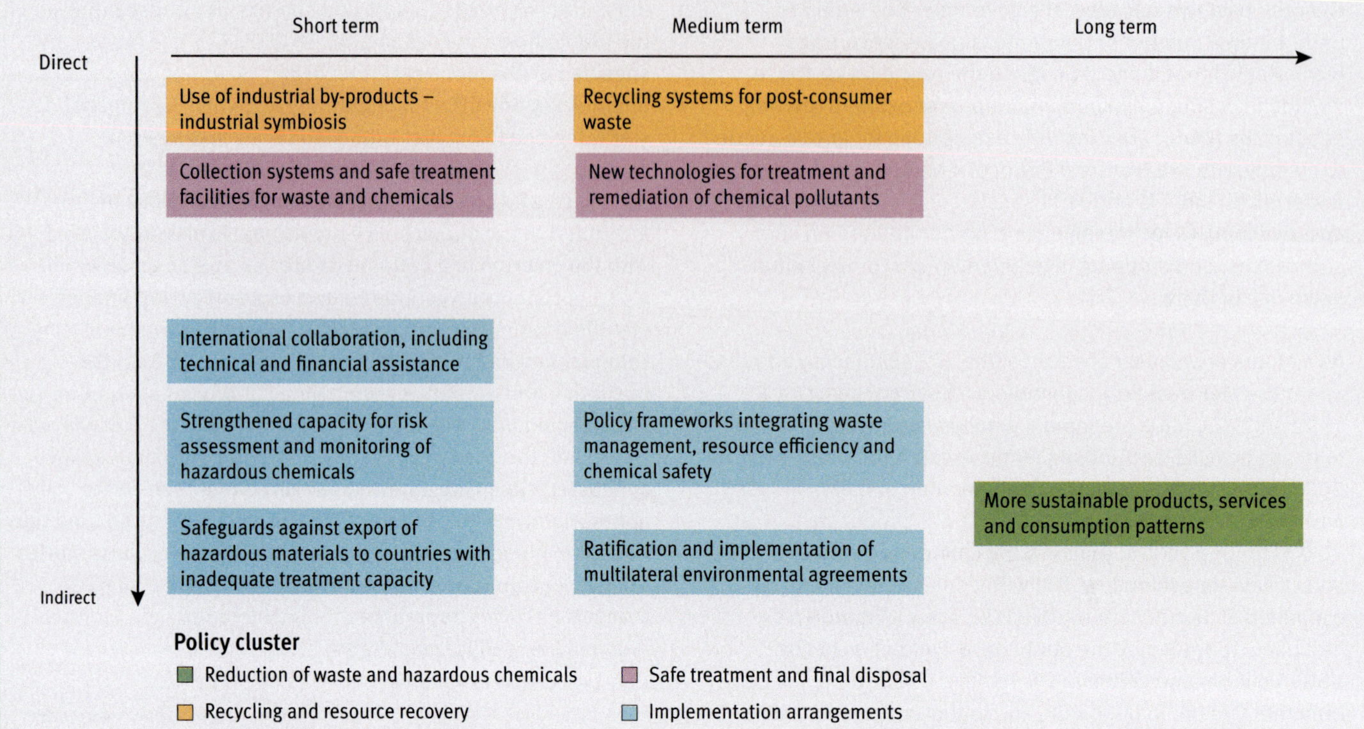

The policy cluster on collection systems and treatment facilities aims to ensure that, once products are in circulation, priority is given to reuse and recycling, with energy recovery and safe disposal as less desirable options. Industrial and municipal solid wastes pose different challenges and require different approaches, although there may be synergies in combined treatment. Effective policy interventions need to cover all stages, including efficient collection systems, safe treatment facilities and functioning markets for recovered materials. The cost of establishing systems of a high environmental standard may seem challenging, but the cost of inaction is also significant. Effective technologies are widely available, including both high-tech solutions and systems based on traditional practices. In Japan, for example, more than 50 million tonnes of solid waste are burned every year, often generating electricity and providing district heating. In 2000, Japan set a target to reduce dioxin emissions from incinerators by about 92 per cent from 1997 levels – a target achieved by 2003, with over 2 000 industrial waste incinerators shut down. The revised target in 2003 was to achieve a further reduction of 30 per cent by 2010 (MOEJ 2007; JFS 2005). The Republic of Korea has been particularly successful in introducing source separation of food waste for separate treatment, a policy that has significantly reduced the emissions of greenhouse gases from waste treatment (Lee 2006). Community-based composting has also been successfully introduced in many cities across Asia and the Pacific, reducing disposal needs and the associated costs for municipalities (ESCAP/IGES 2011).

The principal benefits of these policies include the obvious direct environmental and health benefits as well as the longer-term indirect and/or cumulative impacts of lower concentrations of toxic and hazardous materials. The main limitations of recycling have been associated with finding markets for recovered materials, leading to diminished private-sector enthusiasm after an initial period.

The international collaboration policy cluster addresses the need for a joint approach to meeting challenges as the region's countries become more integrated (Aziz 2010; Nag 2010). This is particularly relevant to waste and chemicals, since end-of-life products and recyclable materials are increasingly traded across borders (Box 10.13), and many chemical pollutants, such as persistent organic pollutants (POPs) and mercury, can spread far from their sources. Developing countries and economies in transition are facing particular challenges in formulating and implementing effective policies to address these problems: international collaboration can provide the technical and financial resources needed. Regional monitoring systems for chemicals in the environment such as the POPs Global Monitoring Programme have been instrumental in identifying pollution sources for the formulation of effective action (UNEP 2008a).

The benefits of these policies depend on the form of collaboration. Strengthening the technical and governance capacity for waste and chemicals has obvious and direct benefits for human health and the environment. Effective and

Box 10.13 Ship breaking in South Asia: implementing a new international environmental agreement

Normally, recycling of materials is regarded as an environmental benefit, but in some cases, such as ship breaking and recycling of e-waste and batteries, the long-term exposure and labour-intensive methods used in developing countries result in negative local impacts on the environment and human health. Since the 1980s, the global centre of ship dismantling and recycling has been South Asia, with Bangladesh, India and Pakistan accounting for 70–80 per cent of the international market. The industry not only provides large volumes of recycled iron and steel, plus other materials, but also creates jobs for thousands of workers from the poorest segments of the population. The direct and indirect beneficiaries in Bangladesh alone are estimated to be half a million people. The majority of workers are young, male and functionally illiterate, often living in cramped shacks near the recycling yards, thus adding to health concerns.

Obsolete ships contain a wide range of hazardous materials for which there are no adequate treatment facilities or occupational health and safety measures in the South Asian yards. Ship breaking is regarded as a "pollution haven" industry, often seeking out jurisdictions with lax environmental controls. However, in 2009, the Bangladesh High Court directed that all ship breaking yards without Department of Environment clearance should close within two weeks, and ordered new rules to be formulated requiring all ship breaking yards to obtain an environmental clearance certificate.

In recognition of these environmental dangers, the International Convention for the Safe and Environmentally Sound Recycling of Ships (the Hong Kong Convention) was adopted in May 2009 and is expected to come into force in 2015. A key requirement is to remove hazardous materials before recycling commences. The Hong Kong Convention will require signatories to ensure that ships are recycled only in countries that are a party to the convention, and in facilities that meet its work safety and hazardous waste handling requirements. An inventory of hazardous materials will need to be completed by the ship owner and provided to the recycler, so that the ability to handle the wastes can be checked. The convention will also specify that certain hazardous materials should not be used in the construction of new ships, thus avoiding these problems when the newer generation of ships is scrapped.

The convention will require South Asian nations to revise their legislation and invest heavily in improved procedures, equipment and facilities if they want to continue this business.

Source: Sarraf *et al.* 2010; EC 2007; Andersen 2001

safe treatment of hazardous materials can be ensured if these operations are carried out in countries with adequate capacity and if shipments to other countries are prevented. Enhanced international information sharing enables traceability and effective preventive action, including strengthened control of transboundary movements.

Environmental governance

The global goal on governance relates to strengthening the multiple dimensions of sustainable development (Box 10.5). Four key policy clusters were identified that could accelerate its achievement (Figure 10.5). Capacity development, access to education and information remain as underlying enabling factors for the effectiveness of each policy cluster.

The policy integration and mainstreaming cluster aims to integrate sustainable development functions, which are commonly fragmented between different ministries and agencies with weak coordination. Increased policy integration and strengthening the capacity of environment and related ministries and agencies at different levels of government can promote win-win opportunities for environment and development. This integration would not only strengthen the organizational capacity and decision-making influence of environment ministries, but also enhance accountability regarding the potential environmental and social impacts of development projects. Integration can be supported through impact assessment and monitoring as well as mainstreaming mitigation measures.

A shipbreaking yard in Chittagong, Bangladesh, where workers are exposed to heavy metals and persistent organic pollutants.

© Pierre Torset /Stillpictures

Figure 10.5 Selected governance policies

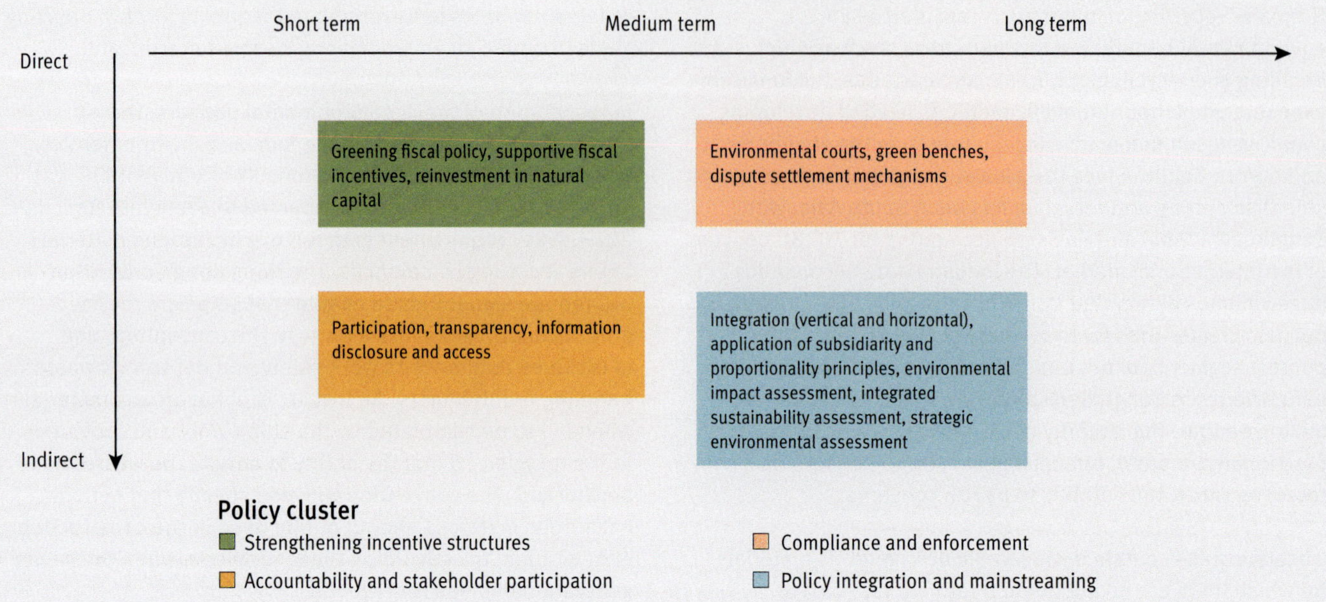

Opportunities for integration include decrees, constitutional amendments, presidential committees and independent peer reviews, or environmental focal points within ministries (Dalal-Clayton and Bass 2009; Jordan and Lenschow 2009). Some degree of mainstreaming is practised by Australia, China, Japan, New Zealand and the Republic of Korea (Box 10.14). Benefits include a more inclusive policy making process, greater coherence of policies and improved implementation. Environmental agendas, however, can be watered down by other stakeholders, and sectoral ministries are not always keen to promote the development projects with the least environmental impacts.

In some cases, achievement of global sustainability goals can be accelerated by allocating authority at more appropriate levels (Berger and Steurer 2009), for example through decentralization and the introduction of participatory approaches in natural resource management (Box 10.15). Delegation of authority needs to be linked with appropriate budget authority, human resources, capacity building and reporting mechanisms to ensure effectiveness and public support (Jordan and Lenschow 2009).

At upstream planning stages, environmental impact assessment for individual projects, cumulative impact assessment for series of projects, and strategic environmental assessment for

Box 10.14 Low-carbon green growth in the Republic of Korea and China

The Republic of Korea has formulated a National Strategy for Green Growth with an overall vision of becoming a global green leader by 2020, focusing on the following broad objectives: mitigation of climate change and energy independence, creation of new engines for economic growth, improvement in quality of life and enhancement of international standing. The strategy is supported by a Green Growth Framework Act and a five-year action plan with targets for greenhouse gas emission reductions, carbon absorption by forests and afforestation.

The National People's Congress has set China on a more sustainable and low-carbon development path through the 12th Five-Year Plan for National Economic and Social Development (2011–2015). Among its binding targets are a 16 per cent reduction in energy intensity, a 17 per cent reduction in carbon intensity, a 6 per cent increase in forest stock volume and a 1.3 per cent increase in forest coverage relative to 2010 levels (Lommen 2011). China's National Development and Reform Commission plays a crucial role in fostering policy integration.

The green growth approach (ESCAP 2005) is an initiative from this region that works to integrate environmental sustainability concerns into overall policy making. It aims to improve the eco-efficiency of economic growth, minimizing resource use and negative environmental impacts. The concept is seen as a key strategy for sustainable development.

Source: ESCAP 2011a

Box 10.15 Participation in the management of natural resources in India and Nepal

In India, about 22 million hectares of forests are under the Joint Forest Management programme, where more than 100 000 committees formed by forest-fringe communities protect state-owned forest patches, receiving in turn a share of forest resources (MOEF 2009a). In conjunction with stringent legislation against the use of forest land for non-forestry purposes, these measures have helped stabilize forest cover after decades of rapid deforestation (MOEF 2009b). Additional incentives for participation have been created by a constitutional amendment that mandates decentralization and devolution of power to local authorities at district, block and village levels (MLJ 2011). In Nepal, over 14 000 community forest user groups have access to fuelwood and fodder, and are additionally provided with income-generating opportunities (DoF 2011).

policies, plans and programmes all provide essential information (World Bank 2006). Ultimately, fully integrated sustainability assessment is desirable (UNEP 2009b). Benefits include reduced risk of harm to human health or livelihoods due to environmental damage, protection of resources and improved cost-effectiveness of environmental protection measures. Moreover, reduced future remediation costs, political acceptance of the need for preventive and remedial measures, and better institutional collaboration are also beneficial. Assessments have some technical limitations as results are not always well used, and stakeholder participation is sometimes weak; impact assessments are only effective if governments commit to ensuring compliance and enforcement of the changes indicated.

The policy cluster for strengthening incentive structures focuses on appropriate pricing mechanisms. The current fiscal incentive structure in many countries does not account for environmental and social externalities and remains a significant barrier to sustainability. Improving the fiscal incentive structure can help accelerate greening of the economy to advance the dual goals of economic development and environmental sustainability (OECD 2011; ESCAP 2010, 2005). Implementation of green taxes can secure a double dividend for the economy and the environment by emphasizing a shift from taxing the "goods" (labour) to taxing the "bads" (unsustainable resource use and pollution) (UNEP 2011; ESCAP 2005). Such a shift would help alter the economic incentive structure in favour of environment-friendly activities, thus reducing trade-offs between economic efficiency and maintenance of environmental quality. To be successful, it needs to be implemented as a package and to include awareness campaigns for political acceptability as well as adequate safeguards for vulnerable groups.

The accountability and stakeholder participation cluster addresses the task of integration, which is underpinned by the collection, storage and sharing of information as well as general access to it, such as requiring industries to self-monitor and self-report on environmental and social performance as well as enabling non-governmental organizations to monitor this information independently. Policy options mandating environmental monitoring provide useful information for decision making. Freedom of information laws can enable people to monitor public- and private- sector performance and demand compliance and enforcement action, such as promoted by the Asian Environmental Compliance and Enforcement Network (AECEN 2011b). Access to information promotes public participation in decision making, helping to avoid mistakes and future costs, and increasing political acceptance of remediation measures. Information disclosure policies provide a sound basis for decision making, identifying priorities and formulating cost-effective strategies. Limitations include the difficulties in collecting relevant and reliable data, potential sensitivities in data sharing and the long-term costs of data management.

Policies that require multi-stakeholder participation are also essential to accelerate achievement of the selected global goals. The benefits of multi-stakeholder participation include improved environmental performance, access by disadvantaged groups to redress environmental harm or eliminate potential risks, a better understanding by all parties of the costs and benefits of specific actions, improved decision making, increased political awareness and active participation, and improved cross-sectoral collaboration. Sometimes institutions accustomed to wielding excessive power resist increased stakeholder participation, necessitating the balancing role of civil society. In the region, the monitoring functions of governments can be usefully supplemented by civil society, especially in situations where government capacity is constrained.

The compliance and enforcement cluster encompasses environmental law and regulation. Imperfect markets, collective

Terraced hillside in the Annapurna village of Ulleri, Nepal where community-based forest management provides a significant contribution to local livelihoods. *© Christoph Achenbach/iStock*

The transition to a low-carbon society will require large-scale investment in renewable energy systems. *© Chinaface/iStock*

action issues and equity concerns characterize many of the environmental issues in the Asia and Pacific region. A final policy cluster, therefore, focuses on access to the environmental judiciary, with policies covering the integrated chain of command-and-control measures, standard setting, monitoring and reporting, investigations and inspection of environmental crimes, establishment of environmental police and "green" courts, and trained environmental judges and prosecutors. Examples include the Environmental Court Act of Bangladesh, the environmental police force in Viet Nam, the green bench of the Supreme and High Courts and the recently established Green Tribunal in India as well as regulatory initiatives to combat corruption in Indonesia and Singapore.

The benefits are to halt or delay potentially environmentally damaging or risky activities, to reach settlements where damage has been caused, to save on future remediation costs if risks are reduced and ultimately to increase social capital. Specialized courts are important as they reduce the burden on enforcement agencies, enable conflict resolution and hold government agencies responsible for implementation. Establishment of green courts could derive from popular environmental movements and international treaties. The limitations are that decisions reached through these procedures are not always enforced, and potential jurisdictional conflicts may emerge between newly formed enforcement agencies and existing sectoral agencies. Court procedures may also be lengthy and costly, thus effectively excluding disadvantaged groups and poor individuals.

Benefits and limitations

The main considerations of benefits and limitations included in the policy analysis are as follows:

- economic and financial;
- social;
- environmental;
- political and institutional; and
- distributional.

In general, the priority policies identified here involve up-front costs, which have been a significant barrier in the developing countries of Asia and the Pacific, where investment priorities have been preferentially directed towards economic growth and poverty reduction rather than environmental goals. The so-called low-hanging fruit, such as energy efficiency measures that can be undertaken relatively easily, are still ripe for harvesting in many countries, but may not achieve the scale of structural change needed to contribute to global goals. For example, the transition to a low-carbon society, which has multiple co-benefits, requires massive investment in new, renewable energy systems which are often not the cheapest option under current conditions (ADB 2009b). In addition, the barriers to adopting energy-efficient technologies can be multi-faceted. The challenge is to change those circumstances, such as through the elimination of fossil fuel subsidies and the accelerated depreciation of existing fossil fuel installations, to create a situation in which renewable energy will be able to compete on equal terms with coal and oil. Continued technological development and cooperation on technology transfer to reduce the costs of this transition are also essential.

The social implications of the priority policies have tended to receive less attention than the economic issues, but a common theme is that improved access to information, stakeholder empowerment, gender equality, access to environmental justice, and equity and benefit sharing will all assist in the transition to sustainable development. Safety nets and compensation for poor households – for example when disadvantaged by removal of subsidies – are also part of the mix. The main barriers are unequal powers and the advantages of elites, conflicting social and private values and the attraction and psychological power of conspicuous consumption. Problems of compliance and enforcement are common from the household to the industry level, while higher household costs and employment impacts on declining industries during policy-initiated transitions can hinder progress. The need to relocate vulnerable communities from areas of risk can also present a challenge.

The political and institutional consequences of the suggested priority policies are critical, as it is unlikely that business-as-usual regimes will enable such a policy mix to be effectively implemented. Political benefits are observed from promoting elements of the green economy that promise economic growth, new employment opportunities and environmental benefits. These benefits are best illustrated in the economic stimulus responses to the global financial crisis, where countries such as the Republic of Korea and China have led the world in green stimulus packages. The political downside, however, is reflected in fears that being a first mover, such as in implementing a carbon tax approach to combat climate change, will lead to unfair advantages for those who lag behind. International coordination, political vision and courage, supported by a willing electorate, are needed to overcome such fears and become a green pioneer.

Institutional changes are commonly observed to be incremental tinkering at the margins rather than wholehearted reform. For example, many countries in Asia and the Pacific have added climate change responsibilities to an existing environmental agency rather than seeking widespread institutional reform that would mainstream climate change across all existing agencies (IGES 2008). Sub-regional agencies such as the ASEAN Centre for Biodiversity and the ASEAN Centre of Energy have been created, but without secured funding to support implementation of their mandates. Overall, it appears that institutional change in Asia and the Pacific has been *ad hoc* rather than systematic.

The viability of PES systems including, REDD+, in large part, hinges on the imperative to support and promote the rights of indigenous peoples and local communities. *© Lee Thomas/iStock*

The distributional impacts of priority policies that determine who wins and who loses have been barely investigated in Asia and the Pacific. In the energy sector, fossil fuel industries stand to lose market share, although leading companies may ultimately make the transition to alternative energy. In relation to biodiversity, new funding arrangements such as payment for ecosystem services including REDD+ suggest that traditional forest-dependent communities should benefit, but there is potential for powerful groups to manipulate such schemes and for traditional users to become losers (Bosetti and Lubowski 2010). In the freshwater sector, agriculture tends to lose out to urban and industrial water users, as they are prepared to pay higher prices (Dinar 2000). In waste recycling, traditional waste collectors and the informal sector are beginning to lose out to more formalized waste management, as the value of material resources embedded in waste rises in response to shortages of raw materials (Medina 2007).

Transfer and replication of successful policies

As most successful policies have had demonstrable success in a limited range of countries, some only in developed countries, others in large economies, the objective of this section is to analyse whether the suggested priority policies are readily transferable and replicable across national boundaries. If they cannot easily be transferred because they depend on local circumstances that are not widely prevalent, they are unlikely to accelerate the achievement of the selected global goals. The factors considered include:

- how many countries have already implemented such policies;
- how quickly the policies have been adopted by multiple countries since their first introduction;
- how easily the private sector has been convinced that the policies are not harmful to their businesses; and
- how the policies have contributed co-benefits that made them even more acceptable.

Internal replication – vertical and horizontal integration

Policy responses in the Asia and Pacific region increasingly need to adopt a cross-sectoral approach. Even though the identification of the recommended priority policies in one sector helps focus attention, policy interventions are needed for simultaneously addressing several of the environmental challenges listed, and packages of complementary policies should combine to achieve desirable environmental outcomes across sectors. For instance, environmental governance for climate change should not be seen as separate from that for water, and environmental policies for water cannot be separated from agriculture and food security because there are interactions between those concerns as well as with climate change. Policies that build a community's capacity to adapt to or deal with water insecurity in general will facilitate their adaptation to the effects of climate change. For example, construction of small reservoirs in upper watershed areas can provide both increased water security for downstream villages and drought and flood control to combat climate change. Climate change never has effects in isolation from those of other drivers or pressures of environmental change, but tends to aggravate the effects of existing ones. What seems like a priority policy intervention today might, over the long run, turn out to be maladapted from a climate change point of view, thus necessitating constant learning from experience and flexible, adaptive management.

External replication – policy transfers across regions

Many of the policies being adopted in the Asia and Pacific region had their origin and initial trials in other regions, often Europe and the United States (Table 10.2). The failure to implement many policies in this region successfully may stem from over-optimism that if a policy already works in a developed country then it should also work in a developing one. For example, the strong command-and-control policy regime to manage air and water pollution in the United States, involving standard setting, permits and prosecution of offenders, tends not to work as well in the developing countries of Asia and the Pacific (AECEN 2004). A policy regime built around voluntary compliance, the social pressures of naming and shaming polluters, and compensation of victims may be more applicable for the socio-cultural context of the region, although measures of effectiveness require further analysis.

Table 10.2 Transferability of priority policies in Asia and the Pacific

Priority policies	Transferability considerations	Case examples of existing or potential policy transfer
Climate change		
Clean energy policy: renewable energy	Resource endowments and local demand may influence transferability; technical capacity may be limiting in some countries	• Maldives: mitigation action to achieve carbon neutrality by 2019 (Box 10.7) • Bangladesh: Renewable Energy Technology Program (Mondal *et al.* 2010) • APEC: Energy Efficiency and Renewable Energy Financing Task Force (APEC 2007); best practice principles for financing energy infrastructure (APEC 2008)
Energy efficiency policy: efficiency improvements in buildings, transport and agriculture	Reducing energy demand is difficult to transfer to economies growing at a fast pace, like China and India	• Japan: energy efficiency and mass transit systems Top Runner programme (UNEP 2011) • Thailand: energy efficiency and mass transit policies (Aumnad 2010) • Singapore: managing motorization in sustainable transport planning (Han 2010)
Technology transfer and diffusion	Poor intellectual property rights and patent protection may limit policy transferability in developing countries	• Japan (Branstetter *et al.* 2006) • Australia: Climate change report under UNFCCC (Government of Australia 2010)
Financial policies: greenhouse gas emissions trading; Clean Development Mechanism (CDM) and Adaptation Fund; amended tariffs and taxes or elimination of fossil fuel subsidies; partnerships	Differing tax regimes and attitudes on subsidies, plus political constraints, may limit transferability	• China: CDM – a proactive and sustainable approach (World Bank 2004); CDM – a value-added approach (World Bank 2011) • Japan: Tokyo's Cap and Trade Program (Nishida and Hua 2011)
Provide incentives for private sector action: feed-in tariffs; private-public partnership	Political views on the role of the private sector may limit transferability in some countries	• Feed-in tariffs have been adopted in Australia, China, Japan, India, Indonesia, Republic of Korea, Mongolia, Pakistan, Philippines, Thailand (IEA/OECD/World Bank 2010; ADB 2009c)
Adaptation policy: integrate climate change adaptation and disaster management; mainstreaming in development planning; mandatory climate proofing of infrastructure	Generally transferable, although difficult to implement, due to poor cross-sectoral coordination	• Samoa: Coastal Infrastructure Management (World Bank 2009) • Tonga: Risk Management and Climate Change Adaptation National Action Plan (Hay 2009)
Land use and forestry management	Generally transferable; major international interest in REDD+; needs local support	• Indonesia: deforestation agreement with Norway (ADB 2010) • China: reforestation programmes (Zhang *et al.* 2010; SDPC 2000) • Australia: West Arnhem land fire abatement (NAILSMA 2011; FAO 2010)
Biodiversity		
Payment for ecosystem services (PES)	Emerging experience suggests that the concept is transferable but would require good governance, including good liaison at various levels between stakeholders and users which is a strong prerequisite for successful establishment of PES	• Viet Nam: payment for forest environmental services (PFES) in Lam Dong Province (Winrock International 2011; George *et al.* 2009); REDD+, transparent, equitable and accountable benefit distribution system (UN-REDD 2011) • Potential and limitations of PES as a means of managing watershed services in mainland South East Asia (George *et al.* 2009)
Ensure that benefits from genetic biodiversity exploitation accrue to traditional stewards of habitats	The potential impacts of the Nagoya Protocol on ABS may not be known yet but it is considered an important agreement that will rectify the issues of equity associated with the commercial use of genetic resources and the associated traditional knowledge	• ASEAN: access and benefit sharing of genetic resources in South East Asian countries (ACB 2011 pp.7–24)
Improve protected area management and minimize land use change, especially deforestation	While these policies can be easily transferred, effective management is harder to transfer than declaration of protected areas	• South East Asia: protected areas and development in the Lower Mekong (ICEM 2003) • East Asia: the CBD Programme of Work on Protected Areas and the East Asian Regional Action Plan (IUCN-WCPA 2011) • Indonesia: transboundary haze management (ADB 2008b)
Freshwater		
Apply adaptive management and integrated water resources management (IWRM)	Transferability depends on institutions in place and strong political will	• Central Asia: IWRM management in Fergana Valley (IWMI 2008)
Promote community-based management for better water allocation	Inclusive water management institutions; bureaucratic/political support for public participation in water sector are needed	• India: Andhra Pradesh (Gupta 2010; Narain 2003; Ballabh 2002; Mollinga 2001; Parthasarathy 2000; Shashidharan 2000)
Encourage rainwater harvesting/ storm water management for increased and improved water storage	Depends on political will to address human basic needs for water as a priority	• Sri Lanka (Government of Sri Lanka 2007) • Australia (Meinzen 2009) • States of India (Rainwater Harvesting Organization 2011)
Support construction of farm dams and renovating existing dams for increased and improved water storage	Access to innovative technologies, water availability, flexible storage system, and an appropriate financing mechanism	• Uzbekistan: rehabilitation of existing reservoirs in Central Asia (Box 10.10) • Cambodia and the Philippines (IFAD 2011)

Table 10.2 Transferability of priority policies in Asia and the Pacific *continued*		
Freshwater *continued*		
Utilize economic instruments and approaches to enhance efficient use of water	Transferability depends on economic context in each country, users' ability and willingness to pay and institutional set-up for payment arrangements	• China: quotas and pricing reform for the Yellow River (Box 10.11) • Cambodia: Phnom Penh (ADB 2009c) • Philippines (IFAD 2011)
Incorporate ecosystem approach/ environmental flow concept to water resources management	Coordination mechanism between line agencies and local people is necessary	• China: Yellow River basin (Box 10.1) • Republic of Korea: Han River Water Use Charge Programme (HREO 2011)
Chemicals and waste		
Adopt policy frameworks promoting waste avoidance, and reduction of the production and use of hazardous chemicals	Transferability depends on institutional capacity for complex management	• Japan: 3Rs strategies and measures (UNEP 2011); Environmental Performance Reviews (OECD 2010); Chemical Substance Control Law • China: Circular Economy Promotion Law (UNEP 2009c)
Establish systems and infrastructure for product reuse and materials recycling, and stimulate markets for recycled materials; include both industrial by-products and post-consumer waste	Transferability depends on political willingness to promote source separation of waste; countries with large informal waste sectors need tailored approaches to avoid negative social impacts	• Japan: Sound Material-Cycle Society (MOEJ 2010) • Republic of Korea: Extended Producer Responsibility System (Government of Korea 2011); hazardous waste management (Yoon *et al.* 2008)
Set up safe disposal facilities for hazardous wastes and chemicals that cannot be recycled, either nationally or sub-regionally, paying special attention to the needs and circumstances of developing countries and countries with economies in transition	Generally transferable, but technical and financial support needed for developing countries	• Indonesia: collective composting in Surabaya (IGES 2008) • Japan: Kityakushu Initiative for a Clean Environment (ESCAP 2011b)
Strengthen international collaboration, including technology transfer and financial support as well as information sharing and policy transfer	Generally transferable; the potential for South-South collaboration is significant but may require external facilitation	• Kazakhstan: pesticide disposal (ADB 2008a)
Reinforce the system to control inappropriate export and import of hazardous chemicals and waste	Transferability depends largely on political will and capacity for effective policy enforcement	• China: Import/export, management and treatment of e-waste (BCRC 2010; UNEP/ SBC 2009)
Governance		
Policy integration and mainstreaming; decentralization according to principle of subsidiarity	Transferability depends on willingness of government agencies to avoid a silo mentality, as well as executive support; decentralization often depends on institutional capacity at local levels	Integration • Samoa, Tonga and possibly other Pacific island countries (ESCAP 2011a) • China: National Development and Reform Commission (ESCAP 2011a) • Republic of Korea: Presidential Committee on Green Growth (ESCAP 2011a) Transferability • Indonesia, Thailand and Philippines have recent experience Decentralization • East Asia: sub-national own-source revenue (World Bank 2005 Chapter 6)
Strengthening incentive structures: greening fiscal policy, supportive fiscal incentives, reinvestment into natural capital	Good financial management mechanisms are generally transferable	• Republic of Korea, Japan, Bhutan, Singapore, Hong Kong
Accountability and stakeholder participation: multi-stakeholder participation, transparency, information disclosure and access	International approach using islands of integrity tends to be effective in combating corruption; information may need to be interpreted and translated into local languages	• Singapore: institutional arrangements to combat corruption (UNDP 2005) • Indonesia: institutional arrangements to combat corruption (UNDP 2005) National Climate Change Councils • Indonesia: National Council on Climate Change (DNPI 2011) • Nepal: Climate Change Council (Government of Nepal 2009) National Councils for Sustainable Development • OECD countries (Japan, Australia, Republic of Korea, New Zealand): national sustainable development strategies (OECD 2006) • Philippines: success stories (PCSD 2011) • Republic of Korea: Presidential Committee on Green Growth (ESCAP 2011a) • Cambodia: Inter-ministerial Green Growth Working Group • India: Planning Commission

In 2010, 12 of the world's 23 mega-cities, comprising 10 million inhabitants, were in Asia. © *Samxmeg/iStock*

Creating an enabling environment for effective policies

Part of the above analysis relates to the enabling and/or impeding factors that have led to the success or otherwise of specific policies. This section starts from the premise that policies are only as good as their implementation. Theoretically brilliant policies that are complex or difficult to implement are unlikely to be easily transferable or replicable across national boundaries.

Moving from successful policies to policies plus enabling conditions

Identifying policies that work, and those that do not, is of little relevance unless the underpinning causal conditions and context are understood. A growing body of policy studies identifies the factors that allow policy change to happen (Huitema and Meijerink 2009). There are many critiques of why policy change in the irrigation sector in South Asia, for instance, has been slow and tardy (Mollinga *et al.* 2003). Accordingly, it is useful to look at what policies have failed to accomplish and what they have achieved (Urs and Whittell 2009).

Filling the policy vacuum: the changing role of markets and civil society

Finally, it is important to look at the changing configuration of actors like the state, markets and civil society. Increasingly, there is a policy vacuum left by the state that the market and/or civil society may intervene to fill. For instance, a common response to the inadequate supply of water in many cities in Asia and the Pacific is the rising popularity of water markets – the sale of safe water by private entrepreneurs – seen in cities like Chennai and Kathmandu, amongst others. Privatization of water markets can increase efficiency, but laws should be put in place to anticipate and limit monopolies.

Given the challenges of balancing growth, poverty alleviation, environmental protection and resource conservation, governance improvements should prioritize addressing environmental concerns at the macro-economic level, specifically focusing on patterns of growth and restructuring the economy (ESCAP/ADB/UNEP 2012; UNEP 2011). Environment remains the foundation of sustainable development and if it is weak or compromised, then the social and economic pillars of sustainable development will also be undermined. By allowing co-benefits, ancillary costs and environmental externalities to be integrated into social and economic decision making, environmental concerns and climate change can be mainstreamed into a sustainable development framework. Carefully monitoring the actual outcomes of policy decisions and making continuous adjustments, with the full input of all stakeholders, are key elements of an adaptive management approach. The recent efforts to promote the green economy and green growth in the region suggest that the time is ripe to begin this transformation.

CONCLUSIONS

The implications of failing to implement the recommended policies or clusters of policies make a strong case for their adoption, as improved environmental governance is necessary to reverse environmental degradation and the unsustainable use of natural resources. This chapter not only identifies some recommended policies but also the changes to governance arrangements necessary for them to be transferred across sectoral and/or national boundaries and replicated quickly. There is adequate experience with several of the priority policies analysed above to justify faster replication, but the differing governance contexts and enabling environments in a region as diverse as Asia and the Pacific may be barriers to adoption. Each of the suggested priority policies or policy clusters has experienced some limitations during implementation, so these challenges also need to be better understood, and additional basic research needs to be undertaken before specific recommendations can be made on the transfer of policies to new jurisdictions. Nevertheless, there is considerable scope for faster uptake of best-practice policies and it is hoped that this chapter will have drawn the attention of the region's decision makers to that potential.

REFERENCES

Abdullaev, I. and Atabaeva, S. (2011). Water sector in Central Asia: slow transformation and potential for cooperation. *International Journal of Sustainable Society* 4(123), 123–129

Abdullaev, I., Kazbekov, J., Jumaboev, K. and Manthrithilake, H. (2009a). Adoption of integrated water resources management principles and its impacts: lessons from Fergana Valley. *Water International* 34(2), 1–12

Abdullaev, I., Kazbekov, J., Manthrithilake, H. and Jumaboev, K. (2009b). Participatory water management at the main canal: a case from South Fergana canal in Uzbekistan. *Journal of Agricultural Water Management* 96(2), 317–329

Abdullaev, I., Ul Hassan, M., Manthrithilake, H. and Yakubov, M. (2006). *The Reliability Improvement in Irrigation Services: Application of Rotational Water Distribution in Tertiary Canals in Central Asia.* International Water Management Institute (IWMI), Research Report 100.

ACB (2011). *Legal and Institutional Development for Promoting Access and Benefit Sharing of Genetic Resources in Southeast Asian Countries.* ASEAN Centre for Biodiversity, Los Baños

ACB (2010). *Technical Report on Gap Analysis on Coverage of Terrestrial and Marine Protected Areas.* ASEAN Centre for Biodiversity, Los Baños

ADB (2010). *Focused Action: Priorities for Addressing Climate Change in Asia and the Pacific.* Asian Development Bank, Manila

ADB (2009a*). Economics of Climate Change in Southeast Asia*. Asian Development Bank, Manila

ADB (2009b). *Investing in Sustainable Infrastructure: Improving Lives in Asia and the Pacific.* Asian Development Bank, Manila

ADB (2009c). *Why is Access to Basic Services Not Inclusive? A Synthesis with a Special Focus on Developing Asia.* ADB Sustainable Development Working Paper Series No. 6. Asian Development Bank, Manila

ADB (2008a). *Partnership on Persistent Organic Pollutants Pesticides Management for Agricultural Production in Central Asian Countries.* Technical Assistance Synthesis Report. Asian Development Bank, Manila. http://www.adb.org/Documents/Reports/Consultant/40040-REG/40040-REG-TACR.pdf (accessed 15 September 2011).

ADB (2008b). *Strengthening Sound Environmental Management in the BIMP-EAGA.* Technical Assistance Consultant's Report. Asian Development Bank, Manila http://www.adb.org/Documents/Reports/Consultant/41075-REG/41075-03-REG-TACR.pdf (accessed 15 September 2011)

AECEN (2011a). *Inventory of Good Practices.* Asian Environmental Compliance and Enforcement Network, Bangkok. http://www.aecen.org/aecen-good-practices (accessed 15 September 2011)

AECEN (2011b). *Principles for Improving Environmental Compliance and Enforcement in Asia.* Asian Environmental Compliance and Enforcement Network, Bangkok. http://www.aecen.org/principles-improving-environmental-compliance-and-enforcement-asia (accessed 15 September 2011)

AECEN (2004). *Environmental Compliance and Enforcement in Thailand: Rapid Assessment.* Asian Environmental Compliance and Enforcement Network, Bangkok. http://www.aecen.org/sites/default/files/TH_Assessmemt.pdf (accessed 6 November 2011)

Andersen, A. (2001). *Worker Safety in the Ship-breaking Industries.* Issue Paper. Internationaol Labour Office, Geneva. http://www.ilo.org/safework/info/publications/WCMS_110357/lang--en/index.htm (accessed 15 September 2011)

APEC (2008). *Priorities for Financing Energy Infrastructure Projects within the APEC Region.* Asia-Pacific Economic Cooperation, Singapore. http://www.ewg.apec.org/documents/BPPFinancingEnergyInfrastructure%282%292008.pdf (accessed 15 September 2011)

APEC (2007). *Progress Report on the APEC Energy Efficiency and Renewable Energy Financing Task Force.* Asia-Pacific Economic Cooperation, Singapore. http://www.ewg.apec.org/documents/063A_EE&RE_Financing2007.pdf (accessed 15 September 2011)

APO (2007). *Solid Waste Management: Issues and Challenges in Asia.* Asian Productivity Organization, Tokyo

ASEAN-WEN (2009). *Illegal Wildlife Trade in Southeast Asia Factsheet.* ASEAN Wildlife Enforcement Network, Bangkok. http://www.asean-wen.org/index.php?option=com_docman&task=doc_details&gid=5&Itemid=80 (accessed 17 September 2011)

Aumnad, P. (2010). Integrated energy and carbon modeling with a decision support system: policy scenarios for low-carbon city development in Bangkok. *Energy Policy* 38(9), 4808–4817

Aziz, Z. (2010). A more integrated and cohesive Asia in the global economy. Speech by Dr. Aziz (Governor of the Central Bank of Malaysia) at the Foreign Bankers' Association of the Netherlands, Amsterdam, 22 June 2010

Ballabh, V. (2002). *Emerging Water Crisis and Political Economy of Irrigation Reforms in India.* Paper prepared for Asian Irrigation in Transition: Responding to Challenges Ahead workshop, 22–23 April 2002. Asian Institute of Technology, Bangkok

Bandaragoda, D. (2006). *Institutional Adaptation for Integrated Water Resources Management: An Effective Strategy for Managing Asian River Basins.* Working Paper 107. International Water Management Institute (IWMI), Colombo

Barnett, J. and Adger, W. (2003). Climate dangers and atoll countries. *Climatic Change* 61: 321–337

Barringer, H. (2003). *A Life Cycle Cost Summary.* International Conference of Maintenance Societies, 20–23 May 2003, Perth

BCRC (2010). *Progress on E-Waste Management and Treatment.* Presentation. Basel Convention Coordinating Centre for Asia and the Pacific (BCRC-China), Tsinghua. http://archive.basel.int/techmatters/ICCM2/PROGRESSS%20ON%20%20E-WASTE%20MANAGEMENT-2009-05_BCRC-China_May%2009.pdf (accessed 19 September 2011)

Berger, G. and Steurer, R. (2009). *Horizontal Policy Integration and Sustainable Development: Conceptual Remarks and Governance Examples.* ESDN Quarterly Report, June 2009. European Sustainable Development Network. http://www.sd-network.eu/?k=quarterly%20reports&report_id=13 (accessed 15 September 2011)

Bosetti, V. and Lubowski, R. (eds.) (2010). *Deforestation and Climate Change: Reducing Carbon Emissions from Deforestation and Forest Degradation.* Edward Elgar Publishing, Cheltenham

Branstetter, L., Fisman, R. and Foley, C. (2006). Do stronger intellectual property rights increase international technology transfer? Empirical evidence from US firm-level data. *Quarterly Journal of Economics* 121(1), 321–349

CBD (2011). *Nagoya Protocol on Access and Benefit-sharing.* Secretariat of Convention on Biological Diversity. http://www.cbd.int/abs/doc/protocol/nagoya-protocol-en.pdfCBD

CBD (2010). *Global Biodiversity Outlook 3.* Secretariat of Convention on Biological Diversity, Montreal

CBD (2002). *Bonn Guidelines on Access to Genetic Resources and Fair and Equitable Sharing of the Benefits Arising out of their Utilization.* Secretariat of the Convention on Biological Diversity. https://www.cbd.int/doc/publications/cbd-bonn-gdls-en.pdf

CBD (1992). *Convention on Biological Diversity.* http://www.cbd.int

CI (2005). *Biodiversity Hotspots.* Conservation International, Arlington, VA. http://www.biodiversityhotspots.org/xp/hotspots/pages/map.aspx (accessed 18 December 2011)

Crooks, S., Herr, D., Tamelander, J., Laffoley, D. and Vandever, J. (2011). *Change through Restoration and Management of Coastal Wetlands and Near-shore Marine Ecosystems: Challenges and Opportunities.* Environment Department Paper 121. World Bank, Washington, DC

Dalal-Clayton, B. and Bass, S. (2009). *The Challenges of Environmental Mainstreaming: Experience of Integrating Environment into Development Institutions and Decisions.* Environmental Governance No. 3. International Institute for Environment and Development, London

De Lopez, T., Tin, P., Iyadomi, K., Santos, S. and McIntosh, B. (2009). Clean Development Mechanism and least developed countries: changing the rules for greater participation. *Journal of Environment and Development* 18(4), 436–452

Dinar, A. (2000). *The Political Economy of Water Pricing Reforms.* World Bank, Washington, DC.

DNPI (2011). Dewan Nasional Perubahan Iklim, Jakarta. http://www.dnpi.go.id/ (accessed 15 September 2011)

DNPI (2010). *Indonesia's Greenhouse Gas Abatement Cost Curve.* Dewan Nasional Perubahan Iklim, Jakarta

DoF (2011). *Status of Community Forest User Groups.* Department of Forests, Government of Nepal.
http://www.dof.gov.np/index.php?option=com_jdownloads&Itemid=102&task=view.download&catid=7&cid=20 (accessed 6 September 2011)

Dukhovny, V., Mirzaev, N. and Sokolov, V. (2008). IWRM implementation: experiences with water sector reforms in Central Asia. In *Central Asian Waters: Social, Economic, Environmental and Governance Puzzle* (eds. Rahaman, M. and Varis, O.). pp.19–35. Water and Development Publications, Helsinki University of Technology, Helsinki

EC (2007). *Ship Dismantling and Pre-cleaning of Ships. Final Report.* European Commission, Directorate General Environment, Brussels

ESCAP (2011a). *Green Growth.* United Nations Economic and Social Commission for Asia and the Pacific, Bangkok. http://www.greengrowth.org/ (accessed 15 December 2011)

ESCAP (2011b). *The Kityakushu Initiative for a Clean Enviornment 2001–2010.* http://www.unescap.org/esd/environment/kitakyushu/ (accessed 15 September 2011)

ESCAP (2010). Ministerial declaration on environment and development in Asia and the Pacific, 2010. E/ESCAP/MCED (6)/11. Ministerial Conference on Environment and Development in Asia and the Pacific, Sixth session, Astana, 27 September-2 October 2010. United Nations Economic and Social Commission for Asia and the Pacific, Bangkok.

ESCAP (2005). *Report of the Ministerial Conference on Environment and Development in Asia and the Pacific, 2005.* United Nations Economic and Social Commission for Asia and the Pacific, Bangkok. http://www.unescap.org/EDC/English/Ministerial/MCED-2005/MCED05_Report.pdf (accessed 11 September 2011)

ESCAP/ADB/UNEP (2012). *Green Growth, Resources and Resilience: Environmental Sustainability in Asia and the Pacific.* United Nations, Bangkok

ESCAP/IGES (2011). *Successful Practices and Policy Database.* United Nations Economic and Social Commission for Asia and the Pacific, Bangkok and Institute for Global Environmental Strategies, Hayama. http://kitakyushu.iges.or.jp/successful_practices/ (accessed 19 September 2011)

FAO (2010) *Forests and Climate Change in the Asia-Pacific Region*. Forests and Climate Change Working Paper 7. Food and Agriculture Organization of the United Nations, Rome. http://www.fao.org/docrep/013/i1759e/i1759e00.pdf (accessed 19 September 2011)

FOEN (2008). *Marrakech Task Force on Sustainable Public Procurement*. Federal Office for the Environment, Swiss Confederation. http://www.unep.fr/scp/marrakech/taskforces/pdf/procurement2.pdf (accessed 15 September 2011)

Frauendorfer, R. and Liemberger, R. (2010). *The Issues and Challenges of Reducing Non-revenue Water*. Asian Development Bank, Mandaluyong City

GEF (2009). *Project Identification Form: Integration of Climate Change Risks into the Maldives Safer Island Development Program*. Global Environment Facility, Washington, DC

George, A., Pierret, A., Boonsaner, A., Christian, V. and Planchon O. (2009). Potential and limitations of payments for environmental services (PES) as a means to manage watershed services in mainland Southeast Asia. *International Journal of the Commons* 3(1), http://www.thecommonsjournal.org/index.php/ijc/article/view/131

GISP (2009). *Global Invasive Species Programme: Annual Report*. Global Invasive Species Programme, Nairobi

Govan, H., Aalbersberg, W., Tawake, A. and Parks, J. (2008). *Locally-Managed Marine Areas: A Guide to Supporting Community-based Adaptive Management*. The Locally-Managed Marine Area Network. http://www.lmmanetwork.org/files/lmmaguide.pdf (accessed 15 September 2011)

Government of Australia (2010). *Australia's Fifth National Communication on Climate Change. A Report under the United Nations Framework Convention on Climate Change 2010*. Department of Climate Change, Commonwealth of Australia. http://www.climatechange.gov.au/~/media/publications/greenhouse-gas/Australia-fifth-national-communication.pdf (accessed 15 September 2011)

Government of Korea (2011). *Extended Producer Responsibility*. Ministry of Environment, Government of Korea. http://eng.me.go.kr/content.do?method=moveContent&menuCode=pol_rec_pol_rec_sys_responsibility (accessed 18 December 2011)

Government of Nepal (2009). *Climate Change Council*. http://www.moenv.gov.np/newwebsite/index.php?view=ccc (accessed 15 September 2011)

Government of Sri Lanka (2007). *Urban Development Authority (Amendment) Act, No. 36 of 2007*. http://www.lankarainwater.org/rwhsl/act_36_2007_e.pdf (accessed 6 November 2011)

Gullison, R., Frumhoff, P., Canadell, J., Field, C., Nepstad, D., Hayhoe, K., Avissar, R., Curran, L., Friedlingstein, P., Jones, C. and Nobre, C. (2007). Tropical forests and climate policy. *Science* 316: 985–986

Gupta, S. (2010). Irrigation governance challenges. Perspectives and initiatives in Andhra Pradesh. *South Asian Water Studies* 2(1), 17–36

Han, S.S. (2010). Managing motorization in sustainable transport planning: the Singapore experience. *Journal of Transport Geography* 18(2), 314–321

Harhay, M., Jason, S., Harhay, S. and Olliaro, P. (2009). Health care waste management: a neglect and growing public health problem worldwide. *Tropical Medicine and International Health* 14, 1–4

Hay, J. (2009). *Institutional and Policy Analysis of Disaster Risk Reduction and Climate Change Adaptation in Pacific Island Countries*. United Nations International System for Disaster Reduction and the United Nations Development Programme, Suva

HREO (2011). Han River Water Use Charge Programme. Han River Environmental Office. http://eng.me.go.kr/docs_hg/tasks/water.jsp (accessed 15 September 2011)

Hughes, G. (2011). *The Myth of Green Jobs*. Report 3. The Global Warming Policy Foundation, London

Huitema, D. and Meijerink, S. (eds.) (2009). *Water Policy Entrepreneurs: A Research Companion to Water Transitions around the Globe*. Edward Elgar Publishing, Cheltenham

ICEM (2003). *Regional Report: Review of Protected Areas and Development of the Lower Mekong River Region*. International Centre for Environmental Management, Indooroopilly, Queensland. http://data.iucn.org/dbtw-wpd/edocs/2003-106-5.pdf (accessed 15 September 2011)

IEA (2010). *World Energy Outlook 2010*. International Energy Agency, Paris

IEA (2007). *Mind the Gap: Quantifying Principal-Agent Problems in Energy Efficiency*. International Energy Agency, Paris

IEA/OECD/World Bank (2010). *The Scope of Fossil Fuel Subsidies in 2009 and Roadmap for Phasing out Fossil-Fuel Subsidies*. Joint Report prepared for G20 Summit, Seoul, 11–12 November 2010

IFAD (2011). *Assessment of the Implementation of the Commune Infrastructure Development Fund*. Project Completion Report. April 2011. International Fund for Agricultural Development, Rome

IGES (2008). *Climate Change Policies in Asia-Pacific: Re-Uniting Climate Change and Sustainable Development*. White Paper. Institute for Global Environmental Strategies, Kanagawa. http://www.iges.or.jp/en/pub/pdf/whitepaper/whitepaper2.pdf (accessed 15 September 2011)

IPCC (2007). *Climate Change 2007: The Physical Science Basis* (eds. Solomon, S., Qi, D. and Manning, M.). Contribution of Working Group I to the Fourth Assessment Report of the Intergovernmental Panel on Climate Change. Cambridge University Press, Cambridge

IUCN-WPA (2011). *Protected Areas of East Asia: Evaluating and Strengthening Implementation of the CBD Programme of Work on Protected Areas and the East Asian Regional Action Plan*. International Union for the Conservation of Nature, Gland, Switzerland

IWMI (2008). *Final Report of IWRM Management in Fergana Valley Project – Phase III May 2005–April 2008*. International Water Management Institute (IWMI), Tashkent. http://publications.iwmi.org/pdf/H041914.pdf (accessed 15 September 2011)

JFS (2005). *Dioxin Emissions from Incinerators Declining throughout Japan*. Japan for Sustainability. http://www.japanfs.org/en/pages/026151.html (accessed 2 November 2011)

Jordan, A., and Lenschow, A. (eds.) (2009). *Innovation in Environmental Policy? Integrating the Environment for Sustainability*. Edward Elgar Publishing, Cheltenham

Khan, T., Quadir, D., Murty, T., Kabir, A., Aktar, F. and Sarker, M. (2002). Relative sea level changes in Maldives and vulnerability of land due to abnormal coastal inundation. *Marine Geodesy* 25, 133–143

Leather, J. and Clean Air Initiative for Asian Cities Center Team (2009). *Rethinking Transport and Climate Change*. ADB Sustainable Development Working Paper Series No. 10. Asian Development Bank, Manila

Lee, D.-H. (2006). *Current Situation and Tasks of Food Waste Recycling in Korea*. Department of Environmental Engineering, University of Seoul

Lommen, Y. (2011). *Towards Sustainable Growth in the People's Republic of China. The 12th Five Year Plan*. ADB Briefs No. 7, May 2011. Asian Development Bank, Manila

Mahmood, K. (1987). *Reservoir Sedimentation: Impact, Extent and Mitigation*. World Bank Technical Report No. 71. World Bank, Washington, DC

Masui, T., Matsumoto, K., Hijioka, Y., Kinoshita, T., Nozawa, T., Ishiwatari, S., Kato, E., Shukla, P., Yamagata, Y. and Kainuma, M. (2011). An emission pathway for stabilization at 6 Wm^{-2} radiative forcing. *Climatic Change* 109(1), 59-76

McKinsey and Company (2009). *Pathways to a Low Carbon Economy*. Version 2 of the Global Greenhouse Gas Abatement Cost Curve. http://www.mckinsey.com/globalGHGcostcurve (accessed 15 September 2011)

Medina, M. (2007). *The World's Scavengers: Salvaging for Sustainable Consumption and Production*. Alta Mira Press, Lanham, MD

Meinzen, S. (2009). *Rainwater Harvesting Policies Throughout the US*. http://www.climateactionplans.com/2009/07/rainwater-harvesting-policies-throughout-the-us/ (accessed 15 September 2011)

Mimura, N., Nurse, L., McLean, R., Agard, J., Briguglio, L., Lefale, P., Payet, R. and Sem, G. (2007). Small islands. In *Climate Change 2007: Impacts, Adaptation and Vulnerability* (eds. Parry, M., Canziani, O., Palutikof, J., van der Linden, P. and Hanson, C.). Contribution of Working Group II to the Fourth Assessment Report of the Intergovernmental Panel on Climate Change. pp.687. Cambridge University Press, Cambridge,

MLJ (2011). *The Constitution (Seventy-Third Amendment) Act, 1992*. Ministry of Law and Justice, Government of India. http://indiacode.nic.in/coiweb/amend/amend73.htm (accessed 15 September 2011)

MOEF (2009a). *India's Fourth Report to the Convention on Biological Diversity*. Ministry of Environment and Forests, Government of India. http://moef.nic.in/downloads/public-information/in-nr-04.pdf (accessed 15 September 2011)

MOEF (2009b). *State of Forests Report 2009*. Ministry of Environment and Forests, Government of India. http://www.fsi.nic.in/india_sfr_2009/india_sfr_2009.pdf (accessed 15 September 2011)

MOEJ (2010). *Annual Report: Establishing a Sound Material-Cycle Society – Milestone toward a Sound Material-Cycle Society through Changes in Business and Life Styles*. Ministry of Environment, Government of Japan

MOEJ (2007). *Technologies to Support a Sound Material-Cycle Society: Development of 3R and Waste Management Technologies*. Ministry of Environment, Government of Japan

Molle, F. and Hoanh, C. (2009). *Implementing Integrated River Basin Management: Lessons from the Red River Basin, Vietnam*. IWMI Research Report No. 131. International Water Management Institute, Colombo

Mollinga, P. (2001). *Power in Motion: A Critical Assessment of Canal Irrigation Reform, with a Focus on India*. Indian PIM Working Paper/Monograph Series No. 1. Indian Network on Participatory Irrigation Management, New Delhi

Mondal A., Kamp, L. and Pachova, N. (2010). Drivers, barriers, and strategies for implementation of renewable energy technologies in rural areas in Bangladesh – an innovation system analysis. *Energy Policy* 38(8), 4626–4634

Nag, R. (2010). Asia's deepening regionalism brings shared prosperity. Special Report. *Development Outreach* October 2010, 45–47. World Bank Institute. http://wbi.worldbank.org/wbi/devoutreach/article/540/asias-deepening-regionalism-brings-shared-prosperity (accessed 15 September 2011)

NAILSMA (2011). *The West Arnhem Land Fire Abatement Project*. North Australian Indigenous Land and Sea Management Alliance. http://www.nailsma.org.au/projects/indigenous_carbon_abatement.html (accessed 19 September 2011)

Narain, V. (2003). *Institutions, Technology and Water Control: Water Users Associations and Irrigation Management Reform in Two Large-scale Systems in India*. Orient Longman, Hyderabad

NDRC (1998a). *Annual Water Use Quota and its Distribution Scheme for the Yellow River*. 14 December 1998. National Development and Reform Commission and Ministry of Water Resources of the People's Republic of China

NDRC (1998b). *The Implementation Regulation for Water Resource Allocation among Provinces in HRB*. 14 December 1998. National Development and Reform Commission and the Ministry of Water Resources of the People's Republic of China

Nicholls, R., Marinova, N., Lowe, J., Brown, S., Vellinga, P., de Gusmao, D., Hinkel, J. and Tol, R. (2011). Sea-level rise and its possible impacts given a 'beyond 4 degrees C world' in the twenty-first century. *Philosophical Transactions of the Royal Society A* 369, 161–181

Nishida, Y. and Hua, Y. (2011). Motivating stakeholders to deliver change: Tokyo's Cap-and-Trade Program. *Building Research and Information* 39(5), 518–533

Nunn, P. (2009). Responding to the challenges of climate change in the Pacific Islands: management and technological imperatives. *Climate Research* 40, 211–231

OECD (2011). *Towards Green Growth*. Organisation for Economic Co-operation and Development, Paris. http://www.oecd.org/dataoecd/37/34/48224539.pdf (accessed 15 September 2011)

OECD (2010). *Environmental Performance Reviews: Japan Highlights*. Organisation for Economic Co-operation and Development, Paris. http://www.oecd.org/dataoecd/51/13/46412900.pdf (accessed 15 September 2011)

OECD (2006). *Good Practices in the National Sustainable Development Strategies of OECD Countries*. Organisation for Economic Co-operation and Development, Paris. http://www.oecd.org/dataoecd/58/42/36655769.pdf (accessed 17 September 2011)

Ozone Cell (1999). *Montreal Protocol: India's Success Story*. Ministry of Environment and Forests, Government of India, New Delhi

Palit, D. and Chaurey, A. (2011). Off-grid rural electrification experiences from South Asia: status and best practices. *Energy for Sustainable Development* 15(3), 266–276

Parthasarathy, R. (2000). Participatory irrigation management programme in Gujarat: institutional and financial issues. *Economic and Political Weekly* XXXV (35) and (36), 3147–3154

PCSD (2011). *Philippines Council for Sustainable Development: Success Stories*. http://pcsd.neda.gov.ph/stories.asp (accessed 15 September 2011)

Rainwater Harvesting Organization (2011). *India – Rainwater Harvesting Policies*. http://www.rainwaterharvesting.org/policy/legislation.htm (accessed 15 September 2011)

Rakhmatullaev, S., Marache, A., Huneau, F., Le Coustumer, P., Bakiev, M. and Motelica-Heino, M. (2010). Geostatistical approach for the assessment of the water reservoir capacity in arid regions: a case study of the Akdarya reservoir, Uzbekistan. *Environmental Earth Sciences* 63(3), 447–460. doi: 10.1007/s12665-010-0711-3

Renner, M. (2008). *Green Jobs: Working for People and the Environment*. Worldwatch Institute, Washington, DC

Sarraf, M., Stuer-Lauridsen, F., Dyoulgerov, M., Bloch, R., Wingfield, S. and Watkinson, R. (2010). *Ship Breaking and Recycling Industry in Bangladesh and Pakistan*. Report No. 58275-SAS. World Bank, Washington, DC

Schaller, G.B. and Vrba, E.S. (1996). Description of the giant muntjac (*Megamuntiacus vuquangensis*) in Laos. *Journal of Mammology* 77(3), 675–683

SDPC (2000). The opinions on further implementing the policy and measures on returning farmland to forests released by the State Council on 25 September 2000, State Issued, No. 24. The State Development and Planning Commission, People's Republic of China (in Chinese). http://www.sdpc.gov.cn/xwfb/t20050708_28195.htm

Shashidharan, E. (2000). Civil society organizations and irrigation management in Gujarat, India. In *Water for Food and Rural Development. Approaches and Initiatives in South Asia* (ed. Mollinga, P.). pp.247–265. Sage Publications, New Delhi

Shekdar, A. (2009). Sustainable solid waste management: an integrated approach for Asian countries. *Waste Management* 29(4), 1438–1448

Srivastava, S. (2011). A framework for regional cooperation on integration of disaster risk reduction and climate change adaptation in South Asia. In *Climate Change and Food Security in South Asia* (eds. Lal, R., Sivakumar, S., Faiz, M. and Islam, K.). pp.569–584. Springer, Netherlands

State Council (2006a). *Water Withdrawal and Water Resource Fee Collection Rules*, 15 April 2006. State Council of China

State Council (2006b). *Yellow River Water Regulating*. 1 August 2006. State Council of China

Tan, X. and Zhang, X. (2010). *Scaling Up Low-Carbon Technology Deployment: Lessons from China*. World Resources Institute, Washington, DC. http://pdf.wri.org/scaling_up_low_carbon_technology_deployment.pdf. (accessed 2 November 2011)

TEEB (2012). The Economics of Ecosystems and Biodiversity. http://www.teebweb.org

Timilsina, G. and Shrestha, A. (2009). Transport sector CO_2 emissions growth in Asia: underlying factors and policy options. *Energy Policy* 37(11), 4523–4539

UNDP (2005). *Institutional Arrangements to Combat Corruption: A Comparative Study*. United Nations Development Programme, Bangkok. http://regionalcentrebangkok.undp.or.th/practices/governance/documents/corruption_comparative_study-200601.pdf (accessed 15 September 2011)

UNEP (2011). *Resource Efficiency: Economics and Outlook for Asia and the Pacific*. United Nations Environment Programme, Bangkok

UNEP (2010). *Scientific Assessment of Ozone Depletion: 2010*. United Nations Environment Programme, Nairobi. http://www.unep.org/PDF/PressReleases/898_ExecutiveSummary_EMB.pdf (accessed 2 November 2011)

UNEP (2009a). *A Case for Climate Neutrality: Case Studies on Moving towards a Low Carbon Economy*. United Nations Environment Programme, Nairobi

UNEP (2009b). *Integrated Assessment: Mainstreaming Sustainability into Policymaking: A Guidance Manual*. United Nations Environment Programme, Nairobi

UNEP (2009c). *UNEP Year Book 2009*. United Nations Environment Programme, Nairobi

UNEP (2008a). *Assessment of Existing Capacity and Capacity Building Needs to Analyse Persistent Organic Pollutants (POPs) in Developing Countries*. Division of Technology, Industry, and Economics, Chemicals Branch, United Nations Environment Programme, Nairobi. http://www.chem.unep.ch/Pops/laboratory/Final%20report%20POPs%20Lab%20Cap_text.pdf (accessed 2 November 2011)

UNEP (2008b). *Freshwater Under Threat: Vulnerability Assessment of Freshwater Resources to Environmental Change – North East Asia 2008*. United Nations Environment Programme, Nairobi

UNEP/SBC (2009). *Report of the Project on the Import/Export Management of E-Waste and Used EEE*. Basel Convention Coordinating Center for Asia and the Pacific, Tsinghua University. http://www.bcrc.cn/col/1257152616046/1276071007264.html (accessed 15 September 2011)

UNESCO-WWAP (2006). *Second United Nations World Water Development Report: Water, a Shared Responsibility*. World Water Assessment Programme, United Nations Educational, Scientific and Cultural Organization, Paris

UNFCCC (1992). *United Nations Framework Convention on Climate Change*. FCCC/INFORMAL/84 GE.05-62220 (E) 200705. http://unfccc.int/resource/docs/convkp/conveng.pdf

UN-REDD (2011). *Support National REDD+ Action: Global Programme Framework Document 2011–2015 Draft*. UN-REDD Programme Sixth Policy Board Meeting, 21–23 March 2011, Da Lat

Urs, K. and Whittell, R. (2009). *Resisting Reform? Water Profits and Democracy*. SAGE Publications, New Delhi

USEPA (2010). *Assessing the Multiple Benefits of Clean Energy: a Resource for States*. US Environmental Protection Agency, Washington, DC

Van der Werf, G., Morton, D., DeFries, R., Olivier, J., Kasibhatla, P., Jackson, R., Collatz, G. and Randerson, J. (2009). CO_2 emissions from forest loss. *Nature Geoscience* 2, 737–738

Van Vliet, N. (2011). *Livelihood Alternatives for the Unsustainable Use of Bushmeat*. CBD Technical Series No. 60. Secretariat of the Convention on Biological Diversity, Montreal

Vörösmarty, C., Meybeck, M., Fekete, B., Sharmad, K., Green, P. and Syvitski, J. (2003). Anthropogenic sediment retention: major global impact from registered river impoundments. *Global and Planetary Change* 39, 169–190

Wang, J. and Zhang. L. (2010). *Water Policy, Management, and Institutions: The Role of Pro-Poor Water Allocation in the Yellow River Basin*. International Food Policy Research Institute (IFPRI), Washington, DC. http://www.ifpri.org/sites/default/files/publications/yrbnote04.pdf (accessed 15 September 2011)

WCD (2000). *Dams and Development: A New Framework for Decision-Making*. World Commission on Dams. Earthscan Publications, London

White, W.R. (2010). *World Water: Resources, Usage and the Role of Man-made Reservoirs*. Foundation for Water Research, Marlow. http://www.fwr.org/wwtrstrg.pdf (accessed 15 September 2011)

Winrock International (2011). *Payment for Forest Environmental Services: A Case Study on Pilot Implementation in Lam Dong Province Vietnam from 2006–2011*. Winrock International, Akansas. http://www.winrock.org/fnrm/files/PaymentForForestEnvironmentalServicesARBCPCaseStudy.pdf (accessed 15 September 2011)

WMO (2010). *Scientific Assessment of Ozone Depletion: 2010*. Global Ozone Research and Monitoring Project Report No. 52. World Meteorological Organization, Geneva

World Bank (2011). *CDM in China: From Taking a Proactive and Sustainable Approach towards a Value Added Approach*. World Bank, Washington, DC. http://web.worldbank.org/WBSITE/EXTERNAL/COUNTRIES/EASTASIAPACIFICEXT/EXTEAPREGTOPENVIRONMENT/0,,contentMDK:21915756~pagePK:34004173~piPK:34003707~theSitePK:502886,00.html (accessed 15 September 2011)

World Bank (2009). *Samoa Second Infrastructure Asset Management Project (Supplemental)*. World Bank, Washington, DC. http://siteresources.worldbank.org/INTPACIFICISLANDS/Resources/SamoaInfrastructureProjectBrief090610.pdf (accessed 15 September 2011)

World Bank (2006). *Where is the Wealth of Nations? Measuring Capital for the 21st Century*. World Bank, Washington, DC

World Bank (2005). Sub-national own-source revenue: getting policy and administration right. In *East Asia Decentralizes: Making Local Government Work*. pp.107–128. World Bank, Washington, DC. http://siteresources.worldbank.org/INTEAPDECEN/Resources/Chapter-6.pdf (accessed 15 September 2011)

World Bank (2004). *Clean Development Mechanism in China: Taking a Proactive and Sustainable Approach*. Former CDM in China Report, World Bank, Washington, DC. http://www.worldbank.org.cn/English/content/cdm-china.pdf (accessed 15 September 2011)

WRI (2005). *Navigating the Numbers, Greenhouse Gas Data and International Climate Policy*. World Resources Institute, Washington, DC

WSSD (2002). *Johannesburg Plan of Implementation*. World Summit on Sustainable Development. http://www.un.org/esa/sustdev/documents/WSSD_POI_PD/English/POIToc.htm

Yang, X. (2003). *Manual on Sediment Management and Measurement*. World Meteorological Organization Operational Hydrology Report No. 47, WMO-No. 948. Secretariat of the World Meteorological Organization, Geneva

Yoon, S., Koo, J., Oh, G., Chung, D. and Yoon, J. (2008). *Current Status and Issues of Hazardous Waste Management in South Korea*. http://www.srcosmos.gr/srcosmos/showpub.aspx?aa=13036 (accessed 15 September 2011)

Young, O. (1992). The effectiveness of international institutions: hard cases and critical variables. In *Governance without Government: Order and Change in World Politics* (eds. Rosenau, J. and Czempiel, E.). Cambridge University Press, Cambridge

Zhang, Q., Bennett, M., Kannan, K. and Jin, L. (2010). *Payments for Ecological Services and Eco-compensation: Practices and Innovations in the People's Republic of China*. Asian Development Bank, Manila

CHAPTER 11

Europe

Coordinating lead authors: Nicolai Dronin and Ruben Mnatsakanian

Lead authors: Thomas Bernauer, Sophie Condé, Karine Danielyan, Lisa Emberson, Anastasia Idrisova, Olena Maslyukivska, Nora Mzavanadze and Alexander Orlov

Contributing authors: Olga Chkanikova (GEO Fellow), Joyeeta Gupta, Naira Harutyunyan, Lawrence Hislop, Pavlos Kassomenos, Jerome Simpson and Åsa Swartling

Principal scientific reviewer: Maria Siwek

Chapter coordinator: Ron Witt

Expert contributions: European Environment Agency staff and Mirjam Schomaker

Main Messages

Europe has strong environmental governance structures and mechanisms in place. In particular, the European Union (EU) has been implementing robust environmental policies over the last four decades. Regular monitoring, reporting and assessment required by legislation is an integral part of EU environmental governance, helping to inform policy makers whether policies are effective, and to identify emerging issues. This concept has already or is being emulated in neighbouring countries and, although to a lesser extent, through the pan-European Environment for Europe ministerial process that was initiated in 1991. Moreover, since the 2002 Earth Summit in Johannesburg, the EU's agenda has been increasingly oriented to external multilateral policies.

Both EU and non-EU European countries are also well on track to meet their own Kyoto targets. European countries are implementing climate-related policies ranging from carbon taxes to emissions trading schemes, stimulating renewable energy systems and local voluntary efforts by municipalities. More recently, climate change adaptation strategies are being developed. Large-scale reductions in anthropogenic greenhouse gas emissions can only be achieved through a tightly coordinated combination of different policies targeting different economic sectors and sources of emissions. The EU, with some of its neighbouring countries, is also a major donor to various global efforts to combat climate change.

Across most of Europe, many aspects of air quality have improved in recent decades, although problems still remain, particularly related to urban air quality, human health and ecosystem degradation. The pan-European scientific monitoring network of the Convention on Long Range Transboundary Air Pollution has been pivotal in building credibility, shaping policies and monitoring air quality trends. A variety of policies, often mutually enforcing, are being applied at regional, national and local scales. Tools cover both obligatory and voluntary implementation mechanisms, and many are being replicated elsewhere in the world, or have the potential to be.

Freshwater policies have been implemented successfully through a mix of policy instruments, often directed through strong umbrella legislation, but challenges such as overuse of water and water pollution persist in parts of Europe. The transboundary nature of most European rivers calls for close international cooperation, and integrated water resources management is increasingly the guiding mechanism for implementation. River basin management plans have shown potential for transfer and use throughout the region, water pollution from non-point sources has been effectively reduced through broad clusters of policies that complement each other, and water metering and water pricing have stimulated more responsible use of water.

Prevention, reuse and recycling of municipal wastes are among the most regulated activities in

the region. Comprehensive legislation supported by monitoring networks is helping to ensure compliance with regulations, but waste volumes continue to grow. In Eastern Europe, a legacy of industrial wastes from the socialist period still poses ecological problems. Policy focus is shifting towards producers' responsibility by encouraging innovative approaches such as ecodesign, new business models and changing life styles. The EU legislation on Registration, Evaluation, Authorisation and Restriction of Chemical substances (REACH), which replaces a patchwork of previous directives and regulations, looks promising for the regulation of chemicals in coming years.

The European region is at the forefront of multinational biodiversity conservation efforts. Networks of protected areas have been successfully established through Natura 2000 and comparable efforts outside the EU, also stimulating an improved knowledge base for preserving and monitoring biodiversity. However, due to landscape, ecosystem and habitat degradation both within and outside protected areas, the overall conservation status of habitats and species is showing no sign of improvement. Through national initiatives for sustainable forest management and payments for ecosystem services, the challenges of biodiversity conservation, climate change and protection of freshwater resources are being addressed in an integrated way, and are already showing positive results.

INTRODUCTION

The pan-European region is very diverse, with its 37 different national languages spoken in the 50 European countries (Table 11.1) (Nations Online 2011) and its range of socio-economic and political systems, as well as in its varied physical environment and means of environmental governance. Europe's land area of 23 million km² (GEO Data Portal 2011; FAO 2010) is characterized by a great variety of (agri)cultural landscapes, urban agglomerations, extensive coastal zones, forests and undisturbed pristine areas. Of the nearly 833 million Europeans, about half live in Western Europe, while some 72 per cent of the entire region's population resides in urban areas (GEO Data Portal 2011; UNDESA 2010).

Conversion to and intensification of agriculture along with increasing demand for greater mobility and urban space have transformed a majority of European landscapes over the past 100 years, causing fragmentation and loss of natural and semi-natural habitats with an associated decline in biodiversity (Chapters 1, 3, 5 and 7) (EEA 2010h; COE 2000). However, exposure of Europe's population to multiple air, water and chemical pollutants has generally declined, and both the European Union (EU) and most non-EU European countries are on track to meet their Kyoto targets (Chapters 1, 2, 4, 6 and 7) (EEA 2010h).

Indeed, considerable progress has been made in meeting environmental targets, with the situation improving in many areas. Nonetheless, concerns about long-term threats to the environment and human health persist, the latter especially for Europe's large urban population (EEA 2010h). Despite some successes in decoupling environmental pressures from economic growth, Europe's environmental footprint remains disproportionately high. This is due to the continued unsustainable use of natural resources both within and outside the region to satisfy the high production and consumption level of its inhabitants (Chapters 1–7) (EEA 2010h).

These trends are increasingly linked and complex, and require an integrated policy approach for which strong governance mechanisms need to be in place. Given that Central and Western Europe in particular have a dense network of political boundaries, a regional focus to tackle environmental issues is necessary. A major attribute of the pan-European region is its economic and political interconnectedness, combined with

Table 11.1 Country groupings used in various environment-related reporting and policy initiatives in Europe

EEA and EU country groups

	Sub-regions	Countries
EEA member countries (EEA-32*)	European Free Trade Association (EFTA) countries	Iceland, Liechtenstein, Norway, Switzerland
EEA member countries (EEA-32*)	European Union member countries (EU-27)	EU-15 Austria, Belgium, Denmark, Finland, France, Germany, Greece, Ireland, Italy, Luxembourg, Netherlands, Portugal, Spain, Sweden, United Kingdom
EEA member countries (EEA-32*)	European Union member countries (EU-27)	EU-12 Bulgaria, Cyprus, Czech Republic, Estonia, Hungary, Latvia, Lithuania, Malta, Poland, Romania, Slovakia, Slovenia
	EU candidate countries	Croatia, Former Yugoslav Republic of Macedonia, Turkey*
	EU potential candidate countries	Albania, Bosnia and Herzegovina, Montenegro, Serbia
	Partner countries of the EU European Neighbourhood Policy	Algeria, Armenia, Azerbaijan, Belarus, Egypt, Georgia, Israel, Jordan, Lebanon, Libya, Republic of Moldova, Morocco, Occupied Palestinian Territory, Syria, Tunisia, Ukraine

UNEP *GEO-5* country groups

Sub-regions	Countries
Western Europe	Andorra, Austria, Belgium, Denmark, Finland, France, Germany, Greece, Iceland, Ireland, Israel, Italy, Liechtenstein, Luxembourg, Malta, Monaco, Netherlands, Norway, Portugal, San Marino, Spain, Sweden, Switzerland, United Kingdom
Central Europe	Albania, Bosnia and Herzegovina, Bulgaria, Croatia, Cyprus, Czech Republic, Estonia, Former Yugoslav Republic of Macedonia, Hungary, Latvia, Lithuania, Montenegro, Poland, Romania, Serbia, Slovakia, Slovenia, Turkey
Eastern Europe	Armenia, Azerbaijan, Belarus, Georgia, Republic of Moldova, Russian Federation, Ukraine

Pan-European Environment for Europe (EfE) process of the United Nations Economic Commission for Europe (UNECE)
Member countries include all those listed above in the *GEO-5* country groups (excluding Holy See) plus Canada, Kazakhstan, Kyrgyzstan, Tajikistan, Turkmenistan, United States of America, Uzbekistan

* Turkey is already an EEA member country (and thus part of EEA-32).

Source: UNECE 2012; EEA 2010h; UNEP 2007b

well-established and strong formal governance mechanisms and structures to address environmental issues at (sub-)regional level. This has made Europe a leader in transboundary as well as global environmental decision-making. In particular, the EU has more than four decades of experience in developing environmental policies: the first Environmental Action Programme (EAP) was adopted in 1972 while the sixth ends in mid-2012. EU legislation is implemented at a national level within EU Member States, with forceful implementation control by EU institutions. The legislation is also used in non-member European states on a case-by-case and voluntary basis.

Policy approaches have evolved over time, using policies and single-issue instruments in the 1970s and 1980s, followed by policy integration and raising public awareness in the 1980s and 1990s and thereafter (EEA 2010h; Hey 2004). An integral part of EU environmental governance is the regular monitoring, reporting and assessment required by EU legislation; these activities help to inform policy makers about effectiveness and also help to identify emerging issues. Since the early 2000s, European environmental policy has increasingly been guided by the fact that well-designed, coherent policies that integrate different sectoral policy domains can provide greater benefits at lower costs than several single policies. As a result, Europe's natural resources are used with increasing efficiency (EEA 2010h).

This concept is already being emulated in EU neighbouring countries and in the pan-European Environment for Europe ministerial process that was initiated in 1991. In September 2011, for example, the Seventh pan-European Environment for Europe Ministerial Conference focused on the sustainable management of water and water-related ecosystems and on greening the economy including mainstreaming the environment into economic development.

The countries of Eastern Europe also have well-developed formal environmental policies and regulations, although the implementation and enforcement of these has often tended to lag. In the early 1990s, following the collapse of industry in Eastern Europe, environmental pressures dropped substantially in many countries, giving the public and authorities a false sense of security. The focus of attention shifted towards more urgent needs related to economic restructuring and development, with an inclination to make the economic transition easier by reducing environmental regulation. At first this strategy worked, but later, when countries regained their economic strength, it began to backfire.

A new wave of improved environmental legislation and policies can now be expected in the non-EU part of Europe, the current global financial crisis notwithstanding. Promising policies include, for example, integrated river basin management and cross-boundary biodiversity conservation. Another example is the Inter-Parliamentary Assembly of the Commonwealth of Independent States (IPA CIS), with its consultative and informative role. It has a Permanent Commission on Agricultural Policy, Natural Resources and Ecology, which advises the parliaments of CIS countries and suggests sample legislation on environmental issues. Practically all aspects of modern environmental policy are covered, ranging from environmental security, environmental insurance and strategic environmental assessments to environmental monitoring, energy conservation and environmental education (IPA CIS 2011).

The Salzach River flowing through Salzburg, Austria, where nature protection, agriculture, energy production and recreational activities are balanced through integrated river basin management. *© Dave Long/iStock*

POLICY APPRAISAL

For this chapter, five key challenges/priority issues were identified for Europe, in no particular order, during a GEO regional consultation held in September 2010:

- climate change;
- air quality;
- freshwater;
- chemicals and waste; and
- biodiversity.

At the GEO Regional Consultation, five international environmental goals related to the key challenges were identified, and regional-level goals were added later where applicable. The group then selected promising policies that have already shown some degree of success in helping to speed up meeting the globally and regionally agreed environmental goals (Table 11.2).

Table 11.2 Selected themes, goals and policy options and examples of success

Themes and international goals	Policy cluster/approach	Regional goals/targets	Policy options	Examples of success
Climate change United Nations Framework Convention on Climate Change (UNFCCC 1992) Articles 2 and 4.8	Combating climate change by creating and using markets	EU 20-20-20 • Reduce EU greenhouse gas emissions to 20% below 1990 levels by 2020 (30% if other industrialized countries make similar commitments and developing countries contribute adequately) • Lower EU energy consumption by 20% compared with projected levels for 2020, through increased energy efficiency • Meet 20% of EU energy needs from renewable sources by 2020 (EC 2009a) • Have EU national adaptation strategies in place by 2015 (ECouncil 2007)	EU Emissions Trading System	Pledges for post-2012 period (Box 11.1)
	Adapting to climate change by working with the public and private sectors, and through command-and-control regulation		Subsidies for renewable energy through feed-in-tariffs	German REFIT scheme (Box 11.2)
			Natural hazard insurance schemes; national adaptation strategies	Transferring innovative climate insurance (Box 11.3)
Air quality Agenda 21 (UNCED 1992) Chapter 9 Paragraph 27	Reducing emission levels through command-and-control regulations and by using markets	EU by 2020 Reduce, compared to 2000: • number of years of life lost due to particulate matter by 47%: • number of premature deaths due to ground-level ozone by at least 10% • forest area affected by ozone by 15% • forest area affected by acidification by 74% • freshwater area affected by acidification by 39% • area affected by eutrophication by 43% (EC 2005)	EU fuel and vehicle standards	Adopting European fuel standards (Figures 11.4 and 11.5)
	Integrated air quality management with the public and private sectors, command-and-control regulations, using markets, creating awareness and voluntary actions		SO_2 reduction	Proof of sulphur dioxide (SO_2) policy benefits (Figure 11.6)
			Local air quality management	Stockholm's air quality management in low-emission zone (Box 11.4)
Freshwater Johannesburg Plan of Implementation (JPOI) (WSSD 2002) Paragraph 26	Integrated water management through command-and-control regulations, with the public and private sectors and by using markets	EU overall goal • Get all water, including lakes, rivers, streams and groundwater aquifers, into a healthy state by 2015 (ECouncil 2000) By end 2012 • Specific targets for 2020 adopted in the EU 2012 Blueprint to Safeguard Europe's Water Resources (EC 2011a)	River basin management plans	Tisza River Basin Management Plan (Box 11.5)
			Policy mixes to reduce non-point sources of water pollution	Denmark's accounting system for nitrogen use in agriculture (Box 11.6)
			Water metering and pricing	Water metering in Armenia (Box 11.7)
Chemicals and waste Johannesburg Plan of Implementation (JPOI) (WSSD 2002) Paragraph 23	Reducing the amount of waste produced through command-and-control regulations and by using markets	• Recycle 50% of annual municipal waste and 70% of annual construction waste in the EU by 2020 • By 31 May 2013: companies must register chemical phase-in substances manufactured or imported in the EU of 100 tonnes or more per year • By 31 May 2018: companies must register chemical phase-in substances manufactured or imported in the EU of 1 tonne or more per year (ETC/SCP 2010)	Waste prevention	Extended producer responsibility (Box 11.8)
			Preparing for reuse Preparing for recycling	Trends in municipal waste processing in the EU (Figure 11.12)
	Comprehensive legislation on chemicals through command-and control-regulations		Registration, Evaluation, Authorisation and Restriction of Chemical substances (REACH)	Too early to report on success
Biodiversity Convention on Biological Diversity (CBD 1992) Articles 8, 10 and 11	Expanding and strengthening ecological networks through legislation and action plans	EU by 2020 • At least 15% of degraded ecosystems restored • Forest management plans in place for all publicly owned forests and forest holdings above a certain size that receive EU Rural Development Policy funding (EC 2011c)	Transboundary EU Natura 2000 network and national non-EU country networks	National ecological network of Ukraine (Box 11.9)
	Integrated management of forest resources and farm land with high nature value through voluntary action and payment for ecosystem services measures, working with both the public and private sector		Mechanisms for payment for ecosystem services Pan-European Forest Europe process	Conserving farmland with a high nature value in Portugal (Box 11.10)

The Introduction to *GEO-5* explains the methodology applied in this appraisal in more detail. It is acknowledged that:

- recent, innovative policies do not have a long enough track record to be selected for this appraisal, though some promising emerging policies are included in the conclusions at the end of this chapter;
- even where actual evidence of policy effectiveness does exist, such environmental improvements cannot usually be directly linked to one single policy or cluster of policies because of impacts from other sectoral policies, economic developments or political restructuring; and
- there are certainly other priority issues in parts of the region – such as the marine and coastal areas surrounding large parts of Europe, a new energy mix, land-use change and land degradation, or developments in Europe's mountainous areas – but these were not among the maximum of five key challenges/priority issues selected for this analysis by the GEO regional consultation.

Climate change

In terms of the total reduction in greenhouse gas emissions, European countries are leading the global climate change mitigation effort by a wide margin. Other large advanced economies have either not ratified the Kyoto Protocol (United States), are failing to meet their Kyoto targets (Canada) or are allowed to increase their emissions (Australia). Japan, with its 6 per cent reduction target, is the main exception. Figure 11.1 presents current emission data and trends for the main sectors in the EU-27, clearly illustrating the dominant role of energy (Chapters 1, 2 and 3).

Figure 11.1 Sectoral trends and projections for EU-27 greenhouse gas emissions, 1990–2020

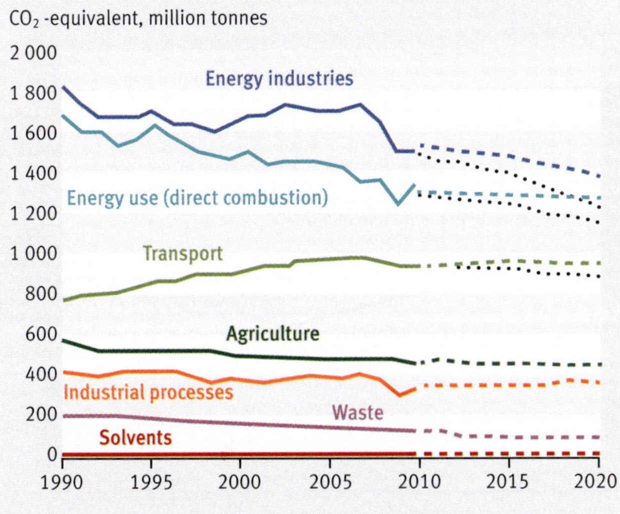

Note: Greenhouse gas emission projections are shown as dashed lines under existing policy measures and black dotted lines with additional measures.

Source: EEA 2011b

The EU-15 is well on track to meet its Kyoto target; indeed, over-compliance may even be achieved when the Clean Development Mechanism, Joint Implementation mechanism, and carbon removals such as forestry activities, are factored in (EEA 2010j). None of the Central and Eastern European countries have faced any problems in meeting their Kyoto Protocol obligations as their targets were set before the fall in emissions associated with the collapse of the Soviet bloc. In addition, regional emission targets for the post-2012 period have been set (Box 11.1).

Box 11.1 Greenhouse gas reduction pledges for the post-2012 period

In March 2007, the EU-27 unilaterally committed itself to reducing its greenhouse gas emissions by at least 20 per cent by 2020 compared to 1990, and to increase this commitment to 30 per cent if other industrialized countries commit to comparable emission reductions and developing countries contribute adequately according to their capabilities. These commitments were renewed by the EU-27 in the Copenhagen Accord in 2009. Similar pledges have been made by other advanced European economies, notably Iceland, Monaco, Norway and Switzerland. The EU has further declared that it will seek to achieve reductions of the order of 80–95 per cent by 2050 (EEA 2010j). In the Copenhagen Accord, the Russian Federation pledged to reduce its emissions by 15–25 per cent by 2020 and 50 per cent by 2050 compared to 1990, and Ukraine by 20 and 50 per cent respectively. Belarus, Croatia, the Former Yugoslav Republic of Macedonia, Moldova and Montenegro have also formally pledged to reduce their emissions. The United Nations Framework Convention on Climate Change (UNFCCC) Conference of the Parties in Cancún in December 2010 formally put those pledges into UN documentation, and the UNFCCC Secretariat will monitor progress (CG 2011).

A recent report by the EU calls for the total elimination of fossil-fuel-powered cars in cities by the year 2050. *© Robert Bremec/iStock*

The building sector will be a key area of opportunity in the EU's ambitious climate and energy package. *© George Clerk/iStock*

Large-scale reductions of greenhouse gas emissions can only be achieved, however, through a tightly coordinated and coherent combination of different policies targeting different economic sectors and sources of emissions. Only then can efficient synergies be achieved.

In 2009, the EU formally adopted its climate and energy package, an integrated approach with binding legislation to implement the EU's three main climate and energy targets:

- reduce EU greenhouse gas emissions to 20 per cent below 1990 levels by 2020;
- reduce EU energy consumption by 20 per cent compared with projected levels by 2020 through increased energy efficiency; and
- meet 20 per cent of EU energy needs from renewable sources, including biofuels, by 2020 (EC 2009a).

These commitments together have been labelled the 20-20-20 targets, which are being implemented through an array of policies ranging from carbon taxes and emissions trading schemes to local voluntary efforts by municipalities (EC 2009b). Two of the most promising policies are discussed below.

European Emissions Trading System

The EU Emissions Trading System (EU ETS) was launched in 2005 as a cornerstone of EU climate policy and the key tool for reducing industrial greenhouse gas emissions in a cost-effective manner. It is the first and largest international scheme for trading emission allowances, and is open to non-EU countries on the condition that they meet the strict standards of the EU ETS.

The EU ETS covers about 40 per cent of EU greenhouse gas emissions. For 2009, the EU carbon trading market was estimated to be worth nearly US$118.5 billion per year, compared to a

Figure 11.2 EU Emissions Trading System cap, 2005–2050

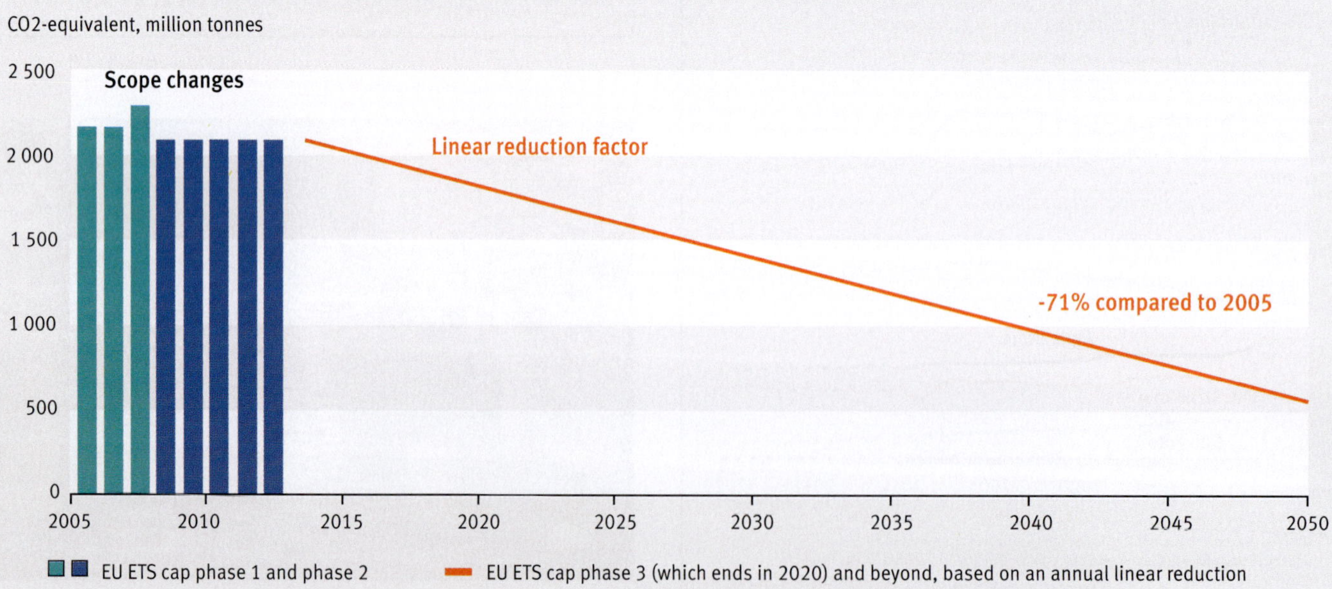

Source: EEA 2011b

global carbon credit market worth an estimated US$143.75 billion. In 2009, the volume of emissions covered by the system reached 6.33 billion tonnes compared to 41 million tonnes covered by the Chicago Climate Exchange (CCX) (Kossoy and Ambrosi 2010).

Falling prices of carbon credits in the first (2005–2007) and second (2008–2013) phases of the EU ETS, although caused by different factors, demonstrate the crucial requirements of the supply of accurate, reliable and constantly updated figures on energy consumption and emissions, verified by strict monitoring (Morris and Worthington 2010; Ellerman 2008). The Third Trading Period (from 1 January 2013) will implement several important changes such as inclusion of airline emissions, increased auctioning of allowances, and an ambitious EU-wide cap instead of national caps. The EU ETS cap will decrease continuously from 2013 onwards using a linear reduction factor (Figure 11.2).

The EU ETS is an attractive option for European countries outside the EU. Iceland, Liechtenstein and Norway are already covered by it through their membership of the European Economic Area Agreement, while Switzerland will be the first non-EU country whose national emissions trading system is linked to the EU ETS (Bart 2007; Ellerman and Buchner 2007) and Australia is exploring this possibility as well (Planet Arc 2011).

Feed-in tariffs for renewable energy systems

Feed-in tariff schemes were elaborated as the main support mechanism for renewable energy systems. Its goal goes far beyond reducing carbon dioxide (CO_2) emissions only; it also takes issues such as energy security, independence from conventional fuel price volatility or decentralization of energy into account (Blanco and Rodrigues 2008).

The policy offers long-term contracts to renewable energy producers, typically based on the cost of generation of each technology with two basic pricing models: a market-independent fixed price, applied by Germany's Renewable Energy Source Act (EEG 2009), and a market-dependent premium price model, used, for example, by Spain (Mendonça *et al.* 2009; Klessmann *et al.* 2008). The German Renewable Energy Feed-in-Tariff (REFIT) scheme, launched as early as 1991, is a successful example (Box 11.3). Spain is another positive example, as the country has established a dynamic, export-oriented and job-creating renewable energy sector even if it has not succeeded in other areas of climate policy (Sills and Roca 2010; Bechberger 2009; del Rio Gonzalez 2008). About two-thirds of EU Member States have now built up renewable energy capacity using feed-in tariffs (Figure 11.3) (Weidner and Mez 2008; Busch 2003).

At least 17 developing countries and emerging economies, including Brazil, China, India, Kenya, Nicaragua, South Africa and the Republic of Tanzania, have feed-in tariff schemes in place, most of which have been implemented within the last five years through, among others, the Global Energy Transfer Feed-in Tariff for Developing Countries (GET FIT) programme (Box 11.2) (REN21 2010). About 60 per cent of projects that have been registered under the Clean Development Mechanism or are in the pipeline for 2012 are for renewable energy, showing that development of this has become the most attractive climate policy option for developing countries (UNFCCC 2011; UNEP Risoe Centre 2010).

Figure 11.3 Electricity capacity in the EU-27 from biomass, on-shore wind and photovoltaic sources, 2005–2010

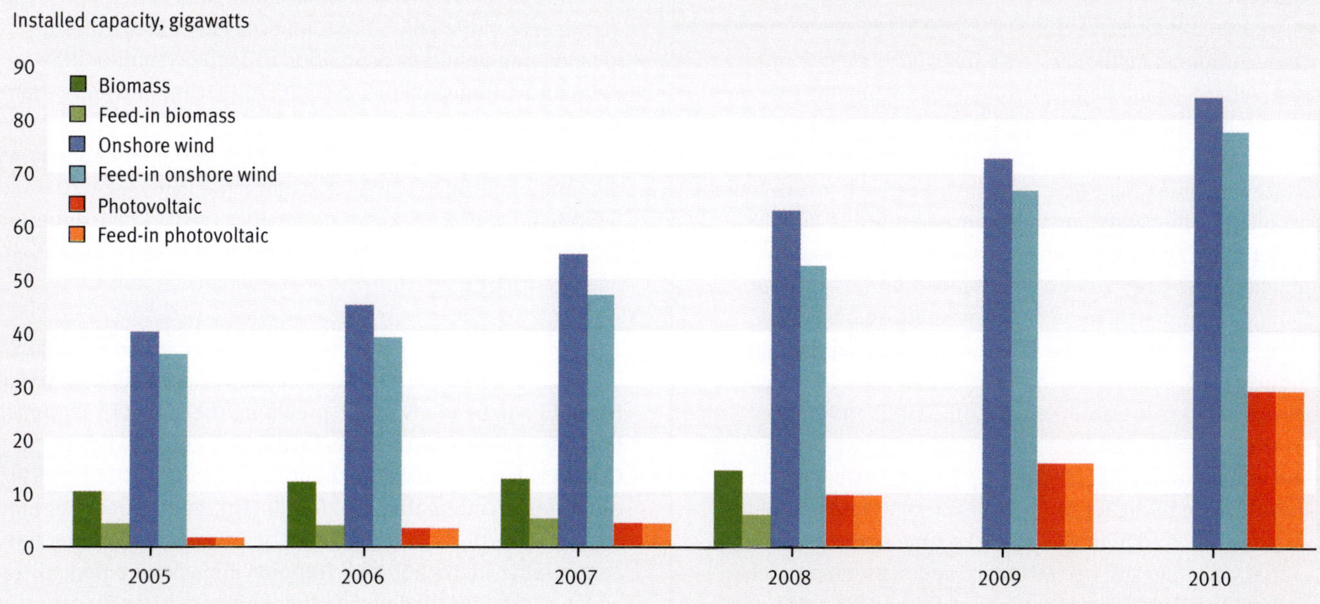

Source: Ragwitz *et al.* 2012

Box 11.2 The German Renewable Energy Feed-in Tariff scheme

Between 2000 and 2010 under the REFIT scheme, the share of electricity in Germany produced from renewable sources increased from 6.3 to about 17 per cent. In 2010, investments in Germany's renewable energy sector amounted to about US$3.5 billion and employed around 370 000 people (Jänicke 2011). The equivalent of 5.8 per cent of Germany's CO_2 emissions in 2009 was thus avoided (AGEE-Stat 2010). The Deutsche Bank Climate Change Advisors established the Global Energy Transfer Feed-in Tariff for Developing Countries (GET FIT) programme, which envisages a REFIT premium for individual producers to be paid by both the national government and the GET FIT fund. The concept of a global fund similar to the GET FIT has already found its way into the footnotes of the climate negotiations in the context of the Nationally Appropriate Mitigation Action (NAMA) programme (UNFCCC 2009).

Climate adaptation policies

When floods caused serious human and material damage in Central Europe in the summer of 2002, the European Commission (EC) reacted immediately by proposing the use of existing funds in a flexible way to respond to the urgent needs of the people affected. By mid-November 2002, an EU Solidarity Fund (EUSF) had been launched to finance short-term responses such as reconstruction of damaged or destroyed infrastructure, and to secure protective infrastructure such as dams and dykes. As the EUSF is restricted to the uninsured sectors of public infrastructure, it should be supplemented with a unified, innovative insurance system, being developed across Europe, which has the power to transfer risks from the local level to national and even global insurance markets through primary insurance and re-insurance (Box 11.3) (EC 2004).

Box 11.3 Transferring innovative climate insurance schemes

Many European insurance companies such as Swiss Re, AXA, Allianz, Munich Re, MicroInsure and Zurich have helped pioneer index-based weather risk transfer instruments in low-income countries. For example, Swiss Re started its index-based weather insurance scheme in India in collaboration with a micro-finance institution and a local insurer in 2004, since when a total of 350 000 policies have been sold to smallholder farmers in India. Similar solutions have been successfully deployed elsewhere: in 2007, Swiss Re designed and implemented index-based weather risk transfer instruments for three village clusters in Sauri (Kenya), Tiby (Mali) and Koraro (Ethiopia), protecting 150 000 farmers against drought risk. The innovation is that insurance pay-outs are based on the performance of the weather index rather than on actual damage incurred or losses suffered. One of the advantages is that pay-outs can be calculated and disbursed quickly and automatically without the need for households to file a claim formally (Warner and Spiegel 2009).

Another tool to assist in preparing for the impacts of climate change is the 2007 EU Floods Directive, under which draft national flood risk maps had to be submitted in 2011, with final versions to be ready by 2013 and final adaptation plans by 2015 (ECouncil 2007). More recently, the EU White Paper on Adaptation to Climate Change (EC 2009b) has moved beyond short-term disaster responses, outlining key steps towards a European framework for long-term adaptation measures and policies to increase resilience, to be implemented at national and local levels. Top-down strategies are envisaged for mainstreaming adaptation into sectoral policies, focusing on sectors such as land-use planning, agriculture, water management and biodiversity/nature conservation. Bottom-up activities focus on building adaptive capacity and implementing action at municipal level (EEA 2010h). In addition, a new EU Clearinghouse on climate change impacts, vulnerability and adaptation was put in place with the first stage of the strategy to run until 2012 (EC 2010c).

Air quality

Although many aspects of air quality across Europe have improved in recent decades due to emission reductions from industry and transport (Chapter 2), air pollution continues to pose a threat to human health, especially in urban areas (EEA 2010h). For example, exposure to fine particulate matter ($PM_{2.5}$) was estimated to have caused 5 million lost life years in 2005 in the EEA-32 (EEA 2010h). Similarly, other air pollutants continue to cause environmental damage to ecosystems, with 10 per cent of the EEA-32 natural ecosystem area still subject to acidifying pollutant deposition caused by sulphur dioxide (SO_2) and nitrogen oxides (NO_x) (EEA 2010h), and more than 40 per cent of sensitive terrestrial and freshwater ecosystems still subject to eutrophying atmospheric nitrogen deposition in the form of nitrogen oxides and ammonia (NH_3) (EEA 2010h). Despite declining peak ground-level ozone (O_3) concentrations, background levels are steadily rising, also leading to ecosystem damage (UNECE 2010).

The Convention on Long Range Transboundary Air Pollution (CLRTAP) of the United Nations Economic Commission for Europe (UNECE) has been pivotal in providing the scientific evidence that has underpinned efforts to shape air quality policy. CLRTAP's flagship 1999 Gothenburg Protocol (UNECE 1999) promotes an integrated multi-pollutant, multi-effect approach to optimize efforts to improve air quality across Europe. It is comparable to the 2001 EU National Emission Ceilings Directive, which establishes legally binding pollutant-specific emission ceilings for nitrogen oxides, non-methane volatile organic compounds, sulphur dioxide and ammonia for the EU-27. The

2008 Clean Air for Europe (CAFE) Directive merges much of the existing air quality legislation to develop long-term, strategic and integrated policy advice.

Such European approaches have been instrumental in providing the impetus for the development of a suite of air quality policies through the establishment of binding emission and air quality standards. Three outstanding environmental success stories are described here: vehicle emissions and fuel standards, the EU Industrial Emissions Directive and local air quality management policies.

European vehicle emission and fuel standards

Historically, road transport has contributed substantially to atmospheric pollution by producing emissions of lead (Pb), nitrogen oxides and particulate matter (Chapter 2). Reduction of these emissions has been achieved through the establishment of EU directives controlling both fuel and vehicle emissions, with fuel policy focused on banning lead and limiting sulphur content (ECouncil 1999, 1998). European Vehicle Emission Standards (Euro standards) control exhaust emissions of nitrogen oxides, non-methane volatile organic compounds and total hydrocarbons, carbon monoxide and particulate matter from new vehicles sold within the EU. Since the establishment of the Euro 1 standards in 1992, more stringent ones have been introduced, tightening controls on different pollutants, vehicle categories, weights and classes, engine volumes and fuel types; Euro 5 standards have been in force since 2007. Figure 11.4 shows the vehicle stock allocated to the Euro standards that have been established to date. Figure 11.5 shows the timeline for the introduction of increasingly stringent Euro standards in the EU and their transferability through the subsequent adoption of these standards in many Asian countries. They have also been adopted in parts of Latin America and Eastern Europe (PCFV 2011b; OECD 2007b).

Figure 11.4 Passenger cars and light-duty trucks meeting Euro standards

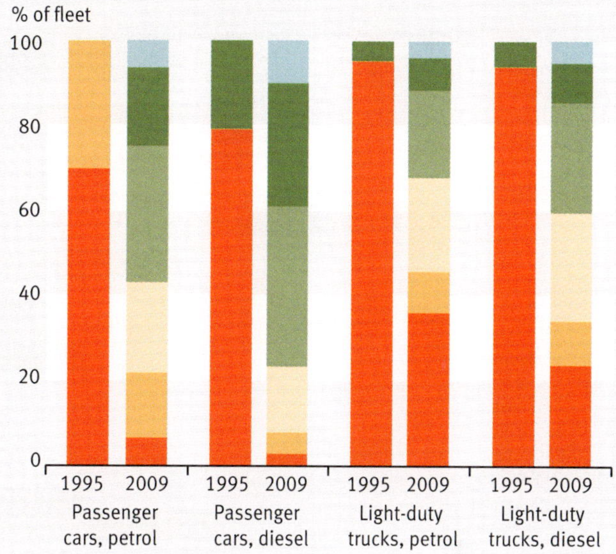

Source: EEA 2010b

Harmful air pollutants can be transported across countries, continents and even oceans, affecting air quality far from the original source.

© Jarek Szymanski/iStock

Figure 11.5 Euro-based standards and their adoption in Asia, 1995–2018

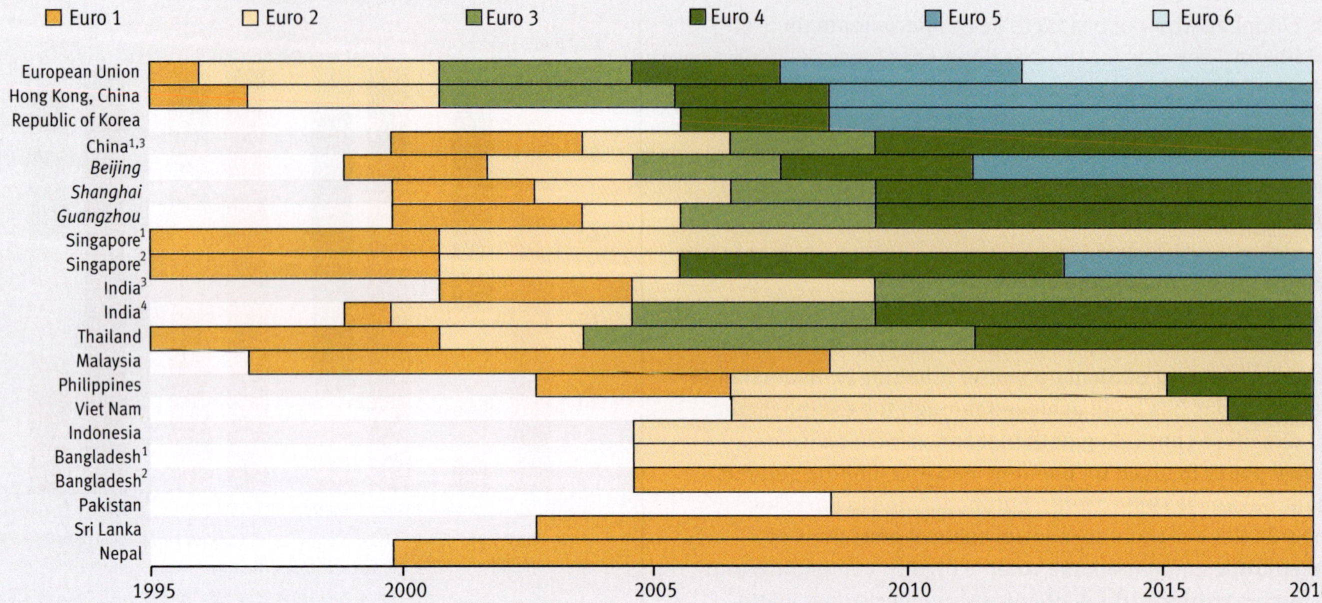

1. Petrol only.
2. Diesel only.
3. Whole country.
4. Agra, Ahmedabad, Bangalore, Chennai, Delhi, Hydrabad, Kanpur, Kolkata, Lucknow, Mumbai, Pune, Sholapur, Surat. All other Indian cities are Euro 2.

Note: The level of adoption varies. If no annotation is given, Euro standards have been adopted country-wide for both petrol- and diesel- fuelled vehicles.

Source: CAI-Asia 2011

Despite a 26 per cent increase in fuel consumption across the entire transport sector between 1990 and 2005, actual pollutant emissions in 2005 were significantly lower than a theoretical no-new-policy scenario assuming conventional technologies and no introduction of Euro standards: in the EEA-32, nitrogen oxide was 40 per cent lower than the scenario figures; carbon monoxide 80 per cent, non-methane volatile organic compounds 68 per cent and particulate matter 60 per cent lower (EEA 2010d). Lead emissions from road transport alone decreased by 99 per cent (EEA 2010c) and sulphur dioxide emissions by 92 per cent between 1990 and 2008 (EEA 2010e). Additional benefits of the Euro standards include increased engine lifetime and lower maintenance costs due to the removal of sulphur (PCFV 2007), better fuel economy and reduced greenhouse gas emissions (ICCT 2007).

The implementation of cleaner fuel policies involves costs related to fuel switching, such as replacing lead with other fuel additives and shifting to producers of low-sulphur crude oil, improving engine technologies and upgrading refineries (PCFV 2007). However, the benefits of lead phase-out and desulphurization in terms of human and ecosystem health in general exceed the costs (Blumberg *et al.* 2004; Lovei 1998). The EU, Japan and the United States lead the world in desulphurization policies (PCFV 2011a) and by 2011, vehicle fuel in the European region was lead free (PCFV 2011c).

The time lag in the effectiveness of vehicle emission policy depends on the average age of the vehicle fleet and the affordability of new vehicles. Awareness raising, product labelling, enforcement and regular control of fuel quality, now considered essential for vehicle policies to reach their full potential, have ensured successful implementation of these policies (PCFV 2007).

EU Industrial Emissions Directive

The 2010 EU Industrial Emissions Directive is designed to consolidate seven existing EU directives that have evolved since the early 1980s and have been instrumental in reducing industrial sulphur dioxide emission. The new directive will combine proven policy measures including technical emission controls, best available techniques, fuel switching and reduced sulphur content in liquid fuels.

Implementation of these measures has resulted in a clear reduction in sulphur dioxide emissions across Europe over recent decades (Figure 11.6), effectively decoupling them from industrial activities, especially in Western Europe (EEA 2010e). To some extent these reductions were aided by socio-political and economic changes between 1990 and 2000 in former socialist Eastern Europe. The reduction of total anthropogenic sulphur dioxide emissions in the EU-27 – by 80 per cent between 1990 and 2009 (EEA 2010e) – has led to substantial declines

Figure 11.6 European sulphur dioxide emission reductions, 1980–2004

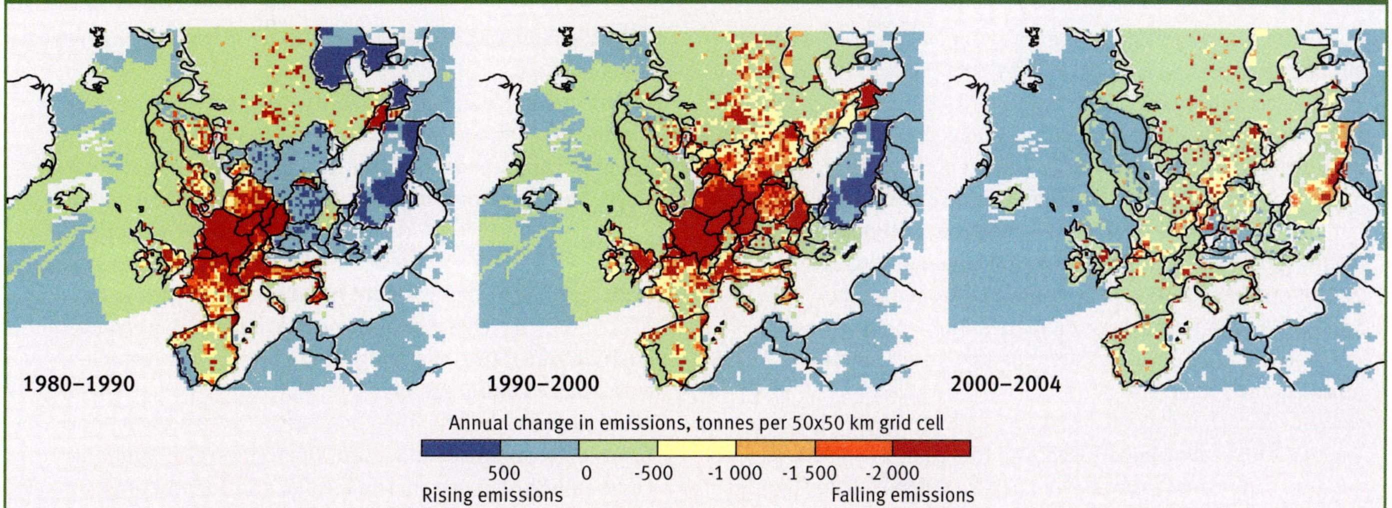

1980s: SO_2 emissions fell dramatically in Western Europe (red and orange), while continuing to increase in the southeast (blue).

1990s: Emission reductions slowed in some Western European countries (yellow and green), but accelerated in Central and Eastern Europe (red and orange), while emissions that had been rising in parts of the southeast also began to fall rapidly.

2000s: SO_2 emissions began to stabilize in most of Europe (green); but began to increase (from green to blue) in the North East Altantic due to shipping activity, and in the Balkans due to recovering industry.

Source: Vestreng *et al.* 2007

in acidification rates as exceedances of critical loads across Europe have been reduced. The implementation of the measures did, however, involve additional costs, requiring investment from the private and public sectors. The new EU Industrial Emissions Directive aims to reduce these costs by streamlining and enhancing cost efficiency and effectiveness (ECouncil 2010). Many control technologies have proven transferability, having been adopted in many Asian countries where they are of particular relevance since 80 per cent of Asia's energy demand is met by coal-fired power. Enhanced penetration of measures across Asia could yield further substantial improvements in sulphur dioxide emission reductions (Klimont *et al.* 2009).

Local air quality management policies

Under the 2008 Clean Air for Europe (CAFE) Directive, local authorities are obliged to prepare air quality management plans to ensure compliance with air quality standards. Many policy measures have focused on urban transport since this sector generates 70 per cent of air pollutants in urban areas (Chapter 1) (EC 2007a). Perhaps the most influential of these policy measures have been low-emission zones that limit or ban the most polluting vehicles from entering urban areas and encourage faster renewal of the vehicle fleet in line with vehicle emission standards. Around 100 low-emission zones in ten European countries have either been established or are in the process of being established (Box 11.4) (LEZ 2011). Other measures include congestion charging, expansion and improvement of public transport and cycling infrastructure, car pooling and cycle sharing systems, renewal or retrofitting of municipal vehicle fleets, and traffic and green areas management. Air quality management plans also require the public dissemination of current information detailing ambient air pollution and exceedances of air quality standards (ECouncil 2008a), with citizens and legal entities having the right to go to court in cases of non-compliance with standards. However, individual lawsuits for breaches of air quality standards are rarely pursued due to costs, time-consuming procedures and low awareness (ECouncil 2008a). Besides, many urban areas in Europe are not compliant with current European air quality legislation (EEA 2010i). To be really successful, local air quality management plans require

A proliferation of designated cycle ways can be seen across several European cities. *© Carsten Madsen/iStock*

Box 11.4 Stockholm's air quality management policies in a low-emission zone

Stockholm's low-emission zone was launched in 1996 and initially targeted heavy-duty vehicles entering the city centre. Vehicles complying with Euro 1 standards were allowed to enter freely while those more than eight years old had to be retrofitted or issued a permit. Enforcement was carried out by police inspections leading to an overall compliance rate of around 90 per cent within a few years (Burman and Johansson 2001). Actual air pollution concentrations in 2000 were down by 0.5–2 per cent for nitrogen oxides and by 0.5–9 per cent for particulate matter compared to the theoretical values calculated for a no-policy situation (Burman and Johansson 2001).

Then in 2007, following a successful trial period in 2006, a variable congestion tax was launched for vehicles entering Stockholm's city centre on weekdays during working hours. Clean vehicles running on electricity and biofuels were exempt from the tax. Burman and Johansson (2010) found that:

- the number of trips and the distance travelled in the inner city decreased in 2006 by 100 000 per day and 8.5 per cent respectively;
- the share of clean vehicles in the private fleet increased from 5 per cent in 2006 to 14 per cent in 2008;
- average pollutant concentrations decreased in the inner city by 10 per cent for nitrogen oxides, 15 per cent for carbon monoxide and 15–20 per cent for particulate matter.

Both air quality management policies were found to be even more effective if supported by additional measures such as green area networks, clean fuels, clean vehicles, extension of public transport and promotion of cycling and walking. Nonetheless, the congestion tax has been shown to generate a net social benefit of around US$95 million (€70 million) per year in the form of shorter and more reliable travel times, reduced greenhouse gas emissions, health and environmental benefits, greater traffic safety, increased public transport and higher government revenue (Eliasson 2009).

adequate monitoring and information systems and appropriate institutional mandates for local authorities.

Freshwater

In large parts of Europe demand for water often exceeds local availability, a trend that is likely to be exacerbated by climate change. In addition, both point and diffuse sources of pollution are still significant in parts of Europe, as a result of which some health risks remain (Chapter 4) (EEA 2010h). Europe's water challenges are driven by competing demands for water by agriculture, industry, public water supply and tourism, and further complicated by the transboundary nature of many European freshwater resources. Addressing these challenges calls for strong environmental governance structures, with a focus on coherent and integrated efforts and regional cooperation (Chapters 1 and 16).

The EU Water Framework Directive (ECouncil 2000) and pan-European UNECE instruments such as the Convention on the Protection and Use of Transboundary Watercourses (Water Convention) provide the basis for solving major water issues in the region. The Water Framework Directive brings the many isolated policies that have been developed in the EU since the mid-1970s together into one coherent legal framework for water policy decision-making within the river basin context. Its main goal is to protect and enhance the status of all EU waters, including groundwater, rivers, lakes and coastal waters, and water-dependent ecosystems, and to ensure long-term sustainable use of water resources. The Water Convention offers a common platform for both EU and non-EU European countries to exchange and transfer knowledge and create common understanding, and is a useful tool for assisting in implementing EU water legislation by non-EU countries.

Three specific policy instruments with some history of effective implementation have been selected for further appraisal: integrated management of transboundary river basins; policy mixes to address non-point sources of pollution; and water metering and volume-based pricing.

Integrated management of transboundary river basins

Water does not stop at administrative or political boundaries, making regional cooperation crucial between countries that share the natural geographical and hydrological unit of a river basin. The overarching approach of integrated water resources management has proven to be an effective policy for valuing, managing and protecting water-related ecosystems (UNECE 2011a). The development of river basin management plans is one of the main tools for implementing the Water Framework Directive, which focuses on pollution prevention and control, greater public participation in water management, and economic analysis of water use. The plans require the integration of industrial, agricultural and rural development, nature conservation and forestry programmes at the river basin scale and, in many cases, transboundary collaboration and coordination through river basin commissions. Progress in cooperation in the region is varied, however.

The first river basin commission in Europe – the International Commission for the Protection of the Rhine – celebrated its 60th anniversary in 2010, and has registered numerous successes over the years. Similar commissions have been established since for many European rivers, gradually moving eastwards despite the fact that in non-EU countries a comprehensive and strong legal basis for cooperation is often still lacking (UNECE 2011b). As many water bodies are shared by EU and non-EU nations, countries are encouraged to jointly prepare river basin management plans: the Tisza River Basin Management Plan provides a recent example of such cooperation across EU borders (Box 11.5).

By sharing the benefits and responsibility of sustainably co-managed water resources, economic development is fostered, establishing a connection between economic activities and the environment. River basin management plans also encourage public participation in working and expert groups. However, this approach still faces serious limitations due to the magnitude and complexity of the problems it seeks to address and the significant number of stakeholders who need to be involved (Figure 11. 7) (Sendzimir *et al.* 2008).

A mix of policies to address diffuse sources of water pollution

Eutrophication, predominantly caused by sewage discharges and agricultural run-off, is a major threat to European freshwater resources. Policies to reduce the flow of nutrients from point sources are well known and have proved to be successful, provided that sufficient funding is allocated to construct and manage water treatment systems. Tackling the problem of diffuse sources of freshwater pollution is much more challenging (DEFRA 2002).

Box 11.5 Integrated Tisza River Basin Management Plan

The Tisza River, which flows through parts of Hungary, Romania, the Republic of Serbia, the Slovak Republic and Ukraine, is the largest tributary of the Danube. The main pressures threatening the region are pollution from nutrient, organic and hazardous substances and both floods and droughts. The countries of the Tisza Basin have prepared an integrated river basin management plan, formally adopted in April 2011, in which the steps and long-term action needed to reach the required improved water status for the basin by 2015 are outlined. The plan attempts to deal with the complex links between different, potentially or actually conflicting, objectives and actors in integrated management (Figure 10.7). Experience gained in developing the plans can be transferred to other basins shared by EU and non-EU countries (UNDP and GEF 2011).

Figure 11.7 Complex links between objectives and actors involved in managing the Tisza Basin

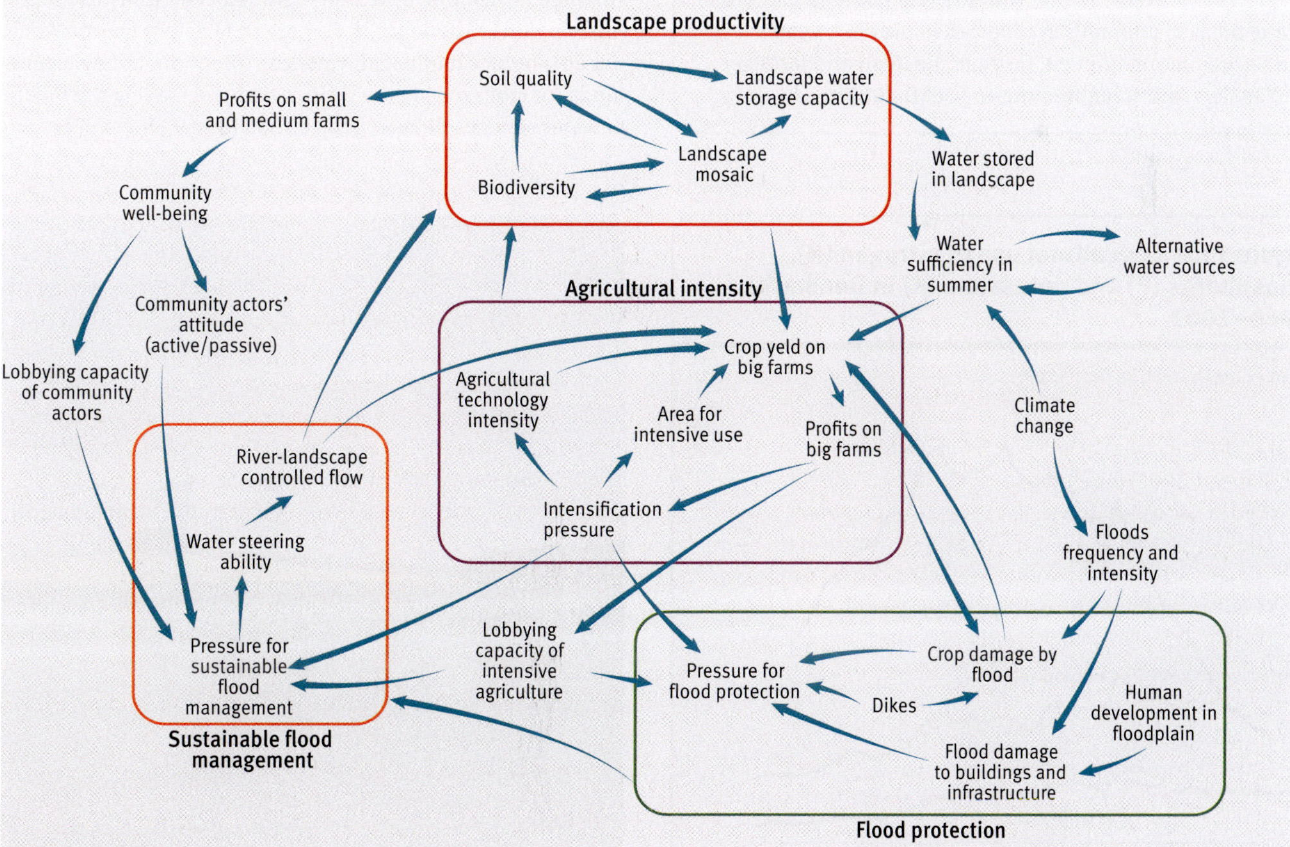

Source: Adapted from Sendzimir *et al.* 2008

There is significant experience in Europe of applying mixes of policies to reduce diffuse nutrient run-off, including accounting systems for the use of nitrogen in agriculture (Box 11.6), regulations on livestock density and the use of animal manure, purchase of nitrogen quotas, taxes on fertilizers, and compensation for converting agricultural land into wetlands or forest (OECD 2007a). Denmark, for example, has applied a large cluster of such mixed policies since the late 1980s, taking their synergistic effects into account, while avoiding disproportionate burdens on any particular stakeholder (Petersen and Knudsen 2010; Jacobsen 2004). As a result, the application of nutrients in Denmark has steadily decreased since the early 1990s (Figure 11.8).

Box 11.6 Nitrogen accounting in Denmark

A cornerstone in Danish policies addressing nutrient run-off is an obligatory, detailed nitrate accounting system introduced in 1993. Under this system, a preliminary, yearly nitrogen quota is calculated for each farm, depending on the area of arable land, the crops planted and the soil type. The accounting system is combined with other obligations, such as regulations on the use of animal manure and limits on livestock density in compliance with the EU Nitrates Directive (91/676/EEC). To reduce administrative costs both for public authorities and for farmers, the Danish authorities developed software in 2005 with pre-filled nitrogen accounts on the basis of information collected in previous years, and further information on, for example, feed and fertilizer wholesalers and slaughterhouses (OECD 2007a).

Figure 11.8 Agricultural use of nitrogen (N), phosphorus (P) and potassium (K) in Denmark, 1960–2007

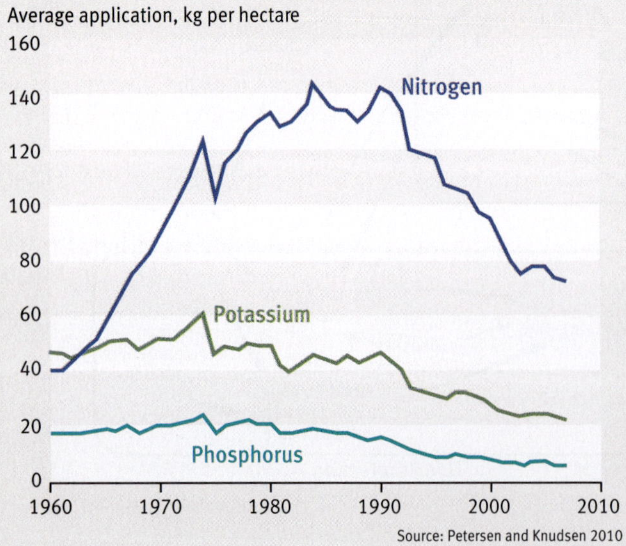

Source: Petersen and Knudsen 2010

Water metering and volume-based pricing

Europe has relatively high water consumption due to high agricultural and industrial demand (Chapter 4). In addition, considerable losses often occur in the supply chain, aggravating shortages in already water-scarce regions. In some countries up to 40 per cent of the total amount of water transported may be lost before it even reaches the consumer, while in others it can be below 10 per cent (EEA 2010h). Metering, cost recovery tariffs and proper pricing structures stimulate more responsible water use, at the same time generating funds for the maintenance of the supply system.

While water metering is a common policy tool in many counties of Western Europe, Central and Eastern European countries are still in the process of transition from a flat-rate price charged per person to a system of volume metering. Various studies reveal that on average, if individual metering systems are in place, reductions of 10–40 per cent can be achieved in household water use (Inman and Jeffrey 2006; Scheuer 2005).

In addition to metering, several Western European countries apply cost recovery tariffs and have introduced site-specific pricing structures. An increasing block rate creates the strongest incentive for conservation, applying a user-pays principle, under which the unit rate for water increases with water use, keeping the price for basic water needs relatively low (Figure 11.9). This system is becoming more common in both households and commercial sectors in Western European countries (OECD 2009). Applying this experience in Central and Eastern Europe would not only reduce inefficient water consumption, but also generate funds for modernizing the water sector, increasing the reliability of water services delivery (Box 11.7).

Farmland in Denmark, where the agricultural use of nitrogen has dropped by more than 50 per cent since 1990. *© BjornRasmussen/iStock*

Figure 11.9 Varying water tariff structures in selected European countries

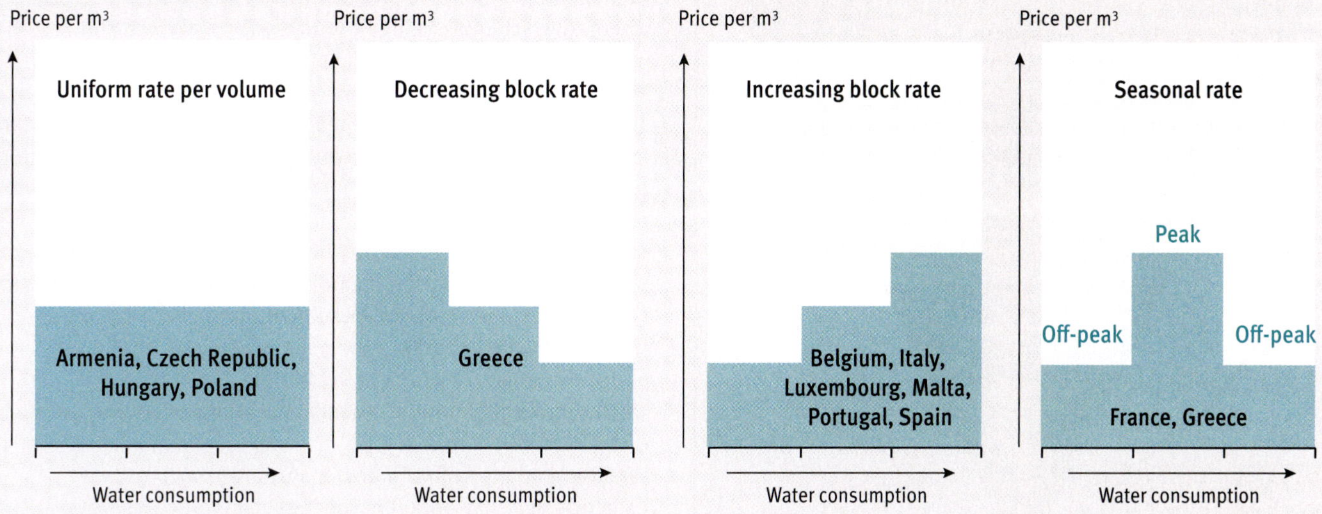

Source: Adapted from Chesnutt *et al.* 1997

This policy, however, faces several limitations. The costs of meter installation could be too heavy a burden for poor households (Melikyan 2003), conflicting with the Millennium Development Goal (MDG) 7c of halving, by 2015, the proportion of the population without sustainable access to safe drinking water and basic sanitation (UN 2000). Also, water pricing should not result in a situation in which personal hygiene and health are compromised in order to pay a water bill (EEA 2009b). To become successful, water pricing and meter installation require a good understanding of relationships between price and use in each sector, taking local conditions into account. Special subsidy schemes could be introduced for providing free meter installation for poor families, gradual repayment terms, and special provisions for writing off vulnerable families' accumulated water debts.

Box 11.7 Water metering in Armenia

By the late 1990s, the water sector in Armenia was severely degraded due to poor maintenance and a failure to invest, with non-revenue water use amounting to around 70 per cent nationwide. Less than 15 per cent of the utility costs were recovered, compared to an average of 30–40 per cent among newly independent countries (OECD 2007a). In 1999, the Armenian government launched the following measures to reform the water supply and sanitation sector:

- reduce sector dependence on state subsidies and donor assistance;
- raise revenues from increased collection of water payments based on metering; and
- restructure water utility debts (OECD 2008).

Soon after the reforms took place, average water use decreased three to four times compared to the use based on flat-rate calculations. The massive process of introducing individual metering became a trigger for a chain of water sector improvements, all backed by a legal, regulatory and institutional framework that enabled private sector involvement accompanied by investment and management efficiencies. As a result, the quality and reliability of water delivery improved.

Chemicals and waste

Both in the EU and in Eastern Europe, issues related to chemicals and waste have always been of primary importance. The EU's waste policy consists of three levels of legislation. The first, so-called horizontal, level defines overall requirements for all waste types and consists of the 2008 Waste Framework Directive, which is the cornerstone of current EU waste policy (ECouncil 2008b), and the 2006 Waste Shipment Regulation. The second level of legislation deals with waste installations and includes the Waste Incineration Directive, the 1999 Landfill Directive and the 2000 Port Facilities Directive. In addition, the 2010 Industrial Emissions Directive also defines requirements for some waste installations. Finally, the third level deals with specific waste streams such as waste containing polychlorinated biphenyls and terphenyls (PCBs/PCTs), waste oils, sewage sludge, electrical accumulators like batteries, and packaging waste. One example of such regulation is the Waste Electrical and Electronic Equipment Directive, which deals with the collection and recycling of such waste (ECouncil 2002b). It also includes the Restriction of Hazardous Substances Directive, which bans the use of certain hazardous substances in electrical and electronic products (ECouncil 2002a).

Figure 11.10 Moving up the waste hierarchy

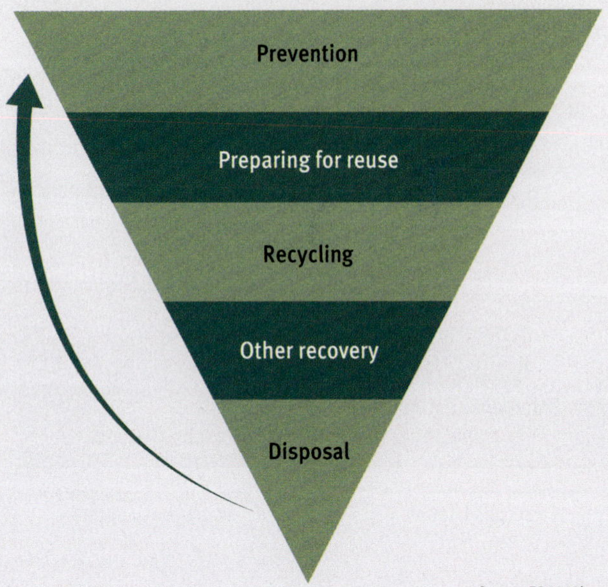

Source: EC 2010b

A basic principle of the EU Waste Framework Directive is the hierarchy of waste management originally established in the 1996 Waste Strategy (Shinn 2005). It states that, in order to better protect the environment, Member States should take measures for their waste treatment in line with the hierarchy shown in Figure 11.10, which is listed in descending order of priority.

One of the EU's major aims has always been waste reduction, but this goal has so far not been achieved (EEA 2010h). On the contrary, the amount of waste has been growing; notable examples including construction and demolition waste, packaging, hazardous and municipal waste, and sewage sludge (EEA 2010h). This trend has to be reversed, particularly as resource efficiency is one of seven flagship initiatives of the EC's Europe 2020 Strategy (EC 2011b, 2011c), which is reflected in the EU goals of decoupling resource use from economic growth, measured as lower resource use per unit of gross domestic product (GDP), and of minimizing waste. In addition to reducing waste generation, it is important to improve waste recycling. Current EU data indicate that only 38 per cent of total waste is reused or recycled (EEA 2010h).

Even though radioactive waste is not a subject of the waste hierarchy, it has important implications for both safety and energy production. On 19 July 2011, the European Council adopted the Radioactive Waste and Spent Fuel Management Directive, which sets standards for the safe disposal of spent fuel and radioactive waste from nuclear power plants as well as from medicine or research. This was a major achievement achievement for nuclear and environmental safety in the EU.

Other non-EU European countries also face significant challenges in waste policy. For example, Belarus, the Russian Federation and Ukraine all have large amounts of industrial waste in landfills as well as mining waste, with few or no financial incentives to recycle them. This is the result of many Soviet-era waste management and reuse practices being abandoned without alternative schemes being introduced (Devyatkin 2009).

Waste prevention

The EU Waste Prevention Directive of 2008 is based on definitions laid down in the Waste Framework Directive, in which prevention has been given the highest priority. Article 3.12 of the directive demands waste prevention through measures taken before a substance, material or product has become waste, by reducing:

- the quantity of waste, including through the reuse of products or the extension of the lifespan of products;
- the adverse impacts of the generated waste on the environment and human health; or
- the content of harmful substances in materials and products.

Waste prevention should also incorporate such aspects as ecodesign, life-cycle approaches, changing business models and consumption patterns (Box 11.8; Figure 11.11).

Box 11.8 Extended producer responsibility

The concept of extended producer responsibility in Europe widens a manufacturer's responsibility across a product's life cycle from its sale to its disposal, creating an incentive to avoid unnecessary waste and encouraging recycling and recovery. One example of this is the Green Dot system, which puts a levy on producers for the collection and recycling of waste components of their packaging (EC 2010b). If well designed, this practice provides significant incentives to introduce waste prevention mechanisms and to consider the entire life cycle of the product (EC 2010b).

Figure 11.11 A life-cycle approach to resource efficiency

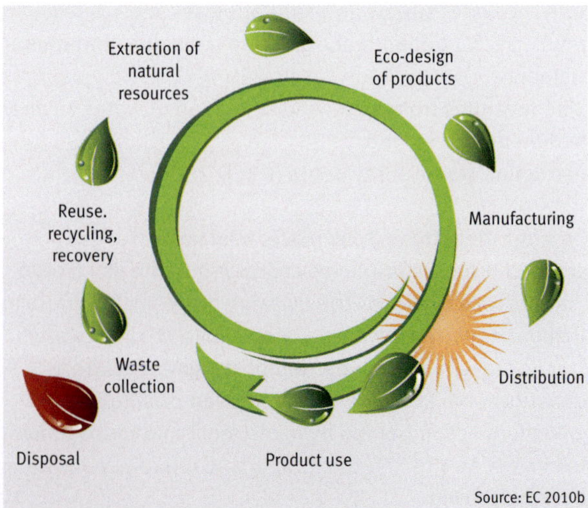

Source: EC 2010b

The practical outcomes of this policy can be achieved through a number of instruments, including legal provisions, voluntary agreements, economic instruments and incentives, and communication strategies.

Reuse and recycling

The Waste Framework Directive also encourages reuse, recycling and recovery, providing a range of options for the recycling of various materials including promoting the establishment of recycling targets, which can be material-specific. The figures show that the average amount of waste per citizen in the EU is approximately 6 tonnes per year. Municipal solid waste alone increased from 468 kg per person in 1995 to 524 kg in 2008, an increase of 12 per cent, caused by the increasing adoption of Western consumption habits in the new Member States (EEA 2011c). However, EU countries have made measurable progress in the efficient use of resources and management of waste, as illustrated by the fact that municipal waste recycling more than doubled between 1995 and 2008, rising from 17 per cent to 40 per cent (Figure 11.12) (EEA 2011c, 2010g).

Recycling has multiple benefits for many areas of the economy.

© Maike Janssen/iStock

Despite such advances, the EU is still not a recycling society, given that as of 2008, the share of municipal waste disposed of in or on land still exceeded 40 per cent (Figure 11.12) (EEA 2010g). Based on various macro-economic scenarios, it is estimated that by 2035 total waste generation in the EU-27 will have increased by 60–84 per cent compared to 2003 levels, although these figures could be significantly revised due to the current economic crisis (EEA 2010h).

Eastern Europe shows quite a different picture. In the Russian Federation with a population of nearly 143 million, the total

Figure 11.12 Trends in municipal solid waste treatment in the EU, 1995–2008

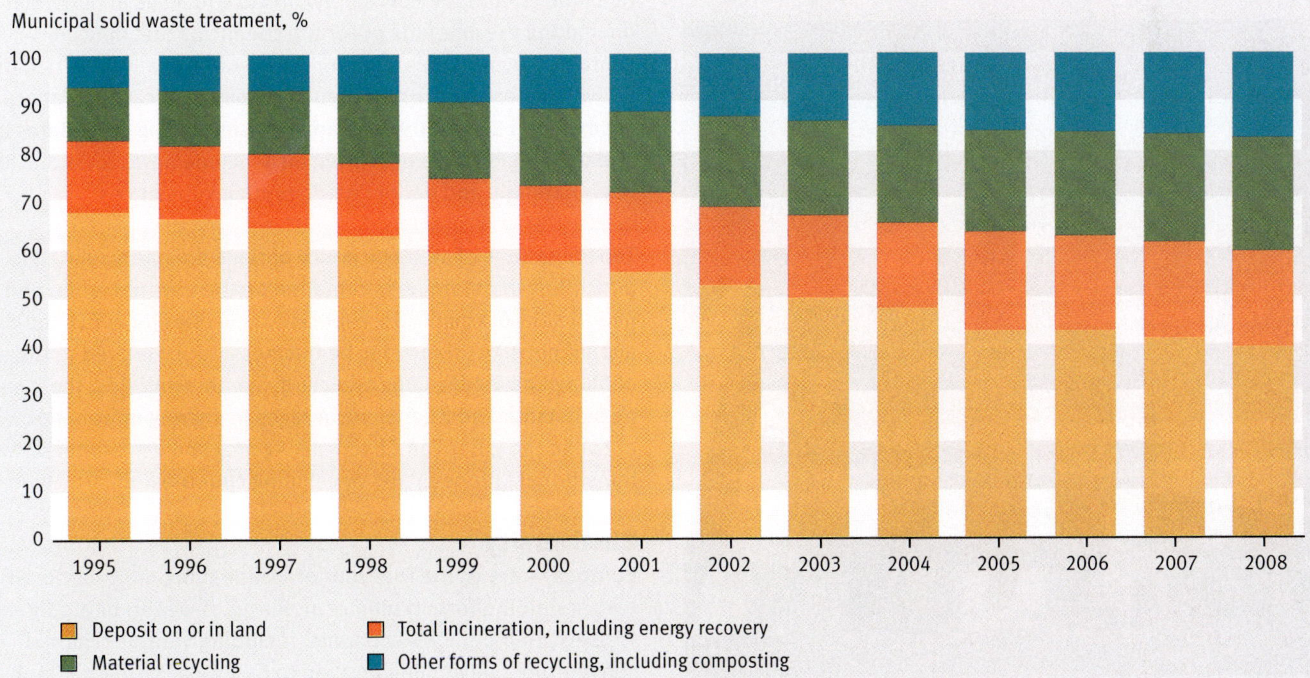

Source: EEA 2011a; Eurostat 2011

amount of waste generated annually is larger than that of the entire EU with a population of 502 million (3.4 billion tonnes and 2.6 billion tonnes respectively) with 90 per cent of waste originating from the mining industry (Eurostat 2011; Devyatkin 2009). On average, however, only about 26 per cent of waste is recycled. Of this recycled waste, 35 per cent is accounted for by industrial waste and only 4–5 per cent by domestic waste. All the other types of waste are effectively not recycled at all (Devyatkin 2009).

A life-cycle approach to waste management could significantly reduce Europe's dependence on imports of raw materials and energy consumption for manufacturing new materials. More significant gains can be made, but only through full implementation of the EU's waste directives and in particular the EU Landfill Directive. Reuse and recycling would also require significant changes in consumer behaviour, which could be helped by information and education campaigns.

Chemicals policy

The most profound and ambitious piece of legislation regulating chemicals in Europe entered into force on 1 June 2007 (EC 2007b). This legislation deals with the Registration, Evaluation, Authorisation and Restriction of Chemical substances (REACH), and replaces a patchwork of previous directives and regulations. The seven objectives that are essential for achieving a sustainable REACH framework are:

- the protection of human health and the environment;
- the maintenance and enhancement of the competitiveness of the EU chemical industry;
- the prevention of fragmentation of the EU's internal market;
- increased transparency;
- the integration with international efforts to regulate the use of chemicals;
- the promotion of non-animal testing; and
- conformity with EU international obligations under the World Trade Organization (WTO) (EC 2007b).

Under the EC regulation REACH, manufacturers are required to gather and register information on the properties of their chemical substances, into a centralized database. *© Carsten Madsen/iStock*

One of the most important elements of REACH is the registration of chemicals. REACH requires companies that make and/or import chemicals to submit registration dossiers to the European Chemicals Agency (ECHA). The 2010 registration deadline was related to bulk chemicals supplied in quantities of more than 1 000 tonnes per year and very hazardous chemicals; by the REACH deadline of 30 November 2010, the agency had received 24 675 registration dossiers for 4 300 substances. Despite significant concerns raised by the chemicals industry about the unprecedented burden REACH placed on companies and some initial technical difficulties, the overall registration process was a success (ECHA 2010). The future deadlines in 2013 and 2018 cover chemicals supplied in smaller quantities (EC 2007b). Additionally, REACH includes some limited provisions for the integrated assessment of cumulative risks from multiple substances and other stressors.

It is expected that implementation and compliance with this legislation will lead to more predictable markets and a reduction in companies' liabilities, especially by providing a level playing field for all market players.

Among new developments is the new EU Toy Safety Directive (2009/48/EC), with Member States expected to have had the new measures under way as of July 2011, and further parts of the directive coming into force in July 2013. Toys come under REACH regulations, and the new safety directive focuses in particular on limiting the amounts of certain chemicals that may be contained in materials used for them. Additionally, in 2013, the EU will implement the new regulation on chemical substances in cosmetics (1223/2009/EF), aimed at simplifying procedures and streamlining terminology. It will also include new provisions for nanomaterials and endocrine-disrupting substances.

The limitations of all these policy options are partly related to difficulties in obtaining information on the environmental and health risks of chemicals, especially new ones for which the risks are unknown. As there may be business issues related to the cost of filling knowledge gaps and clarifying uncertainties, there could be substantial additional advantages in sharing information between the European Chemicals Agency and its counterparts in transitional European and developing countries.

Biodiversity

Europeans are at the forefront of establishing multi-national conservation efforts (Pullin *et al.* 2009). A wealth of biodiversity conservation policies and tools, including various regional conventions, have been applied to European terrestrial and marine ecosystems. At supra-national level, biodiversity conservation is mainly driven by such EU legal instruments as the Nature Directives adopted in 1979 and 1992 (Figure 11.13)

Thanks to improved biodiversity and conservation efforts, the Alpine ibex, once restricted by poaching pressure to Italy's Gran Paradiso National Park, has now recolonized most of the European Alps.

© fotoVoyager/iStock

and the pan-European Biological and Landscape Diversity Strategy adopted at the Third Ministerial Environment for Europe Conference in 1995. Although the EU directives are legally-binding and Pan-European strategy is not, the two are mutually supportive and lead to an improved state of biodiversity in Europe. In 2001, the EU and its Member States committed to halt the loss of biodiversity by 2010 (CBD 2010a), but this target was not met, and the status of biodiversity is still a cause for serious concern (EC 2010d). As a result, a new EU 2020 biodiversity strategy was endorsed in May 2011 (Chapter 5) (EC 2011c; CBD 2010b).

For the purpose of this analysis, three policy clusters were identified as being beneficial in achieving biodiversity conservation goals:

- the establishment of ecological networks as a key means of reducing biodiversity loss (Chapter 5);
- payment for ecosystem services as an instrument for conserving European agro-biodiversity; and
- the sustainable management of forest resources.

Three cases were selected for further appraisal: the EU Natura 2000; agro-environment measures; and the voluntary pan-European Forest Europe process.

Figure 11.13 Sites designated under the Habitats Directive and the Birds Directive, 1995–2009

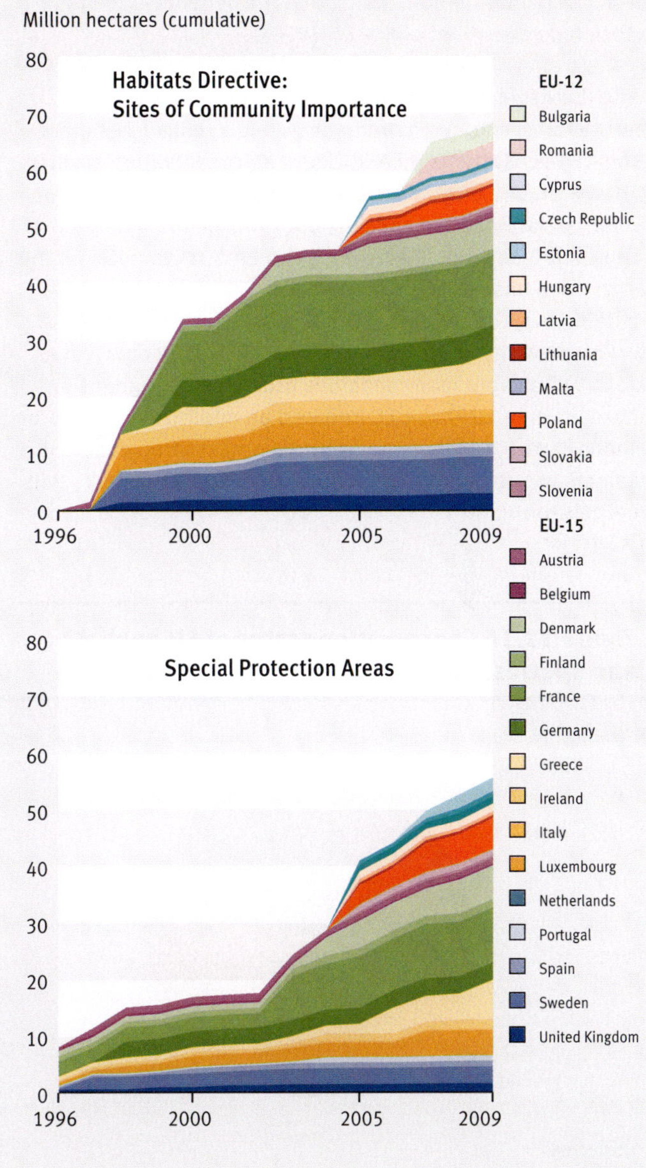

Source: EEA 2010f; ETC/BD *et al.* 2008

The Natura 2000 network

Natura 2000, a tool used by the EU 2020 Biodiversity Strategy, represents the largest supra-national network of protected areas in the world (EEA 2010f). It incorporates sites established under the EU Habitats and Birds Directives and aims to assure the long-term survival of Europe's threatened and most valuable species and habitats (Fock 2011; Watzold *et al.* 2010). It has developed steadily over the last 15 years, and is now made up of more than 26 000 sites covering 18 per cent of the EU's land and sea areas (Figure 11.13) (EC 2010d). Similar network approaches also apply beyond EU borders (Box 11.9).

The Natura 2000 network helps protect vulnerable habitats and species as well as a wide range of ecosystem services, including the regulation of climate (such as mitigation of climate change), purification of water and maintenance of water flows, preservation of landscape and amenity values, and support of tourism and recreation (Gantioler *et al.* 2010; Cliquet *et al.* 2009). Furthermore, it facilitates cooperation beyond national boundaries, contributes to the decentralization of national-level conservation policies, and encourages local and regional economic development by offering job opportunities and helping to attract finance (Ioja *et al.* 2010; Kluvankova-Oravska *et al.* 2009; EC 2008). Even though implementation of the network requires around US$8 billion (€6 billion) annually, there are several examples demonstrating that the benefits exceed the associated costs (Gantioler *et al.* 2010).

While the development of the network has made little headway with marine environments, it is a real success for terrestrial ecosystems (EEA 2010f). The conservation status is, however, still only favourable for less than 20 per cent of terrestrial habitats and species, both within and outside the Natura 2000 network (Figure 11.14) (EEA 2010f). Initially, the designation of sites faced a number of problems, but these are being overcome through the democratization of multi-level biodiversity governance (Beijen 2009; Rauschmayer *et al.* 2009). To avoid many sensitive problems in negotiations, for example, in 1997 the EC initiated an apolitical process for selecting sites in a bio-geographical context through scientific seminars where boundaries were agreed (CEEweb 2011; Papp and Toth 2004).

Figure 11.14 Conservation status of EU habitats and species, 2008

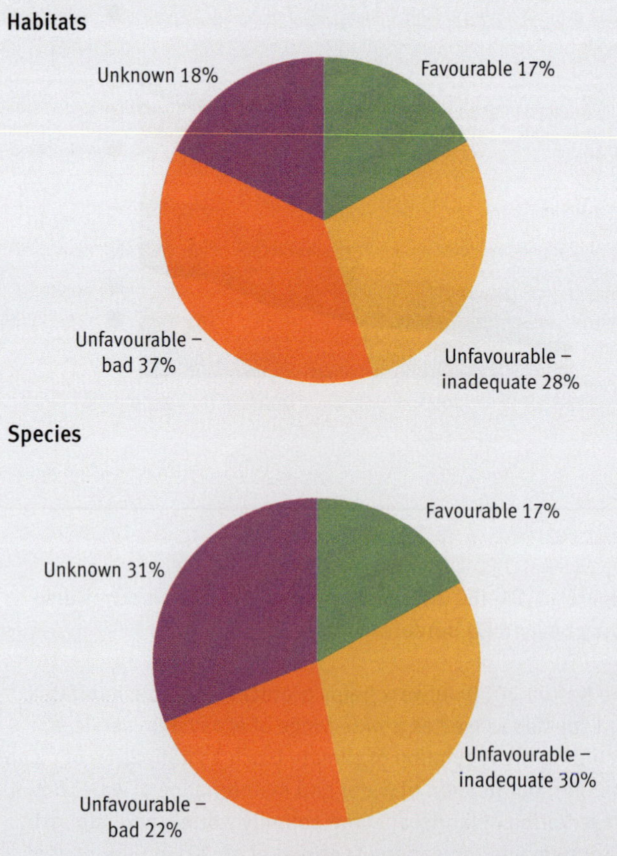

Note: Bulgaria and Romania are not included.

Source: EEA 2010f; ETC/BD *et al.* 2008

Box 11.9 Ukraine's national ecological network

Natura 2000 has been shown to have a significant influence on the development of protected area networks beyond the EU. Similar policies have been adopted both by potential candidate countries and other Central and Eastern European countries (UNEP 2007a). Ukraine, for example, as one of its priority strategic directions for biodiversity conservation, is trying to follow EU policies and has been developing its national ecological network since 2000. Although the creation of this network faces a number of challenges, including a high degree of agricultural expansion and large-scale fragmentation of natural landscapes, it has already resulted in the establishment of transboundary ecological corridors in the Carpathian region. The first corridors were established between 2008 and 2010 as part of a project to realize transboundary ecological connectivity in the Ukrainian Carpathians, linking national parks in Poland, Romania and Ukraine. The establishment of these corridors received full support not only from forest managers and local governments, but also from local communities (Deodatus *et al.* 2010; UNEP 2007a).

Agri-environment measures

The need to preserve high nature-value farmland (Doxa *et al.* 2010; EEA 2009a) in the EU was agreed in 2003 and included in the Kyiv Resolution on Biodiversity (UNECE 2003); it is also highlighted by the EU as a key action to prevent the abandonment or intensification of these lands (EEA 2009a).

Agri-environment measures, an optional policy tool for farmers (Ziolkowska 2009), provide compensation payments covering implementation costs and associated income losses to farmers who commit to preserving the environment and maintaining their farmlands through environmentally friendly practices for at least five years (Box 11.10) (Ziolkowska 2009). Under the EU Common Agricultural Policy (CAP), Member States are obliged to co-finance these measures: between 2007 and 2013, nearly 22 per cent of the expenditure on rural development, some US$27.3 billion (€20 billion), was devoted to them (EC 2010a). Securing financial support and avoiding delay in payments is necessary to ensure farmer commitment (Whittingham 2007; Pinto *et al.* 2005).

In terms of biodiversity conservation, agri-environment measures are at their most successful over large areas (Whittingham 2007), where they also contribute to the maintenance and enhancement of landscapes, protection of the historic environment and of natural resources, and the promotion of public access to the countryside (EEA 2009a). Their high costs, however, may limit their replicability in non-EU European and developing countries. Other limitations to their spread include potential loss of income for farmers and the difficulty predicting their effects on biodiversity (Ziolkowska 2009; Whittingham 2007).

Box 11.10 Conserving high nature-value farmland in Portugal

On the steppe plains of Castro Verde in southern Portugal, the traditional farming system is based on non-irrigated extensive cereal production, with a two- or three-year crop rotation system. These semi-natural mosaic steppe habitats are of value for the conservation of nature, particularly the great bustard (*Otis tarda*). In 1993, a project under EU LIFE – a financial instrument supporting environmental and nature conservation – was implemented to support the preservation of these birds and their habitats by acquiring several farms, leaving the fields fallow, and raising awareness among farmers and landowners. Then, in 1995, an EU agri-environment plan was defined so that the farmers could continue their traditional management practices, rotating crops and maintaining low livestock densities. By 1999 the bird population had improved to such an extent that Castro Verde was included in the Natura 2000 network as a special protected area for birds. Assuring the maintenance and effectiveness of such projects is an essential element in meeting long-term conservation priorities. In this case, however, the agri-environment scheme has not proved popular as delays in subsidy payments led to some farmers withdrawing from the plan (Pinto *et al.* 2005).

Finland is the most forested European country with some 73 per cent of its land covered by forest. *© Samuli Siltanen/iStock*

Forest Europe

Although forests currently (2010) cover 45 and 38 per cent of Europe's and the EU-27's territory respectively, only 26 and 4 per cent of these forests are considered to be undisturbed by humans (Figure 11.15) (Forest Europe *et al.* 2011). Most European forests are heavily exploited and the share of old-growth stands, crucial for forest species, is critically low. Nonetheless, Europe's total forest area is increasing thanks to national policy initiatives coordinated in the Forest Europe framework – a voluntary pan-European policy process for establishing sustainable management of the region's forests.

The Forest Europe process develops common strategies to meet challenges such as climate change and the protection of biodiversity and freshwater, both in Europe and globally (EEA 2010h, 2010a). Since 1990, it has established a collaborative research network on forest ecosystems, a set of pan-European criteria and indicators for sustainable forest management, and a series of action programmes tackling cross-sectoral cooperation and national forest programmes (EEA 2008). Sustainable forest management, as defined by the Ministerial Conference on the Protection of Forests in Europe, has been recognized as a commendable example of the ecosystem approach advocated by the Convention on Biological Diversity (CBD) (EEA 2008).

The benefits of Forest Europe include harmonization of forest policies in European countries that aim to achieve goals for

Figure 11.15 European forest area and status, by region, 2010

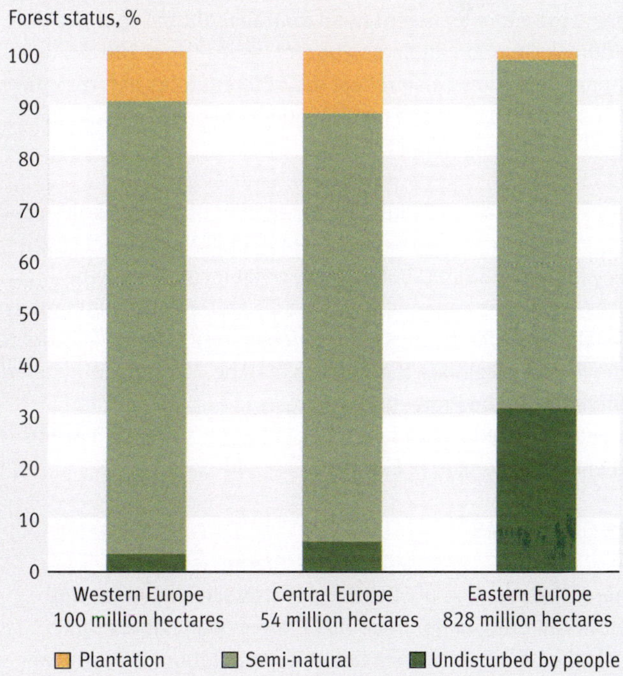

Source: Forest Europe *et al.* 2011

The Forest Europe framework has developed indicators that provide guidance for policy development and assess progress towards sustainable forest management. *© Jens Stolt/iStock*

protecting biodiversity, combating illegal logging and certifying carbon sequestration. Europe has gained 5.1 million hectares of forest since 2005 (Forest Europe *et al.* 2011), and between 2005 and 2010 about 870 million tonnes of CO_2 were removed annually from the atmosphere by photosynthesis and tree biomass growth, about half of it in the EU-27 (Forest Europe *et al.* 2011).

Efforts to enhance the sustainability of forests through management face a lack of national capacity and awareness, and intensifying competition in international forest product markets. There is therefore an urgent need for transnational coordination to address common and cross-border issues (Hogl 2002). National differences also reflect the different roles of forests in various countries and the resulting political need to establish official forest programmes.

The absence of a legally binding agreement on forests at a pan-European level cannot be considered a limitation to successful policy implementation, but at some point it could slow the process down, as common benchmarks and well-defined targets for evaluating effectiveness and efficiency are lacking. In order to improve and accelerate the process, in June 2011 the Ministerial Conference on the Protection of Forests in Europe adopted the Oslo Ministerial Mandate for Negotiating a Legally Binding Agreement on Forests in Europe.

CONCLUSIONS

One of the main conclusions of this chapter is that the coherent application of effective policies across themes and sectors can bring major benefits in terms of an improved physical environment and a healthier population. In European environmental governance, the integration of effective policies under multiple environmental themes and economic sectors is increasingly being taken into account. Even if such policies still have one environmental theme as an entry point, they increasingly cover wide ranges of related aspects. The recent 2009 EU Climate and Energy Package exemplifies such an integrated approach, including binding legislation to achieve three linked targets (the 20-20-20 targets).

It is through such integrated policies that multiple co-benefits can be obtained most cost effectively. Industrial CO_2 emission reduction through emissions trading, for example, will at the same time improve ambient air quality; and promoting renewable energy systems will not only reduce CO_2 emissions, but will also decentralize energy production, potentially improve energy security and provide employment opportunities and economic growth in small and medium-sized companies. Likewise, climate adaptation programmes will increase resilience to climate change effects such as flooding, drought, loss of biodiversity and increased vulnerability to disease, while at the same time improving ambient air quality and reducing greenhouse gas emissions, for example through adjusted agricultural practices, which will also contribute to more sustainable agriculture.

Experience shows that limitations can be overcome if the right enabling regimes are put in place. Common barriers to implementing the policy tools discussed in this chapter are:

- a lack of good data and information to assess impacts and risks and thus support decision making;
- insufficient financial resources from the private and public sectors for dealing with environmental issues both in the EU countries affected by the financial crisis and in non-EU European countries;
- a lack of systematic law enforcement;

- traditional consumption-oriented economic policies that contradict the imperative of more sustainable consumption and hinder the decoupling of human well-being from economic growth;
- intensifying competition in international product markets; and
- increasing egoism, diminished community solidarity and an increasing, though often unjustifiable, sense of community disconnection and insecurity.

Enabling conditions that would increase policy success and replication are:
- more policy coherence, streamlining and simplified procedures that enhance cost efficiency and effectiveness;
- more efficient monitoring systems;
- stronger long-term commitment on the part of politicians and governments;
- stronger enforcement;
- transnational coordination to address common and cross-border issues;
- stronger private-sector involvement by creating and making better use of markets; and
- a more active civil society engaged through awareness raising and strong multi-stakeholder agreements.

Promising emerging innovative policies relevant to the themes covered in this chapter that would help to improve European environmental governance further and would foster scale-up and replication include:

Climate change
- the EU Effort Sharing Decision, which establishes binding emission targets for 2013 –2020 for transport, agriculture, buildings and waste, all sectors that are currently not covered by the EU Emissions Trading System;
- the encouragement of transnational voluntary networks for local action on climate change and air quality, which are actively spreading in Europe and are focusing on more sustainable urban lifestyles, such as Local Governments for Sustainability (ICLEI), Cities for Climate Protection, the Climate Alliance, Energy Cities, CIVITAS and the Aalborg Charter.

Air quality
- devolving responsibility in local air quality management to local administrations, facilitating identification and implementation of policies.

Freshwater
- the EC's expected 2012 Blueprint to safeguard Europe's water resources, which will focus on prevention and preparedness in relation to river basin management, water scarcity and drought, and vulnerability to climate change.

Chemicals and waste
- the proposed mandatory target under the Waste Electrical and Electronic Equipment Directive to recycle 65 per cent of such waste – currently only some 34 per cent is recycled;
- the forthcoming EU regulations on toys to increase protection of the most vulnerable (children);
- the forthcoming EU Cosmetics Directive, to ensure adequate protection from endocrine disruptors and nanomaterials; and

Renewable energies will play a crucial role in the transition to a low-carbon economy; wind alone already accounted for 41 per cent of new power installations in Europe in 2010. *© Mlenny Photography/iStock*

- new measure to address the integrated assessment of cumulative risks from multiple substances and other stressors, filling a main gap in current regulations covering chemicals.

Biodiversity
- the expected adoption of a new Pan-European Biological and Landscape Diversity Strategy 2020, aligned with CBD targets, will reinforce the EU 2020 Biodiversity Strategy;
- the common integrated framework in support of the EU 2020 Biodiversity Strategy, involving a wide range of services and ministries, will create ownership across all relevant policy areas and stakeholders beyond the traditional biodiversity community.

In summary, European examples of regional cooperation on the environment have served as a model for other countries and regions and can potentially serve in the future. Features include current formal institutional structures and the tradition of legislating to improve state and trends in various realms in an integrated way, though as contexts vary adjustments may need to be made in other parts of the world.

Ongoing European attempts strive for continually improving environmental governance, underpinned by strong civil society participation and the recognized right of access to environmental information and justice in environmental matters as laid down in the Aarhus Convention, to date only applied in Europe. These efforts are essential for a proper and robust treatment of the shared environmental space and a healthy future for all.

REFERENCES

AGEE-Stat (2010). *Development of Renewable Energy Sources in Germany 2009: Graphics and Tables.* Version: 15th December 2010. Based on statistical data from the Working Group on Renewable Energies-Statistics (AGEE-Stat). http://www.erneuerbare-energien.de/inhalt/42725/ (accessed 15 December 2011)

Bart, I. (2007). Hungary. In *Allocation in the European Emissions Trading Scheme: Rights, Rents and Fairness* (eds. Ellerman, A.D., Buchner, B.K. and Carraro, C.). pp.246–269. Cambridge University Press, Cambridge

Bechberger, M. (2009). *Renewable Energy in Spain: Conditions for Success and Limitations* (in German). Ibidem-Verlag, Stuttgart

Beijen, B. (2009). The implementation of area protection provisions from European environmental directives in the Member States. *Utrecht Law Review* 5, 101–116

Blanco, M.I. and Rodrigues, G. (2008). Can the future EU ETS support wind energy investments? *Energy Policy* 36, 1509–1520

Blumberg, K., Walsh, M. and Pera, C. (2004). *Low-sulfur Gasoline and Diesel: The Key to Lower Vehicle Emissions.* http://www.unep.org/transport/pcfv/PDF/PubLowSulfurPaper.pdf (accessed 25 May 2011)

Burman, L. and Johansson, C. (2010). *The Effects of the Congestion Tax on Emissions and Air Quality.* SLB-analysis. Stockholm Environment and Health Administration, Stockholm. http://slb.nu/slb/rapporter/pdf8/slb2010_006.pdf (accessed 28 October 2011)

Burman, L. and Johansson, C. (2001). *Stockholm's Low Emissions Zone – Effects on Air Quality in 2000* (in Swedish). SLB-analysis. Stockholm Environment and Health Administration, Stockholm. http://slb.nu/slb/rapporter/pdf8/slb2001_004.pdf (accessed 20 September 2011)

Busch, P.-O. (2003). *The Diffusion of Fixed Feed-in Tariffs and Quotas: Competition of Models in Europe.* FFU-report 03–2003 (in German). Environmental Policy Research Centre, Berlin

CAI-Asia (2011). *Roadmap to Cleaner Fuels and Vehicles in Asia.* CAI-Asia Factsheet No.17. Clean Air Initiative for Asian Cities, Manila

CBD (2010a). *Case Studies Illustrating the Socio-economic Benefits of Ecological Networks.* Secretariat of the Convention on Biological Diversity, Montreal

CBD (2010b). *Strategic Plan for Biodiversity 2011–2020 and the Aichi Targets.* Secretariat of the Convention on Biological Diversity, Montreal

CBD (1992). *Convention on Biological Diversity.* http://www.cbd.int

CEEweb (2011). *What is the New Biogeographic Process?* http://www.ceeweb.org/workingareas/policies/biogeo.htm (accessed 18 September 2011)

CG (2011). *Post-Cancun Analysis.* Policy briefing, January 17, 2011. The Climate Group, London. http://www.theclimategroup.org/_assets/files/Post-Cancun-Analysis_1.pdf (accessed 20 December 2011)

Chesnutt, T.W., Beecher, J.A., Mann, P.C., Clark, D.M., Hanemann, W.M., Raftelis, G.A., McSpadden, C.N., Pekelney, D.M., Christianson, J. and Krop, R. (1997). *Designing, Evaluating, and Implementing Conservation Rate Structures: A Handbook.* California Urban Water Conservation Council, A&N Technical Services, Inc., Santa Monica

Cliquet, A., Backes, C., Harris, J. and Howsam, P. (2009). Adaptation to climate change: legal challenges for protected areas. *Utrecht Law Review* 5, 158–175

COE (2000). *European Landscape Convention.* European Treaty Series No.176. Council of Europe, Strasbourg

DEFRA (2002). *Directing the Flow. Priorities for Future Water Policy.* Department for Environment, Food and Rural Affairs, London

Del Rio Gonzalez, P. (2008). Ten years of renewable electricity policies in Spain: an analysis of successive feed-in tariff reforms. *Energy Policy* 36, 2917–2929

Deodatus, F., Protsenko, L. and Bashta, A. (2010). Introduction. In *Creation of Ecological Corridors in Ukraine. A Manual on Stakeholder Involvement and Landscape-ecological Modelling to Connect Protected Areas, Based on a Pilot in the Carpathians* (eds. Deodatus, F. and Protsenko, L.). pp.11–18. State Agency for Protected Areas of the Ministry of Environmental Protection of Ukraine, Altenburg and Wymenga Ecological Consultants, InterEcoCentre, Kiev

Devyatkin, V. (2009). *Actual Ways of Improving Legislation of Russian Federation Towards Recycling of Industrial Wastes and Other Industrial Outputs.* Report to the Federation Committee of the Russian Parliament on Industrial Policy, 19.02.2009 (in Russian). Federal government-financed agency 'Research Center on resources efficiency and wastes management issues', Moscow

Doxa, A., Bas, Y., Paracchini, M.L., Pointereau, P., Terres, J.-M. and Jiguet, F. (2010). Low-intensity agriculture increases farmland bird abundances in France. *Journal of Applied Ecology* 47, 1348–1356

EC (2011a). *2012 Blueprint to Safeguard Europe's Water Resources.* European Commission, Luxembourg. http://ec.europa.eu/environment/water/pdf/blueprint_leaflet.pdf (accessed 20 December 2011)

EC (2011b). *A Resource-efficient Europe – Flagship Initiative under the Europe 2020 Strategy. Communication from the Commission to the European Parliament, the Council, the European Economic and Social Committee and the Committee of the Regions.* European Commission, Brussels

EC (2011c). *Our Life Insurance, Our Natural Capital: An EU Biodiversity Strategy to 2020.* 3.5.2011 COM(2011) 244 final. European Commission, Brussels. http://ec.europa.eu/environment/nature/biodiversity/comm2006/pdf/2020/1_EN_ACT_part1_v7%5b1%5d.pdf (accessed 15 September 2011)

EC (2010a). *Agriculture and Rural Development: Agri-environmental Measures.* European Commission, Brussels. http://ec.europa.eu/agriculture/envir/measures/index_en.htm (accessed 20 September 2011)

EC (2010b). *Being Wise with Waste: The EU's Approach to Waste Management.* European Commission, Luxembourg

EC (2010c). *EU Climate Change Impacts, Vulnerability and Adaptation Clearinghouse Concept Note and Minimum Requirements for Phase 1.* European Commission, Brussels. http://ec.europa.eu/clima/tenders/2011/208209/clearinghouse_concept_note_en.pdf (accessed 15 December 2011)

EC (2010d). *Nature. Monitoring the Impact of EU Biodiversity Policy.* European Commission, Luxembourg. http://ec.europa.eu/environment/pubs/pdf/factsheets/biodiversity_fsh.pdf (accessed 20 December 2011)

EC (2009a). *The EU Climate and Energy Package.* European Commission, Brussels. http://ec.europa.eu/clima/policies/package/index_en.htm (accessed 15 December 2011)

EC (2009b). *White Paper: Adapting to Climate Change – Towards a European Framework for Action.* COM(2009) 147 final. European Commission, Brussels. http://eur-lex.europa.eu/LexUriServ/LexUriServ.do?uri=COM:2009:0147:FIN:EN:PDF (accessed 20 December 2011)

EC (2008). *The Economics of Ecosystems and Biodiversity (TEEB): An Interim Report.* European Commission, Brussels. http://ec.europa.eu/environment/nature/biodiversity/economics/pdf/teeb_report.pdf (accessed 20 December 2011)

EC (2007a). *Green Paper: Towards a New Culture for Urban Mobility.* COM (2007) 551 Final. European Commission, Brussels

EC (2007b). *REACH in Brief.* European Commission, Brussels. http://ec.europa.eu/environment/chemicals/reach/pdf/2007_02_reach_in_brief.pdf (accessed 26 October 2011)

EC (2005). *Thematic Strategy on Air Pollution. Communication from the Commission to the Council and the European Parliament.* COM (2005) 446 final. European Commission, Brussels

EC (2004). *Proposal for a Directive of the European Parliament and of the Council on Reinsurance and Amending Council Directives 73/239/EEC, 92/49/EEC and Directives 98/78/EC and 2002/83/EC.* European Commission, Brussels. http://ec.europa.eu/internal_market/insurance/docs/reinsurance/directive/com-2004_273-final-en.pdf (accessed 15 December 2011)

ECHA (2010). *The Outcome of the First REACH Registration Deadline.* Press memo. European Chemicals Agency, Helsinki. http://echa.europa.eu/doc/press/press_memo_20101201_en.pdf (accessed 12 December 2011)

ECouncil (2010). *Directive 2010/75/EU of the European Parliament and of the Council of 24 November 2010 on Industrial Emissions (Integrated Pollution Prevention and Control).* European Council, Brussels

ECouncil (2008a). *Directive 2008/50/EC of the European Parliament and of the Council of 21 May 2008 on Ambient Air Quality and Cleaner Air for Europe.* European Council, Brussels

ECouncil (2008b). *Directive 2008/98/EC of the European Parliament and of the Council of 19 November 2008 on Waste.* European Council, Brussels. http://eur-lex.europa.eu/LexUriServ/LexUriServ.do?uri=OJ:L:2008:312:0003:0030:en:PDF (accessed 20 December 2011)

ECouncil (2007). *Directive 2007/60/EC of the European Parliament and of the Council of 23 October 2007 on the Assessment and Management of Flood Risks.* European Council, Brussels

ECouncil (2002a). *Directive 2002/95/EC on the Restriction of the Use of Certain Hazardous Substances in Electrical and Electronic Equipment.* European Council, Brussels. http://europa.eu/legislation_summaries/environment/waste_management/l21210_en.htm (accessed 12 December 2011)

ECouncil (2002b). *Directive 2002/96/EC on Waste Electrical and Electronic Equipment.* European Council, Brussels. http://europa.eu/legislation_summaries/environment/waste_management/l21210_en.htm (accessed 12 December 2011)

ECouncil (2000). *Directive 2000/60/EC of the European Parliament and of the Council of 23 October 2000 Establishing a Framework for Community Action in the Field of Water Policy.* European Council, Brussels

ECouncil (1999). *Council Directive 1999/32/EC of 26 April 1999 Relating to a Reduction in the Sulphur Content of Certain Liquid Fuels and Amending Directive 93/12/EEC.* European Council, Brussels

ECouncil (1998). *Directive 98/70/EC of the European Parliament and of the Council of 13 October 1998 Relating to the Quality of Petrol and Diesel Fuels and Amending Council Directive 93/12/EEC.* European Council, Brussels

EEA (2011a). *Greenhouse Gas Emissions in Europe: A Retrospective Trend Analysis for the Period 1990–2008.* EEA Report No 6/2011. European Environment Agency, Copenhagen

EEA (2011b). *Greenhouse Gas Emission Trends and Projections in Europe 2011. Tracking Progress Towards Kyoto and 2020 targets.* EEA Report No 4/2011. European Environment Agency, Copenhagen

EEA (2011c). *Waste Opportunities. Past and Future Climate Benefits from Better Municipal Waste Management in Europe.* EEA Report No 3/2011. European Environment Agency, Copenhagen

EEA (2010a). *10 Messages for 2010. Forest Ecosystems.* European Environment Agency, Copenhagen

EEA (2010b). *Allocation of Passenger Cars and Light-duty Trucks to the Various Emission Standards. Maps and Graphs.* European Environment Agency, Copenhagen. http://www.eea.europa.eu/data-and-maps/figures/allocation-of-passenger-cars-and (accessed 15 December 2011)

EEA (2010c). *Heavy Metal (HM) Emissions (APE 005) (APE 005).* Assessment published Oct 2010. European Environment Agency, Copenhagen. http://www.eea.europa.eu/data-and-maps/indicators/eea32-heavy-metal-hm-emissions-1/assessment (accessed 15 December 2011)

EEA (2010d). *Impact of Selected Policy Measures on Europe's Air Quality.* Technical Report No 8/2010. European Environment Agency, Copenhagen

EEA (2010e). *Sulphur Dioxide SO_2 Emissions (APE 001) (APE 001).* Assessment published October 2010. European Environment Agency, Copenhagen. http://www.eea.europa.eu/data-and-maps/indicators/eea-32-sulphur-dioxide-so2-emissions-1/assessment (accessed 23 March 2011)

EEA (2010f). *The EU 2010 Biodiversity Baseline.* European Environment Agency, Copenhagen

EEA (2010g). *The European Environment: State and Outlook 2010. Material Resources and Waste.* European Environment Agency, Copenhagen

EEA (2010h). *The European Environment: State and Outlook 2010. Synthesis.* European Environment Agency, Copenhagen

EEA (2010i). *The European Environment: State and Outlook. Air Pollution.* European Environment Agency, Copenhagen

EEA (2010j). *Tracking Progress Towards Kyoto and 2020s Targets in Europe.* EEA Report No 7/2010. European Environment Agency, Copenhagen

EEA (2009a). *Distribution and Targeting of the CAP Budget from a Biodiversity Perspective.* EEA Technical Report No 12. European Environment Agency, Copenhagen

EEA (2009b). *Water Resources Across Europe – Confronting Water Scarcity and Drought.* EEA Report No 2/2009. European Environment Agency, Copenhagen

EEA (2008). *European Forests – Ecosystem Conditions and Sustainable Use.* European Environment Agency, Copenhagen

EEG (2009). *Act Revising the Legislation on Renewable Energy Sources in the Electricity Sector and Amending Related Provisions.* Renewable Energy Sources Act (EEG), Bonn. http://www.bmu.de/english/renewable_energy/downloads/doc/42934.php (accessed 20 December 2011)

Eliasson, J. (2009). A cost-benefit analysis of the Stockholm congestion charging system. *Transportation Research Part A: Policy and Practice* 43, 468–480

Ellerman, A.D. (2008). *The EU Emission Trading Scheme: A Prototype Global System?* Discussion Paper 2008-02. Harvard Project on International Climate Agreements, Cambridge, MA

Ellerman, A.D. and Buchner, B.K. (2007). The European Union emissions trading scheme: origins, allocation, and early results. *Review of Environmental Economics and Policy* 1, 66–87

ETC/BD, EEA and EC-DGENV (2008). *Conservation Status of Habitat Types and Species (Article 17, Habitats Directive 92/43/EEC.* European Topic Centre on Nature Protection and Biodiversity, European Environment Agency and European Commission Directorate-General for Environment. http://www.eea.europa.eu/data-and-maps/data/article-17-database-habitats-directive-92-43-eec (accessed 20 December 2011)

ETC/SCP (2010). *Europe as a Recycling Society. The European Recycling Map.* ETC/SCP working paper 5/2010. European Topic Centre on Sustainable Consumption and Production, Copenhagen. http://eea.eionet.europa.eu/Public/irc/eionetcircle/etc_waste/library?l=/european_recycling/200810_etc-scp-/_EN_1.0_&a=d (accessed 20 December 2011)

Eurostat (2011). *Statistics: Environment and Energy.* http://epp.eurostat.ec.europa.eu/portal/page/portal/statistics/search_database (accessed 12 December 2011)

FAO (2010). *FAO Statistical Databases.* Food and Agriculture Organization of the United Nations, Rome. http://faostat.org (accessed 15 December 2011)

Fock, H. (2011). Natura 2000 and the European Common Fisheries Policy. *Marine Policy* 35, 181–188

Forest Europe, UNECE and FAO (2011). *State of Europe's Forests 2011. Status and Trends in Sustainable Forest Management in Europe.* Ministerial Conference on the Protection of Forests in Europe, Oslo

Gantioler, S., Rayment, M., Bassi, S., Kettunen, M., McConville, A., Landgrebe, R., Gerdes, H. and ten Brink, P. (2010). *Costs and Socio-economic Benefits Associated with the Natura 2000 Network. Final Report to the European Commission, DG Environment on contract ENV.B.2/SER/2008/0038.* Institute for European Environmental Policy, GHK/Ecologic, Brussels

GEO Data Portal. UNEP's online core database with national, sub-regional, regional and global statistics and maps, covering environmental and socio-economic data and indicators. United Nations Environment Programme, Geneva. http://geodata.grid.unep.ch (accessed 15 December 2011)

Hey, C. (2004). EU environmental policies: a short history of the policy strategies. In *EU Environmental Policy Handbook* (ed. Scheuer, S.). European Environmental Bureau, Brussels

Hogl, K. (2002). Patterns of multi-level co-ordination for NFP-processes: learning from problems and success stories of European policy-making. *Forest Policy and Economics* 4, 301–312

ICCT (2007). *Passenger Vehicle Greenhouse Gas and Fuel Economy Standards: A Global Update.* International Council on Clean Transportation, Washington, DC. http://www.theicct.org/sites/default/files/publications/PV_standards_2007.pdf (accessed 15 December 2011)

Inman, D. and Jeffrey, P. (2006). A review of residential water conservation tool performance and influences on implementation effectiveness. *Urban Water Journal* 3, 127–143

Ioja, C., Patroescu, M., Rozylowicz, L., Popescu, V., Verghelet, M., Zotta, M. and Felciuc, M. (2010). The efficacy of Romania's protected areas network in conserving biodiversity. *Biological Conservation* 143, 2468–2476

IPA CIS (2011). *The Inter-Parliamentary Assembly of the Commonwealth of Independent States* (in Russian). http://www.iacis.ru/html/?id=22&str=kom&nid=22 (accessed 9 September 2011)

Jacobsen, B.H. (2004). *Final Economic Evaluation of the Action Plan for the Aquatic Environment II.* Report No.169 (in Danish, with English summary). Danish Research Institute of Food Economics, Copenhagen. http://www.vmp3.dk/Files/Filer/Slutrapporter/Rapport_nr_169.pdf (accessed 15 December 2011)

Jänicke, M. (2011). *The Acceleration of Innovation in Climate Policy. Lessons from Best Practice.* FFU Report. Freie Universität Berlin, Berlin

Klessmann, C., Nabe, C. and Burges, K. (2008). Pros and cons of exposing renewables to electricity market risks – a comparison of the market integration approaches in Germany, Spain, and the UK. *Energy Policy* 36, 3646–3661

Klimont, Z., Cofala, J., Xing, J., Wei, W., Zhang, C., Wang, S., Kejun, J., Bhandari, P., Mathur, R., Purohit, P., Rafaj, P., Chambers, A., Amann, M. and Hao, J. (2009). Projections of SO_2, NO_x and carbonaceous aerosols emissions in Asia. *Tellus B* 61, 602–617

Kluvankova-Oravska, T., Chobotova, V., Banaszak, I., Slavikova, L. and Trifunovova, S. (2009). From government to governance for biodiversity: the perspective of central and Eastern European transition countries. *Environmental Policy and Governance* 19, 186–196

Kossoy, A. and Ambrosi, P. (2010). *State and Trend of the Carbon Market.* The World Bank, Washington, DC

LEZ (2011). Low Emission Zones in Europe website. http://lowemissionzones.eu (accessed 27 May 2011)

Lovei, M. (1998). *Phasing Out Lead from Gasoline. Worldwide Experience and Policy Implications.* World Bank Technical Paper No. 397. World Bank, Washington, DC

Melikyan, L. (2003). Economic and social aspects of reforming water resource management: case of Armenia. In *Drop by Drop: Water Management in the South Caucasus and Central Asia* (ed. O'Hara, S.). pp.29–81. Local Government and Public Service Reform Initiative, Open Society Institute-Budapest, Budapest

Mendonça, M., Jacobs, D. and Sovacool, B. (2009). *Powering the Green Economy. The Feed-in Tariff Handbook.* Earthscan, London

Morris, D. and Worthington, B. (2010). *Cap or Trap? How the EU ETS Risks Locking-in Carbon Emissions.* Sandbag, London. http://sandbag.org.uk/files/sandbag.org.uk/caportrap.pdf (accessed 20 December 2011)

Nations Online (2011). *Official and Spoken Languages of European Countries.* http://www.nationsonline.org/oneworld/european_languages.htm (accessed 19 September 2011)

OECD (2009). *Managing Water for All. An OECD Perspective on Pricing and Financing – Key Messages for Policy Makers.* Organisation for Economic Co-operation and Development, Paris

OECD (2008). *Promoting the Use of Performance-based Contracts Between Water Utilities and Municipalities in EECCA. Case Study No. 2: Armenian Water and Wastewater Company. SAUR Management Contract.* Organisation for Economic Co-operation and Development, Paris. http://www.oecd.org/dataoecd/25/20/40572630.pdf (accessed 15 December 2011)

OECD (2007a). *Financing Water Supply and Sanitation Sector in EECCA Countries, Including Progress in Achieving Water-related Millennium Development Goals (MDGs).* Organisation for Economic Co-operation and Development, Paris. http://www.oecd.org/dataoecd/13/59/39116764.pdf (accessed 15 December 2011)

OECD (2007b). *Policies for a Better Environment: Progress in Eastern Europe, Caucasus and Central Asia. Summary for Policymakers.* Organisation for Economic Co-operation and Development, Paris. http://www.oecd.org/dataoecd/33/27/39271802.pdf (accessed 19 September 2011)

Papp, D. and Toth, C. (2004). *Natura 2000 Site Designation Process with a Special Focus on the Biogeographic Seminars.* CEEweb, Budapest

PCFV (2011a). *Diesel Fuel Sulphur Levels: Global Status, May 2011.* Partnership for Clean Fuels and Vehicles, United Nations Environment Programme, Nairobi. http://www.unep.org/transport/pcfv/PDF/MapWorldSulphur-MAY2011.pdf (accessed 25 May 2011)

PCFV (2011b). *Latin America and the Caribbean. Passenger Vehicle Standards and Fleets.* Partnership for Clean Fuels and Vehicles, United Nations Environment Programme, Nairobi. http://www.unep.org/transport/pcfv/PDF/Maps_Matrices/LAC/matrix/LAC_vehiclestandardsmatrix_august2011.pdf (accessed 19 September 2011)

PCFV (2011c). *Leaded Petrol Phase-out: Global Status, January 2011.* Partnership for Clean Fuels and Vehicles, United Nations Environment Programme, Nairobi. http://www.unep.org/transport/pcfv/regions/global.asp (accessed 25 May 2011)

PCFV (2007). *Opening the Door for Cleaner Vehicles in Developing and Transition Countries: The Role of Lower Sulphur Fuels.* Report of the Sulphur Working Group of the Partnership for Clean Fuels and Vehicles. United Nations Environment Programme, Nairobi. http://www.unep.org/transport/pcfv/PDF/SulphurReport.pdf (accessed 25 May 2011)

Petersen, J. and Knudsen, L. (2010). Accounting nutrients in animal manure. In *Treatment and Use of Organic Residues in Agriculture: Challenges and Opportunities Towards Sustainable Management.* Proceedings of the 14th Ramiran International Conference of the FAO ESCORENA Network on Recycling of Agricultural, Municipal and Industrial Residues in Agriculture, Lisbon, Portugal, 13–15 September 2010. http://www.ramiran.net/ramiran2010/start.html (accessed 15 December 2011)

Pinto, M., Rocha, P. and Moreira, F. (2005). Long-term trends in great bustard (*Otis tarda*) populations in Portugal suggest concentration in single high quality area. *Biological Conservation* 124, 415–423

Planet Arc (2011). EU, Australia to discuss linking carbon trading schemes. http://planetark.org/wen/63170 (accessed 15 December 2011)

Pullin, A., Baldi, A., Can, O.E., Dieterich, M., Kati, V., Livoreil, B., Lovei, G., Mihok, B., Nevin, O., Selva, N. and Sousa-Pinto, I. (2009). Conservation focus on Europe: major conservation policy issues that need to be informed by conservation science. *Conservation Biology* 23, 818–824

Ragwitz, M., Winkler J., Klessmann, C., Gephart, M. and Resch, G. (2012). *Recent Developments of Feed-in Systems in the EU – A Research Paper for the International Feed-In Cooperation.* A report commissioned by the Ministry for the Environment, Nature Conservation and Nuclear Safety (BMU), Bonn

Rauschmayer, F., Berghöfer, A., Omann, I. and Zikos, D. (2009). Examining processes or/and outcomes? Evaluation concepts in European governance of natural resources. *Environmental Policy and Governance* 19, 159–173

REN21 (2010). *Renewables 2010. Global Status Report.* REN21 Secretariat, Paris. http://www.ren21.net/Portals/97/documents/GSR/REN21_GSR_2010 (accessed 15 December 2011)

Scheuer, S. (2005). Water. In *EU Environmental Policy Handbook: A Critical Analysis of EU Environmental Legislation* (ed. Scheuer, S.). pp.125–156. European Environmental Bureau, Brussels

Sendzimir, J., Magnuszewski, P., Flachner, Z., Balogh, P., Molnar, G., Sarvari, A. and Nagy, Z. (2008). Assessing the resilience of a river management regime: informal learning in a shadow network in the Tisza river basin. *Ecology and Society* 13, 1–25

Shinn, M. (2005). Waste. In *EU Environmental Policy Handbook. A Critical Analysis of EU Environmental Legislation* (ed. Scheuer, S.). pp.77–124. European Environmental Bureau, Brussels

Sills, B. and Roca, M. (2010). Spain nearing accord with solar producers on reducing subsidies. Bloomberg, 30 July 2010. http://www.google.com/search?rlz=1C1SVEC_enTJ393TJ394&aq=f&sourceid=chrome&ie=UTF-8&q=Spain+Nearing+Accord+With+Solar+Producers+on+Reducing+Subsidies (accessed 15 December 2011)

UN (2000). *Millennium Development Goals.* United Nations http://www.un.org/millenniumgoals/

UNCED (1992). *Agenda 21.* United Nations Conference on Environment and Development. http://www.un.org/esa/sustdev/documents/agenda21/english/Agenda21.pdf

UNDESA (2010). *World Population Prospects, the 2010 Revision (WPP2010).* Population Division, United Nations Department of Economic and Social Affairs, New York. http://esa.un.org/wpp/unpp/panel_population.htm (accessed 15 December 2011)

UNDP/GEF (2011). *Development and Endorsement of an International River Basin Management Plan.* International Waters Experience Notes. United Nations Development Programme/Global Environment Facility. http://www.icpdr.org/icpdr-files/15503 (accessed 15 December 2011)

UNECE (2012). *Member States and Member States Representatives.* United Nations Economic Commission for Europe, Geneva. http://www.unece.org/oes/nutshell/member_states_representatives.html

UNECE (2011a). *Astana Water Action.* ECE/ASTANA.CONF/2011/5. United Nations Economic Commission for Europe, Geneva. http://www.unece.org/fileadmin/DAM/env/documents/2011/ece/ece.astana.conf.2011.5.e.pdf (accessed 21 December 2011)

UNECE (2011b). *Sustainable Management of Water and Water-related Ecosystems.* ECE/ASTANA.CONF/2011/5. United Nations Economic Commission for Europe, Geneva. http://www.unece.org/fileadmin/DAM/env/documents/2011/ece/ece.astana.conf.2011.3.e.pdf (accessed 21 December 2011)

UNECE (2010). *Hemispheric Transport of Air Pollution. Part A: Ozone and Particulate Matter* (eds. Dentener, F., Keating, T. and Akimoto, H.). Air Pollution Studies No.17. United Nations Economic Commission for Europe, Geneva

UNECE (2003). *Kyiv Resolution of Biodiversity.* Fifth Ministerial Conference Environment for Europe, ECE/CEP/108. United Nations Economic Commission for Europe, Geneva. http://www.unep.ch/roe/documents/biodiv/kiev_conference/documents/biodiv_resolution_e.pdf (accessed 23 February 2012)

UNECE (1999). *Protocol to the 1979 Convention on Long-Range Transboundary Air Pollution to Abate Acidification and Ground-level Ozone.* (The Gothenburg Protocol.) http://www.unece.org/fileadmin/DAM/env/lrtap/full%20text/1999%20Multi.E.Amended.2005.pdf

UNEP (2007a). *Carpathians Environment Outlook 2007.* United Nations Environment Programme Division of Early Warning and Assessment (DEWA)/GRID-Geneva

UNEP (2007b). *Global Environment Outlook 4: Environment for Development.* United Nations Environment Programme, Nairobi

UNEP Risoe Centre (2010). *CDM/JI Pipeline Analysis and Database.* August 2010. United Nations Environment Programme Risoe Centre on Energy, Climate and Sustainable Development http://cdmpipeline.org/ (accessed 15 December 2011)

UNFCCC (2011). *Clean Development Mechanism Methodology Booklet. November 2011.* United Nations Framework Convention on Climate Change, Bonn. https://cdm.unfccc.int/methodologies (accessed 20 December 2011)

UNFCCC (2009). *Report of the Ad Hoc Working Group on Long-term Cooperative Action under the Convention on its Seventh Session.* FCCC/AWGLCA/2009/14, para. 7(g), p.128 and Annex VI. United Nations Framework Convention on Climate Change, Bonn. http://unfccc.int/resource/docs/2009/awglca7/eng/14.pdf (accessed 15 December 2011)

UNFCCC (1992). *United Nations Framework Convention on Climate Change.* FCCC/INFORMAL/84 GE.05-62220 (E) 200705. http://unfccc.int/resource/docs/convkp/conveng.pdf

Vestreng, V., Myhre, G., Fagerli, H., Reis, S. and Tarrasón, L. (2007). Twenty-five years of continuous sulphur dioxide emission reduction in Europe. *Atmospheric Chemistry and Physics* 7, 3663–3681

Warner, K. and Spiegel, A. (2009). Climate change and emerging markets: the role of insurance industry in climate risk management. In *The Geneva Reports – Risk and Insurance Research #2. The Insurance Industry and Climate Change – Contribution to the Global Debate* (ed. Liedtke, P.M.). pp.83–94. The International Association for the Study of Insurance Economics, Geneva

Watzold, F., Mewes, M., Apeldoorn, R., Varjopuro, R., Chmielewski, T. J., Veeneklaas, F. and Kosola, M. (2010). Cost-effectiveness of managing Natura 2000 sites: an exploratory study for Finland, Germany, the Netherlands and Poland. *Biodiversity and Conservation* 19, 2053–2069

Weidner, H. and Mez, L. (2008). German climate change policy. A success story with some flaws. *Journal of Environment and Development* 17, 356–378

Whittingham (2007). Will agri-environment schemes deliver substantial biodiversity gain, and if not why not? *Journal of Applied Ecology* 44, 1–5

WSSD (2002). *Johannesburg Plan of Implementation.* World Summit on Sustainable Development. http://www.un.org/esa/sustdev/documents/WSSD_POI_PD/English/POIToc.htm

Ziolkowska, J. (2009). Environmental benefit, side effects and objective-oriented financing of agri-environmental measures: case study of Poland. *International Journal of Economic Sciences and Applied Research* 2, 71–88

CHAPTER 12

Latin America and the Caribbean

Coordinating lead authors: Keisha Garcia and Joanna Kamiche Zegarra

Lead authors: Ligia Castro, Arturo Flores Martínez, Daniel Fontana Oberling, Elsa Galarza, Alexander Girvan, Ernesto Guhl Nannetti, Gladys Hernández, Paul Hinds, Martha Macedo de Lima Barata, Ana Rosa Moreno, Rodrigo Noriega, Maurice Rawlins (GEO Fellow) and Ernesto Viglizzo

Contributing authors: Dolors Armenteras, Andrea Brusco, Guillermo Castro Herrera, Antonio Clemente (GEO Fellow), Keston Finch, Silvia Giada, Mayte González, Mark Griffith, Martin Obermaier, Mary Otto-Chang, Graciela Metternicht, Keith Nichols, Aida Pacheco, Andrea Salinas, Asha Singh, Michael Taylor, Elisa Tonda, Angel Ureña, Oscar Vallarino, William Wills and Jessica Young

Principal scientific reviewer: Holm Tiessen

Chapter coordinator: Graciela Metternicht

Main Messages

Despite their heterogeneity, countries in Latin America and the Caribbean share a number of common environmental challenges. These include climate change, biodiversity loss and concerns over water and land management. Coastal and marine issues, urbanization, poverty and inequity are also of high priority.

The policies and instruments showcased in this chapter require sound environmental governance to ensure their effectiveness. Strong institutional settings and policy frameworks are the foundation for this, while public participation, monitoring and evaluation, education, and a culture of environmental awareness are fundamental for its efficient functioning.

Policies in the region can only be effective if they succeed in bridging the gap between science and policy making. Robust policies rest on evidence-based research designed to meet the needs of policy makers. Such research should include, where relevant, local and indigenous knowledge, which is an important feature of the region. Researchers and policy makers need to collaborate to acquire the relevant information, knowledge and innovation for environmental decision making.

To be sustainable, the region's natural capital needs to be managed in an integrated fashion across sectors. To respond to the complex nature of the region's environment and its opportunities and challenges, policies should be designed and implemented in ways that transcend the traditional compartmentalized, sectoral approach. This will help the region deal with some of its persistent environmental and associated socio-economic problems, including poverty, inequity and social conflict.

The Latin America and Caribbean region has developed and implemented good examples of transformative policies and approaches. These are usually at national and sub-national levels and offer opportunities for replication both within and outside the region. Their characteristics usually include the effective incorporation of scientific information, knowledge and best practice, links across sectors, and strong governance mechanisms, stakeholder participation, and political will and support.

INTRODUCTION

The 33 countries of Latin America and the Caribbean vary significantly in size and economic development. The region includes both Brazil, the seventh largest economy in the world (*The Economist* 2011) and small island developing states, with their open and fragile economies (Rietbergen *et al.* 2007). Rich in natural resources, the region is home to approximately 23 per cent of the world's forests, 31 per cent of its freshwater resources and six of the world's 17 mega-diverse countries. Although these resources are not evenly distributed, the overall richness and economic importance of the region's ecosystems and its natural capital are undeniable (UNEP 2010b).

Latin American and Caribbean countries face many challenges in managing their rich natural resources. Population growth, as well as unsustainable global and regional production and consumption patterns, drive the increasing demand for, and extraction of raw materials and other natural capital (Chapter 1). This has led to the extensive conversion of natural environments to productive systems, with impacts on the region's biodiversity.

With 79 per cent of its population living in towns and cities (UNEP 2010b), the region is one of the most urbanized in the world. It faces challenges in providing its burgeoning towns and cities with safe water and sanitation, and in addressing air pollution and the contamination of its freshwater, oceans and seas. The associated competition for scarce resources and the inequitable distribution of benefits have led to emerging socio-environmental conflicts and risks to the traditional lifestyles and livelihoods of local and indigenous communities.

Global climate change exacerbates many of the region's existing problems. Extreme weather patterns and climatic events are increasing in frequency and intensity, and sea levels are rising. The impacts are already affecting the region's most vulnerable groups, including its small island developing states and many rural, indigenous and poor populations. Thus, it is ever more important to use water resources efficiently and to conserve and sustain terrestrial, coastal and marine ecosystems. The challenge, however, is great, and the region is far from achieving some of the Millennium Development Goals (MDGs) (UN 2010a). Given the current situation, including poverty throughout Latin America and the Caribbean, there is an urgent need to implement more effective measures to halt and reverse the region's negative environmental trends (UN 2010a).

The region has many laws relating to the environment but, at the same time, the lack of institutional management and capacity to implement and enforce them has constrained their effectiveness (UNEP 2010b). In addition, policies are not keeping pace with production practices or adapting sufficiently to global trends and integration (UNEP 2011a). To address these challenges, governments need a stronger commitment to new policies and to making existing policies and mechanisms more effective. Certain countries are progressing in incorporating new policy mechanisms, such as valuation of ecosystem services, payment for ecosystem services, climate-compatible development, innovative green financing mechanisms, and sound corporate practices, to name a few. Some progress is also being made in developing national environmental/sustainable development strategies that take both cross-sectoral and multi-stakeholder views into account (Bovarnick *et al.* 2010; UNEP 2010b). These positive lessons are a starting point for considering the options available to the region's policy makers.

This chapter highlights policies considered to have the highest potential for increasing environmental sustainability and associated human well-being. A number of interrelated themes have been selected as priorities: environmental governance; water management; biodiversity; soil, land use, land degradation and desertification; and climate change. Sustainable management of oceans and seas is also important, especially to the region's small

Latin America and the Caribbean is the most urbanized region in the developing world. While urbanization exerts great pressure on natural resources and ecosystems, properly managed cities can also be part of the solution to global environmental challenges. © *Aurelio Scetta*

islands (Mahon *et al.* 2011), so coastal and marine issues are also addressed. The following section appraises a number of policy options for the region according to the selected themes, and also addresses key points related to marine and coastal policy.

POLICY APPRAISAL

Environmental governance

Environmental governance (Box 12.1; Figure 12.1) has been identified as a priority theme for the region and is treated as cross-cutting with respect to the other selected environmental themes. This reflects the fact that sound environmental governance will ultimately reverse environmental degradation and help achieve the MDG targets and many multilateral environmental agreements.

Regional context

Governance of the environment and natural resources in Latin America and the Caribbean is a complex mosaic. This stems from the wide diversity of governance systems with different degrees of institutional development and approaches to environmental issues, and different levels of governance mechanisms and performance (Box 12.2). Regional and sub-regional mechanisms play an important role in environmental management, although in many cases the environment is not their main focus.

In recent decades, most Latin American and Caribbean countries have developed national environmental legal and institutional frameworks to formulate strategies and action plans for sustainable natural resource use and environmental protection (UNEP 2010b; Larson 2003). In addition, countries have begun to adopt a more cross-sectoral approach, with other agencies considering environmental issues in addition to those directly responsible for the environment. Despite these achievements, a limited capacity to implement and enforce existing legislation and poor institutional arrangements constrain effectiveness (UNEP 2010b). The weak development of environmental policies in the face of economic, financial, commercial and technological globalization has aggravated the situation (UNEP 2011c). Managing national environmental policies and balancing internal priorities among other sectoral needs, while engaging in multilateral efforts through multilateral environmental agreements, constitutes a major challenge for the region.

Another concern is policy and institutional continuity, which is especially important for environmental issues (Emilsson *et al.* 2004). The timescales over which policies, programmes and projects are realized do not always coincide with those of political terms of office. Options to strengthen the political authority of environmental agencies and maintain essential medium- to long-term efforts include longer terms of office and greater autonomy for technical environmental officers, and creative financing mechanisms to facilitate political independence.

A number of other requirements for good environmental governance must still be met in the region; these are briefly examined in the sections that follow.

Box 12.1 Environmental governance

Environmental policy and institutional frameworks, and the relationships between them, provide the essential foundation of a governance framework. Sound policy frameworks include a set of environmental norms, policies and regulations at various levels – international, hemispheric, regional, sub-regional, national – as well as multilateral and bilateral environmental agreements.

Institutional settings also include civil society and the private sector and the interactions between them. It is important to note that the existence of policies and institutional frameworks alone does not guarantee good environmental governance.

Figure 12.1 The core constituents of environmental governance

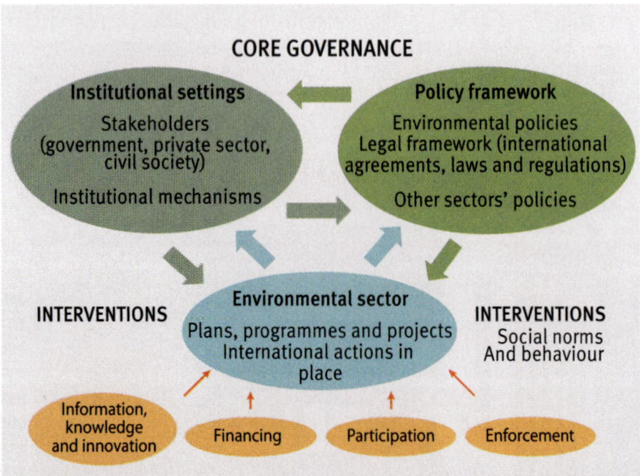

Source: Singh 2008

Box 12.2 Levels of governance in Latin America and the Caribbean

Governance mechanisms occur at a number of levels: the hemispheric level, through the Forum of Ministers of Environment of Latin America and the Caribbean, and at the Wider Caribbean level through the Cartagena Convention on the Protection of the Wider Caribbean Sea; the sub-regional level, as reflected in regional integration movements including the Central America Integration System (SICA), the Caribbean Common Market (CARICOM), the Andean Community (CAN) and the Common Market of the South (MERCOSUR); and at the national level. In addition, there are specific frameworks to manage shared resources, such as the Amazon Cooperation Treaty and many river basin organizations.

Enabling factors for more effective environmental governance

For effective and efficient functioning, a number of enabling conditions should support policy and institutional frameworks, including adequate financial resources, scientific research and information, environmental education and a culture of environmental awareness. In addition, standard governance principles and values such as transparency, accountability, equity, sustainability and inclusive stakeholder participation should underpin any governance framework (Mahon *et al.* 2010; Gaventa and Valderrama 1999).

Policies for generating and disseminating information foster a better understanding of environmental conditions, problems and potential solutions and improve the science-policy interface. Reliable and timely information allows decision makers to respond appropriately and thus improves decision making (Table 12.1a). Where relevant, this information should also incorporate indigenous/local knowledge (Table 12.1b). To influence policy and decision making effectively, environmental information should be transformed into easily understood, scientifically derived indicators to convey clear messages to policy makers

Table 12.1 Environmental governance case studies

Generating environmental information and improving the science-policy interface		
(a)	**Monitoring and evaluation** Central America	The Regional Biodiversity Monitoring and Evaluation Programme (PROMEBIO) has been formulated as a regional programmatic instrument to generate information that enables the management of biodiversity and natural resources, helps the various key stakeholders and decision makers at both regional and national levels to incorporate the issue into their strategic agendas and decisions, and makes a significant contribution to the process of environmental integration in Central America (CCAD-UNDP/GEF 2005).
(b)	**Importance of local knowledge in managing marine protected areas** Belize	Bird Caye National Park and Gladden Spit Marine Reserve are co-managed by Friends of Nature, a local non-governmental organization, through agreements with the government authority – the Forestry and Fisheries Department. Friends of Nature manages the areas, enforcing fishing rules and regulations, which enhances policy effectiveness through local legitimacy and reduces tension between local fishers and federal authorities. Links to fishing communities and international research organizations facilitate the combination of scientific and local knowledge for improved understanding of local environmental conditions (Gray 2008).
Education and environmental culture		
(c)	**Formal environmental education** Mexico	In the Mexican state of Morelos, a new system of formal environmental education has been developed focused on the particularities and features of the natural environment surrounding the public schools. This programme has proved to be of interest to educators, environmental activists and international organizations worldwide (Hurtado Badiola 2008).
(d)	**E-government web** Panama	In 2005, the National Environmental Authority (ANAM) implemented an e-government web-based platform that allows public access to proposed regulations, environmental studies, scientific reports and other key documents including administrative fines and complaints. This is fostering active interaction with non-governmental organizations, the media and the general public (ANAM 2009).
Improving public participation		
(e)	**Co-management in a mangrove ecosystem** Saint Lucia	Co-management of resources is aptly demonstrated in the Mankòtè Mangrove in Saint Lucia, where participatory and collaborative approaches and methods have been used for reconciling economic and other human activities with conservation imperatives (Brown and Renard 2000).
(f)	**Latin America partnership for water funds** Brazil, Colombia, Ecuador, Mexico, Peru	These are private-public, long-term financial partnerships to protect critical watersheds, attracting voluntary contributions from large downstream water users who benefit from upstream water preservation activities including reforestation, ecotourism and water-flow monitoring. They also support green economic opportunities with a positive impact on local communities, such as sustainable farming (Calvache *et al.* 2011).
Environmental economics and market mechanisms		
(g)	**System of Environmental-Economic Accounting (SEEA)** Mexico	This system adjusts national accounts to reflect environmental damage and the depletion of natural resources, such as water and minerals, which cost Mexico nearly US$90 billion annually, or 8% of the country's gross domestic product (GDP) from 2005 to 2009 (INEGI 2011). This is attributed to the globalization of markets, weak and poorly implemented policies and failure to enforce existing laws
Collaboration and coordination		
(h)	**The Caribbean Sea Commission** Wider Caribbean	The Caribbean Sea Commission, established by the Association of Caribbean States (ACS) in 2006, was set up to help advance work on the Caribbean Sea Initiative. This body has the potential to bring greater coherence to the policies and other governance structures associated with the Caribbean Sea (Mahon *et al.* 2011).
Improving environmental justice		
(i)	**A manual for fighting environmental crimes** Peru	In Peru, the Manual for Investigating Environmental Crimes has been developed as a tool for obligatory use by environmental prosecutors. It is intended to guide investigation and punishment of environmental crimes in the Peruvian Amazon and other key ecosystems, as well as to bring greater coherence to approaches aimed at environmental crime prevention (Avina 2011).
(j)	**Environmental justice in practice: the *Mendoza Case*** Argentina	Positive steps were made in resolving a decades-old pollution problem in the Riachuelo watershed of Buenos Aires with the ruling adopted by the Supreme Court of Argentina in the Mendoza case. Residents sued the federal government for damage to their health, resulting in a ruling that held the City of Buenos Aires and the federal government responsible for the damage to and reparation of the watershed, and the setting up of an authority to address the environmental health issues. This authority has embarked on a range of clean-up and restoration efforts (Staveland-Saeter 2011; di Filippo 2000).

Costa Rica has been a pioneer and leader among Latin American countries in the design and development of systems of payment for environmental services. *© Francisco Romero/iStock*

and the public (UNESCO-SCOPE 2006; Cimorelli and Stahl 2005). Importantly, information should not be policy prescriptive but policy relevant, and should provide decision makers with alternatives and associated scenarios (Watson 2005).

Relevant information and indicators also help in monitoring and evaluating the effectiveness of policies and determining if they have allowed management approaches to adapt to new conditions; these are important elements of good environmental governance. Good monitoring and evaluation programmes need to consider appropriate time frames and adequate baselines, and focus on results-based management using appropriate indicators. While well established in internationally sponsored projects (Pasteur and Blauert 2000), planned participatory monitoring and evaluation systems should also be used in government-run initiatives to quantify results and enable adaptive management. It is important that, in addition to quantitative scientific information, monitoring and evaluation regimes include social, political and cultural qualitative data to assess results and develop methods to improve policy effectiveness (Stem *et al.* 2005). Indicators may be process-based to measure progress or outcome-based to measure effectiveness, and should include the evaluation criteria of coverage, effectiveness, sustainability and replication (GEF 2011).

Environmental education gives people a greater sense of responsibility and increases awareness of the consequences of their actions. It promotes an environmentally conscious culture that helps to overcome a general lack of environmental awareness, one of the main causes of adverse change. Furthermore, an environmentally aware culture potentially improves public participation and increases public support for initiatives (Table 12.1c, d). For example, increased environmental awareness is credited with greater public support for developing payment for ecosystem services in Costa Rica (UNDP 2011).

Since the early 1990s, most countries have incorporated provisions for citizen participation in environmental legislation or in thematic or sectoral laws and have created a variety of citizen participation councils (Gaventa and Valderrama 1999). Although national and local regulations have standards for public participation, including those for a variety of instruments such as public hearings and consultations, implementing them effectively continues to be a challenge.

Co-management is one of the most effective and efficient approaches to incorporating public interests in environmental decision making (Moreno-Sánchez and Maldonado 2008). The co-management of protected areas and watersheds by local communities, civil society organizations, indigenous peoples and even the private sector, has become a model of stakeholder participation. This approach has been widely adopted in such areas as biodiversity conservation and forest management (Table 12.1b, e). For example, public-private partnerships used in tandem with economic incentives to protect critical watersheds are evident in a number of countries (Table 12.1f).

In many cases, however, citizens are only consulted at the very end of the decision-making process. This has exacerbated conflicts that integrated water resources management and multi-scale land-use planning are designed to prevent or resolve, including conflicts over water resources and land tenure. It is increasingly clear that there is a need for mechanisms to ensure accountability and transparency to reduce the risk of corruption in decision-making processes and to increase financial flows to environmental programmes (Transparency International 2010).

Negative externalities resulting from market forces are often considered a driving force of adverse environmental change. Thus, in developing future environmental policies in the region, it is of utmost importance to recognize the economic value of ecosystem services. An appreciation of the market value of ecosystem services, which reflects the link between the environment and human well-being in monetary terms, helps promote an environmental culture and improves the political viability of environmental protection. The use of economic incentives encourages citizens and businesses to make decisions based on the true long-term economic value of nature and the services it provides. Examples include Reducing Emissions from Deforestation and Forest Degradation and additionally for conserving and sustainably managing forests and enhancing forest carbon stocks (REDD+); payment for ecosystem services such as the Fund for the Protection of Water in Peru (Table 12.2b); and feed-in tariffs to support renewable energy. Valuing natural assets economically also allows decision makers in the public and private sectors to optimize their cost-benefit analyses and may be used to adjust national accounts and other economic indicators (Table 12.1g). Other tools, such as green funds and environmental taxes, can be used to raise funds for cash-strapped environmental agencies and causes. For example, the Trinidad and Tobago Green Fund couples both sets of tools to fund biodiversity preservation and ecosystem management (Table 12.3j).

Figure 12.2 A governance framework for large marine ecosystems

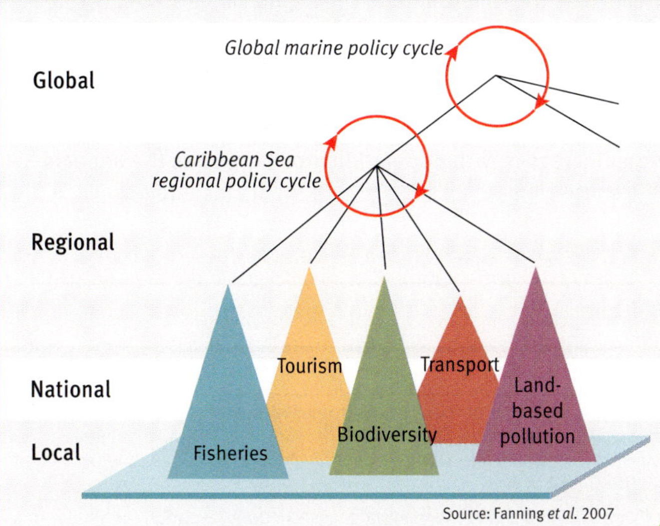

Source: Fanning *et al.* 2007

Effective environmental governance, especially in the context of complex systems, requires that stakeholders collaborate and cooperate; it also requires coordination and harmonization of institutions, policies and other instruments. A number of platforms and mechanisms have been established to facilitate greater collaboration and coordination and improve coherence among governance systems, although these vary in nature, scale and level of success. One such mechanism is the Caribbean Sea Commission (Table 12.1h), which is one of several initiatives under way to strengthen the cohesiveness of the approximately 30 organizations involved from the sub-regional to the international level in coastal and marine management in the Caribbean Sea (Mahon *et al.* 2011). A multi-scale governance framework is proposed for this large marine ecosystem by Fanning *et al.* (2007), which accommodates the diversity of policy cycles at multiple levels and the links between them (Figure 12.2). Such a framework could be adapted for other ecosystems or environmental issues.

Environmental justice is "the fair treatment and meaningful involvement of all people regardless of race, colour, national origin, or income with respect to the development, implementation, and enforcement of environmental laws, regulations, and policies" (USEPA 2011). In recent decades, several Latin American and Caribbean countries have made significant progress in environmental justice, particularly by enacting specialized procedures and mechanisms, as well as by enhancing the capacity of judiciaries; in some cases, this has included establishing specialized tribunals, for example the Tribunal Ambiental Administrativo in Costa Rica, and designating environmental prosecutors (Table 12.1i).

Although there are positive examples of judicial rulings in the region (Table 12.1j), there are still many challenges to improving environmental justice, including institutional and legislative weaknesses, low public participation and lack of awareness and information about people's environmental rights.

The emerging role of the judiciary is also important. In many countries, civil society organizations, prosecutors and individual citizens are using the judicial system to defend environmental rights. This occurs mostly through constitutional courts, but also in criminal and civil courts. In addition, the justice system has been proactive in resolving technically and legally complex disputes by overcoming procedural obstacles and adapting traditional legal institutions to the specifics of environmental law. The judiciary still needs to develop a considerable amount of capacity in addressing environmental issues, however, particularly by training legal professionals, especially lawyers and prosecutors.

Conclusion

Environmental governance should be viewed as a cross-cutting theme across all the other priority issues identified in Latin America and the Caribbean. Despite its complex environmental governance mechanisms, the region has made significant progress in developing national environmental legal and institutional frameworks. Poor institutional arrangements and a limited capacity for implementation and enforcement, among other deficiencies, however, have hampered their effectiveness. A number of enabling conditions need to accompany these frameworks including adequate financial resources, scientific research and information dissemination, environmental education and an improved environmental culture. They also involve the standard governance principles and values of transparency, accountability, equity, sustainability and inclusive stakeholder participation. Such good governance can help reverse the current trend in environmental degradation and help to achieve the targets of the MDGs and many multilateral environmental agreements.

Water

The availability of clean water in sufficient quantities and of sufficient quality was declared a human right by UN Decision 64/292 and is recognized in the constitutions of some Latin American and Caribbean countries. This section describes a set of policy options, presented in clusters, that intend to provide guidance for decisions about water provision and consumption. They were identified as potential options to address the Johannesburg Plan of Implementation Paragraph 26c (WSSD 2002), selected as the region's internationally agreed goal related to water.

Integrated water resources management

Integrated water resources management has been widely acknowledged as a way of achieving long-term solutions to water problems because of its interdisciplinary approach. Its implementation in developing countries, however, has been rather slow (UN-Water 2008). Integrated policies include those associated with:

- strengthened water governance;
- application of economic and financial instruments; and
- improvement of information on water quality and quantity.

Strengthened governance is both a cause and effect of an holistic view of water management because it implies a balance between public interest and the rights of the individual. Economic instruments and information are key tools in managing complex situations such as water scarcity, water-use conflicts and pollution. Economic instruments include mechanisms to change the culture of water use, such as economic valuation and the polluter-pays principle. Information gathering, including outputs from indicators and monitoring processes, supports the management of supply and demand and also helps to sustain traditional knowledge about the links between water, people and the environment. Finally, in the context of climate change, water-related information systems to prevent disasters and manage risk are increasingly important for the region (Table 12.2a, b, c).

Integrated approaches to water management enable resources and capabilities to be used in an efficient, cost-effective and sustainable way, which is ever more important as demand for water increases with population growth, and as the impacts of climate change are felt. Other benefits include fewer

Table 12.2 Water case studies

Integrated water resources management		
(a)	**Strengthened water governance: institutional arrangements** Bolivia	Several community organizations in the peri-urban areas of the city of Cochabamba have engaged in different ways of guaranteeing the water supply for domestic consumption. The result is a network of providers, distributors, committees and various types of arrangements that alleviate basic needs. At the same time, a metropolitan area master plan is under development (GIZ/PROAPAC 2011).
(b)	**Application of economic and financial instruments: payment for ecosystem services** Colombia, Ecuador, Peru	The Fund for the Protection of Water (FONAG) is a trust fund to which water users in Quito, Ecuador, contribute. This fund is used to co-finance activities, projects and programmes for the rehabilitation and conservation of 65 000 hectares of watersheds that supply Quito and surrounding areas. Similar funds have been developed in Colombia and Peru (Cisneros and Lloret 2008).
(c)	**Improved information on water quality and quantity: National System of Environmental Indicators** Mexico	Water is one of the components of the National Environmental Indicators system. This uses indicators for situational assessments of water resources in terms of availability and quality. Indicators reflect pressures on the availability for different uses, its state, and action being taken to preserve water. Also reflected by the indicators are factors that affect water quality, the condition of water bodies in the country, and action being taken to stop their deterioration and allow recovery (SEMARNAT 2009).
Sustainable water provision and consumption		
(d)	**Conservation and restoration of water for supplying ecosystems: rainwater harvesting** Antigua and Barbuda, Bahamas, Brazil, Grenadines, Mexico, Peru	Rainwater harvesting is one of the most successful water management practices in the region because of its relatively low cost and its technical feasibility for multiple uses. In the Caribbean, rainwater harvesting is a source of water for around 500 000 people. Brazil has a programme for the construction of a million rural cisterns; Mexico and Peru have capacity-building and demonstration centres (CEHI/GWP-C 2010; GWP-C 2010; UNEP 2010b; Colegio de Postgraduados 2004).
(e)	**Promotion of water-use efficiency in agriculture** Bolivia	Traditional projects to enhance irrigation systems used to focus mainly on improving the collection, transmission and distribution of water, with water use at the plot level seldom addressed. Recent projects have contributed to the efficiency of irrigation systems, increasing water availability at the plot level by up to 50% by means of best practices in surface irrigation and with adequate technology (PROAGRO/GTZ/DED 2010).
Wastewater treatment and reuse		
(f)	**Lower-cost sewerage systems** Brazil	The condominium sewerage system was developed in Brasilia and Salvador de Bahia as a mechanism to expand sewer services for 1.5 million condominium residents. One of the benefits of this policy is that providers and users need to reach an agreement to facilitate service expansion and adaptation to local needs. Thus, the condominium becomes not only a physical unit of service provision, but a social unit for facilitated collective decisions and the organization of communal action. Additionally, this system provides a discount of 40% on the standard sewerage charge for households (Melo 2005).
Integrated coastal zone management		
(g)	**Coastal Zone Management Unit** Barbados	The Coastal Zone Management Unit in Barbados, established 25 years ago, has two strategic objectives: • sustainable use of the coastal management area by implementing policies that maintain and, where possible, enhance environmental quality while still enabling economic development; and • an effective legal and administrative structure to implement integrated coastal management. Pursuing its objectives, the unit works with the Town and Country Development Planning Office. The unit provides technical expertise to make an informed and just assessment when the site chosen for development falls within the coastal zone management area (CZMU Barbados 2011).
(h)	**Integrated coastal zone management** Colombia	The management plan of the Guapi-Iscuande Integrated Management Unit has enabled ecosystem conservation, support to communities, and rational and alternative use of environmental resources by all involved stakeholders. The plan started in 2000 and has helped to design environmental agendas with local communities including indigenous and ethnic groups, as well as fostering the creation of a local committee for integrated coastal zone management (INVEMAR 2011).
(i)	**Cartagena Convention for the Protection and Development of the Marine Environment** Wider Caribbean	This regional policy framework, which entered into force in 1986, is a comprehensive, umbrella agreement for the protection and development of the marine environment. The convention is supplemented by three protocols: the Protocol Concerning Co-operation in Combating Oil Spills; the Protocol Concerning Specially Protected Areas and Wildlife; and the Protocol Concerning Pollution from Land-Based Sources and Activities (UNEP-CEP 2011a; UNEP 2000).

Paute River in the Andes Mountains of Ecuador, where sustainable water resource management plays a vital role in food security and energy, and in supporting valuable ecosystem services for the Santiago Morona region and beyond. *© Ammit/iStock*

water-related conflicts, such as in managing transboundary basins and other competing uses; increased participation of stakeholders in decision making – including women, indigenous groups and other minorities – that can help reduce marginality and inequity and promotes transparency and accountability; increased water conservation and sustainable distribution; decision making and policy formulation based on evidence and traditional knowledge; and appropriate basin management that contributes to land-use planning policies, helps address issues of food security, ecosystem protection and waste management, and reduces transaction costs in water chains (Dalhuisen and Nijkamp 2002).

Integrated water resources management has only been implemented in Latin America and the Caribbean in a limited way due to fragmented and conflicting institutional mandates, lack of skilled human resources, inadequate mechanisms for effective public participation, lack of sustainable financing and harmonization mechanisms, and a lack of structures and procedures to gather and present data.

Enabling conditions to promote integrated management include:

- water policy reform, including legislation and standards;
- water governance, including institutional frameworks to monitor and enforce legislation, development of institutional capacity to design and implement integrated management plans, projects and long-term programmes at different scales, and greater engagement and use of local knowledge through basin committees;
- land register development, stable governance arrangements, low transaction costs, credible enforcement arrangements and clearly defined rights and/or entitlements for land and water use;
- developing government capacity to collect tax revenues, so that funds can be efficiently and equitably allocated to water programmes and projects; and
- education and information programmes (UNEP 2011c).

Sustainable water provision and consumption

The Latin America and Caribbean region has 31 per cent of the world's freshwater resources (UNEP 2010b). However, given the region's rate of population growth, rapid urbanization and current patterns of water use (UNEP 2010b), sustaining ecosystem services and an adequate water supply for future generations is an increasingly important issue. Investment in infrastructure is necessary, but it alone is not enough to solve

Figure 12.3 Population with access to improved sources of drinking water

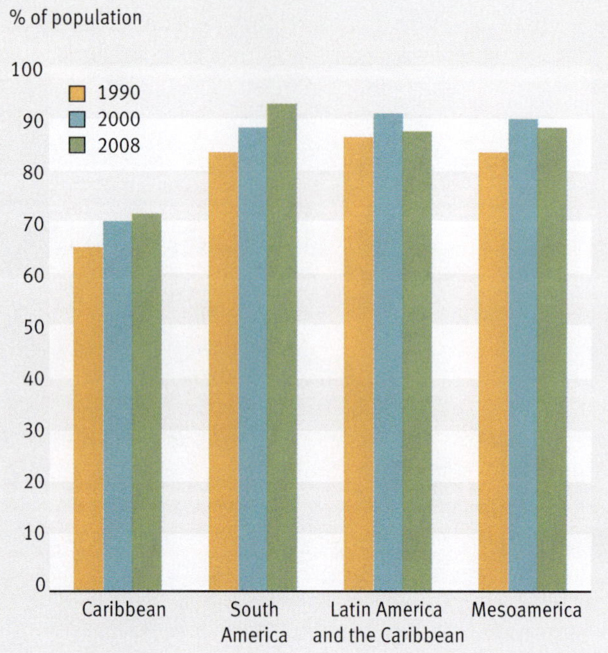

Source: UNEP-EDE 2012

the problem of water supply and demand. There is need for a change in policy making and management approaches, from those based exclusively on managing supply to the inclusion of both supply and demand management (Jouravley 2001). Among users, there is need for a cultural change through education and economic incentives. Two main policy options may be considered:

- conservation and restoration of water-supplying ecosystems;
- promotion of water-use efficiency in human consumption and production activities.

Ecosystems provide a wide range of services within a watershed. Thus, establishing and maintaining the minimum amount of water they require (environmental flows) is vital to ensure a balanced hydrological cycle and a constant water supply. In areas where resources are heavily exploited, improved water-use efficiency is urgently needed through technological developments and by applying traditional and scientific knowledge (Table 12.2d, e). This fosters measures to adapt to climate change as well as reducing costs for water users. Investments in water-use efficiency include the control of unaccounted water at the grid level, installation of water-saving appliances, reuse and recycling systems, rainwater harvesting and water-saving irrigation systems, among others. Although initial investments are high, reduced water use translates into reduced costs in the long term.

Despite notable progress, a significant portion of the region's most vulnerable people – largely in the growing number of informal settlements, or *favelas* – still lack access to clean water and sanitation systems. *© Luoman/iStock*

Figure 12.4 Population with access to improved sanitation

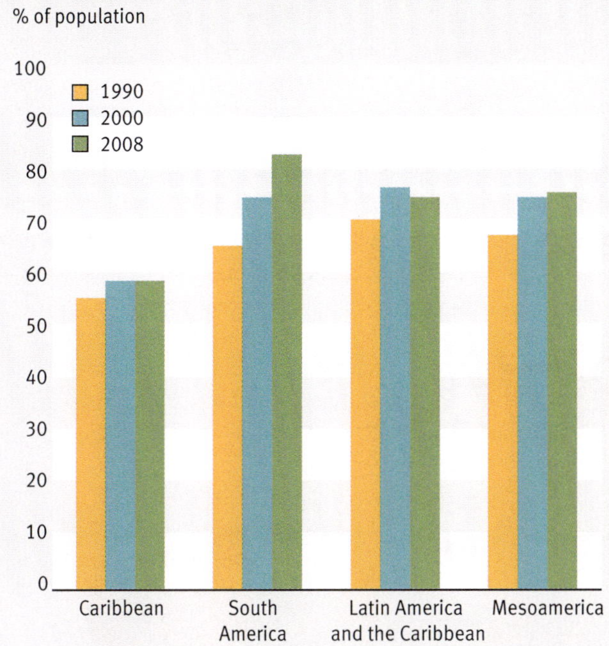

Source: UNEP-EDE 2012

Overall water policy needs reform to ensure that the policies proposed here do not remain isolated projects or campaigns, but have long-term effects. It is thus important to develop the political will to adopt legislation that will effect positive change through encouraging incentives and enforcing penalties. There must be economic incentives such as access to loans with low interest rates and equitable conditions, as well as water-efficiency certification schemes. Management committees, civil society and multi-stakeholder participation are key to success. In summary, sustainable water supply and demand can be achieved when the economic, cultural and social value of water is acknowledged.

Expansion of drinking water and sanitation systems

To achieve MDG 7, 92.5 per cent of the population of Latin America and the Caribbean must have access to safe drinking water and 84.5 per cent to basic sanitation by the end of 2015 (WHO and UNICEF 2010). According to the most recent MDG report, the region has high rates of achievement for the first target and moderate ones for the second. This suggests that the sanitation target will not be met if prevailing trends persist (UN 2010b). Furthermore, there are enormous differences within segments of the population, between urban and rural areas, and between the three sub-regions (Figures 12.3 and 12.4).

The drinking water and sanitation policy cluster includes:

- freshwater augmentation;
- water quality improvement;
- wastewater treatment and reuse; and
- water conservation (UNEP 1997).

These policies are specific to each sub-region on issues such as water use relative to water availability; existing water supply infrastructure including its condition and size, geographical watershed extent, number of people connected and number of people who receive measured water; user characteristics including socio-economic issues, consumption patterns, and essential and non-essential uses; and technical, financial and institutional resources (Sutherland and Fenn 2000). Examples of technological options to expand water availability are rainwater harvesting, water reuse, groundwater recharge and desalination.

These policies require a high level of commitment from governments as well as relatively high financial investments (Trémolet *et al.* 2010). In addition, maintenance costs, lack of technical competency – for example for desalination – and inefficient water-use habits could hinder the expansion of coverage. International cooperation is needed to finance cases requiring special technical or social development skills that governments cannot afford (Melo 2005).

Investing in sanitation is cost effective in terms of the ratio of public cost to estimated health benefits. According to Trémolet *et al.* (2010), the economic benefits range from US$3 to US$34 for each US$1 invested, depending on the region. Household treatment, such as disinfecting water for drinking and food preparation, cuts the primary transmission route for diarrhoeal disease and can pay back up to US$60 for every US$1 invested (WHO and UNICEF 2005). One such example is Brazil, with its innovative mechanisms to promote sewerage systems (Table 12.2f).

Integrated coastal zone management

Population density in the region's coastal zones is significantly greater than in inland areas (Figure 12.5). Coastal infrastructure, urbanization and tourism and land-based pollution are significant pressures on coastal and marine ecosystems. The rise in sea level due to climate change and the increasing frequency of El Niño/La Niña phenomena are also affecting coasts and changing coastline dynamics, ecosystem health, rainfall patterns and river flows, as well as damaging infrastructure.

Integrated coastal zone management is a multidisciplinary and intersectoral approach to land-use planning that promotes effective, meaningful and sustainable management of coastal resources (Ramcharan 2001). Similar to integrated water management, it assimilates the interests and needs of different stakeholders, maintaining ecosystems and their services in a cooperative and rational manner. In the Caribbean, for example, mechanisms have been implemented through the international project for Integrating Watershed and Coastal Area Management in the Small Island Developing States of the Caribbean (IWCAM) and the action plans of Barbados, Belize and Saint Lucia.

Figure 12.5 Estimated population density in Latin America and the Caribbean, 2010

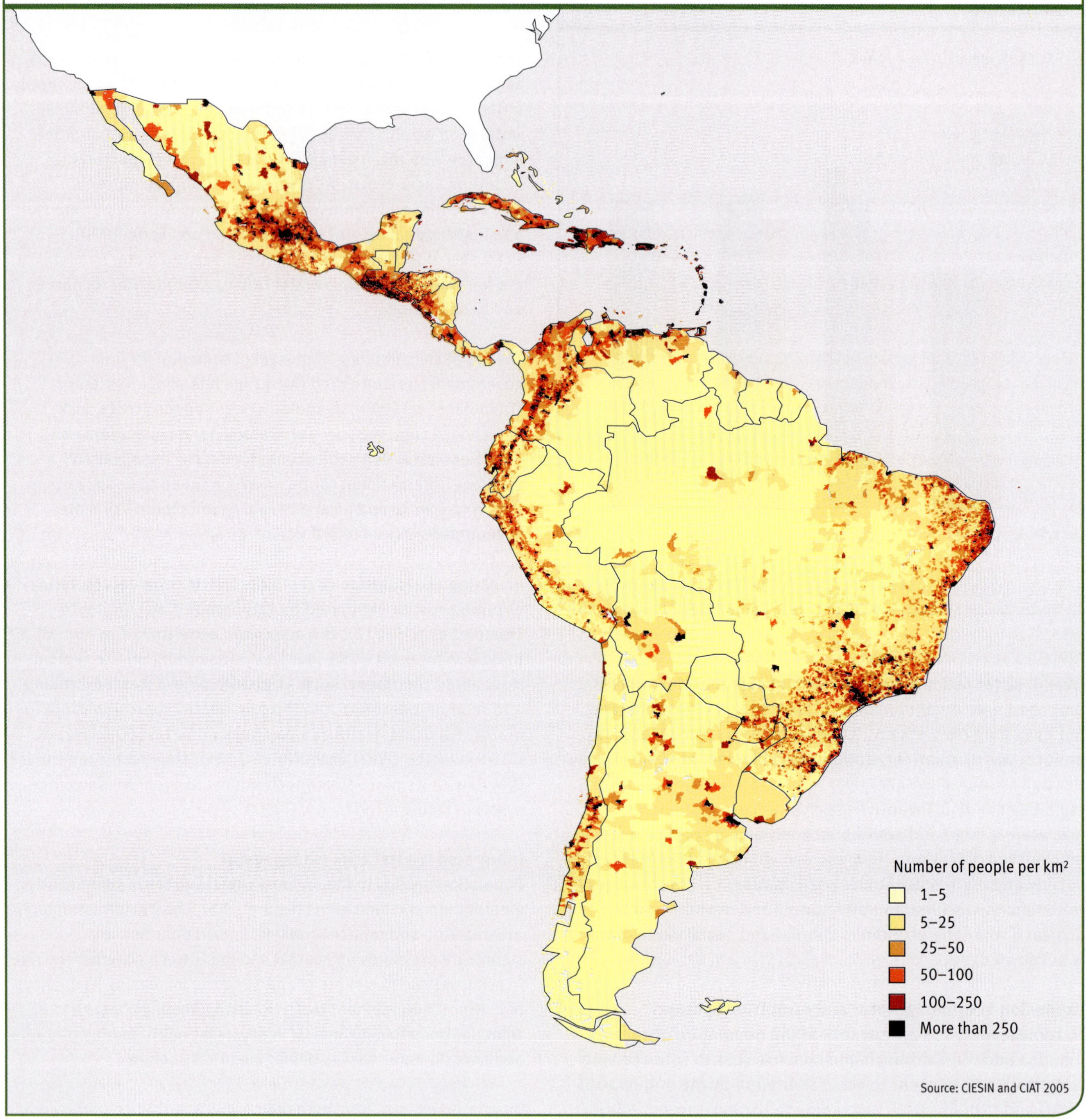

Source: CIESIN and CIAT 2005

The coastal management policy cluster includes the establishment and execution of legislation, regulations, standards and procedures to prevent or minimize environmental degradation, and to protect and restore the quality and function of ecological systems within the coastal zone. It requires an appropriate legal framework, effective institutional structure, and information, data and knowledge for management (Islam and Koudstaal 2003). It also needs a clear and collectively recognized definition of the coastal zone's limits. The foundation for implementing this approach is a coastal zone management action plan, while strengthening monitoring and evaluation capabilities enables progress to be rigorously tracked.

Integrated coastal management promotes the preservation of ecologically sensitive areas such as mangroves, fosters the sustainability of important socio-economic activities such as fisheries and tourism, preserves natural ecosystem functions and services such as coral reefs, and improves the quality of the marine environment, for example by reducing contamination from vessels and in ports. Experience in Barbados, Colombia, Saint Lucia and the wider Caribbean demonstrates these benefits (Table 12.2g, h, i).

Biodiversity

Latin America and the Caribbean is home to approximately 70 per cent of the world's species and almost 20 per cent of its ecoregions (UNEP 2010b). Its economy is highly dependent on this rich biodiversity, yet it is increasingly under threat from human activities (Box 12.3). Although there are numerous biodiversity policies and measures in the region, collectively they do not effectively conserve its biological resources.

Addressing the driving forces that affect biodiversity (Box 12.3) requires equitable, evidence-based, participatory, cross-sectoral policies designed to protect and restore biological resources (Diaz 2010). In the context of the new Aichi Targets (CBD 2010) – 20 targets that form the framework for biodiversity conservation until 2020 (Box 5.1) under the Convention on Biological Diversity (CBD) – and given the region's biodiversity priorities, CBD Article 10 was selected as the internationally agreed biodiversity goal related to this priority issue (CBD 1992). The following four policy options are considered able to help accelerate the region's progress towards meeting this goal.

Increasing and expanding protected areas, improving their management and creating greater connectivity

Latin America and the Caribbean's protected areas, including marine, cover more than 500 million hectares in 4 400 different zones. They are considered to be one of the region's most important policy measures for conserving biological diversity (Bovarnick *et al.* 2010; UNEP 2010b). There is documented evidence that protected areas not only play a role in conserving species and habitats, but also deliver a range of ecosystem services and are considered important in climate change adaptation and mitigation (CBD 2008). If properly managed, they can both contribute to national gross domestic product (GDP) and help to cover their own costs (Table 12.3a, b). Although not often realized, protected areas have the potential to provide a range of social benefits: improving equity and alleviating poverty as well as empowering women, communities and indigenous peoples – all of which are important considerations in the region (Bovarnick *et al.* 2010).

Although protected areas have demonstrated both progress and success in biodiversity conservation in Latin America and the Caribbean, they face a number of challenges. An important one is that isolated areas often offer insufficient protection, but creating biological corridors or improving landscape-scale connectivity can improve protected area performance (Brudvig *et al.* 2009; Dudley and Rao 2008; Bennett 2003). Greater connectivity can also improve species resilience to climate change and provide multiple benefits to humans (Table 12.3c, d) (Harvey *et al.* 2008; Bennett and Mulongoy 2006).

Scarlet macaws, a symbol of Neotropical biodiversity, are now threatened due to habitat loss and poaching. © *Roberto A Sanchez/iStock*

Box 12.3 Threats to biodiversity in Latin America and the Caribbean

Biodiversity in Latin America and the Caribbean is threatened by a number of linked factors including:

- habitat loss, conversion and alteration;
- overharvesting or unsustainable use of terrestrial and aquatic resources;
- unsustainable land management practices;
- contamination of terrestrial and aquatic ecosystems from intensive economic activities;
- the spread of alien invasive species that impact the structure and functioning of ecosystems;
- climate change;
- demographic pressures;
- globalization of markets; and
- weak and poorly implemented policies and failure to enforce laws.

Of all drivers, the conversion of natural environments to productive systems is currently considered the most pressing (Bovarnick *et al.* 2010; UNEP 2010b).

Table 12.3 Biodiversity case studies

Protected areas		
(a)	**Quantifying the contribution of national systems of protected areas to the economy** Mexico	Mexico's federal protected areas contribute at least US$3.5 billion every year to the economy. This represents a return of US$57 for each US$1 invested in protected areas by the federal budget (Bezaury Creel 2009; Bezaury Creel and Pabón Zamora 2009).
(b)	**User fees supporting marine protected area activities** Jamaica	The establishment of marine protected areas is a common way for governments to regulate activities affecting coral and marine resources. In the Montego Bay Marine Park, a tax of 0.1% on tourist equipment helps to cover the park's costs and sustain the park's activities (Reid-Grant and Bhat 2009).
(c)	**Mesoamerican Biological Corridor** Belize, Costa Rica, El Salvador, Guatemala, Honduras, Mexico, Nicaragua, Panama	Established in 1997 by the governments of the eight Central American countries, the Mesoamerican Biological Corridor acts as a primary pathway between large and important areas of habitat, predominantly protected areas. By promoting greater opportunities for local residents to participate in planning and management, the corridor is helping to promote a greater sense of human well-being while ensuring that the biological heritage of the region is protected and enhanced (López and Jiménez 2007; Bennett 2004).
(d)	**Eastern Tropical Pacific Marine Corridor** Colombia, Costa Rica, Ecuador, Panama	A voluntary agreement created the corridor to encourage regional management for the Eastern Tropical Pacific seascape. The initiative has included more than 80 non-governmental organizations, research organizations, local community groups and the private sector in promoting regional cooperation for training, education and the conservation of coastal-marine resources (UNESCO 2011).
(e)	**Amazon Region Protected Areas (ARPA)** Brazil	Brazil's Amazon Region Protected Areas (ARPA) programme is the largest worldwide initiative in tropical forest conservation, aiming to protect 600 000 km² of biologically important areas between 2003 and 2018. ARPA has been innovative in • developing decision support tools in protected area management; • developing financial mechanisms to allow protected areas to be sustainable in the long term; and • engaging a wide range of stakeholders in the decision-making process. ARPA has the potential to avoid 5 billion tonnes of carbon emissions by 2050 (Simpson 2010; Azevedo-Ramos *et al.* 2006).
The ecosystem approach		
(f)	**The ecosystem approach in a forested ecosystem** Bolivia, Brazil, Paraguay	The 240 000-km² Dry Chiquitano Forest ecoregion has adopted an ecosystem-based framework and has highlighted the importance of certain key factors in ecosystem management, including participation of local communities, decentralization of decision making, setting priorities based on ecological integrity, and planning at multiple temporal and spatial scales (Vides-Almonacid *et al.* 2008).
(g)	**Improving water management using the ecosystem approach in South America** Argentina, Brazil, Paraguay, Uruguay	The ecosystem approach has been applied to the management of water resources in the La Plata river basin. This region is considered important for protection because of poverty, the presence of indigenous communities, and likely effects of climate change in the area. Management approaches include integrated water resources management as a key component (Bello *et al.* 2009; Forero 2008).
(h)	**Applying the ecosystem approach in the management of coastal and marine protected areas** Chile	The ecosystem approach is being applied to improve the management of multiple-use marine and coastal protected areas in Chile. Key factors that have been given attention are research, incorporation of local communities in management of the ecosystem, and incorporation of the areas into national land-use plans (De Andrade 2008).
Payment for ecosystem services (PES)		
(i)	**Payment for ecosystem services for watershed protection in Heredia** Costa Rica	To protect the water resources that originate in micro-watersheds in the hills above Heredia, the water authority established a scheme to pay landowners in the upper watershed to reforest the land (US$1 000 per hectare per year under a five-year contract), and to prevent cattle ranching close to the streams (US$100 per hectare per year for ten years). A hydrological tariff, derived from fees charged to water authority customers, supports the scheme. Customers pay approximately US$0.05 per m³ for the ecosystem services delivered by landowners (Jindal and Kerr 2007).
(j)	**Sustainable financing** Trinidad and Tobago	The Green Fund of Trinidad and Tobago was established by the government under the Finance Act 2004, Part XIV – Green Fund Levy – and is capitalized by a turnover tax on corporations operating in the country. The purpose of the fund is to ensure that there is a sustainable source of financing available to support targeted efforts to conserve biodiversity and promote ecosystem management in Trinidad and Tobago (UNEP 2011b).
Access and benefit sharing		
(k)	**National law on access and benefit sharing** Peru	Under the Peruvian Law for the Conservation and Sustainable Use of Biodiversity (Law 26839), the state and the native and peasant communities participate in the protection and promotion of plant genetic resources, and the knowledge, innovations and practices associated with those resources (UNEP 2003).
(l)	**Access and benefit sharing in Central America** Belize, El Salvador, Guatemala, Honduras, Nicaragua, Panama	The Central American Commission on Environment and Development has adopted a Central American Protocol on Access to Genetic and Biochemical Resources and to the Associated Traditional Knowledge (UNEP 2003).

Other ways to enhance protected area effectiveness in the region include:

- advancing conservation in marine and freshwater protected areas that are still largely under-represented;
- effectively integrating indigenous and local communities in protected area management, including, where relevant, by promoting indigenous- and community-conserved areas (Chapter 5);
- promoting the links between conservation and development goals, using land-use planning as a fundamental tool;
- improving research capability and strengthening links between research and decision-making frameworks (Table 12.3e); and

- strengthening the capacity for managing protected areas (Elbers 2011; Mora and Sale 2011; BirdLife International 2009; Cuartas 2008; Guarderas *et al.* 2008; McElhinny 2007; Bennett and Mulongoy 2006; Oviedo 2006; Rivera *et al.* 2006; Burke and Maidens 2004; Geoghegan and Renard 2002).

In addition, key instruments for protected area management include ecotourism and sustainable tourism programmes; balancing the relationship between conservation and development through mechanisms such as payment for ecosystem services, including for carbon dioxide (CO_2) capture and sequestration services and environmental stewardship and usage fees (see below); and the selective extraction of resources (UNEP 2010b; Eguren 2004). Measures such as tax incentives, preservation easements, education, decentralized administration, partnerships with international organizations and outright land purchases may also encourage and promote protected areas and associated corridors and linked landscapes.

In Cuba, the ecosystem approach is being used to restore and improve local crop varieties that are better adapted to the low-input conditions of organic farming. © *Maria Pavlova/iStock*

Applying the ecosystem approach to biodiversity management

The ecosystem, or ecosystem-based, approach is increasingly recognized as an important strategy in biodiversity management, especially in the context of climate change (World Bank 2010). According to the CBD, it is "a strategy for the integrated management of land, water and living resources that promotes conservation and sustainable use in an equitable way" (Box 12.4).

The ecosystem approach is not designed to replace other management and conservation approaches, but rather to complement and support them, for example sustainable forest management, integrated river basin management, integrated marine and coastal area management and sustainable fisheries (Table 12.3f, g, h). In addition, approaches such as creating protected areas, corridors or biosphere reserves and species conservation programmes, as well as action under existing national policy and legislative frameworks, can be integrated to deal with complex ecological situations (Bianchi and Skjoldal 2008; Waltner-Toews *et al.* 2008; CBD 2004).

The ecosystem approach has been identified as a key policy in Latin America and the Caribbean for two main reasons: it is useful for managing water resources, wetlands and land, and in developing payment for ecosystem services; and many pristine ecosystems still exist with high conservation value (Andrade Perez 2008). Because of their size, the Caribbean's small island states also present excellent opportunities for implementing the ecosystem approach, and could serve as case studies for ascertaining its strengths and weaknesses.

Although there are several on-the-ground initiatives applying an ecosystem approach in the region, this has often been done on an *ad hoc*, single-project basis, which remains a challenge. Such initiatives need to be better integrated into institutions, including those concerned with sectors outside biodiversity conservation, such as agriculture, fisheries, forestry and health. More research is also needed (De Freitas *et al.* 2007) to support the development of a monitoring and evaluation framework for each of the principles of the ecosystem approach (CBD 2004). In addition, issues such as illiteracy, land boundaries and the cost of participatory processes all need to be considered in integrating and assessing the impact of the approach in Latin America and the Caribbean (Andrade Perez 2008).

Box 12.4 Key features of the ecosystem approach to biodiversity management

The ecosystem approach goes beyond conservation and promotes the sustainable use of resources with a focus on equity, participation and decentralization. It can be applied in a flexible manner depending on the social, economic, environmental and cultural context. Its key features include:

- emphasizing adaptive management;
- fostering integration by accounting for all usable goods and services and optimizing the mixing of their benefits;
- involving other forms of knowledge, including indigenous and local;
- focusing on people, their society and their culture;
- orientation to environmental and societal conservation;
- applying a two-way approach – top-down and bottom-up;
- a long-term vision; and
- considering goods and services as the product of a healthy ecosystem and not as ends in themselves.

Source: Andrade Perez 2008

Guatemala is recognizing wider land rights for indigenous communities while strengthening access and benefit sharing through recent advances in payment for ecosystem services. *© Holger Mette/iStock*

Enhancing biodiversity conservation through payment for ecosystem services

A number of options grounded in economic theory present promising opportunities for both mainstreaming biodiversity issues and reducing driving forces, while simultaneously supporting development processes and promoting human well-being. Among these is the payment for ecosystem services – or PES – mechanism, which was largely pioneered in Latin America and the Caribbean (Wunder 2007), and which is gaining popularity worldwide as an effective approach to dealing with biodiversity loss (UNEP 2010b; Pfaff *et al.* 2008).

In general terms, PES schemes or systems offer incentives, usually monetary ones, to individuals to protect and ensure the delivery of key ecosystem services at local, national and regional levels. The mechanism can address many of the driving forces of biodiversity loss in the region, especially habitat loss and unsustainable land management, as it usually aims to protect and/or rehabilitate natural vegetation. In addition it can support many existing policies (Box 12.5).

Box 12.5 Payment for ecosystem services (PES) in support of existing policies

PES can be used in conjunction with other Latin American and Caribbean policies, for example:

- protected areas;
- integrated water management;
- conservation and restoration of water-supplying ecosystems;
- sustainable forest management;
- small-scale agro-ecological systems; and
- recovery of degraded lands.

Monetary compensation provides a tangible incentive to protect habitats and their biodiversity by providing sustainable livelihoods (Table 12.3i); it also mitigates the initial needs that drive unsustainable biodiversity resource use (Ferraro 2001). As such, PES has the potential to increase employment and equity (Montagnini and Finney 2011; Bovarnick *et al.* 2010). This reduces poverty, since low-income groups and ecologically sensitive land in the region's developing countries often co-exist (Milder *et al.* 2010). Given that there is a strong link between habitat protection, rehabilitation and a number of ecosystem services – such as water provision and purification, coastal protection, mitigation of greenhouse gas emissions and protection against soil erosion – PES schemes bring multiple co-benefits to a range of sectors (WRI 2009).

Payment for ecosystem services is not without challenges. Its limited application (Redford and Adams 2009) and a lack of information on economic valuation highlight the need to invest more in research and furthering the scientific understanding of local environmental conditions. Certain services cannot be measured, however, and determining the seller of these services is also difficult (Farley and Costanza 2010). Moreover, finding buyers and mobilizing funding is the greatest challenge to implementing PES. Coupling PES with innovative financing mechanisms, however, could address this. Examples include ring-fencing budget allocations for environmental protection, as in the Programme for Forest incentives in Guatemala; earmarking government taxes for environmental protection, such as Brazil's ecological value added tax; providing environmental funds like Trinidad and Tobago's Green Fund (Table 12.3j); and setting up public-private partnerships (FAO 2011; Dijk and Savenije 2009).

Access and benefit sharing

Latin America and the Caribbean's rich genetic resources are important to local communities in sustaining their livelihoods, and especially in providing food security. However, many genetic resources are also the basis of commercial use and production. To promote equity and safeguard the genetic diversity and associated traditional/local knowledge within the region's countries, there has been a growing interest in access and benefit sharing (Chapter 5).

Argentina, Brazil, Costa Rica, Mexico and Peru have developed access and benefit-sharing legislation at the national level (Table 12.3k) and the Andean Community states and Central American Commission on Environment and Development have done so at the sub-regional level (Table 12.3l). At the national level, there are two main groups of associated legislation (Glowka 1998): framework laws on sustainable development, nature conservation and biodiversity (Costa Rica, Mexico, Peru); and dedicated or stand-alone national laws or decrees (Brazil). Access and benefit-sharing considerations can also be incorporated into general environmental framework laws, or existing laws and regulations can be modified to address them, although this has not yet occurred in Latin America and the Caribbean (UNEP 2003).

The CBD's Nagoya Protocol on Access to Genetic Resources and the Fair and Equitable Sharing of Benefits Arising from their

Utilization, adopted in October 2010, now provides a global framework for improving legal certainty and transparency related to access and benefit sharing (CBD 2011), and could help Latin American and Caribbean countries overcome various problems in implementing relevant policy. As of April 2012, 14 countries in the region are signatories to the protocol (CBD 2011). To maximize the advantages of access and benefit-sharing policies, attention needs to be paid to several factors, including:

- undertaking research to better understand and apply access and benefit-sharing principles and the Nagoya Protocol in the regional context;
- enhancing human, technical and financial capacities;
- clarifying legal definitions and interpretations;
- understanding and dealing with the transboundary nature of genetic resources;
- protecting traditional knowledge; and
- negotiating tangible benefits instead of focusing only on access procedures (UNEP 2010a; CBD 2008).

Land use, land degradation and desertification

Pressure on land resources has increased in recent years despite international goals to improve land management (Box 12.6; Chapter 3). To halt and reverse land degradation and ensure renewable resources are used sustainably, policies allowing productive activities with minimal impact on natural ecosystems and their environmental services are required. These include land-use policies that prevent inefficient or inappropriate transformation for agriculture, livestock or illegal crops (Grau and Aide 2008). Examples include sustainable forest management, increasing efficiency and intensifying productivity to reduce environmental impacts, improving waste management, decreasing the amount of new land being cultivated, and helping prevent conflicts due to land, water and other resource shortages. At the same time, environmentally friendly productive activities with tangible economic benefits for land owners and environmental services for society need promotion. Finally, degraded ecosystems need to be rehabilitated and their sustainability ensured to restore the productivity chain that supports ecological balance and social and economic well-being.

Based on a review of current and past experiences in Latin America and the Caribbean, the three land-management policies described in this section are considered the most favourable to attain the goals set out in Paragraph 40b of the Johannesburg Plan of Implementation (WSSD 2002). These are multi-scale land-use planning, sustainable agriculture and livestock production, and the recovery of degraded lands. The climate change and biodiversity sections deal with payment for ecosystem services and sustainable forest management, while the water section discusses policies related to integrated land management and water-use plans.

Multi-scale land-use planning

Land-use plans take account of all the resources and dimensions involved in the development process and help implement integrated land-use management, water resource planning and conservation priorities, while also encouraging inclusive, multi-stakeholder participation. Land-use planning considers a number of inseparable elements: land, renewable and non-renewable resources, and a coherent view of current and historic lands and their uses, existing services, accessibility and cultural influences. Land uses such as agriculture, agroforestry, livestock production, industrial development and mining, among others, must also be considered (Cardenas-Moller and Bianco 2011; ECLAC 2010a; Weber 2009).

Land-use policies involving stakeholder participation, regulations and financial instruments are necessary to prevent a number of

Box 12.6 Key facts about land conditions in Latin America and the Caribbean

- Since the 1960s, arable land has increased by 83 per cent in South America, 46 per cent in Africa and 36 per cent in Asia, coinciding with significant deforestation in all three regions (IPSRM 2010).
- In 2009, more than 280 000 Latin American and Caribbean producers managed 23 per cent of the world's organically farmed land with the highest regional shares in the Dominican Republic and Uruguay (Willer and Kilcher 2011).
- In South America, the area of land used for agriculture increased by 18 per cent during 1970–2009, while livestock production grew by 31 per cent (FAOSTAT 2011).

The pre-Hispanic cultures in the Lake Titicaca region of South America practised weather-adapted agriculture, where *suka kollus* – artificially raised planting mounds separated by canals – receive both moisture and protection from heat. This practice has recently been revived. © *Jim Alfonso Alem*

land-related conflicts, including transboundary controversies over scarce resources between and within sectors and countries; land tenure and titling issues for rural families – where creating cadastral and registration agencies helps create stability; and the rights of minority groups such as indigenous communities and women to land ownership. Furthermore, land-use planning can be an effective mechanism to prevent resource depletion and environmental degradation (Table 12.4a, b, c).

In a wider sense, land-use planning should also include marine and coastal zones because of the interaction between land and aquatic environments (Table 12.4d). According to UNEP-CEP (2011b), land-based activities are the greatest threat to Caribbean coastal and marine habitats. Likewise, the integrity of marine and coastal ecosystems, which is linked to terrestrial well-being, also affects social resilience, especially in terms of public health and livelihoods.

Several countries have either implemented or are preparing land administration plans, including national coastal management acts (Loper *et al.* 2005; Cabeza 2002). Land administration projects in Latin America are focused primarily on facilitating a land market. Although projects have social equity and environmental sustainability goals (Deininger and Bingswanger 1999), these are largely of secondary importance (IDB 2002). In some countries, for example Bolivia, Ecuador and Peru, advances in land administration require improvement in the property market infrastructure (World Bank 2001). Land tenure also needs to be stabilized, especially in post-conflict situations such as those of Colombia, El Salvador, Guatemala and Nicaragua.

Table 12.4 Land case studies in Latin America and the Caribbean

Multi-scale land-use planning		
(a)	**Economic and ecological zoning** Peru	Economic and ecological zoning in the San Martin region contributed to the development of the conceptual and methodological basis for the National Framework for Land Use Planning. Zoning promotes the inclusion of disaster risk management and climate change adaptation in the participatory process of its design and implementation (Castillo 2011).
(b)	**Sustainable mining** Cuba	In 2008, the State Council approved a mining policy that established principles designed to promote the sustainable development of mining in Cuba by creating quality control systems and environmental protection measures, regulating mine closures and determining the recovery of environmental liabilities, among others (ECLAC 2010a).
(c)	**Land policies and community participation** Bolivia, Brazil, Guatemala, Mexico, Peru	Indigenous land management of traditional communal lands of the Bolivian Amazon has been developed to improve the well-being of people living in rural and indigenous communities, and to help protect forest services (Sabogal *et al.* 2008). Other initiatives involve community forest management plans in Oaxaca, Puebla and Quintana Roo, Mexico; in the Maya Biosphere Reserve in El Petén, Guatemala; and in Amazonian communities in Brazil and Peru (UNEP 2010b).
(d)	**Regional marine ecological land-use planning of the Gulf of Mexico and the Caribbean Sea** Mexico	The definition of activities and land uses for this policy were based on three criteria: land suitability; the interests of different stakeholders including tourism, fisheries, agriculture, maritime security and conservation; and other significant issues identified on the basis of their recurrence, intensity or extent. An integrated assessment was prepared that included the most significant interactions between terrestrial and marine ecosystems (SEMARNAT 2011).
Sustainable agriculture and livestock production		
(e)	**Agritourism: connecting sustainable agriculture with tourism** Caribbean	AgroSandals, Jamaica, the Nevis Model of Hotel/Farmer Partnership, and the Tri-Lakes Project in Guyana aim for sustainable agriculture while linking agriculture with tourism and culture, in association with the private sector, community members and government agencies. Programmes have reported good returns: in Jamaica, for example, farmers' sales income increased more than 55 times in the first three years of the initiative, from US$60 000 to US$3.3 million (Harvey 2011).
(f)	**Transition to organic agriculture** Cuba	While ensuring national food security under a trade embargo, Cuba's transition to organic agriculture has also had a positive impact on people's livelihoods by guaranteeing a steady income for a significant proportion of the population. Moreover, the lack of synthetic pesticides in agricultural production is likely to have a positive long-term impact on people's well-being, since such chemicals are often associated with negative health implications including some forms of cancer (UNEP 2011d).
(g)	**Integrated silvo-pastoral ecosystem management** Colombia, Costa Rica, Nicaragua	The Regional Integrated Silvo-pastoral Ecosystem Management Project is piloting the use of payment for ecosystem services to induce the adoption of silvo-pastoral practices in degraded pastures in Colombia, Costa Rica and Nicaragua. In Nicaragua, the area of degraded pasture fell by two-thirds, while pastures with high tree density increased substantially, as did fodder banks and hedges. The project developed an environmental services index and pays participants for net increases in points (Pagiola *et al.* 2007).
(h)	**Sustainable agricultural practices** Argentina	The adoption of conservation tillage and less aggressive pesticides caused, respectively, a dramatic decrease in soil erosion and contamination risk in Argentina throughout 1956–2005. The risk of water and wind-related erosion fell considerably in response to the expansion of zero-tillage agriculture in the last two decades (Viglizzo *et al.* 2011).
(i)	**Integrated crops and livestock with zero tillage** Brazil	Various Brazilian organizations united to develop a project on integrated crop-livestock zero-tillage systems in the Brazilian Cerrado. Rotating such annual crops as maize, soya and rice with no tillage allowed intensification of land use, increased productivity per hectare, and reduced the need for clearing additional land for pasture or arable land. Estimates indicate that this resulted in a reduction in clearance of 0.25–2.5 hectares for every hectare involved in the project. Reported effects of integrating crops and livestock with zero tillage showed less use of leaching herbicides, lower fertilizer use and lower greenhouse gas emissions (Landers 2007).

There are several challenges associated with land-use planning, including the lengthy process of collecting land-use and cadastral data, which requires information about legal titling that is hampered by legal barriers. The often illegal nature of historic land tenure processes, including forced displacement of peasants resulting from civil conflict or corrupt but sophisticated schemes, is another constraint (IDMC 2010; Springer 2006). In addition, transaction costs may be a major obstacle to registration, particularly for the poor (Barnes 2003). Finally, land-use planning policies may discriminate against minority groups such as indigenous and peasant communities, since many land administration projects are based on a simple territorial demarcation and a title issued in the name of the group (Ankersen and Barnes 2003).

The ceding of subsoil rights in indigenous territories to such outside economic interests as oil and mining companies can result in major physical intrusions and habitat damage, generated, for example, by the construction of infrastructure and roads. The Xingú River basin in Brazil is an example of the successful protection of indigenous territory from deforestation through land policies that include community participation (UNEP 2010b; UNEP *et al.* 2009; Sabogal *et al.* 2008).

Regional experiences show that it is much more important to obtain general agreements on land policy direction than to require, *a priori*, a technically perfect legal framework. New legal frameworks have proven ineffective, as insufficient attention has been paid to stakeholder discussions or the dissemination of their rights (Barnes 2003).

Sustainable agriculture and livestock production

In land-use policy making, it is necessary to distinguish between small-scale and large-scale commercial agriculture. Out migration and land sparing, a system under which some land is farmed intensively to maximize yields while other land is protected as a nature reserve, allow more land to be devoted to preserving biodiversity and providing ecosystem services (Green *et al.* 2005), but small-scale agro-ecological systems appear to be a good option for combining hunger alleviation and biodiversity preservation. Perfecto and Vandermeer (2010) suggest using a policy-making matrix that integrates agricultural and conservation elements to boost small-scale agro-ecological options. Policy-making matrices that use a framework built around payment for ecosystem services can significantly strengthen this approach.

Policies that promote organic agriculture, silvo-pastoral practices, ecotourism and sustainable rural tourism, fall within this category. Silvo-pastoral strategies such as planting trees and shrubs in pastures, fodder banks or trees and shrubs as hedges, induce farmers to increase practices that provide ecosystem services – improving biodiversity, sequestering carbon and conserving water resources (Pagiola *et al.* 2007). Policies fostering ecotourism including sustainable rural tourism promote the optimal use of natural resources and respect for socio-cultural diversity, which improves economic viability and distributes benefits more equitably. Well planned rural tourism can promote social development and equity, providing more opportunities for vulnerable groups such as youth, women and indigenous communities (COPLA 2009).

Many Amerindian communities in Columbia are heavily dependent on silvo-pastoral systems for their livelihoods. *© Bob BalestriiStock*

Examples of successful land-use planning are agritourism initiatives in the Caribbean (Table 12.4e) (Harvey 2011); Cuba's transition to organic agriculture (Table 12.4f) (UNEP 2011a); silvo-pastoral practices and PES in Colombia, Costa Rica and Nicaragua (Table 12.4g) (Pagiola *et al.* 2007); and rural-based community tourism in Guatemala and Nicaragua (COPLA 2009).

Land-use policies for large-scale commercial agriculture, which occurs in Argentina and Brazil, should promote sustainability through integrating existing knowledge with input-based farming technology. Policy options include the adoption of agronomic practices such as zero-tillage (Viglizzo *et al.* 2011), minimum tillage, crop diversification, crop rotation and integrated pest management, combined with strategic applications of fertilizers and irrigation water, using low-impact pesticides and the expansion of precision farming procedures (Table 12.4h, i). These practices have had positive impacts in Argentina, where public-private partnerships have been successful (Viglizzo *et al.* 2011); in Paraguay's poultry industry where initiatives for cleaner production have been effective; and in Uruguay's environmentally friendly rice cultivation (UNEP and MercoNet 2011).

Empirical evidence in Latin America and the Caribbean suggests two ways of developing environmentally friendly livestock production systems, regardless of the farming scale: first, by increasing beef productivity through the dilution of maintenance costs; and second, by integrating crops, pastures, fodder and livestock production. The first case results in a significant reduction in land, water, fossil fuels, feed consumption and outputs of manure and greenhouse gases. In the second case, experience with integrated crop rotation, livestock production and zero-tillage operations in the Brazilian Cerrado allowed grain and meat to be produced sustainably on the same lands, thus eliminating the need to deforest more land (Landers 2007) (Table 12.4i).

Coffee cherries on an organic farm in Nicaragua, where there is a burgeoning movement to develop sustainable agricultural value chains that improve farmers' profits. *© Joel Carillet/iStock*

There are examples of successful organic agriculture in most Latin American and Caribbean countries (Table 12.4f), although there is a need to harmonize policies, particularly those related to market access and distribution. Many countries are establishing regulations and standards for organic production while a few are providing limited financial support to pay certification costs during the conversion period (Willer and Kilcher 2011). The current global market for organic production has encouraged the development of standards, certification processes and public-private partnerships to facilitate market access to organic produce.

Access to micro and small-scale credit in poor rural communities is necessary to ensure that land use is managed sustainably. The enabling conditions that facilitate the expansion of sustainable models of large-scale commercial farming generally rely on access to modern technology, for example precision and low-impact farming and information and communication technologies; updated agronomic knowledge; the professional capacity of farmers; good international prices; the financial capacity of individual farmers and investment funds; and credits to farmers' cooperatives.

Box 12.7 Key facts on land degradation in Latin America and the Caribbean

- Land degradation, primarily by water erosion, salinization and reduction of soil fertility, affects approximately 22 per cent of the region's surface area (Bai *et al.* 2008).
- Degradation of cropland in the region's drylands has reached 28 per cent (Zika and Erb 2009).

Restoration of degraded lands

In addition to impacts on biodiversity and the economy, land degradation has social consequences. These include increased vulnerability to floods and dust storms; health risks such as from vector-borne diseases associated with deforestation (Patz and Norris 2004), and illnesses from contaminated sites; loss of environmental services including water source recharge; and decreased carbon sequestration and evapotranspiration. Thus, the region should prioritize the restoration of degraded lands (Box 12.7), which complements conservation and ecosystem management policies oriented to climate change mitigation and adaptation, reduces disaster risk, and helps maintain the hydrologic cycle and water sources.

All available land, particularly degraded or marginal areas, needs to be used efficiently to satisfy the socio-economic and environmental needs of current and future populations as well as to conserve natural ecosystems. Given the environmental, social and economic benefits of land, it is important to institute restoration or rehabilitation policies. Land can simultaneously generate profits through agriculture, livestock husbandry (Aguiar and Roman 2007) or forestry, and maintain and purify water sources, reduce the risk of floods and mudslides and improve people's living conditions (Rees *et al.* 2007). Given the high costs associated with restoration projects, better economic instruments are needed, including government commitments that promote and finance these projects.

Restoring land and environmental services provides new options for productive activities, reduces the vulnerability of populations and decreases the conversion of natural ecosystems to agriculture or pasture. Other commercial activities, such as ecotourism, can also be encouraged. In addition to reclaiming soil and accelerating forest regeneration, for example, the biological corridor project of Nogal-La Selva in Costa Rica (Montagnini 2001) represented an economic incentive for local farmers, while the reforestation project in the Panama Canal watershed reduced the costs of maintaining the canal's infrastructure (ACP 2007; BCEOM-TERRAM 2006). Restoring degraded lands benefits both market-based and non-market-based ecosystem services at multiple spatial scales.

Land restoration policies and action account for the specific conditions of the site and the benefits expected. Effective restoration requires setting specific and clear goals as part of the planning process and ensuring that the parties compromised by the recovery of degraded lands accept them. Thus, implementing the policies requires effective participatory mechanisms that include indigenous and other disadvantaged groups. For this reason, certainty and legitimacy of land tenure are also required.

Climate change

Climate change is exacerbating many of Latin America and the Caribbean's environmental challenges; it also threatens development gains, poverty reduction and economic growth. Although the region accounts for a relatively modest 12 per cent of the world's greenhouse gas emissions, it is already experiencing the adverse consequences of climate change and variability (De la Torre *et al.* 2009). As vulnerability to climate impacts increases, addressing the underlying drivers of risk becomes a top priority. Poverty, marginalization, exclusion from decision-making processes, lack of opportunities, limited access to credit, inadequate education, poor basic infrastructure, inequity, insecure land tenure, and other factors external and internal to the region, continue to exacerbate its vulnerability.

To address climate change, the region needs to commit to the sustained implementation of international and regional agendas, such as the United Nations Framework Convention on Climate Change and its Kyoto Protocol (UNFCCC 1998, 1992), and the Hyogo Framework for Action (ISDR 2005). It should also commit to the sustainable environmental management of forests and key ecosystems; energy efficiency and the development of new, renewable energy sources; ecoagriculture; and the transformation of transport systems, implemented in a socially and environmentally responsible way by respecting people's and communities' rights, and supported by international financial and economic mechanisms (IISD 2010).

With the world's highest percentage of urban dwellers, the Latin America and Caribbean region faces many climate change challenges in its large and growing cities, many of which are located in higher-risk areas on low-lying coastal plains (World Bank 2011b). To build resilience among the segments of the population most in need, municipal policies should be city-specific and work in tandem with national and international efforts for mitigation and adaptation (World Bank 2011b). Although the region's cities have taken many initiatives on policies and activities both to mitigate and adapt to climate change, these have focused mostly on the former. It has been difficult to promote adaptation at the local level without the necessary support from higher levels of government and the international community. This has left a gap in the support and funding of locally determined, locally driven adaptation efforts that serve and work with those most at risk. The best opportunities to adapt to climate change are linked with action to address the underlying causes of vulnerability and respond to more than one problem at a time (Hardoy and Lankao 2011).

The policy clusters described in this section advance progress towards the internationally agreed UNFCCC goal (UNFCCC 1992 Article 3 Paragraph 1–3), selected to address climate change. It emphasizes precautionary measures to anticipate, prevent or minimize the causes of climate change and mitigate its adverse effects.

Reducing the vulnerability of populations through efficient adaptation

Implementing adaptation measures that consider economic, social-ecological and political criteria is an immense challenge. Fostering research programmes on the impacts of climate change, deforestation and land-use change on the natural environment and the social fabric are priorities, as are strengthening evidence-based policy making and the appropriate institutional infrastructure (De la Torre *et al.* 2009).

Box 12.8 Mainstreaming adaptation to climate change in the Caribbean

Mainstreaming adaptation into policy development has proved to be effective in approaching the common goal of increasing resilience. In the Caribbean, for more than 15 years, the Global Environment Facility (GEF), World Bank, CARICOM and other partners have supported a series of policy-oriented adaptation programmes, namely the Caribbean Planning for Adaptation to Climate Change and Mainstreaming Adaptation into Climate Change, both of which have been completed. Currently ongoing are the Special Programme for Adaptation to Climate Change: Implementation of Adaptation Measures in Coastal Zones, and Enhancing Capacity for Adaptation to Climate Change. From this long-term effort, numerous reports from the wider Caribbean have been produced, including sector-specific policy recommendations for water resources, agriculture, forestry, food security, tourism, fisheries and coastal zone management, information and communication, and social services such as health and education (CCCCC 2011).

Policies for adapting to climate change are critical for strengthening natural capital management. This is especially the case for managing changing water flows and improving ecosystem resilience, strengthening direct protection against climate-related threats in cases for which collective action is needed, and strengthening technology transfer and knowledge flows (De la Torre *et al.* 2009).

The following presents a more detailed analysis of the many issues related to adaptation policy development in Latin America and the Caribbean, structured as four policy groupings.

Strengthening ecosystem management for improving resilience: some countries have made significant efforts to provide a more solid methodological and analytical evidence base for understanding the relationship between ecosystem health, resilience and vulnerability. They have also developed economic cost-benefit analyses of ecosystem policy options and their potential in reducing the vulnerability of societies. Innovative policies and financial mechanisms for delivery are also required, as are sustainably resourced and multi-stakeholder capacity building and active participation of local stakeholders in implementing the process. Land-use planning and protected areas are local mechanisms for managing ecosystem services that include the concept of risk reduction (ISDR 2009).

The examples in Table 12.5 provide a combination of economic, environmental and social policy options for climate change adaptation based on improved ecosystem management. Earlier sections on policies for sustaining water provision and consumption, and on payment for ecosystem services are further examples of these policy instruments.

Table 12.5 Climate change case studies

Strengthening ecosystem management		
(a)	**Using the Maya nut tree to increase tropical agro-ecosystem resilience to climate change** Central America, Mexico	The promotion of community-based conservation of the Maya nut tree focuses on rural indigenous women and children, and on increasing agro-ecosystem resilience through the support of different sectors (Buffle and Vohman 2011).
(b)	**An example of synergy: hydrographical basins programme** Cuba	This programme brings together efforts on integrated environmental management; cleaner production, recycling and reuse; reduction of pollution; enterprise management; natural resources protection and biodiversity access and management; and introduction of environmental education at all levels (AMA *et al.* 2009).
(c)	**Mangrove conservation and adaptation** Belize	Efforts have raised local awareness of mangrove habitat as an adaptive climate change management tool and widened the scope of conservation effort through different community initiatives (WWF 2011).
(d)	**Clean energy for marine conservation** Galápagos Marine Reserve, Ecuador	The Japan International Cooperation System Company has helped to plan the introduction of clean solar energy systems in one of 13 islands that form the Galapagos archipelago (UNDP 2010a).
(e)	**Participatory management for an environmentally sensitive area** Trinidad	This project is a participatory process for the protection, development, management and use of the resources of the Aripo Savannas Environmentally Sensitive Area over a 10–15 year period (CANARI 2011).
Towards resilient infrastructure		
(f)	**Sustainable social housing** Brazil	The Sustainable Social Housing Initiative has brought sustainable building practices to social housing programmes, including design criteria and construction practices (UNEP and UNOPS 2011).
(g)	**Coastline protection** Barbados	The Barbados waterfront promenade (boardwalk) was built as an adaptation measure to protect 1.2 km of coastline. The island's coastal resources support diverse ecosystems and the critical tourism industry (Toba 2009).
(h)	**Disaster risk reduction and public policy investments** Costa Rica, Peru	The Ministry of Finance in Peru introduced disaster risk reduction in the assessment of public investments with a strong programme for capacity building. The initiative was replicated by the government of Costa Rica (ISDR 2011).
Strengthening weather monitoring and forecasting tools		
(i)	**Early warning systems** Caribbean, Cuba, Mexico	Forecasts of hydro-meteorological events, volcanoes, tsunamis and droughts have been implemented by Cuba, Mexico's National Centre for Disaster Prevention (CENAPRED) and the Caribbean Disaster Emergency Management Agency (CDEMA), to enable populations to protect themselves from injury and disease (CENAPRED 2011; Rubiera 2010).
(j)	**Early warning, preparedness and risk reduction** Central America	The aim of the Central American Early Warning System (SATCA) is to strengthen early warning systems across disaster-prone Central America, to enhance humanitarian preparedness and build risk reduction capacities among local and regional actors.
Strengthening social protection		
(k)	**Mitigation and adaptation tool for farmers** Central America	The Rainforest Alliance's Climate Module assists farmers identify the risks and impacts of climate change on their farms and communities, and to promote the adoption of good agricultural practices that reduce greenhouse gas emissions, increase carbon sequestration and enhance the capacity of farms to adapt to climate change (Rainforest Alliance 2011).
(l)	**Reducing deforestation in protected areas** Brazil – State of Amazonas	The Bolsa Floresta programme focuses on reducing deforestation in protected areas using four mechanisms: direct payment of grants to smallholder farmers in exchange for forest protection; investments in social improvements in communities; payments to local associations to strengthen local organization and control of the Bolsa Floresta programme; and payments to communities employing sustainable production methods. The programme addresses broad social concerns, which is likely to make the strategy more sustainable in the long term. The programme has reached 32 000 people (Amazonas Sustainable Foundation 2011).
(m)	**Improving community resilience** Nicaragua	The Atención a Crisis pilot project focuses on two interventions: vocational training and a productive investment package to improve the resilience of poor rural households to natural risks and economic downturns (World Bank 2011a).
(n)	**Conservation and sustainable use of biodiversity in the coffee-growing region** Colombia	The Colombian National Coffee Federation has provided local communities with basic infrastructure, improving living conditions on productive agricultural farms. It has also set up a research centre to generate appropriate, competitive and sustainable coffee production technology (Armenteras *et al.* 2005).
(o)	**A model for community-based sustainable ecosystem management** Bolivia	Chalalán Ecolodge is a community business focused on enjoyment of and in-depth learning about the rainforest under the guidance of local indigenous people. It represents a new community business model that integrates environmental issues into design and operation (UNDP 2010b).
(p)	**Adaptation to rapid glacier retreat** Bolivia, Ecuador, Peru	Implemented by the Andean Community of Nations, pilot projects aim to demonstrate the costs and benefits of adapting to rapid glacier retreat in the tropical Andes, and to generate a knowledge base that can then be used to design projects in other vulnerable communities faced with similar challenges.
Encourage diversification of the energy matrix through renewable energy		
(q)	**A tender system for alternative energy** Brazil	Brazil has been encouraging renewable energy through energy auctions, which offer a range of sources such as natural gas and hydro. In 2008, it obtained positive results for biomass, and in 2009 for wind energy, when more than 1 800 megawatts of wind generation was traded (La Rovere *et al.* 2011; Szklo *et al.* 2005).
(r)	**Energy guidelines** Uruguay	In 2006, Uruguay released its Energy Strategy Guidelines to accelerate the transition to renewable energy sources and reduce dependence on oil. These guidelines promote the use of alternative energy sources, especially biofuels, wind energy and biomass as an energy source for industry (PNUMA *et al.* 2008).

Table 12.5 Climate change case studies *continued*

Energy efficiency and low-carbon mobility		
(s)	**Rapid-transit buses** Colombia (also Ecuador, Chile, Mexico)	TransMilenio is a sustainable urban mass-transport system in the metropolitan area of Bogotá, Colombia. The project aims to shift the transport paradigm in urban areas from individual and private to public and inclusive. It replaces the conventional transport system with a bus rapid-transit system, a dedicated track for special buses. First Curitiba and Bogotá, then Río de Janeiro and Porto Alegre have implemented these systems. Six other cities in Colombia, Guayaquil and Quito in Ecuador, Mexico City, and Santiago de Chile have their own bus rapid-transit systems (Grütter Consulting 2006/2010; LIMA 2010; WRI 2010, 2008; Det Norske Veritas 2006).
(t)	**Electricity saving** Brazil	The National Electrical Energy Conservation Programme (PROCEL) is a labelling and reward programme, created in 1985, consisting of a number of sub-programmes in industry, sanitation, education, construction, public buildings, municipal action, promotion of information, technological development and dissemination of experiences and successes (Szklo *et al.* 2005).
(u)	**Transition to energy-efficient lighting** Cuba	Cuba introduced the Cuban Energy Saving Programme (PAEC) in 1997. During the course of just under a decade, less efficient incandescent light bulbs were replaced free of charge with compact fluorescent ones (UNEP 2011c). In 2004, the energy efficiency programme, Energy Revolution, was introduced, under which household appliances throughout the country have been replaced with more efficient ones at subsidized prices (Revolución Cubana 2011).

Towards resilient infrastructure: in light of the risks posed by extreme weather events, reducing the vulnerability of infrastructure systems should be a central objective of climate change adaptation policy. The region has a wide range of potential policy instruments addressing these concerns, the most cost effective and efficient of which rely on enforcing sustainable building standards (Table 12.5f) and relocating vulnerable populations. Large-scale projects to build or replace infrastructure in the coming years present a tremendous opportunity to ensure that physical infrastructure and land-use systems are resilient in a changing climate. The Barbados boardwalk is an example (Table 12.5g).

There is also significant opportunity to improve the cost effectiveness and sustainability of climate-resilient infrastructure investments by more systematically considering ecosystem-based approaches as a component of comprehensive infrastructure adaptation strategies (Henstra and McBean 2009). Another strategy is to integrate disaster risk reduction concepts and methodologies in public investments, as the governments of both Peru and Costa Rica have done (Table 12.5h).

Figure 12.6 Common ground for sustainability

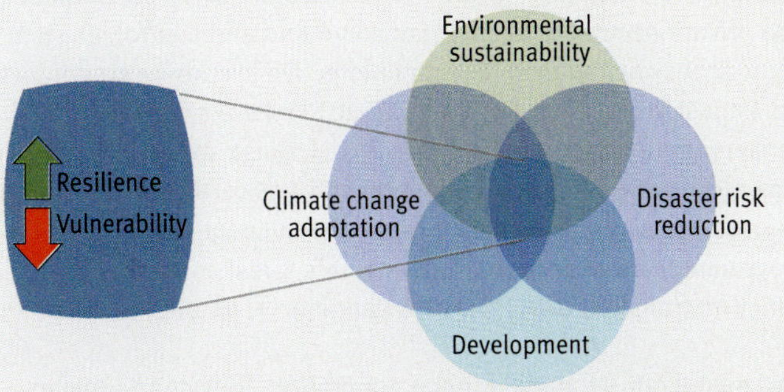

Increasing resilience and decreasing vulnerability are universal goals that lie at the heart of, and are common to, development, environmental sustainability, climate change adaptation and disaster risk reduction. This common ground offers many entry points for integrated intervention.

Strengthening weather monitoring and forecasting tools: early warning systems, one of the main branches of disaster risk reduction, include the monitoring and forecasting of impending events (UNISDR 2006). A number of key intergovernmental organizations work to further early warning policy at the sub-regional level under the Hyogo Framework for Action and through the Regional Platform for Disaster Risk Reduction. Among them are the following: the Centre for Natural Disaster Prevention and Coordination (CEPREDENAC) in Central America; the Andean Committee for Disaster Awareness and Prevention (CAPRADE); and the Caribbean Disaster Emergency Management Agency (CDEMA); as well as humanitarian networks such as the recent REDHU (Humanitarian Assistance of MERCOSUR). Cuba, Mexico, Central America and the Caribbean small islands have implemented weather monitoring and forecasting tools that protect populations from injury and disease (Table 12.5i, j).

Although the region's early warning systems lessen the loss of life, decrease injuries and mitigate against property damage, the World Meteorological Organization stresses the need to re-evaluate national and local emergency preparedness and response plans, which should be based on hazard and vulnerability mapping. It also stresses that countries should strengthen their monitoring and forecasting infrastructure and the skills of technical agencies while improving access to data and technology; strengthen dissemination channels that link national early warning systems to communities focusing on cultural and community needs; and address sustainability issues on the basis of available resources (WMO 2009).

Adaptation policies for social resilience: decreasing vulnerability while increasing resilience is central to development, environmental sustainability, climate change adaptation and

In designing its REDD+ programme, Panama is exploring various scenarios to capture the environmental and social benefits of REDD+ that go beyond carbon. © *Vilainecrevette/iStock*

disaster risk reduction (Figure 12.6). Policy efforts can be integrated around this central challenge (GEF 2011).

Climate change adaptation policies based on social inclusion embrace challenges and opportunities associated with addressing the needs of all segments of the region's population. They are particularly sensitive to the most vulnerable, such as the rural and urban poor and indigenous peoples with traditional lifestyles.

The region's rural households depend heavily on agriculture. Thus, adaptation strategies for coping with the impacts of climate change on agricultural productivity and food security among poor rural households need to include access to such key elements as land, labour, fertilizers, irrigation, infrastructure and financial services (ISDR 2009) as well as technological alternatives. Examples of good policy instruments are the region's agroforestry systems and the Rainforest Alliance's Climate Module, which fosters the adoption of good agricultural practices to reduce greenhouse gas emissions and to enhance the capacity of farms to adapt to climate change in Central America (Table 12.5k) (Rainforest Alliance 2011).

Box 12.9 Brazil's Bolsa Verde

This national-level strategy within the purview of the Brasil Sem Miseria programme envisions the distribution of funds to families living in extreme poverty who promote environmental conservation in the areas where they live and work. Coordinated by the Ministry of Environment, the programme includes environmental capacity building in forest management. Now in its initial stages, the Bolsa Verde has the goal of benefiting nearly 73 000 small-farm families and other traditional communities living in conservation units and land-reform settlements that contain important forest resources (Planalto 2011).

Households, communities and the larger society are increasingly adopting approaches that protect them against the negative impacts of climate change. These include good public policies, such as providing public health services, education, social protection schemes, and supporting active and efficient civil society organizations or government agencies, a solid and well maintained infrastructure, good governance and healthy public finances (Verner 2011).

There are examples of ecologically oriented social policy in Brazil, including Bolsa Verde (Box 12.9), which provides funds to the very poor who work towards environmental conservation, and the State of Amazonas Bolsa Floresta programme (Table 12.5l) (Gebara 2010; May and Millikan 2010). Other examples of policies that increase social resilience come from Bolivia, Colombia, Nicaragua and Peru (Table 12.5m, n, o, p).

Promoting emissions reduction through sustainable forest management, native forest protection and rehabilitation

Maintaining existing forests can be one of the most efficient and cost-effective options for mitigating CO_2 emissions, as can be seen in Brazil, Central America and Mexico (UNEP *et al.* 2010a; Börner and Wunder 2008; Kanninen *et al.* 2007). Protecting and restoring native forests – vitally important in sustaining the livelihoods and cultural heritage of many Latin American and Caribbean peoples – is promoted through such sustainable forest management strategies and results-based payment schemes as REDD+ (Cerbu *et al.* 2011) or the recently created Amazon Fund in Brazil (MMA 2008). Such strategies should focus on carefully integrating and providing benefits to rural and indigenous communities, as there are strong potential synergies with efforts to protect and rehabilitate forest resources (Chhatre and Agrawal 2010; Pereira 2010; Stickler *et al.* 2009). Policy action can rely on a variety of instruments, including payment for ecosystem services, public- and private-sector engagement or command and control, as such approaches can make REDD+ more effective in curbing greenhouse gas emissions (Ezzine-de-Blas *et al.* 2011). Peru's Forest Conservation Programme (MINAM 2011), Bolsa Verde of Brazil (Box 12.9), as well as environmental services certificates, payment for ecosystem services and forest credits in Costa Rica (FONAFIFO programme) are examples of the region's forest conservation policy instruments (UNEP 2010b; Kanninen *et al.* 2007).

Successful efforts at knowledge generation, long-term policies on forest management, and native forest protection and restoration schemes typically strive to provide better information on the value of forest functions and products; strengthen multi-stakeholder involvement; create stronger links between legal, social, environmental, economic and technological tools;

Box 12.10 Energy in Latin America and the Caribbean

- Between 50 and 65 million people live without electricity.
- Electrification rates in areas of Bolivia, Honduras and Nicaragua are below 30 per cent.
- 26 per cent of the region's greenhouse gas emissions is CO_2 from energy generation.
- 23 per cent of the region's energy comes from renewable sources, primarily water, fuelwood and sugarcane products.
- Hydropower generation increased fivefold between 1970 and 2009.
- Several countries have developed regulatory mechanisms for renewable energy.

Source: ECLAC 2011; UNEP and NEF 2010; De la Torre *et al.* 2009; Samaniego *et al.* 2008

and critically evaluate the effectiveness of their objectives by continuously monitoring greenhouse gas emissions reduction and local sustainable development (Corbera and Schroeder 2011; Thompson *et al.* 2011; Cherrington *et al.* 2010; Betts *et al.* 2008; Cortner 2000).

Encouraging diversification of the energy matrix

International prices play a decisive role in defining Latin America and the Caribbean's policies related to fossil fuels. Renewable energy sources have been developed to address growing energy needs (Box 12.10), with hydroelectric projects the preferred energy investment.

Renewable energy sources are a positive alternative to fossil fuels; nevertheless, renewable energy projects can affect the environment and the livelihoods of local communities and, as a result, need to be planned carefully. Given the region's diverse potential for renewable energy – biomass, solar, wind, wave and geothermal – the policy cluster in this section proposes the introduction of renewable sources to the energy matrix.

The benefits of renewable energy sources include:

- the decentralization of investment towards less developed regions, which helps create jobs (for qualified personnel), capacity building and technology transfer (Edenhofer *et al.* 2011; De la Torre *et al.* 2009; Sims *et al.* 2007; Szklo *et al.* 2005);
- a cost-effective alternative to expensive grid extensions (Jacobson and Delucchi 2011); La Rovere *et al.* 2011); and
- great potential to lower emission costs effectively (Sims *et al.* 2007; Szklo *et al.* 2005), thereby reducing energy dependence and positively affecting trade balances.

The suggested policies combine long-term feed-in tariffs with subsidies and tax incentives as ways to provide investment and financial support for the supply chain of electricity generated from renewables, including the transformation of raw materials, manufacturing, and installation of components and systems.

Depending on the policy and regulatory mechanism, renewable energy may increase consumers' energy costs in the short term. However, income-specific tariffs financed by reallocating counter-productive subsidies for non-renewable sources can often help to balance this distortion. If, however, oil prices fall, the opportunity cost may decrease to levels that might not cover costs. This can be addressed by introducing quota-based incentive programmes and long-term contracts with stable prices (De la Torre *et al.* 2009; Guzowsky and Recalde 2008). Adopting policies that subsidize the use of renewable sources in terms of installed capacity (kilowatts), or pay per kilowatt hour generated and sold, can help to improve renewable energy. Likewise, mechanisms such as green certificates, research and development subsides, internalization of external costs and environmental taxes can foster an increase in the share of renewable energy sources in the energy matrix (Guzowsky and Recalde 2008).

Policies involving smart grids and decentralized power generation have the potential to promote greater generation, transport and distribution efficiency and to simultaneously scale up renewables, specifically solar and biomass. Complementarily, transboundary cooperation and integration in the energy sector have been shown to increase electricity supply, widen coverage and enhance system functionality across the region. The Renewable Energy Observatory of Latin America, the Energy and Climate Change Partnership for the Americas and the Mesoamerican Electric Interconnection are examples of policy strategies related to regional cooperation in the energy sector. The Tender System for Alternative Renewable Energies in Brazil and the Energy Strategy Guidelines from Uruguay (Table 12.5q, r) encourage diversification of the energy matrix.

Enhancing efficiency and low-carbon mobility

These policy options aim to reduce energy demand in the residential sector and transport systems while providing more effective and expanded energy distribution to the population. Financial instruments, such as cap-and-trade systems and carbon taxes, funds for research and development and compliance instruments might be adopted as part of the same strategy. A proposed reorganization of public transport systems would improve the efficiency of fossil-fuel use and road space, and change the paradigm from individual and private to public and inclusive.

Related policy strategies foster the use of minimum standards of energy efficiency for electrical appliances (lighting, cooling and heating) and individual vehicles (fuel efficiency standards and the promotion of hybrid cars); and the adoption of energy efficiency stamp programmes and specific nationally appropriate mitigation action. To this end, it is essential to combine public financing instruments, market initiatives and specific policies for research and development, as well as technology transfer to enhance the international transfer of resources associated with new technologies. Table 12.5 (s, t, u) provides examples of associated policy options.

The main benefits of these strategies will be realized over the long term. Some studies reveal that energy efficiency policies have usually reduced implementation costs (Mckinsey and Company 2009; Mckinsey Global Institute 2008a, 2008b, 2007). Furthermore, these policies can help reduce the negative impacts on human health by improving air quality; decrease external energy dependency; increase the reliability of power supply; control demand growth with the potential to reduce energy consumption by 20–25 per cent (ECLAC 2010b); increase productivity and employment; increase the efficiency and competitiveness of domestic energy-intensive industry (Romm 1999); and diminish congestion in cities.

In conjunction with its residential energy efficiency policies, Latin America and the Caribbean has demonstrated the potential to further expand the green design and building market, especially for social housing. One exemplary effort is the government of Mexico's This is Your House initiative and the National Housing Commission's associated Green Mortgage Programme (Comisión Nacional de Vivienda or CONAVI). The government of Brazil is developing initiatives in this vein through the Ministry of Cities' planning instruments within the framework of its Multiannual Plan for Decent Living.

Co-benefits and links between policy options and environmental priorities

To address the complexity of environmental issues, environmental policy making is evolving to transcend traditional, compartmentalized approaches by becoming more integrated and cross-sectoral in nature (UNEP 2009; Persson 2004). This chapter has presented clusters of policies that are believed to have strong co-benefits. An assessment of the policies in the clusters deemed them to be the most co-dependent and mutually supportive policies required to achieve the internationally agreed goals chosen for each thematic issue.

Moreover, in a number of instances the policies or policy clusters associated with certain themes were underscored as benefiting, or being strongly linked with, other policies and environmental themes. Apart from being advantageous to the environment, these policies have positive socio-economic and political impacts. In addition to biodiversity management, policies promoting payment for ecosystem services, for example, are used across a number of issues and in almost all sectors, including land, water and climate change. Policies that focus on integrated management of water resources or the ecosystem approach can also benefit other sectors such as agriculture, fisheries, forestry and land. Many of the climate change policies will ultimately co-benefit the management of land, water and biodiversity resources. Policy makers may find that understanding the links and co-benefits is useful in determining how to maximize the efficiency of existing policies or measures, and in prioritizing the development and implementation of new ones.

It is beyond this chapter's scope to provide an in-depth analysis of all possible links and co-benefits, but policy makers may find

Table 12.6 Links and co-benefits across selected policies

Policy clusters	**Water** Johannesburg Plan of Implementation (JPOI) (WSSD 2002) Paragraph 26c	**Biodiversity** Convention on Biological Diversity (CBD 1992) Article 10	**Land use and land degradation** Johannesburg Plan of Implementation (JPOI) (WSSD 2002) Paragraph 40b	**Climate change** United Nations Framework Convention on Climate Change (UNFCCC 1992) Article 3 Paragraphs 1–3	**Oceans and seas** The Jakarta Mandate of the CBD (CBD 1997)
Integrated water resources management	Efficient, cost-effective and sustainable use of resources and capabilities; reduces water-related conflicts; increases stakeholder participation	Protects watershed habitats; offers potential benefits to biodiversity by protecting freshwater species	Maintains water supply to key ecosystems, reducing the risk of degradation; can help to ensure a sustainable water supply for agriculture	Increases resilience of water supply to climatic changes; reduces the vulnerability of agricultural activities to water-related climate change impacts	Can help to minimize saltwater intrusion in coastal zones; can promote the sustainable supply of water to coastal areas
Enhancing biodiversity conservation through payment for ecosystem services (PES)	Protects watersheds, which can promote a sustainable water supply and provide water purification services	Reduces economic drivers of biodiversity loss; protects ecosystems and species	Reduces economic drivers of land-use change in areas where applied; can be used for rehabilitation of degraded lands	Maintains ecosystems that provide carbon sequestration services; can potentially reduce carbon emissions caused by deforestation	Can help to maintain critical ecosystems such as mangroves and coral reefs; can help to protect coastal and marine species
Multi-scale land-use planning	Protects aquifers and terrestrial areas, ensuring quality delivery of water-related ecosystem services	Reduces the impacts of land-use change in biodiverse regions and habitats	Brings greater coherence to, and management of, different and competing land uses; reduces land degradation	Reduces emissions from deforestation; increases the energy efficiency of urban areas	Can help to reduce the impact of development on coastal ecosystems; very closely linked to integrated coastal zone management
Managing emissions reduction from land-use change: forest management, native forest protection and recovery	May help to improve watershed condition, and contribute to water provision	Forest protection and recovery reduces terrestrial habitat losses of key fauna and maintains supporting ecosystem services for flora	Protecting forest ecosystems reduces the risk of desertification and ensures the delivery of supporting services, such as soil retention, that mitigate the forces of desertification and degradation	Safeguards existing carbon stocks; improves carbon sequestration capacity	Can protect mangroves as existing carbon stocks; can help to reduce emissions from wetlands

Table 12.6 a useful a guide to the type of assessment that may be undertaken to link between and across environmental policies.

CONCLUSIONS

Latin America and the Caribbean's ecosystems and associated natural capital are important to both the region's countries and to the entire planet. However, persistent negative environmental and related socio-economic trends are a clear indication that the measures so far established and implemented to protect them – at national, sub-national or supra-national levels – are insufficient to address either the rate or scale of conversion and consumption prevalent within the region. As a result, Latin American and Caribbean countries continue to face such issues as poverty, inequity and social conflict related to environmental quality.

Throughout this chapter, consideration has been given to policies, approaches and instruments that have demonstrated the potential to improve sustainability in the region, especially for the issues deemed of highest regional priority.

The most salient point arising from the assessment of policy options is that strong environmental governance is a cornerstone for ensuring the success of policies geared at improving sustainability. Without strong governance frameworks to support environmental decision making, efforts to ensure greater environmental sustainability are unlikely to be effective. The following factors were identified as fundamental for strengthening governance frameworks:

- adequate financial resources;
- access to scientific research and information;
- environmental education and the development of an environmental culture;
- the standard governance principles and values of transparency, accountability, equity, sustainability and inclusive participation of all stakeholders; and
- continuity in political systems.

The current limited impact of policies addressing environmental trends also highlights the need to emphasize the root causes that drive change throughout the region. Too often, policies tend to focus on the direct pressures affecting ecosystems and their services, because these are the best understood or they are easiest to deal with. However, until policies begin to address some of the deeper, underlying causes of environmental degradation – or drivers as defined in Chapter 1 – countries are unlikely to meet the goals and targets set out in international, regional and national agreements. There is, thus, a need to invest more in understanding these drivers and the ways they work together. Greater integration of environmental considerations into broader development processes is also needed.

The thematic issues covered have highlighted the interconnectedness of, and links between and among environmental issues. Most of the policy clusters are likely to benefit multiple sectors, once properly implemented. Thus, a close examination of cross-sectoral benefits is an important strategy for policy-makers to apply when considering the priorities and trade-offs associated with implementing a policy or a cluster of policies.

The full effects of decisions taken, or not taken, today will affect the opportunities available to young people and their future families. The Brundtland Commission summed this up in its definition of sustainable development: "satisfying the needs of the present generation without compromising the chance for future generations to satisfy theirs".

© Panama Verde Association, Ennio Arcia

This chapter suggests that existing policies, mechanisms and institutional frameworks at the sub-national, national and regional levels in Latin America and the Caribbean offer a good starting point for strengthening environmental management. In many instances, it is not necessary to reinvent policies and their implementation, or to continue adding to the already saturated landscape of policies. Rather, what is required is a closer examination of existing policies and institutions to see how better to enable and strengthen them to serve more effectively. This approach could help to circumvent the long, sometimes onerous, processes required to build policies and/or new institutions from the ground up, and could accelerate the rate at which countries can work towards meeting internationally agreed goals.

Finally, cooperation is an important element in improving sustainability in the region. Cooperation between and among its countries would facilitate the sharing of information, expertise and technology transfer – the lack of which may currently limit countries in moving to more sustainable paths of development. It could also help to improve the management of ecosystems and species, which commonly cross national boundaries. Cooperation at a global level is also important to ensure that the region's natural capital is maintained and shared in a sustainable and equitable manner.

REFERENCES

ACP (2007). *Programa de Incentivos Económicos Ambientales para la Conservación y Recuperación de los Recursos Hídricos de la Cuenca Hidrográfica del Canal de Panamá 2009–2028*. Autoridad del Canal de Panamá. http://www.acp.gob.pa/esp/pr/press-releases/2009/11/13/pr615.html

Aguiar, M. and Roman, M. (2007). Restoring forage grass to support the pastoral economy of arid Patagonia. In *Restoring Natural Capital: Science, Business and Practice* (eds. Aronson, J., Milton, S.J. and Blignaut, J.N.). pp.112–121. Island Press, Washington, DC

AMA, CITMA and PNUMA (2009). *GEO Cuba: Evaluación del Medio Ambiente Cubano*. Environmental Agency of Cuba, Ministry of Science, Technology and Environment of Cuba and the United Nations Environment Programme

Amazonas Sustainable Foundation (2011). http://www.fas-amazonas.org/en/ (accessed December 2012)

ANAM (2009). *Conservation for Sustainable Development Policy Guidelines of the National Environment Authority*. National Environment Authority of Panama

Andrade Pérez, A. (ed.) (2008). *Applying the Ecosystem Approach in Latin America*. (translator Medina, M.E.). IUCN, Gland

Ankersen, T. and Barnes, G. (2003). Inside the polygon: emerging community tenure systems and forest resource extraction. In *Working Forests in the Latin American Tropics* (eds. Zarin D.J., Alavalapati, J.R.R., Putz, F.E. and Schmink, M., 2004). Columbia University Press, New York

Armenteras, D., Rincón, A. and Ortiz, N. (2005). *Ecological Function Assessment in the Colombian Coffee-growing Region*. Sub-global Assessment Report, Millennium Ecosystem Assessment. http://www.maweb.org/documents_sga/Colombia%20Subglobal%20Report.pdf (accessed 8 December 2011)

Avina (2011). Latin America's Environmental Prosecutors Network produces a manual in Peru. In *Avina Foundation Annual Report*. http://www.informeavina2010.org/english/amazonico.shtml (accessed 30 November 2011)

Azevedo-Ramos, C., Domingues Do Amaral, B., Nepstad, D.C., Soares Filho, B. and Nasi, R. (2006). Integrating ecosystem management, protected areas and mammal conservation in the Brazilian Amazon. *Ecology and Society* 11(2), 17

Bai, Z.G., Dent, D.L., Olsson, L. and Schaepman, M.E. (2008). *Global Assessment of Land Degradation and Improvement. 1 Identification by Remote Sensing*. Report 2008/01. ISRIC World Soil Information, Wageningen

Barnes, G. (2003). Lessons learned: an evaluation of land administration initiatives in Latin America over the past two decades. *Land Use Policy* 20, 367–374

BCEOM-TERRAM (2006). *Valoración Económica de los Recursos Naturales y Diseño de un Sistema de Cuentas Ambientales Satélite en el Marco de las Cuentas Nacionales de Panamá*. http://bdigital.binal.ac.pa/bdp/descarga.php?f=recursosforestalesenpanama.pdf

Bello, E., Rucks, J. and Springer, C. (2009). *Confronting the Challenges of Climate Variability and Change through an Integrated Strategy for the Sustainable Management of the La Plata River Basin*. A United Nations World Water Assessment Programme Dialogue Paper. United Nations Educational, Scientific and Cultural Organization, Paris

Bennett, A.F. (2003). *Linkages in the Landscape: The Role of Corridors and Connectivity in Wildlife Conservation*. Second edition. IUCN, Gland, Switzerland and Cambridge

Bennett, G. (2004). Central America: The Mesoamerican Biological Corridor. In *Integrating Biodiversity Conservation and Sustainable Use: Lessons Learned from Ecological Networks*. IUCN, Cambridge

Bennett, G. and Mulongoy, K.J. (2006). *Review of Experience with Ecological Networks, Corridors and Buffer Zones*. Technical Series No. 23. Secretariat of the Convention on Biological Diversity, Montreal

Betts, R.A., Malhi, Y. and Roberts, J.T. (2008). The future of the Amazon: new perspectives from climate, ecosystem and social sciences. *Philosophical Transactions of the Royal Society B* 363, 1729–1735

Bezaury Creel, J.E. (2009). *El Valor de los Bienes y Servicios que las Áreas Naturales Protegidas Proveen a los Mexicanos*. The Nature Conservancy Programa México – Comisión Nacional de Áreas Naturales Protegidas, México DF

Bezaury Creel, J.E. and Pabón Zamora, L. (2009). *Valuation of Environmental Goods and Services Provided by Mexico's Protected Areas*. The Nature Conservancy-Mexico Program-Comisión Nacional de Áreas Naturales Protegidas, Mexico City

Bianchi, G. and Skjoldal, H.R. (eds.) (2008). *The Ecosystem Approach to Fisheries*. Food and Agriculture Organization of the United Nations, Rome and CABI

BirdLife International (2009). *Ecosystem Profile: The Caribbean Islands Biodiversity Hotspot*. Critical Ecosystem Partnership Fund. Final Draft for submission to the CEPF Donor Council. http://www.cepf.net/Documents/Finaldraft_Caribbean_EP.pdf (accessed 8 December 2011)

Börner, J. and Wunder, S. (2008). Paying for avoided deforestation in the Brazilian Amazon: from cost assessment to scheme design. *International Forestry Review* 10, 496–511

Bovarnick, A., Alpizar, F. and Schnell, C. (eds.) (2010). *The Importance of Biodiversity and Ecosystems in Economic Growth and Equity in Latin America and the Caribbean: An Economic Valuation of Ecosystems*. United Nations Development Programme, New York

Brown, N.A. and Renard, Y. (2000). *Guide to Teaching Participatory and Collaborative Approaches to Natural Resource Management*. CANARI Technical Report 267. Caribbean Natural Resources Institute, Port of Spain

Brudvig, L.A., Damschen, E.I., Tewksbury, J.J., Haddad, N.M. and Levey, D.J. (2009). Landscape connectivity promotes plant biodiversity spillover into non-target habitats. *Proceedings of the National Academy of Sciences of the United States of America* 106, 9328–9332

Buffle, P. and Vohman, E. (2011). *Using the Maya Nut Tree to Increase Tropical Agroecosystem Resilience to Climate Change in Central America and Mexico*. Ecosystems and Livelihoods Adaptation Network (ELAN) Case Study. http://elanadapt.net/sites/default/files/siteimages/maya_nut_0.pdf (accessed 8 December 2011)

Burke, L. and Maidens, J. (2004). *Reefs at Risk in the Caribbean*. World Resources Institute, Washington, DC

Cabeza, A.M. (2002). Ordenación del territorio en América Latina. *Scripta Nova Revista Electrónica De Geografía y Ciencias Sociales* VI (125). http://www.ub.es/geocrit/sn/sn-125.htm (accessed 8 December 2011)

Calvache, A., Benítez, S. and Ramos, A. (2011). *Fondos de Agua, Conservando la infraestructura Verde. Guía de Diseño, Creación y Operación*. Alianza Latinoamericana de Fondos de Agua, The Nature Conservancy, Fundación Femsa y Banco Interamericano de Desarrollo, Bogotá

CANARI (2011). *Participatory Management Planning for the Aripo Savannas, a Protected Area in Trinidad*. Caribbean Natural Resources Institute. http://www.canari.org/fl_ta_1.asp (accessed 14 December 2011)

Cardenas-Moller, M. and Bianco, A. (2011). *Sustainable Development of the Latin American Mining Industry – Its Social Dimension*. Sinclair, Knight and Merz Technical Paper. http://www.skmconsulting.com/cognition/manageddocument.aspx?linkid=633820422135437500 (accessed 8 December 2011)

Castillo, M. (2011). Desarrollo rural reduciendo el riesgo en contextos de cambio climático. In *Sistematización de Experiencias del Programa Desarrollo Rural Sostenible (PDRS-GIZ) en el Perú*. Deutsche Gesellschaft für Internationale Zusammenarbeit (GIZ) and GmbH Programa Desarrollo Rural Sostenible – PDRS, Miraflores. http://www.riesgoycambioclimatico.org/biblioteca/archivos/DC1130.pdf (accessed 9 December 2011)

CBD (2011). *About the Nagoya Protocol*. Convention on Biological Diversity, Montreal. http://www.cbd.int/abs/about/ (accessed 22 November 2011)

CBD (2010). *Aichi Biodiversity Targets*. Convention on Biological Diversity, Montreal. http://www.cbd.int/sp/targets/

CDB (2008). *Access and Benefit-Sharing in Practice: Trends in Partnerships Across Sectors*. Technical Series No. 38. Convention on Biological Diversity, Montreal

CBD (2004). *The Ecosystem Approach (CBD Guidelines)*. Convention on Biological Diversity, Montreal

CBD (1997). *Jakarta Mandate on Marine and Coastal Biological Diversity*. Convention on Biological Diversity. http://www.cbd.int/doc/meetings/mar/jmem-01/official/jmem-01-02-en.pdf

CBD (1992). *Convention on Biological Diversity*. http://www.cbd.int/ (accessed 30 November 2011)

CCAD-UNDP/GEF (2005). *Regional Project to Establish a Program for the Consolidation of the Mesoamerican Biological Corridor (PCCBM)*. Central American Commission for Development and the Environment (CCAD), United Nations Development Programme/Global Environment Facility

CCCCC (2011). *Mainstreaming Adaptation to Climate Change (MACC) Project*. Caribbean Community Climate Change Centre. http://www.caricom.org/jsp/projects/macc%20project/macc.jsp (accessed 9 December 2011)

CEHI and GWP-C (2010). *Toolbox, Rainwater Harvesting in the Caribbean*. Caribbean Environmental Health Institute and Global Water Partnership-Carribean. http://www.cehi.org.lc/Rain/Rainwater%20Harvesting%20Toolbox/about2.htm (accessed 9 December 2011)

CENAPRED (2011). Centro Nacional de Prevención de Desastres, Mexico City. http://www.cenapred.unam.mx/es/ (accessed 9 December 2011)

Cerbu, G.A., Swallow, B.M. and Thompson, D.Y. (2011). Locating REDD: a global survey and analysis of REDD readiness and demonstration activities. *Environmental Science and Policy* 14, 168–180

Cherrington, E.A., Ek, E., Cho, P., Howell, B.F., Hernandez, B.E., Anderson, E.R., Flores, A.I., Garcia, B.C., Sempris, E. and Irwin, D.E. (2010). *Forest Cover and Deforestation in Belize: 1980–2010*. http://www.servir.net/servir_bz_forest_cover_1980-2010.pdf (accessed 9 December 2011)

Chhatre, A. and Agrawal, A. (2010). Trade-offs and synergies between carbon storage and livelihood benefits from forest commons. *Proceedings of the National Academy of Sciences of the United States of America* 106, 17667–17670

CIESIN and CIAT (2005). *Gridded Population of the World, Version 3 (GPWv3)*. Center for International Earth Science Information Network (CIESIN), Columbia University and Centro Internacional de Agricultura Tropical (CIAT). Socioeconomic Data and Applications Center (SEDAC), Columbia University, Palisades, NY. http://sedac.ciesin.columbia.edu/gpw (accessed 20 November 2011)

Cimorelli, A.J. and Stahl, C.H. (2005). Tackling the dilemma of the science-policy interface in environmental policy analysis. *Bulletin of Science Technology Society* 25, 276–284

Cisneros, J. and Lloret, P. (2008). *El Fondo para la Protección del Agua. Mecanismo Financiero para la Conservación y el Cuidado del Agua en Quito, Ecuador*. Seminario Internacional: Cogestión de cuencas hidrográficas experiencias y desafíos. USAID, Quito. http://orton.catie.ac.cr/repdoc/A2983E/A2983E11.PDF (accessed 9 December 2011)

Colegio de Postgraduados (2004). *Centro Internacional de Demonstración y Capacitación en Aprovechamiento del Agua de Lluvia (CIDECALLI)*. Colegio de Postgraduados, Institución de Enseñanza e Investigación en Ciencias Agrícolas, Mexico. http://www.colpos.mx/ircsa/cidecall/odcs/carpeta.pdf (accessed 9 December 2011)

COPLA (2009). *Rural Community-based Tourism In Central America*. Comercio y Pobreza en Latino América (COPLA) (Trade and Poverty in Latin America). http://www.odi.org.uk/resources/docs/5648.pdf (accessed 9 December 2011)

Corbera, E. and Schroeder, H. (2011). Governing and implementing REDD+. *Environmental Science and Policy* 14, 89–99

Cortner, H.J. (2000). Making science relevant to environmental policy. *Environmental Science and Policy* 3, 21–30

Cuartas, M.F. (2008). *State of 101 Protected Areas in Latin America*. Unpublished Masters' project. Duke University, Durham, NC

CZMU Barbados (2011). *Coastal Zone Management Unit Barbados*. http://www.coastal.gov.bb/index.cfm (accessed 9 December 2011)

Dalhuisen, J. and Nijkamp, P. (2002). *Enhancing Efficiency of Water Provision: Theory and Practice of Integrated Water Management Principles*. Tinbergen Institute Discussion Paper, Amsterdam. http://www.tinbergen.nl/ti-publications/discussion-papers.php?paper=303 (accessed 9 December 2011)

De Andrade, R. (2008) The ecosystem approach and the management of marine and coastal protected areas (MCPAs) in Chile. In *Applying the Ecosystem Approach in Latin America* (ed. Andrade Pérez, A.) (translator Medina, M.E.). IUCN, Gland

De Freitas, C.M., de Oliveira, S.G., Schutz, G.E., Freitas, M.B. and Camponovo, M.P.G. (2007). Ecosystem approaches and health in Latin America. *Cadernos Saúde Pública* 23, 283–296

Deininger, K. and Binswanger, H. (1999). The evolution of the World Bank's land policy: principles, experience, and future challenges. *The World Bank Research Observer* 14, 247–276

De La Torre, A., Fajnzylber, P. and Nash, J. (2009). *Low Carbon, High Growth. Latin American Responses to Climate Change: An Overview.* The International Bank for Reconstruction and Development/The World Bank, Latin America and Caribbean Studies, Washington, DC

Det Norske Veritas (2006). *Validation Report* (CDM-UNFCCC). TransMilenio Phase II–IV in Colombia. Veritas, Høvik. http://cdm.unfccc.int/filestorage/U/P/U/UPUWD3ZSZM1IT2Y09EBST6WQT96IJN/DNV_Transmilenio_Validation%20Report_24-09-06ETEL.pdf?t=QmV8bHZ5MTA4fDBP-tgpAfL05u1KjTmJE-U2 (accessed 9 December 2011)

Diaz, S. (2010). *Biodiversity and Human Well-being in Latin America and the Caribbean: A Multi-Sectoral Contribution to the Science-Policy Interface.* Policy Brief. International Council for Science (ICSU)-ROLAC. http://www.icsu.org/icsu-latin-america/publications/policy-briefs/policy-brief-biodiversity/ROLAC_biodiversity_policybrief_en.pdf (accessed 9 December 2011)

Di Filippo, P. (2000). The mists of Riachuelo. *The Argentina Independent*, 11 April 2000. http://www.argentinaindependent.com/socialissues/environment/the-mists-of-riachuelo-/ (accessed 9 December 2011)

Dijk, K. and Savenije, H. (2009). *Towards National Financing Strategies for Sustainable Forest Management in Latin America. Overview of the Present Situation and the Experience in Selected Countries.* Forest Policy and Institutions Working Paper. Food and Agriculture Organization of the United Nations, Rome

Dudley, N. and Rao, M. (2008). *Assessing and Creating Linkages Within and Beyond Protected Areas: A Quick Guide for Protected Area Practitioners.* Quick Guide Series (ed. Ervin, J.). The Nature Conservancy, Arlington, VA

ECLAC (2011). CEPALSTAT: Databases and Statistical Publications. Economic Commission for Latin American and the Caribbean, Santiago. http://websie.eclac.cl/infest/ajax/cepalstat.asp?idioma=i (accessed 9 December 2011)

ECLAC (2010a). *Sustainable Development in Latin America and the Caribbean: Trends, Progress, and Challenges in Sustainable Consumption and Production, Mining, Transport, Chemicals and Waste Management.* Report to the 18th Session of the Commission on Sustainable Development of the United Nations. Economic Commission for Latin American and the Caribbean, Santiago. http://www.un.org/esa/dsd/csd/csd_pdfs/csd-18/rims/LAC_background_eng.pdf (accessed 9 December 2011)

ECLAC (2010b). *Energy Efficiency in Latin America and the Caribbean: Situation and Outlook.* Economic Commission for Latin American and the Caribbean, Santiago

Edenhofer, O., Pichs-Madruga, R., Sokona, Y. and Seyboth, K. (2011). *Summary for Policy Makers. IPCC Special Report on Renewable Energy Sources and Climate Change Mitigation.* Cambridge University Press, Cambridge

Eguren, L. (2004). *El Mercado de Carbono en América Latina y el Caribe: Balance y Perspectivas.* Serie Medio Ambiente y Desarrollo 83. Economic Commission for Latin American and the Caribbean, Santiago. http://www.eclac.org/publicaciones/xml/2/14902/lcl2085e.pdf (accessed 9 December 2011)

Elbers, J. (ed.) (2011). *Las Áreas Protegidas de América Latina: Situación Actual y Perspectivas para el Futuro.* IUCN, Quito

Emilsson, S., Tyskeng, S. and Carlsson, A. (2004). Potential benefits of combining environmental management tools in a local authority context. *Journal of Environmental Assessment Policy and Management* 6, 131–151

Ezzine-de-Blas, D., Börner, J., Violato-Espada, A.-L., Nascimento, N. and Piketty, M.-G. (2011). Forest loss and management in land reform settlements: implications for REDD governance in the Brazilian Amazon. *Environmental Sciences and Policy* 14, 188–200

Fanning, L., Mahon, R., McConney, P., Angulo, J., Burrows, F., Chakalall, B., Gil, D., Haughton, M., Heileman, S., Martínez, S., Ostine, L'O., Oviedo, A., Parsons, S., Phillips, T., Santizo Arroya, C., Simmons, B. and Toro, C. (2007). A large marine ecosystem governance framework. *Marine Policy* 31, 434–443

FAO (2011). *Payments for Ecosystem Services and Food Security.* Office of Knowledge Exchange, Research and Extension, Food and Agriculture Organization of the United Nations, Rome

FAOSTAT (2011). *FAO Statistical Database.* Food and Agriculture Organization of the United Nations, Rome. http://faostat.fao.org/site/377/default.aspx#ancor (accessed 21 March 2012)

Farley, J. and Costanza, R. (2010). Payments for ecosystem services: from global to local. *Ecological Economics* 69, 2060–2068

Ferraro, P. (2001). Global habitat protection: limitations of development interventions and a role for conservation performance payments. *Conservation Biology* 15, 990–1000

Forero, E.G. (2008). The EA and water management: a Latin American perspective. In *Applying the Ecosystem Approach in Latin America* (ed. Andrade Pérez, A.) (translator Medina, M.E.). IUCN, Gland

Gaventa, J. and Valderrama, C. (1999). *Participation, Citizenship and Local Governance.* Background note prepared for Strengthening Participation in Local Governance workshop. Institute of Development Studies, Brighton. http://www.uv.es/~fernandm/Gaventa,%20Valderrama.pdf (accessed 9 December 2011)

Gebara, M.F. (2010). *Benefit-Sharing Mechanisms for REDD: How to Equitably Share Benefits Among Forest Managers?* Oxford Centre for Tropical Forests (OCTF) and Center for International Forestry Research (CIFOR)

GEF (2011). *Tracking Progress for Effective Action – A Framework for Monitoring and Evaluating Adaptation to Climate Change* (Sanahuja, H.). Community of Practice, Global Environment Facility. http://www.climate-eval.org/sites/default/files/file/StudyFrameworksAdaptation_2011_08_20.pdf (accessed 9 December 2011)

Geoghegan, T. and Renard, Y. (2002). Beyond community involvement: lessons from the insular Caribbean. *Parks* 12, 16–25

GIZ/PROAPAC (2011). *Memorias de los Talleres Internacionales sobre Arreglos Institucionales para Provisión de Agua Potable y Gestión de Aguas Residuales.* Programa de Agua Potable y Alcantarillado Sanitario en Pequeñas y Medianas Ciudades, La Paz

Glowka, L. (1998). A guide to designing legal frameworks to determine access to genetic resources. *IUCN Environmental Policy And Law Papers* 34. IUCN Gland, Cambridge and Bonn

Grau, R. and Aide, M. (2008). Globalization and land-use transitions in Latin America. *Ecology and Society* 13, 16

Gray, N.J. (2008). *Producing Success: Co-Management of a Marine Protected Area in Belize.* Presented at Governing Shared Resources: Connecting Local Experience to Global Challenges, 12th Biennial Conference of the International Association for the Study of Commons, Cheltenham, England, July 14–18, 2008

Green, R.E., Cornell, S., Scharlemann, J.P.W. and Balmford, A. (2005). Farming and the fate of wild nature. *Science* 307, 550–555

Grütter Consulting (2006/2010). *Project Design Document: TransMilenio Phase II to IV, Colombia.* http://cdm.unfccc.int/filestorage/E/6/L/E6LUMUUAQA83IUZAPO9XWBMS6BTSAB/PDD%20version%206-09-06.pdf?t=aWN8bHZ5NWhwfDAQChdoLZNFKuUXob3C5eq4 (accessed 9 December 2011)

Guarderas, A.P., Hacker, S.D. and Lubchenco, J. (2008). Current status of marine protected areas in Latin America and the Caribbean. *Conservation Biology* 22, 1630–1640

Guzowski, C. and Recalde, M. (2008). Renewable energy in Argentina: energy policy analysis and perspectives. *International Journal of Hydrogen Energy* 33, 3592–3595

GWP-C (2010). GWP-C creates rainwater harvesting model to help water stricken Caribbean communities. Global Water Partnership, Caribbean. http://www.gwp-caribbean.org/news.aspx?ArticleID=187 (accessed 9 December 2011)

GWSP (2008). Map 72: Environmental water stress indicator (V1.0). In *Digital Water Atlas.* Global Water System Project. http://atlas.gwsp.org (accessed 24 November 2011)

Hardoy, J. and Lankao, P.R. (2011). Latin American cities and climate change: challenges and options to mitigation and adaptation responses. *Current Opinion in Environmental Sustainability* 3, 158–163

Harvey, C., Komar, O., Chazdon, R., Ferguson, B.G., Finegan, B., Griffith, D.M., Martínez-Ramos, M., Morales, H., Nigh, R., Soto-Pinto, L., van Breugel, M. and Wishnie, M. (2008). Integrating agricultural landscapes with biodiversity conservation in the Mesoamerican hotspot. *Conservation Biology* 22, 8–15

Harvey, E. (2011). *Agritourism Development in the Caribbean: Some Experiences and Lessons.* Barbados Agritourism Unit with the Inter-American Institute for Cooperation on Agriculture (IICA). http://agri-tourismlinkages.com/agrosuccess.pdf (accessed 9 December 2011)

Henstra, D. and McBean, G. (2009). *Climate Change and Extreme Weather: Designing Adaptation Policy.* Simon Fraser University, British Columbia. http://act-adapt.org/wp-content/uploads/2011/03/PDF-WeatherSession_BackgroundReport.pdf (accessed 9 December 2011)

Hurtado Badiola, M. (2008). *Environmental Culture.* Editorial Trillas, Mexico

IDB (2002). *Ecuador: Rural Land Regularization and Administration Programme.* Loan Proposal Document EC-0191. Inter-American Development Bank, Washington, DC

IDMC (2010). *Building Momentum for Land Restoration: Towards Property Restitution for IDPs in Colombia.* Internal Displacement Monitoring Centre, Geneva. http://www.internal-displacement.org/8025708F004BE3B1/(httpInfoFiles)/A0CCF5D6CC55525DC12577D600458E97/$file/Colombia_SCR_Nov2010.pdf (accessed 9 December 2011)

IISD (2010). Summit on Latin American and Caribbean unity addresses climate change negotiations. *Climate Change, Policy and Practice, Knowledgebase of UN/Intergovernmental Activities Addressing Global Climate Change 2010.* International Institute for Sustainable Development, Winnipeg. http://climate-l.iisd.org/news/summit-on-latin-american-and-caribbean-unity-addresses-climate-change-negotiations/ (accessed 9 December 2011)

INEGI (2011). *Sistema de Cuentas Económicas y Ecológicas.* Instituto Nacional de Estadística, Geografía e Informática, Mexico. http://www.inegi.org.mx/est/contenidos/proyectos/scn/c_anuales/c_econecol/scee_46.aspx (accessed 9 December 2011)

INVEMAR (2011). *Unidad de Manejo Integrado UMI Guapi-Iscuandé Pacîfico Colombiano.* Instituto Investigaciones Marinas y Costeras, Santa Marta. http://www.invemar.org.co/redcostera1/invemar/docs/2828UMI%20GUAPI.pdf (accessed 9 December 2011)

IPSRM (2010). *Assessing Global Land Use and Soil Management for Sustainable Resource Policies.* International Panel for Sustainable Resource Management (IPSRM/UNEP), Paris

ISDR (2011). *Global Assessment Report on Disaster Risk Reduction: Revealing Risk, Redefining Development.* International Strategy for Disaster Reduction, Geneva. http://www.preventionweb.net/gar (accessed 9 December 2011)

ISDR (2009). *Global Assessment Report on Disaster Risk Reduction.* International Strategy for Disaster Reduction, Geneva

ISDR (2005). *Hyogo Framework for Action 2005–2015: Building the Resilience of Communities to Disaster.* International Strategy for Disaster Reduction, Geneva. http://www.unisdr.org/2005/wcdr/intergover/official-doc/L-docs/Hyogo-framework-for-action-english.pdf

Islam, M.R. and Koudstaal, R. (2003). *Coastal Zone Management: An Analysis of Different Policy Documents.* Working Paper WP009. Program Development Office for Integrated Coastal Zone Management

Jacobson, M.S. and Delucchi M.A. (2011) Providing all global energy with wind, water and solar power, Part I: Technologies, energy resources, quantities and areas of infrastructure, and materials. *Energy Policy* 39, 1154–1169

Jindal, R. and Kerr, J. (2007). *Lessons and Best Practices for Pro-poor Payment for Ecosystem Services.* USAID PES Sourcebook. Sustainable Agriculture and Natural Resource Management Collaborative Research Support Program, Virginia

Jouravley, A. (2001). *Water Management in Latin America and the Caribbean on the Threshold of 21st Century.* Economic Commission for Latin America and the Caribbean, Santiago. http://www.medioambienteonline.com/web/guest/viewpoint-articles/article/-/article/ug2M/21606/-1/859

Kanninen, M., Murdiyarso, D., Seymour, F., Angelsen, A., Wunder, S. and German, L. (2007). *Do Trees Grow on Money? The Implications of Deforestation Research for Policies to Promote REDD.* Center for International Forestry Research (CIFOR), Bogor. http://www.cifor.org/nc/online-library/browse/view-publication/publication/2347.html (accessed 9 December 2011)

Landers, J. (2007). Tropical crop-livestock systems in conservation agriculture: the Brazilian experience. *Integrated Crop Management.* vol. 5. Food and Agriculture Organization of the United Nations, Rome. ftp://ftp.fao.org/docrep/fao/010/a1083e/a1083e.pdf (accessed 9 December 2011)

La Rovere, E.L., do Valle, C., Pereira, A. and Poppe, M.K. (2011). *Projeto "Carta do Sol" – Relatório Técnico: Subsídios para o Planejamento da Promoção da Energia Solar Fotovoltáica no Brasil.* Laboratório Interdisciplinar do Meio Ambiente (LIMA) and Rio de Janeiro Federal University, Rio de Janeiro

Larson, A.M. (2003). Decentralisation and forest management in Brazil: towards a working model. *Public Administration and Development* 23, 211– 226

LIMA (2010). *Inventário de Emissões de Gases do Efeito Estufa da Cidade do Rio de Janeiro.* Laboratório Interdisciplinar do Meio Ambiente/COPPE/UFRJ, Rio de Janeiro

Loper, C.E., Balgos, M.C., Brown, J., Cicin-Sain, B., Edwards, P., Jarvis, C., Lilley, J., Torres de Noronha, I., Skarke, A., Tavares, J.F. and Walker, L. (2005). *Small Islands, Large Ocean States: A Review of Ocean and Coastal Management in Small Island Developing States since the 1994 Barbados Programme of Action for the Sustainable Development of Small Island Developing States (SIDS).* Papers Series No. 2005-1. UNEP/GPA and the Global Forum on Oceans, Coasts, and Islands

López, A. and Jiménez, A. (2007). *Latin American Assessment, Environmental Conflict and Cooperation: The Mesoamerican Biological Corridor as a Mechanism for Transborder Environmental Cooperation.* Report of the Regional Consultation, 4–5 July 2006, Mexico City. United Nations Environment Programme, Nairobi

Mahon, R., Fanning, L. and McConney, P. (2011). *Wider Caribbean Region Ocean Governance Lessons.* Conference on Sustainable Oceans and the Eradication of Poverty in the Context of the Green Economy, Principality of Monaco, 28–30 November 2011

Mahon, R., Fanning, L. and McConney, P. (2010). *Observations on Governance in the Global Environment Facility (GEF) International Waters (IW) Programme.* The GEF Transboundary Waters Assessment Programme (TWAP) Large Marine Ecosystem (LME) Working Group

Maretti, C.C. (2003). *Protected Areas and Indigenous and Local Communities in Brazil.* WCPA Ecosystems, Protected Areas and People (EPP) project. IUCN, Gland

May, P. and Millikan, B. (2010). *The Context of REDD+ in Brazil: Drivers, Agents and Institutions.* Center for International Forestry Research (CIFOR), Bogor

McElhinny, V. (2007). *Information Brief.* Second Latin American Congress of National Parks and Protected Areas, San Carlos de Bariloche, Argentina, 1–4 October 2007. Bank Information Centre (BIC), Washington, DC

McKinsey and Company (2009). *Caminhos para uma Economia de Baixa Emissão de Carbono no Brasil.* McKinsey and Company. http://veja.abril.com.br/40anos/ambiente/pdf/relatorio-mckinsey.pdf (accessed 9 December 2011)

McKinsey Global Institute (2008a). *The Carbon Productivity Challenge: Curbing Climate Change and Sustaining Economic Growth.* McKinsey Global Institute. http://www.mckinsey.com/Insights/MGI/Research/Natural_Resources/The_carbon_productivity_challenge (accessed 9 December 2011)

McKinsey Global Institute (2008b). *The Case for Investing in Energy Productivity.* McKinsey Global Institute. http://www.mckinsey.com/Insights/MGI/Research/Natural_Resources/The_case_for_investing_in_energy_productivity (accessed 9 December 2011)

McKinsey Global Institute (2007). *Curbing Global Energy Demand Growth: The Energy Productivity Opportunity.* McKinsey Global Institute. http://www.mckinsey.com/Insights/MGI/Research/Natural_Resources/Curbing_global_energy_demand_growth (accessed 9 December 2011)

Melo, J.C. (2005). *The Experience of Condominial Water and Sewerage System in Brazil.* World Bank and Water Sanitation Program, Lima

Milder, J.C., Scherr, S.J. and Bracer, C. (2010). Trends and future potential of payment for ecosystem services to alleviate rural poverty in developing countries. *Ecology and Society* 15, 4

MINAM (2011). *Programa Nacional de Conservación de Bosques para la Mitigación del Cambio Climático.* Ministerio del Ambiente, Lima. http://bosques.minam.gob.pe/ (accessed 9 December 2011)

MMA (2008). *Plano Nacional de Mudanças Climáticas.* Ministério do Meio Ambiente, Brasilia

Montagnini, F. (2001). Strategies for the recovery of degraded ecosystems: experiences from Latin America. *Interciencia* 26, 498–503

Montagnini, F. and Finney, C. (2011). Payments for environmental services in Latin America as a tool for restoration and rural development. *Ambio* 40, 285–297

Mora, C. and Sale, P.F. (2011). Ongoing global biodiversity loss and the need to move beyond protected areas: a review of the technical and practical shortcomings of protected areas on land and sea. *Marine Ecology Press Series* 434, 251–266

Moreno-Sánchez, R. and Maldonado, J.H. (2008). *Can Co-management Improve Governance of a Common-Pool Resource? Lessons from a Framed Field Experiment in a Marine Protected Area in the Colombian Caribbean.* Working Paper Series No. 2008-WP5. Latin America and the Caribbean Environmental Economics Program, Turrialba

Oviedo, G. (2006). Community conserved areas in South America. In *Community Conserved Areas* (ed. Goriup, P.). *Parks* 16, 49–55

Pagiola, S., Ramírez, E., Gobbi, J., de Haan, C., Ibrahim, M., Murgueitio, E. and Ruíz, J. (2007). Paying for the environmental services of silvopastoral practices in Nicaragua. *Ecological Economics* 64, 374–385

Pasteur, K. and Blauert, J. (2000). *Participatory Monitoring and Evaluation In Latin America: Overview of Literature with Annotated Bibliography.* The World Bank. http://siteresources.worldbank.org/INTPCENG/1143331-1116505657479/20509244/pme-latam.pdf (accessed 9 December 2011)

Patz, J.A. and Norris, D.E. (2004). Land use change and human health. In *Ecosystems and Land Use Change* (eds. DeFries R., Asner, G. and Houghton, R.). Geophysical Monograph 153. pp.159–167. American Geophysical Union, Washington, DC

Pereira, S.N.C. (2010). Payment for environmental services in the Amazon forest: how can conservation and development be reconciled? *Journal for Environment and Development* 19, 171–190

Perfecto, I. and Vandermeer, J. (2010). The agroecological matrix as alternative to the landsparing/agriculture intensification model. *Proceedings of the National Academy of Sciences of the United States of America* 107, 5787–5791

Persson, A. (2004). *Environmental Policy Integration: An Introduction.* Policy Integration for Sustainability Background Paper. Stockholm Environment Institute, Stockholm

Pfaff, A., Robalino, J.A. and Sanchez-Azofeifa, G.A. (2008). *Payments for Environmental Services: Empirical Analysis for Costa Rica.* Working Paper Series SAN08-05. Terry Sanford Institute of Public Policy, Duke University, Durham, NC

Planalto (2011). *Lei nº 12.512, 2011 – Institui o Programa de Apoio à Conservação Ambiental e o Programa de Fomento às Atividades Produtivas Rurais.* Government of the Republic of Brazil. http://www.planalto.gov.br/ccivil_03/_Ato2011-2014/2011/Lei/L12512.htm

PNUMA, CLAES and DINAMA (2008). *GEO Uruguay 2008.* http://www.pnuma.org/deat1/pdf/GEOUruguay2008.pdf (accessed 16 December 2011)

PROAGRO/GTZ/DED (2010). *Experiencias de la Cooperación Alemana en el Manejo Integral de Cuencas y la Gestión Integral de Recursos Hídricos en Bolivia.* Primera Edición. El Programa de Desarrollo Agropecuario Sostenible (PROAGRO), La Paz

Rainforest Alliance (2011). New tool to help farmers mitigate and adapt to climate change (press release). http://www.rainforest-alliance.org/newsroom/news/san-climate-module-release (accessed 18 December 2011)

Ramcharan, E. (2001). *Elements of Coastal Zone Management: Coastal Zone/Island Systems Management.* CDMC Professional Development Programme, Coastal Infrastructure Design, Construction and Maintenance, Chapter 1. Organization of American States (OAS) and USAID. http://www.oas.org/cdcm_train/courses/course1/chapter%201-eements%20of%20coastal%20management.pdf (accessed 9 December 2011)

Redford, K.H. and Adams, W.M. (2009). Payment for ecosystem services and the challenge of saving nature. *Conservation Biology* 23, 785–787

Rees, W.E., Farley, J., Vesely, É.-T. and de Groot, R. (2007). Valuing natural capital and the costs and benefits of restoration. In *Restoring Natural Capital: Science, Business, and Practice* (eds. Aronson, J., Milton, S.J. and Blignaut, J.N.). pp.227–236. Island Press, Washington, DC

Reid-Grant, K. and Bhat, M.G. (2009). Financing marine protected areas in Jamaica: an exploratory study. *Marine Policy* 33,128–136

Revolución Cubana (2011). *Revolución Energética.* Centro de Información para la Prensa de la Unión de Periodistas de Cuba, Havana. http://revolucioncubana.cip.cu/logros/desafios-del-desarrollo-economico/sector-energetico-1/revolucion-energetica (accessed 9 December 2011)

Rietbergen, S., Hammond, T., Sayegh, C., Hesselink, F. and Mooney, K. (2007). *Island Voices – Island Choices: Developing Strategies for Living with Rapid Ecosystem Change in Small Islands.* IUCN, Gland

Rivera, V.S., Cordero, P.M., Borras, M.F., Govan, H. and Vera, V. (2006). Community conservation areas in Central America: recognising them for equity and good governance. In *Community Conserved Areas* (ed. Goriup, P.). *Parks* 16, 21–27

Romm, J.J. (1999). *Cool Companies: How the Best Businesses Boost Profits and Productivity by Cutting Greenhouse Gas Emissions.* Island Press, Washington, DC

Rubiera, J. (2010). *Early Warning System for Tropical Cyclones in the Republic of Cuba.* Presentation to the DRR Technical Conference, 20–21 September 2010, Bogotá, Colombia. http://www.wmo.int/pages/prog/drr/events/TECORAIII/Session2/Dr.%20Jose%20Rubiera%20(CUBA).pdf (accessed 9 December 2011)

Sabogal, C., de Jong, W., Pokorny, B. and Louman, L. (eds.) (2008). *Manejo Forestal Comunitario en América Latina: Experiencias, Lecciones Aprendidas y Retos Para el Futuro: Resumen Ejecutivo.* CIFOR–CATIE, Turrialba. http://www.cebem.org/cmsfiles/publicaciones/MFC_America_Latina_Resumen_Ejecutivo.pdf (accessed 9 December 2011)

Samaniego, M.R., Garcia-Perez, M., Cortez, L.B., Rosillo-Calle, F. and Mesa, J. (2008). Improvements of Brazilian carbonization industry as part of the creation of a global biomass economy. *Renewable and Sustainable Energy Reviews* 12, 1063–1086

SEMARNAT (2011). *Programa de Ordenamiento Ecológico Marino y Regional del Golfo de México y Mar Caribe.* http://www.semarnat.gob.mx/temas/ordenamientoecologico/Paginas/B_A_GolfoMex_Caribe.aspx (accessed 14 December 2011)

SEMARNAT (2009). *Indicadores Básicos de Desempeño Ambiental de México.* Secretaría de Medio Ambiente y Recursos Naturales, Mexico. http://app1.semarnat.gob.mx/dgeia/indicadores_2010_web/indicadores_2010/02_agua/02_introduccion.html (accessed 23 November 2011)

Simpson, B. (2010) International involvement in preservation of the Brazilian Amazon rainforest: context, constraints and scope. *Asia Pacific Journal of Environmental Law* 13(1), 39–59

Sims, R.E.H., Schock, R.N., Adegbululgbe, A., Fenhann, J., Konstantinaviciute, I., Moomaw, W., Nimir, H.B., Schlamadinger, B., Torres-Martínez, J., Turner, C., Uchiyama, Y., Vuori, S.J.V., Wamukonya, N. and Zhang, X. (2007). Energy supply. In *Climate Change 2007: Mitigation* (eds. Metz, B., Davidson, O.R., Bosch, P.R., Dave, R. and Meyer L.A.). Contribution of Working Group III to the Fourth Assessment Report of the Intergovernmental Panel on Climate Change. Cambridge University Press, Cambridge and New York

Singh, A. (2008). *Governance in the Caribbean Sea: Implications for Sustainable Development.* United Nations – Nippon Foundation Fellowship Programme

Springer, N. (2006). *Colombia: Internal Displacement, Policies and Problems.* A Writenet Report commissioned by United Nations High Commissioner for Refugees, Status Determination and Protection, Information Section (DIPS), UK

Staveland-Sæter, K.I. (2011). *Litigating the Right to a Healthy Environment: Assessing the Policy Impact of "The Mendoza Case".* Chr. Michelsen Institute (CMI) Report, Bergen. http://www.cmi.no/publications/file/4258-litigating-the-right-to-a-healthy-environment.pdf (accessed 28 November 2011)

Stem, C., Margoluis, R., Salafasky, N. and Brown, M. (2005). Monitoring and evaluation in conservation: a review of trends and approaches. *Conservation Biology* 19(2), 295–309

Stickler, C.M., Nepstad, D.C., Coe, M.T., McGrath, D.C., Rodrigues, H.O., Walker, W.S., Soares-Filho, B.S. and Davidson, E.A. (2009). The potential ecological costs and co-benefits of REDD: a critical review and case study from the Amazon region. *Global Change Biology* 15, 2803–2824

Sutherland, D. and Fenn, C. (2000). *Assessments of Water Supply Options.* World Commission on Dams Secretariat, Cape Town

Szklo, A.S., Schaeffer, R., Schuller, M.E. and Chandler, W. (2005). Brazilian energy policies side-effects on CO_2 emissions reduction. *Energy Policy* 33, 343–64

The Economist (2011). Statistics and lies. http://www.economist.com/node/18333018 (accessed 15 November 2011)

Thompson, M.C., Baruah, M. and Carr, E.R. (2011). Seeing REDD+ as a project of environmental governance. *Environmental Science and Policy* 14, 100–110

Toba, N. (2009). Potential economic impacts of climate change in the Caribbean community. In *Assessing the Potential Consequences of Climate Destabilization in Latin America* (ed. Vergara, W.). Latin America and Caribbean Region Sustainable Development Working Paper 32. World Bank Latin America and the Caribbean Region Sustainable Development Department (LCSSD), Washington, DC.
http://irispublic.worldbank.org/85257559006C22E9/All+Documents/85257559006C22E9852575D600577B9B/$File/SDWP%2032%20June%202009.pdf (accessed 9 December 2011)

Transparency International (2010). Cimate governance for a better world. *Transparency International Newsroom: In Focus.* http://www.transparency.org/news_room/in_focus/2010/climate_change (accessed 9 December 2011)

Trémolet, S., Kolsky, P. and Perez, E. (2010). *Financing On-site Sanitation for the Poor: A Six Country Comparative Review and Analysis.* World Bank and Water Sanitation Program, Washington, DC

UN (2010a). *Millennium Development Goals: 2010 Progress Chart.* Statistics Division, Department of Economic and Social Affairs, United Nations, New York. http://unstats.un.org/unsd/mdg/Resources/Static/Products/Progress2010/MDG_Report_2010_Progress_Chart_En.pdf (accessed 9 December 2011)

UN (2010b). *Millennium Development Goals: Advances In Environmentally Sustainable Development in Latin America and the Caribbean.* Economic Commission for Latin America and the Caribbean, Santiago. http://www.eclac.cl/cgi-bin/getProd.asp?xml=/publicaciones/xml/2/38502/P38502.xml&xsl=/dmaah/tpl-i/p9f.xsl&base=/MDG/tpl-i/top-bottom.xsl (accessed 9 December 2011)

UNDP (2011). *Human Development Report 2011. Sustainability and Equity: A Better Future for All.* Palgrave McMillan, New York

UNDP (2010a). *Energía Renovable para la Generación de Energía Eléctrica – Electrificación de Galápagos con Energías Renovables.* Informe de evaluación de medio término. Proyecto GEF/PNUD/MEER. http://erc.undp.org/evaluationadmin/downloaddocument.html?docid=4648 (accessed 16 December 2011)

UNDP (2010b). *Latin America and the Caribbean A Biodiversity Superpower.* http://www.undp.org/latinamerica/biodiversity-superpower/Index.htm (accessed 16 December 2011)

UNEP (2011a). Eficiencia en el uso de recursos en América Latina: perspectivas e implicaciones económicas. *Boletín ONU* 11(263). Centro de Información de Naciones Unidas (CINU). http://www.cinu.mx/comunicados/2011/09/eficiencia-en-el-uso-de-los-re-1/ (accessed 9 December 2011)

UNEP (2011b). *The Green Fund of Trinidad and Tobago.* UNEP Division of Environmental Law and Conventions. http://www.unep.org/dec/onlinemanual/Enforcement/InstitutionalFrameworks/EconomicInstruments/Resource/tabid/1018/Default.aspx (accessed 9 December 2011)

UNEP (2011c). *Toward a Green Economy: Guide to Sustainable Development and Poverty Eradication, Synthesis Responsible For Policy Formulation.* United Nations Environment Programme, Nairobi. http://www.unep.org/greeneconomy/Portals/88/documents/ger/GER_synthesis_sp.pdf (accessed 9 December 2011)

UNEP (2011d). *Green Economy Success Stories.* United Nations Environment Programme, Nairobi. http://www.unep.org/greeneconomy/SuccessStories/OrganicAgricultureinCuba/tabid/29890/Default.aspx (accessed 20 March 2012)

UNEP (2010a). *Access and Benefit Sharing Regional Consultations for Latin America and the Caribbean Countries.* Preparatory meeting of high-level experts. Seventeenth meeting of the forum of Ministers of the Environment of Latin America and the Caribbean, Panama City, Panama, 20–26 April, 2010. United Nations Environment Programme

UNEP (2010b). *Latin America and the Caribbean Environment Outlook: GEO LAC 3.* UNEP, Panamá

UNEP (2009). *Integrated Policy-making for Sustainable Development. A Reference Manual.* UNEP Division of Technology, Industry and Economics (DTIE), Geneva

UNEP (2003). *Legislation on Access to Genetic Resources in Latin America and the Caribbean.* UNEP/ROLAC briefing. UNEP Regional Office for Latin America and the Carribean, Panama

UNEP (2000). *Convention for the Protection and Development of the Marine Environment of the Wider Caribbean Region* and associated protocols. http://www.cep.unep.org/meetings-events/5th-lbs-istac/5th_lbs_istac_documents/cartagena-convention-and-protocols-en.pdf

UNEP (1997). *Source Book of Alternative Technologies for Freshwater Augmentation in Latin America and the Caribbean.* UNEP and International Environmental Technology Centre, Washington, DC

UNEP and MercoNet (2011). *Resource Efficiency in Latin America: Economics and Outlook.* UNEP and the Mercosur Economic Research Network. http://www.uncsd2012.org/rio20/index.php?page=view&type=400&nr=188&menu=45

UNEP and NEF (2010). *Global Trends in Sustainable Energy Investment 2010.* United Nations Environment Programme and New Energy Finance. http://sefi.unep.org/english/globaltrends2010.html. (accessed 9 November 2011).

UNEP and UNOPS (2011). *Construcciones Sostenibles.* United Nations Environment Programme and United Nations Office for Project Services

UNEP-CEP (2011a). Caribbean Environment Programme. http://www.cep.unep.org/

UNEP-CEP (2011b). *Protocol Concerning Pollution From Land-Based Sources and Activities.* United Nations Environment Programme, Caribbean Environment Programme. http://cep.unep.org/cartagena-convention/lbs-protocol/protocol-concerning-pollution-from-land-based-sources-and-activities (accessed 9 December 2011)

UNEP-EDE (2012). UNEP Environmental Data Explorer. http://geodata.grid.unep.ch/

UNEP, ECLAC and GRID Arendal (2010a). *Vital Climate Change Graphics for Latin America and the Caribbean. Special edition for the CoP 16/CMP 6.* http://www.pnuma.org/english/comunicados/061210/LAC_Web_eng_2010-12-07.pdf (accessed 9 December 2011)

UNEP, IUCN and CBD (2010b). *International Payments for Ecosystem Services.* UNEP Division of Technology, Industry and Economics, Economics and Trade Branch. http://www.unep.ch/etb/events/IPES%20Side%20Event%20Bonn/IPES%20SUM%20FINAL.pdf (accessed 9 December 2011)

UNEP, FAO and UNFF (2009). *Vital Forest Graphics.* United Nations Environment Programme, Nairobi

UNESCO (2011). *Eastern Tropical Pacific Seascape Project.* United Nations Educational, Scientific and Cultural Organization. http://whc.unesco.org/en/seascape/ (accessed 9 December 2011)

UNESCO-SCOPE (2006). *How to Improve the Dialogue between Science and Society: The Case of Global Environmental Change.* UNESCO-SCOPE Policy Brief No. 3. United Nations Educational, Scientific and Cultural Organization-Scientific Committee on Problems of the Environment of ICSU, Paris. http://unesdoc.unesco.org/images/0015/001500/150009e.pdf (accessed 9 December 2011)

UNISDR (2006). *Basic Terminology-DRR.* United Nations International Strategy for Disaster Reduction. http://www.unisdr.org/we/inform/terminology (accessed 9 December 2011)

UNFCCC (1998). *Kyoto Protocol to the United Nations Framework Convention on Climate Change.* http://unfccc.int/resource/docs/convkp/kpeng.pdfUN-Water (2008). *UN-Water Annual Report 2008.* United Nations, Geneva

UNFCCC (1992). *United Nations Framework Convention on Climate Change.* FCCC/INFORMAL/84 GE.05-62220 (E) 200705. http://unfccc.int/resource/docs/convkp/conveng.pdf

USEPA (2011). *Environmental Justice.* United States Environmental Protection Agency, Washington, DC. http://www.epa.gov/environmentaljustice/ (accessed 30 November 2011)

Verner, D. (2011). Social implications of climate change in Latin America and the Caribbean. *Economic Premise 61.* World Bank, Washington, DC. http://siteresources.worldbank.org/INTPREMNET/Resources/EP61.pdf (accessed 9 December 2011)

Vides-Almonacid, R., Suarez, H.R.J., Peredo, A.M.L. and Soto, R.V. (2008). The value of the ecosystem approach in the ecoregional management of the Chiquitano Forest in Bolivia and Paraguay. In *Applying the Ecosystem Approach in Latim America* (ed. Andrade Pérez, A.) (translator Medina, M.E.). IUCN, Gland

Viglizzo, E., Frank, F.C., Carreño, L.V., Jobbagy, E.G., Pereyra, H., Clatt, J., Pincen, D. and Ricard, M.F. (2011). Ecological and environmental footprint of 50 years of agricultural expansion in Argentina. *Global Change Biology* 17, 959–973

Waltner-Toews, D., Kay, J.J. and Lister, N.E. (2008). *The Ecosystem Approach: Complexity, Uncertainty, and Managing for Sustainability.* Columbia University Press, New York

Watson R.T. (2005). Turning science into policy: challenges and experiences from the science–policy interface. *Philosophical Transactions of the Royal Society B* 360, 471–477. http://rstb.royalsocietypublishing.org/content/360/1454/471.full (accessed 18 December 2011)

Weber, I. (2009). *Actualizing Sustainable Mining: Whole Mine, Whole Community, Whole Planet Through Industrial Ecology and Community-Based Strategies.* Framework for Responsible Mining. http://www.frameworkforresponsiblemining.org/pubs/ActualizingSustainableMining.pdf (accessed 8 December 2011)

WHO and UNICEF (2010). *Progress on Sanitation and Drinking Water: 2010 Update.* WHO Press, Geneva

WHO and UNICEF (2005). *Water for Life, Making It Happen.* WHO Press, Geneva

Willer, H. and Kilcher, L. (2011). *The World of Organic Agriculture. Statistics and Emerging Trends 2011.* IfOM, Bonn and FiBL, Frick

WMO (2009). *2009 Global Assessment Report on Disaster Reduction: Thematic Progress Review Sub-component on Early Warning Systems.* World Meteorological Organization, Geneva

World Bank (2011a). *Learning from the "Atención a Crisis" Pilot Program in Nicaragua's Drought Region.* World Bank, Washington, DC. http://www.google.com/url?sa=t&rct=j&q=&esrc=s&source=web&cd=1&ved=0CCAQFjAA&url=http 3A 2F 2Fwwwr.worldbank.org 2Fatencionacrisisevaluation&ei=L8jrTrafB8egtwe2h-TaCg&usg=AFQjCNGJO8OG9bbUZMLAugRXxI2nQusPFw (accessed 16 December 2011)

World Bank (2011b). *Urban Development.* World Bank, Washington, DC. http://data.worldbank.org/topic/urban-development (accessed 24 November 2011)

World Bank (2010). *Convenient Solutions to an Inconvenient Truth: Ecosystem-based Approaches to Climate Change.* Environment Department, World Bank, Washington, DC. http://www-wds.worldbank.org/external/default/WDSContentServer/WDSP/IB/2009/07/08/000333037_20090708013334/Rendered/PDF/493130ESW0whit10Box338946B01PUBLIC1.pdf (accessed 8 December 2011)

World Bank (2001). *Land Policy and Administration: Lessons Learned and New Challenges for the Bank's Development Agenda.* World Bank, Washington, DC

WRI (2010). *Modernizing Public Transportation. Lessons Learned from Major Bus Improvements in Latin America and Asia.* World Resources Institute, Washington, DC

WRI (2009). *Stacking Payments for Ecosystem Services.* World Resources Institute, Washington, DC

WRI (2008). *Measuring the Invisible. Quantifying Emissions Reductions from Transport Solutions. Porto Alegre Case Study.* World Resources Institute, Washington, DC

WSSD (2002). *Johannesburg Plan of Implementation.* World Summit on Sustainable Development. http://www.un.org/esa/sustdev/documents/WSSD_POI_PD/English/POIToc.htm

Wunder, S. (2007). Efficiency of payments for environmental services. *Conservation Biology* 21, 48–58

WWF (2011). *Mangrove Conservation and Preserves as Climate Change Adaptation in Belize, Central America: A Case Study.* http://community.eldis.org/.59c095ef/Placencia%20Mangrove%20Reserves%20Case%20Study_final.pdf (accessed 16 December 2011)

Zika, M. and Erb, K. (2009). The global loss of net primary production resulting from human-induced soil degradation in drylands. *Ecological Economics* 69, 310–318

CHAPTER 13

North America

Coordinating lead authors: Jane Barr and James Dobrowolski

Lead authors: John Campbell, Philippe Le Prestre, Lori Lynch and Marc Sydnor

Contributing authors: Robert Adler, José Etcheverry, Alexander Kenny, Catherine Hallmich (GEO Fellow), Jim Lazar, Russell M. Meyer, Robin Newmark, Janet Peace, Julie A. Suhr Pierce and Stephen Yamasaki

Principal scientific reviewer: Dork Sahagian

Chapter coordinators: Jason Jabbour and Ashbindu Singh

Main Messages

Policies and innovative market-based instruments are becoming more successful in improving environmental conditions in North America when they work in concert and in a political environment that allows for sound implementation. For example, the US Clean Air Act includes a cap-and-trade market mechanism to reduce sulphur dioxide (SO_2) emissions that costs less for each unit reduced than traditional environmental regulation. In addition, it is estimated that the Clean Air Act's direct benefit to human health and the environment will reach almost US$2 trillion by 2020, compared to the US$65 billion implementation cost. In Canada, Ontario's Green Energy and Green Economy Act supports a feed-in tariff that has contributed to the growth in renewable energy; wind production in Ontario, for example, increased from 15 megawatts in 2003 to more than 1 100 megawatts in 2009.

Pricing externalities and integrated land management have shown the potential to increase the sustainability of land-use practices in North America. Governments can efficiently diminish environmental impacts by paying land managers for implementing best management practices such as riparian buffers, reduced tillage and reduced fertilizer application. In the United States, taxes and other incentives have increased the total area conserved by local, state and national land trusts to almost 15 million hectares, while payment for ecosystem service programmes, such as farmland preservation programmes that bring together the various benefits to society of farmland and agricultural production, have permanently preserved another 92 million hectares. The US Conservation Reserve Program also pays farmers to withdraw land from production in order to restore soils, providing benefits estimated at US$1.3 billion per year, excluding carbon sequestration, ecosystem protection and other less easily quantified benefits.

The integrated watershed approach, in combination with technological instruments and economic incentives, has proved to be effective in addressing some complex water resource challenges in North America. Currently, the United States and Canada administer this approach through initiatives at the bi-national, regional or state/provincial levels rather than nationally. The Great Lakes and St Lawrence Cities Initiative, a cross-jurisdictional programme, has successfully increased water efficiency and reduced demand across the Great Lakes region.

Increasing renewable energy as a part of the total primary energy supply provides multiple benefits. Case studies from states and provinces in North America indicate that a comprehensive policy approach leads to rapid expansion of renewable energy production. However, development across jurisdictions has been uneven and current policy regimes are not adequate to realize the necessary paradigm shift to achieve a sustainable energy system. This shift will lead to significant reductions in greenhouse gas emissions, and increasing renewable energy production is an integral part of this transition. Research shows that by increasing renewable energy deployment globally, up to 85 per cent of all carbon dioxide (CO_2) emissions could be avoided by 2050 (IPCC 2011).

INTRODUCTION

As previously indicated, *GEO-5* shifts the GEO focus from identifying environmental problems to identifying solutions that governments can then prioritize. This chapter provides examples of a number of policy options and market mechanisms that have shown some success in improving environmental conditions in North America. They are organized by priority environmental theme and their success is related to how they may have helped to speed up the attainment of selected international environmental goals.

The priority themes and related global goals for the North American region of Canada and the United States were chosen during two regional GEO consultations (Table 13.1). In addition to the four priority issues of environmental governance, land use, freshwater and energy, this chapter also addresses the overarching theme of climate change, which is discussed within each of the four thematic sections.

Subsequently, the chapter reviews existing policy approaches, institutional arrangements and market mechanisms related to environmental and natural resources management. The aim is to identify relatively successful policy options that are currently being implemented to address each of the priority issues, and that would also address the related internationally agreed goals. To single out those with most potential to speed up the attainment of the associated goals, the resulting options were subjected to the following criteria, with policies selected if they met some, but not necessarily all, of these criteria:

- responds to, and/or reinforces or fosters interactions and synergies among the priority issues, policy options, regions and/or actors;
- has the potential for replication elsewhere;
- has the potential for scale-up;
- addresses drivers and pressures rather than end-of-pipe solutions;
- focuses on transboundary aspects of the issues and regional solutions;
- can operate as part of a cluster of policies that, if implemented together, are more beneficial than the sum of the separate policies.

Clusters of policy instruments were selected for each of the four areas, and to further refine the assessment, a select number of policies deemed to be the most successful were analysed according to their benefits and drawbacks, the perceived trade-offs of implementing them, and whether their effectiveness could be measured by any specific indicators. In addition, a number of case studies were identified to illustrate how the policies and instruments work in different contexts.

Ultimately, the policies selected are the result of the appraisal process involving a review of the literature and government data, multi-stakeholder consultations and expert opinion. Although the survey was thorough, the policy options were those that could be gleaned by this process and do not represent an exhaustive and comprehensive search; neither do they reflect relative importance compared to those that were not selected. Whether the policy would be effective in a different context and on a different scale is uncertain: for many policies, direct causal evidence of effectiveness is limited. The success of policy instruments depends on their historical, political, cultural, economic and social context. Moreover, each instrument should be evaluated relative to its environmental effectiveness across sectors; its politico-administrative effectiveness in terms of ease of environmental monitoring and validity for decision making;

Table 13.1 Priority themes and related global goals

Environmental governance	
Nusa Dua Declaration (UNEP GC 2010) Paragraph 13	We acknowledge that the advancement of the concept of a green economy in the context of sustainable development and poverty eradication can significantly address current challenges, deliver development opportunities and multiple benefits for all nations.
Johannesburg Plan of Implementation (JPOI) (WSSD 2002) Paragraph 40b	Develop and implement integrated land management and water-use plans that are based on sustainable use of renewable resources and on integrated assessments of socio-economic and environmental potential.
Freshwater	
Johannesburg Plan of Implementation (JPOI) (WSSD 2002) Paragraph 26c	Improve the efficient use of water resources and promote their allocation among competing uses in a way that gives priority to the satisfaction of basic human needs and balances the requirement of preserving or restoring ecosystems and their functions, in particular in fragile environments, with human domestic, industrial and agriculture needs, including safeguarding drinking water quality.
Land use	
Johannesburg Plan of Implementation (JPOI) (WSSD 2002) Paragraph 40b	Develop and implement integrated land management and water-use plans that are based on sustainable use of renewable resources and on integrated assessments of socio-economic and environmental potential.
Energy	
Johannesburg Plan of Implementation (JPOI) (WSSD 2002) Paragraph 40b	With a sense of urgency, substantially increase the global share of renewable energy sources with the objective of increasing its contribution to the total energy supply.

Glacier-fed Portage Lake in south-central Alaska. *© Dave Hughes/iStock*

its contribution to international goals and commitments; and its political feasibility. Whether there are resources adequate to implement a policy, what policy options can be most successfully applied in each region to help accelerate achievement of internationally agreed goals, and the potential for replication, scale-up and transfer are all research questions about which few rigorous studies, if any, exist.

North America is considered a global economic leader, although changes in regional demographics, rapidly emerging global economies and resource constraints all challenge the region's provision of public goods and services. At the same time, fragmented governance, policy instability, lack of clear targets and science policy, and the dilemma of whether to address global issues rather than seeking local solutions hampers the achievement of environmental goals (Chapter 1).

Environmental governance

Environmental governance in North America is best characterized as multi-faceted, partly reflecting the nature of the federal political systems, ideological flux, evolving socio-economic constraints, and the dynamics of environmental issues as well as the knowledge associated with them. Federal governments are no longer the primary leaders in setting the policy agenda or devising innovative policy instruments, yet they remain essential to the ultimate success of those policies, help ensure harmonization across jurisdictions and prevent the development of environmental inequities. In addition, there is a strong tendency to favour market-based instruments because of early successes, and to overlook traditional regulatory instruments. Finally, relative federal disengagement has opened the door to policy initiatives and innovations at the sub-national levels of states and provinces or municipalities, as well as to regional transborder cooperation. The latter is extensive and continues to expand, and its dynamics are further supported by the Commission for Environmental Cooperation, which oversees the environmental accord of the North American Free Trade Agreement (NAFTA).

The examples of policy options on environmental governance given in the following pages are suggested as ways of helping to address these current challenges. In addition, they could help speed up the attainment of Paragraph 13 of the Nusa Dua Declaration, which proposes advancing the green economy, and of paragraph 40b of the Johannesburg Plan of Implementation, which charges decision makers to "develop and implement integrated land management and water-use plans".

Freshwater

The United States and Canada respectively contain 6 and 5 per cent of global renewable water resources, ranking third and fourth overall among nations (FAO 2011). Because of its relative high quality and abundance, water in North America is often taken for granted, although more recently there is recognition of a looming water crisis. Freshwater issues that remain a challenge in some parts of the region include droughts and floods (Cayan *et al.* 2010; Easterling 2000), eutrophication (Smith *et al.* 2006), dams and river fragmentation (Chapter 4), saltwater intrusion (Barlow and Reichard 2010), contamination caused by hydraulic fracturing for natural gas extraction (Kargbo *et al.* 2010), non-point source pollution from agricultural (Ritter and Shirmohammadi 2001; Novotny 1999) and urban run-off (NRC 2008). Climate change may exacerbate these problems by altering both water supply and demand (Vörösmarty *et al.* 2010, 2000; Bates *et al.* 2008).

Since freshwater provides ecosystem services that are central to human health, it is critical to ensure a continued good-quality supply. This chapter provides a number of policy options to illustrate ways of speeding up the achievement of Paragraph 26c of the Johannesburg Plan of Implementation, which stipulates that the efficient use of water resources is to be improved and human needs and ecosystem requirements balanced.

Land use

Land use is a priority issue in North America because it presents both significant environmental concerns and great potential for sustainable development. The natural resource and agricultural sectors contribute significantly to employment and wealth generation; for example, more than 2 million people in the United States work in or support the forestry and agriculture industries (BLS 2011), and in Canada, gross domestic product (GDP) for agriculture, forestry, fishing and hunting totalled US$24.7 billion in 2010 (Industry Canada 2011). In addition, citizen engagement and people's high level of attachment to natural areas have moved land use up on the political agenda. These and other factors, such as fossil fuel extraction and urban development, apply significant pressure on land, often resulting in conflicts over its use. Policy options addressing land use are suggested to help improve progress towards Paragraph 40b of the Johannesburg Plan of Implementation, which urges the development and implementation of integrated land management and water-use plans so as to use renewable resources more sustainably.

Energy

The international goal of urgently increasing renewable energy resources as a part of the total energy supply (Johannesburg Plan of Implementation, Paragraph 20e) was selected to address multiple challenges associated with the current energy system. These challenges include the contribution of fossil fuel combustion to climate change, elevated water consumption and air pollution. However, renewable energy also presents opportunities for sustainable development through increased employment and economic activity and is a necessary element in the transition to a green, sustainable economy. In addition, the processes required for site generation and transmission facilities present opportunities for increasing transparency and cross-agency cooperation, and would also benefit environmental governance and land use. Although the pace of change is still slow, the policy options for increasing renewable energy are examples of current tools applied in North America to speed up the transition to a sustainable energy system.

POLICY APPRAISAL

Environmental governance

North America has used a variety of policy approaches to environmental governance, beginning with regulatory policies, then gradually developing market mechanisms, complemented by measures designed to improve accountability and transparency. The region was a pioneer in cross-border governance, which dates back at least to the 1909 Boundary Waters Treaty, and in developing international environmental law and national parks, including cross-border parks. In the last 20 years, this governance has deepened cross-border ties through the creation of the Conference of New England Governors/Eastern Canadian Premiers on climate change and the Commission for Environmental Cooperation (CEC 2011; Johnson and Beaulieu 1996), and by reinforcing cooperation between provinces and states in managing the Great Lakes and St Lawrence River (Box 13.5), as well as on a variety of other issues, notably protection of waterbirds and sea mammals. The Georgia Basin/Puget Sound International Airshed Strategy in British Columbia and Washington State, for example, is currently the most active bilateral arrangement regarding air quality (Environment Canada 2011). For their part, the proposed creation of watershed boards across the entire Canadian/US border would represent a major leap in the International Joint Commission's regulatory potential (Schwartz 2006). Canada and the United States have also established several jointly protected areas that further harmonize policies.

North America has pioneered the use of many market instruments, now being used with increasing frequency, and there is evidence that some have succeeded in changing behaviour. Command-and-control mechanisms, however, still form the backbone of environmental policy. Because of recent improvements in measures designed to foster accountability and transparency, these increasingly used instruments strengthen the effectiveness of both market instruments and command-and-control mechanisms. Rarely are any of these used exclusively to address a particular environmental issue; it is more common to see a variety of instruments applied. For example, to address littering, many North American municipalities and states or provinces have laws that require a deposit on bottles and cans. This deposit provides a financial incentive – a market instrument – to return the items for recycling. In conjunction, bottles and cans in certain states must clearly display a recycling logo representing the type of material used and providing easy-to-understand and transparent information about recycling. Finally, various regions have banned the inclusion of bottles and cans in solid waste – a command-and-control form of regulation.

Market mechanisms

Market instruments have been used to address a variety of environmental issues in North America. The most recent have targeted air quality and climate change and include an acid rain reduction programme, a greenhouse gas emissions trading programme in the northeastern states and eastern provinces, and a carbon tax in Quebec (2007) and British Columbia (2008) (Box 13.1). Payment for ecosystem services is also gaining wider attention, although such schemes remain limited.

In 1995, the United States instituted a cap-and-trade emissions programme, stemming from amendments to the 1990 Clean Air Act (under Title IV), to reduce sulphur dioxide (SO_2) emissions, the major industrial pollutant responsible for acid rain. This programme is widely credited with reducing sulphur dioxide emissions more cheaply than traditional environmental regulation.

Montreal Metro entrance. In 2007, Quebec became Canada's first province to charge a carbon tax which is being directed towards energy-saving initiatives such as improvements to public transit. *© aetb/iStock*

Box 13.1 The Quebec and British Columbia carbon taxes

In 2007, Quebec became the first North American state or province to introduce a carbon tax. Energy companies are required to pay 0.8 cents for each litre of petrol distributed in Quebec and 0.938 cents for each litre of diesel fuel. Compared to other jurisdictions, however, this tax rate is very low. The revenue-neutral carbon tax in place in British Columbia since 2008 is much more ambitious. Rate increases were phased in, starting at a modest US$10 per tonne of CO_2-equivalent in 2008 and then increasing at a rate of US$5 a year to US$30 a tonne in 2012. The tax's revenue neutrality is achieved by allowing tax reductions for businesses as well as tax reductions for and payments to poorer sections of society. The comprehensive tax applies to all emissions from fossil fuels, accounting for approximately 70 per cent of the province's total emissions. Emissions from fossil fuels exported from British Columbia to other jurisdictions are exempt. In 2010, the tax began to apply to biodiesel as well (BC Ministry of Finance 2008). The new tax did not seem to have significant political repercussions – the provincial party that introduced it was re-elected.

Addressing drawbacks typically associated with carbon taxes may have enhanced its acceptability. This includes mitigating or eliminating the potentially regressive nature of carbon taxation (Metcalf and Weisbach 2008), with comprehensive coverage combined with targeted tax reductions, and reducing potentially large adaptation costs for carbon-intensive industries through a gradual phase-in of the tax (Nordhaus 2010).

Early projections of the average cost for the first phase of the programme ranged from a high of US$307 per tonne of sulphur dioxide removed to US$180 per tonne (1995 dollars). Ellerman *et al.* (2000) estimated that the actual costs were closer to the low end of the projections, in the range of US$186–210 per tonne. In addition, a 2011 US Environmental Protection Agency (EPA) review of the direct benefits to human health and the environment of the Clean Air Act estimates that these will reach almost US$2 trillion by 2020 while implementation costs are US$65 billion – a benefit-cost ratio of 30:1. This was probably due to the flexibility afforded to producers to find low-cost compliance measures, although other factors such as unanticipated technical improvements, lower transport costs and increases in coal production and use efficiencies also played important roles (Chestnut and Mills 2005). Although the costs of many regulatory programmes tend to be overestimated while they are being developed, recent research found that this has been especially the case for market-based programmes (Harrington *et al.* 2008).

The success of the sulphur dioxide trading programme has in part prompted several jurisdictions in Canada to increase the use of market-based instruments. As of 2007, the Alberta greenhouse gas emissions trading system, for example, requires large industrial emitters that have been established more than eight years to reduce the intensity of greenhouse gas emissions by 12 per cent per year relative to a 2003–2005 baseline (Can LII 2011), and purchase carbon offsets or else pay a tax of US$15 per tonne of CO_2-equivalent. While the programme may result in reduced emissions compared to the business-as-usual alternative, it has been heavily criticised for permitting overall increases in carbon emissions by only targeting emissions intensity. In this sense it is not a typical cap-and-trade programme.

A less developed scheme, but one that is emblematic of the readiness of some states and provinces to compensate for perceived federal inaction, is the Western Climate Initiative, which combines seven US states and four Canadian provinces. This has been working since 2007 to develop policies to address climate change, including a regional, economy-wide cap-and-trade programme and forest offset mechanisms (Anderson *et al.* 2010). Only some of the initiative's members – California, Quebec and British Columbia – are currently taking preparatory steps towards implementing this programme in 2012.

Water trading between Canada and the United States and efforts to allocate water efficiently and equitably among various users have triggered considerable political controversy, even before the United Nations acknowledged access to clean water and sanitation as a fundamental human right in 2010. Trading water rights, from farms to cities, for example, can be viewed as making farmland unproductive and favouring urban dwellers over rural residents. In addition, many civil society organizations see the privatization of some water rights as incompatible with the principle of universal and equal access to water.

Water markets, or transferrable water rights, are generally most developed in regions where water allocation is based on first-in-time, first-in-right or the doctrine of prior appropriation (Kenney 2005). In the United States, water markets are prevalent in the arid western states, and in Canada, water trading occurs in Alberta and to a lesser extent in British Columbia and the Territories. The benefits of water trading include the reallocation of water from lower- to higher-value economic uses or from areas where the marginal value is low to where it is high. For instance, where urban users pay much higher rates for water than do rural and agricultural users, trading makes both water buyers and sellers better off economically. There are numerous drawbacks, however. For example, the market value of water may not correspond to its *in situ* environmental value. Moreover, the impact on local water may be externalized to third parties, including changes to the local economy and environmental effects from reduced local water availability (Hanak 2003). Other drawbacks pertain to the very principle promoted by some groups that water should remain a public good and therefore should not be commoditized and traded for profit, the ability of private parties to monopolize the water resources market, and the distortion of the water trading market due to substantial water subsidies for the agricultural sector.

Subsidies and tariffs for clean energy, agricultural production and industrial goods can facilitate the adoption of new, less polluting technologies or projects that enhance energy conservation. Subsidies for installing water-efficient fixtures or the California subsidy programme on residential solar installation, which encourages distributed electrical generation as well as emission-free power production, are two such examples. The Ontario Feed-in Tariff programme, enabled by the 2009 Green Energy and Green Economy Act (Box 13.2), offers stable prices for energy provided by renewable sources and supports Ontario's objective to phase out coal-fired electricity generation by 2014. This programme has contributed to greater reliance on renewable energy sources in Ontario, such as wind power, which increased from 15 megawatts in 2003 to more than 1 100 megawatts in 2009 (Government of Ontario 2009).

While subsidies may help promote technological change, they have also been criticized for increasing the risk of pollution, encouraging overconsumption, and fostering the rapid depletion of natural resources (ten Brink 2011). Agricultural subsidies have come under the greatest scrutiny not only because of their pervasive environmental effects on land use, but also for their negative impact on the agricultural sector and exports of developing countries. Both Canada and the United States also continue to provide large subsidies for the production of non-renewable energy, often in the form of low tax rates for capital investment (Kenny *et al.* 2011; Congressional Budget Office 2005), despite the commitment to the contrary made by the G20 economies in 2009 in Pittsburgh (G20 2009). While some potentially environmentally harmful subsidies may have social or other worthwhile objectives, many may not be equitable, may no longer fulfil their original purpose, or may have unintended outcomes as a result of market distortions. There are many instances where subsidies have either directly or indirectly distorted the market or caused unintended consequences: for example, declining block rate structures for water use, where marginal costs decrease as a function of the total amount of water used, encourage overconsumption.

Payment for ecosystem services, which in one form or another has been used for years but has lately triggered considerable renewed interest, is designed to safeguard or increase the provision of an ecosystem service for which there is high demand but currently no market mechanism. The US Conservation Reserve Program, which provides continuous direct payments to farmers for withdrawing land from production and engaging in soil restoration, is a long-standing and successful example. The US Economic Research Service (ERS) conservatively estimates the programme's benefits to be US$1.3 billion per year, excluding carbon sequestration, ecosystem protection and other less easily quantified benefits (Hellerstein 2010). Other significant ecological benefits include the reversal of landscape fragmentation, maintenance of regional biodiversity, creation

Box 13.2 Ontario: a comprehensive approach to energy

The provincial energy system in Ontario has undergone a number of reforms in the last 30 years. The province had a vertically integrated monopoly until the mid-1990s, but in 1998 moved towards a more market-based model. In 2004, policies were again revised and a hybrid model put in place, in which overall system planning was under one agency; nonetheless, the direction was still towards a market-based model. During this same period, major interruptions occurred with the infrastructure, including the overhaul of seven of 20 nuclear power plants, leading to increased use of coal-fired generation, resulting in major emissions increases with concerns over both the health effects and greenhouse gases. In turn, these concerns led to political pressure and in 2004 the province decided to phase out coal generation as part of a strategy to address climate change and reduce the human impacts and health-care costs of air pollution (Winfield *et al.* 2010).

To achieve the goal, Ontario implemented a variety of conservation and renewable energy initiatives, including the Green Energy and Green Economy Act, a broad-based instrument that enabled the province to implement a comprehensive system of renewable energy feed-in tariffs in 2009. The Ontario feed-in tariff programme provides stable long-term contracts and generation prices specifically tailored for wind, solar, micro-hydro and biomass projects. It also provides a consolidated siting authority, smart grid provisions and additional benefits to attract community energy initiatives and First Nations involvement. The act provided the comprehensive package of policies that created incentives, stimulated new methods to move energy to markets and streamlined the project permission process.

The results of the Green Energy and Green Economy Act have been impressive. The Ontario Power Authority has received supply applications for the production of 10.4 gigawatts of wind power and 6.7 gigawatts of solar photovolatic power; by 2011, there were approximately 3.0 gigawatts of renewable electric power under contract. The provincial power authority also estimated that the renewable energy sector had created 13 000 direct and indirect jobs through the most recent contracts awarded (Mabee *et al.* 2012). The Ontario model is currently being considered in other Canadian provinces including British Columbia and Nova Scotia (Yatchew and Baziliauskas 2011; Ontario Ministry of Energy 2010; Power Authority of Ontario 2010).

of wildlife habitat and favourable changes in regional carbon flux (Gleason *et al.* 2008; Haufler 2005; Dunn *et al.* 1993). The Environmental Quality Incentives Program and the Conservation Security Program of 2002 are two more recent and wide-ranging programmes that seek to reward farmers for sound land management from a multi-functionality perspective. For the same budgetary outlay, the ERS found that environmental performance could improve 12-fold, including an estimated 17 per cent reduction in soil erosion – saving about 36 million tonnes of soil valued at about US$2 per tonne, although the value of reducing sheet and rill erosion alone could be as high as US$332 million when in-stream sediment decreases are included. In addition, nitrogen leaching declined by 14 per cent, nitrogen run-off by 13 per cent, phosphorus run-off by 15 per cent, soil productivity losses by more than 300 per cent, wind erosion by 21 per cent, carbon emissions by 7 per cent, pesticide leaching by 9 per cent, and pesticide run-off by 7 per cent (Cattaneo *et al.* 2005). The US Department of Agriculture has formed an Office of Environmental Markets (previously the Office of Ecosystem Services and Markets formed in 2008) to create guidelines for developing these kinds of market-based policies (USDA 2011).

In Canada, continuous direct payment programmes based on a multi-functionality approach remain uncommon. Some provinces are already using payment for ecosystem services to make it more attractive for farmers to maintain stream habitats, while at the national level efforts are under way to find approaches for comparing the value of services provided by forests (Anderson *et al.* 2010). The implementation of such schemes faces numerous methodological, political and ethical challenges as well as capacity, cost and time constraints, and their long-term impact is still unclear. In general, payment for ecosystem services needs to be complemented with land-use planning frameworks to be effective (Calbick *et al.* 2003).

One innovative and promising economic approach aims to reduce the financial risk of switching to more environmentally sound practices and does not necessarily involve any payment. For instance, in the Canadian province of Prince Edward Island, farmers were offered insurance against the perceived risk that reducing fertilizer use might also reduce yields. In the majority of cases, no payment was needed since reducing fertilizer use did not reduce yields: this was because fertilizer use was already so high that using less had little effect (Cheverie 2009).

Command-and-control mechanisms

The use of public authority to preserve a given resource has a long and successful history. Changing private ownership to public or government ownership and a state-controlled protective regime can eliminate incentives to appropriate the benefits of overexploitation. Indeed, North America pioneered the establishment of the first national parks. This strategy presupposes extensive political and administrative enforcement of the status of these resources, which is more readily available in highly developed economies. Although its effectiveness remains to be seen, the Quebec Water Law of 2009, which considers water a common heritage of the Québécois nation, is a recent and noteworthy example of this type of instrument (Government of Quebec 2009).

Command-and-control mechanisms are often preferred when there are significant threats to human health, when a specific requirement needs to be monitored and enforced, when absolutely no additional environmental harm is permitted, and when simplicity and consistency are desired. In practice, market-based and command-and-control style regulations are often combined to meet an environmental objective. The ban on leaded petrol in the United States, for example, was accompanied by a trading mechanism during the phase-out period so that refineries could meet the declining production allowance in a cost-effective manner.

Although such instruments have become politically challenging to put together, particularly in the United States, there are several noteworthy examples of their successful use, such as standards for drinking water, clean air, toxic chemical releases and fuel; various types of prohibitions including on littering and the introduction of invasive alien species; and requirements on recycling, for example. Canada has the authority to regulate toxic substances, several fuels including diesel and petrol, and a number of fuel quality parameters, including sulphur levels. Greenhouse gas and air pollution regulations have also been implemented in Canada and the United States for new vehicles and engines. With regard to air quality control more generally, Canada monitors and regulates air pollutants through the Canadian Environmental Protection Act and has established National Ambient Air Quality Objectives, although air quality remains the primary responsibility of provinces. In the United States, the Corporate Average Fuel Economy (CAFE) standard regulates the fuel economy of new light-duty vehicles.

One of the drawbacks of these instruments is their weak resilience. When regulations induce changes in behaviour, such as when penalties for failure to obey them are high enough,

Bicycle commuters in San Francisco, California. *© Can Balcioglu/iStock*

these changes usually depend on the continuous enforcement of regulations. Many governments at various levels have tried to green their operations, but the results have often been disappointing and have remained limited as long as they were perceived as top-down mandates and did not change the incentive structure. However, the positive experience of the US Forest Service since 2008, which sought to instil not only a conservation ethic but also a consumption ethic by changing organizational incentives and promoting bottom-up efforts, is instructive in this regard (Jones-Crabtree *et al.* 2008).

Accountability and transparency

Policy instruments designed to increase accountability and transparency seek to make information on environmental performance and the environmental impacts of resource use more widely available to facilitate decision making as well as mobilize a variety of stakeholders. Certainly the best known and most widely disseminated of these policy tools is the requirement for environmental impact assessments, which, when first included in the 1969 US National Environmental Policy Act, mandated preliminary interdisciplinary assessments of the likely environmental impacts of major federal projects (Hironaka 2002). It required US federal officials to include environmental values in a federal decision-making process dominated by technical and economic, if not political, considerations. An environmental impact assessment also requires the identification and evaluation of reasonable alternatives to a proposed federal action, as well as input from concerned stakeholders. Canada adopted its own act in 1992, following previous provincial initiatives. This has since evolved considerably, notably in terms of its target, which goes beyond federal and even publicly funded projects, but also in terms of its scope, with the introduction of sectoral and strategic assessments, and methods that include social variables. Although often criticized for its cost, the delays it can cause, and for ignoring the value of not doing anything at all (null decision), it remains one of the most effective tools for making sounder environmental decisions as well as improving participation.

The requirement to report on polluting emissions is another example of information dissemination that can become an effective policy tool. Canada has a National Pollutant Release Inventory and has implemented a Greenhouse Gas Emissions Reporting Program, while in the United States the EPA requires the reporting of greenhouse gas data and other relevant information from large sources and suppliers. This reporting is now required in 19 US states, and companies will need to report federally on their 2010 emissions in 2011. The US Toxics Release Inventory programme provides stakeholders with information about chemical releases for better decision making. The drawbacks of such instruments include the limited effectiveness of relying on blame-and-shame alone when inventory requirements are not tied to specific obligations. Thus, this instrument is best seen as a complement to market-based or command-and-control approaches.

Providing basic information on the environmental impact of individual citizen behaviour is another useful policy instrument.

A healthy male Peary caribou — listed by COSEWIC as being endangered — stands on guard in the High Arctic. *© Paul Loewen/iStock*

The US EPA and Department of Energy instituted the EnergyStar labelling programme to recognize appliances that perform at or above category benchmarks for energy efficiency. It confers a simple efficiency label to a product, but not detailed information about its energy consumption or the anticipated operating costs. The benefits include its simplicity, which led to the rapid improvement of product efficiency by manufacturers who wanted to qualify for the EnergyStar label (Howarth *et al.* 2000).
In addition, the creation of third-party advisory bodies has proved useful in balancing the needs of science and politics, and provides a means of enhancing policy resilience, that is, the capacity of given policy objectives and means to persist in the face of external challenges. Nationally, the Committee on the Status of Endangered Wildlife in Canada (COSEWIC), where federal and provincial bodies work together with private ones to diagnose problems and recommend action, has protected wildlife conservation from the vagaries of political cycles.

North America has also pioneered the institutionalization of public participation, which helps increase the likelihood of a policy's implementation. Examples include the Great Lakes agreements (Box 13.5), the Commission for Environmental Cooperation's process of citizen submission on enforcement matters, and environmental public hearings, as through Quebec's Environmental Public Hearing Office. Specifically, Articles 14 and 15 of the North American Agreement on Environmental Cooperation (NAAEC) provide a non-adversarial process that allows citizens to file assertions that a Party of the NAAEC (Canada, Mexico or the United States) is failing to enforce its environmental law effectively. In some cases, this process can lead to a record. Political checks and declining funding for the Commission for Environmental Cooperation, which had held constant over the years, have however threatened its effectiveness.

A noteworthy development, reflecting a trend seen in other countries, has been the use of the Office of the Auditor General to evaluate and publicize the degree of implementation of national or sub-national commitments. Canada created a Commissioner of the Environment and Sustainable Development in 1995, and Quebec followed suit in 2006, with both enjoying a fair degree of autonomy. This role will be enhanced by the recent adoption of various sustainable development strategies, both at the federal and provincial/state levels, aiming to make environmental decision making more transparent and accountable. It is too early to assess effectiveness, however, and the lack of uniform sustainable development indicators hampers comparisons between the approaches adopted.

The acceptability, nature and effectiveness of various policy instruments depend on a number of internal and external factors that vary from state to state, province to province and region to region. In the end, successful policies rely on a mix of instruments and incentives. Although market approaches have raised considerable interest and have proved to be efficient in some cases, the enduring value of traditional command-and-control regulation, associated with disclosure requirements, has been most effective in changing the behaviour of major polluters (Harrison and Antweiler 2003).

Land use

One of the most important obstacles to sustainable land use in North America remains the fragmentary nature of land management. Forests, rangelands, croplands and urban, suburban and peri-urban lands are all part of the same landscape mosaic from which people derive survival and quality of life. Often, activities within one land type affect the state of others, as well as other ecosystem services such as air and water quality. Such impacts are often referred to as externalities, as the true costs and benefits of the impact are borne by parties external to the those who control and benefit from the activity. Even within a given land-use type, management responsibilities can be dispersed across several distinct bodies according to the type of activity taking place or to the component that is under consideration – water, fish and wildlife, fossil fuels or recreation, amongst others. In forest planning, for example, forestry, oil and gas, recreation, and the provision of ecosystem services are often managed by entirely separate bodies, even though all activities take place within the forest.

In North America, many land-use policies are gaining support and are now considered to be highly effective in motivating sustainable land use. These policy options work in tandem, providing informational and functional support to achieve the desired goals. This section discusses three policy clusters that have been demonstrated in reality or suggested in theory to be the most promising to coordinate land management and promote sustainable resource use and social, economic and environmental harmonization in North America. These policy clusters are:

- implementing integrated land management plans to encourage and enable sustainable resource use;
- incorporating the true costs and benefits of ecosystem services when developing policy mechanisms; and
- improving planning for and sustainability of public lands.

Implementing integrated land management plans

To speed up achievement of the international goal of sustainably developing and using land in North America, integrated planning is crucial, requiring policies with clear agreed goals and specific targets. Land-use policies need to be set at the appropriate geographic scale – state, province, county and city level – although watersheds or other ecologically relevant geographic scales may be the most logical units for determining a resource-use sustainability plan. Specific targets should be set to obtain the highest benefits for the least social and economic costs. Institutional barriers such as centralized yet fragmented governmental structures should be overcome to allow a regional emphasis, and stakeholders should be allowed to participate in spatial planning. Both regulatory and incentive-based policies can be enacted to encourage target attainment. These policies should motivate individuals and corporations to act in accordance with the established plans. In addition, policies should be developed to encourage the resource sectors to maintain and enhance ecosystem resilience for future generations as well as limit the erosion of ecosystem services.

Jurisdictions throughout North America have adopted many of these policy instruments, to different degrees. For example, in British Columbia, resource companies, environmental groups and coastal First Nations have successfully carried out an ecosystem-based integrated land-use planning exercise, the 2006 Great Bear Forest Agreements, through a collaborative process (McGee *et al.* 2010), although the recent economic downturn has made financing participatory and multi-agency programmes more difficult for state/provincial and local governments. As fiscal issues may become more challenging in the near future, creative financing and regulatory measures coupled with financial incentives could become more important. At the same time, agencies will have more time to develop plans as the pace of industrial, commercial and housing development slows. Planning today may have long-reaching impacts as the economy returns to normal.

States, provinces, counties and cities have taken action to encourage smarter land use through innovative policy mechanisms. These initiatives address many of the challenges related to an optimal land-use pattern while respecting property rights, the need for equity and low-income housing, employment concerns, resource protection and environmental issues. For example, in the United States, the State of Maryland uses a series of incentives in its Smart Growth programme (Box 13.3). The programme rewards people for relocating close to their place of employment, capitalizes on state money for infrastructure by providing it only within planned growth areas (priority funding areas), targets conservation funding to contiguous land and high-conservation-value land within clearly identified Rural Legacy areas, and subsidizes urban redevelopment through its brownfield redevelopment programme. Smart Growth focuses

Vancouver's Sky Train, a light-rail rapid public transit system, contributes towards the achievement of British Columbia's ambitious greenhouse gas reduction targets. *© Wade Jabbour*

on long-term regional considerations of sustainability, valuing community, public transport, employment and housing choices, preserving natural resources and promoting equity.

Similarly in Canada, the Province of Ontario has developed a greenbelt around the City of Toronto (Box 13.4) and protected open space and working lands from further conversion through zoning regulations. Agricultural retention can have economic, cultural and amenity benefits as well as environmental ones. British Columbia has designated an Agricultural Reserve, while Vancouver promotes development near its Sky Train stations. Rather than continue investments in roads and highways that promote an automobile culture, metropolitan areas like Toronto and Vancouver are focusing scarce investment on public transport and transit-oriented development with multiple benefits.

Incorporating the value of ecosystem services in private sector decision making

Market mechanisms, financial incentives and regulatory approaches have moved people to adopt better land-use practices. However, policies intended to benefit society can have unintended consequences. They often require the conversion of forests, grasslands and wetlands to other uses, which results in loss of habitat and biodiversity, impaired water quality, increased flooding, eroded soils and loss of resource-based industries and employment. Governments can help diminish such environmental effects through a number of policy initiatives. The most efficient and least controversial remains the establishment of mechanisms through which users of an ecosystem service, such as water quality, who are willing to pay for the service, compensate land managers for implementing best management practices such as riparian buffers, reduced tillage and reduced fertilizer applications. Taxes and other incentives in the United States have increased the total area conserved by local, state and national land trusts to almost 15 million hectares. Payment for ecosystem service programmes, such as working lands (agricultural and forest) preservation programmes that bring together the various economic and ecological benefits that these lands provide to society, have permanently preserved another 92 million hectares in the United States.

Cap-and-trade systems, such as the one in place for wetlands in the United States (Spieles 2005), can also be established when the users of the ecosystem services are dispersed or even do not yet exist, as in the case of acting in the interests of future generations. Caps need to be established, as in the case of the policy of no net loss of wetlands in the US Clean Water Act, and the magnitude and nature of compensation needs to be determined. While requiring considerable resources of time and effort to establish and implement, the pay-off from a societal point of view is that the market is then able to determine the most efficient means of respecting the cap through a system of trading (Yamasaki *et al.* 2010; Salzman 2005). In the more than 500 wetlands mitigation banking schemes that generate US$3 billion dollars, and the more than 110 habitat banks generating US$370 million in the United States (Madsen *et al.* 2010), land developers include the cost of wetland mitigation when pricing potential land acquisitions. They understand that purchasing land with wetlands will cost more in the end than land without them; either they protect the wetland or are required to restore wetlands elsewhere. Governments can implement programmes to encourage wetland restoration projects that developers can pay for and use to mitigate any wetlands destroyed in their own development projects.

Where potential projects are too fragmented, as is often the case with conservation on working lands, and markets for environmental payments run the risk of remaining excessively

Suburban sprawl outside Austin, Texas. *© Jodi Jacobson/iStock*

thin, governments can opt for more direct financial intervention, such as the Conservation Reserve Program in the United States (as mentioned in the section on market mechanisms, above), under which landowners enter into contracts with the government to implement best management or conservation practices to achieve environmental goals.

Improving sustainability on public lands

In both Canada and the United States, which are endowed with diverse and abundant land resources, the government owns a substantial amount of that land: 89 per cent of the land mass in Canada and 35–40 per cent of it in the United States. While human capital in both countries remains a tremendous asset, many economic sectors continue to generate wealth through natural resource use. Therefore, federal government policies on its own land can have a large impact.

In the United States, principles of multiple use and sustained yield dominated for many years, then in 1993 President Clinton established a goal of achieving sustainable forest management of all US forests by the year 2000. And in 1995, through the Montreal Process and the Santiago Declaration, the United States committed to a process of developing and evaluating national indicators of sustainable forest management. As a result, during the passage of the Federal Ecosystem Management Initiative, its emphasis shifted to ecosystem management with plans focused on long-term sustainability rather than on management to maximize short-term yield (Cortner and Moote 1999; Yaffee *et al.* 1996). However, planning has proven problematic and litigious, and recently, a revised planning rule has been proposed for the nation's public lands. The latest planning rule under consideration stresses the restoration and maintenance of forests and grasslands; the protection of water quality and ecological integrity of riparian areas; habitat provision for plant and animal diversity and species conservation; multiple uses including recreation and industrial applications; public involvement in the planning process including community consultation and all levels of government entities; the use of the best available scientific information to inform the planning process; and the development of a more efficient and adaptive land management planning process (USDA 2012).

While the planning rule is being revised, some groups argue that, instead of a governmental planning approach to help the forest, some type of certification processes for land and management practices should be implemented. Examples of such processes include those used by non-governmental groups including the Forest Stewardship Council and the Marine Stewardship Council's fisheries certification programme (Glickman 2008). Indeed, the province of Quebec's revised Forest Act, which sets the stage for integrated land management with significantly increased responsibilities at the regional level, legislates for wood products from all public forests to be eco-certified by 2013.

Public-private partnerships have become increasingly important as current government funds and staffing are inadequate to assess resources, coordinate sustainable management, and accommodate the increasing demands of multiple users. Public-private partnerships are difficult to foster unless sufficient motivation exists on all sides, as within federal agencies and among their staff, long-term traditions can be difficult to alter without appropriate changes to incentives and reward structures.

Case studies on innovative land-use policies

The policies, underlying conditions and case studies presented here demonstrate that multiple policy instruments can speed up efforts to achieve the internationally agreed goal of implementing integrated land management and water-use plans to ensure the sustainable use of renewable resources (Johannesburg Plan of Implementation Paragraph 40b). In the case of the State of Maryland (Box 13.3), policies leveraged the state's funds to encourage built infrastructure in planned priority areas while providing incentives to create new jobs and develop brownfield sites within the same areas. The planning process involved local communities and used incentives to encourage voluntary participation to achieve the plan's goals, ensuring that it was politically palatable and thus likely to be successful. While encouraging development in and near cities, Maryland also protected valuable resource-rich land from conversion through permanent conservation easements.

In the case of Ontario and British Columbia (Box 13.4), their governments passed regulatory measures to protect environmentally sensitive and working lands while encouraging transit-oriented development within the cities. From a policy perspective, environmentally sensitive and working lands are lumped together, and farming and environmental communities have joined forces on these issues – one of the reasons for so much support for conservation programmes. Conservation practices can be adopted to retain topsoil and prevent erosion

Box 13.3 Maryland's Smart Growth programme: financial incentives and planning

Maryland's Smart Growth programme targets state resources to support development in areas where infrastructure already exists and to avoid the high cost of building infrastructure far from traditional population centres. Priority funding areas are identified within existing communities and other areas where local county and town governments want state investment to support growth and development (Sartori *et al.* 2011; Lewis *et al.* 2009). This approach capitalizes on the influence of state expenditure on economic growth and development. Development is more likely to occur in these planned areas, slowing the conversion of resource-rich land.

In addition, Smart Growth helps protect valuable natural resources, purchasing land and easements in designated Rural Legacy areas that have been selected based on the extent of the development threat and the value of their agricultural, forestry and natural resources. These areas attract both Rural Legacy dollars and money from other preservation and conservation programmes (Lynch and Liu 2007), leading to more contiguous and environmentally beneficial preservation, including retention of interior forests (blocks of trees away from non-forest land or roads), wildlife habitat, groundwater recharge and wetland preservation, as well as agricultural and other productive resource use.

Three additional incentive programmes reward the redevelopment of brownfield sites (Howland 2010), businesses that create jobs in priority funding areas, and citizens who move to live near their place of work. Johns Hopkins University, for example, worked with Baltimore City and the State of Maryland to offer cash grants ranging from US$2 500 to US$17 000 to help university employees buy homes in targeted areas around its campuses (Wiewel and Knaap 2005).

on environmentally sensitive land; wetland protection can be implemented; and streams can be fenced and animals kept out. In many cases, agriculture-related programmes are relatively successful in achieving environmental protection because the opportunity costs to landowners are much lower than for land put to other uses. Programme developers have also come to understand that environmental attributes are often devalued in land markets and have developed new compensation schemes with environmentally sensitive features that benefit landowners. While the use of financial incentives and subsidies differs from regulatory measures, all of these, alone and in combination, can play important roles in addressing land-use issues. Concerns about property rights should be evaluated and addressed. For each policy, decision makers should consider the implied property rights in the existing market structure and how a particular policy will alter this. Regardless of the policy path chosen, cultivating and developing widespread public support and a willingness to plan is essential to the success of any of these policies.

Box 13.4 Canadian land-use reserves in Ontario and British Columbia: command and control

Ontario and British Columbia have protected rural and working lands surrounding major cities through regulatory measures. British Columbia established the Agricultural Land Reserve system, under which agriculture and forestry are the priority uses, and non-agricultural uses are controlled (Cavendish-Palmer 2008; Hanna 1997). The system covers approximately 4.7 million hectares. While it is criticized for not compensating farmland owners sufficiently for the alteration in rights, it has been defended on the grounds that it effectively provides food security and controls urban and peri-urban expansion.

The Ontario Greenbelt protects green space, farmland, forests, wetlands and watersheds around one of Canada's most populated and rapidly growing areas (Ali 2008; Feung and Conway 2007; Taylor *et al.* 2005). It encompasses 730 000 hectares in which limited agricultural uses are permitted, includes environmentally sensitive land and a major aquifer, and contains a UNESCO Biosphere Reserve, the Niagara Escarpment (Cavendish-Palmer 2008; Hanna 1997).

Cross-cutting issues

Implementing the selected land-use policies can provide a number of benefits to support the energy, freshwater and governance goals. Integrated land management may lead to policies that provide co-benefits, such as improving water availability and quality by reducing run-off. This form of planning may also help to identify areas that are most acceptable and best suited for the development of renewable energy, thereby decreasing uncertainty for projects and accelerating implementation. Integrated land management, if it leads to the maintenance of vegetation on a landscape, will also help attain international goals related to climate change.

Freshwater

It is critical to the appropriate use and allocation of freshwater resources that policy instruments designed to meet basic human water needs, as well as water requirements for the production of food and energy, are balanced with the need to maintain other ecosystem services. Three clusters of key policy options identified for North America are integrated watershed management, full-cost pricing and technological solutions.

Integrated watershed management

Integrated watershed planning and management can be applied in combination with other water management measures and has become an indispensible instrument for improving water resources. It is an holistic approach to managing water within drainage areas. This approach is consistent with the broader concept of integrated water resources management discussed in Chapter 4 and aims to achieve optimal and sustainable water availability that will improve human quality of life while maintaining environmental integrity for all species. Integrated watershed planning and management has proved effective in addressing some complex challenges over the last few decades (Heathcote 2009). The method recognizes that water issues cannot be addressed independently but require the balanced consideration of all environmental, social, economic and technical aspects. It may include goals such as flood prevention, enhancement of aquatic habitat and biodiversity, reduction in the loss and degradation of wetlands, pollution control and economic growth. The success of programmes can be assessed through water quality indicators including contaminant concentration, dissolved oxygen and biodiversity, water flow and flood prevention.

Developing and implementing an integrated watershed planning and management policy requires active participation, interaction and collaboration between stakeholders. Currently, this is not administered nationally in the United States and Canada, but through initiatives at the regional or state/provincial level. For example, the Total Maximum Daily Loads programme for pollutant control in the United States is being implemented at the state level as required by the Clean Water Act. States are required to identify impaired waters and calculate the maximum amount of a pollutant a water body can receive and still meet water quality standards, and then develop plans, with public input, to address point and non-point sources of pollutants in an effort to restore and maintain the water quality. Although the programme has shown a varying degree of success across the country – due in part to the differences of each watershed – factors that have been recognized to enhance implementation include a focused watershed plan, active stakeholder involvement, coordination between local and state governments, a diversity of approaches to addressing sources of pollution, and adequate resources for watershed characterization and monitoring (Benham *et al.* 2008). An attractive aspect of integrated watershed planning and management is that it need not require expensive infrastructure such as water treatment and control structures. Therefore, costs do not necessarily restrict implementation, so it can move forward in situations and regions where financial resources are limited. This makes integrated watershed planning and management highly transferable, provided effective coordination and implementation mechanisms can be established. It can also be applied at a diversity of scales ranging from small urban stream restoration projects to large watershed programmes, such as the Great Lakes (Box 13.5), Chesapeake Bay (Hassett *et al.* 2005), the Everglades (Davis and Ogden 1994) and San Francisco Bay (IRWMP 2006). Of its many benefits, perhaps the most notable is that stakeholders are actively involved in selecting the management strategies to solve water resource problems. Active stakeholder involvement, with explicit discussion of issues, improves decision making and acceptance, thus offering advantages over top-down planning, which often lacks public support and understanding.

Integrated watershed planning and management is not without problems, however, and it is often difficult to determine how well it works. In the Chesapeake Bay watershed, it was initiated decades ago in an effort to clean up the estuary and restore coastal fisheries. Projects to improve water quality have largely focused on tributaries, and include re-vegetating riparian areas, improving stream channels and restoring wetlands. Millions of dollars have been spent on thousands of restoration projects within the watershed, yet the success is difficult to gauge, due in part to a lack of comprehensive monitoring of individual projects (Hassett *et al.* 2005). While clear indications of widespread water quality improvements in the Chesapeake Bay have not yet been observed, outcomes in some areas look promising (Ruhl and Rybicki 2010).

In general, integrated watershed planning and management faces serious challenges due largely to the magnitude and complexity of problems as well as socio-political rather than technological or hydrological barriers. The mismatch between watershed boundaries and political boundaries poses a challenge because of the often conflicting needs of the multiple landowners and political entities with jurisdictions in watersheds (Blomquist and Schlager 2005). To overcome this, a watershed authority is typically established to coordinate and implement the plan, and faces the formidable task of bringing together the stakeholders and facilitating agreements to balance the needs of competing interests. Thus, collaboration and public participation are essential. The challenges of creating watershed authorities are magnified when watersheds cross international boundaries. However, these challenges can be met through such efforts as the International Watersheds Initiative, which was conceived by

Smart meters measure residential water consumption. *© Kenneth Cheung*

Box 13.5 Protection and management of the Great Lakes Basin

Canada and the United States share the benefits and responsibility of co-managing the Great Lakes watershed ecosystem, the Earth's largest surface freshwater system (GLIN 2011a) (Figure 13.1). In addition to providing drinking water to nearly 33 million people, this abundant supply of water is at the core of the region's economy. The Great Lakes–St Lawrence River Basin Sustainable Water Resources Agreement of 2005, signed by eight American states and two Canadian provinces, provides a framework for each state and province to manage and protect the basin as a whole. The agreement's principles stem from an ecosystem-based water management approach and include bans on new diversions of water from the basin, with a few exceptions; imposes a consistent standard to review proposed uses of basin water; requires that each state and province develop and implement a water conservation and efficiency programme; strengthens the collection and sharing of technical data among parties; and ensures a strong commitment to continued public involvement in implementing the agreement.

The Great Lakes and St Lawrence Cities Initiative is an example of a successful cross-jurisdictional initiative that has been effective in increasing water efficiency and reducing demand across the region. The initiative includes the objective that, by 2015, all participating cities of the basin reduce water use by 15 per cent relative to levels of use in 2000. By 2010, almost half of the 33 participating cities had collectively achieved a 13 per cent reduction, conserving around 330 million m^3 of water. Policy instruments that helped achieve this reduction include:

- technological instruments such as infrastructure upkeep and water metering;
- economic incentives such as subsidies that promote water efficiency, and reduced water rates for industrial users based on their commitment to implement sustainable water-saving projects; and
- educational outreach (GLSL Cities 2011).

Figure 13.1 The Great Lakes Basin

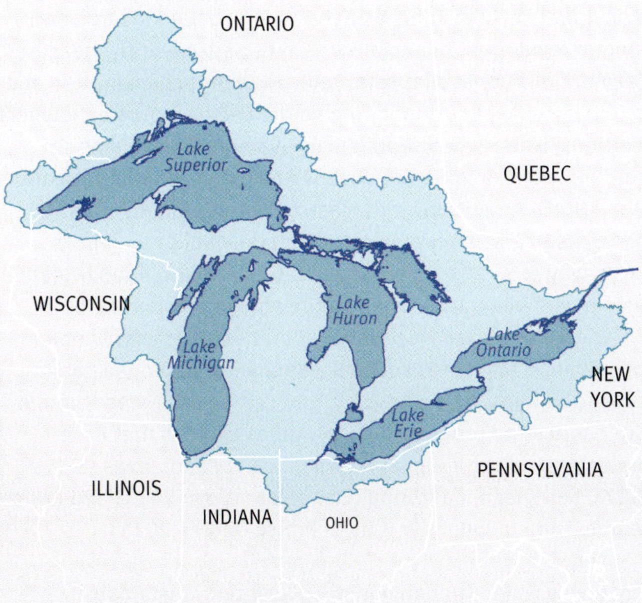

Source: Base from ESRI, 2001; U.S. Army Corps of Engineers,1998; and Environment Canada,1995.

the governments of Canada and the United States to promote the establishment of watershed authorities and facilitate integrated transboundary management (Blaney 2009).

Full-cost pricing

Full-cost pricing of water delivery has been defined by the US EPA as "a pricing structure which fully recovers the cost of providing that service in an economically efficient, environmentally sound, and socially acceptable manner, and which promotes efficient water use by customers" (USEPA 2006). Based on the user-pays and polluter-pays principles, high-volume users pay proportionately more than low-volume users. The aim is to make it possible for all consumers to afford the volume of water necessary for basic human needs while charging increasing prices for consumption beyond that level. Full costs include all public and private costs, both market and non-market values, and account for costs that will be incurred in the future, such as those arising from infrastructure rehabilitation and replacement. In public water provision, once the water delivery infrastructure is in place – such as dams, canals, pumps, pipelines or treatment plants – the marginal cost to the utility company of delivering water to its customers is equal to its variable costs. These costs amount primarily to administrative and maintenance costs, which are near zero compared to the cost of establishing the overall infrastructure. The resulting artificially low market price to customers generally leads to water consumption decisions being based on incomplete information, resulting in overconsumption. In a full-cost pricing model, all infrastructural, environmental and intergenerational costs are included in the delivery price. In practice it is difficult to account for all of these costs accurately; nonetheless, various pricing systems attempt to convey more complete cost information so as to require consumers to pay more of the costs associated with their respective levels of water consumption. One example of how full-cost pricing can be implemented is through increasing block rates, thought to be the most effective in encouraging conservation. In this pricing structure, the amount charged per unit of water consumed increases with the total volume consumed.

Numerous examples of successful implementation of full-cost pricing exist, and are typically evaluated in terms of reductions in water consumption (USEPA 2005). An example is offered by the Marin Municipal Water District (MMWD), a public agency that provides water for 195 000 residents in south and central Marin County, California (MMWD 2011). The MMWD's water rate structure includes a base fee that covers such services as meter reading, billing, meter replacement and repair, customer service, water conservation and administration, and four levels of charge that cover the cost of water transmission, treatment, distribution, watershed maintenance, and importing and recycling water. The MMWD imports a quarter of its water from the Russian River in Sonoma County through an agreement with the Sonoma County Water Agency. The environmental costs of using water from the Russian River stem from the Federal Endangered Species Act, and include expenses related to improving conditions for several fish species that are classified as threatened or endangered, for example by constructing fish ladders, as well as channel maintenance and monitoring. The MMWD is unusual in that customers pay the full cost of water without state and federal subsidies or cost sharing with other water agencies. Rates are comparable to other northern California water agencies, and overall water use has remained relatively stable over the last several decades despite an increasing population (Fryer 2009). These water-saving measures are a result of a better understanding of the true value of water, and have minimized the financial and environmental costs of water supply expansion.

A residential condensing hybrid tankless water heater. This technology produces hot water on demand and is much more energy efficient than a conventional hot water holding tank. *© BanksPhotos/iStock*

Despite successes, there are also some limitations to full-cost pricing, including its complexity compared to the simplicity of traditional marginal-cost pricing structures, making it difficult for consumers to respond to the price information by adjusting their water use. Public outreach campaigns and in-bill information leaflets that describe cost structures are addressing this barrier to some extent. Another limitation is the difficultly in setting prices properly, in particular in identifying and allocating non-market costs such as environmental losses associated with water delivery – for example the environmental effects of constructing new diversion and containment structures. Various formal methods have been developed, however, for assigning market values to non-market costs over time, identifying present and amortized values for those costs, and then adding them to the marginal cost to customers on the basis of water consumed (Renzetti and Kushner 2004; Rogers *et al.* 2002). Implementing full-cost pricing requires adequate institutional support and agreement, as well as the personnel and data necessary for estimating cost components.

Technological solutions and conservation measures

Technological advances and conservation measures can effectively decrease water use in the residential, industrial and agricultural sectors. This has been accomplished in large part through regulation, financial incentives and voluntary measures. Many options are available for reducing water consumption and increasing efficiency depending on the sector, including low-tech solutions, water-saving appliances, water reuse systems and metering. For example, the decline in average residential water use in North America over the last 25 years is largely attributed to increased efficiency standards for household appliances (Rockaway *et al.* 2011). In the agricultural sector, flood irrigation systems are being replaced by more efficient technologies designed to increase crop yield per unit of water use. Simpler conservation measures such as responsible water-use habits go hand-in-hand with efficiency, and can be promoted through water education programmes. Examples of cities that have implemented such programmes include El Paso, Texas (EPWU 2007), San Diego, California (City of San Diego 2011) and Prince George, British Columbia (City of Prince George 2011).

Conserving water through improved long-term sustainable efficiency can lead to a range of economic and environmental benefits. Some of the advantages of this approach include adaptability to site-specific needs, avoidance of more expensive potable water supplies, and the reduced costs of operating and maintaining water distribution and treatment infrastructure, with associated energy savings. For commercial and industrial facilities, savings in water and energy costs realized by implementing efficiency measures can quickly offset the investments made. For instance, in the State of California the average estimated payback period for investing in water-efficient technologies in the commercial, industrial and institutional sector is typically less than two and a half years (Vickers 2001). Obstacles to implementing water efficiency measures include situations in which the capital investment does not justify the water cost savings in the short term, or when a general consensus cannot be reached among stakeholders that the benefits accrued to the water rate payers are worth the investment in the long run. Decisions often depend on the costs associated with water use and water discharge, environmental compliance and production. Other economic incentives may be required in some areas with low water costs, including subsidies, tax credits and grants. In many cases, it will be a combination of sector-specific instruments and incentives appropriate to a region's issues and needs that will allow a variety of innovative and effective water-use efficiency measures to be implemented.

Cross-cutting issues

Policies that promote the integrity of the water cycle and the essential life-supporting services it provides can indirectly help achieve the internationally agreed goals for land use and renewable energy. Successful implementation of integrated watershed planning and management is likely to promote sustainable land use by restoring ecosystem function and enhancing resilience. When the true cost of water supply is assessed, added revenue may be used to fund restoration programmes carried out over the landscape. Water conservation that stems from financial incentives and technological advances will further reduce land degradation and minimize energy requirements for the use and distribution of water. Greater reliance on renewable energy sources will reduce greenhouse gas emissions that cause climate change, which may mitigate projected impacts on the water cycle.

Energy

Canada and the United States are endowed with diverse and abundant renewable energy resources. Transforming that vast potential into a sustainable energy system requires mobilizing political will, behavioural change and smart, comprehensive policies that support renewable energy. There are several environmental issues associated with the current energy system, including climate change, elevated water consumption and air pollution.

Since fossil fuel consumption is the major contributor to increasing atmospheric concentrations of carbon dioxide (CO_2), experts contend that policy interventions should be strengthened, not just to increase renewable energy production, but to substitute renewable energy for the current carbon-emitting energy systems (Delucchi and Jacobson 2011; IPCC 2011; Jacobson and Delucchi 2011; Schneider *et al.* 2000). Renewable electricity technologies offer an effective means of reducing greenhouse gas emissions, thus providing a tool for climate change mitigation (Awerbuch 2006). This section highlights practical lessons learned as well as comprehensive and emerging novel approaches from North America's electricity sector. It has become clear that even partial mitigation of the rate of climate change requires more carbon-free sources of electricity (Schiermeier *et al.* 2008). In addition, policy innovation and technical improvements are rapidly advancing in this sector, thus providing the clearest examples for emulation.

Tehachapi Pass Wind Farm, California, generating clean, renewable energy. *© Patrick Poendl/iStock*

North America's current dependence on fossil fuel resources largely stems from a cycle of pricing effects, partially due to subsidies that favour conventional fossil fuel energy production and that externalize pollution costs. For example, an analysis of all energy subsidies provided in the United States in 2004 shows that 86 per cent went to fossil fuels, 8 per cent to nuclear energy and just 6 per cent to renewables and energy efficiency (Sovacool and Watts 2009). Recently, Energy Secretary Steven Chu announced that the Obama administration intends to repeal US$46.2 billion in subsidies to oil, natural gas and coal companies in the next ten years in order to fund renewable energy spending (Bloomberg 2011). Economists argue that to address these uneven subsidies and other market failures associated with fossil fuels and to accelerate renewable energy deployment, the multiple social and environmental costs of emissions have to be included in the price of conventional energy production (Sovacool 2009a). Smart, novel and comprehensive policies are therefore necessary to provide the incentives, transmission networks, transparency and market space essential to support rapid and sustained renewable energy development and the substitution of fossil fuels.

During the selection process, three policy clusters were identified affecting renewable energy adoption: providing financial support to alter incentives or encourage behavioural change; improving networks and grid flexibility; and decreasing institutional barriers. This section highlights key policies that support current instruments affecting the adoption of renewable energy and discusses the benefits, drawbacks and potential for transfer and scale-up. However, as experts contend and the case studies in this section illustrate, a comprehensive policy approach is important when considering renewable energy support (Sovacool 2009b). Such an approach could accelerate renewable energy development by simultaneously confronting the multiple challenges and barriers that are delaying the transition to a sustainable energy system.

Support to alter incentives or encourage behavioural change

The policy measures described in this section provide market incentives that partially address fossil fuel subsidies and the externalization of the costs of pollution (Sovacool and Watts 2009). Examples already in use in North America include production tax credits, feed-in tariffs and renewable portfolio standards; in addition, governments supply funding for research and development. Production tax credits represent kilowatt-hour tax credits for qualified renewable energy sources while feed-in tariffs typically guarantee grid access and provide long-term contracts for electricity generation at stable prices (DSIRE 2011; Mendonca 2007). Where they are well designed, feed-in tariffs also provide renewable energy premiums using the rate-payer base rather than government funds. Renewable portfolio standard policies also avoid the use of government funds, with the exception of monitoring compliance with the standard, and typically require utilities to procure renewable energy resources as a prescribed percentage of total electricity (Fischer 2010). Investments in research and development help to improve technologies that drive prices down, providing market advantages aimed at increasing the renewable energy market. The close coupling of research and development with investment subsidies has shown to improve policy effectiveness (Soderholm and Klaassen 2007; Klaassen *et al.* 2005).

Improving networks and grid flexibility

Renewable energy sources and current fossil fuel generation facilities are often located in different places, thus requiring networks to transport energy from new source areas to load centres. In addition, fossil fuel generation, which is characterized by long-term capital stock, currently dominates the market, limiting opportunities for new technologies to enter. Several policy measures have been devised that improve the management and characteristics of transmission networks and increase market access and space. These include designating transmission cost recovery and allocation; managing the grid through independent system operators; developing smart grids; and phasing out coal plants. These policies are intended to make it easier to develop infrastructure, open market space and transmit renewable energy from areas of generation to load centres.

Cost recovery and allocation policies provide clear frameworks for developers to recover installation costs from transmission projects, which is necessary to provide an energy transportation network to increase renewable energy use. Currently, it is difficult to finance the development of transmission structures that cross multiple state and provincial jurisdictions, with, in many cases, associated problems in assigning costs and benefit levels. To overcome this, experts have proposed that federal authorities should determine cost allocation (Willrich 2009).

Energy developers also encounter problems with the lack of transparency and access to the grid (Sovacool 2009b) as, traditionally, vertically integrated companies generate, transmit and distribute electricity. In many areas, utility companies still own and operate the transmission assets, leading to a lack of transparency in the availability of transmission. Independent system operators are third-party public institutions responsible for granting access to transmission grids, which could provide desirable conditions for accelerating renewable energy deployment by ensuring transparency and fair access to markets (Joskow 2005). In Texas, where cost allocations are assigned to all supply entities, representing a novel approach for North America (Schumacher *et al.* 2010), the construction of high-voltage electricity transmission is proceeding rapidly (Box 13.6).

Phasing out coal plants is a relatively new policy instrument that decreases greenhouse gas emissions while simultaneously increasing grid flexibility and providing market space for renewable energy. Since coal-fired technology has a limited ability to respond to load fluctuations, these policies typically substitute coal-fired generation with natural gas, which has more responsive technologies that emit lower levels of pollutants and greenhouse gases than coal-fired generation (Dewees 2008). Coal phase-out policies provide public health benefits and accelerate the transition to a sustainable energy system by decreasing emissions that lead to climate change (Winfield *et al.* 2010). This particular policy rapidly internalizes the costs associated with the market failure of fossil fuel energy by targeting concentrated sources of emissions.

Policies for overcoming institutional barriers

The final cluster consists of policies that increase the pace of renewable energy deployment by removing institutional barriers and facilitating long-term planning. One method of removing barriers is by consolidating siting authorities, either by aggregating multiple jurisdictions into one decision-making body or by placing the siting authority in an existing entity; examples are the Province of Ontario (Box 13.2) and the State of Texas (Box 13.6) (Gallant and Fox 2011; Bohn and Lant 2009; Wilson and Stephens 2009).

Agencies may also conduct integrated resource planning, which typically requires involving the public, identifying energy efficiency and resource options, developing action plans, and describing efforts to minimize the environmental effects of resource acquisitions. Experts contend that plans for designing and optimizing systems should now include explicit consideration of grid-connected renewables. They also maintain that including the evaluation of renewable energy sources in integrated resource planning helps develop a cost-effective sustainable energy system (Yilmaz *et al.* 2008).

Benefits of the selected policy measures

Empirical evidence shows that widespread renewable energy results in decreased environmental impacts and increased social benefits (IPCC 2011). Thus, increasing renewable energy production and displacing fossil fuels in the energy system by addressing perverse subsidies, providing paths to markets and market space, and removing institutional barriers could deliver multiple benefits. Environmental benefits include reduced greenhouse gas emissions and air pollutants, lower water use in the case of wind and solar photovoltaics, and decreased water pollution (Sovacool and Watts 2009; Roth and Ambs 2004). Social benefits include enhanced energy security and reliability by diversifying the supply and using indigenous resources, and reduced energy price volatility and disruptions (Awerbuch 2006; Roth and Ambs 2004). In addition, experts maintain that renewable energy developments are associated with enhanced economic development and more jobs (IPCC 2011; Wei *et al.* 2010). Finally, the use of renewable resources also benefits public health through decreased emissions and fewer occupational injuries (Sumner and Layde 2009; Rabl and Spadaro 2000).

Research clearly demonstrates that renewable energy sources generate significantly lower greenhouse gas emissions than fossil fuel options (IPCC 2011; Awerbuch 2006). Scenario analyses indicate that increasing renewable energy deployment from 27 to 77 per cent of the primary energy supply by 2050 may be expected and may achieve savings of up to 85 per cent of global CO_2 emissions for the scenarios with the highest renewable energy shares (IPCC 2011). The majority of the technologies deployed in these scenarios are wind, direct solar and modern biomass, with an annual average cost of less than 1 per cent of global gross domestic product (GDP) per year (Edenhofer *et al.* 2011). Furthermore, experts forecast that by 2030 the production costs, including social costs, of renewable energy would be lower than energy production by fossil fuels (Delucchi and Jacobson 2011; Jacobson and and Delucchi 2011). However, to achieve this transition, existing policies must be significantly strengthened and implemented comprehensively and therefore require additional political will (Jacobson and Delucchi 2011; Sovacool and Watts 2009).

The benefits of improving networks and reducing institutional barriers include lower costs and faster deployment of renewable energy. In the case of transmission, improved networks generally enhance reliability, lower the delivered costs of electricity and restrict the ability of generators to exercise market power (Hirst 2004). Experts commonly call for reducing institutional barriers to expedite the transition to a sustainable energy system (Mitchell *et al.* 2011). Quantitative analysis also shows that reducing siting barriers correlates with increased wind power development (Bohn and Lant 2009).

Potential drawbacks of selected policy measures

The successful implementation of production tax credits or feed-in tariffs requires an in-depth understanding of the various energy prices for all renewable energy sources as well as the costs of externalities. These policies therefore have potential drawbacks. Specifically, production tax credits or feed-in tariffs can be extremely inefficient. Since incentive levels are fixed over time, this may lead to limited innovation and downward price pressures. Likewise, implementing renewable portfolio standards also requires an in-depth knowledge of markets to establish appropriate targets, enforcement mechanisms and sector-specific set-asides. While context dependent, these are usually subsidies aimed at a particular industry (Berry and Jaccard 2001). Inadequately designed renewable portfolio standards may encourage particular technologies and therefore lead to technological lock-ins (Unger and Ahlgren 2005).

A large-scale oil refinery complex in the Alberta oil sands, Canada, near Fort McMurray. *© Dan Barnes/iStock*

In addition, critics argue that implementing renewable energy policies may increase the cost of energy and/or increase tax burdens (Gallant and Fox 2011). These expenses are especially burdensome to lower-income households; however, widespread renewable energy adoption combined with progressive tax design and incentives offers some protection from energy price increases. For example, subsidy programmes already exist to assist low-income households with energy costs in the United States, so expanding existing programmes could provide assistance for vulnerable groups should energy prices rise.

Policies to increase transmission networks and reduce siting barriers also have potential drawbacks. When reallocating the costs of transmission, these policies could result in disproportionate financial burdens on parties who do not benefit. Reducing siting barriers may also decrease public participation.

Replication and transferability of selected policies

The potential for replication and transferability of these policies is not straightforward and is arguably dependent on context and specific instrument design. For example, the North American grid exists in an institutional framework that is highly fragmented, while other countries may have nationally owned networks, in which case fragmentation may not be an issue (Willrich 2009; Joskow 2005). Germany, France, Italy, Japan and Denmark have experience in replicating and transfering feed-in tariffs at the national level, while the United States and Australia have experience with production tax credits and renewable portfolio standards (IEA 2011). Policies on feed-in tariffs and renewable portfolio standards are in force in diverse jurisdictions including Canada, China, Kenya, Portugal and Uganda (IEA 2011). Statistically, correlations demonstrate that the policies are effective, particularly in the case of feed-in tariffs (Haas *et al.* 2011). Direct causal evidence of effectiveness for other policies, however, is limited, as is evidence of the potential for replication and transferability to other jurisdictions (Carley 2009; Doris *et al.* 2009).

Proactive measures to accelerate the use of renewable energy
Achieving the international goal of urgently expanding the share of renewable energy supply in North America's energy mix

Box 13.6 Texas: a rapid expansion of wind energy

Texas has emerged as the leader in the growth of wind energy in the United States, with policies that direct market mechanisms towards achieving the state's energy capacity goals. Policies include customer choice, wholesale electricity markets, and a transmission cost allocation method along with tradable renewable energy credits and federal tax credits (Zarnikau 2011). In addition to these policies, the authority for siting wind farms in Texas is centralized, making it relatively easy to obtain licenses compared to other areas of the country (Bohn and Lant 2009; Wilson and Stephens 2009).

Expanding electricity transmission facilities has been a key component of Texas's package of policy instruments. The state is an unusual jurisdiction in North America because it has a single grid authority, the Electric Reliability Council of Texas (ERCOT). As part of its transmission policies, Texas dispensed with the demonstration of a need and designated Competitive Renewable Energy Zones, thus allowing capacity to be built ahead of need. Texas also allocates costs for these facilities across all companies that provide electric power to consumers within the ERCOT area (Schumacher *et al.* 2010), allowing transmission developers to recover the costs from the installation of new power lines. In addition, the charging of all consumers and not just the beneficiaries provides a consistent framework across the entire grid, eliminating the political dispute over who pays and who benefits from new transmission. These policies, which actively plan for expanding transmission, have been vital in promoting the state's rapid growth in renewable energy production (Figure 13.2).

Figure 13.2 Proposed renewable energy zones, potential transmission expansion and the growth of wind power in Texas

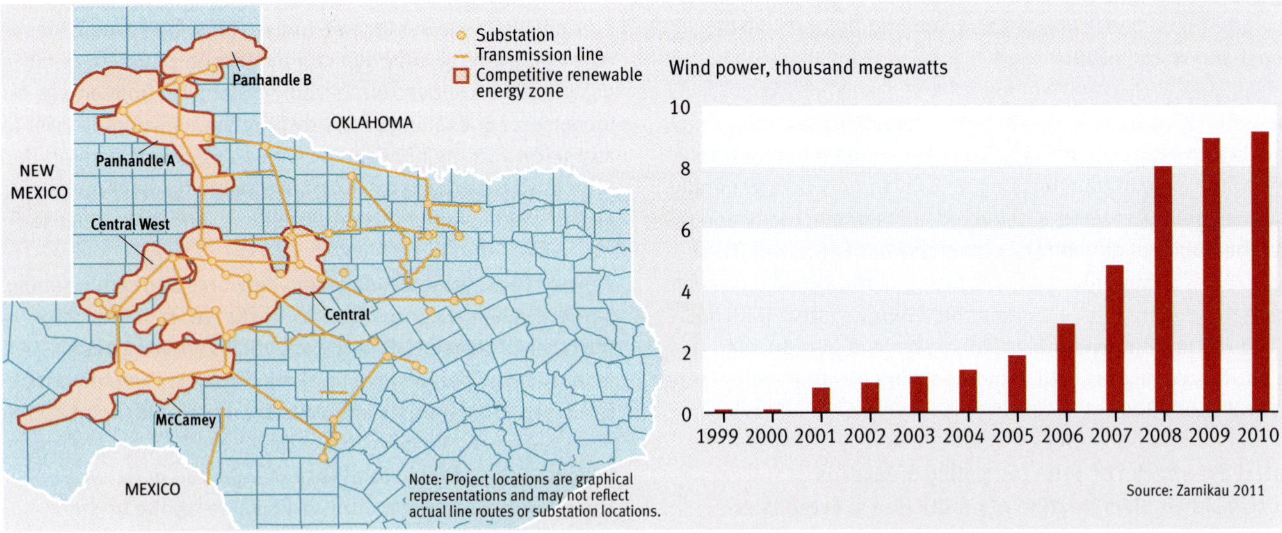

Texas's comprehensive policy package – which mandates renewable energy production, consolidates the siting authority and spreads transmission costs across all consumers – is a novel approach that has provided impressive results. Wind power deployment has grown from a capacity of 50 megawatts in 1999 to more than 9 272 megawatts in early 2010, accounting for 8.4 per cent of the state's total electrical generation in the first quarter of 2010. While there have been challenges in grid integration and additional transmission expansion is currently under way, projections based on the current policies indicate that Texas's wind energy will continue to expand and that solar energy deployment is expected to boom. The achievements and forecasts indicate that if the policy regime is properly designed, market-based initiatives can realize significant and rapid renewable energy development (Zarnikau 2011).

requires mobilizing political will and increasing public support to implement comprehensive renewable energy policies focused on addressing market failures, providing clear market signals, modernizing transmission systems, proving new technologies including energy storage, and streamlining institutional structures. A modernized, clean, reliable and efficient 21st century energy system will provide greater energy security, enhanced price stability and increased economic performance, and may save up to 85 per cent of global greenhouse gas emissions by 2050 (IPCC 2011; Awerbuch 2006).

Current research argues that in accounting for fossil fuel externalities and subsidies, the question appears not to be one of cost, but rather of social and political barriers (Delucchi and Jacobson 2011). Cultivating and developing widespread public participation and support is essential for generating the political will to implement the policies necessary to achieve the internationally agreed goal. The case studies illustrate that comprehensive policy packages that include incentives to offset externality and subsidy advantages afforded to fossil fuels, provide for energy transmission and reduce institutional barriers, can also significantly accelerate the transition to a sustainable energy future.

Cross-cutting issues

Increasing the deployment of renewable energy can provide a number of benefits to support the other internationally agreed goals. Wind and solar photovoltaic renewable energy can decrease water stress since it uses less water than conventional thermo-electric forms of generation (Roth and Ambs 2004). Benefits for land use include relative reductions in greenhouse gas emissions, thereby decreasing potential climate change impacts (Turney and Fthenakis 2011). However, land use for expanding renewable energy systems may require the disturbance of additional areas, depending on the particular technology being deployed (Fthenakis and Kim 2009). At the same time, an integrated approach to siting renewable energy, increased transparency and collaboration between agencies may lead to improvements in environmental governance.

CONCLUSIONS

This chapter has suggested that there are many policies and market instruments that have contributed, however uncertain the causality, towards achieving the internationally agreed goals. It is unlikely the policies were instituted with the global goals in mind, however; rather, the impetus probably came from bi-national, national and sub-regional institutions and governing bodies. It is important for all levels of governance and decision making to set clear short-, medium- and long-term environmental goals and specific targets as a crucial means of inducing a change in behaviour among public and private actors. Performance indicators are necessary to evaluate policy progress and clearly identify successes and shortcomings, and it is also essential to work towards synergy between the goals adopted under climate change and other environmental themes, while keeping in mind the potential contradictions between different environmental goals – at least in the short term, for example in the case of clean air and climate change – as well as between environmental protection and sustainable development, where conservation issues can arise.

Yosemite, one of the largest and least fragmented habitat blocks in the Sierra Nevada, was central to the development of the national park concept in the United States. *© Pgiam/iStock*

Some of the policy examples show how cultivating public will and political support while reducing negative public perceptions has moved the region closer to achieving environmental goals. Public-private partnerships have become increasingly important as government funds and staff have shown to be unable to assess resources, coordinate sustainable management and accommodate the increasing demands of multiple users.

The selected policy options suggest a number of opportunities for future environmental governance in North America. The most efficient and least controversial financial mechanism for ecosystem services focuses on users of an ecosystem service – such as water quality – who are willing to pay for the service and compensate the owners or managers of that resource for implementing best management practices.

Finally, and importantly, the examples reveal that applying successful policy options is complex, often requiring hybrid techniques combining two or more regulatory mechanisms to adjust existing market rules, financial incentives to shift pricing in existing markets, and participatory techniques. Transferring and up-scaling the processes that appear to have contributed to the success of a policy or market instrument will further speed up the achievement of internationally agreed environmental goals. In general, transferring processes is more feasible than replicating policy contents, since more is known about factors that influence the likelihood of transfers. The success of policies and instruments is very context dependent, while processes foster legitimacy and learning. Failing to protect ecosystem services for the generations to come will undoubtedly be more costly – socially, economically and environmentally – than the burden of expanding processes and policies that seem already to have proven successful.

REFERENCES

Ali, A.K. (2008). Greenbelts to contain urban growth in Ontario, Canada: promises and prospects. *Planning, Practice and Research* 23, 533–548

Anderson, J., Gomez W., C., McCarney, G., Adamowicz, W., Chalifour, N., Weber, M., Elgie, S. and Howlett, M. (2010). *Natural Capital: Using Ecosystem Service Valuation and Market-based Instruments as Tools for Sustainable Forest Management: A State of Knowledge Report.* Sustainable Forest Management Network, Edmonton, AB

Awerbuch, S. (2006). Portfolio-based electricity generation planning: policy implications for renewables and energy security. *Mitigation and Adaptation Strategies for Global Change* 11, 693–710

Barlow, P.M. and Reichard, E.G. (2010). Saltwater intrusion in coastal regions of North America. *Hydrogeology Journal* 18, 247–260

Bates, B.C., Kundzewicz, Z.W., Wu, S. and Palutikof, J.P. (eds.) (2008). *Water and Climate Change.* IPCC Technical Paper VI, June 2008. IPCC Secretariat, Geneva

BC Ministry of Finance (2008). *Budget and Fiscal Plan 2008/09–2010/11.* Government of British Columbia. http://www.bcbudget.gov.bc.ca/2008/bfp/2008_Budget_Fiscal_Plan.pdf (accessed 29 November 2011)

Benham, B., Zeckoski, R. and Yagow, G. (2008). Lessons learned from TMDL implementation case studies. *Water Practice* 2, 1–13

Berry, T. and Jaccard, M. (2001). The renewable energy portfolio standard: design considerations and an implementation survey. *Energy Policy* 29, 263–277

Blaney, J.P. (2009). An overview of the International Joint Commission. In *Managing Water Resources in a Time of Global Change: Mountains, Valleys and Flood Plains* (eds. Garrido, A. and Dinar, A.). pp.225–232. Routledge, New York

Blomquist, E. and Schlager, E. (2005). Political pitfalls of integrated watershed management. *Society and Natural Resources* 18, 101–117

Bloomberg (2011). *Obama Seeks to End $46.2 Billion in Energy Tax Breaks in Decade, Chu Says.* http://www.bloomberg.com/news/2011-02-11/obama-seeks-to-end-46-2-billion-in-energy-industry-tax-breaks-over-decade.html

BLS (2011). *Current Employment Statistics.* US Bureau of Labor Statistics. http://www.bls.gov/ces/ (accessed 27 November 2011)

Bohn, C. and Lant, C. (2009). Welcoming the wind? Determinants of wind power development among US states. *The Professional Geographer* 61, 87–100

Calbick, K.S., Day, J.C. and Gunton, T.I. (2003). Land use planning implementation: a 'best practices' assessment. *Environments* 31, 69–82

Can LII (2011). *Specified Gas Emitters Regulation, Alta Reg 139/2007.* Canadian Legal Information Institute, Ottawa, ON. http://www.canlii.org/en/ab/laws/regu/alta-reg-139-2007/latest/alta-reg-139-2007.html (accessed 29 November 2011)

Carley, S. (2009). State renewable energy electricity policies: an empirical evaluation of effectiveness. *Energy Policy* 37, 3071–3081

Cattaneo, A., Claassen, R., Johansson, R. and Weinberg, M. (2005). *Flexible Conservation Measures on Working Land, What Challenges Lie Ahead?* Economic Research Report Number 5. United States Department of Agriculture (USDA) Economic Research Service, Washington, DC

Cavendish-Palmer, H.A. (2008). *Planting Strong Boundaries: Urban Growth, Farmland Preservation, and British Columbia's Agricultural Land Reserve.* MSc thesis. Simon Fraser University, Burnaby, BC

Cayan, D.R., Das, T., Pierce, D.W., Barnett, T.P., Tyree, M. and Gershunov, A. (2010). Future dryness in the southwest US and the hydrology of the early 21st century drought. *National Academy of Sciences of the United States of America* 107, 21271–21276

CEC (2011). Commission for Environmental Cooperation of North America: site map. http://www.cec.org/ (accessed 28 November 2011)

Chestnut, L.G. and Mills, D.M. (2005). A fresh look at the benefits and costs of the US acid rain program. *Journal of Environmental Management* 77, 255

Cheverie, F. (2009). Prince Edward Island ecological goods and services pilot project. In *Proceedings of the Ecological Goods and Services Technical Meeting, Ottawa, Canada. Prairie Habitat Joint Venture.* http://phjv.ca/pdf/090924-EGS-techmeeting-proceedings-final-HR.pdf (accessed 18 December 2011)

City of Prince George (2011). *Water Conservation.* http://princegeorge.ca/cityservices/utilities/Pages/WaterConservation.aspx (accessed 28 May 2011)

City of San Diego (2011). *Water Conservation Program.* http://www.sandiego.gov/water/conservation/consprogram.shtml (accessed 28 May 2011)

Congressional Budget Office (2005). *Taxing Capital Income: Effective Rates and Approaches to Reform.* CBO, Washington, DC (October). http://www.cbo.gov/doc.cfm?index=6792 (accessed 18 December 2011)

Cortner, H. and Moote, M. (1999). *The Politics of Ecosystem Management.* Island Press, Washington, DC

Davis, S.M. and Ogden, J.C. (1994). *Everglades: The Ecosystem and its Restoration.* St Lucie Press, Delray Beach, FL

Delucchi, M.A. and Jacobson, M.Z. (2011). Providing all global energy with wind, water, and solar power. Part II: Reliability, system and transmission costs, and policies. *Energy Policy* 39, 1170–1190

Dewees, D.N. (2008). Pollution and the price of power. *The Energy Journal* 29, 81–100

Doris, E., McLaren, J., Healey, V. and Hockett, S. (2009). *State of the States.* National Renewable Energy Laboratory, US Government Printing Office, Washington, DC

DSIRE (2011). Database of State Incentives for Renewables and Efficiency. http://www.dsireusa.org/ (accessed 19 May 2011)

Dunn, C.P., Stearns, F., Guntenspergen, G.G. and Sharpe, D.M. (1993). Ecological benefits of the Conservation Reserve Program. *Conservation Biology* 7, 132–139

Easterling, D.R., Meehl, G.A., Parmesan, C., Changnon, S.A., Karl, T.R. and Mearns, L.O. (2000). Climate extremes: observations, modeling, and impacts. *Science* 289, 2068–2074

Edenhofer, O., Pichs-Madruga, R., Sokona, Y., Seyboth, K., Matschoss, P., Kadner, S., Zwickel, T., Eickemeier, P., Hansen, G., Schlömer, S. and von Stechow, C. (eds.) (2011). *IPCC Special Report on Renewable Energy Sources and Climate Change Mitigation.* Cambridge University Press, Cambridge and New York

Ellerman, D., Joskow, P., Schmalensee, R., Montero, J.-P., and Bailey, E. (2000). *Markets for Clean Air: The US Acid Rain Program.* Cambridge University Press, Cambridge

Environment Canada (2011). *Georgia Basin-Puget Sound International Airshed Strategy.* http://www.pyr.ec.gc.ca/airshed/index_e.htm (accessed 29 November 2011)

EPWU (2007). *El Paso Water Utilities.* http://www.epwu.org/conservation/education.html?reload (accessed 28 May 2011)

FAO (2011). *AQUASTAT Information System on Water and Agriculture.* Food and Agriculture Organization of the United Nations, Land and Water Development Division, Rome. http://www.fao.org/nr/water/aquastat/dbase/index.stm (accessed 21 March 2011)

Feung, F. and Conway, T. (2007). Greenbelts as an environmental planning tool: a case study of Southern Ontario, Canada. *Journal of Environmental Policy Planning* 9, 101–117

Fischer, C. (2010). Renewable portfolio standards: when do they lower energy prices? *The Energy Journal*, 31, 101–119

Fryer, J. (2009). *Sustaining our Water Future: A Review of the Marin Municipal Water District's Alternatives to Improve Water Supply Reliability.* Food and Water Watch, Washington, DC

Fthenakis, V. and Kim, H.C. (2009). Land use and electricity generation: a life-cycle analysis. *Renewable and Sustainable Energy Reviews* 13, 1465–1474

G20 (2009) *Leaders' Statement: The Pittsburgh Summit.* http://ec.europa.eu/commission_2010-2014/president/pdf/statement_20090826_en_2.pdf

Gallant, P. and Fox, G. (2011). Omitted costs, inflated benefits: renewable energy policy in Ontario. *Bulletin of Science, Technology and Society* 30 September 2011, 1–8

Gleason, R.A., Laubhan, M.K. and Euliss Jr., N. H. (eds.) (2008). *Ecosystem Services Derived from Wetland Conservation Practices in the United States Prairie Pothole Region with an Emphasis on the US Department of Agriculture Conservation Reserve and Wetlands Reserve Programs.* US Geological Professional Paper 1745. USGS, Reston, Virginia, VA

Glicksman, R.L. (2008). Sustainable federal land management: protecting ecological integrity and preserving environmental principal. *Tulsa Law Journal* 44, 147

GLIN (2011a). Great Lakes Information Network. http://www.great-lakes.net/ (accessed 28 May 2011)

GLIN (2011b). Great Lakes Information Network. http://gis.glin.net/maps/ (accessed 21 September 2011)

GLSL Cities (2011). *Great Lakes and St. Lawrence Cities Initiative Annual Report 2010–2011.* http://www.glslcities.org/Reports/Annual%20Report%202011_v8_final.pdf (accessed 27 December 2011)

Government of Ontario (2009). *Ontario's Coal Phase Out Plan.* http://news.ontario.ca/mei/en/2009/09/ontarios-coal-phase-out-plan.html (accessed 29 November 2011)

Government of Quebec (2009). *National Assembly, 39th Legislature, 1st Session: An Act to Affirm the Collective Nature of Water Resources and Provide for Increased Water Resource Protection.* http://www2.publicationsduquebec.gouv.qc.ca/dynamicSearch/telecharge.php?type=5&file=2009C21A.PDF (accessed 29 November 2011)

Haas, R., Resch, G., Panzer, C., Busch, S., Ragwitz, M. and Held, A. (2011). Efficiency and effectiveness of promotion systems for electricity generation from renewable energy sources: lessons from EU countries. *Energy* 36, 2186–2193

Hanak, E. (2003). *Who Should be Allowed to Sell Water in California? Third-Party Issues and the Water Market.* Public Policy Institute of California, San Francisco. http://www.ppic.org/content/pubs/report/r_703ehr.pdf (accessed 27 November 2011)

Hanna, K.S. (1997). Regulation and land-use conservation: a case study of the British Columbia Agricultural Land Reserve. *Journal of Soil and Water Conservation* 52, 166–170

Harrington, W., Morgenstern, R.D. and Nelson, P. (2008). On the accuracy of regulatory cost estimates. *Journal of Policy Analysis and Management* 19, 297–322

Harrison, K. and Antweiler, W. (2003). Incentives for pollution abatement: regulation, regulatory threats, and non-governmental pressures. *Journal of Policy Analysis and Management* 22, 361–382

Hassett, B., Palmer, M., Bernhardt, E., Smith, S., Carr, J. and Hart, D. (2005). Restoring watersheds project by project: trends in Chesapeake Bay tributary restoration. *Frontiers in Ecology and the Environment* 3, 259–267

Haufler, J. B. (2005). Fish and wildlife benefits of Farm Bill conservation programs: 2000–2005 update. *The Wildlife Society Technical Review* 05-2, Bethesda, MD

Heathcote, I.W. (2009). *Integrated Watershed Management: Principles and Practice*. John Wiley & Sons, Inc., Hoboken, NJ

Hellerstein, H. (2010). Challenges facing USDA's Conservation Reserve Program. *Amber Waves* 8

Hironaka, A. (2002). The globalization of environmental protection: the case of environmental impact assessment. *International Journal of Comparative Sociology* 43, 65–78

Hirst, E. (2004). US transmission capacity: a review of transmission plans. *The Electricity Journal* 17, 65–79

Howarth, B.R., Haddad, B.M. and Paton, B. (2000). The economics of energy efficiency: insights from voluntary participation programs. *Energy Policy* 28, 477–486

Howland, M. (2010). The private market for brownfield properties. *Cityscape* 12, 37

IEA (2011). Policies and measures databases. http://www.iea.org/textbase/pm/index.html (accessed 20 May 2011)

Industry Canada (2011). *Gross Domestic Product (GDP): Agriculture, Forestry, Fishing and Hunting*. http://www.ic.gc.ca/cis-sic/cis-sic.nsf/IDE/cis-sic11vlae.html#gdp2a (accessed 29 November 2011)

IPCC (2011). Summary for policymakers. In *IPCC Special Report on Renewable Energy Sources and Climate Change Mitigation* (eds. Edenhofer, O., Pichs-Madruga, R., Sokona, Y., Seyboth, K., Matschoss, P., Kadner, S., Zwickel, T., Eickemeier, P., Hansen, G., Schlömer, S. and von Stechow, C.). Intergovernmental Panel on Climate Change. Cambridge University Press, Cambridge and New York

IRWMP (2006). *Bay Area Integrated Regional Water Management Plan*. http://bairwmp.org/plan/ (accessed 27 May 2011)

Jacobson, M.S. and Delucchi, M.A. (2011). Providing all global energy with wind, water and solar power. Part I: Technologies, energy resources, quantities and areas of infrastructure, and materials. *Energy Policy* 39, 1154–1169

Johnson, P.M. and Beaulieu, A. (1996). *The Environment and NAFTA: Understanding and Implementing the New Continental Law*. Island Press, New York

Jones-Crabtree, A., Wilson, G., McWilliams, R., Patterson, T., Baker, S., Zanowick, M. and Horsch, L. (2008). *Greening from the Ground Up: A Report on the 3-yr Investment Between the Forest Service Washington Office and the Rocky Mountain Region (R2)*. Sustainable Operations WO/R2 Partnership Report. http://www.fs.fed.us/sustainableoperations/documents/200810-GreeningFromTheGroundUpSustainableOperationsInTheForestService.pdf (accessed 29 November 2011)

Joskow, P.A. (2005). Transmission policy in the United States. *Utilities Policy* 13, 95–115

Kargbo, D.M., Wilhelm, R.G. and Campbell, D.J. (2010). Natural gas plays in the Marcellus Shale: challenges and potential opportunities. *Environmental Science and Technology* 44, 5679–5684

Kenney, D.S. (2005). Prior appropriation and water rights reform in the western United States. In *Water Rights Reform: Lessons for Institutional Design* (eds. Bruns, B.R., Claudia Ringler, C. and Meinzen-Dick, R.). pp.167–182. International Food Policy Research Institute, Washington, DC

Kenny, A., Elgie, S. and Sawyer, D. (2011). *Advancing the Economics of Ecosystems and Biodiversity in Canada: A Survey of Economic Instruments for the Conservation and Protection of Biodiversity*. Environment Canada, Ottawa

Klaassen, G., Miketa, A., Larsen, K. and Sundqvist, T. (2005). The impact of R&D on innovation for wind energy development in Denmark, Germany, and the United Kingdom. *Ecological Economics* 54, 227–240

Lewis, R., Knaap, G.-J. and Sohn, J. (2009). Managing growth with priority funding areas: a good idea whose time has yet to come. *Journal of the American Planning Association* 75, 457–478

Lynch, L. and Liu, X. (2007). Impact of designated preservation areas on rate of preservation and rate of conversion. *American Journal of Agricultural Economics* 89, 1205–1210

Mabee, W.E., Mannion, J. and Carpenter, T. (2012). Comparing the feed-in tariff incentives for renewable electricity in Ontario and Germany. *Energy Policy* 40, 480–489

Madsen, B., Carroll, N. and Moore Brands, K. (2010). *State of Biodiversity Markets Report: Offset and Compensation Programs Worldwide*. http://www.ecosystemmarketplace.com/documents/acrobat/sbdmr.pdf (accessed 6 December 2011)

McGee, G., Cullen, A. and Gunton, T. (2010). A new model for sustainable development: a case study of The Great Bear Rainforest regional plan. *Environment, Development and Sustainability* 12, 745–762

Mendonca, M. (2007). *Feed-in Tariffs: Accelerating the Deployment of Renewable Energy*. Earthscan, London

Metcalf, G.E. and Weisbach, D. (2008). The design of a carbon tax. *Harvard Environmental Law Review* 33, 499–556

Mitchell, C., Sawin, J.L., Pokharel, G.R., Kammen, D., Wang, Z., Fifita, S., Jaccard, M., Langniss, O., Lucas, H., Nadai, A., Trujillo Blanco, R., Usher, E., Verbruggen, A., Wüstenhagen, R. and Yamaguchi, K. (2011). Policy, financing and implementation. In *IPCC Special Report on Renewable Energy Sources and Climate Change Mitigation* (eds. Edenhofer, O., Pichs-Madruga, R., Sokona, Y., Seyboth, K., Matschoss, P., Kadner, S., Zwickel, T., Eickemeier, P., Hansen, G., Schlömer, S. and von Stechow, C.). Cambridge University Press, Cambridge and New York

MMWD (2011). Marin Municipal Water District. http://www.marinwater.org/ (accessed 6 December 2011)

Nordhaus, W.D. (2010). Carbon taxes to move toward fiscal sustainability. *The Economists' Voice* 7(3), Article 3

Novotny, V. (1999). Diffuse pollution from agriculture – a worldwide outlook. *Water Science and Technology* 39(3), 1–13

NRC (2008). *Urban Stormwater Management in the United States*. National Research Council of the National Academy of Sciences. The National Academy Press, Washington, DC

Ontario Ministry of Energy (2010). *Green Energy Act*. http://www.energy.gov.on.ca/en/green-energy-act/ (accessed 19 September 2011)

Power Authority of Ontario (2010). *FIT Program microFIT Program*. http:/fit.powerauthority.on.ca (accessed 19 September 2011)

Rabl, A. and Spadaro, J.V. (2000). Public health impacts of air pollution and implications for the energy system. *Annual Review of Energy and the Environment* 25, 601–627

Renzetti, S. and Kushner, J. (2004). Full cost accounting for water supply and sewage treatment: concepts and case application. *Canadian Water Resources Journal* 29, 13–22

Ritter, W.F. and Shirmohammadi, A. (2001). *Agricultural Non-Point Source Pollution: Watershed Management and Hydrology*. Lewis Publishers, New York

Rockaway, T.D., Coomes, P.A., Rivard, J. and Kornstein, B. (2011). Residential water use trends in North America. *Journal of the American Water Works Association* 103, 76–89

Rogers, P., de Silva, R. and Bhatia, R. (2002). Water is an economic good: how to use prices to promote equity, efficiency, and sustainability. *Water Policy* 4, 1–17

Roth, I.F. and Ambs, L.L. (2004). Incorporating externalities into a full cost approach to electric power generation life-cycle costing. *Energy* 29, 2125–2144

Ruhl, H.A. and Rybicki, N.B. (2010). Long-term reductions in anthropogenic nutrients link to improvements in Chesapeake Bay habitat. *Proceedings of the National Acadamy of Sciences of the United States of America* 107(38), 16566–16570

Salzman, J.E. (2005). Creating markets for ecosystem services: notes from the field. *New York University Law Review* 8, 870–961

Sartori, J., Moore, T. and Knaap, G. (2011). *Indicators of Smart Growth in Maryland*. The National Center for Smart Growth Research and Education at the University of Maryland, College Park, MD

Schiermeier, Q., Tollefson, J., Scully, T., Witze, A. and Morton, O. (2008). Energy alternatives: electricity without carbon. *Nature* 454, 816–823

Schneider, H., Easterling, W.E. and Mearms, L.O. (2000). Adaptation: sensitivity to natural variability, assumptions, and dynamic climatic changes. *Climatic Change* 45, 203–221

Schumacher, A., Fink, S. and Porter, K. (2010). Moving beyond paralysis: how states and regions are creating innovative transmission policies for renewable energy projects. *The Electricity Journal* 22, 27–36

Schwartz, A.M. (2006). The management of shared waters: watershed boards past and future. In *Bilateral Ecopolitics: Continuity and Change in Canadian-American Environmental Relations* (eds. Le Prestre, P. and Stoett, P.). pp.133–144. Ashgate Publishing, Aldershot

Smith, V.H., Joye, S.B. and Howarth, R.W. (2006). Eutrophication of freshwater and marine ecosystems. *Limnology and Oceanography* 51, 351–355

Soderholm, P. and Klaassen, G. (2007). Wind power in Europe: a simultaneous innovation-diffusion model. *Environmental and Resource Economics* 36, 163–190

Sovacool, B.K. (2009a). Rejecting renewables: the socio-technical impediments to renewable electricity in the United States. *Energy Policy* 37, 4500–4513

Sovacool, B.K. (2009b). The importance of comprehensiveness in renewable electricity and energy-efficiency policy. *Energy Policy* 37, 1529–1541

Sovacool, B.K. and Watts, C. (2009). Going completely renewable: is it possible (*let alone* desirable)? *The Electricity Journal* 22, 95–111

Spieles, D.J. (2005). Vegetation development in created, restored, and enhanced mitigation wetland banks of the United States. *Wetlands* 25, 51–63

Sumner, S.A. and Layde, P.M. (2009). Expansion of renewable energy industries and implications for occupational health. *Journal of the American Medical Association* 302, 787–789

Taylor, J., Paine, C. and FitzGibbon, J. (2005). From greenbelt to greenways: four Canadian case studies. *Landscape and Urban Planning* 33, 47–64

ten Brink, P. (ed.) (2011). *The Economics of Ecosystems and Biodiversity in National and International Policy Making*. London, Earthscan

Turney, D. and Fthenakis, V. (2011). Environmental impacts from the installation and operation of large-scale solar power plants. *Renewable and Sustainable Energy Reviews* 15(6), 3261–3270

UNEP GC (2010) *Nusa Dua Declaration, Bali, February 2010*. United Nations Environment Programme Governing Council. http://www.unep.org/gc/gcss-xi/Documents/Nusa_Dua_Declaration_Bali_Feb2010.pdf

Unger, T. and Ahlgren, E.O. (2005). Impacts of a common green certificate market on electricity and CO_2-emission markets in the Nordic countries. *Energy Policy* 33, 2152–2163

USDA (2012). *New Forest Planning Rule Seeks to Restore the Nation's Forests through Science and Collaboration*. USDA Forest Service Press Release No. 1158. http://www.fs.fed.us/news/2012/releases/01/planning-rule.shtml (accessed 8 March 2012)

USDA (2011). Office of Environmental Markets (OEM). US Department of Agriculture. http://www.fs.fed.us/ecosystemservices/OEM/ (accessed 6 December 2011)

USEPA (2006). *Expert Workshop on Full Cost Pricing of Water and Wastewater Service: Final Report*. United States Environmental Protection Agency. http://water.epa.gov/infrastructure/sustain/upload/2009_05_26_waterinfrastructures_workshop_si_fullcostpricing.pdf (accessed 29 November 2011)

USEPA (2005). *Case Studies of Sustainable Water and Wastewater Pricing*. EPA 816-R-05-007. Office of Water, United States Environmental Protection Agency. http://www.epa.gov/safewater/smallsystems/pdfs/guide_smallsystems_fullcost_pricing_case_studies.pdf (accessed 29 November 2011)

Vickers, A. (2001). *Handbook of Water Use and Conservation*. WaterPlow Press, Amherst, MA

Vörösmarty, C.J., McIntyre, P.B., Gessner, M.O., Dudgeon, D., Prusevich, A., Green, P., Glidden, S., Bunn, S.E., Sullivan, C.A., Liermann, C.R. and Davies, P.M. (2010). Global threats to human water security and river biodiversity. *Nature* 467, 555–561

Vörösmarty, C.J., Green, P., Salisbury, J. and Lammers, R. (2000). Global water resources: vulnerability from climate change and population growth. *Science* 289, 284–288

Wei, M., Patadia, S. and Kammen, D.M. (2010). Putting renewables and energy efficiency to work: how many jobs can the clean energy industry generate in the US? *Energy Policy* 38, 919–931

Wiewel, W. and Knaap, G. (2005). *Partnerships for Smart Growth: University-Community Collaboration for Better Public Places*. M.E. Sharp, Inc., New York

Willrich, M. (2009). *Electricity Transmission Policy for America: Enabling a Smart Grid, End-to-End*. Energy Innovation Working Paper Series. Industrial Performance Center – Massachusetts Institute of Technology, Cambridge, MA

Wilson, E.J. and Stephens, J.C. (2009). Wind deployment in the United States: resources, policy, and discourse. *Environmental Science and Technology* 43, 9063–9070

Winfield, M., Gibson, R.B., Markvart, T., Gaudreau, K. and Taylor, J. (2010). Implications of sustainability assessment for electric system design: the case of the Ontario power authority's integrated power system plan. *Energy Policy* 38, 4115–4126

WSSD (2002). *Johannesburg Plan of Implementation*. World Summit on Sustainable Development. http://www.un.org/esa/sustdev/documents/WSSD_POI_PD/English/POIToc.htmYaffee, S.L. (1996). *Ecosystem Management in the United States: An Assessment of Current Experience*. Island Press, Washington, DC

Yaffee, S.L., Phillips, A.F., Frentz, I.C., Hardy, P., Maleki, S. and Thorpe, B.E. (1996). *Ecosystem Management in the United States: An Assessment of Current Experience*. Island Press, Washington, DC

Yamasaki, S.H., Guillon, B.M.C., Brand, D. and Patil, A.M. (2010). Market-based payments for ecosystem services: current status, challenges and the way forward. *CAB Reviews: Perspectives in Agriculture, Veterinary Sciences, Nutrition and Natural Resources* 5, 1–13

Yatchew, A. and Baziliauskas, A. (2011). Ontario feed-in tariff programs. *Energy Policy* 39, 3885–3893

Yilmaz, P., Hocaoglu, M.H. and Konukman, A.E.S. (2008). A pre-feasibility case study on integrated resource planning including renewables. *Energy Policy* 36, 1223–1232

Zarnikau, J. (2011). Successful renewable energy development in a competitive electrical market: a Texas case study. *Energy Policy* (Special Section: Renewable energy policy and development) 39, 3906–3913

CHAPTER 14

West Asia

Coordinating lead authors: Amr El-Sammak and Nesreen Ghaddar

Lead authors: Mohamed Abdulrazzak, Anwar Abdu Khalil, Ahmad Fares Asfary, Nesreen Ghaddar, Ibrahim Abdel Gelil, Amr El-Sammak, Mohamed Abdel Raouf Abdel Hamid Aly and Fouad Abousamra

Contributing authors: Abdullah Droubi, Mahmoud Al-Sinai, Asma Abahussain, Mohammad S. Abido, Ahmed Ali Salih, Abdel Hadi Mohamed, Muhyiddine Jradi, Maha Al-Sabbagh, Hashim Al-Sayed, Fouad Abousamra, Ahmed Khalil, Lulwa N Ali, Amir Ibrahim, Mohammad Abdul Rahman Hassan and Mukdad Al-Khateeb

Principal scientific reviewer: Mahmoud Ali

Chapter coordinators: Adel Farid Abdel-Kader and Fouad Abousamra

Main Messages

Initiatives to introduce policy mixes to achieve a higher level of integration at different sectoral levels remain modest. West Asia has, however, made some progress on environmental governance and tends to rely on command-and-control measures rather than market-based instruments.

Financial investment has enabled some countries to make good progress towards Millennium Development Goal targets for water supply and sanitation (MDG 7c), but more efforts are still needed, especially in Yemen. In the past four decades, water policies have focused on supply infrastructure, especially in urban areas, aiming to overcome shortages through technical solutions including desalination. Coordination with other policies that prioritize balancing water supply with demand is crucial. The success of water policies in the region is contingent on political, financial and human commitment, reliable assessment of supply and demand, effective legal and institutional arrangements and active public-private sector partnerships.

National action plans to combat land degradation and desertification should be integrated with the sustainable use of natural resources, biodiversity conservation and plans to reduce the impacts of climate change. Integrated action to reduce land degradation, a significant issue in the region, would also address the regional phenomena of dust storms.

The region needs to strengthen its legislative and institutional frameworks for developing sustainable energy systems if it is to achieve global goals. Policy development to promote energy efficiency and renewable energy is evolving but, despite a wealth of renewable energy sources, the energy sector is still characterized by heavy reliance on fossil fuels, leading to high carbon emissions and adverse environmental impacts. The building sector is a major energy consumer, especially for air-conditioning, though green building practices are now emerging through the adoption of industry energy-efficiency codes.

Countries should confirm their commitments to protect the coastal and marine ecosystem through harmonization of the ecosystem approach with integrated coastal zone management plans and strategies. Strong coastal development plans reflect the implementation of coastal and marine policies in West Asia. Achieving marine biodiversity conservation is progressing through the establishment of marine protected areas and the application of integrated fisheries management.

INTRODUCTION

West Asia is geographically grouped into two sub-regions: the Arabian Peninsula, including Yemen and the Gulf Cooperation Council (GCC) countries of Bahrain, Kuwait, Oman, Qatar, Saudi Arabia and United Arab Emirates; and the Mashriq, which includes Iraq, Jordan, Lebanon, the Occupied Palestinian Territories (OPT) and Syria. The region covers about 4 million km^2 approaching 2.5 per cent of world's total land area. The environment is predominantly arid and semi-arid. Rainfall is scarce but with significant spatial and temporal variability. Water scarcity and frequent and persistent spells of drought are common, making water the region's most precious resource.

The region faces major environmental challenges in the need to address water scarcity; land degradation and desertification; increasing fossil fuel-based energy production and use with high inefficiencies in generation, distribution and end use; and conservation and sustainable use of marine and coastal resources. Climate change is becoming one of the region's main problems with potentially adverse impacts on the economy and human-well being. Water availability is projected to drop in most of the region by 2050, mainly due to rising temperatures and decreased precipitation (UNEP 2010; IPCC 2007). Much of the coast, especially in the GCC countries and Yemen, is vulnerable to sea level rise, which threatens large areas with inundation and saltwater intrusion (AFED 2009).

The drivers of environmental change in the region are linked to peace and security, demography and the state of the economy. The international desire to secure valuable energy resources and disputes including the current political conflict are playing a major role in the ongoing environmental degradation in the region. Environmental damage is escalating and the number of displaced people is increasing, straining the environment and contributing to the degradation of land and water resources (UNEP 2010).

The total population of West Asia was estimated at 134 million in 2010, or 1.94 per cent of the world population. Given an annual growth rate of around 3 per cent, it is expected to reach 205 million by 2030 (UNPD 2008). Although fertility rates in the region are declining, the momentum of population growth is still high, partially due to cultural and religious beliefs and difficulties hindering family planning (UNEP 2010). Urban communities represent more than 90 per cent of the population of the GCC countries, about 75 per cent of the Mashriq sub-region, and 31 per cent in Yemen. These high population growth rates and urbanization together with current consumption patterns compound the pressures on the region's limited land and water resources. Generally speaking, the young and ever-increasing population, as well as its mobility, represent new prospects for development but may also exacerbate pressure on already strained resources and ecosystems. More resources and services are required to support the demand for jobs, housing, health, water, energy and education; hence land use change is expected to be a major issue in the region (UNEP 2010). Furthermore, an influx of expatriates into the GCC countries only adds to pressure on already strained and limited land and water resources (UN ESCWA 2005).

Profits from the export of petroleum have made many countries in the region dependent on a continued oil boom. *© Ryan Lindsay*

Most of the West Asian countries' economies depend on oil and gas export revenues, especially GCC countries. In general, the region holds 52.2 per cent of world oil reserves and 24.6 per cent of world gas resources (OAPEC 2009). Oil and gas exports along with petrochemicals are the main source of income in GCC countries. In the Mashriq sub-region and Yemen, however, agriculture is the main economic activity, contributing 30 per cent of gross domestic product (GDP) and employing more than 40 per cent of the workforce (UN ESCWA 2002), although there are also some extractive industries in countries such as Jordan and Syria. On a per-person basis, the GCC country with the highest GDP is Qatar, earning US$77 000 per person in 2010 (UNDP 2010). These high earnings are reflected in high per-person energy consumption, with many of the GCC countries having carbon dioxide (CO_2) emissions of more than 25 tonnes per person per year in 2006 (UNDP 2010). Furthermore, the concentration of oil and extractive industries in the region strains the environment by polluting the atmosphere and degrading land and water resources. New initiatives are, however, being implemented to reduce the emissions and waste output associated with development, for example at Masdar City in Abu Dhabi (Sgouridis and Kennedy 2010).

The rapid development of the past 30 years has been the main driver of continued degradation of the environment in West Asia. In spite of the progress that has been achieved to meet the MDGs, more effort is needed (UN DESA 2011). Governments of the region are dealing with these challenges by creating suitable conditions and empowered communities, with national environmental policies having been developed in all West Asian countries. The UN Conference on Environment and Development in 1992 – the Rio Earth Summit – accelerated the setting-up and strengthening of environment ministries and authorities, the adoption of national strategies, financial resource mobilization and the creation of partnerships.

Figure 14.1 Priorities for action in West Asia

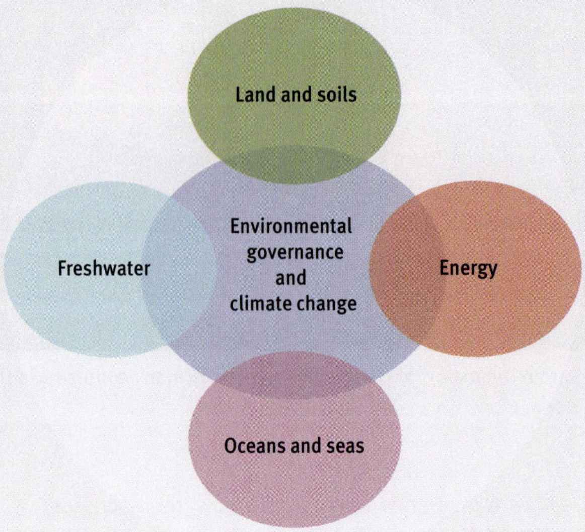

Environmental institutions have been accorded high priority and status in all countries of West Asia (UNEP 2010), and a range of institutions has been established to implement policies, enforce laws and set standards and norms. However, these policies remain sectoral in nature and participation of the major public groups in environmental governance remains weak. There is no clear policy for the integration of these groups in the environmental governance process at either national or regional levels.

The environmental policies of West Asian countries rely mainly on command-and-control mechanisms rather than on economic instruments, though there have recently been various initiatives to use market-based instruments to offer incentives and change behaviour; these include water cost recovery options and a road toll system.

Through a consultative process, the four most pressing environmental challenges identified in West Asia are freshwater; soil, land use, land degradation and desertification; energy; and oceans and seas. Policies and policy considerations relating to the cross-cutting issues of environmental governance and climate change have been incorporated into the four priority areas as appropriate (Figure 14.1).

POLICY APPRAISAL

Freshwater

The water sources of the West Asia region, estimated at 106.5 km^3 (UNEP 2011), consist of renewable surface and shallow groundwater resources supplemented by non-renewable groundwater, desalinated water and treated wastewater. Surface water resources are estimated at 86 km^3 concentrated mainly in the Mashriq sub-region, with 63 km^3 available from mainly shared rivers, the Euphrates, Tigris, Jordan, Yarmouk and Al Kabeer al-Jounbi, and the remaining 13 km^3 supplied by small rivers, springs and intermittent wadi flow (UN ESWCA 2007b; Abdulrazzak *et al.* 2002; Al-Rashed and Sherif 2000; Abdulrazzak 1995, 1994). The total renewable groundwater resources in the region are estimated at 15.5 km^3 (UNEP 2011). Iraq, Lebanon and Syria rely on river flows supplemented by limited groundwater resources, while Jordan, OPT, Yemen and the GCC countries rely on renewable groundwater sources supplemented by extensive non-renewable groundwater reserves and desalinated water (UNEP 2007; Dabour 2006).

Desalinated water, which has become a dependable domestic water supply source, contributes 3.3 km^3 and meets 56 per cent of the domestic water requirements of the GCC countries (Word Bank 2005). The GCC is home to about 44 per cent of world desalination capacity (AFED 2010; UN ESCWA 2007b). Around 2.3 km^3 of treated waste and drainage water is used in urban landscaping and feed-crop production together with 9 km^3 of untreated wastewater. Total water demand in the domestic, industrial and agricultural sectors was estimated at 83.4 km^3 in 1990, rising to 112.8 km^3 in 2000, and is expected to reach 167.4 km^3 in 2025 (UNEP 2011). High population growth and urbanization rates, increased frequency of drought and extreme events, accelerated economic activities and improved standards of living have contributed to the widening gap between supply and demand, and to higher levels of pollution and resource depletion. The region's increasing water scarcity is evident in the reduction in annual per-person renewable water resources from 1 050 m^3 in 1990, to 553 m^3 in 2010; this is expected to fall to 205 m^3 in 2025 compared to a world average of 7 243 m^3 per person per year (CEDARE and AWC 2004).

Water scarcity due to climate change may reduce the available renewable water resources by 15–20 per cent in the next 50 years, leading to decreases in the flow of major rivers and groundwater recharge rates, a higher frequency of flash floods and droughts, and a loss of productivity in rain-fed areas (AFED 2009). The increase in temperature due to climate change is expected to lead to increased water demand, especially for irrigated agriculture; saltwater intrusion from sea level rise; a decline in provisions for tourism, and changes in crop production systems (AFED 2009).

Previous water policies that emphasized the development of supply infrastructure made it possible for most of the countries to be on track to achieve the Millennium Development Goal (MDG) 7c targets for water supply and sanitation. Regional coverage in 2008 reached 92 per cent for water supply and 81 per cent for sanitation, with major achievements in urban areas (UN DESA 2011). Coverage in rural areas is lower, especially in Iraq, Syria, Oman, OPT and Yemen. Higher coverage has been achieved in the GCC countries than in the Mashriq sub-region, corresponding to the availability of financial resources (Figure 14.2). Drinking water coverage in West Asia ranges from 100 per cent in most of the GCC countries to 52 per cent in Yemen. Between 1990 and 2008, the coverage for domestic water supply increased by 4 per cent and for sanitation by 5 per cent. It is estimated that since

Figure 14.2 Domestic water supply and sanitation in West Asia, 1990–2015

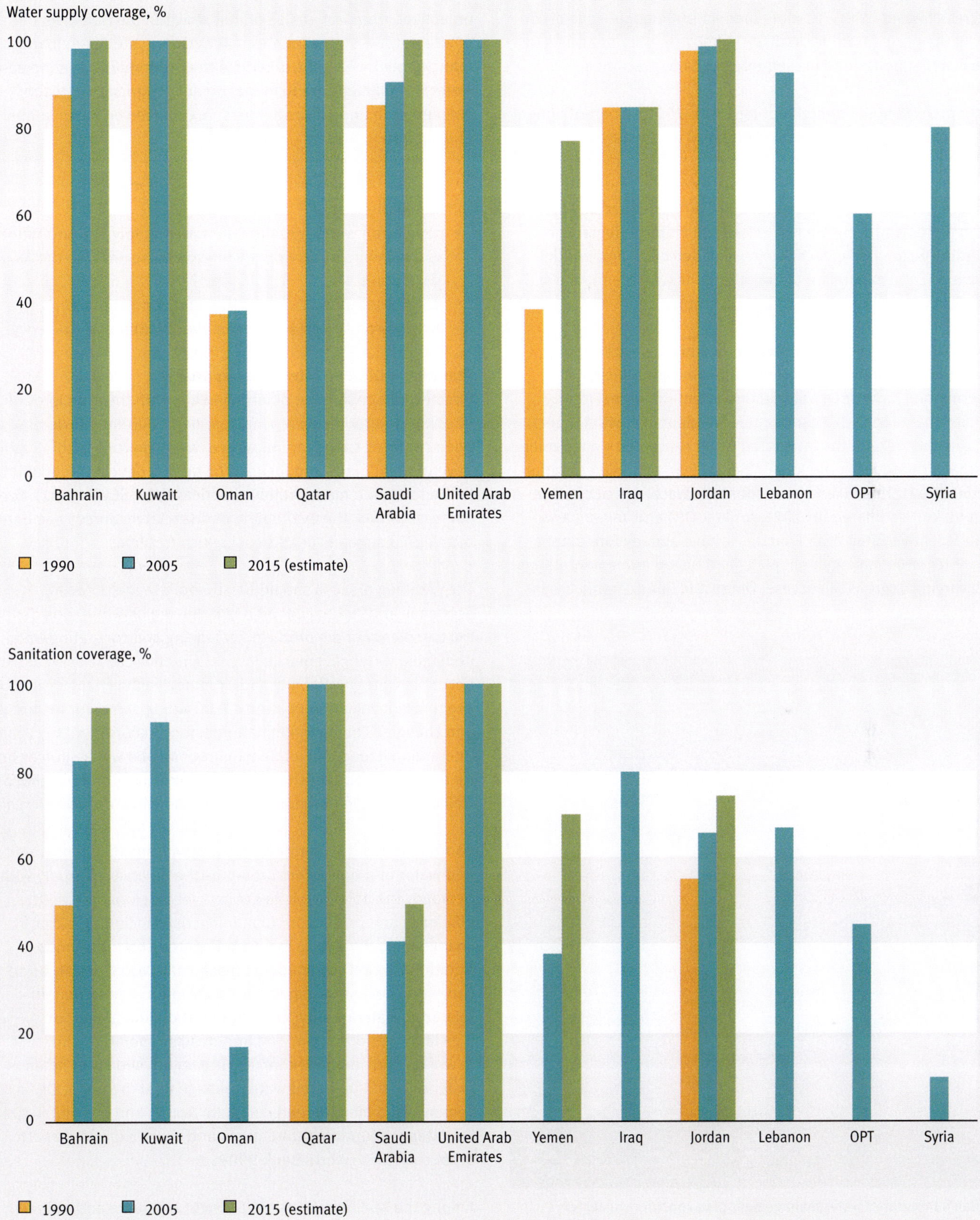

Source: CEDARE and AWC 2004

1990, 47–49 million people have gained access to a drinking water supply and 42–43 million to sanitation (UN DESA 2011). Most countries are expected to meet the MDG targets in 2015, with the exception of Yemen and OPT. Despite the substantial progress towards MDG 7c, more than 41 000 people have died during the period from 1990 to 2008 because of poor access to safe water supplies and inadequate sanitation facilities.

Water policies implemented during 1960–2000 as part of annual or five-year development plans addressed water scarcity by making use of supplies from major rivers, shallow and deep groundwater and desalination. Services were expanded to improve water supply and sanitation coverage, especially in urban areas, and measures were taken to manage demand, including water-saving technology, leak detection and public education, and expansion of irrigation schemes to enhance self-sufficiency in certain commodity food crops.

Since 2000, with the region's water resources – especially renewable resources – exploited to the limit of capacity, governments have been paying more attention to developing policies that emphasize comprehensive planning, with longer horizons in line with the integrated water resources management approach called for in the Johannesburg Plan of Implementation (WSSD 2002). This takes into account the water deficit of more than 50 km^3 estimated for 2025. Jordan, OPT and Yemen, and more recently United Arab Emirates – have already formulated their integrated management plans and begun implementation with varying degrees of success. Oman and Saudi Arabia are in the process of finalizing their plans while others have established timelines (AFED 2010).

Jordan's integrated water management plan considers all water resources in the Lower Jordan Valley, including groundwater, wastewater, saline water and floodwater. *© Miguel Nicolaevsky/iStock*

The region's water policy priorities should focus on three key objectives: comprehensive planning within the framework of integrated water resources management; supply-demand management measures to reduce the water deficit and increase water-use efficiency; and management of agricultural water consumption. Indicators to measure progress on water supply and sanitation are:

- annual per-person water consumption from renewable water sources – or water sustainability – as a measure of scarcity and depletion;
- population with access to safe water supply and sanitation as a measure of service coverage and proximity to the MDG target; and
- water losses from the irrigation and domestic water distribution systems as a measure of water use efficiency.

Planning within an integrated approach

Effective policy calls for planning in line with the principles of integrated water resources management. Current efforts have been confined to the formulation of water policies within national development plans and focused on supply development and limited demand management practices (UN ESCWA 2001). For some countries, the availability of financial resources has been considered a means of addressing the problem.

The planning process should be appropriate to the social, economic and cultural conditions of the region while considering the complexity of the problem: increasing and competing water demands; water governance issues; adaptive capacity in case of uncertain water availability and extreme events; changing socio-economic development patterns – including demographic trends and changing consumption patterns; food security and the volatile international food market; tension over shared water sources both for rivers and aquifers shared between countries of the region as well as other neighbouring countries; and climate change impacts.

Coordinated and integrated planning within and across water and water-related sectors promotes the balance of supply with demand. The objective is to achieve resource sustainability, efficiency and protection; to manage risks including climate change impacts; and to manage disputed shared sources. Additional benefits include increased safe water supply and sanitation coverage, especially for the poor; health benefits related to water quality; compliance and enforcement of legislation; information; and improved cooperation and trust in working on shared resources. The region should build on experience of integrated water management in Jordan, OPT and Yemen – and more recently in Saudi Arabia and the United Arab Emirates – to update future plans and share experience with other countries (World Bank 2009).

Among the key limitations are a lack of adequate and trained technical and managerial capacity to accommodate the highly complex integrated water resources planning process, set well defined objectives, formulate and implement multi-objective measures, and commit the necessary financial and human resources to strengthen governance issues. Difficulties lie in establishing the free dissemination of information and the coordination of different data sources; reliable assessment resources; and identifying water requirements during a period of dynamic socio-economic development and environmental change. For transboundary resources, there are conflicting national interests in forging equitable sharing agreements. All these issues can be addressed through integrated water resources management, supported by strong commitment on the part of decision makers to place water high on the political agenda.

The integrated water resources management framework is not an option but an essential requirement if water management is to be enhanced in the region. The experience gained from its formulation and implementation in Jordan and OPT can inform the planning process of the Mashriq sub-region, while Yemen's plan (Box 14.1) can inform that of the GCC countries, which share similar environmental and social conditions. In addition, documentation of lessons learned through the practical implementation of management measures could help to build national capacity for updating any existing integrated water management plans.

In terms of environmental governance, certain enabling conditions are necessary for the success of the integrated approach. Decision makers and stakeholders must fully understand the relevant policy statements and must define the objectives and mandate of the water and water-related sectors; enable the free dissemination of information; commit the necessary financial and adequately trained human resources; increase reliance on national expertise; adopt a community-based management approach; and enforce coordination mechanisms. This requires comprehensive and enforced legislation.

Supply-demand management to reduce water deficits

On the supply side, appropriate measures include the development of renewable groundwater within sustainable yields; augmentation from water desalination; reuse of adequately treated wastewater; rainwater management and harvesting; artificial groundwater recharge; flood control structures; and a limit on the mining of non-renewable groundwater. Demand measures include economic mechanisms such as partial cost recovery; socially acceptable tariffs; subsidies and incentives for improving water-use efficiency, especially in the irrigation sector; modification of building codes for water saving; leakage control; decentralization of water utilities; groundwater metering; and effective coordination of international funds supplemented by public awareness. This is in addition to supporting non-governmental organizations and stakeholder participation.

Limited management measures have been implemented in Jordan, OPT, Yemen, and recently Saudi Arabia and the United Arab Emirates, mainly through water-saving technology, public education and leak detection in large cities, and incentives such as subsidies and loans.

Expected benefits include coordination in balancing water supply through rational utilization of all sources, with demand reductions achieved within 25 years. Such rationalization includes the use of renewable and non-renewable sources within their sustainable yields; increasing domestic supply from desalination; reusing adequately treated wastewater; establishing strategic groundwater reserves in the Arabian Peninsula sub-region and Jordan; and developing rainfall harvesting infrastructures in Jordan, Lebanon, Oman, Saudi Arabia, Syria, United Arab Emirates and Yemen. Demand management measures aim to reduce water losses from distribution systems from the present levels to 5–20 per cent of the non-renewable resources, especially in the irrigation

Box 14.1 Yemen's integrated water resources management plan

Yemen's progress on water management was supported by an investment plan and the prior establishment of a comprehensive legislative framework. Supply and demand objectives have to a certain extent been achieved, aided by assessment of the water supply along with supply-and-demand management projects in the domestic and irrigation sectors. The supply side includes dams for flood control and recharge, control of groundwater withdrawal in certain areas, reuse of treated wastewater, and rainwater harvesting; while the demand side includes such management tools as the renovation of terraces, decentralization of water utilities, appropriate domestic tariffs, subsidies to improve water-use efficiency in the irrigation sector, incentives, and the creation of water user associations in coordination with the Ministry of Agriculture. Evaluation of climate change impacts was also carried out.

The process involved academics, UN agencies and non-governmental organizations, in addition to coordination of the international funding agencies and identifying the necessary financial and human resources. Benefits have included improved planning at basin level, increased investment in wastewater treatment, increased service coverage in the cities of Sana'a, Aden, Taiz, and Hudadiah, rationalizing the use of groundwater resources especially in the Sana'a basin, and enforcing tariff collection. The principal constraints have been under-commitment of financial and human resources and a lack of public-private sector partnerships (CEDAR and AWC 2004).

sector. Socially acceptable economic tools include gradual cost recovery, and loans and incentives to reduce consumption. Measures also include administrative steps to decentralize functions, change building codes and regulations, encourage stakeholder participation and establish modern agricultural practices, including hydroponics and irrigation systems. These can lead to changes in behaviour and consumption patterns, reduced pollution and depletion especially of non-renewable and shared sources, and improved water productivity, and can contribute to meeting the MDG goals.

The main challenge is to shift water from being heavily regulated and subsidized, which is largely dictated by a strong agricultural lobby, into the realm of partially priced goods and services. In most countries of the region, subsidy policies have contributed to wasteful water consumption, though this is now changing in Jordan, Saudi Arabia and Syria. Challenges also lie in overcoming the reluctance to reuse treated wastewater, providing adequate financial sources in the Mashriq sub-region, and low capacity for integrated and comprehensive planning, especially in the Arabian Peninsula sub-region (AFED 2010). Further, there is reluctance to take appropriate action to reduce the power of the agriculture lobby.

Box 14.2 Leak detection and repair of the distribution system in Bahrain

Water distribution leakage is in the range of 30–50 per cent in certain areas in Bahrain, resulting in the loss of costly desalinated water, contamination with wastewater and changes in the water table that can damage urban infrastructure. Bahrain's management measures achieved a 5–15 per cent reduction in leakage, saving 25 million m^3 of desalinated water and reducing costs by US$18–25 million in 2000 (World Bank 2008). Improvements were seen in water supply reliability and coverage, enhanced technical and managerial staff capacity, and reduced impacts from a high water table such as nuisance odours, soil contamination and damage to urban buildings and roads. In addition, the measures helped in-house water auditing, and increased public awareness and social responsibility for conservation of an already limited resource. Such demand management measures could be replicated in many big cities in the region.

Similar economic and social characteristics across the region provide opportunities to share many supply-and-demand management experiences. Experience of desalination in the GCC countries can be shared with the Mashriq sub-region while taking full consideration of the environmental impacts, especially Jordan, the OPT and Yemen, while measures such as water storage and rainwater harvesting infrastructure are replicable in most of the countries. Other successes include water-saving technologies, leak detection and repair, public awareness campaigns and groundwater metering – introduced, for example, in Bahrain (Box 14.2), Jordan, Saudi Arabia and Syria – and decentralization of water utilities. In addition, water user associations in Jordan (the Local Farmers Association, for example), Oman and Yemen can be replicated in all countries.

Desalination remains the most practical way of meeting rising demand for water in the countries of the Gulf Cooperation Council. © *Tanuki Photography*

Enabling conditions require comprehensive water sector reform and include good governance conducive to inter- and cross-sectoral coordination; adequate investment; financial transparency and accountability; public acceptance of cost recovery tools with socially acceptable tariffs; and application of the polluter-pays principle. Other enabling conditions include commitment to the right to water; ensuring that stakeholders have an active role in decision making; a free flow of information; separation of service providers and regulatory functions; and effective capacity-building programmes.

Management of agricultural water consumption

The agricultural sector, which uses more than 85 per cent of the region's water, has been oriented towards food self-sufficiency in certain commodities and overall food security in light of increasing food prices, rural development and rising incomes. In Lebanon, Jordan, Syria and Yemen, the sector employs 30–40 per cent of the domestic population, while in the GCC countries it depends on foreign labour (UNEP 2010). Agricultural intensification has accelerated groundwater depletion, especially in the Arabian Peninsula, as well as increased agro-pollution and soil salinity. The sector is characterized by low irrigation efficiency of 30–45 per cent and the cultivation of particularly water-thirsty crops, resulting in low water productivity (AOAD 2009). Water scarcity and pollution can be alleviated by increasing the use of adequately treated wastewater; rainwater harvesting on mountain terraces; modern agricultural and irrigation systems; and subsidies, incentives and soft loans to promote the application of water-saving technologies. Water sustainability can also be enhanced by groundwater metering, partial cost-recovery tariffs, application of the virtual water concept, increasing the number of water user associations, market integration between countries, and making use of World Trade Organization (WTO) and other trade agreements.

The benefits of integrated management to the agricultural sector include enhanced water-use efficiency of 15–30 per cent above the current level, resulting in substantial water savings and increasing the water available to meet domestic demand and achieve the MDG targets (UN DESA 2011). Improving water-use efficiency will increase water productivity and farmer income, and conserve non-renewable groundwater for future generations. The current system of subsidies and soft loans available for modern agricultural and irrigation systems provides an effective economic tool to reduce water consumption and prevent groundwater depletion and pollution from agrochemicals. The current secondary and tertiary level wastewater treatment facilities, especially in GCC countries, provide water suitable for a number of crops (UNEP 2010). However, more attention should be given to the monitoring and enforcement of water-saving technologies to reach a defined level of efficiency and achieve appropriate water treatment standards.

Limited commitment to providing the necessary financial resources to implement water-saving irrigation technologies and assess climate change will impact water availability, agricultural productivity and biodiversity. Problems include difficulties in convincing farmers to shift to modern irrigation techniques and use treated wastewater, and in developing adequate human resources to monitor compliance with water treatment standards, along with weak marketing strategies and the impact of foreign labour. Other challenges include overcoming reluctance to move away from the concept of irrigation water as free or heavily subsidized, especially groundwater, to acceptance of cost recovery and pricing of water allocations. There is a need for significant investment in wastewater treatment, awareness campaigns, modernized irrigation and agricultural systems, and the establishment of user associations.

Agricultural policies have to be compatible, coordinated and integrated with broader water policies. Investment is required for wastewater facilities to increase reuse volumes, and for subsidies and loans to increase take-up of water-saving technologies. Enabling conditions must be established for setting efficiency rate targets of 75 per cent for irrigation, and for the gradual phase-out of water-thirsty and low cash-value crops in favour of importing crops (the virtual water concept). Growing wheat, for example, requires large amounts of water. By importing wheat and concentrating on crops that require less water, a country can acquire virtual water and use existing resources more efficiently.

Additional measures include limiting the export of green animal feed, increasing the number of water user associations and taking advantage of WTO and bilateral agreements between Arab countries (UNEP 2010).

Similarities in the irrigation supply, consumption practices and arid environment of most countries provide opportunities to share success stories as well as market and trade complementarities, and take advantage of possibilities for integration at the regional or sub-regional level. The success of large agricultural companies in Saudi Arabia could, for example, help some countries of the Mashriq sub-region to expand their activities and increase water productivity. Use of the virtual water concept and intra-region agricultural policies provide an opportunity for cooperation in agricultural production based on comparative advantage, while conserving local water resources for future generations (Box 14.3).

Box 14.3 Irrigation management in Saudi Arabia

Saudi Arabia's agricultural sector is responsible for more than 85 per cent of the country's water consumption, especially from non-renewable and sometimes shared groundwater resources. During 2005–2007, renewable water resources of 2.5 million m^3 were supplemented with 16.2 million m^3 from non-renewable groundwater sources to satisfy irrigation demands. Even though the 2010 total demand of 18.7 million m^3 is expected to decrease to 12 million m^3 in 2025, the gap between the total irrigation demand and the supply from renewable water sources will still be considerable.

The government recently implemented a number of measures to limit irrigated food production by decreasing the subsidy on diesel fuel and gradually reducing the purchase of local wheat. In 2009, it set a target to eliminate wheat production over an eight-year period, and at the same time increased incentives and loans for modern irrigation systems, provided subsidies for animal feed imports while banning the export of fodder, and established strategic food reserves (AFED 2010). Further measures have been implemented to freeze the amount of land used for agriculture; promote cultivation under glass; improve the coordination of the agricultural sector with other relevant policies; and encourage agricultural investment abroad by forming committees and setting aside funds to encourage the private sector. These measures have contributed to reducing the amount of irrigated land, the production of wheat and groundwater mining, and increased interest in the reuse of treated water (Hussain *et al.* 2010). Future action includes evaluation of irrigation cost-recovery options, groundwater metering and setting limits on water allocation to the various sectors.

Fields in Halabiye, Syria, where scientists are working with farmers to breed more robust crops. *© Joel Carillet/iStock*

Soil, land use, land degradation and desertification

Most of West Asia is characterized by patchy vegetation, sandy soils and arid to hyper-arid conditions. Drylands make up 64 per cent of the total area of 4 million km^2 (Abahussain *et al.* 2002; Al Kassas 1999). Rangelands fall into the largest land-use category, with lands cultivated with annual and permanent crops representing 4.8 per cent, and forests 1.4 per cent (FAOSTAT 2008; AOAD 2007). High and sustained population growth and urbanization rates coupled with rising rates of consumption increase the pressure on limited land resources.

The biophysical characteristics of the region, combined with population growth and socio-economic policies, are the main drivers of land degradation and desertification, one of the main problems facing West Asia. Proximate causes include intensification of crop and livestock production and pastoral activities; development of human settlements and infrastructure; wars; policies that subsidize unsustainable practices such as irrigation with fossil or saline water; overuse of agrochemicals; overstocking of livestock; and lack of appropriate integrated water-land-use planning and management. All these developments have resulted in reduced ecosystem products and services, widespread desertification and land degradation including biodiversity loss, which, in turn, affects human well-being (ACSAD *et al.* 2004).

The impacts of land degradation have been most serious in the countries where the share of agriculture in gross domestic product (GDP) is high, such as Lebanon, Syria and Yemen (UNEP 2010), and are further exacerbated by frequent droughts and climate change. Policies to combat land degradation and desertification must take into consideration the region's multiple challenges of sustained population growth, rapid urbanization rates, increasing demand for natural resources, a declining natural resource base, varying rates of economic growth and increasing incidence of poverty in communities that depend mainly on land resources. A lack of financial resources, appropriate technologies and institutional capacities, as well as limited stakeholder and civil society participation must also be taken into account (SRAP 2007).

Indicators for measuring the progress of selected land-use policies include:

- the proportion of land affected by desertification (erosion and salinization);
- the proportion of land that falls into nationally protected areas and forest;
- livestock numbers relative to the carrying capacity of rangelands;
- land-use change, including the proportion of productive area lost to urbanization;
- the proportion of land under modern irrigation; and
- levels of productivity (tonnes per hectare) and production (tonnes per year).

Successful policy options that show potential to accelerate the achievement of internationally agreed goals can be addressed as three clusters:

- developing rangelands and combating land degradation;
- achieving food security and cropland rehabilitation; and
- adopting integrated policies for improving land and water use with local community participation.

Developing rangelands and combating land degradation

Policies to develop national and regional rangelands help to improve their management by prohibiting cultivation in designated areas while protecting and rehabilitating degraded rangelands (Box 14.4) (Kattach 2008).

The benefits include the protection, conservation, sustainability and improvement of natural vegetation productivity and diversity.

Box 14.4 Protection and rehabilitation of rangelands in Syria

The main objective of Syria's policy is to conserve vegetation density, productivity and biological diversity, improve the livelihoods of local communities, reduce dust and sand storms, and increase carbon sequestration. The aim is to protect and rehabilitate degraded areas in the Al-Bishri rangeland of the Syrian steppe. Implementation involves local community participation in selecting degraded areas, seeding and planting, and the control and reduction of grazing pressure through collaboration between local herders and animal-fattening cooperatives. After three years of rehabilitation and protection, forage production increased from 90 kg to 320 kg per hectare per year and bare soil decreased from 91 to 32 per cent. The diversity of plants increased from 27 species from 23 genera and 13 families to 83 species from 55 genera and 17 families, and the density of edible shrubs increased from 0.02 to 4 plants per m^2 (Kattach 2008). In the long term, vegetation density, productivity and diversity, and carbon sequestration are expected to increase to optimum levels, alongside the prevention of dust and sand storms. This is in addition to providing more forage for livestock, reducing the need for feed and the cost of meat production.

In addition, improved rangelands help to prevent soil erosion, conserve water, increase carbon sequestration, reduce both the frequency and magnitude of dust and sand storms, and provide links to global support for combating desertification. Limitations include reduced open grazing areas for herders, competition with cropping, lower direct financial returns to herders and increasing risk of conflict with local communities.

These policies can be replicated and scaled up for implementation in similar degraded rangelands at regional and global levels.

Achieving food security and cropland rehabilitation

Food security has continued to be the main concern of national governments in the region since the concept was introduced in the 1980s. The world food crisis in 2007, accompanied by soaring food prices, revitalized the need and desire of some countries to become self-sufficient in certain agricultural commodities, and especially to restrict the export of cereals and livestock feed (AOAD 2009). As a result, national agricultural policies were revised to increase agricultural production, and government control of farming systems was relaxed in favour of decentralization. Avenues for offering incentives in the form of price controls, tax breaks and reductions, cereal and animal feed export restrictions, easy loans, and the introduction of efficient techniques in reclamation and irrigation were explored. This is in addition to developing adaptation policies to climate change such as using saline water for agricultural production, developing new local crop varieties tolerant of aridity and drought conditions, and rehabilitating rainwater harvesting systems (Box 14.5).

After introducing a series of incentive measures, benefits have included relative food security with regard to certain commodities, which in turn reduced dependency on food imports and helped to alleviate poverty and hunger.

The limitations to agricultural practices such as flood irrigation have included depletion of water resources in a region that is already water-short. Overuse of aquifers has led groundwater to suffer from saltwater intrusion in coastal regions, and salination has rendered large areas of agricultural land useless and converted landscapes into desert (Hussain *et al.* 2010). However, the governments of the region are left with no choice but to reclaim new areas and salinated fields, and re-cultivate them in order to meet the ever-increasing demands for food. Drought and climate change are working against achieving food security, as persistent drought has continued to affect the region for the past few years.

Policies to improve agricultural productivity are being replicated with modifications to suit each country's economic and social conditions.

Box 14.5 Sustainable agricultural development in Bahrain

The National Strategy for Sustainable Agricultural Development aims to upgrade the agricultural sector in Bahrain. It includes goals and programmes that contribute to agricultural growth and preserve agricultural heritage. The main objectives of the strategy include achieving relative food security, conserving natural resources, protecting agricultural land, using modern technologies to encourage agricultural investment, making the agricultural sector economically efficient, contributing to the needs of citizens and residents for food, and supporting small farmers (Ministry of Municipalities Affairs and Land Use Planning 2010). To attain the strategy's objectives, communication and cooperation were enhanced between the various parties involved in the country's agricultural sector. In addition, the strategy concentrates on involving diverse communities while paying special attention to farmers as central to agricultural development.

Benefits of the strategy include the modernization of production systems, conservation of water and land, increased agricultural productivity, relative food security, reduced groundwater consumption, increased vegetation cover including palm trees, a boost to agricultural trade, promotion of the national economy and a reduction in unemployment.

In order to create enabling conditions, governments of the region eased access to finance and technical services. Agricultural research and extension were strengthened to boost productivity and conserve water and land resources, thus promoting good agricultural practices. New crop varieties suited to drought conditions were introduced as well as new cultivation methods.

Enabling conditions for rangeland protection policies in West Asia have been jointly implemented by various institutions, with assistance from developed countries and research centres. The success of land degradation control depends on the presence of a favourable framework, including organizational, institutional, legal and political structures and processes that promote programme planning and implementation. This included analysis of the factors influencing institutional response capacity, and from there the development of recommendations for capacity building and participatory modalities (UN ESCWA 2007a).

Integrated policies for improving land and water use with local community participation

The 2007 report of the Intergovernmental Panel on Climate Change (IPCC 2007) indicates that the problem of land degradation and desertification prevalent in West Asia will be exacerbated by climate change. The expected increase in temperature, decline in precipitation, and greater intensity and frequency of droughts and dust storms will impact rangelands and rain-fed cropland and contribute to land deterioration, biodiversity loss and the spread and intensification of desertification.

Mindful of this, Jordan's policies deal with the strategic improvement of rain-fed agriculture and the prevention of land degradation and desertification. Achieving these objectives involves long-term mainstreaming of integrated strategies for improving productivity; rehabilitating, conserving and sustaining land and water resources; combating desertification; and mitigating the impacts of drought and climate change.

Tomato crops by the Dead Sea. Here, drip irrigation uses nearly 50 per cent less water than traditional irrigation. *© Ricardo De Mattos*

Implementing these strategies is more effective with the participation of local traditional resource users, recognizing the interrelationship between these and other environmental issues at the local, national, regional and global levels (Box 14.6). The benefits of these policies are the protection, conservation, sustainability and optimization of natural resource productivity and the potential for diversifying income sources with links to global support for improving farmers' livelihoods. Factors that determine success include soil and water conservation, irrigation, forestry, livestock, range management and community-based resource management, enhanced technical capacity of local managers and local institution building. Indicators of success include the long-term rehabilitation of degraded lands, halting of desertification processes and increased resilience to climate change, while short-term benefits include increased agricultural productivity, improvements in individual and family incomes, greater drought resistance of rural production systems and protection of biodiversity (UN ESCWA 2007a).

In semi-arid areas the implementation of policies with an emphasis on agricultural production leads to a reduction in the rangeland available for livestock grazing. In many countries most of the farmers are also livestock owners and their herds graze on lands with low productivity and on crop residues. Farming practices in these areas return very few nutrients and organic matter to the soil and provide little protection from wind erosion. Livestock grazing of virtually all crop residues is particularly problematic (UN ESCWA 2007a). Policy limitations also include the continuous out-migration of the younger members of rural families, creating local labour shortages.

In many countries of the region successful programmes tend to emphasize the significance of comprehensiveness and integration. An excellent policy in one country does not usually stand alone and, as such, cannot easily be transferred or successfully replicated in its original form in a new location (UN ESCWA 2007a). New circumstances, new management and various interdependent problems such as poor and low implementation capacities, lack of financial resources and marginalized local stakeholders, can make many successful programmes lose their effectiveness under replication.

The evaluation and assessment of successful policies have demonstrated that the alleviation of land degradation depends not only on the motivation of individual stakeholders, but also on the creation of enabling conditions for effective collective action by the whole community, and this makes implementation of such policies more challenging. The development of appropriate policy frameworks and incentive structures is key to inducing sustainable management of natural resources. Environmental governance should be incorporated in the activities of social, economic and administrative institutions, with environmental and land-use policies central to the coordination and management of the various sectors of the national economy. Governance encourages the use and application of scientific data and information for the sustainable development of natural resources. On a larger scale, it fosters an understanding of the

Box 14.6 Integrated agricultural management in Al-Karak, Jordan

The main objectives of Jordan's policy were to arrest land degradation, optimize the long-term productive capacity of land and water resources, improve the income of vulnerable farmers, especially women, with their active participation, safeguard and upgrade the productive potential of natural resources and enhance returns, prevent soil degradation, restore soil fertility, promote efficient use of soil and water, strengthen the capacity of local technical and managerial staff, and meet the needs of local farmers. To achieve these objectives, the programme provides technical and financial support aimed at:

- building soil and water conservation structures and improving agricultural production;
- enhancing sustainable land and water management practices;
- promoting rural microfinance to support on- and off-farm activities;
- tree planting;
- building cisterns and dams for water harvesting;
- improving animal husbandry;
- maintaining springs and irrigation canals; and
- constructing small reservoirs, known as *hafira*, to retain run-off for later use.

Local communities have benefited from the newly vibrant agricultural sector through processing local products and having better access to financial services. Some 5 350 households have benefited from the various soil and water conservation measures, while spring protection and/or rehabilitation programmes alone having benefited about 1 000 households (Ministry of Water and Irrigation 2008). The improved agricultural extension services are estimated to have reached about 22 300 households, and the provision of loans and support for developing alternative income-generating activities has benefited more than 5 000 women and landless farmers (UN ESCWA 2007a).

These investments in soil and water conservation have reduced and will continue to reduce the degradation of the fragile ecosystems in the project area. They will also improve vegetative cover, reduce run-off and soil loss, improve soil fertility, and enhance sustainable use of the natural resource base. The project has raised awareness about the impacts of land degradation and desertification while improving farmer's livelihoods, diversifying income sources, and alleviating both poverty and out-migration.

main economic, social and environmental issues among many stakeholders, helping to achieve a balance between governance needs and governance capacity (UN ESCWA 2007a).

Energy

Energy resources

West Asia is one of the major players in the global energy market, having 52.2 per cent of world oil reserves and 24.6 per cent of world gas resources (OAPEC 2009). The region produces nearly 17.3 million barrels of oil a day, accounting for 27.6 per cent of world oil exports. Rapid economic development, population growth, urbanization, and changes in standards of living in West Asian countries have led to increases in energy demand (Figure 14.3) (IEA World Energy Agency 2010). Despite rich renewable resources, the energy sector is characterized by heavy reliance on fossil fuels. In addition, the regional economy is still largely dependent on fossil fuels to fulfill increasing energy demands. The use of fossil fuels is always accompanied by considerable environmental impacts including deteriorating local air quality and rising greenhouse gas concentrations in the atmosphere, contributing to climate change.

Energy consumption rose steadily in most of West Asia between 2004 and 2008, increasing by some 20 per cent over the period (Ruble and Nader 2011). But with accelerating rates of development and rapid urbanization in the majority of the region, energy demand is now increasing drastically in all sectors, including electric power production, domestic energy use and transport. In view of energy security and safety

Figure 14.3. Primary energy consumption in West Asia, 2004–2008

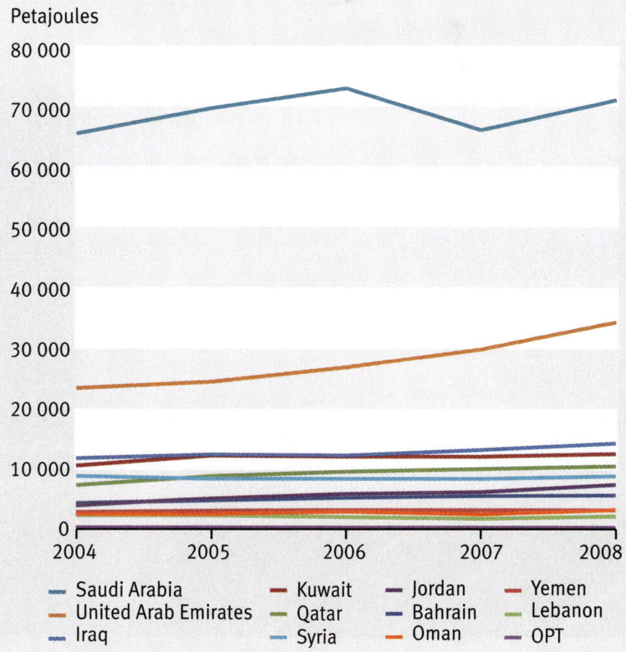

Source: IEA 2010

issues, the sharp increase in oil and gas prices, climate change and environmental considerations, as well as technological advances, energy planning in several countries is now addressing more decentralized energy generation options. The region is characterized by rich renewable resources including solar, wind, geothermal and, to some extent, biomass, and over the past decade has been shifting its policies towards diversification of energy sources and placed energy efficiency and renewable technologies high on national policy agendas. Some examples of renewable energy initiatives include Jordan's objective to generate as much as 7 per cent of its energy from renewable sources by 2015 and 10 per cent by 2020, while solar capacity is expected to reach 300–600 megawatts over the same period; Abu Dhabi's aim of generating up to 7 per cent of its energy from renewable sources, with planned investments reaching US$22 billion; Syria's intention to generate 7.5 per cent of its electrical energy from renewable resources by 2020; and Lebanon's target for renewable energy of 10 per cent of total energy supply by 2013 and 12 per cent by 2020 while also aiming to reduce energy consumption by 6 per cent by the 2013 (Ruble and Nader 2011; Verdeil 2008).

Successful energy policies in the countries of West Asia cluster around two main areas:

- energy efficiency in the building sector including systems for space heating and cooling, and measures for promoting the use of renewable energy resources; and
- energy generation mixes and targets for clean energy production, which require governmental commitment and advanced legislation.

Abu Dhabi's US$600-million solar power plant, expected to be complete in 2012, will be one of the largest concentrated solar power plants in the world. *© Fernando Alonso Herrero*

Indicators for measuring the progress of the selected energy policies are:

- energy savings in percentage terms or cost terms, reduced air conditioning system sizes and impacts on local markets;
- total surface area of solar water heaters installed (market penetration); and
- diversification of energy sources as part of countries' plans, and renewable energy capacity as a proportion of total capacity.

Policies that have been shown to be effective in reducing energy consumption with the participation of the local community address the energy performance of buildings and their cooling and water systems (Hajiah 2010; Maheshwari and Al-Murad 2001), promote renewable energy resources (Shahin 2010; Houri 2006; Kablan 2004) and encourage diversification of energy supply options (Ruble and Nader 2011; Hainoun *et al.* 2010; Reiche 2010). These policies may have high potential for replication in countries with similar climate or socio-economic characteristics and similar regulations to those of West Asia.

Regional policy interventions for improving buildings' energy performance and implementing renewable energy enhancements such as water heating are directly linked to policy formulation regarding population growth, urbanization, and associated economic activities and technological affordability. Further policies that are of equal importance, but have only recently been formulated, address public transport, car fleet age and codes for fuel use.

Building and systems energy performance

Energy efficiency in the building sector has been a primary national target for West Asian countries, and thermal guidelines and codes for buildings have been developed and implemented in most of the region (Ali *et al.* 2008; Alnaser *et al.* 2008; Aftab and Elhadidy 2002). Al-Ajlan *et al.* (2006) reported that improving the efficiency of Saudi Arabia's air-conditioning alone provided a return on investment equivalent to 400–500 megawatts per year of generating capacity, a saving of up to US$250 million per year. The countries' energy performance codes for buildings focused on solutions to improve heating and cooling loads and have, to some extent, addressed the use of efficient systems and processes for heating/cooling and lighting.

More recent codes tackle green building design and performance. Hybrid air-conditioning, for example, has high potential for saving energy either through optimized operation or by integrating renewable energy sources into their function (Farraj *et al.* 2010; Fasiuddin *et al.* 2010; Ghaddar *et al.* 2010; Ghali *et al.* 2008). Building code development has reached an advanced stage, and is now considering smart systems and green designs that meet the American Society of Heating, Refrigeration and Air-conditioning Engineers (ASHRAE) goal of zero-energy building in the coming decade. The development of carbon-neutral Masdar City in Abu Dhabi is a process of "transforming oil wealth into renewable energy leadership", with the long-term goal of a "transition from a 20th century, carbon-based economy into a 21st century sustainable economy" (Reiche 2010).

The introduction of green building codes in some West Asian countries has been successful in reducing electricity consumption in buildings through guiding the selection of building materials and glazing choices, and setting upper limits on lighting intensity and cooling/heating (Al-Temeemi 1995; Kellow 1989). The success of this policy has been due to multiple factors including:

- a rigorous technical methodology for developing a building code responsive to a country's climate and the availability of building materials;
- a short payback period for several of the proposed energy conservation measures;
- the ability to ensure compliance through setting upper limits on the capacity of the electric supply/meter to the building unit;
- the ability to enforce building codes in the public and commercial sectors;
- awareness and understanding among the professional community of best practices to improve building performance; and
- flexibility and room for innovation in the selection and introduction of new energy conservation measures and practices that provide the contractor, owner and operator of the building with choice (Maheshwari and Al-Murad 2001).

The significance of such concepts and products is evident in their incorporation in international standards for green buildings such as the Building Research Establishment Environmental Assessment Method (BREEAM) and Leadership in Energy and Environmental Design (LEED), which use environmental assessment methods and rating systems.

The development of thermal codes and ratings enables countries to reduce operational energy consumption for heating, ventilation, air-conditioning (HVAC) and lighting over a building's life cycle, thus reducing emissions of greenhouse gases. By adopting green building envelopes that provide high insulation and air tightness, energy savings of 30 per cent and higher have been achieved in Bahrain, Jordan and Kuwait, among others (Hajiah 2010; Ministry of Public Work and Housing 2009a, 2009b; Alnaser *et al.* 2008; Maheshwari and Al-Murad 2001). The introduction of building thermal codes and ratings can be implemented through limiting the electric power supply capacity to a building, which forces designers and contractors to follow the code. Energy conservation in buildings extends beyond the envelope to include cooling systems, solar water-heating systems and energy-efficient appliances. It also calls for the greater availability of green market products and technologies associated with building services and materials.

The main challenges hampering the implementation of new building codes include higher capital costs, the need for short and long-term planning, low skill levels and both financial and strategic inadequacy. The economic feasibility of adopting building codes is well established and the costs of some envelope measures have been reduced by the use of local materials, while others such as double glazing remain expensive. Countries with moderate climates such as Lebanon and Syria can use other less expensive cooling methods such as fans and evaporative coolers. Not all GCC countries have implemented building codes, although they can afford them. Nonetheless, the market is open for transfer of green designs and services for buildings.

GCC countries are coordinating their energy regulations and encouraging the exchange of case studies of successful sustainable buildings. Some countries with moderate climates in the Mediterranean, Jordan, for example, have implemented a code or, like Lebanon, are considering guidelines while providing incentives through reduced building fees for those following the code (Chedid and Ghajar 2004). The Kuwait Energy Code for buildings is a good example that could be replicated not only by GCC countries with similar climatic conditions and a need for air-conditioning around the year, but also for West Asian countries facing the problem of drastic increases in energy demand (Box 14.7) (Hajiah 2010; Maheshwari and Al-Murad 2001). In addition, thermal insulation according to the Syrian Thermal Insulation Code is mandatory for new buildings in Syria (Ministry of Electricity 2007a). With climate change resulting in warmer

Box 14.7 Energy conservation in buildings in Kuwait

Demand for electrical power In Kuwait has progressively increased, particularly in the past two decades. Generated capacity was about 11 000 megawatts in 2009, and this is expected to rise to about 22 000 megawatts in 2020 (Hajiah 2010). As all electricity generation depends on fossil fuel resources, power plants consume about 55 per cent of Kuwait's total primary energy. In addition, 85 per cent of electrical peak power and 60 per cent of the country's total annual output is used for air-conditioning and lighting in buildings.

The Ministry of Energy in Kuwait launched its energy code for buildings in 1983 with a set of mandatory standards and regulations to enhance energy conservation and decrease the progressive negative impacts on the climate (Hajiah 2010; Maheshwari and Al-Murad 2001). The main objectives of the building code, which is applied to new and retrofitted air-conditioned buildings, are to decrease the capacity of air-conditioning systems and to reduce the peak power demand by introducing smaller units.

Table 14.1 shows the energy savings and peak-power reductions in some of Kuwait's buildings. The implementation of the energy code has saved Kuwait nearly US$10 billion over the past two decades (Hajiah 2010).

conditions in the Mediterranean region, the adoption of building codes becomes a necessity (UN ESCWA 2008).

Building codes that use local materials, green products and energy-efficient systems for cooling, heating and lighting require cooperative and well-organized planning by the government, by financial, educational and legislative institutions, and by the private sector. It is vital for the government to establish the necessary reforms and then to make them mandatory for all new and retrofitted air-conditioned buildings.

Promoting renewable energy resources

Some West Asian countries have adopted policies to promote the use of solar technologies including solar water heaters, taking advantage of the area's abundance of natural solar energy. These policies particularly address the needs of remote and rural populations with only an unreliable supply of conventional energy or none at all. This has been done in parallel with the adoption of performance standards for solar water-heating systems and awareness campaigns that demonstrate their economic, social and environmental benefits. Policies include subsidies for water heaters and tax exemptions for their manufacture; for example, Syria has made the installation of solar water-heating systems mandatory for new buildings and an assessment has to be carried out and submitted with the application for a building license (Hainoun *et al.* 2010; Kraidy 2007). In Jordan and the OPT, raw materials for manufacturing solar water heaters are tax exempt (PEC 2006; Hrayshat and Al-Soud 2004).

Solar water heaters have multiple benefits. They rely on pollution-free, inexhaustible and safe energy, and are simple, reliable, cheap and easy to install. They reduce consumption of fossil fuels and emissions of greenhouse gases. In summer months, when the entire region enjoys long sunny periods, solar water heaters can meet most of the demand for domestic hot water, drastically reducing consumers' energy use.

Solar water heating has become an increasingly common and cost-effective way of meeting domestic energy demand. © *Igor Bystrov*

The main challenges to the widespread use of solar water-heating systems include fossil fuel or electrical energy subsidies, lack of financing schemes and incentive programmes, low levels of public awareness, limited distribution and the need for a larger number of qualified personnel to design, size, install and maintain the systems. The role of government is indispensible in developing the market through establishing energy standards and labelling programmes, regulatory instruments to mandate installation in

Table 14.1 Energy savings and peak-power reductions in Kuwait

Building	Year of implementation	Energy saving (%)	Peak-power reduction (%)
Kuwait Port Authority	1996	30	20
KISR main building	2000	21	20
MEW and MPW buildings	2004	20	38
Public Authority for Civil Identification	2004	12	5
Al-Fanar shopping mall	2004	8	15
Smart Operation Strategies in eight government buildings	2007	–	40
Building Avenues Mall (Phase 1)	2009	12	2.4

Note: KISR – Kuwait Institute for Scientific Research; MEW – Ministry of Electricity and Water; MPW – Ministry of Public Works.

Source: Hajiah 2010

Sunrise in Bethlehem, which enjoys the extended periods of high solar intensity characteristic of the region. *© Pavel Skopets*

new residential and commercial buildings, innovative financing schemes, and other economic incentives. In addition, testing, certification and accreditation schemes should be introduced to make sure that the quality of the systems is acceptable and meets consumer expectations.

Box 14.8 Solar water heaters in Jordan and the Occupied Palestinian Territories

With economic growth and an increasing population, energy consumption in Jordan is expected to increase by 50 per cent over the next 20 years. Indeed, the primary energy demand of 7.5 million tonnes of oil-equivalent in 2008 is expected to double by 2020. In the OPT, around 96 per cent of the energy requirement is currently met by imports, accounting for up to 19.6 per cent of GDP (Shahin 2010). Around 38 per cent of its energy use is domestic.

The serious scarcity of fossil fuel resources leaves the OPT relying totally on imports, which reached about US$374 million in 2009 (Shahin 2010). The cost of electricity typically represents 10 per cent of household incomes, exceeding levels in neighbouring countries (Abu Hamed *et al.* 2012; Abualkhair 2007).

With Jordan and OPT, like the rest of the region, enjoying long periods of high solar intensity, solar water heating is an effective solution to meeting residential energy demands. By enhancing solar water heating, Jordan aims to increase its share of energy from renewable sources to about 7 per cent in 2015 and 10 per cent in 2020, equivalent to 200–600 megawatts of solar energy (Shahin 2010).

With the same climate as Jordan and OPT (Box 14.8), Lebanon has the potential to promote a range of solar energy applications (Ghaddar *et al.* 2006; Al-Mohamad 2001), and has recently initiated a programme for installation of solar water heaters, providing zero-interest loans to consumers (Houri 2006). Syria has plans for further promotion of solar water heating and GCC countries are currently developing their own plans.

Strengthening the legislative and institutional framework is indispensable in the diffusion of green energy technologies including solar water heaters. To overcome the barrier of high up-front costs, governments can provide financial incentives, as in Jordan, Lebanon and Syria, and/or make concessional funding available to consumers. In addition, marketing campaigns that educate the public about the economic and environmental benefits of renewable energy are vital (Ghaddar *et al.* 2006; Houri 2006; Kablan 2004). All these need to be complemented by building local capacity through training and education programmes.

Diversifying energy supply options

Emerging technologies are expected to accelerate diversification of energy supply options for the region, which has a proven abundance of renewable energy resources, especially solar and wind. Oil-importing countries such as Jordan and Lebanon have already adopted polices to diversify their fuel mix by using renewable energy technologies. The same policies are at the early stages of development in the oil-rich countries of the GCC. As indicated in Table 14.2, many countries of the region have already announced national renewable energy targets.

Table 14.2 Renewable energy targets for selected countries	
Jordan	Wind: 600–1 000 megawatts; solar photovoltaics: 300–600 megawatts; waste-to-energy: 20–50 megawatts
Kuwait	Renewable capacity: 5% by 2020
United Arab Emirates (Abu Dhabi)	Electricity generating capacity: 7% by 2020
Lebanon	Renewable capacity: 12% by 2020
Occupied Palestinian Territories	Renewable capacity: 20% by 2020

Source: Ruble and Nader 2011

The benefits of a diversified energy supply include its contribution to meeting the energy needs of people and stimulating economic growth, which is of particular interest in hydrocarbon-scarce economies. Indigenous renewable energy enables oil-importing countries to secure their supply, avoid the volatility of the global oil market, reduce dependence on imports, and minimize burdens on the state budget. In addition, diversification of sources might encourage West Asian countries to share complementary energy supplies. The region is highly reliant on fossil fuels and has one of the highest carbon footprints in the world (Reiche 2010). Switching to sustainable sources of energy would help improve both environmental quality and public health while reducing greenhouse gas emissions and conserving non-renewable fossil fuel resources for future generations. Promoting renewable energy technologies would, additionally, improve energy access, particularly in remote and rural areas.

Scientists say that a species of coral in the Red Sea could stop growing by 2070 if current warming trends continue. *© Claes Torstensson*

A number of barriers often put renewable energy solutions at an economic, regulatory or institutional disadvantage, and the situation in West Asia is no exception. These barriers include a lack of or weak legal and institutional frameworks; slow and incomplete market-liberalization processes; poor capacity for managing and disseminating information about the opportunities provided by renewable energy technologies; low levels of consumer awareness leading to low demand; lack of national standards, testing and certification schemes; weak capacity in local assembly and manufacturing, distribution, installation and maintenance; and a lack of proper financing schemes together with heavily subsidized prices for oil and gas. To overcome such barriers, different countries have developed a range of policy packages, appropriate to their national circumstances, that combine both regulatory and market-based instruments.

Diversification of the energy supply has high potential for replication in the region. Several countries have already started to develop national energy strategies that include this policy and the rest are planning to do the same.

Governments have a central role to play in setting and developing national energy strategies and master plans. Public-private partnerships are vital for achieving renewable energy objectives, as private-sector investment is often necessary to overcome shortages of the capital needed for expanding energy systems. Governments need to develop an enabling environment conducive to private-sector participation. Reforming the energy sector, allowing independent power producers to enter the market, and formulating regulatory mechanisms to secure fair market competition would be major steps to achieving this.

Oceans and seas

West Asian countries lie within three different major areas: the Regional Organization for the Protection of the Marine Environment (ROPME) area, the Red Sea and Gulf of Aden, and the Eastern Mediterranean Sea. All countries have coastal areas, with the Sultanate of Oman, Saudi Arabia and Yemen having the largest, while Iraq and Kuwait have the smallest (UNEP 2010).

The different coastal and marine environments of West Asia are facing common threats due to pressures resulting from national development plans, including the urbanization of coastal zones, tourism, land use and reclamation (Figure 14.4), maritime and oil traffic, rapid industrialization and overfishing (Sheppard *et al.* 2010). In addition, due to specific socio-economic conditions, the impacts on the marine and coastal environment are more severe in some areas than others. Among the issues are the depletion of living resources, coastal zone degradation and marine pollution, while challenges include integrated coastal zone management, management of marine protected areas and gaps in information and knowledge. Since the main centres of economic activity and population are on the coast in the majority of West Asian countries, sea level rise and related impacts of coastal inundation and increased salinity of aquifers and soil are a real risk. Bahrain, Kuwait, Qatar and the United Arab Emirates are the countries most vulnerable to sea level rise (AFED 2009). Significant warming of seawater due to the outflow of warm water from desalination plants may cause coral mortality, loss of biodiversity, depletion of fisheries, invasion of alien species and other environmental stresses. Given the rapid rates of change and the extent of pressure on the coastal environment, measures of biodiversity are not a clear indication of the background resilience of the system or the overall integrity of ecosystem function (Sheppard *et al.* 2010; Price 2002).

Indicators to measure the progress of the selected policies include:

- marine and coastal biodiversity indices;
- level of compliance with national legislation related to fisheries;
- trends in landings of marine biological stocks;
- funds allocated to research and assessment of marine biodiversity; and
- levels of compliance with measures to protect the coastal and marine environment.

The recommended policies can be grouped into four clusters:

- integrated and ecosystem-based marine planning and management;
- enhancing protection of coastal and marine ecosystems;
- controlling and combating marine pollution; and
- fisheries management.

From within these clusters, three policies were selected that have a history of implementation in most West Asian countries, that show some degree of success in ensuring the sustainable development of the coastal and marine environment, and that can be replicated and transferred. The three selected policies are:

- integrated coastal zone management;
- the establishment of marine protected areas; and
- fish stock enhancement.

Integrated coastal zone management

Integrated coastal zone management is a process of achieving sustainable development goals and objectives in coastal areas within the constraints of physical, social and economic conditions as well as within those of legal, financial and administrative systems and institutions, providing a cooperatively developed framework for the long-term conservation and management of coastal and marine resources (PAP-RAC 2011). One of its fundamental requirements is a strong set of policies for the management of the coastal environment and its resources, backed by appropriate legislation or a similar legal base. Many West Asian countries have developed strong policies and the legal base also exists in Lebanon, Qatar, Saudi Arabia, United Arab Emirates and Yemen (Tortell 2004). However, there appear to be some difficulties in moving on to the next step, which is implementation. Among the different policies and policy tools that form the framework of integrated coastal zone management are an integrated coastal planning process backed by a planning/management authority, or its equivalent, and a coastal planning/management office; a coastal zone monitoring programme; an environmental impact assessment; and implementation of the Global Programme of Action for the Protection of the Marine Environment from Land-based Activities at a regional level.

Kuwait's heavily populated low-lying coast is particularly vulnerable to potential sea level rise. *© Øystein Lund Andersen*

It is important to mention that an integrated approach is essential if national and regional, rather than simply sectoral, objectives and targets are to be achieved. Integration is required between central and local governments; various sectors in government, administration and the community; and governments, civil society and private sectors. In addition, countries sharing the same water body need to adopt regional approaches for policy implementation.

Through the rational planning of activities, integrated coastal zone management facilitates the sustainable development of coastal areas by ensuring that the environment and landscapes are taken into account in harmony with economic, social and cultural development. Benefits include the preservation of coastal zones for current and future generations; ensuring the sustainable use of natural resources, particularly with regard to water; ensuring the preservation of the integrity of coastal ecosystems, landscapes and geomorphology; and preventing and/or reducing the effects of natural hazards and in particular of climate change.

The integrated approach improves coherence between public and private initiatives, and between all decisions made by public authorities at the national and regional levels that affect the use of the coastal zone. Related institutional strengthening can also help communities adapt to the impact of climate change. The region has recently witnessed a new trend in the integration of ecotourism policy in the framework of integrated coastal zone management in Jordan, with a project developed in 2010 by the Regional Organization for the Conservation of the Environment of the Red Sea and Gulf of Aden (PERSGA) and the Aqaba Special Economic Zone Authority, focusing on the enhancement of ecotourism based on coral reefs and other coastal habitats in the Gulf of Aqaba. Such a policy can enhance efforts to protect the coastal and marine environment.

Before an effective integrated resource management system can be established, some barriers need to be faced. The most salient of these can be overcome by the establishment of an appropriate management system and the development of environmentally sound policies for, amongst others, land reclamation, urbanization and fisheries. Many West Asian countries have significant reclamation activities with adverse impacts on coastal and marine ecosystems and their services. These activities sometimes aim to improve land availability, as in Bahrain (Figure 14.4), or to provide vast recreational opportunities. An effective integrated management policy that improves public awareness and enforces existing laws related to the use and protection of natural resources can help to overcome these limitations.

Figure 14.4 Reclaimed land in Bahrain, 1963–2008

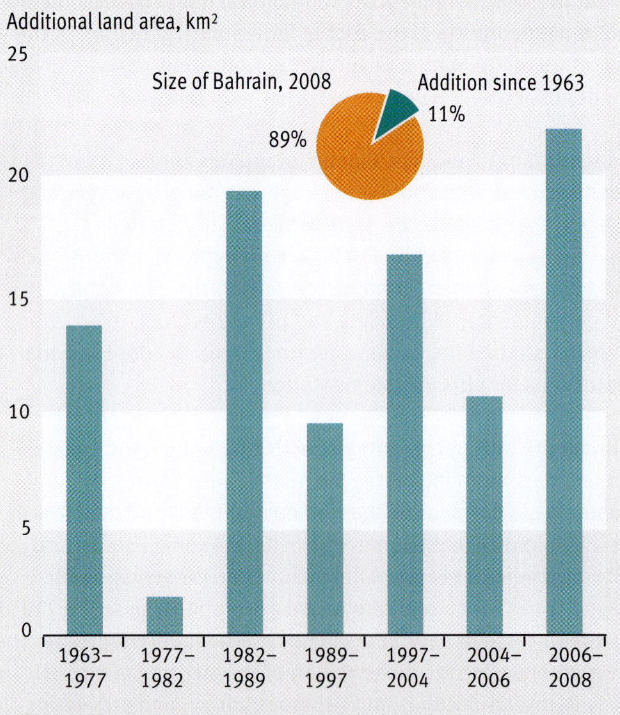

Note: Land reclamation periods vary.

Source: Zainal 2009

Programmes for the integrated management of coastal zones can be replicated either under different sectoral programmes or within the framework of an ecosystem approach, or under a management authority or its equivalent, which would necessitate coastal planning offices. Many regional and international organizations are involved in the transfer of know-how and knowledge between different countries (Box 14.9).

Integrated management of the coastal and marine environment cannot occur under the current arrangements of many West Asian countries, where responsibilities and activities are divided between a number of ministers and organizations. Enabling factors include preparing integrated marine and coastal development plans that incorporate the principles of multiple uses and an ecosystem approach, establishing institutional arrangements for marine and coastal planning, enforcing the outcomes of environmental impact assessments, and building capacity for a better understanding of the marine environment.

Box 14.9 Coastal and Area Management Programme (CAMP) in Lebanon

The CAMP-Lebanon project, part of the Mediterranean Action Plan (MAP), was implemented within UNEP's Coastal Area Management Programme. It aims to improve sustainable coastal management and integrates environmental concerns into development plans (Mehdi 2004). CAMP-Lebanon addresses the conservation of natural coastal resources in the area between Damour and Naqoura, an 8-km-wide strip of land, by applying concepts of sustainable development as well as methods for integrated coastal and marine management alongside economic and social development. The project area was defined at two levels:

- the national coastal area to the south of Beirut; and
- the three municipalities of Damour, Sarafand and Naqoura as the operational area.

Thematic activity on integrated coastal area management was split into several components:

- land-use management;
- cultural heritage and sustainable development;
- the status of environment, agriculture and fishery;
- the socio-economic situation;
- a legal framework; and
- a national strategy.

The most important elements of CAMP-Lebanon have been developed and articulated in the national strategy for integrated coastal area management and the project has developed a crucial legal tool, the proposed law on integrated coastal area management.

Establishment of marine protected areas

Marine biodiversity in the region faces extensive threats including the unprecedented pace of recent construction along and off the coastline. Together with destructive and wasteful fishing, these developments seriously threaten coastal habitats in most West Asian countries (UNEP 2010). Introducing financial incentives through integrating the economic benefits of ecosystem services into development costs is one way to overcome marine biodiversity loss. Marine protected areas are another effective tool for biodiversity conservation, habitat protection and fisheries management, acknowledged at national, regional and international levels. Since 2009, most West Asian countries have had biodiversity strategies in place supported by UNEP's implementation programme and projects funded by the Global Environment Facility (GEF), especially in Lebanon, Iraq, OPT, Syria and Yemen.

Three policy options have been adopted at the national level to speed up the process of reaching agreed goals:

- rehabilitating degraded habitats and conserving biodiversity;
- documenting marine life and biodiversity; and
- establishing multi-purpose protected areas in various marine and coastal ecosystems.

Marine protected areas require a well-defined conservation plan backed by legislation, a monitoring programme to ensure sustainability, and effective partnership between different stakeholders, supported by research into best management practice by such regional marine and coastal programmes as the Regional Organization for Protection of the Marine Environment (ROPME) and the Regional Organization for the Conservation of the Environment of the Red Sea and Gulf of Aden (PERSGA). The benefits of establishing marine protected areas are the conservation and improvement of biological diversity; the maintenance of essential ecological process within the natural system; the sustainable management of marine renewable resources including fisheries; and the protection and restoration of degraded habitats while facilitating the implementation of an ecosystem approach (Box 14.10).

A researcher documents information during a coral reef resilience survey. *© J Tamelander/IUCN*

Fish stock enhancement

Another policy cluster addresses biodiversity conservation through integrated fisheries management, an initiative aimed at tackling the issue of how fish resources can best be shared between competing users within the broad context of ecologically sustainable development (Shing 2001). There are several marine fisheries management policies within the cluster, currently practised to overcome management obstacles and problems facing the fishery sector.

Data clearly show very substantial declines in commercial fish over the past 10–20 years (Sheppard *et al.* 2010), while Bishop (2002) and Sheppard *et al.* (2010) have demonstrated a link between the permanent loss of inter-tidal and shallow sub-tidal nursery grounds with declining fish and shellfish catches. Marine stock enhancement, an approach that addresses these

Box 14.10 Marawah Biosphere Reserve, Abu Dhabi, United Arab Emirates

The Marawah Marine Protected Area, the largest in the region with a total area of 4 255 km², became the first UNESCO Marine Biosphere Reserve in the region in 2007. Marawah itself, just one of 20 islands that make up the protected area, is flanked by Jarnain Island to the north, Abu Al Abyad Island to the east, the mainland to the south and Sir Baniyas to the west. The protected area is a representative example of the Gulf region, containing coastal areas, salt flats (*sabkhas*), shallow waters and shallow islands as well as seagrass habitats. The island hosts a significant population of dugongs, four species of marine turtle, 70 species of fish, and coral reefs and expanses of mangrove (*Avicennia marina*) that are important habitats for many terrestrial and marine species. Resident and migratory bird species such as ospreys, sooty falcons and several tern species, are part of the ecosystem, with bottlenose and humpback dolphins also found in the surrounding waters, making the area important for its biodiversity. Preserving the natural diversity and quality of the coastal and marine environment is what the reserve aims to achieve. It has established a 12-member marine ranger force to carry out surveillance and control programmes, essential infrastructure has been procured and maintained, and the rehabilitation of a mangrove site on Marawah Island has begun. The island is also of great cultural and archaeological significance, with more than 20 sites dating back 7 000 years to the Stone Age (SCENR *et al.* 2008).

issues, involves a set of management measures for releasing farmed organisms to enhance or restore fisheries. Engineered artificial reefs, which help to restore lost or degraded marine and coastal environments, can also enhance re-stocking of depleted commercial fish and shellfish.

Such practices, including sea ranching, stock enhancement and restocking, are widespread, often controversial, and have varying levels of success (Lorenzen *et al.* 2010). Fish stock enhancement, one of the successful policies, involves the annual release of tens of thousands of fish fingerlings in different parts of the territorial waters.

The benefits of this policy are mostly in rehabilitating depleted fisheries, and also include the potential to reduce the time needed to rebuild some severely overexploited ones or improve the productivity of other, healthy fisheries. Only if large-scale fish releases are completed and the effects on the fisheries measured and documented will it be possible to quantify and value the economic benefits of the programme.

Stock enhancement, which can be very costly and absorb major resources, is not a substitute for fisheries management (AFED 2009). Understanding both the biology and the culture technology of a species is critical to the success of stock enhancement programmes, the impacts of which may be insignificant and can be extremely difficult to assess. If not carefully managed, stock enhancement programmes can affect the gene pool of wild stocks and there can be complications when exotic species are introduced as part of the enhancement programme.

The potential for restocking and stock enhancement stems mainly, but not entirely, from the development of technology to produce juveniles of a wide variety of coastal fish and shellfish in hatcheries (Bell *et al.* 2006). Restocking and stock enhancement programmes, in which West Asian countries have developed considerable expertise, are applied in intricate human-environment systems involving dynamic interactions between the resource, the technical intervention and the people who use it, making replication complex (Box 14.11).

A school of striped mackerel feeding in the Red Sea. *© Dirk-Jan Mattaar*

Stock enhancement programmes should be considered within the broader issue of fisheries management (Shams and Uwate

Box 14.11 Fish stock enhancement in Bahrain

Bahrain is an island state and its people have a strong affinity with the sea. Landings of some preferred fish such as grouper have declined dramatically in the last 10–20 years. One successful management policy that is being followed on an annual basis is the enhancement of fish stocks, which have suffered over the years as too many fishers catch too many fish. In 1994, tens of thousands of orange spotted groupers were successfully released; in 1996 and 1997 stock enhancement activity focused on the release of yellow-finned black sea bream and sobaity sea bream.

Several fish species such as white-blotched grouper and striped grouper, parrotfish and sobaity sea bream were rarely seen in fish markets before the start of the release programme in 1994, but the annual release of different species has enabled the refinement of the technology and reduced mortality and costs. The General Directorate of Fisheries, for example, has recently made arrangements with Durat Al-Bahrain, in the southwest of the Kingdom, to deploy different types of artificial reefs to be used for the release of sobaity sea bream and grouper fingerlings.

However, as a result of insufficient funding for the fish release programme, modern fish tagging technology could not be applied to assess its success. Some people, however, have reported large quantities of small groupers and sobaity and shaem sea bream in the market just after fish release. Additionally, in the post-release period, actual grouper landings were high. This suggests that the releases have had a positive impact on the fishery, especially on landings (Zainal and Abdulqader 2009; Shams and Uwate 1996).

1996). Enhancements affect complex fisheries systems and, to be successful, must contribute to a broad set of biological, economic, social and institutional management objectives (Lorenzen 2008), so it is important to compare the costs of the programmes with the benefits. Finally, public cooperation is very helpful.

CONCLUSIONS

Environmental policies in West Asian countries have been developed over the past two decades and continue to progress; however, they need to become proactive rather than reactive. Additionally, environmental governance, rather than merely focusing on environmental policies, needs to take account of societies' common goals and engage with various stakeholders in the design and execution of policies. The integration of sectoral policies is also important. Regional environmental governance is crucial for the West Asia region as the countries share common environmental conditions. There is also a need for clear, integrated policies, not just targets, aimed at shifting the region from current long-established economic systems towards a green economy.

Failure to introduce sectoral policy integration, policy mixes and regional integration will intensify currently unsustainable consumption and production patterns, especially for energy, water, food security and marine resources, with the potentially grave consequences of natural resource depletion and increased pollution, which in turn impact human health and well-being.

The region's marginal biophysical characteristics, population growth, urbanization and socio-economic policies, coupled with high rates of natural resource consumption, are the main drivers of environmental problems. Insecurity and conflicts are also among the regional drivers of environmental degradation. These are further exacerbated by frequent droughts and climate change.

There is a general lack of coherent environmental data and information tools in the West Asia region. The systematic collection, processing, analysis, production, dissemination and exchange of environmental information would lead to more robust decision making and proper policy formulation and implementation. Trends show the need to make use of additional measures to improve enforcement and compliance processes. Moreover, there is a significant need for regular environmental reporting in all West Asian countries as well as greater public and private participation (UNEP 2010).

The involvement of the public in the environmental regulatory systems remains low because people are neither well informed nor encouraged to participate. Although access to general environmental information has recently improved, much effort is still required to achieve real public participation in environmental management.

Several countries in West Asia undertake enabling initiatives to facilitate the implementation of green technologies to reduce pollution and waste, to conserve energy and to rationalize water use. Most of the countries have developed policies to integrate cleaner production concepts in the industrial sector and established centres to help build capacity. However, the effectiveness of policy implementation is unsatisfactory.

Proper allocation of authority within environmental governance needs to be enhanced and environmental institutions empowered. Strengthening the role of various stakeholders, including non-governmental organizations, the private sector and local communities, would improve execution, monitoring, reporting and achieving the collective goals as well as increasing cooperation at national and regional levels, and would lead to better implementation of environmental policies.

Many policy options could be seen as inducing the necessary structural changes to achieve better environmental governance in the region. Such options include the integration of environmental impact assessment in decision-making processes and development plans, the decentralization and development of institutions, and improving access to environmental information to improve public engagement.

The challenge for the water sector lies in achieving sustainable water development through balancing supply and demand, with priority given to meeting the MDG 7c targets. Water resources policy should be coordinated with agriculture, environment, housing, and social and economic policies to achieve internationally agreed goals. The updating of water legislation and strengthening of intra-institutional coordination mechanisms, the free dissemination of information and the enhancement of stakeholder participation are necessary to the comprehensive and integrated management of the water sector. Essential management measures include greater efficiencies in water use, especially in the irrigation sector; protecting water sources from pollution and depletion; and deploying adequate financial and qualified human resources. The integrated water resources management approach provides an innovative planning tool to overcome existing obstacles and address future

Box 14.12 The Council of Arab Ministers Responsible for the Environment (CAMRE)

Within the framework of the League of Arab States (LAS), CAMRE was established as a high-level institution to ensure the proper coordination of environmental policies in the Arab region, which includes all the countries of the West Asia region. CAMRE aims to identify major environmental problems, set priorities and address issues related to a sustainable environment. CAMRE has played, and continues to play, a major role in the coordination of environmental policies of the Arab countries at regional and global levels and has ensured a certain level of replication of environmental policies among West Asian countries. In addition, CAMRE ensures that all LAS institutions are addressing environmental issues in a comprehensive and harmonized manner.

water sector challenges. Successful national formulation and application of an integrated management approach can be replicated across the region given the homogeneity of its natural, physical, social, economic and cultural settings.

Screening and analysis of policies implemented to prevent and mitigate land degradation in West Asia show that the region is in line with the main goal of the Johannesburg Plan of Implementation, Paragraph 40 (WSSD 2002). However, the main challenges are the formulation and implementation of bottom-up policy to encourage community participation, and the enhancement of regional cooperation through projects to conserve natural resources, increase land productivity, and prevent and mitigate soil erosion and dust storms. Integrated land, agriculture and water policies include modern agricultural techniques, sustainable agricultural production systems and afforestation to achieve relative food and water security.

Featuring three integrated wind turbines, the Bahrain World Trade Center boasts the world's most sophisticated skyscraper energy recovery system. *© Klaas Lingbeek- van Kranen*

Policies focused on the energy efficiency of buildings have been successful in the development of green building codes and expanding the market for innovative green services and efficiency-oriented companies, which attract professional participation to improve practices. The public at large understands the policy concepts and has the awareness to implement efficient practices as long as the building fabric options and systems are economically justifiable and mandated by governments. Replication of building and systems energy policies has a high success rate due to the similarity of climates and needs, and the high level of motivation for increased development, innovation, and investment for the greening of the building sector by professional, private and government entities.

Energy policies related to power generation have a more top-down approach that relies on country goals to increase the share of energy production from clean sources. Decision-making tools to decide on the energy generation mix have to include environmental impact assessments. West Asian countries are looking at various options including nuclear energy, as renewable energy generation is still not seen as cost-effective compared with the conventional resources that are abundant in GCC. Energy policies cannot be developed independently of freshwater policies. The challenge lies in how to optimize the development of policies that address the demand of both energy and water at minimal cost to the environment. Cooperation between Arab countries to complement each others' energy needs would have more success where alternative sources of energy generation are used.

Policies for oceans and seas are clustered around the integration of management tools in order to achieve sustainable development of the coastal and marine areas. West Asian countries should confirm their commitment to an ecosystem approach through continued support for the integrated management of the coastal and marine environment. To this end, policy implementation tools such as strategic social and environmental impact assessments should be considered during project planning.

The development and improvement of management systems for marine protected areas and regional networks are important for conservation of the region's biodiversity. Global climate change will exert additional impacts on the coastal and marine environment, and regional adaptation strategies should take into consideration the environmental, social and economic differences between countries. The region has a very high potential for the transfer of several policies to ensure the sustainable development of the coastal and marine environment of the West Asia region.

REFERENCES

Abahussain, A.A., Abdu, A.S., Al-Zubari, W.K., El-Deen, N.A. and Abdul-Raheem, M. (2002). Desertification in the Arab Region: analysis of current status and trends. *Journal of Arid Environments* 51, 521–545

Abdulrazzak, M.J. (1995). Water supplies versus demand in countries of Arabian Peninsula. American Society of Civil Engineering. *Journal of Water Resources Planning and Management* 121, 227–234

Abdulrazzak, M.J. (1994). Review and assessment of water resources of the Gulf Cooperation Council countries. *International Journal of Water Resources Development* 10, 23–37

Abdulrazzak, M., Jurdi, M. and Basma, S. (2002). The role of desalination in meeting water supply demands in Western Asia. *Water International* 27(3), 395–406

Abualkhair, A. (2007). Electricity sector in the Palestinian territories: which priorities for development and peace? *Energy Policy* 35, 2209–2230

Abu Hamed, T., Flamm, H. and Azraq, M. (2012). Renewable energy in the Palestinian territories: opportunities and challenges. *Renewable and Sustainable Energy Reviews* 16(1), 1082–1088

ACSAD, CAMRE and UNEP (2004). *State of Desertification in the Arab World* (updated study). Arab Center for the Study of Arid Zones and Dry Lands, Damascus

AFED (2010). *Arab Environment. Water: Sustainable Management of a Scarce Resource*. 2010 Report of the Arab Forum for Environment and Development, Beirut

AFED (2009). *Executive Summary. Arab Environment. Climate Change: Impact of Climate Change on Arab Countries* (eds. Tolba, M.K. and Saab, N.W.). 2009 Report of the Arab Forum for Environment and Development, Beirut

Aftab, A. and Elhadidy, M.A. (2002). *Energy Conservation Measures for a Typical Detached Single Family House in Dhahran*. Proceedings of the First Symposium on Energy Conservation and Management in Buildings, King Fahd University of Petroleum and Minerals (KFUPM), 5–6 February 2002

Al-Ajlan, S.A., Al-Ibrahim, A.M., Abdulkhaleq, M. and Alghamdi, F. (2006). Developing sustainable energy policies for electrical energy conservation in Saudi Arabia. *Energy Policy* 34(13), 1556–1565

Al-Ajmi, F.F. and Loveday, D.L. (2010). Indoor thermal conditions and thermal comfort in air-conditioned domestic buildings in the dry-desert climate of Kuwait. *Building and Environment* 45, 704–710

Ali, Y., Mustafa, M., Al-Mashaqbah, S., Mashal, K. and Mohsen, M. (2008). Potential of energy savings in the hotel sector in Jordan. *Energy Conversion and Management* 49, 3391–3397

Al-Kassas, M.A. (1999). *Desertification: Degradation of Lands in Arid Areas*. Alam Al-Marefa Series No. 242 (in Arabic). Kuwait

Al-Mohamad, A. (2001). Renewable energy resources in Syria. *Renewable Energy* 24, 365–371

Alnaser, N.W., Flanagan, R. and Alnaser, W.E. (2008). Potential of making over to sustainable buildings in the Kingdom of Bahrain. *Energy and Buildings* 40, 1304–1323

Al-Rashed, M. and Sherif, M.M. (2000). Water resources in the GCC countries: an overview. *Water Resources Management* 14, 59–75

Al-Temeemi, A.S. (1995). Climatic design techniques for reducing cooling energy consumption in Kuwaiti houses. *Energy and Buildings* 23(1), 41–48

AOAD (2009). *Comprehensive Study to Document Agricultural Policies in Arab Countries during the 1st Decade of the 3rd Millennium*. Arab Organization for Agricultural Development, Khartoum. http://www.aoad.org/agrpolicies

AOAD (2007). *Strategy for Sustainable Arab Agricultural Development for the Upcoming Two Decades (2005–2025)*. Arab Organization for Agricultural Development, Khartoum. http://www.aoad.org/El%20strtiga%20

Bell, J.D., Bartley, D.M., Neil, K.L. and Loneragan, R. (2006). Restocking and stock enhancement of coastal fisheries: potential, problems and progress. *Fisheries Research* 8, 1–8

Bishop, J.M. (2002). Fishing and mariculture. In *The Gulf Ecosystem, Health and Sustainability* (eds. Khan, N.Y., Munwar, M. and Price, A.R.G.). pp.253–278. Backhuys Publishers, Leiden

CEDARE and AWC (2004). *Report on the State of the Water in the Arab Region*. Arab Water Council, Cairo. http://www.arabwatercouncil.org/administrator/Modules/CMS/SOW.pdf

Chedid, R.B. and Ghajar, R.F. (2004). Assessment of energy efficiency options in the building sector of Lebanon. *Energy Policy* 32, 647–655

Dabour, N. (2006). Water resources and their use in agriculture in Arab countries. *Journal of Economic Cooperation* 27(1), 1–38. http://www.sesrtcic.org/files/article/25.pdf

EIA (2007). *Country Reports*. US Energy Information Administration. http://www.eia.gov/countries/country-data.cfm?fips=SY

FAOSTAT (2008). *FAO Statistical Databases*. Food and Agriculture Organization of the United Nations, Rome. http://www.faostat.org

Fasiuddin, M., Budaiwi, I. and Abdou, A. (2010). Zero-investment HVAC system operation strategies for energy conservation and thermal comfort in commercial buildings in hot-humid climate. *International Journal of Energy Research* 34(1), 1–19

Ghaddar, N., Ghali, K. and Saadeh, R. (2010). Optimized selection and operation of the combined chilled ceiling system and displacement ventilation. *International Journal of Energy Research* 34(15), 1328–1340

Ghaddar, N., Moukalled, F., Chedid, R., Fadel, M., Mezher, T., Hamzeh, A., Harb, A. and Abdulla, F. (2006). Renewable energies technologies contribution and barriers to poverty alleviation in Jordan, Syria, and Lebanon. Proceedings of The Arab Regional Solar Energy Conference (ARSEC), 5–7 November 2006, University of Bahrain, Bahrain. *Journal of the Association of Arab Universities for Basic and Applied Sciences* 358–371

Ghali, K., Othmani, M. and Ghaddar, N. (2008). Integration of desiccant dehumidification wheel with air-conditioning system in Beirut: performance and energy savings. *International Journal of Green Energy* 5(5), 360–372

Hainoun, A., Seif Aldin, M. and Almoustafa, S. (2010). Formulating an optimal long-term energy supply strategy for Syria using MESSAGE model. *Energy Policy* 38, 1701–1714

Hajiah, A. (2010). *Sustainable Energy in Kuwait – Challenges and Opportunities*. UNDP Regional Consultation Meeting: Climate Change Impacts in the Arab Region: Towards Sustainable Energy Resources, Challenges and Opportunities, 6 October 2010. http://www.arabclimatewatch.org/knowledge/sustainable_energy/ALi%20Ebraheem%20Hajiah-Energy%20Efficient%20Building.pdf

Houri, A. (2006). Solar water heating: current status and future prospects. *Renewable Energy* 31, 663–675

Hrayshat, E.S. and Al-Soud, M.S. (2004). Solar energy in Jordan: current state and prospects. *Renewable and Sustainable Energy Reviews* 8, 193–200

Hussain, G., Alquwaizany, A. and Al-Zarah, A. (2010). Guidelines for irrigation water quality and water management in the Kingdom of Saudi Arabia: an overview. *Journal of Applied Sciences* 10, 79–96

IEA (2010). *World Energy Statistics 2010*. International Energy Agency, Paris. http://www.iea.org/stats/index.asp

IPCC (2007) *Climate Change 2007: Synthesis Report* (eds. Pachauri, R.K. and Reisinger, A.). Contribution of Working Groups I, II and III to the Fourth Assessment Report of the Intergovernmental Panel on Climate Change. IPCC, Geneva

Kablan, M.M. (2004). Techno-economic analysis of the Jordanian solar water heating system. *Energy* 29(7), 1069–1079

Kattach, G. (2008). The use of forage plants for landscape management and soil conservation in dry areas. In *Conservation Agriculture for Sustainable Land Management to Improve the Livelihood of People in Dry Areas* (eds. Stewart, B.A., Asfary, A.F., Belloum, A., Steiner, K. and Friedrich, T.). pp.219–26. Proceedings of the International Workshop, Damascus, 7–9 May 2007, organised by the Arab Center for the Study of Arid Zones and Dry Lands (ACSAD) and GTZ

Kellow, M. (1989). Kuwait's approach to mandatory energy-conservation standards for buildings. *Energy* 14(8), 491–502

Kraidy, A. (2007). *Energy Efficiency and Renewable Energy, Syria – National Study*. Mediterranean and National Strategies for Sustainable Development. Priority Field of Action 2: Energy and Climate Change. Plan Bleu Regional Activity Centre, Sophia Antipolis. http://www.planbleu.org/publications/atelier_energie/SY_national_study_final.pdf.

Lorenzen, K. (2008). Understanding and managing enhancement fisheries systems. *Reviews in Fisheries Science* 16(1–3),10–23

Lorenzen, K., Leber, K.M. and Blankenship, H.L. (2010). Responsible approach to marine stock enhancement: an update. *Reviews in Fisheries Science* 18(2), 189–210

Maheshwari, G.P. and Al-Murad, R. (2001). Impact of energy-conservation measures on cooling load and air-conditioning plant capacity. *Applied Energy* 69(1), 59–67

Mehdi, S. (2004). *Coastal Area Management Programme (CAMP) Lebanon: Final Integrated Report*. Priority Action Programme, Coastal Management Center, Split. http://www.pap-thecoastcentre.org

Ministry of Electricity (2007a). *Building Thermal Insulation Code in Syria*. National Energy Research Center, Government of Syria

Ministry of Electricity (2007b). *Syria's Master Plan for Renewable Energy*. Government of Syria

Ministry of Municipalities Affairs and Land Use Planning (2010). The national strategy for sustainable agricultural development of the Kingdom of Bahrain. In *Seeds for OUR Future*. Manama

Ministry of Public Work and Housing (2009a). *Energy Efficient Building Code*. Government of Jordan

Ministry of Public Work and Housing (2009b). *Thermal Insulation Code*. Government of Jordan

Ministry of Water and Irrigation (2008). *A National Water Demand Management Policy*. Government of Jordan

OAPEC (2009). *Annual Statistical Report 2009*. Organization of Arab Petroleum Exporting Countries. http://www.oapecorg.org/publications/ASR/A%20S%20R%202009.pdf
PAP-RAC (undated) Mediterranean Action Plan Priority Actions Programme-Regional Activity Centre. www.pap-thecoastcentre.org (accessed 2011)

PEC (2006). *SOLATERM Project, Country Report 2006*. Palestinian Energy and Environment Research Center

Price, A.R.G. (2002). Simultaneous 'hot spots' and 'cold spots' of marine biodiversity and implications for global conservation. *Marine Ecology Progress Series* 24, 23–27

Reiche, D. (2010). Renewable energy policies in the Gulf countries. A case study of the carbon-neutral "Masdar City" in Abu Dhabi. *Energy Policy* 38, 378–382

Ruble, E. and Nader, P. (2011). Transforming shortcomings into opportunities: can market incentives solve Lebanon's energy crisis? *Energy Policy* 39(5), 2467–2474

SCENR, EAD, NCRI and EWS-WWF (2008). *Conservation and Management Plan for Abu Dhabi and Eastern Qatar Coral Reefs*. Prepared by Supreme Council for the Environment and Natural Reserves (SCENR) of the State of Qatar, Environment Agency of Abu Dhabi (EAD), National Coral Reef Institute (NCRI) and Emirates Wildlife Society in association with the World Wide Fund for Nature (EWS-WWF) and with support from Dolphin Energy Ltd

Sgouridis, S. and Kennedy, S. (2010). Tangible and fungible energy: hybrid energy market and currency system for total energy management. A Masdar City case study. *Energy Policy* 38(4), 1749–1758

Shahin, W. (2010). *Jordan's Energy Efficiency Strategy*. National Efficiency Plan for Regional Energy Challenges: The Arab EE Directive. National Energy Research Center, Amman

Shams, A.J. and Uwate, K.R. (1996). *Bahrain Fish Release Activities: 1994 to Present*. Directorate of Fisheries, Ministry of Works and Agriculture, State of Bahrain

Sheppard, C., Al-Husiani, M., Al-Jamali, F., Al-Yamani, F., Baldwin, R. , Bishop, J. , Benzoni, F., Dutrieux, E., Dulvy, N.K., Durvasula, S.R.V., Jones, D.A., Loughland, R., Medio, D., Nithyanandan, M. , Pilling, G.M., Polikarpov, I., Price, A.R.G., Purkis, S., Riegl, B., Saburova, M., Namin, K.S., Taylor, O., Wilson, S. and Zainal, K. (2010). The Gulf: a young sea in decline. *Marine Pollution Bulletin* 60, 13–38

Shing, C.C.A. (2001). *Case Study of the Integrated Coastal Fisheries Management Project: A Pilot Project for the Gulf of Paria, Trinidad*. Caribbean Natural Resources Institute (CANARI) Technical Report No. 280. http://canari.org/chanashing.pdf

SRAP (2007). *Integrated Natural Resource Management for Combating Desertification in West Asia*. UNCCD/SRAP Pilot Projects in Jordan, Lebanon, Syria and Yemen 2003–2006, Final Report. United Nations Convention to Combat Desertification/Sub-Regional Action Programme

Tortell, P. (2004). *Thoughts on Integrated Coastal Zone Management (ICZM) in Saudi Arabia*. The Regional Organization for the Conservation of the Environment of the Red Sea and Gulf of Aden (PERSGA), Jeddah

UN DESA (2011). *The Millennium Development Goals Report 2011*. United Nations Department for Economic and Social Affairs, New York

UNDP (2010). *Human Development Report 2010*. United Nations Development Programme, New York

UNEP (2011). *Towards a Green Economy: Pathways to Sustainable Development and Poverty Eradication*. United Nations Environment Programme. http://www.unep.org/greeneconomy

UNEP (2010). *The Environment Outlook for the Arab Region*. UNEP Regional Office for West Asia, League of Arab States and CEDARE. http://eoar.cedare.int/report/EOAR%20Full.pdf

UNEP (2007). Freshwater of the West Asia region. In *Global Environmental Outlook: Environment for Development*. United Nations Environment Programme, Nairobi. http:www.unep.org/geo

UN ESCWA (2008). *Promoting Sustainable Energy Production and Consumption in the Arab Region*. United Nations Economic and Social Commission for Western Asia. http://esa.un.org/marrakechprocess/pdf/ESCWA_SEPC_paper_15march2008.pdf

UN ESCWA (2007a). *Land Degradation Assessment and Prevention: Selected Case Studies from the ESCWA Region*. United Nations Economic and Social Commission for Western Asia. United Nations, New York. http://www.arab-hdr.org/publications/other/escwa/landdegradation-07e.pdf

UN ESCWA (2007b). *State of Water Resources in the ESCWA Region*. ESCWA Water Development Report 2. ESCWA/SDPD/2007/6. United Nations Economic and Social Commission for Western Asia

UN ESCWA (2005). *Promoting IWRM Plans in ESCWA Member Countries*. E/ESCWA/SDPD/2005/10. United Nations Economic and Social Commission for Western Asia.

UN ESCWA (2002). *World Summit on Sustainable Development: Assessment Report for the ESCWA Region*. E/ESCWA/ENR/2002/19. Economic and Social Commission for Western Asia. United Nations, New York. http://www.escwa.un.org/divisions/sdpd/wssd/pdf/assess.pdf

UN ESCWA (2001). *Enhancing the Application of Integrated Water Resources Management in the ESCWA Region*. ESCWA/SDPD/2004/6/Summary. United Nations Economic and Social Commission for Western Asia. United Nations, New York

UNPD (2008). *World Population Prospects: The 2008 Revision*. Population Division, Department of Economic and Social Affairs, United Nations, New York

Verdeil, É. (2008). Electricity in Middle East policy. *Maghreb Machrek* 195(1), 109–128

World Bank (2009). *Management's Discussion and Analysis and Condensed Quarterly Financial Statements September 30 2009*. http://treasury.worldbank.org/web/BRD_MDA_and_Condensed_Quarterly_Financial_Statements_Sep_2009.pdf (accessed 20 December 2011)

World Bank (2008). *World Development Report 2009: Reshaping Economic Geography*. World Bank, Washington, DC

World Bank (2005). *A Water Sector Assessment Report on the Countries of the Gulf Cooperation Council*. World Bank, Washington, DC

WSSD (2002). *Johannesburg Plan of Implementation*. World Summit on Sustainable Development. http://www.un.org/esa/sustdev/documents/WSSD_POI_PD/English/POIToc.htm

Zainal, K. (2009). *The Cumulative Impacts of Reclamation and Dredging Activities*. Report for Regional Organization for the Protection of the Marine Environment (ROPME), Kuwait

Zainal, K. and Abdulqader, A. (2009). Fisheries. In *Marine Atlas of Bahrain* (eds. Loughland, R. and Zainal, A.J.). Geomatec Bahrain Centre for Studies and Research, Miracle Publishing

CHAPTER 15

Policy Options: Regional Summary

Coordinating lead authors: Asha Singh and Renat Perelet

Contributing authors: Jane Barr, Ludgarde Coppens, Nicolai Dronin, Jose Etcheverry, Amr El-Sammak, Lailai Li, Clever Mafuta, Catherine P. McMullen, Flavia Rovia (GEO Fellow) and Joanna Kamiche Zegarra

Principal scientific reviewer: Ada Ignaciuk

Chapter coordinators: Matthew Billot and Ludgarde Coppens

Main Messages

The selection of freshwater, climate change and environmental governance as priorities by all regions suggests a recognition that these issues have reached a point of global importance requiring responses that could have relevance worldwide. Climate change exerts extreme pressure on ecological systems, including on freshwater by exacerbating problems of water supply and demand. Two regions considered climate change to be cross-cutting, and assessed how policies in each theme help to attain international goals related to climate change.

There are common elements in successful policies across the regions. Tools such as integrated water resources and coastal zone management; the removal of environmentally harmful subsidies, especially on fossil fuels and/or carbon taxes; renewable energy, marine protected areas, and cross-boundary biodiversity conservation, are all examples of policies used in more than one region, but customized to each context. Formal, robust, and well-established governance mechanisms and structures at all governance levels are a necessary foundation for successful implementation of environmental policies.

The policies selected by the regions are successful because of some underlying principles. These include policies that are mutually reinforcing and have benefits across sectors, address drivers, invest in monitoring and evaluation to allow revision and enhance accountability, or involve multi-stakeholder participation at local, national and regional level.

There is sufficient experience to enable faster transfer and replication of several of the priority policies. This, however, would be greatly improved by the sharing of experiences between donor and recipient practitioners and stakeholders, learning the specific skills of how to assess potential policies for particular needs and how to adapt these to the selected situation, and establishing capacity and institutional development to support the enhancement and propagation of these skills.

While many of these policies are long-standing management concepts, their application can be innovative if certain principles are adhered to. This includes policies that are mutually reinforcing with positive impacts in more than one thematic domain, and policies that address drivers – as defined in Chapter 1. Concentrating on these deeper, underlying causes of environmental degradation will allow the goals and targets set out in international, regional and national agreements to be met in a more effective way.

Transboundary cooperation is important when natural areas are shared. It promotes understanding and the transfer of knowledge between neighbours, and leads to a collective response to shared problems, allowing new opportunities and ways of overcoming these common problems to be identified.

Improved environmental governance is needed if environmental degradation and the unsustainable use of natural resources are to be reversed. Critical components include multi-stakeholder support, raising public awareness among all stakeholders, stronger mechanisms for financial sustainability, enhanced institutional capacity, adequate legal frameworks and strong compliance mechanisms. Community leadership demonstrated, for example, in the formation of water maintenance trust funds or wetland management schemes provides local services, helps resolve inter-community conflicts, demonstrates the value of participation and learning, and provides income-generating opportunities.

Policies that have proven successful can be analysed for their ability to leverage societal transformation. Understanding the potential of these policies, alone or in combination, could help facilitate transformative change and enhance the effect that policy makers have on reaching sustainable development objectives at local, national, regional and international levels.

INTRODUCTION

Humans have long been aware of the effects on their local environment of resource use, waste production and land use, but only in the last few decades has it been realized how such activities affect the global environment. In the past, when there were fewer humans and each used natural resources less intensively, the capacities of the atmosphere, land and water could carry the load of human consumption and production. But a significant proportion of the 7 billion humans alive today are actively exploiting the planet's resources at accelerating rates and intensities that surpass the carrying capacity of the Earth's systems (Krausmann *et al.* 2009; Liu *et al.* 2003; McNeill 2000). As mentioned in Chapter 1, the scale, spread and rate of change of global drivers are without precedent.

The concept of planetary boundaries was introduced by Rockström *et al.* (2009) to identify those key environmental processes that provide humanity with a safe operating space for well-being. Scientific analysis established nine planetary boundaries with approaching thresholds beyond which humans could not thrive in the Earth System. Of these nine thresholds, three may already have been passed: climate change, biodiversity loss, and the removal of nitrogen from the atmosphere (fixing) for use in fertilizers and weapons (Rockström *et al.* 2009). The concept of impending thresholds, tipping points and crossing boundaries are familiar to those who study complex systems (Limburg *et al.* 2002). Biologists and marine scientists term abrupt changes in the species that populate an ecosystem regime shifts (Kraberg *et al.* 2011; Rodionov and Overland 2005). Earth System scientists are currently debating the imminence of destabilization in various tipping elements affecting global climate (Lenton *et al.* 2008). A more in-depth discussion of the role of planetary boundaries can be found in Chapter 7.

This chapter provides a summary of Chapters 9–14 and aims to determine those approaches and policies that show promise for adoption and adaptation elsewhere. The outcome of this appraisal is to offer policy options that can help meet internationally agreed goals efficiently and effectively and also lead to international, regional and local co-benefits. It is useful to identify at which level the policies identified as promising are best suited to apply leverage (Chapter 16).

REGIONAL SUMMARY

The selection of themes

The *GEO-5* Introduction describes the process by which each region selected priority themes and internationally agreed goals, as well as the policy appraisal methodology by which promising policies were identified. During the consultations, some regions decided that certain themes cut across the priority environmental challenges that were selected for the region.

The selection of priority themes and goals, which was limited to a maximum of five or six for each region, provides a first indication of what is considered important (Table 15.1).

Different regions focused on different aspects of the same theme. For example, while only two regions selected energy as a priority theme (Table 15.1), three others – Asia and the Pacific, Europe, and Latin America and the Caribbean – included energy

Table 15.1 Priority themes by region

	Africa	Asia and the Pacific	Europe	Latin America and the Caribbean	North America	West Asia
Environmental governance	Cross-cutting	Theme	Cross-cutting	Cross-cutting	Theme	Cross-cutting
Climate change	Theme	Theme	Theme	Theme	Cross-cutting	Cross-cutting
Energy					Theme	Theme
Air pollution			Theme			
Land	Theme			Theme	Theme	Theme
Freshwater	Theme	Theme	Theme	Theme	Theme	Theme
Oceans and seas	Theme					Theme
Biodiversity	Theme	Theme	Theme	Theme		
Chemicals and waste		Theme	Theme			

Selected as a cross-cutting theme   Selected as a theme

in their selection of policies that hold promise for meeting the climate change goal. In Latin America and the Caribbean, policies on wastewater treatment and coastal zone management are included in the water theme, while policies on coastal and marine protected areas are included in the biodiversity theme.

The selection by all regions of freshwater, climate change and environmental governance as priorities suggests recognition that these issues have reached a point of global importance requiring responses that could have relevance worldwide.

Africa

For the first time in 2009, Africa's total population exceeded 1 billion, of which 395 million (or almost 40 per cent) lived in urban areas. Africa should prepare for a total population increase of about 60 per cent between 2010 and 2050, with the urban population tripling to 1.23 billion during this period (UN-Habitat 2010). Africa's cities are characterized by extremes of prosperous centres and poor, informal settlements, and many governments struggle to provide social services including access to water, food and energy security, and management of environmental risks. Climate change and other adverse environmental change may accelerate urbanization (UN-Habitat 2010) and further strain governments' ability to cope (Mohamed-Katerere 2009).

The importance of transboundary natural resource policies is increasingly recognized in Africa as an essential strategy to enable integrated environmental management, economic integration, conflict resolution, and equitable socio-economic development. The integrated water resources management of the Senegal River Basin provides a practical example of transboundary cooperation that strengthens political links while enhancing agricultural productivity and food security. This initiative also improves transport, allowing year-round navigation, and ensures a reliable source of hydroelectricity.

Much of the growth in many of Africa's fastest-growing cities is a result of the expansion of informal settlements, which offer its residents limited access to basic infrastructure. *© iStock/Steven Allan*

The achievement of policy goals in a given sector or region is now understood to be closely linked to potential improvements and co-benefits in additional areas (Chapter 9). For example, under the Sustainable Land Management Programme in Ethiopia, 177 local watersheds are being protected to enhance land productivity and rural development, simultaneously strengthening resilience in downstream communities and countries (TerrAfrica 2009). For nations faced with limited resources, maximizing policy synergies helps deliver social, ecological and economic benefits, reduces trade-offs, and provides multiple paths for addressing common drivers and pressures.

The establishment of networks of marine protected areas creates additional benefits and provides more effective management opportunities than reliance on isolated ones. Regional cooperation is essential to sustain effective marine protected area networks. Addressing challenges – such as disparities in governance, institutional structures, wealth distribution, social capital and the collection of ecological data – and strengthening enabling conditions help to establish networks.

The effective use of tools and mechanisms to track and monitor environmental performance and change increases the capacity to respond effectively and efficiently to new challenges, including risks such as extreme events. For example, East Africa's Intergovernmental Authority on Development established a Conflict and Early Warning and Response Mechanism that helps communities plan their pastoral activities and food production better, building resilience against threats from famine and inter-community conflicts over grazing and crop production.

Regional cooperation, community-driven strategies and public-private partnerships can support learning, improve sustainability and encourage ecosystem approaches. The recently adopted Mangrove Charter for West Africa, which is complemented by country-specific action plans, is an example of regional cooperation. The relative success of a community-based mangrove management project in Cameroon demonstrates the value of participation and learning for successful adaptation (Ajonina *et al.* 2009). The restoration of 5 hectares of locally significant mangroves in Mauritius, financed by a local bank with technical assistance from government and implemented by a non-governmental organization and the local community, illustrates how partnerships between government, the private sector and civil society can help preserve natural resources and provide a practical climate change adaptation strategy that helps local people cope better with extreme events such as storm surges (ADD 2011, 2009). An island-wide survey has been undertaken to identify potential areas for replication.

Human rights protection is increasingly recognized as critical in strengthening human well-being while delivering environmental benefits. South Africa's Free Basic Water Policy, for example, benefits many impoverished households by guaranteeing access to 25 litres of water per person per day for domestic use within 200 metres of their homes. This strategy reduces the burden on women as well as providing health benefits (Mehta 2005).

The policy also strengthens municipal institutions and helps to reduce surface water pollution from domestic sources by placing authority at the local level.

Asia and the Pacific

The Asia and Pacific region has become a global engine of economic growth, but with large intra-regional differences. China is the world's largest carbon dioxide (CO_2) emitter, while most Pacific island nations are among the smallest. Water endowments range from the highly arid temperate zones and water-stressed small island states to Himalayan snowfields and abundant tropics. There is a wide diversity of environmental governance systems and mechanisms. The region's challenges include lifting millions of people out of poverty, navigating the turbulence of globalization, and dealing with some of the most polluted landscapes on Earth.

As the region is the fastest growing source of greenhouse gas emissions in the world, decisions to implement policies supporting carbon neutrality, renewable energy, conservation, and efficiency are crucial to the success of global efforts to address climate change. Countries in Asia and the Pacific, such as China, India and Indonesia, are reducing and removing fossil fuel subsidies with the objectives of reducing state budget burdens, preventing the use of public funds to support the wealthiest and those responsible for the greatest energy consumption, ensuring fairness for alternative energy development, and reducing environmental damage and contributions to climate change (IEA *et al.* 2010).

Of the ten countries in the world that are most at risk from climate change impacts, six are in the Asia and Pacific region. Key areas for action include integrating climate change adaptation and disaster risk reduction, mainstreaming adaptation concerns into development policies and plans, promoting ecosystem-based adaptation, and developing climate-proofed infrastructure. In the Maldives, policy research continues on possible relocation sites for populations displaced by sea level rise. At the same time, policy implementation increases resilience of individual islands by offering support with measures for afforestation, improving drainage, supplementing natural ridges, replenishing beaches, cultivating mangroves and fostering coral reef health (GEF 2009).

The Asia and Pacific region faces significant water-related challenges. Integrated water resource management planning, balancing water supply and demand through coordination among users, improved water quality management, appropriate pricing and multi-stakeholder participation, represent essential strategies to solve water problems in the region. In China, the Yellow River was cut off from the sea for 226 days in 1997, after episodic flows with shorter cut-offs in preceding years. In 1998, China initiated a programme restricting water withdrawals from the river and assigning quotas to users, with compliance enforcement measures including fines for exceeding withdrawal quotas (NDRC 1998). Since 2000, the river has flowed all the way to the sea.

For the Maldives, adaptation is a multi-dimensional goal that aims to increase the resilience of its vulnerable island systems against climate hazards and risks, and to achieve sustainable development.

© iStock/Tuomas Kujansuu

The emerging economies of Asia and the Pacific are exerting immense pressure on natural resources and ecosystem services. Although progress has been achieved through expanding protected areas, conserving species, addressing direct drivers of biodiversity loss, implementing community-based management and innovative financing, the scale of these efforts is insufficient to address current biodiversity and habitat losses. However, some successes can be replicated. In Viet Nam, forest ecosystem service payment schemes charge tourism operators and downstream water and power utilities for upstream water regulation, soil conservation and landscape preservation. Payments for the protection of 210 000 hectares of forest went to forestry businesses and forestry management boards, as well as to 9 870 households comprising mostly ethnic minorities (Winrock International 2011).

Many of the policy successes observed in the region are context specific. Therefore, policy transfer and emulation initiatives require careful analysis of the underlying political, cultural, economic and social contexts and their influence on policy implementation and success. Creating the necessary enabling environment is as important as selecting the right combination of policies.

Europe

The pan-European region is very diverse, with 37 national languages spoken in its 50 European countries (Nations Online 2011), a range of socio-economic and political systems, a variety of physical environments and differing means of environmental governance. Europe's land area of 23 million km^2 (GEO Data Portal 2011; FAO 2010) is characterized by a diversity of (agri)cultural landscapes, urban agglomerations, extensive coastal zones, forests and undisturbed pristine areas. Of the nearly 833 million Europeans, about half live in Western Europe, while some 72 per cent of the entire region's population lives in urban areas (GEO Data Portal 2011; UNDESA 2010).

This region has formal, robust, and well-established governance mechanisms and structures to address environmental issues. For more than four decades, the European Union (EU) has developed and implemented environmental policies. That process has evolved from targeted policies and single-issue instruments in the 1970s and 1980s, through diffuse policy integration and public awareness in the 1980s and 1990s, to policy coherence and other systemic approaches since the late 1990s.

At the pan-European scale, the Environment for Europe ministerial process, initiated in 1991, reflects this process of environmental governance. Both EU and non-EU European countries are on schedule to meet their Kyoto targets on cutting greenhouse gas emissions, and are among the major donors to international efforts to address climate change worldwide. The EU's Emissions Trading System (ETS), which covers about 40 per cent of EU emissions, provides valuable implementation and design lessons for other regions. For 2009, the EU carbon trading market was estimated to be worth more than US$118 billion per year, part of a global carbon-credit market estimated at US$144 billion, with the volume of emissions covered reaching 6.3 billion tonnes (EC 2009a, 2009b; Ellerman and Buchner 2007).

Two other promising policy options in the European region concern the development of renewable energy and adaptation strategies. Feed-in tariffs (FIT) for renewable energy systems were established in Germany more than 20 years ago and are being emulated successfully both throughout the EU and globally (Jänicke 2011). In its adaptation strategies, the EU is moving away from short-term disaster responses, aiming for long-term adaptation measures and policies that will be implemented at national and local levels focusing on land-use planning, agriculture, water management and biodiversity/nature conservation, as well as building adaptive capacity and taking action to increase resilience to climate change.

Although air quality in Europe has improved in recent decades, there are still some outstanding issues, particularly in relation to urban air quality, human health, air pollutants and ecosystem degradation. The efforts of the Convention on Long Range Transboundary Air Pollution (CLRTAP) and its pan-European scientific network have been pivotal in documenting air quality issues and in building credibility, shaping policies and ultimately monitoring trends in air quality improvement. Many of these policies and initiatives have strong replication potential – particularly where rapid industrialization is degrading air quality.

European governmental institutions also play a vital role in solving freshwater issues, while integrated water resources management is becoming a key guiding mechanism for decision-making. The transboundary nature of most European rivers calls for close cooperation between user countries through the creation of river basin management plans (UNECE 2011). Also, information provision and market-based instruments, such as water metering and incentive-based water pricing, show potential as policy approaches that can yield 20–40 per cent reductions in household water use.

Waste volumes continue to grow despite strong regulation. In Eastern Europe, a legacy of industrial wastes from the Soviet era still poses significant ecological problems (Devyatkin 2009). Policy focus is evolving to make producers responsible for waste reduction, reuse, and recycling and by encouraging the development of new technologies and greater reliance on life-cycle approaches. Legislation has been developed for specific wastes such as electrical and electronic equipment, various chemical substances, and toxic and radioactive by-products.

Europe is at the forefront of multi-national conservation efforts. Through Natura 2000, a coherent network of protected areas, biodiversity monitoring, and conservation activities has been established. However, biodiversity loss remains a problem due to continuing landscape, ecosystem, and habitat degradation. Nevertheless, initiatives such as Forest Europe are addressing biodiversity conservation, climate change, and the protection of freshwater resources and have already contributed to the increase of the total forest area in European countries (Forest Europe *et al.* 2011).

Latin America and the Caribbean

The 33 countries of Latin America and the Caribbean vary significantly in size and economic development. The region includes both Brazil, the seventh largest economy in the world (The Economist 2011) and small island developing states with open and fragile economies (Rietbergen *et al.* 2007). Rich in natural resources, the region is home to approximately 23 per cent of the world's forests and 31 per cent of its freshwater resources. Although these resources are not evenly distributed, the overall richness and economic importance of the region's ecosystems and its natural capital are undeniable (UNEP 2010). With 79 per cent of its population living in towns and cities (UN-Habitat 2010), the region is one of the most urbanized in the world. It faces challenges in providing its burgeoning towns and cities with safe water and sanitation, and in addressing air pollution and the contamination of its freshwater, oceans and seas.

The performance of the region's environmental institutions is currently constrained not by the absence of laws but by a paucity of political will, limited procedural continuity and inadequate enforcement instruments. The region also requires greater financial resources to facilitate sustainable management efforts and to ensure conservation of biological resources.

Achieving a more sustainable model of development requires improved national and regional strategies to facilitate cross-sectoral policies, at relevant scales, that can address environmental and economic issues simultaneously. To improve governance, active community participation and a high level of inter-institutional cooperation are also needed. The combination of these approaches can help improve environmental problems while enhancing human well-being. These types of initiatives are also crucial to address the most serious challenges faced in the region: poverty and inequality.

Careful analysis and evaluation of social needs, at local and regional levels, facilitates the implementation of more effective environmental initiatives that can also address social development. For example, the innovative transport initiatives implemented first in Curitiba, Brazil, and subsequently in Bogotá, Colombia, illustrate that well-designed projects can yield multiple environmental and social benefits, such as climate mitigation and improved mobility options (WRI 2010).

Integrated water resources management promotes the coordinated development and management of water, land and related resources. If carefully designed and implemented, it can also maximize economic and social welfare in an equitable manner without compromising the sustainability of vital ecosystems.

Green financing options and mechanisms to ensure better inclusion of key stakeholders are also recognized as important approaches to help reduce environmental degradation and threats to the region's biodiversity. For example, the Fund for the Protection of Water (FONAG), a trust fund to which water users contribute, is used to co-finance the rehabilitation and conservation of 65 000 hectares of watersheds that supply water in Quito, Ecuador, and surrounding areas. Similar funds have been developed in Colombia and Peru (Cisneros and Lloret 2008).

Since the 1960s, arable land has increased by 83 per cent in South America, 46 per cent in Africa and 36 per cent in Asia, coinciding with significant deforestation in all three regions (IPSRM 2010). In 2009, more than 280 000 Latin American and Caribbean producers managed 23 per cent of the world's organically farmed land, with the highest regional shares in Dominican Republic and Uruguay (Willer and Kilcher 2011). The overall area of land used for agriculture in South America increased by 20 per cent between 1970 and 2008, while livestock production grew by 37 per cent (FAO 2010).

Based on a review of Latin America and the Caribbean's current and past experiences, the three land management policies considered the most favourable to attain the goals set out in the Johannesburg Plan of Implementation (WSSD 2002) include multi-scale land-use planning, sustainable agriculture and livestock production, and the recovery of degraded lands. Payment for ecosystem services, sustainable forest management, and policies related to integrated land management and water-use plans address multiple themes including climate change, biodiversity and water goals.

Efforts to strengthen technology networks and knowledge flows are needed to improve the management of natural capital and the use of land-based and coastal and marine resources. Such capacity-building efforts are also important to help improve policies that promote the conservation of biodiversity and water, and climate change mitigation and adaptation, while helping to yield more effective development results (CCCCC 2011).

North America

North America is considered a global economic leader, although changes in regional demographics, rapidly emerging global economies and resource constraints all challenge the country's provision of public goods and services. At the same time, fragmented governance, policy instability, lack of clear targets and science policy, and the dilemma of whether to address global issues rather than seeking local solutions, hamper the achievement of environmental goals (Chapter 1).

In North America, there has been a recent trend emphasizing the success of market-based instruments over the use of regulations and standards in environmental management. However, there is a need for additional empirical data to show the actual impacts of these market-based instruments. Such policy options are best approached in a complementary manner, as market-based instruments need a clear and strong regulatory framework to work well. Furthermore, accountability and transparency can increase their environmental effectiveness while helping to ensure fair and equitable social outcomes.

Integrated water resources management in combination with technological instruments and economic incentives has proved effective in addressing complex water resource challenges. Integrated strategies have great potential for replication, provided that effective coordination and implementation mechanisms accompany them. They require a coordinated effort between stakeholders at multiple geographic and political scales, as well as adequate scientific evidence and monitoring to ensure that appropriate action is taken in response to altered water regimes resulting from climate change and increasing demand. The adoption of policies for maintaining the availability, sustainable use and fair allocation of freshwater to meet the needs of both humans and nature must be embedded into an holistic approach at the scale of a hydrological basin.

The Great Lakes-St. Lawrence River Basin Sustainable Water Resources Agreement of 2005, involving eight US states and two Canadian provinces, provides a framework for each state and province to manage and protect the basin as a whole. The Great Lakes and St. Lawrence Cities Initiative includes the objective that by 2015, all participating cities of the basin reduce water use by 15 per cent below 2000 levels. By 2010, almost half of the 33 participating cities had collectively achieved a 13 per cent reduction conserving a total of 330 million m^3 of water.

The pricing of externalities and integrated land management show potential to increase the sustainability of land-use practices in North America. Jurisdictions throughout the region have adopted many of these policy instruments to different degrees. For example, in British Columbia, Canada, resource companies, environmental groups and coastal First Nations have successfully carried out an ecosystem-based integrated land-use planning exercise, the 2006 Great Bear Forest Agreements, through a collaborative process (McGee *et al.* 2010). Taxes and other incentives in the United States have increased the total area conserved by local, state and national land trusts to around 15 million hectares. Programmes on payment for ecosystem services have permanently preserved another 92 million hectares in the United States.

At the state or provincial level, North America's growth of renewable energy technologies and the associated rise in the proportion of renewable energy in the region's energy mix can be traced to policy goals supported by market mechanisms such as feed-in tariffs. Greatly increasing the use of renewable energy in North America is technically feasible and can provide multiple benefits including reduction of greenhouse gas emissions, lower energy prices and decreased market volatility, while creating new employment and economic opportunities.

Carbon taxes can also contribute to energy efficiencies. The Canadian provinces of Quebec and British Columbia imposed carbon taxes in 2007 and 2008, respectively. Quebec's tax is very low but British Columbia's revenue-neutral scheme is more ambitious, starting at US$10 per tonne emitted in 2008 and increasing to US$30 per tonne in 2012. Addressing drawbacks typically associated with carbon taxes such as comprehensive coverage combined with targeted tax reductions, and reducing potentially large adaptation costs for carbon-intensive industries, appears to have helped it gain citizen acceptance.

West Asia

The predominantly arid and semi-arid region of West Asia covers about 4 million km^2. Rainfall is scant but with significant spatial and temporal variability. Water scarcity and frequent and persistent droughts are common, making water the region's most precious resource. The region also faces major environmental challenges in the need to address land degradation and desertification; increasing fossil-fuel-based energy production and use with high inefficiencies in generation, distribution and end use; and the conservation and sustainable use of marine and coastal resources. Climate change is becoming one of the region's main problems, with potentially adverse impacts on the economy and human well-being.

West Asia has made considerable progress in environmental governance. For instance, the League of Arab States established the Council of Arab Ministers Responsible for the Environment (CAMRE) as a high-level institution to ensure proper coordination of environmental policies across the region. CAMRE aims to identify major environmental problems, set priorities and address issues related to sustainable environment. However, deteriorating environmental trends indicate a need for using additional policy instruments such as incentives, monitoring mechanisms, economic and environmental assessment tools, environmental education, and public awareness strategies.

Countries in West Asia rely heavily on regulatory strategies rather than market-based instruments. Although new initiatives to introduce policy mixes aimed at achieving higher levels of integration between different sectors do exist, they remain modest. As an illustration, for the last four decades water policies have mostly focused on supply strategies aimed at overcoming shortages through technical solutions including desalination. This supply-focused approach, enabled by the availability of strong financial resources within key countries, specifically the Gulf Cooperation Council States, has resulted in good progress in meeting the Millennium Development Goal on water supply and sanitation, particularly in urban areas (UNDESA 2011).

The ongoing prevalence of unsustainable patterns of demand and consumption is nevertheless resulting in depletion of water resources and widespread deterioration in water quality, a situation that also increases regional tension over shared sources. Integrated water resource management represents a promising approach to achieve resource sustainability (CEDARE and AWC 2004). In addition, fair pricing of water services is becoming widely understood as crucial for achieving better demand management. In Saudi Arabia, the government has launched a number of measures to reduce consumption by the agricultural sector. It initially limited domestic food production by decreasing the subsidy on diesel fuel and is gradually reducing the government's own purchase of local wheat. In 2009, it set a target to gradually eliminate domestic wheat production over an eight-year period, while increasing incentives and loans for installing modern irrigation systems, providing subsidies for animal feed imports, banning the export of fodder and establishing strategic food reserves (AFED 2010; Hussain *et al.* 2010).

Land degradation and desertification are linked with a number of challenges including food production, biodiversity loss, deterioration of water resources and climate change. National action plans to combat land degradation and desertification therefore need to be better integrated with sustainable approaches to natural resources, biodiversity conservation efforts and climate change initiatives (Ministry of Municipalities Affairs and Land Use Planning 2010).

Many West Asian nations are currently implementing strong coastal development plans. However, regional authorities still need to confirm their commitment to protect coastal and marine ecosystems through the application of ecosystem-based management. The protection of marine and coastal areas from climate change, oil spills and land-based sources of pollution still represent important challenges for this region. The development and implementation of strategies such as crisis management and risk assessment represent very significant tools for climate change adaptation and for the protection of marine environments. The establishment of marine protected areas and the application of integrated fisheries management represent promising solutions to enhance marine biodiversity conservation (Sheppard *et al.* 2010; Price 2002).

West Asia has vast and valuable renewable energy resources but the energy sector is still characterized by heavy reliance on fossil fuels, which results in adverse environmental impacts and

In an effort to reduce damage to coastal and marine ecosystems, Oman has established two marine protected areas. *© iStock/Steven Allan*

high carbon intensity. Policy efforts to promote energy efficiency and renewable energy to mitigate climate change are evolving. However, to achieve global goals and to develop sustainable energy systems, the region still needs to strengthen its legislative and institutional frameworks. In particular, the building sector in West Asia is a major energy consumer, largely due to a significant demand for air-conditioning. A switch to green building practices is emerging through the adoption of building energy efficiency codes, which, along with renewable energy development, represents great opportunities for the entire region (Ministry of Public Work and Housing 2009).

COMMONALITIES

GEO-5's regional assessments, presented in Chapters 9–14, identify policy responses and instruments based on best practice. Common threads can be traced between and among the regions with particular policy approaches proving successful in a number of cases. Those policy responses adopted successfully in more than one region have a greater likelihood of accelerating achievement of internationally agreed goals.

Successful policy tools and instruments

Environmental governance

At the regional and global levels, environmental governance has evolved into a set of organizations, policy instruments, financing mechanisms, rules, procedures and norms that regulate the processes of environmental protection.

Absent or inadequate governance is one of the major issues in sustainable development, and many proactive efforts are being made to overcome these barriers, including multi-level/multi-stakeholder participation; increased introduction of the principle of subsidiarity; governance at local levels; policy synergy and removal of conflict; strategic environmental assessment; accounting systems that value natural capital and ecosystem services; improved access to information, public participation and environmental justice; capacity development; and improved goal setting and monitoring systems.

Climate change

A major concern for many countries is how to build resilience, especially in the most vulnerable communities, to climate change impacts already set in motion by past greenhouse gas emissions. Policies are oriented to climate change mitigation and adaptation and disaster risk reduction.

Promising climate change policies already under implementation include the removal of environmentally harmful subsidies, especially on fossil fuels; carbon taxes; forestry incentives for carbon sequestration; emissions trading schemes; climate insurance; capacity building and financing; and climate change preparedness and adaptation such as climate proofing infrastructure.

Energy

Legislative and institutional frameworks for developing sustainable energy systems are needed to achieve global goals.

Successful policies include increased international cooperation in the area of transfer and application of energy-saving technologies; promotion of energy efficiency; increased use of renewable energy; feed-in tariffs; restriction on fossil fuel subsidies; low emission zones within cities; and research and development, especially on batteries and other forms of energy storage.

Air pollution

Europe was the only region to select air pollution as a priority theme and perform an appraisal of policy options.

Successful policies include fuel and vehicle emission standards; control of industrial pollution through technical emission controls, best available techniques, fuel switching and reduced sulphur content in liquid fuels; and local air quality management plans including adequate monitoring and information systems and appropriate institutional mandates for local authorities.

Land

Land policy has a role in preventing environmental degradation and its social and economic costs.

Clear and protected rights, and effective rules defining access and regulating land, water and other natural resource use, are all essential means of ensuring long-term sustainable land and resources management. Successful policy options include integrated watershed (catchment) management; resource-efficient urban growth; protecting prime agricultural land; improved forest management; payment for ecosystem services and REDD+; and agroforestry and silvo-pastoral practices.

Freshwater

The equitable and sustainable management of freshwater is a major challenge for all water users, with most governments, from the local to the international level, facing the need to realign the availability of water with human and economics-based demand at levels that also maintain ecosystem integrity and environmental sustainability. In large part, this realignment requires the integration of environmental considerations, alongside domestic, agricultural and industrial requirements, into the drafting and implementation of national and international policies and legislation. Given that environmental considerations have historically been deemed secondary or even of no priority in decision-making related to the allocation and management of freshwater resources, the realignment will have to focus, at least initially, on expanding the attention paid to environmental concerns in the existing people-centred processes (UNEP 2010).

Policies identified as successful across the regions include integrated water resources management; the conservation and sustainable use of wetlands; promotion of water-use efficiency; water metering and volumetric-based tariffs implemented at a national or sub-national level; recognizing safe drinking water and sanitation as a basic human right/need; and effluent charges.

Oceans and seas

Polices such as integrated coastal zone management and marine protected areas, and economic instruments such as user fees have provided a level of success in terms of management.

Biodiversity

Biodiversity policies promote the protection, conservation and sustainable use of biologically diverse ecosystems and habitats. In doing so, they create significant public benefits and contribute to social well-being.

Successful policy instruments include market-based instruments for ecosystem services, including payment for ecosystem services and Reducing Emissions from Deforestation and Forest Degradation (REDD+); increasing and improving the management of protected areas; establishing transboundary biodiversity and wildlife corridors; community-based participation and management; and sustainable agricultural practices.

Chemicals and waste

Important international legal instruments and frameworks have been adopted with regard to the sound management of hazardous chemicals and wastes, including such policies as the registration of chemicals; extended producer responsibility; product redesign and design for the environment; life cycle analysis; reduce, reuse and recycle – the 3Rs – alongside cleaner production; national and regional hazardous waste treatment systems; and control of inappropriate export and import of hazardous chemicals and waste.

Applying policies in a more effective way

Many of the policies selected as promising are based on well-studied and accepted management concepts such as integrated water resources and coastal zone management, and protected areas. However, some common conclusions were identified across the regions, indicating that the application of these management concepts can be innovative if certain principles are adhered to.

Wastewater treatment aims to remove as much of the suspended solids as possible before the remaining water – or effluent – is discharged back to the environment. *© Christian Uhrig/iStock*

Cross-cutting policies across themes and sectors

The selected policy options were often identified as being promising because they are mutually reinforcing, with positive impacts in more than one thematic domain.

- "It is important to maximize opportunities by focusing on options that are mutually reinforcing and cross-cutting." (Chapter 9 – Africa)
- "The coherent application of effective policies across themes and sectors can bring major benefits in terms of an improved physical environment and a healthier population." (Chapter 11 – Europe)
- "To be sustainable, the region's natural capital needs to be managed in an integrated fashion across sectors." (Chapter 12 – Latin America and the Caribbean)
- "Failure to introduce sectoral policy integration, policy mixes and regional integration will intensify currently unsustainable consumption and production patterns, especially for energy, water, food security and marine resources, with the potentially grave consequences of natural resource depletion and increased pollution, which in turn impact human health and well-being." (Chapter 14 – West Asia)

Addressing drivers

There is increasing awareness among practitioners of a need to shift attention away from the effects of environmental degradation and instead focus on the underlying drivers.

- "Policy responses are beginning to shift from a focus on environmental impacts to addressing the key drivers through market- and information-based approaches." (Chapter 10 – Asia and the Pacific)
- "Until policies begin to address some of the deeper, underlying causes of environmental degradation – or drivers

– countries are unlikely to meet the goals and targets set out in international, regional and national agreements." (Chapter 12 – Latin America and the Caribbean)

Monitoring, evaluation and accountability

Monitoring and evaluation can be used to improve policy design, increase the accountability or ownership of stakeholders and identify promising practices that can be applied subsequently or in other country settings.

- "Investing in monitoring and evaluation, as well as social learning, supports the revision and modification of policy responses. Strong accountability helps secure government and private-sector commitment to implementation and to achieve agreed outcomes (Najam and Halle 2010). Developing performance indicators rather than effort-based indicators, such as the number of meetings held, improves clarity about how and to what extent the purpose of the policy is being achieved (Najam and Halle 2010). Strong and effective national and sub-regional reporting systems help hold implementing agencies to account and provide an opportunity to document successes, which in turn set the basis for up-scaling and replication." (Chapter 9 – Africa)
- "Improved monitoring and data collection, and access to information and legal redress, have the potential to alter the drivers of environmental change and unsustainable development." (Chapter 10 – Asia and the Pacific)
- "Enabling conditions that would increase policy success and replication include more efficient monitoring systems." (Chapter 11 – Europe)
- "Performance indicators are necessary to evaluate policy progress and clearly identify successes and shortcomings." (Chapter 13 – North America)
- "The systematic collection, processing, analysis, production, dissemination and exchange of environmental information would lead to sound decision making and proper policy formulation and implementation." (Chapter 14 – West Asia)

Transboundary cooperation

Natural areas shared by neighbouring countries are not only a common treasure, but also a common responsibility. They can be either a source of conflict or a source of cooperation and prosperity.

- "Cooperation has been shown to be effective for achieving sustainable management, including policy options for transboundary coastal and land-based resource management, and where there are multiple stakeholders. This has improved equity, enhanced skills sharing and reduced conflict." (Chapter 9 – Africa)
- "Efforts to enhance the sustainability of forests through management face a lack of national capacity and awareness, and intensifying competition in international forest product markets. There is therefore an urgent need for transnational coordination to address common and cross-border issues (Hogl 2002)." (Chapter 11 – Europe)
- "Cooperation is an important element in improving sustainability in the region. Cooperation between its countries will facilitate the sharing of information, expertise and technology transfer, the lack of which may currently limit countries in moving to more sustainable paths of development. It could also help to improve the management of transboundary ecosystems and species." (Chapter 12 – Latin America and the Caribbean)

Multi-stakeholder participation at local and national levels

The benefits of involving stakeholders in decision-making processes have been acknowledged. They include opportunities to share views, needs and knowledge; build consensus; enable participants to influence outcomes; and build commitment and a feeling of ownership to enhance and ensure implementation.

- "Several of the options presented, including sustainable land management, show that a high degree of participation at local and government levels helps to ensure relevance, with good outcomes for strengthening sustainability. Decentralization and devolution policies, including in community-based resource management, have achieved positive outcomes for communities and for the environment." (Chapter 9 – Africa)
- "Successful implementation of policies requires the establishment of a planning framework for adaptive and integrated management of water resources, under which appropriate pricing and multi-stakeholder participation are essential. Governance improvements are critical to enhanced accountability as a means of achieving sustainable development." (Chapter 10 – Asia and the Pacific)
- "Enabling conditions that would increase policy success and replication include a more active civil society engaged through awareness raising and strong multi-stakeholder agreements." (Chapter 11 – Europe)
- "The standard governance principles and values of transparency, accountability, equity, sustainability and inclusive participation of all stakeholders are fundamental for strengthening governance frameworks." (Chapter 12 – Latin America and the Caribbean)
- "Active stakeholder involvement, with explicit discussion of issues, improves decision making and acceptance, offering advantages over top-down planning, which often lacks public support and understanding." (Chapter 13 – North America)
- "Environmental governance, rather than merely focusing on environmental policies, needs to take account of societies' common goals and engage with various stakeholders in the design and execution of policies." (Chapter 14 – West Asia)

CHALLENGES AND OPPORTUNITES

Traditionally, policy analyses have been conducted in the context of a specific plan, programme or project at the local or national level, focusing on economic and social costs and benefits and involving specific stakeholder groups. But policy analysis now faces new challenges as rules and norms adapt to changing expectations (Hajer 2003). The *GEO-5* policy mandate aimed to extend the scope of analysis and appraisal to identify successful environmental policies in the regions and to highlight their potential to speed up the achievement of internationally agreed goals. The analysis is intended to provide policy makers with promising avenues for exploration.

It is not certain whether the policies identified and appraised are the optimal choices with respect to the selected international goal, although there is evidence of their effectiveness. Furthermore, clusters of policies were identified, not just individual ones, acknowledging that most policies are implemented as part of a complementary package. Whether a policy could be effective in a different context or on a different scale is uncertain: for many policies, direct evidence of the specific reasons for their effectiveness is limited, as is evidence of the potential for their transferability and replication. Undoubtedly, political will remains an essential ingredient for success. However, direct causality is always difficult to isolate in dynamic systems such as societies. Inconvenient variables can be difficult to eliminate, and convenient ones usually cannot be introduced spontaneously. So experimentation and observation must continue despite inherent problems.

Information and indicators

Investing in and generating environmental knowledge and translating it into information that can be used in governance and policy development are essential for management success (Adger *et al.* 2005), and require a better interface between science, policy makers and communities. To influence policy and decision making effectively, environmental information should be transformed into scientifically derived, easily understood indicators that convey clear messages to policy makers and the public (UNESCO-SCOPE 2006; Cimorelli and Stahl 2005).

Policy instruments designed to increase accountability and transparency seek to make information on environmental performance and the impacts of resource use more widely available, facilitating decision making and mobilizing a variety of stakeholders. Relevant information and indicators also help in monitoring and evaluating the effectiveness of policies and determining whether they allowed management approaches to adapt to new conditions – these are important elements of good environmental governance. At regional and sub-regional levels mechanisms for sharing information and knowledge could be better utilized.

Environmental impact assessment for individual projects, cumulative impact assessment for series of projects, and strategic environmental assessment for policies, plans and programmes all provide essential information (World Bank 2006). Documented cases of policies that appear to have made a contribution to achieving environmental goals can also be valuable. The promising policies described are illustrated by case studies, which provide insights into the context within which they were successful.

Transboundary and regional cooperation

Environmental problems do not respect borders. The transboundary nature of environmental degradation is the result of scale, as pollution or damage affect larger and larger areas. Solving transboundary environmental problems can provide an opportunity to establish regional cooperation.

Addressing shared environmental problems through regional cooperation can support the transfer of innovative solutions based on common characteristics of terrain, climate, economic activities and history, among others, which increases the likelihood of success. Transboundary approaches often enhance cooperation and reduce conflict by facilitating dialogue, establishing networks and encouraging learning and knowledge sharing. This helps create the political stability needed for economic and development cooperation.

Other benefits of cooperation are an augmentation of national efforts, the transfer of capacities and conservation efforts involving several stakeholders across borders. The key challenges are sustainability, the differing capacities of the institutions involved and the political nature of cooperation whenever sensitive sovereign issues arise.

Chapters 9–14 offer many examples of successful transboundary initiatives:

- In Africa, although there are significant variations in focus, structure, delivery and scope, the rapid increase in transboundary natural resource management demonstrates that this policy, despite some challenges, has high potential for replication and for managing Africa's diverse shared ecosystems.
- In Asia and the Pacific, transboundary collaboration fosters the cooperation of national institutions to the benefit of multiple countries, as demonstrated by several examples involving cross-boundary interest in protecting areas with high levels of biodiversity such as the Greater Mekong sub-region, the Terai Arc landscape in India and Nepal, Sulu-Sulawesi marine areas and the Coral Triangle.
- The transboundary nature of most European rivers calls for close international cooperation, and integrated water resources management is increasingly the guiding mechanism for implementation.
- In Latin America and the Caribbean, transboundary cooperation and integration in the energy sector have been shown to increase electricity supply, widen coverage and enhance system functionality.
- The International Watersheds Initiative, which was conceived by the governments of Canada and the United States, promotes the establishment of watershed authorities and facilitates integrated transboundary watershed management.
- In West Asia, there are conflicting national interests in forging equitable sharing agreements for transboundary resources. These issues, however, could be addressed through integrated water resources management supported by strong commitment by decision makers to place water high on the political agenda.

Challenges specific to each region influence policy approaches

Each region reflects varying characteristics – population, area, level of internal cohesiveness regarding shared history and culture, language, wealth distribution and education. Questions of political will, economic capacity, history and other intangibles persist, with the regions weighing these differently.

In Africa, population growth, rapid urbanization, climate change, unsustainable development choices and weak governance persist as critical challenges to achieving both the environmental and the social aspects of important regional goals. Addressing human well-being was taken as a point of departure for strengthening environmental policy and implementation. The support of donors has been crucial to the implementation of some policies. The principles of the Paris Declaration on Aid Effectiveness – ownership, harmonization, alignment, management for results and mutual accountability – define collaboration with donors and are designed to ensure that aid supports agreed government priorities and uses, and strengthens, government systems rather than developing parallel institutions.

The Asia and Pacific region has become a global engine of economic growth, but this success has come at the cost of some of the planet's most threatened ecosystems. Many of the policies being adopted in the region had their origin and initial trials in other regions, often Europe and the United States. The failure to implement many of these policies successfully may stem from the assumption that if a policy works in a developed country then it should also work in a developing one. For example, the strong command-and-control policy regime to manage air and water pollution in the United States, involving standard setting, permits and prosecution of offenders, tends not to work as well in the developing countries of Asia and the Pacific (AECEN 2004). A policy regime built around voluntary compliance, the social pressures of naming and shaming polluters, and compensation where appropriate may be more applicable for the socio-cultural context of the region, although measures of effectiveness require further analysis.

In Europe, concerns about long-term threats to the environment and human health persist, the latter especially for its large urban population (EEA 2010). Despite some successes in decoupling environmental pressures from economic growth, Europe's environmental footprint remains disproportionately high due to the continued unsustainable use of natural resources, both within and outside the region, to satisfy the high consumption and production level of its inhabitants (Chapters 1–7) (EEA 2010). To deal with these trends an integrated policy approach is required, for which strong governance mechanisms need to be in place. Given that Central and Western Europe in particular have a dense network of political boundaries, a regional focus to tackle environmental issues is necessary, with focus on transboundary as well as global environmental decision making. Regular monitoring, reporting and assessment required by legislation are an integral part of EU environmental governance.

Latin American and Caribbean countries face many challenges in managing their rich natural resources. Population growth, as well as unsustainable global and regional production and consumption patterns, drives the increasing demand for and extraction of raw materials and other natural capital (Chapter 1). This has led to the extensive conversion of natural environments to productive systems, with impacts on the region's biodiversity. To be sustainable, the region's natural capital needs to be managed in an integrated fashion across sectors. To respond to the complex nature of the region's environment, its opportunities and challenges, policies should be designed and implemented in ways that transcend the traditional compartmentalized, sectoral approach. This will help the region deal with some of its persistent environmental and associated socio-economic problems, including poverty, inequity and social conflict.

In North America, changes in regional demographics, rapidly emerging global economies and resource constraints all challenge the countries' provision of public goods and services. At the same time, fragmented governance, policy instability, lack of clear targets and science policy, and the dilemma of whether to address global issues rather than seeking local solutions, hampers the achievement of environmental goals (Chapter 1). Federal governments are no longer the primary leaders in setting the policy agenda or devising innovative policy instruments, yet they remain essential to the ultimate success of those policies, help ensure harmonization across jurisdictions and prevent the development of environmental inequities. In addition, there is a strong tendency to favour market-based instruments because of early successes, and to overlook traditional regulatory instruments. Finally, relative federal disengagement has opened the door to policy initiatives and innovations at sub-national levels, by states and provinces or municipalities, as well as to regional transboundary cooperation. The latter is extensive and continues to expand, and its dynamics are further supported by the Commission for Environmental Cooperation, which oversees the environmental accord of the North American Free Trade Agreement (NAFTA).

The drivers of environmental change in West Asia are linked to peace and security, demography and the state of the economy. The international desire to secure valuable energy resources and disputes, including current political conflict, play a major role in ongoing environmental degradation. Environmental damage is escalating and the number of displaced people is increasing, straining the environment and contributing to the degradation of land and water resources (UNEP 2010). Although environmental policies in the region have developed over the past two decades and continue to progress, they need to become proactive rather than reactive. Additionally, environmental governance, rather than merely focusing on environmental policies, needs to take account of societies' common goals and engage with various stakeholders in the design and execution of policies. The integration of sectoral policies is also important. Regional environmental governance is crucial for the region's countries share many common environmental conditions.

Challenges in policy replication

There is a degree of scepticism as to whether a policy can be successfully reproduced and applied in different circumstances and for different stakeholders with different needs and expectations. The differing governance contexts and enabling environments in a region as diverse as Asia and the Pacific, for example, may be barriers to adoption. However, as suggested in Chapter 10, there is sufficient experience with several of the priority policies analysed to justify faster replication.

The Asia and Pacific region (Chapter 10) considered the following factors when evaluating the potential for replication of policies:

- how many countries have already implemented such policies;
- how quickly the policies have been adopted by multiple countries since their first introduction;
- how easily the private sector has been convinced that the policies are not harmful to their businesses; and
- how the policies have contributed co-benefits that made them even more acceptable.

Part of the above analysis relates to the enabling and/or impeding factors that have led to the success or otherwise of specific policies. Europe (Chapter 11) identified the following enabling conditions for policy success and replication:

- more policy coherence, streamlining and simplified procedures that enhance cost efficiency and effectiveness;
- more efficient monitoring systems;
- stronger long-term commitment on the part of politicians and governments;
- stronger enforcement;
- transnational coordination to address common and cross-border issues;
- stronger private-sector involvement by creating and making better use of markets; and
- a more active civil society engaged through awareness raising and strong multi-stakeholder agreements.

The Latin America and Caribbean region (Chapter 12) has developed and implemented good examples of policies and approaches, usually at national and sub-national levels, that offer opportunities for replication both within and outside the region. Their characteristics usually include the effective incorporation of scientific information, knowledge and best practice, links across sectors, strong governance mechanisms, stakeholder participation, and political will and support.

The potential for transferability and replication of policies identified in North America (Chapter 13) is not straightforward and is dependent on context and specific instrument design. For example, the institutional framework of the North American electricity grid is highly fragmented, while many other countries have nationally owned networks (Willrich 2009). Denmark, France, Germany, Italy and Japan all have experience with feed-in tariffs at the national levels, while the United States and Australia have experience of production tax credits and renewable portfolio standards (IEA 2011). Policies on feed-in tariffs and renewable energy portfolio standards are in force in a variety of jurisdictions including Canada, China, Kenya, Portugal and Uganda (IEA 2011). Statistically, correlations demonstrate that the policies are effective, particularly in the case of feed-in tariffs (Haas *et al.* 2011). Direct causal evidence of effectiveness of other policies, however, is limited, as is evidence of their potential for transfer to and replication in other jurisdictions (Carley 2009; Doris *et al.* 2009).

In West Asia (Chapter 14), an excellent policy in one country does not usually stand alone and, as such, cannot easily be transferred or successfully replicated in its original form (UN ESCWA 2007). New circumstances, new management and various interdependent problems such as poor and low implementation capacities, lack of financial resources and marginalized local stakeholders, can make many successful programmes lose their effectiveness when replicated. The Council of Arab Ministers Responsible for the Environment (CAMRE) played, and continues to play, a major role in the coordination of environmental policies of the Arab countries at regional and global levels and has ensured a certain level of replication of environmental policies among West Asian countries.

Enforcement officers on a survey mission at the Myanmar-Thailand border discuss joint action to combat transnational crime through regionally organized law enforcement, October 2011. *© UNODC*

On the other hand, examples of successful replication from within and across regions provide possible tools for ensuring further success. Sharing experiences between practitioners and stakeholders may be a first step to a better understanding of the specific conditions in which a policy has been successful, and whether it could be replicated in another context, and to what extent.

Compliance and enforcement

However admirably designed, sensitively implemented and wisely administered, policies emerging from internationally agreed goals must be enforced to ensure continuity and accountability, but decision makers rarely give sufficient priority to this. Political will and leadership are required at all levels of implementation and enforcement. Compliance regimes can be more effective if the authority with responsibility is clearly identified and transparency is maintained to allow both higher levels of government and citizen-stakeholders to understand where the enforcement falls short. In this way, policy success is achievable.

Future work

Efforts to make the greatest possible use of quantitative evidence of policy effectiveness have highlighted opportunities for future work. In particular, there is a great need for governments and organizations to increase policy monitoring, for further research into policy effectiveness, and for the development of assessment methodologies that take greater account of synergistic policy effects and feedbacks.

PLANETARY POLICY PERSPECTIVES

The policies presented in *GEO-5* can be analysed for their utility to leverage societal transformation. Understanding the potential of these policies, alone or in combination, could help facilitate transformative change and enhance the effect that policy makers have on reaching sustainable development objectives at local, national, regional and international levels.

Feedback and adjustments – managing markets

These include price adjustment and other market instruments that are intended to reduce burdens and provide signals that correct or reinforce patterns of behaviour.

Example 1: Volumetric-based tariffs and water metering in Armenia (Chapter 11)

- Metering, cost-recovery tariffs and proper pricing structures stimulate more responsible water use while generating funds for the maintenance of the supply system.
- Various studies reveal that, on average, if individual metering systems are in place, reductions of 10–40 per cent can be achieved in household water use (Inman and Jeffrey 2006; Scheuer 2005).
- Soon after the reforms took place in Armenia, average water use decreased three to four times compared to use based on flat-rate calculations. The massive process of introducing individual metering became a trigger for a chain of water sector improvements, all backed by a legal, regulatory and institutional framework that enabled private-sector involvement accompanied by investment and management efficiencies. As a result, the quality and reliability of water delivery improved.

Example 2: Carbon taxes in British Columbia (Chapter 13)

- The revenue-neutral carbon tax in British Columbia phased in rate increases starting at a modest US$10 per tonne of CO_2-equivalent in 2008, increasing at a rate of US$5 a year thereafter to US$30 a tonne in 2012.
- The tax's revenue neutrality is achieved by allowing tax reductions for businesses as well as for poorer sections of society who also receive payments.
- The tax applies to emissions from fossil fuels, accounting for approximately 70 per cent of the province's total emissions (emissions from fossil fuels exported from British Columbia to other jurisdictions are exempt).
- Addressing the drawbacks typically associated with carbon taxes may have enhanced the acceptability of British Columbia's policy. This includes mitigating or eliminating the potentially regressive nature of carbon taxation by its comprehensive coverage combined with targeted tax reductions, and reducing potentially large adaptation costs for carbon-intensive industries through a gradual phase-in of the tax.

Rules and incentives – state action

The creation and administration of rules is a key policy leverage point, as these have direct influence on and power over the actions of individuals and groups.

Example 1: The Free Basic Water Policy of South Africa, allowing households to benefit from free, secure access to water (Chapter 9)

- The South African constitution provides a right of access to sufficient water, implemented through the Free Basic Water Policy. Many impoverished households benefit from secure access to at least 25 litres of water per person per day within 200 metres of the household for domestic use (Mehta 2005). This is in line with the World Health Organization's recommendation for minimum consumption, though it does not cover broader health and livelihood needs.
- Positive outcomes include a saving of the time and effort women and girls spend collecting water, freeing them to engage in other activities, removing the need to resort to unprotected water sources and reducing vulnerability to water-borne disease (Mehta 2005). In addition, citizens directly attribute such policies to good governance, and this in turn supports long-term political stability.
- A major challenge for the policy is to strike a balance between the human benefits and the cost (DWAF 2002). However, improvements in human well-being are seen as outweighing the associated costs (Stalk 2004).
- Failure to provide the legally guaranteed quantity has resulted in citizens instituting litigation.
- Critical enabling factors include addressing cost recovery, identifying target groups, ensuring financing, managing demand and facilitating the expansion of infrastructure.

Example 2: Energy conservation in buildings in Kuwait (Chapter 14)

- Demand for electrical power in Kuwait has progressively increased, particularly in the past two decades. As all electricity generation depends on fossil fuels, power plants consume about 55 per cent of Kuwait's total primary energy. In addition, 85 per cent of electrical peak power and 60 per cent of the country's total annual output is used for air-conditioning and lighting in buildings.
- The Ministry of Energy in Kuwait launched its energy code for buildings in 1983 with a set of mandatory standards and regulations to enhance energy conservation and decrease the progressive negative impacts on the climate.
- The main objectives of the building code, which is applied to new and retrofitted air-conditioned buildings, are to decrease the capacity of air-conditioning systems and to reduce peak power demand by introducing smaller units.
- Implementation of the energy code has saved Kuwait nearly US$10 billion over the past two decades.

Mindsets – civil society

Measures that shift the paradigms guiding individual behaviour create shared mindsets that translate into vision, goals and collective action.

Example 1: Payment for ecosystems services – including the costs of environmental and resource use in the value of ecosystems (Chapter 12)

- In general terms, payment for ecosystem services schemes offer incentives, usually monetary, to individuals to protect and ensure the delivery of key ecosystem services at the local, national and regional levels.
- The mechanism can address many of the driving forces of biodiversity loss, especially habitat loss and unsustainable land management, as it usually aims to protect and/or rehabilitate natural vegetation.
- Payment for ecosystem services can be used in conjunction with other policies such as protected areas, integrated water resources management, conservation and restoration of water-supplying ecosystems, sustainable forest management, small-scale agro-ecological systems and the recovery of degraded lands.
- Lack of information on economic valuation highlights the need to invest more in research and further the scientific understanding of local environmental conditions.

Example 2: Participation in the management of natural resources in India and Nepal (Chapter 10)

- In India, about 22 million hectares of forests are included in the Joint Forest Management programme, under which more than 100 000 committees formed by forest-fringe communities protect patches of state-owned forest, receiving a share of forest resources in return (MOEF 2009a).
- In conjunction with stringent legislation prohibiting the use of forest land for non-forestry purposes, forest cover has stabilized after decades of rapid deforestation (MOEF 2009b).
- Additional incentives for participation have been created by a constitutional amendment that mandates decentralization and devolution of power to local authorities at village, intermediate and district levels (MLJ 2011).
- In Nepal, more than 14 000 community forest user groups have access to fuelwood and fodder, and are additionally provided with income-generating opportunities (DoF 2011).

CONCLUSION

This summary shows that in response to common concerns in the regions over freshwater, climate change and environmental governance, there are a variety of successful responses ranging from local actions to guarantee water quality, through transboundary agreements to tackle concerns affecting shared resources, to national programmes designed to shift the behaviour of whole economic sectors.

Improvements in governance are crucial if global goals are to be achieved, requiring the integration of sustainability concerns across all policy areas. Key elements, highlighted by the regions, include better enforcement at all appropriate levels of government, improved monitoring and data collection, greater access to information, increased multi-stakeholder participation and capacity building.

Whether a policy is replicable in a different context and on a different scale remains uncertain. Direct evidence of specific reasons for effectiveness is limited, as is evidence of the potential for transferability. The effectiveness of a specific policy or instrument and its potential to be scaled up or implemented elsewhere depends on significant context-specific variables, on the nature of the specific environmental issue, and on the existence of detailed and rigorous analyses of factors associated with its effectiveness in terms of behavioural change.

Nonetheless, common elements can be seen among the regions, with particular policy approaches proving successful in a number of places. Some of these, such as integrated approaches to water resources and coastal zone management or the creation of protected areas, draw on well-established concepts. But significantly, where policies have been successfully replicated, there is evidence that their application has taken significant account of local cultures, conditions and needs.

Responses at the local, national, regional and international levels interact and generate incremental, structural and transformational change. There is no universal solution to environmental degradation and a range of responses is required to address the diversity of regional needs. However, on issues of common global concern, coordination, participation, and cooperation are vital for meeting jointly internationally agreed goals and targets, while also addressing the capacity deficits of a range of countries.

Better progress in achieving internationally agreed goals can be made if policies apply leverage at the most advantageous place. Examples of promising policies identified in the regional chapters can be found at all leverage levels, namely feedback and adjustments in managing markets; rules and incentives in state action; and mindsets at the level of civil society.

REFERENCES

ADD (2011). *Mangrove Propagation at Le Morne with the Active Participation of the Vulnerable Local Communities and Preparation of a GIS Map Highlighting Potential Sites for an Island-wide Mangrove Restoration Programme.* ADD/MCB-FF Project Third Interim Quarterly Report. Association pour le Développement Durable, Mauritius. http://www.addmauritius.org/GEO%20 5%20Third%20MCB%20FF%20report.doc (accessed 11 November 2011)

ADD (2009). *Improving the Livelihood and Welfare of Artisanal Fishermen and Other Coastal Communities in Le Morne Village.* ADD/DCP/EU Project Final Report. Association pour le Développement Durable, Mauritius. http://www.addmauritius.org/FINAL%20NARRATIVE%20 REPORT_sgw%201.doc (accessed 11 November 2011)

Adger, W.N., Huges, T.P., Folke, C., Carpenter, S.R. and Rockstrom, J. (2005). Social-ecological resilience to coastal disasters. *Science* 309, 1036–1039

AECEN (2004). *Environmental Compliance and Enforcement in Thailand: Rapid Assessment.* Asian Environmental Compliance and Enforcement Network, Bangkok. http://www.aecen.org/ sites/default/files/TH_Assessmemt.pdf (accessed 6 November 2011)

AFED (2010). *Report on the Arab Water Sustainable Management of Scarce Water Sources.* Report published by the Arab Forum for Environment and Development (AFED), Beirut

Ajonina, G., Tchikangwa, B., Chuyong, G. and Tchamba, M. (2009). The challenges and prospects of developing a community based generalizable method to assess mangrove ecosystems vulnerability and adaptation to climate change impacts: experience from Cameroon. In *The Relevance of Mangrove Forests to African Fisheries, Wildlife and Water Resources. Nature and Faune* (eds. Bojang, F. and Ndeso-Atanga, A.). Vol. 24 pp.16–25. Food and Agriculture Organization of the United Nations, Accra. ftp://ftp.fao.org/docrep/fao/012/ak995e/ ak995e00.pdf (accessed 29 November 2011)

Carley, S. (2009). State renewable energy electricity policies: an empirical evaluation of effectiveness. *Energy Policy* 37, 3071–3081

CCCCC (2011). *Mainstreaming Adaptation to Climate Change (MACC) Project.* Caribbean Community Climate Change Centre. http://www.caricom.org/jsp/projects/macc%20project/ macc.jsp (accessed 9 December 2011)

CEDARE and AWC (2004). *Report on the Status of the Water in the Arab Region.* Cairo Egypt. http://www.arabwatercouncil.org/administrator/Modules/CMS/SOW.pdf

Cimorelli, A.J. and Stahl, C.H. (2005). Tackling the dilemma of the science-policy interface in 14 environmental policy analyses. *Bulletin of Science Technology Society* 25, 276–284

Cisneros, J. and Lloret, P. (2008). El Fondo para la protecciòn del agua. Mecanismo financiero para la conservaciòn y el cuidado del agua en Quito, Ecuador. In *Seminario Internacional Cogestión de cuencas hidrográficas experiencias y desafíos USAID.* Quito http://orton.catie. ac.cr/repdoc/A2983E/A2983E11.PDF (accessed 9 December 2011)

Devyatkin, V. (2009). *Actual Ways of Improving Legislation of Russian Federation Towards Recycling of Industrial Wastes and Other Industrial Outputs.* Report to the Federation Committee of the Russian Parliament on Industrial Policy, 19.02.2009 (in Russian). Federal government-financed agency 'Research Center on resources efficiency and wastes management issues', Moscow

DoF (2011). *Status of Community Forest User Groups.* Department of Forests, Government of Nepal. http://www.dof.gov.np/index.php?option=com_jdownloads&Itemid=102&task=view. download&catid=7&cid=20 (accessed 6 September 2011)

Doris, E., McLaren, J., Healey, V. and Hockett, S. (2009). *State of the States.* National Renewable Energy Laboratory, US Government Printing Office, Washington, DC

DWAF (2002). *Free Basic Water: Tap into Life. Regulations and guidelines.* Department of Water Affairs and Forestry, Directorate of Interventions and Operations Support, Pretoria

EC (2009a). *The EU Climate and Energy Package.* European Commission, Brussels. http:// ec.europa.eu/clima/policies/package/index_en.htm (accessed 15 December 2011)

EC (2009b). *White Paper: Adapting to Climate Change – Towards a European Framework for Action.* COM(2009) 147 final. European Commission, Brussels. http://eur-lex.europa.eu/ LexUriServ/LexUriServ.do?uri=COM:2009:0147:FIN:EN:PDF (accessed 20 December 2011)

EEA (2010). *The European Environment: State and Outlook 2010. Synthesis.* European Environment Agency, Copenhagen

Ellerman, A.D. and Buchner, B.K. (2007). The European Union emissions trading scheme: origins, allocation, and early results. *Review of Environmental Economics and Policy* 1, 66–87

FAO (2010). *FAO Statistical Databases.* Food and Agriculture Organization of the United Nations, Rome. http://faostat.org (accessed 15 December 2011)

Forest Europe, UNECE and FAO (2011). State of Europe's Forests 2011. Status and Trends in Sustainable Forest Management in Europe. Ministerial Conference on the Protection of Forests in Europe, Oslo

GEF (2009). *Project Identification Form: Integration of Climate Change Risks into the Maldives Safer Island Development Program.* Global Environment Facility, Washington, DC

GEO Data Portal (2011). UNEP's online core database with national, sub-regional, regional and global statistics and maps, covering environmental and socio-economic data and indicators. United Nations Environment Programme, Geneva. http://geodata.grid.unep.ch (accessed 15 December 2011) (now called Environmental Data Explorer)

Haas, R., Resch, G., Panzer, C., Busch, S., Ragwitz, M. and Held, A. (2011). Efficiency and effectiveness of promotion systems for electricity generation from renewable energy sources: lessons from EU countries. *Energy* 36, 2186–2193

Hajer, M., (2003). Policy without polity? Policy analysis and the institutional void. *Policy Sciences* 36, 175–195

Hogl, K. (2002). Patterns of multi-level co-ordination for NFP-processes: learning from problems and success stories of European policy-making. *Forest Policy and Economics* 4, 301–312

Hussain, G., Alquwaizany, A. and Al-Zarah, A. (2010). Guidelines for irrigation water quality and water management in the Kingdom of Saudi Arabia: an overview. *Journal of Applied Sciences* 10, 79–96

IEA (2011). *Policies and Measures Databases.* http://www.iea.org/textbase/pm/index.html (accessed 20 May 2011)

IEA, OECD and World Bank (2010). *The Scope of Fossil Fuel Subsidies in 2009 and Roadmap for Phasing Out Fossil-Fuel Subsidies.* Joint Report prepared for G20 Summit, Seoul, 11–12 November 2010

Inman, D. and Jeffrey, P. (2006). A review of residential water conservation tool performance and influences on implementation effectiveness. *Urban Water Journal* 3, 127–143

IPSRM (2010). *Assessing Global Land Use and Soil Management for Sustainable Resource Policies.* International Panel for Sustainable Resource Management (IPSRM/UNEP), Paris, France

Jänicke, M. (2011). *The Acceleration of Innovation in Climate Policy. Lessons from Best Practice.* FFU Report. Freie Universität Berlin, Berlin

Kraberg, A.C., Wasmund, N. Vanaverbeke, J., Schiedek, D., Wiltshire, K.H. and Mieszkowska, N. (2011) Regime shifts in the marine environment: the scientific basis and political context. *Marine Pollution Bulletin* 62(1), 7–20

Krausmann, F., Gingrich, S., Eisenmenger, N., Erb, K.-H., Haberl, H. and Fischer-Kowalski, M. (2009). Growth in global materials use, GDP and population during the 20th century. *Ecological Economics* 68(10), 2696–2705

Lenton, T.M., Held, H., Kriegler, E., Hall, J.W., Lucht, W., Rahmstorf, S. and Schellnhuber, H.J. (2008). Tipping elements in the Earth's climate system. *Proceedings of the National Academy of Sciences of the United States of America* 105(6), 1786–1793

Limburg, J.E., O'Neill, R.V., Costanza, R.C. ABD Farber, S. (2002). Complex systems and valuation. *Ecological Economics* 41, 409– 420

Liu, J., Daily, G.C., Ehrlich, P.R. and Luck, G.W. (2003). Effects of household dynamics on resource consumption and biodiversity. *Nature* 421, 530–533

McGee, G., Cullen, A. and Gunton, T. (2010). A new model for sustainable development: a case study of The Great Bear Rainforest regional plan. *Environment, Development and Sustainability* 12, 745–762

McNeill , J.R. (2000). *Something New Under the Sun: An Environmental History of the Twentieth Century.* Norton, New York

Mehta, L. (2005). *Unpacking Rights and Wrongs: Do Human Rights Make a Difference? The Case of Water Rights in India and South Africa.* IDS Working Paper 260. Institute of Development Studies, Brighton

Ministry of Municipalities Affairs and Land Use Planning (2010). The national strategy for sustainable agricultural development of the Kingdom of Bahrain. In *Seeds for OUR Future.* Manama

Ministry of Public Work and Housing (2009). *Energy Efficient Building Code.* Government of Jordan

MLJ (2011). *The Constitution (Seventy-Third Amendment) Act, 1992.* Ministry of Law and Justice, Government of India. http://indiacode.nic.in/coiweb/amend/amend73.htm (accessed 15 September 2011)

MOEF (2009a). *India's Fourth Report to the Convention on Biological Diversity.* Ministry of Environment and Forests, Government of India. http://moef.nic.in/downloads/public-information/in-nr-04.pdf (accessed 15 September 2011)

MOEF (2009b). *State of Forests Report 2009.* Ministry of Environment and Forests, Government of India. http://www.fsi.nic.in/india_sfr_2009/india_sfr_2009.pdf (accessed 15 September 2011)

Mohamed-Katerere, J.C. (2009). Climate change, natural resource governance and human security in Africa. Charting new paths. In *Natural Resource Governance and Human Security in Africa. Emerging Issues and Trends* (eds. Kesselman, B., Hughes, T., Kabemba, C., Matose, F. and Rocha, J.). Pax-Africa, Johannesburg

Najam, A. and Halle, M. (2010). Global environmental governance: the challenge of accountability. *Sustainable Development Insights* 005. Frederick S. Pardee Center for the Study of the Longer-Range Future, Boston University

Nations Online (2011). *Official and Spoken Languages of European Countries.* http://www. nationsonline.org/oneworld/european_languages.htm (accessed 19 September 2011)

NDRC (1998). *Annual Water Use Quota and its Distribution Scheme for the Yellow River.* 14 December 1998. National Development and Reform Commission and Ministry of Water Resources of the People's Republic of China

Price, A.R.G. (2002). Simultaneous 'hot spots' and 'cold spots' of marine biodiversity and implications for global conservation. *Marine Ecology Progress Series* 24, 23–27

Rietbergen, S., Hammond, T., Sayegh, C., Hesselink, F. and Mooney, K. (2007). *Island Voices – 38 Island Choices: Developing Strategies for Living with Rapid Ecosystem Change in Small Islands.* IUCN, Gland

Rockström, J., Steffen, W., Noone, K., Persson, Ö., Chapin, F.S., Lambin, E.F., Lenton, T.M., Scheffer, M., Folke, C., Schellnhuber, H.J., Nykvist, B., De Wit, C.A., Hughes, T., Van Der Leeuw, S., Rodhe, H., Sörlin, S., Snyder, P.K., Costanza, R., Svedin, U., Falkenmark, M., Karlberg, L., Corell, R.W., Fabry, V.J., Hansen, J., Walker, B., Liverman, D., Richardson, K., Crutzen, P. and Foley, J.A. (2009). A safe operating space for humanity. *Nature* 461(7263), 472–475

Rodionov, S., and J. Overland. (2005). Application of a sequential regime shift detection method to the Bering Sea ecosystem. *ICES Journal of Marine Science* 62(3), 328–32

Scheuer, S. (2005). Water. In *EU Environmental Policy Handbook: A Critical Analysis of EU Environmental Legislation* (ed. Scheuer, S.). European Environmental Bureau, Brussels

Sheppard, C., Al-Husiani, M., Al-Jamali, F., Al-Yamani, F., Baldwin, R. , Bishop, J. , Benzoni, F., Dutrieux, E., Dulvy, N.K., Durvasula, S.R.V., Jones, D.A., Loughland, R., Medio, D., Nithyanandan, M. , Pilling, G.M., Polikarpov, I., Price, A.R.G., Purkis, S., Riegl, B., Saburova, M., Namin, K.S., Taylor, O., Wilson, S. and Zainal, K. (2010). The Gulf: a young sea in decline. *Marine Pollution Bulletin* 60, 13–38

Stalk, A. (2004). *Management of the Free Basic Water Policy in South Africa.* Master project. Roskilde University, Roskilde

TerrAfrica (2009). *Enhancing the TerrAfrica Partnership.* http://www.unep.org/south-south-cooperation/exchangeplatform/Publications/GlobalMechanismTeamPublications/EnhancingtheTerrAfricapartnership/tabid/5780/Default.aspx. (accessed 11 September 2011)

The Economist (2011). Statistics and lies. http://www.economist.com/node/18333018. (accessed 15 November 2011)

UNDESA (2011). *The Millennium Development Goals Report – 2011.* United Nations, New York. http://www.un.org/millenniumgoals/11_MDG%20Report_EN.pdf (accessed 14 April 2012)

UNDESA (2010). *World Population Prospects, the 2010 Revision (WPP2010).* Population Division, United Nations Department of Economic and Social Affairs, New York. http://esa.un.org/wpp/unpp/panel_population.htm (accessed 15 December 2011)

UNECE (2011). *Sustainable Management of Water and Water-related Ecosystems.* ECE/ASTANA.CONF/2011/5. United Nations Economic Commission for Europe, Geneva. http://www.unece.org/fileadmin/DAM/env/documents/2011/ece/ece.astana.conf.2011.3.e.pdf (accessed 21 December 2011)

UNEP (2010). *Global Environment Outlook: Latin America and the Caribbean 3.* United Nations Environment Programme, Panama City

UNESCO-SCOPE (2006). *How to Improve the Dialogue between Science and Society: The Case of Global Environmental Change.* UNESCO-SCOPE Policy Brief No. 3. United Nations Educational, Scientific and Cultural Organization-Scientific Committee on Problems of the Environment, Paris. http://unesdoc.unesco.org/images/0015/001500/150009e.pdf (accessed 9 December 2011)

UN ESCWA (2007). *Land Degradation Assessment and Prevention: Selected Case Studies from the ESCWA Region.* United Nations Economic and Social Commission for Western Asia. United Nations, New York. http://www.arab-hdr.org/publications/other/escwa/landdegradation-07e.pdf

UN-Habitat (2010). *The State of African Cities 2010: Governance, Inequality and Urban Land Markets.* United Nations Human Settlements Programme, Nairobi

Willer, H. and Kilcher, L. (2011). *The World of Organic Agriculture. Statistics and Emerging Trends 2011.* IfOM, Bonn and FiBL, Frick

Willrich, M. (2009). *Electricity Transmission Policy for America: Enabling a Smart Grid, End-to-End.* Energy Innovation Working Paper Series. Industrial Performance Center – Massachusetts Institute of Technology, Cambridge

Winrock International (2011). *Payment for Forest Environmental Services: A Case Study on Pilot Implementation in Lam Dong Province Vietnam from 2006-2011.* Arkansas and Virginia

World Bank (2006). *Where is the Wealth of Nations? Measuring Capital for the 21st Century.* World Bank, Washington, DC

WRI (2010). *Modernizing Public Transportation: Lessons Learned from Major Bus Improvements in 27 Latin America and Asia.* World Resources Institute, Washington, DC

WSSD (2002). Johannesburg Plan of Implementation. World Summit on Sustainable Development. http://www.un.org/esa/sustdev/documents/wssd_poi_pd/English/poitoc.htm

Part 3: Global Responses

Chapter 16:
Scenarios and Sustainability Transformation

Chapter 17:
Global Responses

"Imagine our descendants in the year 2200 or 2500. They might liken us to aliens who have treated the Earth as if it were a mere stopover for refueling, or even worse, characterize us as barbarians who would ransack their own home. Living up to the Anthropocene means building a culture that grows with Earth's biological wealth instead of depleting it. Remember, in this new era, nature is us."

Paul J. Crutzen, Nobel Lauriate

CHAPTER 16

Scenarios and Sustainability Transformation

Coordinating lead authors: Begum Ozkaynak, Laszlo Pinter and Detlef P. van Vuuren

Lead authors: Livia Bizikova, Villy Christensen, Martina Floerke, Marcel Kok, Paul Lucas, Diane Mangalagiu, Rob Alkemade, Trista Patterson, John Shilling and Darren Swanson

Contributing authors: Andrea Bassi, Fabio Feldmann, Jill Jäger, Washington Ochola, Weishuang Qu, Kilaparti Ramakrishna, Claudia Ringler, Pinar Ertor (GEO Fellow) and Natalia Pervushina (GEO Fellow)

Chapter coordinators: Matthew Billot and Nalini Sharma

Main Messages

Meeting an ambitious set of sustainability targets by the middle of the century is possible but current supporting policies and strategies are not adequate to achieve this. Scenario studies show that without greater efforts to implement appropriate short-term policies, to shift investments to achieve necessary long-term structural changes, and to introduce behavioural transformations, it will not be possible to meet sustainability targets. These relate to international agreements on environmental protection and human development for issues like atmosphere and climate change, land and food security, water and biodiversity.

Transforming both consumption and production is important. Scenario studies suggest that targets can be met, but only if measures are taken to influence the levels and patterns of consumption and production. Most current policies focus on changes in production processes to achieve targets, but fail to address consumption. However, changes in consumption levels and patterns have great – but as yet unrealized – potential to reduce environmental pressures.

Effective implementation of wide-ranging technical and policy measures needs to be supported by a shift in underlying motivations and value patterns. Changes need to be both short and long term, and to combine technology, investment and governance measures along with lifestyle modifications grounded in a mindset shift towards values based on sustainability and equity. They also need to reflect regional differences and priorities. Technical measures alone are unlikely to be enough, and will not have the required level of societal support if not accompanied by transformations at all leverage points.

Accomplishing such complex transformations requires a gradual but steadily accelerating transition process. Some successful policy innovations are already happening, but need to be mainstreamed to be more effective. There is also a need to stop doing the things that pull the Earth System towards unsustainability. At the same time, it is important to provide resources, build capacity and create an enabling environment in a way that is consistent with visions of a sustainable world.

Broad-based social contracts grounded in jointly developed visions of a sustainable future would help to bring key stakeholders on board. The transition requires a high degree of consensus and coordination of action between social actors – governments, the private sector and civil society.

To ensure coherence, contextually sensitive transition pathways could be developed as joint visions of the future. These can be agreed on as informal or formal social contracts that respect the requirement to assure sustainable access to the resources necessary for human well-being.

The transition process needs to be based on adaptive management. Uncertainties play a key role in the problems of the Earth System. As a result, management should be based on learning-by-doing processes, periodic reassessment based on new learning, and a great diversity of measures. This will provide better insurance against wholesale failure on critical issues – due either to inherent uncertainties or inadequate implementation – and be mutually strengthening as well.

There is a need for clear long-term environment and development targets and for stronger accountability in international agreements. Given that environmental and societal Earth System changes can be slow, long-term visions and goals – expressed as social contracts – could help focus investments and technology development, induce societal change, and engage other actors in society.

INTRODUCTION

The nature and scale of changes described in Part 1 indicate that, without additional policies, the global environment will degrade further – from a situation that already raises considerable concern. A crucial question, therefore, is how to halt and reverse such trends.

While previous *Global Environment Outlook* (GEO) reports have explored several scenarios looking at very different futures (UNEP 2002, 2007), the emphasis of *GEO-5* is on the choices and strategies that could, from 2012, lead to a sustainable future. This is advanced by looking at two very different storylines based on a review of existing scenario studies:

- a view of the world in 2050 assuming business-as-usual paths and behaviours – "conventional world" scenarios; and
- an alternative that leads to results consistent with our current understanding of sustainability and agreed-upon goals and targets on the road to 2050 – "sustainable world" scenarios.

A key difference between the two is how deeply transformation occurs, supporting the emergence of alternative development trajectories (Figure 16.1).

The ambitious goals of this systemic transformation require increasing the power of collective thinking, creativity and coordination. Cultivating profound long-term change is neither a linear nor a simple process, especially in complex dynamic systems that often exhibit non-linear behaviour or tipping points (Lenton *et al.* 2008; Folke *et al.* 2002; Levin 1998). For this reason, knowledge of the component parts of a system, their relationships, interactions and emergent behaviour can help policy makers understand, anticipate and strategize outcomes for the longer run, even when evidence of those changes may not be immediately apparent. Sustainable world scenarios represent and require many fundamental shifts in society-environment interactions. These scenario outcomes are designed to be consistent with the best available science on the Earth System and the aspirations for environment and sustainable development manifested in multilateral agreements. They combine the effects of mainstreaming promising measures that already exist with change at progressively deeper structural levels.

Figure 16.1 Conventional world and sustainable world scenarios in *GEO-5*

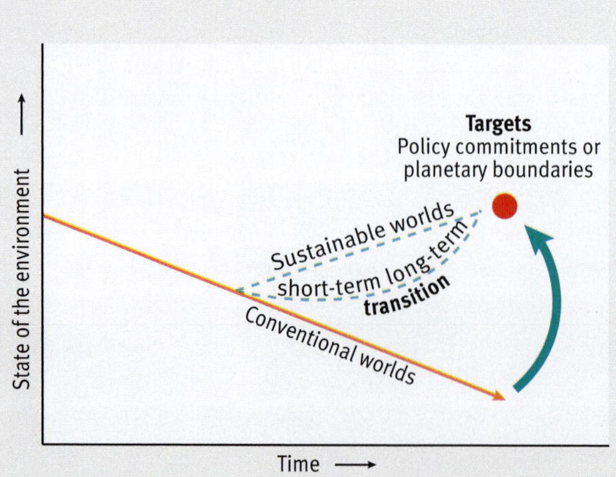

Source: Adapted from PBL 2012

A system is a set of things – here, people and ecosystems in the Earth System – that interact with one another within a defined boundary, and that produce their own pattern of behaviour over time. Complexity theory shows that small, calculated, strategically applied actions can effect great change. In a complex system, leverage points are those where the outcome is disproportionate to the input. Identifying and acting on effective leverage points is especially difficult in cases in which casting off old paradigms is involved, but once a leverage point has been pushed in the right direction, the resultant change can be particularly lasting and profound (Meadows 1999).

> *"Magical leverage points are not easily accessible, even if we know where they are and which direction to push them on. There are no cheap tickets to mastery. You have to work at it, whether that means rigorously analyzing a system or rigorously casting off your own paradigms and throwing yourself into the humility of Not Knowing."*
>
> (Meadows 1999)

Figure 16.2 depicts the transformative layers where leverage points may be found. The more embedded drivers are in the system, the more enduring and profound the change needs to be: shifting mindsets is at the core of transformation because it translates into visions, goals and collective action. Changing rules and incentives is literally a game changer in the pursuit of transformation, because the right incentives can create structural change and influence key drivers. The outer layer of transformation involves creating and listening to feedback and making adjustments to environmental pressures to maintain progress towards sustainability.

Shepherding a complex dynamic system through transition is not a linear or unidirectional process, and progress or setbacks occurring in each layer continuously influence the others. Thus a policy approach that diversifies strategy across all the layers of transformation presents a diverse and resilient portfolio for using leverage to its greatest advantage. The result is an integrated policy that produces both short-term and profound long-term system shifts by cultivating and targeting ever deeper system change over time, while monitoring it as well as feeding it with successes at more superficial levels. Clearly, producing these outcomes requires an ability to accept some uncertainty, while also holding a strong focus on integrating and documenting progress towards a clearly articulated vision of success. The section on visions, goals and targets on the road to 2050 articulates a vision of a desired state of the environment, with goals and targets based on existing international agreements. The section on sustainability pathways reviews existing

Figure 16.2 Layers of transformation

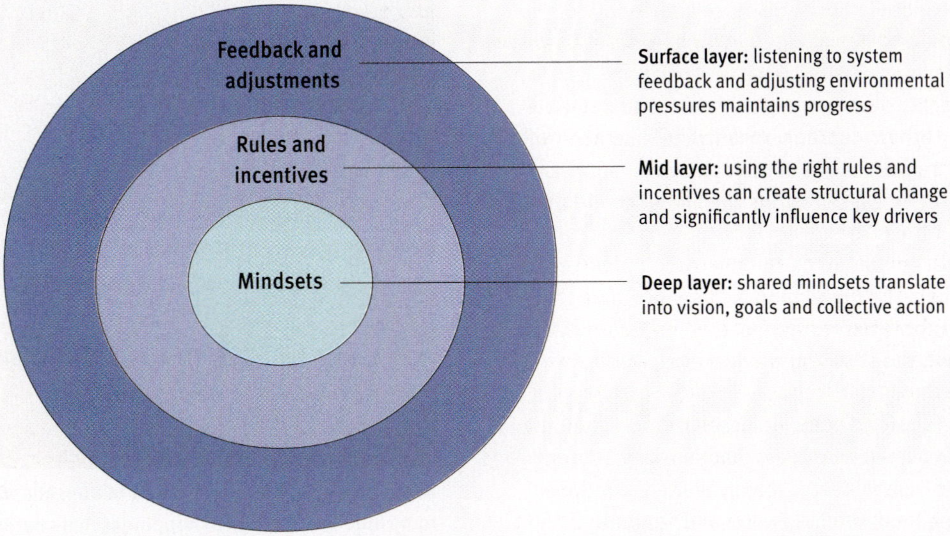

Source: Meadows 1999

scenarios to provide an outlook for the drivers of environmental change and the pathways that society could follow on the road to 2050 to achieve goals and targets. Over the last few years, a large number of scenario-based assessments of global environmental problems and human development have been published, including the Intergovernmental Panel on Climate Change (IPCC) climate assessments (Nakicenovic and Swart 2000), the *Global Environment Outlook* reports (UNEP 2007, 2002), the *Millennium Ecosystem Assessment* (MA 2005a), the *International Assessment of Agricultural Science and Technology Development* (IAASTD 2009b) and the *World Water Development Reports* (UNESCO 2009, 2006). Most of these used an explorative approach with widely diverging scenarios that assess what might happen in the future. Van Vuuren *et al.* (2011a) discuss many of the assessments and highlight emerging commonalities. Generally, the scenarios explore a wide range of possible outcomes but, importantly and by design, almost none involves meeting sustainability targets – or sets them out as an objective. In contrast, some visioning exercises carried out by various organizations, such as the World Business Council on Sustainable Development (WBCSD 2010), the International Geosphere Biosphere Programme (Jäger and Cornell 2011) and UNEP's *Green Economy Report* (UNEP 2011c), have tried to do so.

The section on advancing sustainability explores strategic elements that have the potential to advance the transition onto pathways consistent with visions of a sustainable world. Changing the current unsustainable course demands a scale of effort without precedent in human history (Steffen *et al.* 2005). Meeting this challenge will require a diverse portfolio of strategies and measures, partly as a form of insurance against failure, but also to reflect the different and dynamically changing condition of individual countries and ecosystems around the world (Innes *et al.* 2005; Speth 2005). Taking into account recent scientific advances in understanding the functioning and governance of closely coupled human systems and ecosystems, this section provides guidance for developing response measures and strategies at the sub-global level. A strategic element identifies visioning and the ability to build social and political consensus around visions of sustainable-future outcomes as essential but underrated aspects of sustainable development governance (Costanza 2000; Meadows 1996). Society-wide relearning and the phase-out of unsustainable policies and practices are discussed, coupled with the redirection of resources towards high-leverage intervention points, including those that better align peoples' mindsets with sustainability and redefine the common meaning of progress as something broader and more meaningful than gross domestic product (GDP). Finally, the need for approaching the transition as an adaptive learning process to build resilience (Loorbach 2007; Holling 2001; Lee 1993) is identified. While providing guidance at the sub-global level, these strategic elements also serve as a starting point for the discussion of responses within international institutions.

VISIONS, GOALS AND TARGETS ON THE ROAD TO 2050

This section introduces a sustainable world vision for 2050 with specific goals and targets derived mainly from existing international agreements. Visioning activates awareness, emotion and imagination with the intention of bringing new systems into being, while the 40-year perspective allows enough space for societies to identify policy options and initiate the structural transformations required.

The challenge of meeting both human needs and human aspirations within the planet's carrying capacity makes the overall ambition complex (UNEP 2011c, 2007; WBCSD 2010; MA 2005b; WCED 1987). A number of countries have attained high levels of human development, but this has often been at the expense of the global natural resource base and environmental quality, and has resulted in high levels of greenhouse gas emissions (Figure 16.3). Given the analysis in Part 1, it is clear that this development path is not sustainable in the long run. Many other countries, meanwhile, are faced with prioritizing the basic human needs of their citizens – such as energy, food and water – over protecting the global commons. In general, these countries currently exert a lower per-person pressure on the global environment, but the overall pressure can still be significant where the population is large, or where there are local environmental problems. Furthermore, when future dynamics are taken into account, the situation may get even worse.

A vision of a sustainable world could be based on the simultaneous achievement of broad overarching goals that take into account the fulfilment of basic human needs, mainly related to reliable and affordable energy, food, drinking water and sanitation, and the achievement of environmental sustainability at the global, national, regional and local levels. Such a vision was the basis of the 1992 Rio Declaration (UNCED 1992) and was advanced further in the Millennium Development Goals (MDGs) (UN 2000).

A sustainable world cannot be realized until it is widely and actively envisioned. Of the available tools to address the challenges ahead,

Farmers thresh their rice harvest in Punakha, Bhutan, the first country to include the concept of happiness in its national measurements of development. *© Gill Fickling/UN Photo*

visioning is critical to crafting profound and enduring change. Vision statements have a specific form: they describe the future, yet they are phrased in the present tense as if the desired change had already occurred. Box 16.1 presents a possible vision for

Figure 16.3 Twin challenge

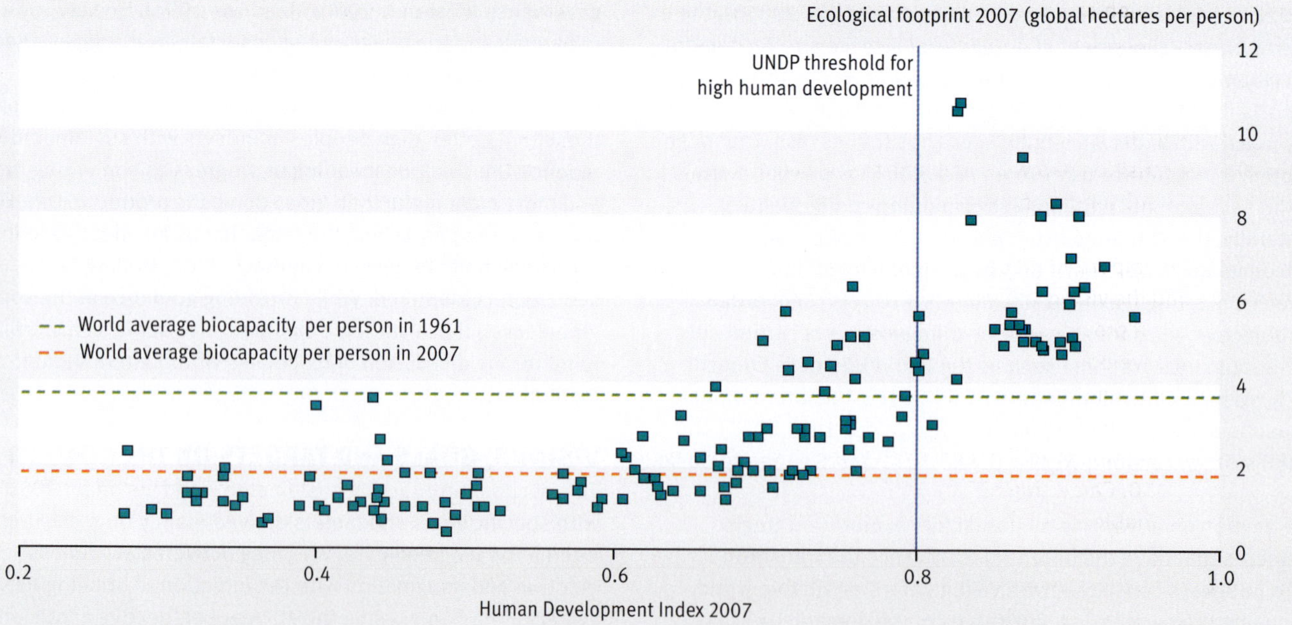

This figure plots countries on the basis of two indicators: the Human Development Index (HDI) and the ecological footprint per person. In order to achieve sustainability, countries must move towards the bottom right corner and as such decouple human development from natural resource use and environmental impacts (UNEP 2011c). The figure shows that worldwide, no country held that position in 2007.

Note: A global hectare is a hypothetical area equivalent to 1 hectare of globally averaged productivity.

Source: Global Footprint Network 2010; UNDP 2009

Box 16.1 A possible vision of the world on a path towards sustainability in 2050

The year is 2050. What appeared to be so improbable at the start of the second decade of the century is turning out to be possible, after all. Changes have been great, and there have been deep losses. Although people expect and are prepared for far greater changes than any yet experienced, a sense of possibility abounds as there have been so many successes.

Climate change is still a problem, but emissions have nearly halved compared to four decades ago. Basic drinking water and sanitation needs of even the poorest have been met. Learning and mimicking nature's resilience has helped restore ecological function in areas once considered irretrievably lost. The most devastating projections for ocean acidification, groundwater salinity, desertification and land degradation have not materialized – with real implications for the food system worldwide. An eco-efficient, highly diversified agricultural system ensures that food shortages are infrequent, local, and mostly due to extreme weather events. Civil instability and conflict over resources, food and water is now rare. More humans enjoy a higher quality of life for longer than ever before, without denying future generations the same possibility.

Most of the world's citizens are actively and personally engaged with humanity's goal of living within planetary limits. Peak oil and peaks in the supply of some other natural resources have come and gone, but thanks to radical changes in lifestyle and resource productivity, have not led to the disruption of absolute scarcity. Leadership is everywhere; as a result, diverse, innovative, bottom-up initiatives abound and are spread through social networks, faster than ever before.

Governance systems, more than ever, are creating synergistic impacts. There has been a tangible shift in the willingness to look for truly sustainable alternatives, a consensus to aim for prosperity rather than continued economic growth at all costs, a commitment to redirect investments to green entrepreneurship and innovation. Knowledge of nature, species and ecosystems is used as a measure and model for humanity's greatest challenges. Indigenous and traditional knowledge, women's access to education, governance and decision making, and a successful balance in North-South and developed-developing perspectives are the forums for pursuing these goals: each shows clearly that human systems esteem diversity as a form of wisdom.

How did this happen? Perhaps things had to get worse before they got better. Perhaps each problem that the financial, social, and ecological debt crises presented had a positive outcome. Ironically, the lynchpin in this transition was an element that had previously been all but overlooked in international governance. A generation of young people emerged, by nature more comfortable with visioning, social networking and truth-telling than its elders. The resulting intergenerational contract, building on momentum already present in society supported a generation of problem-solvers who had never learned values and behaviours that undermined planetary life-support systems, and who could envision solutions and success previously unforeseen.

Achieving these outcomes was the result of a major global effort that began at the 2012 World Summit in Rio.

2050, in line with the internationally agreed goals summarized in Table 16.1. Its transition pathways and key strategic elements are described in the subsequent sections of this chapter.

Obviously, other important global sustainable development targets exist, and the vision and goals outlined here (Box 16.1; Table 16.1) cannot provide a complete picture of a sustainable world. A vision develops through evolution and must have contributions from many people before it is mature and compelling. Thus, the vision captured here is only a start: it represents an invitation to individuals to envision the world they really want in 2050. Catalysing human imagination is integral to realizing a sustainable, desirable future.

The analysis that follows reflects the thematic structure of Part 1: first the global drivers of environmental change, followed by the environmental themes of atmosphere, land, water and biodiversity. As there has been little scenario building on the theme of Chapter 6 – chemicals and waste – it is not included in this analysis, although for the sake of completeness it appears in Table 16.1. Since meeting basic human needs is crucial to a sustainable development strategy, related human well-being objectives are addressed, where appropriate, under each theme. In the selection process, care was taken to ensure that targets properly address the desired state of the global environment and basic human needs.

At this point, it is also important to note that internationally agreed goals and targets are, by definition, a result of political compromise. They take into account scientifically established thresholds to a varying degree, but they may not be fully in agreement with them, particularly as scientific understanding of where such thresholds may lie is also evolving. Therefore, in some cases, selected targets only include a qualitative description of the objective, without further specification, either because of lack of data or because there is still divergence over quantifiable targets. For the climate change problem, for instance, what level of global mean temperature increase is seen as dangerous – and therefore must be avoided according to the qualitative objective agreed in the United Nations Framework Convention on Climate Change (UNFCCC) – is highly debated. Yet, the risks associated with temperature rise are clear, and some

Table 16.1 Goals and targets on the road to 2050

Themes	Goals	Targets
Atmosphere		
United Nations Framework Convention on Climate Change (UNFCCC 1992) Article 2 **Cancun Agreements (UNFCCC 2010) Article 1 Paragraph 4**	Prevent dangerous anthropogenic interference with the climate system	Stabilizing greenhouse gas emissions at a level that would hold the increase in global average temperature below 2°C above pre-industrial levels
Convention on Long-range Transboundary Air Pollution (CLRTAP 1979) Article 2 **World Health Organization guidelines (WHO 2006)**	Reduce and prevent air pollution	Limiting the concentration of pollutants (such as $PM_{2.5}$, PM_{10}, SO_2, NO_2, O_3, CO, Pb) in line with WHO guidelines
Johannesburg Plan of Implementation (JPOI) (WSSD 2002) Paragraph 9a **Energy for a Sustainable Future (AGECC 2010)**	Improve access to reliable, affordable, economically viable and environmentally sound energy supplies	Achieve universal access to modern energy supplies by 2030
Land		
FAO World Food Summit Plan of Action (FAO 1996) Paragraph 33g	Conservation and sustainable use of land	Reduce salinization, combat desertification, reduce cropland expansion and prevent soil pollution and degradation
FAO World Food Summit Plan of Action (FAO 1996) Paragraph 33g **Agenda 21 (UNCED 1992b) Chapter 11.12a**	Sustain forest cover	Reduce the deforestation rate and expand forest areas
UN Millennium Declaration (UN 2000) MDG 1 Target 1c	Eradicate hunger	Halve, between 1990 and 2015, the proportion of people who suffer from hunger, and eradicate hunger by 2050
Water		
Johannesburg Plan of Implementation (JPOI) (WSSD 2002) Paragraph 25d **UN Millennium Declaration (UN 2000) Paragraph 23**	Sustain water resources, protect water quality and aquatic ecosystems	Intensify water pollution prevention to reduce health hazards and protect ecosystems Stop the unsustainable exploitation of water resources by developing water management strategies at the regional, national and local levels, which promote both equitable access and adequate supplies
UN Millennium Declaration (UN 2000) MDG 7 Target 7c	Universal provisioning of safe drinking water and improved sanitation	Halve, by 2015, the proportion of the population without sustainable access to safe drinking water and basic sanitation and ensure full access by 2050
Biodiversity		
Convention on Biological Diversity (CBD) Aichi Biodiversity Targets (CBD 2010) Target 5 **CBD Aichi Biodiversity Targets (CBD 2010a) Target 12**	Improve the status of biodiversity by safeguarding ecosystems, species and genetic diversity and promote its sustainable use and fair and equitable benefit sharing	By 2020, at least halve and where feasible bring close to zero the rate of loss of all natural habitats, including forests, and significantly reduce degradation and fragmentation By 2020, prevent the extinction of known threatened species, and improve and sustain their conservation status, particularly of those most in decline
United Nations Convention on the Law of the Sea (UNCLOS 1982) Article 192 **Convention on Biological Diversity Decision II/10 (Jakarta Mandate 1995)** **FAO Code of Conduct for Responsible Fisheries (FAO 1995) Paragraph 6.2**	Protect and preserve the marine environment	Promote conservation and sustainable use of the coastal and marine ecosystems as well as their natural resources Promote the maintenance of the quality, diversity and availability of fishery resources in sufficient quantities for present and future generations
Chemicals and waste		
Johannesburg Plan of Implementation (JPOI) (WSSD 2002) Paragraph 23 **Stockholm Convention on Persistent Organic Pollutants (2009)**	Reduce chemical pollution to protect human health and the environment	By 2020, use and produce chemicals in ways that lead to the minimization of significant adverse effects on human health and the environment Protect human health and the environment from persistent organic pollutants
Rotterdam Convention on Certain Hazardous Chemicals in International Trade (Rotterdam Convention 1998) Article 1	Monitor and control the trade in certain hazardous chemicals	Promote shared responsibility in the international trade of certain hazardous chemicals in order to protect human health and the environment from potential harm and to contribute to their environmentally sound use
Johannesburg Plan of Implementation (JPOI) (WSSD 2002) Paragraph 22	Minimize the amount of waste and promote reuse and recycling	Prevent and minimize waste and maximize reuse, recycling and use of environmentally friendly alternative materials

are becoming more acute, so preventative and precautionary action is necessary as a form of sustainability insurance (UNFCCC 2010). Strengthening the long-term globally agreed target on the basis of the best available scientific knowledge, rather than of political consensus, is still an issue under consideration at UNFCCC meetings. Overall, the targets are based on the latest multilateral agreements associated with the goals that inspire and guide humanity towards a specific destination.

PATHWAYS TO ACHIEVE LONG-TERM SUSTAINABILITY GOALS

This section examines the existing literature on quantitative scenarios to outline how sustainability targets might be met. The chapter looks at the scenarios used in earlier assessments and published in the scientific literature and summarizes the findings for the conventional world projections, scenarios that describe the consequences of continuing current policies, and compares them to sustainable world projections, those that aim to reach the long-term targets envisioned above. The purpose of this comparison is to assess the gap between these different pathways and to discuss how it might be closed. The two scenario categories have some general characteristics. Conventional world scenarios typically extrapolate historical trends, assuming no new policy direction, also described in the literature as business-as-usual. They also usually assume a continuing increase in the use of material goods and services, driven by the same market dynamics dominant in the world today. These scenarios thus tend to ignore the risks associated with environmental degradation and resource scarcity. Sustainable world scenarios, in contrast, explore the changes required if sustainable development goals are to be met. Clearly, this category includes a wide range of scenarios based on the use of advanced technologies, increased efficiency and/or lifestyle changes. In some cases new calculations have been performed to respond to gaps in the literature.

Drivers

Population and income

The global population is projected to grow to 8–10.5 billion people by 2050 (Figure 16.4) (UNDESA 2011; Lutz *et al.* 2008). By far the largest share of population growth is expected in countries that are currently low-income, mainly in sub-Saharan Africa, Northern Africa and West Asia, and South Asia. Generally, low population scenarios are more likely to lead to lower environmental pressure than high ones, although high population scenarios that result in low emissions can also be found in the literature (van Vuuren *et al.* 2012). Still, the importance of population growth in the context of sustainability targets has been recognized at the highest UN level (ICPD 1994). Investing in women's education is one of the most effective methods of reducing population growth, as women with higher levels of education have fewer children. Scenario analysis by Lutz and Samir (2011) shows a global population ranging from 8.9 to 10.0 billion people in 2050 as a result of a high or low education scenario alone.

Nearly all scenarios project a further increase in GDP as an indicator of economic development, although there is variation between scenarios, with global average per-person growth rates ranging between 1.2 and 2.2 per cent annually (Figure 16.4). The relationship between income and environmental change is ambiguous. On the one hand, high income tends to coincide with high consumption levels, leading to further environmental degradation. On the other, an increase in income can also coincide with lower population levels, an increasing appreciation of a clean environment and rapid technological change. These

Figure 16.4 Population and income projections in the scenario literature, 2000–2050

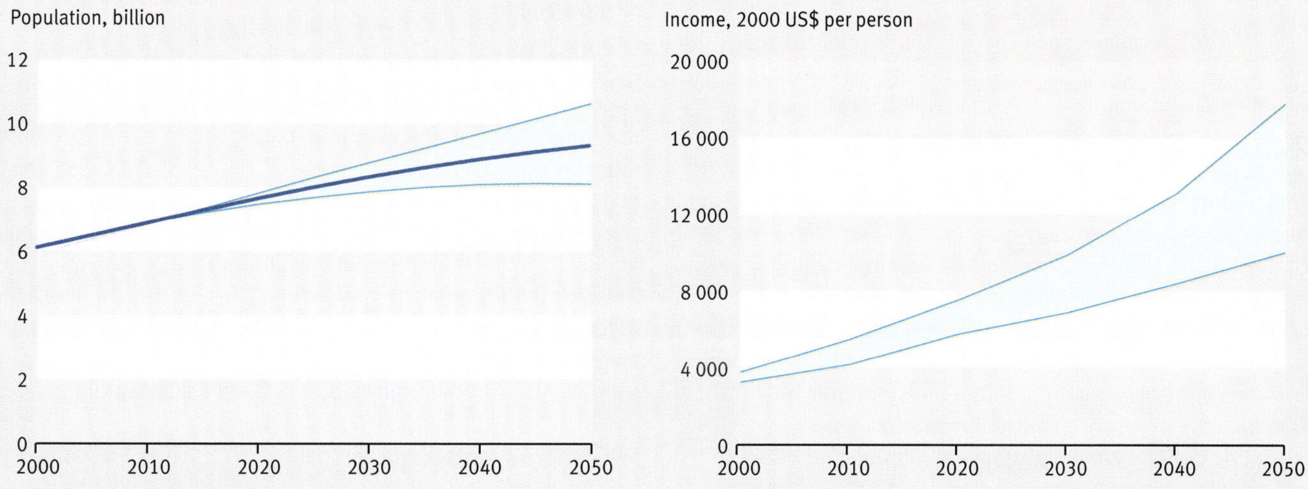

Note: The shaded areas indicate the 10–90th percentile literature range.

Source: (GDP) van Vuuren *et al.* 2012; (population) UNDESA 2009

A young girl receiving training in trade in the Kapisa Province of Afghanistan, where the vast majority of primary and secondary students are boys. *© Eskinder Debebe/UN Photo*

trends may lead to a decrease in environmental pressure as incomes rise, as observed, for instance, for local air pollution, the so-called environmental Kuznets curve (Chapter 1) (van Ruijven *et al.* 2008; Riahi *et al.* 2007; Smith 2005; Stern 2003). This effect has, however, not been observed for many global environmental problems, including carbon dioxide (CO_2) emissions, and there have been reports of displacement effects – with production shifting to low-income countries – underlying some Kuznets observations (Luzzati and Orsini 2009). It is important to note that it is not the level or rate of economic growth that determines environmental impacts, but its structure. For example, a focus on services rather than material goods could reduce the pressure on natural resources. Consistent with this scientific debate, no straightforward relationship between income projections and achievement of sustainability targets can be found in global scenarios (van Vuuren *et al.* 2012). Some scholars emphasize the positive feedbacks between economic growth and achieving sustainability goals, the green growth paradigm for example (WCED 1987), while others highlight trade-offs and the correlation between consumption rates and the flow of material goods, as in the steady-state economy paradigm (Czech and Daly 2004; Daly 1974, 1971). The difference depends on factors such as technological development, macro-economic feedbacks and avoided environmental damage (UNEP 2011b).

Consumption

Global average consumption levels have risen significantly during recent decades, outpacing simultaneous improvements in efficiency. Increases in both the number and size of cars have been much greater than improvements in fuel efficiency, leading to a rapid increase in overall transport fuel consumption (Girod *et al.* 2012). In fact, efficiency improvement itself may induce higher consumption levels by decreasing the cost of consumption – the rebound effect. Changing consumption patterns could form an important part of a sustainable development strategy as it often leads to multiple benefits and brings environmental concerns closer to consumers. Historically, however, campaigns aimed at changing consumption patterns have not always been successful. The effectiveness of changes in consumption can be illustrated by scenario studies on dietary change, discussed below in the section on land.

There is increasing recognition of the importance of the environment for human development and quality of life (World Bank 2008; UNEP 2007; MA 2005b). Currently, an estimated 24 per cent of the global disease burden and 23 per cent of all deaths can be attributed to environmental factors (Prüss-Üstün and Corvalán 2006). For child mortality, in particular, low levels of food intake, unsafe drinking water, a lack of basic sanitation and the use of solid fuels for cooking and heating are important drivers (Black *et al.* 2010). Analysis shows that, despite some progress, in conventional world scenarios, full access to sufficient food, water, and energy will not be reached for many countries in South Asia and sub-Saharan Africa by 2030, or even 2050 (World Bank/IMF 2011; Hilderink *et al.* 2009).

Atmosphere

Conventional world scenarios

Almost all scenarios that assume no major policy changes expect energy consumption to continue to grow worldwide. On average, they project energy consumption to increase by a factor of three over the 21st century, with a range of 2.5–5.5 (van Vuuren *et al.* 2012; Clarke *et al.* 2010; Fisher *et al.* 2007). Moreover, such conventional world scenarios project fossil fuels to retain a large market share as their price, especially for coal, is expected to be lower than that of alternative fuels. Despite the domination of fossil fuels, however, most scenarios project a significant increase in non-fossil energy production, including biomass, solar, wind and other renewables, as well as nuclear.

The projections also indicate that by 2030 nearly 3 billion people, mostly in rural areas in sub-Saharan Africa and Asia, will still rely on traditional biomass for cooking and heating, while about 1 billion people will have no access to electricity (GEA 2011; IEA *et al.* 2010). Conventional world scenarios also

Figure 16.5 Emissions and temperature scenarios

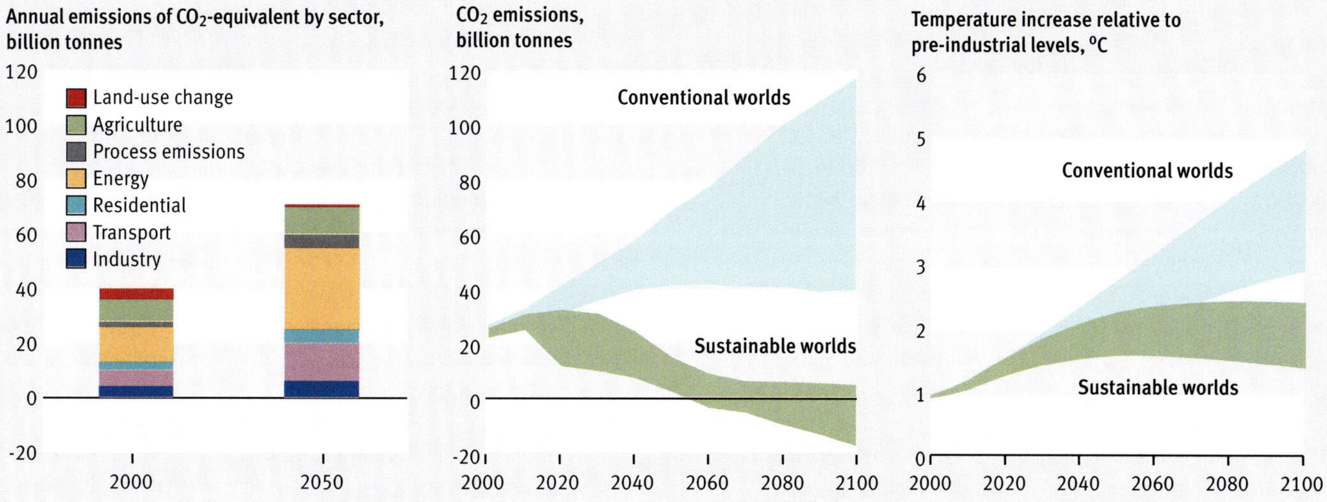

Note: Emission and temperature scenarios cover a longer time period than other scenarios in this chapter because of inertia in the climate system. The shaded areas indicate the 10-90th percentile literature range.

Source: van Vuuren *et al.* 2008a, 2008b; Fisher *et al.* 2007

consistently project high numbers of people suffering from the health consequences of traditional fuel use, with around 1.5 million annual premature deaths resulting from indoor air pollution in 2030 (IEA *et al.* 2010). The use of traditional fuels in inefficient stoves can also have serious implications for deforestation and local and regional air pollution (FAO 2006a; IEA 2006; Arnold *et al.* 2003). It should be noted that fuelwood use could actually increase in response to rising prices for modern fuels (Easterling *et al.* 2007).

Increasing fossil fuel use implies increasing emissions of greenhouse gases. On average, conventional world scenarios project greenhouse gas emissions to more or less double in the next 50 years (van Vuuren *et al.* 2012; PBL 2009; Fisher *et al.* 2007). Scientific knowledge leaves little doubt that a consequence of the increase will be a steady rise in global mean temperature (Figure 16.5) (van Vuuren *et al.* 2008a, 2008b; IPCC 2007), of 3–5°C relative to pre-industrial levels by the end of the century. There is considerable uncertainty regarding both climate change and its impacts. The IPCC's Fourth Assessment Report indicates that a warming of 4°C is likely to have negative effects on agricultural yields in most parts of the world (Easterling *et al.* 2007), with sensitive systems – coral reefs, some mountain ecosystems, polar sea ice and many of the world's glaciers – likely to be lost and sea levels possibly rising by more than 1 metre by the end of the century. Moreover, there is a risk of passing critical thresholds for the functioning of the Amazon rainforest (IPCC 2007), as well as an increase in the frequency of storms, droughts and other extreme weather events.

Historically, people have tended to invest more in the control of air pollution as they become more affluent. Typically, emissions increase in the early stages of development, but may diminish as incomes rise. In conventional world scenarios, emissions are usually shown to decrease slowly in the first decades of the 21st century in high-income countries, but to increase in low-income ones (van Ruijven *et al.* 2008). Globally, for many air pollutants

Figure 16.6 Scenarios for sulphur emissions

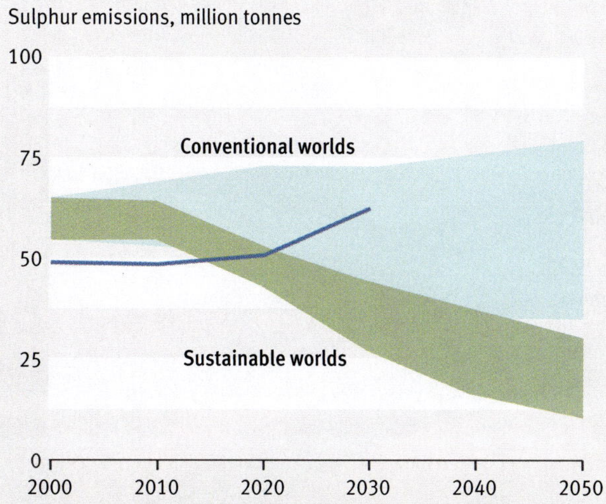

Conventional world scenarios include those without climate policy; sustainable world scenarios include those with stringent climate policy. The blue line represents the scenario under current legislation.

Note: The shaded area indicates the 10-90th percentile literature range.

Source: Van Vuuren *et al.* 2008a; Cofala *et al.* 2007

this results in a pattern of stable or slightly decreasing emissions (Figure 16.6); although with considerable uncertainty. During most of the century, therefore, these scenarios suggest that targets for health standards are unlikely to be met in many parts of the world.

Sustainable world scenarios

Several scenario studies assess the sustainability target of providing universal access to modern energy (Table 16.1) (GEA 2011; Pachuari *et al.* 2011; van Ruijven *et al.* 2012; IEA 2010). Improving access to electricity requires accelerating the pace of electrification in the least developed countries either by grid expansion or by the development of decentralized mini-grids or off-grid systems (AGECC 2010). For cooking, to increase energy efficiency and decrease health impacts, the main strategy is to promote the use of advanced-combustion biomass stoves or to make a full transition to cleaner fuels (Venkataraman *et al.* 2010). Scenario analysis shows that such strategies could avoid more than 1 million premature deaths per year up to 2030 (GEA 2011). Estimates of the annual investments to implement such strategies range from US$10 billion to 140 billion per year (GEA 2011; Bazilian *et al.* 2010; IEA *et al.* 2010). Scenario analysis also suggests that the climate impact of ensuring access to modern energy to all is small: the increase in CO_2 emissions from fossil fuels could be around 1 per cent of global emissions in 2030 compared to a conventional world scenario, but this would be compensated for by reduced fuelwood demand and thereby reduced deforestation (GEA 2011; IEA *et al.* 2010).

Based on current estimates of the uncertainty in climate sensitivity (IPCC 2007), targets for greenhouse gas concentrations of 450 parts per million (ppm) and 400 ppm CO_2-equivalent would ensure a median chance of 50 per cent and 70 per cent, respectively, of staying below the UNFCCC's agreed limit of a 2°C temperature increase (Table 16.1) (Meinshausen *et al.* 2006). Sustainable world scenarios show that to reach such targets, global greenhouse gas emissions would need to peak in just one to two decades, then fall to around half the current levels, or even less, by 2050, and by the end of century would need to reach zero or even result in a net absorption – which could, for instance, be done by afforestation or by the combination of bio-energy and carbon capture and storage (van Vuuren and Riahi 2011; UNEP 2010a). There are four basic ways to reduce emissions:

- changing the structure of economic growth;
- increasing energy efficiency through technology or lifestyle changes;

Masdar City, under construction near Abu Dhabi, United Arab Emirates, will rely entirely on solar and other renewable energy sources, with the ambition of becoming the first zero-carbon, zero-waste city in the world. © *www.masdar.ae*

- changing energy supply including using zero-carbon energy options; and
- implementing end-of-pipe measures such as carbon capture and storage.

Reaching the sustainable world emission reduction targets (Table 16.1) requires a broad portfolio of measures. Figure 16.7 provides an indication of the size of the transition involved. Very many scenarios show that a low-carbon economy can be achieved with currently identifiable technologies, and they share a number of common features (Clarke *et al.* 2010; ECF 2010; Fisher *et al.* 2007; van Vuuren *et al.* 2007):

- Energy efficiency improvement is a robust option under all scenarios.
- Energy generation provides very significant emission reduction potential by introducing combinations of renewable energy, nuclear power and/or carbon capture and storage, although each of these has limitations and drawbacks. Indeed, the centralized energy sector could even achieve net negative emissions if it were based on bio-energy use with carbon capture and storage.
- Reducing non-CO_2 greenhouse gas emissions such as methane, nitrous oxide, and black carbon and ozone precursors could contribute significantly to mitigating climate change at relatively low cost, for example by reducing methane emissions from energy production and some of the methane emissions from livestock and rice cultivation.
- Lifestyle changes are not often explicitly accounted for in scenarios, but may achieve considerable reductions such as in transport or food consumption.
- The use of bio-energy is very common in low-emission scenarios (van Vuuren *et al.* 2010). Bio-energy production may, however, have serious consequences for biodiversity, food production and greenhouse gas emissions (Dornburg *et al.* 2010; Searchinger *et al.* 2008; Bringezu *et al.* 2009; Fargione *et al.* 2008), so needs to be carefully monitored and the available potential used as efficiently as possible. Moreover, it would be important to focus the use of bio-energy on sectors where it would provide the greatest benefit.
- Access to modern energy sources can be improved by grid, mini-grid or off-grid expansion, subsidies and grants, or micro-loans for stoves.

Although air quality could be significantly improved in low-income countries through the rapid introduction of state-of-the-art technologies, it should be noted that air pollutant emissions are also strongly influenced by structural changes in the energy sector. In addition to the importance of pollutants such as methane and black carbon for climate change, the fact that they also negatively affect health and crop growth (via ground-level ozone) might be a much more important trigger for their reduction than climate change alone. The sustainable world scenarios show that stringent climate policies and existing air pollution control measures could significantly reduce emissions. The adoption of such strategies could be successful in achieving World Health Organization air quality targets, and their combined benefit would be delivered at much lower costs than the sum of separate strategies to meet climate and air pollution goals (UNEP 2011a; GEA 2011; Bollen 2008).

There are several possible consequences to the movement towards a low-carbon society. Some of these are co-benefits, for instance for greenhouse gases and air pollution, or emission reductions and improved energy security. There are some trade-

Figure 16.7 An example of primary energy use and annual change in CO_2 emissions in sustainable world scenarios

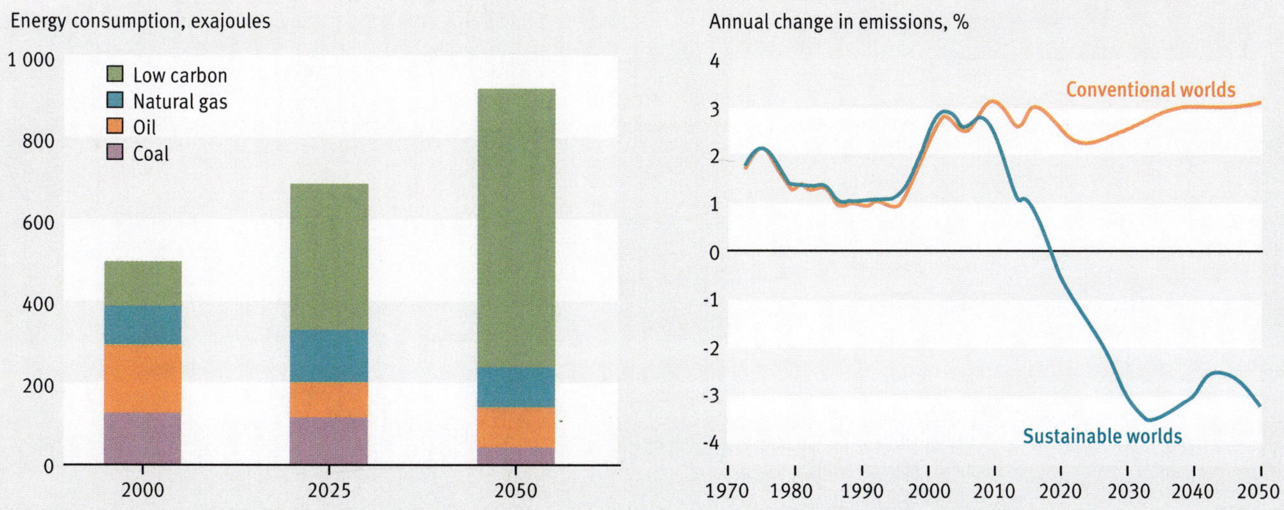

The category Low carbon refers to renewable energy, nuclear power and fossil fuels in combination with carbon capture and storage and efficiency, and illustrates the level of transition required. Different models and studies suggest different combinations.

Source: PBL 2009

offs as well. For instance, the reduction in aerosol emissions that results from measures to reduce greenhouse gas emissions will initially partly offset the climate benefits from current aerosol cooling effects. An important trade-off concerns bio-energy, but other technologies also have side effects. Hydropower infrastructure can have several impacts, such as loss of agricultural land, replacement of settlements, biodiversity loss and ongoing greenhouse gas emission episodes from water bodies (Fearnside 2011; St. Louis *et al.* 2000). Wind turbines regularly face opposition from local communities and carbon capture and storage (once it can be applied on a large scale) may entail risks of CO_2 release. Climate policy will also interact with forestry management, which could lead to both positive and negative impacts on biodiversity.

Land

Conventional world scenarios

A wide range of models, from economic to biophysical, has been used to explore future trends in land use (Smith *et al.* 2010). All the studies project a strong increase in the demand for food up to 2050, driven by a rising population and dietary change spurred by economic growth. Between 2000 and 2050, global cereal demand is projected to increase by 70–75 per cent while meat consumption is expected to double (Thornton 2010; IAASTD 2009a; FAO 2006b). Meeting these needs, while avoiding a large expansion of agricultural land and protecting biodiversity, will be a major challenge. Ensuring food security will also be an issue as world food markets are likely to be influenced by increasing resource scarcity. A key mitigating factor will be continued investments in yield growth and intensification on existing cultivated land (FAO 2011; UNEP 2011b; Rosegrant *et al.* 2009).

Food security exists when all people, at all times, have physical, social and economic access to enough safe and nutritious food to meet their dietary needs and food preferences for an active and healthy life (FAO 1996). Looking at per-person food availability in the conventional world scenarios, by 2050 average calorie availability per person per day will be around 3 000–3 500 calories, but projected food availability in sub-Saharan Africa is much lower, in the range of 2 100–3 350 calories. Environmental degradation, lack of investment and competition for land are expected to push world food prices upwards, causing additional stress to the poor, especially in urban areas (OECD/FAO 2011; IAASTD 2009a). In other words, the conventional world scenarios suggest that it is very unlikely that malnutrition will be fully eradicated by 2050 without major policy shifts (IAASTD 2009a; UNEP 2007; FAO 2006b; MA 2005a). The prevalence of child malnutrition for 2050 in developing countries is projected to range between 13 per cent and 25 per cent (Figure 16.8; Box 16.2). The highest levels of undernourishment are projected in countries that currently suffer from hunger, have high population growth rates and poor prospects for rapid economic growth, and possess limited agricultural resources (FAO 2006b).

Population growth and dietary change have contributed to a rising demand for agricultural products. In the past four decades,

Tea plantation in Limuru, Kenya. The overall productivity of Kenya's tea plantations is considered among the highest in the world. © *Jason Jabbour*

Figure 16.8 Food consumption and child undernourishment under different scenarios

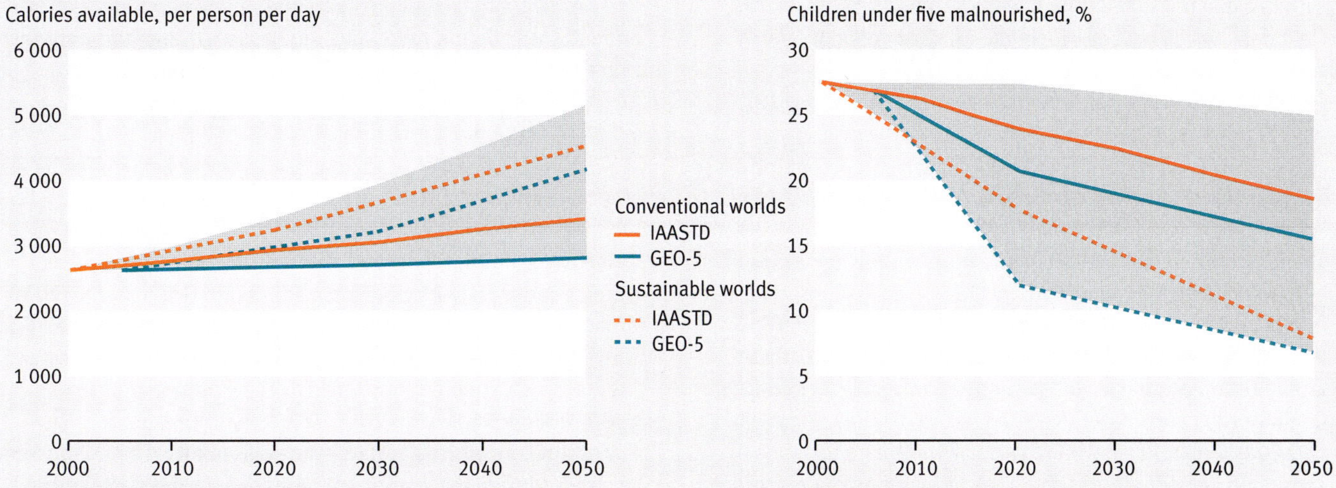

Note: Scenarios selected from GEO-4, the IAASTD study and the Millennium Ecosystem Assessment are those that most reflect the sustainable and conventional worlds on the basis of storyline or quantitative elaboration. The shaded area shows the range in the literature. *GEO-5* results as discussed in box 16.2.

Source: Hughes *et al.* 2011; IAASTD 2009a; UNEP 2007; FAO 2006b; MA 2005a

around 78 per cent of the global increase in agricultural supply has been achieved through increases in yield and greater efficiency in the supply chain; a further 7 per cent has come from increased cropping intensity; while a mere 15 per cent has come from expansion of the arable area (Smith *et al.* 2010; Bruinsma 2003). At the regional level, however, large differences can be seen. In sub-Saharan Africa, for instance, only 34 per cent of the rise in output was derived from yield increases, and the remaining 66 per cent came from area expansion (Mery *et al.* 2010; Smith *et al.* 2010). These factors are expected to continue to be important in the future, although trends will differ over time and across regions.

Yield growth has slowed over the last several decades (FAOSTAT 2012). Moreover, environmental pressures, including the effects of climate change and ground-level ozone, could also have a negative impact on yields in the future. IPCC estimates of the potential global impact of climate change on crops under high temperature increase scenarios (4°C compared to pre-industrial), although uncertain, suggest that if no adaptation occurs climate change could have a substantial negative impact on yields, of 10–35 per cent at all latitudes for crops like maize and wheat. Adaptation measures could, in aggregate, ward off negative impacts in temperate regions, but could not avoid an average yield reduction of around 10 per cent in tropical zones (Easterling *et al.* 2007).

When focusing on crop production, scenarios show some variation in terms of expected land use (Figure 16.9) (Smith *et al.* 2010). The 2050 projections for cropland increase range from as low as 6 per cent, through an average increase of around 10–20 per cent (van Vuuren *et al.* 2008b) to more than 30 per cent as suggested by the IPCC's A2 scenario, which is based on high population growth. Regional results can be very different: while a considerable expansion of arable land is expected in Africa, Asia and Latin America, this is compensated for by a decrease in harvested area in the temperate zone (van Vuuren *et al.* 2008b; UNEP 2007). As land degradation (CBD 2010b) is typically not accounted for in scenario analysis, the real impacts could be worse.

Figure 16.9 Trends in land use, 1970–2050

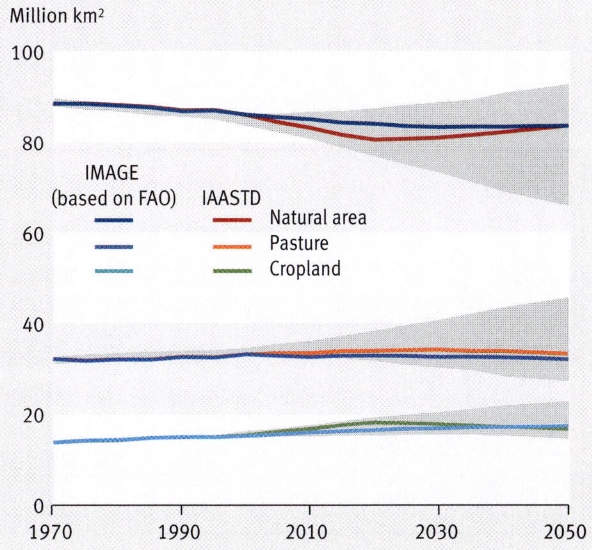

Note: Shaded area indicates 10–90th percentile literature range

Source: Rose *et al.* 2012; Hurtt et al. 2011; Smith *et al.* 2010; IAASTD 2009a; OECD 2008a; UNEP 2007; FAO 2006; MA 2005a

For animal products, existing scenarios indicate that most of the increases in global livestock production will occur in developing countries (Bouwman *et al.* 2005). In grazing systems, strong growth is expected for confined livestock production systems, while most studies show an increase of 10 per cent or less in pasture areas.

Sustainable world scenarios

Given the strong connection between agricultural production and the ecosystem services that provide food, forage, fibre, energy and biodiversity, achieving sustainable development as it affects agriculture and land resources requires an integrated approach (Smith *et al.* 2010). Such an integrated approach would take into

Box 16.2 Integrated simulation of the 2050 targets for climate, food and land

Can very high investments in agriculture and water productivity help to achieve the sustainability objectives discussed earlier in this chapter? Here, this question is explored using the International Food Policy Research Institute (IFPRI) IMPACT model (International Model for Policy Analysis of Agricultural Commodities and Trade) (Nelson *et al.* 2010; Rosegrant *et al.* 2008). Previous analyses have shown the importance of economic development in reducing hunger and malnutrition (Nelson *et al.* 2010).

Compared to a conventional world scenario, economic growth in developing countries is assumed to be higher and population growth to be lower overall (Nelson *et al.* 2010). Additional investments in agricultural research and development will lead to rapid increases in agricultural production: as a result, by 2030, grain yields are 15 per cent greater than in the corresponding conventional world scenario and by 2050, they are 35 per cent greater. Furthermore, livestock numbers are up by 30 per cent. It is also assumed that the UNFCCC's agreed limit of a 2°C temperature rise relative to pre-industrial levels has been achieved, and that there is full access to safe drinking water by 2050, and that all girls have access to secondary schooling by 2030. Finally, the water efficiency improvements suggested by the sustainable water withdrawal scenarios are also included (with the exception of a constant irrigated area) (Box 16.3).

The changes outlined above result in average cereal prices being respectively 21 per cent and 39 per cent lower by 2030 and 2050 compared to the conventional world scenario. Under the conventional world scenario, the global harvested crop area is expected to grow at 0.23 per cent per year or 169 million hectares for 2005–2050, with contractions in some OECD countries and Asia more than offset by increases in sub-Saharan Africa and Latin America. In the sustainable world scenario, on the other hand, crop area contracts by 116 million hectares by 2030 and by 201 million hectares by 2050. The lower food prices suggested by the sustainable world scenario are expected to boost affordability and thus access to food, increasing daily calorie availability in the developing world by 496 kilocalories by 2030 and 1 336 kilocalories by 2050. As a result, about 50 million children would be malnourished, a fall of 66 million or 57 per cent. The model calculations, however, also show that eradicating hunger in 2050 is a complex, multifaceted challenge: importantly, significant steps can be made through changes in investment and policies. Key factors that can make a difference for childhood malnutrition include increasing the availability of food, access to safe drinking water and increased female secondary school enrolment. Moreover, mitigation of and adaptation to climate change will make a positive difference for agricultural production.

Table 16.2 Selected indicators for the conventional and sustainable world scenarios

	2005	2030 conventional worlds	2050 conventional worlds	2030 sustainable worlds	2050 sustainable worlds
Area-weighted grain prices, US$ per tonne	150	202	253	160	154
Total crop harvested area, thousand hectares	1 520 811	1 684 798	1 689 758	1 569 207	1 489 230
Developing country calorie availability, per person per day	2 637	2 717	2 823	3 213	4 159
Malnourished children worldwide, million	153	136	115	78	50
Proportion of children malnourished in India, %	46	41	39	30.7	27.4

Source: New calculations IMPACT model; Nelson *et al.* 2010; Rosegrant *et al.* 2008

account the relationship between competing demands for limited land resources and the environmental impact on agricultural production (UNEP 2010b).

One key to ensuring future access to food is investment in agricultural research to improve productivity (Rosegrant *et al.* 2009). Another is to reduce food wastage and loss – currently around 10–40 per cent of agricultural production is wasted (Parfitt *et al.* 2010). Through lifestyle changes, technological development and investment in infrastructure, it is possible to significantly remedy this wastage (Jäger and Cornell 2011; Parfitt *et al.* 2010). Alterations in people's diets could also help reduce the need for additional production. Scenario studies have looked into the consequences of reducing consumption of livestock products through substitution of vegetable alternatives. Different results have been published on the land consequences of such dietary changes. Some studies suggest considerable reductions in land-use (Ten Brink *et al.* 2010; Stehfest *et al.* 2009), while others highlight the risk of a rebound effect, with reduced meat consumption in developed countries leading to increased meat and cereal consumption in the rest of the world (Rosegrant *et al.* 1999). There is also some discussion of the health implication of low-meat diets: although studies claim benefits in high-income countries from reducing overconsumption of meat, low-meat diets need to be well designed.

Very few projections can be found that actually result in food security for all by 2050. According to the Food and Agriculture Organization of the United Nations (FAO 2009), however, it could be accomplished if global food supplies expand by about 70 per cent above current levels. The conditions under which this could be achieved include political stability, good governance, food security strategies, integration of world markets and strong economic growth, all on the basis of an increase in agricultural production in Africa, Asia and Latin America. This implies ensuring that agricultural trade liberalization – integration of world markets – does not lead to negative impacts on vulnerable communities (Jäger and Cornell 2011). The International Assessment of Agricultural Knowledge, Science and Technology for Development (IAASTD 2009a) includes a scenario with increased investments in agricultural technology combined with increased investment in water infrastructure and female secondary education. This scenario shows a significant increase in food availability but still leaves 8 per cent of children malnourished, mostly in sub-Saharan Africa. Clearly, access to food needs to be viewed in the context of poverty reduction both to promote rural development and to provide direct and immediate access to food for the most needy (Broca 2002).

Figure 16.9 shows that some scenarios lead to little expansion and even contraction of the agricultural area. One contributing factor is reduced rates of population growth leading to smaller increases in the demand for food. If historical yield growth rates can be maintained, the global agricultural area could be stabilized or even reduced. Indeed, the analyses by IAASTD (2009a) and Thornton (2010) show that it is possible to increase yields significantly on the basis of better use of agricultural knowledge, science and technology – although this will be far from easy. Policies that lead to higher agricultural yields need to be combined with policies that reduce or avoid soil degradation and other negative environmental trends such as the loss of crop resilience to pests (Killham 2010; Petermann *et al.* 2008; Kaiser *et al.* 2007; Paulitz *et al.* 2002). Finally, it is also necessary to effectively and fairly define property rights and develop local institutions that foster longer-term investment in the use of water, soil and biological resources in agriculture and forestry (FAO 2011; Von Braun and Meinzen-Dick 2009; Hazell and Wood 2008). The crop yields in the sustainable world scenarios also benefit from reduced climate change impacts. The IPCC assessment indicates that the combination of agricultural adaptation and an increase in temperature of less than 2°C above pre-industrial levels may even result, in aggregate, in positive impacts on globally averaged yields (Easterling *et al.* 2007).

Policy options that could improve sustainable agricultural production include:

- supporting investments in increasing crop yields in developing countries to limit the expansion of agricultural land and to close the yield gap between developed and developing countries;
- promoting adaptation to climate change by encouraging crops and crop varieties that are more resilient under changing climatic conditions;
- investing in infrastructure, food processing and storage techniques to reduce food waste;
- better utilizing urban-rural landscapes for food systems and natural resource use;
- reducing consumption of livestock products; and
- strengthening land-use policies and planning by promoting integrated land and resource management.

While it is technically possible to achieve zero expansion of agricultural area and still produce sufficient or more food, implementing this presents a large number of challenges. Key among these are the continuing degradation of land and water, climate change and the rising demand for biofuels (FAO 2009). For example, the current trend of increasing emissions and water pollution from nitrogen fertilizers is expected to increase by 2050, despite potential for greater efficiency in fertilizer use (Power 2010; Bruinsma 2003). Furthermore, the predicted growth in animal production in developing countries will lead to more methane and nitrous oxide emissions from manure, although the projected improvement in husbandry will somewhat reduce emissions per animal (Smeets *et al.* 2009; Bouwman *et al.* 2006). While climate change mitigation is designed to reduce negative climate impacts, in specific regions such as the Russian Federation, such policy will result in forestalling potentially positive yield changes. Furthermore, improvements will also depend on potential trade-offs between allocations of agricultural land for crops or for biofuels, which could further threaten food production and security.

Water

Conventional world scenarios

Chapter 4 shows that many regions are seriously affected by an imbalance between the availability and withdrawal of water, water stress and water pollution from different sources. River systems are considered the most endangered ecosystems on the planet and their loss of biodiversity has been faster over the past 30 years than in any other terrestrial or marine ecosystem (Jenkins and Lowe 2003). The Johannesburg Plan of Implementation (WSSD 2002) Paragraph 26c calls not only for an efficient and well-balanced use of freshwater resources but also for the safeguarding of drinking water quality. Important drivers of water scarcity include population growth, increasing water consumption, pollution and climate change. Increasing water use, river and reservoir regulation, or non-treated return flows lead to alterations in flow regimes that contribute to intensified and complex conflicts between ecosystem requirements and the management of rivers for human water supply and energy generation. Climate change can affect freshwater problems in many ways through changes in precipitation, discharge rates, extreme events, reduced dilution capacity of rivers and salinization due to sea level rise (Schneider *et al.* 2011; Bates *et al.* 2008).

Several scenario assessments exist, showing large variations in projected water withdrawals – the total volume of water extracted from surface or groundwater sources for various uses – based on different assumptions on such factors as population, consumption patterns and technology availability (Figure 16.10). Most of the estimates of water withdrawals indicate a large global net increase, but with significant regional differences. The most important factor for this increase is the growth of household water use, followed by industrial and agricultural use (Alcamo *et al.* 2007). As a result of increasing withdrawals, effluent returns are also likely to increase, many of which remain untreated in low-income regions. For the *GEO-4* "markets first" scenario, for instance, the volume of untreated wastewater was reported to steadily increase despite improvement in treatment. In the *GEO-4* "sustainability first" scenario, in contrast, the volume of untreated wastewater fell as a result of an overall decrease in wastewater due to greater efficiency (UNEP 2007). The *GEO-5* calculations are discussed in more detail in Box 16.3.

Increases in water withdrawals are projected to lead to an increase in water stress (Arnell *et al.* 2011; Alcamo *et al.* 2007, 2003; Cosgrove and Rijsberman 2000; Vörösmarty *et al.* 2000). More than 2 billion people currently live in severely water-stressed areas, primarily in Asia (Figure 16.12). Model simulations show that under the conventional world scenarios, the number of people living in water-stressed areas, as well as the extent of these areas, is expected to rise substantially due to population growth, increased use of water and climate change. The ratio of future to current numbers is given in Figure 16.12, which also demonstrates the uncertainty in the scenarios as a result of different assumptions about climate and other global change. Increases are highest in Africa, where the number of people living in severe water-stressed areas is expected to grow fourfold (median). In South America and Asia significant changes may also occur. In many river basins under severe water stress, there will be competition between domestic, industrial and agricultural users. It should be noted that changes in the use and natural condition of water somewhere in a river catchment area will affect the availability and quality of water downstream.

Figure 16.10 Water withdrawals under different scenarios, 2000–2050

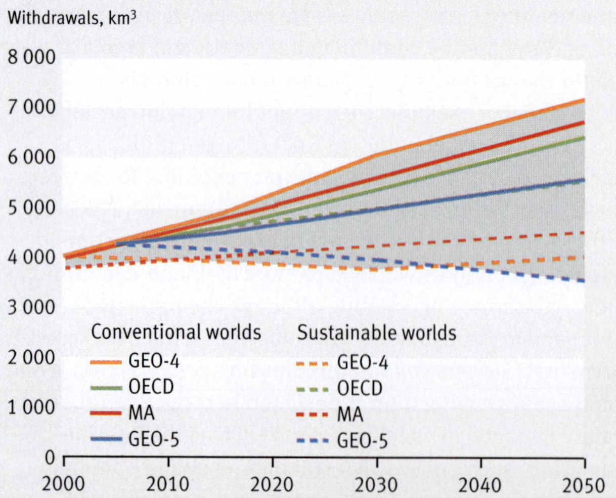

Note: The shaded area shows the range in the literature.

Source: Bakkes *et al.* 2008; OECD 2008a; UNEP 2007; Alcamo *et al.* 2005b; MA 2005a

A woman on her way to the water distribution site in Tora, Northern Darfur. The closest water source is more than an hour's walk from the village. *© Olivier Chassot/UN Photo*

Currently nearly 1 billion people lack access to clean drinking water and 2.6 billion lack access to improved sanitation services (WHO/UNICEF 2010). In 2004, unsafe water and inadequate sanitation were responsible for around 1.6 million deaths and 6.3 per cent of worldwide disability adjusted life-years (DALYs), mainly due to diarrhoea (WHO/UNICEF 2010). Scenario analysis suggests that by 2015, 627 million people will still live without access to clean drinking water and 2.7 billion people will live without access to improved sanitation. Different studies project long-term developments in drinking water and sanitation, either by assuming a continuation of the 1990–2000 improvement rates (Prüss-Üstün *et al.* 2004) or by using cross-sectional relationships with socio-economic indicators (OECD 2012; Hughes *et al.* 2011). These studies project that the proportion of the world's population without access to safe drinking water will diminish from 23 per cent in 2000 to 3–5 per cent in 2050. For sanitation, the proportion will diminish from 51 per cent in 2000 to 15–18 per cent in 2050. This would lead to a significant reduction in the numbers of children suffering from related ill health (OECD 2012; Hughes *et al.* 2011).

Sustainable world scenarios

The goal for water is to reduce global water stress. Several scenarios have explored the scope for achieving this, such as the "technogarden" scenario from the *Millennium Ecosystem Assessment* (Alcamo *et al.* 2005b), and three alternative

Box 16.3 The sustainable world scenario for water withdrawals

Simulations were carried out to calculate future water availability, water withdrawals (Figures 16.10 and 16.11) and water stress (Figure 16.12) under conventional and sustainable world scenario conditions, assuming the same socio-economic development. In these calculations, for the sustainable world scenario the following assumptions were made:

- stringent efficiency measures are taken in industry and residential water use;
- the irrigation area remains constant;
- climate policies lead to a reduced demand for thermal cooling in power generation as fossil-fuel-powered plants are partly replaced by renewable energy sources; and
- the climate patterns were assumed to be consistent with limiting global temperature rise to 2°C above pre-industrial levels.

As a result, global water withdrawals after 2015 are projected to decrease substantially compared to the conventional world scenario (Figure 16.11). Nevertheless, regions affected by severe water stress still exist and the number of people living in river basins suffering from water stress could reach 3.9 billion in 2050 (Figure 16.12). An important lesson is that improvements in efficiency of water use are necessary to reduce water withdrawals, but are insufficient to avoid water scarcity. The core issue is the amount of water used for irrigation and the high concentration of water demand in urban areas. In other words, if water scarcity is to be reduced further, fundamental changes in agricultural practices are needed – improving irrigation efficiency, shifting irrigated areas from water-scarce to water-rich basins, moving from irrigated crops to rain-fed crops, or relying on imports from other regions.

Figure 16.11 Water withdrawals under conventional world and sustainable world scenarios, 2005–2050

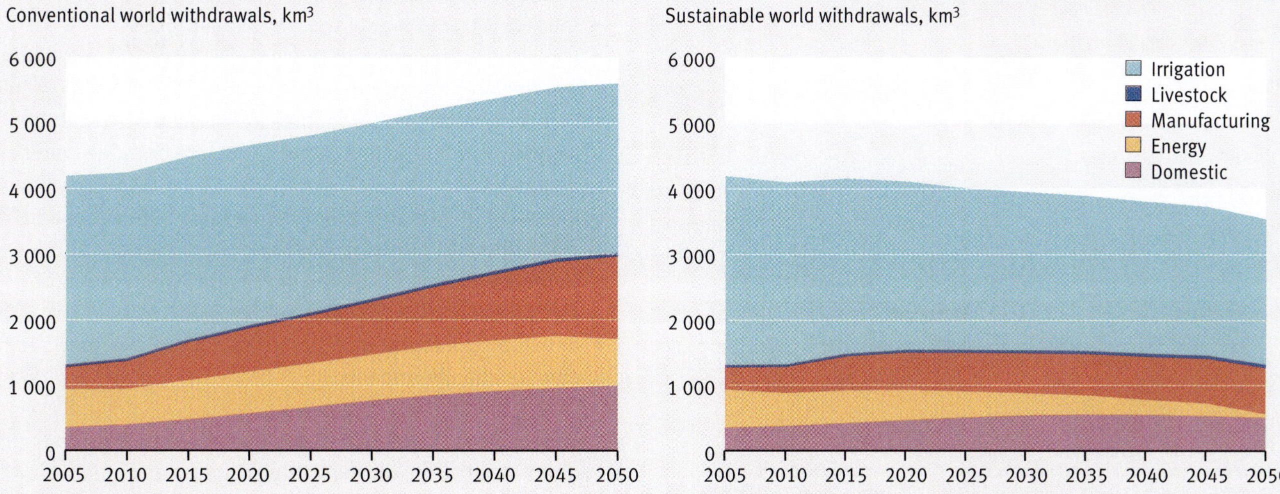

Source: New calculations for *GEO-5*; WaterGap model from Alcamo *et al.* 2003 and Flörke and Alcamo 2004

Figure 16.12 Water stress under current conditions and for 2050 under conventional and sustainable world scenarios

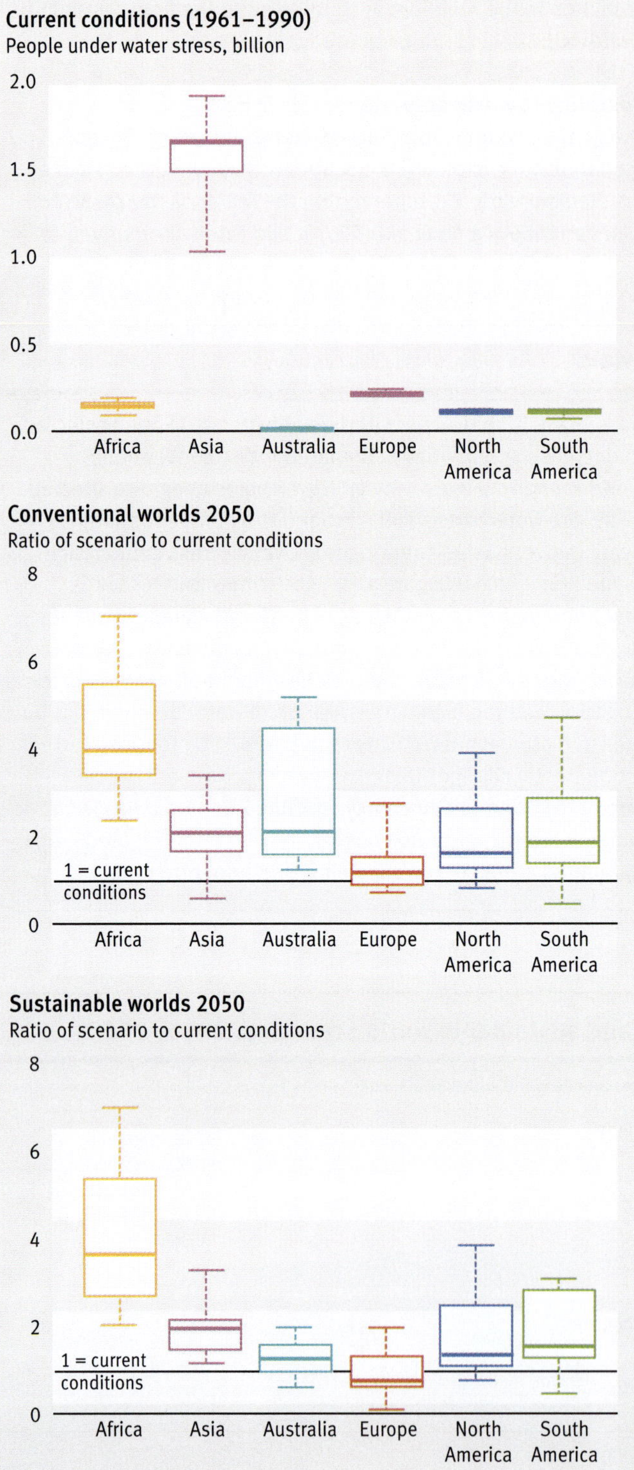

Note: By using a Box-Whisker plot the five-number-summary can be depicted, i.e. the minimum, lower quartile, median, upper quartile, and the maximum are presented in the same graph. The uncertainty ranges expressed in the plot represent different model runs categorized as "baseline" and "challenge" scenarios by two global hydrological/water models taking into account different conditions.

Source: Arnell *et al.* 2011; Alcamo *et al.* 2007, 2005b; UNEP 2007

transition scenarios developed by the WBCSD (WBCSD 2006). The main measure, increased efficiency, is discussed in Box 16.3. In general, the sustainable world scenarios lead to lower numbers of people living with severe water stress, mostly as a result of a reduction in water withdrawals due to behavioural and technological change. However, even under sustainable world scenario assumptions, some regions experience a doubling (median) in the number of people living with water stress compared with current conditions (Figure 16.12). Nevertheless, due to regional population growth and spatial variation of climate change impacts, an increase in water stress is apparent compared to current conditions. This implies that competition between water-related sectors would still be important.

The OECD's *Environmental Outlook to 2050* (OECD 2012) assesses the costs and benefits of halving the number of people without access to safe drinking water by 2030 compared to 2005 levels, and of full access by 2050. The study indicates that, to reach such goals, significant additional infrastructure investments and operations and maintenance resources would be needed. The average required was estimated at US$1.9 billion globally each year between 2010 and 2030, and US$7.6 billion each year between 2031 and 2050. The improved access to drinking water and sanitation would also lead to other major benefits: Hutton and Haller (2004) estimate that every dollar spent on drinking water and sanitation creates economic returns of US$12–34, depending on the region and the technology. Three-quarters of these benefits stem from decreased collection time, particularly for women and especially when water is piped to premises, while the other benefits are mostly linked to a reduction in water-borne disease and death, such as from diarrhoea. The OECD (2012) projects total avoided deaths of around 81 000 per year in 2050. Possible policy levers to improve access to safe drinking water and reduce water stress include:

- investing in research, development and training to increase irrigation efficiency;
- controlling the extent of irrigated areas;
- using waste water and desalinated water to conserve freshwater resources;
- reusing water in manufacturing industries;
- investing in devices and processes for reusing grey water (wastewater from domestic activities);
- investing in education to raise awareness of the need to save water and the link between unsafe drinking water and disease;
- investing in infrastructure for accessing safe water and for collecting and purifying waste water;
- reducing the use of cooling water in the generation of electricity; and
- developing adaptation and mitigation policies to reduce climate change impacts.

As agriculture is the largest user of water worldwide, water stress and food security are strongly interrelated. Lack of sufficient water, possibly caused by alternative water use, can limit food production; at the same time, water consumption for agricultural production could limit residential and industrial water supply.

These relationships further emphasize the need for integrated resource planning at the level of water basins.

Terrestrial biodiversity

Conventional world scenarios

Different scenario studies and assessments have considered biodiversity loss (CBD 2010b, 2006; UNEP 2007; van Vuuren *et al.* 2006; MA 2005a; Sala *et al.* 2000), including information on extinction rates, changes in forest cover and changes in species abundance and distribution (Leadley *et al.* 2010). An important element of future biodiversity loss is land-use change (Figure 16.13), with particularly important trends including the loss of mangroves, wetlands and tropical forests (CBD 2010b). Scenarios indicate a further decline of tropical forests, while the area of temperate forests is likely to expand. Another factor affecting biodiversity is natural asset exploitation: an increasing population with rising affluence in the conventional world scenarios implies, for example, greater demand for forest and tree products. Currently, most timber products are extracted from natural forests, while plantations provide some 35 per cent of harvested wood (Sohngen *et al.* 2001). The increased timber demand is expected to lead to a further expansion of managed forest in tropical zones, at the expense of unmanaged forests (Gibson *et al.* 2011).

Projections of extinction rates vary widely between studies. Those based on species/area relationships lead to relatively high extinction rates, whereas species distribution models that allow for migration generate much lower rates. However, all estimates of future extinction rates are considerably higher than is considered sustainable. Conservative estimates of future rates of extinction expect them to be similar to the currently high rates (Pereira *et al.* 2010).

Sustainable world scenarios

It has been shown that the expansion of agricultural land could be avoided by increasing crop yields combined with policies to reduce food waste, control biofuel use, conserve resources and shift diets (Figure 16.14) (Ten Brink *et al.* 2010; Stehfest *et al.* 2009; Wise *et al.* 2009). Analysis has also shown that the 2020 target to prevent the extinction of known threatened species could be achieved by a well chosen network of protected areas in combination with reduced habitat loss (Butchart *et al.* 2012; Ricketts *et al.* 2005). In a recent study, a combination of policy options – including the expansion of protected areas into a well-chosen network covering 29 per cent of the world's surface, an increase in agricultural productivity and reduced post-harvest losses, dietary change, improved forest management, and climate mitigation – suggested a significant restoration of natural areas and reduced biodiversity loss (Ten Brink *et al.* 2010). A great impact on land-use change could also be achieved through financial mechanisms: Wise *et al.* (2009) demonstrated that a policy that provided equal monetary incentives to reduce greenhouse gas emissions to all emission sources, including land use, could lead to the preservation of both managed and

Figure 16.13 Changes in the extent of forest up to 2050 in different global scenarios, and estimated rates of species loss

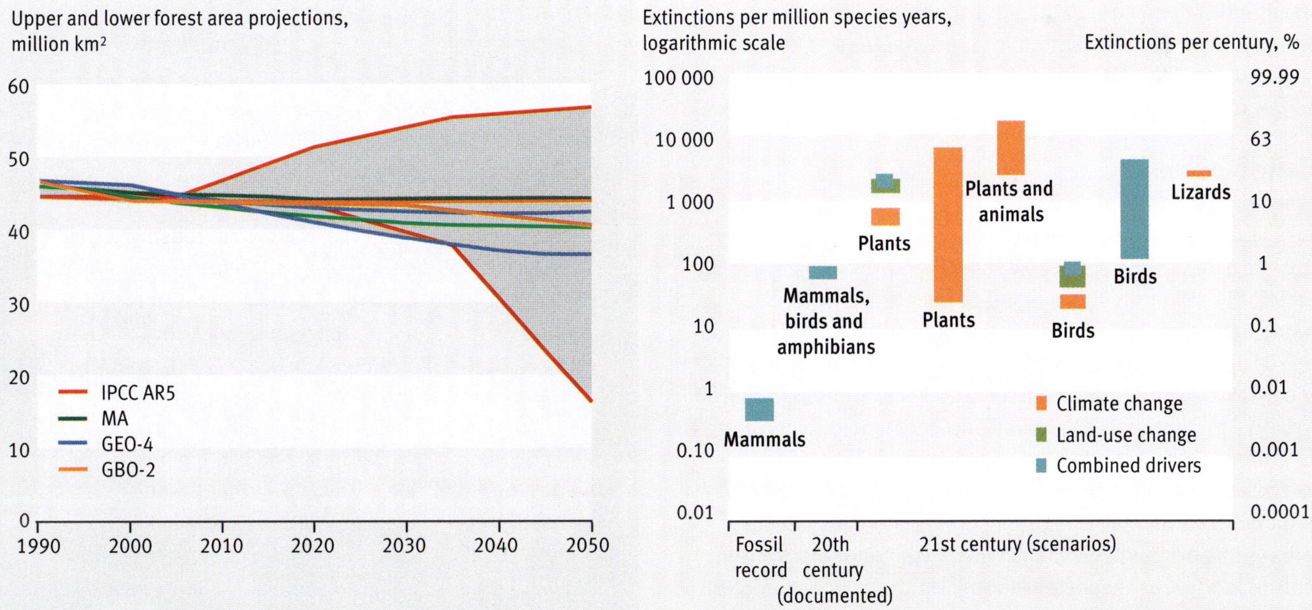

The graph offers a comparison of extinction rates in the distant and recent past, with projections of species committed to extinction during the 21st century according to different global scenarios. The extinction rate caused by each driver and the total extinction rates are differentiated when possible.

Note: For 20th-century extinctions, mammals fall into the upper bound, and birds and amphibians into the lower bound.

Source: CBD 2010b; Pereira *et al.* 2010a

Figure 16.14 Options for reducing biodiversity loss by 2050

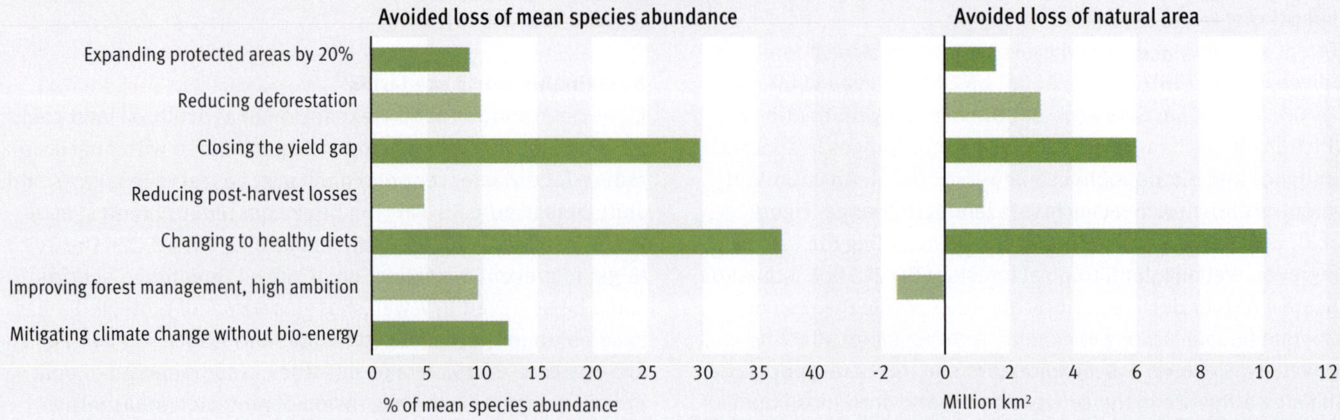

Note: Mean species abundance is the average relative to the original state.

Source: Adapted from Ten Brink *et al.* 2010

unmanaged forest. Other measures that could also be considered include agroforestry and the certification of sustainable wood products (Angelsen 2010).

A critical threat to biodiversity is rising demand for agricultural land leading to conversion of natural habitats, whereas a substantial increase in yields reduces the demand for land and is considered necessary to reduce habitat loss. This may, however, cause a decline in the biodiversity and ecosystem services of farmlands (Robinson and Sutherland 2002; Tilman *et al.* 2002) but an increase in biodiversity across the landscape (Phalan *et al.* 2011). An expansion of protected areas may increase competition for land, decreasing the potential for agricultural production, which could in turn lead to higher food prices.

Aquatic biodiversity

Conventional world scenarios

Conventional world projections of changes in the biodiversity of aquatic ecosystems are scarce; however, pressures are expected to remain high as a result of increasing water scarcity, climate change, pollution and exploitation (CBD 2010b; Rands *et al.* 2010). For freshwater systems, organic pollution and dam construction are important threats. For marine systems, destructive and intensive fisheries and ocean acidification are among the main factors that could reduce biodiversity (Halpern *et al.* 2008; Pinnegar *et al.* 2006; Pauly *et al.* 2003). Ocean acidification may transform coral reefs into systems dominated by other species, and will cause major disruptions in marine food webs, especially in the Southern Ocean (McNeil and Matear 2008).

About 32 per cent of wild marine fish populations are classified as collapsed, depleted or recovering (FAO 2010), and while there has been some rebuilding in areas with strong fisheries management (Worm *et al.* 2009), the majority of the world's fisheries are operating with severe overcapacity (Anticamara *et al.* 2011) resulting in significant economic losses (Arnason *et al.* 2009; Srinivasan *et al.* 2012). Global assessments for exploited marine invertebrates indicate similar trends (Purcell *et al.* 2011). Overexploitation has already depleted fishery yields, reduced

Certain marine fish populations have been depleted to such an extent that they may not be able to recover. *© J Tamelander/IUCN*

the abundance of large fish and caused local extinctions. Analysis indicates that global wild fish catches will decrease in the future unless fishing effort and catch rates are reduced to sustainable levels (Figure 16.15). Projections also show that if current trends persist, the populations of medium and larger fish in the world's oceans will continue to decrease while small fish may increase in abundance due to release from predation (Ten Brink *et al.* 2010; Pauly *et al.* 2003).

Sustainable world scenarios

Reducing fishing effort, even to the level of maximum sustainable yields for all fish populations, could make an important difference for a sustainable world. It would require a stringent reduction, but only temporarily while larger- and medium-sized fish populations recover (Ten Brink *et al.* 2010; Pauly *et al.* 2003). After this period, fishing could return to a level that can be sustained in the long run.

Increasing the protected area of land and sea would reduce the availability of suitable land and fishing grounds for food production. Increasing protection will only therefore be effective if it is done in combination with more efficient production methods on the land under cultivation and the establishment of sustainable fisheries. It should also be noted that reduced fishing effort initially leads to reduced fish landings. This implies that while populations rebuild, demand for wild-caught fish needs to be replaced by demand for aquaculture – which itself has specific environmental impacts – or crop and animal products. As fish populations rebuild, the supply of fish would, however, grow to a long-term sustainable level that is comparable to the peak 1980s catch levels (Ten Brink *et al.* 2010; Pauly *et al.* 2003).

Synthesis: gaps and sustainability pathways

The review of conventional world and sustainable world scenarios in relation to the strategic goals discussed earlier in the chapter illustrates that continuing on the current trajectory would lead to major environmental damage and a serious loss of ecosystem services by 2050. It would also leave many people without sustainable access to food, water and energy. In contrast, the sustainable world scenarios show how societies could meet some of the 2050 targets or, at least, join a trajectory that would make meeting such targets more feasible. The changes suggested in the sustainable world scenarios include all kinds of measures related to greater technological implementation, changes in consumption patterns and improved management. In general, major shifts from current trends are required for each specific issue. It should be noted that, due to inertia in the human-environment system, several targets still imply significant environmental change, such as the target to limit temperature increase to 2°C above pre-industrial levels. Sustainable world scenarios, as well as requiring mitigating measures, call for measures to cope with or adapt to these adverse effects.

Table 16.3 presents an overview of the key measures suggested in the thematic sections of *GEO-5*, including changes in consumption and production levels and patterns. On the production side, changes include improving efficiency and using fewer inputs; input switching and producing with low- or non-polluting inputs; end-of-pipe measures; and integrated production systems. On the consumption side, changes consist of alterations in lifestyle, such as dietary shifts and greater use of public transport. A focus on education, including awareness raising, investment in infrastructure and the creation or

Figure 16.15 Marine catches with and without a reduction in fishing effort, by region, 1950–2050

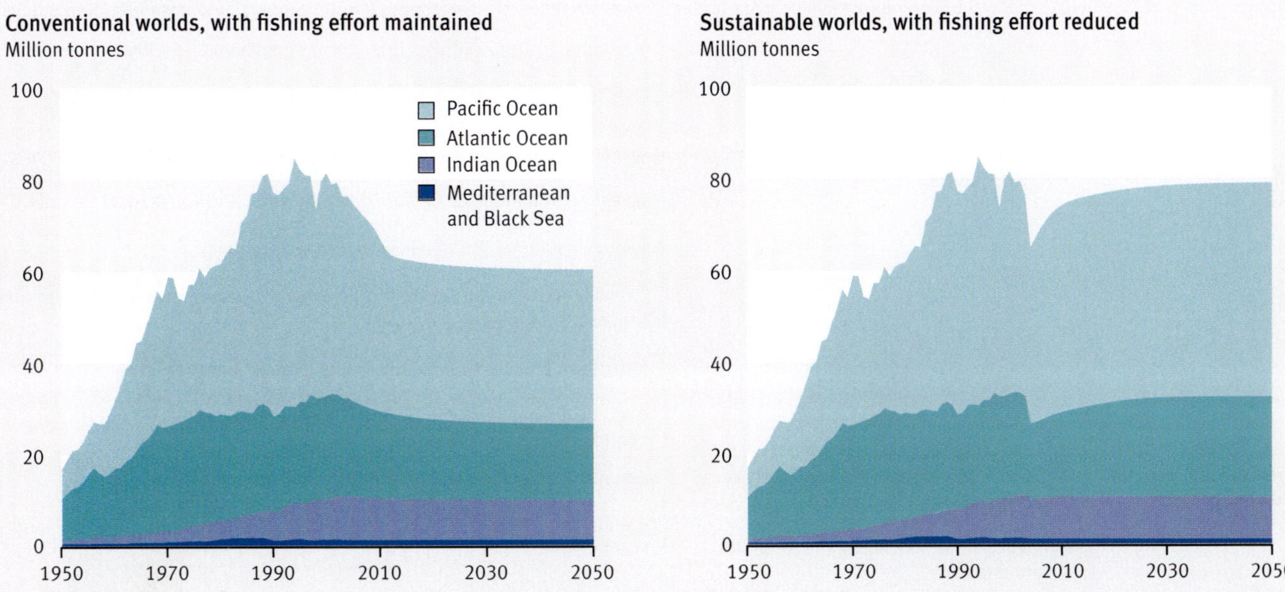

Source: Ten Brink *et al.* 2010

strengthening of markets, and adaptation to unmitigated change are other measures included in the sustainable world scenarios. Measures can also be related to the layers of transformation in Figure 16.2: while many measures address the outer layer, others deal with the mid or even deep layers, such as changes in behaviour, mainly the result of education and awareness raising.

Table 16.3 Overview of the gap between the conventional and sustainable world scenarios and important measures to close the gap

Theme	Gap between the conventional and sustainable world scenarios	Examples of important measures to close the gap
Atmosphere and energy	By 2050 greenhouse gas emissions have increased by 70% compared to now, while the vision requires a 50% reduction By 2030 1 billion people live without access to electricity and nearly 3 billion rely on traditional biomass for cooking and heating Air quality levels are still above WHO guidelines in most developing countries	Reduce carbon intensity by 4–5% per year compared to a baseline increase of 2% per year, partly by increasing the contribution of non-carbon energy options to more than 50% and by significantly increasing energy efficiency Increase investment in electrification Create a smart subsidy and micro-financing system to provide modern fuels for cooking and heating to the poorest Promote less energy-intensive lifestyles and material consumption Increase investment in research and development Use technology to reduce air pollution
Land and food	By 2050 13–25% of all children are undernourished By 2050 cropland has increased by 10–20% compared to 2010 By 2050 pasture area has increased by 10% compared to 2010	Increase crop yields and overall agricultural productivity by, for example, closing the yield gap between deveped and developing countries Encourage planting crops and crop varieties that are better suited under changing climatic conditions Reduce food waste Improve use of urban-rural landscapes for food systems and natural resources Strengthen land-use policies and planning Make appropriate social, technological and economic investments in infrastructure and regulate agriculture, including increasing efficiency of irrigation, nutrient recycling and pest management Reduce consumption of livestock products
Water	In 2050 6.5 billion people live in areas under water stress By 2030 5–8% of the population live without safe drinking water By 2030 17–28% of the population live without improved sanitation	Invest in research, development and training to increase irrigation efficiency Control the extent of irrigated areas Use wastewater and desalinated water to save freshwater resources Invest in education to raise commitment and awareness for water saving Invest in infrastructure for safe drinking water and wastewater treatment Reduce cooling water needs with new technologies Increase reuse of water in manufacturing industries; invest in devices and processes for reusing grey water Mitigation policies to prevent climate change impacts Adaptation measures for climate change such as rainwater harvesting, flood control for rivers and water transfer
Terrestrial and aquatic biodiversity	By 2050 forest area has further decreased compared to 2010 Extinction rates are clearly above fossil record rates Global fish are exploited above sustainable levels	Conserve key terrestrial and aquatic biodiversity Reduce the pressure on land, mainly through the options under the land theme Reduce global fishing effort Improve forest management

Box 16.4 Integrated global analysis of sustainability scenarios

The scenario assessments address gaps and sustainability pathways by theme, as well as the potential trade-offs and co-benefits with the other themes. It is, however, important to further analyse the links between themes in an integrated manner. Here, the global Threshold 21 (T21) model (Bassi *et al.* 2010) is used to address how to achieve the goals set out in Table 16.1 by focusing on the investments needed and the trade-offs and synergies of the interventions across various sectors. Two alternative sustainable world scenarios to 2050 are compared to a conventional world scenario. Scenario A focuses entirely on additional investments in transforming technology and production to achieve the goals. Scenario B focuses on how adding lifestyle change reduces those investments.

Scenario A shows that investments of about 2 per cent of GDP between 2011 and 2050 will make the necessary shift to sustainable development. The interventions include efforts to reduce energy demand in buildings, industry and transport (38 per cent of investments); shifting to more renewable energy sources (31 per cent); increasing food production through ecological agricultural practices (10 per cent); rebuilding fishery stocks (8 per cent); and sustainable management of forests (3 per cent) and water (10 per cent). Thus, achieving the climate goal emerges as requiring the highest investment. The measures in the energy and agriculture sector reduce greenhouse gas emissions to a level that limits atmospheric concentrations below 450 ppm. In T21, managing natural resources more sustainably also helps restore key natural resources or greatly mitigate their depletion (Table 16.4). Scenario B shows that lifestyle changes that reduce demand for energy, including for transport, heating and cooling, water and biomass, will reduce the need for investments in production and technology to about 1.2 per cent of GDP. It should be noted that the costs of the lifestyle changes are not accounted for and will at least involve the costs of transmission of information.

In both scenarios, the synergies from the additional investments focused on environmental sustainability and on reducing the stress that natural resource depletion puts on economic productivity. Overall, the macro-economic results of the T21 model show that investing to reach the sustainability goals creates more jobs than a conventional world scenario and leads to higher GDP growth. The shift to greener investment will lead to slightly lower growth rates in initial years, as is usual for transition investments, but A and B scenarios' GDP will pass business-as-usual projections well before 2030 (UNEP 2011c). This is a significant, but uncertain, result as some other models show negative GDP impacts, depending on the assumed cost of fossil fuels and renewable alternatives, the impacts of environmental change and responses to increases in investments (Bassi and Eaton 2011; Clarke *et al.* 2010; Fisher *et al.* 2007). Taking account of the cross-sectoral impacts of an integrated approach, such as synergies between agriculture and forestry, enables further understanding of the complexity underlying mutual socio-economic and environmental dependence and the need for coordinated programmes and investments to achieve the *GEO-5* goals.

Table 16.4 Threshold 21 scenario results for key indicators

	2011	2050		
	Conventional worlds	**Conventional worlds**	**Sustainable worlds A**	**Sustainable worlds B**
Economic sector				
Real GDP, US$ billion per year	69 363	155 192	190 428	181 762
GDP, US$ per person per year	9 996	17 472	21 166	20 217
Additional investment, US$ billion per year	0	0	3 712	2 133
Social sector				
Total population, billion	7.0	8.9	9.0	9.0
Calories per person per day	2 787	2 981	3 348	3 234
Population below US$2 per day, %	19.5	11.1	8.9	9.4
Human Development Index	0.60	0.67	0.71	0.70
Total employment, million people	3 186	4 624	4 689	4 612
Environmental sector				
Forest area, billion hectares	3.9	3.7	4.5	4.5
Waste generation, million tonnes per year	11 242	13 855	14 497	14 338
Ratio of footprint to biocapacity	1.5	2.1	1.1	1.2
Primary energy demand, million tonnes of oil-equivalent per year	12 956	19 733	13 421	12 470
Renewable energy share of primary demand, %	13	13	27	26
Fossil fuel greenhouse gas emissions, billion tonnes per year	32.1	52.0	18.9	20.6

There are important synergies between the different goals. For example, reduced climate change would improve water availability and crop yields as well as relieve pressure on biodiversity. Decreased consumption of food, water and fossil fuels would reduce the mitigation requirements to achieve the biodiversity, water stress and climate change objectives, while increased agricultural yields would lessen the pressure on biodiversity. In some cases, however, options for a specific theme may induce important trade-offs with other themes. Policies that combat environmental degradation may have impacts on human development, and vice versa: creating bio-reserves, for example, may increase land and food prices, while desalinating water would significantly increase the demand for energy. Ignoring such cross-sectoral links might jeopardize the success of the sustainability transition effort and lead to significant delays in reaching the targets. Strategies must therefore go beyond theme-oriented conventional thinking and take a broader systemic view that reflects these links. Central to this is how the measures introduced for the different themes would work together. Box 16.4 describes an example of an analysis in which integrated scenarios are explored. This shows that it is indeed possible to identify pathways that would meet multiple sustainable development targets.

ADVANCING SUSTAINABILITY

Given the massive gap between conventional and sustainable worlds in 2050, it is clear that inertia is a principal obstacle – in the form of dominant unsustainable processes, structures and habits. Moving off the current path will require a transformation without precedent in human history (Steffen *et al.* 2005; Takács-Sánta 2004). Guiding changes of such magnitude and complexity will require time and patience to facilitate a sometimes steady, sometimes fitful, transition process. During this process, structures and underlying mental models need to be evaluated – some phased out and others energetically phased in. These underlying models should be consistent with the desired trajectory for targets relating to atmosphere and climate, land, water and biodiversity, resource efficiency and waste management. The changes must effectively transform society's material metabolism, especially those elements and dynamics currently locking countries into trajectories that are not the ones they would prefer. The changes needed to realize sustainable world trajectories have to be diverse and should combine demographic, technological, governance and investment measures, along with lifestyle changes resulting from shifting mindsets towards dematerialized values. They also need to provide sufficient leverage to break the inertia of unsustainable trends.

Diversity is key to understanding the structure and function of complex adaptive systems and enhancing their resilience to stress (Innes *et al.* 2005). A diversity of potential responses is required because effective interventions must be sensitive to socio-cultural and environment-development contexts. Diversity also helps strengthen resilience and provides a form of insurance if some of the responses fail, as many have over the last few decades with regard to biodiversity, climate and other key environmental issues (Speth 2005).

Ecuador, one of the UN-REDD programme's partner countries, is prioritizing social and environmental co-benefits in its REDD+ readiness preparations.

© Elena Kalistratova/iStock

Grounding responses in local processes and experience not only builds diversity, it can also tap into knowledge that has evolved in direct connection with a particular environment. This section focuses on the sub-global level, where countries, communities or other entities, when addressing environmental problems from the perspective of a particular place, self-organize their response mechanisms. The self-organizing potential of communities, businesses, civil society and other actors can be strengthened by building networks, by matching the scale of governance to the scale of the ecosystem in question, and by promoting innovation and action (Berkes *et al.* 2003).

Paying attention to how response measures interact with, support or constrain each other is difficult but increasingly important. The significance of this in the case of ecosystem services has been recognized and may involve the reduction of one to enhance another, for example sacrificing the potential value of mining to optimize a forest's carbon sequestration or biodiversity value – or vice versa (Rodríguez *et al.* 2006). At the same time, a diversity of measures offers increased opportunities for finding synergies and supporting measures that produce multiple benefits (UNEP 2011b), such as the carbon sequestration and biodiversity co-benefits of forest conservation.

While responses at the sub-global level are likely to be diverse and grounded in local conditions, it is important to identify some common strategic elements for advancing sustainability. This section discusses four strategic elements:

- compelling visions and social contracts;
- reversing the unsustainable;
- leverage points; and
- adaptive management and governance.

Such strategic elements inform and guide the development and implementation of specific policy measures under the widest range of geographic and thematic conditions, across all scales from the local to the national and regional. In some cases they also apply at the global level, with implications for the international priority responses discussed in Chapter 17.

Social consensus around compelling visions of sustainability

Initiating and managing the transition towards internationally agreed goals and targets requires setting well-defined directions through a clearly articulated vision. Governments and other organizations at all levels should develop the capacity to engage society in expressing cohesive visions of an environmentally sustainable future around which new social consensus – in the form of deliberately agreed social contracts, sectoral and thematic strategies and policies – can be developed and implemented.

Based on the experience of public- and private-sector organizations, developing visions of an environmentally sustainable future is an effective mechanism for achieving progress in a desired direction (Costanza 2000). The generation of shared visions is not only an essential element, but also an underrated element of environmental strategic policy development and management (Meadows 1996). Current public institutions often have limited capacity to construct legitimate and credible visions relying on stakeholders' inputs (Walker *et al.* 2006). This has also been recognized by the Global Survey on Sustainable Lifestyles, which concluded that the missing links between global environmental challenges and individual action are pragmatic, holistic and compelling visions of what sustainable societies could look like (UNEP 2011c).

In the absence of a clear, coherent vision that reflects the links between social, economic and environmental issues, policies may lead to disadvantageous trade-offs, often sacrificing environmental or social goals in the face of more quantifiable economic objectives. The result is often addressing one sustainability problem while passing the true cost to another sector, community, region or even another generation, creating even more persistent and complex risks in the long term (Loorbach and Rotmans 2005).

Visioning is important both for exploring the broader implications of the effort needed for simultaneously meeting internationally agreed global goals and targets, and for discussing environmental futures from strategic sub-global points of view under different economic, social and ecological conditions. The power of visions to provide direction, navigate pathways, recognize solutions and explore uncertainty has been recognized and illustrated by the growing number of sub-global efforts that involve articulating a vision or outlook for the future. Some prominent examples follow.

- *Regional projection of options for resource efficiency improvements.* UNEP and regional partners prepared an outlook to explore how resource efficiency improvements can help maintain ecosystem health and contribute to the provision of essential ecosystem services in the rapidly industrializing economies of Asia and the Pacific (UNEP 2011b).
- *Scenario visioning at the state and city level.* A combination of qualitative visioning and quantitative modelling methods, involving extensive participation, has been used in Minnesota, United States. The purpose of the process was to help regional leaders make strategic decisions about sustainability, to identify related knowledge and research gaps and to introduce systems thinking into policy making and planning (Schmitt-Olabisi *et al.* 2010). An innovative 100-year vision has been developed for Panjim, the capital of the Indian state of Goa and surrounding areas, around the concept of "RUrbanism", an integration of urban and rural development in terms of resource use and the convergence of human well-being (Revi *et al.* 2006).
- *Visions focused on solutions to specific environmental problems.* A vision has been developed to tackle the acute water shortages in Kuwait (Al-Damkhi *et al.* 2009). Pathways of possible emission reduction and low-carbon futures have been developed for specific regions in Europe (Matthes *et al.* 2006) and cities in North America (Metro City of Vancouver 2011; Danish Architecture Centre 2011).

The development of visions involves creative tensions between quantitative model-based projections and qualitative normative articulation of what is desired for the future (Schmitt-Olabisi *et al.* 2010; Patel *et al.* 2007; Strauss 1987). A vision refers to a moving target, guiding the self-organizing and innovative forces of a society – forces that otherwise would remain diffuse. A vision differs from a goal; it is a tangible image of a future that is not subjected to the exact definitions involved in setting and achieving goals (Jaeger *et al.* 2000).

When developing visions, it is essential to ensure integration across policy themes to capture the intricately linked ecological, social, economic, ethical and institutional dimensions of sustainability problems, while reflecting on uncertainty – including surprise, critical thresholds and abrupt change, which are inherent in non-linear natural and social systems such as the Earth System (Swart *et al.* 2004). The visioning process should also account for human decisions as a key conditioning factor – the constitution, reproduction and reformulation of human needs, wants, vulnerabilities and values is essential for illuminating consumption, social goals, institutional innovation, social learning and the prospects for alternative futures (Robinson 2004; Swart *et al.* 2004).

Participatory, integrated visioning processes are most useful when they are iterative, support policy development and adaptation, and are embedded in institutional cultures with adequate capacity to manage the process. Embedding elements of a vision in institutions can take place through legal and administrative means. However, creating and maintaining the political will that makes such measures feasible requires more – as does recruiting society to adopt lifestyle changes included in the sustainability scenarios elaborated above.

According to the German Advisory Council on Global Change, formalizing the transition agenda to a low-carbon economy can be achieved in the form of new social contracts (WBGU 2011). The council argues that the need for such contracts is grounded in the joint responsibility of states and their global communities – business, science, civil society and even individuals – for tackling threats to the stability of the Earth System, with changes going beyond technical and bureaucratic reforms. Formulated around positive visions of the future, a new form of interaction between politics, society, the economy and science would need to be defined to bring creativity, resources, capacity, legitimacy and political will together in the interest of navigating the transition and achieving tangible progress towards outcomes consistent with such visions.

A social contract is a contract or agreement between people to form a society that determines their moral and/or political obligations. With Socrates being its earliest known proponent in antiquity, the concept is almost as old as philosophy. Social contracts can exist in different forms at different levels, and can outline different obligations for specific stakeholders. For example, a social contract for science would need to focus on the commitment of the scientific community to systematically apply its creativity to addressing the fundamental problems of the Earth System and the public's interests therein (Lubchenco 1998). But social contracts could take other forms in other sectors. Standards with a focus on the sustainability of products and processes in sectors such as forestry or agriculture have been proposed as a form of social contract, with the state providing an overall operating framework but with non-govenmental organizations, businesses and consumer associations working out and codifying the details (Giovannucci and Ponte 2005).

Social contracts would need to address both short-term and long-term issues. The Stockholm Memorandum of Nobel Prize Winners called for a dual-track approach: short-term actions, emergency solutions to address the most pressing environmental trends and their drivers within today's faulty institutional framework; and long-term changes focused on transforming the institutional framework itself, to create an environment for innovation, learning and action without the barriers of today. In the short term, they call for focus on achieving the Millennium Development Goals (MDGs), and in the long run for a new agreement between developed and developing countries to scale up the investment and capacity building necessary to reach those goals (Royal Swedish Academy of Sciences 2011).

Reversing the unsustainable

The transition towards internationally agreed sustainability goals and targets requires not only introducing innovative new measures and policies, but also the rapid phasing out of policies and practices that reinforce vested interests that prevent the transition from happening.

Learning what not to do is a necessary, yet often neglected, precondition to framing the improvement of sustainability and, in particular, to the integrated sustainable governance of complex social-ecological systems. Understanding the constraints and opportunities for bringing such framing into science, education and policy debate while avoiding simple ideological discussions is essential in advancing to a sustainable world (Tàbara and Pahl-Wostl 2007). However, phasing out existing policies and practices is not always easy. Policies create dependencies, and discontinuing a policy may hurt economic and other vested interests and result in adjustments. Yet discontinuing unsustainable policies and practices may also free up resources and create new niches that innovative measures, consistent with the sustainable world trajectory, can fill.

One important area for the phasing out of existing unsustainable measures is counterproductive government subsidies. These are a widespread phenomenon encouraging unsustainable activities in sectors such as agriculture, energy and transport (van Beers and van den Bergh 2009). A subsidy is a "payment by a government to an individual or firm, the intent of which, theoretically, is to decrease the divergence between social costs and benefits – to internalize externalities" (Myers and Kent 2001).
Subsidies can produce socially desirable outcomes, such as

provision of a public good that would be undersupplied by the market. When properly designed and applied, they can also provide investment to help green industry or technology start-ups become economically competitive (Bagstad *et al.* 2007). However, not all subsidies help progress towards agreed goals and targets. There are so-called "perverse" subsidies that increase the divergence between private and social costs and benefits (Myers and Kent 2001), typically by increasing the size and pollution intensity of economic activities, and often without clear, compensating social benefits but with economic benefits channelled to a small minority of entrenched interests in business-as-usual (van Beers and van den Bergh 2009, 2001). Examples include:

- agricultural subsidies that encourage the intensification of production or the expansion of farmland at the expense of natural ecosystems; without precautions, such measures often lead to significant negative impacts on biodiversity and habitats (Robin *et al.* 2003);
- fossil fuel subsidies increase greenhouse gas emissions and thus contribute to climate change; while these subsidies are often designed to keep energy costs lower for the poor, they typically end up benefiting medium- to high-income households: in 2008, for example, the Indonesian Ministry of Economic Affairs concluded that the top 40 per cent of high-income families benefited from 70 per cent of the subsidies, while the bottom 40 per cent of low-income families from only 15 per cent of them (IISD GSI 2011; IEA 2008);
- subsidies for road transport, in which subsidized road construction directly destroys habitat, and burning subsidized fossil fuels in motor vehicles is a major contributor to air pollution and climate change (Myers and Kent 2001).

Discontinuing existing unsustainable practices may be a long and uneasy process, but does lead to a shift to sustainable behaviour. An example of reversing the unsustainable can be observed in marine fisheries through the implementation of the Wellington Convention signed in 1989, which prohibits fishing with long driftnets in the South Pacific. This type of fishing is destructive because of its non-selectiveness and the high level of by-catch. Driftnet fishing was widely used, especially during the 1980s, and posed a threat to fish stocks, in particular to albacore tuna. Although the convention initially created tension among the distant-water fishing nations, it managed to balance the economic interests of fisheries with the pressure on the marine environment and led to the adoption of a global moratorium on full driftnet fishing on the high seas (Techera 2011; Hewison 1993). The International Convention for the Regulation of Whaling (ICRW), which originally aimed to prevent the oversupply of whale products but turned into a key instrument of whale conservation (Maffei 1997), can stand as another example. The governance regime for whales has contributed to more sustainable practices and a change in mindsets, allowing a transition from predominantly consumptive exploitation of a natural resource (whaling) to non-consumptive use such as whale watching and related tourism.

Sustainability not only requires being aware of complexities and uncertainties, but also developing normative patterns of knowledge creation and collective behaviour that render action possible (Mangalagiu *et al.* 2011). A transition to sustainability demands profound changes in understanding, interpretative frameworks and broader cultural values, just as it requires transformations in the practices, institutions and social structures that regulate and coordinate individual behaviour. In this context, it is essential to get to the position where people, industry and governments can easily distinguish between objective facts and opinions that are presented as facts to advance particular interests, and rely on the former to make informed decisions. This is where education becomes paramount, and involves raising awareness of the challenges and solutions as well as harnessing and communicating diverse knowledge.

A road penetrates a rainforest in the Chiriqui Highlands of Panama. The growing networks of roads through tropical rainforest are of serious environmental concern. *© Alfredo Maiquez*

Reversing or phasing out unsustainable policies and practices is essential. But it is only a first step that must be accompanied by investment in investigating solutions that incorporate traditional knowledge and novel forms of sustainability science as well as engaging with broader civil society (Bäckstrand 2003). Reversing unsustainable practices must be accompanied by providing society with knowledge and practices that are consistent with management within planetary means.

Applying leverage

Achieving internationally agreed goals and targets will require that policy makers look for advantageous places to intervene and apply leverage through the design and implementation of a diverse array of policies and instruments that:

- facilitate a mindset shift to align with sustainability principles;
- change the rules and incentives to advance sustainable practices; and
- create feedback and make adjustments to keep environmental pressures in check.

The comparison of conventional and sustainable world scenarios brings into focus the challenge at hand: an urgent need for a societal system-wide shift in the way energy is generated and used, in consumption patterns, and in natural resource management – to quickly adjust the pathway towards sustainability – in the context of an unstable global economy and present underachievement of most of the MDGs. In short, there is neither time nor money to waste. However, the crisis of sustainability may help increase awareness and understanding of the problem, its underlying causes and the relationship between them, and create momentum. Future policy efforts must be as efficient and effective as humanly possible: by specifically addressing the deeply embedded underlying drivers and targeting and coordinating progress towards long-term change.

Progress toward deep transformation will not be rapid, but evidence of change and progress at levels less deeply embedded in the system will help create the conditions that support and assist deep level change. What is emphasized here is the importance of constructing a diversified portfolio of policy intervention at levels that are fairly easy to access and influence, in conjunction with other interventions which involve deeper analysis, coordination and structural change.

Three layers of transformation were introduced in Figure 16.2 where leverage points might be found, providing practical guidance for policy makers in managing the sustainability transition. Reflecting on the presence or absence of intervention at each of these leverage points can help broaden and diversify policy emphasis, leading to a more resilient and responsive overall strategy. Several examples follow, beginning with those that are at the core of the sustainability transition.

Shift mindsets

At the core of the sustainability transition lie critical reflection and changes in the mindsets and goals that determine how issues themselves are framed. Mindsets in this respect refers both to those held by individuals and to those shared mindsets that define how social groups, cultures, nations or the human species at large approach things. Recognizing the importance of mindsets in the sustainability transition allows an opportunity to reflect and examine underlying assumptions, identify shared values and cultivate common ground. Each of these contributes to defining the shared goals and the compelling visions necessary to bring these changes about. Consider the following examples:

- *Youth education for sustainable development.* Introducing the principles of sustainability at an early and formative age in primary and secondary school supports the power of complex problem solving and can instil the belief that change is possible and preferable, and that a sustainable world is indeed achievable. The United Nations Decade of Education for Sustainable Development (UNESCO 2011) is an example of a global response meant to facilitate national and sub-national efforts in this regard. Policy interventions that help schools integrate sustainability into curriculums have the potential to change mindsets.
- *Social marketing*. While the world is replete with advertisements for products and services, there is a dearth of public messaging to promote the principles of sustainability and non-material aspects of well-being such as health, leisure or time with friends and family. Changes in the rules and ethics of advertising and marketing that match the audience-reach of product and service advertising, but that communicate critical behavioural change for a sustainable future – such as water and energy conservation or the use of green products – have the potential to change mindsets.
- *Beyond GDP.* Nobel prize-winning economists Joseph Stiglitz and Amartya Sen were commissioned by President Nicholas Sarkozy of France to examine the measurement of economic performance and social progress. The Stiglitz-Sen Commission called for a shift in emphasis from measuring economic production to measuring people's well-being and the sustainability of that well-being (Stiglitz *et al.* 2009). They stressed the importance of measuring such aspects as the state of health, education, personal activities, environmental conditions, social connections and political voice, as well as insecurity, inequalities and proximity to dangerous levels of environmental damage. An example of the type of transformation that such a change in mindset represents is Bhutan's Gross National Happiness Index and the national priority to focus on increasing happiness rather than merely focusing on perpetual GDP growth (Government of Bhutan 2011).

Change the rules and incentives

Coordinating deep and enduring system change is neither a single pathway nor a linear process. For example, the rules of a system often arise from a change in mindset, but in turn help support mindset shifts. At this level of system intervention the emphasis is on getting the signals right. Because rules and incentives can institute structural change, they represent the game changers that can catalyse and retain a strong influence on system behaviour over time.

The Rio Branco sawmill in the Brazilian Amazon, working under Forest Stewardship Council certification. Third-party forest certification systems are being used as a way to promote sustainable practices and reduce the carbon footprint of products while improving market access for community-based forest enterprises. © *Antoine Lorgnier*

Policy instruments such as laws, taxes, subsidies and market mechanisms can be directed to shift specific drivers that affect the state of the environment, and the more influential the driver, the more systemic the change can be. Some instruments that are already in place in many jurisdictions, such as pollution charges, can create direct incentives for the reduction of emissions. Other measures, however, with potentially more far-reaching impacts, may require and represent a deeper mind-shift. For example, payment for ecosystem services schemes include a suite of approaches that attempt to attach value to ecological functions that are usually left out of cost-benefit calculations and mainstream economic models, including direct public and private payments (Milder *et al.* 2010). Payment for ecosystem services has been advanced for example in Latin America and the Caribbean (Wunder 2007), and China has implemented some of the largest schemes in the world (ADB 2010).

Create feedback and make adjustments

Interventions aimed at less structurally embedded parts of a system can contribute to the sustainability transition, especially when they catalyse mass action. Interventions that strengthen feedback are designed to "deliver information to a place where it wasn't going before and therefore causing people to behave differently" (Meadows 1999). Such feedback provides the evidence base for the mitigation of environmental pressures. Examples include:

- *Household water and energy metering.* A digital household meter for electricity and water consumption can have a significant impact on individual behaviour. In Armenia, for example, studies on water consumption soon after meter installation revealed that on average water use decreased three to four times (OECD 2008b).
- *Product labelling.* Providing information such as the carbon footprint of products, or forest or marine stewardship certification, can influence consumer behaviour and lead to transformation across sectors. For example, the Marine Stewardship Council (MSC) provides certification standards and requirements for sustainable seafood. As of 2011 there were 133 MSC-certified fisheries representing almost 6 per cent of the total wild fish catch (MSC 2011).
- *Community indicator systems.* Communities that undertake multi-stakeholder processes to identify priority aspects of quality of life and indicators to track progress over time, create important feedback loops that can influence collaborative action and transformation in communities. In a 2011 study (US-GAO 2011), the US Government Accountability Office highlighted that community indicator systems are a "vehicle for encouraging civic engagement both through the system's development process and through action once the indicator system is in place"; "help address community or national challenges by facilitating collaboration of various parties inside and outside of government"; and "provide solutions to long-term challenges". One example from Nevada's Reno and Sparks communities is *Truckee Meadows Tomorrow, Quality of Life Indicators* (TMT 2011).

Adaptive governance

Recognizing that humanity is encroaching on critical planetary boundaries, new modes of adaptive governance are needed to initiate transition management and achieve internationally agreed goals and targets.

This analysis has demonstrated that the transition path towards a sustainable world scenario is feasible but requires navigating a wide range of highly complex and interrelated issues simultaneously. Contextually, society's pursuit of well-being and the requisite use of natural resources is a complex adaptive system, where different systems interact and adapt to one another, giving rise to the emergent ability of both people and ecosystems to self-organize in response to sudden shocks and more slowly changing stresses (Liu *et al.* 2007). Such a system is never at a standstill, but rather is in a constant process of incremental adaptation, reconfiguration, modification, revision and re-ordering, where long periods of stability or equilibrium are visited by short periods of radical change (Grin *et al.* 2010; Loorbach 2007).

In such unpredictable settings, it is nearly impossible to create a fail-proof blueprint or to formulate optimal policies. What is required instead is an inclusive, learn-by-doing process with careful monitoring of policy effects, and an ability to make critical choices and improvements consistent with the trajectories leading to established goals. Society has already experienced the inadequacies of inflexible blueprint-style approaches and is gaining both experience and insights into alternative strategies and policies that are more adaptive and that help build resilience. Resilience thinking puts three aspects of social-ecological systems at the centre: resilience, adaptability and transformability. Resilience refers to the capacity of a system – such as a country or an ecosystem – to adapt to change, deal with surprise, and retain its basic function and structure while remaining within critical boundaries. Adaptability – part of resilience – represents the capacity to adjust responses to changing external drivers and internal processes, and thereby channel development along the preferred trajectory in what is called a stability domain. Transformability is the capacity to cross thresholds, enter new development trajectories, abandon unsustainable actions and chart better pathways to established targets (Folke *et al.* 2010).

There are different related approaches emerging that put these concepts into practice, including adaptive management, sustainability transition management, adaptive governance and adaptive policy making. They each have common features as well as different niches and scales in which they are relevant. The adaptive management approach, pioneered in the 1980s and 1990s (Lee 1993; Holling 1978), offers practical and experience-based guidance for the type of skilful navigation that would be necessary for local and regional place-based natural resource management efforts.

Recommendations for managing the transition to sustainability, resilience thinking and intervening in complex adaptive systems all provide governance-level insights (Grin *et al.* 2010; Loorbach 2007; Berkes *et al.* 2003; Rihani 2002; Ruitenbeek and Cartier 2001; Axelrod and Cohen 2000). In relation to the design and implementation of adaptive policy instruments, there is also an array of research and experience with practical applications to draw from that embody many of the same principles as adaptive management and managing sustainability transitions (Swanson *et al.* 2010; Walker and Marchau 2003; Bankes 2002; Dewey 1927). An example of what adaptive governance and policy-making entail in relation to watershed management in India is shown in Box 16.5. The range of research and experience cited in this paragraph reveals a consistent set of critical functions for adaptive governance and managing the transition to sustainability, here largely adopting the terminology of Loorbach (2007):

- *Multi-actor deliberation and agenda building.* Many stakeholders influence societal change. Governance must, therefore, be participatory to recognize advantageous leverage points, the levers for change and the correct direction to move them; to achieve coherent coalitions for creating shared notions of goals and ambitions; and to strengthen policy design and implementation.
- *Futures analysis and long-term collective goal setting.* Integrated and forward-looking assessments are critical tools that inform ongoing processes of change by systematically reflecting upon the future and developing shared notions of future goals and targets.

Box 16.5 India's National Watershed Development Project for Rainfed Areas (NDWPRA) – adaptive governance and policy-making at the sub-national level

The objectives of India's NWDPRA project include:

- sustainable enhancement of agricultural productivity and production;
- restoration of the ecological balance in degraded and fragile rain-fed ecosystems by greening these areas through an appropriate mix of trees, shrubs and grasses;
- reduction in regional disparity between irrigated and rain-fed areas; and
- creation of sustained employment opportunities for the rural poor.

In Maharashtra, the NDWPRA project launched in 1990/91 continued through India's Ninth Five-Year Plan, when it was considerably restructured. Greater emphasis was placed on decentralization and community participation. In the Tenth Five-Year Plan (2002–2007), the state of Maharashtra continued to implement the scheme with a participatory approach, extending it to 433 micro-watersheds across 33 districts, with a targeted treatment area of 203 000 hectares. The shift towards a decentralized approach contributed to improvements in water-management and represents an example of adaptive governance.

Source: Swanson and Bhadwal 2009

- *Enabling self-organization and networking*. Creating opportunities for cooperation and replicating successes, ensuring that social capital remains intact, and guaranteeing that members of the population are free and able to interact, are all fundamental elements of building the capacity of actors and policy itself to plan for and adapt to surprises.
- *Variation, experimentation and innovation*. Diversity of responses forms a common risk-management approach, and continuous reflection and improvement helps to develop a context in which innovation for desired change can thrive.
- *Reflexivity and adaptation*. Systemic review of past, present and future sustainability conditions and policy performance through interaction and cooperation with a range of stakeholders is critical for continuous improvement and social learning.

These critical functions of sustainability transition management and adaptive governance, together with the other strategic elements described earlier – social consensus for achieving compelling visions, discarding the unsustainable, and applying leverage in policy making – provide practical guidance for advancing sustainability and achieving internationally agreed targets.

In its simplest expression, a green economy can be thought of as one which is low carbon, resource efficient and socially inclusive. © *Ilias Kordelakos*

CONCLUSIONS

The envisioned sustainable world aims simultaneously to achieve universal human well-being and environmental sustainability at global, national, regional and local levels. The vision assumes that, by 2050, all people have access to food, safe drinking water, improved sanitation and modern sources of energy, all within the ecological limits of the planet.

Without major course correction, however, continuing on the current trajectory would lead, by 2050, to major environmental damage, a serious loss of ecosystem services, depletion of natural resources and many people left without sustainable access to food, water or energy. As a consequence, most internationally agreed goals and targets would be missed, some by a wide margin, particularly those related to climate change, biodiversity, water and food security.

The review of sustainable world scenarios suggests that measures can be put in place to help achieve these targets and reduce the risk of Earth System changes and negative impacts on future human development. Measures at the mid layer of transformation, such as rule changes, will not be enough to move to a sustainable world pathway. Structural measures and stronger policy action are needed to influence both production and consumption patterns. Such changes should be both short- and long-term, and combine technology, investment and governance measures along with lifestyle modifications grounded in a mindset shift towards sustainability and equity-based values.

A transformation of such complexity requires a gradual, but steadily strengthening, transition process. During this, there is a need to stop activities that pull the Earth System towards unsustainability. At the same time, it is important to provide resources, build capacity and create an enabling environment for all in a way that is consistent with the vision of a sustainable world.

Such a transition requires a high degree of consensus and coordination between societal actors with diverse interests and working environments. As a first step, broad-based social contracts, grounded in jointly developed visions of a sustainable future, would be needed to focus people's attention on the future. To ensure coherence among all societal actors, contextually sensitive transition pathways can be developed and agreed as joint visions of the future that respect social responsibility and ensure that the rest of society has sustainable access to the resources necessary for well-being. Given that systemic changes, both environmental and societal, are often slow, long-term goals would help focus investment and technological development, induce societal change and engage other actors in society.

The transition process would necessarily be based on adaptive management, since uncertainties play a key role in the problems of the Earth System. A diversity of measures would provide better insurance against wholesale failure on critical issues – due either to inherent uncertainties or inadequate implementation – and be mutually strengthening. Of course, strategies to achieve the targets will necessarily differ between developed and developing countries, and between regions, countries, and communities within those categories. The momentum of green economy initiatives to integrate environmental imperatives into major sectors also needs to be sustained and turned into detailed work on policy, innovation and practice. Reversing ecosystem degradation while meeting the increasing demand for resources may seem challenging, but policy measures that help achieve environmental goals and targets also have the potential to deliver benefits to human well-being.

All this requires political determination and strong governance. Questions of how to do it and what kind of global responses and institutions would be needed are discussed in Chapter 17.

REFERENCES

ADB (2010). *An Eco-Compensation Policy Framework for the People's Republic of China: Challenges and Opportunities*. Asian Development Bank, Manila. http://www.adb.org/documents/reports/eco-compensation-prc/eco-compensation-prc.pdf

AGECC (2010). *Energy for a Sustainable Future: Summary Report and Recommendations*. The Secretary-General's Advisory Group on Energy and Climate Change. United Nations, New York. http://www.unido.org/fileadmin/user_media/Publications/download/AGECCsummaryreport.pdf (accessed 24 September 2011)

Alcamo, J., Floerke, M. and Maerker, M. (2007). Future long-term changes in global water resources driven by socio-economic and climatic changes. *Hydrological Sciences Journal* 52, 247–275

Alcamo, J., van Vuuren, D.P. and Cramer, W. (2005a). Change in ecosystem services and their drivers across the scenarios. In *Ecosystems and Human Well-being: Scenarios. Volume 2* (eds. Carpenter, S.R., Pingali, P., Bennett, E.M. and Zurek, M.B.). Island Press, Washington

Alcamo, J., van Vuuren, D., Ringler, C., Cramer, W., Masui, T., Alder, J. and Schulze, K. (2005b). Changes in nature's balance sheet: model-based estimates of future worldwide ecosystem services. *Ecology and Society* 10(2)

Alcamo, J., Döll, P., Henrichs, T., Kaspar, F., Lehner, B., Rösch, T. and Siebert, S. (2003). Development and Testing of the WaterGAP 2 Global Model of Water Use and Availability. *Hydrological Sciences Journal*, 48(3), 317–337

Al-Damkhi, A.M., Al-Fares, R.A., Al-Khalifa, K.A. and Abdul-Wahab, S.A. (2009). Water issues in Kuwait: a future sustainability vision. *International Journal of Environmental Studies* 66(5), 619–636

Angelsen, A. (2010). Policies for reduced deforestation and their impact on agricultural production. *Proceedings of the National Academy of Sciences of the United States of America* 107(46), 19639–19644

Anticamara, J.A., Watson, R., Gelchu, A. and Pauly, D. (2011). Global fishing effort (1950–2010): trends, gaps, and implications. *Fisheries Research* 107, 131–136

Arnason, R., Kelleher, K. and Willman, R. (2009). *The Sunken Billions: The Economic Justification for Fisheries Reform*. World Bank, Washington, DC and Food and Agriculture Organization of the United Nations, Rome

Arnell, N.W., van Vuuren, D.P. and Isaac, M. (2011). The implications of climate policy for the impacts of climate change on global water resources. *Global Environmental Change* 21(2), 592–603

Arnold, M., Kohlin, G., Persson, R. and Shepherd, G. (2003). *Fuel Wood Revisited: What Has Changed in the Last Decade?* Center for International Forestry Research, Jakarta

Axelrod, R. and Cohen, M.D. (2000). *Harnessing Complexity: Organizational Implications of a Scientific Frontier*. Basic Books, New York

Bäckstrand, K. (2003). Civic science for sustainability: reframing the role of experts, policy-makers and citizens in environmental governance. *Global Environmental Politics* 3(4), 24–41

Bagstad, K.J., Stapleton, K. and D'Agostino, J.R. (2007). Taxes, subsidies, and insurance as drivers of United States coastal development. *Ecological Economics* 6(3), 285–298

Bakkes, J.A. and Bosch, P.R. (eds.) (2008). *Background Report to the OECD Environmental Outlook to 2030. Overviews, Details, and Methodology of Model-based Analysis*. Netherlands Environmental Assessment Agency, Bilthoven

Bankes, S.C. (2002). Tools and techniques for developing policies for complex and uncertain systems. *Proceedings of the National Academy of Sciences of the United States of America* 99(3), 7263–7266

Bassi, A.M. and Eaton, D. (2011). In defence of green economy report. *Nature* 475, 454

Bassi, A.M., Pedercini, M., Ansah, J.P. and Tan, Z. (2010). *T21-World Model Documentation, Modeling the Green Economy*. Millennium Institute, Arlington, VA

Bates, B., Kundzewicz, Z.W., Shaohong, W. and Palutikof, J. (2008). *Climate Change and Water*. IPCC Technical Paper VI. Intergovernmental Panel on Climate Change. IPCC Secretariat, Geneva

Bazilian, M., Nussbaumer, P., Haites, E., Levi, M., Howells, M. and Yumkella, K.K. (2010). Understanding the scale of investment for universal energy access. *Geopolitics of Energy* 32, 10–11

Berkes, F., Colding, J. and Folke, C. (2003). *Navigating Social-Ecological Systems: Building Resilience for Complexity and Change*. Cambridge University Press, Cambridge

Black, R.E., Cousens, S., Johnson, H.L., Lawn, J.E., Rudan, I., Bassani, D.G., Jha, P., Campbell, H., Walker, C.F., Cibulskis, R., Eisele, T., Liu, L. and Mathers, C. (2010). Global, regional, and national causes of child mortality in 2008: a systematic analysis. *Lancet* 375, 1969–1987

Bollen, J.C. (2008). *Energy Security, Air Pollution, and Climate Change: An Integrated Cost Benefit Approach*. Netherlands Environmental Assessment Agency, Bilthoven

Bouwman, A.F., van der Hoek, K.W., Drecht, G.V. and Eickhout, B. (2006). World livestock and crop production systems, land use and environment between 1970 and 2030. In *Rural Lands, Agriculture and Climate beyond 2015: A New Perspective on Future Land Use Patterns* (eds. Brouwer, F. and McCarl, B.). pp.75–89. Springer, Dordrecht

Bouwman, A.F., van der Hoek, K.W., Eickhout, B. and Soenario, I. (2005). Exploring changes in world ruminant production systems. *Agricultural Systems* 84(2), 121–153. doi:110.1016 j.agsy 2004.1005.1006

Bringezu, S., Schütz, H., O'Brien, M., Kauppi, L., Howarth, R.W. and McNeely, J. (2009). *Towards Sustainable Production and Use of Resources: Assessing Biofuels*. International Panel for Sustainable Resource Management. United Nations Environment Programme, Division of Technology, Industry and Economics (UNEP DTIE), Paris

Broca, S.S. (2002). *Food Insecurity, Poverty and Agriculture: A Concept Paper*. Agricultural and Development Economics Division, Food and Agriculture Organization of the United Nations (FAO), Rome

Bruinsma, J. (ed.) (2003). *World Agriculture: Towards 2015/2030. An FAO Perspective*. Earthscan, London

Butchart, S.H.M., Scharlemann, J.P.W., Evans, M.I., Quader, S., Aricò, S., Arinaitwe, J., Balman, M., Bennun, L.A., Besançon, C., Boucher, T.M., Bertzky, B., Brooks, T.M., Burfield, I.J., Burgess, N.D., Chan, S., Clay, R.P., Crosby, M.J., Davidson, N.C., De Silva, N., Devenish, C., Dutson, G.C.L., Díaz Fernández, D.F., Fishpool, L.D.C., Fitzgerald, C., Foster, M., Heath, M.F., Hockings, M., Hoffmann, M., Knox, D., Larsen, F.W., Lamoreux, J.F., Loucks, C., May, I., Millett, J., Molloy, D., Morling, P., Parr, M., Ricketts, T.H., Seddon, N., Skolnik, B., Stuart, S.N., Upgren, A. and Woodley, S. (2012). Protecting important sites for biodiversity contributes to meeting global conservation targets. *PLoS ONE* 7(3)

CBD (2010a). *Aichi Biodiversity Targets*. Tenth Meeting of the Conference of the Parties to the Convention on Biological Diversity (COP 10), Nagoya. Secretariat of the Convention on Biological Diversity, Montreal

CBD (2010b). *Global Biodiversity Outlook 3*. Secretariat of the Convention on Biological Diversity, Montreal

CBD (2006). *Global Biodiversity Outlook 2*. Secretariat of the Convention on Biological Diversity, Montreal

Clarke, L., Edmonds, J., Krey, V., Richels, R., Rose, S. and Tavoni, M. (2010). International climate policy architectures: overview of the EMF 22 international scenarios. *Energy Economics* 31(2), S64–S81

CLRTAP (1979). *Convention on Long-range Transboundary Air Pollution (CLRTAP)*. United Nations Economic Commission for Europe (UNECE), Geneva

Cofala, J., Amann, M., Klimont, Z., Kupiainen, K., Höglund-Isaksson, L. (2007). Scenarios of global antropogenic emissions of air pollutants and methane until 2030. *Atmospheric Environment* 41, 8468–8499

Cosgrove, W. and Rijsberman, F. (2000). *World Water Vision: Making Water Everybody's Business*. World Water Council, Earthscan Publications, London

Costanza, R. (2000). Visions of alternative (unpredictable) futures and their use in policy analysis. *Conservation Ecology* 4(1), 5

Czech, B. and Daly, H.E. (2004). The steady state economy – what it is, entails and connotes. *Wildlife Society Bulletin* 32(2), 598–605

Daly, H.E. (1974). The economics of the steady state. *American Economic Review* 64(2), 15–21

Daly, H.E. (1971). *The Stationary-State Economy: Toward a Political Economy of Biophysical Equilibrium and Moral Growth*. Distinguished Lecture Series No. 2. University of Alabama, Alabama

Danish Architecture Centre (2011). *Lyon: An Overall Vision for Transport – Urban Mobility Master Plan*. Danish Architecture Centre, Copenhagen. http://sustainablecities.dk/en/city-projects/cases/lyon-an-overall-vision-for-transport-urban-mobility-master-plan

Dewey, J. (1927). *The Public and its Problems*. Holt and Company, New York

Dornburg, V., van Vuuren, D., van de Ven, G., Langeveld, H., Meeusen, M., Banse, M., van Oorschot, M., Ros, J., van den Born, G.J., Aiking, H., Londo, M., Mozaffarian, H., Verweij, P., Lysen, E. and Faaij, A. (2010). Bioenergy revisited: key factors in global potentials of bioenergy. *Energy and Environment Science* 3, 258–267

Easterling, W., Aggarwal, P., Batima, P., Brander, K., Erda, L., Howden, M., Kirilenko, A., Morton, J., Soussana, J.-F., Schmidhube, S. and Tubiello, F. (2007). Food, fibre and forest products. In *Climate Change 2007: Impacts, Adaptation and Vulnerability* (eds. Parry, M.L., Canziani, O.F., Palutikof, J.P., van der Linden, P.J. and Hanson, C.E.). Contribution of Working Group II to the Fourth Assessment Report of the Intergovernmental Panel on Climate Change. Cambridge University Press, Cambridge

ECF (2010). *Roadmap 2050. A Practical Guide to a Prosperous, Low Carbon Europe*. European Climate Foundation, The Hague

FAO (2011). *The State of the World's Land and Water Resources for Food and Agriculture (SOLAW): Managing Systems at Risk*. Food and Agriculture Organization of the United Nations, Rome and Earthscan, London

FAO (2010). *State of World Fisheries and Aquaculture 2010*. Food and Agriculture Organization of the United Nations, Rome.

FAO (2009). *High-level Expert Forum: How to Feed the World in 2050*. Food and Agriculture Organization of the United Nations, Rome

FAO (2006a). WISDOM – *East Africa: Woodfuel Integrated Supply/Demand Overview Mapping (WISDOM) Methodology. Spatial Woodfuel Production And Consumption Analysis of Selected African Countries*. Forestry Department, Food and Agriculture Organization of the United Nations, Rome

FAO (2006b). *World Agriculture: Towards 2030/2050. Prospects for Food, Nutrition, Agriculture and Major Commodity Groups*. Food and Agriculture Organization of the United Nations, Rome

FAO (1996). *Rome Declaration on World Food Security and World Food Summit Plan of Action*. Adopted at the World Food Summit, November 13–17, Rome. Food and Agriculture Organization of the United Nations, Rome

FAO (1995). *Code of Conduct for Responsible Fisheries*. Food and Agriculture Organization of the United Nations, Rome

FAOSTAT (2012). *FAO Statistical Databases*. Food and Agriculture Organization of the United Nations, Rome. http://faostat.org

Fargione, J., Hill, J., Tilman, D., Polasky, S. and Hawthorne, P. (2008). Land clearing and the biofuel carbon debt. *Science* 319, 1235–1238

Fearnside P.M. (2011). Methane emissions from hydroelectric dams. *Science* 28 July 2011, 50. http://www.sciencemag.org/content/331/6013/50/reply

Fisher, B., Nakicenovic, N., Alfsen, K., Corfee Morlot, J.C., de la Chesnaye, F., Hourcade, J.-C., Jiang, K., Kainuma, M., La Rovere, E., Matysek, A., Rana, A., Riahi, K., Richels, R., Rose, S., van Vuuren, D. and Warren, R. (2007).) Issues related to mitigation in the long-term context. In *Climate Change 2007. Mitigation of Climate Change* (eds. Metz, B., Davidson, O.R, Bosch, P.R., Dave, R. and Meyer, L.). pp.169–250. Contribution of Working Group III to the Fourth Assessment Report of the Intergovernmental Panel on Climate Change. Cambridge University Press, New York

Flörke, M. and Alcamo, J. (2004). *European Outlook on Water Use*. European Environment Agency, Copenhagen. http://scenarios.ewindows.eu.org/reports/fol949029

Folke, C., Carpenter, S.R., Walker, B., Scheffer, M., Chapin, T. and Rockström, J. (2010). Resilience thinking: integrating resilience, adaptability and transformability. *Ecology and Society* 15(4), 20. http://www.ecologyandsociety.org/vol15/iss4/art20/

Folke, C., Carpenter, S., Elmqvist, T., Gunderson, L., Holling, C.S. and Walker, B. (2002). Resilience and sustainable development: building adaptive capacity in a world of transformations. *Ambio* 31(5), 437–440

GEA (2011). *Global Energy Assessment*. Cambridge University Press, Cambridge

Gibson, L., Ming Lee, T., Pin Koh, L., Brook, B.W., Gardner, T.A., Barlow, J., Peres, C.A., Bradshaw, C.J.A., Laurance, W.F., Lovejoy, T.E. and Sodhi, N.S. (2011). Primary forests are irreplaceable for sustaining tropical biodiversity. *Nature* 478, 378–381

Giovannucci, D. and Ponte, S. (2005). Standards as a new form of social contract? Sustainability initiatives in the coffee industry. *Food Policy* 30(3), 284–301

Girod, B., van Vuuren, D.P. and Deetman, S. (2012). Global travel within the 2°C climate target. *Energy Policy* 45, 152–166

Global Footprint Network (2010). *The Ecological Wealth of Nations: Earth's Biocapacity as a New Framework for International Cooperation*. http://www.footprintnetwork.org/images/uploads/Ecological_Wealth_of_Nations.pdf (accessed 24 September 2011)

Government of Bhutan (2011). *Gross National Happiness*. National Portal of Bhutan, Government of Bhutan. http://www.bhutan.gov.bt/government/gnh.php

Grin, J., Rotmans, J. and Schot, J. (2010). *Transitions to Sustainable Development. New Directions in the Study of Long Term Transformative Change*. Routledge, New York, London

Halpern, B.S., Walbridge, S., Selkoe, K.A., Kappel, C.V., Micheli, F., D'Agrosa, C., Bruno, J.F., Casey, K.S., Ebert, C., Fox, H.E., Fujita, R., Heinemann, D., Lenihan, H.S., Madin, E.M.P., Perry, M.T., Selig, E.R., Spalding, M., Steneck, R. and Watson, R. (2008). A global map of human impact on marine ecosystems. *Science* 319, 948

Hazell, P. and Wood, S. (2008). Drivers of change in global agriculture. *Philosophical Transactions of the Royal Society B* 363, 495–515

Hewison, G.J. (1993). The Convention for the Prohibition of Fishing with Long Driftnets in the South Pacific. *Case Western Reserve Journal of International Law* 25, 449

Hilderink, H.B.M., Lucas, P.L. and Kok, M. (eds.) (2009). *Beyond 2015: Long-term Development and the Millennium Development Goals*. Netherlands Environmental Assessment Agency, Bilthoven

Holling, C.S. (2001). Understanding the complexity of economic, ecological and social systems. *Ecosystems* 4(5), 390–405

Holling, C.S. (1978). *Adaptive Environmental Assessment and Management*. John Wiley, New York

Hughes, B.B., Kuhn, R., Peterson, C.M., Rothman, D.S. and Solórzano, J.R. (2011). *Improving Global Health. Patterns of Potential Human Progress Vol.3*. Oxford University Press, New Delhi

Hutton, G. and Haller, L. (2004). *Evaluation of the Costs and Benefits of Water and Sanitation Improvements at the Global Level*. World Health Organization, Geneva

Hurtt, G., Chini, L., Frolking, S., Betts, R., Edmonds, J., Feddema, J., Fisher, G., Goldewijk, K.K., Hibbard, K.A., Houghton, R., Janetos, A., Jones, C.D., Kindermann, G. Kinoshita, T., Goldewijk, K.K., Riahi, K., Shevliakova, E., Smith, S., Stehfest, E., Thomson, A., Thornton, P., van Vuuren D.P., and Wang, Y.P. (2011). Harmonization of land-use scenarios for the period 1500–2100: 600 years of global gridded annual land-use transitions, wood harvest, and resulting secondary lands. *Climatic Change* 109(1), 117–161

IAASTD (2009a) *High-level Expert Forum: How to Feed the World in 2050*. International Assessment of Agricultural Science and Technology for Development: Global Report. Island Press, Washington, DC and Rome

IAASTD (2009b) *Synthesis Report: A Synthesis of the Global and Sub-Global IAASTD Reports*. International Assessment of Agricultural Science and Technology for Development: Global Report. Island Press, Washington, DC and Rome. http://www.agassessment.org/reports/IAASTD/EN/Agriculture%20at%20a%20Crossroads_Synthesis%20Report%20(English).pdf

ICPD (1994). *Report of the International Conference on Population and Development, Cairo, 5–13 September 1994*. UN Population Fund

IEA (2010). *World Energy Outlook 2010*. International Energy Agency, Paris

IEA (2008). *Energy Policy Review of Indonesia*. International Energy Agency, Paris. http://www.iea. org/textbase/nppdf/free/2008/Indonesia2008.pdf

IEA (2006). *Angola, Towards an Energy Strategy*. International Energy Agency, Paris

IEA/UNDP/UNIDO (2010). Energy Poverty: How To Make Modern Energy Access Universal? Special early excerpt of the *World Energy Outlook 2010* for the UN General Assembly on the Millennium Development Goals. International Energy Agency, United Nations Development Programme and United Nations Industrial Development Organization. OECD/IEA, Paris

IISD GSI (2011). *A High-Impact Initiative for Rio+20: A Pledge to Phase out Fossil-Fuel Subsidies*. Global Subsidies Initiative of the International Institute for Sustainable Development (IISD), Geneva and Winnipeg

Innes, A.D., Campion, P.D. and Griffith, F.E. (2005). Complex consultations and the "edge of chaos". *British Journal of General Practice* 55(510), 47–52

IPCC (2007). *Climate Change 2007: Synthesis Report* (eds. Pachauri, R.K. and Reisinger, A.). Contribution of Working Groups I, II and III to the Fourth Assessment Report of the Interngovernmental Panel on Climate Change. Cambridge University Press, Cambridge

Jaeger, C.C., Kasemir, B., Stoll-Kleemann, S., Schibli, D. and Dahinden, U. (2000). Climate change and the voice of the public. *Integrated Assessment Journal* 1, 339–349

Jäger, J. and Cornell, S.E. (eds.) (2011). *The Planet in 2050: The Lund Discourse of the Future*. Routledge

Jakarta Mandate (1995). *The Jakarta Mandate on the Conservation and Sustainable Use of Marine and Coastal Biological Diversity*. http://www.ngo.grida.no/wwfneap/Projects/Reports/jakmand.pdf

Jenkins, G. and Lowe, J. (2003). *Handling Uncertainties in the UKCIP02 Scenarios of Climate Change*. Hadley Centre Technical Note 44. Met Office, Exeter

Kaiser, M., Ellerbrock, R.H. and Gerke, H.H. (2007). Long-term effects of crop rotation and fertilization on soil organic matter composition. *European Journal of Soil Science* 58, 1460–1470

Killham, K. (2010). Integrated soil management – moving towards globally sustainable agriculture. Foresight Project on Global Food and Farming Futures. *Journal of Agricultural Science* 149, 29–36

Leadley, P., Pereira, H.M., Alkemade, R., Fernandez-Manjarres, J.F., Proenca, V., Scharlemann, J.P.W. and Walpole, M.J. (2010). *Biodiversity Scenarios: Projections of 21st Century Change in Biodiversity and Associated Ecosystem Services*. Technical Series No. 50. Secretariat of the Convention on Biological Diversity, Montreal

Lee, K. (1993). *Compass and Gyroscope: Integrating Science and Politics for the Environment*. Island Press, Washington, DC

Lenton, T.M., Held, H., Kriegler, E., Hall, J.W., Lucht, W., Rahmstorf, S. and Schellnhuber, H.J. (2008). Tipping elements in the Earth's climate system. *Proceedings of the National Academy of Sciences of the United States of America* 105(6), 1786–1793

Levin, S.A. (1998). Ecosystem and the biosphere as complex adaptive systems. *Ecosystems* 1(5), 431–436

Liu, J., Dietz, T., Carpenter, S.R., Alberti, M., Folke, C., Moran, E., Pell, A.N., Deadman, P., Kratz, T., Lubchenco, J., Ostrom, E., Ouyang, Z., Provencher, W., Redman, C.L., Schneider, S.H. and Taylor, W.W. (2007). Complexity of coupled human and natural systems. *Science* 317(5844), 1513–1516

Loorbach, D. (2007). *Transition Management: New Mode of Governance for Sustainable Development*. International Books, Utrecht

Loorbach, D. and Rotmans, J. (2005). Managing transitions for sustainable development. In *Industrial Transformation – Disciplinary Approaches Towards Transformation Research* (eds. Wieczorek, A.J. and Olshoorn, X.). pp.187–206. Kluwer Academic Publishers Dordrecht

Lubchenco, J. (1998). Entering the century of the environment: a new social contract for science. *Science* 279(5350), 491–497

Lutz, W. and Samir, K.C. (2011). Global human capital: integrating education and population. *Science* 333(6042), 587

Lutz, W., Sanderson, W. and Scherbov, S. (2008). The coming acceleration of global population ageing. *Nature* 451, 716–719

Luzzati, T. and Orsini, M. (2009). Investigating the energy-environmental Kuznets curve. *Energy* 34, 291–300

MA (2005a). *Ecosystems and Human Health: Scenarios*. Millennium Ecosystem Assessment. Island Press, Washington, DC

MA (2005b). *Ecosystems and Human Health: Synthesis*. Millennium Ecosystem Assessment. Island Press, Washington, DC

Maffei, M.C. (1997). The International Convention for the Regulation of Whaling. *International Journal of Coastal and Marine Law* 12(3), 287–305

Mangalagiu, D., Wilkinson, A. and Kupers, R. (2011). When futures lock in the present. In *Reframing the Problem of Climate Change: From Zero Sum Game to Win-win Solutions* (eds. Jaeger, C.C., Hasselmann, K., Leipold, G, Mangalagiu, D. and Tàbara, J.D.). Earthscan, London and Washington DC

Matthes, F.C., Gores, S., Graichen, V., Repenning, J. and Zimmer, W. (2006). *The Vision Scenario for the European Union*. Öko-Institut e.V., Berlin and Freiburg

McNeil, B.I. and Matear, R.J. (2008). Southern Ocean acidification: a tipping point at 450-ppm atmospheric CO_2. *Proceedings of the National Academy of Sciences of the United States of America* 105, 18860–18864

Meadows, D. (1999). *Leverage Points: Places to Intervene in a System*. The Sustainability Institute, Hartland. http://www.sustainabilityinstitute.org/pubs/Leverage_Points.pdf

Meadows, D.H. (1996). *Envisioning a Sustainable World*. Prepared for the Third Biennial Meeting of the International Society for Ecological Economics, 24-28 October 1994, San Jose, Costa Rica. http://www.infoark.org/InfoArk/Sustainability/Envisioning%20a%20Sustainable%20World%20-%20Meadows_1994-10-24.pdf

Meinshausen, M., Hare, B., Wigley, T.M.L., van Vuuren, D., den Elzen, M.G.J. and Swart, R. (2006). Multi-gas emissions pathways to meet climate targets. *Climatic Change* 75(1–2), 151–194

Mery, G., Katila, P., Galloway, G., Alfaro, R.I., Kanninen, M., Lobovikov, M. and Varjo, J. (eds.) (2010). *Forests and Society – Responding to Global Drivers of Change*. IUFRO World Series Volume 25. IUFRO – The Global Network for Forest Science Cooperation, Vienna

Metro City of Vancouver (2011). *Climate Smart*. Metro Vancouver Program Information. Metro Vancouver, Vancouver. https://climatesmartbusiness.com/metrovancouver/#overview

Milder, J.C., Scherr, S.J. and Bracer, C. (2010). Trends and future potential of payment for ecosystem services to alleviate rural poverty in developing countries. *Ecology and Society* 15(2), 4. http://www.ecologyandsociety.org/vol15/iss2/art4/

MSC (2011). *Marine Stewardship Council: Certified Sustainable Seafood*. http://www.msc.org

Myers, N. and Kent, J. (2001). *Perverse Subsidies. How Tax Dollars Can Undercut the Environment and the Economy*. Island Press, Washington, DC

Nakicenovic, N. and Swart, R. (eds.) (2000). *Emissions Scenarios*. IPCC Special Report. Intergovernmental Panel on Climate Change. Cambridge University Press, Cambridge

Nelson, G.C., Rosegrant, M.W., Palazzo, A., Gray, I., Ingersoll, C., Robertson, R., Tokgoz, S., Zhu, T., Sulser, T.B., Ringler, C., Msangi, S. and You, L. (2010). *Food Security, Farming, and Climate Change to 2050*. International Food Policy Research Institute (IFPRI), Washington, DC

OECD (2012). *OECD Environmental Outlook to 2050*. Organisation for Economic Co-operation and Development, Paris

OECD (2008a). *OECD Environmental Outlook to 2030*. Organisation for Economic Co-operation and Development, Paris

OECD (2008b). *Promoting the Use of Performance-Based Contracts between Water Utilities and Municipalities in EECCA. Case Study No. 2: Armenian Water and Wastewater Company, SAUR Management Contract*. Organisation for Economic Co-operation and Development, Paris

OECD/FAO (2011). *Agricultural Outlook 2011–2020*. Organisation for Economic Co-operation and Development, Paris and Food and Agriculture Organization of the United Nations, Rome

Parfitt, J., Barthel, M. and Macnaughton, S. (2010). Food waste within food supply chains: quantification and potential for change to 2050. *Philosophical Transactions of the Royal Society B* 365, 3065–3081

Patel, M., Kok, K. and Rothman, D.S. (2007). Participatory scenario construction in land use analysis: an insight into the experiences created by stakeholder involvement in the Northern Mediterranean. *Land Use Policy* 24(3), 546–561

Paulitz, T., Smiley, R.W. and Cook, R.J. (2002). New insights into the make-up and management of soilborne cereal pathogens under direct seeding in the Pacific Northwest. *Canadian Journal of Phytopathology* 24, 416–428

Pauly, D., Alder, J., Bennett, E., Christensen, V., Tyedmers, P. and Watson, R. (2003). The future for fisheries. *Science* 302(5649), 1359–1361

PBL (2012). *Roads from Rio+20: Pathways to achieve global sustainability goals by 2050*. Van Vuuren, DP and Kok, MTJ (eds.). Den Haag/Bilthoven, the Netherlands, PBL Netherlands Environmental Assessment Agency.

PBL (2009). *Growing within Limits*. Netherlands Environmental Assessment Agency, Bilthoven

Pereira, H.M., Leadley, P.W., Proença, V., Alkemade, R., Scharlemann, J.P.W., Fernandez-Manjarrés, J.F., Araújo, M.B., Balvanera, P., Biggs, R., Cheung, W.W.L., Chini, L., Cooper, D., Gilman, E.L., Guénette, S., Hurtt, G.C., Huntington, H.P., Mace, G.M., Oberdorff, T., Revenga, C., Rodrigues, P., Scholes, R.J., Sumaila, U.R. and Walpole, M. (2010). Scenarios for global biodiversity in the 21st century. *Science* 330(6010), 1496–1501

Petermann, J.S., Fergus, A.J.F., Turnbull, L.A. and Schmid, B. (2008). Janzen-Connell effects are widespread and strong enough to maintain diversity in grasslands. *Ecology* 89, 2399–2406

Phalan, B., Onial, M., Balmford, A. and Green, R.E. (2011). Reconciling food production and biodiversity conservation: land sharing and land sparing compared. *Science* 333, 1289–1291

Pinnegar, J.K., Viner, D., Hadley, D., Dye, S., Harris, M., Berkout, F. and Simpson, M. (2006). *Alternative Future Scenarios for Marine Ecosystems: Technical Report*. Cefas, Lowestoft

Power, A.G. (2010). Ecosystem services and agriculture: tradeoffs and synergies. *Philosophical Transactions of the Royal Society B* 365, 2959–2971

Prüss-Üstün, A. and Corvalán, C. (2006). *Preventing Diseases Through Healthy Environments: Towards an Estimate of the Environmental Burden of Disease*. World Health Organization, Geneva

Prüss-Üstün, A., Kay, D., Fewtrell, L. and Bartram, J. (2004). Unsafe water, sanitation and hygiene. In *Comparative Quantification of Health Risks: Global and Regional Burden of Disease Attributable to Selected Major Risk Factors* (eds. Ezzati, M., Lopez, A.D., Rodgers, K.B. and Murray, C.J.L.). World Health Organization, Geneva

Purcell, S.W., Mercier, A., Conand, C., Hamel, J.-F., Toral-Granda, M.V., Lovatelli, A. and Uthicke, S. (2011). Sea cucumber fisheries: global analysis of stocks, management measures and drivers of overfishing. *Fish and Fisheries* (forthcoming)

Rands, M.R.W., Adams, W.M., Bennun, L., Butchart, S.H.M., Clements, A., Coomes, D., Entwistle, A., Hodge, I., Kapos, V., Scharlemann, J.P.W. and Vira, B. (2010). Biodiversity conservation: challenges beyond 2010. *Science* 329, 1298–1303

Revi, A., Prakash, S., Mehrotra, R., Bhat, G.K., Gupta, K. and Gore, R. (2006). Goa 2100: the transition to a sustainable RUrban design. *Environment and Urbanization* 18(1), 51–65

Riahi, K., Grübler, A. and Nakicenovic, N. (2007). Scenarios of long-term socio-economic and environmental development under climate stabilization. *Technological Forecasting and Social Change* 74(7), 887–935

Ricketts, T.H., Dinerstein, E., Boucher, T., Brooks, T.M., Butchart, S.H.M., Hoffman, M., Lamoreux, J.F., Morrison, J., Parr, M., Pilgrim, J.D., Rodrigues, A.S.L., Secrest, W., Wallace, G.E., Berlin, K., Bielby, J., Burgess, N.D., Church, D.R., Cox, N., Knox, D., Loucks, C., Luck, G.W., Master, L.L., Moore, R., Naidoo, R., Ridgely, R., Schatz, G.E., Shire, G., Strand, H., Wettengel, W. and Wikramanayake, E. (2005). Pinpointing and preventing imminent extinctions. *Proceedings of the National Academy of Sciences of the United States of America* 102(51), 18497–18501

Rihani, S. (2002). *Complex Systems Theory and Development Practice: Understanding Non-Linear Realities*. Zed Books, New York

Robin, S., Wolcott, R. and Quintela, C.E. (2003). *Perverse Subsidies and the Implications for Biodiversity: A Review of Recent Findings and the Status of Policy Reforms*. Proceeding of the 5th World Parks Congress: Sustainable Finance Stream, September 2003, Durban, South Africa. http://www.conservationfinance.org/guide/WPC/WPC_documents/Overview_PanB_Wolcott_v2.pdf

Robinson, J. (2004). Squaring the circle? Some thoughts on the idea of sustainable development. *Ecological Economics* 48(4), 369–384

Robinson, R.A. and Sutherland, W.J. (2002). Post-war changes in arable farming and biodiversity in Great Britain. *Journal of Applied Ecology* 39, 157–176

Rodríguez, J.P., Beard, T.D., Bennett Jr., E.M., Cumming, G.S., Cork, S., Agard, J., Dobson, A.P. and Peterson, G.D. (2006). Trade-offs across space, time, and ecosystem services. *Ecology and Society* 11(1). http://www.ecologyandsociety.org/vol11/iss1/art28/

Rose, S.K, Ahammad, H, Eickhout, B., Fisher, B., Kurosawa, A., Rao, S., Riahi, K. and van Vuuren, D.P. (2012). Land-based mitigation in climate stabilization. *Energy Economics* 34(1), 365–380

Rosegrant, M.W., Ringler, C., Sulser, T.B., Ewing, M., Palazzo, A., Zhu, T., Nelson, G.C., Koo, J., Robertson, R., Msangi, S. and Batka, M. (2009). *Agriculture and Food Security under Global Change: Prospects for 2025/2050*. Prepared for the Strategy Committee of the CGIAR. International Food Policy Research Unit (IFPRU), Washington, DC. http://alliance.cgxchange.org/documentation-for-the-development-of-the-cgiar-strategy-and-mega-programs/SRF_IMPACT10-10-09c.pdf

Rosegrant, M.W., Ringler, C., Msangi, S., Sulser, T.B., Zhu, T. and Cline, S.A. (2008). *International Model for Policy Analysis of Agricultural Commodities and Trade (IMPACT): Model Description*. International Food Policy Research Unit (IFPRU), Washington, DC. http://www.ifpri.org/sites/default/files/publications/impactwater.pdf

Rosegrant, M.W., Leach, N. and Gerpacio, R.V. (1999). Alternative futures for world cereal and meat consumption. *Proceedings of the Nutrition Society* 58, 219–234

Rotterdam Convention (1998). *Rotterdam Convention on the Prior Informed Consent Procedure for Certain Hazardous Chemicals and Pesticides in International Trade*. http://www.pic.int/TheConvention/Overview/TextoftheConvention/tabid/1048/language/en-US/Default.aspx

Royal Swedish Academy of Sciences (2011). *The Stockholm Memorandum. Tipping the Scales towards Sustainability*. 3rd Nobel Laureate Symposium on Global Sustainability, Stockholm, 16–19 May 2011. Royal Swedish Academy of Sciences, Stockholm

Ruitenbeek, J. and Cartier, C. (2001). *The Invisible Wand: Adaptive Co-Management as an Emergent Strategy in Complex Bio-economic Systems*. Centre for International Forestry Research, Bogor

Sala, O.E., Chapin III, F.S., Armesto, J.J., Berlow, E., Bloomfield, J., Dirzo, R., Huber-Sanwald, E., Huenneke, L.F., Jackson, R.B., Kinzig, A., Leemans, R., Lodge, D.M., Mooney, H.A., Oesterheld, M., LeRoy Poff, N., Sykes, M.T., Walker, B.H., Walker, M. and Wall, D.H. (2000). Global biodiversity scenarios for the year 2100. *Science* 287(5459), 1770–1774

Schmitt-Olabisi, L.K., Kapuscinski, A.R., Johnson, K.A., Reich, P.B., Stenquist, B. and Draeger, K.J. (2010). Using scenario visioning and participatory system dynamics modeling to investigate the future: lessons from Minnesota 2050. *Sustainability* 2(8), 2686–2706

Schneider, C., Flörke, M., Geerling, G., Duel, H., Grygoruk, M. and Okruszko, T. (2011). The future of European floodplain wetlands under a changing climate. *Journal of Water and Climate Change* 2(2–3), 106–122

Searchinger, T., Heimlich, R., Houghton, R.A., Dong, F., Elobeid, A., Fabiosa, J., Tokgoz, S., Hayes, D. and Yu, T.-H. (2008). Use of US croplands for biofuels increases greenhouse gases through emissions from land-use change. *Science* 319, 1238–1240

Smeets, E.W.M., Bouwman, A.F., Stehfest, E., van Vuuren, D.P. and Posthuma, A. (2009). The contribution of N_2O emissions to the greenhouse gas balance of first-generation biofuels. *Global Change Biology* 15, 1–23. doi: 10.1111/j.1365-2486.2008.01704.x

Smith, S.J. (2005). Income and pollutant emissions in the ObjECTS MiniCAM model. *Journal of Environment and Development* 14(1), 175–196

Smith, P., Gregory, P.J., van Vuuren, D., Obersteiner, M., Havlík, P., Rounsevell, M., Woods, J., Stehfest, E. and Bellarby, J. (2010). Competition for land. *Philosophical Transactions of the Royal Society B* 365(1554), 2941–2957

Sohngen, B., Mendelsohn, R. and Sedjo, R. (2001). A global model of climate change impacts on timber markets. *Journal of Agricultural and Resource Economics* 26(2), 326–343

Speth, J.G. (2005). *Red Sky at Morning: America and the Crisis of the Global Environment*. Yale University Press, New Haven, CT

Srinivasan, U.T., Watson, R. and Sumaila, U.R. (2012). Global fisheries losses at the exclusive economic zone level, 1950 to present. *Marine Policy* 36, 544–549

Steffen, W., Sanderson, R.A., Tyson, P.D., Jäger, J., Matson, P.A., Moore III, B., Oldfield, F., Richardson, K., Schellnhuber, H.-J., Turner, B.L. and Watson, R.J. (2005). *Global Change and the Earth System*. Springer, Berlin

Stehfest, E., Bouwman, L., van Vuuren, D.P., den Elzen, M.G.J., Eickhout, B. and Kabat, P. (2009). Climate benefits of changing diet. *Climatic Change* 95(1–2), 83–102

Stern, D.I. (2003). *The Environmental Kuznets Curve*. International Society for Ecological Economics/ Internet Encyclopedia of Ecological Economics. http://www.ecoeco.org/pdf/stern.pdf

Stiglitz, J.E., Sen, A. and Fitoussi, J.-P. (2009). *Report by the Commission on the Measurement of Economic Performance and Social Progress*. Technical Report September 2009. http://www.stiglitz-sen-fitoussi.fr

St. Louis, V.L., Kelly, C.A., Duchemin, E., Rudd. J.W.M. and Rosenberg, D.M. (2000). Reservoir surfaces as sources of greenhouse gases to the atmosphere: a global estimate. *BioScience* 50, 766–775

Stockholm Convention (2009). *Stockholm Convention on Persistent Organic Pollutants (POPs) – as Amended in 2009*. http://chm.pops.int/Convention/Media/Publications/tabid/506/language/en-US/Default.aspx (accessed 20 November 2011)

Strauss, A.L. (1987). *Qualitative Analysis for Social Scientists*. Cambridge University Press, Cambridge

Swanson, D.A. and Bhadwal, S. (eds). (2009). *Creating Adaptive Policies: A Guide for Policy-making in an Uncertain World*. Sage Publications, New Delhi/IDRC, Ottawa

Swanson, D.A., Barg, S., Tyler, S., Venema, H.D., Tomar, S., Bhadwal, S., Nair, S., Roy, D. and Drexhage, J. (2010). Seven tools for creating adaptive policies. *Technological Forecasting and Social Change* 77, 924–939

Swart, R.J., Raskin, P. and Robinson, J. (2004). The problem of the future: sustainability science and scenario analysis. *Global Environmental Change* 14(2004), 137–146

Tabara J.D. and Pahl-Wostl, C. (2007). Sustainability learning in natural resource use and management. *Ecology and Society* 12(2), 3

Takács-Sánta, A. (2004). The major transitions in the history of human transformation of the biosphere. *Human Ecology Review* 11(1), 51–66

Techera, E.J. (2011). Convention for the Prohibition of Fishing with Long Drift Nets in the South Pacific. In *Encyclopaedia of Sustainability. Vol. 3: The Law and Politics of Sustainability* (eds. Bossellman, K., Fogel, D. and Ruhl, J.B.). Berkshire Publishing, Great Barrington

Ten Brink, B., van der Esch, S., Kram, T., van Oorschot, M., Alkemade, J.R.M., Ahrens, R., Bakkenes, M., Bakkes, J.A., van den Berg, M., Christensen, V., Janse, J., Jeuken, M., Lucas, P., Manders, T., van Meijl, H., Stehfest, E., Tabeau, A., van Vuuren, D. and Wilting, H. (2010). *Rethinking Global Biodiversity Strategies: Exploring Structural Changes in Production and Consumption to Reduce Biodiversity Loss*. Netherlands Environmental Assessment Agency (PBL), Bilthoven

Thornton, P.K. (2010). Livestock production: recent trends, future prospects. *Philosophical Transactions of the Royal Society B* 365, 2853–2867

Tilman, D., Cassman, K.G., Matson, P.A., Naylor, R. and Polasky, S. (2002). Agricultural sustainability and intensive production practices. *Nature* 418, 671–677

TMT (2011). *Truckee Meadows Tomorrow: Engaging the Community, Measuring Our Progress*. http://www.truckeemeadowstomorrow.org/

UN (2000). *United Nations Millennium Declaration*. Resolution adopted by the General Assembly. United Nations, New York

UNCED (1992). *Agenda 21*. United Nations Convention on Environment and Development http://www.un.org/esa/sustdev/documents/agenda21/english/Agenda21.pdf

UNCLOS (1982). *The United Nations Convention on the Law of the Sea*. Montego Bay

UNDESA (2011). *World Population Prospects: The 2010 Revision*. Population Division, United Nations Department of Economic and Social Affairs. http://esa.un.org/unpd/wpp/index.htm

UNDESA (2009). *World Population Prospects: The 2008 Revision*, Population Division of the Department of Economic and Social Affairs of the United Nations Secretariat.

UNDP (2009). *Human Development Report 2009. Overcoming Barriers: Human Mobility and Development*. United Nations Development Programme, New York

UNEP (2011a). *Integrated Assessment of Black Carbon and Tropospheric Ozone: Summary for Decision Makers*. http://www.unep.org/dewa/Portals/67/pdf/BlackCarbon_SDM.pdf

UNEP (2011b). *Resource Efficiency: Economics and Outlook for Asia and the Pacific. Key Messages and Highlights*. United Nations Environment Programme, Nairobi. http://www.unep.org/roap/Portals/96/REEO_AP_Key.pdf

UNEP (2011c). *Towards a Green Economy: Pathways to Sustainable Development and Poverty Eradication*. United Nations Environment Programme, Nairobi. http://www.unep.org/greeneconomy/GreenEconomyReport/tabid/29846/Default.aspx (accessed 17 November 2011)

UNEP (2010a). *Are the Copenhagen Accord Pledges Sufficient to Limit Global Warming to 1.5 or 2 Degrees C? Emissions Gap Report*. United Nations Environmental Programme, Nairobi

UNEP (2010b). *Global Environment Outlook: Latin America and the Caribbean – GEO LAC 3*. United Nations Environment Programme, Regional Office for Latin America and the Caribbean, Panama City

UNEP (2007). *Global Environment Outlook 4*. United Nations Environment Programme, Nairobi

UNEP (2002). *Global Environment Outlook 3*. United Nations Environment Programme, Nairobi

UNESCO (2011). *UN Decade of Education for Sustainable Development (2005–2014)*. United Nations Educational, Scientific and Cultural Organization, Paris. http://www.unesco.ca/en/interdisciplinary/ESD/default.aspx

UNESCO (2009). *Water in a Changing World. The United Nations World Water Development Report 3*. United Nations Educational, Scientific and Cultural Organization, Paris. http://webworld.unesco.org/water/wwap/wwdr/wwdr3/pdf/WWDR3_Water_in_a_Changing_World.pdf

UNESCO (2006). *Water – A Shared Responsibility. The United Nations World Water Development Report 2*. United Nations Educational, Scientific and Cultural Organization, Paris. http://www.unesco.org/water/wwap/wwdr/wwdr2/pdf/wwdr2_front_matter.pdf

UNFCCC (2010). *Report of the Conference of the Parties on Its Sixteenth Session*. United Nations Framework Convention on Climate Change, Cancun. http://unfccc.int/resource/docs/2010/cop16/eng/07a01.pdf#page=2

UNFCCC (1992). *United Nations Framework Convention on Climate Change*. FCCC/INFORMAL/84 GE.05-62220 (E) 200705. http://unfccc.int/resource/docs/convkp/conveng.pdf

UNCED (1992a). *Rio Declaration on Environment and Development*. United Nations Convention on Environment and Development, Rio de Janeiro

US-GAO (2011). *Key Indicator Systems: Experiences of Other National and Subnational Systems Offer Insights for the United States*. United States Government Accountability Office. http://www.gao.gov/new.items/d11396.pdf

Van Beers, C. and van den Bergh, J.C.J.M. (2009). Environmental harm of hidden subsidies: global warming and acidification. *Ambio* 38(6), 339–341

Van Beers, C. and van den Bergh, J.C.J.M. (2001). Perseverance of perverse subsidies and their impact on trade and environment. *Ecological Economics* 36(3), 475–486

Van Ruijven, B.J., Schers, J. and van Vuuren, D.P. (2012). Model-based scenarios for rural electrification in developing countries. *Energy* 38, 386–397

Van Ruijven, B., Urban, F., Benders, R.M.J., Moll, H.C., van der Sluijs, J.P., de Vries, B. and van Vuuren, D.P. (2008). Modeling energy and development: an evaluation of models and concepts. *World Development* 36(12), 2801–2821

Van Vuuren, D.P., Riahi, K., Moss, R., Edmonds, J., Thomson, A., Nakicenovic, N., Kram, T., Berkhout, F., Swart, R., Janetos, A., Rose, S.K. and Arnell, N. (2012). A proposal for a new scenario framework to support research and assessment in different climate research communities. *Global Environmental Change* 22, 21–35

Van Vuuren, D.P. and Riahi, K. (2011). The relationship between short-term emissions and long-term concentration targets – a letter. *Climatic Change* 104(3–4), 793–801

Van Vuuren, D.P., Kok, M., Girod, B., Lucas, P., de Vries, H.J.M. and (2011a). Scenarios in global environmental assessments: key characteristics and lessons for future use. *Global Environmental Change* (submitted)

Van Vuuren, D.P., Bellevrat, E., Kitous, A. and Isaac, M. (2010). Bio-energy use and low stabilization scenarios. *The Energy Journal* 31 (Special Issue 1), 193–222

Van Vuuren, D.P., Meinshausen, M., Plattner, G.K., Joos, F., Strassmann, K.M., Smith, S.J., Wigley, T.M.L., Raper, S.C.B., Riahi, K., de la Chesnaye, F., den Elzen, M.G.J., Fujino, J., Jiang, K.,

Nakicenovic, N., Paltsev, S. and Reilly, J.M. (2008a). Temperature increase of 21st century mitigation scenarios. *Proceedings of the National Academy of Sciences of the United States of America* 105(40), 15258–15262

Van Vuuren, D.P., Ochola, W.O., Riha, S., Giampietro, M., Ginzo, H., Henrichs, T., Hussain, S., Kok, K., Makhura, M., Mirza, M., Palanisami, K.P., Ranganathan, C.R., Ray, S., Ringler, C., Rola, A., Westhoek, H., Zurek, M., Avato, P., Best, G., Birner, R., Cassman, K., de Fraiture, C., Easterling, B., Idowu, J., Pongali, P., Rose, S., Thornton, P.K. and Wood, S. (2008b). Outlook on agricultural change and its drivers. In *Agriculture at a Crossroads* (eds. McIntyre, B.D., Herren, H.R., Wakhungu, J. and Watson, R.T.). pp.255–305. Island Press, Washington, DC

Van Vuuren, D.P., den Elzen, M.G.J., Lucas, P.L., Eickhout, B., Strengers, B.J., van Ruijven, B., Wonink, S. and van Houdt, R. (2007). Stabilizing greenhouse gas concentrations at low levels: An assessment of reduction strategies and costs. *Climatic Change* 81, 119–159

Van Vuuren, D.P., Sala, O.E. and Pereira, H.M. (2006). The future of vascular plant diversity under four global scenarios. *Ecology and Society* 11(2), 25

Venkataraman, C., Sagar, A.D., Habib, G., Lam, N. and Smith, K. (2010). The Indian National Initiative for Advanced Biomass Cookstoves: the benefits of clean combustion. *Energy for Sustainable Development* 14, 63–72

Von Braun, J. and Meinzen-Dick, R. (2009). *Land Grabbing by Foreign Investors in Developing Countries – Risks and Opportunities.* IFPRI Policy Brief No. 13. International Food Policy Research Institute, Washington, DC

Vörösmarty, C.J., Green, P., Salisbury, J. and Lammers, R. (2000). Global water resources: vulnerability from climate change and population growth. *Science* 289, 284–288

Walker, W.E. and Marchau, V.A.W.J. (2003). Dealing with uncertainty in policy analysis and policy-making. *Integrated Assessment Journal* 4(1), 1–4

Walker, B.H., Gunderson, L.H., Kinzig, A.P., Folke, C., Carpenter, S.R. and Schultz, L. (2006). A handful of heuristics and some propositions for understanding resilience in social-ecological systems. *Ecology and Society* 11(1), 13. http://www.ecologyandsociety.org/vol11/iss1/art13/

WBCSD (2010). *Vision 2050*. World Business Council for Sustainable Development, Geneva

WBCSD (2006). *Business in the World of Water: WBCSD Water Scenarios to 2025*. World Business Council for Sustainable Development, Geneva

WBGU (2011). *World in Transition. A Social Contract for Sustainability*. Summary for Decision-Makers. German Advisory Council on Global Change (WBGU). WBGU Secretariat, Berlin

WCED (1987). *Our Common Future: Report of the World Commission on Environment and Development*. World Commission on Environment and Development. Oxford University Press, New York

WHO (2006). WHO *Air Quality Guidelines for Particulate Matter, Ozone, Nitrogen Dioxide and Sulfur Dioxide: Global Update 2005*. World Health Organization, Geneva

WHO/UNICEF (2010). *Progress on Sanitation and Drinking Water: 2010 Update*. WHO/UNICEF Joint Monitoring Programme for Water Supply and Sanitation. World Health Organization, Geneva and United Nations Children's Fund, New York

Wise, M., Calvin, K., Thomson, A., Clarke, L., Bond-Lamberty, B., Sands, R., Smith, S.J., Janetos, A. and Edmonds, J. (2009). Implications of limiting CO_2 concentrations for land use and energy. *Science* 324(5931), 1183–1186

World Bank (2008). *Global Monitoring Report – MDGs and the Environment: Agenda for Inclusive and Sustainable Development.* World Bank, Washington, DC

World Bank/IMF (2011). *Global Monitoring Report 2011: Improving the Odds of Achieving the MDGs*. World Bank, Washington, DC

WSSD (2002). *Johannesburg Plan of Implementation*. World Summit on Sustainable Development. http://www.un.org/esa/sustdev/documents/WSSD_POI_PD/English/POIToc.htm

Worm, B., Hilborn, R., Baum, J., Branch, T., Collie, J., Costello, C., Fogarty, M., Fulton, E., Hutchings, J., Jennings, S., Jensen, O., Lotze, H., Mace, P., McClanahan, T., Minto, C., Palumbi, S., Parma, A., Ricard, D., Rosenberg, A., Watson, R. and Zeller, D. (2009). Rebuilding global fisheries. *Science* 325, 578

Wunder, S. (2007). The efficiency of payments for environmental services in tropical conservation. *Conservation Biology* 21(1), 48–58

CHAPTER 17

Global Responses

Coordinating lead authors: Ivar Baste, Maria Ivanova and Bernice Lee

Lead authors: Satishkumar Belliethathan, Ibrahim Abdel Gelil, Joyeeta Gupta, Peter M. Haas, Zerisenay Habtezion, Achim Halpaap, Jennifer Clare Mohamed-Katerere, Peter King, Marcel Kok, Marcus Lee and Trista Patterson

Contributing authors: Vivien Campal, Bradnee Chambers, Melissa Goodall (GEO Fellow), Slobodan Milutinovic and Felix Preston (GEO Fellow)

Chapter coordinators: Matthew Billot and Nalini Sharma

Main Messages

Environmental degradation heightens risks and reduces opportunities for the advancement of human well-being, especially for poor and vulnerable populations. Harmful environmental changes are taking place in an increasingly globalized, industrialized and interconnected world, with a growing global population and unsustainable production and consumption patterns. The degradation of ecosystem services is narrowing development opportunities and could threaten future human well-being.

The prospect for improving human well-being is dependent on the capacity of individuals, institutions, countries and the global community to respond to environmental change. Innovative and transformative policies and technologies could assist society to overcome current barriers to achieving sustainable development. A more balanced approach to addressing environmental, economic and social concerns could also help.

Even though national and regional responses have begun to address environmental challenges, a polycentric governance approach is needed to attain effective, efficient and equitable outcomes. This approach recognizes a diversity of settings and assumes multiple centres of activity and authority, which, given the range of capacity needs, are critical to generate adequate responses to environmental challenges.

Environmental responses are attracting greater financial flows but these still fall short of the resources needed. The Organisation for Economic Co-operation and Development countries' aid commitments to the three UN conventions on biodiversity, climate and desertification grew from US$5.1 billion in 1999 to US$17.4 billion in 2009. The same countries allocated US$22.9 billion to official development assistance for climate change mitigation and adaptation in 2010. Yet, the cost for developing countries to adapt to climate change alone has been estimated at US$70–US$100 billion a year for 2010–2050.

Global responses have a key role to play in promoting coordination, integration and systemic considerations. They can help set goals and develop metrics, support capacity enhancement, generate financial resources and facilitate the sharing of best practices. At the global level, a results-based approach to advancing human well-being and environmental sustainability could be anchored in the strategies and associated response options that follow below. The United Nations Conference on Sustainable Development (Rio+20) provides an opportunity to take stock, assess achievements and shortcomings, and begin to stimulate transformative global responses. The suggested strategies are part of a systemic approach, which could highlight barriers and inform adjustments, learning and continuous improvement.

Global response options

Framing environmental goals in the context of sustainable development, and monitoring outcomes. A process could be initiated to revisit and extend the Millennium Development Goals in the form of sustainable development goals centred on human well-being, with measurable metrics, keeping in mind the need for the coherent and balanced integration of environmental, economic and social dimensions.

Enhancing the effectiveness of global institutions. The sustainable development agenda could be elevated and mainstreamed into the core of decision making within the UN system, supported by enhanced cooperation with and between environmental, economic and social institutions.

Investing in enhanced capacities for addressing environmental change. Delivering results will require strengthened national capacities to develop, deliver and implement strategies to combat environmental degradation. A UN system-wide framework for capacity building could strengthen the national capacities required to implement specific multilateral environmental agreements.

Supporting technological innovation and development. Mechanisms from collaborative research and development (R&D), knowledge platforms and global prize funds for environmentally sound technologies could be scaled up to accelerate the innovation and diffusion of technologies critical to the transition to a global green economy.

Strengthening rights-based approaches and access to environmental justice through recognition, enforcement and implementation in global and regional institutions. Principle 10 of the Rio Declaration on Environment and Development recognizes the importance of procedural environmental rights. For the past 20 years regional experience has demonstrated that such rights provide a basis for citizens to participate in safeguarding both human and environmental well-being.

Deepening and broadening stakeholder engagement. The private sector and civil society could be invited to explore the use of new information and communications technology to build a stakeholder web to enhance access to information and stakeholder engagement, and to mobilize new partnerships. An intergenerational assembly could provide an opportunity for future leaders and sustainability champions to interact and foster a joint vision for a sustainable future.

INTRODUCTION

Global environmental change such as climate change and the degradation of ecosystem services is heightening risks and reducing opportunities, especially for poor and vulnerable populations. Such change is taking place in an increasingly globalized, urbanized, interconnected and fast-moving world amidst shifting geopolitical power balances. Burgeoning flows of goods and services, capital and technology, information and labour all fuel a growing global population with implications for patterns of consumption and production. The scale and persistence of global environmental problems require sustained collective efforts to meet internationally agreed goals. Responses at national and regional levels are already available, but addressing the underlying drivers of global environmental degradation, rather than the pressures or symptoms, would require the sustained evolution of rules, institutions, economic systems and values to transform the current approach to environmental management. In addition, adequate and stable financial resources, political commitment, knowledge and operational capacity are also imperative. But these enabling conditions and the requisite governance mechanisms and structures vary considerably between regions and countries.

There is no single, overarching solution to environmental challenges. Yet many environmental problems, particularly those pertaining to the global commons, can only be addressed through collective action. Global responses are also critical for enhancing national capacity and facilitating the uptake of solutions among nations with regional commonalities. Responses at national and global levels interact and generate incremental, structural and transformational change (Putnam 1988). The engagement of non-state actors at different levels has, for example, fostered knowledge exchange and strengthened capacities. Policy changes adopted by individual governments can transmit normative signals, exert peer pressure or encourage learning and replication – providing incentives for the collective adoption of international norms, rules, laws or policies. In several areas – climate, biodiversity, chemicals – global environmental treaties have set new goals, standards and expectations for state performance. The embodiment of these goals and standards in national laws, regulations and action plans in turn induces member states to comply. Global responses integrating a mix of strategies, values, principles, investments and measures supported by a diverse range of capacities can enable national and regional choices.

The capacity of the international community to deliver solutions to environmental problems is a function of its ability to establish and maintain flexible and holistic governance and management frameworks at global and national levels. Ideally, frameworks to accomplish these objectives should be based on clear and measurable goals, verifiable strategies, and strong monitoring and evaluation mechanisms. Adaptive governance is an emerging approach for addressing multi-dimensional, ever-evolving environmental and socio-economic challenges that exhibit a high degree of uncertainty (Gunderson *et al.* 2010; Dietz *et al.* 2003); and it facilitates decision making in complex systems under the circumstances of abrupt, disorganizing or turbulent change (Folke *et al.* 2005). In addition, trust building, involving stakeholder participation and mechanisms for feedback, could help ensure that change is both sustainable and equitable (Kydd 2005; Levi-Faur 2005; Braithwaite and Drahos 2000).

Although results-based management is most commonly used to manage internal organizational processes, it is a perspective that also enhances transparency and accountability when focused on international processes. Modified to reflect the global scale,

Secretary-General Ban Ki-moon addresses the opening high-level segment of the UN Climate Change Conference (COP16) in Cancún, Mexico, urging governments to mobilize the highest level of political will, and to deliver progress towards an eventual international treaty. *© Paulo Filgueiras/UN Photo*

a systematic and comprehensive results-based global approach could be anchored in six response options:

- framing environmental goals in the context of sustainable development;
- enhancing the effectiveness of global institutions;
- investing in enhanced capacities for addressing environmental change;
- supporting technological innovation and development;
- strengthening rights-based approaches and access to environmental justice; and
- deepening and broadening stakeholder engagement.

These strategies build on the lessons learned and opportunities identified in *GEO-5*. Aligning goals and strategies within a common vision lies at the basis of an integrated approach, which could be instrumental in improving the effectiveness and efficiency of responses at global, national and local levels. The following section assesses the state of global responses to date and highlights gaps and barriers that have hindered the collective ability to manage environmental change. Proposals are then made for results-based responses anchored in the many different centres of governance for addressing global environmental challenges and advancing human well-being.

STATE OF GLOBAL RESPONSES

Over the past 40 years a wide range of responses to environmental problems has been implemented as a set of interacting systems with multiple actors at different scales. Conventional responses at national and global levels include the creation of rules, laws and institutions, with international organizations established to serve as conveners at the global scale; as arbiters for exchange, sharing experiences, articulating interests and aggregating preferences; as sources of expertise; and as enablers of a broader social dialogue (Bearce and Bondanella 2007; Esty and Ivanova 2002; Bartlett *et al.* 1995). The public sector accounts for about 30 per cent of gross domestic product (GDP) worldwide (World Bank 2011) and is an essential tool in creating enabling conditions for societal change, with public-private partnerships and social networks bringing new opportunities for engagement. Notwithstanding all the good intentions and efforts, however, the Earth and its sub-systems are showing signs of considerable degradation.

The global response framework: from isolation to integration

Currently, environmental problems are rarely tackled in an integrated fashion. The connectedness of climate change, water resources, desertification and biodiversity loss, for example, makes isolated governance responses inadequate and potentially counterproductive. A more integrated approach to substantive issues and spatial scale demands a new adaptive governance framework.

Global environmental issues can be divided into those that are common to many or most countries, including pollution of water bodies or solid waste disposal, and those that affect the global commons such as pollution of the global atmosphere or the open seas. Not all environmental issues require a global scale of governance. Some can be addressed through cooperation between a few countries, for example the transboundary water concerns of the Mekong or Zambezi rivers or networks of protected areas for endangered marine species with limited ranges. However, problems of the global commons – ones that cumulatively lead to negative global trends and/or whose drivers are essentially global – often require international treaties to ensure collective global action. Relationships between international and national scales of governance tend to be structured as:

- bottom-up: countries harmonize national policy, with the policy space created determining the ability to accommodate international commitments;
- top-down: thresholds, targets and principles are decided at global level and then translated into national-scale implementation;
- multi-level: policy development that addresses the complex relationships between the different levels of governance and the actors involved.

Governments have used international institutions as key instruments when global action is required. Cooperative behaviour is enhanced by changing the environment within which collective agreements arise, increasing awareness about particular issues, and enhancing national capacity to deal with the problems in question (Haas *et al.* 1993). In the environmental field, international institutions have channelled information, created norms and principles, provided training and financial resources to affected countries, and catalysed actions at multiple levels of governance (Young 2010, 2002). The UN General Assembly formally initiated the international environmental agenda through Resolution 2398 (XXIII) on 3 December 1968, calling for the convening of the United Nations Conference on the Human Environment, the 1972 Stockholm Conference. The assembly framed the environmental challenge as an integral part of economic and social development, with UNEP established as the institutional mechanism to ensure follow-up to the environmental dimension of the conference outcome. The promotion and coordination of environmental activities within the whole UN system was one of the core functions governments delegated to UNEP. With environmental awareness on the rise, the new programme also initiated a number of new international agreements aiming to address emerging environmental issues.

Environmental activities have become an integral component of the UN system, represented by its programmes, agencies, secretariats and coordinating mechanisms. Table 17.1 identifies core instruments anchored within the UN system's multi-sectoral environmental response regime as they relate to the environmental issues assessed in Part 1 of *GEO-5*. The many instruments at the regional level, such as international agreements on transboundary waters, are not included. Nonetheless, the table illustrates that the competence and capacities of the environmental institutions in the UN system are distributed among its different entities and policy sectors, reflecting the growing importance of various governing bodies across the system.

Table 17.1 Core elements of the UN system-wide environmental response regime

Theme	Instruments
Sustainable development	**International soft law instruments and institutions:** Rio Declaration on Environment and Development and Agenda 21 from the United Nations Conference on Environment and Development (UNCED, 1992); Johannesburg Plan of Implementation (JPOI); Millennium Development Goals (MDGs); Commission on Sustainable Development (CSD); and UN system entities **Science processes:** Group on Earth Observations and its Global Earth Observation System of Systems (GEOSS); Inter-Agency and Expert Group (IAEG) on MDG indicators coordinated by the United Nations Statistics Division **Inter-agency bodies:** Chief Executive Board for Coordination (CEB); High Level Committee on Policy (HLCP); Executive Committee on Economic and Social Affairs (ECESA)
Environment broadly defined	**International soft law instruments and institutions:** Declaration and Programme of Action from the Stockholm Conference on the Human Environment; UNEP; Global Environment Facility (GEF); the environment-related portfolio of 44 UN system entities including the United Nations Development Programme (UNDP), Food and Agriculture Organization of the United Nations (FAO), United Nations Educational, Scientific and Cultural Organization (UNESCO) and the World Bank Group. **Science processes:** Global Environment Outlook (GEO) (UNEP); International Panel for Sustainable Resource Management (UNEP); Millennium Ecosystem Assessment (MA) **Funds:** Environment Fund (UNEP); GEF Trust Fund; World Bank environmental and natural resource management (ENRM) lending portfolio; the environment portfolio of other UN Multi-Donor Trust Funds (MDTF) administered by UNDP **Inter-agency bodies:** Environment Management Group (EMG)
Atmosphere	**Multilateral environmental agreements:** Vienna Convention (1985) and Montreal Protocol (1987); United Nations Framework Convention on Climate Change (UNFCCC, 1992) and Kyoto Protocol (1997) **International soft law instruments and institutions:** a broad range of UN system entities, including FAO, the United Nations Conference on Trade and Development (UNCTAD), UNDP, UNEP and the World Meteorological Organization (WMO) as well as the World Bank have programme activities related to atmosphere **Science processes:** Technology and Economic Assessment Panel (TEAP) under the Montreal Protocol (UNEP); Intergovernmental Panel on Climate Change (IPCC) administered by WMO and UNEP; World Climate Research Programme (WCRP) **Funds:** Montreal Fund (UNEP); GEF is the financial mechanism for the UNFCCC; Clean Development Mechanism (CDM); Special Climate Change Trust Fund (SCCF), Adaptation Trust Fund and Least Developed Countries Trust Fund (LDCF) administered by the GEF; Environment Fund (UNEP) **Inter-agency bodies:** Working Group on Climate Change under the High-Level Committee on Policy (HLCP) of the Chief Executives Board for Coordination (CEB), and UN Energy
Land	**Multilateral environmental agreements:** United Nations Convention to Combat Desertification (UNCCD, 1994) **International soft law instruments and institutions:** A broad range of UN system entities, including FAO, the International Fund for Agricultural Development (IFAD), UNDP, UNEP, UN-Habitat, World Health Organization (WHO), World Food Programme (WFP) and the World Bank, have programme activities on land-related issues **Science processes:** Covered by the Global Environment Outlook (GEO) and Millennium Ecosystem Assessment (MA). **Funds:** GEF as the financial mechanism for UNCCD; Global Mechanism (UNCCD); Environment Fund (UNEP) **Inter-agency bodies:** EMG Issue Management Group on Land
Water	**Multilateral environmental agreements:** United Nations Convention on the Law of the Sea (UNCLOS, 1994); International Convention for the Prevention of Pollution from Ships (MARPOL, 1973); International Convention on Oil Pollution Preparedness, Response and Cooperation (OPRC, 1990); Convention on the Prevention of Marine Pollution by Dumping of Wastes and Other Matter (LDC, 1972); International Convention for the Control and Management of Ships' Ballast Water and Sediments (2004); United Nations Convention on the Law of the Non-navigable Uses of International Water Courses (1997) (not entered into force) **International soft law instruments and institutions:** International Oceanographic Commission, administered by UNESCO; Global Programme of Action for the Protection of the Marine Environment from Land-based Activities (GPA) administered by UNEP; FAO Code of Conduct for Responsible Fisheries; a broad range of UN system entities including FAO, the International Maritime Organization (IMO), UNDP, UNEP, UNESCO, WMO and the World Bank, have programme activities related to oceans and water **Science processes:** Regular process for global reporting and assessment of the state of the marine environment (UNCLOS); Group of Experts on Scientific Aspects of Marine Environmental Protection (GESAMP); World Water Development Report (UNESCO); and covered by the GEO and the MA **Funds:** GEF international waters focal area; Environment Fund (UNEP) **Inter-agency bodies:** UN-Oceans and UN-Water
Biodiversity	**Multilateral environmental agreements:** Ramsar Convention on Wetlands (1971); World Heritage Convention (WHC, 1972); Convention on International Trade in Endangered Species of Wild Fauna and Flora (CITES, 1973); Convention on Migratory Species (CMS, 1979); Convention on Biological Diversity (CBD, 1992) and the Cartagena Protocol (2000); Treaty on Plant Genetic Resources (ITPGRFA, 2001) **International soft law instruments and institutions:** Commission on Genetic Resources for Food and Agriculture under FAO; United Nations Forum on Forests (UNFF); a broad range of UN system entities, including FAO, IMO, UNDP, UNEP, UNESCO, United Nations University (UNU), United Nations World Tourist Organization (UNWTO), WHO, WMO, World Trade Organization (WTO) and the World Bank, have programme activities related to biodiversity **Science processes:** Intergovernmental science-policy platform on biodiversity and ecosystem services (IPBES); Global Biodiversity Outlook (CBD); Global Forest Resource Assessment; State of the World Fisheries and Aquaculture; State of the World's Plant Genetic Resources for Food and Agriculture; State of the World's Animal Genetic Resources for Food and Agriculture; International Assessment on Agricultural Science and Technology for Development (IAASTD), and covered by GEO and the Millennium Ecosystem Assessment (MA) **Funds:** GEF as the financial mechanism for CBD; Environment Fund (UNEP) **Inter-agency bodies:** EMG Issue Management Group on Biodiversity; Biodiversity Liaison Group

Theme	Instruments
Table 17.1 Core elements of the UN system-wide environmental response regime *continued*	
Chemicals and waste	**Multilateral environmental agreements** Basel Convention on the Control of Transboundary Movements of Hazardous Wastes and their Disposal (1989); Rotterdam Convention on the Prior Informed Consent Procedure for Certain Hazardous Chemicals and Pesticides in International Trade (1998); Stockholm Convention on Persistent Organic Pollutants (POPs, 2001). **International soft law instruments and institutions:** Negotiation of a convention on mercury (UNEP); Strategic Approach to International Chemicals Management (SAICM); a range of UN entities, including FAO, International Labour Organization (ILO), UNDP, UNEP, United Nations Industrial Development Organization (UNIDO), United Nations Institute for Training and Research (UNITAR), WHO and the World Bank, have programme activities related to chemicals **Science processes:** Covered by GEO **Funds:** GEF is the financial mechanism for the POPs Convention; Environment Fund (UNEP) **Inter-agency bodies:** Inter-Organization Programme for the Sound Management of Chemicals (IOMC)

While the level of environmental integration both within and outside the UN system is significant – representing an important source of competence and capacity – the integrated governance of the diverse and multi-sectoral environmental field is complex and, at times, problematic (Oberthür and Stokke 2011), especially for sustainability goals.

At times it seems that calls to simplify are at odds with a need to capture system complexity: on one hand, governments have requested that the United Nations foster synergies between compatible multilateral environmental agreements and identify guiding elements for realizing such synergies while respecting the autonomy of the conferences of the parties (UNEP 2011e). There have been several attempts to integrate these diverse efforts, including clustering the chemical conventions by convening simultaneous Extraordinary Conferences of the Parties to the Basel, Rotterdam and Stockholm Conventions in February 2010 in Bali, Indonesia. There is also coordination between the three Rio conventions – on climate, biodiversity, and desertification – under the Joint Liaison Group and the 44 UN entities of the Environment Management Group, the UN-wide coordination body for the environment (UNEP 2011f).

On the other hand, interrelationships and interdependencies exist between all levels of governance and intervention – from the individual and community scale to the global level. Multiple causal mechanisms are at work, including normative influences, prices and markets, political pressure and incentives, persuasion, social learning and the science-policy interface (Simmons *et al.* 2006). Each may operate in isolation or with others, build pressures over time and in combination. Multi-scale interventions can be both counterproductive and mutually enforcing. In such interventions, countries can adopt policies that they then encourage others to adopt as international norms and/or law, yet those policies may influence others adversely. Once in place, the norms and economic incentives may affect behaviour more broadly. They may also engender future changes to legal regimes, normative signals, social learning and resource transfers. In addition, diverse actors including members of civil society organizations, scientific networks and research institutions, international organizations, faith communities and the private sector are engaged not only in demanding but also in providing global responses (Slaughter 2004; Commission on Global Governance 1995; Rosenau and Czempiel 1991; Keohane and Nye 1971).

The late Hon Indira Gandhi, then Prime Minister of India, addressing the UN Conference on the Human Environment in Stockholm on 5 June 1972.

© Yutaka Nagata/UN Photo

Legal and policy framework

As Part 1 of this report illustrates, it is difficult to measure success in achieving environmental goals, especially if those goals are not quantified. Environmental treaties form the core legal and policy framework for the global environment and aim to set appropriate goals for the international community. While environmental laws are legally binding, the lack of specific targets and timetables often implies that these are, in effect, soft law guidelines rather than hard law frameworks. Some treaties are also difficult to implement because of the lack of capacity in individual countries. In addition, verifying change in environmental quality and attributing changes to specific policy measures is challenging without robust and comparable data, especially at the global level.

Environmental treaties

Today, there are more than 500 international treaties and other agreements that relate to the environment, of which 323 are regional and 302 date from 1972 and the early 2000s. The core of the global environmental legal framework, however, is made up of a more limited number of treaties with a growing number of ratifications (Figure 17.1). Most of the new agreements have established new, independent bureaucracies and this proliferation has fragmented authority in international environmental governance. Thus, while the creation of the various environmental conventions and protocols can be viewed as an achievement, it also raises the need for continuing support in developing countries when national administrations become overloaded with reporting requirements and countless international meetings (Najam 2005; Biermann 2004).

A distinguishing feature of the more effective treaties is their development through the interplay of organized scientific communities (Haas and Stevens 2011) and a moderate to strong international institution (Biermann and Siebenhüner 2009; Haas 2007). The scientific community informs treaties that reflect an understanding of the problems and their solutions, while the institutions integrate the science into draft treaties, help promote the ideas of the scientists, coordinate meetings, compile information repositories, provide incentives to states to participate in negotiations and assist member states in complying with their obligations. Innovation in technology, networking, coordination and knowledge management can help this process. Chapter 16 has also pointed to the critical role of planning in creating the conditions suitable to coordinate integrated, complex or multi-procedural outcomes.

The Montreal Protocol on Substances that Deplete the Ozone Layer is one example of such a success. Under this, countries have almost totally eliminated the production of chlorofluorocarbons (CFCs) in just 20 years. The protocol's success resulted from:

- scientific consensus on the problem;
- public awareness and pressure;
- existence of a cost-effective substitute;
- private sector buy-in;
- leadership from both an international institution, UNEP, and a national agency, the US Environmental Protection Agency;
- a concrete plan of action; and
- mobilization of financial resources in developed countries to assist developing countries and economies in transition.

Unfortunately, one of the replacement chemicals, hydrofluorocarbons, has a high global warming potential and now needs to be phased out to address climate change – illustrating the interconnectivity of environmental problems. Similarly, cold temperatures in the upper atmosphere, possibly due to climate change, are resulting in increased loss of ozone, particularly over the Arctic.

Capacity building and diffusion of policy tools

To ensure a responsive and cohesive approach to meeting country needs and achieving environmental results and

Figure 17.1 Growth in ratification of environmental treaties, 1971–2011

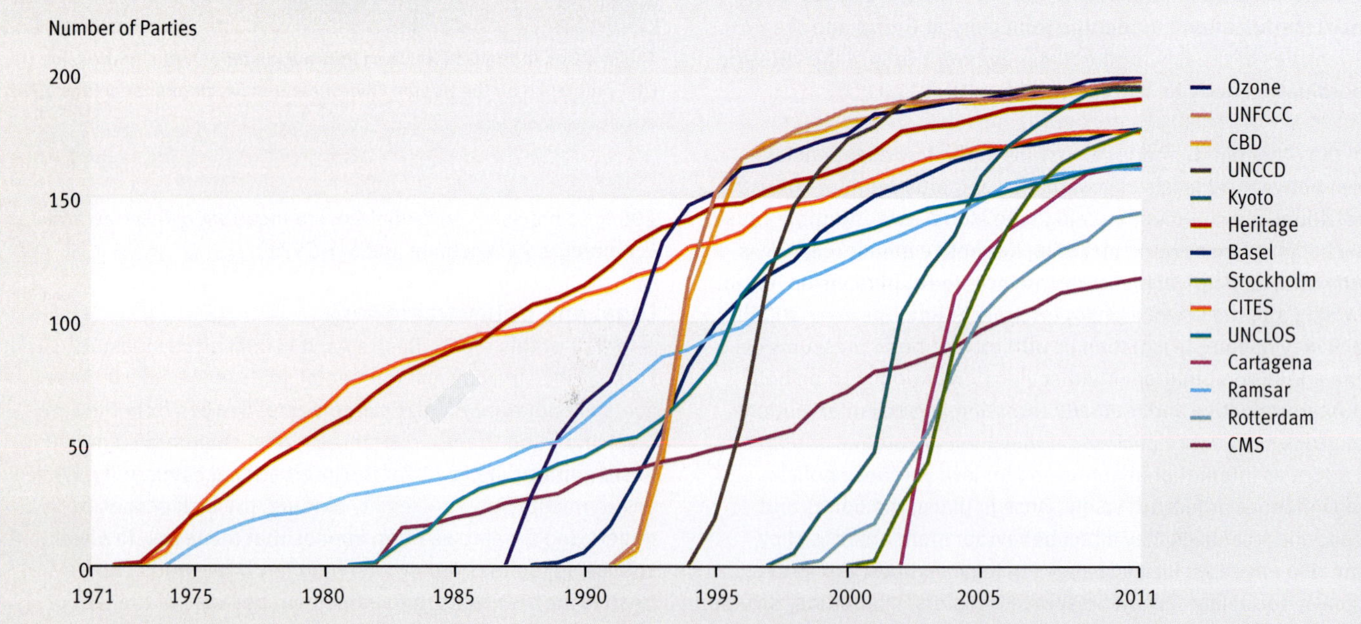

Source: UNEP Environmental Data Explorer, compiled from various MEA Secretariats http://geodata.grid.unep.ch

Box 17.1 Diffusion of policy tools worldwide – the case of strategic environmental assessment

One widely used policy tool across all regions is strategic environmental assessment, which helps to integrate national environmental policies. This goes beyond environmental impact assessments to ensure that environmental, social and economic information is incorporated into decision making in a unified manner. The process involves analysing the likely impacts of decisions; fostering public participation; developing and comparing alternatives fostering; recording the impact, options and comments from the public in a report; ensuring that the report is taken into account when making final decisions; and informing the public about the decisions.

Strategic environmental assessments were initially implemented in Europe, but have since spread to many countries, increasingly becoming an obligatory part of national legislation (OECD 2012). Experiences in Africa, for example, show that this type of assessment has to go beyond the project level and be carried out on the level of policy plans to become effective. In Guinea, it has been used to develop the co-management of reserved forests; in Zambia and Zimbabwe, the International Union for the Conservation of Nature (IUCN) supported its use for planning developments around Victoria Falls – a World Heritage site; Morocco used it to provide an analysis of legal, regulatory, and institutional aspects of environmental impacts in the large-scale irrigation sector (Economic Commission of Africa 2005); while in Ghana it has improved the management of mangrove forests (Sampong 2004).

outcomes, developing and implementing a system-wide capacity-building framework is crucial (OECD 2011b). Studies from international organizations (Baser and Morgan 2008), academics (Eyben 2006), non-governmental organizations (Lipson and Warren 2006) and other practitioners (James and Wrigley 2007) suggest that capacity building:

- is a complex human process based on values, emotions and beliefs;
- involves the main actors taking responsibility for the process of change;
- involves shifts in power and identity;
- involves changes in relationships between elements of human systems;
- is uncertain and unpredictable in its outcomes; and
- is strongly shaped by culture and values (Woodhill 2010).

This implies greater consideration and recognition of the less visible aspects of capacity building, such as values, legitimacy, identity and self-confidence, as well as other non-monetary forms of motivation (Aragón and Macedo 2010). It also involves improving access to key resources such as finance, technology and knowledge that underpin capacity and capabilities. Capacity building can also be advanced by building on lessons learned from a diffusion of policy tools. Strategic environmental assessments are an example of diffusion of policy instruments where timing, public participation and credibility of policy analysis stand out as important determinants of success (Runhaar and Driesen 2007).

Financial flows

Expanding the donor base, increasing the availability and accessibility of funds, and ensuring stable and predictable financial flows are among the top priorities in international environmental governance (UNEP 2010). The first financing mechanism designed explicitly and exclusively for global environmental purposes was the Environment Fund. Created in 1972 through UN General Assembly Resolution 2997 as one of the core elements of the new environment programme, the Environment Fund was intended to finance new environmental initiatives within the UN system and to assist developing countries. Today, environmental financing comes in the form of environmental aid from bilateral and multilateral donors, including through funds dedicated to specific environmental concerns such as the Montreal Fund for the Implementation of the Montreal Protocol to support ozone-related work, climate funds to support mitigation and adaptation, funds to combat deforestation, and others. The Global Environment Facility (GEF) is the largest funder of projects that specifically seek to improve the global environment through support for the additional costs of transforming projects with national benefits into projects with global environmental benefits.

Rapid developments in information technology over the past 20 years have revolutionized many aspects of life, including the development of truly global financial markets. *© Robert Churchill/iStock*

Box 17.2 Identifying financial flows for environmental response

It is currently difficult to get a complete picture of the amount of resources invested in environmental activities at normative and operational levels. Financial resources are often counted several times as funding flows from one organization to another or between funding categories. This double counting is also due to lack of distinct definitions and the inherent overlaps between categories of spending. Reported financial figures are often not fully comparable because the financial year and budget procedures vary among institutions. Furthermore, since much of the investment in environmental activities happens through integrating environmental perspectives and issues into policies, programmes and projects, it is often difficult to distinguish environmental activities from sectoral ones. For example, as much as 85 per cent of the World Bank's environmental and natural resource management (ENRM) projects are currently being managed by non-environmental sectors in the bank (UNEP 2011c). Several important developments illustrate annual financial flows for responses to environmental challenges.

- The carbon market stalled in 2010 at US$142 billion after rapid growth, partly because of lack of regulatory clarity. The figure includes the value of the primary and secondary Clean Development Mechanism (CDM) markets, which respectively stood at US$1.5 billion and US$18.3 billion (World Bank 2011).
- Countries of the Organisation for Economic Co-operation and Development (OECD) allocated the following official development assistance (ODA):
  - up to US$22.9 billion, 15 per cent of total ODA, to climate change mitigation and adaptation in 2010 (OECD 2011c);
  - US$4.3 billion to biodiversity in 2009 (OECD 2011a);
  - US$1.9 billion to desertification in 2009 (OECD 2011a).
- UNEP (2011c) reports the following indicative levels of financial environmental flows:
  - annual pledges to the GEF for the fifth replenishment agreed in 2010 amount to US$1.1 billion;
  - the World Bank's environmental and natural resource management (ENRM) portfolio, including the GEF, in 2008 reached US$3 billion;
  - the United Nations Development Programme's (UNDP) expenditures on environmental activities, including the GEF, in 2009 were US$1.1 billion;
  - UNEPs budget for 2010 was US$0.5 billion including the GEF, the Environment Fund and the Multilateral Fund for the implementation of the Montreal Protocol;
  - the combined annual budget for the three Rio conventions in the period 2008–2011 was of the order of US$0.1 billion.

Yet, long-standing commitments from developed countries to improve access to finance for developing countries remain largely unfulfilled, and insufficient and unpredictable financial resources continue to constrain effective environmental governance at all levels (OECD 2011b). It is currently difficult to identify the financial flows for environmental responses (Box 17.2) as there is no tracking system to monitor resources invested in environmental activities by the United Nations and other international institutions (UNEP 2011e). A review of the existing data shows that while there are significant financial investments in climate change and other environmental initiatives, they fall well short of the scale required to address the challenges (Behrens 2009; Müller 2009; UNDP 2007). For example, the World Bank estimates that the price tag between 2010 and 2050 for adapting to an approximately 2°C warmer world by 2050 will be in the range of US$70 billion to US$100 billion a year (World Bank 2010a).

The Environment Fund

The Environment Fund is the principal source of financing for the implementation of UNEP's programme, and was established by the UN General Assembly in 1972. Altogether, 181 countries have made at least one voluntary contribution in the period between 1973 and 2011, with 12 countries having maintained their regular annual contributions over the whole period (UNEP 2012). The four-decade trend depicted in Figure 17.2, however, shows that the original intention to grow the fund proportionally

Figure 17.2 The Environment Fund, 1973–2009

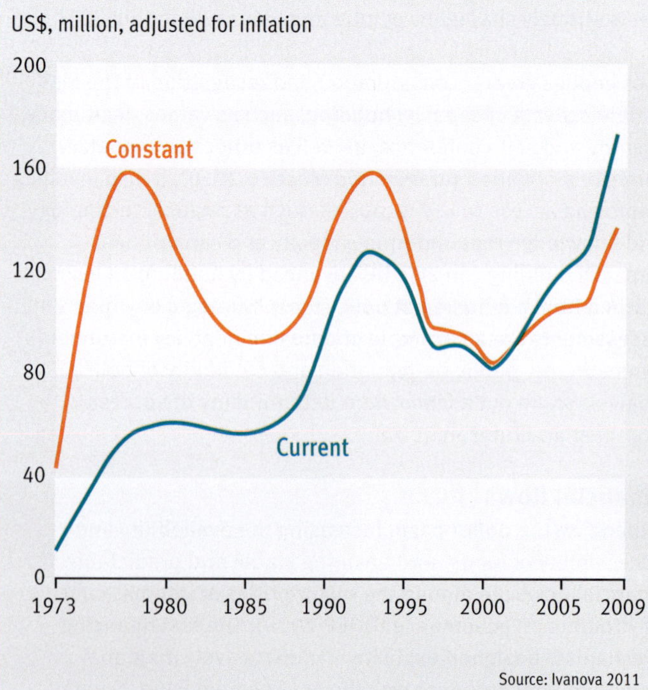

Source: Ivanova 2011

Table 17.2: Financial resources available to selected global multilateral environmental agreements, 2010

Cluster: Atmosphere	**US$, million**
Convention on Long-range Transboundary Air Pollution (CLRTAP)	3.62
Vienna Convention on Substances that Deplete the Ozone Layer	4.84
United Nations Framework Convention on Climate Change (UNFCCC)	107.90
Total	**116.36**
Cluster: Biodiversity	
Convention on the Conservation of Migratory Species of Wild Animals (CMS)	0.33
Cartagena Protocol on Biosafety	2.76
Convention on Wetlands of International Importance, especially as Waterfowl Habitat (Ramsar Convention)	4.67
Convention on International Trade in Endangered Species of Wild Fauna and Flora (CITES)	5.07
Convention on Biological Diversity (CBD)	12.36
Total	**25.19**
Cluster: Chemicals and wastes	
Convention on the Prior Informed Consent Procedure for Certain Hazardous Chemicals and Pesticides in International Trade (Rotterdam Convention)	0.93
Convention on Persistent Organic Pollutants (Stockholm/POPs Convention)	5.47
Convention on the Control of Transboundary Movements of Hazardous Wastes and Their Disposal (Basel Convention)	5.84
Total	**12.24**
Other agreements	
Convention Concerning the Protection of the World Cultural and Natural Heritage (World Heritage Convention)	1.95
United Nations Convention to Combat Desertification in Countries Experiencing Serious Drought and/or Desertification, Particularly in Africa (UNCCD)	5.90
Total	**7.85**
TOTAL	**161.64**

Source: Ivanova and Delina *forthcoming in 2012.*

with intensifying environmental problems has not been truly realized. While it shows some growth in current terms, in real terms (adjusted for inflation), the fund fell by 44 per cent between 1977 and 1987 and is only now reaching the US$160 million per biennium that UNEP attracted in the 1970s and again in the run-up to the 1992 Rio Earth Summit (Ivanova 2011).

Multilateral environmental agreements

As discussed, governments created multilateral environmental agreements when new environmental issues emerged. Table 17.2 offers an overview of financial flows for the secretariats of MEAs by cluster – the GEF provides funding for project-level work in these clusters.

The Multilateral Fund for the Implementation of the Montreal Protocol

Funding for implementation of most multilateral environmental agreements comes through special funds, the largest of which is the Multilateral Fund for the Implementation of the Montreal Protocol. Created in 1990 and administered by UNEP, it helps developing countries comply with the protocol's control measures. It has been replenished eight times since the beginning of its operation in 1991, with contributions from the industrialized countries, including countries in transition, assessed according to the UN scale of assessment. The significant financial resources devoted to the ozone treaty – during 1991–2011 governments pledged US$2.8 billion to the Montreal Protocol (UNEP 2011f) – can be seen both as a reason for and an indicator of the treaty's effectiveness in eliminating the production and consumption of most CFCs. Significant initial investment was critical to the fund's success, and this initial success stimulated sustained investment in the longer-term.

The Global Environment Facility

Established as a US$1 billion pilot programme in the World Bank in 1992, the GEF has evolved to become the financial mechanism for several multilateral environmental agreements, including UNFCCC, the Convention on Biodiversity (CBD), the United Nations Convention to Combat Desertification (UNCCD) and the Stockholm Convention. Over the past 20 years, the GEF has allocated US$10 billion for more than 2 800 projects in more than 168 developing countries and economies in transition, and more than 13 000 small grants totalling US$634 million have been made directly to civil society and community-based organizations (GEF 2011). Although the GEF was initially a partnership between the World Bank, UNDP and UNEP, today it is in partnership with ten UN agencies, 182 civil society organizations and the private sector. This diversity of participation is directly related to the GEF's co-financing requirement, through which it has leveraged more than US$47 billion of funds additional to those available through UN and World Bank channels since 1992. The GEF's operational arrangements have also evolved, with a new system for the transparent allocation of resources (STAR) implemented in 2010. That same year, donors pledged more than US$4.3 billion in replenishment for the period 2010–2014 (GEF-5), representing a 55 per cent increase in resources over GEF-4 (GEF 2010). During 1991–2010, the GEF invested more than US$50 billion, US$40.7 billion of which came from co-financing from development partners – almost half of these funds were used for climate change mitigation and adaptation (Figure 17.3). In 2010 alone, the GEF disbursed a little over US$5 billion for its work; 81 per cent of this amount was from co-financing (GEF 2010).

Figure 17.3: GEF portfolio and co-financing allocations by focal area, 1991–2010

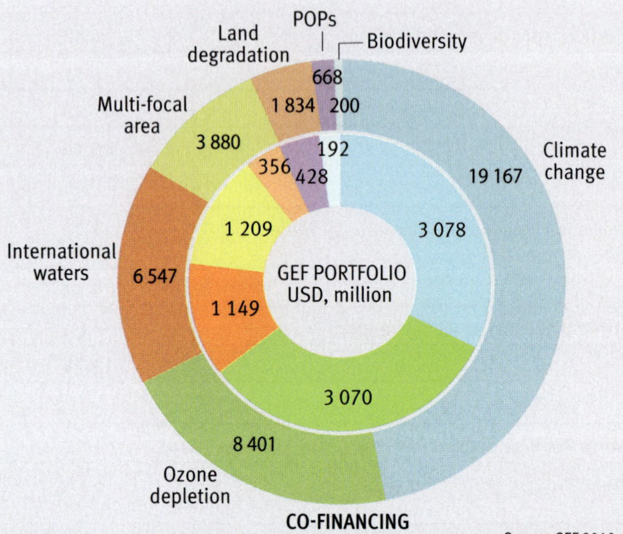

Source: GEF 2010

Environmental official development assistance

Close to US$100 billion of aid, an average of 15 per cent of the global total, was committed to the environment in 1998–2007 (Castro and Hammond 2009), with the most significant source of environmental financing being official development assistance (ODA) from the OECD countries. OECD countries' aid commitments targeted at the objectives of the three Rio conventions combined grew from US$5.1 billion in 1999 to US$17.4 billion in 2009 (Figure 17.4), largely because of increases in funds targeted at climate change. The challenges of proliferation of institutional mechanisms, however, are acute in the environmental field.

Box 17.3 International aid for the environment

Environmental aid is no different from aid to other sectors: there are too many actors, adding to the administrative burden on countries and donors, and impeding aid effectiveness. The average partner country has 17 donors out of the 23 members and 10 major multilateral agencies that report to the OECD's Development Assistance Committee (DAC). Adding up the number of donors in each country for the 153 countries that receive ODA shows that there are 2 617 donor/recipient partnerships, all of which need to be maintained through policy dialogue, planning, coordination, accounting and reporting. In 1 571 of these partnerships – 60 per cent – donors are providing environment aid.

The individual size of the projects and programmes that constitute these partnerships varies enormously. At one extreme, four donors annually provide over US$10 million of core environmental aid on average per partner, while at the other extreme there are 11 donors whose average aid per partner is less than US$1 million. Moreover, the number of small partnerships has increased faster in the past decade than the larger ones. This plethora of partnerships is just a glimpse of the complexity of the architecture of environmental aid. Each donor operates through an average of three agencies using literally thousands of channels. There are also 30 or more bilateral donors that are not members of the DAC, and dozens of small multilateral agencies operating environmental aid programmes.

Official development finance for the environment is big business, with thousands of actors and annual commitments exceeding US$15 billion. But, in common with the health sector, the number of actors and channels needs to be rationalized through a better division of labour. Otherwise, as aid to the environment is scaled up in response to the challenges of climate change, there is a danger that developing countries will be further overburdened with a plethora of competing actors, funds and initiatives, which will undermine the effectiveness of the aid being provided and limit developmental and environmental results.

Source: Castro and Hammond 2009

Figure 17.4 OECD countries' aid commitments to UNCCD, CBD and UNFCCC, 1998–2009

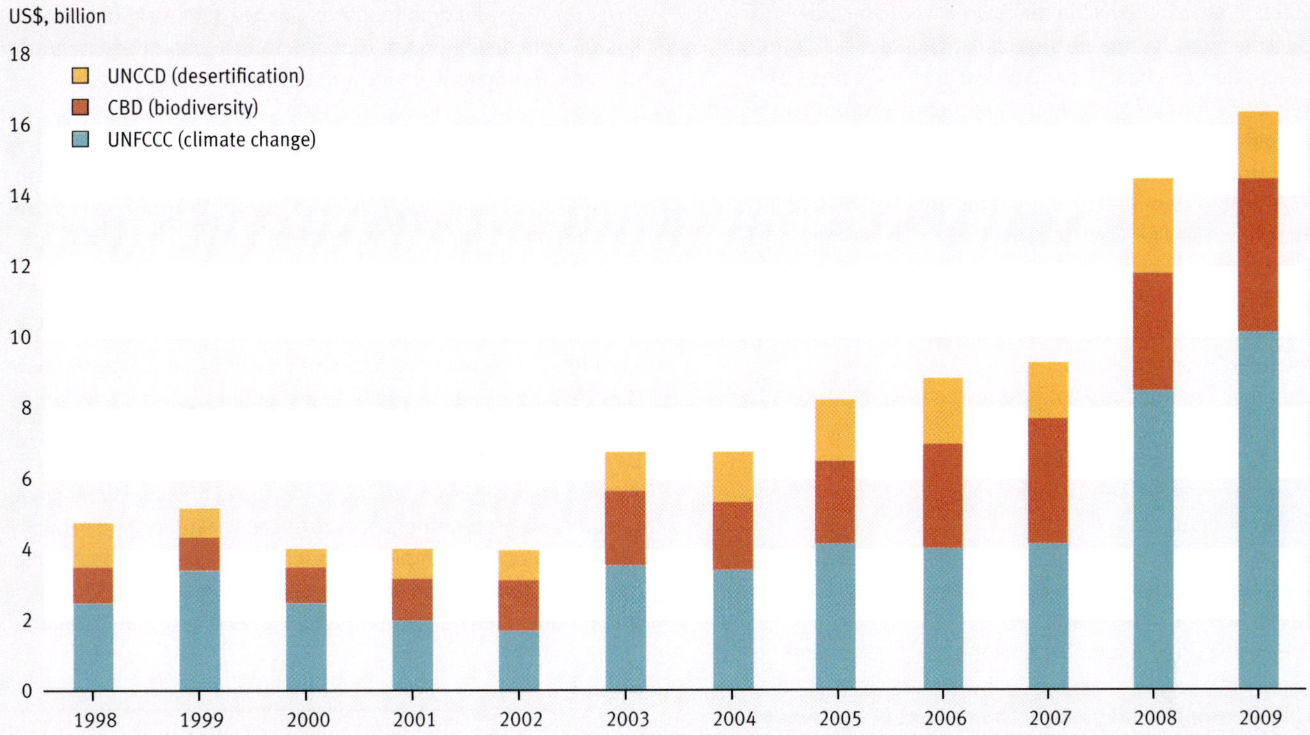

Source: OECD 2011a

Stakeholder engagement: from consultation to participation

Traditionally reserved for governments, the global arena is now open to a broader range of civil society actors, including non-governmental organizations, businesses, faith groups and academic institutions (Willetts 2011; Gemmill and Bamidele-Izu 2002). Over the years, the role of stakeholders in global governance has shifted from simply being consulted in the 1960s, to serving as back-stage managers in the 1970s, to being protected and empowered in the 1980s, to being invited as partners in the 1990s, to the present state as practitioners carrying out initiatives on the ground. This transition has been described as one of increasing engagement and empowerment (Gupta 2003). Stakeholders have had significant impact, from their role in requesting advisory opinions from the International Court of Justice on the legality of nuclear weapons (Yamin 2001), to involvement in environmental litigation (Beyerlin 2001). In global water governance, non-state actors have even taken the lead, filling an institutional void and responding to an emerging need for a global response to water problems (Varady and Iles-Shih 2009; Pahl-Wostl *et al.* 2008).

National and global responses to environmental challenges require effective engagement of multiple stakeholders – at different levels of governance – in the collective definition, adoption and implementation of solutions (Parts 1 and 2). At the global level, collaboration between actors comes into play at different stages of policy intervention, including agenda setting and framing; rule-making; enforcement; and assessing resilience (Underdal 1998; Haas 2000). By participating in global governance, civil society groups have the opportunity to communicate concerns from local stakeholders to international organizations. Additionally, civil society groups facilitate informed public debate by collecting and disseminating information about, and critical evaluations of, international governance (Steffek and Nanz 2008).

Participatory approaches can be transformational (Hickey and Mohan 2005; Chambers 1997; Mohan 2002) or instrumental (Neef 2008; Hooper 2005; Mohan 2002; Mayo and Craig 1995). In the information age, decision makers possess a plethora of new means to engage stakeholders. Social networks, for example, can be valuable if unpredictable assets for engaging citizens in active, emergent and functional communities of practice. Crowdsourcing, the act of sourcing tasks normally performed by individuals by issuing an open call, is increasingly used to encourage community-based design and democratic participation. Iceland recently crowd-sourced a wiki/open-source version of its constitution and the result was a proposal for several changes with broad public support (Constitutional Council of Iceland 2011). In fact, non-governmental organizations have been critical to the development and existence of an open, publicly accessible internet conducive to democratic global engagement (Willetts 2011).

OPTIONS FOR GLOBAL RESPONSES

Effective responses to global environmental problems require a management framework that embodies a holistic and adaptive approach at all levels. Such a framework would include clear and measurable goals, verifiable strategies, and sound monitoring and evaluation mechanisms to address the root causes of emerging environmental problems, reducing environmental and social vulnerability, and accommodating multiple perspectives and solutions. At each stage, a multi-stakeholder interactive and iterative process would take place. This approach would facilitate the adoption of more realistic plans that can be continuously monitored, as well as promoting ownership and accountability (FAO 2010; UNDG 2010). Against this backdrop, this section assesses the rationale for the following six linked and mutually reinforcing response options:

- framing environmental goals in the context of sustainable development;
- enhancing the effectiveness of global institutions;
- investing in enhanced capacities for addressing environmental change;
- supporting technological innovation and development;
- strengthening rights-based approaches and access to environmental justice; and
- deepening and broadening stakeholder engagement.

Framing environmental goals in the context of sustainable development

Findings from *GEO-5* reinforce the importance of setting measurable goals and targets to effectively monitor progress and advance the sustainability agenda. Goal-setting arenas at the international level include not only public institutions such as the UN system but also civil society groups and private-sector associations, among others. Global goals need to be complemented by synergized regional, national and local goals, as well as concrete national action plans.

The Millennium Development Goals (MDGs) represent a results-based approach to advancing human well-being by setting and monitoring global development outcomes. Lessons should be learnt from the MDGs to implement a results-based framework with global goals for sustainable development, including the environment, and provide clear metrics for measuring and tracking progress. The MDG addressing environmental sustainability, MDG 7, for example, has proven challenging to implement in most countries, not least because of a lack of measurable indicators (World Bank 2005). The MDG 7 target of integrating the principles of sustainable development into country policies and programmes, and reversing the loss of environmental resources, is the only non-quantitative target in the MDG framework. As a result, according to the OECD (2008), MDG 7 "often gets pushed aside in the programmes of bilateral donors and international financing institutions".

A more balanced set of goals for sustainable development could more effectively help address the risks that environmental change may pose to development. This can be illustrated by the scenario analysis from the *Human Development Report 2011* (UNDP 2011), which shows that countries with a very high Human Development Index (HDI) score are projected to be less affected by environmental risks than those in all other categories (Figure 17.5), and demonstrates the need for a set of goals for sustainable development that promotes a balanced integration of its environmental, social and economic dimensions.

Effective monitoring of environmental outcomes requires establishing quantifiable metrics or conditional states that can be measured, such as the nitrate concentration in a body of water or the number of species inhabiting a specific area (Jordan *et al.* 2010). Methodological techniques – such as gap analysis, distance-to-target comparative analysis and benchmarking – can provide valuable insights into how countries perform relative to each other. Common indicators can facilitate knowledge transfer as they help governments at all levels to identify and share successful implementation strategies (Strange and Bayley 2008). The MEAs have been developing global goals that focus on articulating the desired state of the environment, reducing pressures and creating joint measures, together with technical assistance and capacity building to scale up implementation. Increasingly, these goals have been specified so that results can be identified through quantifiable metrics or conditional states that can be measured.

Figure 17.5 Scenarios projecting the impacts of environmental risks on human development, 1980–2050

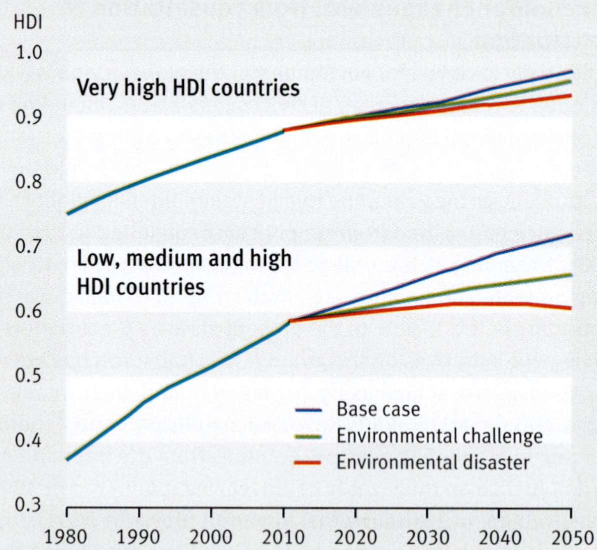

The HDI's base case scenario assumes limited changes in equality, environmental threats and risks, while the environmental challenges scenario envisions intensified environmental risks. The biophysical and human systems in the environmental disaster scenario are put under severe stress by, for example, the overuse of fossil fuels and falling water tables, glacial melting, progressive deforestation and land degradation, dramatic declines in biodiversity, greater frequency of extreme weather and increased civil conflict.

Source: UNDP 2011

Without clear metrics for measuring progress towards sustainable development, achieving internationally agreed goals will remain elusive. In bringing sustainability to the core of decision making, rethinking the way economic development and human well-being are currently measured and monitored becomes crucial (Pintér *et al.* 2011; Stiglitz *et al.* 2009). This requires a broader set of indicators for measuring economic, social and environmental dimensions of sustainable development that go beyond GDP, currently the most widely used indicator of economic development. This has long been called for, and measurement reform has recently received greater attention in political agendas, as illustrated by such efforts as the on-going review and revision of the framework for environmental accounts led by the UN Statistics Division (United Nations 2011), the adjusted net national savings methods of the World Bank (World Bank 2010b), the European Commission's Beyond GDP programme (Stiglitz *et al.* 2009), the OECD's Measuring the Progress of Societies initiative (Hall *et al.* 2010) and UNEP's Green Economy Initiative (UNEP 2011d). These have resulted in the development of environmental and social indicators and aggregate indices to complement GDP and traditional national accounts that are now beginning to be applied.

Governments, academia, civil society and the private sector could collaborate in the development of environmental goals for the global, regional and national levels within the sustainable development framework. Examples of such collaborations include internationally agreed goals to:

- stabilize greenhouse gas concentrations in the atmosphere at a level that would keep the increase in global temperature below 2°C relative to pre-industrial levels, and enhance long-term cooperative action to combat climate change on the basis of equity (UNFCCC 2010);
- halt the loss of biodiversity in order to ensure that ecosystems are resilient and continue to provide essential services, thereby securing the planet's variety of life and contributing to human well-being and poverty eradication (CBD 2010);
- reverse and prevent desertification and land degradation and mitigate the effects of drought in affected areas in order to support poverty reduction and environmental sustainability (UNCCD 2008).

Any internationally agreed sustainable development goals would need to be customized and translated into national targets in order to facilitate measurement of progress towards meeting those targets and facilitate support for their implementation. The development and implementation of internationally agreed indicators could then be coupled with piloting, capacity building in statistical offices, and collaboration with the private sector, research institutions and non-governmental organizations. The data collected and utilized through the monitoring of indicators could be maintained and shared through collaborative databases open to the public. Such goal-setting processes could draw on regional experiences and be informed by and draw inspiration from various schemes that address Earth System challenges, such as the Earth Charter Initiative (2011) and the *Stockholm Memorandum: Tipping the Scales Towards Sustainability* (Nobel Laureate Symposium 2011). Furthermore, incentive and accountability mechanisms would need to be put in place for monitoring the progress towards achieving goals, acknowledging and supporting successes and thereby facilitating progress.

Enhancing the effectiveness of global institutions

Successful global responses to environmental challenges require accurate data and rigorous analysis, agreement on any course of action, and effective execution and implementation of agreed strategies at all levels. The science-policy interface has been strengthened in recent years, particularly in the field of indicators, assessments and early warning systems, supported by developments in research, modelling, monitoring and observations, and especially by advances in information and communication technologies. Significant attention has been given to the design and governance structure of these processes to ensure their scientific independence and credibility as well as their legitimacy and relevance to the full and meaningful participation of developing countries (UNEP 2011e). The interface

Box 17.4 Response option 1: Framing environmental goals in the context of sustainable development and monitoring outcomes

Establish a sustainable development goals framework that integrates the contribution of the environment to development and poverty reduction. A process could be initiated to revisit and extend the MDGs in the form of Sustainable Development Goals (SDGs) with clear, measurable indicators, keeping in mind the need for a coherent and balanced integration of the environmental, economic and social pillars of sustainability.

These goals could serve as a common reference point for action and accountability for a wide range of actors, including intergovernmental institutions, the private sector, civil society and individuals. The framework could articulate a vision for enhancing human well-being – as it relates to health, material needs, social relations and security – in an inter- and intra-generational, equitable manner.

Such a framework could be complemented by targets and measurable indicators building on initiatives that go beyond GDP. Placed within such a framework, a sub-set of global environmental goals could draw on an assessment of existing international environmental agreements and instruments, including the three Rio conventions – UNFCCC, CBD and UNCCD.

could be further strengthened by addressing inequalities in scientific capacity through scaling up support for science-policy capacity in developing countries. Additional efforts could include strengthening data-gathering systems, cooperation on enhancing the connectivity and efficiency of existing international environmental assessments, scientific panels and information networks, and targeting the communication of scientific findings to various audiences. Technology can enable resource users to make better decisions and can give decision makers access to better and timelier information about environmental conditions, helping them organize effective responses. The diffusion of global positioning systems (GPS), mobile telephones and other decentralized technologies, including social media, can strengthen citizen engagement and involvement, which can help create a more robust foundation for coherent decision making. This sort of dynamic approach would strengthen institutions by adding wider society into the science-policy interface – further broadening the depth of stakeholder engagement and integrating the concept of human well-being into action and implementation.

Agreement on a global course of action requires effective coordination, which is especially difficult at the global level. Within the United Nations, the environment falls within the portfolio of organizations with thematic and functional focuses directly related to the environment, and of other institutions that have integrated the environment as a priority area (UNEP 2011c). Consultations on how to strengthen international environmental governance have been active since 2006, when the UN General Assembly agreed to explore the possibility of a more coherent institutional framework to address environmental activities efficiently. The negotiations, including consultations under UNEP's Governing Council, revealed that regularized processes for developing, implementing, assessing and revising a UN system-wide environmental strategy and the division of labour are necessary. A new strategy, developed through an inclusive process involving governments and seeking input from civil society, including the private sector, could facilitate inter-agency cooperation and clarify the division of labour within the UN system (UNEP 2011e). A recent review of the evolution of environmental activities in the United Nations has demonstrated that there are substantive environmental resources and capacities within the system (UNEP 2011f), which could be mobilized and better utilized through a regularized process of developing, implementing and revising an overarching strategy.

While the system for global environmental governance has grown rapidly, there has been no systematic assessment of the performance of international organizations in relation to their mandates or to their impact on environmental quality. The absence of a scientifically credible and politically legitimate assessment of the institutional aspects and options for strengthening international environmental governance could be a reason why it has proved difficult for countries to agree on a way forward. The United Nations Conference on Sustainable Development in 2012 (Rio+20) is an opportunity to initiate an assessment of institutional performance and a strategic analysis of options for strengthening international environmental governance. The process could follow the procedures of the GEO process and take the form of a special GEO report that builds on and deepens the analysis of the global response. The intergovernmental aspects of the process could

The five successive Executive Directors of UNEP gathered together in Glion, Switzerland, for the Global Environmental Governance Forum in June 2009, where they lent their voices and support for an international agreement on climate change. *© Satishkumar Belliethathan/Global Environmental Governance Project*

Box 17.5 Response option 2: Enhancing the effectiveness of global institutions

Elevate and mainstream the sustainable development agenda into the core of decision making within the UN system, supported by enhanced cooperation with and between environmental, economic and social institutions. Cooperation between key institutions could be strengthened, including by building on and further enhancing the work of the Chief Executives Board on Coordination (CEB) and the Environmental Management Group (EMG) as currently guided by the Economic and Social Council (ECOSOC) and the UNEP Governing Council respectively.

Within the institutional framework of sustainable development:

- **convene a science-policy interface forum with representatives from existing environmental assessments, scientific panels and information networks** to advance their connectivity and efficiency, facilitate ways of meeting the science-policy capacity needs of developing countries, strengthen data gathering and target the communication of scientific findings to various audiences;
- **launch a consultative process for the development of a system-wide strategy for the environment in the UN system,** built around those environmental goals that already have international agreement. The UNEP Governing Council, the principle subsidiary body of the General Assembly on environmental matters, could set up a process by tasking the EMG as the principle inter-agency environment body to draft the strategy and then devise a process for review by and consultation with the governing bodies of members of the EMG and other inter-agency bodies and stakeholders;
- **initiate a strategic review of entities in the international environmental system** that compares actual performance to expected results, identifies key constraints and opportunities and outlines ways to measure impact. An independent review would help clarify the environmental mandates of existing organizations, elaborate a substantive vision for global environmental governance and outline ways to address priority issues. It would also collate reports on the status of reform efforts, set short- and long-term goals, and establish timeframes to complete reforms. It could provide a replicable template for similar assessments of other global public goods and help build a foundation for continued United Nations reforms.

be ensured through an initial and concluding intergovernmental consultation and government peer review. Scientific credibility could be ensured by appointing leading scientific experts both from within and beyond the UN system, and through extensive scientific and system-wide peer review.

Investing in enhanced capacities for addressing environmental change

Enhancing capacity requires multi-dimensional and systemic approaches. The capacity of individuals, institutions and organizations as well as societies and communities to implement effective policies is tied to a complex set of tangible and intangible attitudes, resources, strategies and skills (Aragón and Macedo 2010). Part 2 of *GEO-5* highlights the inadequacy of purely technical assistance and emphasizes the importance of governance systems, knowledge systems, technology and shared value systems in reducing vulnerability and strengthening resilience to environmental change. Limited capacity for designing, implementing and reviewing the effectiveness of policies can be a significant barrier to successful policy replication, scaling and learning, especially in developing countries. More focused policies are needed on the less visible aspects of capacity enhancement, such as values, legitimacy, identity and self-confidence, as well as other non-monetary forms of motivation (Aragón and Macedo 2010).

Effective environmental governance is made more challenging by the range of sectoral agencies whose decisions can have an environmental impact. Organizational design at the international and national level rests on the functional division of authority into isolated decision-making units. While governments and the international system have, since the 1972 Stockholm Conference, striven to remedy gaps in information flow and authority, environment ministries remain relatively weak within national governments and within the international system. Economic ministries have maintained their influence, and thus efforts to develop policies to internalize the ecological externalities of economic development have continued to be weak.

Numerous countries and international organizations have experimented with institutional designs to improve the flow of information between functional authorities. For instance, France, Spain and the United States created coordinative environmental councils to work with other government agencies to complement the regulatory authority of environment ministries. At the international level, the United Nations has tried to encourage inter-agency cooperation and to internalize environmental considerations into the policies of other functional agencies (Haas and Haas 1995; Ivanova and Roy 2007).

Policy experience and best practice at a range of scales can also provide lessons for policy development and strengthening capacities. Part 2 of *GEO-5* offers several examples of relatively successful regional policy design and implementation, such as the Maldives adopting a goal of carbon neutrality by 2020, and the European Union Industrial Emissions Directive that resulted in significant reductions in sulphur dioxide emissions across Europe. In addition to these, the use of strategic environmental

Aerial view of Malè, capital of the Republic of the Maldives. In September 2011, the country launched an online campaign seeking help from the world's top experts on how to achieve carbon neutrality by 2020. *© Lucyna Koch/iStock*

assessment has generated examples of how environmental goals can be integrated into and addressed through national development policy (Box 17.1).

Another key capacity challenge lies in the dearth of financial resources. Insufficient predictability and availability of funds has been a key constraint on effective environmental governance at all levels. Yet, global foreign direct investment flows in 2010 were US$1.2 trillion (UNCTAD 2010), far exceeding the value of development funding from international organizations or ODA-related flows. Innovative financial instruments that leverage private investments and improve environmental performance could help bridge the finance gap (Girishankar 2009) – through, for example, linking financing to environmental outcomes (World Bank 2010c). Such instruments include debt-for-nature swaps, payment for ecosystem services, emissions trading and carbon finance, as well as tools from development finance such as green bonds, microcredit, insurance and other risk management instruments, and performance derivatives (Sander and Cranford 2010). More recent ideas include advanced market commitments that guarantee revenues to companies for a limited time to stimulate markets, and prize funds for environmentally sound technologies.

At the national level, targeted policies and instruments are needed to facilitate large-scale green investments, generate necessary resources for public expenditure on environmental priorities and encourage green consumer choices. These could include eco-taxes, performance standards, public procurement strategies, green financing instruments such as green bonds, and green accounting mechanisms (UNEP 2010). Income from taxes related to environmental outcomes – on electricity, heating fuels, transport fuels, greenhouse gas emissions, air pollution, water and waste – raised 2–3 per cent of GDP in European countries in 2007, US$400 billion (€304 billion) in revenues or 6.2 per cent of total taxes and social contributions (Georgescu 2010). In addition, some countries, such as the United Kingdom, are establishing green infrastructure banks or greening existing investment institutions, while at the international level there are proposals to generate large-scale additional revenues through coordinated levies on aviation and shipping and financial transactions (Barbier 2012; Steckhan 2009).

The term green economy was coined some 20 years ago in the publication *Blueprint for a Green Economy* (Pearce *et al.* 1989). The authors argued that a green economy that values environmental assets, employs pricing policies and regulatory changes to translate these values into market incentives, and adjusts the economy's measure of GDP for environmental loss was needed to ensure the well-being of current and future generations. A renewed focus on a green economy has resulted in reports on how to promote public and private investments in different sectors of the economy to help address unprecedented levels of environmental change and advance the sustainable

Box 17.6 Response option 3: Investing in enhanced capacities for addressing environmental change

Establish a UN system-wide framework for capacity building. Such a framework would strengthen the national capacity required to implement environmental policies and could be an integral part of a system-wide strategy on the environment established within the wider institutional framework for sustainable development.

Adopt a green economy roadmap, possibly within the context of a sustainable development goals framework. A roadmap would set out how human well-being can be enhanced through public and private investments in the sectors of the economy that cover demand for, and supply of the goods, services and technologies needed to address unprecedented levels of environmental change, and that advance the sustainable use of natural resources. A combination of market-based mechanisms and regulatory structures might be needed to create employment and economic activities, but the appropriate policy mix would depend on national circumstances and contexts. The full spectrum of available measures includes public investments, green accounting, subsidies, taxes, charges, sustainable trade, creation of new markets, planning, standards, regulations, technological innovation, technology transfer and capacity building.

Establish policy banks. This would enable the sharing of examples of sound environmental policy, design and implementation from different regions, including a green economy roadmap. This could provide opportunities for learning, adaptation or replication at an unprecedented scale, with the early involvement of multiple stakeholders facilitating the development and uptake of relevant reforms. Matching needs to the right policy tools is unlikely to occur organically without facilitation or brokerage. Governments and other actors could consider establishing decentralized, possibly open-access, web-based sustainability policy banks to:

- act as a repository of good practices for learning and replication;
- assist governments and stakeholders in identifying good practices for their priority areas for intervention;
- provide a forum for discussions on tailoring policies to national needs; and
- provide rosters of experts to assist in realistic applications in particular countries and at sub-national levels.

Principles for greener investments. A financial strategy could be built on a set of common principles, and a renewed commitment could be made to meeting existing obligations and creating sufficient and predictable funding to promote a green economy and sustainable living. These norms would serve to guide:

- the greening of existing and new investments to improve the environmental impacts and outcomes of all investments, including mainstreaming the environment in development expenditure;
- raising additional resources for green investments through new mechanisms such as green taxation; and
- public-private partnerships that leverage private funding sources while also addressing environmental objectives.

Establish a system for financial tracking. Regular reviews and renewal of funding commitments would further evolve private-public partnerships and scale up direct budget support to ensure mainstreaming of the environment in development, more effective participation in global processes and improved analytical capacity.

use of natural resources (UNEP 2011c, 2011d). There have been concerns that a green economy might create unsustainable jobs, lead to inequity, create trade distortions or promote new forms of green protectionism (UNEP 2011c). Such concerns would have to be addressed through existing mechanisms such as trade agreements and a balanced integration of the three pillars of sustainable development. A calibrated framework of sustainable development goals could guide an investment roadmap to the green economy and help ensure that such investments are socially and financially sustainable (Bina and Camera 2011).

In addition to increasing financing for the environment, a related, overarching priority would be to make all investment decisions – both public and private – greener. With US$24–30 trillion to be invested in infrastructure globally over the next 20 years (CG/LA Infrastructure 2008), the challenge to integrate environmental considerations into investment decisions is daunting. A set of principles for green investment by institutional investors, governments and international bodies could accelerate the growing commitment to making investments greener. Many of the policy instruments mentioned above would deliver opportunities and benefits for both the environment and the economy (Part 2).

Supporting technological innovation and development

Technology has a substantial role to play in the effort to meet the most pressing global environmental challenges. Advanced and environmentally sustainable technologies can help developing economies leapfrog the resource-intensive, highly polluting growth phase. This is not just about technologically advanced solutions but also other adaptive ones. As technology systems include not only the deployment of hardware but also knowledge and know-how, lessons from traditional knowledge and practices can also be shared and adapted (IPCC 2001). Increased domestic capacity for innovation, including adapting existing technologies to local conditions, is an important goal for many countries.

Technologies can help improve environmental performance along the supply chain from resource extraction to manufacture and transportation and more efficient, greener end-use equipment for consumers. System-level technology links are often crucial for transformative change. For example, the smart grid concept aims to integrate electric vehicles, the power sector, information management and consumers into a single network. Technologies are also essential for successful adaptation to changing environments, from drought-resilient seeds through efficient methods of irrigation to flood defences.

But technologies and technological systems have a much broader role in green transformation than direct mitigation and adaptation. They play a key enabling role in such areas as remote and onsite monitoring of environmental change; early warning systems and new types of collaborative problem solving including crowdsourcing. Social networks are also having a significant but unpredictable impact on the environmental activity of governments, non-governmental organizations and communities.

In all of these areas, the ability of individuals, companies and institutions to absorb both new technologies and available finance varies in the different national contexts of developing countries (Ruggie 2008; Puustjarvi *et al.* 2003), making capacity building and demonstration projects key enabling factors (WBCSD 2010).

Addressing the technology gap is at the fore of international negotiations on responses to environmental challenges. Since 1990, developed countries have agreed to take all practicable steps to encourage the transfer of green technologies and know-how to developing countries. But this agenda has seen slow progress, with persistent disagreements even about what constitutes a technology transfer. The processes through which large-scale transfers should occur are not straightforward, given that most technologies are owned by the private sector rather than governments.

***Tûranor PlanetSolar*, the world's largest solar powered boat – topped by 500 m² of solar panels – was the first solar electric vehicle to circumnavigate the globe.** © *Tatiana Kakhill/iStock*

Some developing countries are critical of existing technology transfer regimes due to the high transaction costs of obtaining information or negotiating and acquiring technologies protected by intellectual property rights, and a lack of clarity in defining

Box 17.7 Technology Mechanism at the UNFCCC

At the 2010 UN Climate Change Conference (COP 16) in Cancún, Mexico, governments agreed to establish a Technology Mechanism to facilitate technology cooperation and transfer. It comprises an executive committee and a climate technology centre and network.

The priorities for this mechanism include the development and enhancement of the capacities and technologies of developing countries; deployment and diffusion of environmentally sound technologies and know-how; increased public and private investment in technology development, deployment, diffusion and transfer; strengthening of national systems of innovation and technology innovation centres; and development and implementation of national technology plans for mitigation and adaptation. Further, it is hoped that the technology mechanism will stimulate and encourage – through collaboration with the private sector, public institutions, academia and research institutions – the development and transfer of existing and emerging environmentally sound technologies and opportunities for North-South as well as South-South technology cooperation.

In December 2011, governments adopted the modalities for the Technology Mechanism at the 17th Conference of the Parties in Durban. Since funding has always been a barrier to technology transfer to developing countries, the Green Climate Fund (GCF), newly established at Durban, could also help speed up implementation of the goals set by the international community to combat climate change, and promote a paradigm shift towards low-emission and climate-resilient development pathways. Though the UNFCCC process has been a useful forum to initiate global intergovernmental procedures to foster global technology transfer, collaborative work with the other multilateral environmental agreements, such as CBD, is also imperative to ensure the development and transfer of technologies for achieving other global environmental goals.

Box 17.8 Response option 4: Supporting technological innovation and development

Accelerating the innovation and diffusion of technologies is a critical element of any holistic support framework that seeks to encourage the uptake of environmentally sound technologies in the transition to a global green economy. This includes:

Collaborative R&D. Collaborative research for environmentally sustainable technologies could be coordinated between governments and the private sector for early pre-competitive stages of technology system development, before particular standards or industry value chains become embedded in national economies and the global industrial system, as happened in the semi-conductor market. Model technology cooperation agreements could take different levels of development and jurisdictional requirements into account to limit the potential of patent-related conflicts and to encourage joint development. National laboratories could be twinned, or new ones set up that are multilaterally managed and funded in pursuit of agreed long-term technology objectives, ideally with the participation of industry.

Support for knowledge-sharing platforms. Collaborative initiatives on agriculture and the environment, such as the Consultative Group on International Agricultural Research (CGIAR), demonstrate the potential of stakeholder advice platforms and provide support for knowledge-sharing structures at the regional level. These initiatives could be emulated to scale up much needed environmentally sustainable technologies. Existing and potential barriers to the development and diffusion of such technologies could be assessed at the sectoral level to create the most appropriate incentives.

Global prize funds to stimulate innovation on green technologies. Prize funds and similar awards could be an effective means of bridging innovation gaps, including technologies to improve sustainability for the poor, as demonstrated by some successes in the public health and energy sectors. A range of global technology prizes could be established to promote innovation in all areas that support sustainability, especially for developing economies. Such prize funds could function as a patent pool and/or a repository for cross-licensing environmental technologies.

what is protected and what is not (Li and Correa 2009; Barton 2007; Hutchison 2006; Commission for Intellectual Property Rights 2002). The impact of intellectual property (IP) rights on the technological advancement of developing countries varies according to the sector (Barton 2007), with countries like China and India making significant advances in technological development and acquisition despite the barriers (Puustjarvi *et al.* 2003). Maskus (2010) argues that although patents and IP rights may not in fact restrict access to environmentally sustainable technologies, there may be needs for beneficial differentiation in patent rights such as "*ex ante* extensions of patent terms tied to licensing commitments, expedited patent examinations in environmentally sustainable technologies, investments in patent transparency and landscaping efforts, and facilitation of voluntary patent pools".

Technological innovation has the potential to reduce the cost of achieving global environmental objectives (OECD 2010). The costs of implementing green policies have often turned out to be far lower than those projected, in part due to technological advances. Investment in research and development (R&D) is mostly undertaken by the private sector and is increasingly global in nature, but government actions and public policy can help leverage the power of markets to solve environmental challenges through innovation. Efforts to increase the flow of technology to developing countries and economies in transition include the UNFCCC decision to create a new Technology Mechanism (Box 17.7).

International cooperation is needed to build and strengthen innovation links between different sectors, especially between developed and developing economies. This is not least because many transformative approaches involve complex changes to technology systems and new forms of industrial models that are yet to be demonstrated at scale. International cooperative research could help pool development risks, share information (OECD 2011b) and overcome barriers to private-sector investment. Yet innovation cooperation is primarily a national activity, not an international one. A study of six clean energy sectors points out that only 1.5 per cent of patents are co-assigned, listing more than one company or institution as co-owners, and only 2 per cent of these joint patents are shared between developed and developing economy companies and institutions (Lee *et al.* 2009).

Strengthening rights-based approaches and access to environmental justice

Human and environmental rights can play a valuable role in ensuring that governments stay on track in meeting environmental goals and in providing safeguards against the adoption of environmental policies that reduce human and ecological well-being. Several important developments are evident in environmental rights. Adverse impacts on human health from environmental misuse are increasingly seen as violations of the human right to life (Kravchenko and Bonine 2008). Further, globally agreed human rights frameworks increasingly emphasize the intersection between human well-being and environmental health, as well as social-ecological resilience (Campese *et al.* 2009; ICHRP 2008; Jeffery 2005; Hunter *et al.* 2001), setting the basis for sustainability in environmental decision making.

Aerial view of oil rigs in the Niger Delta, close to a village. Over recent decades, the delta has faced extensive environmental degradation, undermining sustainable environmental management and the right to access to a clean environment. *© Eric Miller/Still Picturesock*

The environmental aspects of the current human rights framework are still too weak, however, to ensure that citizens are able to protect their well-being and hold governments accountable. In part, this is because environmental rights law at the global level is predominately soft law, making it easy for states to avoid their responsibility, with regional courts and judicial bodies not always able to ensure that their decisions are put into effect. For example, the decision of the African Commission on Human Rights that pollution from oil exploration in the Niger Delta that affects environmental quality and human health constitutes a violation of the right to a clean environment in the African Charter has never been put into effect. In contrast, implementation of the 1998 United Nations Economic Commission for Europe (UNECE) Convention on Access to Information, Public Participation in Decision-making and Access to Justice in Environmental Matters – the Aarhus Convention – demonstrates that effective procedural rights and state acceptance can be effective in protecting people and the environment. Replicating this approach regionally or globally is one option for giving effect to Rio Principle 10 by both state (UNECE 2011) and non-governmental organizations (Barreira 2012; UN-NGLS 2007). In 2011, the Meeting of the Parties to the Convention adopted a decision encouraging accession by states outside the UNECE region and a simplified procedure for doing so, creating a pathway for propagating the protection offered by this international environmental rights treaty on a global scale (UNECE 2010).

Although environmental rights are widely recognized, full access to justice at the national level can be difficult to implement. The effectiveness of legal systems has been hindered by local inability to access the courts, with a lack of financial resources, distance from courts, and language barriers being key challenges. Further, state entities do not always understand the nature of their obligations under environmental and human rights law (Serra and Tanner 2008). Global and regional investment in strengthening these national systems by enhancing citizen and state capacity could improve access to justice.

Despite its limitations, soft law can play a valuable role in shifting the culture of environmental practice by providing a basis for citizen advocacy, including claims for access to vital livelihood resources such as land and water; for governments to review their current practice; and for strengthening public participation in environmental decision making, particularly where citizens' rights are affected. For example, with the adoption of the United Nations Declaration on the Rights of Indigenous People (UNDRIP) (United Nations 2007), the United Nations agreed that all its activities must be based on the recognition of these rights. For example, the UN programme for Reducing Emissions from Deforestation and Forest Degradation (UN-REDD) seeks to incorporate the UNDRIP rights into its practice and policy by, among other action, giving effect to the right to free prior informed consent. The recognition by the UN General Assembly of the human right to water and sanitation (Gupta *et al.* 2010) is another clear step forward to promote human rights. At the national level, the 1992 Rio Declaration on Environment and Development is widely accepted as a framework for allocating rights and responsibilities between states and citizens. Importantly, soft law can catalyse hard law agreements – the Aarhus Convention was negotiated in response to Principle 10 of the Rio Declaration (UNCED 1992).

Given current limitations within the human rights system, there is a renewed debate on the merits and drawbacks of a

Box 17.9 Response option 5: Strengthening rights-based approaches and access to environmental justice

Recognize the links between human rights, environmental rights and the responsibilities of states, which can set the basis for better environmental performance where these rights are incorporated in decision making. Improving understanding of how this can be achieved through learning from best practice at inter-state and inter-regional levels should be facilitated. Existing human rights platforms could provide the basis for dialogue between diverse actors, including states, academics and communities, and strengthen and clarify understanding.

Develop a global legal instrument, or a series of regional instruments, to strengthen access to information, public participation and access to justice in environmental matters, based on Principle 10 of the Rio Declaration and taking into consideration experience gained with the UNECE Aarhus Convention (1998).

Create a shared legal normative basis for action. Globally, a number of soft law norms have emerged to protect the environment in an equitable and responsible manner. There is, however, a need for a legal process to promote the hardening of these norms into legally binding rights and responsibilities that would provide a shared legal normative basis for action.

Recognize and support the different kinds of dispute resolution systems, including indigenous systems to ensure that justice is delivered. Although a number of formal and alternative dispute resolution systems are being developed in different forums, and national courts are offering non-nationals the right to seek adjudication on their environmental claims, there is need for a process that recognizes and supports these kinds of dispute resolution processes.

Establish a process for creating an international environmental court to address violations of environmental standards. Agreeing to a process for considering the establishment of an international environmental court is an important first step in improving dispute resolution. It is important to build on the experience of existing judicial systems at the regional level and within the human rights field, avoid duplication and ensure sufficient human capacity and finance.

judicial system for the environment at the international level. Options discussed range from an international court for the environment to enhanced complaint mechanisms to turning environmental and related equity rights into law (Klabbers *et al.* 2009). A number of models have been proposed as a template for an International Court for the Environment (ICE), which could function as a Court of First Instance for the International Court of Justice, rendering decisions and/or advisory opinions, or act as a specialized environmental tribunal in the spirit of the Permanent Court of Arbitration, or could provide a combination of negotiation, conciliation, arbitration and adjudication similar to the Dispute Settlement Body of the WTO. Furthering the ICE, the International Court for the Environment Coalition (2011) proposed three characteristics for it:

- the court or tribunal should have specialized environmental judges or a process that addresses the current gaps between international law and environmental science;
- its standing should be offered to non-state actors, provided the cases satisfied a materiality threshold, i.e. what is considered relevant or material to a case; and
- the court should incorporate the common law principle of *stare decisis*, which would establish precedence in the international environmental legal order.

Traditional adjudication, however, faces some significant constraints, which could reduce the court's effectiveness in resolving international resource disputes, such as those over the use and sharing of natural resources. Analysis of the use of adjudication by international courts and tribunals reveals four categories of limitation:

- parties may refuse to submit to adjudication;
- the judicial decision might not address the merits of the dispute;
- non-compliance is not punishable; and
- recurrence of the dispute or conflict (Spain 2011).

These constraints can be overcome through the use of integrated methods for dispute settlement and resolution. Ultimately, however, successful resolution of international resource dispute hinges on the availability of mechanisms – judicial or otherwise – that allow for the active engagement of non-state actors and for addressing the concerns of all parties with legitimacy, fairness and speed.

Deepening and broadening stakeholder engagement

The complexity and varied nature of global environmental challenges facing the international community today, as demonstrated in Part 1, require a range of interventions beyond action by public institutions. Many of the solutions described in Part 2 also demand collective action by civil society, private sector actors, the media and academic and research institutions.

The role of civil society actors in supporting global environmental governance has evolved over the past 40 years to create groups that operate from the local to the global level, offering means by which to connect global policy to local action. Non-governmental organizations tend to be more flexible than governments and intergovernmental agencies, and can therefore offer swift support for seeking and implementing solutions. They frequently have the capacity to conduct in-depth research, collect and disseminate

Box 17.10 Social learning

Social learning comprises formal or informal processes to share knowledge and lessons, at different levels and across different communities, to support innovative problem solving required for addressing unprecedented environmental change. Social learning is as much about changes in relationships and in individual and collective attitudes and mindsets as it is about practical tools and institutional change to deal with new challenges (Pahl-Wostl 2006). Platforms for social learning include, for example, biotechnology citizens' juries (Pimbert 2011), Oxfam's climate dialogues and the United Nations Human Rights Council's Social Forum.

Access to technology and information is vital, but not sufficient, for effective social learning. Governance and the dynamics of interaction between actors are important factors that determine what knowledge and experience is shared and how it is used. In order to be effective, collaborative learning requires open communication, engagement outside established decision-making circles, consideration of multiple kinds of knowledge, unrestrained thinking and self-reflection (Woodhill 2010; Keen *et al.* 2005; Schulster *et al.* 2003).

At the global scale, social learning can be fostered by facilitating institutional openness, multi-level governance both horizontal and vertical, and dialogue between sectors as well as between different communities. Specific options for a strong global response to foster social learning include:

- promoting actor and stakeholder learning networks – South-South, global inter-generational dialogues and private-public;
- cross- and multi-actor participation in international decision making such as Conferences of the Parties;
- improved transparency and access to information;
- support of experimentation and variation; and
- improvement of monitoring and regular reviews of policies and experiments using rigorous analysis and providing rapid feedback on success or failure.

data, and support assessment and monitoring (Gemmill and Bamidele-Izu 2002), together with awareness raising and public mobilization. Similarly, academic institutions can offer unique support for global responses by granting credibility through scientific and technical support. Non-governmental organizations and academic institutions together contribute to increasing public participation, creating and maintaining knowledge networks and facilitating the diffusion of knowledge and ideas (Ramos 2009; Eriksson and Sundelius 2005; Stone and Maxwell 2005). The recently launched Global University Partnership on Environment and Sustainability (GUPES), a UNEP-planned consultative, sharing and learning platform for university leaders from developed and developing countries, offers a possible model for collaboration between international organizations and universities (UNEP 2011b).

As attempts to transition to a green economy advance, business engagement in a variety of ways and at multiple scales can also bring added value to global responses. The Montreal Protocol offers an example of a successful international environmental treaty, in which a critical element in the negotiations was the inclusion of businesses and NGOs in drafting the treaty

The future in our hands: access to information and technology is vital; open communication and diffusion of knowledge with ever increasing public participation can lead to collective action from global to local level and vice versa. *© Peeter Viisimaa/iStock*

and supporting its implementation. In treating businesses as collaborators rather than constituents, they can become engaged in the problem, the strategy and the implementation (Ivanova *et al.* 2007). Some also benefit from being first movers in commercial terms. While the Montreal Protocol was relatively limited in its scope and fairly straightforward in terms of policy decisions, this strategy could offer useful lessons for other agreements and initiatives.

Businesses have also taken the lead in developing private certification schemes, which are an emerging approach to environmental governance. Supply-chain management guidelines have been effective in promoting sustainable practices in forestry through the Forest Stewardship Council, and fisheries by the Marine Stewardship Council (Auld *et al.* 2008; Cashore *et al.* 2004) and for establishing broader global standards of corporate social responsibility through the United Nations Global Compact (Ruggie 2001). Such efforts depend on proper institutional design that includes legitimate third-party verification, supportive government institutions at the national level, clear relationships between the private sector and civil society and a public awareness of the meaning of the codes. Schemes initially focused on one sector can lead to similar approaches being applied in others, as public and corporate awareness and experience develop. Similarly, national schemes are sometimes scaled up to the regional or international level. The dangers of voluntary approaches for environmental policy, however, include their non-enforceability, poor monitoring and a lack of transparency (OECD 1999).

Collaboration and engagement of sub-national authorities is another important element of public participation. Cities, for example, have embarked on environmental and sustainability

Vancouver, Canada, used the 2010 Winter Olympics to boost its efforts to become a greener, more sustainable and more resilient city.

© Amanda Mitchell

action of their own (Box 17.11). While bottom-up initiatives such as these might not deliver the necessary degree of change, the proactive measures provide channels for implementation, engagement and feedback on the efficacy of policy (Otto-Zimmerman 2011).

Box 17.11 Cities and climate action

Many cities worldwide have begun to take climate action, illustrating the important role that sub-national actors can play in addressing global environmental problems. Most city efforts thus far have focused on mitigation rather than adaptation (Hoornweg *et al.* 2011), with more than 2 000 cities now committed to reducing their greenhouse gas emissions (ICLEI 2010). Their motivations for taking climate action is complex and varied, typically reflecting frustration with the limited progress in international negotiations and the desire of city leaders to respond to citizens' concerns.

Climate action by cities and sub-national regions has also taken on a global dimension. Cities are increasingly acting in concert and learning from one another, with little distinction between North and South. Globally, the landscape of networks and entities active in climate change adaptation and mitigation at the city level is emerging but fragmented: these include the C40 Cities Climate Leadership Group, ICLEI – Local Governments for Sustainability, the World Mayors Council on Climate Change, the Covenant of Mayors and the Climate Alliance in Europe, and the Climate Protection Agreement of the United States Conference of Mayors.

Climate action by cities has increasingly been recognized by national governments and at the international level. Although the UNFCCC and its Kyoto Protocol did not originally include any explicit role for cities, this has been changing. The 16th Conference of the Parties in Cancun recognized local governments as key governmental stakeholders in global climate change efforts, with numerous references in Decision CP.16. More recently, ICLEI – Local Governments for Sustainability has called for a framework of global environmental governance that includes local and sub-national governments as part of a multi-stakeholder system of collaborative policy setting, implementation and accountability (Otto-Zimmerman 2011). The European Union has also called for new governance modes that foster social innovation and that adopt an holistic approach to environmental and energy issues in cities (EC 2011).

Box 17.12 **Response option 6:** Deepening and broadening stakeholder engagement

Build a stakeholder web for sustainability. Non-state actors and the private sector could be invited to explore how the modalities of a stakeholder web could evolve by building on existing structures and utilizing modern information and communication technologies including social media. The web could help identify issues on which the public sector may need to act in terms of implementation of the Rio Declaration's Principle 10 as it relates to access to information and stakeholder engagement. Principle 10 could also act as a platform to mobilize new partnerships for action towards implementing internationally agreed goals and targets, such as a possible framework of sustainable development goals and the transition towards an inclusive green economy.

Establish an inter-generational assembly. An inter-generational assembly could provide an opportunity for future leaders and sustainability champions to interact and foster a joint vision for a sustainable future. The idea of the assembly could be taken up as part of the current reform process, building on discussions at the UN Conference on Sustainable Development. In terms of concrete outputs, the assembly could also promote access to information and shared accountability through innovative tools that would support decision making, including a global database of conceptual innovations in environmental governance and management.

While the public sector is an essential agent in creating enabling conditions for societal change, the private sector and civil society are also core agents. The implementation of the Rio Declaration's Principle 10 could help further empower individuals, the private sector and non-state actors in responding to environmental problems. The principle, amongst others, recognizes that each individual should have appropriate access to information about the environment that is held by public authorities, and that states should facilitate and encourage public awareness and participation by making information widely available. Although stakeholder participation in intergovernmental affairs and public-private partnerships has evolved through, for example, the Commission on Sustainable Development, a greater deepening and broadening of stakeholder engagement, through using modern information and communication technologies for example, could make society better prepared to respond to the scale of environmental change. Civil society and the private sector could be invited to develop a stakeholder web for sustainability, building on existing structures aimed at mobilizing action to implement internationally agreed goals and targets.

Current decision-making processes tend to focus on the short term – to the probable detriment of future generations. Explicit future orientation is an important element of adaptive governance strategies for sustainable development and, while foresight processes are a regular part of decision-making processes (de Lattre-Gasquet 2009; Green and Stewart 2004), broader mechanisms to help strengthen the voice of future generations could be considered.

Governments possess various options to strengthen the voice of future generations at various levels (Brown Weiss 1992). They can install an office that has responsibility for ensuring that the interests of future generations are considered, for investigating complaints and for providing warnings of emerging problems. States could also give standing in their national courts and administrative bodies to a representative of future generations, who might function as a guardian. Another approach is to designate an ombudsman for future generations or to appoint commissioners for future generations who could operate internationally, nationally or locally. This was advocated by the World Commission on Environment and Development and in some countries, Hungary for example, experiments are now under way with ombudsmen – who in national law have the responsibility to safeguard social and environmental conditions to the benefit of future generations (JNO 2010).

CONCLUSION: RESPONDING TO EARTH SYSTEM CHALLENGES

When the international community last took stock of the state of the environment in 2007 as part of the *GEO-4* process, promises and recommendations were made to tackle the environmental challenges. But neither the scope of environmental policy nor the speed of its implementation has been sufficient. Efforts to reduce the pressures from the underlying drivers – including enhanced resource efficiency and climate mitigation measures – may have resulted in moderate successes but have fundamentally failed to reduce environmental problems on a global scale.

Five years on, it is clearer than ever that there is no global panacea or single, overarching solution to environmental challenges. Rather, collective action built around strategies, values, principles, investments and measures, supported by a diverse range of competencies and capacities, needs to be woven into the fabric of nations, international society and its institutions. Ultimately, the prospect for improving human well-being is critically dependent on the capacity of individuals and countries as well as the global community to respond – through mitigation and adaptation – to environmental change. While the modalities of multilateral cooperation need to be kept under review to ensure their effectiveness, the key challenge of addressing capacity issues in the developed and developing world remains.

As *GEO-5* demonstrates, however, notwithstanding the enormous challenges, there are great opportunities to scale up policies that could help take the world's citizens along trajectories that begin to reverse negative environmental

Rio de Janeiro, host to the 2012 United Nations Conference on Sustainable Development. *© Zxvisual/iStock*

trends, and that address the inequalities and inadequacies of the institutional frameworks within which human society operates. It is also imperative for the international community to invest in solutions that will help tackle the root causes, not merely the symptoms, of environmental degradation, from fundamental shifts in values through the design and structure of institutions to innovative policy frameworks. Modified to reflect the global scale, a systematic and comprehensive results-based global approach could be anchored in the six response options outlined in this chapter.

The 2012 United Nations Conference on Sustainable Development (Rio+20) provides an opportunity for the international community to take stock, assess achievements and shortcomings, and stimulate transformative global responses. It is also an opportunity for the international community, from individual member states to the United Nations, to demonstrate political leadership in tackling these complex challenges. This chapter has identified a number of response options that together could help society address the problems of global environmental change. While these do not guarantee success, they expose, clearly and systematically, whether or not progress is being made. In addition, evaluation and collective learning could enable the identification of barriers to implementation. This, in turn, could inform adjustments and adaptive management as part of a larger, systemic approach to global governance.

Integrated governance of socio-ecological systems must be cross-sectoral, cross-scale, and across time. Authority and accountability must be dispersed to the appropriate level of decision-making – subsidiarity – while including a broad set of actors beyond the state and enhancing their capacity.

At the global level, it remains a daunting challenge to design and implement effective measures that can motivate citizens, companies, institutions, networks and governments to cooperate and deliver ambitious policies and action. Highlighting the rewards of cooperation and shared purpose could embolden efforts to overcome barriers and past trajectories, reversing unsustainable trends that that were once considered insurmountable. The rewards of progress are often obscured against a landscape fraught with challenges and inequities. In the end, openness to possibility – reflecting the optimism, creativity and potential of young people around the world – and investing in an environment in which multiple sustainable and desirable solutions can emerge, would probably be the most effective, and meaningful, global response.

REFERENCES

Aragón, A.O. and Macedo, J.C.G. (2010). A 'systemic theories of change' approach for purposeful capacity development. Institute of Development Studies. *IDS Bulletin* 41(3), 87–99

Auld, G., Bernstein, S. and Cashore, B. (2008). The new corporate social responsibility. *Annual Review of Environment and Resources* 33, 413–435

Barbier, E. (2012). Sustainability: Tax 'societal ills' to save the planet. *Nature* 483, 30

Barreira, A. (2012). *Public Participation in MEAs Compliance: A Proposal to Rio +20 to Improve the Institutional Framework for Sustainable Development.* Instituto Internacional de Derecho y Medio Ambiente (IIDMA), Madrid

Bartlett, R.V., Priya, A.K. and Madhu, M. (1995). *International Organizations and Environmental Policy.* Greenwood Press, Westport

Barton, J.H. (2007). *Intellectual Property and Access to Clean Energy Technologies.* International Centre for Trade and Sustainable Development, Geneva

Baser, H. and Morgan, P. (2008). *Capacity, Change and Performance: Study Report.* Discussion Paper 59B. European Centre for Development Policy Management (ECDPM), Maastricht

Bearce, D.H. and Bondanella, S. (2007). Intergovernmental organizations, socialization and member-state interest convergence. *International Organization* 61(4), 703–733

Behrens, A. (2009). Financial impacts of climate change mitigation. *Climate Change Law Review* 3(2), 179–87

Beyerlin, U. (2001). The role of NGOs in international environmental litigation. *Heidelberg Journal of International Law* 61, 358–378

Biermann, F. (2004). *Ecological Interdependence and State Power: Explaining the Bargaining Success of Developing Countries in Global Environmental Negotiations.* 45th Annual Convention of the International Studies Association, Montreal

Biermann, F. and Siebenhüner, B. (2009). *Managers of Global Change.* MIT Press, Cambridge, MA

Bina, O. and La Camera, F. (2011). Promise and shortcomings of a green turn in recent policy responses to the "double crisis". *Ecological Economics* 70, 2308–2316

Botes, L. and van Rensburg, D. (2000). Community participation in development: nine plagues and twelve commandments. *Community Development Journal* 35(1), 41–58

Braithwaite, J. and Drahos, P. (2000). *Global Business Regulation.* Cambridge University Press, Cambridge

Brown Weiss, E. (1992). Intergenerational equity: a legal framework for global environmental change. In *Environmental Change and International Law: New Challenges and Dimensions* (ed. Brown Weiss, E.). Chapter 12. United Nations University Press, Tokyo

Campese, J., Sunderland, T., Greiber, T. and Oviedo, G. (2009). *Rights-based Approaches: Exploring Issues and Opportunities for Conservation.* Center for International Forestry Research (CIFOR) and IUCN, Bogor

Cashore, B., Auld, G. and Newsom, D. (2004). *Governing through Markets: Forest Certification and the Emergency of Non-State Authority.* Yale University Press, New Haven

Castro, R. and Hammond, B. (2009). *The Architecture of Aid for the Environment: A Ten Year Statistical Perspective.* CFP Working Paper Series No. 3. Concessional Finance and Global Partnerships Vice Presidency. World Bank, Washington, DC. http://siteresources.worldbank.org/CFPEXT/Resources/Aid_Architecture_for_the_Environment.pdf iclei

CBD (2010). *Decision X/2 of the Tenth Meeting of the Conference of Parties of the Convention on Biological Diversity on the Strategic Plan for Biodiversity.* http://www.cbd.int/decision/cop/?id=12268 (accessed 23 December 2011)

CG/LA Infrastructure (2008). *The Global Infrastructure Marketplace: The Next Twenty Years.* http://cg-la.com/en/products/global-infra-market-2030 (accessed 7 May 2011)

Chambers, R. (1997). *Whose Reality Counts? Putting the First Last.* Intermediate Technology, London

Commission for Intellectual Property Rights (2002). *Integrating Intellectual Property Rights and Development Policy.* Commission for Intellectual Property Rights, London

Commission on Global Governance (1995). *Our Global Neighbourhood.* Oxford University Press, Oxford

Constitutional Council of Iceland (2011). The Constitutional Council hands over the bill for a new constitution. http://stjornlagarad.is/english (accessed 24 December 2011)

De Lattre-Gasquet, M. (2009). *Foresight.* http://knowledge.cta.int/en/Dossiers/S-T-Issues-in-Perspective/Foresighting/Articles/Foresight (accessed 27 September 2011)

Dietz, T.E., Ostrom, E. and Stern, P.C. (2003). The struggle to govern the commons. *Science* 302, 1907–1912

Earth Charter Initiative (2011). *The Earth Charter.* http://www.earthcharterinaction.org/content/pages/Read-the-Charter.html (accessed 25 December 2011)

EC (2011). *Cities of Tomorrow: Challenges, Visions, Ways Forward.* European Commission, Directorate General for Regional Policy, Brussels

Economic Commission for Africa (2005). *Review of the Application of Environmental Impact Assessment in Selected African Countries.* United Nations Economic Commission for Africa, Addis Ababa

Eriksson, J. and Sundelius, B. (2005). Molding minds that form policy: how to make research useful. *International Studies Perspectives* 6(1), 51–7

Esty, D. and Ivanova, M. (2002). Revitalizing global environmental governance: a function-driven approach. In *Global Environmental Governance: Options and Opportunities* (eds. Esty, D. and Ivanova, M.). Yale School of Forestry and Environmental Studies, New Haven

Eyben, R. (2006). The road not taken: international aid's choice of Copenhagen over Beijing. *Third World Quarterly* 27(4), 595–608

FAO (2010). *Results-Based Management.* Food and Agriculture Organization of the United Nations, Rome. http://www.fao.org/about/57743/en/ (accessed 6 June 2011)

Folke, C., Hahn, T., Olsson, P. and Norberg, J. (2005). Adaptive governance of social-ecological systems. *Annual Review of Environment and Resources* 30, 441–73

GEF (2011). *Annual Report on Impact.* GEF/ME/C.41/inf.01. http://www.thegef.org/gef/sites/thegef.org/files/documents/GEF_ME_C.41_Inf.01_%20GEF_Annual_Report_on_Impact.pdf (accessed 25 December 2011)

GEF (2010). *Annual Report 2010.* Global Environmental Facility. http://www.thegef.org/gef/sites/thegef.org/files/publication/WBAnnualReportText.revised.pdf (accessed 22 December 2011)

Gemmill, B. and Bamidele-Izu, A. (2002). The role of NGOs and civil society in global environmental governance. In *Global Environmental Governance: Options and Opportunities* (eds. Esty, D. and Ivanova, M.). Yale School of Forestry and Environmental Studies, New Haven

Georgescu, M.A. (2010). *Distribution of Environmental Taxes in Europe by Tax Payers in 2007.* Eurostat Report. European Commission

Girishankar, N. (2009). *Innovating Development Finance: From Financing Sources to Financial Solutions.* CFP Working Paper Series No. 1. Concessional Finance and Global Partnerships Vice Presidency. World Bank, Washington, DC

Green, D. and Stewart, D. (2004). *The Foresight Process in Practice.* http://www.busi.mun.ca/irishchair/Foresight_process.doc. (accessed 7 May 2011)

Gunderson, L., Allen, C. and Holling, C. (2010). *Foundations of Ecological Resilience.* Island Press, Washington, DC

Gupta, J. (2003). The role of non-state actors in international environmental affairs. *Heidelberg Journal of International Law* 63(2), 459–486

Gupta, J., Ahlers, R. and Ahmed, L. (2010). The human right to water: moving towards consensus in a fragmented world. *Review of European Community and International Environmental Law* 19(3), 294–305

Haas, P.M. (2007). Epistemic communities and international environmental law. In *Oxford Handbook of International Environmental Law.* (eds. Bodansky, D., Hey, E. and Brunnee, J.). Oxford University Press, Oxford

Haas, P.M. (2000). International institutions and social learning in the management of environmental risks. *Policy Studies Journal* 28(3) 558–575

Haas, P.M. and Stevens, C. (2011). Organized science, usable knowledge and multilateral environmental governance. In *Governing the Air* (eds. Lidskog, R. and Sundqvist, G.). MIT Press, Cambridge, MA

Haas, P.M. and Haas, E.B. (1995). Learning to learn: improving international governance. *Global Governance* 1, 255

Haas, P.M., Keohane, R.O. and Levy, M.A. (1993). Institutions for the Earth: sources of effective international environmental protection. In *Global Environmental Accords Series* (ed. Levy, M.A.). MIT Press, Cambridge, MA

Hall, J., Giovanni, E., Morrone, A. and Ranuzzi, G. (2010). *A Framework to Measure the Progress of Societies.* Organisation for Economic Co-operation and Development (OECD), Paris

Hickey, S. and Mohan, G. (2005). Relocating participation within a radical politics of development. *Development and Change* 36(2), 237–262

Hooper, B. (2005). *Integrated River Basin Governance: Learning from International Experience.* IWA Publishing, London

Hoornweg, D., Freire, M., Lee, M.J., Bhada-Tata, P. and Yuen, B. (2011). *Cities and Climate Change: Responding to an Urgent Agenda.* World Bank, Washington, DC

Hunter, D., Salzman, J. and Zaelke, D. (2001). *International Environmental Law and Policy.* Foundation Press

Hutchison, C. (2006). Does TRIPS facilitate or impede climate change technology transfer into developing countries? *University of Ottawa Law and Technology Journal* 3(2), 517–537

ICE Coalition (2011). *Creating the International Court for the Environment.* http://icecoalition.com/wp-content/uploads/2011/11/ICE-Coalition-Rio-contribution.pdf (accessed 18 December 2011)

ICHRP (2008). *Climate Change and Human Rights: A Rough Guide.* International Council on Human Rights Policy, Geneva

ICLEI (2010). *Cities in a Post-2012 Climate Policy Framework.* Local Governments for Sustainability (ICLEI), Bonn

IPCC (2001). *Setting the Stage: Climate Change and Sustainable Development.* Agenda 21, Paragraph 34.3. Intergovernmental Panel on Climate Change. Cambridge University Press, Cambridge

Ivanova, M. (2011). *Financing Environmental Governance: Lessons from the United Nations Environment Programme.* Governance and Sustainability Issue Brief Series: Brief 1. Center for Governance and Sustainability, University of Massachusetts Boston, Boston

Ivanova, M. and Delina, L. (forthcoming in 2012). *Financing Environmental Governance: Survey of the Financial Landscape.* Governance and Sustainability Issue Brief Series: Brief 5. Center for Governance and Sustainability. University of Massachusetts Boston, Boston

Ivanova, M. and Roy, J. (2007). The architecture of global environmental governance: pros and cons of multiplicity. In *Global Environmental Governance: Perspectives on the Current Debate.* (eds. Swart, L. and Perry, E.) Center for UN Reform Education, New York

Ivanova, M., Gordon, D., and Roy, J. (2007). Towards Institutional Symbiosis: Business and the United Nations in Environmental Governance. *Review of European Community and International Environmental Law* (RECIEL) 16 (2), 123–134

James, R. and Wrigley, R. (2007). *Investigating the Mystery of Capacity Building.* Praxis Paper 18. International NGO Training and Research Centre (INTRAC), Oxford

Jeffery, M. (2005). Environmental governance: a comparative analysis of public participation and access to justice. *Journal of South Pacific Law* 9 (2), 1–31

JNO (2010). Parliamentary Commissioner for Future Generations. http://www.jno.hu/en

Jordan, S.J., Sharon, E.H., Yoskowitz, D., Smith, L.M., Summers, J.K., Russell, M. and Benson, W.H. (2010). Accounting for natural resources and environmental sustainability: linking ecosystem services to human well-being. *Environmental Science Technology* 44(5), 1530–1536

Keen, M., Brown, V.A. and Dyball, R. (2005). Social learning: a new approach to environmental management. In *Social Learning in Environmental Management.* Earthscan, London

Keohane, R.O. and Nye, J.S. (1971). *Transnational Relations and World Politics.* Harvard University Press, Cambridge

Klabbers, J., Peters, A. and Ulfstein, G. (2009). *The Constitutionalization of International Law.* Oxford University Press, Oxford

Kravchenko, S. and Bonine, J.E. (2008). *Human Rights and the Environment.* Carolina Academic Press, Durham

Kydd, A.H. (2005). *Trust and Mistrust in International Relations.* Princeton University Press, Princeton

Lee, B., Iliev, I. and Preston, F. (2009). *Who Owns our Low Carbon Future? Intellectual Property and Energy Technologies.* Chatham House Report, London

Levi-Faur, D. (2005). The global diffusion of regulatory capitalism. *The Annals of the American Academy of Political and Social Science* 598, 12–34

Li, X. and Correa, C. (2009). *How Developing Countries Can Manage Intellectual Property Rights to Maximize Access to Knowledge.* South Centre, Geneva

Lipson, B. and Warren, H. (2006). *International Non-Governmental Organizations' Approaches to Civil Society and Capacity Building: Overview Survey.* Paper for Capacity Building Conference. International NGO Training and Research Centre (INTRAC), Oxford

Maskus, K. (2010). *Differentiated Intellectual Property Regimes for Environmental and Climate Technologies.* OECD Environment Working Papers No. 17. Organisation for Economic Co-operation and Development (OECD), Paris

Mayo, M. and Craig, G. (1995). Community participation and empowerment: the human face of structural adjustment or tools for democratic transformation? In *Community Empowerment: A Reader in Participation and Development* (eds. Craig, G. and Mayo, M.). Zed Books, London

Mohan, G. (2002). Participatory development. In *The Companion to Development Studies* (eds. Desai, V. and Potter, R.B.). Arnold, London

Müller, B. (2009). *International Adaptation Finance: The Need for an Innovative and Strategic Approach.* http://iopscience.iop.org/1755-1315/6/11/112008/pdf/1755-1315_6_11_112008.pdf (accessed 25 December 2011)

Najam, A. (2005). Developing countries and global environmental governance: from contestation to participation to engagement. *International Environmental Agreements: Politics, Law and Economics* 5(3), 303–321

Neef, A. (2008). Lost in translation: the participatory imperative and local water governance in North Thailand and southwest Germany. *Water Alternatives* 1(1), 89–110

Njoh, A.J. (2002). Barriers to community participation in development planning: lessons from Mutengene (Camaroon) self-help water project. *Community Development Journal* 37(3), 233–48

Nobel Laureate Symposium (2011). *Third Nobel Laureate Symposium on Global Sustainability: Transforming the World in an Era of Global Change.* http://globalsymposium2011.org/ (accessed 25 December 2011)

Oberthür, S. and Stokke, O.S. (2011). *Managing Institutional Complexity: Regime Interplay and Global Environmental Change.* MIT Press, Cambridge

OECD (2012). *Strategic Environmental Assessment in Development Practice: A Review of Recent Experience.* OECD Publishing, Organisation for Economic Co-operation and Development, Paris. doi: 10.1787/9789264166745-en

OECD (2011a). *Aid Commitments Targeted at the Objectives of the Rio Conventions.* Organisation for Economic Co-operation and Development, Paris. http://www.oecd.org/dataoecd/2/9/48707955.xls (accessed 22 December 2011)

OECD (2011b). *A Country System Approach to Capacity Development for Environment.* Organisation for Economic Co-operation and Development, Paris

OECD (2011c). Environment: climate change aid up to USD 22.9 billion in 2010, says OECD's Gurría. OECD News Room. Organisation for Economic Co-operation and Development, Paris. http://www.oecd.org/document/4/0,3746,en_21571361_44315115_49170628_1_1_1_1,00.html (accessed 22 December 2011)

OECD (2010). *The Influence of Regulation and Economic Policy in the Water Sector on the Level of Technology Innovation in the Sector and its Contribution to the Environment: The Case Study of Israel.* Organisation for Economic Co-operation and Development, Paris

OECD (2008). *OECD Environmental Outlook to 2030.* Organisation for Economic Co-operation and Development, Paris

OECD (1999). *Voluntary Approaches for Environmental Policy: An Assessment.* Organisation for Economic Cooperation and Development, Paris

Otto-Zimmerman, K. (2011). *Embarking on Global Environmental Governance.* ICLEI Paper 2011-1. Local Governments for Sustainability (ICLEI), Bonn

Pahl-Wostl, C. (2006). The importance of social learning in restoring the multifunctionality of rivers and floodplains. *Ecology and Society* 11(1), 10

Pahl-Wostl, C., Gupta, J. and Petry, D. (2008). Governance and the Global Water System: Towards a Theoretical Exploration. *Global Governance* 14, 419–436

Pearce, D.W., Markandya, A. and Barbier, E. (1989). *Blueprint for a Green Economy.* Earthscan, London

Pimbert, M. (2011). *Participatory Research and On-farm Management of Agricultural Biodiversity in Europe.* International Institute for Environment and Development (IIED), London

Pintér, L., Hardi, P., Martinuzzi, A. and Hall, J. (2011). Bellagio STAMP: principles for sustainability assessment and measurement. *Ecological Indicators* (forthcoming)

Putnam, R.D. (1988). Diplomacy and domestic politics: the logic of two-level games. *International Organization* 42, 429–460

Puustjärvi, E., Katila, M. and Simula, M. (2003). *Transfer of Environmentally Sound Technologies from Developed Countries to Developing Countries.* Indufor, Helsinki

Ramos, T.B. (2009). Development of regional sustainability indicators and the role of academia in this process: the Portuguese practice. *Journal of Cleaner Production* 17(12), 1101–1115

Rosenau, J.N. and Czempiel, E.O. (1991). *Governance without Government: Change and Order in World Politics.* Cambridge Studies in International Relations. Cambridge University Press, New York

Ruggie, J.G. (2008). *Embedding Global Markets: An Enduring Challenge.* Ashgate Publishing, London

Ruggie, J.G. (2001). Global-governance.net: the global compact as learning network. *Global Governance* 7, 371

Runhaar, H. and Driessen, P.P.J. (2007).What makes strategic environmental assessment successful environmental assessment? The role of context in the contribution of SEA to decision-making. *Impact Assessment and Project Appraisal* 25(1), 2–14

Sampong, E. (2004). *A Review of the Application of Environmental Impact Assessment (EIA) in Ghana.* United Nations Economic Commission for Africa, Addis Ababa

Sander, K. and Cranford, M. (2010). *Financing Environmental Services in Developing Countries.* 2010 Environment Strategy Analytical Background Papers. The World Bank Group. http://siteresources.worldbank.org/EXTENVSTRATEGY/Resources/6975692-1289855310673/20101201-Financing-Environmental-Investments.pdf (accessed 23 May 2012)

Schulster, T.A., Decker, D.J. and Pfeffer, M.J. (2003). Social learning for collaborative natural resource management. *Society and Natural Resources* 15, 309–326

Serra, C. and Tanner C. (2008). Legal empowerment to secure and use land and resource rights in Mozambique. In *Legal Empowerment in Practice: Using Legal Tools to Secure Land Rights in Africa* (eds. Cotula, L. and Matheiu, P.). International Institute for Environment and Development (IIED) and Food and Agriculture Organiation of the United Nations (FAO), London

Simmons, B.A., Dobbin, F. and Garrett, G. (2006). *International Organization.* The International Organization Foundation and Cambridge University Press, Cambridge

Slaughter, A.-M. (2004). *A New World Order.* Princeton University Press, Princeton

Spain, A. (2011). Beyond adjudication. *Stanford Environmental Law Journal* 30, 343

Steckhan, O. (2009). *Financial Flows for Environment.* World Bank, United Nations Development Programme (UNDP) and United Nations Environment Programme (UNEP). http://bit.ly/vOXPDS (accessed 20 December 2011)

Steffek, J. and Nanz, P. (2008). *Emergent Patterns of Civil Society Participation in Global and European Governance*. https://www.palgrave.com/PDFs/0230006396.Pdf (accessed 23 December 2011)

Stiglitz, J.E., Sen, A. and Fitoussi, J.P. (2009). *Report by the Commission on the Measurement of Economic Performance and Social Progress*. Commission on the Measurement of Economic Performance and Social Progress, Paris

Stone, D. and Maxwell, S. (2005). *Global Knowledge Networks and International Development: Bridges Across Boundaries*. Psychology Press, London

Strange, T. and Bayley, A. (2008). *Sustainable Development: Linking Economy, Society, Environment*. Organisation for Economic Co-operation and Development (OECD), Paris

UNCCD (2008). *Decision 3/COP.8 of the Eighth Meeting of the Conference of Parties of the UN Convention to Combat Desertification on the 10-year Strategic Plan and Framework to Enhance the Implementation of the Convention*. http://www.cbd.int/decision/cop/?id=12268 (accessed 23 December 2011)

UNCED (1992). *Rio Declaration on Environment and Development*. United Nations Conference on Environment and Development. http://www.unep.org/Documents.Multilingual/Default.asp?documentid=78&articleid=1163 (accessed 23 May 2012)

UNCTAD (2010). *World Investment Report 2010: Investing in a Low-Carbon Economy*. United Nations Conference on Trade and Development, Geneva. http://www.unctad.org/en/docs/wir2010_en.pdf (accessed 19 December 2011)

Underdal, A. (1998). *The Politics of International Environmental Management*. Kluwer Academic Publishers, Dordrecht

UNDG (2010). *Millennium Development Goals Thematic Papers: Thematic Paper on MDG 7 Environmental Sustainability*. United Nations Development Group, New York

UNDP (2011). *Human Development Report 2011. Sustainability and Equity: A Better Future for All*. United Nations Development Programme (UNDP), New York. http://hdr.undp.org/en/reports/global/hdr2011/download/ (accessed 24 December 2011)

UNDP (2007). *Human Development Report 2007–2008. Fighting Climate Change: Human Solidarity in a Divided World*. United Nations Development Programme. Palgrave Macmillan, New York

UNECE (2011). *Landmark meeting of Aarhus Convention welcomes global accession*. http://www.unece.org/press/pr2011/11env_p32e.html (accessed 18 April 2012)

UNEP Data Explorer, http://geodata.grid.unep.ch/

UNEP (2012). Environment Fund: Resource mobilization. http://www.unep.org/rms/en/Financing_of_UNEP/Environment_Fund/index.asp (accessed 19 May 2012)

UNEP (2011a). *Decisions Adopted by the Governing Council/Global Ministerial Environment Forum at its Twenty-Sixth Session*. United Nations Environment Programme, Nairobi

UNEP (2011b). *Enhanced Coordination across the United Nations System, Including the Environment Management Group*. Report of the Executive Director: UNEP/GC.26/15. United Nations Environment Programme, Nairobi

UNEP (2011c). *Environment in the UN System: Note by the Executive Director*. UNEP/GC.26/INF/23. United Nations Environment Programme, Nairobi. http://www.unep.org/gc/gc26/information-docs.asp (accessed 22 December 2011)

UNEP (2011d). *Global Green New Deal Policy Brief*. United Nations Environment Programme, Nairobi. http://www.unep.org/pdf/A_Global_Green_New_Deal_Policy_Brief.pdf (accessed 25 December 2011)

UNEP (2011e). *Outcome of the Work of the Consultative Group of Ministers or High-level Representatives on International Environmental Governance*. Note by the Executive Director. UNEP/GC.26/18. United Nations Environment Programme, Nairobi

UNEP (2011f). *Status of Contributions and Disbursements*. UNEP/OzL.Pro/ExCom/64/3. United Nations Environment Programme, Nairobi

UNEP (2010). *Advancing the Biodiversity Agenda: A UN System-wide Contribution*. United Nations Environment Programme, Nairobi

UNFCCC (2010). *Decision 2/CP.15 of the Fifteenth Meeting of the Conference of Parties of the UN Framework Convention on Climate Change on the Copenhagen Accord*. http://unfccc.int/resource/docs/2009/cop15/eng/11a01.pdf#page=4 (accessed 23 December 11)

UN-NGLS (2007). UNEO: A Champion for environment in the 21st Century, but what role for stakeholders? A multi-stakeholder conversation. United Nations Non- Governmental Liaison Service (NGLS), Stakeholder Forum and ANPED http://www.un-ngls.org/IMG/pdf/ReformingInternationalEnvironmentalGovernance-mtg_report.pdf (accessed 18 April 2012)

United Nations (2011). *Report of the Committee of Experts on Environmental-Economic Accounting*. Note by the Secretary-General. E/CN.3/2011/7. http://unstats.un.org/unsd/statcom/doc11/2011-7-UNCEEA-e.pdf (accessed 29 December 2011)

United Nations (2007). *General Assembly Resolution 61/295*. United Nations Declaration on the Rights of Indigenous Peoples, New York

Varady, R.G. and Iles-Shih, M. (2009). Global water initiatives: what do the experts think? In *Impacts of Megaconferences on the Water Sector: Water Resources Development and Management* (eds. Biswas, A.K. and Tortajada, C.). Springer, Berlin

WBCSD (2010). *The Business Case for Sustainable Development: Making a Difference Towards the Johannesburg Summit 2002 and Beyond*. World Business Council for Sustainable Development, Geneva

Willetts, P. (2011). *Non-Governmental Organizations in World Politics: The Construction of Global Governance*. Routledge, Global Institutions Series, London

Woodhill, J. (2010). Capacities for institutional innovation: a complexity perspective. *Institute of Development Studies Bulletin* 41(3) *Special Issue: Reflecting Collectively on Capacities for Change*, 47–59

World Bank (2011). *State and Trends of the Carbon Market 2011*. World Bank, Washington, DC. http://siteresources.worldbank.org/INTCARBONFINANCE/Resources/State_and_Trends_Updated_June_2011.pdf (accessed 22 December 2011)

World Bank (2010a). *The Cost to Developing Countries of Adapting to Climate Change: New Methods and Estimates*. World Bank, Washington, DC

World Bank (2010b). *Innovative Finance for Development Solutions: Initiatives of the World Bank Group*. World Bank, Washington, DC

World Bank (2010c). *World Development Report 2010: Development and Climate Change*. World Bank, Washington, DC

World Bank (2005). *Ensuring Environmental Sustainability: Measuring Progress Toward the 7th Millennium Development Goal*. World Bank, Washington, DC

Yamin, F. (2001). NGOs and international environmental law: a critical evaluation of their roles and responsibilities. *Review of European Community and International Environmental Law* 10(2), 149–162

Young, O.R. (2010). *Institutional Dynamics: Emergent Patterns in International Environmental Governance*. http://mitpress.mit.edu/catalog/item/default.asp?ttype=2&tid=12318 (accessed 22 December 2011)

Young, O.R. (2002). *The Institutional Dimensions of Environmental Change: Fit, Interplay, and Scale*. http://mitpress.mit.edu/catalog/item/default.asp?ttype=2&tid=8725 (accessed 22 December 2011)

"There is a huge amount to be done if we are to reach a state of sustainability. Do not despair, do not be weighed down by it. All I ask of you is that you go home and do what you can."

Wangari Maathai (1940-2011), Nobel Lauriate.

The GEO-5 Process

MANDATE

In February 2009, as part of UNEP's overall mandate to keep the global environment under review, the 25th Session of the UNEP Governing Council/Global Ministerial Environment Forum reconfirmed the *Global Environmental Outlook (GEO)* mandate, requesting the Executive Director to:

> *"continue to conduct a comprehensive, integrated and scientifically credible global environmental assessment, avoiding duplication and building on ongoing assessment work, to support decision-making processes at all levels, in the light of the continuing need for up-to-date, scientifically credible, policy-relevant information on environmental change worldwide, including analyses of cross-cutting issues and indicator-based components"*

and to:

> *"strengthen the policy relevance of* GEO-5 *by including an analysis of case studies of policy options, that incorporates environmental, economic, social and scientific data and information and their indicative costs and benefits to identify promising policy options to speed up achievement of the internationally agreed goals such as those agreed at the Millennium Summit in 2000 and in Multilateral Environmental Agreements"* (UNEP/GC.25/2/III) (http://www.unep.org/gc/gc25/Docs/Proceedings-English.pdf).

Support for the fifth *Global Environment Outlook* (*GEO-5*) was further approved in November 2011 by the United Nations General Assembly 2nd Committee (Economic and Financial) (resolution A/C.2/66/L.57) (http://daccess-dds-ny.un.org/doc/UNDOC/LTD/N11/601/65/PDF/N1160165.pdf).

OBJECTIVES, SCOPE AND PROCESS

The objectives, scope and process for *GEO-5* were defined and adopted in a Final Statement by the Global Intergovernmental and Multi-stakeholder Consultation that took place in March 2010 and included 91 governmental representatives and 55 other main stakeholders.

Objectives

The consultation reaffirmed the above mandate by identifying the following objectives for the assessment:

- provide a comprehensive, integrated and scientifically credible global environmental assessment to support decision-making processes at appropriate levels;
- engage all governments, relevant UN bodies, and other stakeholders in *GEO-5* in order to support and strengthen its scientific credibility, policy relevance and legitimacy;
- strengthen the on-going process of capacity building for developing countries and countries with economies in transition to conduct environmental monitoring and assessments at all levels, in collaboration with relevant on-going activities of UNEP and other initiatives, including through south-south and triangular cooperation;
- inform, as appropriate, the strategic directions of UNEP and other relevant UN bodies;
- strengthen the policy relevance of *GEO-5* by including an analysis of case studies of policy options, that incorporates environmental, economic, social and scientific data and information and their indicative costs and benefits to identify promising policy options to speed up achievement of the internationally agreed goals such as those agreed at the Millennium Summit in 2000 and in multilateral environmental agreements;
- inform and learn from relevant global and regional processes and meetings where progress towards these agreed goals is discussed; and
- identify data gaps in the thematic issues considered by *GEO-5*.

Scope

GEO-5 builds on previous GEO reports and continues to provide an analysis of the state, trends and outlook of the global environment. It differs from previous GEO reports in its emphasis on internationally agreed goals and in providing possible means of accelerating achievement of those goals. *GEO-5* is made up of three distinct but closely linked parts.

Part 1 assesses the state and trends of the global environment in relation to key internationally agreed goals such as the Millennium Development Goals and those of various multilateral environmental agreements. The assessment is based on national, regional and global analyses and datasets.

Part 2 prioritizes a number of environmental themes per region, selected through a consultative process in light of relevant internationally agreed goals. The regional assessments identify and appraise promising policy responses that could help speed up the achievement of these goals.

Part 3 identifies options with potential to assist the transition towards sustainable development and suggests possibilities for global response.

The consultation proposed ten key questions for UNEP to address. To a great extent, these questions helped define the scope of the GEO-5 assessment and to guide the process.

Key questions for Part 1

i. What are the current drivers, state and trends and outlook for the global environment?
ii. Do the current drivers, state and trends of the environment reflect progress towards meeting internationally agreed goals?
iii. What are the main challenges for the life-supporting functions of the Earth System and the drivers that cause them?

iv. To what extent do existing monitoring and observation activities and institutional arrangements meet the need to keep the state and trends of the environment under review?
v. What are the main gaps and barriers to meeting the agreed goals?

Key questions for Part 2

vi. Which internationally agreed goals are high priorities for each region?
vii. What policy options can be most successfully applied in each region to help speed up meeting internationally agreed goals?
viii. What policy options facilitate environmental monitoring and its use in decision making?

Key questions for Part 3

ix. What policy approaches could be suitable for scaling up in order to accelerate meeting internationally agreed goals?
x. What types of sustainable change and innovation are needed over the long term?

Process

The March 2010 consultation also provided direction for strengthening the process of the *GEO-5* assessment through:

- engaging the best available scientific and policy expertise;
- ensuring scientific credibility, policy relevance, and legitimacy of the assessment by engaging a wide range of stakeholders;
- constituting multidisciplinary groups of experts nominated by governments and other stakeholders using a transparent process;
- establishing three overarching advisory groups: a High-Level Intergovernmental Advisory Panel to provide guidance to experts; a Science and Policy Advisory Board to ensure the scientific credibility of the process; and a Data and Indicators Working Group to provide core data support to the process;
- subjecting the assessment to extensive scientific expert peer-review and government review;
- continuing to target institutional capacity building by engaging developing country experts; and
- communicating key messages and findings to target audiences in an accessible manner.

PARTNERSHIPS AND COLLABORATION

The development of *GEO-5* involved extensive collaboration both within UNEP and between UNEP and a network of multidisciplinary experts, research institutions and GEO collaborating centres, all of whom made their valuable time and knowledge available to the process.

The consultation requested that experts for content development, including reviewers and advisory groups, be nominated by governments and other main stakeholders including GEO collaborating centres and other partners, based on their expertise and using a transparent process drawing from the nomination process of the Intergovernmental Panel on Climate Change (IPCC). The nominated experts were then engaged by the UNEP Secretariat on the basis of their expertise, with due consideration of gender and regional balances.

Figure 1 The development of GEO-5: milestones in the production process

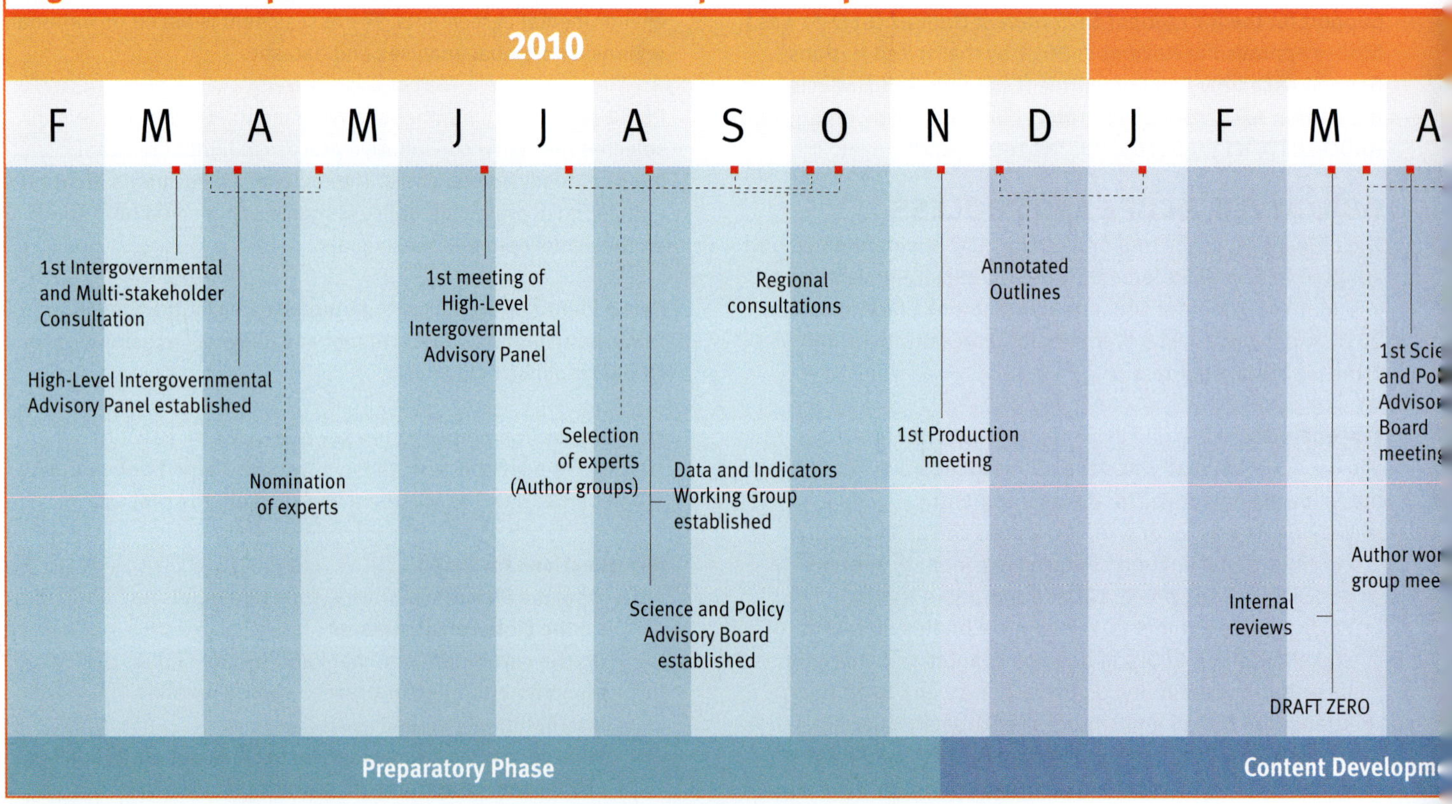

Chapter expert groups

The *GEO-5* report has 17 chapters. An expert working group was established for each chapter to conceptualize, research, draft, revise and finalize the manuscripts. More than 310 authors were involved in content development. Each chapter expert group included 5–38 individuals under the leadership of two or three coordinating lead authors and supported by a UNEP chapter coordinator. Other members of the chapter expert groups comprised lead authors and contributing authors.

GEO-5 fellows

GEO-5 continued to pursue the Fellowship initiative established during the *GEO-4* process in 2005. This engages early career professionals in the GEO process so that they can gain experience from participating in a major global environmental assessment. A total of 21 fellows from 18 countries participated in *GEO-5*.

Outreach working group

An outreach working group was established that included one member of each chapter expert group as well as UNEP experts. The group prepared the overarching outreach strategy for *GEO-5* and identified target audiences and relevant meetings to disseminate findings.

REVIEW PROCESS

The *GEO-5* assessment underwent three rounds of review involving more than 300 experts. The first was an internal one within UNEP; the second was an external review by governments and UNEP's extensive network of science and policy experts, including those nominated by governments and other stakeholders. The final review was undertaken by governments and well-known scientific experts from both the natural and social science communities. The final round of expert review was an independent peer-review process facilitated by the Earth System Science Partnership (ESSP). The ESSP sent a call for reviewers to its global expert network and then selected interested experts based on their field as well as gender and geographical balance. In the final expert peer review, each chapter had three to four expert scientific reviewers with extensive experience in the subject area covered by the respective chapter. The content development process and all review stages were supported by the Science and Policy Advisory Board who provided guidance to chapter authors, reviewers and the UNEP Secretariat to ensure that the process was scientifically credible and robust.

GEO-5 advisory groups

Three external specialized advisory bodies were established to support the assessment process.

High-Level Intergovernmental Advisory Panel

The panel comprised 20 high-level government representatives from all six UNEP regions. The panel, using the Global Environmental Goals framework (for more detail see http:/geg.informea.org/goals), identified the internationally agreed goals for *GEO-5* to assess, and formulated strategic advice to *GEO-5* authors and other groups to assist them in their assessment of the goals. They also provided initial guidance on the structure and content of the *GEO-5 Summary for Policy Makers* and further guidance

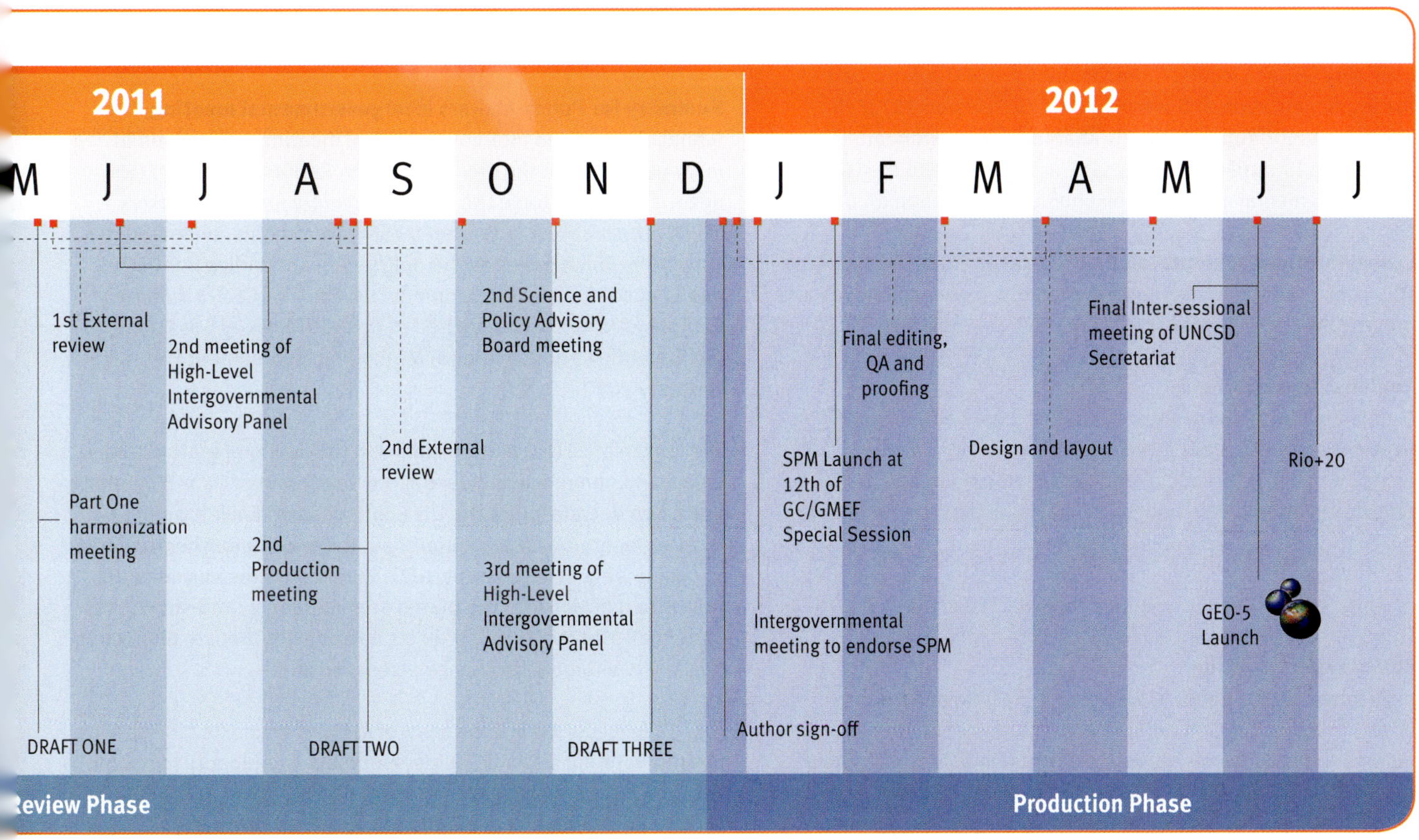

Participants at the intergovernmental meeting to endorse the GEO-5 Summary for Policy Makers in Gwangju, Republic of Korea.

to the experts in finalizing the draft in preparation for the final intergovernmental negotiation. In addition, *ad-hoc* guidance was provided to UNEP throughout the assessment process, in particular in aligning the *GEO-5* process with the relevant processes of the 2012 United Nations Conference on Sustainable Development (Rio+20). The panel met three times in 2010 and 2011.

Science and Policy Advisory Board

The board comprised 18 distinguished scientists and senior representatives from the policy community and met twice in 2011. The board was responsible for strengthening the scientific credibility and policy relevance of the assessment by providing guidance throughout the process. They provided high-level strategic advice; standards and guidelines for the assessment and review process; and undertook a mid-term and final evaluation of the assessment process.

Data and Indicators Working Group

The group met once in March 2011 and provided support to the assessment process on the use of core datasets and indicators. They consulted with experts to identify priority environmental indicators and identified available datasets, as well as data gaps and related issues.

CONSULTATION PROCESS

UNEP organized global and regional consultations and meetings throughout the assessment process. The following are some of the key meetings convened since the inception in November 2009.

GEO-5 planning meetings

Two planning meetings were convened with experts familiar with the GEO process, including UNEP's GEO specialists, in November 2009 and January 2010. The meetings focused on reviewing lessons learned from the previous GEO processes and the implementation of Governing Council Decision 25/2/III. The experts formulated UNEP's analytical framework and a vision of the future global assessment to propose to the Global Intergovernmental and Multi-stakeholder Consultation on *GEO-5*.

Global Intergovernmental and Multi-stakeholder Consultation

This consultation defined and adopted the scope, objectives and process for *GEO-5* in March 2010.

Regional consultations

A series of seven regional consultations were held between September and October 2010. The consultations, which engaged many different stakeholders, determined five or six priority environmental challenges in each region and selected internationally agreed goals of relevant concern, as well as identifying potential policy options in the region, implementation of which could speed up meeting the selected goals.

Policy expert meeting

In October 2010 a policy expert group, including one policy expert from each region nominated to participate in the *GEO-5* regional policy analysis, as well as several independent policy experts, was convened to discuss the challenge of policy analysis in the context of identifying policies that help speed up the attainment of internationally agreed goals. The policy expert group provided guidelines for the regional policy analysis.

Global production and authors' meetings

Two global production and authors' meetings were convened in November 2010 and in September 2011 to discuss and develop *GEO-5* chapter content and outlines, to address review comments, and to harmonize different approaches and presentation styles.

Chapter working group meetings

More than 30 meetings were convened to prepare, review and revise the drafts for individual chapters.

Summary for Policy Makers intergovernmental meeting

A final open-ended intergovernmental meeting was convened in January 2012 in the City of Gwangju, Republic of Korea, to negotiate and endorse the *GEO-5 Summary for Policy Makers* (SPM). The meeting, attended by 53 governments, endorsed the summary, which presents the policy-relevant findings of *GEO-5* and is published as a separate document. The *GEO-5 Summary for Policy Makers* was launched at the 12th Special Session of the Governing Council/Global Ministerial Environment Forum in February 2012.

The launch of *GEO-5* will coincide with the final preparatory stages of the UN Conference on Sustainable Development (Rio+20), taking place two decades after the UN Conference on Environment and Development (Rio Earth Summit) set the agenda for the transition to sustainable development. *GEO-5* highlights the current state, trends and outlook for the planet and its people, and showcases more than 100 initiatives, projects and policies that are pioneering positive environmental change across the globe.

GEO-5 highlights not just the perils of delaying action, but the options for transforming sustainable development from theory to reality.

Further information is available at www.unep.org/geo

Acronyms and Abbreviations

3Rs	reduce, reuse, recycle
4Rs	reduce, reuse, recycle and re-think
ABC	atmospheric brown cloud
ABS	access and benefit sharing
ACC	adaptation to climate change
ACCOBAMS	Agreement on the Conservation of Cetaceans of the Black Sea, Mediterranean Sea and contiguous Atlantic Area
ACP	Panama Canal Authority
ACS	Association of Caribbean States
ACSAD	Arab Center for Studies of Arid Zones and Dry Lands
ACTO	Amazon Cooperation Treaty Organization
ADB	Asian Development Bank
ADFEC	Abu Dhabi Future Energy Company
AEM	agri-environment measures
AEWA	African-Eurasian Migratory Waterbird Agreement
AHTEG	*Ad Hoc* Technical Expert Group
AICS	Australian Inventory of Chemical Substances
AIDS	acquired immune deficiency syndrome
ALR	Agricultural Land Reserve (Canada)
AMAP	Arctic Monitoring and Assessment Programme
AMCs	advanced market commitments
AMCEN	African Ministerial Conference on the Environment
ANAM	National Environmental Authority of Panama
AOAD	Arab Organization for Agricultural Development
APVMA	Australian Pesticides and Veterinary Medicines Authority
AQG	air quality guidelines
ASCLME	Agulhas and Somali Current Large Marine Ecosystems
ASCOBANS	Agreement on the Conservation of Small Cetaceans of the Baltic, North East Atlantic, Irish and North Seas
ASEAN	Association of Southeast Asian Nations
ATS	Antarctic Treaty System
AZEs	Alliance for Zero Extension sites
BBOP	Business and Biodiversity Offsets Programme
BC	black carbon
BCLME	Benguela Current Large Marine Ecosystem
BFP	Bolsa Floresta Programme
BPA	bisphenol-A
BRIC	Brazil, Russia, India and China
CAA	Clean Air Act (United States)
CAC	command and control
CAFE	Corporate Average Fuel Economy (United States)
CAN	Andean Community
CAP	Common Agricultural Policy of the EU
CAPRADE	Andean Committee for Disaster Awareness and Prevention
CAR	1) Central African Republic, or 2) Central Albertine Rift
CARICOM	Caribbean Common Market
CAS	1) complex adaptive systems, or 2) Chemicals Abstract Service
CBD	Convention on Biological Diversity (UN)
CBNRM	Community Based Natural Resources Management
CBR	crude birth rate
CCAD	Central American Commission on Environment and Development
CCCCC	Caribbean Community Climate Change Centre
CDC	Centers for Disease Control and Prevention (United States)
CDEMA	Caribbean Disaster Emergency Management Agency
CDM	Clean Development Mechanism
CEB	Chief Executive Board for Coordination (UN)
CEC	Commission for Environmental Cooperation (under NAFTA)
CEPA	Canadian Environmental Protection Act
CEHI	Caribbean Environmental Health Institute
CEPREDENAC	Centre for Natural Disaster Awareness and Prevention
CFC	chlorofluorocarbon
CFU	community forest unit
CGIAR	Consultative Group on International Agricultural Research
CH_4	methane
CITES	Convention on International Trade in Endangered Species of Wild Fauna and Flora
CLRTAP	Convention on Long-range Transboundary Air Pollution
CMC	Chemical Management Center
CMP	Chemicals Management Plan
CMS	Convention on the Conservation of Migratory Species of Wild Animals
CO	carbon monoxide
CO_2	carbon dioxide
CONAVI	Comisión Nacional de Vivienda
COP	conference of the parties
COSEWIC	Status of Endangered Wildlife in Canada
CRED	Centre for Research on the Epidemiology of Disasters
CRP	Conservation Reserve Program (United States)
CSA	environmental services certificates
CSCL	Chemical Substance Control Law
CSD	Commission on Sustainable Development
CSP	Conservation Security Program (United States)
CSIRO	Commonwealth Scientific and Industrial Research Organisation (Australia)
CSO	civil society organisation
CSRP	Sub-regional Fisheries Commission
CZMU	Coastal Zone Management Unit (Barbados)
DAC	Development Assistance Committee (OECD)
DALY	disability adjusted life year
DDT	dichlorodiphenyltrichloroethane
DESA	Department of Economic and Social Affairs (UN)
DEWA	1) Division of Early Warning and Assessment (UNEP), or 2) Dubai Electricity and Water Authority

DPSIR	drivers, pressures, state, impacts, responses
DRC	Democratic Republic of the Congo
DRR	disaster risk reduction
EA	ecosystem approach
EAC	East African Community
EAF	ecosystem approach to fisheries
EAP	Environmental Action Programme of the EU
EBA	ecosystem based adaptation
EC	European Commission
ECESA	Executive Committee on Economic and Social Affairs (UN)
ECHA	European Chemicals Agency
ECLAC	Economic Commission for Latin America and the Caribbean of the United Nations
ECOWAS	Economic Community of West African States
EE	energy efficiency
EEA	European Environment Agency
EU	European Union
EIA	1) Energy Information Administration, or 2) environmental impact assessment
EIONET	European Environment Information and Observation Network
EKC	environmental Kuznets curve
EM-DAT	Emergency Events Database
EMEP	European Monitoring and Evaluation Programme
EMG	Environment Management Group
ENRM	Environmental and Natural Resources Management (World Bank)
EPA	1) environmental performance assessment, or 2) Environmental Protection Agency (United States)
EQIP	Environmental Quality Incentives Program (United States)
ERMA	Environmental Risk Management Authority
ERS	Economic Research Service (United States)
ES	Earth System
ESA	1) environmentally sensitive area, or 2) European Space Agency
ESI	environmental services index
ESS	Earth system science
ETS	emissions trading scheme
EU	European Union
EUROBATS	Agreement on the Conservation of Populations of European Bats
Ex-COPs	Extraordinary Conferences of the Parties to the Basel, Rotterdam and Stockholm Conventions
EWS	early warning system
FAO	Food and Agriculture Organization of the United Nations
FDI	foreign direct investment
FIBA	Fondation Internationale du Banc d'Arguin
FIT	feed-in tariff
FIT-FIR	first-in-time, first-in-right (or the Doctrine of Prior Appropriation)
FLORES	Forest Land Oriented Resources Envisioning System
FON	Friends of Nature
FONAFIFO	Fondo Nacional de Financiamiento Forestal
FONAG	Fund for the Protection of Water
FSC	Forest Stewardship Council
G7	Group of Seven (Canada, France, Germany, Italy, Japan, United Kingdom, United States)
G8	Group of Eight (Canada, France, Germany, Italy, Japan, Russian Federation, United Kingdom, United States)
GAPS	Global Atmospheric Passive Sampling
GATT	General Agreement on Tariffs and Trade
GCC	Gulf Cooperation Council
GCF	Green Climate Fund
GCLME	Guinea Current Large Marine Ecosystem
GCM	general circulation models
GCP	gross cell product
GDP	gross domestic product
GEF	Global Environment Facility
GEMS	Global Environmental Monitoring System
GEO	Global Environment Outlook
GEOSS	Global Earth Observation System of Systems
GESAMP	Group of Experts on Scientific Aspects of Marine Environmental Protection
GHG	greenhouse gas
GIS	geographical information systems
GISS	Goddard Institute for Space Studies
GLASOD	Global Assessment of Human-Induced Soil Degradation
GM	Global Mechanism
GMO	genetically modified organism
GNP	gross national product
GPA	Global Programme of Action for the Protection of the Marine Environment from Land-based Activities
GPCP	Global Precipitation Climatology Project
GPI	genuine progress indicator
GPW	Gridded Population of the World
GUPES	Global University Partnership on Environment and Sustainability
GW	gigawatt
GWP	1) Global Water Partnership, or 2) global warming potential
GWSP	Global Water System Project
HAB	harmful algal blooms
HCFC	hydrochlorofluorocarbon
HCH	hexachlorocyclohexane
HDI	Human Development Index
HFA	Hyogo Framework for Action
HFC	hydrofluorocarbon
HIV	human immunodeficiency virus
HKHT	Hindu Kush-Himalayan-Tibetan
HLCP	High Level Committee on Policy
HLIAP	High-Level Intergovernmental Advisory Panel
HS	Harmonized System
HTAP	hemispheric transport of air pollution
HWS	human water security
IAEG	Inter-agency and Expert Group
IATTC	Inter-American Tropical Tuna Commission
IBA	important bird area
ICARM	integrated coastal and river management
ICCA	indigenous and community-conserved areas

ICE International Court for the Environment
ICHRP International Council on Human Rights Policy
ICLEI Local Governments for Sustainability
ICLZT integrated rotating crops, livestock production and zero-tillage operations
ICRISAT International Crop Research Institute for the Semi-Arid Tropics
ICT information and communication technology
ICZM integrated coastal zone management
IDB Inter-American Development Bank
IDMC Internal Displacement Monitoring Centre
IEA 1) International Energy Agency, or 2) integrated environmental assessment
IFAD International Fund for Agricultural Development
IFPRI International Food Policy Research Institute
IGRAC International Groundwater Resources Assessment Centre
IIASA International Institute for Applied System Analysis
IISD International Institute for Sustainable Development
IJC International Joint Commission
ILBM integrated lake basin management
ILC indigenous and local communities
ILEC International Lake Environment Committee
ILM 1) integrated land management, or 2) indigenous land management
ILO International Labour Organization
IMO International Maritime Organization
IMPACT International Model for Policy Analysis of Agricultural Commodities and Trade
INBO International Network of Basin Organizations
INVERMAR Invertec Pesquera Mar de Chiloé
IOC Intergovernmental Oceanographic Commission of UNESCO
IOMC Inter-organizational Programme for the Sound Management of Chemicals
IP intellectual property
IPA indigenous protected area
IPA CIS Inter-Parliamentary Assembly of the Commonwealth of Independent States
IPAT Impact = Population x Affluence x Technology
IPBES Intergovernmental Science-Policy Platform on Biodiversity and Ecosystem Services
IPCC Intergovernmental Panel on Climate Change
IPR intellectual property rights
IPSI International Partnership on Satoyama Initiative
IPSRM International Panel for Sustainable Resource Management
IRP integrated resource planning
ISDR International Strategy for Disaster Reduction
ISEW Index of Sustainable Economic Welfare
ITPGRFA International Treaty on Plant Genetic Resources for Food and Agriculture
ITF International Transport Forum
IUCN International Union for Conservation of Nature
IWI International Watersheds Initiative (North America)
IWM integrated watershed planning and management
IWRM integrated water resources management
JHU Johns Hopkins University (United States)
JPOI Johannesburg Plan of Implementation
JPoI Joint Plan of Implementation
JRC European Commission Joint Research Centre
LAC Latin America and the Caribbean
LAS League of Arab States
LECZ low elevation coastal zone
LDC 1) least developed country, or 2) London Dumping Convention: Convention on the Prevention of Marine Pollution by Dumping of Wastes and Other Matter
LDCF Least Developed Countries Trust Fund
LEZ low emission zone
LIFDC low-income food deficit countries
LME large marine ecosystem
LPG liquefied petroleum gas
LRTAP long-range transboundary air pollution
MA Millennium Ecosystem Assessment
MAP Mediterranean Action Plan for the Barcelona Convention
MARPOL International Convention for the Prevention of Pollution From Ships
M&E monitoring and evaluation
MDG Millennium Development Goal
MDTF Multi-Donor Trust Funds (UN)
MEA multilateral environmental agreement
MERCOSUR Mercado Común del Sur
MFA material flow accounting
MINAM Portal del Ministerio del Ambiente del Perú
MMAs marine managed area
MMWD Marin Municipal Water District
MPA marine protected area
MSC Marine Stewardship Council
MSW municipal solid waste
N_2O nitrous oxide
NAAEC North American Agreement on Environmental Cooperation
NAFA National Forest Authority
NAFTA North American Free Trade Agreement
NAMA nationally appropriate mitigation actions
NASA National Aeronautics Space Administration (United States)
NBI Nile Basin Initiative
NBSAP national biodiversity strategies and action plans
NEG/ECP New England Governors/Eastern Canadian Premiers
NEPA National Environment Policy Act (United States)
NEPA National Environmental Protection Agency (China)
NEPAD New Partnership for Africa's Development
NEPAD CAADP NEPAD Comprehensive Africa Agriculture Development Programme
NERC 1) National Energy Research Center (Jordan; Syria), or 2) Natural Environment Research Council (United Kingdom)
NPP net primary productivity
NGO non-governmental organization
NH_3 ammonia
NH_x ammonia and ammonium

NHANES National Health and Nutrition Examination Survey
NICNAS National Industrial Chemicals Notification and Assessment Scheme
NOAA National Oceanic and Atmospheric Administration (United States)
NOWPAP Action Plan for the Protection, Management and Development of the Marine and Coastal Environment of the Northwest Pacific Region
NO_2 nitrogen dioxide
NO_x nitrogen oxides
NPRI National Pollutant Release Inventory (Canada)
NRTEE National Roundtable on the Economy and the Environment
NMVOCs non-methane volatile organic compounds
O_3 ozone
OCP organochlorine pesticides
ODA official development assistance
ODS ozone-depleting substance
OECD Organisation for Economic Co-operation and Development
OP obsolete pesticide
OPRC International Convention on Oil Pollution Preparedness, Response and Cooperation
OPT Occupied Palestinian Territories
OSPAR Convention for the Protection of the Marine Environment of the North-East Atlantic
PA protected area
PAEC Cuban Energy Saving Programme
PAH polycyclic aromatic hydrocarbons
PBDE polybrominated diphenyl ethers
PCB polychlorinated biphenyls
PCT polychlorinated terphenyls
PERI Political Economy Research Institute, Univeristy of Massachusetts (United States)
PES payment for ecosystem services
PM particulate matter
$PM_{2.5}$ particulate matter with a diameter of 2.5 micrometres (0.0025 millimetre) or less
PM_{10} particulate matter with a diameter of 10 micrometres (0.01 millimetre) or less
POPs persistent organic pollutants
PPCDAm Action Plan for Protection and Control of Deforestation in the Amazon
PPP purchasing power parity
PROCEL Programa Nacional de Conservação de energia eléctrica (National Electrical Conservation Programme) (Brazil)
PSP paralytic shellfish poisoning
PTC production tax credit
QSAR quantitative structure-activity relationships
R&D research and development
RAFNET Rwanda Agro-forestry Network
RCP representative concentration pathways
REACH Registration, Evaluation, Authorisation and Restriction of Chemical substances programme (EU)
RE renewable energy
REC renewable energy credits
REDD Reducing Emissions from Deforestation and Forest Degradation
REFIT renewable energy feed-in-tariff
REMP renewable energy master plan
REMPEC Regional Marine Pollution Emergency Response Centre for the Mediterranean Sea
RES renewable energy systems
ROPME Regional Organization for the Protection of the Marine Environment of the sea area surrounded by Bahrain, Iran, Iraq, Kuwait, Oman, Qatar, Saudi Arabia and the United Arab Emirates
RPBR Río Plátano Biosphere Reserve (Honduras)
RPS Renewable Portfolio Standard
RWH rainwater harvesting
SADC Southern African Development Community
SAICM Strategic Approach to International Chemicals Management
SCBD Secretariat of the Convention on Biological Diversity
SCCF Special Climate Change Trust Fund
SEA strategic environmental assessment
SEEA System of Environmental-Economic Accounting
SEMARNAT Secretaría de Medio Ambiente y Recursos Naturales
SFM sustainable forest management
SICA Sistema de la Integración Centroamericana (Central America Integration System)
SIDS small island developing states
SLCF short-lived climate forcer
SLM sustainable land management
SNACs Significant New Activity Controls
SNS sacred natural sites
SNURs Significant New Use Rules
SOE state owned enterprises
SoE state of the environment
SOER State of the Environment Report of the EEA
SO_x sulphur oxides
SO_2 sulphur dioxide
SPB sustainability policy banks
STAR System for the Transparent Allocation of Resources
SST sea surface temperature
SWF Sovereign Wealth Funds
TCO traditional communal lands
TBNRM transboundary natural resources management
TEAP Technology and Economic Assessment Panel (the Montreal Protocol)
TEEB The Economics of Ecosystems and Biodiversity
TEK traditional ecological knowledge
TEU twenty-foot-equivalent units
TFCA transfrontier conservation areas
TK traditional knowledge
TM technology mechanism
TMDL total maximum daily load
TRI Toxics Release Inventory (United States)
TRIPs trade-related aspects of international property rights

TSCA Toxic Substances Control Act (United States)
UK United Kingdom
UN United Nations
UNCCD United Nations Convention to Combat Desertification
UNCED United Nations Conference on Environment and Development
UNCLOS United Nations Convention on the Law of the Sea
UNCSD United Nations Commission on Sustainable Development
UNCTAD United Nations Conference on Trade and Development
UNDG United Nations Development Group
UNDP United Nations Development Programme
UNDRIP United Nations Declaration on the Rights of Indigenous Peoples
UNECE United Nations Economic Commission for Europe
UNEP United Nations Environment Programme
UNEP-CEP United Nations Environment Programme – Caribbean Environment Programme
UNEP-PCFV United Nations Environment Programme – Partnership for Clean Fuels and Vehicles
UNEP-WCMC United Nations Environment Programme – World Conservation Monitoring Centre
UNESCO United Nations Educational, Scientific and Cultural Organization
UNFCCC United Nations Framework Convention on Climate Change
UNFF United Nations Forum on Forests
UNHCR The United Nations Refugee Agency
UNICEF United Nations Children's Fund
UNIDO United Nations Industrial Development Organization
UNITAR United Nations Institute for Training and Research
UN-REDD United Nations collaborative initiative on Reducing Emissions from Deforestation and forest Degradation in Developing Countries
UNSD United Nations Statistics Division
UNU United Nations University
UNWTO United Nations World Tourism Organization
USA United States of America
USAID United States Agency for International Development
US EPA United States Environmental Protection Agency
UV ultraviolet
VITEK vitality of traditional ecological knowledge
VOC volatile organic compound
vPvB very persistent and very bioaccumulative
WAIS West Antarctic ice sheet
WBCSD World Business Council for Sustainable Development
WCI Western Climate Initiative (North America)
WCRP World Climate Research Programme
WFD Waste Framework Directive of the EU
WFP World Food Programme (United Nations)
WHC World Heritage Convention
WHO World Health Organization
WIO Western Indian Ocean
WMO World Meteorological Organization
WRI World Resources Institute
WSSD World Summit on Sustainable Development
WTO World Trade Organization
WTP willingness to pay
WUE water-use efficiency
WWAP World Water Assessment Programme
WWDR World Water Development Report
WWF World Wide Fund for Nature
ZZE economic and ecological zoning

Contributors

***GEO-5* Author teams**

Chapter 1 Drivers: Susana B. Adamo, Columbia University, USA; Jane Barr, independent expert, Canada; David Laborde Debucquet, International Food Policy Research Institute, USA; Elizabeth R. Desombre, Wellesley College, USA; Thomas Dietz, Michigan State University, USA; Matthew Gluschankoff, University of California, Santa Barbara, USA; Konstadinos Goulias, University of California, Santa Barbara, USA; Jason Jabbour, UNEP, Kenya; Yeojoo Kim, Korea Environment Institute, Republic of Korea; Marc A. Levy, Center for International Earth Science Information Network, USA; David López-Carr, University of California, Santa Barbara, USA; Catherine P. McMullen, independent consultant, Canada; Alexandra C. Morel, Centre for International Earth Science Information Network, USA; Ana Rosa Moreno, National Autonomous University of Mexico, Mexico; Siwa Msangi, International Food Policy Research Institute, USA; Matthew Paterson, University of Ottawa, Canada; Batimaa Punsalmaa, Water Authority, Ministry of Nature, Environment and Tourism, Mongolia; Eugene A. Rosa, Washington State University, USA; Paul F. Steinberg, Harvey Mudd College, USA; Ray Tomalty, McGill University, Canada; Craig Townsend, Johns Hopkins University, USA.

Chapter 2 Atmosphere: May Antoniette Ajero, Clean Air Initiative-Asia Center, Philippines; Susan Casper Anenberg, US Environmental Protection Agency, USA; Paulo Artaxo, University of São Paulo, Brazil; Geir Braathen, World Meteorological Organization, Switzerland; Luis Abdon Cifuentes, Pontificia Universidad Catolica de Chile, Chile; Lisa Emberson, Stockholm Environment Institute, UK; Sara Feresu, University of Zimbabwe, Zimbabwe; Kevin Hicks, Stockholm Environment Institute, UK; Msafiri Jackson, Ardhi University, Tanzania; Johan C. I. Kuylenstierna, Stockholm Environment Institute, UK; Yousef Meslmani, Atomic Energy Commission, Syria; Nicholas Muller, Middlebury College, USA; Frank Murray, Murdoch University, Australia; Seydi Ababacar Ndiaye, Labo de Physique et de l'Atmosphère et de l'Océan, Senegal; Emily Nyaboke (GEO Fellow), Intergovernmental Authority on Development Climate Prediction and Applications Centre, Kenya; Nguyen Thi Kim Oanh, Asian Institute of Technology, Thailand; T.S. Panwar, The Energy and Resources Institute, India; Linn Persson, Stockholm Environment Institute, Sweden; Drew Shindell, NASA Goddard Institute for Space Studies, USA; Sara Terry, US Environmental Protection Agency, USA; Eric Zusman, Institute for Global Environmental Strategies, Japan.

Chapter 3 Land: Magdi T. Abdelhamid, National Research Centre, Egypt; T. Mitchell Aide, University of Puerto Rico, USA; Björn Alfthan, UNEP/GRID-Arendal, Norway; Fethi Ayache, Université de Sousse, Tunisia; Asmeret Asefaw Berhe, University of California, Merced, USA; Saturnino (Jun) M. Borras Jr., Erasmus University Rotterdam, Philippines; Chizoba Chinweze, Nnamdi Azikiwe University, Nigeria; Tahia Devisscher, Stockholm Environment Institute, UK; Tom P. Evans, Indiana University, USA; Jana Frélichová, Charles University, Prague, Czech Republic; Lawrence Hislop, UNEP/GRID-Arendal, Norway; Carol A. Hunsberger, Carleton University, Canada; Jason Jabbour, UNEP, Kenya; Shashi Kant, University of Toronto, Canada; David López-Carr, University of California, Santa Barbara, USA; Hillary Masundire, University of Botswana, Botswana; Juan Albaladejo Montoro, Centro de Edafología y Biología Aplicada del Segura, Consejo Superior de Investigaciones Científicas, Spain; William K. Pan, Johns Hopkins University, USA; Narcisa G. Pricope (GEO Fellow), University of Florida, USA; Roberto Sánchez-Rodríguez, University of California, Riverside, USA; Björn Schulte-Herbrüggen, UNEP-WCMC, UK; Jessica Smith, UNEP-WCMC, UK; Carlos Souza Jr., Amazon Institute of People and the Environment, Brazil; Tracy L. Timmins (GEO Fellow), University of Calgary, Canada; Héctor Francisco del Valle, Centro Nacional Patagónico, Argentina; Joris de Vente Centro de Edafología y Biología Aplicada del Segura, Consejo Superior de Investigaciones Científicas, Spain; Leo C. Zulu, Michigan State University, USA.

Chapter 4 Water: Maite Aldaya, Botín Foundation, Spain; Hermanni Backer, Helsinki Commission, Finland; Erica Brown Gaddis, SWCA Environmental Consultants, USA; Paul Roger Glennie, UNEP-DHI Centre for Water and Environment, Denmark; Yi Huang, Peking University, China; Hans Günter Brauch, Freie University of Berlin, Germany; Peter Koefoed Bjørnsen, UNEP-DHI Centre for Water Environment, Denmark; Salif Diop, UNEP, Kenya; Mariele Evers, Leuphana University of Lueneburg, Germany; Carlo Giupponi, University of Venice Ca' Foscari, Italy; Sherry Heileman, independent consultant, France; Gensuo Jia, Chinese Academy of Sciences, China; Ljubomir Jeftic, independent consultant, Croatia; Alioune Kane, Universite Cheikh Anta Diop de Dakar, Senegal; Tiina Kurvits, UNEP/GRID-Arendal, Norway; Robin Mahon, University of West Indies, Barbados; Walter Rast, Texas State University, USA; Santiago Reyna, National University of Cordoba, Argentina; Lisa Speer, Natural Resources Defense Council, USA; Jaap van Woerden, UNEP, Switzerland; Roy Victor Watkinson, Roy Watkinson Environmental Consulting Ltd, UK; Judith Weis, Rutgers University, USA.

Chapter 5 Biodiversity: John Agard, University of West Indies, Trinidad and Tobago; Dolors Armenteras, Universidad Nacional de Colombia, Colombia; Mario Baudoin, San Andres University, Bolivia; Kabir Bavikatte, Natural Justice, South Africa; Bastian Bertsky, UNEP-WCMC, UK; Neil Burgess, University of Copenhagen, Denmark; Stuart H.M. Butchart, Birdlife International, UK; Joji Carino, International Indigenous Forum on Biodiversity, Philippines; William W.L. Cheung, University of East Anglia, UK; Ben Collen, Zoological Society of London, UK; Nigel Dudley, Equilibrium, UK; C. Max Finlayson, Charles Sturt University, Australia; Leslie G. Firbank, University of Leeds, UK; Rodrigo Fuentes, ASEAN Centre for Biodiversity, Philippines; Alessandro Galli, Global Footprint Network, Italy; Yogesh Gokhal, The Energy and Resources Institute, India; Simon Hales, University of Otago, New Zealand; Marc Hockings, University of Queensland, Australia; Robert Höft, Secretariat of the UN Convention on Biological Diversity, Canada; J. Carter Ingram, Wildlife Conservation Society, USA; Valerie Kapos, UNEP-WCMC, UK; Justin Kitzes, University of California, Berkeley, USA; Ashish Kothari Kalpavriksh, Environment Action Group, India; Linda Krueger, Wildlife Conservation Society, USA; Melodie A. McGeoch, South Africa National Parks, South Africa; Thomasina E.E. Oldfield, Traffic International, UK; Christian Prip, Ministry of Environment, Denmark; Camilo Garcia Ramirez, National University of Colombia, Colombia; Kent H. Redford, Wildlife Conservation Society, USA; Monica Marcela Morales Rivas (GEO Fellow), Universidad Nacional de Colombia, Colombia; John G. Robinson, Wildlife Conservation Society, USA; Alison M. Rosser, UNEP-WCMC, UK; Jörn P.W. Scharlemann, UNEP-WCMC, UK; Holly Shrumm, Natural Justice, South Africa; Damon Stanwell-Smith, UNEP-WCMC, UK; Heikki Toivonen, Finnish Environment Institute, Finland; Bas Verschuuren, WCPA Specialist Group on Cultural and Spiritual Values of Protected Areas, Netherlands; Johanna von Braun, Natural Justice, South Africa; Matt Walpole, UNEP-WCMC, UK.

Chapter 6 Chemicals and Waste: Ricardo Barra, University of Concepción, EULA Environmental Sciences Centre, Chile; Borislava Batandjieva, Consultancy Services, Bulgaria; Arthur Russell Flegal Jr., University of California, Santa Cruz, USA; Walter Giger, Giger Research Consulting, Switzerland; Ivan Holoubek, Research Centre for Toxic Compounds in the Environment, Masaryk University, Czech Republic; Heather Jones-Otazo, Health Canada, Canada; Liu Lili, Basel Convention Coordinating Center for Asia and the Pacific, China; Philip Edward Metcalf, Independent Consultant, British/South African; Karina Silvia Beatriz Miglioranza, National Council of Scientific and Technological Research, Mar del Plata University, Argentina; Mónica Patricia Montory Gonzalez (GEO Fellow), University of Concepción, Chile; Adebola A. Oketola (GEO Fellow), University of Ibadan, Nigeria; Oladele Osibanjo, Basel Convention Coordinating Centre for Training and Technology Transfer for the African Region, University of Ibadan, Nigeria; Pierre Portas, Waste Environment Cooperation Centre, Switzerland; Ian Rae, University of Melbourne, Australia; Martin Scheringer, Institute for Chemical and Bioengineering, Eidgenössische Technische Hochschule Zürich, Switzerland; Claudia ten Have, UNEP, Kenya; Roy Victor Watkinson, Roy Watkinson Environmental Consulting Ltd, UK.

Chapter 7 An Earth System Perspective: Genrikh Alekseev, Arctic and Antarctic Research Institute, Russia; Opha Pauline Dube, University of Botswana, Botswana; Niki Frantzeskaki, Dutch Research Institute for Transitions, Netherlands; Benjamin Gaddis, SWCA Environmental Consultants, USA; Andrew Githeko, Medical Research Institute, Kenya; Jill Jäger, independent expert, UK; Pushker Kharecha, NASA Goddard Institute for

Space Studies, USA; Derk Loorbach, Dutch Research Institute for Transitions, Netherlands; Neeyati Patel, UNEP, Kenya; James Reynolds, Duke University, USA; Johan Rockström, Stockholm Environment Institute, Sweden; Jan Rotmans, Dutch Research Institute for Transitions, Netherlands; Vladimir Ryabinin, World Meteorological Organization, Switzerland; Jiansheng Ye (GEO Fellow), Lanzhou University, China.

Chapter 8 Data Needs: Charles Davies, UNEP, Kenya; Ashbindu Singh, UNEP, USA; Jaap van Woerden, UNEP, Switzerland.

Chapter 9 Africa: Ameer Abdulla, International Union for Conservation of Nature, Centre for Mediterranean Cooperation, Spain; Osman Mirghani M. Ali, University of Khartoum, Sudan; Adnan A. Awad, University of the Western Cape, South Africa; Habtemariam Kassa Belay, Center for International Forestry Research, Ethiopia Office, Ethiopia; Kerry W. Bowman, University of Toronto, Canada; Rannveig K. Formo, UNEP/GRID-Arendal, Norway; Marina Gomei, World Wildlife Fund, Italy; Charlotte Karibuhoye, Foundation Internationale du Banc d'Arguin, Senegal; Winnie Lau, Forest Trends, USA; Masego Madzwamuse, independent consultant, South Africa; Clever Mafuta, UNEP/GRID-Arendal, Norway; Jennifer Clare Mohamed-Katerere, independent expert, South Africa; Francis Mwaura, University of Nairobi, Kenya; Valerie Rabesahala, independent consultant, Madagascar; Sachooda Ragoonaden, Indian Ocean Commission, Mauritius; Bevlyne Sithole, Shanduko Centre for Agrarian Research, Zimbabwe.

Chapter 10 Asia and the Pacific: Iskandar Abdullaev, Deutsche Gesellschaft für Internationale Zusammenarbeit (GIZ) GmbH, Uzbekistan; Raquibul Amin, International Union for Conservation of Nature, Thailand; Yumiko Asayama, National Institute for Environmental Studies, Japan; Magnus Bengtsson, Institute for Global Environmental Strategies, Japan; Robert Dobias, USAID/Climate Change Adaptation Project Preparation Facility for Asia-Pacific, Thailand; Mark Elder, Institute for Global Environmental Strategies, Japan; Rodrigo Fuentes, ASEAN Biodiversity Centre, Philippines; Anirban Ganguly, The Energy and Resources Institute, India; Prodipto Ghosh, The Energy and Resources Institute, India; Guibin Jiang, Research Center for Eco-Environmental Sciences, Chinese Academy of Sciences, China; Mikiko Kainuma, National Institute for Environmental Studies, Japan; Yatsuka Kataoka, Institute for Global Environmental Strategies, Japan; Peter N. King, Institute for Global Environmental Studies, Thailand; Robert Kipp, Institute for Global Environmental Strategies, Japan; Marie Leroy, Institute for Political Studies, Science Po, France; Keping Ma, Institute of Botany, Chinese Academy of Sciences, China; Vishal Narain, Management Development Institute, India; Simon Hoiberg Olsen (GEO Fellow), Institute for Global Environmental Strategies, Japan; Shavkat Rakhmatullaev, Deutsche Gesellschaft für Internationale Zusammenarbeit (GIZ) GmbH, Uzbekistan; Nilapha Ratanavong (GEO Fellow), Regional Resource Centre for Asia and the Pacific, Thailand; Jianbo Shi, Research Center for Eco-Environmental Sciences, Chinese Academy of Sciences, China; Diana Suhardiman, International Water Management Institute – Southeast Asia, Indonesia; Poh Poh Wong, University of Adelaide, Australia; Shiqiu Zhang, Peking University, China.

Chapter 11 Europe: Thomas Bernauer, Swiss Federal Institute of Technology, Switzerland; Olga Chkanikova (GEO Fellow), Lund University, Sweden; Sophie Condé, National Museum of Natural History, France; Karine Danielyan, Yerevan State University, Armenia; Nicolai Dronin, Moscow State University, Russia; Lisa Emberson, Stockholm Environment Institute, UK; Joyeeta Gupta, Vrije Universiteit Amsterdam, Netherlands; Naira Harutyunyan, Central European University, Hungary; Anastasia Idrisova, Central European University, Hungary; Pavlos Kassomenos, University of Ioannina, Greece; Olena Maslyukivska, National University of Kyiv-Mohyla Academy, Ukraine; Ruben Mnatsakanian, Central European University, Hungary; Nora Mzavanadze, Central European University, Hungary; Alexander Orlov, The State University of New York, Stony Brook, USA; Mirjam Schomaker, independent consultant, Switzerland; Jerome Simpson, The Regional Environmental Center for Central and Eastern Europe, Hungary; Åsa Swartling, Stockholm Environment Institute, Sweden.

Chapter 12 Latin America and the Caribbean: Andrea Brusco, UNEP, Panama; Ligia Castro, CAF – Development Bank of Latin America, Panama; Antonio Clemente (GEO Fellow), Water Center for the Humid Tropics of Latin America and the Caribbean, Panama; Keston Finch, The Cropper Foundation, Trinidad and Tobago; Elsa Galarza, Universidad del Pacifico, Peru; Silvia Giada, UNEP, Panama; Alexander Girvan, The Cropper Foundation, Trinidad and Tobago; Mayte González, The Nature Conservancy, Panama; Keisha Garcia, The Cropper Foundation, Trinidad and Tobago; Mark Griffith, UNEP, Panama; Gladys Hernández, Centro de Investigaciones de la Economía Mundial, Cuba; Guillermo Castro Herrera, International Sustainable Development Center, Panama; Paul Hinds, College of Science, Technology and Applied Arts of Trinidad and Tobago, Trinidad and Tobago; Martha Macedo de Lima, Barata Instituto Oswaldo Cruz, Brazil; Arturo Flores Martínez, Secretaría de Medio Ambiente y Recursos Naturales – SEMARNAT, Mexico; Graciela Metternicht, UNEP, Panama; Ana Rosa Moreno, Universidad Nacional Autónoma de México, Mexico; Ernesto Guhl Nannetti, Institute for Sustainable Development – CIDES, Colombia; Keith Nichols, Organisation of the Eastern Caribbean States, St. Lucia; Rodrigo Noriega, International Sustainable Development Center – CIDES, Panama; Daniel Fontana Oberling, Federal University of Rio de Janeiro, Brazil; Martin Obermaier, Federal University of Rio de Janeiro, Brazil; Mary Otto-Chang, independent consultant, Jamaica; Aida Pacheco, Universidad del Pacifico, Peru; Maurice Rawlins (GEO Fellow), The Cropper Foundation, Trinidad and Tobago; Andrea Salinas, UNEP, Panama; Asha Singh, Caribinvest (West Indies) Limited, Guyana; Michael Taylor, University of West Indies, Jamaica; Elisa Tonda, UNEP, Panama; Angel Ureña, Panama Canal Authority, Panama; Oscar Vallarino, Panama Canal Authority, Panama; Ernesto Viglizzo, National Institute of Agricultural Technology, Argentina; Jessica Young, MarViva Foundation, Panama; William Wills, Federal University of Rio de Janeiro, Brazil; Joanna Noelia Kamiche Zegarra, Universidad del Pacifico, Peru.

Chapter 13 North America: Robert Adler, University of Utah, USA; Jane Barr, independent expert, Canada; John Campbell, US Forest Service, USA; James Dobrowolski, US Department of Agriculture, USA; José Etcheverry, York University, Toronto, Canada; Catherine Hallmich (GEO Fellow), Commission for Environmental Cooperation, Canada; Jim Lazar, The Regulatory Assistance Project, USA; Philippe Le Prestre, Université Laval, Canada; Lailai Li, Stockholm Environment Institute, Thailand; Alexander Kenny, Center for International Sustainable Development Law, Canada; Lori Lynch, University of Maryland, USA; Russell M. Meyer, Pew Center on Global Climate Change, USA; Robin Newmark, US Department of Energy, USA; Janet Peace, Pew Center on Global Climate Change, USA; Julie A. Suhr Pierce, US Department of Agriculture Natural Resources Conservation Service, USA; Marc Sydnor, University of Denver, USA; Stephen Yamasaki, EcoTerra Solutions, Canada.

Chapter 14 West Asia: Asma Abahussain, Arabian Gulf University, Bahrain; Ibrahim Abdel Gelil, Arabian Gulf University, Egypt; Mohamed Abdulrazzak, Independent Expert, Saudi Arabia; Anwar Abdu Khalil, Arabian Gulf University, Bahrain; Mohammad S. Abido, Damascus University, Syria; Fouad Abousamra, UNEP, Syria; Mukdad Al-Khateeb, Environment Research Center, Iraq; Maha Al-Sabbagh, Arabian Gulf University, Bahrain; Lulwa N Ali, Kuwait Institute for Scientific Research, Kuwait; Mahmoud Al-Sibai, Arab Center for the Studies of Arid Zones and Dry Lands, Syria; Hashim Al-Sayed, University of Bahrain, Bahrain; Abdullah Droubi, Arab Center for the Studies of Arid Zones and Dry Lands, Syria; Amr El-Sammak, Arabian Gulf University, Egypt; Ahmad Fares Asfary, Independent Expert, Syria; Nesreen Ghaddar, American University of Beirut, Lebanon; Mohamed Abdel Raouf Abdel Hamid Aly, Gulf Research Center, Egypt; Amir Ibrahim, Tishreen University, Syria; Mohammad Abdul Rahman Hassan, Dubai Municipality, UAE, Muhyiddine Jradi, American University of Beirut, Lebanon; Ahmed Khalil, Regional Organization for the Conservation of the Environment of the Red Sea and Gulf of Aden, Sudan; Abdel Hadi Mohamed, Arabian Gulf University, Sudan; Amr El-Sammak and Ahmed Ali Salih, Arabian Gulf University, Sudan.

Chapter 15 Regional Summary: Jane Barr, independent expert, Canada; Ludgarde Angèle Elisa Coppens, UNEP, Kenya; Nicolai Dronin, Moscow

State University, Russia; Amir El-Sammak, Arabian Gulf University, Bahrain; Jose Etcheverry, York University, Toronto, Canada; Lailai Li, Stockholm Environment Institute, Thailand; Clever Mafuta, UNEP/GRID-Arendal, Norway; Catherine P. McMullen, independent consultant, Canada; Renat Perelet, Institute for Systems Analysis, Russia; Flavia Rovira (GEO Fellow), Centro de Investigaciones Económicas, Uruguay; Asha Singh, Cariblnvest (West Indies) Limited, Guyana; Joanna Noelia Kamiche Zegarra, Universidad del Pacifico, Peru.

Chapter 16 Scenarios Sustainability Transformation: Pinar Ertör Akyazi (GEO Fellow), Boğaziçi University, Turkey; Rob Alkemade, PBL Netherlands Environmental Assessment Agency, Netherlands; Andrea Bassi, Millennium Institute, USA; Livia Bizikova, International Institute for Sustainable Development, Canada; Villy Christensen, University of British Columbia, Canada; Fabio Feldmann, consultant, Brazil; Martina Floerke, University of Kassel, Germany; Jill Jäger, independent expert, UK; Marcel Kok, PBL Netherlands Environmental Assessment Agency, Netherlands; Paul Lucas, PBL Netherlands Environmental Assessment Agency, Netherlands; Diane Mangalagiu, University of Oxford, UK; Washington Ochola, Regional University Forum for Capacity Building, Kenya; Begum Ozkaynak, Boğaziçi University, Turkey; Trista Patterson, US Department of Agriculture, Forest Service, USA; Natalia Pervushina (GEO Fellow), Central European University, Hungary; Laszlo Pinter, Central European University/International Institute for Sustainable Development, Hungary/Canada; Weishuang Qu, Millennium Institute, USA; Kilaparti Ramakrishna, Woods Hole Research Center, USA; Claudia Ringler, International Food Policy Research Institute, Germany; John Shilling, Millennium Institute, USA; Darren Swanson, International Institute for Sustainable Development, Canada; Detlef van Vuuren, PBL Netherlands Environmental Assessment Agency, Netherlands.

Chapter 17 Global Responses: Ibrahim Abdel Gelil, Arabian Gulf University, Bahrain; Ivar Baste, Directorate for Nature Management, Norway; Satishkumar Belliethathan, Horn of Africa – Regional Environment Centre/ Network, Ethiopia; Vivien Campal, Secretary of State for Environment and Sustainable Development, Guinea-Bissau; Bradnee Chambers, UNEP, Kenya; Melissa Goodall (GEO Fellow), Yale University, USA; Joyeeta Gupta, Vrije Universiteit Amsterdam, Netherlands; Peter M. Haas, University of Massachusetts Amherst, USA; Zerisenay Habtezion, Harvard University, USA; Achim Halpaap, UNITAR, Switzerland; Maria Ivanova, University of Massachusetts Boston, USA; Peter N. King, Institute for Global Environmental Strategies, Thailand; Marcel Kok, PBL Netherlands Environmental Assessment Agency, Netherlands; Bernice Lee, Chatham House, UK; Marcus Lee, The World Bank, USA; Slobodan Milutinovic, University of Nis, Serbia; Jennifer Clare Mohamed-Katerere, independent expert, South Africa; Trista Patterson, US Department of Agriculture, Forest Service, USA; Felix Preston (GEO-Fellow), Chatham House, UK.

Scientific Peer-Reviewers (*Coordinated by ESSP*): Keigo Akimoto, Research Institute of Innovative Technology for the Earth, Japan; Mahmoud Ali, Arab Organization for Agricultural Development, Syria; Erik Ansink, Vrije Universiteit Amsterdam, Netherlands; Masroor Ellahi Babar, University of Veterinary and Animal Sciences, Pakistan; David Barkin, Universidad Autónoma Metropolitana, Mexico; Janos Bogardi, University of Bonn, Germany; Philippe Bourdeau, Director (ret.) DG Research, European Commission; Josep Canadell, Marine and Atmospheric Research, Commonwealth Scientific and Industrial Research Organization, Australia; Graciela Ana Canziani, Universidad Nacional del Centro de la Provincia de Buenos Aires, Argentina; Andrea Birgit Chavez Michaelesen, Universidad Nacional Amazonica de Madre de Dios, Peru; Kevin Cheung, Macquarie University, Australia; Antonio Cruzado, Oceans Catalonia International SL, Spain; Shobhakar Dhakal, National Institute for Environmental Studies, Japan; Serigne Faye, Universite Cheikh Anta Diop de Dakar, Senegal; Marina Fischer-Kowalski, Alpen Adria Universitaet, Austria; Amadou Thierno Gaye, Universite Cheikh Anta Diop de Dakar, Senegal; Mark O. Gessner, Berlin Institute of Technology, Germany; Evgeny Gordov, Institute of Monitoring of Climatic and Ecological Systems, Russia; Dagmar Haase, Helmholtz Centre for Environmental Research, Germany; Itsuki Handoh, Research Institute for Humanity and Nature, Japan; Nick Harvey, University of Adelaide, Australia; Lars Hein, Wageningen University, Netherlands; Gerhard J. Herndl, Royal Netherlands Institute for Sea Research, Netherlands; Shu-Li Huang, National Taipei University, Taiwan Province of China; Falk Huettmann, University of Alaska-Fairbanks, USA; Ada Ignaciuk, Earth System Sciences Partnership, France; Muhammad Mohsin Iqbal, Global Change Impact Studies Centre, Pakistan; Louise Jackson, University of California, Davis, USA; Sharad Jain, Indian Institute of Technology Roorkee, India; Ian Jenkinson, Agency for Consultation and Research in Oceanography, France; Rainer Krug, Stellenbosch University, South Africa; Nelson Lourenco, IGBP-International Geosphere Biosphere Programme/Global Change, Portugal; Angela M. Maharaj, Macquarie University, Australia; Miyuki Nagashima, Research Institute of Innovative Technology for the Earth, Japan; Daiju Narita, Kiel Institute for the World Economy, Germany; Isabelle Niang, University of Dakar, Senegal; Patrick Nunn, University of New England, Australia; Jay O'Keeffe, Rhodes University, South Africa; Jean-Pierre Ometto, Brazilian National Institute for Space Research, Brazil; Ursula Oswald Spring, National University of Mexico, Mexico; Claudia Pahl-Wostl, Institute for Environmental Systems Research, Germany; Nirmalie Pallewatta, University of Colombo, Sri Lanka; Henrique M. Pereira, University of Lisbon, Portugal; Erika Pires Ramos, Brazilian Institute for the Environment and Renewable Natural Resources, Brazil; Germán Poveda, Universidad Nacional de Colombia, Colombia; Francesc Prenafeta, Institute of Agrifood Research and Technology, Spain; Seema Purushothaman, Centre for Conservation Governance and Policy, ATREE, India; Dork Sahagian, Lehigh University, USA; Galia Selaya, Madre de Dios-Pando Consortium, Bolivia; Mika Sillanpaa, Lappeenranta University of Technology, Finland; Maria Siwek, University of Technology and Life Sciences, Poland; Erika Techera, University of Western Australia, Australia; Holm Tiessen, Inter-American Institute for Global Change Research, Brazil; Klement Tockner, Institute of Freshwater Ecology and Inland Fisheries, Germany; Aysun Uyar, Research Institute for Humanity and Nature, Japan; Emma Archer van Garderen, Council for Scientific and Industrial Research, South Africa; Tracy Van Holt, East Carolina University, USA; Stefano Vignudelli, National Research Council, Italy; Hassan Virji, International START Secretariat, USA; Angela Wagener, Pontifical Catholic University of Rio de Janeiro, Brazil; Hong Yang, Swiss Federal Institute for Aquatic Science and Technology, Switzerland.

High-Level Intergovernmental Advisory Panel: Rajender Ahlawat, Ministry of Environment and Forests, India; Hussein A. Al-Gunied, Ministry of Water and Environment, Yemen; Wahid Al-Shuely, Ministry of Environment and Climate Change, Oman; Liana Bratasida, Ministry of Environment, Indonesia; Burcu Bursali, Ministry of Environment and Forestry, Turkey; Sandra De Carlo, Ministry of Environment, Brazil; Mantang Cai, Peking University, China; Jorge Laguna Celis, Ministry of Foreign Affairs, Mexico; Guilherme da Costa, Secretariat of State for Environment and Sustainable Development, Guinea Bissau; Raouf Dabbas, Ministry of Environment, Jordan; Martijn Dadema, Ministry of Foreign Affairs, Netherlands; Idunn Eidheim, Ministry of Environment, Norway; Prudence Galega, Ministry of Environment and Protection of Nature, Cameroon; Nilkanth Ghosh, Ministry of Environment and Forests, India; Rosario Gomez, Ministry of Environment, Peru; Xia Guang, Ministry of Environmental Protection, China; Han Huiskamp, Ministry of Foreign Affairs, Netherlands; Jos Lubbers, Ministry of Foreign Affairs, Netherlands; John Michael Matuszak, US Department of State, USA; Samira Nateche, Ministry of Land and Planning, Environment and Tourism, Algeria; Kim Thi Thuy Ngoc, Ministry of Natural Resources and Environment, Vietnam; Jose Rafael Almonte Perdomo, Ministry of Environment and Natural Resources, Dominican Republic; Majid Shafie-Pour-Motlagh, Department of Environment, Iran; Van Tai Nguyen, Institute of Strategy and Policy on Natural Resources and Environment, Vietnam; Jiang Wei, Ministry of Environmental Protection, China; Albert Williams, Department of the Environment, Vanuatu; Daniel Ziegerer, Federal Office of Environment, Switzerland.

Science and Policy Advisory Board: Asma Ali Abahussain, Arabian Gulf University, Bahrain; Pinhas Alpert, Tel Aviv University, Israel; Torkil Jonch Clausen, UNEP-DHI Centre for Water and Environment, Denmark; Ahmed Djoghlaf, Secretariat of the UN Convention on Biological Diversity, Canada; Susanne Dröge, German Institute for International and Security Affairs, Germany; Kejun Jiang, Energy Research Institute, China; Nicholas King, Global Biodiversity Information Facility, Denmark; Filipo Lansigan, University of Los Banos, Philippines; Anne Larigauderie, DIVERSITAS, France; Emilio Lèbre La Rovere, Laboratório Interdisciplinar de Meio Ambiente/COPPE/UFRJ, Brazil; Jacqueline McGlade, European Environment Agency, Denmark; Luisa T. Molina, Massachusetts Institute of Technology, USA; Toral Patel-Weynand, US Department of Agriculture Forest Service, USA; Nicolas Perritaz, Federal Office for the Environment, Switzerland; Carlos A. Quesada, University of San Jose,

Costa Rica; Chirapol Sintunawa, Mahidol University, Thailand; Sandra Torrusio, National Commission of Space Activities, Argentina; George Varughese, Development Alternatives Group, India; Robert Watson, Department for Environmental, Food and Rural Affairs, UK.

Data and Indicators Working Group: Asma Ali Abahussain, Arabian Gulf University, Bahrain; Ezgi Akpinar-Ferrand, University of Cincinnati, Turkey; Barbara Clark, European Environment Agency, Denmark; Sandra de Carlo, Ministry of Environment, Brazil; Volodymyr Demkine, UNEP, Kenya; Alexander Gorobets, Sevastopol National Technical University, Ukraine; Eszter Horvath, United Nations Statistics Division, USA; Koffi Kouadio, Ministry of Environment, Water and Forest, Cote d'Ivoire; Murari Lal, University of the South Pacific, Fiji; Samwiri Musisi-Nkambwe, University of Botswana, Botswana; Ambinistoa Lucie Noasilalaonomenjanahary, Ministry of Environment and Forest, Madagascar; Toral Patel-Weynand, US Forest Service, USA; Muhammad Munir Sheikh, Global Change Impact Studies Center, Pakistan; Ashbindu Singh, UNEP, USA; Anil Kumar Thanappan, Environmental Agency – Abu Dhabi, UAE; Susan Tumwebaze, Makerere University, Uganda; Héctor Tuy, University Raphael Landivar, Guatemala; Jaap van Woerden, UNEP, Switzerland.

Extended UNEP Team: Henry Aguilar, Mozaharul Alam, Jacqueline Alder, Jacqueline Alvarez, Meryem C. Amar, Neville Ash, Margarita Astrálaga, Mario Boccucci, Vivienne Caballero, Christopher Corbin, Mara Angélica Murillo Correa, Artie Dubrie, Heidelore Fiedler, Alex Forbes, Amy Fraenkel, Sandor Frigyik, Joanna Granados, Julie Greenwalt, Moustapha Kamal Gueye, Niklas Hagelberg, Jonathan Gilman, Silja Halle, Ampai Harakunarak, Arab Hoballah, Melanie Hutchinson, David Jensen, Bob Kakuyo, Khaled Klaly, Alexander Koch, Fanina R. Kodre-Alexander, Nicolas Kosoy, Angela Lusigi, Janet Macharia, Kaj Madsen, Katarina Magulova, Isabel Martínez, Patricia Miranda, David H.W. Morgan, Richard Munang, Masa Nagai, Theodore Oben, Young-Woo Park, Wahida Patwa-Shah, Alex Pires, Ravi Prabhu, Purna Rajbhandari, Jean Jacob Sahou, Andrea Salinas, John Scanlon, Yasmin Shehata, Gemma Shepherd, Guido Sonnemann, Tunnie Srisakulchairak, Angele Lu Sy, Claudia ten Have, Dechen Tshering, Stephen Twomlow, Carla Valle-Klann, James Vener, Kamar Yousuf, Massimiliano Zandomeneghi, Max Zieren.

Other United Nations Bodies: Russel Arthuton, IOC of UNESCO; Magaran Bagayoko, WHO; Juan Carlos Belausteguigoitia, The World Bank; Ruhiza Jean Boroto, FAO; Christopher Briggs, UNDP; Seon-Mi Choi, UNDP; Henrik Oksfeldt Enevoldsen, UNESCO; José Escamilla, PAHO; José Javier Gómez, UN ECLAC; Jacob Gyamfi-Aidoo, UNDP; Peter Holmgren, FAO; Mahendra Joshi, UNFF; Mikhail Kokine, UN ECE; Lars Gunnar Marklund, FAO; Johnson Nkem, UNDP; Emilio Pinto, PAHO; Hitomi Rankine, UN ESCAP; Mukundan Pillay, WHO; Paul Steele, UNDP; Terrence Thompson, WHO; María Noel Vaeza, UNOPS; Walter Vergara, The World Bank; Margarita Zambrano, UNHCR.

External Reviewers: Asma Ali Abahussain, Arabian Gulf University, Bahrain; Mohammad Abido, Arabian Gulf University, Bahrain; Mariam Akhtar-Schuster, Desertnet International, Germany; Stephanie Aktipis, US Department of State, USA; Dhari Al-Ajmi, Kuwait Institute for Scientific Research, Kuwait; Jean Albergel, Institut de Recherche pour le Développement, France; Mukdad Al-Khateeb, University of Technology, Iraq; Habiba Al Marashi, Emirates Environmental Group, UAE; Sergio Alvarez, Ministry of Environment Rural and Marine Affairs, Spain; Li An, San Diego State University, USA; Matheus Marques Andreozzi, Ministry of Environment, Brazil; Michelle Andriamahazo, Ministry of Agriculture, Madagascar; Fabio Franca Silva Araujo, Ministry of Environment, Brazil; Fethi Ayache, Université de Sousse, Tunisia; Julio Cesar Baena, Ministry of Environment, Brazil; Robert Bakiika, Environmental Management for Livelihood Improvement Bwaise Facility, Uganda; Jan Bakkes, PBL Netherlands Environmental Assessment Agency, Netherlands; Zoltan Balint, FAO, Hungary; Martha Macedo de Lima Barata, Instituto Oswaldo Cruz, Brazil; Alisson Barbieri, Universidade Federal de Minas Gerais, Brazil; Garfield Barnwell, Caribbean Community Secretariat, Guyana; Stephen Bates, Department of Sustainability Environment Water Population and Communities, Australia; Adriana Panhol Bayma, Ministry of Environment, Brazil; Douglas Beard, US Geological Survey, USA; Asmeret Asefaw Berhe, University of California Merced, USA; Martial Bernoux, Institut de Recherche pour le Développement, France; Alka Bharat, Maulana Azad National Institute of Technology, India; Janos Bogardi, University of Bonn, Germany; Hans-Georg Bohle, University of Bonn, Germany; Marcel Bovy, Sustainability Guidance, Netherlands; Andreas Brink, Joint Research Center – European Commission, Italy; Carmen Burghelea, University of Vigo, Romania; Nadia Bystriakova, Natural History Museum, UK; Jillian Campbell, United Nations Secretariat, USA; Rita Cerutti, Environment Canada, Canada; Antony Challenger, Ministry of Environment and Natural Resources, Mexico; Hung Chak Ho, Mississippi State University, USA; Ge Chazhong, Chinese Academy for Environmental Planning, China; Marion Cheatle, independent expert, UK; Mbow Cheikh, Université Cheikh Anta Diop de Dakar, Senegal; Weixue Cheng, Chinese Academy for Environmental Planning, China; Norma Cherry-Fevrier, Ministry of Finance, Economic Affairs and National Development, Saint Lucia; Barthod Christian, Ministry of Ecology, Sustainable Development, Transport and Housing, France; Adriano Ciani, Perugia University, Italy; Barbara Clark, European Environment Agency, Denmark; Petru Cocirta, Institute of Ecology and Geography of the Academy of Sciences, Moldova; Ana Corado, US Environmental Protection Agency, USA; Sérgio Ferreira Cortizo, Ministry of Environment, Brazil; Sylvie Côté, Environment Canada, Canada; Sandra De Carlo, Ministry of Environment, Brazil; Nathalie Delrue, OECD, France; Xiangzheng Deng, Institute of Geographic Sciences and Natural Resources Research, Chinese Academy of Sciences, China; Alvaro Aguilar Díaz, Ministerio de Ambiente, Energía y Telecomunicaciones, Costa Rica; Kelly Rain Dodge, US Department of State, USA; Ida Edwertz, Ministry of Environment, Sweden; Kassem El-Saddik, Développement Sans Frontieres, Lebanon; Karlheinz Erb, Institute of Social Ecology, USA; Keston Finch, The Cropper Foundation, Trinidad and Tobago; Jeff Fox, US Department of State, USA; Teodoro Georgiadis, Institute of Biometeorology of the National Research Council, Italy; Matthew Gerdin, US Department of State, USA; Anju Ghoorah, Ministry of Environment and Sustainable Development, Mauritius; Alexander Girvan, The Cropper Foundation, Trinidad and Tobago; Kees Klein Goldewijk, PBL Netherlands Environmental Assessment Agency, Netherlands; Liza Grandia, Clark University, USA; Xia Guang, Ministry of Environmental Protection, China; Andres Guhl, Universidad de los Andes, Colombia; Rodrigo Afonso Guimarães, Ministry of Environment, Brazil; Slayde Hawkins, Forest Trends, USA; Hans-Joachim Hermann, Federal Environmental Agency, Germany; Jeff Herrick, US Department of Agriculture, USA; Vicki Hird, World Society for the Protection of Animals, UK; Yi Huang, Peking University, China; Lloyd C. Irland, University of Maine, USA; Klaus Jacob, Freie Universität Berlin, Germany; Ljubomir Jeftic, independent consultant, Croatia; Gensuo Jia, Chinese Academy of Sciences, China; Li Jinhui, Basel Convention Coordinating Centre for Asia and the Pacific, China; Liu Jinyuan, Institute of Geographic Sciences and Natural Resources Research, Chinese Academy of Sciences, China; Daniel Jones, Department of Environment, Food and Rural Affairs, UK; Heather Jones-Otazo, Health Canada, Canada; Muhyiddine Jradi, American University of Beirut, Lebanon; Wilfred Kadewa, University of Malawi, Malawi; Douglas Karlen, US Department of Agriculture, USA; Jiang Kejun, Energy Research Institute, China; Martin Kijazi, University of Toronto, Canada; Nicolas King, Global Biodiversity Information Facility, Denmark; Barbara Knox-Seith, US Agency for International Development, USA; Noriko Kobayashi, Ministry of Foreign Affairs, Japan; Murari Lal, University of the South Pacific, Fiji; Greg Liknes, US Department of Agriculture, USA; Ronald Macfarlane, Toronto Public Health, Canada; Mazen Malkawi, WHO, Jordan; Cai Mantang, Peking University, China; Ney Maranhão, Ministry of Environment, Brazil; Saskia Marijnissen, UNDP/GEF Project on Partnership Interventions for the Implementation of the Strategic Action Programme for Lake Tanganyika, Burundi; Bernado Marke, Ministry of Foreign Affairs, Brazil; Mike McGahuey, US Agency for International Development, USA; Elizabeth McLanahan, National Oceanic and Atmospheric Administration, USA; Carlos Mena, Universidad San Francisco de Quito, Ecuador; Alexander Metcalf, US Environmental Protection Agency, USA; Frank Müller, Asia Pacific Roundtable for Sustainable Consumption and Production, Thailand; Michele Muniz, Ministry of Environment, Brazil; John K. Musingi, University of Nairobi, Kenya; Mark Nelson, US Department of Agriculture Forest Service, USA; Keith E. Nichols, Organisation of Eastern Caribbean States, Saint Lucia; Itzchel Nieto, Ministry of Environment and Natural Resources, Mexico; Taina Nikula, Ministry of the Environment, Finland; Theophile Niyonzima, National University of Rwanda, Rwanda; Ambinintsoa Lucie Noasilalaonomenjanahary, Ministry of Environment and Forest, Madagascar; Patrick Nussbaumer, United Nations Industrial Development Organization, Austria; Htwe Nyo, National Commission for Environmental Affairs, Myanmar; Alice Oluoko-Odingo, University of Nairobi, Kenya; Andréa Oncala, Ministry of Environment, Brazil; Konrad Otto-Zimmermann, ICLEI – Local Governments for Sustainability, Germany; Dawn Parker, University of Waterloo, Canada; Toral Patel-Weynand, US Department of Agriculture Forest Service, USA; Netatua Pelesikoti, Secretariat of the Pacific Regional Environment Programme, Samoa; Maria Pena, University of the West Indies, Barbados; Monica Peres, Ministry of Environment, Brazil; Nicolas Perritaz, Federal Office for the Environment, Switzerland; Rebecca L. Powell, University of Denver, USA; Narcisa G. Pricope, University of Florida, USA; Kaushalya Ramachandran, Central Research Institute for Dryland Agriculture, Indian Council of Agricultural Research, India; Maurice Rawlins, The Cropper Foundation, Trinidad and Tobago; Richard Roseman, US Department of State, USA; Kurt Riitters, US Department of Agriculture Forest Service, USA; Taeho Ro, Korea Environment Institute, Republic of Korea; John Romankiewicz, US Department of State, USA; Dale Rothman, University of Denver, USA; Najib Saab, Arab Forum for Environment and Development, Lebanon; Nurhuda Binti Salam, Department of Environment, Malaysia; Neil Sampson, Vision Forestry, LLC, USA; Henriette Schweizerhof, Federal Ministry for the Environment, Nature Conservation and Nuclear Safety, Germany; Richard Sigman, OECD, France; Juliana Simões, Ministry of Environment, Brazil; Benjamin Sleeter, US Geological Survey, USA; Stephan Slingerland, PBL Netherlands Environmental Assessment Agency, Netherlands; William Sonntag, US Environmental Protection Agency, USA; Anand Sookun, Central Statistics Office, Mauritius; Mary Beth Steisslinger, Global Commons Trust, USA; Karen Regina Suassuna, Ministry of Environment, Brazil; Danling Tang, South China Sea Institute of Oceanology, Chinese Academy of Sciences, China; Tracy Timmins, University of Calgary, Canada; Mary Andy Rowen Tobiason, US Agency for International Development, USA; Bella Tonkonogy, Department of Treasury, USA; Darin Tooney, US Department of State, USA; Jerry Touval, The Nature Conservancy, USA; Nathalie Unterstell, Ministry of Environment, Brazil; Niko Urho, Ministry of the Environment, Finland; Ingrid Verstraeten, US Geological Survey, USA; Anne Wein, US Geological Survey, USA; Judith S. Weis, Rutgers University, USA; Mona M. Westergaard, Environmental Protection Agency, Denmark; Dano Wilusz, US Department of State, USA; Maria Witmer, PBL Netherlands Environmental Assessment Agency, Netherlands; Kerstin Wortman, Federal Ministry for the Environment, Nature Conservation and Nuclear Safety, Germany; Lesley Woudberg, Ministry for the Environment, New Zealand; H. E. Mohammadi Zadeh, Department of Environment, Iran; Jieqing Zhang, Ministry of Environmental Protection, China; Daniel Ziegerer, Federal Office for the Environment, Switzerland.

Individuals and institutions – from governments, partner institutions, the scientific community and the private sector – who contributed to the *GEO-5* assessment process in a variety of ways, including *GEO-5* regional and intergovernmental consultations:

Africa: Ahmed Abdelrehim, Centre for Environment and Development for the Arab Region and Europe, Egypt; Ali Adan Ali, National Museums of Kenya, Kenya; Jonathan Addo Allotey, Environmental Protection Agency, Ghana; Marie-Laetitia Busokeye, Rwanda Environment Management Authority, Rwanda; Lizete Marina Firmino, Minister of Environment, Angola; Osman Mirghani Mohammed Ali, University of Khartoum, Sudan; Ayman Tharwat Amin, Ministry of Foreign Affairs, Egypt; Daniel S. Amlalo, Environmental Protection Agency, Ghana; Michelle Andriamahazo, Ministry of Agriculture, Madagascar; Samuel Ndonwi Ayonghe, University of Buea, Cameroon; Adnan A. Awad, University of the Western Cape, South Africa; Robert Bakiika, Environmental Management for Livelihood Improvement Bwaise Facility, Uganda; Ndey Sireng Bakurin, National Environment Agency, The Gambia; Philip O. Bankole, Federal Ministry of Environment, Nigeria; Mohammed El Bouch, Ministry of Water and Environment, Morocco; Viriatú Cassamá, Secretariat of State for Environment and Sustainable Development, Guinea-Bissau; Oliver Chapeyama, independent consultant, Botswana; Mbow Cheikh, Université Cheikh Anta Diop de Dakar, Senegal; Thandiwe Chikomo, Birdlife International, Kenya; Tabeth Chiuta, World Fish Center, Zambia; Famara Drammeh, Daily Observer News Paper, The Gambia; Scopas Jibi Dima, Ministry of Environment, South Sudan; Mathieu Ducrocq, International Union for Conservation of Nature, Mauritania; Nadia Makram Ebeid, Center for Environment and Development for the Arab Region and Europe, Egypt; Abu Bakr Elsiddig Ahmed Eltohami, Omdurman Ahlia University, Sudan; Thiyu Kohoga Essobiyou, Ministry of Environment and Forest Resources, Togo; Serigne Faye, Universite Cheikh Anta Diop de Dakar, Senegal; Michael Vosa Flyman, Department of Environmental Affairs, Botswana; Cheikh Fofana, Secrétariat Intérimaire du Volet Environnement du NEPAD, Senegal; Louis Gachimbi, National Environment Management Authority, Kenya; Tesfaye Woldeyes Gammo, Ethiopia; Brad Garanganga, SADC Drought Monitoring Centre, Zimbabwe; Jean Paul Gaudechoux, Indian Ocean Commission, Mauritius; Noha Ekram Abdel Gawad, Centre for Environment and Development for the Arab Region and Europe, Egypt; Amadou Thierno Gaye, Universite Cheikh Anta Diop de Dakar, Senegal; Mercy Wamukore Gichora, Kenya Forestry Research Institute, Kenya; John Githaiga, University of Nairobi, Kenya; Sives Govender, Network for the Co-operative Management of Environmental Information in Africa (EIS Africa), South Africa; Youssouf Hamadi, Ministry of Production, Fisheries, Environment, Industry, Energy and Handicraft, Comoros; Mamoudou Hamadou, Ministère de l'Hydraulique et de l'Environnement, Niger; Mohamed Salem Hamouda, Environment General Authority, Libya; Pascal Houenou, Network for Environment and Sustainable Development in Africa, Cote d'Ivoire; Issa Ibro, Ministère de l'Environnement et de la Lutte Contre la Désertification, Niger; I. A. Jaiyeoba, Ahmadu Bello University, Nigeria; Remi Jiagho, Union Internationale pour la Conservation de la Nature, Cameroon; Marie Rose Kabura, Ministry of Water, Environment, Land Use and Urban Planning, Burundi; Wilfred Kadewa, University of Malawi, Malawi; Adjakouma Kakou, Radio des Nations Unies, Cote d'Ivoire; Timothy Kaluma, Ministry of Foreign Affairs, Kenya; Mona Mohamed Kamal, Egyptian Environmental Affairs Agency, Egypt; Macharia Kamau, Ministry of Foreign Affairs, Kenya; Alioune Kane, Universite Cheikh Anta Diop de Dakar, Senegal; Samuel Kanyamibwa, independent consultant, Rwanda; Lydia Karanja, National Environment Management Authority, Kenya; Charlotte Karibuhoye, Foundation Internationale du Banc d'Arguin, Senegal; Habtemariam Kassa, Center for International Forestry Research – Ethiopia Office, Ethiopia; Norah M. Kendeli, Ministry of Foreign Affairs, Kenya; Ahmed Khalil, Regional Organization for the Conservation of the Environment of the Red Sea and Gulf of Aden, Sudan; Mamadou Khouma, International Development Consulting, Senegal; John Kiringe, University of Nairobi, Kenya; Boniface Kiteme, Centre for Training and Research in ASAL Development, Kenya; Yao Bernard Koffi, Ministry of Environment, Water and Forest, Cote d'Ivoire; Kassim Kulindwa, Norwegian University of Life Sciences/University of Dar es Salaam, Tanzania; Christian Padingani Kunkadi, Ministry of Environment, Nature Conservation and Tourism, Democratic Republic of Congo; Winnie Lau, Forest Trends, USA; Robert Lewis Lettington, legal consultant, Kenya; Everlyn Macharia, Ministry of Foreign Affairs, Kenya; Lapologang Magole, University of Botswana, Botswana; Amadou Maiga, Ministère de l'Environnement et de l'Assainissement, Mali; Willy R. Makundi, independent consultant, Tanzania; Joel Celestin Mamboundou, Croissance Saine Environnement, Gabon; Anna Mampye, Department of Environmental Affairs and Tourism, South Africa; Paul Stephen Maro, University of Dar es Salaam, Tanzania; Isabelle Masinde, African Wildlife Foundation, Kenya; Klaus Mithoefer, African Insect Science for Food and Health, Kenya; Nosiku S. Munyinda, University of Zambia, Zambia; Telly Eugene Muramira, National Environment Management Authority, Uganda; John K. Musingi, University of Nairobi, Kenya; Mukundi Mutasa, Topline Research Solutions, Zimbabwe; Nyawira Muthiga, Wildlife Conservation Society, Kenya; Francis Mwaura, University of Nairobi, Kenya; Richard Mwendandu, Ministry of Environment and Mineral Resources, Kenya; David Melchisédéck Yangbondo, Central African Republic; Jacques Andre Ndione, Centre de Suivi Ecologique, Senegal; Parkinson Ndonye, Ministry of Environment and Mineral Resources, Kenya; Alleta R. Nenguke, Environment Management Agency, Zimbabwe; Tcharbuahbokengo Nfinn, Federation of Environmental and Ecological Diversity for Agricultural Revampment and Human Rights, Cameroon; Erasmo Roberto Nhachungue, Ministry of Environmental Affairs, Mozambique; Musisi Nkambwe, University of Botswana, Botswana; Isabelle Niang, Universite Cheikh Anta Diop de Dakar, Senegal; Betty Nzioka, National Environment Management Authority, Kenya; Jorge Rafael Jora Obiamo, Ministry of Fisheries and the Environment, Equatorial Guinea; David Obura, Coastal Oceans Research and Development in the Indian Ocean, Kenya; Ochieng Ogodo, Science and Development Network, Kenya; Jay O'Keeffe, Rhodes University, South Africa; Olukayode Oladipo, Bells University of Technology, Nigeria; Alice Oluoko-Odingo, University of Nairobi, Kenya; David Ongare, National Environment Management Authority, Kenya; Alfred Opere, University of Nairobi, Kenya; George Olago Owuor, Ministry of Foreign Affairs, Kenya; Chedly Rais, Regional Activity Centre for Specially Protected Areas, Tunisia; Belinda Reyers, Council for Scientific and Industrial Research, South Africa; John L. Roberts, independent consultant, Mauritius; Houssein Rirache Roble, Direction de l'Environnement du Territoire et de l'Énvironnement, Djibouti; Mayar Sabet, Centre for Environment and Development for the Arab Region and Europe, Egypt; Jefter Sakupwanya, ORGUT Consulting AB, Mozambique; Camille Flore Jepang Sandjong, Programme Regional Eau et Zones Humides, Cameroon;

Gerald Musoke Sawula, National Environment Management Authority, Uganda; Ashraf Nour Shalaby, League of Arab States, Egypt; Constantine Shayo, Tanzania; Cletus Ignace Shengena, Vice-President's Office, Tanzania; Gift Sikaundi, Environment Council of Zambia, Zambia; Nouri Soussi, Ministry of Environment and Sustainable Development, Tunisia; Sokhna Sy Diallo, Direction de l'Environnement et des Etablissements Classés, Senegal; Egline Tawuya, Southern African Research and Documentation Centre/Musokotwane Environment Resource Centre for Southern Africa, Zimbabwe; Ben Wandago, International Union for Conservation of Nature Eastern Africa Regional Office, Kenya; Baraza Wangwe, National Environment Management Authority, Kenya; Harun Warui, Kenya Agricultural Research Institute, Kenya.

Asia and the Pacific: Joseph Aitaro, Ministry of Natural Resources, Environment and Tourism, Palau; Chamina Priyankari Alexander, South Asia Cooperative Environment Programme, Sri Lanka; Cholpon Alibakieva, State Agency on Environment Protection and Forestry, Kyrgyz Republic; Chonchinee Amawatana, Asian Development Bank, Thailand; Saikia Anshuman, International Union for Conservation of Nature Asia Regional Office, Thailand; Kamil Ashimov, State Agency on Environment Protection and Forestry, Kyrgyzstan; Uddhav Prasad Baskota, Ministry of Environment, Nepal; Henry Bastaman, Ministry of Environment, Indonesia; Mirza Salman Babar Beg, Ministry of Foreign Affairs, Pakistan; Mantang Cai, Peking University, China; Akihiro Chiashi, Ritsumeikan University, Japan; Kanchan Chopra, University of Delhi Enclave, India; Munir Chowdhury, Ministry of Environment and Forests, Bangladesh; Yoo Yeon Chul, Ministry of Environment, Republic of Korea; Nicholas T. Dammen, Ministry of Foreign Affairs, Indonesia; Ashish Despande, Maulana Azud National Institute of Technology, India; Laksmi Dhewanthi, Ministry of Environment, Indonesia; Chazhong Ge, Chinese Academy of Environment Planning, China; Manuel D. Gerochi, Department of Environment and Natural Resources, Philippines; Abbas Golriz, Department of International Economic Affairs and Specialized Agencies, Iran; Xia Guang, Ministry of Environmental Protection, China; Toshiaki Ichinose, National Institute for Environmental Studies, Japan; Dahe Jiang, Tongji University, China; Galiya Karibzhanova, Ministry of Environment Protection, Kazakhstan; Soudavee Keopaseuth, Water Resources and Environment Administration, Lao PDR; Cheol Hee Kim, Pusan National University, Republic of Korea; Masanori Kobayashi, Institute for Global Environmental Strategies, Japan; Peter Kouwenhoven, CLIMsystems, New Zealand; D. Johny Kusumo, Ministry of Environment, Indonesia; Murari Lal, University of the South Pacific, Fiji; Kosimiki Latu, Secretariat of the Pacific Regional Environment Programme, Samoa; Byoung Yoon Lee, National Institute of Biological Resources, Republic of Korea; Dong Li, Tongji University, China; Daniela Liggett, University of Canterbury, New Zealand; Demetrio Jr. Luciano, Department of Environment and Natural Resources, Philippines; Nguyen Hung Minh, Vietnam Environment Administration, Vietnam; Arabindra Mishra, The Energy and Resources Institute, India; Khieu Muth, Ministry of Environment, Cambodia; Seul-ki Myoung, UNEP National Committee, Republic of Korea; Hasnun Nahar, Ministry of Environment and Forests, Bangladesh; Somrudee Nicrowattanayingyong, Thailand Environment Institute, Thailand; Nuradi Noeri, Ministry of Foreign Affairs, Indonesia; Rahul Pandey, Integrated General Systems Analysis Labs, India; Majid Shafie-Pour-Motlagh, Department of Environment, Iran; Meera Pandit Pattni, Ministry of Foreign Affairs, Thailand; Batimaa Punsalmaa, Ministry of Nature, Environment and Tourism, Mongolia; Atiq Rahman, Bangladesh Centre for Advanced Studies, Bangladesh; Bakhodir Rakhmanov, State Committee for the Nature Protection, Uzbekistan; Neelam Rana, Development Alternatives Group, India; Kim Sanghoon, Ministry of Environment, Republic of Korea; Vivek Saxena, Ministry of Environment and Forests, India; Heinz Schandl, Commonwealth Scientific and Industrial Research Organization, Australia; M. I. Sharif, Bangladesh Centre for Advanced Studies, Bangladesh; Keshav Prasad Sharma, Ministry of Environment, Nepal; Leena Srivastava, The Energy and Resources Institute, India; Anond Snidvongs, Southeast Asia START Regional Center, Thailand; Laska Sophal, Ministry of Environment, Cambodia; Nguyen Van Tai, Institute of Strategy and Policy on Natural Resources and Environment, Vietnam; Kiyoshi Takahashi, National Institute for Environmental Studies, Japan; Eiji Tanaka, Ministry of Foreign Affairs, Japan; Abhimuk Tantihabhakul, Office of National Resources and Environmental Policy and Planning, Thailand; Tshering Tashi, National Environment Commission, Bhutan; Nenenteiti Teariki-Ruatu, Ministry of Environment, Lands and Agriculture Development, Kiribati; Maung Maung Than, Ministry of Environment Conservation and Forestry, Myanmar; Somsak Triamjangarun, Ministry of Foreign Affairs, Thailand; Karma Tshering, National Environment Commission, Bhutan; Hoang Duong Tung, Ministry of Natural Resources and Environment, Vietnam; Ahmed Ashan Uddin, Center for Global Change, Bangladesh; Jinnan Wang, Chinese Academy for Environmental Planning, China; Supat Wangwongwatana, Ministry of Natural Resources and Environment, Thailand; Yohpy Ichsan Wardana, Ministry of Foreign Affairs, Indonesia; Albert Abel Williams, Department of Environment and Conservation, Vanuatu; Huang Yi, Peking University, China; Hai Yu, Ministry of Environmental Protection, China; Zhang Yutian, Ministry of Foreign Affairs, China; Tshewang Zangmo, National Environment Commission, Bhutan.

Europe: Rashad Allahverdiyev, Ministry of Ecology and Natural Resources, Azerbaijan; Ros Almond, World Wildlife Fund IUCN/SSC Sustainable Use Specialist Group, UK; Valentine Altmater, Ministry of Foreign Affairs, France; Markus Amann, International Institute for Applied Systems Analysis, Austria; Erik Ansik, Vrije Universiteit Amsterdam, Netherlands; John Barrett, University of York, UK; Heike Baumueller, Chatham House, UK; Volodymyr Bilokon, Ministry of Ecology and Natural Resources, Ukraine; Bastian Bomhard, UNEP-WCMC, UK; Ninni Maud Christina Lundblad Borén, Swedish Environmental Protection Agency, Sweden; Daniela Breidler, Ministry of Foreign Affairs, Austria; Andreas Michael Burger, Federal Environmental Agency, Germany; Olga Butko, Ministry of Ecology and Natural Resources, Ukraine; Francisco Cadarso, Ministry of Agriculture, Food and Environment, Spain; Sophie Condé, National Museum of Natural History, France; William Darwall, International Union for Conservation of Nature, UK; Nicolas Dasnois, Ministry of Foreign Affairs, France; David Dent, CABI Bioscience, UK; Erdoğan Erturk, Ministry of Forestry and Water Affairs, Turkey; Joan Fabres, UNEP/GRID-Arendal, Norway; Jon Geddes, Ministry of Foreign Affairs, UK; Luminita Guminita Ghita, Ministry of Environment and Forests, Romania; Nino Gokhelashvili, Ministry of Environment Protection, Georgia; Richard Gregory, Royal Society for the Protection of Birds, UK; Charles Hieronymi, Federal Office for the Environment, Switzerland; Peter Hooda, Kingston University London, UK; Monika Kaczyńska, Ministry of the Environment, Poland; Larisa Kharatova, Ministry of Nature Protection, Armenia; Richard Klein, Stockholm Environment Institute, Sweden; Natalija Koprivanac, University of Zagreb, Croatia; Hratch Kouyoumjian, Regional Science, Technology and Innovation Observatory, UK; Fred Langeweg, PBL Netherlands Environmental Assessment Agency, Netherlands; Evgeny Lazarev, Ministry of Foreign Affairs, Belarus; Melissa Leach, Institute for Development Studies, UK; Rik Leemans, Earth Systems Science Partnership, Netherlands; Roger Levett, Levett-Therivel, UK; Georgina Mace, Imperial College London, UK; Tural Mammadov, Ministry of Ecology and Natural Resources, Azerbaijan; Tom Manders, PBL Netherlands Environmental Assessment Agency, Netherlands; Peter P. Mollinga, University of London, UK; Davut Oguz, Ministry of Forest and Water Affairs, Turkey; Nebojsa D. Redzic, Environmental Protection Agency, Serbia; Lisa Schipper, Stockholm Environment Institute, Sweden; Rima Mekdaschi Studeer, University of Bern, Switzerland; George Dragos Zaharescu, Vigo University, Spain; Dalia Maier, Ministry of Environment and Forests, Romania; Juliet Migwi, Ministry of Foreign Affairs, United Kingdom; Marketa Mohn, Ministry of the Environment, Czech Republic; Markus Ohndorf, Institut für Umweltentscheidungen (ETH Zürich), Switzerland; Véronique Plocq-Fichelet, Scientific Committee on Problems of the Environment, France; John Laing Roberts, independent expert, UK; Renate Schubert, Swiss Federal Institute of Technology, Switzerland; David Stanners, European Environment Agency, Denmark; Wendelin Stark, Swiss Federal Institute of Technology, Switzerland; Thomas Stratenwerth, Federal Ministry for the Environment, Nature Conservation and Nuclear Safety, Germany; Petra Tacheci, Ministry of the Environment, Czech Republic; Victoria Thoresen, Partnership for Education and Research about Responsible Living, Norway; Jurjen van der Vlugt, Ministry of Foreign Affairs, Netherlands.

Latin America and the Caribbean: Gherda Barreto, Ministerio del Ambiente y Recursos Naturales, Nicaragua; Marcela Bonilla, Ministry of the Environment, Colombia; Ralph Carnegie, University of West Indies, Barbados; Mónica Castillo, Comisión Centroamericana de Ambiente y Desarrollo, El Salvador; Candy Degracia, Asociación Panamá Verde, Panama; Randolph Antonio Edmead, Ministry of Sustainable Development, Saint Kitts and Nevis; Edgar Ek, Department of the Environment, Belize; Kenneth Fearon, Panama; Jose Feres, Institute of Applied Economic Research, Brazil; Argelia Estela Fernández, Agencia de Medio Ambiente, Cuba; Edwin Giovanni Tobar Guzman, Ministerio de Ambiente y Recursos Naturales, Guatemala; Arica Marianne Hill, Ministry of Agriculture, Land,

Housing, and the Environment, Antigua and Barbuda; Kenrick Leslie, Caribbean Community Climate Change Centre, Belize; Patricia Maccagno, Secretaría de Medio Ambiente y Desarrollo Sustentable, Argentina; Mirella Martinez, Florida State University, USA; Diana Martucci, Ministerio del Ambiente, Ecuador; Anthony McKenzie, National Environment and Planning Agency, Jamaica; Marcelo Núñez, Ministry of the Environment, Ecuador; Gabriel Rodríguez Márquez, Instituto Interamericano de Cooperación para la Agricultura, Costa Rica; César E. Rodríguez Ortega, Secretaría de Medio Ambiente y Recursos Naturales – SEMARNAT, Mexico; Luis Javier Campuzano Pina, Ministry of Foreign Affairs, Mexico; Jose-Manuel Sandoval, Ministry of Environment and Sustainable Development, Colombia; Sealy Sean, Ministry of Environment, Water Resources and Drainage, Barbados; Rodrigo Tarté, Fundación Ciudad del Saber, Panama; Vaitoti Tupa, National Environment Service, Cook Islands; Malena Sarlo, Fundacion Mar Viva, Panama; Jessica Young, Fundación Mar Viva, Panama.

North America: Karen Bakker, University of British Columbia, Canada; Scott Barclay, National Science Foundation, USA; M. Bruce Beck, University of Georgia, USA; Luc Bouthilier, Université Laval, Canada; Paula Brand, Environment Canada, Canada; Edward Carr, University of South Carolina, USA; Richard Connor, Unisféra International Centre, Canada; Tooney Darin, US Department of State, USA; Ligia Castro de Doens, Land Eco Services, USA; Stewart Elgie, University of Ottawa, Canada; James Galloway, Nitrogen Initiative, USA; Kathryn Harrison, University of British Columbia, Canada; David Houle, University of Toronto, Canada; Giorgios Kallis, University of California, Berkeley, USA; Douglas Macdonald, University of Toronto, Canada; Ronald Macfarlane, Toronto Public Health, Canada; Jerry Melillo, The Ecosystems Center, USA; Jean Mercier, Université Laval, Canada; Tim Morris, Walter and Duncan Gordon Foundation, USA; Adil Najam, Boston University, USA; Daniel Pauly, University of British Columbia, Canada; Jim Perry, University of Minnesota, USA; Rebecca L. Powell, University of Denver, USA; Carmen Revenga, The Nature Conservancy, USA; Andrew Rosenberg, Conservation International, USA; Roberto Sanchez-Rodriguez, University of California, Riverside, USA; Beverly Sithole, Management Consulting, USA; John D. Shilling, Millennium Institute, USA; Sarah Ryker, Science and Technology Policy Institute, USA; Liana Talaue-Mcmanus, University of Miami, USA; Tim Weis, The Pembina Institute, Canada.

West Asia: Hesham Abd-El Rasol, Arabian Gulf University, Bahrain; Yousef Attallah Ibrahim Abu-Safieh, Palestinian Environment Quality Authority, Palestine; Mohammad Mosa Afaneh, Ministry of Environment, Jordan; Saif Saad Abdaljabbar Al-Aany, Ministry of Foreign Affairs, Iraq; Ahmed Hammodi Hamdi Al-Husseini, Ministry of Environment, Iraq; Lulwa N. Ali, Kuwait Institute for Scientific Research, Kuwait; Sabah Aljenaid, College of Graduate Studies, Bahrain; Khalil Ismail Abdulsahib Al-Mosawi, Ministry of Foreign Affairs, Iraq; Bara Al-Nakeeb, Ministry of Environment, Iraq; Maha Al-Sabbagh, Arabian Gulf University, Bahrain; Hashim Al-Sayed, University of Bahrain, Bahrain; Waleed Al-Zubari, Arabian Gulf University, Bahrain; Yahia Awaidah, Consultants for Sustainable Development, Syria; Mohammad Badran, Regional Organization for the Conservation of the Environment of the Red Sea and Gulf of Aden, Saudi Arabia; Abdullah Droubi, The Arab Center for the Studies of Arid Zones, Syria; Alaa El-Sadek, Arabian Gulf University, Bahrain; Anwar Abdu Khalil, Arabian Gulf University, Bahrain; Mohammed Alaa Abdel Moati, Ministry of Environment, Qatar; Abdel Hadi Mohamed, Arabian Gulf University, Bahrain; Riad Sadek, American University of Beirut, Lebanon; Mohammed Saidam, Environment Monitoring and Research Central Unit, Jordan; Ahmed Salih, Arabian Gulf University, Bahrain; Walid Rajab Shahin, National Energy Research Center, Jordan; Batir Wardam, Ministry of Environment, Jordan.

In-kind support for GEO Fellows: American University of Beirut, Lebanon; Antioch University, USA; Regional Resource Centre for Asia and the Pacific, Thailand; Bogazici University, Turkey; University of Calgary, Canada; Central European University, Hungary; Concepción University, Chile; UNEP/GRID-Arendal, Norway; Institute for Global Environmental Strategies, Japan; IGAD Climate Prediction and Application Centre, Kenya; Lanzhou University, China; Lund University, Sweden; McGill University, Canada; Peking University, China; Red Mercosur, Uruguay; Chatham House, UK; The Cropper Foundation, Trinidad and Tobago; Universidad Nacional de Colombia; University of Florida, USA; Water Center for the Humid Tropics of Latin America and the Caribbean, Panama.

Contributing institutions and organizations: Abu Dhabi Global Environmental Data Initiative (AGEDI); Arabian Gulf University, Bahrain; Alexandria University, Egypt; American University of Beirut, Lebanon; ASEAN Centre for Biodiversity, Philippines; Central European University, Hungary; Centre de Suivi Ecologique, Senegal; Centre for Environment and Development for the Arab Region and Europe, Egypt; Centre for International Earth Science Information Network, Columbia University, USA; Centro de Investigaciones de la Economía Mundial, Cuba; College of Science, Technology and Applied Arts of Trinidad and Tobago, Trinidad and Tobago; Columbia University, USA; Commission for Environmental Cooperation, Canada; Deutsche Gesellschaft für Internationale Zusammenarbeit (GIZ) GmbH, Uzbekistan; DIVERSITAS – International Programme of Biodiversity Science, France; Earth System Science Partnership, France; European Environment Agency, Denmark; Federal University of Rio de Janeiro Interdisciplinary Environment Laboratory, Brazil; Higher Institute for Water Management, Syria; ICLEI – Local Governments for Sustainability, Germany; Indiana University, USA; Indian Ocean Commission, Mauritius; Institut für Umweltentscheidungen (ETH Zürich), Switzerland; Institute for Global Environmental Strategies, Japan; Institute Oswaldo Cruz, Brazil; Institute for Sustainable Development, Colombia; International Institute for Sustainable Development, Canada; International Sustainable Development Center, Panama; International Union for Conservation of Nature, Cameroon and Thailand; International Water Management Institute – Southeast Asia, Lao PDR; Kuwait Institute for Scientific Research, Kuwait; Moscow State University, Russia; National Environment Management Authority, Uganda; National Institute for Environmental Studies, Japan; National Institute of Agricultural Technology, Argentina; Network for Environment and Sustainable Development in Africa, Côte d'Ivoire; PBL Netherlands Environmental Assessment Agency, Netherlands; Regional Organization for the Conservation of the Environment of the Red Sea and Gulf of Aden, Kingdom of Saudi Arabia; Research Center for Eco-Environmental Science, Chinese Academy of Sciences, China; Secretaria de Medio Ambiente y Recursos Naturales, Mexico; Secretariat of the UN Convention on Biological Diversity, Canada; Secretariat of the UN Convention to Combat Desertification in Countries Experiencing Serious Drought and/or Desertification, Particularly in Africa, Germany; Secretariat of the UN-REDD Programme, Switzerland; Southern African Research and Documentation Centre/Musokotwane Environment Resource Centre for Southern Africa, Zimbabwe; Stockholm Environment Institute, Sweden, Thailand and United Kingdom; The Arab Center for the Studies of Arid Zones, Syria; The Cropper Foundation, Trinidad and Tobago; The Energy and Resources Institute, India; Tishreen University, Syria; Universidad del Pacífico, Peru; Universidad Nacional Autónoma de México, Mexico; University of Bahrain, Bahrain; University of Technology, Iran; University of Toronto, Faculty of Forestry, Canada; World Resources Institute, USA.

Note: The affiliations of experts are provided for identification purposes only. The name of the country usually refers to the location of the institution the expert is affiliated with.

Glossary

This glossary is compiled from citations in different chapters, and draws from glossaries and other resources available on the websites of the following organizations, networks and projects:

American Meteorological Society; Asian Development Bank; Center for Transportation Excellence (United States); Charles Darwin University (Australia); Consultative Group on International Agricultural Research; Convention on Wetlands of International Importance especially as Waterfowl Habitat (Ramsar); Edwards Aquifer Website (United States); Encyclopedia of Earth; Europe's Information Society; European Commission Environment A to Z; European Environment Agency; European Nuclear Society; Food and Agriculture Organization of the United Nations; Foundation for Research; Science and Technology (New Zealand); Global Earth Observation System of Systems; Global Footprint Network; GreenFacts Glossary; Illinois Clean Coal Institute (United States); Intergovernmental Panel on Climate Change; International Centre for Research in Agroforestry; International Comparison Program; International Federation of Organic Agriculture Movements; International Research Institute for Climate and Society at Columbia University (United States); International Strategy for Disaster Reduction; Lyme Disease Foundation (United States); Millennium Ecosystem Assessment; Ministerial Conference on the Protection of Forests in Europe; National Safety Council (United States); Natsource (United States); Organisation for Economic Co-operation and Development; Professional Development for Livelihoods (United Kingdom); Redefining Progress (United States); SafariX eTextbooks Online; TheFreeDictionary.com; United Nations Convention to Combat Desertification in Countries Experiencing Serious Drought and/or Desertification, Particularly in Africa; United Nations Development Programme; United Nations Educational, Scientific and Cultural Organization; United Nations Framework Convention on Climate Change; United Nations Industrial Development Organization; United Nations Statistics Division; US Department of Agriculture; US Department of the Interior; US Department of Transportation; US Energy Information Administration; US Environmental Protection Agency; US Geological Survey; USLegal.com; Water Footprint Network, the Netherlands; Water Quality Association (United States); Wikipedia; World Bank; World Health Organization and World Intellectual Property Organization.

Abundance
The number of individuals or related measure of quantity (such as biomass) in a population, community or spatial unit.

Abrupt change
The change that takes place so rapidly and unexpectedly that human or natural systems have difficulty adapting to it.

Acidification
Change in natural chemical balance caused by an increase in the concentration of acidic elements.

Acidity
A measure of how acid a solution may be. A solution with a pH of less than 7.0 is considered acidic.

Adaptation
Adjustment in natural or human systems to a new or changing environment, including anticipatory and reactive adaptation, private and public adaptation, and autonomous and planned adaptation.

Adaptive capacity
The potential or ability of a system, region or community to adapt to the effects or impacts of a particular set of changes. Enhancement of adaptive capacity is a practical means of coping with changes and uncertainties, reducing vulnerabilities and promoting sustainable development.

Adaptive governance
A governance approach that incorporates methods of adaptive management, adaptive policy making and transition management for addressing complex, uncertain and dynamic issues. Adaptive governance relies on polycentric institutional arrangements for decision making at multiple scales. Spanning the local and global levels, this form of governance provides for collaborative, flexible, learning-based approaches to ecosystem management.

Adaptive environmental governance
The process of design and execution of policy based on contemporary understanding and ongoing analysis of evolving environmental problems. Spanning the local and global levels, it relies on polycentric institutional arrangements for decision-making at multiple scales and provides for evidence-based, consultative and participatory ecosystem management able to evolve along with the problems it aims to address.

Adaptive management
A systematic management paradigm that assumes natural resource management policies and actions are not static, but are adjusted based on the combination of new scientific and socio-economic information.

Aerosols
A collection of airborne solid or liquid particles, with a typical size between 0.01 and 10 micrometres (μm), that reside in the atmosphere for at least several hours. Aerosols may be of either natural or anthropogenic origin.

Afforestation
Establishment of forest plantations on land that is not classified as forest.

Alien species (also non-native, non-indigenous, foreign, exotic)
Species accidentally or deliberately introduced outside its normal distribution.

Anthropocene
A term used by scientists to name a new geologic epoch (following the most recent Holocene) characterized by significant changes in the Earth's atmosphere, biosphere and hydrosphere due primarily to human activities.

Anthroposphere
The total human presence throughout the Earth System including its culture, technology, built environment, and activities associated with these. The anthroposphere complements the term Anthropocene.

Aquatic ecosystem
Basic ecological unit composed of living and non-living elements interacting in water.

Arable land
Land under temporary crops (double-cropped areas are counted only once), temporary meadows for mowing or pasture, land under market and kitchen gardens, and land temporarily fallow (less than five years). The abandoned land resulting from shifting cultivation is not included in this category.

Billion
10^9 (1 000 000 000)

Bioaccumulation
The increase in concentration of a chemical in organisms. Also used to describe the progressive increase in the amount of a chemical in an organism resulting from rates of absorption of a substance in excess of its metabolism and excretion.

Biocapacity
The capacity of ecosystems to produce useful biological materials and to absorb waste materials generated by humans, using current management schemes and extraction technologies. The biocapacity of an area is calculated by multiplying the actual physical area by the yield factor and the appropriate equivalence factor. Biocapacity is usually expressed in units of global hectares.

Biodiversity (a contraction of biological diversity)
The variety of life on Earth, including diversity at the genetic level, among species and among ecosystems and habitats. It includes diversity in abundance, distribution and behaviour. Biodiversity also incorporates human cultural diversity, which can both be affected by the same drivers as biodiversity, and itself has impacts on the diversity of genes, other species and ecosystems.

Biofuel
Fuel produced from dry organic matter or combustible oils from plants, such as alcohol from fermented sugar or maize, and oils derived from oil palm, rapeseed or soybeans.

Biogas
Gas, rich in methane, which is produced by the fermentation of animal dung, human sewage or crop residues in an airtight container.

Biogeochemical cycles
The flow of chemical elements and compounds between living organisms (biosphere) and the physical environment (atmosphere, hydrosphere, lithosphere).

Biological corridor
A section of habitat designated to restore or conserve the connection between habitats that have been fragmented by human or natural causes.

Biomass
Organic material, above and below ground and in water, both living and dead, such as trees, crops, grasses, tree litter and roots.

Biomagnification
The build up of certain substances in the bodies of organisms at higher trophic levels of food webs. Organisms at lower trophic levels accumulate small amounts. Organisms at the next higher level of the food chain eat many of these lower-level organisms and hence accumulate larger amounts. The tissue concentration increases at each trophic level in the food web when there is efficient uptake and slow elimination

Biome
The largest unit of ecosystem classification that is convenient to recognize below the global level. Terrestrial biomes are typically based on dominant vegetation structure (such as forest or grassland). Ecosystems within a biome function in a broadly similar way, although they may have very different species composition. For example, all forests share certain properties regarding nutrient cycling, disturbance and biomass that are different from the properties of grasslands.

Biosphere
The part of the Earth and its atmosphere in which living organisms exist or that is capable of supporting life.

Biotechnology (modern)
The application of *in vitro* nucleic acid techniques, including recombinant deoxyribonucleic acid (DNA) and direct injection of nucleic acid into cells or organelles, or fusion of cells beyond the taxonomic family, that overcome natural physiological, reproductive or recombination barriers and that are not techniques used in traditional breeding and selection.

Black carbon
Operationally defined aerosol based on measurement of light absorption and chemical reactivity and/or thermal stability. Black carbon is formed through the incomplete combustion of fossil fuels, biofuel and biomass, and is emitted as part of anthropogenic and naturally occurring soot. It consists of pure carbon in several linked forms. Black carbon warms the Earth by absorbing sunlight and re-emitting heat to the atmosphere and by reducing albedo (the ability to reflect sunlight) when deposited on snow and ice.

Blue water
Fresh surface and groundwater, in other words, the water in freshwater lakes, rivers and aquifers. The blue water footprint

is the volume of surface and groundwater consumed as a result of the production of a good or service. Blue water consumption refers to the volume of freshwater used and then evaporated or incorporated into a product. It also includes water abstracted from surface or groundwater in a catchment and returned to another catchment or the sea. It is the amount of water abstracted from groundwater or surface water that does not return to the catchment from which it was withdrawn.

Bleaching (of coral reefs)
A phenomenon occurring when corals under stress expel their mutualistic microscopic algae, called zooxanthellae. This results in a severe decrease or even total loss of photosynthetic pigments. Since most reef-building corals have white calcium carbonate skeletons, these then show through the corals' tissue and the coral reef appears bleached.

Capacity development
The process through which individuals, organizations and societies obtain, strengthen and maintain the capabilities to set and achieve their own development objectives over time.

Cap and trade (system)
A regulatory or management system that sets a target level for emissions or natural resource use, and, after distributing shares in that quota, lets trading in those permits determine their price.

Capital
Resource that can be mobilized in the pursuit of an individual's goals. Thus, natural capital (natural resources such as land and water), physical capital (technology and artefacts), social capital (social relationships, networks and ties), financial capital (money in a bank, loans and credit), human capital (education and skills).

Carbon dioxide equivalent (CO_2-equivalent or CO_2e)
The universal unit of measurement used to indicate the global warming potential of the different greenhouse gases. Carbon dioxide – a naturally occurring gas that is a by-product of burning fossil fuels and biomass, land-use changes and other industrial processes – is the reference against which other greenhouse gases are measured.

Carbon sequestration
The process of increasing the carbon content of a reservoir other than the atmosphere.

Carbon stock
The quantity of carbon contained in a "pool", meaning a reservoir or system which has the capacity to accumulate or release carbon.

Catchment (area)
The area of land from which precipitation drains into a river, basin or reservoir. See also *Drainage basin*.

Certified emission reductions (CERs)
Certification of greenhouse gas emission reductions issued pursuant to the Clean Development Mechanism of the Kyoto Protocol, and measured in tonnes of carbon dioxide equivalent.

Civil society
The aggregate of non-governmental organizations and institutions representing the interests and will of citizens.

Clean Development Mechanism (CDM)
The mechanism provided by Article 12 of the Kyoto Protocol, designed to assist developing countries achieve sustainable development by permitting industrialized countries to finance projects for reducing greenhouse gas emissions in developing countries and receive carbon credits for doing so.

Climate change
The UN Framework Convention on Climate Change defines climate change as "a change of climate which is attributed directly or indirectly to human activity that alters the composition of the global atmosphere and which is in addition to natural climate variability observed over comparable time periods."

Climate proofing
A shorthand term for identifying risks to a development project, or any other specified natural or human asset, as a consequence of climate variability and change, and ensuring that those risks are reduced to acceptable levels through long-lasting and environmentally sound, economically viable, and socially acceptable changes implemented at one or more of the following stages in the project cycle: planning, design, construction, operation, and decommissioning.

Climate variability
Variations in the mean state and other statistics (such as standard deviations and the occurrence of extremes) of the climate on all temporal and spatial scales beyond that of individual weather events. Variability may be due to natural internal processes in the climate system (internal variability), or to variations in natural or anthropogenic external forcing (external variability).

Chlorofluorocarbons (CFCs)
A group of chemicals, consisting of chlorine, fluorine and carbon, highly volatile and of low toxicity, widely used in the past as refrigerants, solvents, propellants and foaming agents. Chlorofluorocarbons have both ozone depletion and global warming potential.

Congener
A term in chemistry that refers to one of many variants or configurations of a common chemical structure.

Conservation tillage
Breaking the soil surface without turning over the soil.

Consumptive water use
The use or removal of water from a water basin that renders it unavailable for further use.

Cross-cutting issue
An issue that cannot be adequately understood or explained without reference to the interactions of several of its dimensions that are usually defined separately.

Crowd-sourcing
A problem-solving and production process that involves outsourcing tasks to a network of people, also known as the crowd. This process can occur both online and offline.

Cultural services
In the context of ecosystems, the non-material benefits for people, including spiritual enrichment, cognitive development, recreation and aesthetic experience.

Datum
A single piece of information used for reference or analysis.

Dataset
A collection of data on a particular issue.

DDT (dichlorodiphenyltrichloroethane)
A synthetic organochlorine insecticide, one of the persistent organic pollutants listed for control under the Stockholm Convention on Persistent Organic Pollutants.

Dead zone
A part of a water body so low in oxygen that normal life cannot survive. The low-oxygen conditions usually result from eutrophication caused by fertilizer run-off from land.

Deforestation
Conversion of forested land to non-forest areas.

Desertification
Land degradation in arid, semi-arid and dry sub-humid areas resulting from various factors, including climatic variations and human activities. It involves crossing thresholds beyond which the underpinning ecosystem cannot restore itself, but requires ever-greater external resources for recovery.

Disability-adjusted life years (DALYS)
The sum of years of potential life lost due to premature mortality and the years of productive life lost due to disability.

Disaster risk reduction
The conceptual framework of elements intended to minimize vulnerability to disasters throughout a society, to avoid (prevention) or limit (mitigation and preparedness) the adverse impacts of hazards, within the broad context of sustainable development.

Drainage basin (also called watershed, river basin or catchment)
Land area where precipitation runs off into streams, rivers, lakes and reservoirs. It is a land feature that can be identified by tracing a line along the highest elevations between different areas, often a ridge.

Driver
The overarching socio-economic forces that exert pressures on the state of the environment.

Drylands
Areas characterized by lack of water, which constrain two major, linked ecosystem services: primary production and nutrient cycling. Four dryland sub-types are widely recognized: dry sub-humid, semi-arid, arid and hyper-arid, showing an increasing level of aridity or moisture deficit.

Early warning
The provision of timely and effective information, through identified institutions, that allows individuals exposed to a hazard to take action to avoid or reduce their risk and prepare an effective response.

Earth System
The Earth System is a complex social-environmental system of interacting physical, chemical, biological and social components and processes that determine the state and evolution of the planet and life on it.

Ecoagriculture
An approach to landscape management that simultaneously advances agricultural production, conservation of biodiversity and ecosystem services, and sustainable rural livelihoods.

Ecological footprint
A measure of the area of biologically productive land and water an individual, population or activity uses to produce all the resources it consumes and to absorb the corresponding waste (such as carbon dioxide emissions from fossil fuel use), using prevailing technology and resource management practices. The ecological footprint is usually measured in global hectares.

Ecosystem
A dynamic complex of plant, animal and micro-organism communities and their non-living environment, interacting as a functional unit.

Ecosystem approach
A strategy for the integrated management of land, water and living resources that promotes conservation and sustainable use in an equitable way. An ecosystem approach is based on the application of appropriate scientific methods, focused on levels of biological organization that encompass the essential structure, processes, functions and interactions among and between organisms and their environment. It recognizes that humans, with their cultural diversity, are an integral component of many ecosystems.

Ecosystem function
An intrinsic ecosystem characteristic related to the set of conditions and processes whereby an ecosystem maintains its integrity (such as primary productivity, food chain and biogeochemical cycles). Ecosystem functions include such processes as decomposition, production, nutrient cycling, and movements of nutrients and energy.

Ecosystem health
The degree to which ecological factors and their interactions are reasonably complete and function for continued resilience, productivity and renewal of the ecosystem.

Ecosystem management
An approach to maintaining or restoring the composition, structure, function and delivery of services of natural and modified ecosystems for the goal of achieving sustainability. It is based on an adaptive, collaboratively developed vision of desired future conditions that integrates ecological, socio-economic, and institutional perspectives, applied within a geographic framework, and defined primarily by natural ecological boundaries.

Ecosystem resilience
The level of disturbance that an ecosystem can withstand without crossing a threshold to become a different structure or deliver different outputs. Resilience depends on ecological dynamics as well as human organizational and institutional capacity to understand, manage and respond to these dynamics.

Ecosystem services
The benefits of ecosystems. These include provisioning services, such as food and water regulating services, such as flood and disease control cultural services, such as spiritual, recreational and cultural benefits and supporting services, such as nutrient cycling, that maintain the conditions for life on Earth. Sometimes called ecosystem goods and services.

Ecotourism
Travel undertaken to witness the natural or ecological quality of particular sites or regions, including the provision of eco-friendly services to facilitate such travel.

Effluent
In issues of water quality, refers to liquid waste (treated or untreated) discharged to the environment from sources such as industrial process and sewage treatment plants.

El Niño (also El Niño-Southern Oscillation (ENSO))
In its original sense, it is a warm water current that periodically flows along the coast of Ecuador and Peru, disrupting the local fishery. This oceanic event is associated with a fluctuation of the inter-tropical surface pressure pattern and circulation in the Indian and Pacific Oceans, called the Southern Oscillation. This atmosphere-ocean phenomenon is collectively known as El Niño-Southern Oscillation, or ENSO. During an El Niño event, the prevailing trade winds weaken and the equatorial countercurrent strengthens, causing warm surface waters in the Indonesian area to flow eastward to overlie the cold waters of the Peru current off South America. This event has great impact on the wind, sea surface temperature and precipitation patterns in the tropical Pacific. It has climatic effects throughout the Pacific region and in many other parts of the world. The opposite of an El Niño event is called La Niña.

Emission inventory
Details the amounts and types of pollutants released into the environment.

Endangered species
A species is endangered when the best available evidence indicates that it meets any of the criteria A to E specified for the endangered category of the IUCN Red List, and is therefore considered to be facing a very high risk of extinction in the wild.

Endocrine disruptor
An external substance that interferes (through mimicking, blocking, inhibiting or stimulating) with function(s) of the hormonal system and consequently causes adverse health effects in an intact organism, or its progeny, or (sub) populations.

Energy intensity
Ratio of energy consumption to economic or physical output. At the national level, energy intensity is the ratio of total domestic primary energy consumption or final energy consumption to gross domestic product or physical output. Lower energy intensity shows greater efficiency in energy use.

Environmental education
The process of recognizing values and clarifying concepts in order to develop skills and attitudes necessary to understand and appreciate the interrelatedness of humans, their culture and biophysical surroundings. Environmental education also entails practice in decision-making and self-formulation of a code of behaviour about issues concerning environmental quality.

Environment statistics
Statistics that describe the state of and trends in the environment, covering the media of the natural environment (air/climate, water, land/soil), the living organisms within the media, and human settlements.

Environmental assessment
The entire process of undertaking an objective evaluation and analysis of information designed to support environmental decision making. It applies the judgement of experts to existing knowledge to provide scientifically credible answers to policy -relevant questions, quantifying where possible the level of confidence. It reduces complexity but adds value by summarizing, synthesizing and building scenarios, and identifies consensus by sorting out what is known and widely accepted from what is not known or not agreed. It sensitizes the scientific community to policy needs and the policy community to the scientific basis for action.

Environmental flows
Quantity, timing and quality of water flows required to sustain freshwater and estuarine ecosystems and the human livelihoods and well-being that depend on these ecosystems. Through implementation of environmental flows, water managers strive to achieve a flow regime, or pattern, that provides for human uses and maintains the essential processes required to support healthy river ecosystems.

Environmental impact assessment (EIA)
An analytical process or procedure that systematically examines the possible environmental consequences of a given activity or project. The aim is to ensure that the environmental implications are taken into account before the decisions are made.

Environmental health
Those aspects of human health and disease that are determined by factors in the environment. It also refers to the theory and practice of assessing and controlling factors in the environment that can potentially affect health. Environmental health includes both the direct pathological effects of chemicals, radiation and some biological agents, and the effects, often indirect, on health and well-being of the broad physical, psychological, social and aesthetic environment. This includes housing, urban development, land use and transport.

Environmental monitoring
Regular, comparable measurements or time series of data on the environment.

Environmental policy
A policy aimed at addressing environmental problems and challenges.

Equity
Fairness of rights, distribution and access. Depending on context, this can refer to access to resources, services or power.

Eutrophication
The degradation of water or land quality due to enrichment by nutrients, primarily nitrogen and phosphorous, which results in excessive plant (principally algae) growth and decay. Eutrophication of a lake normally contributes to its slow evolution into a bog or marsh and ultimately to dry land. Eutrophication may be accelerated by human activities that speed up the ageing process.

Evapotranspiration
Combined loss of water by evaporation from the soil or surface water, and transpiration from plants and animals.

E-waste (electronic waste)
A generic term encompassing various forms of electrical and electronic equipment that has ceased to be considered of value and is disposed of.

External cost (also externality)
A cost that is not included in the market price of the goods and services produced. In other words, a cost not borne by those who create it, such as the cost of cleaning up contamination caused by discharge of pollution into the environment.

Feedback
Where non-linear change is driven by reactions that either dampen change (negative feedbacks) or reinforce change (positive feedbacks).

Floods (river, flash and storm surge)
Usually classified into three types: river flood, flash flood and storm surge. River floods result from intense and/or persistent rain over large areas. Flash floods are mostly local events resulting from intense rainfall over a small area in a short period of time. Storm surge floods occur when floodwater from the ocean or large lakes is pushed on to land by winds or storms.

Food security
Physical and economic access to food that meets people's dietary needs as well as their food preferences.

Forest
Land spanning more than 0.5 hectares with trees higher than 5 metres and a canopy cover of more than 10 per cent, or trees able to reach these thresholds *in situ*. It does not include land that is predominantly under agricultural or urban use.

Forest degradation
Changes within the forest that negatively affect the structure or function of the stand or site, and thereby lower the capacity to supply products and/or services.

Forest management
The processes of planning and implementing practices for the stewardship and use of forests and other wooded land aimed at achieving specific environmental, economic, social and/or cultural objectives.

Forest plantation
Forest stands established by planting and/or seeding in the process of afforestation or reforestation. They are either of introduced species (all planted stands), or intensively managed stands of indigenous species, which meet all the following criteria: contain one or two species, are of similar age and regularly spaced. "Planted forest" is another term used for plantation.

Fossil fuel
Coal, natural gas and petroleum products (such as oil) formed from the decayed bodies of animals and plants that died millions of years ago.

Fuel cell
A device that converts the energy of a chemical reaction directly into electrical energy. It produces electricity from external supplies of fuel (such as hydrogen) and oxidant (such as oxygen). A fuel cell can operate as long as the necessary flows are maintained. Fuel cells differ from batteries in that they consume reactant, which must be replenished, while batteries store electrical energy chemically in a closed system. One great advantage of fuel cells is that they generate electricity with very little pollution: much of the hydrogen and oxygen used in generating electricity ultimately combine to form water. Fuel cells are being developed as power sources for motor vehicles, as well as stationary power sources.

Genetic diversity
The variety of genes within a particular species, variety or breed.

GEO Data Portal (now Environmental Data Explorer)
The source for datasets used by UNEP and its partners in the Global Environment Outlook report and other integrated environmental assessments. Its online database holds more than 500 different variables, including national, sub-regional, regional and global statistics as well as geospatial datasets (maps), covering themes such as freshwater, population, forests, emissions, climate, disasters, health and GDP. geodata.grid.unep.ch

Global commons
Natural un-owned assets such as the atmosphere, oceans, outer space and the Antarctic.

Global Earth Observation System of Systems (GEOSS)
A network aiming to link existing and planned Earth observing systems (e.g., satellites and networks of weather stations and ocean buoys) around the world, support the development of new systems where gaps currently exist, and promote common technical standards so that data from the thousands of different instruments can be combined into coherent datasets. It aims to provide decision support tools to policy makers and other users in areas such as health, agriculture and disasters.

Global hectare
A hypothetical hectare with world-average ability to produce resources and absorb wastes.

Global (international) environmental governance
The assemblage of laws and institutions that regulate society-nature interactions and shape environmental outcomes.

Global observation system
A set of coordinated monitoring activities that would collect much needed data at a global scale on a variety of indicators such as biodiversity, water quality and quantity, atmospheric pollution, land degradation and chemical release.

Global public good
Public goods that have universal benefits, covering multiple groups of countries and all populations.

Global warming
Increase in surface air temperature, referred to as the global temperature, induced by emissions of greenhouse gases into the air.

Globalization
The increasing integration of economies and societies around the world, particularly through trade and financial flows, and the transfer of culture and technology.

Governance
The act, process, or power of governing for the organization of society/ies. For example, there is governance through the state, the market, or through civil society groups and local organizations. Governance is exercised through institutions: laws, property-rights systems and forms of social organization.

Greenhouse effect
A process by which thermal radiation from a planetary surface is absorbed by atmospheric greenhouse gases, and is re-radiated in all directions. Since part of this re-radiation is back towards the surface and the lower atmosphere, it results in an elevation of the average surface temperature above what it would be in the absence of the gases.

Greenhouse gases (GHGs)
Gaseous constituents of the atmosphere, both natural and anthropogenic, that absorb and emit thermal radiation. This property causes the greenhouse effect. Water vapour (H_2O), carbon dioxide (CO_2), nitrous oxide (N_2O), methane (CH_4) and ozone (O_3) are the primary greenhouse gases in the Earth's atmosphere. There are human-made greenhouse gases in the atmosphere, such as halocarbons and other chlorine- and bromine-containing substances. Beside CO_2, N_2O and CH_4, the Kyoto Protocol deals with sulphur hexafluoride (SF_6), hydrofluorocarbons (HFCs), perfluorocarbons (PFCs) and nitrogen trifluoride (NF_3).

Green water
The precipitation on land that does not run off or recharge the groundwater but is stored in the soil or temporarily stays on top of the soil or vegetation. Eventually, this part of precipitation evaporates or transpires through plants. The green water footprint is the volume of rainwater consumed during the production process. This is particularly relevant for agricultural and forestry products (products based on crops or wood), where it refers to the total rainwater evapotranspiration (from fields and plantations) plus the water incorporated into the harvested crop or wood.

Grey water
Water the quality of which has been adversely affected by human use, in industrial, agriculture or domestically. The grey water footprint of a product is an indicator of freshwater pollution that can be associated with the production of a product over its full supply chain. It is defined as the volume of freshwater that is required to assimilate the load of pollutants based on natural background concentrations and existing ambient water quality standards. It is calculated as the volume of water that is required to dilute pollutants to such an extent that the quality of the water remains above agreed water quality standards.

Gross domestic product (GDP)
The value of all final goods and services produced in a country in one year. GDP can be measured by adding up all of an economy's incomes – wages, interest, profits, and rents – or expenditures – consumption, investment, government purchases, and net exports (exports minus imports).

Ground-truthing
A process by which the content of satellite images, aerial photographs – or maps based on them – is compared with the reality on the ground through site visits and field surveys. It is used to verify the accuracy of the images or the way they have been interpreted to produce maps.

Groundwater
Water that flows or seeps downward and saturates soil or rock, supplying springs and wells. The upper surface of the saturate zone is called the water table.

Gyres
A large system of rotating ocean currents, primarily driven by wind movement. Large gyres exist in the Indian Ocean, North Atlantic, North Pacific, South Atlantic and South Pacific.

Habitat
(1) The place or type of site where an organism or population occurs naturally.
(2) Terrestrial or aquatic areas distinguished by geographic, living and non-living features, whether entirely natural or semi-natural.

Habitat fragmentation
Alteration of habitat resulting in spatial separation of habitat units from a previous state of greater continuity.

Hard law
Legally binding obligations that are precise (or can be made precise through adjudication or the issuance of detailed regulations) and that delegate authority for interpreting and implementing the law. In the context of international law, hard law refers to treaties or international agreements, as well as customary laws. These documents create enforceable obligations and rights for states and other international entities. See also *Soft law*.

Hazard
A potentially damaging physical event, phenomenon or human activity that may cause the loss of life or injury, property damage, social and economic disruption or environmental degradation.

Hazardous waste
A used or discarded material that can damage human health and the environment. Hazardous wastes may include heavy metals, toxic chemicals, medical wastes or radioactive material.

Heavy metals
A subset of elements that exhibit metallic properties, including transitional metals and semi-metals (metalloids), such as arsenic, cadmium, chromium, copper, lead, mercury, nickel and zinc, that have been associated with contamination and potential toxicity.

High seas
The oceans outside national jurisdictions, lying beyond each nation's exclusive economic zone or other territorial waters.

Human well-being
The extent to which individuals have the ability to live the kinds of lives they have reason to value; the opportunities people have to pursue their aspirations. Basic components of human well-being include: security, meeting material needs, health and social relations.

Hydrochlorofluorocarbons (HCFCs)
Organic and human-made substances composed of hydrogen, chlorine, fluorine and carbon atoms. As the ozone-depleting potential of HCFCs is much lower than that of CFCs, HCFCs were considered acceptable interim substitutes for CFCs.

Hydrological cycle
Succession of stages undergone by water in its passage from the atmosphere to the Earth's surface and its return to the atmosphere. The stages include evaporation from land, sea or inland water, condensation to form clouds, precipitation, accumulation in the soil or in water bodies, and re-evaporation.

Hydrosphere
All of the Earth's water, including surface water (water in oceans, lakes and rivers), groundwater (water in soil and beneath the Earth's surface), snow, ice and water in the atmosphere, including water vapour.

Hypoxia
Lack of oxygen. In the context of eutrophication and algal blooms, hypoxia is the result of a process that uses up dissolved oxygen in the water. Algal blooms cause water to become more opaque, thereby reducing light availability to submerged aquatic vegetation, and interfering with beneficial human water uses. When the bloom dies off, algae sink to the bottom and are decomposed by bacteria using up the available oxygen. Hypoxia is particularly severe in the late summer, and can be so severe in some areas that they are referred to as "dead zones" because only bacteria can survive there.

Improved drinking water
"Improved" sources of drinking water include piped water into dwellings; piped water into yards/plots; public taps or standpipes; tube wells or boreholes; protected dug wells; protected springs; rainwater.

Improved sanitation
"Improved" sanitation includes flush lavatories; piped sewer systems; septic tanks; flush/pour flush to pit latrines; ventilated improved pit latrines (VIP); pit latrines with slab; composting lavatories.

Integrated coastal zone management (ICZM)
Approaches that integrate economic, social and ecological perspectives for the management of coastal resources and areas.

Integrated water resources management (IWRM)
A process which promotes the coordinated development and management of water, land and related resources, in order to maximize the resultant economic and social welfare in an equitable manner without compromising the sustainability of vital ecosystems.

Institutions
Regularized patterns of interaction by which society organizes itself: the rules, practices and conventions that structure human interaction. The term is wide and encompassing, and could be taken to include law, social relationships, property rights and tenurial systems, norms, beliefs, customs and codes of conduct as much as multilateral environmental agreements, international conventions and financing mechanisms. Institutions could be formal (explicit, written, often having the sanction of the state) or informal (unwritten, implied, tacit, mutually agreed and accepted). Formal institutions include law, international environmental agreements, bylaws and memoranda of understanding. Informal institutions include unwritten rules, codes of conduct and value systems. The term "institutions" should be distinguished from organizations.

IPCC scenarios
Six future-emission scenarios based on four scenario families A1, A2, B1 and B2, where A represents globalized development,

B represents regionalized development, while 1 refers to economic growth and 2 refers to environmental stewardship.

IPAT formulation
Impact = Population x Affluence x Technology. An equation developed in the 1970s to describe humanity's influence/impact on the environment.

Jevons paradox
The proposition that technological progress that increases the efficiency with which a resource is used tends to increase (rather than decrease) the rate of consumption of that resource.

Joint Implementation
A mechanism provided by Article 6 of the Kyoto Protocol, whereby a country included in Annex I of the UNFCCC may acquire emission reduction units when it helps to finance projects that reduce net emissions in another industrialized country.

Kuznets curve (environmental)
A relationship between economic development and environmental pollution. Based on empirical evidence, some forms of local pollution (airborne lead, sulphur) declined significantly in industrialized countries despite robust economic growth. This follows a general pattern of poor countries being relatively unpolluted, middle-income countries more polluted, and rich countries clean again.

Kyoto Protocol
A protocol to the 1992 United Nations Framework Convention on Climate Change (UNFCCC) adopted at the Third Session of the Conference of the Parties to the UNFCCC in 1997 in Kyoto, Japan. It contains legally binding commitments in addition to those included in the UNFCCC. Countries included in Annex B of the protocol (most OECD countries and countries with economies in transition) agreed to control their national anthropogenic emissions of greenhouse gases (CO_2, CH_4, N_2O, HFCs, PFCs, SF_6 and NF_3) so that the total emissions from these countries would be at least 5 per cent below 1990 levels in the commitment period, 2008 to 2012.

Land cover
The physical coverage of land, usually expressed in terms of vegetation cover or lack of it. Influenced by but not synonymous with land use.

Land degradation
The reduction or loss of the biological or economic productivity and complexity in croplands, pastures, forest or woodlands resulting from climate variability, natural processes, and unsustainable human activities.

Land use
The functional dimension of land for different human purposes or economic activities. Examples of land use categories include agriculture, industrial use, transport and protected areas.

Land-use planning
The systematic assessment of land and water potential, alternative patterns of land use and other physical, social and economic conditions, for the purpose of selecting and adopting land-use options which are most beneficial to land users.

Legal regime
A system of principles and rules governing something, and which is created by law. It is framework of legal rules.

Legitimacy
Measure of political acceptability or perceived fairness. State law has its legitimacy in the state; local law and practices work on a system of social sanction, in that they derive their legitimacy from a system of social organization and relationships.

Leverage point
A place in a system's structure where a relatively small amount of force can effect change. It is a low leverage point if a small amount of force causes a small change in system behaviour, or a high leverage point if a small amount of force causes a large change.

Life-cycle analysis
A technique to assess the environmental impacts associated with all the stages of the life of a product – from raw material extraction through materials processing, manufacture, distribution, use, repair and maintenance, and disposal or recycling (cradle-to-grave).

Lifetime (in the atmosphere)
The approximate amount of time it takes for concentrations of an atmospheric pollutant to return to the background level (assuming emissions cease) as a result of either being converted to another chemical compound or being taken out of the atmosphere through a sink. Atmospheric lifetimes can vary from hours or weeks (sulphate aerosols) to more than a century (CFCs).

Lithosphere
The outer part of the Earth, consisting of the crust and upper mantle. It is about 55 km thick beneath the oceans and up to about 200 km thick beneath the continents. The solid part of the Earth, as contrasted with the atmosphere and hydrosphere.

Low emission zone (LEZ)
Urban areas where travel by polluting vehicles is limited or banned.

Low-impact pesticides
Pesticides considered to be of minimal risk compared to other pesticides. In order for a pesticide to be truly low impact, considerations beyond the choice of pesticide product must be considered, including the timing, method and site of application.

Lyme disease
A multi-system bacterial infection caused by the spirochaete *Borrelia burgdoferi*. These spirochaetes are maintained in nature in the bodies of wild animals, and transmitted from one animal to another through the bite of an infected tick. People and pets are incidental hosts to ticks.

Mainstreaming
Taking into consideration as an integral part of the issue in question.

Material flow accounting
The quantification of all materials used in economic activities. It accounts for the total material mobilized during the extraction of materials and for the materials actually used in economic processes measured in terms of their mass.

Marine protected area (MPA)
A geographically defined marine area that is designated or regulated and managed to achieve specific conservation objectives.

Mega-cities
Urban areas with more than 10 million inhabitants.

Mega-heatwave
An event with regional mean temperature anomalies (over an area of ~1 million km^2) of extraordinary amplitude (approximately ≥3 SD (standard deviations)) at sub-seasonal scales, of at least 7 days.

Millennium Development Goals (MDGs)
The eight Millennium Development Goals – which range from halving extreme poverty to halting the spread of HIV/AIDS and providing universal primary education, all by the target date of 2015 – form a blueprint agreed to by all the world's countries and all the world's leading development institutions.

Morphology
The branch of biology that deals with the form of living organisms, and with relationships between their structures.

Multilateral environmental agreements (MEAs)
Treaties, conventions, protocols and contracts between several states regarding specified environmental problems.

Nanomaterial
A natural, incidental or manufactured material containing particles, in an unbound state, as an aggregate or as an agglomerate and where, for 50 per cent or more of the particles in the number size distribution, one or more external dimension is in the size range 1–100 nanometres (a nanometre is one billionth of a metre). Such particles/materials are generally termed as nanoparticles (NPs), nanochemicals or nanomaterials (NMs).

Natural capital
Natural assets in their role of providing natural resource inputs and environmental services for economic production. Natural capital includes land, minerals and fossil fuels, solar energy, water, living organisms, and the services provided by the interactions of all these elements in ecological systems.

Net primary productivity (NPP)
The rate at which all the plants in an ecosystem produce net useful chemical energy. Some net primary production goes toward growth and reproduction of primary producers, while some is consumed by herbivores.

Nitrogen deposition
The input of reactive nitrogen, mainly derived from nitrogen oxides and ammonia emissions, from the atmosphere into the biosphere.

Non-state actors
Non-state actors are categorized as entities that (i) participate or act in the sphere of international relations; organizations with sufficient power to influence and cause change in politics which (ii) do not belong to or exist as a state-structure or established institution of a state; do not have the characteristics of this, these being legal sovereignty and some measure of control over a country's people and territories.

No-till (zero tillage)
A technique of drilling (sewing) seed with little or no prior land preparation, which has a positive impact on soil erosion.

Nutrient pollution
Contamination of water resources by excessive inputs of nutrients.

Nutrients
The approximately 20 chemical elements known to be essential for the growth of living organisms, including nitrogen, sulphur, phosphorus and carbon.

Oil sands
A complex mixture of sand, water and clay trapping very heavy oil, known as bitumen.

Organic agriculture
A production system that sustains the health of soils, ecosystems and people. It relies on ecological processes, biodiversity and cycles adapted to local conditions, rather than the use of synthetic inputs.

Organic carbon (OC)
Organic carbon, as used in climate research, usually refers to the carbon fraction of the aerosol that is not black. This term is an oversimplification because organic carbon may contain hundreds or thousands of different organic compounds with varying atmospheric behaviour. It is the quantity that results from thermal analysis of carbon aerosols.

Organizations
Bodies of individuals with a specified common objective. Organizations could be political organizations, political parties, governments and ministries; economic organizations, federations of industry; social organizations (non-governmental organizations (NGOs) and self-help groups) or religious organizations (church and religious trusts). The term organizations should be distinguished from institutions.

Organochlorine compounds
Any of a class of organic chemical compounds containing carbon, hydrogen and chlorine, such as dioxins, poly-chlorinated-biphenyls (PCBs) and some pesticides such as DDT.

Overexploitation
The excessive use of raw materials without considering the long-term ecological impacts of such use.

Overshoot
The situation that occurs when humanity's demand on the

biosphere exceeds supply or regenerative capacity. At the global level, ecological deficit and overshoot are the same, since there is no net import of resources to the planet.

Ozone layer
A region of the atmosphere situated at an altitude of 10–50 km above the Earth's surface (called the stratosphere) which contains diluted ozone.

Participatory approach
Securing an adequate and equal opportunity for people to place questions on an agenda and to express their preferences about a final outcome during decision making to all group members. Participation can occur directly or through legitimate representatives. Participation may range from consultation to the obligation of achieving a consensus.

Particulate matter (PM)
Tiny solid particles or liquid droplets suspended in the air.

Pastoralism, pastoral system
The husbandry of domestic animals as a primary means of obtaining resources.

Pasture
Ground covered with grass or herbage, used or suitable for the grazing of livestock.

Payment for environmental services/payment for ecosystem services (PES)
Appropriate mechanisms for matching the demand for environmental services with incentives for land users whose actions modify the supply of those environmental services.

Permafrost
Soil, silt and rock located in perpetually cold areas, and that remains frozen year-round for two or more years.

Persistent organic pollutants (POPs)
Chemical substances that persist in the environment, bioaccumulate through the food web, and pose a risk of causing adverse effects to human health and the environment.

Phytoplankton
Microscopically small plants that float or swim weakly in fresh or saltwater bodies.

Planetary boundaries
A framework designed to define a safe operating space for humanity for the international community, including governments at all levels, international organizations, civil society, the scientific community and the private sector, as a precondition for sustainable development.

Policy
Any form of intervention or societal response. This includes not only statements of intent, but also other forms of intervention, such as the use of economic instruments, market creation, subsidies, institutional reform, legal reform, decentralization and institutional development. Policy can be seen as a tool for the exercise of governance. When such an intervention is enforced by the state, it is called public policy.

Policy bank
A repository of good practices in policy development and implementation, supported by facilitation services that help stakeholders to identify relevant policy lessons and tools and adapt them to local needs.

Policy dialogue
A platform for major stakeholders, such as government authorities and non-governmental organizations, for awareness raising, capacity building, policy-preparation and policy-implementation.

Policy diffusion
The process of a policy being taken up, copied, implemented in other areas, fields, regions or sectors.

Pollutant
Any substance that causes harm to the environment when it mixes with soil, water or air.

Pollution
The presence of minerals, chemicals or physical properties at levels that exceed the values deemed to define a boundary between good or acceptable and poor or unacceptable quality, which is a function of the specific pollutant.

Polycentric
Having many centres, especially of authority or control.

Poverty
The state of one who lacks a defined amount of material possessions or money. Absolute poverty refers to a state of lacking basic human needs, which commonly include clean and fresh water, nutrition, health care, education, clothing and shelter.

Precautionary approach/principle
The precautionary approach or precautionary principle states that if an action or policy has a suspected risk of causing harm to the public or to the environment, in the absence of scientific consensus that the action or policy is harmful, the burden of proof that it is not harmful falls on those taking the action.

Prediction
The act of attempting to produce a description of the expected future, or the description itself, such as "it will be 30°C tomorrow, so we will go to the beach".

Premature deaths
Deaths occurring earlier due to a risk factor than would occur in the absence of that risk factor.

Primary energy
Energy embodied in natural resources (such as coal, crude oil, sunlight or uranium) that has not undergone any anthropogenic conversion or transformation.

Projection
The act of attempting to produce a description of the future subject to assumptions about certain preconditions, or the description itself, such as "assuming it is 30°C tomorrow, we will go to the beach."

Propagation of effects
An impact at one level in a system, even a very small one, may lead to larger changes as this impact moves up (or down) through a system.

Protected area
A clearly defined geographical space, recognized, dedicated and managed, through legal or other effective means, to achieve the long-term conservation of nature with associated ecosystem services and cultural values.

Provisioning services
The products obtained from ecosystems, including, for example, genetic resources, food and fibre, and freshwater.

Public-private partnership
A contractual agreement between a public agency (federal, state or local) and a private sector entity. Through such an agreement, the skills and assets of each sector (public and private) are shared in delivering a service or facility.

Public sector
The portion of society that comprises the general government sector plus all public corporations including the central bank.

Purchasing power parity (PPP)
The number of currency units required to purchase an amount of goods and services equivalent to what can be bought with one unit of the currency of the base country, for example, the US$.

Radiative forcing
A measure of the net change in the energy balance of the Earth with space, that is, the change in incoming solar radiation minus outgoing terrestrial radiation.

REDD/REDD+
Reducing Emissions from Deforestation and Forest Degradation in Developing Countries. REDD+ involves enhancing existing forests and increasing forest cover. In order to meet these objectives, policies need to address enhancement of carbon stocks by providing funding and investments in these areas.

Reforestation
Planting of forests on lands that have previously contained forest, but have since been converted to some other use.

Regulating services
The benefits obtained from the regulation of ecosystem processes, including, for example, the regulation of climate, water and some human diseases.

Remote sensing
Collection of data about an object from a distance. In the environmental field, it normally refers to aerial or satellite data for meteorology, oceanography or land cover assessment.

Renewable energy source
An energy source that does not rely on finite stocks of fuels. The most widely known renewable source is hydropower; other renewable sources are biomass, solar, tidal, wave and wind.

Resilience
The capacity of a system, community or society potentially exposed to hazards to adapt by resisting or changing in order to reach and maintain an acceptable level of functioning and structure.

Resistance
The capacity of a system to withstand the impacts of drivers without displacement from its present state.

Results-based management
A managerial approach that builds on defining realistic expected results, monitoring progress towards their achievement, integrating lessons learned into management decisions and reporting on performance.

Riparian
Related to or located on the bank of a natural watercourse, usually a river, but sometimes a lake, tidewater or enclosed sea.

River fragmentation
Degree to which river connectivity and flow regimes have been altered, usually by dams and reservoirs.

Regional Organization for the Protection of the Marine Environment (ROPME) Sea Area
The sea area surrounded by the eight Member States of the Regional Organization for the Protection of the Marine Environment (ROPME): Bahrain, Iran, Iraq, Kuwait, Oman, Qatar, Saudi Arabia and the United Arab Emirates.

Run-off
A portion of rainfall, melted snow or irrigation water that flows across the ground's surface and is eventually returned to streams. Run-off can pick up pollutants from air or land and carry them to receiving waters.

RUrbanism
An integration of urban and rural development in terms of sustainable resource use and the convergence of human well-being.

Sahel
A loosely defined strip of transitional vegetation that separates the Sahara desert from the tropical savannahs to the south. The region is used for farming and grazing, and because of the difficult environmental conditions at the border of the desert, the region is very sensitive to human-induced land-cover change. It includes parts of Senegal, the Gambia, Mauritania, Mali, Niger, Nigeria, Burkina Faso, Cameroon and Chad.

Salinization/salination
The process by which water-soluble salts accumulate in the soil. Salinization may occur naturally or because of conditions resulting from management practices.

Scale
The spatial, temporal (quantitative or analytical) dimension used to measure and study any phenomena. Specific points on a scale can thus be considered levels (such as local, regional, national and international).

Scenario
A description of how the future may unfold based on if-then propositions, typically consisting of a representation of an initial situation, a description of the key drivers and changes that lead to a particular future state. For example, "given that we are on holiday at the coast, if it is 30°C tomorrow, we will go to the beach".

Seagrass bed
Profusion of grass-like marine plants, usually on shallow, sandy or muddy areas of the seabed.

Secondary pollutant
Not directly emitted as such, but forms when other pollutants (primary pollutants) react in the atmosphere.

Security
Relates to personal and environmental security. It includes access to natural and other resources, and freedom from violence, crime and war, as well as security from natural and human-caused disasters

Sediment
Solid material that originates mostly from disintegrated rocks and is transported by, suspended in or deposited from water, wind, ice and other organic agents

Sedimentation
Strictly, the act or process of depositing sediment from suspension in water or ice. Broadly, all the processes whereby particles of rock material are accumulated to form sedimentary deposits. Sedimentation, as commonly used, involves transport by water, wind, ice and organic agents.

Sequestration
In *GEO-5*, sequestration refers to the capture of carbon dioxide in a manner that prevents it from being released into the atmosphere for a specified period of time.

Service advertising
Advertising with a central focus on public welfare.

Sex-disaggregated data
Data separated by sex/gender in order to allow differential impacts on men and women to be measured.

Shared waters
Water resources shared by two or more governmental jurisdictions.

Short-lived climate forcers
Substances such as methane, black carbon, tropospheric ozone, and many hydrofluorocarbons, which have a significant impact on climate change, and a relatively short lifespan in the atmosphere compared to carbon dioxide and other longer-lived gases.

Siltation
The deposition of finely divided soil and rock particles on the bottom of stream and riverbeds and reservoirs.

Silvopastoral strategy
The integration of trees and shrubs in pastures with animals for economic, ecological and social sustainability.

Siting authority
A clearly defined and legitimate agency that authorizes construction of, for example, electricity transmission equipment.

Social contract
A contract or agreement between people to form a society that determines their moral and/or political obligations. Social contracts can be formal or informal and define the relationship between individuals and their government on the basis of mutual consent.

Social learning
Process in which individuals observe the behaviour of others and its consequences, and modify their own behaviour accordingly.

Social network
A social structure made up of a set of actors, such as individuals or organizations, and the ties between these actors, such as relationships, connections or interactions.

Soft law
Rules that are neither strictly binding in nature nor completely lacking legal significance. They are weakened along one or more of the dimensions of obligation, precision and delegation. In the context of international law, soft law refers to guidelines, policy declarations or codes of conduct which set standards of conduct. However, they are not directly enforceable.

Soil acidification
A naturally occurring process in humid climates that has long been the subject of research, whose findings suggest that acid precipitation affects the productivity of terrestrial plants.

Species (biology)
An interbreeding group of organisms that is reproductively isolated from all other organisms, although there are many partial exceptions to this rule. A generally agreed fundamental taxonomic unit that, once described and accepted, is associated with a unique scientific name.

Species diversity
Biodiversity at the species level, often combining aspects of species richness, their relative abundance and their dissimilarity.

Species richness/abundance
The number of species within a given sample, community or area.

Strategic environmental assessment (SEA)
A range of analytical and participatory approaches that aim to integrate environmental considerations into policies, plans and programmes and evaluate the links with economic and social considerations. An SEA is undertaken for plans, programmes and policies. It helps decision makers reach a better understanding of how environmental, social and economic considerations fit together.

Stratospheric ozone depletion
Chemical destruction of the stratospheric ozone layer, particularly by substances produced by human activities.

Structural adjustment
A process of market-oriented economic reform aimed at reducing inflation and creating conditions for economic growth.

Supporting services
Ecosystem services that are necessary for the production of all other ecosystem services. Some examples include biomass production, production of atmospheric oxygen, soil formation and retention, nutrient cycling, water cycling and provisioning of habitat.

Surface water
All water naturally open to the atmosphere, including rivers, lakes, reservoirs, streams, impoundments, seas and estuaries. The term also covers springs, wells or other collectors of water that are directly influenced by surface waters.

Sustainability
A characteristic or state whereby the needs of the present population can be met without compromising the ability of future generations or populations in other locations to meet their needs.

Sustainable agriculture and livestock production
Management of agricultural and livestock resources to satisfy human needs while maintaining or enhancing environmental quality and conserving natural resources for future generations.

Sustainable development
Development that meets the needs of the present generation without compromising the ability of future generations to meet their own needs.

Sustainable forest management (SFM)
The stewardship and use of forests and forest lands in a way, and at a rate, that maintains their biodiversity, productivity, regeneration capacity, vitality and potential to fulfill, now and in the future, relevant ecological, economic and social functions, at local, national and global levels, and that does not cause damage to other ecosystems.

Symbiotic relationship
A relationship between two different organisms living in close physical association, typically to the advantage of both

Synergies
These arise when two or more processes, organizations, substances or other agents interact in such a way that the outcome is greater than the sum of their separate effects.

System
A system is a collection of component parts that interact with one another within some boundary.

Taxonomy
A system of nested categories (taxa) reflecting evolutionary relationships or morphological similarities.

TechnoGarden
The TechnoGarden scenario depicts a globally connected world relying strongly on technology and highly managed, often engineered ecosystems, to deliver ecosystem services.

Technology
Physical artefacts or the bodies of knowledge of which they are an expression. Examples are water extraction structures, such as tube wells, renewable energy technologies and traditional knowledge. Technology and institutions are related. Any technology has a set of practices, rules and regulations surrounding its use, access, distribution and management.

Technology effect
Reducing the net or at least per-person impact of resource consumption due to technological innovations.

Technology transfer
A broad set of processes covering the flows of know-how, experience and equipment among different stakeholders.

Temperate region
The region in which the climate undergoes seasonal change in temperature and moisture. Temperate regions of the Earth lie primarily between 30° and 60° latitude in both hemispheres.

Thermohaline circulation (THC)
Large-scale density-driven circulation in the ocean, caused by differences in temperature and salinity. In the North Atlantic, the thermohaline circulation consists of warm surface water flowing northward and cold deep water flowing southward, resulting in a net poleward transport of heat. The surface water sinks in highly restricted sinking regions located in high latitudes. Also referred to as the (global) ocean conveyer belt or the meridional overturning circulation (MOC).

Threshold
The level of magnitude of a system process at which sudden or rapid change occurs. A point or level at which new properties emerge in an ecological, economic or other system, invalidating predictions based on mathematical relationships that apply at lower levels.

Tipping point
The critical point in an evolving situation that leads to a new and sometimes irreversible development.

Tokenism
The policy or practice of making only a symbolic effort.

Topography
The study or detailed description of the surface features of a region.

Toxic pollutants
Pollutants that cause death, disease or birth defects in organisms that ingest or absorb them.

Traditional or local ecological knowledge
A cumulative body of knowledge, know-how, practices or representations maintained or developed by peoples with extended histories of interaction with the natural environment.

Transformation
State of being transformed. In the context of *GEO-5*, transformation refers to a series of actions that explores opportunities to stop doing the things that pull the Earth System in the wrong direction and at the same time provide resources, capacity and an enabling environment for all that is consistent with the sustainable-world vision.

Transitions
Non-linear, systematic and fundamental changes of the composition and functioning of a societal system with changes in structures, cultures and practices.

Transpiration
The loss of water vapour from parts of plants, especially in leaves but also in stems, flowers and roots.

Trillion
10^{12} (1 000 000 000 000)

Trophic level
Successive stages of nourishment as represented by the links of the food chain. Put simply, the primary producers (phytoplankton) constitute the first trophic level, herbivorous zooplankton the second and carnivorous organisms the third trophic level.

Tropospheric ozone
Ozone at the bottom of the atmosphere, and the level at which humans, crops and ecosystems are exposed. Also known as ground-level ozone.

Urban sprawl
The decentralization of the urban core through the unlimited outward extension of dispersed development beyond the urban fringe, where low density residential and commercial development exacerbates fragmentation of powers over land use.

Urbanization
An increase in the proportion of the population living in urban areas.

Virtual water trade
The idea that when goods and services are traded, the water needed to produce them (embedded) is traded as well.

Vulnerability
An intrinsic feature of people at risk. It is a function of exposure, sensitivity to impacts of the specific unit exposed (such as a watershed, island, household, village, city or country), and the ability or inability to cope or adapt. It is multi-dimensional, multi-disciplinary, multi-sectoral and dynamic. The exposure is to hazards such as drought, conflict or extreme price fluctuations, and also to underlying socio-economic, institutional and environmental conditions.

Wastewater treatment
Any of the mechanical, biological or chemical processes used to modify the quality of wastewater in order to reduce pollution levels.

Water conflict
A confrontation between countries, states, or groups over water resources.

Water footprint
An indicator of water use that looks at both direct and indirect water use of a consumer or producer. The water footprint of an individual, community, nation or business is defined as the total volume of freshwater that is used to produce the goods and services consumed by the individual, community or nation, or produced by the business.

Water quality
The chemical, physical and biological characteristics of water, usually in respect to its suitability for a particular purpose.

Water scarcity
Occurs when annual water supplies drop below 1 000 m^3 per person, or when more than 40 per cent of available water is used.

Water security
A term that broadly refers to the sustainable use and protection of water systems, the protection against water related hazards (floods and droughts), the sustainable development of water resources and the safeguarding of (access to) water functions and services for humans and the environment.

Water stress
Occurs when low water supplies limit food production and economic development, and affect human health. An area is experiencing water stress when annual water supplies drop below 1 700 m^3 per person.

Wetland
Area of marsh, fen, peatland, bog or water, whether natural or artificial, permanent or temporary, with water that is static or flowing, fresh, brackish or salt, including areas of marine water to a depth, at low tide, that does not exceed 6 metres.

Woodland
Wooded land, which is not classified as forest, spanning more than 0.5 hectares, with trees higher than 5 metres and a canopy cover of 5–10 per cent, or trees able to reach these thresholds in situ, or with a combined cover of shrubs, bushes and trees above 10 per cent. It does not include areas used predominantly for agricultural or urban purposes.

Index

Note: Page numbers in **bold** refer to illustrations; those in *italic* refer to tables and boxes

S

T

U